PERIODIC TABLE OF THE ELEMENTS

KEY

| 79 — Atomic number |
| Au — Symbol |
| Gold — Name |
| 196.9665 — Atomic weight |

An element

Metals
Metalloids
Nonmetals, noble gases

Period number

Group number, U.S. system
Group number, IUPAC system

Numbers in parentheses are mass numbers of radioactive isotopes.

Main Table

1A (1)	2A (2)	3B (3)	4B (4)	5B (5)	6B (6)	7B (7)	8B (8)	8B (9)	8B (10)	1B (11)	2B (12)	3A (13)	4A (14)	5A (15)	6A (16)	7A (17)	8A (18)
1 **H** Hydrogen 1.0079																	2 **He** Helium 4.0026
3 **Li** Lithium 6.941	4 **Be** Beryllium 9.0122											5 **B** Boron 10.811	6 **C** Carbon 12.011	7 **N** Nitrogen 14.0067	8 **O** Oxygen 15.9994	9 **F** Fluorine 18.9984	10 **Ne** Neon 20.1797
11 **Na** Sodium 22.9898	12 **Mg** Magnesium 24.3050											13 **Al** Aluminum 26.9815	14 **Si** Silicon 28.0855	15 **P** Phosphorus 30.9738	16 **S** Sulfur 32.066	17 **Cl** Chlorine 35.4527	18 **Ar** Argon 39.948
19 **K** Potassium 39.0983	20 **Ca** Calcium 40.078	21 **Sc** Scandium 44.9559	22 **Ti** Titanium 47.88	23 **V** Vanadium 50.9415	24 **Cr** Chromium 51.9961	25 **Mn** Manganese 54.9380	26 **Fe** Iron 55.847	27 **Co** Cobalt 58.9332	28 **Ni** Nickel 58.693	29 **Cu** Copper 63.546	30 **Zn** Zinc 65.39	31 **Ga** Gallium 69.723	32 **Ge** Germanium 72.61	33 **As** Arsenic 74.9216	34 **Se** Selenium 78.96	35 **Br** Bromine 79.904	36 **Kr** Krypton 83.80
37 **Rb** Rubidium 85.4678	38 **Sr** Strontium 87.62	39 **Y** Yttrium 88.9059	40 **Zr** Zirconium 91.224	41 **Nb** Niobium 92.9064	42 **Mo** Molybdenum 95.94	43 **Tc** Technetium (98)	44 **Ru** Ruthenium 101.07	45 **Rh** Rhodium 102.9055	46 **Pd** Palladium 106.42	47 **Ag** Silver 107.8682	48 **Cd** Cadmium 112.411	49 **In** Indium 114.82	50 **Sn** Tin 118.710	51 **Sb** Antimony 121.757	52 **Te** Tellurium 127.60	53 **I** Iodine 126.9045	54 **Xe** Xenon 131.29
55 **Cs** Cesium 132.9054	56 **Ba** Barium 137.327	57 **La** Lanthanum 138.9055	72 **Hf** Hafnium 178.49	73 **Ta** Tantalum 180.9479	74 **W** Tungsten 183.85	75 **Re** Rhenium 186.207	76 **Os** Osmium 190.2	77 **Ir** Iridium 192.22	78 **Pt** Platinum 195.08	79 **Au** Gold 196.9665	80 **Hg** Mercury 200.59	81 **Tl** Thallium 204.3833	82 **Pb** Lead 207.2	83 **Bi** Bismuth 208.9804	84 **Po** Polonium (209)	85 **At** Astatine (210)	86 **Rn** Radon (222)
87 **Fr** Francium (223)	88 **Ra** Radium 227.0278	89 **Ac** Actinium (227)	104 **Rf** Rutherfordium (261)	105 **Db** Dubnium (262)	106 **Sg** Seaborgium (263)	107 **Bh** Bohrium (262)	108 **Hs** Hassium (265)	109 **Mt** Meitnerium (266)	110 (269)	111 (272)	112 (277)		114 (285)		116 (289)		

Lanthanides (6)

| 58 **Ce** Cerium 140.115 | 59 **Pr** Praseodymium 140.9076 | 60 **Nd** Neodymium 144.24 | 61 **Pm** Promethium (145) | 62 **Sm** Samarium 150.36 | 63 **Eu** Europium 151.965 | 64 **Gd** Gadolinium 157.25 | 65 **Tb** Terbium 158.9253 | 66 **Dy** Dysprosium 162.50 | 67 **Ho** Holmium 164.9303 | 68 **Er** Erbium 167.26 | 69 **Tm** Thulium 168.9342 | 70 **Yb** Ytterbium 173.04 | 71 **Lu** Lutetium 174 |

Actinides (7)

| 90 **Th** Thorium 232.0381 | 91 **Pa** Protactinium 231.0359 | 92 **U** Uranium 238.0289 | 93 **Np** Neptunium (237) | 94 **Pu** Plutonium (244) | 95 **Am** Americium (243) | 96 **Cm** Curium (247) | 97 **Bk** Berkelium (247) | 98 **Cf** Californium (251) | 99 **Es** Einsteinium (252) | 100 **Fm** Fermium (257) | 101 **Md** Mendelevium (258) | 102 **No** Nobelium (259) | 103 **Lr** Lawrencium |

P9-BUG-705

FoodPix/Karl Petzke

Introduction to Organic and Biochemistry

FIFTH EDITION

Frederick A. Bettelheim
Adelphi University

William H. Brown
Beloit College

Jerry March
Adelphi University

THOMSON

BROOKS/COLE

Australia • Canada • Mexico • Singapore
Spain • United Kingdom • United States

DEDICATION: *This book is dedicated to the memory of John Vondeling, visionary editor and friend.*

THOMSON
BROOKS/COLE

CHEMISTRY EDITOR: John Holdcroft

DEVELOPMENT EDITOR: Sandra Kiselica

ASSISTANT EDITOR: Alyssa White

EDITORIAL ASSISTANT: Lauren Raike

TECHNOLOGY PROJECT MANAGER: Ericka Yeoman-Saler

MARKETING MANAGER: Jeffrey L. Ward

MARKETING ASSISTANT: Mona Weltmer

ADVERTISING PROJECT MANAGER: Stacey Purviance

PROJECT MANAGER, EDITORIAL PRODUCTION: Lisa Weber

PRINT/MEDIA BUYER: Barbara Britton

PERMISSIONS EDITOR: Joohee Lee

PRODUCTION SERVICE: Thompson Steele, Inc.

TEXT DESIGNER: Baugher Design Inc.

PHOTO RESEARCHER: Jane Sanders

COPY EDITOR: Thompson Steele, Inc.

ILLUSTRATOR: J/B Woolsey and Thompson Steele, Inc.

COVER DESIGNER: Baugher Design Inc.

COVER IMAGE: ©FoodPix/Karl Petzke

COVER PRINTER: Transcontinental Printing/Interglobe

COMPOSITOR: Thompson Steele, Inc.

PRINTER: Transcontinental Printing/Interglobe

For more information about our products, contact us at:
Thomson Learning Academic Resource Center
1-800-423-0563
For permission to use material from this text, contact us by:
Phone: 1-800-730-2214
Fax: 1-800-730-2215
Web: http://www.thomsonrights.com

Library of Congress Control Number: 2003100305

Student Edition with InfoTrac College Edition: ISBN 0-534-40176-7

Student Edition without InfoTrac College Edition: ISBN-0-534-40189-9

Instructor's Edition: ISBN 0-534-40177-5

International Student Edition: ISBN 0-534-40211-9 (not for sale in the United States)

Brooks/Cole—Thomson Learning
10 Davis Drive
Belmont, CA 94002
USA

Asia
Thomson Learning
5 Shenton Way #01-01
UIC Building
Singapore 068808

Australia/New Zealand
Thomson Learning
102 Dodds Street
Southbank, Victoria 3006
Australia

Canada
Nelson
1120 Birchmount Road
Toronto, Ontario M1K 5G4
Canada

Europe/Middle East/Africa
Thomson Learning
High Holborn House
50/51 Bedford Row
London WC1R 4LR
United Kingdom

Latin America
Thomson Learning
Seneca, 53
Colonia Polanco
11560 Mexico D.F.
Mexico

Spain/Portugal
Paraninfo
Calle/Magallanes, 25
28015 Madrid, Spain

FREDERICK A. BETTELHEIM is a Distinguished University Research Professor at Adelphi University, and a Visiting Scientist at the National Eye Institute NIH. He has coauthored every edition of *Introduction to General, Organic & Biochemistry* and several laboratory manuals, including *Laboratory Experiments for General, Organic & Biochemistry* and *Experiments for Introduction to Organic Chemistry*. He is the author of *Experimental Physical Chemistry* and coauthor of numerous monographs and research articles. Professor Bettelheim received his Ph.D. from the University of California, Davis, and his areas of specialization have included the biochemistry of proteins and carbohydrates and the physical chemistry of polymers. He was Fulbright Professor at the Weizman Institute, Israel, and Visiting Professor at the University of Uppsala, Sweden, at Technion, Israel, and at the University of Florida. He was keynote lecturer at the 16th International Conference on Chemical Education, August 2000, in Budapest, Hungary, where his lecture topic was "Modern Trends in Teaching Chemistry to Nurses and Other Health-Related Professionals."

WILLIAM H. BROWN, a new coauthor for the sixth edition of *Introduction to General, Organic and Biochemistry,* is Professor of Chemistry at Beloit College, where he has twice been named Teacher of the Year. He is also the author of two best-selling undergraduate texts, *Introduction to Organic Chemistry*, sixth edition, and *Organic Chemistry*, third edition. His regular teaching responsibilities include organic chemistry, advanced organic chemistry, and special topics in pharmacology and drug synthesis. He was the leader of a team revising the organic chemistry material appearing in the upcoming *Encyclopaedia Britannica CD-ROM*. Professor Brown received his Ph.D. from Columbia University under the direction of Gilbert Stork and did postdoctoral work at the California Institute of Technology and the University of Arizona.

JERRY MARCH (d. 1997) was the coauthor of *Introduction to General, Organic & Biochemistry* for its first five editions. He was the sole author of the best-selling *Advanced Organic Chemistry* text, now in its fifth edition, which has been translated into many languages, including Russian and Japanese. He was a member of the Physical Organic Chemistry Commission of the International Union of Pure and Applied Chemistry, and as such was instrumental in the development of several nomenclature systems. Professor March received his Ph.D. at Pennsylvania State University and was Professor of Chemistry at Adelphi University, specializing in organic chemistry.

BRIEF CONTENTS

PART 1 ORGANIC CHEMISTRY 1

CHAPTER 1 Organic Chemistry 2
2 Alkanes 18
3 Alkenes 46
4 Benzene and Its Derivatives 75
5 Alcohols, Ethers, and Thiols 91
6 Chirality: The Handedness of Molecules 114
7 Acids and Bases 134
8 Amines 167
9 Aldehydes and Ketones 183
10 Carboxylic Acids, Anhydrides, Esters, and Amides 202

PART 2 BIOCHEMISTRY 229

CHAPTER 11 Carbohydrates 230
12 Lipids 258
13 Proteins 289
14 Enzymes 317
15 Chemical Communications: Neurotransmitters and Hormones 336
16 Nucleotides, Nucleic Acids, and Heredity 360
17 Gene Expression and Protein Synthesis 384
18 Bioenergetics: How the Body Converts Food to Energy 409
19 Specific Catabolic Pathways: Carbohydrate, Lipid, and Protein Metabolism 430
20 Biosynthetic Pathways 455
21 Nutrition 468
22 Immunochemistry 486
23 Body Fluids 508

Appendix I Exponential Notation A-1
Appendix II Significant Figures A-5
Answers to In-Text and Odd-Numbered End-of-Chapter Problems A-8

CONTENTS

PART 1
ORGANIC CHEMISTRY   1

CHAPTER 1
Organic Chemistry   2

1.1 Introduction   2
1.2 Sources of Organic Compounds   4
1.3 Structure of Organic Compounds   6
1.4 Functional Groups   8

Summary   14
Problems   14

CHEMICAL CONNECTIONS
1A Taxol—A Story of Search and Discovery   5
1B Combinatorial Chemistry   6

CHAPTER 2
Alkanes   18

2.1 Introduction   18
2.2 Structure of Alkanes   19
2.3 Constitutional Isomerism in Alkanes   21
2.4 Nomenclature of Alkanes   23
2.5 Cycloalkanes   27
2.6 Shapes of Alkanes and Cycloalkanes   28
2.7 Cis-Trans Isomerism in Cycloalkanes   32
2.8 Physical Properties   34
2.9 Reactions of Alkanes   37
2.10 Important Haloalkanes   38
2.11 Sources of Alkanes   39

Summary   41
Key Reactions   41
Problems   41

CHEMICAL CONNECTIONS
2A The Poisonous Puffer Fish   31
2B The Environmental Impact of Freons   39
2C Octane Rating: What Those Numbers at the Pump Mean   40

CHAPTER 3
Alkenes   46

3.1 Introduction   46
3.2 Structure   47
3.3 Nomenclature   48
3.4 Physical Properties   55
3.5 Naturally Occurring Alkenes: The Terpenes   55

3.6 Addition Reactions of Alkenes   57
3.7 Polymerization of Ethylene and Substituted Ethylenes   64

Summary   69
Key Reactions   69
Problems   70

CHEMICAL CONNECTIONS
3A Ethylene, a Plant Growth Regulator   47
3B The Case of the Iowa and New York Strains of the European Corn Borer   52
3C Cis-Trans Isomerism in Vision   54
3D Recycling Plastics   68

CHAPTER 4
Benzene and Its Derivatives   75

4.1 Introduction   75
4.2 The Structure of Benzene   76
4.3 Nomenclature   77
4.4 Reactions of Benzene and Its Derivatives   81
4.5 Phenols   83

Summary   87
Key Reactions   87
Problems   88

CHEMICAL CONNECTIONS
4A DDT—A Boon and a Curse   79
4B Carcinogenic Polynuclear Aromatics and Smoking   80
4C Iodide Ion and Goiter   81
4D The Nitro Group in Explosives   82
4E FD & C No. 6 (a.k.a. Sunset Yellow)   83
4F Capsaicin, for Those Who Like It Hot   85

CHAPTER 5
Alcohols, Ethers, and Thiols   91

5.1 Introduction   91
5.2 Alcohols   92
5.3 Reactions of Alcohols   96
5.4 Ethers   102
5.5 Thiols   104
5.6 Important Alcohols   107

Summary   108

Tom and Pat Leeson/Photo Researchers, Inc.

Key Reactions 109
Problems 109

CHEMICAL CONNECTIONS
5A Nitroglycerin, an Explosive
and a Drug 94
5B Breath-Alcohol Screening 101
5C Ethylene Oxide: A Chemical Sterilant 103
5D Ethers and Anesthesia 105

CHAPTER 6
Chirality: The Handedness
of Molecules 114

6.1 Introduction 114
6.2 Enantiomerism: A New Kind of
Stereoisomerism 114
6.3 Naming Stereocenters: The *R,S* System 119
6.4 Molecules with Two or More
Stereocenters 123
6.5 Optical Activity: How Chirality Is Detected in
the Laboratory 127
6.6 The Significance of Chirality in the
Biological World 128

Summary 129
Problems 130

CHEMICAL CONNECTIONS
6A Chiral Drugs 126

CHAPTER 7
Acids and Bases 134

7.1 Introduction 134
7.2 Acid and Base Strength 136
7.3 Brønsted-Lowry Acids and Bases 138
7.4 The Position of Equilibrium in Acid–Base
Reactions 140
7.5 Acid Ionization Constants 142
7.6 Some Properties of Acids and Bases 144
7.7 Self-Ionization of Water 147
7.8 pH and pOH 148
7.9 The pH of Aqueous Salt Solutions 150
7.10 Titration 152
7.11 Buffers 154
7.12 The Henderson-Hasselbalch Equation 158

Summary 160
Problems 161

CHEMICAL CONNECTIONS
7A Some Important Acids and Bases 137
7B Acid and Base Burns of the Cornea 141
7C Drugstore Antacids 145
7D Acidosis and Alkalosis 159

CHAPTER 8
Amines 167

8.1 Introduction 167
8.2 Structure and Classification 167

8.3 Nomenclature 170
8.4 Physical Properties 172
8.5 Basicity 173
8.6 Reaction with Acids 175
8.7 Epinephrine: A Prototype for the
Development of New Bronchodilators 177

Summary 178
Key Reactions 178
Problems 179

CHEMICAL CONNECTIONS
8A Amphetamines (Pep Pills) 168
8B Alkaloids 169
8C Tranquilizers 172
8D The Solubility of Drugs in Body Fluids 176

CHAPTER 9
Aldehydes and Ketones 183

9.1 Introduction 183
9.2 Structure and Bonding 183
9.3 Nomenclature 184
9.4 Physical Properties 187
9.5 Oxidation 188
9.6 Reduction 189
9.7 Addition of Alcohols 192
9.8 Keto-Enol Tautomerism 194

Summary 195
Key Reactions 195
Problems 196

CHEMICAL CONNECTIONS
9A Some Naturally Occurring Aldehydes
and Ketones 186

CHAPTER 10
Carboxylic Acids, Anhydrides, Esters,
and Amides 202

10.1 Introduction 202
10.2 Carboxylic Acids 203
10.3 Anhydrides, Esters, and Amides 208
10.4 Preparation of Esters—Fischer
Esterification 212
10.5 Preparation of Amides 213
10.6 Hydrolysis of Anhydrides, Esters,
and Amides 215
10.7 Phosphoric Anhydrides
and Esters 219
10.8 Step-Growth Polymerizations 219

Summary 222
Key Reactions 223
Problems 223

CHEMICAL CONNECTIONS
10A The Pyrethrins: Natural Insecticides of
Plant Origin 210
10B The Penicillins and Cephalosporins: β-Lactam
Antibiotics 211

10C From Willow Bark to Aspirin—and Beyond 214
10D Ultraviolet Sunscreens and Sunblocks 216
10E Barbiturates 217
10F Stitches That Dissolve 221

PART 2
BIOCHEMISTRY 229

CHAPTER 11
Carbohydrates 230

11.1 Introduction 230
11.2 Monosaccharides 231
11.3 The Cyclic Structure of Monosaccharides 235
11.4 Reactions of Monosaccharides 240
11.5 Disaccharides and Oligosaccharides 245
11.6 Polysaccharides 249
11.7 Acidic Polysaccharides 250

Summary 252
Key Reactions 253
Problems 254

CHEMICAL CONNECTIONS
11A Galactosemia 235
11B L-Ascorbic Acid (Vitamin C) 237
11C Testing for Glucose 243
11D A, B, AB, and O Blood Types 246
11E High-Fructose Corn Syrup 249

CHAPTER 12
Lipids 258

12.1 Introduction 258
12.2 The Structure of Triglycerides 259
12.3 Properties of Triglycerides 261
12.4 Complex Lipids 264
12.5 Membranes 265
12.6 Glycerophospholipids 267
12.7 Sphingolipids 269
12.8 Glycolipids 270
12.9 Steroids 273
12.10 Steroid Hormones 277
12.11 Bile Salts 281
12.12 Prostaglandins, Thromboxanes, and Leukotrienes 282

Summary 285
Problems 285

CHEMICAL CONNECTIONS
12A Rancidity 261
12B Waxes 262
12C Soaps and Detergents 264
12D Transport Across Cell Membranes 266
12E The Myelin Sheath and Multiple Sclerosis 270
12F Lipid Storage Diseases 272
12G Anabolic Steroids 278
12H Oral Contraception 281
12I Action of Anti-inflammatory Drugs 282

CHAPTER 13
Proteins 289

13.1 Introduction 289
13.2 Amino Acids 290
13.3 Zwitterions 292
13.4 Cysteine: A Special Amino Acid 294
13.5 Peptides and Proteins 294
13.6 Some Properties of Peptides and Proteins 296
13.7 The Primary Structure of Proteins 298
13.8 The Secondary Structure of Proteins 301
13.9 The Tertiary and Quaternary Structures of Proteins 304
13.10 Glycoproteins 310
13.11 Denaturation 310

Summary 313
Problems 313

CHEMICAL CONNECTIONS
13A Glutathione 296
13B AGE and Aging 298
13C The Use of Human Insulin 300
13D Sickle Cell Anemia 301
13E Protein/Peptide Conformation–Dependent Diseases 303
13F Proteomics, Ahoy! 307
13G The Role of Quaternary Structures in Mechanical Stress and Strain 308
13H Laser Surgery and Protein Denaturation 312

CHAPTER 14
Enzymes 317

14.1 Introduction 317
14.2 Naming and Classifying Enzymes 319
14.3 Common Terms in Enzyme Chemistry 320
14.4 Factors Affecting Enzyme Activity 321
14.5 Mechanism of Enzyme Action 323
14.6 Enzyme Regulation 327
14.7 Enzymes in Medical Diagnosis and Treatment 330

Summary 331
Problems 332

CHEMICAL CONNECTIONS
14A Muscle Relaxants and Enzyme Specificity 319
14B Acidic Environment and *Helicobacter* 223
14C Active Sites 325
14D Sulfa Drugs as Competitive Inhibitors 326
14E Noncompetitive Inhibition and Heavy Metal Poisoning 327
14F A Zymogen and Hypertension 328
14G One Enzyme, Two Functions 330

Lester Lefkowitz/Tony Stone Worldwide

CHAPTER 15
Chemical Communications:
Neurotransmitters and Hormones 336

15.1 Introduction 336
15.2 Chemical Messengers: Neurotransmitters and Hormones 338
15.3 Cholinergic Messenger: Acetylcholine 341
15.4 Amino Acids as Messengers: Neurotransmitters 346
15.5 Adrenergic Messengers: Neurotransmitters/ Hormones 347
15.6 Peptidergic Messengers: Neurotransmitters/ Hormones 353
15.7 Steroid Messengers: Hormones 356

Summary 357
Problems 357

CHEMICAL CONNECTIONS
15A Calcium as a Signaling Agent (Secondary Messenger) 342
15B Nerve Gases and Antidotes 343
15C Botulism and Acetylcholine Release 344
15D Alzheimer's Disease and Acetylcholine Transferase 345
15E Parkinson's Disease: Depletion of Dopamine 350
15F Nitric Oxide as a Secondary Messenger 352
15G Diabetes 354
15H Tamoxifen and Breast Cancer 356

CHAPTER 16
Nucleotides, Nucleic Acids, and Heredity 360

16.1 Introduction 360
16.2 Components of Nucleic Acids 361
16.3 Structure of DNA and RNA 365
16.4 RNA 371
16.5 Genes, Exons, and Introns 372
16.6 DNA Replication 373
16.7 DNA Repair 377
16.8 Cloning 380

Summary 381
Problems 382

CHEMICAL CONNECTIONS
16A Anticancer Drugs 365
16B Telomers, Telomerase, and Immortality 374
16C DNA Fingerprinting 375
16D Pharmacogenomics: Tailoring Medication to an Individual's Predisposition 377
16E Apoptosis: Programmed Cell Death 379

CHAPTER 17
Gene Expression and Protein Synthesis 384

17.1 Introduction 384
17.2 Transcription 385
17.3 The Role of RNA in Translation 388
17.4 The Genetic Code 389
17.5 Translation and Protein Synthesis 390
17.6 Gene Regulation 397
17.7 Mutations, Mutagens, and Genetic Diseases 400
17.8 Recombinant DNA 403

Summary 406
Problems 407

CHEMICAL CONNECTIONS
17A "Antisense" Makes Sense 387
17B Viruses 394
17C AIDS 395
17D The Promoter: A Case of Targeted Expression 398
17E Mutations and Biochemical Evolution 401
17F Oncogenes 402
17G p53: A Central Tumor Suppressor Protein 402

CHAPTER 18
Bioenergetics: How the Body Converts Food to Energy 409

18.1 Introduction 409
18.2 Cells and Mitochondria 411
18.3 The Principal Compounds of the Common Catabolic Pathway 413
18.4 The Citric Acid Cycle 417
18.5 Electron and H^+ Transport 420
18.6 Phosphorylation and the Chemiosmotic Pump 422
18.7 The Energy Yield 424
18.8 Conversion of Chemical Energy to Other Forms of Energy 425

Summary 426
Problems 427

CHEMICAL CONNECTIONS
18A The Importance of Iron–Sulfur Cluster Assembly 421
18B 2,4-Dinitrophenol as an Uncoupling Agent 422
18C Protection Against Oxidative Damage 423

CHAPTER 19
Specific Catabolic Pathways: Carbohydrate, Lipid, and Protein Metabolism 430

19.1 Introduction 430
19.2 Glycolysis 431
19.3 The Energy Yield from Glucose 436
19.4 Glycerol Catabolism 438
19.5 β-Oxidation of Fatty Acids 438
19.6 The Energy Yield from Stearic Acid 439
19.7 Ketone Bodies 441

Douglas Brown

19.8 Catabolism of the Nitrogen of Amino
Acids 443
19.9 Catabolism of the Carbon Skeleton of Amino
Acids 448
19.10 Catabolism of Heme 450

Summary 451
Problems 451

CHEMICAL CONNECTIONS

19A Lactate Accumulation 435
19B Effects of Signal Transduction on Metabolism 438
19C Ketoacidosis in Diabetes 443
19D Lipomics 443
19E Ubiquitin and Protein Targeting 446
19F Hereditary Defects in Amino Acid Catabolism: PKU 449
19G Jaundice 450

CHAPTER 20
Biosynthetic Pathways 455

20.1 Introduction 455
20.2 Biosynthesis of Carbohydrates 456
20.3 Biosynthesis of Fatty Acids 460
20.4 Biosynthesis of Membrane Lipids 461
20.5 Biosynthesis of Amino Acids 464

Summary 465
Problems 466

CHEMICAL CONNECTIONS

20A Photosynthesis 458
20B Prenylation of Ras Protein and Cancer 464
20C Amino Acid Transport and Blue Diaper Syndrome 465

CHAPTER 21
Nutrition 468

21.1 Introduction 468
21.2 Nutrition 468
21.3 Calories 471
21.4 Carbohydrates in the Diet and Their Digestion 472
21.5 Fats in the Diet and Their Digestion 473
21.6 Proteins in the Diet and Their Digestion 474
21.7 Vitamins, Minerals, and Water 475

Summary 483
Problems 483

CHEMICAL CONNECTIONS

21A Parenteral Nutrition 469
21B Too Much Vitamin Intake? 480
21C Dieting and Artificial Sweeteners 480
21D Food for Performance Enhancement 482

CHAPTER 22
Immunochemistry 486

22.1 Introduction 486
22.2 Organs and Cells of the Immune System 489

22.3 Antigens and Their Presentation by Major
Histocompatibility Complex 491
22.4 Immunoglobulin 493
22.5 T Cells and Their Receptors 498
22.6 Control of Immune Response: Cytokines 500
22.7 How to Recognize "Self" 501

Summary 504
Problems 505

CHEMICAL CONNECTIONS

22A Antibodies and Cancer
Therapy 497
22B Immunization 499
22C Mobilization of Leukocytes:
Tight Squeeze 500
22D Myasthenia Gravis: An
Autoimmune Disease 503
22E Glycomics 504

CHAPTER 23
Body Fluids 508

23.1 Introduction 508
23.2 Functions and
Composition of
Blood 509
23.3 Blood as a Carrier of
Oxygen 513
23.4 Transport of Carbon Dioxide
in the Blood 515
23.5 Blood Cleansing: A Kidney Function 515
23.6 Buffer Production: Another Kidney
Function 518
23.7 Water and Salt Balance in Blood
and Kidneys 518
23.8 Blood Pressure 519

Summary 520
Problems 521

CHEMICAL CONNECTIONS

23A Using the Blood–Brain Barrier to Eliminate
Undesirable Side Effects of Drugs 509
23B Blood Clotting 512
23C Breathing and Dalton's Law 514
23D Sex Hormones and Old Age 519
23E Hypertension and Its Control 520

Appendix I
Exponential Notation A-1
Appendix II
Significant Figures A-5
Answers to In-Text and Odd-Numbered
End-of-Chapter Problems A-8
Glossary G-1
Credits C-1
Index I-1

NSF Oasis Project/Norbett Wu Photography

*To see the world in a grain of sand
And heaven in a wild flower
Hold infinity in the palm of your hand
And eternity in an hour.*

WILLIAM BLAKE,
AUGURIES OF INNOCENCE

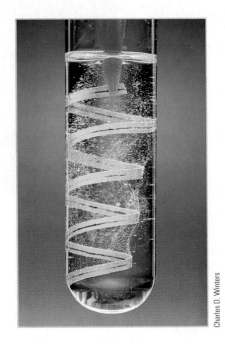

Charles D. Winters

PERCEIVING ORDER in the nature of the world is an ontological—not just pedagogical—need. It is our primary aim to convey the relationship among facts, thereby presenting a totality of the scientific edifice built over the centuries. In this process we marvel at the unity of laws that govern everything in the ever-exploding dimensions: from photons to protons, from hydrogen to water, from carbon to DNA, from genome to intelligence, from our planet to the galaxy and to the known universe. Unity in all diversity.

In writing the preface of the fifth edition of our textbook, we cannot help but ponder the paradigm shift that has occurred during the last 25 years. From the seventies slogan of "Better living through chemistry" to today's epitaph of "Life by chemistry," one is able to discern the change in focus. Chemistry not only helps to provide the amenities of good life but also lies at the heart of our concept and is the preoccupation of life itself. This shift in emphasis demands that our textbook, which is designed primarily for the education of future practitioners of health sciences, should attempt to provide both the basics and the scope of the horizon within which chemistry touches our life.

The increasing use of our textbook made this new edition possible, and we wish to thank our colleagues who adopted the previous editions for their courses. Testimony from colleagues and students indicates that we managed to convey our enthusiasm for the subject to students, who find this book great help in studying difficult concepts.

In the new edition, we strive further to present an easily readable and understandable text. At the same time, we emphasize the inclusion of new relevant concepts and examples in this fast-growing discipline, especially in the Biochemistry part. We maintain an integrated view of chemistry. From the very beginning in the Organic Chemistry part, we include biochemical substances to illustrate the principles. The progress is an ascension from the simple to the complex. We urge our colleagues to advance to the chapters of biochemistry as fast as possible, because there lies most of the material that is relevant to the future professions of our students.

Dealing with such a giant field in one course—and possibly the only course in which students get an exposure to chemistry—makes the selection of the material an overarching enterprise. Even though we tried to keep the book to a manageable size and proportion, we inevitably included more topics than could possibly be covered in a two-semester course. Our aim was to provide a variety of material, from which the instructor can select the topics he or she deems most important. We organized the sections so that each can stand independently; consequently, leaving out sections or even entire chapters will not cause fundamental cracks in the total edifice. We have increased the number of topics covered and provided 25 percent more new problems, many of them challenging and thought-provoking.

Audience

As with the previous editions, we developed this book for nonchemistry majors, mainly those entering health sciences and related fields, such as nursing, medical technology, physical therapy, and nutrition. Students in environmental studies can also benefit from it.

While teaching the chemistry of the human body is our ultimate goal, we try to show that each subsection of chemistry is important in its own right, besides being required for future understanding.

Chemical Connections (Medical and Other Applications of Chemical Principles)

The Chemical Connections features contain applications of the principles discussed in the text. Comments from users of the earlier editions indicate that these boxes have been especially well received and provide a much-requested relevance to the text. In the fifth edition, we have updated some of these features and added a number of new ones.

The majority of Chemical Connections boxes deal with health-related applications. Ones new to this edition focus on topics such as COX-2 inhibitors and their use in chemoprevention, amyloid structures in prion diseases including systematic amyloidosis and Alzheimer's disease, proteomics, zymogen and hypertension, botulinum toxin (Botox), phamacogenomics, "antisense drugs," four modes of combating AIDS, fusion inhibitors, iron–sulfur cluster assemblies, lipomics, structures of photosystems I and II, prenylation of Ras protein, potential anticancer vaccines, and glycomics.

The presence of these features allows for a considerable degree of flexibility. If an instructor wants to assign only the main text, the Chemical Connections do not interrupt continuity, and the essential material will be covered. However, because they enhance the core material, most instructors will probably wish to assign at least some of the Chemical Connections. In our experience, students are eager to read the relevant Chemical Connections, even without assignments, and they do so with discrimination. From such a large number of boxes, an instructor can judiciously select those that best fit the particular needs of the course. To enhance the material in the Chemical Connections, we provide problems at the end of each chapter covering all of the boxes.

Metabolism: Color Code

The biological functions of chemical compounds are explained in each of the biochemistry chapters and in many of the organic chapters. Throughout the book, emphasis is placed in chemistry rather than physiology. In the past, we have received much positive feedback regarding the way in which we have organized the topic of metabolism (Chapters 17, 18, and 19). We have maintained this organization in the current edition.

First we introduce the common metabolic pathway through which all food will be utilized (the citric acid cycle and oxidative phosphorylation). Only then do we discuss the specific pathways leading to the common pathway. We find this organization to be a useful pedagogic device, and it enables us to sum the caloric value of each type of food because its utilization through the common pathway has already been learned. Finally, we separate the catabolic pathways from the anabolic pathways by treating them in different chapters, emphasizing the different ways in which the body breaks down and builds up different molecules.

The topic of metabolism is a difficult one for most students, and we have tried to explain it as clearly as possible. As in the previous edition, we enhance the clarity of presentation by the use of a color code for the most important biological compounds discussed in Chapters 17, 18, and 19. Each

type of compound is screened in a specific color, which remains the same throughout the three chapters. These colors are as follows:

ATP and other nucleoside triphosphates

ADP and other nucleoside diphosphates

The oxidized coenzymes NAD$^+$ and FAD

The reduced coenzymes NADH and FADH$_2$

Acetyl coenzyme A

In figures showing metabolic pathways, we display the numbers of the various steps in yellow.

In addition to this major use of a color code, other figures in various parts of the book are color coded so that the same color is used for the same entity throughout. For example, in all Chapter 13 figures that show enzyme–substrate interactions, enzymes are always shown in blue and substrates in orange.

Features

Chemical Connections are a series of 100 in-depth and intriguing essays that describe applications of chemical concepts presented in the text. Examples include "Nitroglycerin, an Explosive and a Drug" (Chapter 5), "The Solubility of Drugs in Body Fluids" (Chapter 7), and "AGE and Aging" (Chapter 21).

Worked examples and corresponding problems are included throughout each chapter to help students develop sound problem-solving skills. Each of the 120 examples includes a detailed solution and is accompanied by a similar problem for students to try on their own.

 CD-ROM icons in the text refer to **Interactive General, Organic and Biochemistry CD-ROM, version 2.0,** which accompanies this book. Students are encouraged to use the CD for additional help or information whenever they see this icon.

Margin notes offer additional bits of information, such as historical notes, and reminders, which complement nearby text.

Molecular models, including ball-and-stick models, space-filling models, and electron density maps, appear throughout the text and serve as appropriate aids to visualizing molecular properties and interactions.

A dynamic art program enhances the book's clarity and makes the text visually exciting. A number of new annotated art and macro/micro art pieces have been added. Many new photographs have been added to the book as well, illustrating reactions, chemical procedures, and applications.

Margin definitions of many new terms reinforce key terminology.

Key Reactions, which appear at the ends of many organic chapters, are an annotated list of new reactions introduced in that chapter (Chapters 1–10), along with an example of the reaction and the section in which it is introduced. These lists help students to summarize the organic reactions they have learned.

End-of-chapter problems are paired to give students two tries at solving a particular problem type. Problem numbers that appear in color indicate difficult problems; ones marked with this tetrahedral icon are application-type problems. Problems are broken up by section topics, followed by an additional set of unclassified problems. Problems based on the Chemical Connections help to reinforce the applications of the chemical principles.

Chapter Summaries are section-by-section synopses of the key terms and concepts introduced in each chapter.

The Glossary at the end of the book gives a definition of each new term along with the number of the section in which it is introduced.

Answers to all in-text and odd-numbered end-of-chapter problems are provided at the end of the book. Detailed, worked-out solutions to these same problems are provided in the *Student Solutions Manual.*

Style

Feedback from colleagues and students alike indicates that one of the book's major assets is its style, which addresses students directly using simple and clear phrasing. In the fifth edition, we continue to make special efforts to provide clear and concise writing. Our hope is that this style will facilitate the understanding and absorption of difficult concepts.

Problems

Approximately 25 percent of the problems in this edition are new. The number of more challenging, thought-provoking questions has also been increased. The end-of-chapter problems are grouped and given subheads in order of topic coverage. Problem numbers in color specify more challenging ideas; those marked with a tetrahedral icon are application-type problems; and a group of problems are based on the Chemical Connections boxes. In the last group, called "Additional Problems," problems are not arranged in any specific order.

The answers to all in-text and odd-numbered end-of-chapter problems appear at the end of the book. Answers to the even-numbered end-of-chapter problems are supplied in the *Instructor's Manual.*

Ancillaries

The following ancillaries are available to qualified adopters. Please contact your local Thomson • Brooks/Cole sales representative for details.

- **Interactive General, Organic and Biochemistry CD-ROM, version 2.0,** is an unparalleled multimedia presentation of GOB topics. Created by William Vining of the University of Massachusetts, and Susan Young of Hartwick College, it features extensive use of videos and animations to create a strong link between observable macroscopic properties and molecular-scale chemical interpretations. It is supported by an interactive database of three-dimensional molecular models.

- **Student Study Guide,** by William Scovell of Bowling Green State University, includes reviews of chapter objectives, important terms and comparisons, focused reviews of concepts, and self-tests.

- **Student Solutions Manual,** by William Brown of Beloit College and Rodney Boyer of Hope College. This ancillary contains complete worked-out solutions to all in-text and end-of-chapter problems.

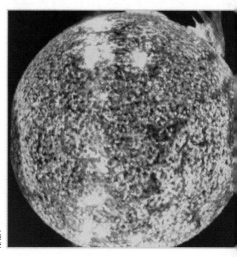

NASA

- **Instructor's Manual,** by William Brown of Beloit College and Rodney Boyer of Hope College, contains worked-out solutions to all even-numbered end-of-chapter problems. This manual also contians suggested course outlines.

- **InfoTrac® College Edition** Every new copy of *General, Organic and Biochemistry,* seventh edition, comes packaged with four months of free access to **InfoTrac College Edition.** This online resource features a comprehensive database of reliable, full-length articles (not abstracts) from thousands of top academic journals and popular sources—updated daily and spanning 22 years! Just some of the journals available 24 hours a day, seven days a week: *American Scientist, Science, Science News, Science Weekly,* and thousands more.

- **MyCourse 2.0** is our new, free, online course builder. Whether you want only the easy-to-use tools to build it or the content to furnish it, MyCourse 2.0 offers you a simple solution for creating a custom course Web site that allows you to assign, track, and report on student progress. Contact your Thomson•Brooks/Cole sales representative for details or visit **http://mycourse.thomsonlearning.com** for a free demo.

Charles D. Winters

- **Multimedia Manager for Organic Chemistry: A Microsoft® PowerPoint® Link Tool** is a digital library and presentation tool that is available on one convenient multiplatform CD-ROM. Thanks to its easy-to-use interface, you can readily take advantage of Brooks/Cole's presentations, which consist of art and tables from the text, as well as more than 700 PowerPoint presentation lecture slides especially created for this book by William E. Brown. The presentations are available in a variety of e-formats that are easily exported into presentation software or used on Web-based course support materials. You can even customize your own presentation by importing your personal lecture slides or any other material you choose. The result is an interactive and fluid lecture that truly engages your students.

- **ExamView® Computerized Testing** allows you to enhance your range of assessment and tutorial activities—and save yourself time in the process. You can create, deliver, and customize tests and study guides (both print and online) in minutes with this user-friendly assessment and tutorial system. **ExamView** offers both a *Quick Test Wizard* and an *Online Test Wizard* that guide you step-by-step through the process of creating tests. Using **ExamView**'s complete word processing capabilities, you can enter an unlimited number of new questions or edit existing questions. And **ExamView** is the only test generator that offers a "what-you-see-is-what-you-get" feature that allows you to see the test you are creating on the screen exactly as it will print! You can explore **ExamView** online at **http://www.examview.com/product_info/tour.htm**.

- **Printed Test Bank** contains approximately 50 multiple-choice questions per chapter for each of the 22 chapters in this text.

- **WebTutor™ Advantage on WebCT and Blackboard** Students have access to study tools that correspond chapter by chapter with the book, while professors can use **WebTutor Advantage** to provide virtual office hours, post syllabi, track student progress on the practice quizzes, and more. Visit **http://webtutor.thomsonlearning.com** for more information and to try WebTutor Advantage.

- **Overhead Transparencies** The 120 full-color overhead transparencies include figures and tables taken directly from the text.

- **Brooks/Cole Chemistry Book Companion Web Site** at **http://www.brookscole.com/chemistry** includes a rich array of teaching and learning tools for students and instructors. It includes online quizzes, tutorials, and more.

- **Laboratory Experiments for General, Organic & Biochemistry, fifth edition,** by Frederick A. Bettelheim and Joseph M. Landesberg, includes 52 experiments that illustrate important concepts and principles in general, organic, and biochemistry. Three new experiments are featured: (1) fermentation of a carbohydrate: ethanol from sucrose; (2) isolation of DNA from onions; and (3) neurotransmission: an example of enzyme specificity. In addition, many experiments have been revised. All experiments have new pre- and post-lab questions. The large number of experiments allows sufficient flexibility for the instructor.

- **Instructor's Manual to Accompany Laboratory Experiments** will help instructors in grading the answers to questions and in assessing the range of experimental results obtained by students. The *Instructor's Manual* also contains reminders for professors to pass on to students and details on how to handle disposal of waste chemicals.

Charles D. Winters

Acknowledgments

The publication of a book such as this one requires the efforts of many more people than merely the authors. We would like to thank the following professors who offered many valuable suggestions for this new edition:

JOHN T. BARBAS, Valdosta State University
MARK BENVENUTO, University of Detroit Mercy
VICKY BEVILACQUA, Kennesaw State University
ERWIN BOSCHMANN, Indiana University
SETH ELSHEIMER, University of Central Florida
MARK ERICKSON, Hartwick College
JOHN FOLKROD, University of Minnesota, Duluth
DONALD HARRISS, University of Minnesota, Duluth
JACK HEFLEY, Blinn College
JESSE W. JONES, Baylor University
LAURA KIBLER-HERZOG, Georgia State University
PAMELA MARKS, Arizona State University
THOMAS NALLI, Winona State University
ELVA MAE NICHOLSON, Eastern Michigan University
KIMBERLY PACHECO, University of Northern Colorado
DAVID REINHOLD, Western Michigan University
KAREN SCHUSTER, Florida Community College at Jacksonville
ERIC SIMANEK, Texas A&M University
BOBBY STANTON, University of Georgia
RICHARD TREPTOW, Chicago State University
BURTON TROPP, Queens College, CUNY

We thank John Holdcroft, Chemistry Editor, for promoting and facilitating the production of this edition. We have special thanks for Sandi Kiselica, our Senior Development Editor, who has been a rock of support through

Charles D. Winters

the entire revision process. We greatly appreciate her ability to set challenging but manageable schedules for us and her constant encouragement as we worked to meet those deadlines. Sandi has also been a valuable resource person with whom we could discuss everything from pedagogy to details of the art program. We also thank Nicole Barone of Thompson Steele, Inc. for her determination and iron-hand coordination of the production processes.

Organic Chemistry

Chapter 1, Organic Chemistry, is little changed except for an added Chemical Connections box on combinatorial chemistry.

In **Chapter 2, Alkanes,** we introduce the concept of line-angle formulas and continue using these formulas throughout the organic chapters (Chapters 2–10). They are easier to draw than the usual condensed structural formulas as well as easier to visualize. In previous editions, line-angle formulas were used only to represent cycloalkane and aromatic rings. A new Section 2.9B describes the halogenation of alkanes. The following section treats chlorofluorocarbons (the Freons) and their replacements, plus haloalkane solvents. A new Chemical Connections box describes the environmental impact of Freons. Section 2.5, covering the terpenes, has been shortened considerably. A new Chemical Connections box in **Chapter 3,** treats ethylene, a plant growth regulator.

The positions of Chapters 4 and 5 have been reversed from the sixth edition. **Chapter 4, Benzene and Its Derivatives,** now follows immediately after the treatment of alkenes and alkynes. We have expanded the discussion of phenols to include phenols and antioxidants. In this subsection, we treat auto-oxidation of unsaturated fatty acid hydrocarbon chains as a radical-chain mechanism leading to hydroperoxidation. We also discuss how substituted phenols such as BHT and BHA act as antioxidants. **Chapter 5, Alcohols, Ethers, and Thiols,** has been reorganized so that the material now discusses alcohols first (Sections 5.2 and 5.3), then ethers (Section 5.4), and finally thiols (Section 5.5).

In **Chapter 6, Chirality: The Handedness of Molecules,** the introduction of the concept of a stereocenter in Section 6.2 is done much more slowly and carefully using 2-butanol as a prototype. Section 6.4, which treats Molecules with Two or More Stereocenters, has been shortened to concentrate on recognizing a stereocenter by the placement of four different groups on a carbon. The relationship in cyclic compounds between cis-trans isomers and enantiomers has been reduced to a minimum.

Chapter 7, Acids and Bases, now introduces the use of curved arrows to show the flow of electrons in organic reactions. Specifically, we use them here to show the flow of electrons in proton-transfer reactions. We have dropped the discussion of normality and equivalents from the discussion of the stoichiometry of acid–base reactions. The major revision in this chapter is an expansion of the discussion of acid–base buffers and the Henderson-Hasselbalch equation.

In **Chapter 8, Amines,** a new Section 8.7, "Epinephrine: A Prototype for the Development of New Bronchodilators," traces the development of new asthma medications such as albuterol (Proventil) from epinephrine as a lead drug. To **Chapter 9, Aldehydes and Ketones,** we have added a discussion of $NaBH_4$ as a carbonyl-reducing agent with emphasis on it as a hydride transfer reagent. We then make the parallel to NADH as a carbonyl-reducing agent and as a hydride transfer agent.

To **Chapter 10, Carboxylic Acids, Anhydrides, Esters, and Amides,** we have added a brief discussion of the preparation of amides by treating a carboxylic acid with an amine (which gives a salt) and then heating this salt to eliminate water. A more useful preparation of amides is the

treatment of an amine with an anhydride. We have added two new Chemical Connections: "Pyrethrins—Natural Insecticides of Plant Origin," and "Ultraviolet Sunscreens and Sunblocks."

Biochemistry

The most significant revision in **Chapter 11, Carbohydrates,** is a new Chemical Connections box on high-fructose corn syrup.

In **Chapter 12, Lipids,** new topics include a description of potassium ion channels and the mode of transportation of K^+. The discussion of steroids in Section 12.9 has been reorganized and expanded. Detailed discussions on the structure, transport, and concentrations of the different lipoproteins have been added. The functions of steroidal and nonsteroidal anti-inflammatory agents are discussed, including the COX-2 inhibitors and their use in chemoprevention.

In **Chapter 13, Proteins,** we present an expanded Chemical Connections on protein conformation-dependent diseases that summarizes amyloid structures in prion-related diseases such as mad cow disease, Creutzfeld-Jacob disease, systematic amyloidosis, and Alzheimer's disease. New discussions focus on the quaternary structures of hemoglobin, collagen, and integral membrane proteins. A new Chemical Connections highlights the new discipline of proteomics and its function.

The discussion of enzymes now includes details of the chemistry of active sites in **Chapter 14, Enzymes.** In 65 percent of enzyme-catalyzed reactions, the active site operates with acid–base reactions. A new Chemical Connections, "Zymogen and Hypertension," describes the specificity of different ACE inhibitors.

New features in **Chapter 15, Chemical Communications: Neurotransmistters and Hormones,** are the updating of the list of drugs affecting neurotransmission, inclusion of the role of the phosphatidylinositol phosphates as secondary messengers, and the recognition of neurosteroids synthesized in the brain. New Chemical Connections describe the use of the botulinum toxin (Botox) in the cosmetics industry and the neurodegenerative role played by beta amyloid versus tau protein in Alzheimer's disease.

The ever-increasing developments in molecular biology necessitated rewriting of **Chapter 16, Nucleotides, Nucleic Acids, and Heredity.** It now inludes a new section on quaternary and higher structures of DNA leading to the superstructures of chromosomes. The section on DNA replication now includes material on the opening up of superstructures and the components of replisomes. A new section describes the different mechanisms of DNA repair. A new Chemical Connections feature deals with phamacogenomics' ability to tailor medication to an individual's predisposition.

Chapter 17, Gene Expression and Protein Synthesis, underwent major reorganization and expansion. We now concentrate on gene expression and protein synthesis as they operate in eukaryotes. Gene regulation has been expanded to cover different entries on the transcriptional, translational, and post-translational levels. New Chemical Connections cover antisense drugs, four modes of combating AIDS including the new "fusion inhibitors," and the role of the tumor suppression protein p53 in aging.

Chapter 18, Bioenergetics: How the Body Converts Food to Energy, has a new Chemical Connections on iron–sulfur cluster assemblies.

The only new feature in **Chapter 19, Specific Catabolic Pathways: Carbohydrate, Lipid, and Protein Metabolism,** is a Chemical Connections on lipomics, the collective lipid components of a cell.

In **Chapter 20, Biosynthetic Pathways,** there is extended coverage of photosynthesis, including the newly discovered structures of photosystems I and II. Stepwise synthesis of cholesterol up to the C10 and C15 units are detailed. A new Chemical Connections features prenylation of the Ras protein and its effect on signal transduction.

Chapter 21, Nutrition, has been reorganized to present separate sections on carbohydrates in diet and their digestion and similarly on fats and proteins. References to the latest Dietary Reference Intakes are included as well as discussion of current fad diets such as the Atkins diet and raw food diet.

Chapter 22, Immunochemistry, has been expanded to include a new section on how to recognize "self," which includes coverage of autoimmune diseases and their treatment. The existing sections were subdivided to facilitate the comprehension of difficult concepts. A new subsection was written to explain how the body acquires the diversity to react to different antigens. Among the Chemical Connections is a new feature on potential anticancer vaccines and on glycomics (a distant cousin of genomics and proteomics).

In **Chapter 23, Body Fluids,** we moved the Chemical Connections "Breathing and Dalton's Law" here from Chapter 5 to provide a better survey of gas transports in the blood.

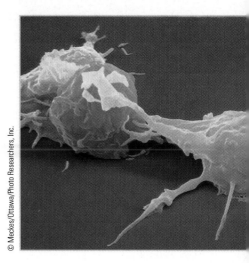

© Meckes/Ottawa/Photo Researchers, Inc.

Key:

ChemConn = Chemical Connections Box number
Sect. = Section number
Prob. = Problem number

Acid and base burns to the cornea	ChemConn 7B
Acidic polysaccharides in the body	Sect. 11.8
Acidosis and Alkalosis	ChemConn 7D
Active sites	ChemConn 14C
AGE and aging	ChemConn 13B
AIDS	ChemConn 17C
Albuterol (Proventil)	Sect. 8.7
Alkaloids	ChemConn 8B
Alzheimer's disease	ChemConn 23D,
	ChemConn 15D
Amphetamines	ChemConn 8A
Anabolic steroids	ChemConn 12G
Angiotensin	ChemConn 14F
Antacids	ChemConn 7C
Antibodies and cancer therapy	ChemConn 22A
Anticancer drugs	ChemConn 15A
Antidepressants	Sect. 15.5F
Antidepressants	Prob. 6.33
Antigens	Sect. 22.3
Antihistamines	Sect. 15.5G
Anti-inflammatory drugs	ChemConn 12I
Antioxidants	Sect. 4.5C
Antisense drugs	ChemConn 17A
Apoptosis, programmed cell death	ChemConn 16E
Artificial sweeteners	ChemConn 21C
Ascorbic acid (vitamin C)	ChemConn 11B
Aspartame	Prob. 10.56
Aspirin and other NSAIDs	ChemConn 10C
Asthma	Sect. 12.12
Atherosclerosis: levels of LDL and HDL	Sect. 12.9E
Atropine	Prob. 8.50
Attention deficit disorder (ADD)	ChemConn 15F
Autoimmune diseases	Sect. 22.7C
Autoxidation	Sect. 4.5C
B cells	Sect. 22.2C
Barbiturates	ChemConn 10E
Basal caloric requirement	Sect. 21.3
Basic excision repair (BER) of DNA	Sect. 16.7
BHT, an antioxidant in foods	Sect. 4.5C
Bile salts	Sect. 12.11
Blood buffers	Sect. 7.11D
Blood cleansing	Sect. 23.5
Blood clotting	ChemConn 23B
Blood pressure	Sect. 23.8
Blood types	ChemConn 11D
Blue diaper syndrome	ChemConn 20C
Botox	ChemConn 15C
Botulism	ChemConn 15C
Breath-alcohol screening	ChemConn 5B
Breathing and Dalton's law	ChemConn 23C
Bronchodilators and asthma	Sect. 8.7
Brown fat and hybernation	ChemConn 18B
Calcium, as a signaling agent	ChemConn 15A
Capsaicin	ChemConn 4F
Carbohydrates as signals of virulence and malignance	ChemConn 22E
Carcinogenic PAHs and smoking	ChemConn 4B
Carcinogens	ChemConn 17E
β-Carotene	Prob. 3.60
Cephalosporins	ChemConn 10B
Chiral drugs	ChemConn 6A
Chirality, in biomolecules	Sect. 6.6A
Cholera	Sect. 15.5D
Cis-trans isomerism, in vision	ChemConn 3C
Cocaine	ChemConn 8B
Cocaine addiction	ChemConn 15C
Coniine	ChemConn 8B
Cox-2 inhibitor drugs	ChemConn 12I
Creatine, performance enhancement	ChemConn 21D
Cystic fibrosis	ChemConn 19E
Cytochrome P-450, in detoxification	ChemConn 18C
Cytokines	Sect. 22.6
DDT	ChemConn 4A
Diabetes	ChemConn 15G
Dichloroacetic acid	Sect. 10.2D
Dietary Reference Intakes (DRI)	Sect. 21.2
Diets	Sect. 21.2
2,4-Dinitrophenol, as an uncoupling agent	ChemConn 18B
DNA fingerprinting	ChemConn 16C
Enzymes, in diagnosis	Sect. 14.7
Enzymes, in therapy	Sect. 14.7
Ephedrine	Prob. 6.26
Epibatadine	Prob. 8.51
Epinephrine	Sect. 8.7
Esters, as flavoring agents	Prob. 10.47

Ethers and anesthesia — ChemConn 5D

Ethylene, a plant growth regulator — ChemConn 3A

Ethylene oxide, a chemical sterilant — ChemConn 5C

Fluid mosaic model of membranes — Sect. 12.5

Freons — ChemConn 2B

Galactosemia — ChemConn 11A

Gallstones — Sect. 12.9A

Gene therapy — Sect. 17.8

Genetic code — Sect. 17.4

Genetic engineering — Sect. 17.8

Glutathione — ChemConn 13A

G-protein/cAMP cascade — Sect. 15.5C

Heart enzymes — Sect. 14.7

Heavy Metal poisoning — ChemConn 14E, Sect. 13.10

Helicobacter — ChemConn 14B

Heme products, in bruises — Sect. 19.10

Heparin — Sect. 11.7B

High-fructose corn syrup — ChemConn 11E

Hormones — Sect. 15.2

Human insulin — ChemConn 13C

Hyaluronic acid — Sect. 11.8A

Hypertension and its control — ChemConn 23E

Hypoglycemic awareness — Sect. 13C

Ibuprofen — Sect. 6.3

Immune system — Sect. 22.1B

Immunization — ChemConn 22B

Immunoglobulin — Sect. 22.4

Innate immunity — Sect. 22.1A

Insulin — Sect. 13.7

Iodide ion and goiter — ChemConn 4C

Iron–sulfur cluster assembly — ChemConn 18A

Jaundice — ChemConn 19G

Ketoacidosis, in diabetes — ChemConn 19C

Ketone bodies — Sect. 19.7

Lactate accumulation — ChemConn 19A

Laser in situ keratomileusis (LASIK) — ChemConn 13H

Laser surgery and protein denaturation — ChemConn 13H

Lipid storage diseases — ChemConn 12F

Lycopene — Prob. 3.61

Mad cow disease — ChemConn 13E

Menstrual cycle — Sect. 12.10B

Mobilization of leukocytes — ChemConn 22C

Monoclonal antibodies — Sect. 22.4C

Morphine and enkephalins — Sect. 15.6

Mucins — Sect. 13.10

Multiple sclerosis — ChemConn 12E

Muscle relaxants — ChemConn 14A, Sect. 15.3E

Mutagens — ChemConn 17E

Mutations and biochemical evolution — ChemConn 17E

Myasthenia Gravis — ChemConn 22D

Naproxen — Sect. 6.3

Nerve gases and antidotes — ChemConn 15B

Neurotransmitters — Sect. 15.2

Nicotine — ChemConn 8B

Nitric oxide — ChemConn 3C, 15F

Nitroglycerin, an explosive and a drug — ChemConn 5A

Oncogenes — ChemConn 17F

Oral contraception — ChemConn 12H

Oxidative damage — ChemConn 18C

p53, a central tumor suppressor protein — ChemConn 17G

Parenteral Nutrition — ChemConn 21A

Parkinson's disease — ChemConn 15E

Penicillins — ChemConn 10B

pH of some common materials — Sect. 7.8

Phenols, as antioxidants — Sect. 4.5C

Phenylcyclidine (PCP) — Sect. 15.4B

Phenylketonurea (PKU) — ChemConn 19F

Photorefractive keratectomy (PRK) — ChemConn 13H

Photosynthesis — ChemConn 20A

Poison ivy — Sect. 4.5A

Poisonous puffer fish — ChemConn 2A

Polynuclear aromatic hydrocarbons (PAHs) — Sect. 4.3D

Prenylation of Ras protein and cancer — ChemConn 20B

Promoter, a case of targeted expression — ChemConn 17D

Propofol — Prob. 5.36

Prostaglandin endoperoxide synthase, two functions — ChemConn 14G

Protein/peptide conformation-dependent diseases — ChemConn 13E

Proteoglycans of extracellular matrix — Sect. 13.10

Proteomics — ChemConn 13F

Proventil — Sect. 8.7

Pyrethrins — ChemConn 10A

Quaternary structures, in mechanical stress — ChemConn 13F

Rancidity — ChemConn 12A

Recommended Daily Allowances (RDA) — Sect. 21.2

Relative sweetness — Sect. 11.5D

Senile systematic amyloidosis — ChemConn 13E

Sex hormones and old age — ChemConn 23D

Sickle cell anemia — ChemConn 13D

Side effects of drugs — ChemConn 23A

Sideroblastic anemia — ChemConn 18A

Signal transduction   ChemConn 19B
Smoking and carcinogenesis   ChemConn 4B
Solubility of drugs in body fluids   ChemConn 8D
Statins   Sect. 20.4
Stitches that dissolve   ChemConn 10F
Sulfa drugs   ChemConn 22D,
                                            ChemConn 14D
Sunscreens and sunblocks   ChemConn 10D
Synthetic food dyes   ChemConn 4E

T cells   Sect. 22.2C
Tailoring medication to an   ChemConn 16D
    individual's predisposition
Tamoxifen and breast cancer   ChemConn 15H
Taxol   ChemConn 1A
Telomers and immortality   ChemConn 24B,
                                            ChemConn 16B

Testing for glucose   ChemConn 10C
Tetrodotoxin   ChemConn 2A
Tranquilizers   ChemConn 8C
Transport across cell membranes   ChemConn 12D

Ubiquitin and protein targeting   ChemConn 19E

Viagra and blood vessel dilation   ChemConn 15F
Viruses   ChemConn 17B
Vitamin A and vision   ChemConn 3C
Vitamin excess   ChemConn 21B
Vitamins in diet   Sect. 21.7

Xeroderma pigmentosa   Sect. 16.7

Zymogen and hypertension   ChemConn 14F

Christy Carter/Brand Hollman Photography, Inc.

PART 1 ORGANIC CHEMISTRY

Chapter 1
Organic Chemistry

Chapter 2
Alkanes

Chapter 3
Alkenes

Chapter 4
Benzene and Its Derivatives

Chapter 5
Alcohols, Ethers, and Thiols

Chapter 6
Chirality: The Handedness
of Molecules

Chapter 7
Acids and Bases

Chapter 8
Amines

Chapter 9
Aldehydes and Ketones

Chapter 10
Carboxylic Acids, Anhydrides,
Esters, and Amides

CHAPTER **1**

1.1   *Introduction*

1.2   *Sources of Organic Compounds*

1.3   *Structure of Organic Compounds*

1.4   *Functional Groups*

The bark of the Pacific yew contains paclitaxel, a substance that has proved effective in treating certain types of ovarian and breast cancer.

Tom & Pat Leeson/Photo Researchers Inc.

Organic Chemistry

1.1 Introduction

Organic chemistry is the chemistry of the compounds of carbon. As you study Chapters 1–10 (organic chemistry) and 11–31 (biochemistry), you will see that organic compounds are everywhere around us. They are in our foods, flavors, and fragrances; in our medicines, toiletries, and cosmetics; in our plastics, films, fibers, and resins; in our paints, varnishes, and glues; and, of course, in our bodies and the bodies of all other living organisms.

Perhaps the most remarkable feature of organic compounds is that they involve the chemistry of carbon and only a few other elements—chiefly, hydrogen, oxygen, and nitrogen. While the majority of organic compounds contain carbon and just these three elements, many also contain sulfur, a halogen (fluorine, chlorine, bromine, or iodine), or phosphorus.

When this book was written, there were 114 known elements. Organic chemistry concentrates on carbon, just one of the 114. The chemistry of the other 113 elements comes under the field of inorganic chemistry. As we see in Figure 1.1, carbon is far from being among the most abundant elements in Earth's crust. In terms of elemental abundance, approximately 75% of

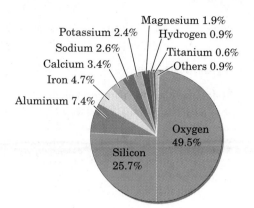

Figure 1.1 Abundance of the elements in the Earth's crust.

Earth's crust is composed of just two elements: oxygen and silicon. These two elements are the components of silicate minerals, clays, and sand. In fact, carbon is not even among the ten most abundant elements. Instead, it is merely one of the components making up the remaining 0.9% of Earth's crust. Why, then, do we pay this special attention to just one element from among 114?

The first reason is largely historical. In the early days of chemistry, scientists thought organic compounds were those produced by living organisms, and that inorganic compounds were those found in rocks and other nonliving matter. At the time, they believed that a "vital force," possessed only by living organisms, was necessary to produce organic compounds. In other words, chemists believed that they could not synthesize any organic compound starting with only inorganic compounds. This theory was very easy to disprove if, indeed, it was wrong: It required only one experiment in which an organic compound was made from inorganic compounds. In 1828, Friedrich Wöhler (1800–1882) carried out such an experiment. He heated an aqueous solution of ammonium chloride and silver cyanate, both inorganic compounds and, to his surprise, obtained urea, an "organic" compound found in urine.

Organic and inorganic compounds differ in properties because they differ in structure, not because they obey different natural laws. One set of natural laws applies to all compounds.

$$NH_4Cl + AgNCO \xrightarrow{\text{heat}} H_2N-\overset{\overset{\displaystyle O}{\|}}{C}-NH_2 + AgCl$$

Ammonium   Silver   Urea   Silver
chloride   cyanate   chloride

Although this single experiment of Wöhler's sufficed to disprove the "doctrine of vital force," it took several years and a number of additional experiments for the entire scientific community to accept the fact that organic compounds could be produced in the laboratory. This discovery meant that the terms "organic" and "inorganic" no longer had their original meaning because, as Wöhler demonstrated, organic compounds could be obtained from inorganic materials. A few years later, August Kekulé (1829–1896) put forth a new definition—organic compounds are those containing carbon— and his definition has been accepted ever since.

A second reason for the study of carbon compounds as a separate discipline is the sheer number of organic compounds. Chemists have discovered or made more than 10 million of them, and an estimated 10,000 new ones are discovered or prepared in the laboratory each year. By comparison,

Table 1.1   A Comparison of Properties of Organic and Inorganic Compounds

Organic Compounds	Inorganic Compounds
Bonding is almost entirely covalent.	Most have ionic bonds.
May be gases, liquids, or solids with low melting points (less than 360°C).	Most are solids with high melting points.
Most are insoluble in water.	Many are soluble in water.
Most are soluble in organic solvents such as diethyl ether, toluene, and dichloromethane.	Almost all are insoluble in organic solvents.
Aqueous solutions do not conduct electricity.	Aqueous solutions conduct electricity.
Almost all burn.	Very few burn.
Reactions are usually slow.	Reactions are often very fast.

chemists have discovered or made an estimated 1.7 million inorganic compounds. Thus approximately 85% of all known compounds are organic compounds.

A third reason—and one particularly important for those of you who plan to study biochemistry—is that carbohydrates, lipids, proteins, enzymes, nucleic acids (DNA and RNA), hormones, vitamins, and almost all other important chemicals in living systems are organic compounds. Furthermore, their reactions are often strikingly similar to those observed occurring in test tubes. For this reason, knowledge of organic chemistry is essential for an understanding of biochemistry.

One final point about organic compounds. They generally differ from inorganic compounds in many of their properties, some of which are shown in Table 1.1. Most of these differences stem from the fact that the bonding in organic compounds is almost entirely covalent, while most inorganic compounds have ionic bonds.

Of course, there are many exceptions to the generalizations in Table 1.1. Some organic compounds behave like inorganic compounds, and vice versa, but the generalizations remain largely true for the vast majority of compounds of both types.

1.2 Sources of Organic Compounds

Chemists obtain organic compounds in two principal ways: isolation from nature and synthesis in the laboratory.

A Isolation from Nature

■ **Sugar cane, Hawaii.**

D. E. Cox/Stone/Getty Images

Living organisms are chemical "factories." Each plant and animal—even microorganisms such as bacteria—makes thousands of organic compounds by a process called biosynthesis. One way, then, to get organic compounds is to extract them from biological sources. In this book, we will encounter many compounds that are or have been obtained in this way. Some important examples include vitamin E, the penicillins, table sugar, insulin, quinine, and the anticancer drug Taxol (see Chemical Connections 1A). Nature also supplies us with three other important sources of organic compounds: natural gas, petroleum, and coal; we discuss them in Section 2.11.

CHEMICAL CONNECTIONS 1A

Taxol—A Story of Search and Discovery

In the early 1960s, the National Cancer Institute undertook a program to analyze samples of native plant materials in the hope of discovering substances that would prove effective in the fight against cancer. Among the materials tested was an extract of the bark of the Pacific yew, *Taxus brevifolia,* a slow-growing tree found in the old-growth forests of the Pacific Northwest. This extract proved to be remarkably effective in treating certain types of ovarian and breast cancer, even in cases where other forms of chemotherapy failed. The structure of the cancer-fighting component of yew bark was determined in 1962, and the compound was named paclitaxel (Taxol).

Paclitaxel
(Taxol)

■ **Pacific yew bark being stripped for Taxol extraction.**

Unfortunately, the bark of a single 100-year-old tree yields only about 1 g of Taxol, not enough for effective treatment of even one cancer patient. Furthermore, getting Taxol means stripping the bark from trees, which kills them. In 1994, chemists succeeded in synthesizing Taxol in the laboratory, but the drug's cost was far too high to be economical. Fortunately, an alternative natural source of the drug was found. Researchers in France discovered that the needles of a related plant, *Taxus baccata,* contain a compound that can be converted to Taxol in the laboratory. Because the needles can be gathered without harming the plant, it is not necessary to kill trees to obtain the drug.

Taxol inhibits cell division by acting on microtubules, a key component of the scaffolding of cells. Before cell division can take place, the cell must disassemble these microtubule units, and Taxol prevents this disassembly. Because cancer cells are the fastest dividing cells, the drug effectively controls their spreading.

The remarkable success of Taxol in the treatment of breast and ovarian cancer has stimulated research efforts to discover and/or synthesize other substances that work the same way in the body and that may be even more effective anticancer agents than Taxol.

B Synthesis in the Laboratory

Ever since Wöhler synthesized urea, organic chemists have sought to develop more ways to make the same compounds found in nature. In recent years, the methods for doing so have become so sophisticated that there are few natural organic compounds, no matter how complicated, that chemists cannot synthesize in the laboratory.

Compounds made in the laboratory are identical to those found in nature, assuming, of course, that each is 100% pure. There is no way that anyone can tell whether a sample of a compound was made by chemists or obtained from nature. If, however, a compound is not 100% pure, we can usually tell from the identity of the impurities whether the compound was isolated from nature or made in a laboratory. As a consequence, pure ethanol made by chemists has exactly the same physical and chemical properties as the pure ethanol prepared by distilling wine. The same is true for

■ **Vitamin C in an orange is identical to its synthetic tablet form.**

ascorbic acid (vitamin C). There is no advantage, therefore, in paying more money for vitamin C obtained from a natural source than for synthetic vitamin C, because the two are identical in every way.

Organic chemists, however, have not been satisfied with just duplicating nature's compounds. They have also synthesized compounds not found in nature. In fact, the majority of the more than 10 million known organic compounds are purely synthetic and do not exist in living organisms. For example, many modern drugs—Valium, Vasotec, Prozac, Zantac, Cardizem, Lasix, Viagra, and Enovid—are synthetic organic compounds not found in nature. Even the over-the-counter drugs aspirin and ibuprofen are synthetic organic compounds not found in nature.

1.3 Structure of Organic Compounds

A structural formula shows all the atoms present in a molecule as well as the bonds that connect the atoms to each other. The structural formula for ethanol, for example, shows all nine atoms and the eight bonds that connect them:

$$
\begin{array}{c}
\quad\;\; \text{H} \quad\; \text{H} \\
\quad\;\; | \qquad | \\
\text{H}-\text{C}-\text{C}-\text{O}-\text{H} \\
\quad\;\; | \qquad | \\
\quad\;\; \text{H} \quad\; \text{H}
\end{array}
$$

The Lewis model of bonding enables us to see how carbon forms four covalent bonds that may be various combinations of single, double, and triple bonds. Furthermore, the valence-shell electron-pair repulsion (VSEPR) model tells us that the most common bond angles about carbon atoms in covalent compounds are approximately 109.5°, 120°, and 180°.

CHEMICAL CONNECTIONS 1 B

Combinatorial Chemistry

Since the beginning of synthetic organic chemistry about 200 years ago, the goal has been to produce pure organic compounds with high percentage yields. More often than not, a particular use for a compound was discovered accidentally. Such was the case, for example, with aspirin and the barbiturates. Later, when researchers discovered the mode of action of certain drugs, it became possible to design new compounds that would have a reasonable probability of being more effective or having fewer side effects than some existing drug. Such was the case with ibuprofen and albuterol (see Section 8.7). Even in such a "rational" approach (as compared to the "shotgun" approach, in which compounds were synthesized at random), the goal was still to obtain a single pure compound, no matter how many steps were involved in the synthesis or how difficult the steps were. In the last few years this strategy has changed, mostly in pharmaceutical research.

Combinatorial chemistry is a process in which chemists try to produce as many compounds from as few building blocks as possible. For example, if we allow 20 compounds (building blocks) to react at random, and could find a method to restrict the combination in such a way that only molecules containing three of these building blocks are produced, then a mixture of 8000 different compounds (20^3) would result. Perhaps this mixture might include two or three candidates with the potential to be useful drugs. The task, then, is to select from this synthetic soup those compounds that promise to be biologically active. Researchers accomplish this goal by screening the mixture for a desired biological activity and then separating promising candidates for further study. In essence, it is cheaper and faster to produce a large "library" of combinatorial synthetic compounds and then "fish out" the biologically active ones than to pursue the old way of randomly synthesizing new compounds and screening them one at a time.

Table 1.2 Single, Double, and Triple Bonds in Compounds of Carbon (bond angles are predicted using the VSEPR model)

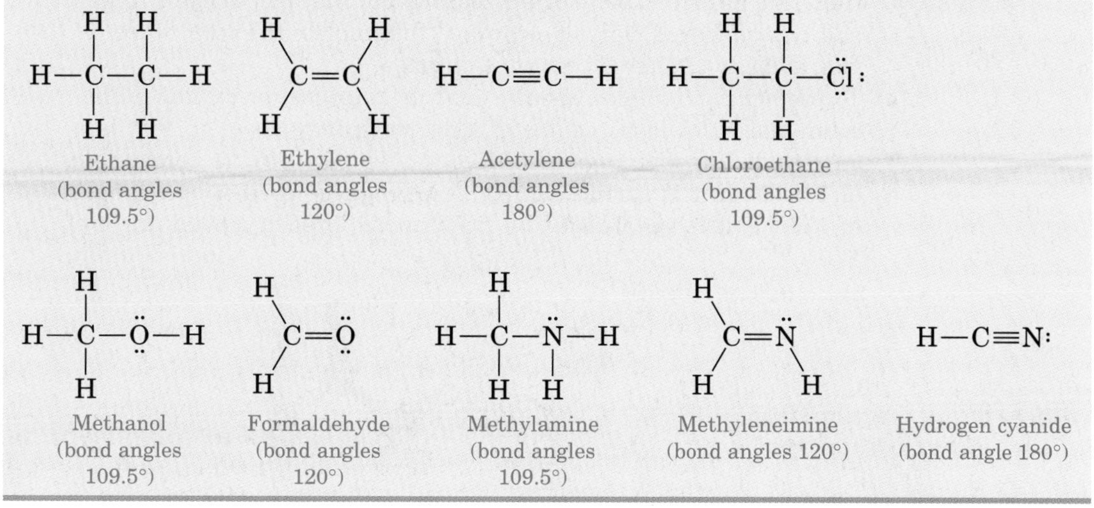

Ethane
(bond angles
109.5°)

Ethylene
(bond angles
120°)

Acetylene
(bond angles
180°)

Chloroethane
(bond angles
109.5°)

Methanol
(bond angles
109.5°)

Formaldehyde
(bond angles
120°)

Methylamine
(bond angles
109.5°)

Methyleneimine
(bond angles 120°)

Hydrogen cyanide
(bond angle 180°)

Table 1.2 shows several covalent compounds containing carbon bonded to hydrogen, oxygen, nitrogen, and chlorine. From these examples, we see the following:

- Carbon normally forms four covalent bonds and has no unshared pairs of electrons.
- Nitrogen normally forms three covalent bonds and has one unshared pair of electrons.
- Oxygen normally forms two covalent bonds and has two unshared pairs of electrons.
- Hydrogen forms one covalent bond and has no unshared pairs of electrons.
- Chlorine (and fluorine, bromine, and iodine) normally forms one covalent bond and has three unshared pairs of electrons.

EXAMPLE 1.1

The structural formulas for acetic acid, CH_3COOH, and ethylamine, $CH_3CH_2NH_2$, are

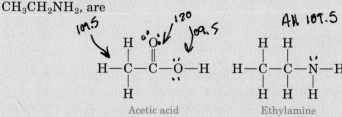

Acetic acid   Ethylamine

(a) Complete the Lewis structure for each molecule by adding unshared pairs of electrons so that each atom of carbon, oxygen, and nitrogen has a complete octet.
(b) Using the VSEPR model, predict all bond angles in each molecule.

Solution

(a) Each carbon atom is surrounded by eight valence electrons and therefore has a complete octet. To complete the octet of each oxygen, add two unshared pairs of electrons. To complete the octet of nitrogen, add one unshared pair of electrons.

(b) To predict bond angles about a carbon, nitrogen, or oxygen atom, count the number of regions of electron density about it. If it is surrounded by four regions of electron density, predict bond angles of 109.5°. If it is surrounded by three regions, predict bond angles of 120°. If it is surrounded by two regions, predict a bond angle of 180°.

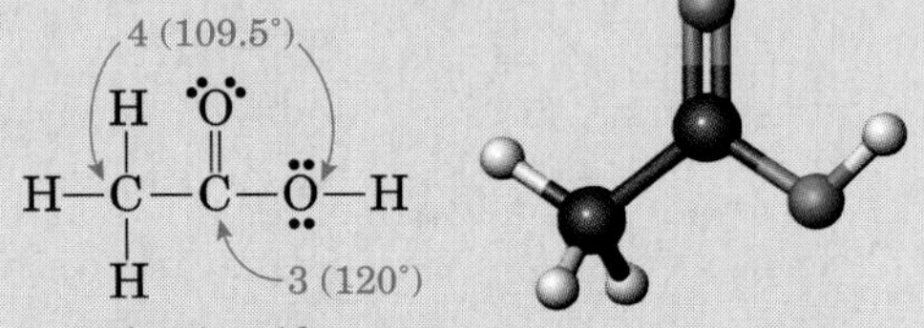

Acetic acid

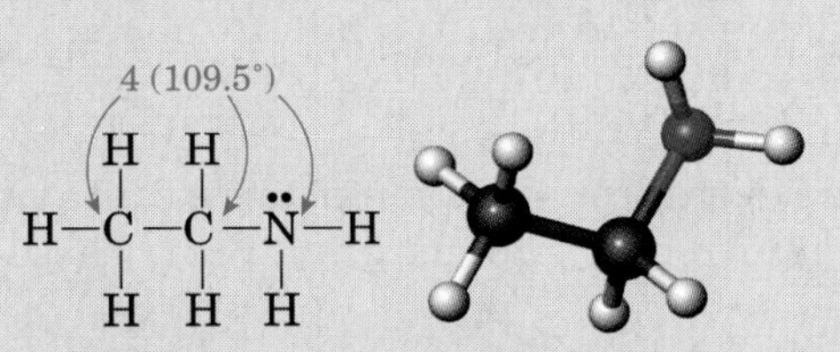

Ethylamine

Problem 1.1

The structural formulas for ethanol, CH_3CH_2OH, and propene, $CH_3CH=CH_2$, are

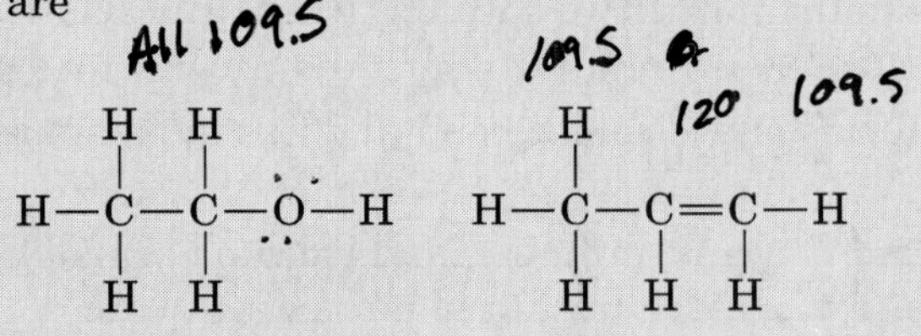

All 109.5

Ethanol

109.5 120 109.5

Propene

(a) Complete the Lewis structure for each molecule showing all valence electrons.

(b) Using the VSEPR model, predict all bond angles in each molecule.

1.4 Functional Groups

Functional group An atom or group of atoms within a molecule that shows a characteristic set of physical and chemical properties

As noted in Section 1.1, more than 10 million organic compounds have been discovered or made by organic chemists. It might seem an almost impossible task to learn the physical and chemical properties of so many compounds. Fortunately, the study of organic compounds is not as formidable a task as you might think. While organic compounds can undergo a wide variety of chemical reactions, only certain portions of their structures are changed in any particular reaction. We call the part of an organic molecule that undergoes chemical reactions a **functional group.** As we will see, the

Table 1.3 Five Common Functional Groups

Family	Functional Group	Example	Name
Alcohol	—OH	CH_3CH_2OH	Ethanol
Amine	—NH_2	$CH_3CH_2NH_2$	Ethanamine
Aldehyde	$\overset{\displaystyle O}{\overset{\|}{-C-H}}$	$CH_3\overset{\displaystyle O}{\overset{\|}{C}}H$	Ethanal
Ketone	$\overset{\displaystyle O}{\overset{\|}{-C-}}$	$CH_3\overset{\displaystyle O}{\overset{\|}{C}}CH_3$	Acetone
Carboxylic acid	$\overset{\displaystyle O}{\overset{\|}{-C-OH}}$	$CH_3\overset{\displaystyle O}{\overset{\|}{C}}OH$	Acetic acid

See the **Interactive General, Organic, and Biochemistry CD-ROM, version 2.0,** for further exploration on this topic.

same functional group, in whatever organic molecule it occurs, undergoes the same types of chemical reactions. Therefore, we do not have to study the chemical reactions of even a fraction of the 10 million known organic compounds. Instead, we need to identify only a few characteristic functional groups and then study the chemical reactions that each undergoes.

Functional groups are also important because they are the units by which we divide organic compounds into families of compounds. For example, we group those compounds that contain an —OH (hydroxyl) group bonded to a tetrahedral carbon into a family called alcohols; compounds containing a —COOH group belong to a family called carboxylic acids. Table 1.3 introduces five of the most common functional groups. A complete list of all functional groups we will study appears on the inside back cover of the text.

At this point, our concern is simply pattern recognition—that is, how to recognize one of these five functional groups when you see it, and how to draw structural formulas of molecules containing them. We will have more to say about the physical and chemical properties of these and several other functional groups in Chapters 1–8.

Functional groups also serve as the basis for naming organic compounds. Ideally, each of the 10 million or more organic compounds must have a name different from the name of every other compound. We will show how these names are derived in Chapters 1–8 as we study individual functional groups in detail.

To summarize, functional groups

- Are sites of chemical reaction—a particular functional group, in whatever compound it is found, undergoes the same types of chemical reactions;
- Determine in large measure the physical properties of a compound;
- Are the units by which we divide organic compounds into families; and
- Serve as a basis for naming organic compounds.

A Alcohols

The functional group of an **alcohol** is an **—OH (hydroxyl) group** bonded to a tetrahedral carbon atom (a carbon having bonds to four atoms). In the general formula of an alcohol (shown on the next page on the left), the symbol R— indicates either a hydrogen or another carbon group. The im-

Alcohol A compound containing an —OH (hydroxyl) group bonded to a tetrahedral carbon atom

Hydroxyl group An —OH group bonded to a tetrahedral carbon atom

■ **Table wine contains about 10 to 13% ethanol.**

Charles D. Winters

portant point of the general structure is the —OH group bonded to a tetra-hedral carbon atom.

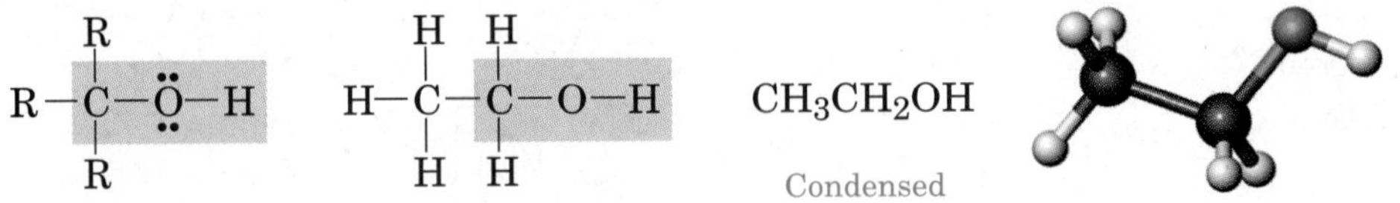

$$R-\overset{\displaystyle R}{\underset{\displaystyle R}{C}}-\overset{..}{\overset{..}{O}}-H \qquad H-\overset{\displaystyle H}{\underset{\displaystyle H}{C}}-\overset{\displaystyle H}{\underset{\displaystyle H}{C}}-O-H \qquad CH_3CH_2OH$$

Functional group   Structural formula   Condensed structural formula

R = H or carbon group   An alcohol (Ethanol)

Here we represent the alcohol as a **condensed structural formula,** CH_3CH_2OH. In a condensed structural formula, CH_3 indicates a carbon bonded to three hydrogens, CH_2 indicates a carbon bonded to two hydrogens, and CH indicates a carbon bonded to one hydrogen. Unshared pairs of electrons are generally not shown in a condensed structural formula.

Alcohols are classified as **primary (1°), secondary (2°),** or **tertiary (3°),** depending on the number of carbon atoms bonded to the carbon bearing the —OH group.

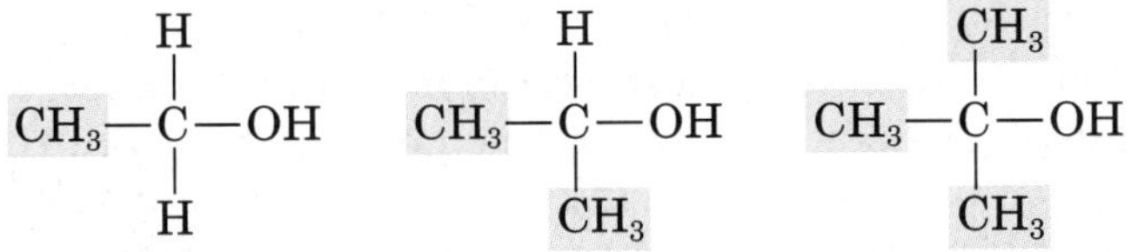

$$CH_3-\overset{\displaystyle H}{\underset{\displaystyle H}{C}}-OH \qquad CH_3-\overset{\displaystyle H}{\underset{\displaystyle CH_3}{C}}-OH \qquad CH_3-\overset{\displaystyle CH_3}{\underset{\displaystyle CH_3}{C}}-OH$$

A 1° alcohol   A 2° alcohol   A 3° alcohol

EXAMPLE 1.2

Draw Lewis structures and condensed structural formulas for the two alcohols with molecular formula C_3H_8O. Classify each as primary, secondary, or tertiary.

Solution
Begin by drawing the three carbon atoms in a chain. The oxygen atom of the hydroxyl group may be bonded to the carbon chain at two different positions on the chain: either to an end carbon or to the middle carbon.

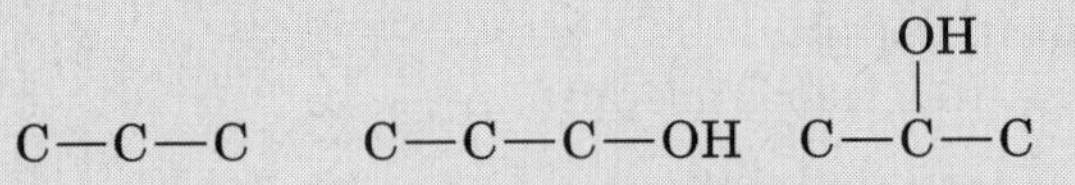

$$C-C-C \qquad C-C-C-OH \qquad C-\overset{\displaystyle OH}{\overset{|}{C}}-C$$

Carbon chain   The two locations for the —OH group

Finally, add seven more hydrogens, giving a total of eight as shown in the molecular formula. Show unshared electron pairs on the Lewis structures but not on the condensed structural formulas.

Lewis structures   Condensed structural formulas

$$H-\overset{\displaystyle H}{\underset{\displaystyle H}{C}}-\overset{\displaystyle H}{\underset{\displaystyle H}{C}}-\overset{\displaystyle H}{\underset{\displaystyle H}{C}}-\overset{..}{\overset{..}{O}}-H \qquad CH_3CH_2CH_2OH$$

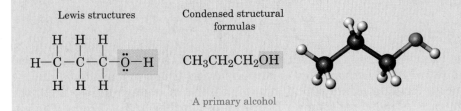

A primary alcohol

H :O: H
H—C—C—C—H
H H H

OH
CH₃CHCH₃
A secondary alcohol

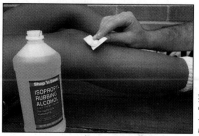

Charles D. Winters

■ **2-Propanol (isopropyl alcohol) is used to disinfect cuts and scrapes.**

The secondary alcohol, whose common name is isopropyl alcohol, is the cooling, soothing component in rubbing alcohol.

Problem 1.2

Draw Lewis structures and condensed structural formulas for the four alcohols with molecular formula $C_4H_{10}O$. Classify each alcohol as primary, secondary, or tertiary. (*Hint:* First consider the order of attachment of the four carbon atoms; they can be bonded either four in a chain or three in a chain with the fourth carbon as a branch on the middle carbon. Then consider the points at which the —OH group can be bonded to each carbon chain.)

B Amines

The functional group of an **amine** is an **amino group**—a nitrogen atom bonded to one, two, or three carbon atoms. In a **primary (1°) amine,** nitrogen is bonded to one carbon group. In a **secondary (2°) amine,** it is bonded to two carbon groups, and in a **tertiary (3°) amine,** it is bonded to three carbon groups. The second and third structural formulas can be written in a more abbreviated form by collecting the CH_3 groups and writing them as $(CH_3)_2NH$ and $(CH_3)_3N$.

Amine An organic compound in which one, two, or three hydrogens of ammonia are replaced by carbon groups: RNH_2, R_2NH, or R_3N

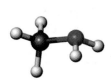

CH_3NH_2

Methylamine
(a 1° amine)

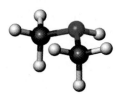

CH_3NH or $(CH_3)_2NH$
|
CH_3

Dimethylamine
(a 2° amine)

CH_3NCH_3 or $(CH_3)_3N$
|
CH_3

Trimethylamine
(a 3° amine)

EXAMPLE 1.3

Draw condensed structural formulas for the two primary amines with molecular formula C_3H_9N.

Solution

For a primary amine, draw a nitrogen atom bonded to two hydrogens and one carbon.

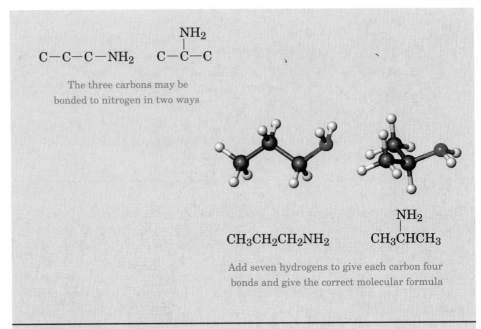

NH₂ · · ·

C—C—C—NH₂   C—C—C
                    |
                   NH₂

The three carbons may be
bonded to nitrogen in two ways

$CH_3CH_2CH_2NH_2$ CH_3CHCH_3
                        |
                       NH₂

Add seven hydrogens to give each carbon four
bonds and give the correct molecular formula

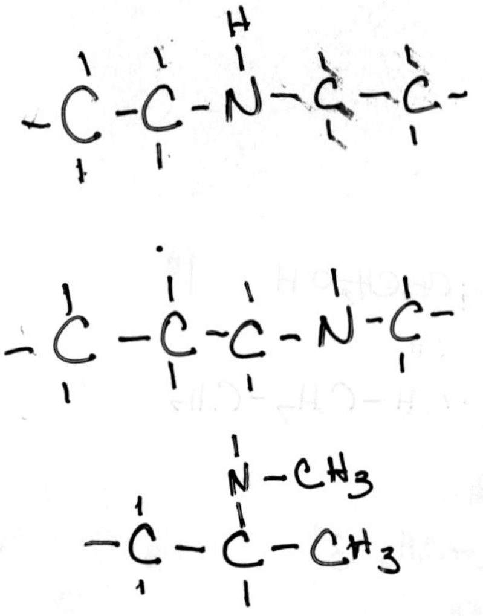

Problem 1.3

Draw structural formulas for the three secondary amines with molecular formula $C_4H_{11}N$.

C Aldehydes and Ketones

Both aldehydes and ketones contain a C=O (**carbonyl**) **group.** The **aldehyde** functional group contains a carbonyl group bonded to a hydrogen. In formaldehyde, CH_2O, the simplest aldehyde, two hydrogens are bonded to its carbonyl carbon. In a condensed structural formula, the aldehyde group may be written showing the carbon–oxygen double bond as CH=O, or, alternatively, it may be written —CHO. The functional group of a **ketone** is a carbonyl group bonded to two carbon atoms. In the general structural formula of each functional group, we use the symbol R to represent other groups bonded to carbon to complete the tetravalence of carbon.

Carbonyl group A C=O group

Aldehyde A compound containing a carbonyl group bonded to a hydrogen; a —CHO group

Ketone A compound containing a carbonyl group bonded to two carbon groups

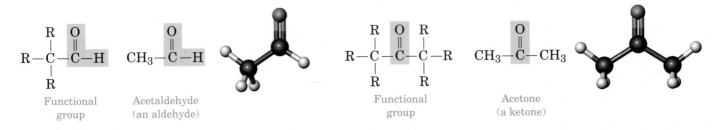

| Functional group | Acetaldehyde (an aldehyde) | | Functional group | Acetone (a ketone) |

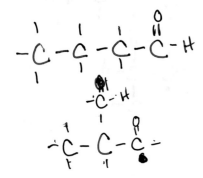

EXAMPLE 1.4

Draw condensed structural formulas for the two aldehydes with molecular formula C_4H_8O.

Solution

First draw the functional group of an aldehyde, and then add the remaining carbons. These may be bonded in two ways. Then, add seven hydrogens to complete the four bonds of each carbon.

$$\overset{O}{\overset{\parallel}{CH_3CH_2CH_2CH}}$$

or

$$CH_3CH_2CH_2CHO$$

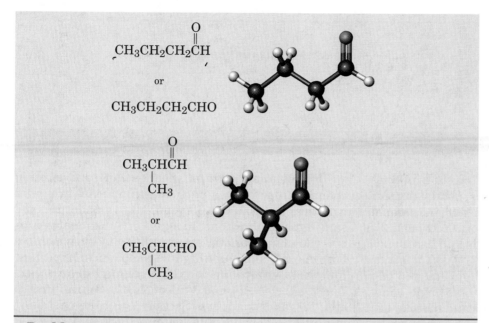

$$\overset{O}{\overset{\parallel}{CH_3CHCH}}$$
$$\overset{|}{CH_3}$$

or

$$CH_3CHCHO$$
$$\overset{|}{CH_3}$$

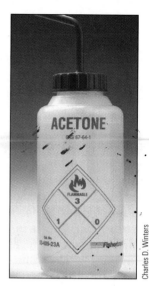

■ **Acetone is a ketone.**

Problem 1.4

Draw condensed structural formulas for the three ketones with molecular formula $C_5H_{10}O$.

D Carboxylic Acids

The functional group of a **carboxylic acid** is a **—COOH (carboxyl: carbonyl + hydroxyl) group.** In a condensed structural formula, a carboxyl group may also be written —CO₂H.

> **Carboxylic acid** A compound containing a —COOH group

> **Carboxyl group** A —COOH group

$$\overset{O}{\overset{\parallel}{RCOH}}$$
Functional group

$$\overset{O}{\overset{\parallel}{CH_3COH}}$$
Acetic acid (a carboxylic acid)

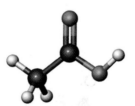

EXAMPLE 1.5

Draw a condensed structural formula for the single carboxylic acid with molecular formula $C_3H_6O_2$.

Solution
The only way the carbon atoms can be written is three in a chain, and the —COOH group must be on an end carbon of the chain.

$$\overset{O}{\overset{\parallel}{CH_3CH_2COH}} \quad \text{or} \quad CH_3CH_2COOH$$

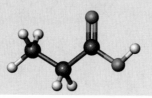

Problem 1.5

Draw condensed structural formulas for the two carboxylic acids with molecular formula $C_4H_8O_2$.

S U M M A R Y

Organic chemistry (Section 1.1) is the study of compounds containing carbon. Chemists obtain organic compounds either by isolation from plant and animal sources or by synthesis in the laboratory (Section 1.2).

Carbon normally forms four bonds and has no unshared pairs of electrons (Section 1.3). Its four bonds may be four single bonds, two single bonds and one double bond, or one triple bond and one single bond. Nitrogen normally forms three bonds and has one unshared pair of electrons. Its bonds may be three single bonds, one single bond and one double bond, or one triple bond. Oxygen normally forms two bonds and has two unshared pairs of electrons. Its bonds may be two single bonds or one double bond.

Functional groups (Section 1.4) are sites of chemical reactivity; a particular functional group, in whatever compound it is found, always undergoes the same types of reactions. In addition, functional groups are the characteristic structural units by which we both classify and name organic compounds. Important functional groups include the **hydroxyl group** (Section 1.4A) of 1°, 2°, and 3° alcohols; the **amino group** of 1°, 2°, and 3° amines (Section 1.4B); the **carbonyl group** of aldehydes and ketones (Section 1.4C); and the **carboxyl group** of carboxylic acids (Section 1.4D).

P R O B L E M S

Numbers that appear in color indicate difficult problems.
▶ designates problems requiring application of principles.

Review of Lewis Structures

1.6 How many electrons are in the valence shell of each of the following atoms? Write a Lewis dot structure for an atom of each element.

(a) Carbon .Ċ. 4   (b) Oxygen :Ö: 6
(c) Nitrogen .Ṅ. 5   (d) Fluorine :F̈: 7

1.7 What is the relationship between the number of electrons in the valence shell of each of the following atoms and the number of covalent bonds it forms?

(a) Carbon 4   (b) Oxygen 2
(c) Nitrogen 3   (d) Hydrogen 1

1.8 Write Lewis structures for these compounds. Show all valence electrons. None of them contains a ring of atoms. (*Hint:* Remember that carbon has four bonds, nitrogen has three bonds and one unshared pair of electrons, oxygen has two bonds and two unshared pairs of electrons, and each halogen has one bond and three unshared pairs of electrons.)

(a) H_2O_2, hydrogen peroxide
(b) N_2H_4, hydrazine
(c) CH_3OH, methanol
(d) CH_3SH, methylethiol
(e) CH_3NH_2, methanamine
(f) CH_3Cl, chloromethane

1.9 Write Lewis structures for these compounds. Show all valence electrons. None of them contains a ring of atoms.

(a) CH_3OCH_3, dimethyl ether
(b) C_2H_6, ethane
(c) C_2H_4, ethylene
(d) C_2H_2, acetylene
(e) CO_2, carbon dioxide
(f) CH_2O, formaldehyde
(g) H_2CO_3, carbonic acid
(h) CH_3COOH, acetic acid

1.10 Write Lewis structures for these ions.

(a) HCO_3^-, bicarbonate ion
(b) CO_3^{2-}, carbonate ion
(c) CH_3COO^-, acetate ion
(d) Cl^-, chloride ion

1.11 Why are the following molecular formulas impossible?

(a) CH_5 (b) C_2H_7

Review of the VSEPR Model

1.12 Explain how to use the valence-shell electron-pair repulsion (VSEPR) model to predict bond angles about atoms of carbon, oxygen, and nitrogen.

1.13 Suppose you forget to take into account the presence of the unshared pair of electrons on nitrogen in the molecule NH_3. What would you predict for the H—N—H bond angles?

1.14 Suppose you forget to take into account the presence of the two unshared pairs of electrons on the oxygen atom of ethanol, CH_3CH_2OH. What would you predict for the C—O—H bond angle?

1.15 Use the VSEPR model to predict the bond angles about each highlighted atom. (*Hint:* Remember to take into account the presence of unshared pairs of electrons.)

(a) H—C—C—O—H (with H atoms above and below each carbon)

(b) H—C=C—Cl (with H atoms below each carbon)

(c) H—C—C≡C—H (with H above, below, and H on carbon)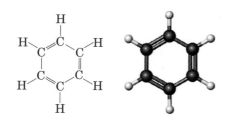

1.16 Use the VSEPR model to predict the bond angles about each highlighted atom.

(a) H—C—Ö—H (with double-bonded O above C)

(b) H—C—N̈—H (with H atoms above and below C and N)

(c) H—Ö—N̈=Ö

1.17 Following is a structural formula and a ball-and-stick model of benzene, C_6H_6.

(structural formula of benzene)

(a) Predict each H—C—C and C—C—C bond angle in benzene.

(b) Predict the shape of a benzene molecule.

Sources of Organic Compounds

1.18 Is there any difference between vanillin made synthetically and vanillin extracted from vanilla beans?

1.19 Suppose that you are told that only organic substances are produced by living organisms. How would you rebut this assertion?

1.20 What important experiment did Wöhler carry out in 1828?

Organic Compounds

1.21 List the four principal elements that make up organic compounds and give the number of bonds each typically forms.

1.22 Think about the types of substances in your immediate environment, and make a list of those that are organic—for example, textile fibers. We will ask you to return to this list later in the course and to refine, correct, and possibly expand it.

Functional Groups

1.23 What is meant by the term "functional group"?

1.24 List three reasons why functional groups are important in organic chemistry.

1.25 Draw Lewis structures for each of the following functional groups. Show all valence electrons.

(a) Carbonyl group
(b) Carboxyl group
(c) Hydroxyl group
(d) Primary amino group

1.26 Complete the following structural formulas by adding enough hydrogens to complete the tetravalence of each carbon. Then write the molecular formula of each compound.

(a) C—C=C—C—C (with C above the third carbon)

(b) C—C—C—C—OH (with O double-bonded above the third carbon)

(c) C—C—C—C (with O double-bonded above the third carbon)

(d) C—C—C—H (with O double-bonded above the second carbon, and C below the second carbon)

(e) C—C—C—C—NH₂ (with C above and C below the second carbon)

(f)
$$\underset{\underset{NH_2}{|}}{C}-\overset{\overset{O}{\|}}{C}-OH$$

(g)
$$C-\underset{\underset{OH}{|}}{C}-C-C-C$$

(h)
$$C-\underset{\underset{OH}{|}}{C}-C-\overset{\overset{O}{\|}}{C}-OH$$

(i) $C{=}C{-}C{-}OH$

1.27   Some of the following structural formulas are incorrect (that is, they do not represent real compounds) because they have atoms with an incorrect number of bonds. Which structural formulas are incorrect, and which atoms in them have an incorrect number of bonds?

(a) $H{-}C{\equiv}C{-}\underset{\underset{H}{|}}{\overset{\overset{H}{|}}{C}}{-}H$

(b) $H{-}C{=}\underset{\underset{H}{|}}{\overset{\overset{Cl}{|}}{C}}{-}H$
     with H below the first C

(c) $H{-}\underset{\underset{H}{|}}{N}{-}\underset{\underset{H}{|}}{\overset{\overset{H}{|}}{C}}{-}\underset{\underset{H}{|}}{\overset{\overset{H}{|}}{C}}{-}O{-}H$

(d) $H{-}\underset{\underset{H}{|}}{\overset{\overset{H}{|}}{C}}{-}\underset{\underset{H}{|}}{\overset{\overset{H}{|}}{C}}{=}O{-}H$

(e) $H{-}O{-}\underset{\underset{H}{|}}{\overset{\overset{H}{|}}{C}}{-}\underset{\underset{H}{|}}{\overset{\overset{H}{|}}{C}}{-}\overset{\overset{O}{\|}}{C}{-}O{-}H$

(f) $H{-}\underset{\underset{H}{|}}{\overset{\overset{H}{|}}{C}}{-}\underset{\underset{H}{|}}{\overset{\overset{H}{|}}{C}}{-}\overset{\overset{O}{\|}}{C}{-}H$

1.28   What is the meaning of the term "tertiary (3°)" when it is used to classify alcohols?

1.29   Draw a structural formula for the one tertiary (3°) alcohol with molecular formula $C_4H_{10}O$.

1.30   What is the meaning of the term "tertiary (3°)" when it is used to classify amines?

1.31   Draw a structural formula for the one tertiary (3°) amine with molecular formula $C_4H_{11}N$.

1.32   Identify the functional group(s) in each compound.

(a) $CH_3{-}\underset{\underset{OH}{|}}{CH}{-}\overset{\overset{O}{\|}}{C}{-}OH$
     Lactic acid
     (produced in actively respiring muscle; gives sourness to sour cream)

(b) $HO{-}CH_2{-}CH_2{-}OH$
     Ethylene glycol
     (a component of antifreeze)

(c) $CH_3{-}\underset{\underset{NH_2}{|}}{CH}{-}\overset{\overset{O}{\|}}{C}{-}OH$
     Alanine
     (one of the building blocks of proteins)

(d) $HOCH_2\underset{\underset{HO}{|}}{CH}\overset{\overset{O}{\|}}{CH}$
     Glyceraldehyde
     (the simplest carbohydrate)

(e) $H_2NCH_2CH_2CH_2CH_2CH_2CH_2NH_2$
     1,6-Hexanediamine
     (one component of nylon-66)

1.33   Identify the functional group(s) in each compound.

(a) $CH_3CH_2\overset{\overset{O}{\|}}{C}CH_3$
     2-Butanone
     (a solvent for paints and lacquers)

(b) $HO\overset{\overset{O}{\|}}{C}CH_2CH_2CH_2CH_2\overset{\overset{O}{\|}}{C}OH$
     Hexanedioic acid
     (the second component of nylon-66)

(c) $H_2NCH_2CH_2CH_2CH_2\underset{\underset{NH_2}{|}}{CH}\overset{\overset{O}{\|}}{C}OH$
     Lysine
     (one of the 20 amino acid building blocks of proteins)

(d) $HOCH_2\overset{\overset{O}{\|}}{C}CH_2OH$
     Dihydroxyacetone
     (a component of several artificial tanning lotions)

1.34   Draw condensed structural formulas for all compounds with molecular formula C_4H_8O that contain a carbonyl group (there are two aldehydes and one ketone).

1.35 Draw structural formulas for the following:

(a) The four primary (1°) alcohols with molecular formula $C_5H_{12}O$

(b) The three secondary (2°) alcohols with molecular formula $C_5H_{12}O$

(c) The one tertiary (3°) alcohol with molecular formula $C_5H_{12}O$

1.36 Draw structural formulas for the six ketones with molecular formula $C_6H_{12}O$.

1.37 Draw structural formulas for the eight carboxylic acids with molecular formula $C_6H_{12}O_2$.

1.38 Draw structural formulas for the following:

(a) The four primary (1°) amines with molecular formula $C_4H_{11}N$

(b) The three secondary (2°) amines with molecular formula $C_4H_{11}N$

(c) The one tertiary (3°) amine with molecular formula $C_4H_{11}N$

Chemical Connections

1.39 (Chemical Connections 1A) How was Taxol discovered?

1.40 (Chemical Connections 1A) In what way does Taxol interfere with cell division?

1.41 (Chemical Connections 1B) What is the goal of combinatorial chemistry?

Additional Problems

1.42 Use the VSEPR model to predict the bond angles about each atom of carbon, nitrogen, and oxygen in these molecules. (*Hint:* First add unshared pairs of electrons as necessary to complete the valence shell of each atom and then make your predictions of bond angles.)

(a) $CH_3CH_2CH_2OH$

(b) $CH_3CH_2\overset{\overset{\displaystyle O}{\|}}{C}H$

(c) $CH_3CH{=}CH_2$

(d) $CH_3C{\equiv}CCH_3$

(e) $CH_3\overset{\overset{\displaystyle O}{\|}}{C}OCH_3$

(f) $CH_3\overset{\overset{\displaystyle CH_3}{|}}{N}CH_3$

1.43 Silicon is immediately below carbon in the Periodic Table. Predict the C—Si—C bond angle in tetramethylsilane, $(CH_3)_4Si$.

1.44 Phosphorus is immediately below nitrogen in the Periodic Table. Predict the C—P—C bond angle in trimethylphosphine, $(CH_3)_3P$.

1.45 Draw the structure for a compound with molecular formula

(a) C_2H_6O that is an alcohol

(b) C_3H_6O that is an aldehyde

(c) C_3H_6O that is a ketone

(d) $C_3H_6O_2$ that is a carboxylic acid

1.46 Draw structural formulas for the eight aldehydes with molecular formula $C_6H_{12}O$.

1.47 Draw structural formulas for the three tertiary (3°) amines with molecular formula $C_5H_{13}N$.

1.48 Which of these covalent bonds are polar and which are nonpolar?

(a) C—C   (b) C=C   (c) C—H

(d) C—O   (e) O—H   (f) C—N

(g) N—H   (h) N—O

1.49 Of the bonds in Problem 1.48, which is the most polar? Which is the least polar?

1.50 Using the symbol $\delta+$ to indicate a partial positive charge and $\delta-$ to indicate a partial negative charge, indicate the polarity of the most polar bond (or bonds if two or more have the same polarity) in each of the following molecules.

(a) CH_3OH

(b) CH_3NH_2

(c) $HSCH_2CH_2NH_2$

(d) $CH_3\overset{\overset{\displaystyle O}{\|}}{C}CH_3$

(e) $H\overset{\overset{\displaystyle O}{\|}}{C}H$

(f) $CH_3\overset{\overset{\displaystyle O}{\|}}{C}OH$

InfoTrac College Edition

For additional readings, go to InfoTrac College Edition, your online research library, at

http://infotrac.thomsonlearning.com

CHAPTER 2

2.1 Introduction

2.2 Structure of Alkanes

2.3 Constitutional Isomerism in Alkanes

2.4 Nomenclature of Alkanes

2.5 Cycloalkanes

2.6 Shapes of Alkanes and Cycloalkanes

2.7 Cis-Trans Isomerism in Cycloalkanes

2.8 Physical Properties

2.9 Reactions of Alkanes

2.10 Important Haloalkanes

2.11 Sources of Alkanes

A petroleum refinery. Petroleum, along with natural gas, provides nearly 90 percent of the organic raw materials for the synthesis and manufacture of synthetic fibers, plastics, detergents, drugs, dyes, adhesives, and a multitude of other products.

J. L. Bohin/Photo Researchers, Inc.

Alkanes

2.1 Introduction

Alkane A saturated hydrocarbon whose carbon atoms are arranged in an open chain.

Hydrocarbon A compound that contains only carbon and hydrogen atoms

Saturated hydrocarbon A hydrocarbon that contains only carbon–carbon single bonds

Aliphatic hydrocarbon An alkane

In this chapter, we examine the physical and chemical properties of **alkanes,** the simplest type of organic compounds. Actually, alkanes are members of a larger class of organic compounds called hydrocarbons. A **hydrocarbon** is a compound composed of only carbon and hydrogen. Figure 2.1 shows the four classes of hydrocarbons, along with the characteristic type of bonding between carbon atoms in each class. Alkanes are **saturated hydrocarbons;** that is, they contain only carbon–carbon single bonds. Saturated in this context means that each carbon in the hydrocarbon has the maximum number of hydrogens bonded to it. A hydrocarbon that contains one or more carbon–carbon double bonds, triple bonds, or benzene rings is classified as an **unsaturated hydrocarbon.** We study alkanes (saturated hydrocarbons) in this chapter and alkenes, alkynes, and arenes (unsaturated hydrocarbons) in Chapters 3 and 4.

We commonly refer to alkanes as **aliphatic hydrocarbons** because the physical properties of the higher members of this class resemble those of the long carbon-chain molecules we find in animal fats and plant oils (Greek: *aleiphar,* fat or oil).

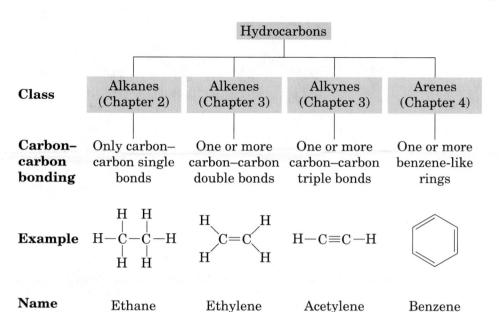

Figure 2.1 The four classes of hydrocarbons.

Class	Alkanes (Chapter 2)	Alkenes (Chapter 3)	Alkynes (Chapter 3)	Arenes (Chapter 4)
Carbon–carbon bonding	Only carbon–carbon single bonds	One or more carbon–carbon double bonds	One or more carbon–carbon triple bonds	One or more benzene-like rings
Name	Ethane	Ethylene	Acetylene	Benzene

2.2 Structure of Alkanes

Methane, CH_4, and ethane, C_2H_6, are the first two members of the alkane family. Figure 2.2 shows Lewis structures and ball-and-stick models for these molecules. The shape of methane is tetrahedral, and all H—C—H bond angles are 109.5°. Each carbon atom in ethane is also tetrahedral, and the bond angles in it are all approximately 109.5°. Although the three-dimensional shapes of larger alkanes are more complex than those of methane and ethane, the four bonds about each carbon atom are still arranged in a tetrahedral manner, and all bond angles are still approximately 109.5°.

The next members of the alkane family are propane, butane, and pentane. In the representations on page 20, these hydrocarbons are drawn first as condensed structural formulas, which show all carbons and hydrogens. They can also be drawn in a more abbreviated form called a **line-angle formula.** In this type of representation, a line represents a carbon–carbon bond and an angle represents a carbon atom. A line ending in space represents a —CH₃ group. To count hydrogens from a line-angle formula, you simply add enough hydrogens in your mind to give each carbon its required four bonds.

Structural formulas for alkanes can be written in yet another condensed form. For example, the structural formula of pentane contains three CH_2 (**methylene**) groups in the middle of the chain. We can group them

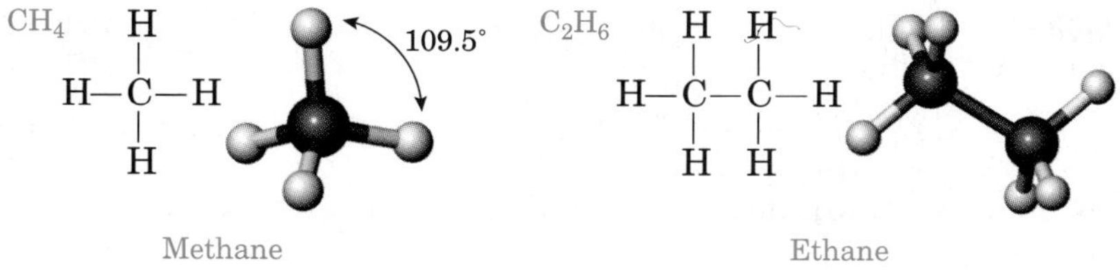

> **Line-angle formula** An abbreviated way to draw structural formulas in which each angle and line terminus represents a carbon atom and each line represents a bond

Figure 2.2 Methane and ethane.

Ball-and-stick model

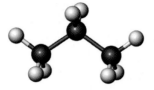

Line-angle formula

Condensed structural
formula

$CH_3CH_2CH_3$

Propane

$CH_3CH_2CH_2CH_3$

Butane

$CH_3CH_2CH_2CH_2CH_3$

Pentane

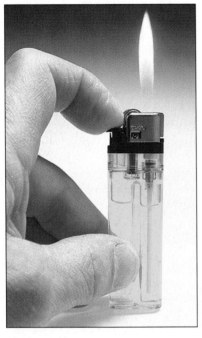

Charles D. Winters

■ **Butane, $CH_3CH_2CH_2CH_3$, is the fuel in this lighter. Butane molecules are present in the liquid and gaseous states in the lighter.**

together and write the structural formula $CH_3(CH_2)_3CH_3$. Table 2.1 gives the first ten alkanes with unbranched chains. Note that the names of all these alkanes end in "-ane." We will have more to say about naming alkanes in Section 2.4.

EXAMPLE 2.1

Table 2.1 gives the condensed structural formula for hexane. Draw a line-angle formula for this alkane, and number the carbons on the chain beginning at one end and proceeding to the other end.

Solution
Hexane contains six carbons in a chain. Its line-angle formula is

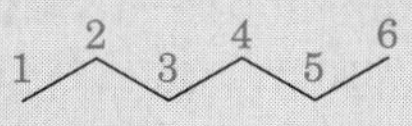

Problem 2.1
Following is a line-angle formula for an alkane. What is the name and molecular formula of this alkane?

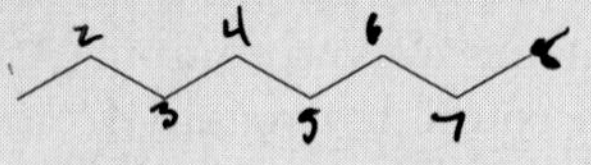

octane

Table 2.1 The First Ten Alkanes with Unbranched Chains

Name	Molecular Formula	Condensed Structural Formula	Name	Molecular Formula	Condensed Structural Formula
methane	CH_4	CH_4	hexane	C_6H_{14}	$CH_3(CH_2)_4CH_3$
ethane	C_2H_6	CH_3CH_3	heptane	C_7H_{16}	$CH_3(CH_2)_5CH_3$
propane	C_3H_8	$CH_3CH_2CH_3$	octane	C_8H_{18}	$CH_3(CH_2)_6CH_3$
butane	C_4H_{10}	$CH_3(CH_2)_2CH_3$	nonane	C_9H_{20}	$CH_3(CH_2)_7CH_3$
pentane	C_5H_{12}	$CH_3(CH_2)_3CH_3$	decane	$C_{10}H_{22}$	$CH_3(CH_2)_8CH_3$

2.3 Constitutional Isomerism in Alkanes

Constitutional isomers are compounds that have the same molecular formula but different structural formulas. By "different structural formulas," we mean that they differ in the kinds of bonds (single, double, or triple) and/or in their connectivity (the order of attachment among their atoms). For the molecular formulas CH_4, C_2H_6, and C_3H_8, only one order of attachment of their atoms is possible so there are no isomers for these molecular formulas. For the molecular formula C_4H_{10}, two structural formulas are possible: In butane, the four carbons are bonded in a chain; in 2-methylpropane, three carbons are bonded in a chain with the fourth carbon as a branch on the chain. The two constitutional isomers with molecular formula C_4H_{10} are drawn here both as condensed structural formulas and as line-angle formulas. Also shown are ball-and-stick models of each.

> **Constitutional isomers**
> Compounds with the same molecular formula but a different order of attachment of their atoms

Constitutional isomers have also been called structural isomers, an older term that is still in use.

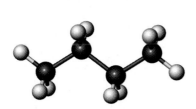

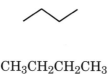

$$CH_3CH_2CH_2CH_3$$

Butane
(bp −0.5°C)

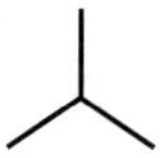

$$\begin{array}{c} CH_3 \\ | \\ CH_3CHCH_3 \end{array}$$

2-Methylpropane
(bp −11.6°C)

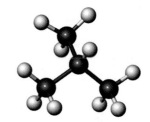

We refer to butane and 2-methylpropane as constitutional isomers; they are different compounds and have different physical and chemical properties. Their boiling points, for example, differ by approximately 11°C.

In Section 1.4, we encountered several examples of constitutional isomers, although we did not call them that at the time. We saw, for example, that there are two alcohols with molecular formula C_3H_8O, two primary amines with molecular formula C_3H_9N, two aldehydes with molecular formula C_4H_8O, and two carboxylic acids with molecular formula $C_4H_8O_2$.

To determine whether two or more structural formulas represent constitutional isomers, we write the molecular formula of each and then compare them. All compounds that have the same molecular formula but different structural formulas are constitutional isomers.

EXAMPLE 2.2

Do the structural formulas in each set represent the same compound or constitutional isomers? You will find it helpful to redraw each molecule as a line-angle formula, which will make it easier for you to see similarities and differences in molecular structure.

(a) $CH_3CH_2CH_2CH_2CH_2CH_3$ and $\underset{\underset{CH_2CH_2CH_3}{|}}{CH_3CH_2CH_2}$ (Each is C_6H_{14})

(b) $\underset{\underset{CH_3}{|}}{\overset{\overset{CH_3}{|}}{CH_3CHCH_2CH}}$ and $\underset{\underset{CH_3}{|}}{\overset{\overset{CH_3}{|}}{CH_3CH_2CHCHCH_3}}$ (Each is C_7H_{16})

Solution

First, find the longest chain of carbon atoms. Note that it makes no difference whether the chain is drawn straight or bent. As structural formulas are drawn in this problem, there is no attempt to show three-dimensional shapes. Second, number the longest chain from the end nearest the first branch. Third, compare the lengths of the two chains and the sizes and locations of any branches. Structural formulas that have the same molecular formula and the same order of attachment of atoms represent the same compound; those that have the same molecular formula but different orders of attachment of their atoms represent constitutional isomers.

(a) Each structural formula has an unbranched chain of six carbons; they are identical and represent the same compound.

$$\underset{CH_3CH_2CH_2CH_2CH_2CH_3}{\overset{1\quad 2\quad 3\quad 4\quad 5\quad 6}{}} \quad \text{and} \quad \underset{CH_3CH_2CH_2}{\overset{1\quad 2\quad 3}{}}$$

$$\underset{CH_2CH_2CH_3}{\overset{4\quad 5\quad 6}{}}$$

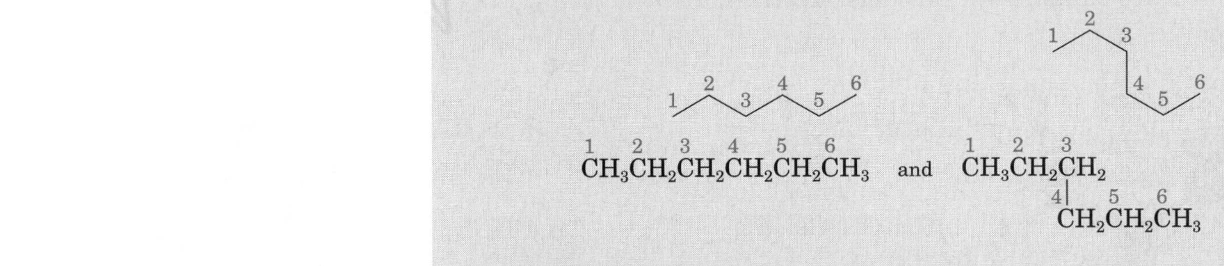

(b) Each structural formula has the same molecular formula, C_7H_{16}. In addition, each has a chain of five carbons with two CH_3 branches. Although the branches are identical, they are found at different locations on the chains. Therefore, these structural formulas represent constitutional isomers.

$$\underset{\underset{CH_3}{|}}{\overset{\overset{5}{CH_3}}{CH_3CHCH_2CH}} \quad \text{and} \quad \underset{\underset{CH_3}{|}}{\overset{\overset{CH_3}{|}}{CH_3CH_2CHCHCH_3}}$$

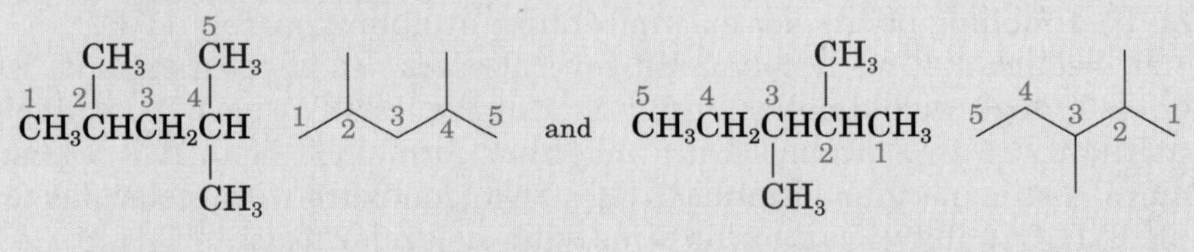

Problem 2.2

Do the structural formulas in each set represent the same compound or constitutional isomers?

(a) 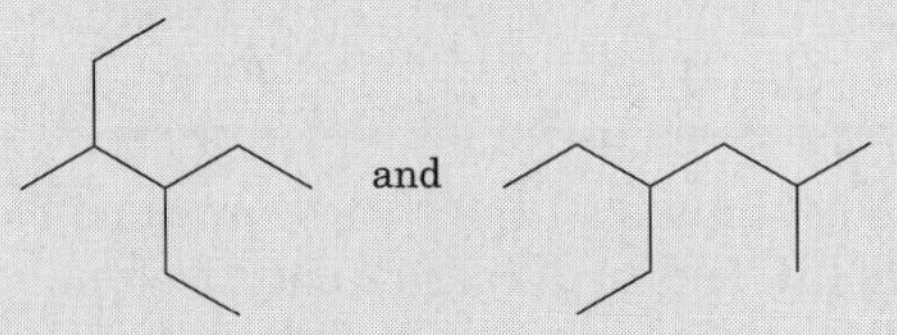   and

(b) 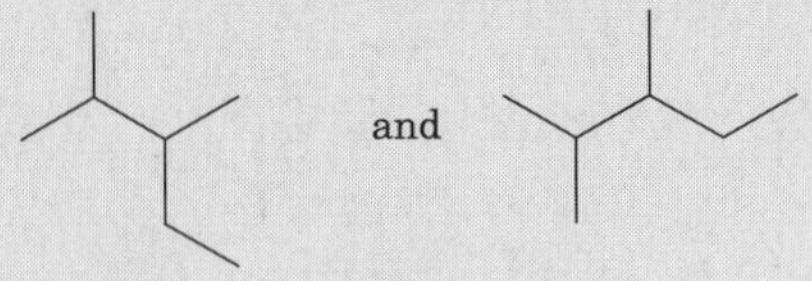   and

EXAMPLE 2.3

Draw structural formulas for the five constitutional isomers with molecular formula C_6H_{14}.

Solution

In solving problems of this type, you should devise a strategy and then follow it. Here is one possible strategy. First, draw a line-angle formula for the constitutional isomer with all six carbons in an unbranched chain. Then, draw line-angle formulas for all constitutional isomers with five carbons in a chain and one carbon as a branch on the chain. Finally, draw line-angle formulas for all constitutional isomers with four carbons in a chain and two carbons as branches. No constitutional isomers for C_6H_{14} having only three carbons in the longest chain are possible.

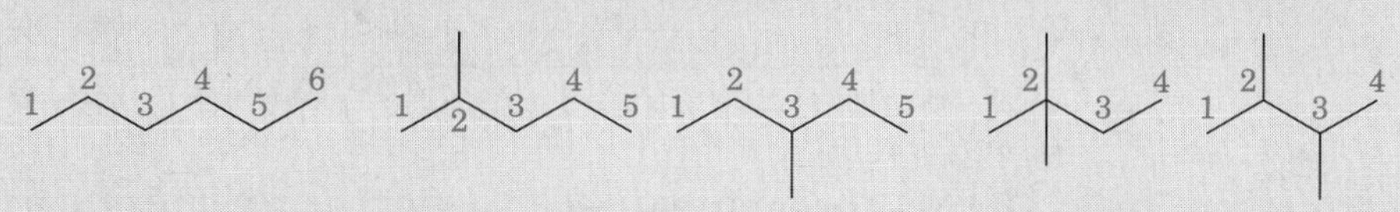

Six carbons in an
unbranched chain

Five carbons in a chain;
one carbon as a branch

Four carbons in a chain;
two carbons as branches

Problem 2.3

Draw structural formulas for the three constitutional isomers with molecular formula C_5H_{12}.

The ability of carbon atoms to form strong, stable bonds with other carbon atoms results in a staggering number of constitutional isomers, as the following table shows:

Molecular Formula	Number of Constitutional Isomers
CH_4	1
C_5H_{12}	3
$C_{10}H_{22}$	75
$C_{15}H_{32}$	4,347
$C_{25}H_{52}$	36,797,588
$C_{30}H_{62}$	4,111,846,763

Thus, for even a small number of carbon and hydrogen atoms, a very large number of constitutional isomers is possible. In fact, the potential for structural and functional group individuality among organic molecules made from just the basic building blocks of carbon, hydrogen, nitrogen, and oxygen is practically limitless.

2.4 Nomenclature of Alkanes

A The IUPAC System

Ideally, every organic compound should have a name from which its structural formula can be drawn. For this purpose, chemists have adopted a set of rules established by the **International Union of Pure and Applied Chemistry (IUPAC).**

The IUPAC name for an alkane with an unbranched chain of carbon atoms consists of two parts: (1) a prefix to show the number of carbon atoms in the chain and (2) the suffix **-ane** to show that the compound is a saturated

See the **Interactive General, Organic, and Biochemistry CD-ROM, version 2.0,** for further exploration on this topic.

Table 2.2	Prefixes Used in the IUPAC System to Show the Presence of One to Ten Carbons in an Unbranched Chain		
Prefix	**Number of Carbon Atoms**	**Prefix**	**Number of Carbon Atoms**
meth-	1	hex-	6
eth-	2	hept-	7
prop-	3	oct-	8
but-	4	non-	9
pent-	5	dec-	10

hydrocarbon. Table 2.2 gives the prefixes used to show the presence of one to ten carbon atoms.

The IUPAC chose the first four prefixes listed in Table 2.2 because they were well established long before the nomenclature was systematized. For example, the prefix *but-* appears in the name butyric acid, a compound of four carbon atoms formed by the air oxidation of butter fat (Latin: *butyrum*, butter). The prefixes to show five or more carbons are derived from Latin numbers. Table 2.1 gives the names, molecular formulas, and condensed structural formulas for the first ten alkanes with unbranched chains.

IUPAC names of alkanes with branched chains consist of a parent name that shows the longest chain of carbon atoms and substituent names that indicate the groups attached to the parent chain. For example,

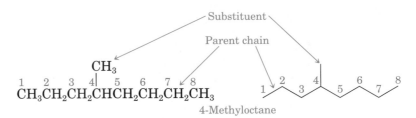

4-Methyloctane

A substituent group derived from an alkane by removal of a hydrogen atom is called an **alkyl group** and is commonly represented by the symbol **R—**. Alkyl groups are named by dropping the "-ane" from the name of the parent alkane and adding the suffix **-yl.** Tables 2.3 gives the names and condensed structural formulas for eight of the most common alkyl groups. The prefix *sec-* is an abbreviation for "secondary," meaning a carbon bonded to two other carbons. The prefix *tert-* is an abbreviation for "tertiary," meaning a carbon bonded to three other carbons.

The rules of the IUPAC system for naming alkanes are as follows:

1. The name for an alkane with an unbranched chain of carbon atoms consists of a prefix showing the number of carbon atoms in the chain and the ending *-ane.*

2. For branched-chain alkanes, take the longest chain of carbon atoms as the parent chain and its name becomes the root name.

3. Give each substituent on the parent chain a name and a number. The number shows the carbon atom of the parent chain to which the substituent is bonded. Use a hyphen to connect the number to the name.

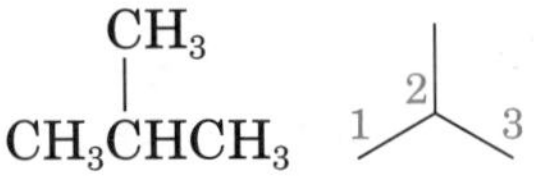

CH₃
|
CH₃CHCH₃

2-Methylpropane

Alkyl group A group derived by removing a hydrogen from an alkane; given the symbol R—

R— A symbol used to represent an alkyl group

Table 2.3   Names of the Most Common Alkyl Groups

Name	Condensed Structural Formula	Name	Condensed Structural Formula
methyl	$-CH_3$	isobutyl	$-CH_2CHCH_3$ $\quad\quad\ \ \mid$ $\quad\quad\ \ CH_3$
ethyl	$-CH_2CH_3$	sec-butyl	$-CHCH_2CH_3$ $\ \ \mid$ $\ \ CH_3$
propyl	$-CH_2CH_2CH_3$	tert-butyl	$\quad\quad CH_3$ $\quad\quad\mid$ $-CCH_3$ $\quad\quad\mid$ $\quad\quad CH_3$
isopropyl	$-CHCH_3$ $\ \ \mid$ $\ \ CH_3$		
butyl	$-CH_2CH_2CH_2CH_3$		

4. If there is one substituent, number the parent chain from the end that gives the substituent the lower number.

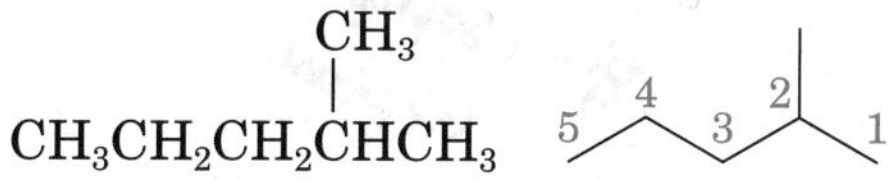

$$\begin{array}{c} CH_3 \\ \mid \\ CH_3CH_2CH_2CHCH_3 \end{array}$$

2-Methylpentane
(not 4-methylpentane)

5. If the same substituent occurs more than once, number the parent chain from the end that gives the lower number to the substituent encountered first. Indicate the number of times the substituent occurs by a prefix di-, tri-, tetra-, penta-, hexa-, and so on. Use a comma to separate position numbers.

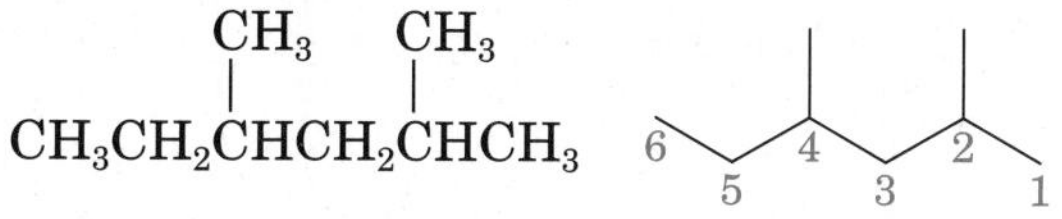

$$\begin{array}{c} CH_3 \quad\ CH_3 \\ \mid \quad\quad \mid \\ CH_3CH_2CHCH_2CHCH_3 \end{array}$$

2,4-Dimethylhexane
(not 3,5-dimethylhexane)

6. If there are two or more different substituents, list them in alphabetical order and number the chain from the end that gives the lower number to the substituent encountered first. If there are different substituents in equivalent positions on opposite ends of the parent chain, give the substituent of lower alphabetical order the lower number.

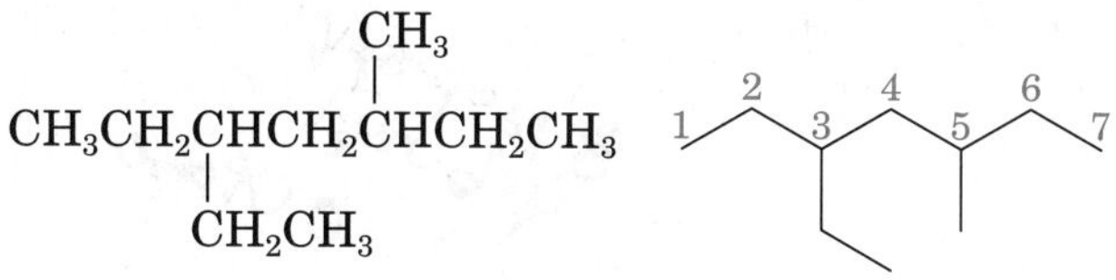

$$\begin{array}{c} CH_3 \\ \mid \\ CH_3CH_2CHCH_2CHCH_2CH_3 \\ \mid \\ CH_2CH_3 \end{array}$$

3-Ethyl-5-methylheptane
(not 3-methyl-5-ethylheptane)

7. Do not include the prefixes di-, tri-, tetra-, and so on, or the hyphenated prefixes *sec-* and *tert-* in alphabetizing. Alphabetize the names of substituents first, and then insert these prefixes. In the following example, the alphabetizing parts are ethyl and methyl, not ethyl and dimethyl.

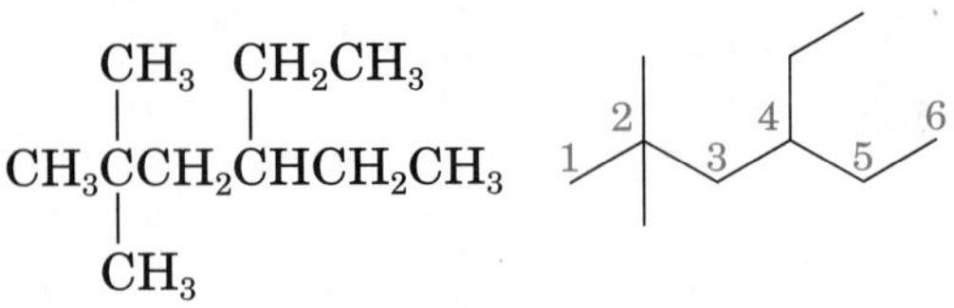

$$CH_3 \quad CH_2CH_3$$
$$CH_3CCH_2CHCH_2CH_3$$
$$CH_3$$

4-Ethyl-2,2-dimethylhexane
(not 2,2-dimethyl-4-ethylhexane)

EXAMPLE 2.4

Write the molecular formula and IUPAC name for each alkane.

(a) 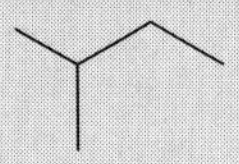   *2-methyl butane*

(b) 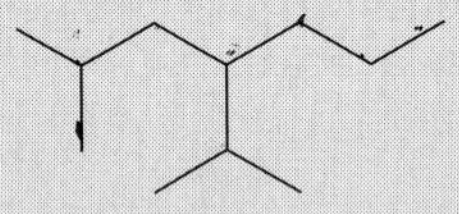   *2-methyl 4-isopropyl heptane*

Solution

The molecular formula of (a) is C_5H_{12}, and that of (b) is $C_{11}H_{24}$. In (a), number the longest chain from the end that gives the methyl substituent the lower number (rule 4). In (b), list the isopropyl and methyl substituents in alphabetical order (rule 6).

(a) 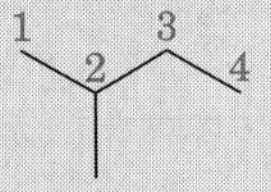   2-Methylbutane

(b) 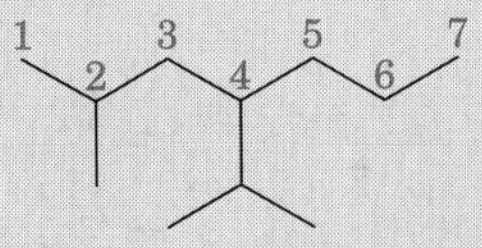   4-Isopropyl-2-methylheptane

---

Problem 2.4

Write the molecular formula and IUPAC name for each alkane.

(a) 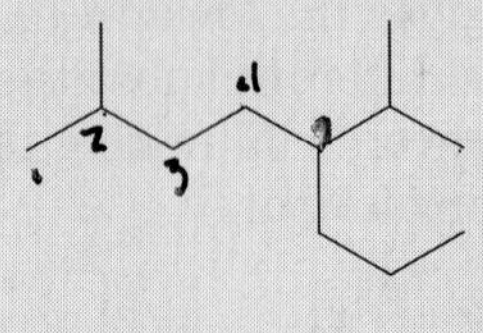   *2 methyl 5 isopropyl octane*

(b) 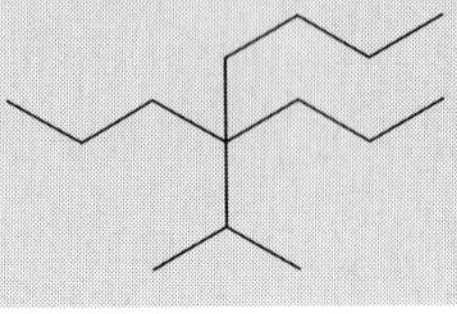   *8 isopropyl 4 butyl octane*

B Common Names

In the older system of **common nomenclature,** the total number of carbon atoms in an alkane, regardless of their arrangement, determines the name. The first three alkanes are methane, ethane, and propane. All alkanes of formula C_4H_{10} are called butanes, all those of formula C_5H_{12} are called pentanes, and all those of formula C_6H_{14} are called hexanes. For alkanes beyond propane, **iso** indicates that one end of an otherwise unbranched chain terminates in a $(CH_3)_2CH—$ group. Following are examples of common names:

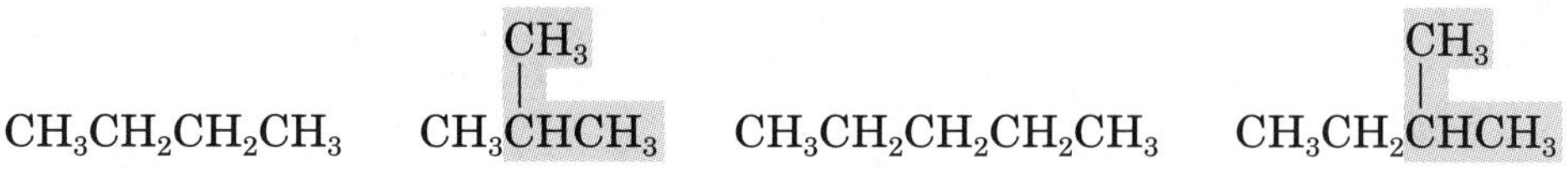

$$CH_3CH_2CH_2CH_3 \qquad CH_3\overset{\overset{\displaystyle CH_3}{|}}{C}HCH_3 \qquad CH_3CH_2CH_2CH_2CH_3 \qquad CH_3CH_2\overset{\overset{\displaystyle CH_3}{|}}{C}HCH_3$$

Butane   Isobutane   Pentane   Isopentane

This system of common names has no way of handling other branching patterns; for more complex alkanes, we must use the more flexible IUPAC system.

In this book, we concentrate on IUPAC names. From time to time we also use common names, especially when chemists and biochemists use them almost exclusively in their everyday discussions. When the text uses both IUPAC and common names, we always give the IUPAC name first, followed by the common name in parentheses. In this way, you should have no doubt about which name is which.

2.5 Cycloalkanes

A hydrocarbon that contains carbon atoms joined to form a ring is called a **cyclic hydrocarbon.** When all carbons of the ring are saturated, the hydrocarbon is called a **cycloalkane.** Cycloalkanes of ring sizes ranging from 3 to more than 30 carbon atoms are found in nature, and in principle there is no limit to ring size. Five-membered (cyclopentane) and six-membered (cyclohexane) rings are especially abundant in nature; for this reason, we concentrate on them in this book.

Organic chemists rarely show all carbons and hydrogens when writing structural formulas for cycloalkanes. Rather, we use line-angle formulas to represent cycloalkane rings. We represent each ring by a regular polygon having the same number of sides as there are carbon atoms in the ring. For example, we represent cyclobutane by a square, cyclopentane by a pentagon, and cyclohexane by a hexagon (Figure 2.3).

To name a cycloalkane, prefix the name of the corresponding open-chain alkane with *cyclo-,* and name each substituent on the ring. If there is only one substituent on the ring, there is no need to give it a location number. If there are two substituents, number the ring beginning with the substituent of lower alphabetical order.

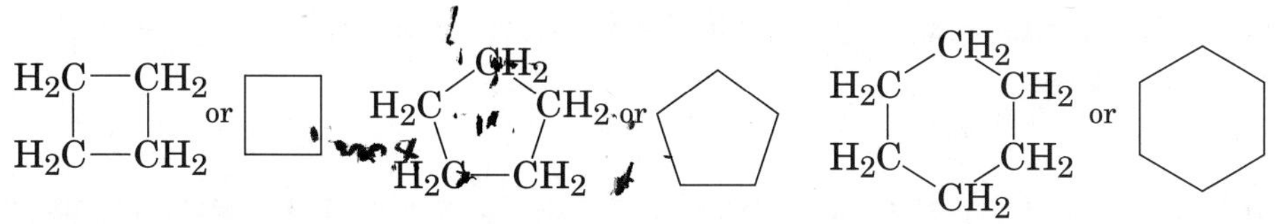

> **Cycloalkane** A saturated hydrocarbon that contains carbon atoms bonded to form a ring

$$\begin{matrix} H_2C—CH_2 \\ | \quad\quad | \\ H_2C—CH_2 \end{matrix} \text{ or } \square \qquad \begin{matrix} CH_2 \\ H_2C \quad CH_2 \\ H_2C—CH_2 \end{matrix} \text{ or } \pentagon \qquad \begin{matrix} CH_2 \\ H_2C \quad CH_2 \\ H_2C \quad CH_2 \\ CH_2 \end{matrix} \text{ or } \hexagon$$

Cyclobutane   Cyclopentane   Cyclohexane

Figure 2.3 Examples of cycloalkanes.

EXAMPLE 2.5

Write the molecular formula and IUPAC name for each cycloalkane.

(a)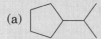

(b)

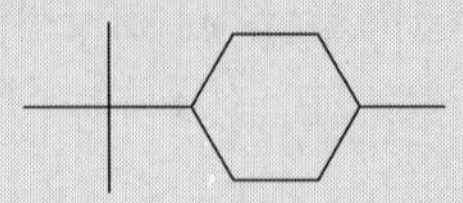

Solution

(a) The molecular formula of this compound is C_8H_{16}. Because there is only one substituent on the ring, there is no need to number the atoms of the ring. The IUPAC name of this cycloalkane is isopropyl-cyclopentane.

(b) The molecular formula is $C_{11}H_{22}$. To name this compound, first number the atoms of the cyclohexane ring beginning with *tert*-butyl, the substituent of lower alphabetical order (remember, alphabetical order here is determined by the b of butyl, and not by the t of *tert*-). The name of this cycloalkane is 1-*tert*-butyl-4-methylcyclohexane.

Problem 2.5

Write the molecular formula and IUPAC name for each cycloalkane.

(a)

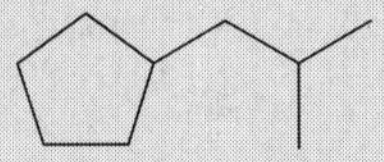

(b)

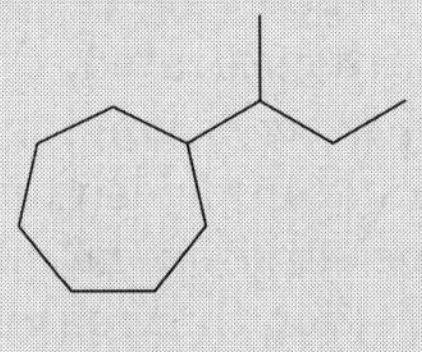

(c)

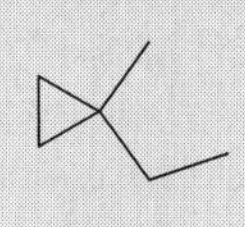

2.6 Shapes of Alkanes and Cycloalkanes

In this section, we concentrate on ways to visualize molecules as three-dimensional objects and to visualize bond angles and relative distances between various atoms and functional groups within a molecule. We urge you to build molecular models of these compounds and to study and manipulate the models we have prepared for you on the CD-ROM supplied with this text. Organic molecules are three-dimensional objects, and it is essential that you become comfortable in dealing with them as such.

A Alkanes

Although the VSEPR model gives us a way to predict the geometry about each carbon atom in an alkane, it provides no information about the three-dimensional shape of an entire molecule. The fact is that there is free rota-

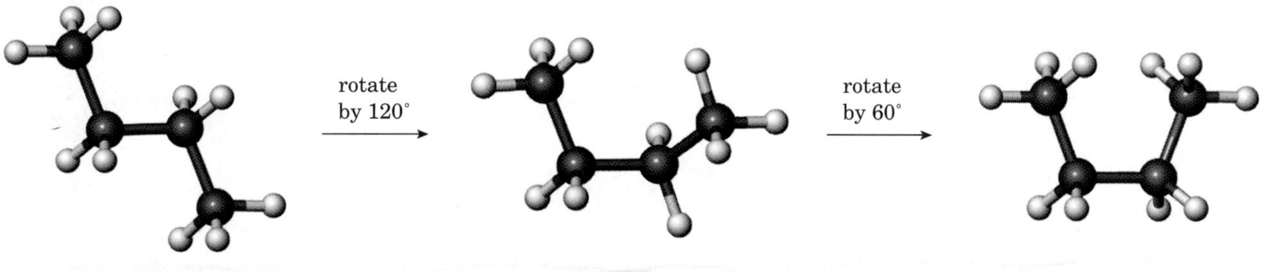

(a) Least crowded conformation; methyl groups are farthest apart.

rotate by 120°

(b) Intermediate crowding; methyl groups are closer to each other.

rotate by 60°

(c) Most crowded conformation; methyl groups are closest to each other.

Figure 2.4 Three conformations of a butane molecule.

tion about each carbon–carbon bond in an alkane and, as a result, even a molecule as simple as ethane has an infinite number of possible three-dimensional shapes, or **conformations.**

Figure 2.4 shows three conformations for a butane molecule. Conformation (a) is the most stable because the methyl groups at the ends of the four-carbon chain are farthest apart. Conformation (b) is formed by a rotation of 120° about the single bond joining carbons 2 and 3. In this conformation, there is some crowding of groups because the two methyl groups are closer together. Rotation about the C_2—C_3 single bond by another 60° gives conformation (c), which is the most crowded because the two methyl groups face each other.

Figure 2.4 shows only three of the possible conformations for a butane molecule. In fact, there are an infinite number of possible conformations that differ only in the angles of rotation about the various C—C bonds within the molecule. Within an actual sample of butane, the conformation of each molecule constantly changes as a result of collisions with other butane molecules and with the walls of the container. Even so, at any given time, a majority of butane molecules are in the most stable, fully extended conformation. There are the fewest butane molecules in the most crowded conformation.

To summarize, for any alkane (except, of course, for methane), there are an infinite number of conformations. The majority of molecules in any sample will be in the least crowded conformation; the fewest will be in the most crowded conformation.

> **Conformation** Any three-dimensional arrangement of atoms in a molecule that results from rotation about a single bond

B Cycloalkanes

We limit our discussion to the conformations of cyclopentanes and cyclohexanes, because they are the most common carbon rings found in nature.

Cyclopentane

The most stable conformation of cyclopentane is the **envelope conformation** shown in Figure 2.5. In it, four carbon atoms are in a plane, and the fifth is bent out of the plane, like an envelope with its flap bent upward. All bond angles in cyclopentane are approximately 109.5°.

Cyclohexane

The most stable conformation of cyclohexane is the **chair conformation** (Figure 2.6), in which all bond angles are approximately 109.5°.

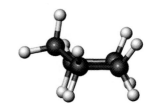

Figure 2.5 Cyclopentane. The most stable conformation is an envelope conformation.

> **Chair conformation** The most stable conformation of a cyclohexane ring; all bond angles are approximately 109.5°

Figure 2.6 Cyclohexane. The most stable conformation is a chair conformation.

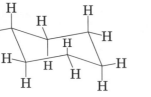

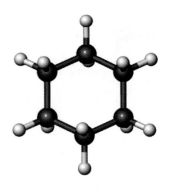

(*a*) Skeletal model

(*b*) Ball-and-stick model viewed from the side

(*c*) Ball-and-stick model viewed from above

Axial position A position on a chair conformation of a cyclohexane ring that extends from the ring parallel to the imaginary axis of the ring

Equatorial position A position on a chair conformation of a cyclohexane ring that extends from the ring roughly perpendicular to the imaginary axis of the ring

In a chair conformation, the twelve C—H bonds are arranged in two different orientations. Six of them are **axial bonds,** and the other six are **equatorial bonds.** One way to visualize the difference between these two types of bonds is to imagine an axis running through the center of the chair, (Figure 2.7). Axial bonds are oriented parallel to this axis. Three of the axial bonds point up; the other three point down. Notice also that axial bonds alternate, first up and then down, as you move from one carbon to the next.

Equatorial bonds are oriented approximately perpendicular to our imaginary axis and also alternate first slightly up and then slightly down as you move from one carbon to the next. Notice also that if the axial bond on a carbon points upward, the equatorial bond on that carbon points slightly downward. Conversely, if the axial bond on a particular carbon points downward, the equatorial bond on that carbon points slightly upward.

Finally, notice that each equatorial bond is oriented parallel to two ring bonds on opposite sides of the ring. A different pair of parallel C—H bonds is shown in each of the following structural formulas, along with the two ring bonds to which each pair is parallel:

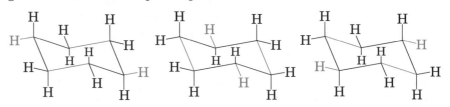

Figure 2.7 Chair conformation of cyclohexane showing equatorial and axial C—H bonds.

Axis through the center of the ring

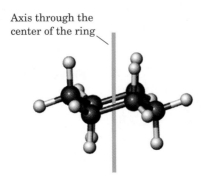

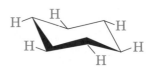

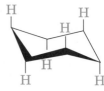

(*a*) Ball-and-stick model showing all 12 hydrogens

(*b*) The six equatorial C—H bonds shown in red

(*c*) The six axial C—H bonds shown in blue

The Poisonous Puffer Fish

Nature is by no means limited to carbon in six-membered rings. Tetrodotoxin, one of the most potent toxins known, is composed of a set of interconnected six-membered rings, each in a chair conformation. All but one of these rings contain atoms other than carbon.

Tetrodotoxin is produced in the liver and ovaries of many species of *Tetraodontidae,* especially the puffer fish, so called because it inflates itself to an almost spherical spiny ball when alarmed. It is evidently a species highly preoccupied with defense, but the Japanese are not put off by its prickly appearance. They regard the puffer, called *fugu* in Japanese, as a delicacy. To serve it in a public restaurant, a chef must be registered as sufficiently skilled in removing the toxic organs so as to make the flesh safe to eat.

Tetrodotoxin blocks the sodium ion channels, which are essential for neurotransmission (Section 15.3). This blockage pre-

Tim Rock/Animals Animals

■ **A puffer fish with its body inflated.**

vents communication between neurons and muscle cells and results in weakness, paralysis, and eventual death.

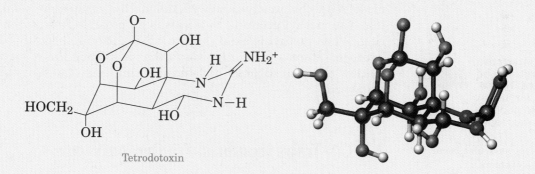

Tetrodotoxin

EXAMPLE 2.6

Following is a chair conformation of methylcyclohexane showing a methyl group and one hydrogen. Indicate by a label whether each is equatorial or axial.

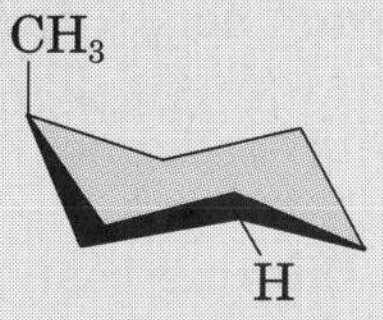

Solution

The methyl group is axial, and the hydrogen is equatorial.

Problem 2.6

Following is a chair conformation of cyclohexane with carbon atoms numbered 1 through 6. Draw methyl groups that are equatorial on carbons 1, 2, and 4.

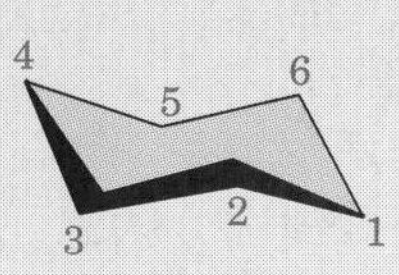

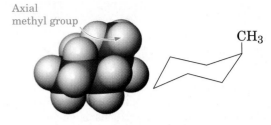

Equatorial
methyl group

Axial
methyl group

CH₃

CH₃

(a) Equatorial methylcyclohexane

(b) Axial methylcyclohexane

Figure 2.8 Methylcyclohexane. The three hydrogens of the methyl group are shown in green to make them stand out more clearly.

See the **Interactive General, Organic, and Biochemistry CD-ROM, version 2.0,** for further exploration on this topic.

Cis-trans isomers have also been called geometric isomers.

Cis-trans isomers Isomers that have the same order of attachment of their atoms but a different arrangement of their atoms in space due to the presence of either a ring or a carbon–carbon double bond

Planar representations of five- and six-membered rings are not spatially accurate because these rings normally exist as envelope and chair conformations. Planar representations, however, are adequate for showing cis-trans isomerism.

Suppose that —CH₃ or another group on a cyclohexane ring may occupy either an equatorial or axial position. Chemists have discovered that a six-membered ring is more stable when the maximum number of substituent groups are equatorial. Perhaps the simplest way to see that this is so is to examine molecular models. Figure 2.8(a) shows a space-filling model of methylcyclohexane with the methyl group equatorial. In this position, the methyl group is as far away as possible from other atoms of the ring. When methyl is axial [Figure 2.8(b)], it quite literally bangs into two hydrogen atoms on the top side of the ring. Thus the more stable conformation of a substituted cyclohexane ring has the substituent group(s) equatorial.

2.7 Cis-Trans Isomerism in Cycloalkanes

Cycloalkanes with substituents on two or more carbons of the ring show a type of isomerism called **cis-trans isomerism.** Cycloalkane cis-trans isomers have (1) the same molecular formula and (2) the same order of attachment of their atoms, but (3) a different arrangement of their atoms in space because of restricted rotation around the carbon–carbon single bonds of the ring. We study cis-trans isomerism in cycloalkanes in this chapter and in alkenes in Chapter 3.

We can illustrate cis-trans isomerism in cycloalkanes by using 1,2-dimethylcyclopentane as an example. In the following structural formulas, the cyclopentane ring is drawn as a planar pentagon viewed edge-on. (In determining the number of cis-trans isomers in a substituted cycloalkane, it is adequate to draw the cycloalkane ring as a planar polygon.) Carbon– carbon bonds of the ring that project forward are shown as heavy lines. When viewed from this perspective, substituents attached to the cyclopentane ring project above and below the plane of the ring. In one isomer of 1,2-dimethylcyclopentane, the methyl groups are on the same side of the ring (either both above or both below the plane of the ring); in the other isomer, they are on opposite sides of the ring (one above and one below the plane of the ring). The prefix **cis** (Latin, on the same side) indicates that the substituents are on the same side of the ring; the prefix **trans** (Latin, across) indicates that they are on opposite sides of the ring.

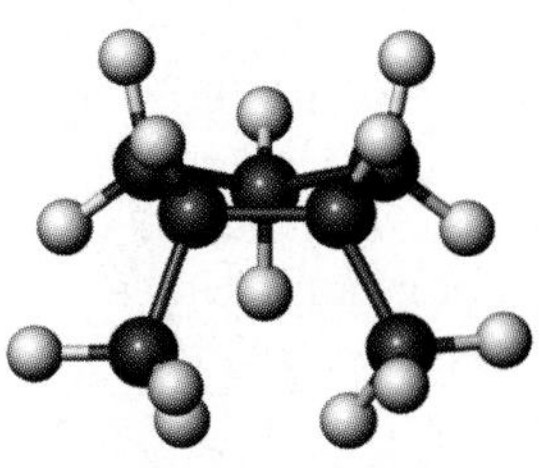

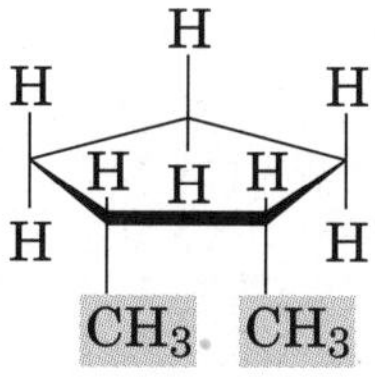

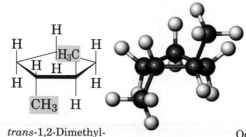

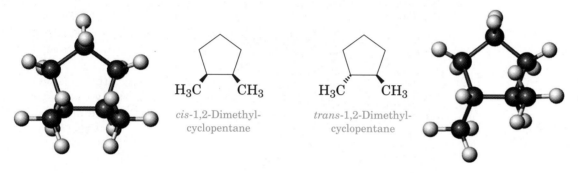

cis-1,2-Dimethyl-
cyclopentane

trans-1,2-Dimethyl-
cyclopentane

Occasionally hydrogen atoms are
written before the carbon, H_3C—,
to avoid crowding or to emphasize
the C—C bond, as in H_3C—CH_3.

Alternatively, we can view the cyclopentane ring from above, with the
ring in the plane of the paper. Substituents on the ring then either project
toward you (that is, they project up above the page) and are shown by solid
wedges, or they project away from you (project down below the page) and
are shown by broken wedges. In the following structural formulas, we show
only the two methyl groups; we do not show hydrogen atoms of the ring.

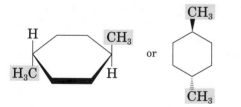

H_3C CH_3

cis-1,2-Dimethyl-
cyclopentane

H_3C CH_3

trans-1,2-Dimethyl-
cyclopentane

Similarly, we can draw a cyclohexane ring as a planar hexagon and view
it through the plane of the ring. Alternatively, we can view it from above
with substituent groups pointing toward you shown by solid wedges, and
those pointing away from you shown by broken wedges.

H

CH_3

H_3C

H

or

CH_3

CH_3

trans-1,4-Dimethylcyclohexane

H

H

H_3C

CH_3

or

CH_3

CH_3

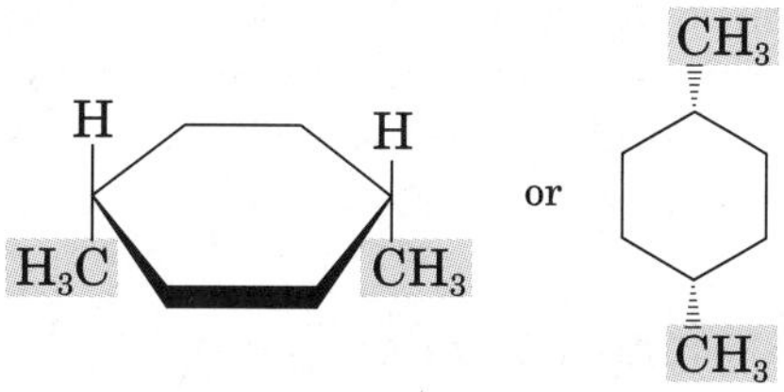

cis-1,4-Dimethylcyclohexane

Because cis-trans isomers differ in the orientation of their atoms in
space, they are **stereoisomers.** Cis-trans isomerism is one type of
stereoisomerism. We will study another type of stereoisomerism, called
enantiomerism, in Chapter 6.

Stereoisomers Isomers that
have the same connectivity (the
same order of attachment of their
atoms) but a different orientation
of their atoms in space

EXAMPLE 2.7

Which of the following cycloalkanes show cis-trans isomerism? For each
that does, draw both isomers.
(a) Methylcyclopentane   (b) 1,1-Dimethylcyclopentane
(c) 1,3-Dimethylcyclobutane

Solution

(a) Methylcyclopentane does not show cis-trans isomerism; it has only one substituent on the ring.

(b) 1,1-Dimethylcyclobutane does not show cis-trans isomerism because only one arrangement is possible for the two methyl groups. Because both groups are bonded to the same carbon, they must be trans to each other—one above the ring, the other below it.

(c) 1,3-Dimethylcyclobutane shows cis-trans isomerism. The two methyl groups may be cis or they may be trans.

cis-1,3-Dimethylcyclobutane   *trans*-1,3-Dimethylcyclobutane

Problem 2.7

Which of the following cycloalkanes show cis-trans isomerism? For each that does, draw both isomers.

(a) 1,3-Dimethylcyclopentane
(b) Ethylcyclopentane
(c) 1,3-Dimethylcyclohexane

2.8 Physical Properties

The most important property of alkanes and cycloalkanes is their almost complete lack of polarity. The electronegativity difference between carbon and hydrogen is $2.5 - 2.1 = 0.4$ on the Pauling scale. Given this small difference, we classify a C—H bond as nonpolar covalent. Therefore, alkanes are nonpolar compounds and the only interactions between their molecules are the very weak London dispersion forces.

A Melting and Boiling Points

The boiling points of alkanes are lower than those of almost any other type of compound of the same molecular weight. In general, both boiling and melting points of alkanes increase with increasing molecular weight (Table 2.4).

Alkanes containing 1 to 4 carbons are gases at room temperature. Alkanes containing 5 to 17 carbons are colorless liquids. High-molecular-weight alkanes (those containing 18 or more carbons) are white, waxy solids. Several plant waxes are high-molecular-weight alkanes. The wax found in apple skins, for example, is an unbranched alkane with molecular formula $C_{27}H_{56}$. Paraffin wax, a mixture of high-molecular-weight alkanes, is used for wax candles, in lubricants, and to seal home-canned jams, jellies, and other preserves. Petrolatum, so named because it is derived from petroleum refining, is a liquid mixture of high-molecular-weight alkanes. It is sold as mineral oil and Vaseline, used as an ointment base in pharmaceuticals and cosmetics, and as a lubricant and rust preventive.

Alkanes that are constitutional isomers are different compounds and have different physical and chemical properties. Table 2.5 lists the boiling

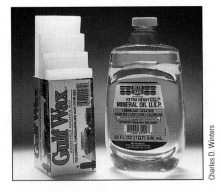

Charles D. Winters

■ **Paraffin wax and mineral oil are mixtures of alkanes.**

See the **Interactive General, Organic, and Biochemistry CD-ROM, version 2.0,** for further exploration on this topic.

Table 2.4 Physical Properties of Some Unbranched Alkanes

Name	Condensed Structural Formula	Molecular Weight (amu)	Melting Point (°C)	Boiling Point (°C)	Density of Liquid (g/mL at 0°C)*
methane	CH_4	16.0	−182	−164	(a gas)
ethane	CH_3CH_3	30.1	−183	−88	(a gas)
propane	$CH_3CH_2CH_3$	44.1	−190	−42	(a gas)
butane	$CH_3(CH_2)_2CH_3$	58.1	−138	0	(a gas)
pentane	$CH_3(CH_2)_3CH_3$	72.2	−130	36	0.626
hexane	$CH_3(CH_2)_4CH_3$	86.2	−95	69	0.659
heptane	$CH_3(CH_2)_5CH_3$	100.2	−90	98	0.684
octane	$CH_3(CH_2)_6CH_3$	114.2	−57	126	0.703
nonane	$CH_3(CH_2)_7CH_3$	128.3	−51	151	0.718
decane	$CH_3(CH_2)_8CH_3$	142.3	−30	174	0.730

*For comparison, the density of H_2O is 1 g/mL at 4°C.

points of the five constitutional isomers with molecular formula C_6H_{14}. The boiling point of each branched-chain isomer is lower than that of hexane itself; the more branching there is, the lower the boiling point. These differences in boiling points are related to molecular shape in the following way. As branching increases, the alkane molecule becomes more compact, and its surface area decreases. As surface area decreases, London dispersion forces act over a smaller surface area. Hence the attraction between molecules decreases, and boiling point decreases. Thus, for any group of alkane constitutional isomers, the least branched isomer generally has the highest boiling point, and the most branched isomer generally has the lowest boiling point.

B Solubility: A Case of "Like Dissolves Like"

Because alkanes are nonpolar compounds, they are not soluble in water, which dissolves only ionic and polar compounds. Water is a polar substance, and its molecules associate with one another through hydrogen bonding. Alkanes do not dissolve in water because they cannot form hydrogen bonds

Table 2.5 Boiling Points of the Five Isomeric Alkanes with Molecular Formula C_6H_{14}

Name	bp (°C)
hexane	68.7
3-methylpentane	63.3
2-methylpentane	60.3
2,3-dimethylbutane	58.0
2,2-dimethylbutane	49.7

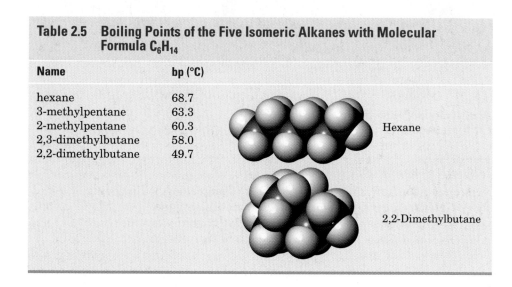

Hexane

2,2-Dimethylbutane

with water. Alkanes, however, are soluble in each other, an example of "like dissolves like." Alkanes are also soluble in other nonpolar organic compounds, such as toluene and diethyl ether.

C Density

The average density of the liquid alkanes listed in Table 2.4 is about 0.7 g/mL; that of higher-molecular-weight alkanes is about 0.8 g/mL. All liquid and solid alkanes are less dense than water (1.0 g/mL) and, because they are insoluble in water, they float on water.

EXAMPLE 2.8

Arrange the alkanes in each set in order of increasing boiling point.
(a) Butane, decane, and hexane
(b) 2-Methylheptane, octane, and 2,2,4-trimethylpentane

Solution
(a) All three compounds are unbranched alkanes. As the number of carbon atoms in the chain increases, London dispersion forces between molecules increase, and boiling points increase. Decane has the highest boiling point, and butane has the lowest boiling point.

Butane
bp –0.5°C

Hexane
bp 69°C

Decane
bp 174°C

(b) These three alkanes are constitutional isomers with the molecular formula C_8H_{18}. Their relative boiling points depend on the degree of branching. 2,2,4-Trimethylpentane is the most highly branched isomer and, therefore, has the smallest surface area and the lowest boiling point. Octane, the unbranched isomer, has the largest surface area and the highest boiling point.

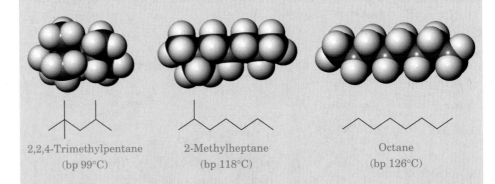

2,2,4-Trimethylpentane
(bp 99°C)

2-Methylheptane
(bp 118°C)

Octane
(bp 126°C)

Problem 2.8
Arrange the alkanes in each set in order of increasing boiling point.
(a) 2-Methylbutane, pentane, and 2,2-dimethylpropane
(b) 3,3-Dimethylheptane, nonane, and 2,2,4-trimethylhexane

2.9 Reactions of Alkanes

The most important chemical property of alkanes and cycloalkanes is their inertness. They are quite unreactive toward any of the normal ionic reaction conditions. Under certain conditions, however, alkanes do react with oxygen, O_2. By far their most important reaction with oxygen is oxidation (combustion) to form carbon dioxide and water. They also react with bromine and chlorine to form halogenated hydrocarbons.

A Reaction with Oxygen: Combustion

Oxidation of hydrocarbons, including alkanes and cycloalkanes, is the basis for their use as energy sources for heat [natural gas, liquefied petroleum gas (LPG), and fuel oil] and power (gasoline, diesel fuel, and aviation fuel). Following are balanced equations for the complete combustion of methane, the major component of natural gas, and propane, the major component of LPG or bottled gas. The heat liberated when an alkane is oxidized to carbon dioxide and water is called its heat of combustion.

$$CH_4 \ + \ 2O_2 \longrightarrow CO_2 + 2H_2O \ + \ 212 \text{ kcal/mol}$$
Methane

$$CH_3CH_2CH_3 + 5O_2 \longrightarrow 3CO_2 + 4H_2O \ + \ 530 \text{ kcal/mol}$$
Propane

B Reaction with Halogens: Halogenation

If we mix methane with chlorine or bromine in the dark at room temperature, nothing happens. If, however, we heat the mixture to 100°C or higher or expose it to light, a reaction begins at once. The products of the reaction between methane and chlorine are chloromethane and hydrogen chloride. What occurs is a substitution reaction—in this case, the substitution of a chlorine for a hydrogen in methane.

$$CH_4 + Cl_2 \xrightarrow{\text{heat or light}} CH_3Cl + HCl$$
Methane   Chloromethane
                            (Methyl chloride)

If chloromethane is allowed to react with more chlorine, further chlorination produces a mixture of dichloromethane, trichloromethane, and tetrachloromethane.

$$CH_3Cl + Cl_2 \xrightarrow{\text{heat}} CH_2Cl_2 + HCl$$
Chloromethane   Dichloromethane
(Methyl chloride)   (Methylene chloride)

$$CH_2Cl_2 \xrightarrow[\text{heat}]{Cl_2} CHCl_3 \xrightarrow[\text{heat}]{Cl_2} CCl_4$$
Dichloromethane   Trichloromethane   Tetrachloromethane
(Methylene chloride)   (Chloroform)   (Carbon tetrachloride)

In the last equation, the reagent Cl_2 is placed over the reaction arrow and the equivalent amount of HCl formed is not shown. Placing reagents over reaction arrows and omitting by-products is commonly done to save space.

We derive IUPAC names of haloalkanes by naming the halogen atom as a substituent (fluoro-, chloro-, bromo-, iodo-) and alphabetizing it along with other substituents. Common names consist of the common name of the alkyl group followed by the name of the halogen (chloride, bromide, and so forth) as a separate word. Dichloromethane (methylene chloride) is a widely used solvent for organic compounds.

EXAMPLE 2.9

Write a balanced equation for the reaction of ethane with chlorine to form chloroethane, C_2H_5Cl.

Solution
The reaction of ethane with chlorine results in the substitution of one of the hydrogen atoms of ethane by a chlorine atom.

$$CH_3CH_3 + Cl_2 \xrightarrow{\text{heat or light}} CH_3CH_2Cl + HCl$$
Ethane   Chloroethane
                 (Ethyl chloride)

Problem 2.9

The reaction of propane with chlorine gives two products with the molecular formula C_3H_7Cl. Draw structural formulas for these two compounds, and give each an IUPAC and common name.

2.10 Important Haloalkanes

One of the major uses of haloalkanes is as intermediates in the synthesis of other organic compounds. Just as we can replace a hydrogen atom of an alkane by a halogen, we can replace the halogen atom, in turn, by a number of other functional groups. In this way, we can construct more complex molecules. In contrast, alkanes that contain several halogens are often quite unreactive, a fact that has proved especially useful in the design of several classes of consumer products.

A The Chlorofluorocarbons

Of all the haloalkanes, the **chlorofluorocarbons (CFCs)** manufactured under the trade name Freons are the most widely known. CFCs are nontoxic, nonflammable, odorless, and noncorrosive. Originally, they seemed to be ideal replacements for the hazardous compounds such as ammonia and sulfur dioxide formerly used as heat-transfer agents in refrigeration systems. Among the CFCs most widely used for this purpose were trichlorofluoromethane (CCl_3F, Freon-11) and dichlorodifluoromethane (CCl_2F_2, Freon-12). The CFCs also found wide use as industrial cleaning solvents to prepare surfaces for coatings, to remove cutting oils and waxes from millings, and to remove protective coatings. In addition, they were employed as propellants for aerosol sprays.

B Solvents

Several low-molecular-weight haloalkanes are excellent solvents in which to carry out organic reactions and to use as cleaners and degreasers. Carbon tetrachloride ("carbon tet") was the first of these compounds to find wide ap-

CHEMICAL CONNECTIONS 2B

The Environmental Impact of Freons

Concern about the environmental impact of CFCs arose in the 1970s when researchers found that more than 4.5×10^5 kg/yr of these compounds were being emitted into the atmosphere. In 1974, Sherwood Rowland and Mario Molina, both of the United States, announced their theory, which has since been amply confirmed, that these compounds destroy the stratospheric ozone layer. When released into the air, CFCs escape to the lower atmosphere. Because of their inertness, however, they do not decompose there. Slowly, they find their way to the stratosphere, where they absorb ultraviolet radiation from the sun and then decompose. As they do so, they set up a chemical reaction that leads to the destruction of the stratospheric ozone layer, which shields Earth against short-wavelength ultraviolet radiation from the sun. An increase in short-wavelength ultraviolet radiation reaching Earth is believed to promote the destruction of certain crops and agricultural species, and even to increase the incidence of skin cancer in light-skinned individuals.

This concern prompted two conventions, one in Vienna in 1985 and one in Montreal in 1987, held by the United Nations Environmental Program. The 1987 meeting produced the Montreal Protocol, which set limits on the production and use of ozone-depleting CFCs and urged the complete phase-out of their production by the year 1996. This phase-out resulted in enormous costs for manufacturers and is not yet complete in developing countries.

Rowland, Molina, and Paul Crutzen (a Dutch chemist at the Max Planck Institute for Chemistry in Germany) were awarded the 1995 Nobel Prize for chemistry. As noted in the award citation by the Royal Swedish Academy of Sciences, "By explaining the chemical mechanisms that affect the thickness of the ozone layer, these three researchers have contributed to our salvation from a global environmental problem that could have catastrophic consequences."

The chemical industry responded to the crisis by developing replacement refrigerants that have much lower ozone-depleting potential. The most prominent of these replacements are the hydrofluorocarbons (HFCs) and hydrochlorofluorocarbons (HCFCs). These compounds are much more chemically reactive in the atmosphere than the Freons and are destroyed before they reach the stratosphere. However, they cannot be used in air conditioners in 1994- and earlier-model cars.

HFC-134a   HCFC-141b

plication, but its use for this purpose has since been discontinued because it is now known that carbon tet is both toxic and carcinogenic. Today, the most widely used haloalkane solvent is dichloromethane, CH_2Cl_2.

2.11 Sources of Alkanes

The two major sources of alkanes are natural gas and petroleum. **Natural gas** consists of approximately 90 to 95% methane, 5 to 10% ethane, and a mixture of other relatively low-boiling alkanes, chiefly propane, butane, and 2-methylpropane.

Petroleum is a thick, viscous, liquid mixture of thousands of compounds, most of them hydrocarbons, formed from the decomposition of marine plants and animals. Petroleum and petroleum-derived products fuel automobiles, aircraft, and trains. They provide most of the greases and lubricants required for the machinery of our highly industrialized society. Furthermore, petroleum, along with natural gas, provides nearly 90 percent of the organic raw materials for the synthesis and manufacture of synthetic fibers, plastics, detergents, drugs, dyes, and a multitude of other products.

The fundamental separation process in refining petroleum is fractional distillation (Figure 2.9). Practically all crude petroleum that enters a refinery goes to distillation units, where it is heated to temperatures as high as 370 to 425°C and separated into fractions. Each fraction contains a mixture of hydrocarbons that boils within a particular range.

■ A petroleum fractional distillation tower.

Ashland Oil Company

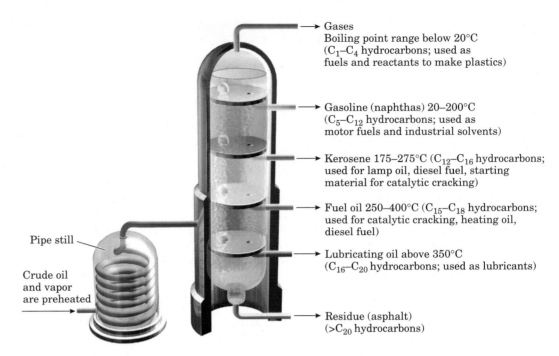

Gases
Boiling point range below 20°C
(C_1–C_4 hydrocarbons; used as
fuels and reactants to make plastics)

Gasoline (naphthas) 20–200°C
(C_5–C_{12} hydrocarbons; used as
motor fuels and industrial solvents)

Kerosene 175–275°C (C_{12}–C_{16} hydrocarbons;
used for lamp oil, diesel fuel, starting
material for catalytic cracking)

Fuel oil 250–400°C (C_{15}–C_{18} hydrocarbons;
used for catalytic cracking, heating oil,
diesel fuel)

Lubricating oil above 350°C
(C_{16}–C_{20} hydrocarbons; used as lubricants)

Residue (asphalt)
(>C_{20} hydrocarbons)

Pipe still

Crude oil
and vapor
are preheated

Figure 2.9 Fractional distillation of petroleum. The lighter, more volatile fractions are removed from higher up the column; the heavier, less volatile fractions are removed from lower down.

 CHEMICAL CONNECTIONS 2C

Octane Rating: What Those Numbers at the Pump Mean

Gasoline is a complex mixture of C_6 to C_{12} hydrocarbons. Its quality as a fuel for internal combustion engines depends on how much it makes an engine "knock." Engine knocking occurs when some of the air–fuel mixture explodes prematurely and independently of ignition by the spark plug.

Two compounds were selected for rating gasoline quality. 2,2,4-Trimethylpentane (isooctane) has very good antiknock properties (the fuel–air mixture burns smoothly in the combustion chamber) and was assigned an octane rating of 100. Heptane, in contrast, has poor antiknock properties and was assigned an octane rating of 0.

■ **Typical octane ratings of commonly available gasolines.**

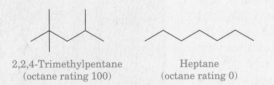

2,2,4-Trimethylpentane
(octane rating 100)

Heptane
(octane rating 0)

The **octane rating** of a particular gasoline is a number equal to the percentage of 2,2,4-trimethylpentane in a mixture of 2,2,4-trimethylpentane and heptane that has the same antiknock properties as the gasoline being rated. For example, the antiknock properties of 2-methylhexane are the same as those of a mixture of 42% 2,2,4-trimethylpentane and 58% heptane; therefore, the octane rating of 2-methylhexane is 42. Ethanol, which is added to gasoline to produce gasohol, has an octane rating of 105. Octane itself has an octane rating of −20.

SUMMARY

A **hydrocarbon** is a compound that contains only carbon and hydrogen (Section 2.1). A **saturated hydrocarbon** contains only single bonds. An **alkane** is a saturated hydrocarbon whose carbon atoms are arranged in an open chain (Section 2.2). **Constitutional isomers** have the same molecular formula but a different order of attachment of their atoms (Section 2.3).

Alkanes are named according to a set of rules developed by the **International Union of Pure and Applied Chemistry** (Section 2.4). The IUPAC name of an alkane consists of two parts: a prefix that tells the number of carbon atoms in the parent chain, and the ending **-ane.** Substituents derived from alkanes by removal of a hydrogen atom are called **alkyl groups** and are given the symbol **R—.**

An alkane that contains carbon atoms bonded to form a ring is called a **cycloalkane** (Section 2.5). To name a cycloalkane, we prefix the name of the open-chain alkane with **cyclo-.**

A **conformation** is any three-dimensional arrangement of the atoms of a molecule that results from rotation about a single bond (Section 2.6). The lowest-energy conformation of cyclopentane is an **envelope conformation.** The lowest-energy conformation of cyclohexane is a **chair conformation.** In a chair conformation, six bonds are **axial** and six bonds are **equatorial.** A substituent on a six-membered ring is more stable when it is equatorial than when it is axial.

Cis-trans isomers of cycloalkanes have (1) the same molecular formula and (2) the same order of attachment of their atoms, (3) but a different orientation of their atoms in space because of the restricted rotation around the C—C bonds of the ring (Section 2.7). With cis-trans isomers of cycloalkanes, **cis** means that substituents are on the same side of the ring; **trans** means that they are on opposite sides of the ring.

Alkanes are nonpolar compounds, and the only forces of attraction between their molecules are London dispersion forces (Section 2.8). At room temperature, low-molecular-weight alkanes are gases, higher-molecular-weight alkanes are liquids, and very-high-molecular-weight alkanes are solids. For any group of alkane constitutional isomers, the least branched isomer generally has the highest boiling point, and the most branched isomer generally has the lowest boiling point.

Alkanes are insoluble in water but soluble in each other and in other nonpolar organic solvents such as toluene (Section 2.8B). All liquid and solid alkanes are less dense than water (Section 2.8C).

Natural gas consists of 90 to 95 percent methane with lesser amounts of ethane and other lower-molecular-weight hydrocarbons (Section 2.11). **Petroleum** is a liquid mixture of thousands of different hydrocarbons.

KEY REACTIONS

1. **Oxidation of Alkanes (Section 2.9A)** Oxidation of alkanes to carbon dioxide and water is the basis for their use as sources of heat and power.

$$CH_3CH_2CH_3 + 5O_2 \longrightarrow 3CO_2 + 4H_2O + 530 \text{ kcal/mol}$$
Propane

2. **Halogenation of Alkanes (Section 2.9B)** Reaction of an alkane with chlorine or bromine results in the substitution of a halogen atom for a hydrogen.

$$CH_3CH_3 + Cl_2 \xrightarrow{\text{heat or light}} CH_3CH_2Cl + HCl$$
Ethane   Chloroethane
                     (Ethyl chloride)

PROBLEMS

Numbers that appear in color indicate difficult problems.
◢ designates problems requiring application of principles.

Alkanes

2.10 Define:
   (a) Hydrocarbon   (b) Alkane
   (c) Saturated hydrocarbon

2.11 Why is it not accurate to describe unbranched alkanes as "straight chain" alkanes?

2.12 What is meant by the term "line-angle formula" as applied to alkanes and cycloalkanes?

2.13   For each condensed structural formula, write a line-angle formula.

(a) CH₃CH₂CHCHCH₂CHCH₃
with CH₂CH₃ and CH₃ above, CH(CH₃)₂ below

(b) CH₃CCH₃
with CH₃ above and CH₃ below

(c) (CH₃)₂CHCH(CH₃)₂

(d) CH₃CH₂CCH₂CH₃
with CH₂CH₃ above and CH₂CH₃ below

(e) (CH₃)₃CH

(f) CH₃(CH₂)₃CH(CH₃)₂

2.14   Write the molecular formula for each alkane.

(a)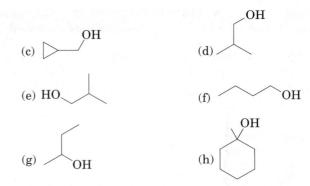

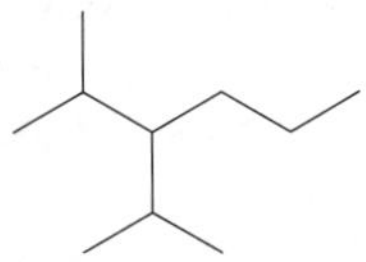

(b)

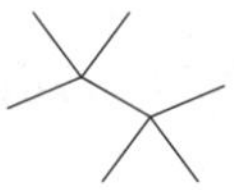

(c)

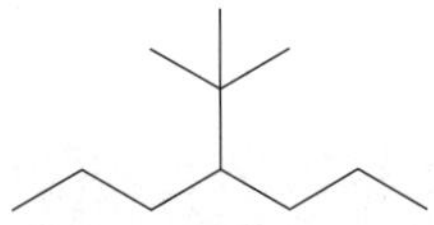

Constitutional Isomerism

2.15   Which statements are true about constitutional isomers?
(a) They have the same molecular formula.
(b) They have the same molecular weight.
(c) They have the same order of attachment of atoms.
(d) They have the same physical properties.

2.16   Each member of the following set of compounds is an alcohol; that is, each contains an —OH (hydroxyl group, Section 1.4A). Which structural formulas represent the same compound, and which represent constitutional isomers?

(a) 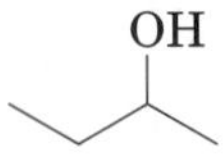   OH

(b) 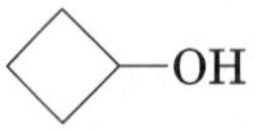   —OH

2.17   Each member of the following set of compounds is an amine; that is, each contains a nitrogen bonded to one, two, or three carbon groups (Section 1.4B). Which structural formulas represent the same compound, and which represent constitutional isomers?

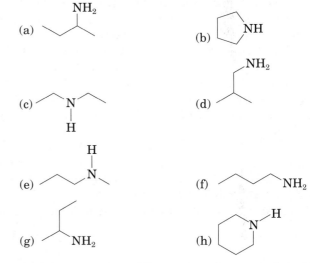

2.18   Each member of the following set of compounds is either an aldehyde or ketone (Section 1.4C). Which structural formulas represent the same compound, and which represent constitutional isomers?

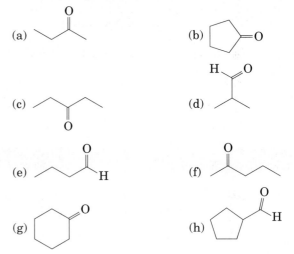

2.19 For which parts do the pairs of structural formulas represent constitutional isomers?

(a)  and

(b) 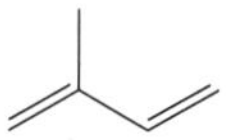 and

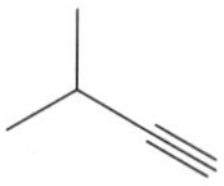

(c) 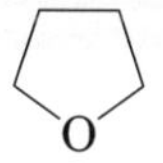 and

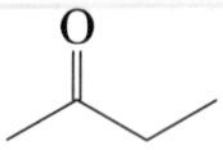

(d) 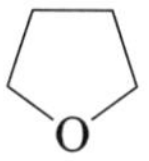 and

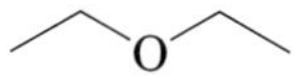

(e) 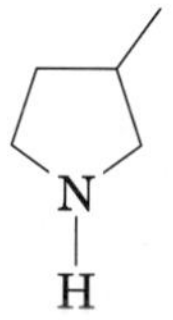 and

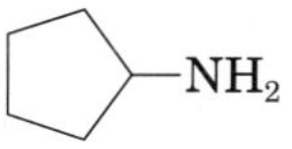

(f) 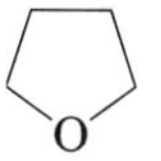 and

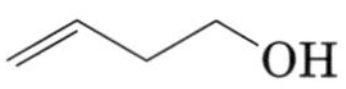

2.20 Draw line-angle formulas for the nine constitutional isomers with molecular formula C_7H_{16}.

Nomenclature of Alkanes and Cycloalkanes

2.21 Name these alkyl groups:

(a) CH_3CH_2—

(b) $CH_3\overset{\displaystyle CH_3}{\underset{\displaystyle |}{CH}}$—

(c) $CH_3\overset{\displaystyle CH_3}{\underset{\displaystyle |}{CH}}CH_2$—

(d) $CH_3\overset{\displaystyle CH_3}{\underset{\displaystyle \underset{\displaystyle CH_3}{|}}{\overset{\displaystyle |}{C}}}$—

2.22 Write the IUPAC names for isobutane and isopentane.

2.23 Write the IUPAC names for these alkanes and cycloalkanes

(a) $CH_3\underset{\displaystyle \underset{\displaystyle CH_3}{|}}{CH}CH_2CH_2CH_3$

(b) $CH_3\underset{\displaystyle \underset{\displaystyle CH_3}{|}}{CH}CH_2CH_2\underset{\displaystyle \underset{\displaystyle CH_3}{|}}{CH}CH_3$

(c) $CH_3(CH_2)_4\underset{\displaystyle \underset{\displaystyle CH_2CH_3}{|}}{CH}CH_2CH_3$

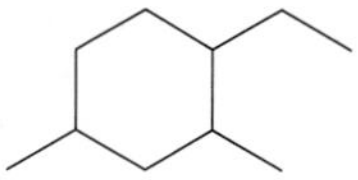

2.24 Write line-angle formulas for these alkanes and cycloalkanes.
  (a) 2,2,4-Trimethylhexane
  (b) 2,2-Dimethylpropane
  (c) 3-Ethyl-2,4,5-trimethyloctane
  (d) 5-Butyl-2,2-dimethylnonane
  (e) 4-Isopropyloctane
  (f) 3,3-Dimethylpentane
  (g) *trans*-1,3-Dimethylcyclopentane
  (h) *cis*-1,2-Diethylcyclobutane

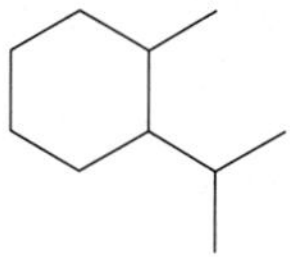

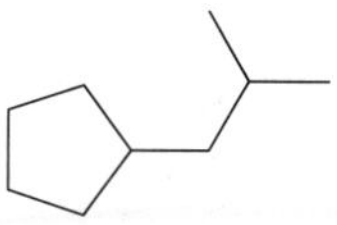

Conformations of Alkanes and Cycloalkanes

2.25 Define conformation.

2.26 The condensed structural formula of butane is $CH_3CH_2CH_2CH_3$. Explain why this formula does not show the geometry of the real molecule.

2.27 Draw a conformation of ethane in which hydrogen atoms on adjacent carbons are as far apart as possible. Also draw a conformation in which they are as close together as possible. In a sample of ethane molecules at room temperature, which conformation is the more likely?

Cis-Trans Isomerism in Cycloalkanes

2.28 What structural feature of cycloalkanes makes cis-trans isomerism in them possible?

2.29 Is cis-trans isomerism possible in alkanes?

2.30 Name and draw structural formulas for the cis and trans isomers of 1,2-dimethylcyclopropane.

2.31 Name and draw structural formulas for the six cycloalkanes with molecular formula C_5H_{10}. Include cis-trans isomers as well as constitutional isomers.

2.32 Why is equatorial methylcyclohexane more stable than axial methylcyclohexane?

2.33 Following is a structural formula for 2-isopropyl-5-methylcyclohexanol.

2-Isopropyl-5-methylcyclohexanol

Using a planar hexagon representation for the cyclohexane ring, draw a structural formula for the cis-trans isomer with isopropyl trans to —OH and methyl cis to —OH. If you answered this part correctly, you have drawn the isomer found in nature and given the name menthol.

2.34 On the left is a representation of the glucose molecule. Convert this representation to the alternative representation using the ring on the right. (We discuss the structure and chemistry of glucose in Chapter 2.)

2.35 On the left is a representation for 2-deoxy-D-ribose. This molecule is the "D" of DNA. Convert this representation to the alternative representation using the ring on the right. (We discuss the structure and chemistry of this compound in more detail in Chapter 2.)

2-Deoxy-D-ribose

Physical Properties of Alkanes and Cycloalkanes

2.36 In Problem 2.20, you drew structural formulas for the nine constitutional isomers with molecular formula C_7H_{16}. Predict which isomer has the lowest boiling point and which has the highest boiling point.

2.37 What unbranched alkane (Table 2.4) has about the same boiling point as water? Calculate the molecular weight of this alkane, and compare it with the molecular weight of water.

2.38 What generalizations can you make about the densities of alkanes relative to the density of water?

2.39 What generalization can you make about the solubility of alkanes in water?

2.40 Suppose that you have samples of hexane and octane. Could you tell the difference by looking at them? What color would they be? How could you tell which is which?

2.41 As you can see from Table 2.4, each CH_2 group added to the carbon chain of an alkane increases its boiling point. This increase is greater going from CH_4 to C_2H_6 and from C_2H_6 to C_3H_8 than it is going from C_8H_{18} to C_9H_{20} or from C_9H_{20} to $C_{10}H_{22}$. What do you think is the reason for this difference?

2.42 How are the boiling points of hydrocarbons during petroleum refining related to their molecular weight?

Reactions of Alkanes

2.43 Write balanced equations for the combustion of the following hydrocarbons. Assume that each is converted completely to carbon dioxide and water.

(a) Hexane

(b) Cyclohexane

(c) 2-Methylpentane

2.44 The heat of combustion of methane, a component of natural gas, is 212 kcal/mol. That of propane, a component of LP gas, is 530 kcal/mol. On a gram-for-gram basis, which hydrocarbon is the better source of heat energy?

2.45 Draw structural formulas for these haloalkanes.

(a) Bromomethane

(b) Chlorocyclohexane

(c) 1,2-Dibromoethane

(d) 2-Chloro-2-methylpropane

(e) Dichlorodifluoromethane (Freon-12)

2.46 The reaction of chlorine with pentane gives a mixture of three chloroalkanes with molecular formula $C_5H_{11}Cl$. Write a line-angle formula and IUPAC name for each.

Chemical Connections

2.47 (Chemical Connections 2A) How many rings in tetrodotoxin contain only carbon atoms? How many contain nitrogen atoms? How many contain two oxygen atoms?

2.48 (Chemical Connections 2B) What are Freons? Why were they considered ideal compounds to use as heat-transfer agents in refrigeration systems? Give structural formulas of two Freons used for this purpose.

2.49 (Chemical Connections 2B) In what way do Freons negatively affect the environment?

2.50 (Chemical Connections 2B) What are HFCs and HCFCs? How does their use in refrigeration systems avoid the environmental problems associated with the use of Freons?

2.51 (Chemical Connections 2C) What is an "octane rating"? What two reference hydrocarbons are used for setting the scale of octane ratings?

2.52 (Chemical Connections 2C) Octane has an octane rating of −20. Will it produce more or less engine knocking than heptane does?

2.53 (Chemical Connections 2C) When ethanol is added to gasoline to produce "gasohol," it promotes more complete combustion of the gasoline and is an octane booster. Compare the heats of combustion of 2,2,4-trimethylpentane (1304 kcal/mol) and ethanol (327 kcal/mol). Which has the higher heat of combustion in kcal/mol? In kcal/g?

Additional Problems

2.54 Tell whether the compounds in each set are constitutional isomers.

(a) CH_3CH_2OH and CH_3OCH_3

(b) $CH_3\overset{\overset{\textstyle O}{\|}}{C}CH_3$ and $CH_3CH_2\overset{\overset{\textstyle O}{\|}}{C}H$

(c) $CH_3\overset{\overset{\textstyle O}{\|}}{C}OCH_3$ and $CH_3CH_2\overset{\overset{\textstyle O}{\|}}{C}OH$

(d) $CH_3\overset{\overset{\textstyle OH}{|}}{C}HCH_2CH_3$ and $CH_3\overset{\overset{\textstyle O}{\|}}{C}CH_2CH_3$

(e) 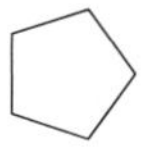 ⬠ and $CH_3CH_2CH_2CH_2CH_3$

(f) 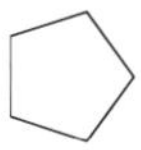 ⬠ and $CH_2{=}CHCH_2CH_2CH_3$

2.55 Explain why each of the following is an incorrect IUPAC name. Write the correct IUPAC name for the compound.

(a) 1,3-Dimethylbutane
(b) 4-Methylpentane
(c) 2,2-Diethylbutane
(d) 2-Ethyl-3-methylpentane
(e) 2-Propylpentane
(f) 2,2-Diethylheptane
(g) 2,2-Dimethylcyclopropane
(h) 1-Ethyl-5-methylcyclohexane

2.56 Which compounds can exist as cis-trans isomers? For each that does, draw both isomers using solid and

dashed wedges to show the orientation in space of the —OH and —CH₃ groups.

(a) 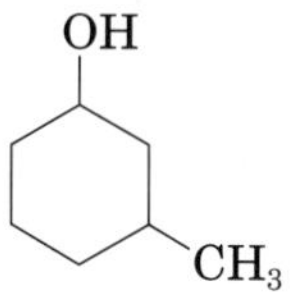 [cyclohexane with OH and CH₃ substituents]

(b) 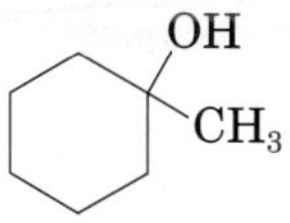 [cyclohexane with OH and CH₃ substituents]

(c) 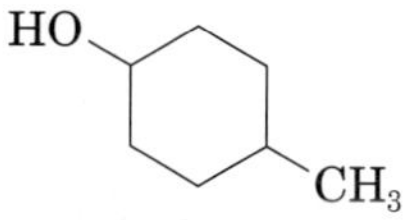 [cyclohexane with HO and CH₃ substituents]

2.57 Tetradecane, $C_{14}H_{30}$, is an unbranched alkane with a melting point of 5.9°C and a boiling point of 254°C. Is tetradecane a solid, liquid, or gas at room temperature?

2.58 Dodecane, $C_{12}H_{26}$, is an unbranched alkane. Predict the following:

(a) Will it dissolve in water?
(b) Will it dissolve in hexane?
(c) Will it burn when ignited?
(d) Is it a liquid, solid, or gas at room temperature and atmospheric pressure?
(e) Is it more or less dense than water?

2.59 As stated in Section 2.8, the wax found in apple skins is an unbranched alkane with molecular formula $C_{27}H_{56}$. Explain how the presence of this alkane prevents the loss of moisture from within an apple.

InfoTrac College Edition

For additional readings, go to InfoTrac College Edition, your online research library, at

http://infotrac.thomsonlearning.com

CHAPTER 3

3.1 Introduction

3.2 Structure

3.3 Nomenclature

3.4 Physical Properties

3.5 Naturally Occurring Alkenes: The Terpenes

3.6 Addition Reactions of Alkenes

3.7 Polymerization of Ethylene and Substituted Ethylenes

Carotene is a naturally occurring polyene in carrots and tomatoes (Problem 3.61).

Charles D. Winters

Alkenes

3.1 Introduction

In this chapter, we begin our study of unsaturated hydrocarbons. Recall from Section 2.1 that these compounds contain one or more carbon–carbon double bonds, triple bonds, or benzene-like rings. In this chapter we study **alkenes.** The simplest alkene is ethylene.

Alkene An unsaturated hydrocarbon that contains a carbon–carbon double bond

Alkyne An unsaturated hydrocarbon that contains a carbon–carbon triple bond

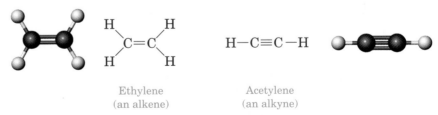

Ethylene
(an alkene)

Acetylene
(an alkyne)

Alkynes are unsaturated hydrocarbons that contain one or more carbon–carbon triple bonds. The simplest alkyne is acetylene. Because alkynes are not widespread in nature and have little importance in biochemistry, we will not study their chemistry.

Ethylene, a Plant Growth Regulator

As noted below, ethylene occurs only in trace amounts in nature. True enough, but scientists have discovered that this small molecule is a natural ripening agent for fruits. Thanks to this knowledge, fruit growers can now pick fruit while it is still green and less susceptible to bruising. When they are ready to pack the fruit for shipment, the growers can treat it with ethylene gas. Alternatively, fruit can be treated with ethephon (Ethrel), which slowly releases ethylene and initiates fruit ripening.

$$Cl-CH_2-CH_2-\overset{\overset{\displaystyle O}{\|}}{\underset{\underset{\displaystyle OH}{|}}{P}}-OH$$

Ethephon

The next time you see ripe bananas in the market, you might wonder when they were picked and whether their ripening was artificially induced.

Compounds containing carbon–carbon double bonds are especially widespread in nature. Furthermore, several low-molecular-weight alkenes, including ethylene and propene, have enormous commercial importance in our modern, industrialized society. The organic chemical industry worldwide produces more pounds of ethylene than any other chemical. Annual production in the United States alone exceeds 55 billion pounds.

What is unusual about ethylene is that it occurs only in trace amounts in nature. The enormous amounts of it required to meet the needs of the chemical industry are derived the world over by thermal cracking of hydrocarbons. In the United States and other areas of the world with vast reserves of natural gas, the major process for the production of ethylene is thermal cracking of the small quantities of ethane extracted from natural gas. In **cracking,** a saturated hydrocarbon is converted to an unsaturated hydrocarbon plus H_2. Ethane is cracked by heating it in a furnace to 800–900°C for a fraction of a second.

$$CH_3CH_3 \xrightarrow[\text{(thermal cracking)}]{800-900°C} CH_2{=}CH_2 + H_2$$

Ethane   Ethylene

Europe, Japan, and other parts of the world with limited supplies of natural gas depend almost entirely on thermal cracking of petroleum for their ethylene.

From the perspective of the chemical industry, the single most important reaction of ethylene and other low-molecular-weight alkenes is polymerization, which we discuss in Section 3.7. The crucial point to recognize is that ethylene and all of the commercial and industrial products made from it are derived from either natural gas or petroleum—both nonrenewable natural resources!

3.2 Structure

A Alkenes

Using the VSEPR model, we predict bond angles of 120° about each carbon in a double bond. The observed H—C—C bond angle in ethylene, for example, is 121.7°, close to the predicted value. In other alkenes, deviations from the predicted angle of 120° may be somewhat larger because of interaction between alkyl groups bonded to the doubly bonded carbons. The C—C—C bond angle in propene, for example, is 124.7°.

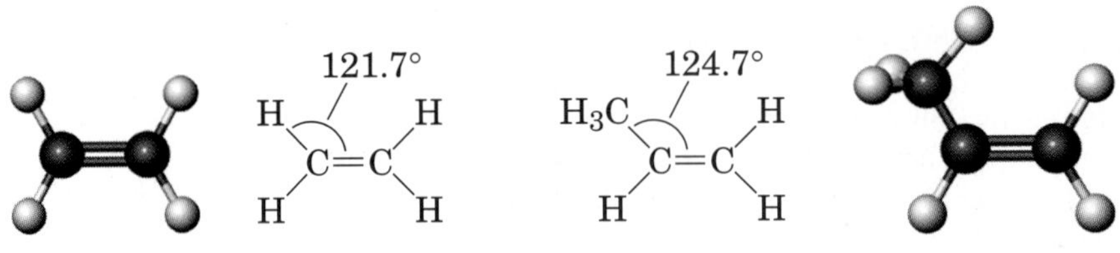

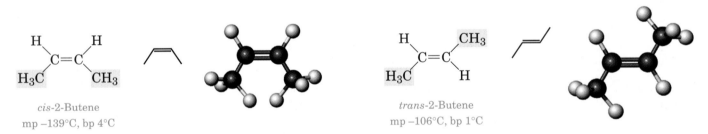

Ethylene   Propene

If we look at a molecular model of ethylene, we see that the two carbons of the double bond and the four hydrogens bonded to them all lie in the same plane; that is, ethylene is a flat or planar molecule. Furthermore, chemists have discovered that, under normal conditions, no rotation is possible about the carbon–carbon double bond of ethylene or, for that matter, of any other alkene. Whereas free rotation occurs about each carbon–carbon single bond in an alkane (Section 2.6A), rotation about the carbon–carbon double bond in an alkene does not normally take place. For an important exception to this generalization about carbon–carbon double bonds, however, see Chemical Connections 3C: Cis-Trans Isomerism in Vision.

B Cis-Trans Isomerism in Alkenes

Cis-trans isomerism Isomers that have the same order of attachment of their atoms but a different arrangement of their atoms in space due to the presence of either a ring or a carbon–carbon double bond

Because of the restricted rotation about a carbon–carbon double bond, an alkene in which each carbon of the double bond has two different groups bonded to it shows **cis-trans isomerism.** For example, 2-butene has two cis-trans isomers. In *cis*-2-butene, the two methyl groups are located on one side of the double bond and the two hydrogens are on the other side. In *trans*-2-butene, the two methyl groups are on opposite sides of the double bond. *Cis*-2-butene and *trans*-2-butene are different compounds and have different physical and chemical properties.

cis-2-Butene
mp −139°C, bp 4°C

trans-2-Butene
mp −106°C, bp 1°C

3.3 Nomenclature

Alkenes and alkynes are named using the IUPAC system. As we will see, some are still referred to by their common names.

A IUPAC Names

See the **Interactive General, Organic, and Biochemistry CD-ROM, version 2.0,** for further exploration on this topic.

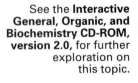

The key to the IUPAC names of alkenes is the ending **-ene.** Just as the ending "-ane" tells us that a hydrocarbon chain contains only carbon–carbon single bonds, so the ending "-ene" tells us that it contains a carbon–carbon double bond. To name an alkene:

  1. Find the longest carbon chain that includes the double bond. Indicate the length of the parent chain by using a prefix that tells the number of carbon atoms in it (Table 2.2).

2. Number the chain from the end that gives the lower set of numbers to the carbon atoms of the double bond. Designate the position of the double bond by the number of its first carbon.

3. Branched alkenes are named in a manner similar to alkanes; substituent groups are located and named.

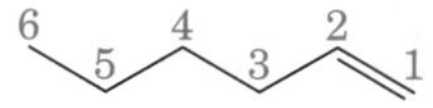

$$CH_3CH_2CH_2CH_2CH\!=\!CH_2$$
1-Hexene

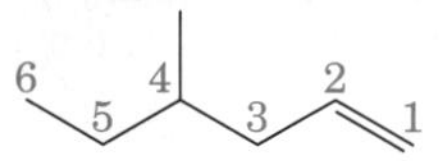

$$CH_3CH_2CHCH_2CH\!=\!CH_2$$
4-Methyl-1-hexene

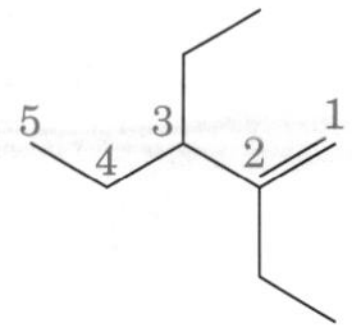

$$CH_3CH_2CHC\!=\!CH_2$$
2,3-Diethyl-1-pentene

Note that, although 2,3-diethyl-1-pentene has a six-carbon chain, the longest chain that contains the double bond has only five carbons. The parent alkane is, therefore, a pentane rather than a hexane, and the molecule is named as a disubstituted 1-pentene.

The key to the IUPAC name of an alkyne is the ending **-yne,** which shows the presence of a carbon–carbon triple bond. Thus $HC\!\equiv\!CH$ is ethyne and $CH_3C\!\equiv\!CH$ is propyne. In higher alkynes, number the longest carbon chain that contains the triple bond from the end that gives the lower set of numbers to the triply bonded carbons. Indicate the location of the triple bond by the number of its first carbon atom.

$$CH_3CHC\!\equiv\!CH$$
3-Methyl-1-butyne

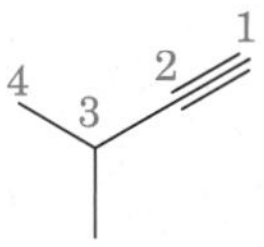

$$CH_3CH_2C\!\equiv\!CCH_2CCH_3$$ 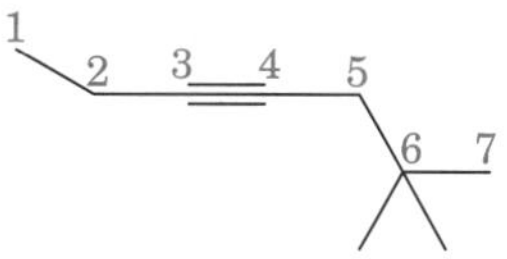
6,6-Dimethyl-3-heptyne

EXAMPLE 3.1

Write the IUPAC name of each unsaturated hydrocarbon.

(a) $CH_2\!=\!CH(CH_2)_5CH_3$

(b)
$$\underset{H_3C}{\overset{H_3C}{\diagdown}}C\!=\!C\overset{CH_3}{\underset{H}{\diagup}}$$

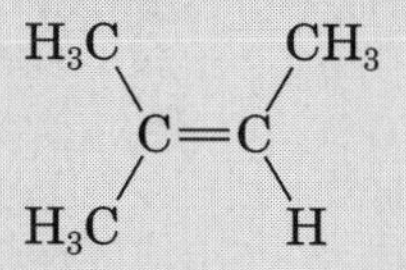

(c) $CH_3(CH_2)_2C\!\equiv\!CCH_3$

Solution

(a) The parent chain contains eight carbons; thus the parent alkane is octane. To show the presence of the carbon–carbon double bond, change -ane to -ene. Number the chain beginning with the first carbon of the double bond. This alkene is 1-octene.

(b) Because there are four carbon atoms in the chain containing the
carbon–carbon double bond, the parent alkane is butane. The double
bond is between carbons 2 and 3 of the chain, and there is a methyl
group on carbon 2. This alkene is 2-methyl-2-butene.

(c) There are six carbons in the parent chain, with the triple bond between carbons 2 and 3. This alkyne is 2-hexyne.

Problem 3.1

Write the IUPAC name of each unsaturated hydrocarbon.

(a)

(b)

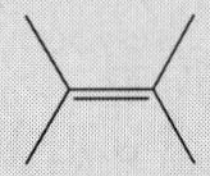

(c)

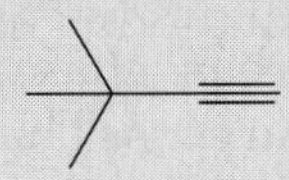

B Common Names

Despite the precision and universal acceptance of IUPAC nomenclature,
some alkenes and alkynes—particularly those of low molecular weight—
are known almost exclusively by their common names. Three examples
are:

	$CH_2{=}CH_2$	$CH_3CH{=}CH_2$	$CH_3\overset{\displaystyle CH_3}{\underset{\displaystyle |}{C}}{=}CH_2$
IUPAC name:	Ethene	Propene	2-Methylpropene
Common name:	Ethylene	Propylene	Isobutylene

We derive common names for alkynes by prefixing the names of the substituents on the carbon–carbon triple bond to the name *acetylene:*

	$HC{\equiv}CH$	$CH_3C{\equiv}CH$	$CH_3C{\equiv}CCH_3$
IUPAC name:	Ethyne	Propyne	2-Butyne
Common name:	Acetylene	Methylacetylene	Dimethylacetylene

C Cis and Trans Configurations of Alkenes

The orientation of the carbon atoms of the parent chain determines whether
an alkene is cis or trans. If the carbons of the parent chain are on the same
side of the double bond, the alkene is cis; if they are on opposite sides, it is a
trans alkene. In the first example on the next page, they are on opposite
sides and the compound is a trans alkene. In the second example, they are
on the same side and the compound is a cis alkene.

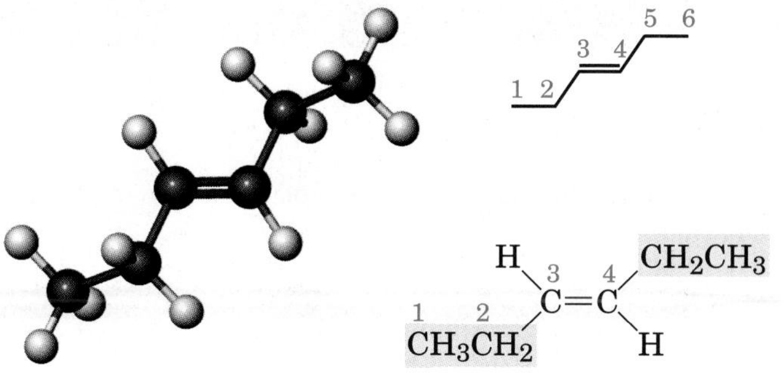

5 6
3 4
1 2

$$
\begin{array}{cc}
H & CH_2CH_3 \\
\underset{3}{\diagdown} & \underset{4}{\diagup} \\
C=C \\
\underset{CH_3CH_2}{\overset{1}{\diagup}} \underset{2}{} & \overset{}{\diagdown} H
\end{array}
$$

trans-3-Hexene

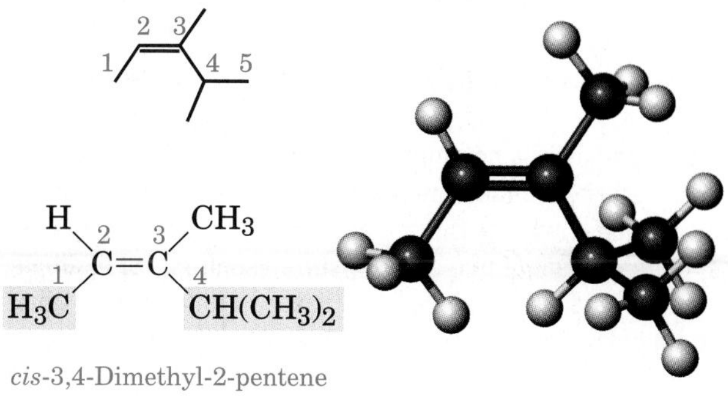

2 3
1 4 5

$$
\begin{array}{cc}
H & CH_3 \\
\underset{2}{\diagdown} & \diagup \\
C=C \\
\underset{H_3C}{\overset{1}{\diagup}} & \underset{4}{\diagdown} CH(CH_3)_2
\end{array}
$$

cis-3,4-Dimethyl-2-pentene

EXAMPLE 3.2

Name each alkene and specify its configuration.

(a) 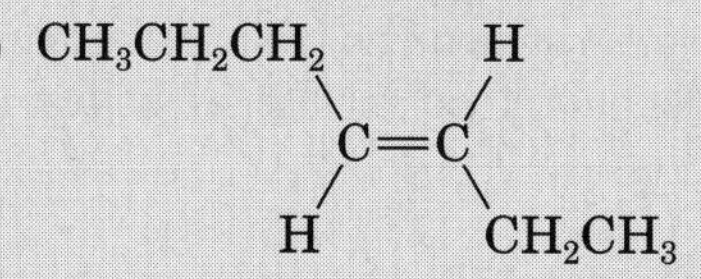
$$
\begin{array}{cc}
CH_3CH_2CH_2 & H \\
\diagdown & \diagup \\
C=C \\
\diagup & \diagdown \\
H & CH_2CH_3
\end{array}
$$

(b) 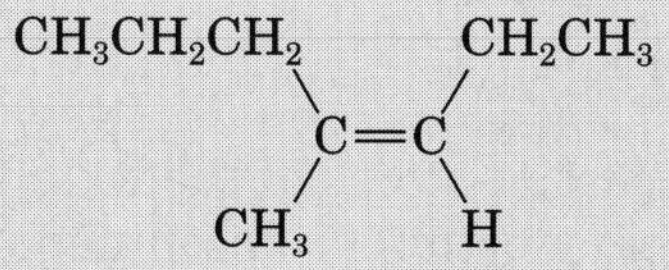
$$
\begin{array}{cc}
CH_3CH_2CH_2 & CH_2CH_3 \\
\diagdown & \diagup \\
C=C \\
\diagup & \diagdown \\
CH_3 & H
\end{array}
$$

Solution

(a) The chain contains seven carbon atoms and is numbered from the right to give the lower number to the first carbon of the double bond. The carbon atoms of the parent chain are on opposite sides of the double bond. This alkene is *trans*-3-heptene.

(b) The longest chain contains seven carbon atoms and is numbered from the right so that the first carbon of the double bond is carbon 3 of the chain. The carbon atoms of the parent chain are on the same side of the double bond. This alkene is *cis*-4-methyl-3-heptene.

Problem 3.2

Name each alkene and specify its configuration.

(a)

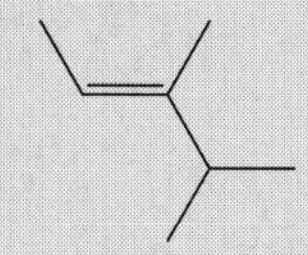

(b)

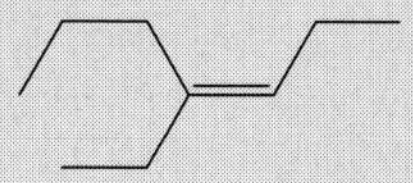

CHEMICAL CONNECTIONS 3B

The Case of the Iowa and New York Strains of the European Corn Borer

Although humans communicate largely by sight and sound, the vast majority of other species in the animal world communicate by chemical signals. Often, communication within a species is specific for one of two or more stereoisomers. For example, a member of a given species may respond to the cis isomer of a chemical but not to the trans isomer. Alternatively, it might respond to a precise blend of cis and trans isomers but not to other blends of these same isomers.

Several groups of scientists have studied the components of the sex **pheromones** of both the Iowa and New York strains of the European corn borer. Females of these closely related species secrete the sex attractant 11-tetradecenyl acetate. Males of the Iowa strain show maximum response to a mixture containing 96% of the cis isomer and 4% of the trans isomer. When the pure cis isomer is used alone, males are only weakly attracted. Males of the New York strain show an entirely different response pattern. They respond maximally

to a mixture containing 3% of the cis isomer and 97% of the trans isomer.

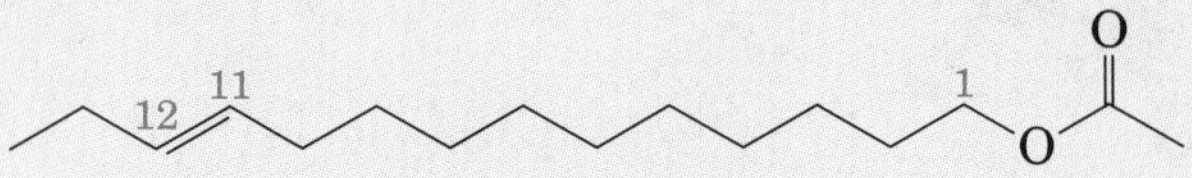

trans-11-Tetradecenyl acetate

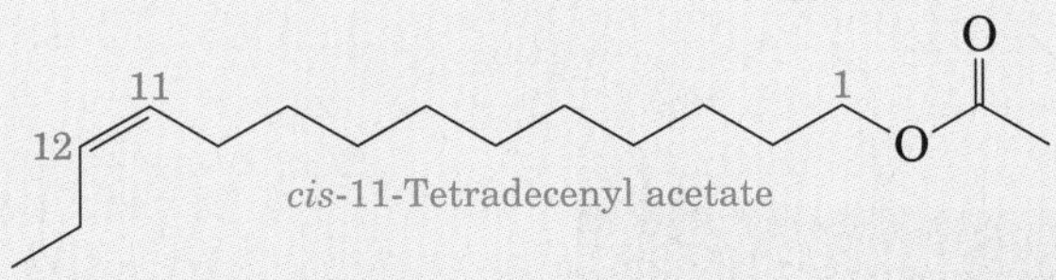

cis-11-Tetradecenyl acetate

Evidence suggests that an optimal response to a narrow range of stereoisomers as we see here is widespread in nature and that most insects maintain species isolation for mating and reproduction by the stereochemistry of their pheromones.

D Cycloalkenes

In naming **cycloalkenes,** we number the carbon atoms of the ring double bond 1 and 2 in the direction that gives the substituent encountered first the lower number. Therefore, it is not necessary to use a location number for the carbons of the double bond. Number substituents and list them in alphabetical order.

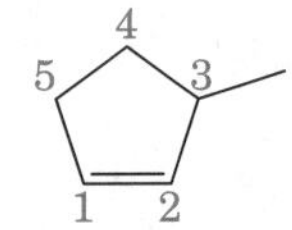

3-Methylcyclopentene
(not 5-methylcyclopentene)

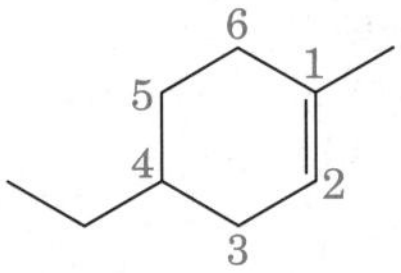

4-Ethyl-1-methylcyclohexene
(not 5-ethyl-2-methylcyclohexene)

EXAMPLE 3.3

Write the IUPAC name for each cycloalkene.

(a)

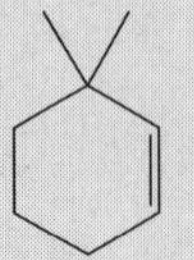

(b)

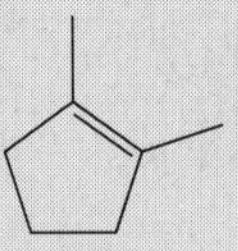

(c)

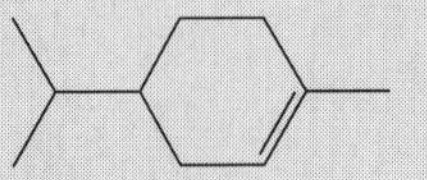

Solution
(a) 3,3-Dimethylcyclohexene
(b) 1,2-Dimethylcyclopentene
(c) 4-Isopropyl-1-methylcyclohexene

Problem 3.3

Write the IUPAC name for each cycloalkene.

(a)

(b)

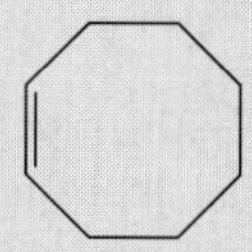

(c)

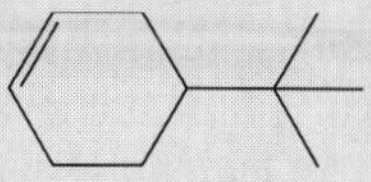

E Dienes, Trienes, and Polyenes

We name alkenes that contain more than one double bond as alkadienes, alkatrienes, and so on. We refer to those that contain several double bonds more generally as polyenes (Greek: *poly*, many). Following are three dienes:

$$CH_2=CHCH_2CH=CH_2$$
1,4-Pentadiene

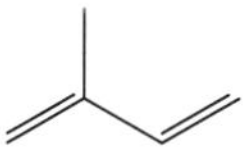

$$CH_2=CCH=CH_2$$ with CH₃ substituent
2-Methyl-1,3-butadiene
(Isoprene)

1,3-Cyclopentadiene

We saw earlier that for an alkene with one carbon–carbon double bond that can show cis-trans isomerism, two cis-trans isomers are possible. For an alkene with n carbon–carbon double bonds, each of which can show cis-trans isomerism, 2^n cis-trans isomers are possible.

EXAMPLE 3.4

How many cis-trans isomers are possible for 2,4-heptadiene?

$$CH_3-CH=CH-CH=CH-CH_2-CH_3$$
2,4-Heptadiene

Solution

This molecule has two carbon–carbon double bonds, each of which shows cis-trans isomerism. As shown in the following table, $2^2 = 4$ cis-trans isomers are possible. Line-angle formulas for two of these dienes are shown below.

Double-Bond	
C_2-C_3	C_4-C_5
trans	trans
trans	cis
cis	trans
cis	cis

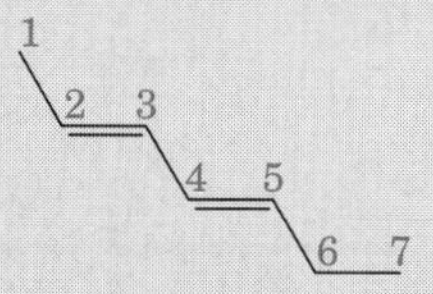

trans,trans-2,4-Heptadiene

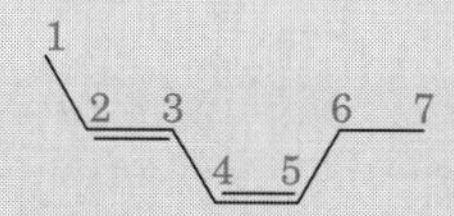

trans,cis-2,4-Heptadiene

Problem 3.4

Draw structural formulas for the other two cis-trans isomers of 2,4-heptadiene.

CHEMICAL CONNECTIONS 3C

Cis-Trans Isomerism in Vision

The retina, the light-detecting layer in the back of our eyes, contains reddish compounds called visual pigments. Their name, rhodopsin, is derived from the Greek word meaning "rose-colored." Each rhodopsin molecule is a combination of one molecule of a protein called opsin and one molecule of 11-*cis*-retinal, a derivative of vitamin A in which the CH₂OH group of carbon 15 is converted to an aldehyde group, —CH=O.

When rhodopsin absorbs light energy, the less stable 11-cis double bond, is converted to the more stable 11-trans double bond. This isomerization changes the shape of the rhodopsin molecule, which in turn causes the neurons of the optic nerve to fire and produce a visual image.

The retina of vertebrates contain two kinds of cells that contain rhodopsin: rods and cones. Cones function in bright light and are used for color vision; they are concentrated in the central portion of the retina, called the macula, and are responsible for the greatest visual acuity. The remaining area of the retina consists mostly of rods, which are used for peripheral and night vision. 11-*cis*-Retinal is present in both cones and rods. Rods have one kind of opsin, whereas cones have three kinds—one for blue, one for green, and one for red color vision.

15 CH₂OH

$\xrightarrow{\text{enzyme-catalyzed oxidation}}$

11 12 15 CH=O

Vitamin A (retinol)

Vitamin A aldehyde
(11-*trans*-retinal)

11 12 CH=O

11-*cis*-retinal

$\xrightarrow{\substack{\text{H}_2\text{N}-\text{opsin} \\ -\text{H}_2\text{O}}}$

11 12 CH=N—opsin

Rhodopsin
(visual purple)

enzyme-catalyzed isomerization of the 11-trans double bond to 11-cis

1. light strikes rhodopsin
2. the 11-cis double bond is isomerized to 11-trans
3. a nerve impulse is sent via the optic nerve to the visual cortex

11 12 CH=O $\xleftarrow[\substack{\text{opsin} \\ \text{removed}}]{\text{H}_2\text{O}}$

11-*trans*-retinal

11 12 CH=N—opsin

EXAMPLE 3.5

Draw all cis-trans isomers possible for the following unsaturated alcohol:

$$\underset{}{\text{CH}_3\text{C}}{=}\text{CHCH}_2\text{CH}_2\underset{}{\text{C}}{=}\text{CHCH}_2\text{OH}$$

with CH₃ groups on the two marked carbons

Solution

Cis-trans isomerism is possible only about the double bond between carbons 2 and 3 of the chain. It is not possible for the other double bond because carbon 7 has two identical groups on it (review Section 3.2B). Thus, $2^1 = 2$ cis-trans isomers are possible. The trans isomer of this alcohol, named geraniol, is a major component of the oils of rose, citronella, and lemon grass.

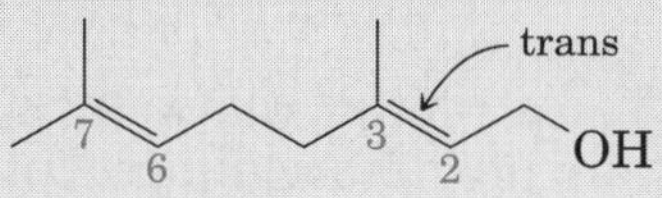

The trans isomer

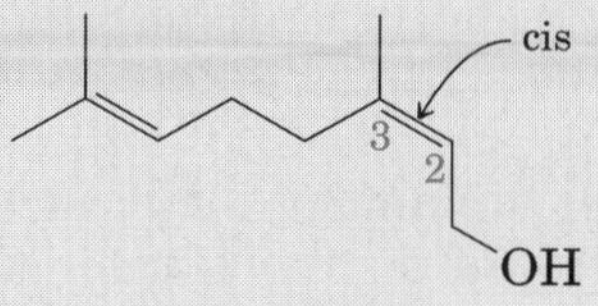

The cis isomer

■ **Lemon grass from Shelton Herb Farm, North Carolina.**

Problem 3.5

How many cis-trans isomers are possible for the following unsaturated alcohol?

$$CH_3C{=}CHCH_2CH_2C{=}CHCH_2CH_2C{=}CHCH_2OH$$

An example of a biologically important polyunsaturated alcohol for which a number of cis-trans isomers are possible is vitamin A. Each of the four carbon–carbon double bonds in the chain of carbon atoms bonded to the substituted cyclohexene ring has the potential for cis-trans isomerism. Therefore, $2^4 = 16$ cis-trans isomers are possible for this structural formula. Vitamin A, the isomer shown here, is the all-trans isomer.

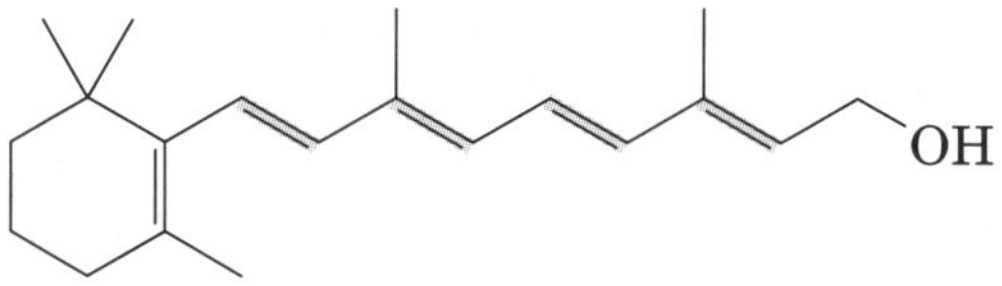

Vitamin A (retinol)

3.4 Physical Properties

Alkenes and alkynes are nonpolar compounds, and the only attractive forces between their molecules are very weak London dispersion forces. Their physical properties, therefore, are similar to those of alkanes having the same carbon skeletons. Alkenes and alkynes that are liquid at room temperature have densities less than 1.0 g/mL (they float on water). They are insoluble in water but soluble in one another and in other nonpolar organic liquids.

3.5 Naturally Occurring Alkenes: The Terpenes

Among the compounds found in the essential oils of plants are a group of substances called **terpenes,** all of which have in common the fact that their carbon skeletons can be divided into two or more carbon units that are iden-

Terpene A compound whose carbon skeleton can be divided into two or more units identical to the carbon skeleton of isoprene

tical with the carbon skeleton of isoprene. Carbon 1 of an **isoprene unit** is called the head, and carbon 4 is called the tail. A terpene is a compound in which the tail of one isoprene unit becomes bonded to the head of another isoprene unit.

$$CH_2{=}\overset{\overset{\displaystyle CH_3}{|}}{C}{-}CH{=}CH_2$$

2-Methyl-1,3-butadiene
(Isoprene)

Head · · · C · · · Tail

$$\overset{1}{C}{-}\overset{2}{C}{-}\overset{3}{C}{-}\overset{4}{C}$$

Isoprene unit

Terpenes are among the most widely distributed compounds in the biological world, and a study of their structure provides a glimpse into the wondrous diversity that nature can generate from a simple carbon skeleton. Terpenes also illustrate an important principle of the molecular logic of living systems: In building large molecules, small subunits are bonded together by a series of enzyme-catalyzed reactions and then chemically modified by additional enzyme-catalyzed reactions. Chemists use the same principles in the laboratory, but their methods do not have the precision and selectivity of the enzyme-catalyzed reactions of cellular systems.

Probably the terpenes most familiar to you, at least by odor, are components of the so-called essential oils obtained by steam distillation or ether extraction of various parts of plants. Essential oils contain the relatively low-molecular-weight substances that are largely responsible for characteristic plant fragrances. Many essential oils, particularly those from flowers, are used in perfumes.

One example of a terpene obtained from an essential oil is myrcene (Figure 3.1), a component of bayberry wax and oils of bay and verbena. Myrcene is a triene with a parent chain of eight carbon atoms and two one-carbon branches. The two isoprene units in myrcene are joined by bonding the tail of one unit to the head of the other unit. Figure 3.1 also shows three other terpenes, each of ten carbon atoms. In limonene and menthol, nature has formed an additional bond between two carbons to form a six-membered ring.

Farnesol, a terpene with molecular formula $C_{15}H_{26}O$, includes three isoprene units. Derivatives of both farnesol and geraniol are intermediates in the biosynthesis of cholesterol (Section 20.4).

■ **California laurel,** *Umbelluria californica,* **one source of myrcene.**

Farnesol
(Lily-of-the-valley)

Figure 3.1 Four terpenes, each derived from two isoprene units bonded from the tail of the first unit to the head of the second unit. In limonene and menthol, formation of an additional carbon–carbon bond creates a six-membered ring.

Head

Tail

Myrcene
(Bay oil)

Geraniol
(Rose and
other flowers)

Forming this
bond makes
the ring

Limonene
(Lemon
and orange)

Menthol
(Peppermint)

Vitamin A (Section 3.3E), a terpene with molecular formula $C_{20}H_{30}O$, consists of four isoprene units bonded head-to-tail and cross-linked at one point to form a six-membered ring.

3.6 Addition Reactions of Alkenes

The most characteristic reaction of alkenes is addition to the carbon–carbon double bond: The double bond is broken and in its place single bonds form to two new atoms or groups of atoms. Table 3.1 shows several examples of alkene addition reactions, along with the descriptive name(s) associated with each reaction.

At this point, you might ask why a carbon–carbon double bond is a site of chemical reactivity, whereas carbon–carbon single bonds are quite unreactive under most experimental conditions. One way to answer this question is to focus on the changes in bonding that occur as a result of an alkene addition reaction. Consider, for example, the addition of hydrogen to ethylene. As a result of this addition, one double bond and one single bond (the H—H bond of H_2) are replaced by three single bonds, giving a net conversion of one bond of the double bond to two single bonds.

$$
\underset{\substack{\text{one double bond}\\\text{and one single bond}}}{\overset{\displaystyle H}{\underset{\displaystyle H}{}}\!\!C\!=\!C\!\!\overset{\displaystyle H}{\underset{\displaystyle H}{}}} + H\!-\!H \xrightarrow{\text{are replaced by}} \underset{\text{three single bonds}}{H\!-\!\overset{\displaystyle H}{\underset{\displaystyle H}{C}}\!-\!\overset{\displaystyle H}{\underset{\displaystyle H}{C}}\!-\!H} + \text{heat}
$$

This and almost all other addition reactions of alkenes are exothermic, which means that the products are more stable (have lower energy) than the reactants. Remember, though, that just because an alkene addition reaction is exothermic doesn't mean that it occurs rapidly. The rate of a chem-

Table 3.1 Characteristic Addition Reactions of Alkenes	
Reaction	**Descriptive Name(s)**
$\text{C}=\text{C} + \text{HCl} \longrightarrow \overset{\displaystyle H}{-}\!\overset{\vert}{\text{C}}\!-\!\overset{\displaystyle Cl}{\underset{\vert}{\text{C}}}\!-$	hydrochlorination
$\text{C}=\text{C} + \text{H}_2\text{O} \longrightarrow \overset{\displaystyle H}{-}\!\overset{\vert}{\text{C}}\!-\!\overset{\displaystyle OH}{\underset{\vert}{\text{C}}}\!-$	hydration
$\text{C}=\text{C} + \text{Br}_2 \longrightarrow \overset{\displaystyle Br}{-}\!\overset{\vert}{\text{C}}\!-\!\overset{\displaystyle Br}{\underset{\vert}{\text{C}}}\!-$	bromination
$\text{C}=\text{C} + \text{H}_2 \longrightarrow \overset{\displaystyle H}{-}\!\overset{\vert}{\text{C}}\!-\!\overset{\displaystyle H}{\underset{\vert}{\text{C}}}\!-$	hydrogenation (reduction)

See the **Interactive General, Organic, and Biochemistry CD-ROM, version 2.0,** for further exploration on this topic.

ical reaction depends on its activation energy, not on how exothermic or endothermic it is. In fact, the addition of hydrogen to an alkene is immeasurably slow room temperature but, as we will see soon, proceeds quite rapidly in the presence of a suitable catalyst.

A Addition of Hydrogen Halides

The hydrogen halides HCl, HBr, and HI add to alkenes to give haloalkanes (alkyl halides). Addition of HCl to ethylene, for example, gives chloroethane (ethyl chloride):

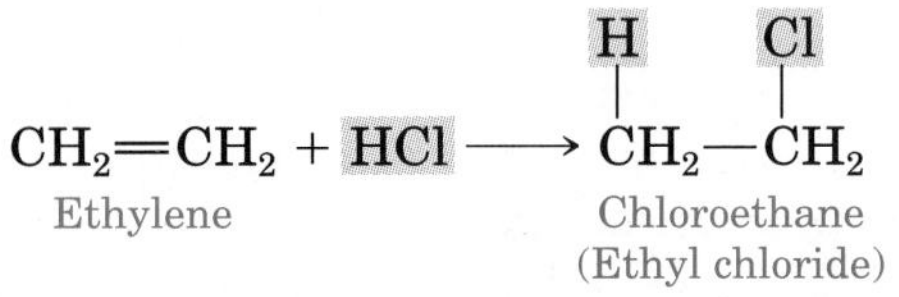

$$CH_2{=}CH_2 + \boxed{HCl} \longrightarrow \underset{\substack{\text{Chloroethane}\\\text{(Ethyl chloride)}}}{\overset{\substack{H\quad\ Cl\\|\qquad\ |}}{CH_2{-}CH_2}}$$

Ethylene

Addition of HCl to propene gives 2-chloropropane (isopropyl chloride); hydrogen adds to carbon 1 of propene and chlorine adds to carbon 2. If the orientation of addition were reversed, 1-chloropropane (propyl chloride) would form. The observed result is that almost no 1-chloropropane forms. Because 2-chloropropane is the observed product, we say that addition of HCl to propene is **regioselective.**

$$\underset{\text{Propene}}{\overset{2\qquad\ 1}{CH_3CH{=}CH_2}} + \boxed{HCl} \longrightarrow \underset{\text{2-Chloropropane}}{\overset{\substack{Cl\quad\ H\\|\qquad\ |}}{CH_3CH{-}CH_2}} \qquad \underset{\substack{\text{1-Chloropropane}\\\text{(not formed)}}}{\overset{\substack{H\quad\ Cl\\|\qquad\ |}}{CH_3CH{-}CH_2}}$$

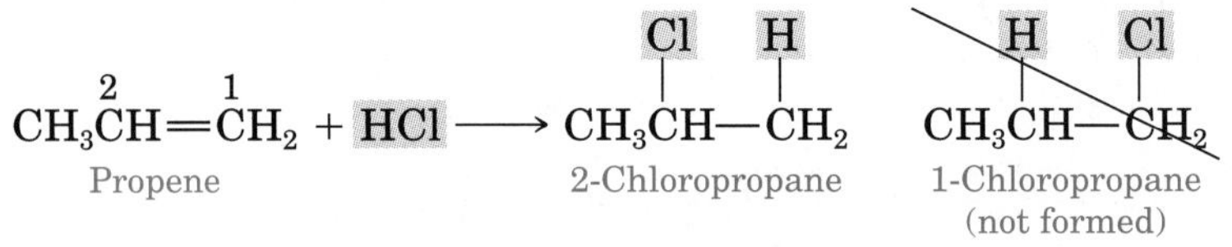

This regioselectivity was noted by Vladimir Markovnikov (1838–1904), who made the following generalization known as **Markovnikov's rule:** In the addition of HX (where X = halogen) to an alkene, hydrogen adds to the doubly bonded carbon that has the greater number of hydrogens already bonded to it; halogen adds to the other carbon.

EXAMPLE 3.6

Draw a structural formula for the product of each alkene addition reaction.

(a) 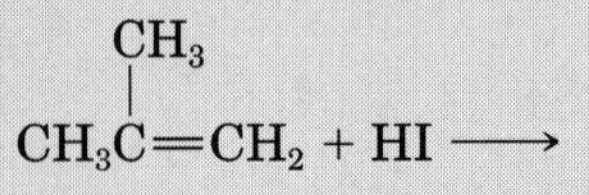 $\underset{}{\overset{\overset{\displaystyle CH_3}{|}}{CH_3C}}{=}CH_2 + HI \longrightarrow$

(b) 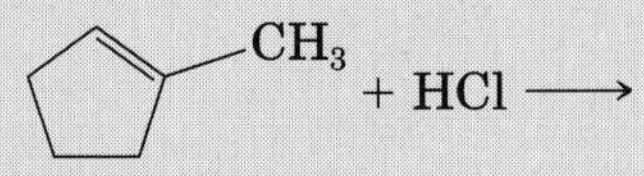 cyclopentene with $-CH_3$ + HCl $\longrightarrow$

Solution
(a) Markovnikov's rule predicts that the hydrogen of HI adds to carbon 1 and the iodine adds to carbon 2 to give 2-iodo-2-methylpropane.
(b) H adds to carbon 2 of the ring and Cl adds to carbon 1 to give 1-chloro-1-methylcyclopentane.

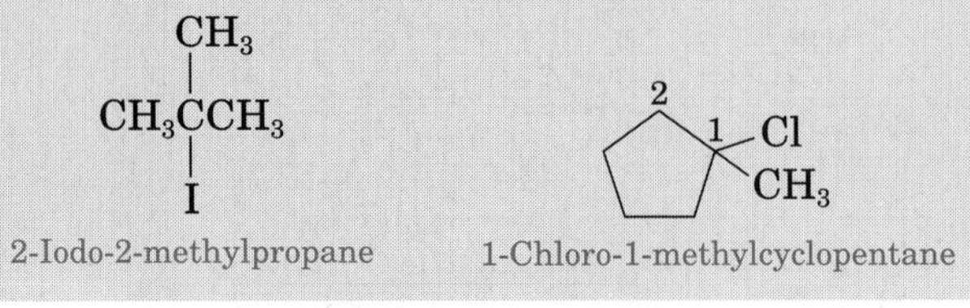

$$\underset{\text{2-Iodo-2-methylpropane}}{\overset{\substack{CH_3\\|}}{\underset{\substack{|\\I}}{CH_3CCH_3}}} \qquad \underset{\text{1-Chloro-1-methylcyclopentane}}{\text{(cyclopentane ring with } ^2,\ ^1{-}Cl,\ CH_3)}$$

Problem 3.6

Draw a structural formula for the product of each alkene addition reaction.

(a) $CH_3CH{=}CH_2 + HBr \longrightarrow$

(b) 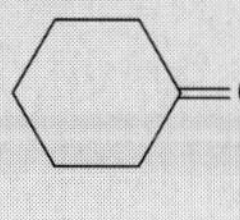 ⬡$={=}CH_2 + HBr \longrightarrow$

Markovnikov's rule tells us what happens when we add HCl, HBr, or HI to a carbon–carbon double bond. We know that in the addition of HCl or other halogen acid, one bond of the double bond and the H—Cl bond are broken, and that new C—H and C—Cl bonds form. But chemists also want to know *how* this conversion happens. Are the C=C and H—X bonds broken and both new covalent bonds formed all at the same time? Or does this reaction take place in a series of steps? If the latter, what are these steps, and in what order do they take place?

Chemists account for the addition of HX to an alkene by a two-step **reaction mechanism,** which we illustrate for the reaction of 2-butene with hydrogen chloride to give 2-chlorobutane. Step 1 is the addition of H^+ to 2-butene. To show this addition, we use a curved arrow that shows the repositioning of an electron pair from its origin (the tail of the arrow) to its new location (the head of the arrow). Recall that we used curved arrows in Section 7.1 to show bond breaking and bond formation in proton-transfer reactions. We now use curved arrows in the same way to show bond breaking and bond formation in a reaction mechanism.

Step 1 results in the formation of an organic cation. One carbon atom in this cation has only six electrons in its valence shell and so carries a charge of +1. A species containing a positively charged carbon atom is called a **carbocation** (carbon + cation). Carbocations are classified as primary (1°), secondary (2°), or tertiary (3°), depending on the number of carbon groups bonded to the carbon bearing the positive charge.

> **Reaction mechanism** A step-by-step description of how a chemical reaction occurs

> **Carbocation** A species containing a carbon atom with only three bonds to it and bearing a positive charge

Mechanism: Addition of HCl to 2-Butene

Step 1: Reaction of the carbon–carbon double bond of the alkene with H^+ forms a 2° carbocation intermediate. In forming this intermediate, one bond of the double bond breaks and its pair of electrons forms a new covalent bond with H^+. One carbon of the double bond is left with only six electrons in its valence shell and, therefore, has a positive charge.

$$CH_3CH{=}CHCH_3 + H^+ \longrightarrow CH_3\overset{+}{C}H\overset{\overset{\textstyle H}{|}}{-}CHCH_3$$

A 2° carbocation intermediate

Step 2: Reaction of the 2° carbocation intermediate with chloride ion completes the valence shell of carbon and gives 2-chlorobutane.

$$:\!\ddot{Cl}\!:^- + CH_3\overset{+}{C}HCH_2CH_3 \longrightarrow CH_3\overset{\overset{\textstyle :\ddot{Cl}:}{|}}{C}HCH_2CH_3$$

Chloride ion   A 2° carbocation intermediate

EXAMPLE 3.7

Propose a two-step mechanism for the addition of HI to methylenecyclohexane to give 1-iodo-1-methylcyclohexane.

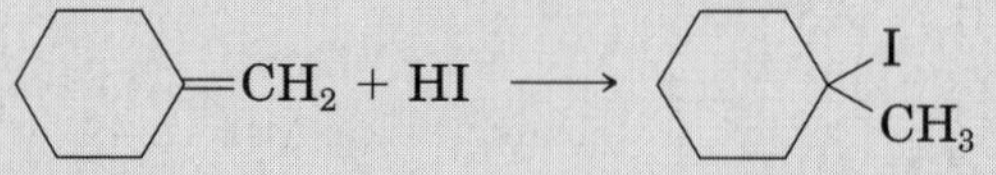

$$
\text{(cyclohexane)}=\text{CH}_2 + \text{HI} \longrightarrow \text{(cyclohexane)}\overset{\displaystyle\text{I}}{\underset{\displaystyle\text{CH}_3}{}}
$$

Methylenecyclohexane   1-Iodo-1-methylcyclohexane

Solution

The mechanism is similar to that proposed for the addition of HCl to propene.

Step 1: Reaction of H^+ with the carbon–carbon double bond forms a new C—H bond to the carbon bearing the greater number of hydrogens and gives a 3° carbocation intermediate.

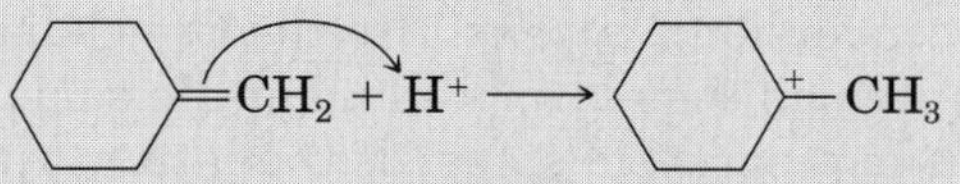

$$
\text{(cyclohexane)}=\text{CH}_2 + \text{H}^+ \longrightarrow \text{(cyclohexane)}\overset{+}{-}\text{CH}_3
$$

A 3° carbocation
intermediate

Step 2: Reaction of the 3° carbocation intermediate with iodide ion completes the valence shell of carbon and gives the product.

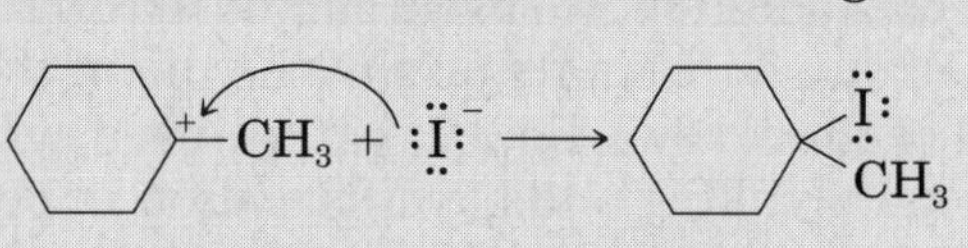

$$
\text{(cyclohexane)}\overset{+}{-}\text{CH}_3 + \text{:}\ddot{\text{I}}\text{:}^- \longrightarrow \text{(cyclohexane)}\overset{\displaystyle\ddot{\text{I}}\text{:}}{\underset{\displaystyle\text{CH}_3}{}}
$$

Problem 3.7

Propose a two-step mechanism for the addition of HBr to 1-methylcyclohexene to give 1-bromo-1-methylcyclohexane.

B Addition of Water: Acid-Catalyzed Hydration

Hydration Addition of water

In the presence of an acid catalyst, most commonly concentrated sulfuric acid, water adds to the carbon–carbon double bond of an alkene to give an alcohol. Addition of water is called **hydration.** In the case of simple alkenes, hydration follows Markovnikov's rule: H adds to the carbon of the double bond with the greater number of hydrogens and OH adds to the carbon with the fewer hydrogens.

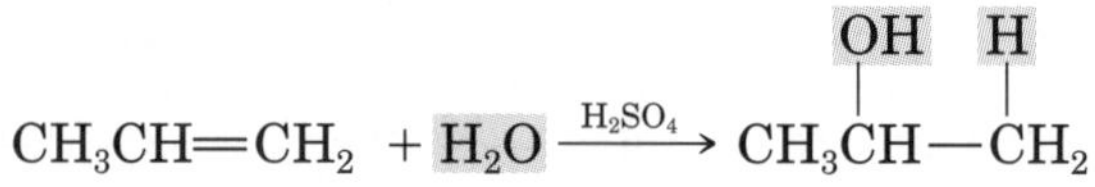

$$
\text{CH}_3\text{CH}{=}\text{CH}_2 + \text{H}_2\text{O} \xrightarrow{\text{H}_2\text{SO}_4} \overset{\displaystyle\text{OH}\quad\text{H}}{\text{CH}_3\text{CH}{-}\text{CH}_2}
$$

Propene   2-Propanol

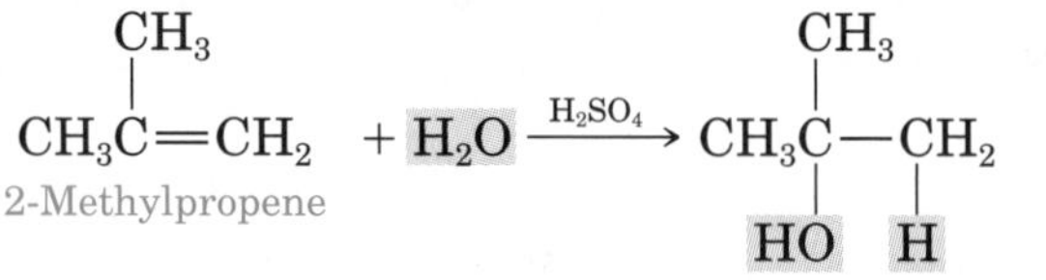

Most industrial ethanol is made by the acid-catalyzed hydration of ethylene.

$$
\overset{\displaystyle\text{CH}_3}{\text{CH}_3\text{C}{=}\text{CH}_2} + \text{H}_2\text{O} \xrightarrow{\text{H}_2\text{SO}_4} \overset{\displaystyle\text{CH}_3}{\underset{\displaystyle\text{HO}\quad\text{H}}{\text{CH}_3\text{C}{-}\text{CH}_2}}
$$

2-Methylpropene   2-Methyl-2-propanol

EXAMPLE 3.8

Draw a structural formula for the alcohol formed by the acid-catalyzed hydration of 1-methylcyclohexene.

Solution

Markovnikov's rule predicts that H adds to the carbon with the greater number of hydrogens, which, in this case, is carbon 2 of the cyclohexene ring. OH then adds to carbon 1.

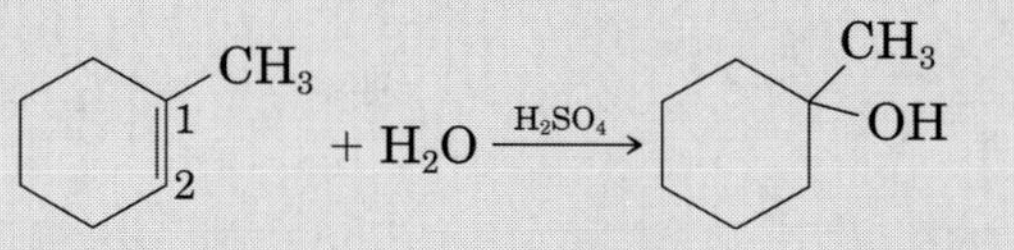

$$\underset{\text{1-Methylcyclohexene}}{\text{(ring with CH}_3\text{, positions 1,2)}} + H_2O \xrightarrow{H_2SO_4} \underset{\text{1-Methylcyclohexanol}}{\text{(ring with CH}_3\text{ and OH)}}$$

Problem 3.8

Draw a structural formula for the alcohol formed from acid-catalyzed hydration of each alkene:
(a) 2-Methyl-2-butene
(b) 2-Methyl-1-butene

The mechanism for the acid-catalyzed hydration of an alkene is similar to what we proposed for the addition of HCl, HBr, and HI to an alkene and is illustrated by the hydration of propene. This mechanism is consistent with the fact that acid is a catalyst. One H^+ is consumed in Step 1, but another is generated in Step 3.

Mechanism: Acid-Catalyzed Hydration of Propene

Step 1: Addition of H^+ to the carbon of the double bond with the greater number of hydrogens gives a 2° carbocation intermediate.

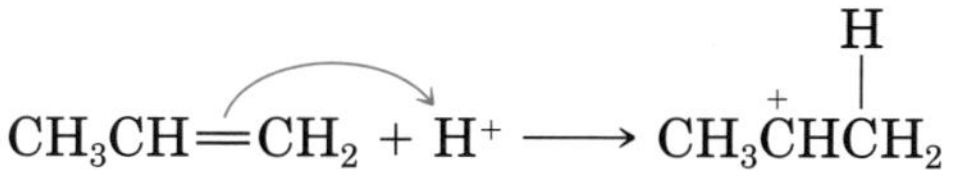

$$CH_3CH{=\!=}CH_2 + H^+ \longrightarrow CH_3\overset{+}{C}HCH_2{-}H$$

A 2° carbocation intermediate

Step 2: The carbocation intermediate completes its valence shell by forming a new covalent bond with an unshared pair of electrons of the oxygen atom of H_2O to give an **oxonium ion.**

$$CH_3\overset{+}{C}HCH_3 + \overset{..}{:}\!O{-}H \longrightarrow CH_3CHCH_3$$
$$\text{(with H below O)} \qquad \text{(with } H\!-\!\overset{..}{O}{}^+\!\!-\!H \text{ above)}$$

An oxonium ion

> **Oxonium ion** An ion in which oxygen is bonded to three other atoms and bears a positive charge

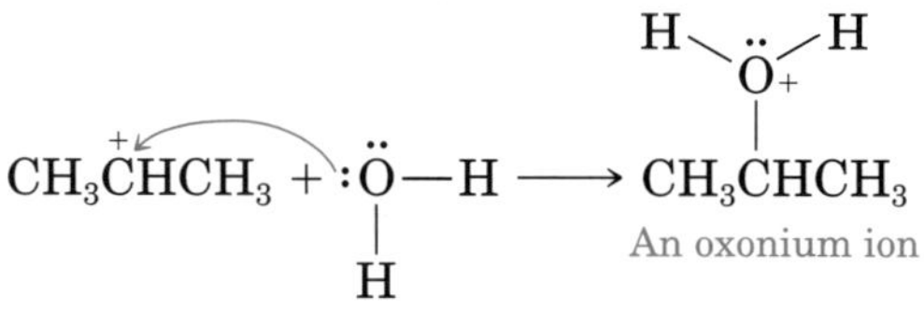

Step 3: Loss of H^+ from the oxonium ion gives the alcohol and generates a new H^+ catalyst.

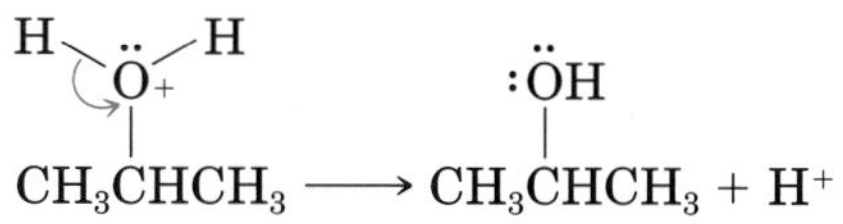

$$\underset{CH_3CHCH_3}{H\!-\!\overset{..}{O}{}^+\!\!-\!H} \longrightarrow \underset{CH_3CHCH_3}{:\!\overset{..}{O}H} + H^+$$

EXAMPLE 3.9

Propose a three-step reaction mechanism for the acid-catalyzed hydration of methylenecyclohexane to give 1-methylcyclohexanol.

Solution

The reaction mechanism is similar to that for the acid-catalyzed hydration of propene.

Step 1: Reaction of the carbon–carbon double bond with H^+ gives a 3° carbocation intermediate.

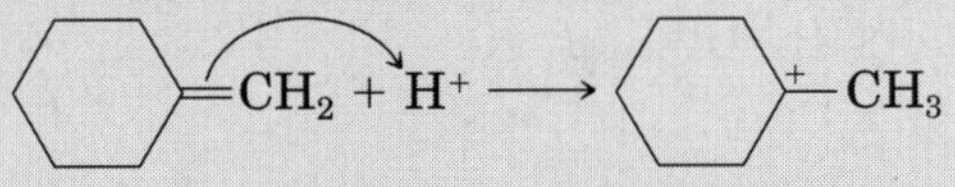

A 3° carbocation intermediate

Step 2: Reaction of the carbocation intermediate with water completes the valence shell of carbon and gives an oxonium ion.

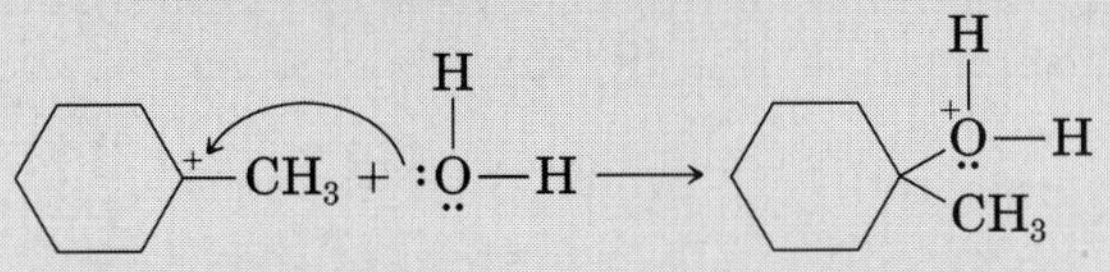

An oxonium ion

Step 3: Loss of H^+ from the oxonium ion completes the reaction and generates a new H^+ catalyst.

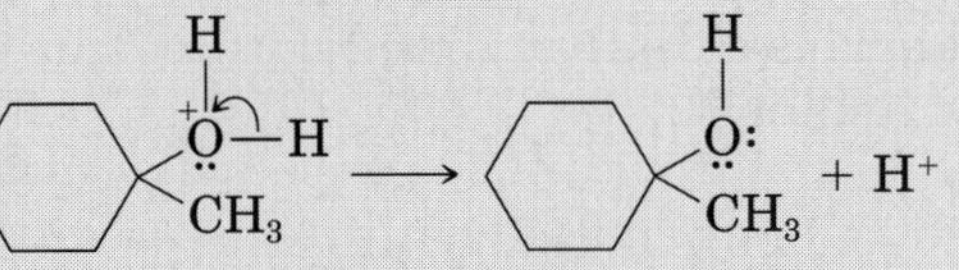

Problem 3.9

Propose a three-step reaction mechanism for the acid-catalyzed hydration of 1-methylcyclohexene to give 1-methylcyclohexanol.

C Addition of Bromine and Chlorine

Chlorine, Cl_2, and bromine, Br_2, react with alkenes at room temperature by addition of halogen atoms to the carbon atoms of the double bond. This reaction is generally carried out either by using the pure reagents or by mixing them in an inert solvent, such as dichloromethane, CH_2Cl_2.

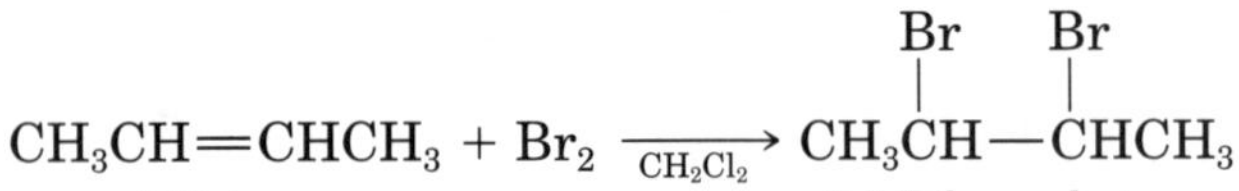

$$CH_3CH{=}CHCH_3 + Br_2 \xrightarrow{CH_2Cl_2} CH_3\overset{|}{\underset{}{C}}H{-}\overset{|}{\underset{}{C}}HCH_3$$

2-Butene   2,3-Dibromobutane

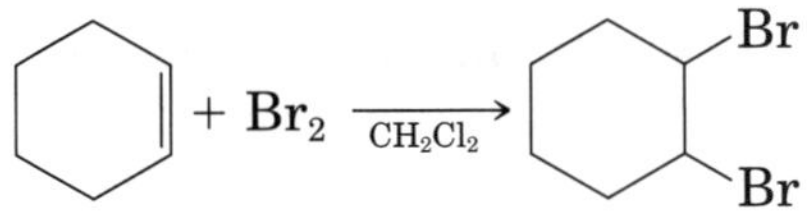

Cyclohexene   1,2-Dibromocyclohexane

Addition of bromine is a useful qualitative test for the presence of an alkene. If we dissolve bromine in carbon tetrachloride, the solution is red. In

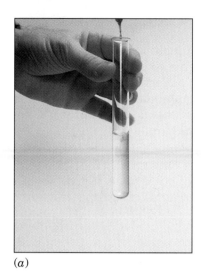

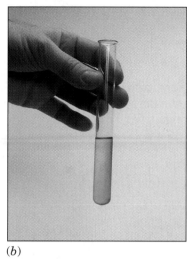

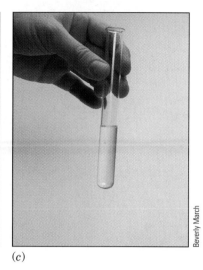

Beverly March

(a)   (b)   (c)

■ (*a*) **A drop of Br₂ dissolved in CCl₄ is added to an unknown liquid. If the color remains after stirring (*b*), it indicates the absence of unsaturation. If the color disappears (*c*), the unknown is unsaturated.**

contrast, alkenes and dibromoalkanes are colorless. If we mix a few drops of the red bromine solution with an unknown sample suspected of being an alkene, disappearance of the red color as bromine adds to the double bond tells us that an alkene is, indeed, present.

EXAMPLE 3.10

Complete these reactions.

(a) [cyclopentene] + Br₂ $\xrightarrow{\text{CH}_2\text{Cl}_2}$

(b) [1-methylcyclohexene] + Cl₂ $\xrightarrow{\text{CH}_2\text{Cl}_2}$

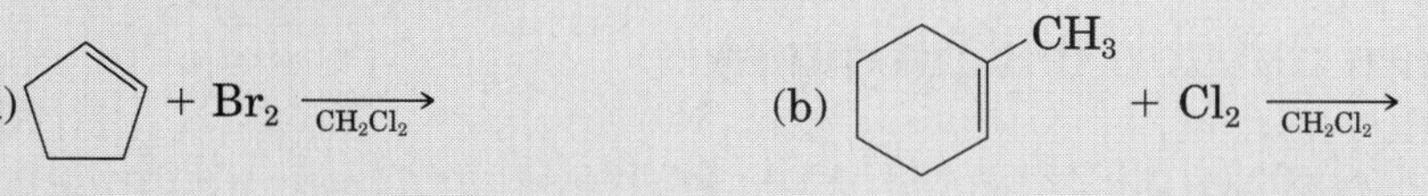

Solution
In addition of Br₂ or Cl₂ to a cycloalkene, one halogen adds to each carbon of the double bond.

(a) [cyclopentene] + Br₂ $\xrightarrow{\text{CH}_2\text{Cl}_2}$ [1,2-dibromocyclopentane]

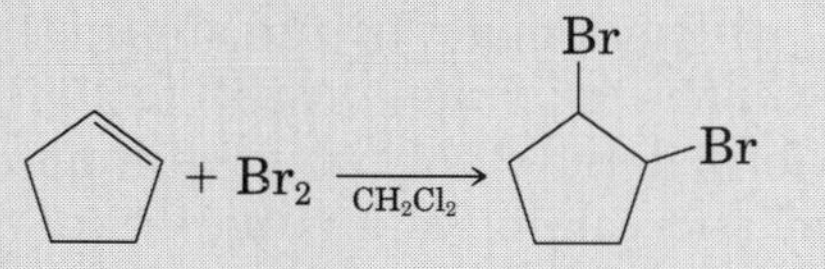

(b) [1-methylcyclohexene] + Cl₂ $\xrightarrow{\text{CH}_2\text{Cl}_2}$ [1,2-dichloro-1-methylcyclohexane]

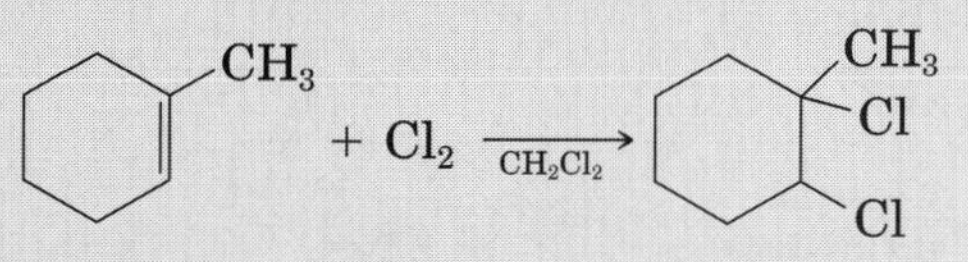

Problem 3.10
Complete these reactions.

(a) $\text{CH}_3\underset{\underset{\text{CH}_3}{|}}{\overset{\overset{\text{CH}_3}{|}}{\text{C}}}\text{CH}{=}\text{CH}_2$ + Br₂ $\xrightarrow{\text{CH}_2\text{Cl}_2}$

(b) [methylenecyclohexane] + Cl₂ $\xrightarrow{\text{CH}_2\text{Cl}_2}$

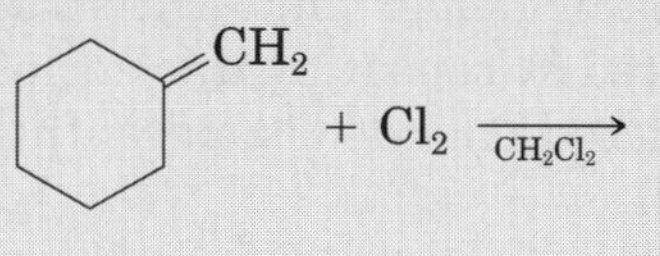

D Addition of Hydrogen: Reduction

Virtually all alkenes react quantitatively with molecular hydrogen, H_2, in the presence of a transition metal catalyst to give alkanes. Commonly used transition metal catalysts include platinum, palladium, ruthenium, and nickel. Because the conversion of an alkene to an alkane involves reduction by hydrogen in the presence of a catalyst, the process is called **catalytic reduction** or, alternatively, **catalytic hydrogenation.**

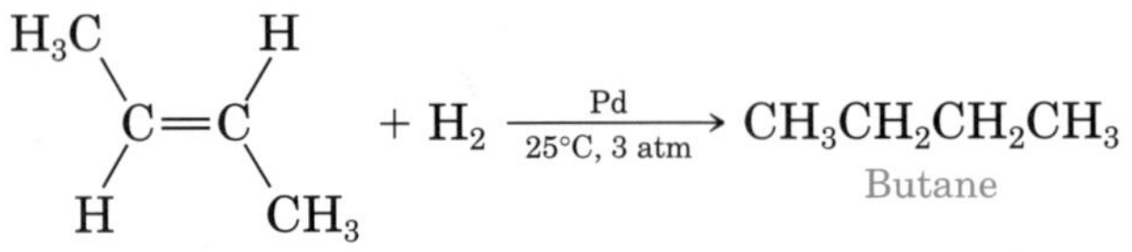

In Section 12.3 we will see how catalytic hydrogenation is used to solidify liquid vegetable oils to margarines and semisolid cooking fats.

$$
\begin{array}{c}
H_3C \qquad\qquad H \\
\diagdown \qquad\quad \diagup \\
C{=}C \qquad + H_2 \xrightarrow[\text{25°C, 3 atm}]{\text{Pd}} CH_3CH_2CH_2CH_3 \\
\diagup \qquad\quad \diagdown \\
H \qquad\qquad CH_3 \\
\textit{trans}\text{-2-Butene} \qquad\qquad\qquad\qquad\qquad \text{Butane}
\end{array}
$$

$$
\bigcirc\!\!\!\!\parallel \quad + H_2 \xrightarrow[\text{25°C, 3 atm}]{\text{Pd}} \quad \bigcirc
$$

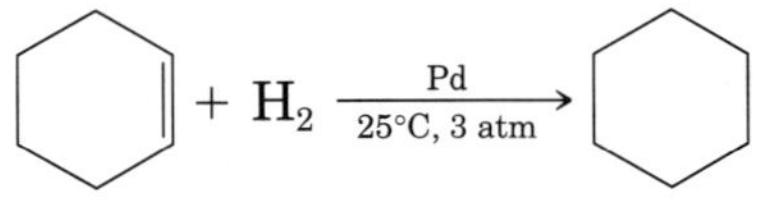

Cyclohexene   Cyclohexane

The metal catalyst is used in the form of a finely powdered solid. The reaction is carried out by dissolving the alkene in ethanol or another nonreacting organic solvent, adding the solid catalyst, and exposing the mixture to hydrogen gas at pressures ranging from 1 to 100 atm.

3.7 Polymerization of Ethylene and Substituted Ethylenes

A Structure of Polyethylenes

From the perspective of the chemical industry, the single most important reaction of alkenes is the formation of **chain-growth polymers** (Greek words: *poly*, many, and *meros*, part). In the presence of certain compounds called initiators, many alkenes form polymers made by the stepwise addition of **monomers** (Greek words: *mono*, one, and *meros*, part) to a growing polymer chain, as illustrated by the formation of polyethylene from ethylene. In alkene polymers of industrial and commercial importance, n is a large number, typically several thousand.

Polymer From the Greek *poly*, many, and *meros*, parts; any long-chain molecule synthesized by bonding together many single parts (monomers)

Monomer From the Greek *mono*, single, and *meros*, parts; the simplest nonredundant unit from which a polymer is synthesized

$$
nCH_2{=}CH_2 \xrightarrow[\text{(polymerization)}]{\text{initiator}} \left(\!CH_2CH_2\!\right)_{\!n}
$$

Ethylene   Polyethylene

To show the structure of a polymer, we place parentheses around the repeating monomer unit. The structure of an entire polymer chain can be reproduced by repeating this enclosed structure in both directions. A subscript n is placed outside the parentheses to indicate that this unit is repeated n times.

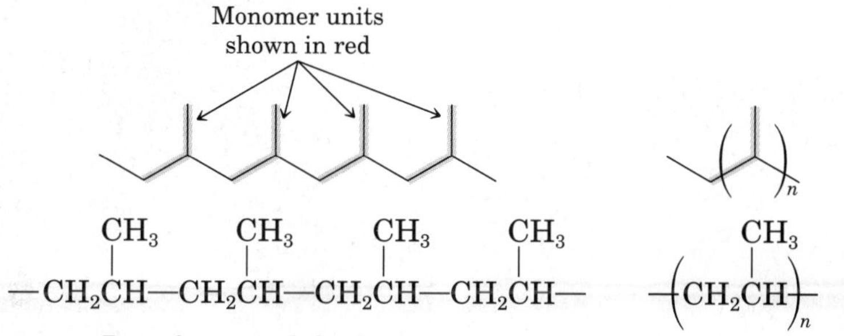

Monomer units
shown in red

$$\begin{array}{c} CH_3 \qquad CH_3 \qquad CH_3 \qquad CH_3 \\ | \qquad\quad | \qquad\quad | \qquad\quad | \\ -CH_2CH-CH_2CH-CH_2CH-CH_2CH- \end{array}$$

Part of an extended polymer chain

$$\begin{array}{c} CH_3 \\ | \\ \left(CH_2CH\right)_n \end{array}$$

The repeating unit

The most common method of naming a polymer is to attach the prefix **poly-** to the name of the monomer from which the polymer is synthesized— for example, polyethylene and polystyrene. Where the name of the monomer consists of two words (for example, the monomer vinyl chloride), its name is enclosed in parentheses.

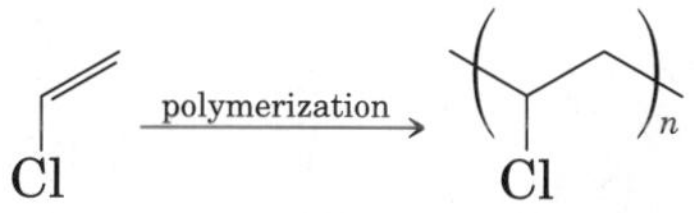

$$\overset{\displaystyle \diagup}{\underset{Cl}{\diagdown}} \xrightarrow{\text{polymerization}} \left(\overset{\diagup}{\underset{Cl}{\diagdown}}\right)_n$$

Vinyl chloride   Poly(vinyl chloride)
(PVC)

Table 3.2 lists several important polymers derived from ethylene and substituted ethylenes along with their common names and most important uses.

Table 3.2 Polymers Derived from Ethylene and Substituted Ethylenes

Monomer Formula	Common Name	Polymer Name(s) and Common Uses
$CH_2{=}CH_2$	ethylene	polyethylene, Polythene; break-resistant containers and packaging materials
$CH_2{=}CHCH_3$	propylene	polypropylene, Herculon; textile and carpet fibers
$CH_2{=}CHCl$	vinyl chloride	poly(vinyl chloride), PVC; construction tubing
$CH_2{=}CCl_2$	1,1-dichloroethylene	poly(1,1-dichloroethylene); Saran Wrap is a copolymer with vinyl chloride
$CH_2{=}CHCN$	acrylonitrile	polyacrylonitrile, Orlon; acrylics and acrylates
$CF_2{=}CF_2$	tetrafluoroethylene	polytetrafluoroethylene, PTFE; Teflon, nonstick coatings
$CH_2{=}CHC_6H_5$	styrene	polystyrene, Styrofoam; insulating materials
$CH_2{=}CHCOOCH_2CH_3$	ethyl acrylate	poly(ethyl acrylate), latex paint
$CH_2{=}\underset{\underset{CH_3}{\mid}}{C}COOCH_3$	methyl methacrylate	poly(methyl methacrylate), Lucite; Plexiglas; glass substitutes

■ Some articles made from chain-growth polymers. (*a*) Saran Wrap, a copolymer of vinyl chloride and 1,1-dichloroethylene. (*b*) Plastic containers for various supermarket products, made mostly from polyethylene and polypropylene. (*c*) Teflon-coated kitchenware. (*d*) Articles made from polystyrene.

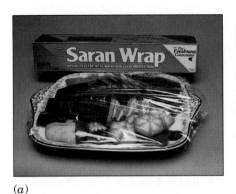

(*a*)

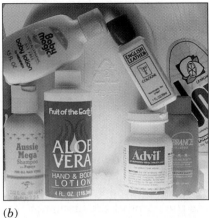

(*b*)

(*c*)

(*d*)

a, c, Beverly March; b, d, Charles D. Winters

A peroxide is any compound that contains an —O—O—bond, for example, hydrogen peroxide, H—O—O—H.

■ Polyethylene films are produced by extruding molten plastic through a ringlike gap and inflating the film into a balloon.

The Stock Market

B Low-Density Polyethylene

The first commercial process for ethylene polymerization used peroxide initiators at 500°C and 1000 atm and yielded a tough, transparent polymer known as **low-density polyethylene (LDPE).** At the molecular level, LDPE chains are highly branched, with the result that they do not pack well together and London dispersion forces between them are weak. LDPE softens and melts at about 115°C, which means that it cannot be used in products that are exposed to boiling water.

Today, approximately 65% of all LDPE is used for the manufacture of films by the blow-molding technique illustrated in Figure 3.2. LDPE film is inexpensive, which makes it ideal for packaging such consumer items as baked goods and vegetables and for the manufacture of trash bags.

C High-Density Polyethylene

An alternative method for polymerization of alkenes, and one that does not involve peroxide initiators, was developed by Karl Ziegler of Germany and Giulio Natta of Italy in the 1950s. Polyethylene from Ziegler-Natta systems, termed **high-density polyethylene (HDPE),** has little chain branching. Consequently, its chains pack together more closely than those of LDPE, with the result that London dispersion forces between chains of HDPE are

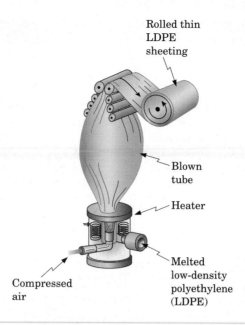

Rolled thin
LDPE
sheeting

Blown
tube

Heater

Compressed
air

Melted
low-density
polyethylene
(LDPE)

Figure 3.2 Fabrication of LDPE film. A tube of melted LDPE along with a jet of compressed air is forced through an opening and blown into a gigantic, thin-walled bubble. The film is then cooled and taken up onto a roller. This double-walled film can be slit down the side to give LDPE film or sealed at points along its length to make LDPE bags.

stronger than those in LDPE. HDPE has a higher melting point than LDPE and is three to ten times stronger.

Approximately 45% of all HDPE products are made by the blow-molding process shown in Figure 3.3. HDPE is used for consumer items such as milk and water jugs, grocery bags, and squeezable bottles.

Figure 3.3 Blow molding an HDPE container. (*a*) A short length of HDPE tubing is placed in an open die and the die is closed, sealing the bottom of the tube. (*b*) Compressed air is forced into the hot polyethylene/die assembly, and the tubing is literally blown up to take the shape of the mold. (*c*) After the assembly cools, the die is opened, and there is the container!

(*a*)   (*b*)   (*c*)

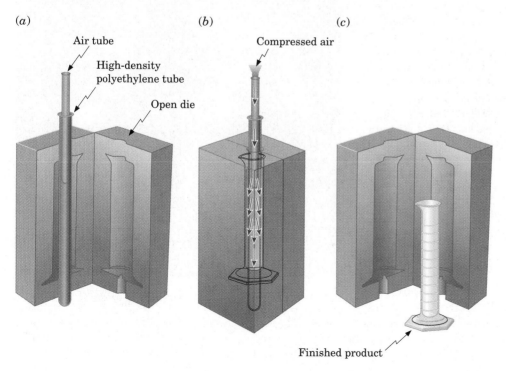

Air tube

High-density
polyethylene tube

Open die

Compressed air

Finished product

CHEMICAL CONNECTIONS 3D

Recycling Plastics

Plastics are polymers that can be molded when hot and that retain their shape when cooled. Because they are durable and lightweight, plastics are probably the most versatile synthetic materials in existence. In fact, the current production of plastics in the United States exceeds the U.S. production of steel. Plastics have come under criticism, however, for their role in the trash crisis. They account for approximately 21% of the volume and 8% of the weight of solid wastes, with most plastic waste consisting of disposable packaging and wrapping.

Six types of plastics are commonly used for packaging applications. In 1988, manufacturers adopted recycling code letters developed by the Society of the Plastics Industry as a means of identifying them.

Code	Polymer	Common Uses
1 PET	poly(ethylene terephthalate)	soft drink bottles, household chemical bottles, films, textile fibers
2 HDPE	high-density polyethylene	milk and water jugs, grocery bags, squeezable bottles
3 V	poly(vinyl chloride), PVC	shampoo bottles, pipes, shower curtains, vinyl siding, wire insulation, floor tiles
4 LDPE	low-density polyethylene	shrink wrap, trash and grocery bags, sandwich bags, squeeze bottles
5 PP	polypropylene	plastic lids, clothing fibers, bottle caps, toys, diaper linings
6 PS	polystyrene	Styrofoam cups, egg cartons, disposable utensils, packaging materials, appliances
7	all other plastics	various

Currently, only poly(ethylene terephthalate) (PET) and high-density polyethylene are recycled in large quantities. The synthesis and structure of PET, a polyester, is described in Section 18.8B.

The process for recycling most plastics is simple, with separation of the plastic from other contaminants being the most labor-intensive step. For example, PET soft drink bottles usually have a paper label and adhesive that must be removed before the PET can be reused. Recycling begins with hand or machine sorting, after which the bottles are shredded into small chips. An air cyclone then removes paper and other lightweight materials. After any remaining labels and adhesives are eliminated with a detergent wash, the PET chips are dried. The PET produced by this method is 99.9% free of contaminants and sells for about half the price of the virgin material.

■ **These students are wearing jackets made from recycled PET soda bottles.**

Charles D. Winters

SUMMARY

An **alkene** is an unsaturated hydrocarbon that contains a carbon–carbon double bond (Section 3.1). An **alkyne** is an unsaturated hydrocarbon that contains a carbon–carbon triple bond.

The structural feature that makes **cis-trans isomerism** possible in alkenes is restricted rotation about the two carbons of the double bond (Section 3.2B). The cis or trans configuration of an alkene is determined by the orientation of the atoms of the parent chain about the double bond. If atoms of the parent chain are located on the same side of the double bond, the configuration of the alkene is cis; if they are located on opposite sides, the configuration is trans.

In IUPAC names, the presence of a carbon–carbon double bond is indicated by a prefix showing the number of carbons in the parent chain and the ending **-ene.** (Section 3.3A). Substituents are numbered and named in alphabetical order. The presence of a carbon–carbon triple bond is indicated by a prefix showing the number of carbons in the parent chain and the ending **-yne.** The carbon atoms of a cycloalkene are numbered 1 and 2 in the direction that gives the smaller number to the first substituent (Section 3.3D). Compounds containing two double bonds are called di-

enes, those with three double bonds are trienes, and those containing four or more double bonds are called polyenes (Section 3.3E).

Because alkenes and alkynes are nonpolar compounds and the only interactions between their molecules are London dispersion forces, their physical properties are similar to those of alkanes with similar carbon skeletons (Section 3.4).

The characteristic structural feature of a **terpene** is a carbon skeleton that can be divided into two or more **isoprene units** (Section 3.5).

A characteristic reaction of alkenes is addition to the double bond (Section 3.6). In addition, the double bond breaks and bonds to two new atoms or groups of atoms form in its place. A **reaction mechanism** is a step-by-step description of how a chemical reaction occurs, including the role of the catalyst (if one is present). A **carbocation** contains a carbon with only six electrons in its valence shell and bears a positive charge.

Polymerization is the process of bonding together many small **monomers** into large, high-molecular-weight **polymers** (Section 3.7).

KEY REACTIONS

1. **Addition of HX (Section 3.6A)** Addition of HX to the carbon–carbon double bond of an alkene follows Markovnikov's rule. The reaction occurs in two steps and involves formation of a carbocation intermediate.

2. **Acid-Catalyzed Hydration (Section 3.6B)** Addition of H_2O to the carbon–carbon double bond of an alkene follows Markovnikov's rule. The reaction occurs in three steps and involves formation of carbocation and oxonium ion intermediates.

3. **Addition of Bromine and Chlorine (Section 3.6C)** Addition to a cycloalkene gives a 1,2-dihalocycloalkane.

4. **Reduction—Formation of Alkanes (Section 3.6D)** Catalytic reduction involves addition of hydrogen to form two new C—H bonds.

5. **Polymerization of Ethylene and Substituted Ethylenes (Section 3.7)** In polymerization of alkenes, monomer units bond together without the loss of any atoms.

P R O B L E M S

Numbers that appear in color indicate difficult problems.
▶ designates problems requiring application of principles.

Structure of Alkenes and Alkynes

3.11 What is the difference in structure between a saturated hydrocarbon and an unsaturated hydrocarbon?

3.12 Each carbon atom in ethane and in ethylene is surrounded by eight valence electrons and has four bonds to it. Explain how the VSEPR model (Section 3.9) predicts a bond angle of 109.5° about each carbon in ethane but an angle of 120° about each carbon in ethylene.

3.13 Predict all bond angles about each highlighted carbon atom.

(a)

(b) 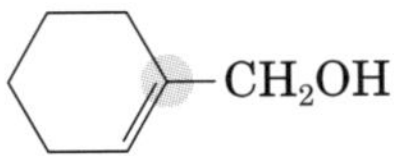 ——CH_2OH

(c) $HC\equiv C—CH\equiv CH_2$

(d)

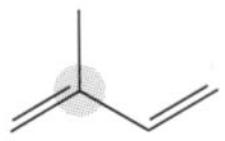

3.14 Predict all bond angles about each highlighted carbon atom.

(a)

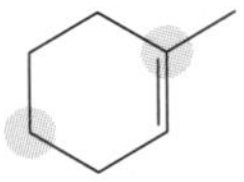

(b)

(c)

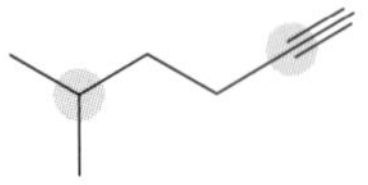

(d)

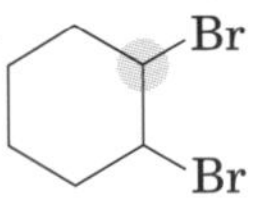

Nomenclature of Alkenes and Alkynes

3.15 Draw a structural formula for each compound.
   (a) *trans*-2-Methyl-3-hexene
   (b) 2-Methyl-3-hexyne
   (c) 2-Methyl-1-butene
   (d) 3-Ethyl-3-methyl-1-pentyne
   (e) 2,3-Dimethyl-2-pentene
   (f) *cis*-2-Hexene

3.16 Draw a structural formula for each compound.
   (a) 3-Chloropropene
   (b) 3-Methylcyclohexene
   (c) 1,2-Dimethylcyclohexene
   (d) *trans*-3,4-Dimethyl-3-heptene
   (e) Cyclopropene
   (f) 3-Hexyne

3.17 Write the IUPAC name for each unsaturated hydrocarbon.
   (a) $CH_2\!\!=\!\!CH(CH_2)_4CH_3$

   (b)

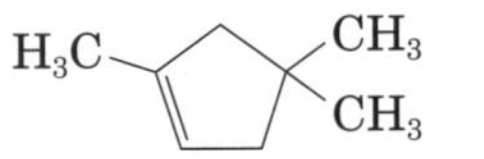

   (c)

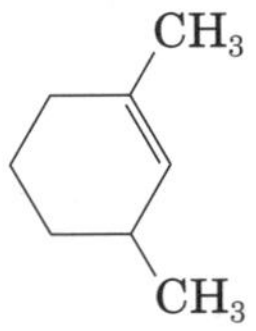

   (d) $(CH_3)_2CHCH\!\!=\!\!C(CH_3)_2$
   (e) $CH_3(CH_2)_5C\!\!\equiv\!\!CH$
   (f) $CH_3CH_2C\!\!\equiv\!\!CC(CH_3)_3$

3.18 Write the IUPAC name for each unsaturated hydrocarbon.

(a) 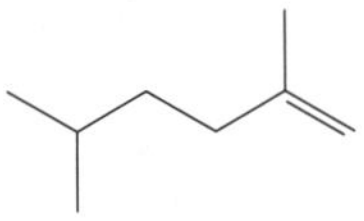   (b)

(c) 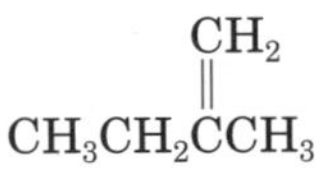 $CH_3CH_2\overset{\overset{\displaystyle CH_2}{\|}}{C}CH_3$ (d)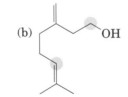

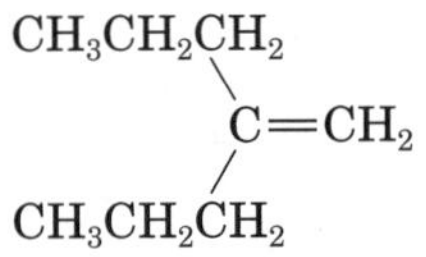

3.19 Explain why each name is incorrect, and then write a correct name.
   (a) 1-Methylpropene
   (b) 3-Pentene
   (c) 2-Methylcyclohexene
   (d) 3,3-Dimethylpentene
   (e) 4-Hexyne
   (f) 2-Isopropyl-2-butene

3.20 Explain why each name is incorrect, and then write a correct name.
   (a) 2-Ethyl-1-propene
   (b) 5-Isopropylcyclohexene
   (c) 4-Methyl-4-hexene
   (d) 2-*sec*-Butyl-1-butene
   (e) 6,6-Dimethylcyclohexene
   (f) 2-Ethyl-2-hexene

Cis-Trans Isomerism in Alkenes and Cycloalkenes

3.21 Which of these alkenes show cis-trans isomerism? For each that does, draw structural formulas for both isomers.
(a) 1-Hexene
(b) 2-Hexene
(c) 3-Hexene
(d) 2-Methyl-2-hexene
(e) 3-Methyl-2-hexene
(f) 2,3-Dimethyl-2-hexene

3.22 Which of these alkenes shows cis-trans isomerism? For each that does, draw structural formulas for both isomers.
(a) 1-Pentene
(b) 2-Pentene
(c) 3-Ethyl-2-pentene
(d) 2,3-Dimethyl-2-pentene
(e) 2-Methyl-2-pentene
(f) 2,4-Dimethyl-2-pentene

3.23 Name and draw structural formulas for all alkenes with molecular formula C_5H_{10}. As you draw these alkenes, remember that cis and trans isomers are different compounds and must be counted separately.

3.24 In Chapter 12 on the biochemistry of lipids, we will study the three long-chain unsaturated carboxylic acids shown below. Each has 18 carbons and is a component of animal fats, vegetable oils, and biological membranes. Because of their presence in animal fats, they are called fatty acids. How many cis-trans isomers are possible for each fatty acid?

Oleic acid $CH_3(CH_2)_7CH{=}CH(CH_2)_7COOH$
Linoleic acid $CH_3(CH_2)_4(CH{=}CHCH_2)_2(CH_2)_6COOH$
Linolenic acid $CH_3CH_2(CH{=}CHCH_2)_3(CH_2)_6COOH$

3.25 The fatty acids in Problem 3.24 occur in animal fats, vegetable oils, and biological membranes almost exclusively as the all-cis isomers. Draw line-angle formulas for each fatty acid showing the cis configuration about each carbon–carbon double bond.

3.26 In Chapter 12, we also encounter this 20-carbon fatty acid. How many cis-trans isomers are possible for it?

$$CH_3(CH_2)_4(CH{=}CHCH_2)_4CH_2CH_2COOH$$
Arachidonic acid

3.27 ▶ Draw a line-angle formula for arachidonic acid showing the cis configuration about each double bond.

3.28 What structural feature in alkenes makes cis-trans isomerism in them possible? What structural feature in cycloalkanes makes cis-trans isomerism in them possible? What do these two structural features have in common?

3.29 For each molecule that shows cis-trans isomerism, draw the cis isomer.

3.30 Draw structural formulas for all compounds with molecular formula C_5H_{10} that are
(a) Alkenes that do not show cis-trans isomerism.
(b) Alkenes that do show cis-trans isomerism.
(c) Cycloalkanes that do not show cis-trans isomerism.
(d) Cycloalkanes that do show cis-trans isomerism.

3.31 ▶ β-Ocimene, a triene found in the fragrance of cotton blossoms and several essential oils, has the IUPAC name *cis*-3,7-dimethyl-1,3,6-octatriene. (Cis refers to the configuration of the double bond between carbons 3 and 4, the only double bond in this molecule about which cis-trans isomerism is possible). Draw a structural formula for β-ocimene.

Terpenes

3.32 Which of these terpenes (Section 3.5) show cis-trans isomerism?
(a) Myrcene   (b) Geraniol
(c) Limonene   (d) Farnesol

3.33 Show that the structural formula of vitamin A (Section 3.3E) can be divided into four isoprene units joined by head-to-tail linkages and cross-linked at one point to form the six-membered ring.

Addition Reactions of Alkenes

3.34 Define "alkene addition reaction." Write an equation for an addition reaction of propene.

3.35 What reagent and/or catalysts are necessary to bring about each conversion?

(a) $CH_3CH{=}CHCH_3 \longrightarrow CH_3CH_2\overset{\displaystyle Br}{\underset{\displaystyle |}{C}}HCH_3$

(b) $CH_3\overset{\displaystyle CH_3}{\underset{\displaystyle |}{C}}{=}CH_2 \longrightarrow CH_3\overset{\displaystyle CH_3}{\underset{\displaystyle |}{\underset{\displaystyle |}{\underset{\displaystyle OH}{C}}}}CH_3$

(c) ⬠ $\longrightarrow$ ⬠—I

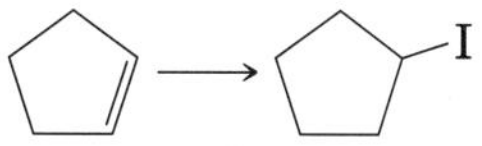

(d) 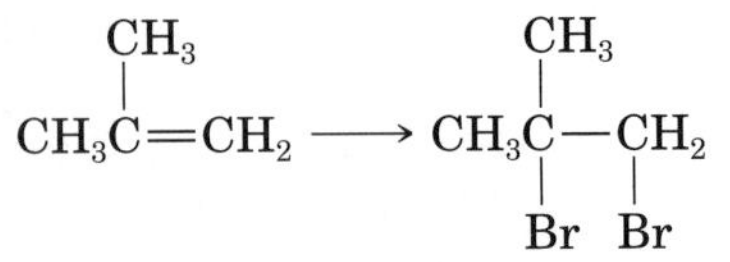 $CH_3\underset{\underset{CH_3}{|}}{C}=CH_2 \longrightarrow CH_3\underset{\underset{Br}{|}}{\overset{\overset{CH_3}{|}}{C}}-\underset{\underset{Br}{|}}{CH_2}$

3.36 Complete these equations.

(a) 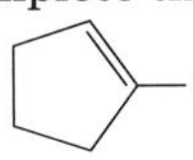 ⬡—CH_2CH_3 + HCl ⟶

(b) 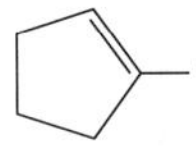 ⬡—CH_2CH_3 + H_2O $\xrightarrow{H_2SO_4}$

(c) $CH_3(CH_2)_5CH=CH_2$ + HI ⟶

(d) 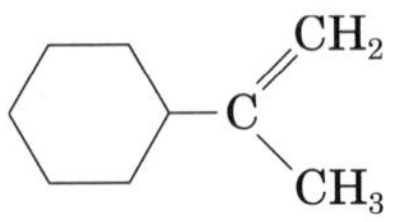 ⬡—$\underset{\underset{CH_3}{|}}{\overset{\overset{CH_2}{||}}{C}}$ + HCl ⟶

(e) $CH_3CH=CHCH_2CH_3$ + H_2O $\xrightarrow{H_2SO_4}$

(f) $CH_2=CHCH_2CH_2CH_3$ + H_2O $\xrightarrow{H_2SO_4}$

3.37 Draw structural formulas for all possible carbocations formed by the reaction of each alkene with HCl. Label each carbocation as primary, secondary, or tertiary.

(a) 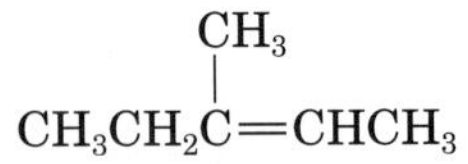 $CH_3CH_2\underset{\underset{CH_3}{|}}{C}=CHCH_3$

(b) $CH_3CH_2CH=CHCH_3$

(c) 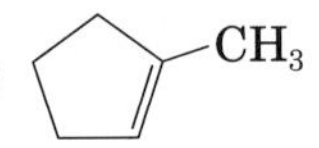 ⬠—CH_3

(d) 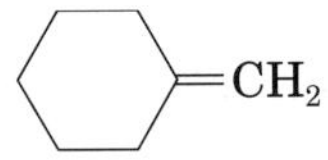 ⬡=CH_2

3.38 Draw a structural formula for the product formed by treatment of 2-methyl-2-pentene with each reagent.
  (a) HCl
  (b) H_2O in the presence of H_2SO_4

3.39 Draw a structural formula for the product of each reaction.
  (a) 1-Methylcyclohexene + Br_2
  (b) 1,2-Dimethylcyclopentene + Cl_2

3.40 Draw a structural formula for an alkene with the indicated molecular formula that gives the compound shown as the major product. Note that more than one alkene may give the same compound as the major product.

(a) C_5H_{10} + H_2O $\xrightarrow{H_2SO_4}$ $CH_3\underset{\underset{OH}{|}}{\overset{\overset{CH_3}{|}}{C}}CH_2CH_3$

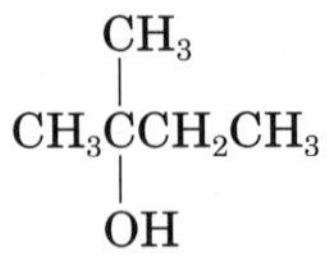

(b) C_5H_{10} + Br_2 ⟶ $CH_3\overset{\overset{CH_3}{|}}{CH}\underset{\underset{Br}{|}}{CH}\underset{\underset{Br}{|}}{CH_2}$

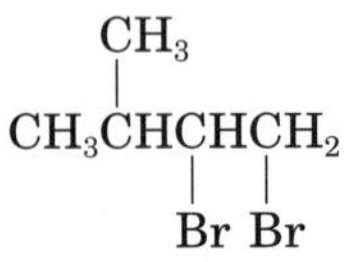

(c) C_7H_{12} + HCl ⟶ ⬡$\overset{CH_3}{\underset{Cl}{}}$

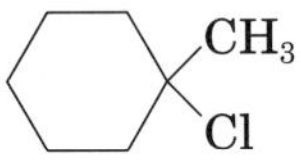

3.41 Draw a structural formula for an alkene with molecular formula C_5H_{10} that reacts with Br_2 to give each product.

(a) 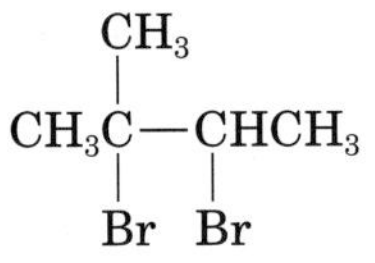 $CH_3\underset{\underset{Br}{|}}{\overset{\overset{CH_3}{|}}{C}}-\underset{\underset{Br}{|}}{CH}CH_3$

(b) $CH_2\underset{\underset{Br}{|}}{\overset{\overset{CH_3}{|}}{C}}CH_2CH_3$

(c) $CH_2\underset{\underset{Br}{|}}{CH}\underset{\underset{Br}{|}}{CH_2}CH_2CH_3$

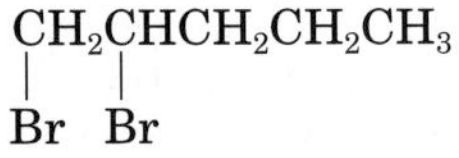

3.42 Draw a structural formula for an alkene with molecular formula C_5H_{10} that reacts with HCl to give the indicated chloroalkane as the major product. More than one alkene may give the same compound as the major product.

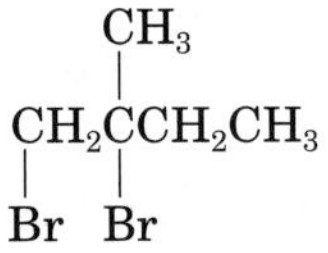

(a) 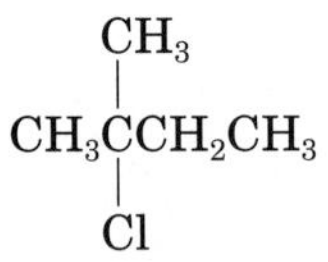 $CH_3\underset{\underset{Cl}{|}}{\overset{\overset{CH_3}{|}}{C}}CH_2CH_3$

(b) $CH_3\underset{\underset{Cl}{|}}{CH}\overset{\overset{CH_3}{|}}{CH}CH_3$

(c) $CH_3\underset{\underset{Cl}{|}}{CH}CH_2CH_2CH_3$

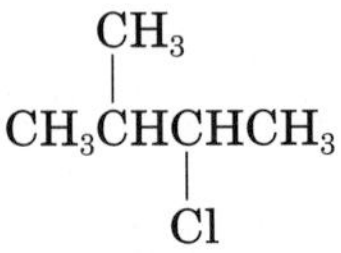

3.43 Draw the structural formula of an alkene that undergoes acid-catalyzed hydration to give the indicated alcohol as the major product. More than one alkene may give each alcohol as the major product.

(a) 3-Hexanol

(b) 1-Methylcyclobutanol

(c) 2-Methyl-2-butanol

(d) 2-Propanol

3.44 Terpin, $C_{10}H_{20}O_2$, is prepared commercially by the acid-catalyzed hydration of limonene (Figure 3.1).

(a) Propose a structural formula for terpin.

(b) How many cis-trans isomers are possible for the structural formula you propose?

(c) Terpin hydrate, the isomer of terpin in which the methyl and isopropyl groups are trans to each other, is used as an expectorant in cough medicines. Draw a structural formula for terpin hydrate showing the trans orientation of these groups.

3.45 Draw the product formed by treatment of each alkene with H_2/Ni.

(a)

$$H_3C \quad\quad H$$
$$C{=}C$$
$$H \quad\quad CH_2CH_3$$

(b)

$$H \quad\quad H$$
$$C{=}C$$
$$H_3C \quad\quad CH_2CH_3$$

(c)

(d)

3.46 Hydrocarbon A, C_5H_8, reacts with 2 moles of Br_2 to give 1,2,3,4-tetrabromo-2-methylbutane. What is the structure of hydrocarbon A?

3.47 Show how to convert ethylene to these compounds.

(a) Ethane

(b) Ethanol

(c) Bromoethane

(d) 1,2-Dibromoethane

(e) Chloroethane

3.48 Show how to convert 1-butene to these compounds.

(a) Butane

(b) 2-Butanol

(c) 2-Bromobutane

(d) 1,2-Dibromobutane

Chemical Connections

3.49 (Chemical Connections 3A) What is one function of ethylene as a plant growth regulator?

3.50 (Chemical Connections 3B) What is the meaning of the term "pheromone"?

3.51 (Chemical Connections 3B) What is the molecular formula of 11-tetradecenyl acetate? What is its molecular weight?

3.52 (Chemical Connections 3B) Assume that 1×10^{-12} g of 11-tetradecenyl acetate is secreted by a single corn borer. How many molecules is this?

3.53 (Chemical Connections 3C) What different functions are performed by the rods and cones in the eye?

3.54 (Chemical Connections 3C) In which isomer of retinal is the end-to-end distance longer, the all-trans isomer or the 11-cis isomer?

3.55 (Chemical Connections 3D) What types of consumer products are made of high-density polyethylene? What types of products are made of low-density polyethylene? One type of polyethylene is currently recyclable and the other is not. Which is which?

3.56 (Chemical Connections 3D) In recycling codes, what do these abbreviations stand for?

(a) V   (b) PP   (c) PS

Additional Problems

3.57 Write line-angle formulas for all compounds with molecular formula C_4H_8. Which are sets of constitutional isomers? Which are sets of cis-trans isomers?

3.58 Knowing what you do about the meaning of the terms "saturated" and "unsaturated" as applied to alkanes and alkenes, what do you think these same terms mean when they are used to describe animal fats such as those found in butter and animal meats? What might the term "polyunsaturated" mean in this same context?

3.59 Name and draw structural formulas for all alkenes with molecular formula C_6H_{12} that have these carbon skeletons. Remember to consider cis and trans isomers.

(a)

$$C$$
$$|$$
$$C-C-C-C-C$$

(b)

$$C \quad C$$
$$| \quad\; |$$
$$C-C-C-C$$

(c)

$$C$$
$$|$$
$$C-C-C-C$$
$$|$$
$$C$$

3.60 Below is the structural formula of lycopene, $C_{40}H_{56}$, a deep-red compound that is partially responsible for the red color of ripe fruits, especially tomatoes. Approximately 20 mg of lycopene can be isolated from 1 kg of fresh, ripe tomatoes.

(a) Show that lycopene is a terpene—that is, its carbon skeleton can be divided into two sets of four isoprene units with the units in each set joined head-to-tail.

(b) How many of the carbon–carbon double bonds in lycopene have the possibility for cis-trans isomerism? Lycopene is the all-trans isomer.

3.61 As you might suspect, β-carotene, $C_{40}H_{56}$, precursor to vitamin A, was first isolated from carrots. Dilute solutions of β-carotene are yellow—hence its use as a food coloring. In plants, this compound is almost always present in combination with chlorophyll to assist in the harvesting of the energy of sunlight. As tree leaves die in the fall, the green of their chlorophyll molecules is replaced by the yellows and reds of carotene and carotene-related molecules. Compare the carbon skeletons of β-carotene and lycopene. What are the similarities? What are the differences?

3.62 Draw the structural formula for a cycloalkene with molecular formula C_6H_{10} that reacts with Cl_2 to give each compound.

(a)

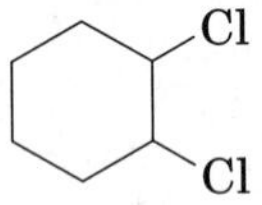

(b)

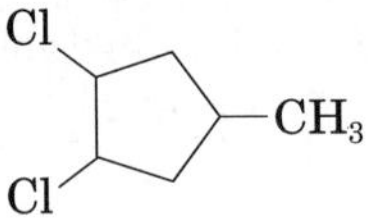

(c)

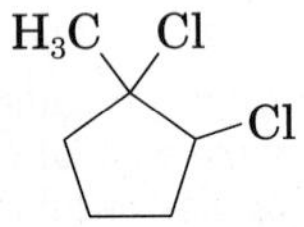

(d)

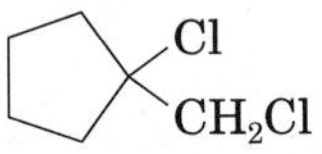

3.63 Propose a structural formula for the product(s) when each of the following alkenes is treated with H_2O/H_2SO_4. Why are two products formed in part (b), but only one in parts (a) and (c)?

(a) 1-Hexene gives one alcohol with molecular formula $C_6H_{14}O$.

(b) 2-Hexene gives two alcohols, each with molecular formula $C_6H_{14}O$.

(c) 3-Hexene gives one alcohol with molecular formula $C_6H_{14}O$.

3.64 *cis*-3-Hexene and *trans*-3-hexene are different compounds and have different physical and chemical properties. Yet, when treated with H_2O/H_2SO_4, each gives the same alcohol. What is the alcohol and how do you account for the fact that each alkene gives the same one?

3.65 Draw the structural formula of an alkene that undergoes acid-catalyzed hydration to give each of the following alcohols as the major product. More than one alkene may give each compound as the major product.

(a)

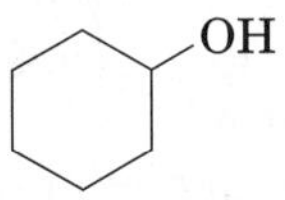

(b)

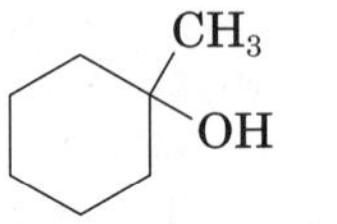

(c)

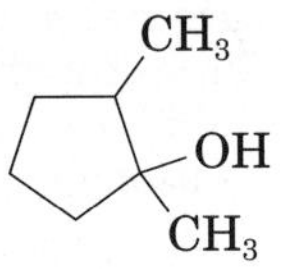

(d)

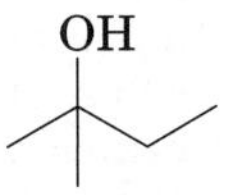

3.66 Show how to convert cyclopentene into these compounds.

(a) 1,2-Dibromocyclopentane

(b) Cyclopentanol

(c) Iodocyclopentane

(d) Cyclopentane

InfoTrac College Edition

For additional readings, go to InfoTrac College Edition, your online research library, at

http://infotrac.thomsonlearning.com

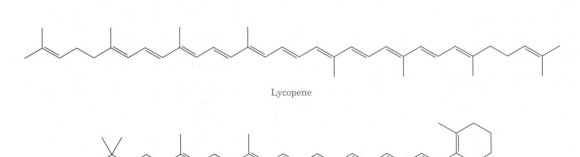

Lycopene

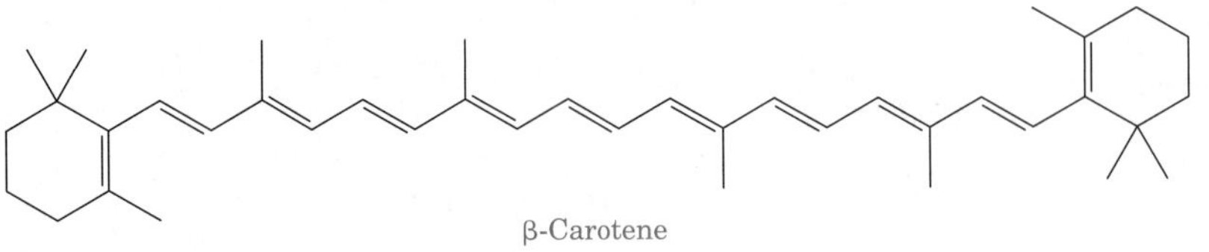

β-Carotene

CHAPTER 4

4.1 Introduction

4.2 The Structure of Benzene

4.3 Nomenclature

4.4 Reactions of Benzene and Its Derivatives

4.5 Phenols

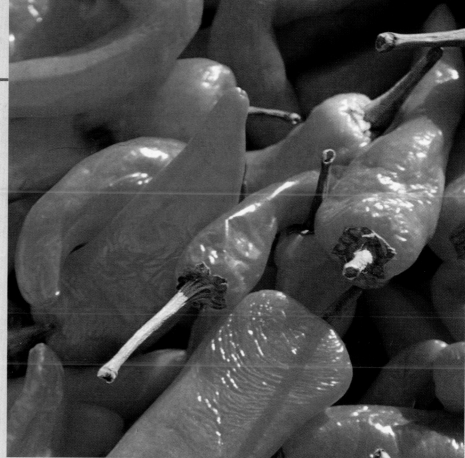

Peppers of the capsicum family (see Chemical Connections 4F).

Douglas Brown

Benzene and Its Derivatives

4.1 Introduction

So far we have described the three classes of hydrocarbons—alkanes, alkenes, and alkynes—called aliphatic hydrocarbons. More than 150 years ago, organic chemists realized that yet another class of hydrocarbons existed—one whose properties are quite different from those of aliphatic hydrocarbons. Because some of these new hydrocarbons have pleasant odors, they were called **aromatic compounds.** Today we know that not all aromatic compounds share this characteristic. Some do have a pleasing odors, but some have no odor at all, and others have downright unpleasant odors. A more appropriate definition of an aromatic compound is any compound that has one or more benzene-like rings.

We use the term **arene** to describe aromatic hydrocarbons. Just as a group derived by removal of an H from an alkane is called an alkyl group and given the symbol R—, so a group derived by removal of an H from an arene is called an **aryl group** and given the symbol **Ar—**.

Aromatic compound Benzene or one of its derivatives

Arene A compound containing one or more benzene rings

4.2 The Structure of Benzene

Benzene is an important compound in both the chemical industry and the laboratory, but it must be handled carefully. Not only is it poisonous if ingested in liquid form, but its vapor is also toxic and can be absorbed either by breathing or through the skin. Long-term inhalation can cause liver damage and cancer.

Benzene, the simplest aromatic hydrocarbon, was discovered by Michael Faraday (1791–1867) in 1825. Its structure presented an immediate problem to chemists of the day. Benzene has the molecular formula C_6H_6, and a compound with so few hydrogens for its six carbons (compare hexane, C_6H_{14}), chemists argued, should be unsaturated. But benzene does not behave like an alkene (the only class of unsaturated hydrocarbons known at that time). Whereas 1-hexene, for example, reacts instantly with Br_2 (Section 12.6C), benzene does not react at all with this reagent. Nor does benzene react with HBr, H_2O/H_2SO_4, or H_2/Pd, reagents that normally add to carbon–carbon double bonds.

A Kekulé's Structure of Benzene

The first structure for benzene was proposed by Friedrich August Kekulé in 1872 and consisted of a six-membered ring with alternating single and double bonds, with one hydrogen attached to each carbon.

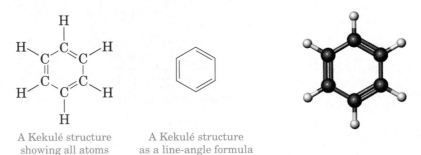

A Kekulé structure
showing all atoms

A Kekulé structure
as a line-angle formula

Although Kekulé's proposal was consistent with many of the chemical properties of benzene, it was contested for years. The major objection was its failure to account for the unusual chemical behavior of benzene. If benzene contains three double bonds, Kekulé's critics asked, why doesn't it undergo reactions typical of alkenes?

B Resonance Structure of Benzene

Resonance hybrid A molecule best described as a composite of two or more Lewis structures

The concept of resonance, developed by Linus Pauling in the 1930s, provided the first adequate description of the structure of benzene. According to the theory of resonance, certain molecules and ions are best described by writing two or more Lewis structures and considering the real molecule or ion to be a **resonance hybrid** of these structures. Each individual Lewis structure is called a **contributing structure.** To show that the real molecule is a resonance hybrid of the two Lewis structures, we position a double-headed arrow between them.

The two contributing structures for benzene are often called Kekulé structures.

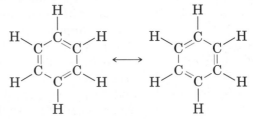

Alternative Lewis contributing structures for benzene

The resonance hybrid has some of the characteristics of each Lewis contributing structure. For example, the carbon–carbon bonds are neither single nor double but something intermediate. It has been determined experimentally that the length of the carbon–carbon bond in benzene is not as long as a carbon–carbon single bond nor as short as a carbon–carbon double bond, but rather midway between the two. The closed loop of six electrons (two from the second bond of each double bond) characteristic of a benzene ring is sometimes called an **aromatic sextet.**

Wherever we find resonance, we find stability. The real structure is generally more stable than any of the fictitious Lewis contributing structures. The benzene ring is greatly stabilized by resonance, which explains why it does not undergo the addition reactions typical of alkenes.

4.3 Nomenclature

A One Substituent

Monosubstituted alkylbenzenes are named as derivatives of benzene—as for example, ethylbenzene. The IUPAC system retains certain common names for several of the simpler monosubstituted alkylbenzenes, including **toluene** and **styrene.**

Ethylbenzene   Toluene   Styrene

The IUPAC system also retains the common names for the following compounds:

Phenol   Anisole   Aniline   Benzaldehyde   Benzoic acid

The substituent group derived by loss of an H from benzene is called a **phenyl group,** C_6H_5—, for which the common symbol is Ph—. In molecules containing other functional groups, phenyl groups are often named as substituents.

Phenyl group (C_6H_5—; Ph—)   1-Phenylcyclohexene

$$C_6H_5\overset{4}{C}H_2\overset{3}{C}H_2\overset{2}{C}H=\overset{1}{C}H_2$$

4-Phenyl-1-pentene

Phenyl group C_6H_5—, the aryl group derived by removing a hydrogen atom from benzene

■ **Household products containing benzene derivatives.**

See the **Interactive General, Organic, and Biochemistry CD-ROM, version 2.0,** for further exploration on this topic.

B Two Substituents

When two substituents occur on a benzene ring, three isomers are possible. We locate the substituents either by numbering the atoms of the ring or by using the locators *ortho* (*o*), *meta* (*m*), and *para* (*p*). The numbers 1,2- are equivalent to *ortho* (Greek: straight); 1,3- to *meta* (Greek: after); and 1,4- to *para* (Greek: beyond).

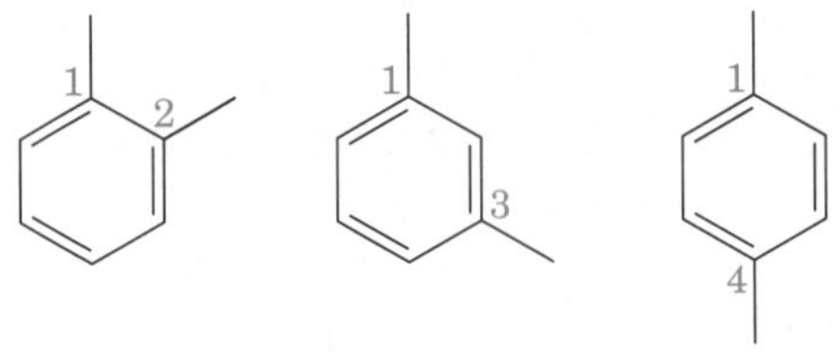

1,2- or *ortho-*   1,3- or *meta*   1,4- or *para*

When one of the two substituents on the ring imparts a special name to the compound (for example, $-CH_3$, $-OH$, $-NH_2$, or $-COOH$), we name the compound as a derivative of that parent molecule, and assume that the substituent occupies ring position number 1. The IUPAC system retains the common name **xylene** for the three isomeric dimethylbenzenes. Where neither substituent imparts a special name, we locate the two substituents and list them in alphabetical order before the ending "benzene." The carbon of the benzene ring with the substituent of lower alphabetical ranking is numbered C-1.

p-Xylene is a starting material for the synthesis of poly(ethylene terephthalate). Consumer products derived from this polymer include Dacron polyester and Mylar films (Section 10.8).

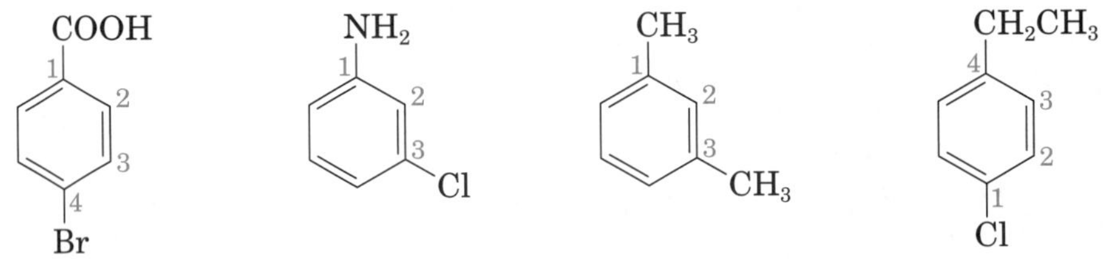

4-Bromobenzoic acid
(*p*-Bromobenzoic acid)

3-Chloroaniline
(*m*-Chloroaniline)

1,3-Dimethylbenzene
(*m*-Xylene)

1-Chloro-4-ethylbenzene
(*p*-Chloroethylbenzene)

C Three or More Substituents

When three or more substituents are present on a benzene ring, we specify their locations by numbers. If one of the substituents imparts a special name, then we name the molecule as a derivative of that parent molecule. If none of the substituents imparts a special name, then we locate the substituents, number them to give the smallest set of numbers, and list them in alphabetical order before the ending "benzene." In the following examples, the first compound is a derivative of toluene, and the second is a derivative of phenol. Because no substituent in the third compound imparts a special name, we list its three substituents in alphabetical order followed by the word "benzene."

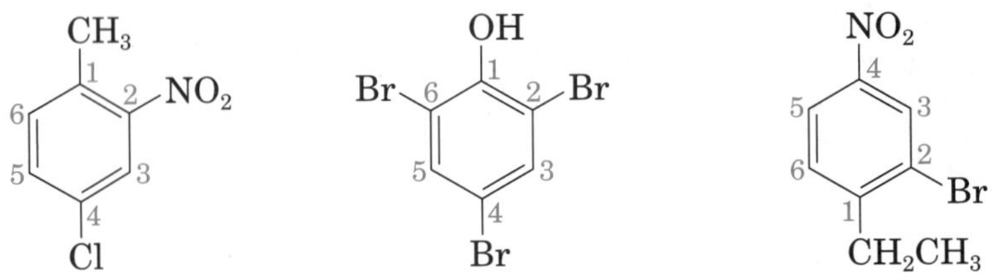

4-Chloro-2-nitrotoluene   2,4,6-Tribromophenol   2-Bromo-1-ethyl-4-nitrobenzene

CHEMICAL CONNECTIONS 4A

DDT: A Boon and a Curse

Probably the best-known insecticide worldwide is *di*chloro*di*phenyl*tri*chloroethane (not an IUPAC name), commonly abbreviated DDT.

Cl—⬡—CH—⬡—Cl
        |
       CCl₃

Dichlorodiphenyltrichloroethane
(DDT)

This compound was first prepared in 1874, but it was not until the late 1930s that its potential as an insecticide was recognized. First used for this purpose in 1939, it proved extremely effective in ridding large areas of the world of the insect hosts that transmit malaria and typhus. In addition, it was so effective in killing crop-destroying insect pests that crop yields in many areas of the world increased dramatically.

Widespread use of DDT, however, has been a double-edged sword. Despite DDT's well-known benefits, it has an enormous disadvantage. Because it resists biodegradation, it remains in the soil for years—and this persistence in the environment creates the problem. Scientists estimate that the tissues of adult humans contain, on average, five to ten parts per million of DDT.

The dangers associated with the persistence of DDT in the environment were dramatically portrayed by Rachel Carson in her 1962 book *The Silent Spring*, which documented the serious

Courtesy of the U.S. Department of Agriculture

■ **DDT was sprayed on crops in the United States until its use was banned in the 1970s.**

decline in the population of eagles and other raptors as well as many other kinds of birds. Scientists discovered soon thereafter that DDT inhibits the mechanism by which these birds incorporate calcium into their egg shells, with the result that shells become so thin and weak that they break during incubation.

Because of these problems, almost all nations have now banned the use of DDT for agricultural purposes. It is still used, however, in some areas to control the population of disease-spreading insects.

EXAMPLE 4.1

Write names for these compounds.

(a) H₃C—⬡—I

(b) [structure with COOH, Br, Br]

(c) [structure with NH₂, Cl]

Solution

(a) The parent is toluene, and the compound is 3-iodotoluene or *m*-iodotoluene.
(b) The parent is benzoic acid, and the compound is 3,5-dibromobenzoic acid.
(c) The parent is aniline, and the compound is 4-chloroaniline or *p*-chloroaniline.

Problem 4.1

Write names for these compounds.

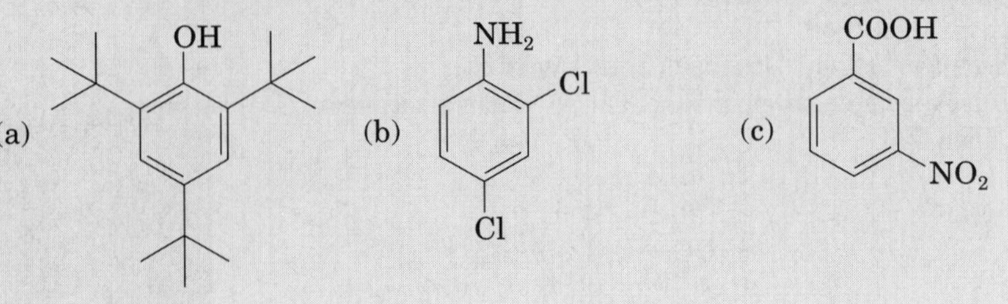

(a)   (b)   (c)

D Polynuclear Aromatic Hydrocarbons

Polynuclear aromatic hydrocarbon A hydrocarbon containing two or more benzene rings, each of which shares two carbon atoms with another benzene ring

Polynuclear aromatic hydrocarbons (PAHs) contain two or more benzene rings, with each pair of rings sharing two adjacent carbon atoms. Naphthalene, anthracene, and phenanthrene, the most common PAHs, and substances derived from them are found in coal tar and high-boiling petroleum residues. At one time, naphthalene was used as moth balls and an insecticide in preserving woolens and furs, but its use decreased after the introduction of chlorinated hydrocarbons such as *p*-dichlorobenzene.

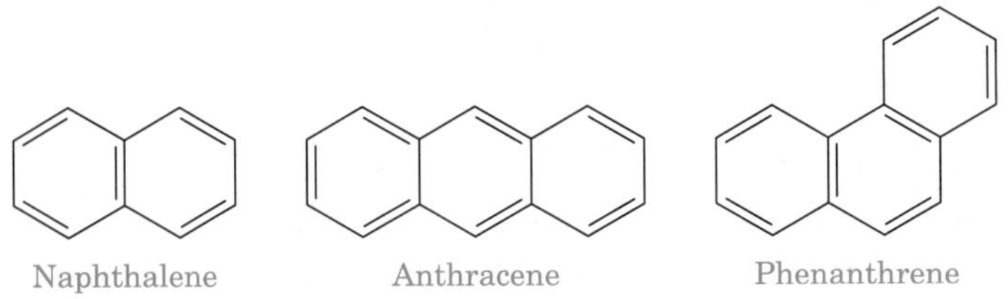

Naphthalene   Anthracene   Phenanthrene

CHEMICAL CONNECTIONS 4B

Carcinogenic Polynuclear Aromatics and Smoking

A **carcinogen** is a compound that causes cancer. The first carcinogens to be identified were a group of polynuclear aromatic hydrocarbons, all of which have at least four aromatic rings. Among them is benzo[a]pyrene, one of the most carcinogenic of the aromatic hydrocarbons. It is formed whenever there is incomplete combustion of organic compounds. Benzo[a]pyrene is found, for example, in cigarette smoke, automobile exhaust, and charcoal-broiled meats.

Benzo[a]pyrene causes cancer in the following way. Once it is absorbed or ingested, the body attempts to convert it into a more soluble compound that can be excreted easily. By a series of enzyme-catalyzed reactions, benzo[a]pyrene is transformed into a **diol** (two —OH groups) **epoxide** (a three-membered ring, one atom of which is oxygen). This compound can bind to DNA by reacting with one of its amino groups, thereby altering the structure of DNA and producing a cancer-causing mutation.

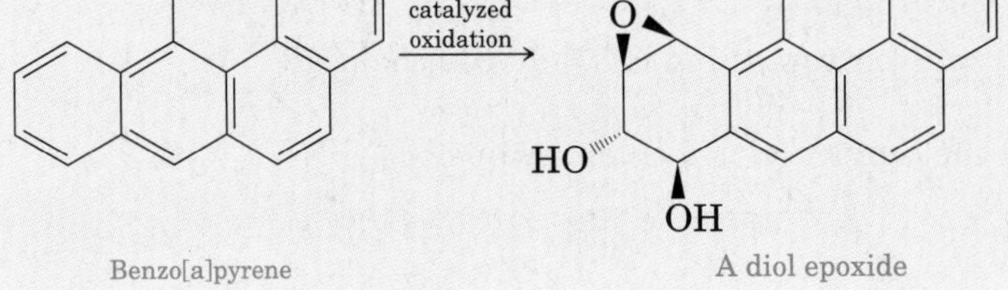

Benzo[a]pyrene   A diol epoxide

Iodide Ion and Goiter

One hundred years ago, goiter, an enlargement of the thyroid gland caused by iodine deficiency, was common in the central United States and central Canada. This disease results from underproduction of thyroxine, a hormone synthesized in the thyroid gland. Young mammals require this hormone for normal growth and development. Its lack during fetal development results in mental retardation. Low levels of thyroxine in adults result in hypothyroidism, commonly called goiter, the symptoms of which are lethargy, obesity, and dry skin.

Iodine is an element that comes primarily from the sea. Rich sources of it, therefore, are fish and other seafoods. The iodine in our diets that doesn't come from the sea most commonly is derived from food additives. Most of the iodide ion in the North American diet comes from table salt fortified with sodium iodide, commonly referred to as iodized salt. Another source is dairy products, which accumulate iodide because of the iodine-containing additives used in cattle feeds and the iodine-containing disinfectants used on milking machines and milk storage tanks.

Thyroxine

4.4 Reactions of Benzene and Its Derivatives

By far the most characteristic reaction of aromatic compounds is substitution at a ring carbon, which we give the name **aromatic substitution.** Groups we can introduce directly on the ring include the halogens, the nitro ($-NO_2$) group, and the sulfonic acid ($-SO_3H$) group.

See the **Interactive General, Organic, and Biochemistry CD-ROM, version 2.0,** for further exploration on this topic.

A Halogenation

As noted in Section 4.2, chlorine and bromine do not react with benzene, in contrast to their instantaneous reaction with cyclohexene and other alkenes (Section 3.6C). In the presence of an iron catalyst, however, chlorine reacts rapidly with benzene to give chlorobenzene and HCl:

$$\text{Benzene}-H + Cl_2 \xrightarrow{FeCl_3} \text{Chlorobenzene}-Cl + HCl$$

Treatment of benzene with bromine in the presence of $FeCl_3$ results in formation of bromobenzene and HBr.

B Nitration

When we heat benzene or one of its derivatives with a mixture of concentrated nitric and sulfuric acids, one of the hydrogen atoms bonded to the ring is replaced by a nitro ($-NO_2$) group.

$$\text{}-H + HNO_3 \xrightarrow{H_2SO_4} \text{Nitrobenzene}-NO_2 + H_2O$$

CHEMICAL CONNECTIONS 4D

The Nitro Group in Explosives

Treatment of toluene with three moles of nitric acid in the presence of sulfuric acid as a catalyst results in nitration of toluene three times to form the explosive 2,4,6-trinitrotoluene, TNT. It is the presence of these three nitro groups that confers the explosive properties to TNT. Similarly, it is the presence of three nitro groups that confers the explosive properties to nitroglycerin.

In recent years, several new explosives have been discovered, all of which contain multiple nitro groups. Among them are RDX and PETN. The plastic explosive Semtex, for example, is a mixture of RDX and PETN. It was used in the destruction of Pan Am flight 103 over Lockerbie, Scotland, in December 1988.

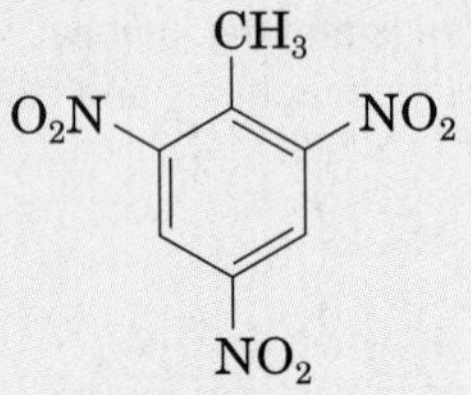

2,4,6-Trinitrotoluene
(TNT)

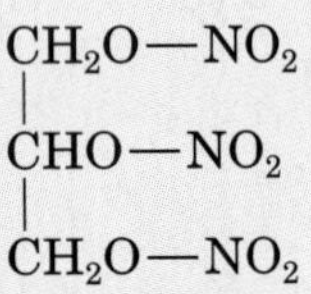

Trinitroglycerin
(Nitroglycerin)

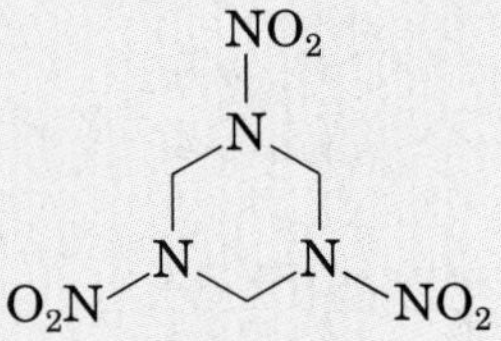

Cyclonite
(RDX)

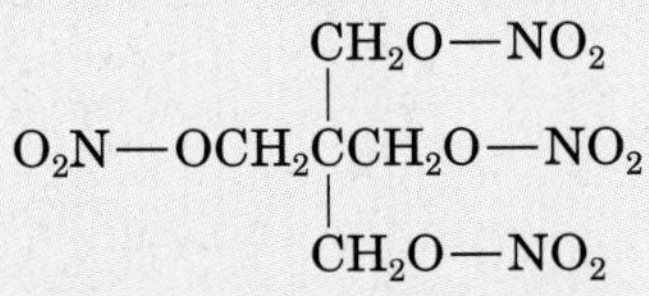

Pentaerythritol tetranitrate
(PETN)

A particular value of nitration is that we can reduce the resulting $-NO_2$ group to a primary amino group, $-NH_2$, by catalytic reduction using hydrogen in the presence of a transition-metal catalyst. In the following example, neither the benzene ring nor the carboxyl group is affected by these experimental conditions:

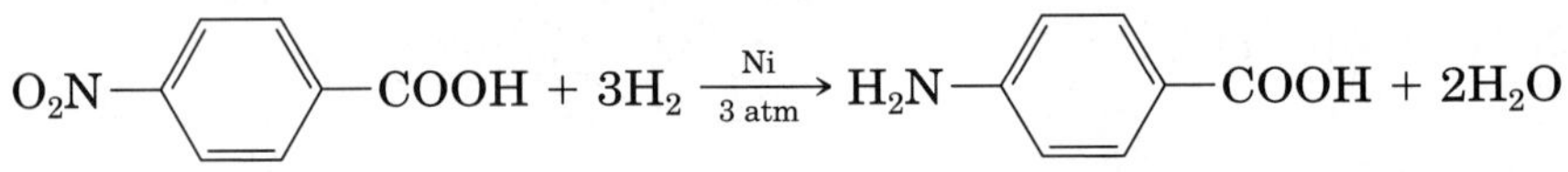

$$O_2N-\!\!\!\bigcirc\!\!\!-COOH + 3H_2 \xrightarrow[\text{3 atm}]{\text{Ni}} H_2N-\!\!\!\bigcirc\!\!\!-COOH + 2H_2O$$

4-Nitrobenzoic acid
(*p*-Nitrobenzoic acid)

4-Aminobenzoic acid
(*p*-Aminobenzoic acid, PABA)

Bacteria require *p*-aminobenzoic acid to synthesize folic acid (Section 21.5), which is in turn required for the synthesis of the heterocyclic aromatic amine bases of nucleic acids (Section 16.2). Whereas bacteria can synthesize folic acid from *p*-aminobenzoic acid, folic acid is a vitamin for humans and must be obtained in the diet. Chemical Connections 14D describes how sulfa drugs inhibit the synthesis of this vitamin.

C Sulfonation

Heating an aromatic compound with concentrated sulfuric acid results in formation of a sulfonic acid, all of which are strong acids, comparable in strength to sulfuric acid.

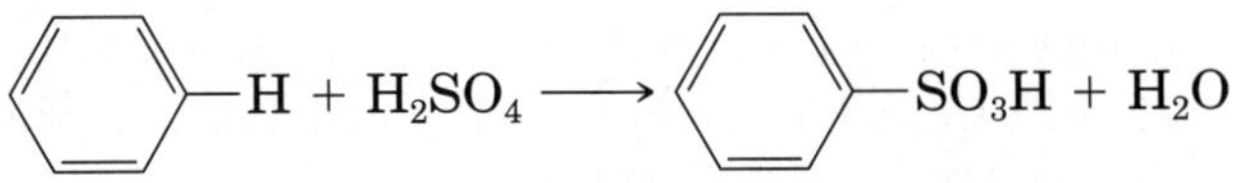

$$\bigcirc\!\!\!-H + H_2SO_4 \longrightarrow \bigcirc\!\!\!-SO_3H + H_2O$$

Benzenesulfonic acid

A major use of sulfonation is in the preparation of synthetic detergents, an important example of which is sodium 4-dodecylbenzenesulfonate. To prepare this type of detergent, a linear alkylbenzene such as dodecylben-

CHEMICAL CONNECTIONS 4E

FD & C No. 6 (a.k.a. Sunset Yellow)

Did you ever wonder what gives gelatin desserts their red, green, orange, or yellow color? Or what gives margarines their yellow color? Or what gives maraschino cherries their red color? If you read the content labels, you will see code names such as FD & C Yellow No. 6 and FD & C Red No. 40.

At one time, the only colorants for foods were compounds obtained from plant or animal materials. Beginning as early as the 1890s, however, chemists discovered a series of synthetic food dyes that offered several advantages over natural dyes, such as greater brightness, better stability, and lower cost. Opinion remains divided on the safety of their use. No synthetic food colorings are allowed, for example, in Norway and Sweden. In the United States, the Food and Drug Administration (FDA) has certified seven synthetic dyes for use in foods, drugs, and cosmetics (FD & C). They include two yellows, two reds, two blues, and one green. When these dyes are used alone or in combinations, they can approximate the color of almost any natural food.

Following are structural formulas for Allura Red (Red No. 40) and Sunset Yellow (Yellow No. 6). These and the other five food dyes certified in the United States have in common three or more benzene rings and two or more ionic groups, either the sodium salt of a carboxylic acid group, —COO⁻Na⁺, or the sodium salt of sulfonic acid group, —SO₃⁻Na⁺. These ionic groups make the dyes soluble in water.

To return to our original questions, maraschino cherries are colored with FD & C Red No. 40, and margarines are colored

with FD & C Yellow No 6. Gelatin desserts use either one or a combination of the seven certified food dyes to produce their color.

Allura Red
(FD & C Red No. 40)

Sunset Yellow
(FD & C Yellow No. 6)

---

zene is treated with concentrated sulfuric acid to give an alkylbenzenesulfonic acid. The sulfonic acid is then neutralized with sodium hydroxide.

$$CH_3(CH_2)_{10}CH_2 - \bigcirc \xrightarrow[\text{2. NaOH}]{\text{1. H}_2\text{SO}_4} CH_3(CH_2)_{10}CH_2 - \bigcirc - SO_3^-Na^+$$

Dodecylbenzene   Sodium 4-dodecylbenzenesulfonate, SDS
(an anionic detergent)

Alkylbenzenesulfonate detergents were introduced in the late 1950s, and today they claim nearly 90% of the market once held by natural soaps. Chemical Connections 12C discusses the chemistry and cleansing action of soaps and detergents.

4.5 Phenols

A Structure and Nomenclature

The functional group of a **phenol** is a hydroxyl group bonded to a benzene ring. Substituted phenols are named either as derivatives of phenol or by common names.

> **Phenol** A compound that contains an —OH bonded to a benzene ring

Charles D. Winters

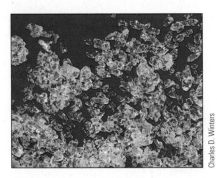

■ **Phenol in crystalline form.**

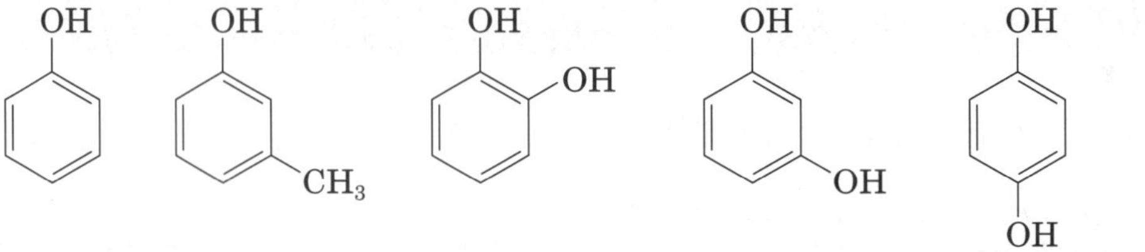

Phenol | 3-Methylphenol (*m*-Cresol) | 1,2-Benzenediol (Catechol) | 1,3-Benzenediol (Resorcinol) | 1,4-Benzenediol (Hydroquinone)

Phenols are widely distributed in nature. Phenol itself and the isomeric cresols (*o*-, *m*-, and *p*-cresol) are found in coal tar. Thymol and vanillin are important constituents of thyme and vanilla beans, respectively. Urushiol is the main component of the irritating oil of poison ivy. It can cause severe contact dermatitis in sensitive individuals.

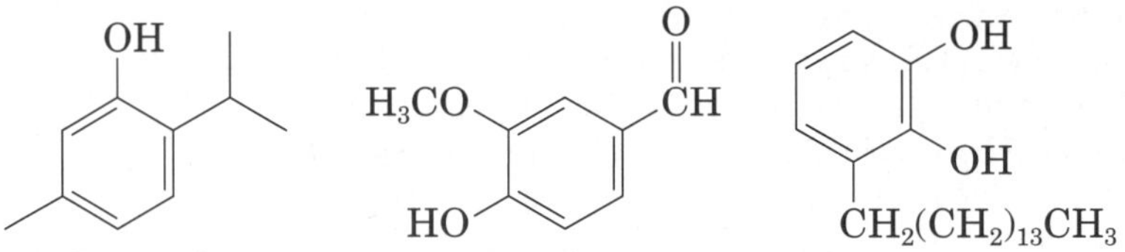

©John D. Cunningham/Visuals Unlimited

■ **Vanilla beans.**

2-Isopropyl-5-methylphenol (Thymol) | 4-Hydroxy-3-methoxybenzaldehyde (Vanillin) | Urushiol

B Acidity of Phenols

Phenols are weak acids, with pK_a values of approximately 10 (Table 8.3). Most phenols are insoluble in water, but they react with strong bases, such as NaOH and KOH, to form water-soluble salts.

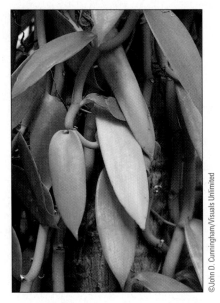

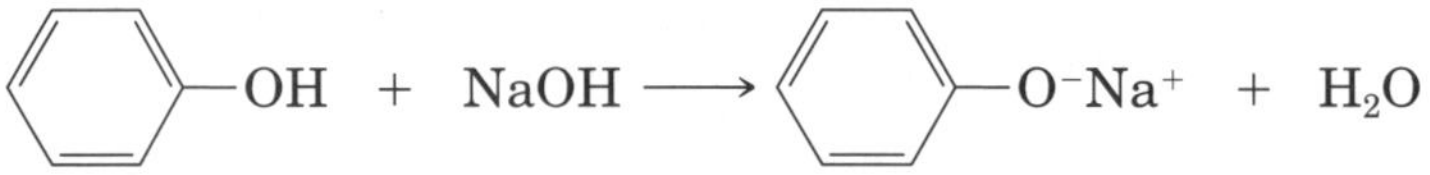

Phenol $pK_a = 9.95$ (stronger acid) | Sodium hydroxide (stonger base) | Sodium phenoxide (weaker base) | Water $pK_a = 15.7$ (weaker acid)

Most phenols are such weak acids that they do not react with weak bases such as sodium bicarbonate; that is, they do not dissolve in aqueous sodium bicarbonate.

Charles D. Winters

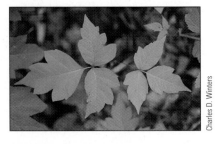

■ **Poison ivy.**

C Phenols as Antioxidants

An important reaction for living systems, foods, and other materials that contain carbon–carbon double bonds is **autoxidation**—that is, oxidation requiring oxygen and no other reactant. If you open a bottle of cooking oil that has stood for a long time, you will notice a hiss of air entering the bottle. This sound occurs because the consumption of oxygen by autoxidation of the oil creates a negative pressure inside the bottle.

Cooking oils contain esters of polyunsaturated fatty acids. You need not worry now about what esters are; we will discuss them in Chapter 10. The important point here is that all vegetable oils contain fatty acids with long hydrocarbon chains, many of which have one or more carbon–carbon double bonds (see Problems 3.24 and 3.26 for the structures of four of these fatty acids). Autoxidation takes place adjacent to these double bonds.

CHEMICAL CONNECTIONS 4 F

Capsaicin, for Those Who Like It Hot

Capsaicin, the pungent principal from the fruit of various species of peppers (*Capsicum* and *Solanaceae*), was isolated in 1876, and its structure was determined in 1919. Capsaicin contains both a phenol and a phenol ether.

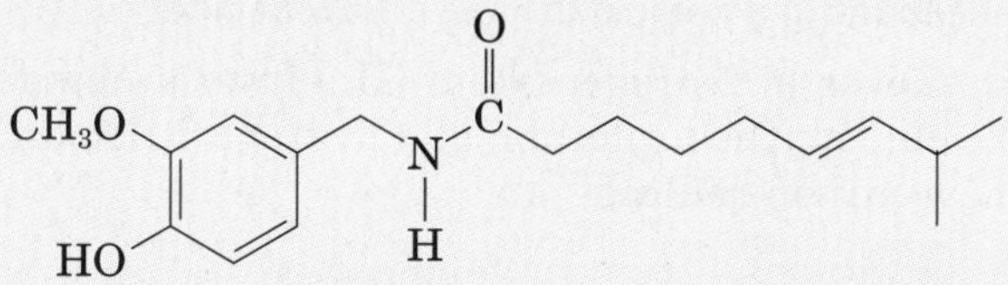

Capsaicin
(from various types of peppers)

The inflammatory properties of capsaicin are well known; the human tongue can detect as little as one drop in 5 L of water. We all know of the burning sensation in the mouth and sudden tearing in the eyes caused by a good dose of hot chili peppers. For this reason, capsaicin-containing extracts from these flaming foods are used in sprays to ward off dogs or other animals that might nip at your heels while you are running or cycling.

Ironically, capsaicin is able to both cause and relieve pain. Currently, two capsaicin-containing creams, Mioton and Zostrix, are prescribed to treat the burning pain associated with postherpetic neuralgia, a complication of the disease known as shingles. They are also prescribed for diabetics to relieve persistent foot and leg pain.

Chuck Pefley/Stone/Getty Images

■ **Red chilies being dried in the sun.**

Autoxidation is a radical chain process that converts an R—H group to an R—O—O—H group, called a hydroperoxide. This process begins when a hydrogen atom with one of its electrons (H·) is removed from a carbon adjacent to one of the double bonds in a hydrocarbon chain. The carbon losing the H· has only seven electrons in its valence shell, one of which is unpaired. An atom or molecule with an unpaired electron is called a **radical.**

Step 1: Chain Initiation—Formation of a Radical from a Nonradical Compound

Removal of a hydrogen atom (H·) may be initiated by light or heat. The product formed is a carbon radical; that is, it contains a carbon atom with one unpaired electron.

$$—CH_2CH=CH—\overset{\overset{\displaystyle H}{|}}{CH}— \xrightarrow[\text{or heat}]{\text{light}} —CH_2CH=CH—\overset{\displaystyle \cdot}{C}H—$$

Section of a fatty
acid hydrocarbon chain

A carbon radical

Step 2a: Chain Propagation—Reaction of a Radical to Form a New Radical

The carbon radical reacts with oxygen, itself a diradical, to form a hydroperoxy radical. The new covalent bond of the hydroperoxy radical forms by the combination of one electron from the carbon radical and one electron from the oxygen diradical.

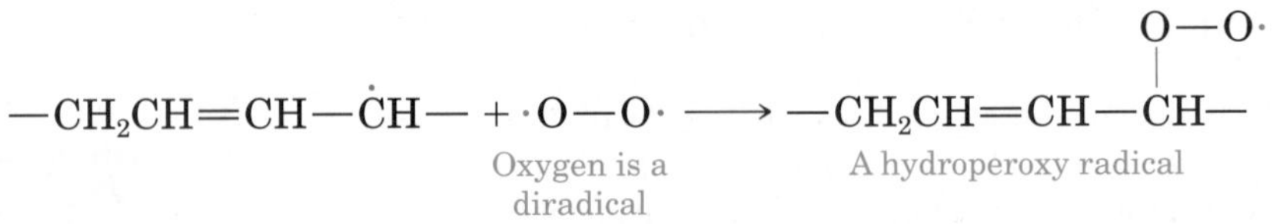

$$—CH_2CH{=}CH—\overset{\cdot}{C}H— \;+\; \cdot O{-}O\cdot \;\longrightarrow\; —CH_2CH{=}CH—\overset{\displaystyle O{-}O\cdot}{\overset{|}{C}H}—$$

Oxygen is a diradical   A hydroperoxy radical

Step 2b: Chain Propagation—Reaction of a Radical to Form a New Radical

The hydroperoxy radical removes a hydrogen atom (H·) from a new fatty acid hydrocarbon chain to complete the formation of a hydroperoxide and at the same time produce a new carbon radical.

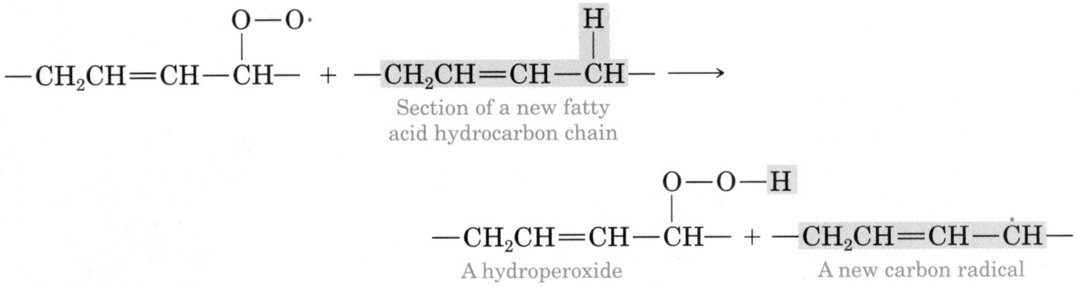

$$—CH_2CH{=}CH—\overset{\displaystyle O{-}O\cdot}{\overset{|}{C}H}— \;+\; —CH_2CH{=}CH—\overset{\displaystyle H}{\overset{|}{C}H}— \;\longrightarrow$$

Section of a new fatty acid hydrocarbon chain

$$—CH_2CH{=}CH—\overset{\displaystyle O{-}O{-}H}{\overset{|}{C}H}— \;+\; —CH_2CH{=}CH—\overset{\cdot}{C}H—$$

A hydroperoxide   A new carbon radical

The most important point about the pair of chain propagation steps is that they form a continuous cycle of reactions. The new radical formed in Step 2b next reacts with another molecule of O_2 by step 2a to give a new hydroperoxy radical. This new hydroperoxy radical then reacts with a new hydrocarbon chain to repeat Step 2b, and so forth. This cycle of propagation steps repeats over and over in a chain reaction. Thus, once a radical is generated in Step 1, the cycle of propagation steps may repeat many thousands of times, generating thousands and thousands of hydroperoxide molecules. The number of times the cycle of chain propagation steps repeats is called the **chain length.**

Hydroperoxides themselves are unstable and, under biological conditions, degrade to short-chain aldehydes and carboxylic acids with unpleasant "rancid" smells. These odors may be familiar to you if you have ever smelled old cooking oil or aged foods that contain polyunsaturated fats or oils. Similar formation of hydroperoxides in the low-density lipoproteins (Section 12.4) deposited on the walls of arteries leads to cardiovascular disease in humans. In addition, many effects of aging are thought to result from the formation and subsequent degradation of hydroperoxides.

Fortunately, nature has developed a series of defenses against the formation of these and other destructive hydroperoxides, including the phenol vitamin E (Section 21.5). This compound is one of nature's "scavengers." It inserts itself into either Step 2a or 2b, donates an H· from its —OH group to the carbon radical, and converts the carbon radical back to its original hydrocarbon chain. Because the vitamin E radical is stable, it breaks the cycle of chain propagation steps, thereby preventing further formation of destructive hydroperoxides. While some hydroperoxides may form, their numbers are very small and they are easily decomposed to harmless materials by one of several enzyme-catalyzed reactions.

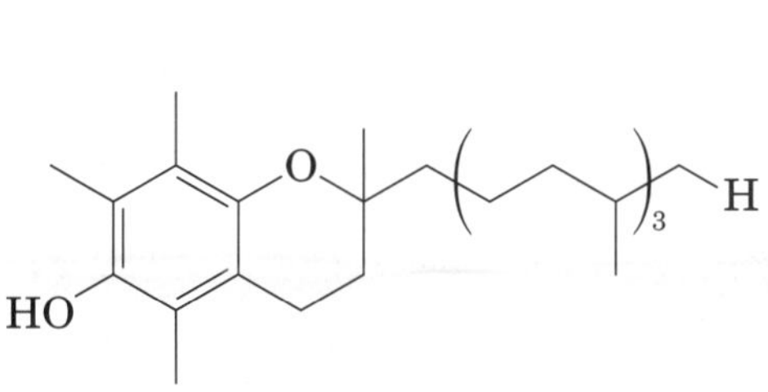

OH

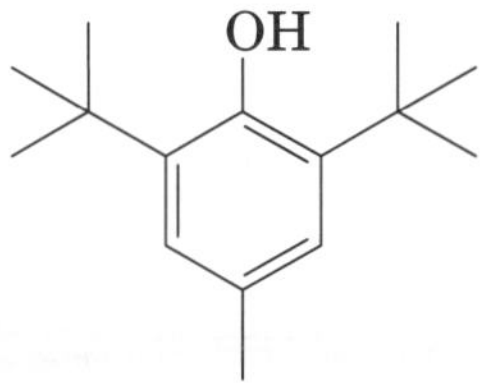

Butylated *hydroxy-toluene*
(BHT)

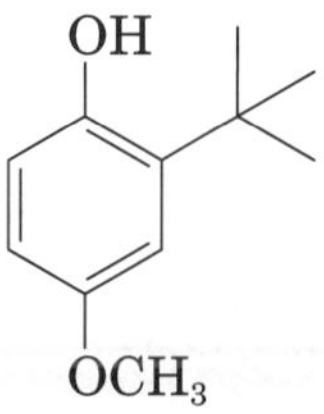

OH

OCH$_3$

Butylated *hydroxy-anisole*
(BHA)

HO

Vitamin E

Unfortunately, vitamin E is removed in the processing of many foods and food products. To make up for this loss, phenols such as BHT and BHA are added to foods to "retard spoilage" (as they say on the packages) by autoxidation. Likewise, similar compounds are added to other materials, such as plastics and rubber, to protect them against autoxidation.

S U M M A R Y

Benzene and its alkyl derivatives are classified as **aromatic hydrocarbons** or **arenes** (Section 4.1). The first structure for benzene was proposed by Friedrich August Kekulé in 1872 (Section 4.2A). The theory of **resonance,** developed by Linus Pauling in the 1930s, provided the first adequate structure for benzene (Section 4.2B).

Aromatic compounds are named according to the IUPAC system (Section 4.3). The C$_6$H$_5$— group is named **phenyl.** Two substituents on a benzene ring may be located by either numbering the atoms of the ring or by using the locators **ortho (o), meta (m),** and **para (p)** (Section 4.3B). **Polynuclear aromatic hydrocarbons** contain two or more benzene rings, each

sharing two adjacent carbon atoms with another ring (section 4.3D).

A characteristic reaction of aromatic compounds is **aromatic substitution,** in which another atom or group of atoms is substituted for a hydrogen atom of the aromatic ring (Section 4.4). Typical aromatic substitution reactions are halogenation, nitration, and sulfonation.

The functional group of a **phenol** is an —OH group bonded to a benzene ring (Section 4.5A). Phenol and its derivatives are weak acids, with pK_a equal to approximately 10.0 (Section 4.5B). Vitamin E, a phenolic compound, is an antioxidant (Section 4.5C). Phenolic compounds such as BHT and BHA are synthetic antioxidants.

K E Y   R E A C T I O N S

1. Halogenation (Section 4.4A) Treatment of an aromatic compound with Cl$_2$ or Br$_2$ in the presence of an FeCl$_3$ catalyst substitutes a halogen for an H.

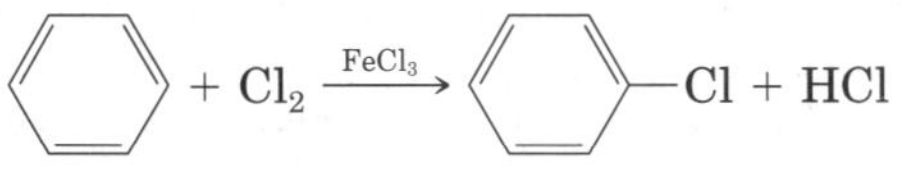

$$\text{+ Cl}_2 \xrightarrow{\text{FeCl}_3} \text{—Cl + HCl}$$

2. Nitration (Section 4.4B) Treatment of an aromatic compound with a mixture of concentrated nitric and sulfuric acids substitutes a nitro group for an H.

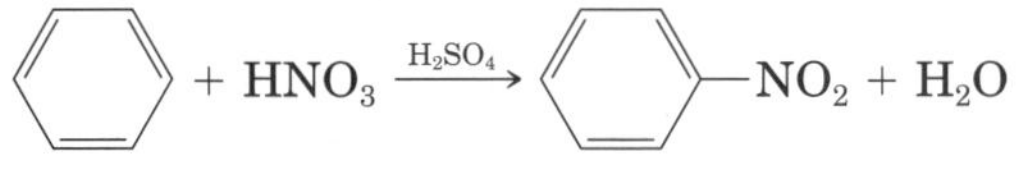

$$\text{+ HNO}_3 \xrightarrow{\text{H}_2\text{SO}_4} \text{—NO}_2 \text{ + H}_2\text{O}$$

3. Sulfonation (Section 4.4C) Treatment of an aromatic compound with concentrated sulfuric acid substitutes a sulfonic acid group for an H.

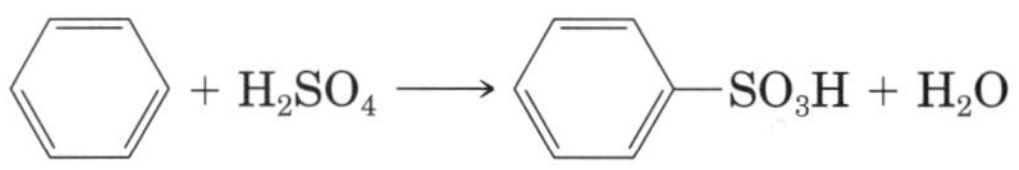

$$\text{+ H}_2\text{SO}_4 \longrightarrow \text{—SO}_3\text{H + H}_2\text{O}$$

4. Reaction of Phenols with Strong Bases (Section 4.5B) Phenols are weak acids and react with strong bases to form water-soluble salts.

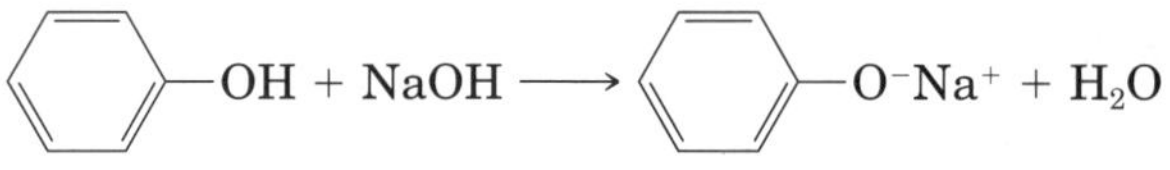

$$\text{—OH + NaOH} \longrightarrow \text{—O}^-\text{Na}^+ \text{ + H}_2\text{O}$$

P R O B L E M S

Numbers that appear in color indicate difficult problems.
◀ designates problems requiring application of principles.

Aromatic Compounds

4.2 What is the difference in structure between a saturated and an unsaturated compound?

4.3 Define *aromatic compound.*

4.4 Why are alkenes, alkynes, and aromatic compounds said to be unsaturated?

4.5 Do aromatic rings have double bonds? Are they unsaturated? Explain.

4.6 Can an aromatic compound be a saturated compound?

4.7 Draw at least two structural formulas for the following. (Several constitutional isomers are possible for each part.)
(a) An alkene of six carbons
(b) A cycloalkene of six carbons
(c) An alkyne of six carbons
(d) An aromatic hydrocarbon of eight carbons

4.8 Write a structural formula and the name for the simplest (a) alkane, (b) alkene, (c) alkyne, and (d) aromatic hydrocarbon.

4.9 Account for the fact that the six-membered ring in benzene is planar but the six-membered ring in cyclohexane is not.

4.10 The compound 1,4-dichlorobenzene (*p*-dichlorobenzene) has a rigid geometry that does not allow free rotation. Yet no cis-trans isomers exist for this structure. Explain why it does not show cis-trans isomerism.

4.11 One analogy often used to explain the concept of a resonance hybrid is to relate a rhinoceros to a unicorn and a dragon. Explain the reasoning in this analogy and how it might relate to a resonance hybrid.

Nomenclature and Structural Formulas

4.12 Name these compounds.

(a)

(b)

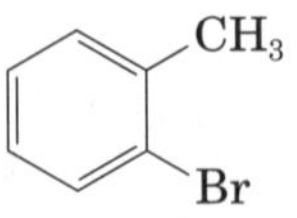

(c) $C_6H_5CH_2CH_2CH_2Cl$

(d)

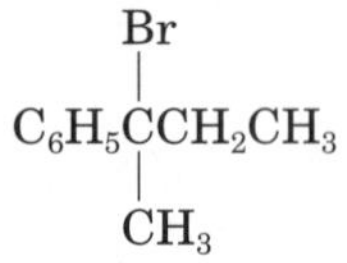

(e)

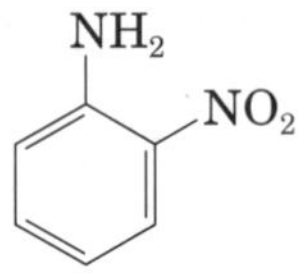

(f)

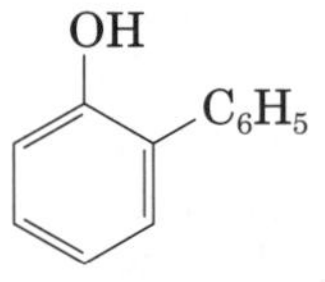

(g)

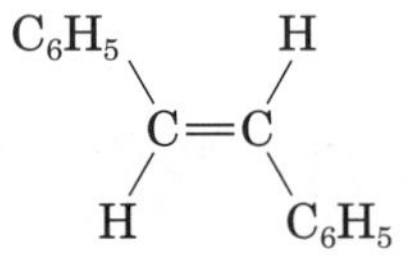

(h)

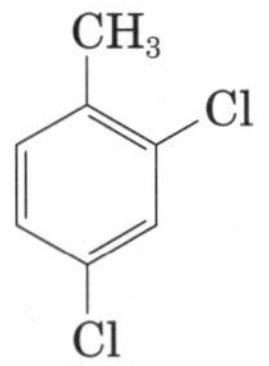

4.13 Draw structural formulas for these compounds.
(a) 1-Bromo-2-chloro-4-ethylbenzene
(b) 4-Bromo-1,2-dimethylbenzene
(c) 2,4,6-Trinitrotoluene
(d) 4-Phenyl-1-pentene
(e) *p*-Cresol
(f) 2,4-Dichlorophenol

4.14 We say that naphthalene, anthracene, phenanthrene, and benzo[a]pyrene are polynuclear aromatic hydrocarbons. In this context, what does "polynuclear" mean? What does "aromatic" mean? What does "hydrocarbon" mean?

Reactions of Aromatic Compounds

4.15 Suppose you have unlabeled bottles of benzene and cyclohexene. What chemical reaction could you use to tell which bottle contains which chemical? Explain what you would do, what you would expect to see, and how you would explain your observations.

4.16 Three possible products with molecular formula C_6H_4BrCl can form when bromobenzene is treated with chlorine, Cl_2, in the presence of $FeCl_3$ as a catalyst. Name and draw a structural formula for each product.

4.17 The reaction of bromine with toluene in the presence of $FeCl_3$ gives a mixture of three products, all with the molecular formula C_7H_7Br. Name and draw a structural formula for each product.

4.18 What reagents and/or catalysts are necessary to carry out each conversion? Each conversion requires only one step.

(a) Benzene to nitrobenzene

(b) 1,4-dichlorobenzene to 2-bromo-1,4-dichlorobenzene

4.19 What reagents and/or catalysts are necessary to carry out each conversion? Each conversion requires two steps.

(a) Benzene to 3-nitrobenzenesulfonic acid

(b) Benzene to 1-bromo-4-chlorobenzene

4.20 Aromatic substitution can be done on naphthalene. Treatment of naphthalene with concentrated H_2SO_4 gives two (and only two) different sulfonic acids. Draw a structural formula for each.

Phenols

4.21 Both phenol and cyclohexanol are only slightly soluble in water. Account for the fact that phenol dissolves in aqueous sodium hydroxide but cyclohexanol does not.

4.22 Define *autoxidation*.

4.23 Autoxidation is described as a radical chain reaction. What is meant by the term "radical" in this context? By the term "chain"? By the term "chain length"?

4.24 Show that, if you add Steps 2a and 2b of the radical chain mechanism for the autoxidation of a fatty acid hydrocarbon chain, you arrive at the following net equation:

$$
\begin{array}{c}
\quad\quad\quad\quad\text{H} \\
\quad\quad\quad\quad | \\
-\text{CH}_2\text{CH}{=}\text{CH}-\text{CH}- \;+\; \cdot\text{O}-\text{O}\cdot
\end{array}
$$

Section of a fatty acid   Oxygen
hydrocarbon chain

$$
\begin{array}{c}
\quad\quad\quad\quad\quad\quad\quad\quad\quad\text{O}-\text{O}-\text{H} \\
\quad\quad\quad\quad\quad\quad\quad\quad\quad | \\
\longrightarrow -\text{CH}_2\text{CH}{=}\text{CH}-\text{CH}-
\end{array}
$$
A hydroperoxide

4.25 How does vitamin E function as an antioxidant?

4.26 What structural features are common to vitamin E, BHT, and BHA (the three antioxidants presented in Section 4.5C)?

Chemical Connections

4.27 (Chemical Connections 4A) From what parts of its common name are the letters *DDT* derived?

4.28 (Chemical Connections 4A) What are the advantages and disadvantages of using DDT as an insecticide?

4.29 (Chemical Connections 4A) Would you expect DDT to be soluble or insoluble in water? Explain.

4.30 (Chemical Connections 4A) One of the degradation products of DDT is dichlorodiphenyldichloroethylene (DDE), which is formed by the loss of HCl from DDT. DDE inhibits the enzyme responsible for the incorporation of calcium ion into bird egg shells. Draw a structural formula for DDE.

4.31 (Chemical Connections 4A) What is meant by the term "biodegradable"?

4.32 (Chemical Connections 4B) What is a carcinogen? What kind of carcinogens are found in cigarette smoke?

4.33 (Chemical Connections 4C) In the absence of iodine in the diet, goiter develops. Explain why goiter is a regional disease.

4.34 (Chemical Connections 4D) Calculate the molecular weight of each explosive shown in this Chemical Connection. In which explosive do the nitro groups contribute the largest percentage of molecular weight?

4.35 (Chemical Connections 4E) What are the differences in structure between Allura Red and Sunset Yellow?

4.36 (Chemical Connections 4E) Which feature of Allura Red and Sunset Yellow makes them water-soluble?

4.37 (Chemical Connections 4E) What color would you get if you mix Allura Red and Sunset Yellow? (*Hint:* Remember the color wheel.)

4.38 (Chemical Connections 4E) Problems 3.60 and 3.61 gave structural formulas for lycopene and β-carotene, two natural food-coloring agents. Compare the structural formulas for these compounds with those of Allura Red and Sunset Yellow.

(a) What structural feature or features do these four compounds have in common that might account for the fact that each is colored?

(b) Why are these two food and cosmetic dyes soluble in water, while lycopene and β-carotene are not?

4.39 (Chemical Connections 4F) From what types of plants is capsaicin isolated?

4.40 (Chemical Connections 4F) Identify the phenol group in capsaicin.

4.41 (Chemical Connections 4F) How many cis-trans isomers are possible for capsaicin? Is the structural formula shown the cis isomer or the trans isomer?

4.42 (Chemical Connections 4F) In what ways is capsaicin used in medicine?

Additional Problems

4.43 The structure for naphthalene given in Section 4.3D is only one of three possible resonance structures. Draw the other two.

4.44 Draw structural formulas for these compounds.

(a) 1-Phenylcyclopropanol

(b) Styrene

(c) *m*-Bromophenol

(d) 4-Nitrobenzoic acid

(e) Isobutylbenzene

(f) *m*-Xylene

4.45 2,6-Di-*tert*-butyl-4-methylphenol (BHT, Section 4.5C) is an antioxidant added to processed foods to "retard spoilage." How does BHT accomplish this goal?

4.46 Write the structural formula for the product of each reaction.

(a)  ⬡ + HNO₃ $\xrightarrow{\text{H}_2\text{SO}_4}$

(b) 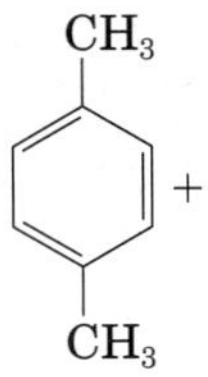 [structure with CH₃ groups] + Br₂ $\xrightarrow{\text{FeCl}_3}$

(c) 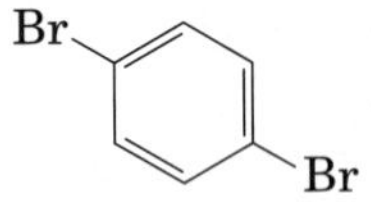 [structure with Br groups] + H₂SO₄ $\longrightarrow$

4.47 Styrene reacts with bromine to give a compound with molecular formula $C_8H_8Br_2$. Draw a structural formula for this compound.

InfoTrac College Edition

For additional readings, go to InfoTrac College Edition, your online research library, at

http://infotrac.thomsonlearning.com

CHAPTER **5**

5.1 Introduction

5.2 Alcohols

5.3 Reactions of Alcohols

5.4 Ethers

5.5 Thiols

5.6 Important Alcohols

Fermentation vats of wine grapes at the Beaulieu Vineyards, California.

Alcohols, Ethers, and Thiols

5.1 Introduction

In this chapter, we study the physical and chemical properties of alcohols and ethers, two classes of oxygen-containing organic compounds. We also study thiols, a class of sulfur-containing organic compounds. Thiols are like alcohols in structure, except that they contain an —SH group rather than an —OH group.

$$CH_3CH_2OH \qquad CH_3CH_2OCH_2CH_3 \qquad CH_3CH_2SH$$
Ethanol   Diethyl ether   Ethanethiol
(An alcohol)   (An ether)   (A thiol)

These three compounds are certainly familiar to you. Ethanol is the fuel additive in gasohol, the alcohol in alcoholic beverages, and an important laboratory and industrial solvent. Diethyl ether was the first inhalation anesthetic used in general surgery. It too is an important laboratory and industrial solvent. Ethanethiol, like other low-molecular-weight thiols, has a stench. Traces of ethanethiol are added to natural gas so that gas leaks can be detected by the smell of the thiol.

Figure 5.1   Methanol, CH_3OH. (*a*) Lewis structure and (*b*) ball-and-stick model. The H—C—O bond angle is 108.6°, very close to the tetrahedral angle of 109.5°.

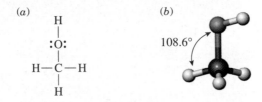

5.2 Alcohols

A Structure

The functional group of an **alcohol** is an **—OH (hydroxyl) group** bonded to a tetrahedral carbon atom (Section 1.4A). Figure 5.1 shows a Lewis structure and a ball-and-stick model of methanol, CH_3OH, the simplest alcohol.

B Nomenclature

IUPAC names of alcohols are derived in the same manner as those for alkenes and alkynes, with the exception that the ending of the parent alkane is changed from -*e* to -*ol*. The ending -*ol* tells us that the compound is an alcohol.

1. Select the longest carbon chain that contains the —OH group as the parent alkane and number it from the end that gives the —OH the lower number. In numbering the parent chain, the location of the —OH group takes precedence over alkyl and aryl groups, and halogens.

2. Change the ending of the parent alkane from **-e** to **-ol,** and use a number to show the location of the —OH group. For cyclic alcohols, numbering begins at the carbon bearing the —OH group; this carbon is automatically carbon-1.

3. Name and number substituents, and list them in alphabetical order.

To derive common names for alcohols, we name the alkyl group bonded to —OH and then add the word "alcohol." Following are IUPAC names and, in parentheses, common names for eight low-molecular-weight alcohols:

Solutions in which ethanol is the solvent are called tinctures.

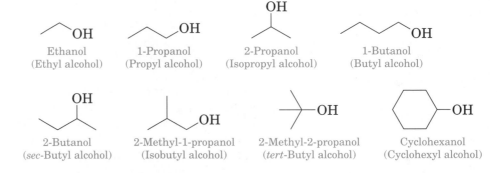

EXAMPLE 5.1

Write the IUPAC name for each alcohol.

(a) 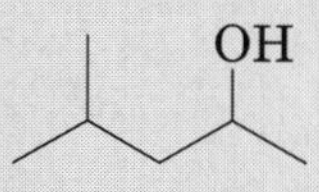
(b)

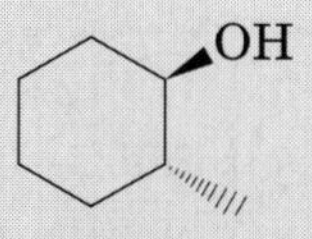

Solution

(a) The parent alkane is pentane. We number the parent chain from the direction that gives the lower number to the carbon bearing the —OH group. This alcohol is 4-methyl-2-pentanol.

(b) The parent cycloalkane is cyclohexane. We number the atoms of the ring beginning with the carbon bearing the —OH group as carbon 1. This alcohol is *trans*-2-methylcyclohexanol.

Problem 5.1

Write the IUPAC name for each alcohol.

(a) 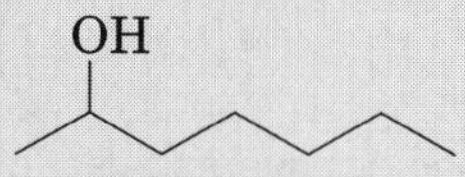
(b) 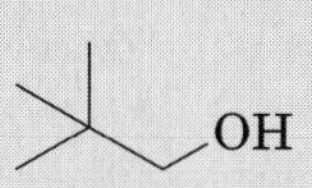
(c)

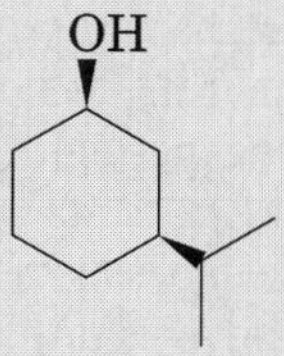

We classify alcohols as **primary (1°), secondary (2°), or tertiary (3°),** depending on the number of carbon groups bonded to the carbon bearing the —OH group (Section 1.4A).

EXAMPLE 5.2

Classify each alcohol as primary, secondary, or tertiary.

(a)

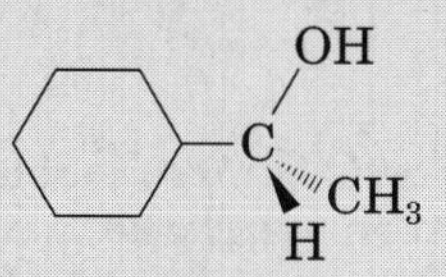

$$
\begin{array}{c}
OH \\
| \\
C\text{\tiny\textbf{w}}CH_3 \\
| \\
H
\end{array}
$$

(b)

$$
\begin{array}{c}
CH_3 \\
| \\
CH_3COH \\
| \\
CH_3
\end{array}
$$

(c) —CH_2OH

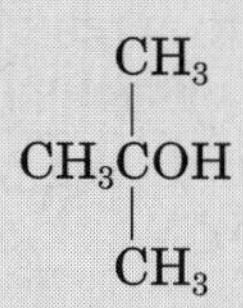

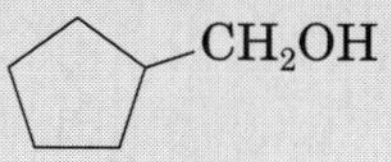

Solution
(a) Secondary (2°)   (b) Tertiary (3°)
(c) Primary (1°)

Problem 5.2

Classify each alcohol as primary, secondary, or tertiary.

(a) [structure with OH]   (b) [structure with OH]

(c) [structure with OH]   (d) [structure with OH]

CHEMICAL CONNECTIONS 5A

Nitroglycerin, an Explosive and a Drug

In 1847, Ascanio Sobrero (1812–1888) discovered that treatment of 1,2,3-propanetriol, more commonly named glycerin, with nitric acid in the presence of sulfuric acid gives a pale yellow, oily liquid called nitroglycerin. Sobrero also discovered the explosive properties of this compound; when he heated a small quantity of it, it exploded!

$$\begin{array}{c} CH_2-OH \\ | \\ CH-OH \\ | \\ CH_2-OH \end{array} + 3HNO_3 \xrightarrow{H_2SO_4} \begin{array}{c} CH_2-ONO_2 \\ | \\ CH-ONO_2 \\ | \\ CH_2-ONO_2 \end{array} + 3H_2O$$

1,2,3-Propanetriol   1,2,3-Propanetriol trinitrate
(Glycerol, Glycerin)   (Nitroglycerin)

Nitroglycerin very soon became widely used for blasting in the construction of canals, tunnels, roads, and mines and, of course, for warfare.

One problem with the use of nitroglycerin was soon recognized: It was difficult to handle safely, and accidental explosions occurred frequently. This problem was solved by the Swedish chemist Alfred Nobel (1833–1896), who discovered that a clay-like substance called diatomaceous earth absorbs nitroglycerin so that it will not explode without a fuse. He gave the name *dynamite* to this mixture of nitroglycerin, diatomaceous earth, and sodium carbonate.

Surprising as it may seem, nitroglycerin is used in medicine to treat angina pectoris, the symptoms of which are sharp chest pains caused by reduced flow of blood in the coronary artery. Nitroglycerin, which is available in liquid form (diluted with alcohol to render it nonexplosive), tablet form, or paste form, relaxes the smooth muscles of blood vessels, causing dilation of the coronary artery. This dilation, in turn, allows more blood to reach the heart.

When Nobel became ill with heart disease, his physicians advised him to take nitroglycerin to relieve his chest pains. He refused, saying he could not understand how the explosive could relieve chest pains. It took science more than 100 years to find the answer. We now know that it is nitric oxide, NO, derived from the nitro groups of nitroglycerin, that relieves the pain. (See Chemical Connections 15F.)

■ (*a*) Nitroglycerin is more stable if absorbed onto an inert solid, a combination called dynamite. (*b*) The fortune of Alfred Nobel, 1833–1896, built on the manufacture of dynamite, now funds the Nobel Prizes.

(a)

(b)

a, Charles D. Winters; b, The Bettmann Archive/Corbis

In the IUPAC system, we name a compound containing two hydroxyl groups as a **diol,** one containing three hydroxyl groups as a **triol,** and so on. We retain the final *-e* in the name of the parent alkane—for example, 1,2-ethanediol.

As with many other organic compounds, common names for certain diols and triols have persisted. We often refer to compounds containing two hydroxyl groups on adjacent carbons as **glycols.** Ethylene glycol and propylene glycol are synthesized from ethylene and propylene, respectively; hence their common names.

■ **Ethylene glycol is a polar molecule and dissolves readily in the polar solvent water.**

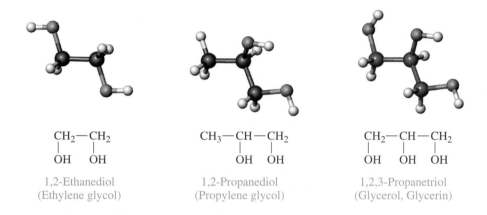

CH₂—CH₂
| |
OH   OH

1,2-Ethanediol
(Ethylene glycol)

CH₃—CH—CH₂
        | |
       OH   OH

1,2-Propanediol
(Propylene glycol)

CH₂—CH—CH₂
| | |
OH   OH   OH

1,2,3-Propanetriol
(Glycerol, Glycerin)

C Physical Properties of Alcohols

The most important physical property of alcohols is the polarity of their —OH groups. Because of the large difference in electronegativity between oxygen and carbon ($3.5-2.5 = 1.0$) and between oxygen and hydrogen ($3.5-2.1 = 1.4$), both the C—O and O—H bonds of an alcohol are polar covalent, and alcohols are polar molecules, as illustrated in Figure 5.2 for methanol.

Alcohols have higher boiling points than do alkanes, alkenes, and alkynes of similar molecular weight (Table 5.1), because alcohol molecules associate with one another in the liquid state by **hydrogen bonding**. The strength of hydrogen bonding between alcohol molecules is approximately 2 to 5 kcal/mol, which means that extra energy is required to separate hydrogen-bonded alcohols from their neighbors (Figure 5.3).

Because of increased London dispersion forces between larger molecules, the boiling points of all types of compounds, including alcohols, increase with increasing molecular weight.

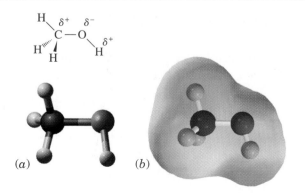

(a)   (b)

Figure 5.2 Polarity of the C—O—H bonds in methanol. (*a*) There are partial positive charges on carbon and hydrogen and a partial negative charge on oxygen. (*b*) An electron density map showing a partial negative charge (in red) around oxygen and a partial positive charge (in blue) around hydrogen of the OH group.

Figure 5.3 The association of ethanol molecules in the liquid state. Each O—H can participate in up to three hydrogen bonds (one through hydrogen and two through oxygen). Only two of these three possible hydrogen bonds per molecule are shown here.

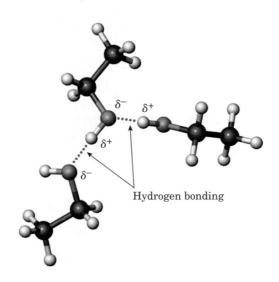

Hydrogen bonding

Table 5.1	Boiling Points and Solubilities in Water of Four Sets of Alcohols and Alkanes of Similar Molecular Weight			
Structural Formula	**Name**	**Molecular Weight (amu)**	**Boiling Point (°C)**	**Solubility in Water**
CH_3OH	methanol	32	65	infinite
CH_3CH_3	ethane	30	−89	insoluble
CH_3CH_2OH	ethanol	46	78	infinite
$CH_3CH_2CH_3$	propane	44	−42	insoluble
$CH_3CH_2CH_2OH$	1-propanol	60	97	infinite
$CH_3CH_2CH_2CH_3$	butane	58	0	insoluble
$CH_3CH_2CH_2CH_2OH$	1-butanol	74	117	8 g/100 g
$CH_3CH_2CH_2CH_2CH_3$	pentane	72	36	insoluble

Alcohols are much more soluble in water than are hydrocarbons of similar molecular weight (Table 5.1), because alcohol molecules interact by hydrogen bonding with water molecules. Methanol, ethanol, and 1-propanol are soluble in water in all proportions. As molecular weight increases, the water solubility of alcohols becomes more like those of hydrocarbons of similar molecular weight. Higher-molecular-weight alcohols are much less soluble in water because the size of the hydrocarbon portion of their molecules (which decreases water solubility) becomes so large relative to the size of the —OH group (which increases water solubility).

5.3 Reactions of Alcohols

In this section, we study the acidity of alcohols, their dehydration to alkenes, and their oxidation to aldehydes, ketones, or carboxylic acids.

A Acidity of Alcohols

Alcohols have about the same pK_a values as water which means that aqueous solutions of alcohols have the same pH as that of pure water. In Section 4.5B, we studied phenols, another class of compounds that contains an

—OH group. Phenols are weak acids and react with aqueous sodium hydroxide to form water-soluble salts.

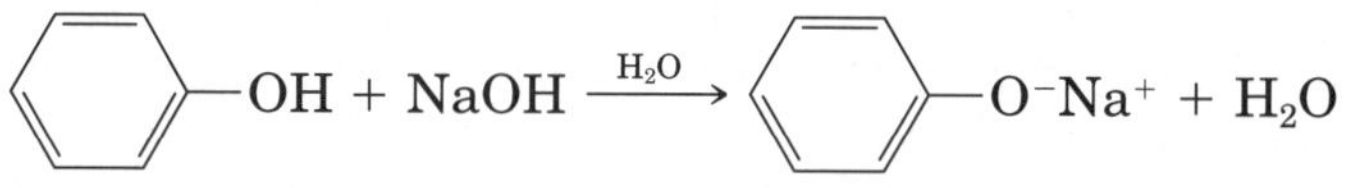

Phenol — Sodium phenoxide (a water-soluble salt)

Alcohols are considerably weaker acids than phenols and do not react in this manner.

B Acid-Catalyzed Dehydration Alkenes

We can convert an alcohol to an alkene by eliminating a molecule of water from adjacent carbon atoms in a reaction called **dehydration.** In the laboratory, dehydration of an alcohol is most often brought about by heating it with either 85% phosphoric acid or concentrated sulfuric acid. Primary alcohols—the most difficult to dehydrate—generally require heating in concentrated sulfuric acid at temperatures as high as 180°C. Secondary alcohols undergo acid-catalyzed dehydration at somewhat lower temperatures. Tertiary alcohols generally undergo acid-catalyzed dehydration at temperatures only slightly above room temperature.

Dehydration Elimination of a molecule of water from an alcohol; an OH is removed from one carbon, and an H is removed from an adjacent carbon

$$\underset{\text{Ethanol}}{CH_3CH_2OH} \xrightarrow[180°C]{H_2SO_4} \underset{\text{Ethylene}}{CH_2{=}CH_2} + H_2O$$

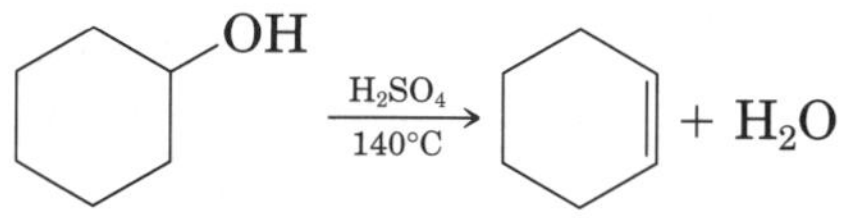

Cyclohexanol — Cyclohexene

See the **Interactive General, Organic, and Biochemistry CD-ROM, version 2.0,** for further exploration on this topic.

$$\underset{\substack{\text{2-Methyl-2-propanol}\\(\textit{tert}\text{-Butyl alcohol})}}{CH_3\overset{\overset{\large CH_3}{|}}{\underset{\underset{\large OH}{|}}{C}}CH_3} \xrightarrow[50°C]{H_2SO_4} \underset{\substack{\text{2-Methylpropene}\\\text{(Isobutylene)}}}{CH_3\overset{\overset{\large CH_3}{|}}{C}{=}CH_2} + H_2O$$

Thus the ease of acid-catalyzed dehydration of alcohols follows this order:

1° alcohols 2° alcohols 3° alcohols

Ease of dehydration of alcohols →

When the acid-catalyzed dehydration of an alcohol yields isomeric alkenes, the alkene having the greater number of alkyl groups on the double bond generally predominates. In the acid-catalyzed dehydration of 2-butanol, for example, the major product is 2-butene, which has two alkyl groups (two methyl groups) on its double bond. The minor product is 1-butene, which has only one alkyl group (an ethyl group) on its double bond.

$$\underset{\text{2-Butanol}}{CH_3CH_2\overset{\overset{\large OH}{|}}{C}HCH_3} \xrightarrow[heat]{H_3PO_4} \underset{\substack{\text{2-Butene}\\(80\%)}}{CH_3CH{=}CHCH_3} + \underset{\substack{\text{1-Butene}\\(20\%)}}{CH_3CH_2CH{=}CH_2} + H_2O$$

EXAMPLE 5.3

Draw structural formulas for the alkenes formed by the acid-catalyzed dehydration of each alcohol. For each reaction, predict which alkene will be the major product.

(a) 3-Methyl-2-butanol
(b) 2-Methylcyclopentanol

Solution

(a) Elimination of H_2O from carbons 2–3 gives 2-methyl-2-butene; elimination of H_2O from carbons 1–2 gives 3-methyl-1-butene. 2-Methyl-2-butene has three alkyl groups (three methyl groups) on its double bond and is the major product. 3-Methyl-1-butene has only one alkyl group (an isopropyl group) on its double bond and is the minor product.

$$\overset{4}{C}H_3\overset{3}{C}H(CH_3)\overset{2}{C}H(OH)\overset{1}{C}H_3 \xrightarrow[\text{acid-catalyzed dehydration}]{H_2SO_4} CH_3C(CH_3){=}CHCH_3 + CH_3CH(CH_3)CH{=}CH_2 + H_2O$$

3-Methyl-2-butanol; 2-Methyl-2-butene (major product); 3-Methyl-1-butene

(b) The major product, 1-methylcyclopentene, has three alkyl groups on its double bond. The minor product, 3-methylcyclopentene, has only two alkyl groups on its double bond.

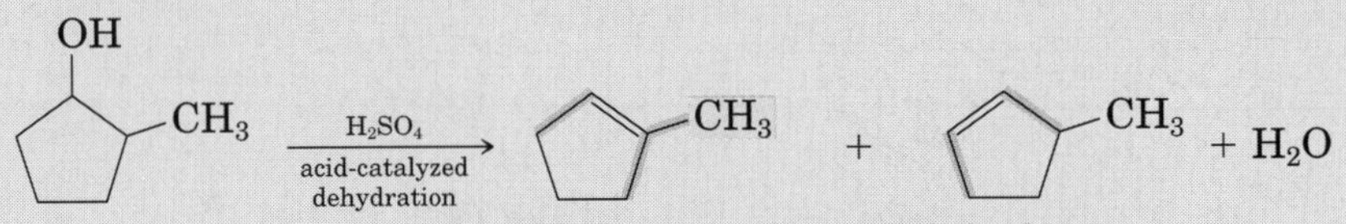

2-Methylcyclopentanol; 1-Methylcyclopentene (major product); 3-Methylcyclopentene

Problem 5.3

Draw structural formulas for the alkenes formed by the acid-catalyzed dehydration of each alcohol. For each reaction, predict which alkene will be the major product.

(a) 2-Methyl-2-butanol
(b) 1-Methylcyclopentanol

In Section 3.6B, we discussed the acid-catalyzed hydration of alkenes to give alcohols. In this section, we discuss the acid-catalyzed dehydration of alcohols to give alkenes. In fact, hydration–dehydration reactions are reversible. Alkene hydration and alcohol dehydration are competing reactions, and the following equilibrium exists:

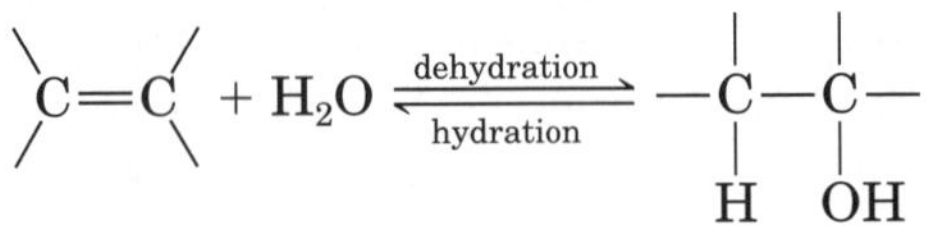

An alkene; An alcohol

In accordance with Le Chatelier's principle, large amounts of water (in other words, using dilute aqueous acid) favor alcohol formation, whereas scarcity of water (using concentrated acid) or experimental conditions where water is removed (heating the reaction mixture above 100°C) favor alkene formation. Thus, depending on the experimental conditions, we can use the hydration–dehydration equilibrium to prepare either alcohols or alkenes, each in high yields.

EXAMPLE 5.4

In part (a), acid-catalyzed dehydration of 2-methyl-3-pentanol gives Compound A. Treatment of Compound A with water in the presence of sulfuric acid in part (b) gives Compound B. Propose structural formulas for Compounds A and B.

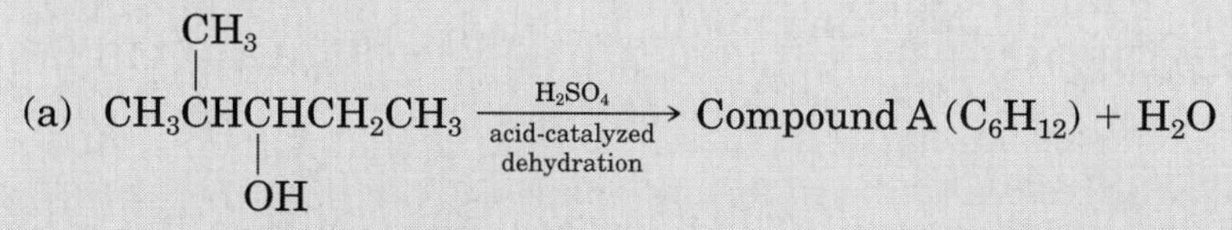

$$\text{(a)} \quad \underset{\underset{\displaystyle \text{OH}}{|}}{\overset{\overset{\displaystyle CH_3}{|}}{CH_3CHCHCH_2CH_3}} \xrightarrow[\substack{\text{acid-catalyzed} \\ \text{dehydration}}]{H_2SO_4} \text{Compound A } (C_6H_{12}) + H_2O$$

(b) Compound A (C_6H_{12}) + H_2O $\xrightarrow{H_2SO_4}$ Compound B $(C_6H_{14}O)$

Solution

(a) Acid-catalyzed dehydration of 2-methyl-3-pentanol gives 2-methyl-2-pentene, an alkene with three substituents on its double bond: two methyl groups and one ethyl group.

(b) Acid-catalyzed addition of water to this alkene gives 2-methyl-2-pentanol in accord with Markovnikov's rule (Section 12.6B).

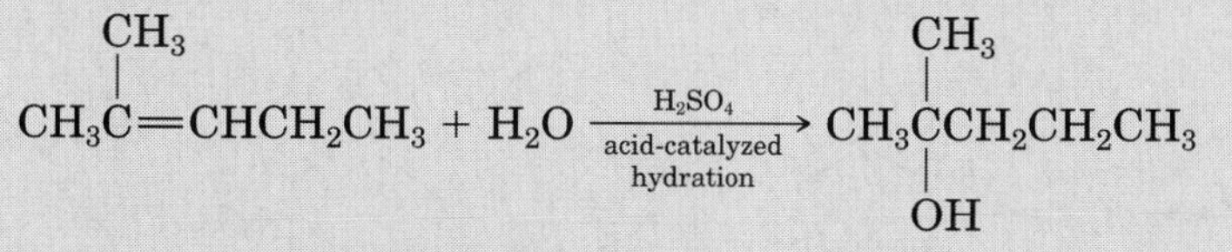

$$\underset{\underset{\text{Compound A } (C_6H_{12})}{}}{\overset{\overset{CH_3}{|}}{CH_3C}{=}CHCH_2CH_3} + H_2O \xrightarrow[\substack{\text{acid-catalyzed} \\ \text{hydration}}]{H_2SO_4} \underset{\underset{\underset{\text{Compound B } (C_6H_{14}O)}{}}{\underset{\displaystyle OH}{|}}}{\overset{\overset{CH_3}{|}}{CH_3CCH_2CH_2CH_3}}$$

Problem 5.4

Acid-catalyzed dehydration of 2-methylcyclohexanol gives Compound C (C_7H_{12}). Treatment of Compound C with water in the presence of sulfuric acid gives Compound D $(C_7H_{14}O)$. Propose structural formulas for Compounds C and D.

C Oxidation of Primary and Secondary Alcohols

Oxidation of a primary alcohol can give an aldehyde or a carboxylic acid, depending on the experimental conditions. Following is a series of transformations in which a primary alcohol is oxidized first to an aldehyde and then to a carboxylic acid. The fact that each transformation involves oxidation is indicated by the letter O in brackets over the reaction arrow.

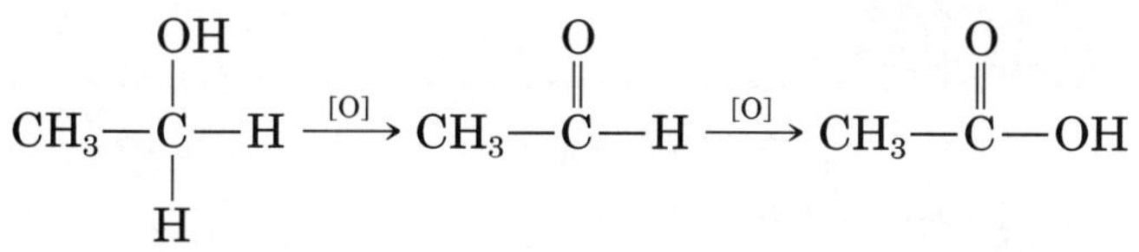

$$\underset{\underset{\text{H}}{|}}{\overset{\overset{OH}{|}}{CH_3{-}C{-}H}} \xrightarrow{[O]} \overset{\overset{\displaystyle O}{\|}}{CH_3{-}C{-}H} \xrightarrow{[O]} \overset{\overset{\displaystyle O}{\|}}{CH_3{-}C{-}OH}$$

A primary alcohol   An aldehyde   A carboxylic acid

According to one definition, oxidation is either the loss of hydrogens or the gain of oxygens. Using this definition, conversion of a primary alcohol to an aldehyde is an oxidation because the alcohol loses hydrogens. Conversion of an aldehyde to a carboxylic acid is an oxidation because the aldehyde gains an oxygen.

The reagent most commonly used in the laboratory for the oxidation of a primary alcohol to a carboxylic acid is potassium dichromate, $K_2Cr_2O_7$, dissolved in aqueous sulfuric acid. Using this reagent, oxidation of 1-octanol, for example, gives octanoic acid. This experimental condition is more than sufficient to oxidize the intermediate aldehyde to a carboxylic acid.

■ **The peppermint plant,** *Mentha piperita,* **is a perennial herb with aromatic qualities used in candies, gums, hot and cold beverages, and garnishes for punch and fruit.**

$$CH_3(CH_2)_6CH_2OH \xrightarrow[\text{H}_2\text{SO}_4]{\text{K}_2\text{Cr}_2\text{O}_7} CH_3(CH_2)_6\overset{\overset{\displaystyle O}{\|}}{C}H \xrightarrow[\text{H}_2\text{SO}_4]{\text{K}_2\text{Cr}_2\text{O}_7} CH_3(CH_2)_6\overset{\overset{\displaystyle O}{\|}}{C}OH$$

1-Octanol   Octanal   Octanoic acid

Although the usual product of oxidation of a primary alcohol is a carboxylic acid, it is often possible to stop the oxidation at the aldehyde stage by distilling the mixture. We remove the aldehyde, which usually has a lower boiling point than either the primary alcohol or the carboxylic acid, from the reaction mixture before it can be oxidized further.

Oxidation of a secondary alcohol using potassium dichromate as the oxidizing agent gives a ketone. Menthol, a secondary alcohol that is present in peppermint and other mint oils, is used in liqueurs, cigarettes, cough drops, perfumery, and nasal inhalers. Its oxidation product, menthone, is also used in perfumes and artificial flavors.

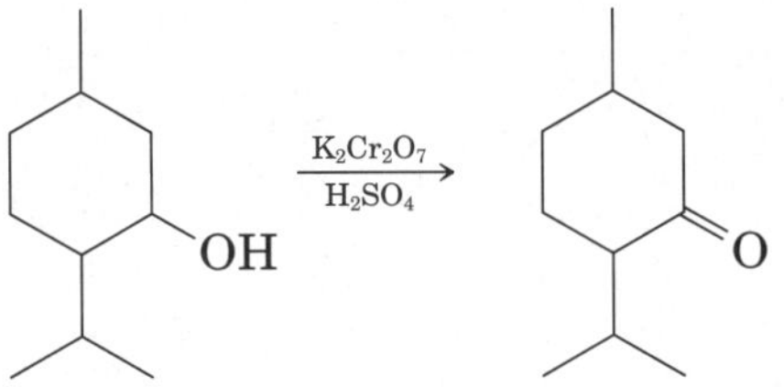

$\xrightarrow[\text{H}_2\text{SO}_4]{\text{K}_2\text{Cr}_2\text{O}_7}$

2-Isopropyl-5-methyl-cyclohexanol (Menthol)   2-Isopropyl-5-methyl-cyclohexanone (Menthone)

Tertiary alcohols are resistant to oxidation because the carbon bearing the —OH is bonded to three carbon atoms and, therefore, cannot form a carbon–oxygen double bond.

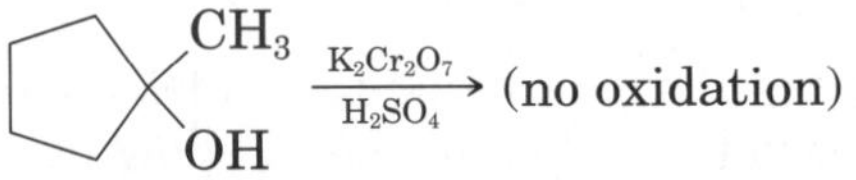

$\xrightarrow[\text{H}_2\text{SO}_4]{\text{K}_2\text{Cr}_2\text{O}_7}$ (no oxidation)

1-Methylcyclopentanol

EXAMPLE 5.5

Draw a structural formula for the product formed by oxidation of each alcohol with potassium dichromate.
(a) 1-Hexanol
(b) 2-Hexanol

CHEMICAL CONNECTIONS 5B

Breath-Alcohol Screening

Potassium dichromate oxidation of ethanol to acetic acid is the basis for the original breath-alcohol screening test used by law enforcement agencies to determine a person's blood alcohol content (BAC). The test is based on the difference in color between the dichromate ion (reddish orange) in the reagent and the chromium(III) ion (green) in the product.

$$CH_3CH_2OH + Cr_2O_7^{2-}$$

Ethanol   Dichromate ion
(reddish orange)

$$\xrightarrow[H_2O]{H_2SO_4} \ \underset{\text{Acetic acid}}{CH_3\overset{O}{\overset{\|}{C}}OH} \ + \ \underset{\substack{\text{Chromium(III) ion}\\\text{(green)}}}{Cr^{3+}}$$

In its simplest form, breath-alcohol screening uses a sealed glass tube that contains a potassium dichromate–sulfuric acid reagent impregnated on silica gel. To administer the test, the ends of the tube are broken off, a mouthpiece is fitted to one end, and the other end is inserted into the neck of a plastic bag. The person being tested then blows into the mouthpiece to inflate the plastic bag.

As breath containing ethanol vapor passes through the tube, reddish orange dichromate ion is reduced to green chromium(III) ion. To estimate the concentration of ethanol in the breath, one measures how far the green color extends along the length of the tube. When it extends beyond the halfway point, the person is judged as having a sufficiently high blood alcohol content to warrant further, more precise testing.

The test described here measures the alcohol content of the breath. The legal definition of being under the influence of alcohol, however, is based on alcohol content in the blood, not in the breath. The correlation between these two measurements is based on the fact that air deep within the lungs is in equilibrium with blood passing through the pulmonary arteries, and an equilibrium is established between blood alcohol and breath alcohol. Based on tests in persons drinking alcohol, researchers have determined that 2100 mL of breath contains the same amount of ethanol as 1.00 mL of blood.

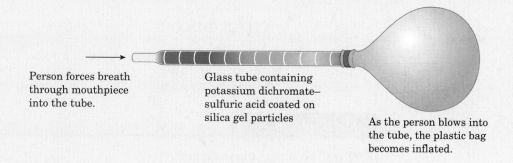

Person forces breath through mouthpiece into the tube.

Glass tube containing potassium dichromate–sulfuric acid coated on silica gel particles

As the person blows into the tube, the plastic bag becomes inflated.

Solution

Oxidation of 1-hexanol, a primary alcohol, gives either hexanal or hexanoic acid, depending on the experimental conditions. Oxidation of 2-hexanol, a secondary alcohol, gives 2-hexanone.

(a) [structure] Hexanal   or   [structure] Hexanoic acid   (b) [structure] 2-Hexanone

Problem 5.5

Draw the product formed by oxidation of each alcohol with potassium dichromate.

(a) Cyclohexanol

(b) 2-Pentanol

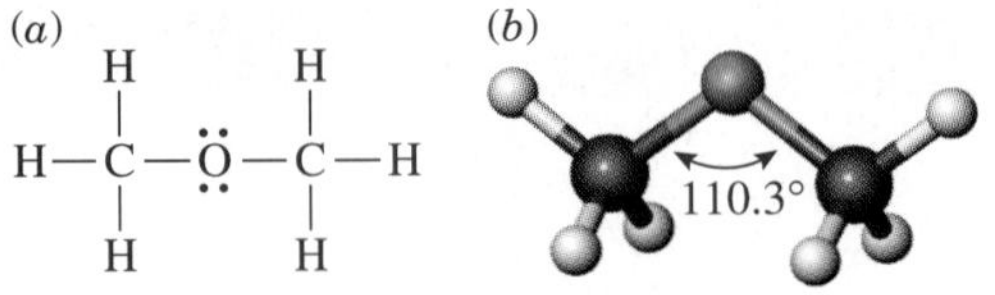

Figure 5.4 Dimethyl ether, CH_3OCH_3. (*a*) Lewis structure and (*b*) ball-and-stick model. The C—O—C bond angle is 110.3°, close to the tetrahedral angle of 109.5°.

(*a*)

$$H-\overset{\overset{\displaystyle H}{|}}{C}-\overset{\cdot\cdot}{\underset{\cdot\cdot}{O}}-\overset{\overset{\displaystyle H}{|}}{C}-H$$

(*b*)

110.3°

Ether A compound containing an oxygen atom bonded to two carbon atoms

5.4 Ethers

A Structure

The functional group of an **ether** is an atom of oxygen bonded to two carbon atoms. Figure 5.4 shows a Lewis structure and a ball-and-stick model of dimethyl ether, CH_3OCH_3, the simplest ether.

B. Nomenclature

Although we can use the IUPAC system to name ethers, chemists almost invariably use common names for low-molecular-weight ethers. We derive common names by listing the alkyl groups attached to oxygen in alphabetical order and adding the word ether. Alternatively, we name one of the groups on oxygen as an alkoxy group. The —OCH_3 group, for example, is named "methoxy" to indicate a <u>meth</u>yl group bonded to <u>oxygen</u>.

$$CH_3CH_2OCH_2CH_3$$

Diethyl ether

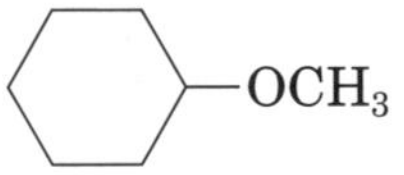

⬡—OCH_3

Cyclohexyl methyl ether
(Methoxycyclohexane)

EXAMPLE 5.6

Write the common name for each ether.

(a) 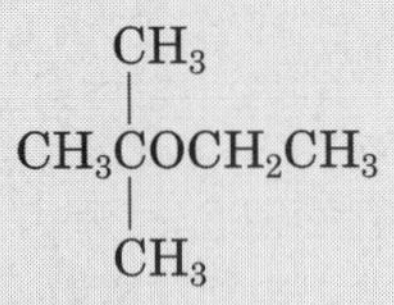 $CH_3\overset{\overset{\displaystyle CH_3}{|}}{\underset{\underset{\displaystyle CH_3}{|}}{C}}OCH_2CH_3$

(b) 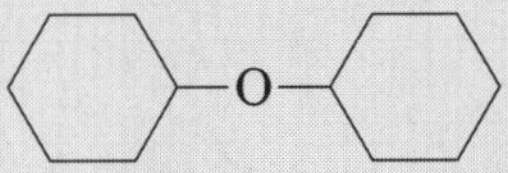 ⬡—O—⬡

Solution

(a) The groups bonded to the ether oxygen are *tert*-butyl and ethyl. The compound's common name is *tert*-butyl ethyl ether.
(b) Two cyclohexyl groups are bonded to the ether oxygen. The compound's common name is dicyclohexyl ether.

Problem 5.6

Write the common name for each ether.

(a) 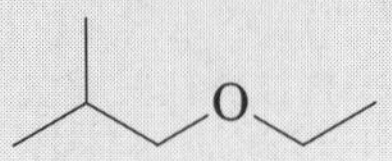 ⟍⟋⟍O⟍

(b) 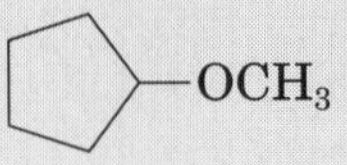 ⬠—OCH_3

Ethylene Oxide: A Chemical Sterilant

Ethylene oxide is a colorless, flammable gas, with a boiling point of 11°C. Because it is such a highly strained molecule (the normal tetrahedral bond angles of both C and O are compressed to approximately 60° in this molecule), ethylene oxide reacts with the amino ($-NH_2$) and sulfhydryl ($-SH$) groups present in biological materials.

At sufficiently high concentrations, it reacts with enough molecules in cells to cause the death of microorganisms. This toxic property is the basis for ethylene oxide's use as a fumigant in foodstuffs and textiles and its use in hospitals to sterilize surgical instruments.

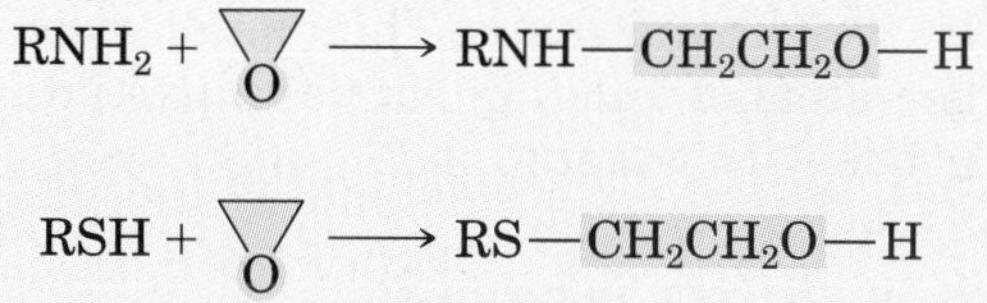

$$RNH_2 + \underset{O}{\triangledown} \longrightarrow RNH-CH_2CH_2O-H$$

$$RSH + \underset{O}{\triangledown} \longrightarrow RS-CH_2CH_2O-H$$

Cyclic ethers are ethers in which one of the atoms in a ring is oxygen. These ethers are also known by their common names. Ethylene oxide is an important building block for the organic chemical industry (Section 5.6). Tetrahydrofuran is a useful laboratory and industrial solvent.

> **Cyclic ether** An ether in which the ether oxygen is one of the atoms of a ring

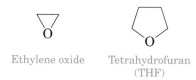

Ethylene oxide   Tetrahydrofuran
(THF)

C Physical Properties

Ethers are polar compounds in which oxygen bears a partial negative charge and each attached carbon bears a partial positive charge (Figure 5.5). However, only weak forces of attraction exist between ether molecules in the pure liquid; consequently, the boiling points of ethers are close to those of hydrocarbons of similar molecular weight.

We can illustrate the effect of hydrogen bonding on physical properties by comparing the boiling points of ethanol (78°C) and its constitutional isomer dimethyl ether (−24°C). The difference in boiling point between these

Figure 5.5 Ethers are polar molecules, but only weak attractive interactions exist between ether molecules in the liquid state.

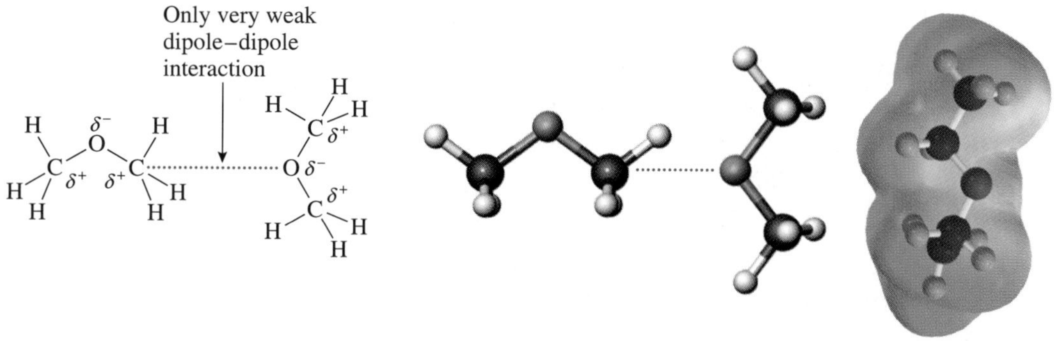

two compounds is due to the presence in ethanol of a polar O—H group, which is capable of forming hydrogen bonds. This hydrogen bonding increases intermolecular associations and thus gives ethanol a higher boiling point than dimethyl ether.

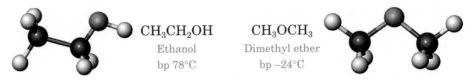

CH_3CH_2OH CH_3OCH_3
Ethanol   Dimethyl ether
bp 78°C   bp −24°C

Ethers are more soluble in water than hydrocarbons of similar molecular weight and shape. Their greater solubility is due to the fact that the oxygen atom of an ether carries a partial negative charge and forms hydrogen bonds with water.

D Reactions of Ethers

Ethers resemble hydrocarbons in their resistance to chemical reaction. They do not react with oxidizing agents, such as potassium dichromate. Likewise, they do not react with reducing agents such as H_2 in the presence of a metal catalyst (Section 3.6D). Furthermore they are not affected by most acids or bases at moderate temperatures. Because of their general inertness to chemical reaction and their good solvent properties, ethers are excellent solvents in which to carry out many organic reactions. The most important ether solvents are diethyl ether and tetrahydrofuran.

5.5 Thiols

The most outstanding property of low-molecular-weight **thiols** is their stench. They are responsible for unpleasant odors such as those from skunks, rotten eggs, and sewage. The scent of skunks is due primarily to two thiols:

$$CH_3CH=CHCH_2SH$$
2-Butene-1-thiol

$$CH_3\overset{\underset{\displaystyle |}{CH_3}}{CH}CH_2CH_2SH$$
3-Methyl-1-butanethiol

A Structure

The functional group of a thiol is an **—SH (sulfhydryl) group** bonded to a tetrahedral carbon atom. Figure 5.6 shows a Lewis structure and a ball-and-stick model of methanethiol, CH_3SH, the simplest thiol.

B Nomenclature

The sulfur analog of an alcohol is called a thiol (*thi-* from the Greek: *theion*, sulfur) or, in the older literature, a **mercaptan,** which literally means "mercury capturing." Thiols react with Hg^{2+} in aqueous solution to give sulfide salts as insoluble precipitates. Thiophenol, C_6H_5SH, for example, gives $(C_6H_5S)_2Hg$.

In the IUPAC system, we name thiols by selecting the longest carbon chain that contains the —SH group as the parent alkane. To show that the

---

■ **The scent of the spotted skunk, native to the Sonoran Desert, is a mixture of two thiols, 2-butene-1-thiol and 3-methyl-1-butanethiol.**

Thiol A compound containing an —SH (sulfhydryl) group bonded to a tetrahedral carbon atom

Mercaptan A common name for any molecule containing an —SH group

Figure 5.6 Methanethiol, CH_3SH. (*a*) Lewis structure and (*b*) ball-and-stick model. The H—S—C bond angle is 100.3°, somewhat smaller than the tetrahedral angle of 109.5°.

(*a*)

$$H-\overset{\underset{\displaystyle |}{\overset{\displaystyle |}{H}}}{C}-\overset{..}{\underset{..}{S}}-H$$

(*b*)

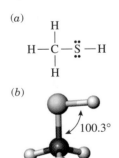

100.3°

CHEMICAL CONNECTIONS 5D

Ethers and Anesthesia

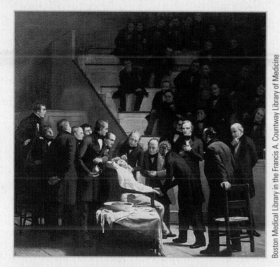

Boston Medical Library in the Francis A. Countway Library of Medicine

■ **This painting by Robert Hinckley shows the first use of ether as an anesthetic in 1846. Dr. Robert John Collins was removing a tumor from the patient's neck, and dentist W. T. G. Morton—who discovered its anesthetic properties—administered the ether.**

Before the mid–1800s, surgery was performed only when absolutely necessary, because no truly effective general anesthetic was available. More often than not, patients were drugged, hypnotized, or simply tied down.

In 1772, Joseph Priestly isolated nitrous oxide, N_2O, a colorless gas. In 1799, Sir Humphry Davy demonstrated this compound's anesthetic effect, naming it "laughing gas." In 1844, an American dentist, Horace Wells, introduced nitrous oxide into general dental practice. One patient awakened prematurely, however, screaming with pain; another died during the procedure. Wells was forced to withdraw from practice, became embittered and depressed, and committed suicide at age 33. In the same period, a Boston chemist, Charles Jackson, anesthetized himself with diethyl ether; he also persuaded a dentist, William Morton, to use it. Subsequently, they persuaded a surgeon, John Warren, to give a public demonstration of surgery under anesthesia. The operation was completely successful, and soon general anesthesia by diethyl ether became routine practice for general surgery.

Diethyl ether was easy to use and caused excellent muscle relaxation. Blood pressure, pulse rate, and respiration were usually only slightly affected. The gas's chief drawbacks were its irritating effect on the respiratory passages and its after-effect of nausea.

Among the inhalation anesthetics used today are several halogenated ethers, the most important being enflurane and isoflurane.

Enflurane
(Ethrane)

Isoflurane
(Forane)

compound is a thiol, we add the suffix *-thiol* to the name of the parent alkane. We number the parent chain in the direction that gives the —SH group the lower number.

We derive common names for simple thiols by naming the alkyl group attached to —SH and adding the word mercaptan.

$$CH_3CH_2SH$$
Ethanethiol
(Ethyl mercaptan)

$$\underset{\text{2-Methyl-1-propanethiol}}{\overset{\overset{\textstyle CH_3}{\textstyle |}}{CH_3CHCH_2SH}}$$
2-Methyl-1-propanethiol
(Isobutyl mercaptan)

■ **Mushrooms, onions, garlic, and coffee all contain sulfur compounds. This thiol is present in the aroma of coffee.**

Charles D. Winters

EXAMPLE 5.7

Write the IUPAC name for each thiol.

(a) ⟍⟋⟍⟋⟍SH

(b) SH

Solution

(a) The parent alkane is pentane. We show the presence of the —SH group by adding *thiol* to the name of the parent alkane. The IUPAC name of this thiol is 1-pentanethiol. Its common name is pentyl mercaptan.

(b) The parent alkane is butane. The IUPAC name of this thiol is 2-butanethiol. Its common name is *sec*-butyl mercaptan.

Problem 5.7

Write the IUPAC name for each thiol.

(a)

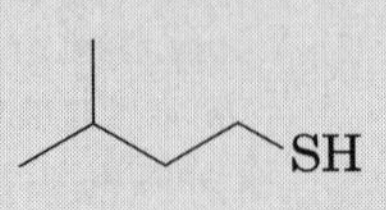

(b)

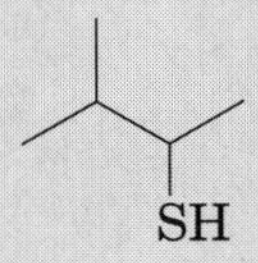

C Physical Properties

Because of the small difference in electronegativity between sulfur and hydrogen $(2.5 - 2.1 = 0.4)$, we classify the S—H bond as nonpolar covalent. Because of this lack of polarity, thiols show little association by hydrogen bonding. Consequently, they have lower boiling points and are less soluble in water and other polar solvents than alcohols of similar molecular weight. Table 5.2 gives boiling points for three low-molecular-weight thiols. Shown for comparison are boiling points of alcohols with the same number of carbon atoms.

Earlier we illustrated the importance of hydrogen bonding in alcohols by comparing the boiling points of ethanol ($78°C$) and its constitutional isomer dimethyl ether ($-24°C$). By comparison, the boiling point of ethanethiol is $35°C$ and that of its constitutional isomer dimethyl sulfide is $37°C$. Because the boiling points of these constitutional isomers are almost identical, we know that little or no association by hydrogen bonding occurs between thiol molecules.

$$CH_3CH_2SH \qquad CH_3SCH_3$$

Ethanethiol   Dimethyl sulfide
bp 35°C   bp 37°C

D Reactions of Thiols

Thiols are weak acids (pK_a 10), and are comparable in strength to phenols (Section 4.5B). Thiols react with strong bases such as NaOH to form thiolate salts.

$$CH_3CH_2SH + NaOH \xrightarrow{H_2O} CH_3CH_2S^-Na^+ + H_2O$$

Ethanethiol   Sodium
(pK_a 10)   ethanethiolate

The most common reaction of thiols in biological systems is their oxidation to **disulfides,** the functional group of which is a disulfide (—S—S—) bond. Thiols are readily oxidized to disulfides by molecular oxygen. In fact, they are so susceptible to oxidation that they must be protected from con-

Disulfide A compound containing an —S—S— group

Table 5.2	Boiling Points of Three Thiols and Alcohols with the Same Number of Carbon Atoms		
Thiol	**Boiling Point (°C)**	**Alcohol**	**Boiling Point (°C)**
methanethiol	6	methanol	65
ethanethiol	35	ethanol	78
1-butanethiol	98	1-butanol	117

tact with air during storage. Disulfides, in turn, are easily reduced to thiols by several reducing agents. This easy interconversion between thiols and disulfides is very important in protein chemistry, as we will see in Chapters 13 and 14.

$$2HOCH_2CH_2SH \underset{\text{reduction}}{\overset{\text{oxidation}}{\rightleftarrows}} HOCH_2CH_2S\!-\!SCH_2CH_2OH$$
$$\text{A thiol} \qquad\qquad\qquad \text{A disulfide}$$

We derive the common names of disulfides by listing the names of the groups attached to sulfur and adding the word "disulfide."

5.6 Important Alcohols

As you study the alcohols described in this section, you should pay particular attention to two key points. First, they are derived almost entirely from petroleum, natural gas, or coal—all nonrenewable resources. Second, many are themselves starting materials for the synthesis of valuable commercial products, without which our modern industrial society could not exist.

At one time **methanol** was derived by heating hard woods in a limited supply of air—hence the name "wood alcohol." Today methanol is derived entirely from the catalytic reduction of carbon monoxide. Methanol, in turn, is the starting material for the preparation of several important industrial and commercial chemicals, including acetic acid and formaldehyde. Treatment of methanol with carbon monoxide in the presence of a rhodium catalyst gives acetic acid. Partial oxidation of methanol gives formaldehyde. An important use of this one-carbon aldehyde is in the preparation of phenol-formaldehyde and urea-formaldehyde glues and resins, which are used as molding materials and as adhesives for plywood and particle board for the construction industry.

■ **Methanol, CH$_3$OH, is the fuel used in cars of the type that race in Indianapolis.**

D. Young/Tom Stack & Associates

$$\text{Coal or methane} \xrightarrow{[O]} \underset{\substack{\text{Carbon}\\\text{monoxide}}}{CO} \xrightarrow{2H_2} \underset{\text{Methanol}}{CH_3OH} \Big\langle \begin{array}{l} \xrightarrow[\text{catalyst}]{CO} \underset{\text{Acetic acid}}{CH_3COOH} \\ \xrightarrow[\text{oxidation}]{O_2} \underset{\text{Fomaldehyde}}{CH_2O} \end{array}$$

The bulk of the **ethanol** produced worldwide is prepared by acid-catalyzed hydration of ethylene, itself derived from the cracking of the ethane separated from natural gas (Section 2.11). Ethanol is also produced by fermentation of the carbohydrates in plant materials, particularly corn and molasses. The bulk of the fermentation-derived ethanol is used as an "oxygenate" additive to produce gasohol, which is a blend containing as much as 10% ethanol in gasoline. Combustion of gasohol produces less air pollution than combustion of gasoline itself.

$$CH_2{=}CH_2 \quad \xrightarrow[]{} \quad \begin{array}{l} \xrightarrow{H_2O,\ H_2SO_4} CH_3CH_2OH \xrightarrow[180°C]{H_2SO_4} CH_3CH_2OCH_2CH_3 \\ \text{Ethanol} \qquad\qquad\qquad \text{Diethyl ether} \\[2em] \xrightarrow[\text{catalyst}]{O_2} H_2C{-}CH_2 \xrightarrow{H_2O,\ H_2SO_4} HOCH_2CH_2OH \\ \text{Ethylene} \qquad\qquad \text{Ethylene glycol} \\ \text{oxide} \end{array}$$

Ethylene

Acid-catalyzed dehydration of ethanol gives **diethyl ether,** an important laboratory and industrial solvent. Ethylene is also the starting material for the preparation of **ethylene oxide.** This compound has few direct uses. Rather, ethylene oxide's importance is as an intermediate for the production of **ethylene glycol,** a major component of automobile antifreeze. Ethylene glycol freezes at −12°C and boils at 199°C, which makes it ideal for this purpose. In addition, reaction of ethylene glycol with the methyl ester of terephthalic acid gives the polymer poly(ethylene terephthalate), abbreviated PET or PETE (Section 10.8B). Ethylene glycol is also used as a solvent in the paint and plastics industry, and in the formulation of printer's inks, ink pads, and inks for ball-point pens.

Isopropyl alcohol, the alcohol in rubbing alcohol, is made by acid-catalyzed hydration of propene. It is also used in hand lotions, after-shave lotions, and similar cosmetics. A multistep process also converts propene to epichlorohydrin, one of the key components in the production of epoxy glues and resins.

Glycerin is a by-product of the manufacture of soaps by saponification of animal fats and tropical oils (Section 12.3). The bulk of the glycerin used for industrial and commercial purposes, however, is prepared from propene. Perhaps the best-known use of glycerin is for the manufacture of nitroglycerin. Glycerin is also used as an emollient in skin care products and cosmetics, in liquid soaps, and in printing inks.

$$CH_2{=}CHCH_3 \quad \xrightarrow[]{} \quad \begin{array}{l} \xrightarrow[\text{steps}]{\text{several}} ClCH_2CH{-}CH_2 \xrightarrow[\text{other reagents}]{\text{several steps and}} \text{Epoxy glues and resins} \\ \qquad\qquad \text{Epichlorohydrin} \\[1.5em] \qquad\qquad OH \\ \xrightarrow{H_2O,\ H_2SO_4} CH_3CHCH_3 \\ \qquad\qquad \text{Isopropyl alcohol} \\[1.5em] \qquad HO\ HO\ OH \\ \xrightarrow[\text{steps}]{\text{several}} CH_2CHCH_2 \\ \qquad\qquad \text{Glycerin, Glycerol} \end{array}$$

Propene

S U M M A R Y

The functional group of an **alcohol** is an **—OH (hydroxyl) group** bonded to a tetrahedral carbon atom (Section 5.2A). IUPAC names of alcohols are derived by changing the **-e** of the parent alkane to **-ol** (Section 5.2B). The parent chain is numbered from the end that gives the carbon bearing the —OH group the lower number. Common names for alcohols are derived by naming the alkyl group bonded to the —OH group and adding the word *alcohol*. Alcohols are classified as **1°, 2°,** or **3°,** depending on the number of carbon groups bonded to the carbon bearing the —OH group. Compounds containing two hydroxyl groups on adjacent carbons are called **glycols.**

Alcohols are polar compounds with oxygen bearing a partial negative charge and both the carbon and hydrogen bonded to it bearing partial positive charges (Section 5.2C). They associate in the liquid state by hydrogen bonding; as a consequence, their boiling points are higher than those of hydrocarbons of similar molecular weight. Because of increased London dispersion forces, the boiling points of alcohols increase with increasing molecular weight. Alcohols interact with water by hydrogen bonding and are more soluble in water than are hydrocarbons of similar molecular weight.

Alcohols have about the same pK_a value as pure water. For this reason, aqueous solutions of alcohols have the same pH as that of pure water (Section 5.3A).

The functional group of an **ether** is an atom of oxygen bonded to two carbon atoms (Section 5.4A). Common names for ethers are derived by naming the two groups attached to oxygen followed by the word *ether* (Section 5.4B). In **cyclic ethers,** oxygen is one of the atoms in a ring. Ethers are weakly polar compounds. Their boiling points are close to those of hydrocarbons of similar molecular weight (Section 5.4C). Because they form hydrogen bonds with water, however, ethers are more soluble in water than are hydrocarbons of similar molecular weight.

A **thiol** contains an **—SH (sulfhydryl) group** (Section 5.5A). Thiols are named in the same manner as alcohols, but the suffix **-e** of the parent alkane is retained, and **-thiol** is added (Section 5.5B). Common names for thiols are derived by naming the alkyl group bonded to —SH and adding the word *mercaptan.* The S—H bond is nonpolar, and the physical properties of thiols resemble those of hydrocarbons of similar molecular weight (Section 5.4C).

KEY REACTIONS

1. Acid-Catalyzed Dehydration of an Alcohol (Section 5.3B) When isomeric alkenes are possible, the major product is generally the more substituted alkene.

OH
|
$CH_3CH_2CHCH_3$

$\xrightarrow[\text{heat}]{H_3PO_4}$ $CH_3CH=CHCH_3$ + $CH_3CH_2CH=CH_2$ + H_2O
                                                         Major product

2. Oxidation of a Primary Alcohol (Section 5.3C) Oxidation of a primary alcohol by potassium dichromate gives either an aldehyde or a carboxylic acid, depending on the experimental conditions.

$CH_3(CH_2)_6CH_2OH \xrightarrow[H_2SO_4]{K_2Cr_2O_7} CH_3(CH_2)_6\overset{O}{\overset{\|}{C}}H$

$\xrightarrow[H_2SO_4]{K_2Cr_2O_7} CH_3(CH_2)_6\overset{O}{\overset{\|}{C}}OH$

3. Oxidation of a Secondary Alcohol (Section 5.3C) Oxidization of a secondary alcohol by potassium dichromate gives a ketone.

$CH_3(CH_2)_4\overset{OH}{\overset{|}{C}}HCH_3 \xrightarrow[H_2SO_4]{K_2Cr_2O_7} CH_3(CH_2)_4\overset{O}{\overset{\|}{C}}CH_3$

4. Acidity of Thiols (Section 5.5D) Thiols are weak acids, with pK_a equal to approximately 10. They react with strong bases to form water-soluble thiolate salts.

CH_3CH_2SH + NaOH $\xrightarrow{H_2O}$ $CH_3CH_2S^-Na^+$ + H_2O
Ethanethiol   Sodium
(pK_a 10)   ethanethiolate

5. Oxidation of a Thiol to a Disulfide (Section 5.5D) Oxidation of a thiol gives a disulfide. Reduction of a disulfide gives two thiols.

$2HOCH_2CH_2SH \underset{\text{reduction}}{\overset{\text{oxidation}}{\rightleftharpoons}} HOCH_2CH_2S-SCH_2CH_2OH$
A thiol   A disulfide

PROBLEMS

Numbers that appear in color indicate difficult problems. ▶ designates problems requiring applications of principles.

Structure and Nomenclature of Alcohols, Ethers, and Thiols

5.8 What is the difference in structure between a primary, secondary, and tertiary alcohol?

5.9 Which are secondary alcohols?

(a)

(b) $(CH_3)_3COH$

(c)

(d)

5.10 Which of the alcohols in Problem 5.9 are primary? Which are tertiary?

5.11 Write the IUPAC name of each compound.

(a) 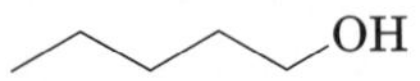 OH

(b) 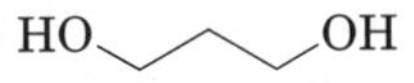 HO OH

(c)

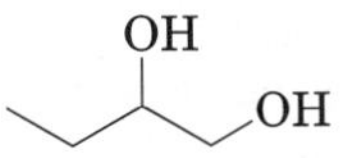

(d)

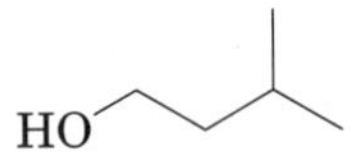

(e)

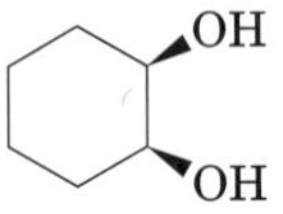

(f)

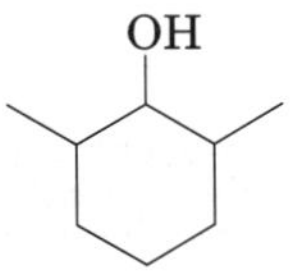

5.12 Draw a structural formula for each alcohol.
(a) Isopropyl alcohol
(b) Propylene glycol
(c) 5-Methyl-2-hexanol
(d) 2-Methyl-2-propyl-1,3-propanediol
(e) 1-Octanol
(f) 3,3-Dimethylcyclohexanol

5.13 Draw a structural formula for each alcohol.
(a) Isobutyl alcohol
(b) 1,4-Butanediol
(c) 5-Methyl-1-hexanol
(d) 1,3-Pentanediol
(e) *trans*-1,4-Cyclohexanediol
(f) 1-Chloro-2-propanol

5.14 Write the common name for each ether.

(a)

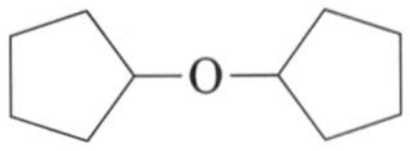

(b) $[CH_3(CH_2)_4]_2O$

(c) $CH_3\overset{\displaystyle CH_3}{\underset{\displaystyle |}{CH}}O\overset{\displaystyle CH_3}{\underset{\displaystyle |}{CH}}CH_3$

5.15 Write the common name for each ether.

(a)

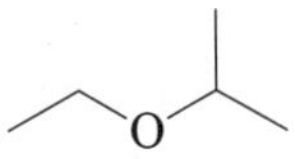

(b)

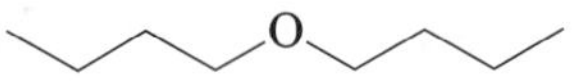

(c)

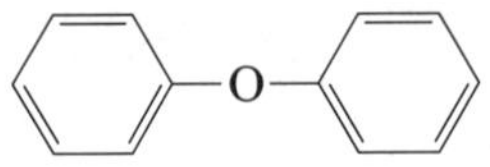

5.16 Write the IUPAC name for each thiol.

(a) $CH_3CH_2\overset{\displaystyle SH}{\underset{\displaystyle |}{CH}}CH_3$

(b) $CH_3CH_2CH_2CH_2SH$

(c)

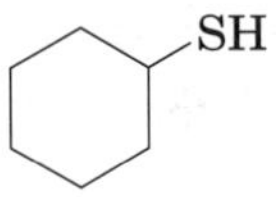

5.17 Write the common name for each thiol in Problem 5.16.

5.18 Both alcohols and phenols contain an —OH group. What structural feature distinguishes these two classes of compounds? Illustrate your answer by drawing the structural formulas of a phenol of six carbon atoms and an alcohol of six carbon atoms.

5.19 Name the functional groups in each compound.

(a)

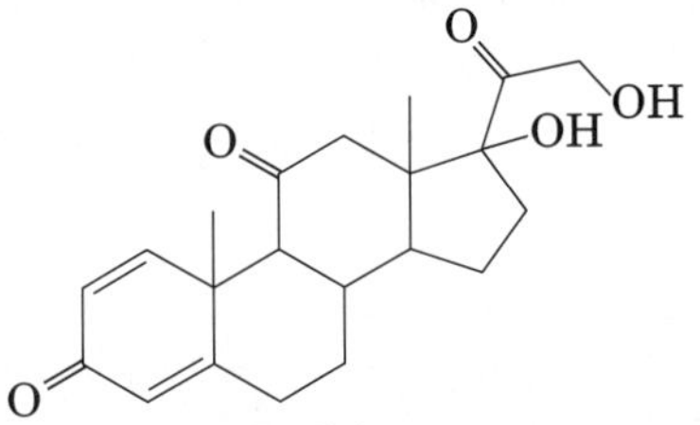

Prednisone
(a synthetic anti-
inflammatory steroid)

(b)

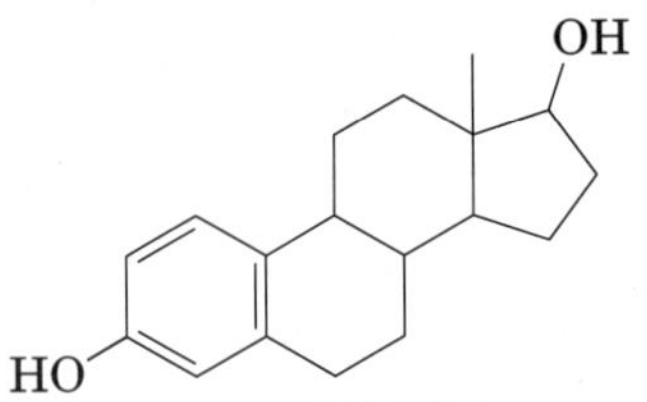

Estradiol
(a female sex hormone;
Section 20.10)

Physical Properties of Alcohols, Ethers, and Thiols

5.20 Explain in terms of noncovalent interactions why the low-molecular-weight alcohols are soluble in water but the low-molecular-weight alkanes and alkenes are not.

5.21 Explain in terms of noncovalent interactions why the low-molecular-weight alcohols are more soluble in water than the low-molecular-weight ethers.

5.22 Why does the water solubility of alcohols decrease as molecular weight increases?

5.23 Show hydrogen bonding between methanol and water in the following ways:
(a) Between the oxygen of methanol and a hydrogen of water
(b) Between the hydrogen of methanol's OH group and the oxygen of water

5.24 Show hydrogen bonding between the oxygen of diethyl ether and a hydrogen of water.

5.25 Arrange these compounds in order of increasing boiling point. Values in °C are −42, 78, 117, and 198.
(a) $CH_3CH_2CH_2CH_2OH$
(b) CH_3CH_2OH
(c) $HOCH_2CH_2OH$
(d) $CH_3CH_2CH_3$

5.26 Arrange these compounds in order of increasing boiling point. Values in °C are 0, 35, and 97.
(a) $CH_3CH_2CH_2OH$
(b) $CH_3CH_2OCH_2CH_3$
(c) $CH_3CH_2CH_2CH_3$

5.27 Following are structural formulas for 1-butanol and 1-butanethiol. One of these compounds has a boiling point of 98°C, the other has a boiling point of 117°C. Which compound has which boiling point?

$CH_3CH_2CH_2CH_2OH$ $CH_3CH_2CH_2CH_2SH$
  1-Butanol   1-Butanethiol

5.28 Explain why methanethiol, CH_3SH, has a lower boiling point (6°C) than methanol, CH_3OH (65°C), even though methanethiol has a higher molecular weight.

5.29 2-Propanol (isopropyl alcohol) is commonly used as rubbing alcohol to cool the skin. 2-Hexanol, also a liquid, is not suitable for this purpose. Why not?

5.30 Explain why glycerol is much thicker (more viscous) than ethylene glycol, which in turn is much thicker than ethanol.

5.31 From each pair, select the compound that is more soluble in water.
(a) CH_3OH or CH_3OCH_3

(b) $\overset{\text{OH}}{\underset{|}{CH_3CHCH_3}}$ or $\overset{\text{CH}_2}{\underset{||}{CH_3CCH_3}}$

(c) $CH_3CH_2CH_2SH$ or $CH_3CH_2CH_2OH$

5.32 Arrange the compounds in each set in order of decreasing solubility in water.
(a) Ethanol, butane, and diethyl ether
(b) 1-Hexanol, 1,2-hexanediol, and hexane

Synthesis of Alcohols

5.33 Give the structural formula of an alkene or alkenes from which each alcohol can be prepared.
(a) 2-Butanol
(b) 1-Methylcyclohexanol
(c) 3-Hexanol
(d) 2-Methyl-2-pentanol
(e) Cyclopentanol

Reactions of Alcohols

5.34 Show how to distinguish between cyclohexanol and cyclohexene by a simple chemical test. Tell what you would do, what you would expect to see, and how you would interpret your observation.

5.35 Compare the acidity of alcohols and phenols, both classes of organic compounds that contain an —OH group.

5.36 Both 2,6-diisopropylcyclohexanol and the intravenous anesthetic Propofol are insoluble in water. Show how these two compounds can be distinguished by their reaction with aqueous sodium hydroxide.

2,6-Diisopropylcyclohexanol   2,6-Diisopropylphenol
                                      (Propofol)

5.37 Write equations for the reaction of 1-butanol, a primary alcohol, with these reagents.
(a) H_2SO_4, heat
(b) $K_2Cr_2O_7$, H_2SO_4

5.38 Write equations for the reaction of 2-butanol with these reagents.
(a) H_2SO_4, heat
(b) $K_2Cr_2O_7$, H_2SO_4

5.39 Write the formula for the product formed by treating each of the following compounds with $K_2Cr_2O_7 / H_2SO_4$.
(a) 1-Octanol
(b) 1,4-Butanediol

5.40 Show how to convert cyclohexanol to the following compounds.
(a) Cyclohexene
(b) Cyclohexane
(c) Cyclohexanone
(d) Bromocyclohexane

5.41 Show reagents and experimental conditions to synthesize each compound from 1-propanol. More than one step is required for some parts of this problem.

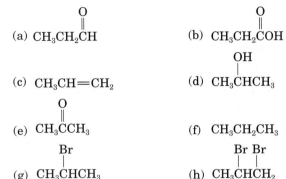

(a) $CH_3CH_2\overset{\displaystyle O}{\overset{\|}{C}}H$

(b) $CH_3CH_2\overset{\displaystyle O}{\overset{\|}{C}}OH$

(c) $CH_3CH{=}CH_2$

(d) $CH_3\overset{\displaystyle OH}{\underset{|}{C}}HCH_3$

(e) $CH_3\overset{\displaystyle O}{\overset{\|}{C}}CH_3$

(f) $CH_3CH_2CH_3$

(g) $CH_3\overset{\displaystyle Br}{\underset{|}{C}}HCH_3$

(h) $CH_3\overset{\displaystyle Br}{\underset{|}{C}}H\overset{\displaystyle Br}{\underset{|}{C}}H_2$

5.42 Show how to convert the given alcohol to compounds (a) and (b).

[cyclopentane]—CH_2OH

(a) [cyclopentane]—$\overset{\displaystyle O}{\overset{\|}{C}}H$

(b) [cyclopentane]—$\overset{\displaystyle O}{\overset{\|}{C}}OH$

5.43 Name two important alcohols derived from ethylene and give two important uses of each.

5.44 Name two important alcohols derived from propene and give an important use of each.

Reaction of Thiols

5.45 Following is a structural formula for the amino acid cysteine:

$$HS-CH_2-\underset{\underset{\displaystyle NH_2}{|}}{CH}-\overset{\overset{\displaystyle O}{\|}}{C}-OH$$

(a) Name the three functional groups in cysteine.

(b) In the human body, cysteine is oxidized to a disulfide. Draw a structural formula for this disulfide.

5.46 Lipoic acid is a growth factor for many bacteria and protozoa and an essential component of several enzymes involved in human metabolism.

[structural formula of lipoic acid with S–S ring and COOH]

Lipoic acid

(a) Name the two functional groups in lipoic acid.

(b) At one stage in its function in human metabolism, the disulfide bond of lipoic acid is reduced to two thiol groups. Draw a structural formula for this reduced form of lipoic acid.

Chemical Connections

5.47 (Chemical Connections 5A) When was nitroglycerin discovered? Is this substance a solid, liquid, or gas?

5.48 (Chemical Connections 5A) What was Alfred Nobel's discovery that made nitroglycerin safer to handle?

5.49 (Chemical Connections 5A) What is the relationship between the medical use of nitroglycerin to relieve the sharp chest pains (angina) associated with heart disease and the gas nitric oxide, NO?

5.50 (Chemical Connections 5B) What is the color of dichromate ion? Of chromium(III) ion? Explain how the conversion of one to the other is used in breath-alcohol screening.

5.51 (Chemical Connections 5B) The legal definition of being under the influence of alcohol is based on blood alcohol content. What is the relationship between breath alcohol content and blood alcohol content?

5.52 (Chemical Connections 5C) What does it mean to say that ethylene oxide is a highly strained molecule?

5.53 (Chemical Connections 5D) What are the advantages and disadvantages of using diethyl ether as an anesthetic?

5.54 (Chemical Connections 5D) Show that enflurane and isoflurane are constitutional isomers.

5.55 (Chemical Connections 5D) Would you expect enflurane and isoflurane to be soluble in water? Would you expect them to be soluble in organic solvents such as hexane?

Additional Problems

5.56 Write a balanced equation for the complete combustion of ethanol, the alcohol added to gasoline to produce gasohol.

5.57 Write a balanced equation for the complete combustion of methanol, the alcohol used in the engines of formula racing cars.

5.58 Knowing what you do about electronegativity, the polarity of covalent bonds, and hydrogen bonding, would you expect an N—H----N hydrogen bond to be stronger than, the same strength as, or weaker than an O—H----O hydrogen bond?

5.59 Draw structural formulas and write IUPAC names for the eight isomeric alcohols with molecular formula $C_5H_{12}O$.

5.60 Draw structural formulas and write common names for the six isomeric ethers with molecular formula $C_5H_{12}O$.

5.61 Explain why the boiling point of ethylene glycol, (198°C) is so much higher than that of 1-propanol, (97°C), even though their molecular weights are about the same.

5.62 1,4-Butanediol, hexane, and 1-pentanol have similar molecular weights. Their boiling points, arranged from lowest to highest, are 69°C, 138°C, and 230°C. Which compound has which boiling point?

5.63 Of the three compounds given in Problem 5.62, one is insoluble in water, another has a solubility of 2.3 g/100 g water, and one is infinitely soluble in water. Which compound has which solubility?

5.64 Each of the following compounds is a common organic solvent. From each pair of compounds, select the solvent with the greater solubility in water.
(a) CH_2Cl_2 or CH_3CH_2OH
(b) $CH_3CH_2OCH_2CH_3$ or CH_3CH_2OH

5.65 Show how to prepare each compound from 2-methyl-1-propanol.
(a) 2-Methylpropene
(b) 2-Methyl-2-propanol
(c) 2-Methylpropanoic acid, $(CH_3)_2CHCOOH$

5.66 Show how to prepare each compound from 2-methylcyclohexanol.

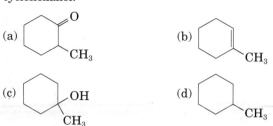

Info Trac College Edition

For additional readings, go to InfoTrac College Edition, your online research library, at

http://infotrac.thomsonlearning.com

CHAPTER **6**

6.1 Introduction

6.2 Enantiomerism: A New Kind of Stereoisomerism

6.3 Naming Stereocenters: The (R,S) System

6.4 Molecules with Two or More Stereocenters

6.5 Optical Activity: How Chirality Is Detected in the Laboratory

6.6 The Significance of Chirality in the Biological World

Median cross section through the shell of a chambered nautilus found in the deep waters of the Pacific Ocean. The shell shows handedness; this cross section is a left-handed spiral.

Lester Lefkowitz/Tony Stone Worldwide

Chirality: The Handedness of Molecules

6.1 Introduction

In this chapter, we study the relationship between objects and their mirror images; that is, we study isomers called enantiomers and diastereomers. Figure 6.1 summarizes the relationship among these isomers and those we studied in Chapters 2 through 5.

The significance of enantiomers is that, except for inorganic and a few simple organic compounds, the vast majority of molecules in the biological world show this type of isomerism. Furthermore, approximately one-half of all medications used to treat humans show it as well.

6.2 Enantiomerism: A New Kind of Stereoisomerism

As an example of a molecule that shows enantiomerism, let us consider 2-butanol. As we discuss this molecule, we will focus on carbon 2, the carbon

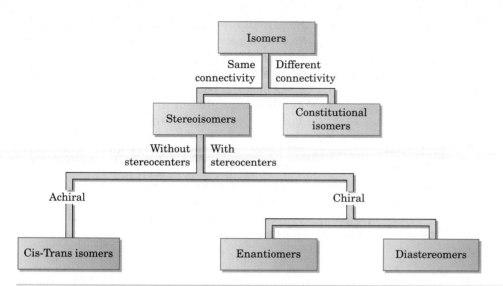

Figure 6.1 Relationships among isomers. In this chapter, we study enantiomers and diastereomers.

bearing the —OH group. What makes this carbon of interest is the fact that it has four different groups bonded to it.

$$\text{OH}$$
$$\text{CH}_3\text{CHCH}_2\text{CH}_3$$
2-Butanol

This structural formula does not show the shape of 2-butanol or the orientation of its atoms in space. To do so, we must consider the molecule as a three-dimensional object.

Original molecule   Mirror image

On the left is what we will call the "original molecule" and a ball-and-stick model of it. In this drawing, the —OH and —CH₃ groups are in the plane of the paper, the —H is behind it, and the —CH₂CH₃ group is in front of it. In the middle is a mirror. On the right is a **mirror image** of the original molecule along with a ball-and-stick model of the mirror image. Every molecule—and, in fact, every object in the world around us—has a mirror image.

The question we now need to ask is, "What is the relationship between the original molecule of 2-butanol and its mirror image?" To answer this question, you need to imagine that you can pick up the mirror image and move it in space in any way you wish. If you can move the mirror image in space and find that it fits over the original so that every bond, atom, and de-

See the **Interactive General, Organic, and Biochemistry CD-ROM, version 2.0,** for further exploration on this topic.

The terms "superposable" and "superimposable" mean the same thing, and both are used currently.

tail of the mirror image matches exactly the bonds, atoms, and details of the original molecule, then the two are **superposable.** In other words, the mirror image and the original represent the same molecule; they are merely oriented differently in space. If, no matter how you turn the mirror image in space, it will not fit exactly on the original with every detail matching, then the two are **nonsuperposable;** they are different molecules.

One way to see that the mirror image of 2-butanol is not superposable on the original molecule is illustrated in the following drawings. Imagine that you hold the mirror image by the C—OH bond and rotate the bottom part of molecule by 180° about this bond. The —OH group retains its position in space, but the $—CH_3$ group, which was to the right and in the plane of the paper, remains in the plane of the paper but is now to the left. Similarly, the $—CH_2CH_3$ group, which was in front of the plane of the paper and to the left, is now behind the plane and to the right.

Original molecule — Mirror image of original molecule — Mirror image rotated by 180°

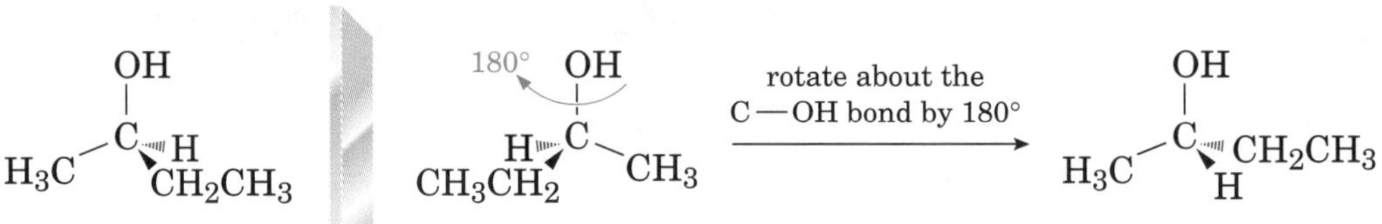

Now move the mirror image in space and try to fit it on the original molecule so that all bonds and atoms match.

Mirror image turned by 180° →

Original molecule →

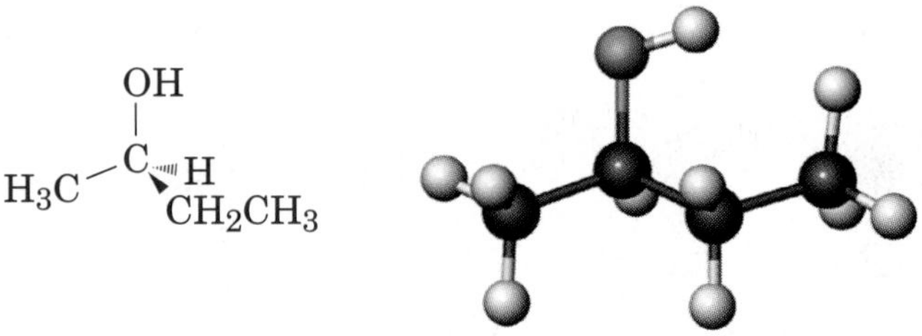

■ **The threads of a drill or screw twist along the axis of the helix, and some plants climb by sending out tendrils that twist helically.**

Charles D. Winters

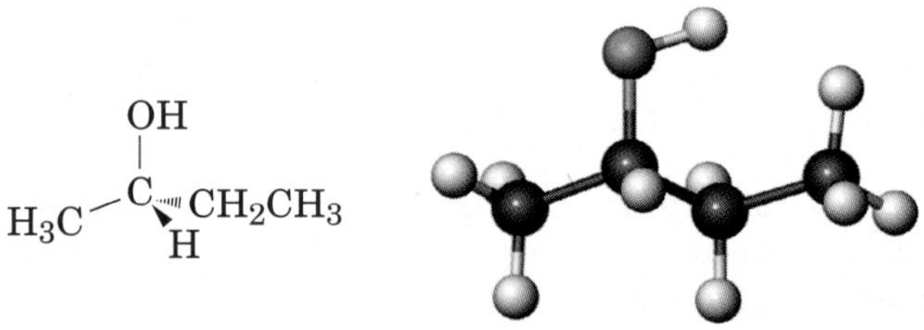

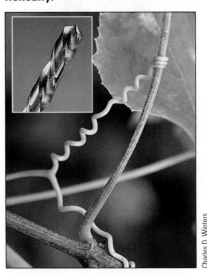

By turning the mirror image as we did, its —OH and $—CH_3$ groups now fit exactly on top of the —OH and $—CH_3$ groups of the original molecule. However, the —H and $—CH_2CH_3$ groups of the two do not match. The —H is away from you in the original but toward you in the mirror image; the $—CH_2CH_3$ group is toward you in the original but away from you in the mirror image. We conclude that the original molecule of 2-butanol and its mirror image are nonsuperposable and, therefore, different compounds.

To summarize, we can turn and rotate the mirror image of 2-butanol in any direction in space, but as long as no bonds are broken and rearranged, we can make only two of the four groups bonded to carbon 2 of the mirror image coincide with the groups on the original molecule. Because 2-butanol

(a)   (b)

■ (a) Mirror images of two wood carvings. The mirror image cannot be superposed on the actual model. The man's right arm rests on the camera in the mirror image; in the actual statue, however, the man's left arm rests on the camera. (b) Left- and right-handed sea shells. If you cup a right-handed shell in your right hand with your thumb pointing from the narrow end to the wide end, the opening will be on your right.

and its mirror image are not superposable, they are isomers. Isomers such as these are called **enantiomers.** Enantiomers, like gloves, always occur in pairs.

Objects that are not superposable on their mirror images are said to be **chiral** (pronounced "ki-ral," rhymes with "spiral;" from the Greek: *cheir,* hand); that is, they show handedness. We encounter chirality in three-dimensional objects of all sorts. Your left hand is chiral, and so is your right hand. A spiral binding on a notebook is chiral. A machine screw with a right-handed twist is chiral. A ship's propeller is chiral. As you examine the objects in the world around you, you will undoubtedly conclude that the vast majority of them are chiral.

As we said after we examined the original and mirror image of 2-butanol, the most common cause of enantiomerism in organic molecules is the presence of a carbon with four different groups bonded to it. Let us examine this statement further by considering 2-propanol, which has no such carbon atom. In this molecule, one carbon atom is bonded to three different groups, but none is bonded to four different groups.

On the left is a three-dimensional representation of 2-propanol; on the right is its mirror image. Also shown are ball-and-stick models of each molecule.

Enantiomers Stereoisomers that are nonsuperposable mirror images; refers to a relationship between pairs of objects

Chiral From the Greek *cheir,* hand; an object that is not superposable on its mirror image

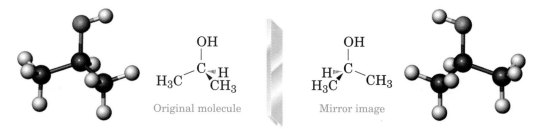

Original molecule   Mirror image

The question we now ask is, "What is the relationship of the mirror image to the original?" This time, let us rotate the mirror image by 120° about the C—OH bond, and then compare it to the original. After performing this rotation, we see that all atoms and bonds of the mirror image fit exactly on the original. Thus the structures we first drew for the original

See the **Interactive General, Organic, and Biochemistry CD-ROM, version 2.0,** for further exploration on this topic.

Figure 6.2 Three-dimensional representations of lactic acid and its mirror image.

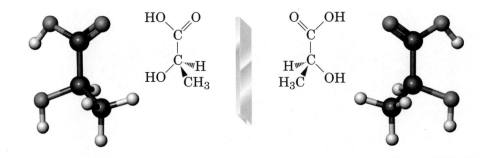

molecule and its mirror image are, in fact, the same molecule—just viewed from different perspectives.

Original molecule | Mirror image of original molecule | Rotate about the C—OH bond by 120° | Mirror image rotated by 120°

Achiral Objects that lack chirality; an object that is superposable on its mirror image

Stereocenter A tetrahedral carbon atom that has four different groups bonded to it

Racemic mixture A mixture of equal amounts of two enantiomers

If an object and its mirror image are superposable, then they are identical and no enantiomerism is possible. We say that such an object is **achiral** (without chirality). Examples of achiral objects include an undecorated cup, an unmarked baseball bat, a regular tetrahedron, a cube, and a sphere.

To repeat, the most common cause of chirality in organic molecules is a tetrahedral carbon atom with four different groups bonded to it. We call such a carbon atom a **stereocenter.** 2-Butanol has one stereocenter; 2-propanol has none. As another example of a molecule with a stereocenter, let us consider 2-hydroxypropanoic acid, more commonly named lactic acid. Lactic acid is a product of anaerobic glycolysis. (See Section 19.2 and also Chemical Connections 19A.) It is also what gives sour cream its sour taste.

Figure 6.2 shows three-dimensional representations for lactic acid and its mirror image. In these representations, all bond angles about the central carbon atom are approximately 109.5°, and the four bonds from it are directed toward the corners of a regular tetrahedron. Lactic acid shows enantiomerism; that is, the original molecule and its mirror image are not superposable but rather are different molecules.

An equimolar mixture of two enantiomers is called a **racemic mixture,** a term derived from the name racemic acid (Latin: *racemus,* a cluster of grapes). Racemic acid is the name originally given to the equimolar mixture of the enantiomers of tartaric acid that forms as a by-product during the fermentation of grape juice.

EXAMPLE 6.1

Each of the following molecules has one stereocenter marked by an asterisk. Draw three-dimensional representations for its enantiomers.

(a) $CH_3\overset{*}{C}HCH_2CH_3$ with Cl substituent

(b) phenyl—$\overset{*}{C}HCH_3$ with NH_2 substituent

Solution

First, draw the carbon stereocenter showing the tetrahedral orientation of its four bonds. One way to do so is to draw two bonds in the plane of the paper, a third bond toward you in front of the plane, and the fourth bond away from you behind the plane. Now place the four groups bonded to the stereocenter on these positions. How you position them is purely arbitrary; there are 12 equivalent ways to do it. To draw an original of (a), for example, place the —Cl group in the vertical position in the plane of the paper, and the —H to the left also in the plane. Place the —CH₂CH₃ away from you and the —CH₃ toward you; this orientation gives the enantiomer of (a) on the left. Its mirror image is on the right.

(a)

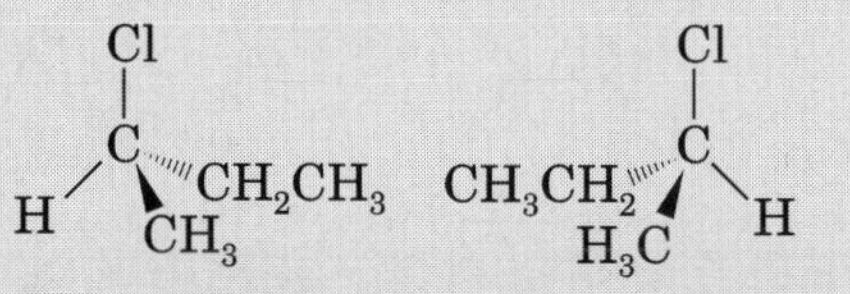

(b)

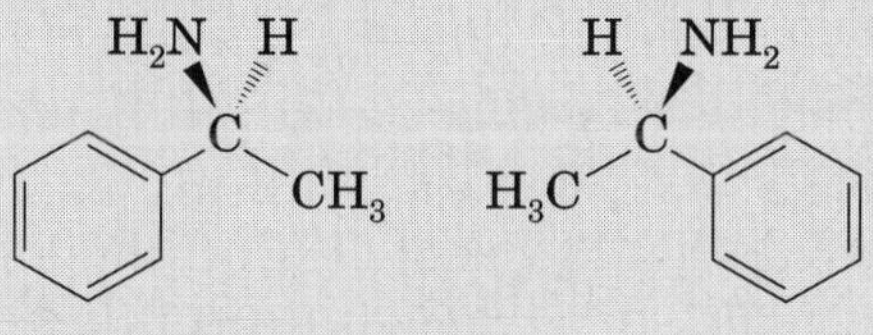

Problem 6.1

Each of the following molecules has one stereocenter marked by an asterisk. Draw three-dimensional representations for its enantiomers.

(a)

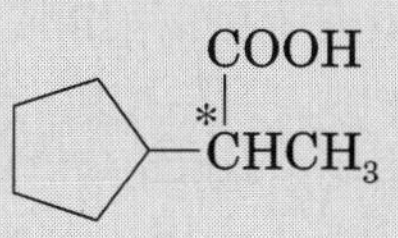

(b) 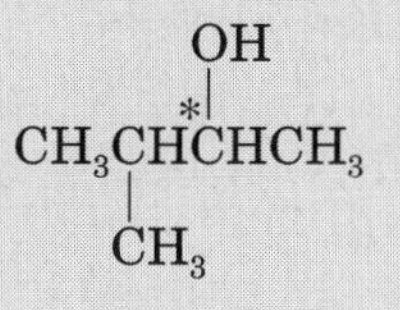 $CH_3CHCHCH_3$

6.3 Naming Stereocenters: The *R,S* System

Because enantiomers are two different compounds, each must have a different name. The over-the-counter drug ibuprofen, for example, shows enantiomerism and can exist as the pair of enantiomers shown here:

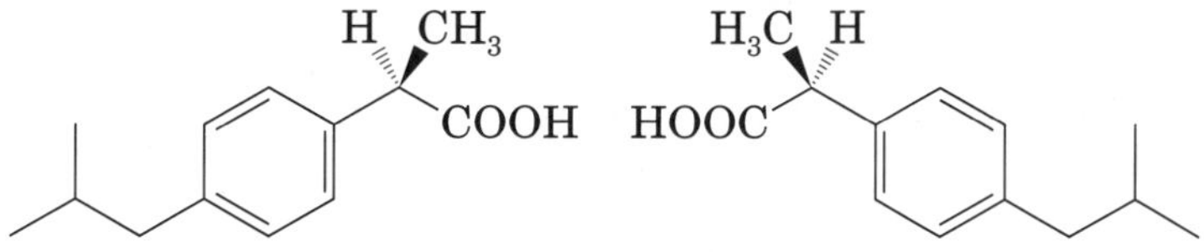

The inactive enantiomer of ibuprofen

The active enantiomer

Only one enantiomer of ibuprofen is biologically active. It reaches therapeutic concentrations in the human body in approximately 12 minutes, whereas the racemic mixture takes approximately 30 minutes. However, in this case, the inactive enantiomer is not wasted. The body converts it to the active enantiomer, but that process takes time.

What we need is a way to designate which enantiomer of ibuprofen (or one of any other pair of enantiomers, for that matter) is which without hav-

See the **Interactive General, Organic, and Biochemistry CD-ROM, version 2.0,** for further exploration on this topic.

R,S system A set of rules for specifying configuration about a stereocenter

ing to draw and point to one or the other of the enantiomers. To do so, chemists have developed the **R,S system.** The first step in assigning an R or S configuration to a stereocenter is to arrange the groups bonded to it in order of priority. Priority is based on atomic number: The higher the atomic number, the higher the priority. If a priority cannot be assigned on the basis of the atoms bonded directly to the stereocenter, look at the next atom or set of atoms and continue to the first point of difference; that is, continue until you can assign a priority.

Table 6.1 shows the priorities of the most common groups encountered in organic and biochemistry. In the R,S system, a $C=O$ is treated as if carbon were bonded to two oxygens by single bonds; thus $CH=O$ has a higher priority than $-CH_2OH$, in which carbon is bonded to only one oxygen.

EXAMPLE 6.2

Assign priorities to the groups in each set.
(a) $-CH_2OH$ and $-CH_2CH_2OH$
(b) $-CH_2CH_2OH$ and $-CH_2NH_2$

Solution
(a) The first point of difference is O of the $-OH$ group compared to C of the $-CH_2OH$ group.

First point of
difference

$-CH_2\overset{}{O}H$ $-CH_2\overset{}{C}H_2OH$
Higher priority   Lower priority

(b) The first point of difference is C of the $-CH_2OH$ group compared to N of the $-NH_2$ group.

First point of
difference

$-CH_2\overset{}{C}H_2OH$ $-CH_2\overset{}{N}H_2$
Lower priority   **Higher priority**

Problem 6.2

Assign priorities to the groups in each set.

                                            O
                                            ∥
(a) $-CH_2OH$ and $-CH_2CH_2COH$

                                         O
                                         ∥
(b) $-CH_2NH_2$ and $-CH_2COH$

To assign an R or S configuration to a stereocenter:

1. Assign a priority from 1 (highest) to 4 (lowest) to each group bonded to the stereocenter.

Table 6.1   R,S Priorities of Some Common Groups

Atom or Group	Reason for Priority: First Point of Difference (Atomic Number)
—I	iodine (53)
—Br	bromine (35)
—Cl	chlorine (17)
—SH	sulfur (16)
—OH	oxygen (8)
—NH_2	nitrogen (7)
$-\overset{\displaystyle O}{\overset{\|}{C}}OH$	carbon to oxygen, oxygen, then oxygen (6 $\longrightarrow$ 8, 8, 8)
$-\overset{\displaystyle O}{\overset{\|}{C}}NH_2$	carbon to oxygen, oxygen, then nitrogen (6 $\longrightarrow$ 8, 8, 7)
$-\overset{\displaystyle O}{\overset{\|}{C}}H$	carbon to oxygen, oxygen, then hydrogen (6 $\longrightarrow$ 8, 8, 1)
—CH_2OH	carbon to oxygen (6 $\longrightarrow$ 8)
—CH_2NH_2	carbon to nitrogen (6 $\longrightarrow$ 7)
—CH_2CH_3	carbon to carbon (6 $\longrightarrow$ 6)
—CH_2H	carbon to hydrogen (6 $\longrightarrow$ 1)
—H	hydrogen (1)

2. Orient the molecule in space so that the lowest-priority group (4) is directed away from you as would be, for instance, the steering column of a car. The three higher-priority groups (1–3) then project toward you, as would the spokes of a steering wheel.

3. Read the three groups projecting toward you in order from highest (1) to lowest (3) priority.

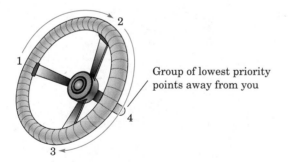

Group of lowest priority points away from you

> **R** Used in the *R,S* system to show that, when the lowest-priority group is oriented away from you, the order of priority of groups on a stereocenter is clockwise

4. If reading the groups 1-2-3 proceeds in a clockwise direction, the configuration is designated as **R** (Latin: *rectus,* straight); if reading the groups 1-2-3 proceeds in a counterclockwise direction, the configuration is **S** (Latin: *sinister,* left). You can also visualize this system as follows: Turning the steering wheel to the right equals R and turning it to the left equals S.

> **S** Used in the *R,S* system to show that, when the lowest-priority group is oriented away from you, the order of priority of groups on a stereocenter is counterclockwise

EXAMPLE 6.3

Assign an R or S configuration to each stereocenter.

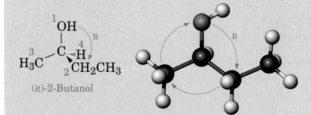

(a) 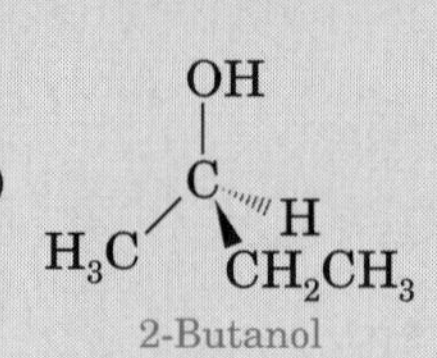 2-Butanol

(b) 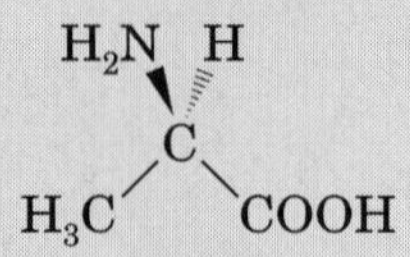 Alanine

Solution
View each molecule through the stereocenter and along the bond from
the stereocenter to the group of lowest priority.
(a) The order of decreasing priority is
$-OH > -CH_2CH_3 > -CH_3 > -H$. Therefore, view the molecule
along the C—H bond with the H pointing away from you. Reading
the other three groups in the order 1-2-3 occurs in the clockwise direction. Therefore, the configuration is R and this enantiomer is
(R)-2-butanol.

(R)-2-Butanol

With —H, the lowest-priority
group, pointing away from you,
this is what you see

(b) The order of decreasing priority is
$-NH_2 > -COOH > -CH_3 > -H$. View the molecule along the
C—H bond with H pointing away from you. Reading the groups in
the order 1-2-3 occurs in the clockwise direction; therefore, the configuration is R and this enantiomer is (R)-alanine.

(R)-Alanine

With —H, the lowest-priority
group, pointing away from you,
this is what you see

Problem 6.3
Assign an R or S configuration to each stereocenter.

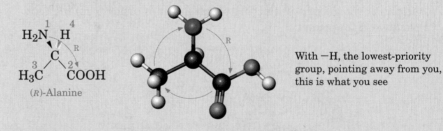

(a) 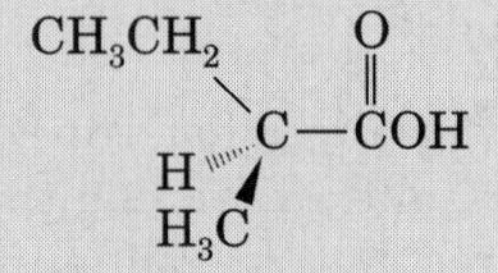 2-Methylbutanoic acid

(b) 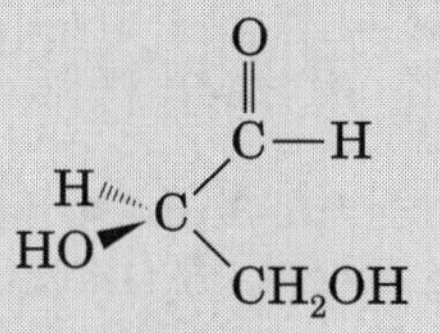 Glyceraldehyde

Now let us return to our three-dimensional drawing of the enantiomers
of ibuprofen and assign each an R or S configuration. In order of decreasing
priority, the groups bonded to the stereocenter are —COOH (1) >

—C_6H_5 (2) > —CH_3 (3) > —H (4). In the enantiomer on the left, reading the groups on the stereocenter in order of priority is clockwise and, therefore, this enantiomer is (R)-ibuprofen. Its mirror image is (S)-ibuprofen.

(R)-Ibuprofen
(the inactive enantiomer)

(S)-Ibuprofen
(the active enantiomer)

The R,S system can be used to specify the configuration of any stereocenter in any molecule. It is not, however, the only system used for this purpose. There is also a D,L system, which is used primarily to specify the configuration of carbohydrates (Chapter 11) and amino acids (Chapter 13).

6.4 Molecules with Two or More Stereocenters

For a molecule with n stereocenters, the maximum number of stereoisomers possible is 2^n. We have already verified that, for a molecule with one stereocenter, $2^1 = 2$ stereoisomers (one pair of enantiomers) are possible. For a molecule with two stereocenters, a maximum of $2^2 = 4$ stereoisomers (two pairs of enantiomers) is possible; for a molecule with three stereocenters, a maximum of $2^3 = 8$ stereoisomers (four pairs of enantiomers) is possible; and so forth.

A Molecules with Two Stereocenters

We begin our study of molecules with two stereocenters by considering 2,3,4-trihydroxybutanal, a molecule with two stereocenters.

$$
\begin{array}{c}
\text{CHO} \\
|\\
*\text{CHOH} \\
|\\
*\text{CHOH} \\
|\\
\text{CH}_2\text{OH}
\end{array}
$$

2,3,4-Trihydroxybutanal

The maximum number of stereoisomers possible for this molecule is $2^2 = 4$, each of which is drawn in Figure 6.3.

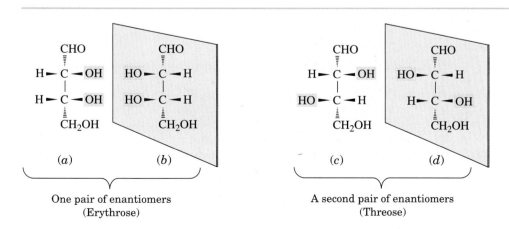

(a)   (b)

One pair of enantiomers
(Erythrose)

(c)   (d)

A second pair of enantiomers
(Threose)

Figure 6.3 The four stereoisomers of 2,3,4-trihydroxybutanal.

Stereoisomers (a) and (b) are nonsuperposable mirror images and are, therefore, a pair of enantiomers. Stereoisomers (c) and (d) are also nonsuperposable mirror images and are a second pair of enantiomers. We describe the four stereoisomers of 2,3,4-trihydroxybutanal by saying that they consist of two pairs of enantiomers. Enantiomers (a) and (b) are named **erythrose.** Erythrose is synthesized in erythrocytes (red blood cells); hence the derivation of its name. Enantiomers (c) and (d) are named **threose.** Erythrose and threose belong to the class of compounds called carbohydrates, which we discuss in Chapter 11.

We have specified the relationship between (a) and (b) and that between (c) and (d). What is the relationship between (a) and (c), between (a) and (d), between (b) and (c), and between (b) and (d)? The answer is that they are **diastereomers**—stereoisomers that are not mirror images.

Diastereomers Stereoisomers that are not mirror images of each other

See the **Interactive General, Organic, and Biochemistry CD-ROM, version 2.0,** for further exploration on this topic.

EXAMPLE 6.4

1,2,3-Butanetriol has two stereocenters (carbons 2 and 3); thus $2^2 = 4$ stereoisomers are possible for it. Following are three-dimensional representations for each.

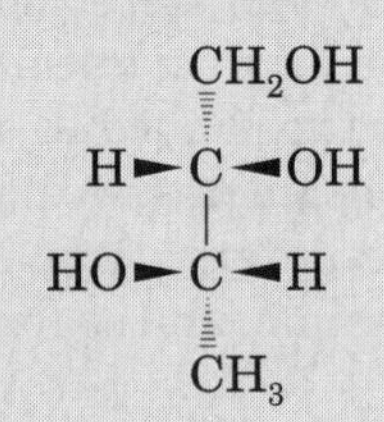

$$
\begin{array}{cccc}
CH_2OH & CH_2OH & CH_2OH & CH_2OH \\
H\!-\!C\!-\!OH & H\!-\!C\!-\!OH & HO\!-\!C\!-\!H & HO\!-\!C\!-\!H \\
HO\!-\!C\!-\!H & H\!-\!C\!-\!OH & HO\!-\!C\!-\!H & H\!-\!C\!-\!OH \\
CH_3 & CH_3 & CH_3 & CH_3 \\
(1) & (2) & (3) & (4)
\end{array}
$$

(a) Which stereoisomers are pairs of enantiomers?
(b) Which stereoisomers are diastereomers?

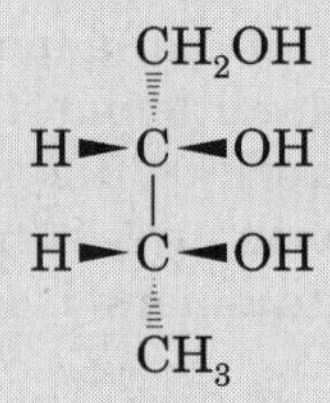

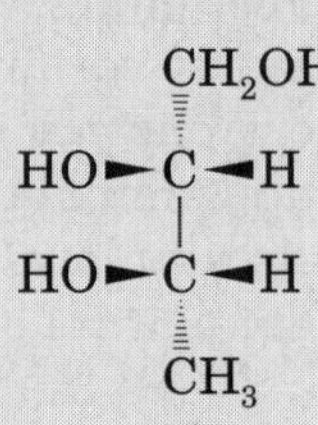

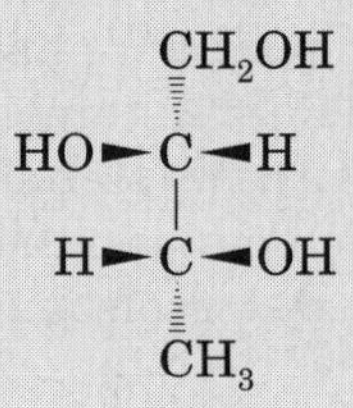

Solution

(a) Enantiomers are stereoisomers that are nonsuperposable mirror images. Compounds (1) and (4) are one pair of enantiomers, and compounds (2) and (3) are a second pair of enantiomers.
(b) Diastereomers are stereoisomers that are not mirror images. Compounds (1) and (2), (1) and (3), (2) and (4), and (3) and (4) are diastereomers.

Problem 6.4

3-Amino-2-butanol has two stereocenters (carbons 2 and 3); thus $2^2 = 4$ stereoisomers are possible for it.

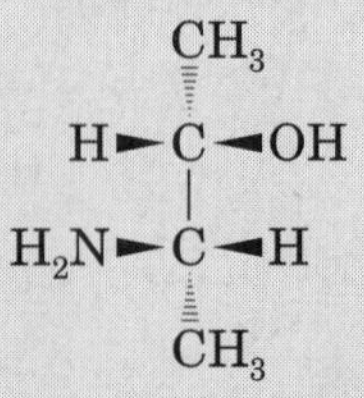

$$
\begin{array}{cccc}
CH_3 & CH_3 & CH_3 & CH_3 \\
H\!-\!C\!-\!OH & H\!-\!C\!-\!OH & HO\!-\!C\!-\!H & HO\!-\!C\!-\!H \\
H_2N\!-\!C\!-\!H & H\!-\!C\!-\!NH_2 & H\!-\!C\!-\!NH_2 & H_2N\!-\!C\!-\!H \\
CH_3 & CH_3 & CH_3 & CH_3 \\
(1) & (2) & (3) & (4)
\end{array}
$$

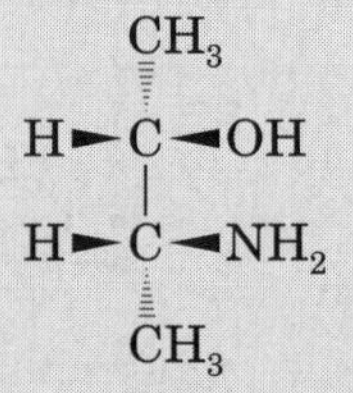

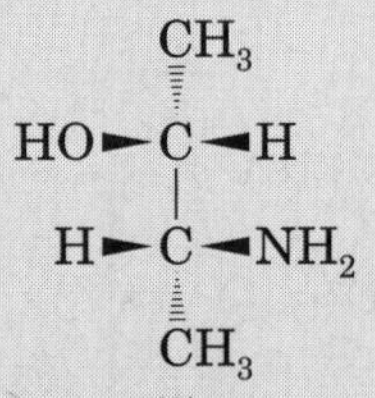

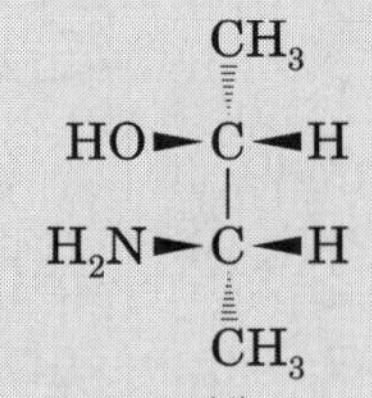

(a) Which stereoisomers are pairs of enantiomers?
(b) Which sets of stereoisomers are diastereomers?

We can analyze chirality in cyclic molecules with two stereocenters in the same way we analyzed it in acyclic compounds.

EXAMPLE 6.5

How many stereoisomers are possible for 3-methylcyclopentanol?

Solution

Carbons 1 and 3 of this compound are stereocenters. There are, therefore, $2^2 = 4$ stereoisomers possible for this molecule.

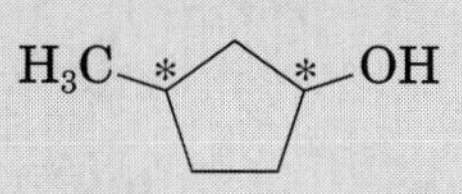

$$H_3C \underset{*}{\diagup}\diagdown \underset{*}{\diagup} OH$$

Problem 6.5

How many stereoisomers are possible for 3-methylcyclohexanol?

EXAMPLE 6.6

Mark the stereocenters in each compound with an asterisk. How many stereoisomers are possible for each?

(a)

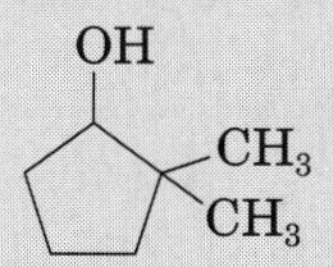

$$\begin{array}{c} OH \\ | \\ \diagup\diagdown CH_3 \\ CH_3 \end{array}$$

(b)

OH

(c)
$$\begin{array}{ccc} OH & & O \\ | & & \| \\ CH_3-CH-CH-COH \\ & | \\ & NH_2 \end{array}$$

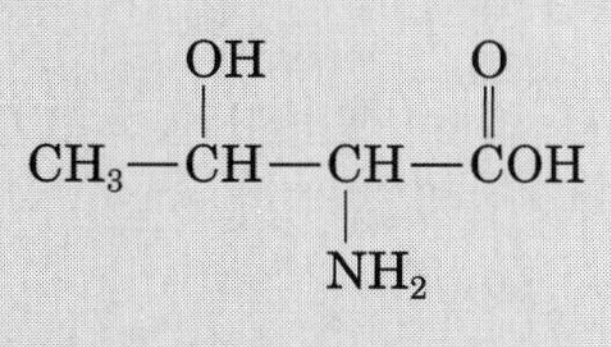

Solution

Each stereocenter is marked with an asterisk, and under each compound is given the number of stereoisomers possible for it. In (a), the carbon bearing the two methyl groups is not a stereocenter; this carbon has only three different groups bonded to it.

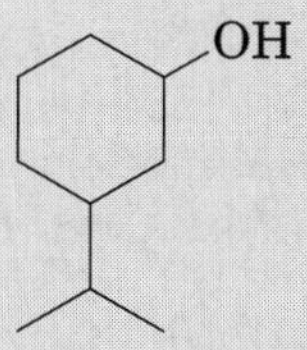

(a)

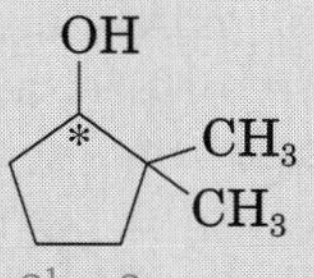

$$\begin{array}{c} OH \\ | \\ *\diagup\diagdown CH_3 \\ CH_3 \end{array}$$

$2^1 = 2$

(b)

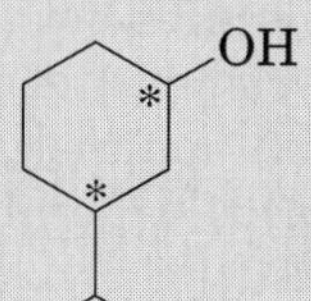

OH
*

*

$2^2 = 4$

(c) 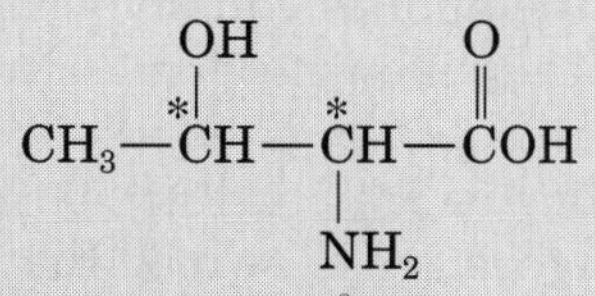
$$\begin{array}{ccc} OH & & O \\ *| & *| & \| \\ CH_3-CH-CH-COH \\ & | \\ & NH_2 \end{array}$$

$2^2 = 4$

Problem 6.6

Mark all stereocenters in each compound with an asterisk. How many stereoisomers are possible for each?

(a)

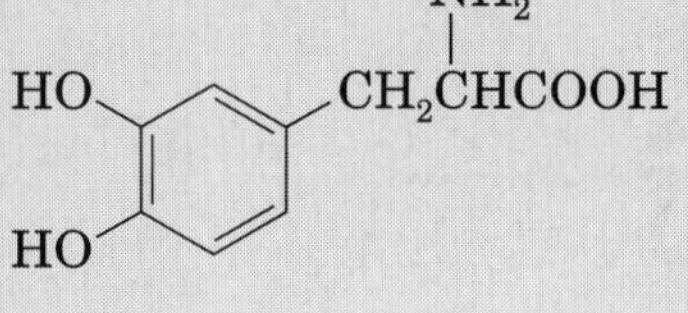

$$\begin{array}{c} NH_2 \\ | \\ HO\diagdownCH_2CHCOOH \\ HO\diagup \end{array}$$

(b)
$$\begin{array}{c} OH \\ | \\ CH_2{=}CHCHCH_2CH_3 \end{array}$$

(c)

OH

NH_2

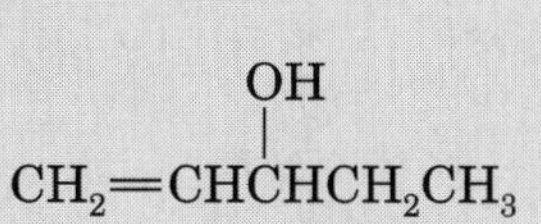

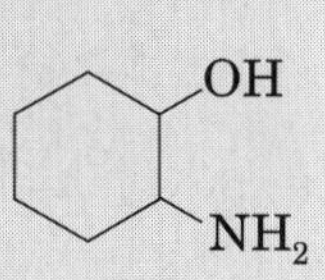

CHEMICAL CONNECTIONS 6A

Chiral Drugs

Some common drugs used in human medicine—for example, aspirin (Chemical Connections 10C)—are achiral. Others, such as the penicillin and erythromycin classes of antibiotics and the drug Captopril, are chiral and are sold as single enantiomers. Captopril is very effective for the treatment of high blood pressure and congestive heart failure (See Chemical Connections 4F). It is manufactured and sold as the (s,s)-stereoisomer.

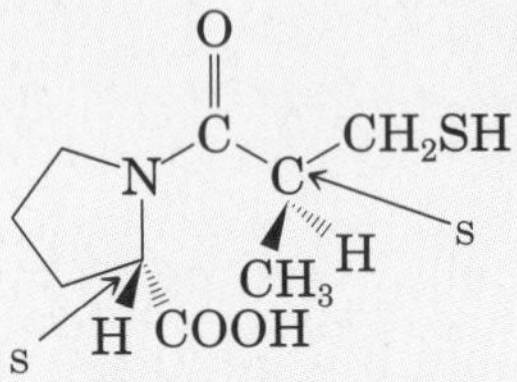

Captopril

A large number of chiral drugs, however, are sold as racemic mixtures. The popular analgesic ibuprofen (the active ingredient in Motrin, Advil, and many other nonaspirin analgesics) is an example.

Recently, the U.S. Food and Drug Administration established new guidelines for the testing and marketing of chiral drugs. After reviewing these guidelines, many drug companies have decided to develop only single enantiomers of new chiral drugs.

In addition to regulatory pressure, pharmaceutical developers must deal with patent considerations. If a company has a patent on a racemic mixture of a drug, a new patent can often be taken out on one of its enantiomers.

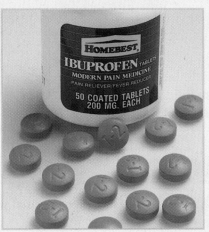

■ **Ibuprofen (for example, Advil) is sold as a racemic mixture, and naproxen (for example, Aleve) is sold as the s enantiomer.**

B Molecules with Three or More Stereocenters

The 2^n rule applies equally well to molecules with three or more stereocenters. In the following disubstituted cyclohexanol with three stereocenters, each stereocenter is marked with an asterisk. A maximum of $2^3 = 8$ stereoisomers are possible for this molecule. Menthol, one of the eight, has the configuration shown on the right. The configuration at the carbon bearing the —OH group is R. Menthol is present in peppermint and other mint oils.

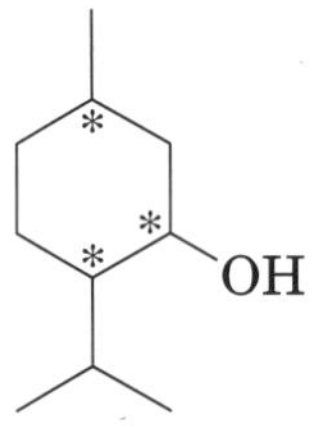

2-Isopropyl-5-methyl-cyclohexanol

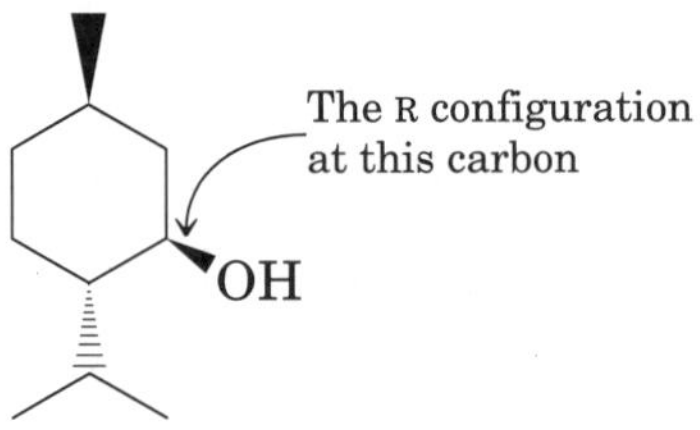

Menthol

The R configuration at this carbon

Cholesterol is a more complicated molecule with eight stereocenters. To identify them, remember to add an appropriate number of hydrogens to complete the tetravalence of each carbon you think might be a stereocenter.

Cholesterol
(has 8 stereocenters; 256
stereoisomers are possible)

Cholesterol
(this is the stereoisomer
present in human metabolism)

6.5 Optical Activity: How Chirality Is Detected in the Laboratory

A Plane-Polarized Light

As we have already established, the two members of a pair of enantiomers are different compounds, and we must expect, therefore, that some of their properties differ. One such property relates to their effect on the plane of polarized light. Each member of a pair of enantiomers rotates the plane of polarized light; for this reason, each enantiomer is said to be **optically active.** To understand how optical activity is detected in the laboratory, we must first understand plane-polarized light and how a polarimeter, the instrument used to detect optical activity, works.

Ordinary light consists of waves vibrating in all planes perpendicular to its direction of propagation. Certain materials, such as a Polaroid sheet (a plastic film like that used in polarized sunglasses), selectively transmit light waves vibrating only in parallel planes. Electromagnetic radiation vibrating in only parallel planes is said to be **plane polarized.**

> **Optically active** Showing that a compound rotates the plane of polarized light

> **Plane-polarized light** Light with waves vibrating in only parallel planes

B A Polarimeter

A **polarimeter** consists of a light source emitting unpolarized light, a polarizer, an analyzer, and a sample tube (Figure 6.4). If the sample tube is empty, the intensity of light reaching the detector (in this case, your eye) is at its maximum when the axes of the polarizer and analyzer are parallel to

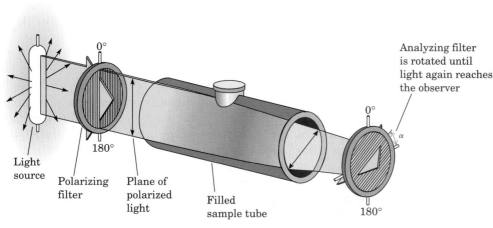

Analyzing filter
is rotated until
light again reaches
the observer

0°

180°

Light
source

Polarizing
filter

Plane of
polarized
light

Filled
sample tube

0°

180°

Figure 6.4 Schematic diagram of a polarimeter with its sample tube containing a solution of an optically active compound. The analyzer has been turned clockwise by α degrees to restore the light field.

each other. If the analyzer is turned either clockwise or counterclockwise, less light is transmitted. When the axis of the analyzer is at right angles to the axis of the polarizer, the field of view is dark.

When a solution of an optically active compound is placed in the sample tube, it rotates the plane of the polarized light. If it rotates the plane clockwise, we say it is **dextrorotatory;** if it rotates the plane counterclockwise, we say it is **levorotatory.** Each member of a pair of enantiomers rotates the plane of polarized light by the same number of degrees, but in opposite directions. If one enantiomer is dextrorotatory, the other is levorotatory.

The number of degrees by which an optically active compound rotates the plane of polarized light is called its **specific rotation** and is given the symbol $[\alpha]$. Specific rotation is defined as the observed rotation of an optically active substance at a concentration of 1 g/mL in a sample tube 10 cm long. A dextrorotatory compound is indicated by a plus sign in parentheses, $(+)$, and a levorotatory compound is indicated by a minus sign in parentheses, $(-)$. Common practice is to report the temperature (in °C) at which the measurement is made and the wavelength of light used. The most common wavelength of light used in polarimetry is the sodium D line, the same wavelength responsible for the yellow color of sodium-vapor lamps.

Following are specific rotations for the enantiomers of lactic acid measured at 21°C and using the D line of a sodium-vapor lamp as the light source:

> **Dextrorotatory** Clockwise (to the right) rotation of the plane of polarized light in a polarimeter

> **Levorotatory** Counterclockwise (to the left) rotation of the plane of polarized light in a polarimeter

The $(+)$ enantiomer of lactic acid is produced by muscle tissue in humans. The $(-)$ enantiomer is found in sour milk.

$$
\begin{array}{cc}
\text{COOH} & \text{COOH} \\
| & | \\
\text{H}_3\text{C}\underset{\text{OH}}{\overset{\text{C}\cdots}{\diagup}}\text{H} & \text{H}\overset{\text{C}}{\underset{\text{HO}}{\diagdown}}\text{CH}_3
\end{array}
$$

(S)-$(+)$-Lactic acid (R)-$(-)$-Lactic acid
$[\alpha]_D^{21} = +2.6°$ $[\alpha]_D^{21} = -2.6°$

6.6 The Significance of Chirality in the Biological World

Except for inorganic salts and a few low-molecular-weight organic substances, the molecules in living systems—both plant and animal—are chiral. Although these molecules can exist as a number of stereoisomers, almost invariably only one stereoisomer is found in nature. Of course, instances do occur in which more than one stereoisomer is found, but these isomers rarely exist together in the same biological system.

A Chirality in Biomolecules

Perhaps the most conspicuous examples of chirality among biological molecules are the enzymes, all of which have many stereocenters. Consider chymotrypsin, an enzyme in the intestines of animals that catalyzes the digestion of proteins (Chapter 14). Chymotrypsin has 251 stereocenters. The maximum number of stereoisomers possible is 2^{251}—a staggeringly large number, almost beyond comprehension. Fortunately, nature does not squander its precious energy and resources unnecessarily; any given organism produces and uses only one of these stereoisomers.

■ **The horns of this African gazelle show chirality; one horn is the mirror image of the other.**

Courtesy of William Brown

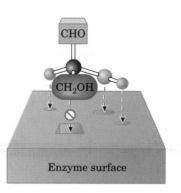

Figure 6.5 A schematic diagram of an enzyme surface that can interact with (*R*)-glyceraldehyde at three binding sites, but with (*s*)-glyceraldehyde at only two of these sites.

(*R*)-Glyceraldehyde
fits the three binding
sites on surface

(*s*)-Glyceraldehyde
fits only two of the
three binding sites

B How an Enzyme Distinguishes
Between a Molecule and Its Enantiomer

An enzyme catalyzes a biological reaction of a molecule by first positioning it at a **binding site** on its surface. An enzyme with specific binding sites for three of the four groups on a stereocenter can distinguish between a molecule and its enantiomer or one of its diastereomers. Assume, for example, that an enzyme involved in catalyzing a reaction of glyceraldehyde has three binding sites: one specific for —H, a second specific for —OH, and a third specific for —CHO. Assume further that the three sites are arranged on the enzyme surface as shown in Figure 6.5. The enzyme can distinguish (*R*)-glyceraldehyde (the natural or biologically active form) from its enantiomer because the natural enantiomer can be absorbed with three groups interacting with their appropriate binding sites; for the s enantiomer, at best only two groups can interact with these binding sites.

Because interactions between molecules in living systems take place in a chiral environment, it should come as no surprise that a molecule and its enantiomer or one of its diastereomers elicit different physiological responses. As we have already seen, (*s*)-ibuprofen is active as a pain and fever reliever while its R enantiomer is inactive. The s enantiomer of the closely related analgesic naproxen is also the active pain reliever of this compound, but its R enantiomer is a liver toxin!

(*s*)-Ibuprofen

(*s*)-Naproxen

S U M M A R Y

A **mirror image** is the reflection of an object in a mirror (Section 6.2). **Enantiomers** are a pair of stereoisomers that are nonsuperposable mirror images of each other. A **racemic mixture** contains equal amounts of two enantiomers. An object that is not superposable on its mirror image is said to be **chiral.** An **achiral** object is one that lacks chirality; that is, it has a superposable mirror image (Section 6.2). The most common cause of chirality in organic molecules is the presence of a tetrahedral carbon atom with four different groups bonded to it. Such a carbon is called a **stereocenter.** The configuration at a stereocenter can be specified using the **R,S system** (Section 6.3). For a molecule with n stereocenters, the maximum number of stereoisomers possible is 2^n.

Light with waves that vibrate in only parallel planes is said to be **plane polarized** (Section 6.5A). To measure optical activity, we use a **polarimeter** (Section 6.5B). A compound is said to be **optically active** if it rotates the plane of polarized light. If it rotates the plane clockwise, it is **dextrorotatory;** if it

rotates the plane counterclockwise, it is **levorotatory.** Each member of a pair of enantiomers rotates the plane of polarized light by an equal number of degrees, but in opposite directions.

An enzyme catalyzes biological reactions of molecules by first positioning them at binding sites on its surface (Section 6.6). An enzyme with binding sites specific for three of the four groups on a stereocenter can distinguish between a molecule and its enantiomer or one of its diastereomers.

P R O B L E M S

Numbers that appear in color indicate difficult problems.
▶ designates problems requiring application of principles.

Chirality

6.7 What does the term "chiral" mean? Give an example of a chiral molecule.

6.8 What does the term "achiral" mean? Give an example of an achiral molecule.

6.9 Define the term "stereoisomer." Name three types of stereoisomers.

6.10 In what way do constitutional isomers differ from stereoisomers? In what way are they the same?

6.11 Which of the following objects are chiral (assume that there is no label or other identifying mark)?

(a) Pair of scissors
(b) Tennis ball
(c) Paper clip
(d) Beaker
(e) The swirl created in water as it drains out of a sink or bathtub

6.12 2-Pentanol is chiral but 3-pentanol is not. Explain.

6.13 2-Butene exists as a pair of cis-trans isomers. Is *cis*-2-butene chiral? Is *trans*-2-butene chiral? Explain.

6.14 Explain why the carbon of a carbonyl group cannot be a stereocenter.

Enantiomers

6.15 Which of the following compounds contain stereocenters?

(a) 2-Chloropentane
(b) 3-Chloropentane
(c) 3-Chloro-1-butene
(d) 1,2-Dichloropropane

6.16 Which of the following compounds contain stereocenters?

(a) Cyclopentanol
(b) 1-Chloro-2-propanol
(c) 2-Methylcyclopentanol
(d) 1-Phenyl-1-propanol

6.17 Using only C, H, and O, write structural formulas for the lowest-molecular-weight chiral molecule of each of the following.

(a) Alkane
(b) Alkene
(c) Alcohol
(d) Aldehyde
(e) Ketone
(f) Carboxylic acid

6.18 Draw the mirror image for each molecule.

(a)

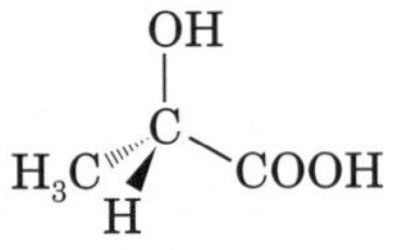

$$\underset{H}{\overset{OH}{\underset{|}{\overset{|}{C}}}}$$
$$H_3C \quad COOH$$

(b) 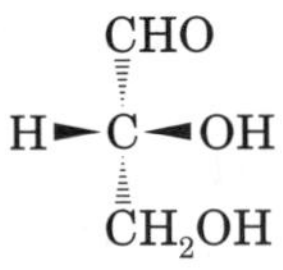
$$\overset{CHO}{\underset{CH_2OH}{\overset{|}{H-C-OH}}}$$

(c)

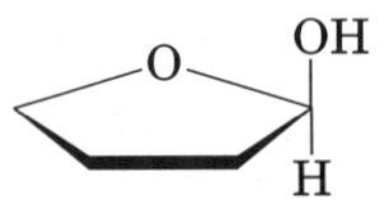

(d)

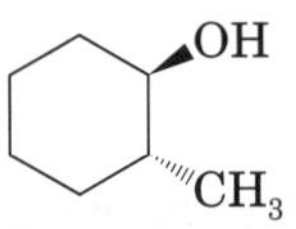

6.19 Draw the mirror image for each molecule.

(a) 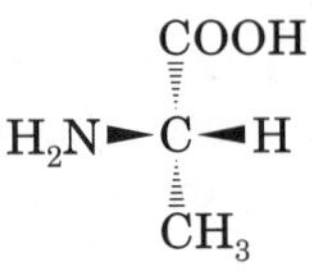 $H_2N - C - H$, with COOH above and CH_3 below

(b)

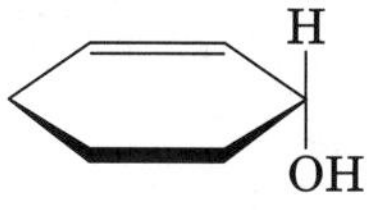

(c)

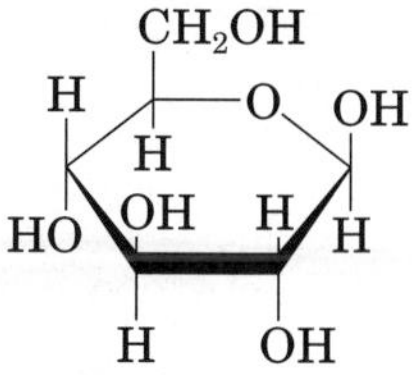

(d) H₃C⟍⟋NH₂

(c)

(d)

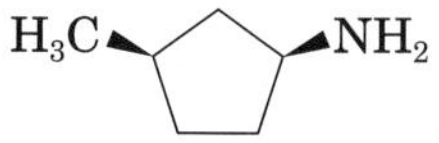

6.20 Mark each stereocenter in these molecules with an asterisk. Note that not all contain stereocenters.

(a)

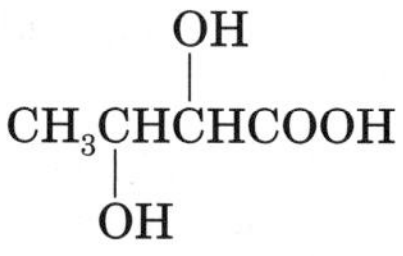

(b)

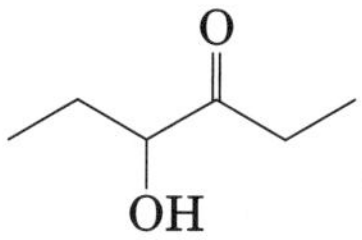

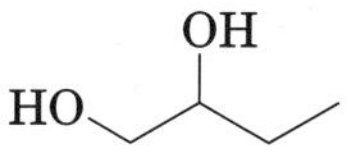

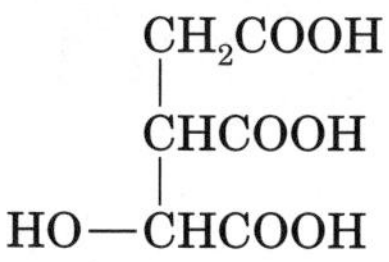

(c)

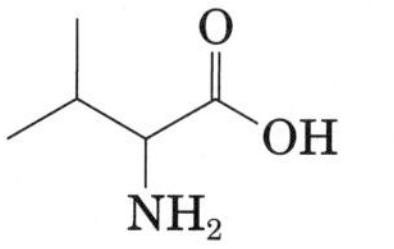

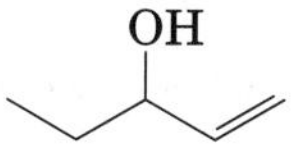

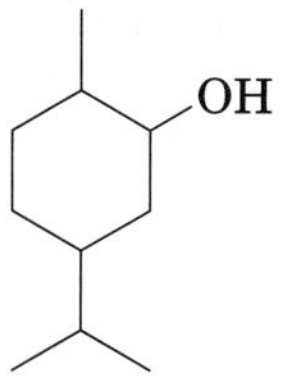

(d)

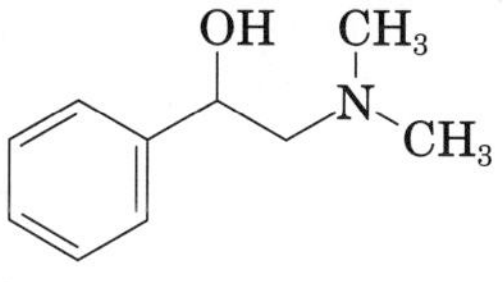

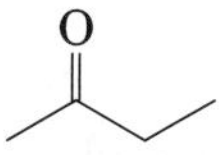

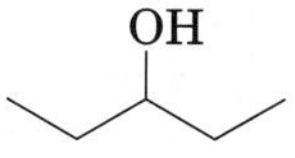

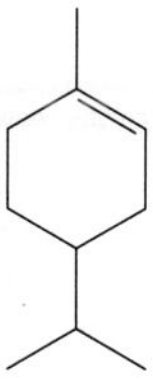

6.23 Label all stereocenters in each molecule with an asterisk. How many stereoisomers are possible for each molecule?

(a)

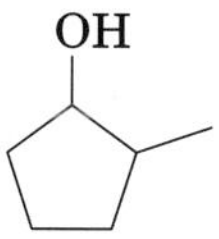

(b)

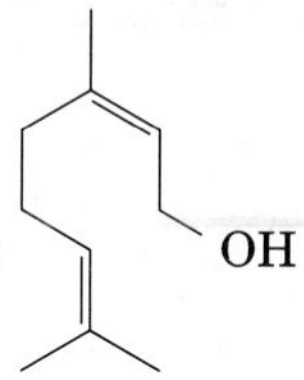

(c)

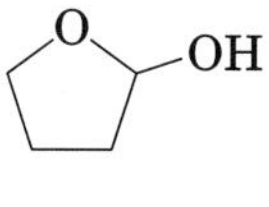

(d) 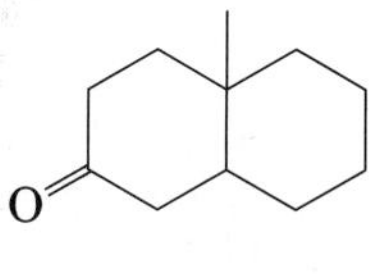 O

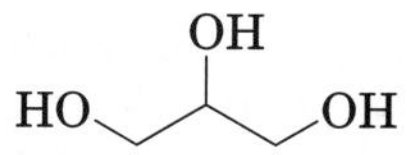

6.21 Mark each stereocenter in these molecules with an asterisk. Note that not all contain stereocenters.

(a) HO⟍⟋OH

(b) HO⟍⟋

(c)

(d)

The R,S System

6.24 Assign priorities to the groups in each set.
(a) —H, —CH₃, —OH, —CH₂OH
(b) —CH₃, —H, —COOH, —NH₂
(c) —CH₃, —CH₂SH, —NH₂, —COOH

Molecules with Two or More Stereocenters

6.22 Label all stereocenters in each molecule with an asterisk. How many stereoisomers are possible for each molecule?

(a) CH₃CHCHCOOH with OH, OH

(b) HO—CHCOOH with CH₂COOH, CHCOOH

6.25 Which molecules have R configurations?

(a) 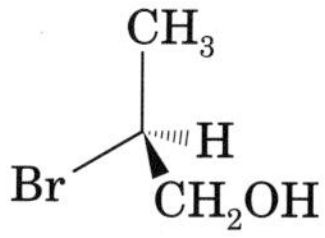 Br, CH₃, H, CH₂OH

(b) 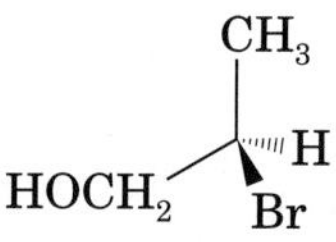 HOCH₂, CH₃, H, Br

(c) 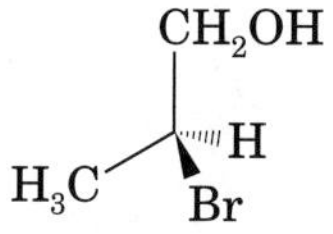 H₃C, CH₂OH, H, Br

(d) 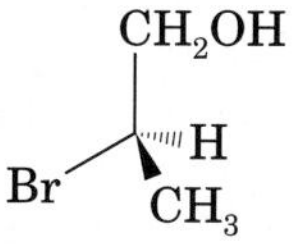 Br, CH₂OH, H, CH₃

6.26 For centuries, Chinese herbal medicine has used extracts of *Ephedra sinica* to treat asthma. The asthma-relieving component of this plant is ephedrine, a very potent dilator of the air passages of the lungs. The naturally occurring stereoisomer is levorotatory and has the following structure. Assign an R or S configuration to each stereocenter.

Ephedrine $[\alpha]_D^{21}$ −41

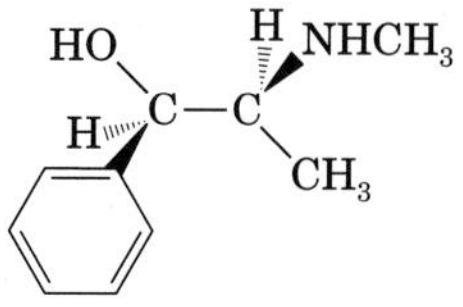

6.27 The specific rotation of naturally occurring ephedrine, shown in Problem 6.26, is $-41°$. What is the specific rotation of its enantiomer?

6.28 What is a racemic mixture? Is a racemic mixture optically active—that is, will it rotate the plane of polarized light?

Chemical Connections

6.29 (Chemical Connection 6A) What does it mean to say that a drug is chiral? If a drug is chiral, will it be optically active—that is, will it rotate the plane of polarized light?

Additional Problems

6.30 Which of the eight alcohols with molecular formula $C_5H_{12}O$ are chiral?

6.31 Write the structural formula of an alcohol with molecular formula $C_6H_{14}O$ that contains two stereocenters.

6.32 Which carboxylic acids with molecular formula $C_6H_{12}O_2$ are chiral?

6.33 Following are structural formulas for three of the most widely prescribed drugs to treat depression. Label all stereocenters in each, and state the number of stereoisomers possible for each.

(a)

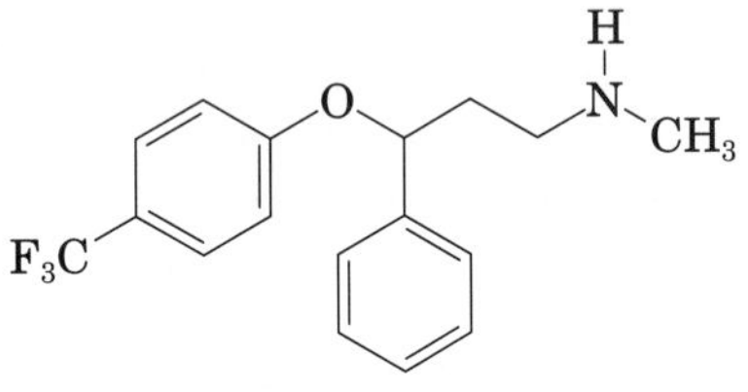

Fluoxetine
(Prozac)

(b)

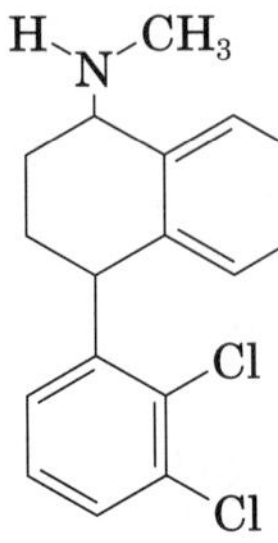

Sertraline
(Zoloft)

(c)

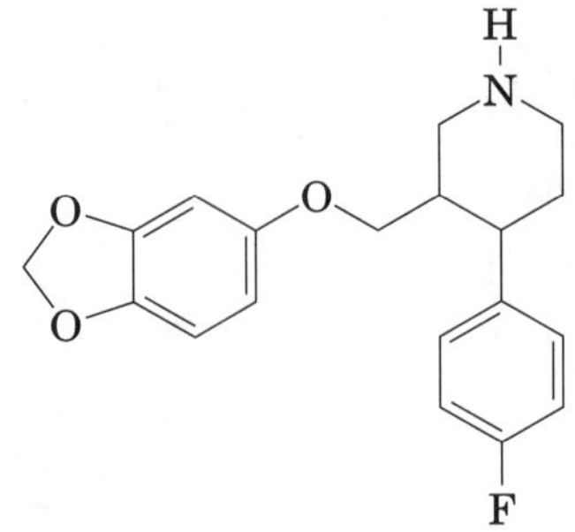

Paroxetine
(Paxil)

6.34 Label the four stereocenters in amoxicillin, which belongs to the family of semisynthetic penicillins.

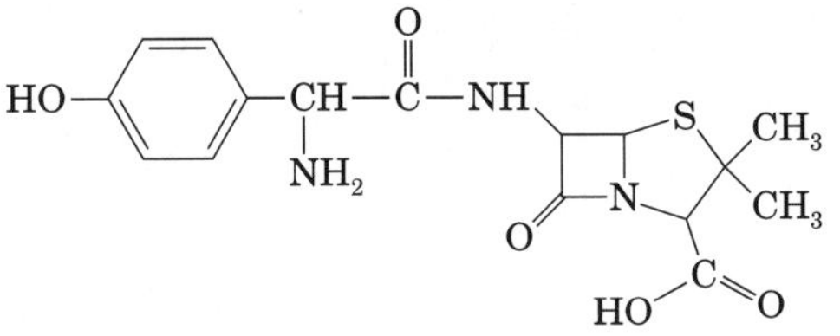

Amoxicillin

6.35 Following is a chair conformation of glucose:

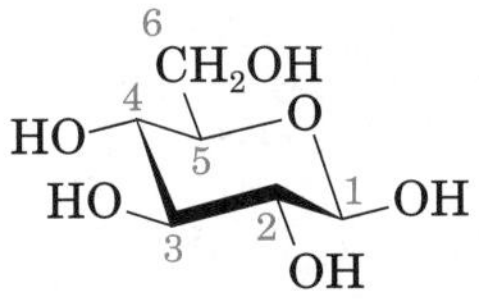

(a) Identify the five stereocenters in this molecule.
(b) How many stereoisomers are possible?
(c) How many pairs of enantiomers are possible?

6.36 Consider a cyclohexane ring substituted with one hydroxyl group and one methyl group. Draw a structural formula for a compound of this composition that
(a) Does not show cis-trans isomerism and has no stereocenters.
(b) Shows cis-trans isomerism but has no stereocenters.
(c) Shows cis-trans isomerism and has two stereocenters.

6.37 Triamcinolone acetonide, the active ingredient in Azmacort Inhalation Aerosol, is a steroid used to treat bronchial asthma.

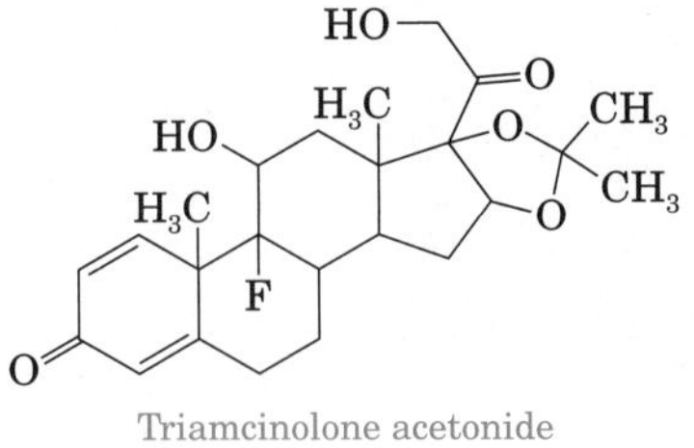

Triamcinolone acetonide

(a) Label the eight stereocenters in this molecule.

(b) How many stereoisomers are possible for it? (Of these, only one is the active ingredient in Azmacort.)

6.38 The next time you have the opportunity to view a collection of sea shells that have a helical twist, study the chirality (handedness) of their twists. For each kind of shell, do you find an equal number of left-handed and right-handed twists or, for example, do they all have the same handedness? What about the handedness among various kinds of shells with helical twists?

6.39 The next time you have an opportunity to examine any of the seemingly endless varieties of blond spiral pasta (rotini, fusilli, radiatori, tortiglione, and so forth), examine their twists. Do the twists of any one kind all have a right-handed twist or a left-handed twist, or are they a racemic mixture?

6.40 Think about the helical coil of a telephone cord or the spiral binding on a notebook. Suppose that you view the spiral from one end and find that it has a left-handed twist. If you view the same spiral from the other end, does it have a left-handed twist from that end as well or does it have a right-handed twist?

InfoTrac College Edition

For additional readings, go to InfoTrac College Edition, your online research library, at

http://infotrac.thomsonlearning.com

CHAPTER 7

7.1   Introduction

7.2   Acid and Base Strength

7.3   Brønsted-Lowry Acids
       and Bases

7.4   The Position of
       Equilibrium in Acid–Base
       Reactions

7.5   Acid Ionization Constants

7.6   Some Properties of Acids
       and Bases

7.7   Self-Ionization of Water

7.8   pH and pOH

7.9   The pH of Aqueous
       Salt Solutions

7.10   Titration

7.11   Buffers

7.12   The Henderson-
       Hasselbalch Equation

These U.S runners have just won the gold medal in the 4 x 400 m relay race at the 1996 Olympic games. The buildup of an organic acid called lactic acid has caused severe muscle pain. In a short time the pain will be gone.

RB-GM-J.O. Atlanta/Liason Agency/Getty Images

Acids and Bases

7.1 Introduction

We frequently encounter acids and bases in our daily lives. Oranges, lemons, and vinegar are examples of acidic foods, and sulfuric acid is in our automobile batteries. As for bases, we take antacid tablets for heartburn and use household ammonia as a cleaning agent. What do these substances have in common? Why are acids and bases usually discussed together?

In 1884, a young Swedish chemist named Svante Arrhenius (1859–1927) answered the first question by proposing what was then a new definition of acids and bases. According to the Arrhenius definition, an **acid** is a substance that produces H_3O^+ ions in aqueous solution, and a **base** is a substance that produces OH^- ions in aqueous solution.

This definition of acid is a slight modification of the original Arrhenius definition, which was that an acid produces hydrogen ions, H^+. Today we know that H^+ ions cannot exist in water. A H^+ ion is a bare proton, and a charge of $+1$ is too concentrated to exist on such a tiny particle. Therefore, a H^+ ion in water immediately combines with an H_2O molecule to give a **hydronium ion,** H_3O^+.

Hydronium ion The H_3O^+ ion

$$H^+(aq) + H_2O(\ell) \longrightarrow H_3O^+(aq)$$
Hydronium ion

Apart from this modification, the Arrhenius definitions of acid and base are still valid and useful today, as long as we are talking about aqueous solutions.

When an acid dissolves in water, it reacts with the water to produce H_3O^+. For example, hydrogen chloride, HCl, in its pure state is a poisonous, choking gas. When HCl dissolves in water, it reacts with a water molecule to give hydronium ion and chloride ion:

$$H_2O(\ell) + HCl(aq) \longrightarrow H_3O^+(aq) + Cl^-(aq)$$

Thus a bottle labeled aqueous "HCl" is actually not HCl at all, but rather an aqueous solution of H_3O^+ and Cl^- ions in water.

We can show the transfer of a proton from an acid to a base by using a symbol called a **curved arrow.** First we write the Lewis structure of each reactant and product. Then we use curved arrows to show the change in position of electron pairs during the reaction. The tail of the curved arrow is located at the electron pair. The head of the curved arrow shows the new position of the electron pair.

$$H-\overset{\cdot\cdot}{\underset{|}{O}}: + H-\overset{\cdot\cdot}{\underset{\cdot\cdot}{Cl}}: \longrightarrow H-\overset{\cdot\cdot}{\underset{|}{O}}{}^+-H + :\overset{\cdot\cdot}{\underset{\cdot\cdot}{Cl}}:{}^-$$
$$\;\;\;\;\;H \; H$$

In this equation, the curved arrow on the left shows that an unshared pair of electrons on oxygen forms a new covalent bond with hydrogen. The curved arrow on the right shows that the pair of electrons of the H—Cl bond is given entirely to chlorine to form a chloride ion. Thus, in the reaction of HCl with H_2O, a proton is transferred from HCl to H_2O and, in the process, an O—H bond forms and an H—Cl bond is broken.

With bases, the situation is slightly different. Many bases are metal hydroxides, such as KOH, NaOH, $Mg(OH)_2$, and $Ca(OH)_2$. These compounds are ionic solids and, when they dissolve in water, their ions merely separate, and each ion is solvated by water molecules. For example,

$$NaOH(s) \xrightarrow{H_2O} Na^+(aq) + OH^-(aq)$$

Other bases are not hydroxides. Instead, they produce OH^- ions in water by reacting with water molecules. The most important example of this kind of base is ammonia, NH_3, a poisonous, choking gas. When ammonia dissolves in water, it reacts with water to produce ammonium ions and hydroxide ions.

$$NH_3(aq) + H_2O(\ell) \rightleftharpoons NH_4^+(aq) + OH^-(aq)$$

As we will see in Section 7.2, ammonia is a weak base and the position of the equilibrium for its reaction with water lies considerably toward the left. In a 1.0 *M* solution of NH_3 in water, for example, only about 4 molecules of NH_3 out of every 1000 react with water to form NH_4^+ and OH^-. Thus, when ammonia is dissolved in water, it exists primarily as NH_3 molecules. Nevertheless, some OH^- ions are produced and, therefore, NH_3 is a base.

Although we know that acidic aqueous solutions do not contain H^+ ions, we frequently use the terms "H^+" and "proton" when we really mean "H_3O^+." The three terms are generally interchangeable.

 See the **Interactive General, Organic, and Biochemistry CD-ROM, version 2.0,** for further exploration on this topic.

The boiling point of liquid ammonia is −33°C.

We show the equation with equilibrium arrows because it is reversible.

Bottles of NH_3 in water are sometimes labeled "ammonium hydroxide" or "NH_4OH," but this gives a false impression of what is really in the bottle. Most of the NH_3 molecules have not reacted with the water, so the bottle contains mostly NH_3 and H_2O and only a little NH_4^+ and OH^-.

We indicate how the reaction of ammonia with water takes place by using curved arrows to show the transfer of a proton from a water molecule to an ammonia molecule. Here, the curved arrow on the left shows that the unshared pair of electrons on nitrogen forms a new covalent bond with a hydrogen of a water molecule. At the same time as the new N—H bond forms, an O—H bond of a water molecule breaks and the pair of electrons forming the H—O bond moves entirely to oxygen forming OH^-.

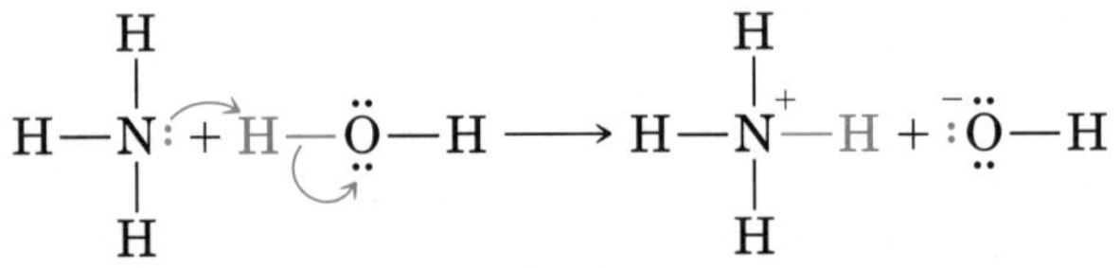

$$
\begin{array}{c}
\mathrm{H} \\
| \\
\mathrm{H-\overset{\cdot\cdot}{N}\: + H - \overset{\cdot\cdot}{\underset{\cdot\cdot}{O}} - H} \\
| \\
\mathrm{H}
\end{array}
\longrightarrow
\begin{array}{c}
\mathrm{H} \\
| \\
\mathrm{H - \overset{+}{N} - H} \\
| \\
\mathrm{H}
\end{array}
+ \;\; {}^{-}\!\overset{\cdot\cdot}{\underset{\cdot\cdot}{O}} - \mathrm{H}
$$

Thus ammonia produces an OH^- ion by taking H^+ from a water molecule and leaving OH^- behind.

7.2 Acid and Base Strength

Strong acid An acid that ionizes completely in aqueous solution

Weak acid An acid that is only partially ionized in aqueous solution

Strong base A base that ionizes completely in aqueous solution

Weak base A base that is only partially ionized in aqueous solution

All acids are not equally strong. According to the Arrhenius definition, a **strong acid** is one that reacts completely or almost completely with water to form H_3O^+ ions. Table 7.1 gives the names and molecular formulas for six of the most common strong acids. These six acids are strong acids because, when they dissolve in water, they dissociate completely to give H_3O^+ ions.

Weak acids produce a much smaller concentration of H_3O^+ ions. Acetic acid, for example, is a weak acid. In water it exists primarily as acetic acid molecules; only a few acetic acid molecules (4 out of every 1000) are converted to acetate ions.

$$CH_3COOH(aq) + H_2O(\ell) \rightleftharpoons CH_3COO^-(aq) + H_3O^+(aq)$$

Acetic acid   Acetate ion

There are only four common **strong bases** (Table 7.1), all of which are metal hydroxides. They are strong bases because, when they dissolve in water, they ionize completely to give OH^- ions. As we saw in Section 7.1, ammonia is a **weak base** because the equilibrium for its reaction with water lies far to the left.

Table 7.1	Six Strong Acids and Four Strong Bases		
Formula	**Name**	**Formula**	**Name**
HCl	Hydrochloric acid	LiOH	Lithium hydroxide
HBr	Hydrobromic acid	NaOH	Sodium hydroxide
HI	Hydroiodic acid	KOH	Potassium hydroxide
HNO_3	Nitric acid	$Ba(OH)_2$	Barium hydroxide
H_2SO_4	Sulfuric acid		
$HClO_4$	Perchloric acid		

CHEMICAL CONNECTIONS 7A

Some Important Acids and Bases

STRONG ACIDS Sulfuric acid, H_2SO_4, is used in many industrial processes. In fact, more tons of sulfuric acid are manufactured in the United States than any other chemical, organic or inorganic.

Hydrochloric acid, HCl, is an important acid in chemistry laboratories. Pure HCl is a gas, and the HCl in laboratories is an aqueous solution. HCl is the acid in the gastric fluid in your stomach, where it is secreted at a strength of about 5%.

Nitric acid, HNO_3, is a strong oxidizing agent. A drop of it causes the skin to turn yellow because the acid reacts with proteins on the skin. A yellow color upon contact with nitric acid has long been a test for proteins.

WEAK ACIDS Acetic acid, CH_3COOH, is present in vinegar (about 5%). Pure acetic acid is called glacial acetic acid because of its melting point of 17°C, which means that it freezes on a moderately cold day.

Boric acid, H_3BO_3, is a solid. Solutions of boric acid in water were once used as antiseptics, especially for eyes. Boric acid is toxic when swallowed.

Phosphoric acid, H_3PO_4, is one of the strongest of the weak acids. The ions produced from it—$H_2PO_4^-$, HPO_4^{2-}, and PO_4^{3-}—are important in biochemistry (see also Section 26.3).

STRONG BASES Sodium hydroxide, NaOH, also called lye, is the most important of the strong bases. It is a solid whose aqueous solutions are used in many industrial processes, including the

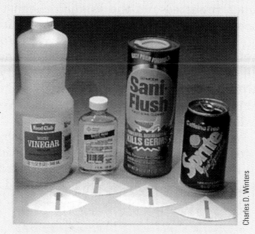

■ **Weak acids are found in many common materials. In the foreground are strips of litmus paper that have been dipped into solutions of these materials. Acids turn litmus paper red.**

■ **Weak bases are also common in many household products. These cleaning agents all contain weak bases. Bases turn litmus paper blue.**

manufacture of glass and soap. Potassium hydroxide, KOH, also a solid, is used for many of the same purposes as NaOH.

WEAK BASES Ammonia, NH_3, the most important weak base, is a gas with many industrial uses. One of its chief uses is for fertilizers. A 5% solution is sold in supermarkets as a cleaning agent, and weaker solutions are used as "spirits of ammonia" to revive people who have fainted.

Magnesium hydroxide, $Mg(OH)_2$, is a solid that is insoluble in water. A suspension of about 8% $Mg(OH)_2$ in water is called milk of magnesia and is used as a laxative. $Mg(OH)_2$ is also used to treat waste water in metal-processing plants and as a flame retardant in plastics.

It is important to understand that the strength of an acid or base is not related to its concentration. HCl is a strong acid, whether it is concentrated or dilute, because it dissociates completely in water to chloride ions and hydronium ions. Acetic acid is a weak acid, whether it is concentrated or dilute, because the equilibrium for its reaction with water lies far to the left. When acetic acid dissolves in water, most of it is present as undissociated CH_3COOH molecules.

$$HCl(aq) + H_2O(\ell) \longrightarrow Cl^-(aq) + H_3O^+(aq)$$

$$\underset{\text{Acetic acid}}{CH_3COOH(aq)} + H_2O(\ell) \rightleftharpoons \underset{\text{Acetate ion}}{CH_3COO^-(aq)} + H_3O^+(aq)$$

We saw that electrolytes (substances that produce ions in aqueous solution) can be strong or weak. The strong acids and bases in Table 7.1 are strong electrolytes. Almost all other acids and bases are weak electrolytes.

7.3 Brønsted-Lowry Acids and Bases

The Arrhenius definitions of acid and base are very useful in aqueous solutions. But what if water is not involved? In 1923 the Danish chemist Johannes Brønsted and the English chemist Thomas Lowry independently proposed the following definitions: An **acid** is a proton donor, a **base** is a proton acceptor, and an **acid–base reaction** is a proton transfer reaction. Furthermore, according to the Brønsted-Lowry definitions, any pair of molecules or ions that can be interconverted by transfer of a proton is called a **conjugate acid–base pair.** When an acid transfers a proton to a base, the acid is converted to its **conjugate base.** When a base accepts a proton, it is converted to its **conjugate acid.**

We can illustrate these relationships by examining the reaction of hydrogen chloride with water to form chloride ion and hydronium ion. The acid HCl donates a proton and is converted to its conjugate base Cl^-. The base H_2O accepts a proton and is converted to its conjugate acid H_3O^+.

> **Conjugated acid–base pair** A pair of molecules or ions that are related to one another by the gain or loss of a proton

> **Conjugate base** In the Brønsted-Lowry theory, a substance formed when an acid donates a proton to another molecule or ion

> **Conjugate acid** In the Brønsted-Lowry theory, a substance formed when a base accepts a proton

$$\underset{\substack{\text{Hydrogen} \\ \text{chloride} \\ \text{(Acid)}}}{HCl(aq)} + \underset{\substack{\text{Water} \\ \text{(Base)}}}{H_2O(\ell)} \longrightarrow \underset{\substack{\text{Chloride} \\ \text{ion} \\ \text{(Conjugate} \\ \text{base of HCl)}}}{Cl^-(aq)} + \underset{\substack{\text{Hydronium} \\ \text{ion} \\ \text{(Conjugate} \\ \text{acid of water)}}}{H_3O^+(aq)}$$

We have illustrated the application of the Brønsted-Lowry definitions using water as a reactant. The Brønsted-Lowry definitions, however, are more general than the Arrhenius definitions and do not require water as a reactant. Consider the following reaction between acetic acid and ammonia:

$$\underset{\substack{\text{Acetic acid} \\ \text{(Acid)}}}{CH_3COOH} + \underset{\substack{\text{Ammonia} \\ \text{(Base)}}}{NH_3} \rightleftharpoons \underset{\substack{\text{Acetate} \\ \text{ion} \\ \text{(Conjugate} \\ \text{base of} \\ \text{acetic acid)}}}{CH_3COO^-} + \underset{\substack{\text{Ammonium} \\ \text{ion} \\ \text{(Conjugate} \\ \text{acid of} \\ \text{ammonia)}}}{NH_4^+}$$

We can use curved arrows to show how this reaction takes place. The curved arrow on the right shows that the unshared pair of electrons on nitrogen becomes shared to form a new H—N bond. At the same time that the H—N bond forms, the O—H bond breaks and the electron pair of the O—H bond moves entirely to oxygen to form —O^- of the acetate ion. The result of these two electron-pair shifts is the transfer of a proton from an acetic acid molecule to an ammonia molecule.

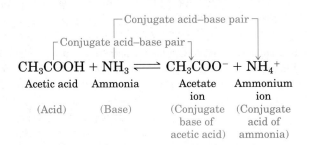

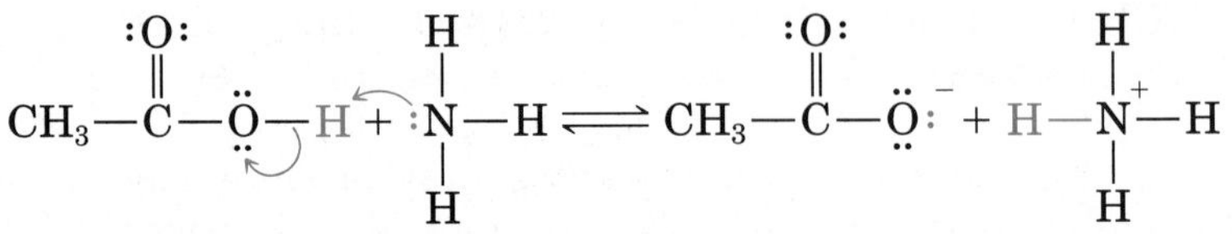

Acetic acid (Proton donor)   Ammonia (Proton acceptor)   Acetate ion   Ammonium ion

Table 7.2   Some Acids and Their Conjugate Bases

	Acid	Name	Conjugate Base	Name	
Strong Acids	HI	Hydroiodic acid	I^-	Iodide ion	Weak Bases
	HCl	Hydrochloric acid	Cl^-	Chloride ion	
	H_2SO_4	Sulfuric acid	HSO_4^-	Hydrogen sulfate ion	
	HNO_3	Nitric acid	NO_3^-	Nitrate ion	
	H_3O^+	Hydronium ion	H_2O	Water	
	HSO_4^-	Hydrogen sulfate ion	SO_4^{2-}	Sulfate ion	
	H_3PO_4	Phosphoric acid	$H_2PO_4^-$	Dihydrogen phosphate ion	
	CH_3COOH	Acetic acid	CH_3COO^-	Acetate ion	
	H_2CO_3	Carbonic acid	HCO_3^-	Bicarbonate ion	
	H_2S	Hydrogen sulfide	HS^-	Hydrogen sulfide ion	
	$H_2PO_4^-$	Dihydrogen phosphate ion	HPO_4^{2-}	Hydrogen phosphate ion	
	NH_4^+	Ammonium ion	NH_3	Ammonia	
	HCN	Hydrocyanic acid	CN^-	Cyanide ion	
	C_6H_5OH	Phenol	$C_6H_5O^-$	Phenoxide ion	
	HCO_3^-	Bicarbonate ion	CO_3^{2-}	Carbonate ion	
	HPO_4^{2-}	Hydrogen phosphate ion	PO_4^{3-}	Phosphate ion	
Weak Acids	H_2O	Water	OH^-	Hydroxide ion	Strong Bases
	C_2H_5OH	Ethanol	$C_2H_5O^-$	Ethoxide ion	

Table 7.2 gives examples of common acids and their conjugate bases. As you study the examples of conjugate acid–base pairs in Table 7.2, note the following points:

1. An acid can be positively charged, neutral, or negatively charged. Examples of these charge types are H_3O^+, H_2CO_3, and $H_2PO_4^-$, respectively.

2. A base can be negatively charged or neutral. Examples of these charge types are Cl^- and NH_3, respectively.

3. Acids are classified as monoprotic, diprotic, or triprotic depending on the number of protons each may give up. Examples of **monoprotic acids** include HCl, HNO_3, and CH_3COOH. Examples of **diprotic acids** include H_2SO_4 and H_2CO_3. An example of a **triprotic acid** is H_3PO_4. Carbonic acid, for example, loses one proton to become bicarbonate ion, and then a second proton to become carbonate ion.

$$H_2CO_3 + H_2O \rightleftharpoons HCO_3^- + H_3O^+$$
Carbonic   Bicarbonate
acid   ion

$$HCO_3^- + H_2O \rightleftharpoons CO_3^{2-} + H_3O^+$$
Bicarbonate   Carbonate
ion   ion

Monoprotic acid An acid that can give up only one proton

Diprotic acid An acid that can give up two protons

Triprotic acid An acid that can give up three protons

Amphiprotic A substance that can act as either an acid or a base

4. Several molecules and ions appear in both the acid and conjugate base columns; that is, each can function as either an acid or a base. The bicarbonate ion, HCO_3^-, for example, can give up a proton to become CO_3^{2-} (in which case it is an acid) or it can accept a proton to become H_2CO_3 (in which case it is a base). A substance that can act as either an acid or a base is called **amphiprotic.** The most important amphiprotic substance in Table 7.2 is water, which can accept a proton to become H_3O^+ or lose a proton to become OH^-.

5. A substance cannot be a Brønsted-Lowry acid unless it contains a hydrogen atom, but not all hydrogen atoms can be given up. For example,

■ A box of Arm & Hammer baking soda (sodium bicarbonate). Sodium bicarbonate is composed of Na^+ and HCO_3^-, the amphiprotic bicarbonate ion.

acetic acid, CH_3COOH, has four hydrogens but is monoprotic; it gives up only one of them. Similarly, phenol, C_6H_5OH, gives up only one of its six hydrogens:

$$C_6H_5OH + H_2O \rightleftharpoons C_6H_5O^- + H_3O^+$$
$$\text{Phenol} \qquad\qquad \text{Phenoxide} \atop \text{ion}$$

6. There is an inverse relationship between the strength of an acid and the strength of its conjugate base: The stronger the acid, the weaker its conjugate base. HI, for example, is the strongest acid listed in Table 7.2 and I^-, its conjugate base, is the weakest base. As another example, CH_3COOH (acetic acid) is a stronger acid than H_2CO_3 (carbonic acid); conversely, CH_3COO^- (acetate ion) is a weaker base than HCO_3^- (bicarbonate ion).

7.4 The Position of Equilibrium in Acid–Base Reactions

We know that HCl reacts with H_2O according to the following equilibrium:

$$HCl + H_2O \longrightarrow Cl^- + H_3O^+$$

We also know that HCl is a strong acid, which means that the position of this equilibrium lies very far to the right. In fact, this equilibrium lies so far to the right that out of every 10,000 HCl molecules dissolved in water, all but one react with water molecules to give Cl^- and H_3O^+.

As we have also seen, acetic acid reacts with H_2O according to the following equilibrium:

$$CH_3COOH + H_2O \rightleftharpoons CH_3COO^- + H_3O^+$$
$$\text{Acetic acid} \qquad\qquad \text{Acetate ion}$$

Acetic acid is a weak acid. Only a few acetic acid molecules react with water to give acetate ions and hydronium ions, and the major species present in equilibrium in aqueous solution is CH_3COOH. The position of this equilibrium, therefore, lies very far to the left.

In these two acid–base reactions, water is the base. But what if we have a base other than water as the proton acceptor? How can we determine which are the major species present at equilibrium? That is, how can we determine if the position of equilibrium lies toward the left or toward the right?

As an example, let us examine the acid–base reaction between acetic acid and ammonia to form acetate ion and ammonium ion. As indicated by the question mark over the equilibrium arrow, we want to determine if the position of this equilibrium lies toward the left or toward the right.

$$CH_3COOH + NH_3 \overset{?}{\rightleftharpoons} CH_3COO^- + NH_4^+$$
$$\underset{\text{(Acid)}}{\text{Acetic acid}} \quad \underset{\text{(Base)}}{\text{Ammonia}} \quad \underset{\substack{\text{(Conjugate base} \\ \text{of } CH_3COOH)}}{\text{Acetate ion}} \quad \underset{\substack{\text{(Conjugate acid} \\ \text{of } NH_3)}}{\text{Ammonium ion}}$$

In this equilibrium, there are two acids present: acetic acid and ammonium ion. There are also two bases present: ammonia and acetate ion. One way to analyze this equilibrium is to view it as a competition of the two bases, ammonia and acetate ion, for a proton. Which is the stronger base? The information we need to answer this question is found in Table 7.2. We first

CHEMICAL CONNECTIONS 7B

Acid and Base Burns of the Cornea

The cornea is the outermost part of the eye. This transparent tissue is very sensitive to chemical burns, whether caused by acids or by bases. Acids and bases in small amounts and in low concentrations that would not seriously damage other tissues, such as skin, may cause severe corneal burns. If not treated promptly, these burns can lead to permanent loss of vision.

For acid splashed in the eyes, immediately wash the eyes with a steady stream of cold water. Quickly removing the acid is of utmost importance. Time should not be wasted looking for a solution of some mild base to neutralize it because removing it is more helpful than neutralizing it. The extent of the burn can be assessed after all the acid has been washed out.

Concentrated ammonia or sodium hydroxide of the "liquid plumber" type can also cause severe damage when splattered in the eyes. Again, immediately washing the eyes is of utmost importance; however, burns by bases can later develop ulcerations in which the healing wound deposits scar tissue that is not transparent, thereby impairing vision. For this reason, first aid is not sufficient for burns by bases, and a physician should be consulted immediately.

Charles D. Winters

■ **Cleaning products like these contain bases. Care must be taken to make sure that they are not splashed into anyone's eyes.**

Fortunately, the cornea is a tissue with no blood vessels and, therefore, can be transplanted without immunological rejection problems. Today, corneal grafts are common.

determine which conjugate acid is the stronger acid and then use this information along with the fact that the stronger the acid, the weaker its conjugate base. From Table 7.2, we see that CH_3COOH is the stronger acid, which means that CH_3COO^- is the weaker base. Conversely, NH_4^+ is the weaker acid, which means than NH_3 is the stronger base. We can now label the relative strengths of each acid and base in this equilibrium:

$$CH_3COOH \; + \; NH_3 \; \underset{}{\overset{?}{\rightleftharpoons}} \; CH_3COO^- \; + \; NH_4^+$$

| Acetic acid | Ammonia | Acetate ion | Ammonium ion |
| (Stronger acid) | (Stronger base) | (Weaker base) | (Weaker acid) |

In an acid–base reaction, the position of equilibrium always favors reaction of the stronger acid and stronger base to form the weaker acid and weaker base. Thus, at equilibrium, the major species present are the weaker acid and weaker base. In the reaction between acetic acid and ammonia, therefore, the equilibrium lies to the right and the major species present are acetate ion and ammonium ion:

$$CH_3COOH \; + \; NH_3 \; \rightleftharpoons \; CH_3COO^- \; + \; NH_4^+$$

| Acetic acid | Ammonia | Acetate ion | Ammonium ion |
| (Stronger acid) | (Stronger base) | (Weaker base) | (Weaker acid) |

To summarize, the four steps we use in determining the position of an acid–base equilibrium are as follows:

1. Identify the two acids in the equilibrium; one is on the left side of the equilibrium, and the other on the right side.

2. Using the information in Table 7.2, determine which acid is the stronger acid and which is the weaker acid.

3. Identify the stronger base and weaker base in each equilibrium. Remember that the stronger acid gives the weaker conjugate base and the weaker acid gives the stronger conjugate base.

4. The stronger acid and stronger base react to give the weaker acid and weaker base. The position of equilibrium, therefore, lies on the side of the weaker acid and weaker base.

EXAMPLE 7.1

For each acid–base equilibrium, label the stronger acid, the stronger base, the weaker acid, and the weaker base. Then predict whether the position of equilibrium lies toward the right or toward the left.
(a) $H_2CO_3 + OH^- \rightleftharpoons HCO_3^- + H_2O$
(b) $HPO_4^{2-} + NH_3 \rightleftharpoons PO_4^{3-} + NH_4^+$

Solution
Arrows over each equilibrium show the conjugate acid–base pairs. The position of equilibrium in (a) lies toward the right. In (b) it lies toward the left.

(a) $H_2CO_3 + OH^- \rightleftharpoons HCO_3^- + H_2O$
  Stronger   Stronger   Weaker   Weaker
    acid   base   base   acid

(b) $HPO_4^{2-} + NH_3 \rightleftharpoons PO_4^{3-} + NH_4^+$
   Weaker   Weaker   Stronger   Stronger
    acid   base   base   acid

Problem 7.1
For each acid–base equilibrium, label the stronger acid, the stronger base, the weaker acid, and the weaker base. Then predict whether the position of equilibrium lies toward the right or toward the left.
(a) $H_3O^+ + I \rightleftharpoons H_2O + HI$
(b) $CH_3COO^- + H_2S \rightleftharpoons CH_3COOH + HS^-$

7.5 Acid Ionization Constants

In Section 7.2, we learned that acids vary in the extent to which they produce H_3O^+ when added to water. Because the ionizations of weak acids in water are all equilibria, we can use equilibrium constants to tell us quantitatively just how strong any weak acid is. The reaction that takes place when a weak acid, HA, is added to water is

HA is the general formula for an acid.

$$HA + H_2O \rightleftharpoons A^- + H_3O^+$$

The equilibrium expression for this ionization is

$$K_{eq} = \frac{[A^-][H_3O^+]}{[HA][H_2O]}$$

Table 7.3 K_a and pK_a Values for Some Weak Acids

Formula	Name	K_a	pK_a
H_3PO_4	Phosphoric acid	7.5×10^{-3}	2.12
$HCOOH$	Formic acid	1.8×10^{-4}	3.75
$CH_3CH(OH)COOH$	Lactic acid	8.4×10^{-4}	3.08
CH_3COOH	Acetic acid	1.8×10^{-5}	4.75
H_2CO_3	Carbonic acid	4.3×10^{-7}	6.37
$H_2PO_4^-$	Dihydrogen phosphate ion	6.2×10^{-8}	7.21
H_3BO_3	Boric acid	7.3×10^{-10}	9.14
NH_4^+	Ammonium ion	5.6×10^{-10}	9.25
HCN	Hydrocyanic acid	4.9×10^{-10}	9.31
C_6H_5OH	Phenol	1.3×10^{-10}	9.89
HCO_3^-	Bicarbonate ion	5.6×10^{-11}	10.25
HPO_4^{2-}	Hydrogen phosphate ion	2.2×10^{-13}	12.66

Increasing acid strength →

Notice that this expression contains the concentration of water. Because water is the solvent and its concentration changes very little when we add HA to it, we can treat the concentration of water, $[H_2O]$, as a constant equal to 1000 g/L or approximately 55.5 mol/L. We can then combine these two constants (K_{eq} and $[H_2O]$) to define a new constant called an **acid ionization constant, K_a.**

$$K_a = K_{eq}[H_2O] = \frac{[A^-][H_3O^+]}{[HA]}$$

> **Acid ionization constant (K_a)**
> An equilibrium constant for the ionization of an acid in aqueous solution to H_3O^+ and its conjugate base; also called an acid dissociation constant

The value of the acid ionization constant for acetic acid, for example, is 1.8×10^{-5}. Because acid ionization constants for weak acids are numbers with negative exponents, acid strengths are often expressed as pK_a, where p$K_a = -\log K_a$. The pK_a of acetic acid is 4.75. Table 7.3 gives names, molecular formulas, and values of K_a and pK_a for some weak acids. As you study the entries in this table, note the inverse relationship between the values of K_a and pK_a. The weaker the acid, the smaller its K_a, but the larger its pK_a.

One reason for the importance of K_a is that it immediately tells us how strong an acid is. For example, Table 7.3 shows us that although acetic acid, formic acid, and phenol are all weak acids, their strengths as acids are not the same. Formic acid, with a K_a of 1.8×10^{-4}, is stronger than acetic acid whereas phenol, with a K_a of 1.3×10^{-10}, is much weaker than acetic acid.

EXAMPLE 7.2

The K_a for benzoic acid is 6.5×10^{-5}. What is the pK_a of this acid?

Solution

Take the logarithm of 6.5×10^{-5} on your scientific calculator. The answer is -4.19. Because pK_a is equal to $-\log K_a$, you must multiply this value by -1 to get pK_a. The pK_a of benzoic acid is 4.19.

Problem 7.2

The K_a for hydrocyanic acid, HCN, is 4.9×10^{-10}. What is its pK_a?

■ **All these fruits and fruit drinks contain organic acids.**

EXAMPLE 7.3

Which is the stronger acid:
(a) Benzoic acid with a K_a of 6.5×10^{-5}, or hydrocyanic acid with a K_a of 4.9×10^{-10}?
(b) Boric acid with a pK_a of 9.14 or carbonic acid with a pK_a of 6.37?

Solution
(a) Benzoic acid is the stronger acid; it has the greater K_a value.
(b) Carbonic acid is the stronger acid; it has the smaller pK_a value.

Problem 7.3

Which is the stronger acid:
(a) Carbonic acid, pK_a = 6.37, or ascorbic acid (vitamin C), pK_a = 4.1?
(b) Aspirin, pK_a = 3.49, or acetic acid, pK_a = 4.75?

7.6 Some Properties of Acids and Bases

Chemists of today do not taste the substances they work with, but 150 and 200 years ago they routinely did so. That is how we know that acids taste sour and bases taste bitter. The sour taste of lemons, vinegar, and many other foods, for example, is due to the acids they contain.

A Neutralization

The most important reaction of acids and bases is that they react with each other in a process called neutralization. This name is appropriate because, when a strong corrosive acid such as hydrochloric acid reacts with a strong corrosive base such as sodium hydroxide, the product (a solution of ordinary table salt in water) has neither acidic nor basic properties. We call such a solution *neutral*. Section 7.10 discusses neutralization reactions in detail.

Figure 7.1 A ribbon of magnesium metal reacts with aqueous HCl to give H_2 gas and aqueous $MgCl_2$.

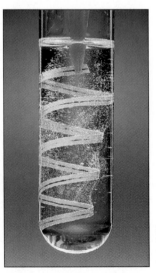

B Reactions with Metals

Strong acids react with certain metals (called active metals) to produce hydrogen gas, H_2, and a salt. Hydrochloric acid, for example, reacts with magnesium metal to give the salt magnesium chloride and hydrogen gas (Figure 7.1):

$$Mg(s) \; + \; 2HCl(aq) \; \longrightarrow \; MgCl_2(aq) \; + \; H_2(g)$$
Magnesium   Hydrochloric   Magnesium   Hydrogen
acid   chloride

The reaction of an acid with an active metal to give a salt and hydrogen gas is a redox reaction. The metal is oxidized to a metal ion and H^+ is reduced to H_2.

C Reaction with Metal Hydroxides

Acids react with metal hydroxides to give a salt and water.

$$HCl(aq) \; + \; KOH(aq) \; \longrightarrow \; H_2O(\ell) \; + \; KCl(aq)$$
Hydrochloric   Potassium   Water   Potassium
acid   hydroxide   chloride

C H E M I C A L CONNECTIONS **7 C**

Drugstore Antacids

Stomach fluid is normally quite acidic because of its HCl content. At some time, everyone has undoubtedly gotten "heartburn" caused by excess stomach acidity. At that time, you may have taken an antacid, which, as the name implies, is a substance that neutralizes acids—in other words, it is a base.

The word "antacid" is a medical term, not one used by chemists. It is, however, found on the labels of many medications available in drugstores and supermarkets. Almost all of them use bases such as $CaCO_3$, $Mg(OH)_2$, $Al(OH)_3$, and $NaHCO_3$ to decrease the acidity of the stomach. Simethicone is often added as an antiflatulant.

Also in drugstores and supermarkets are nonprescription drugs labeled "acid reducers." Among these are the brands Zantac, Tagamet, Pepcid, and Axid. Instead of neutralizing acidity, these compounds reduce the secretion of acid into the stomach. In larger doses (sold only with a prescription), some of these drugs are used in the treatment of stomach ulcers.

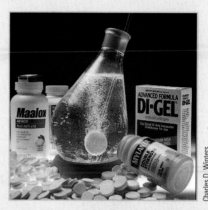

■ **Commercial remedies for excess stomach acid.**

Charles D. Winters

Both the acid and metal hydroxide are ionized in aqueous solution. Furthermore, the salt formed is an ionic compound that is present in aqueous solution as anions and cations. Therefore, the actual equation for the reaction of HCl and KOH is more accurately written showing all of the ions present:

$$H_3O^+ + Cl^- + K^+ + OH^- \longrightarrow 2H_2O + Cl^- + K^+$$

We can simplify this equation by omitting the spectator ions, which gives the following equation for the net ionic reaction of any strong acid and strong base to give a salt and water:

$$H_3O^+ + OH^- \longrightarrow 2H_2O$$

D Reaction with Metal Oxides

Strong acids react with metal oxides to give water and a salt as shown in the following net ionic equation:

$$2H_3O^+(aq) + CaO(s) \longrightarrow 3H_2O(\ell) + Ca^{2+}(aq)$$
$$\text{Calcium}$$
$$\text{oxide}$$

E Reactions with Carbonates and Bicarbonates

When a strong acid is added to a carbonate such as sodium carbonate, bubbles of carbon dioxide gas are rapidly given off. The overall reaction is a summation of two reactions. In the first reaction, carbonate ion reacts with H_3O^+ to give carbonic acid. Almost immediately, in the second reaction, carbonic acid decomposes to carbon dioxide and water. The following equations show the individual reactions and then the overall reaction:

$$2H_3O^+(aq) + CO_3^{2-}(aq) \longrightarrow H_2CO_3(aq) + 2H_2O(\ell)$$
$$H_2CO_3(aq) \longrightarrow CO_2(g) + H_2O(\ell)$$
$$\overline{2H_3O^+(aq) + CO_3^{2-}(aq) \longrightarrow CO_2(g) + 3H_2O(\ell)}$$

Strong acids also react with bicarbonates such as potassium bicarbonate to give carbon dioxide and water:

$$H_3O^+(aq) + HCO_3^-(aq) \longrightarrow H_2CO_3(aq) + H_2O(\ell)$$
$$H_2CO_3(aq) \longrightarrow CO_2(g) + H_2O(\ell)$$
$$\overline{H_3O^+(aq) + HCO_3^-(aq) \longrightarrow CO_2(g) + 2H_2O(\ell)}$$

To generalize, any acid stronger than carbonic acid will react with a carbonate or bicarbonate to give CO_2 gas.

The production of CO_2 is what makes bread and biscuit doughs and cake batters rise. The earliest method used to generate CO_2 for this purpose involved the addition of yeast, which catalyzes the fermentation of carbohydrates to produce carbon dioxide and ethanol (Chapter 19):

$$C_6H_{12}O_6 \xrightarrow{\text{Yeast}} 2CO_2 + 2C_2H_5OH$$
Glucose Ethanol

■ **Baking powder contains the weak acid dihydrogen phosphate ion and the weak base bicarbonate ion. When mixed together in water, the acid and base react to produce carbon dioxide gas. Bubbles of the gas are seen in the picture.**

The production of CO_2 by fermentation, however, is slow. Sometimes it is desirable to have its production take place more rapidly, in which case bakers use the reaction of $NaHCO_3$ (sodium bicarbonate, also called **baking soda**) and a weak acid. But which weak acid? Vinegar (a 5% solution of acetic acid in water) would work, but it has a potential disadvantage—it imparts a particular flavor to foods. For a weak acid that imparts little or no flavor, bakers use either sodium dihydrogen phosphate, NaH_2PO_4, or potassium dihydrogen phosphate, KH_2PO_4. The two salts do not react with $NaHCO_3$ when they are dry but, when mixed with water in a dough or batter, they react quite rapidly to produce CO_2. The production of CO_2 is even more rapid in an oven!

$$H_2PO_4^-(aq) + H_2O(\ell) \rightleftharpoons HPO_4^{2-}(aq) + H_3O^+(aq)$$
$$HCO_3^-(aq) + H_3O^+(aq) \longrightarrow CO_2(g) + 2H_2O(\ell)$$
$$\overline{H_2PO_4^-(aq) + HCO_3^-(aq) \longrightarrow HPO_4^{2-}(aq) + CO_2(g) + H_2O(\ell)}$$

F Reactions with Ammonia and Amines

Any acid stronger than NH_4^+ (Table 7.2) is strong enough to react with NH_3 to form a salt. In the following reaction, the salt formed is ammonium chloride, NH_4Cl, which is shown as it would be ionized in aqueous solution:

$$HCl(aq) + NH_3(aq) \rightarrow NH_4^+(aq) + Cl^-(aq)$$

In Chapter 8 we will meet a family of compounds called amines, which are similar to ammonia except that one or more of the three hydrogen atoms of ammonia are replaced by carbon groups. A typical amine is methylamine, CH_3NH_2. The base strength of most amines is similar to that of NH_3, which means that amines also react with acids to form salts. The salt formed in the reaction of methylamine with HCl is methylammonium chloride, shown here as it would be ionized in aqueous solution:

$$HCl(aq) + CH_3NH_2(aq) \longrightarrow CH_3NH_3^+(aq) + Cl^-(aq)$$

Methylamine   Methylammonium ion

The reaction of ammonia and amines with acids to form salts is very important in the chemistry of the body, as we will see in later chapters.

7.7 Self-Ionization of Water

We have seen that an acid produces H_3O^+ ions in water and that a base produces OH^- ions. Suppose that we have absolutely pure water, with no added acid or base. Surprisingly enough, even pure water contains a very small number of H_3O^+ and OH^- ions. They are formed by the transfer of a proton from one molecule of water (the proton donor) to another (the proton acceptor).

$$H_2O + H_2O \rightleftharpoons OH^- + H_3O^+$$

Acid   Base   Conjugate   Conjugate
base of H_2O acid of H_2O

What is the extent of this reaction? We know from the information in Table 7.2 that, in this equilibrium, H_3O^+ is the stronger acid and OH^- is the stronger base. Therefore, as shown by the arrows, the equilibrium for this reaction lies far to the left. We shall soon see exactly how far, but first let us write the equilibrium expression:

$$K_{eq} = \frac{[H_3O^+][HO^-]}{[H_2O]^2}$$

Because the degree of self-ionization of water is so slight, we can treat the concentration of water, $[H_2O]$, as a constant equal to 1000 g/L or approximately 55.5 mol/L, just as we did in Section 7.5 in developing the K_a for a weak acid. We can then combine these two constants (K_{eq} and $[H_2O]^2$) to define a new constant called the **ion product of water, K_w.** In pure water at room temperature, K_w has a value of 1.0×10^{-14}.

$$K_w = K_{eq}[H_2O]^2 = [H_3O^+][OH^-]$$

$$K_w = 1.0 \times 10^{-14}$$

In pure water H_3O^+ and OH^- form in equal amounts (see the balanced equation for the self-ionization of water), so their concentrations must be equal. That is, in pure water,

$$[H_3O^+] = 1.0 \times 10^{-7} \text{ mol/L}$$
$$[OH^-] = 1.0 \times 10^{-7} \text{ mol/L}$$
In pure water

These are very small concentrations, not enough to make pure water a conductor of electricity. Pure water is not an electrolyte.

The equation for the ionization of water is important because it applies not only to pure water but also to any water solution. The product of $[H_3O^+]$ and $[OH^-]$ in any aqueous solution is equal to 1.0×10^{-14}. If, for example, we add 0.010 mol of HCl to 1 L of pure water, it reacts completely to give H_3O^+ ions and Cl^- ions. The concentration of H_3O^+ will be 0.010 *M*, or 1.0×10^{-2} *M*. This means that the $[OH^-]$ must be $1.0 \times 10^{-14}/1.0 \times 10^{-2} = 1.0 \times 10^{-12}$ *M*.

EXAMPLE 7.4

The $[OH^-]$ of an aqueous solution is 1.0×10^{-4} *M*. What is its $[H_3O^+]$?

Solution
We substitute into the equation:

$$[H_3O^+][OH^-] = 1.0 \times 10^{-14}$$

$$[H_3O^+] = \frac{1.0 \times 10^{-14}}{1.0 \times 10^{-4}} = 1.0 \times 10^{-10} \, M$$

Problem 7.4
The $[OH^-]$ of an aqueous solution is 1.0×10^{-12} *M*. What is its $[H_3O^+]$?

Aqueous solutions can have a very high $[H_3O^+]$ but the $[OH^-]$ must then be very low, and vice versa. Any solution with a $[H_3O^+]$ greater than 1.0×10^{-7} *M* is acidic. In such solutions, of necessity $[OH^-]$ must be less than 1.0×10^{-7} *M*. The higher the $[H_3O^+]$, the more acidic the solution. Similarly, any solution with an $[OH^-]$ greater than 1.0×10^{-7} *M* is basic. Pure water, in which $[H_3O^+]$ and $[OH^-]$ are equal (they are both 1.0×10^{-7} *M*), is neutral—that is, neither acidic nor basic.

7.8 pH and pOH

See the **Interactive General, Organic, and Biochemistry CD-ROM, version 2.0,** for further exploration on this topic.

Because hydronium ion concentrations for most solutions are numbers with negative exponents, these concentrations are more commonly expressed as pH, where

$$pH = -\log[H_3O^+]$$

In Section 7.7, we saw that a solution is acidic if its $[H_3O^+]$ is greater than 1.0×10^{-7}, and that it is basic if its $[H_3O^+]$ is less than 1.0×10^{-7}. We can now state the definitions of acidic and basic solutions in terms of pH:

> A solution is acidic if its pH is less than 7.0
> A solution is basic if its pH is greater than 7.0
> A solution is neutral if its pH is equal to 7.0

Although pure water has a pH of 7, tap water usually has a pH of approximately 6 because it contains dissolved CO_2 (from the atmosphere), which reacts with water to give dissolved carbonic acid, H_2CO_3.

EXAMPLE 7.5

(a) The $[H_3O^+]$ of a certain liquid detergent is 1.4×10^{-9} *M*. What is its pH?
(b) The pH of black coffee is 5.3. What is its $[H_3O^+]$?

Solution

(a) On your calculator, take the log of 1.4×10^{-9}. The answer is -8.85. Multiply this value by -1 to give the pH of 8.85.

(b) Enter 5.3 into your calculator and then press the $+/-$ key to change the sign to minus and give -5.3. Then take the antilog of this number. The $[H_3O^+]$ of black coffee is 5×10^{-6}.

Problem 7.5

(a) The $[H_3O^+]$ of an acidic solution is $3.5 \times 10^{-3}\,M$. What is its pH?

(b) The pH of tomato juice is 4.1. What is its $[H_3O^+]$?

Just as pH is a convenient way to designate the concentration of H_3O^+, pOH is a convenient way to designate the concentration of OH^-:

$$pOH = -\log\,[OH^-]$$

In aqueous solutions, the ion product of water, K_w, is 1×10^{-14} and thus $pK_w = -\log K_w = 14$. Therefore, we can write

$$14 = pH + pOH$$

Thus, once we know the pH of a solution, we can easily calculate the pOH.

EXAMPLE 7.6

The $[OH^-]$ of a strongly basic solution is 1.0×10^{-2}. What are the pOH and pH of this solution?

Solution

The pOH is 2, and the pH is $14 - 2 = 12$.

Problem 7.6

The $[OH^-]$ of a solution is 1.0×10^{-4}. What are the pOH and pH of this solution?

All fluids in the human body are aqueous; that is, the only solvent present is water. Consequently, all body fluids have a pH value. Some of them have a narrow pH range; others have a wide pH range. The pH of blood, for example, must be between 7.35 and 7.45 (slightly basic). If it goes outside these limits, illness and even death may result (see Chemical Connections 7D). In contrast, the pH of urine can vary from 5.5 to 7.5. Table 7.4 gives the pH values of some common materials.

One thing you must remember when you see a pH value is that, because pH is a logarithmic scale, an increase (or decrease) of one pH unit means a tenfold decrease (or increase) in the $[H_3O^+]$. For example, a pH of 3 does not sound very different from a pH of 4. The first, however, means a $[H_3O^+]$ of $10^{-3}\,M$, whereas the second means a $[H_3O^+]$ of $10^{-4}\,M$. The $[H_3O^+]$ of the pH 3 solution is ten times the $[H_3O^+]$ of the pH 4 solution.

There are two ways to measure the pH of an aqueous solution. One way is to use a pH paper, which is made by soaking plain paper with a mixture of pH indicators. A pH **indicator** is a substance that changes color at a certain pH. When we place a drop of solution on this paper, the paper turns a certain color. To determine the pH, we compare the color of the paper with the colors on a chart supplied with the paper.

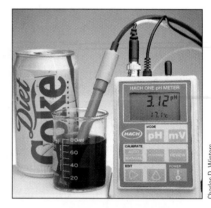

Charles D. Winters

■ The pH of this soft drink is 3.12. Soft drinks are often quite acidic.

Charles D. Winters

■ The pH of three household substances. The colors of the acid–base indicators in the flasks show that vinegar is more acidic than club soda, and the cleaner is basic.

Litmus paper turns red in acidic solution and blue in basic solution. It is often used as an approximate guide to acidity or basicity.

■ **Strips of paper impregnated with indicator are used to find an approximate pH.**

Charles D. Winters

Table 7.4   pH Values of Some Common Materials

Material	pH	Material	pH
Battery acid	0.5	Saliva	6.5–7.5
Gastric juice	1.0–3.0	Pure water	7.0
Lemon juice	2.2–2.4	Blood	7.35–7.45
Vinegar	2.4–3.4	Bile	6.8–7.0
Tomato juice	4.0–4.4	Pancreatic fluid	7.8–8.0
Carbonated beverages	4.0–5.0	Sea water	8.0–9.0
Black coffee	5.0–5.1	Soap	8.0–10.0
Urine	5.5–7.5	Milk of magnesia	10.5
Rain (unpolluted)	6.2	Household ammonia	11.7
Milk	6.3–6.6	Lye (1.0 M NaOH)	14.0

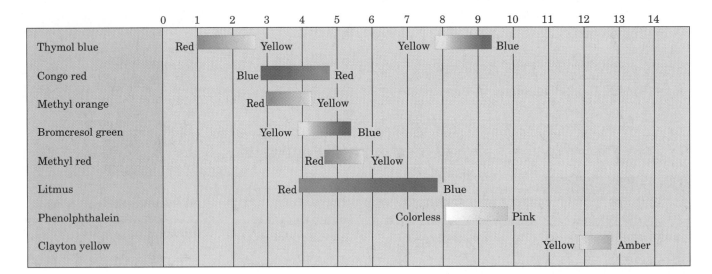

Figure 7.2   Some acid–base indicators. Note that some indicators have two color changes.

One acid–base indicator is the compound methyl orange. When a drop of methyl orange is added to an aqueous solution with a pH of 3.2 or lower, this indicator turns red and the entire solution becomes red. When added to an aqueous solution with a pH of 4.4 or higher, this indicator turns yellow. These particular limits and colors apply only to methyl orange. Other indicators have other limits and colors (Figure 7.2).

The second way for determining pH is more accurate. In this method, we use a pH meter (Figure 7.3). We dip the electrode of the pH meter into the solution whose pH is to be measured, and then read the pH on a dial. The most commonly used pH meters read pH to the nearest hundredth of a unit.

7.9 The pH of Aqueous Salt Solutions

When an acid is added to water, the solution becomes acidic (pH less than 7); when a base is added to water, the solution becomes basic (pH greater than 7). Suppose we add a salt to water? What happens to the pH then? At first thought, you might suppose that all salts are neutral and that adding any salt to pure water leaves the pH at 7. This is, in fact, true for some salts. Many salts, however, are acidic or basic; as a result, they cause the pH of pure water to change. To understand how this change in pH occurs, let us look at sodium acetate, a typical basic salt.

Figure 7.3 Copper(II) sulfate, $CuSO_4$, is an acidic salt. Because copper(II) sulfate is the salt of a strong acid and a weak base, aqueous solutions of it is acidic. The pH of the 0.1 M solution shown here is 4.33.

Like almost all salts, sodium acetate is 100% ionized in water so that, when we add CH_3COONa to water, we get a solution containing Na^+ ions and CH_3COO^- ions. The Na^+ ions do not react with the water, but the CH_3COO^- ions do according to the following equilibrium:

$$CH_3COO^- + H_2O \rightleftharpoons CH_3COOH + OH^-$$

Acetate ion	Water	Acetic acid	Hydroxide ion
(Weaker acid)	(Weaker base)	(Stronger acid)	(Stronger base)

Using the reasoning we developed in Section 7.4 to predict the position of equilibrium in an acid–base reaction, we see that this equilibrium lies to the left—that is, toward the weaker acid and the weaker base. It has been determined experimentally that at equilibrium in a 0.1 M solution of sodium acetate, only about 1 out of every 13,500 acetate ions is converted to acetic acid and OH^- ions. Nevertheless, enough OH^- ions are produced to increase the pH to about 8.88. Because this pH is higher than 7, the solution is basic, and sodium acetate is called a basic salt. Although acetic acid is also being produced, the pH of an aqueous solution depends only on the concentration of H_3O^+ and OH^- and not on the concentration of any other un-ionized acid or base.

We can predict whether a salt will be acidic, neutral, or basic in water solution if we know something about the strength of the acid and base. We have four cases:

1. A salt of a strong base and a weak acid raises the pH of pure water and, therefore, is called a **basic salt.** An example is sodium acetate, which we just discussed.

2. A salt of a strong acid and a weak base lowers the pH of pure water and, therefore, is called an **acidic salt.** An example is ammonium chloride, NH_4Cl. The Cl^- ion does not react with H_2O, but the NH_4^+ ion does to a slight extent to form NH_3 and H_3O^+. The presence of H_3O^+ lowers the pH.

$$NH_4^+ + H_2O \rightleftharpoons NH_3 + H_3O^+$$

Ammonium ion	Water	Ammonia	Hydronium ion
(Weaker acid)	(Weaker base)	(Stronger base)	(Stronger acid)

Copper(II) sulfate (Figure 7.3) is another example of an acidic salt.

3. A salt of a strong acid and a strong base does not change the pH of pure water and, therefore, is called a **neutral salt.** Examples of neutral salts are $NaCl$, KNO_3, and Na_2SO_4.

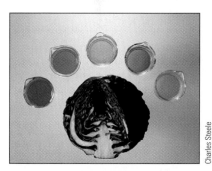

■ **Red cabbage juice is a naturally occurring acid–base indicator. From left to right are solutions of pH 1, 4, 7, 10, and 13.**

Basic salt The salt of a weak acid and a strong base; an aqueous solution of a basic salt is acidic

Acidic salt The salt of a strong acid and a weak base; an aqueous solution of an acidic salt is basic

Neutral salt A salt whose aqueous solution is neither acidic nor basic

■ Because CH_3COONa, sodium acetate, is the salt of a weak acid and a strong base, aqueous solutions of this salt are basic. The pH of the solution shown here is 8.88. An inert, insoluble solid has been added to the flask to enhance the visibility of the liquid in the flask.

Titration An analytical procedure whereby we react a known volume of a solution of known concentration with a known volume of a solution of unknown concentration

End point The point at which there is an equal amount of acid and base in a neutralization reaction

See the **Interactive General, Organic, and Biochemistry CD-ROM, version 2.0,** for further exploration on this topic.

4. A salt of a weak acid and a weak base may or may not change the pH of pure water. Both ions react with water. Sometimes the effects cancel each other, and sometimes they do not. We will not try to make predictions in such cases.

7.10 Titration

Laboratories, whether medical, academic, or industrial, are frequently asked to determine the exact concentration of a particular substance in solution—for example, the concentration of acetic acid in a given sample of vinegar, or the concentrations of iron, calcium, and magnesium ions in a sample of "hard" water. Determinations of solution concentrations can be made using an analytical technique called a **titration.**

In a titration, we react a known volume of a solution of known concentration with a known volume of a solution of unknown concentration. The solution of unknown concentration may contain an acid (such as stomach acid), a base (such as ammonia), an ion (such as Fe^{2+} ion), or any other substance whose concentration we are asked to determine. If we know the titration volumes and the mole ratio in which the solutes react, we can then calculate the concentration of the second solution.

Titrations must meet several requirements:

1. We must know the equation for the reaction so that we can determine the stoichiometric ratio of reactants to use in our calculations.

2. The reaction must be rapid and complete.

3. When the reactants have combined exactly, there must be a clear-cut change in some measurable property of the reaction mixture. We call the point at which the reactants combine exactly the **end point** of the titration.

4. We must have accurate measurements of the amount of each reactant.

Let us apply these requirements to the titration of a solution of sulfuric acid of known concentration with a solution of sodium hydroxide of unknown concentration. We know the balanced equation for this acid–base reaction, so requirement 1 is met.

$$2NaOH(aq) + H_2SO_4(aq) \longrightarrow Na_2SO_4(aq) + 2H_2O(\ell)$$
$$\underset{\text{(Concentration not known)}}{} \quad \underset{\text{(Concentration known)}}{}$$

Sodium hydroxide ionizes in water to form sodium ions and hydroxide ions; sulfuric acid ionizes to form hydronium ions and sulfate ions. The reaction between hydroxide and hydronium ions is rapid and complete, so requirement 2 is met.

To meet requirement 3, we must be able to observe a clear-cut change in some measurable property of the reaction mixture at the end point. For acid–base titrations, we use the sudden pH change that occurs at this point. Suppose we add the sodium hydroxide solution slowly. As it is added, it reacts with hydronium ions to form water. As long as any unreacted hydronium ions are present, the solution is acidic. When the number of hydroxide ions added exactly equals the original number of hydronium ions, the solution becomes neutral. Then, as soon as any extra hydroxide ions are added, the solution becomes basic. We can observe this sudden change in pH by reading a pH meter.

Another way to observe the change in pH at the end point is to use an acid–base indicator (Section 7.8). Such an indicator changes color when the

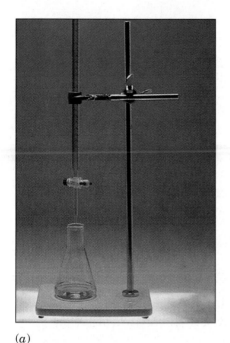
(a)

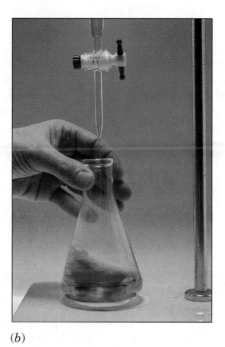
(b)

(c)

Charles D. Winters

Figure 7.4 An acid–base titration. (*a*) An acid of known concentration is in the Erlenmeyer flask. (*b*) When a base is added from the buret, the acid is neutralized. (*c*) The end point is reached when the color of the indicator changes from colorless to pink.

solution changes from acidic to basic. Phenolphthalein, for example, is colorless in acid solution and pink in basic solution. If this indicator is added to the original sulfuric acid solution, the solution remains colorless as long as excess hydronium ions are present [Figure 7.4(*a*)]. After enough sodium hydroxide solution has been added to react with all of the hydronium ions, the next drop of base provides excess hydroxide ions, and the solution turns pink (Figure 7.4(*c*)). Thus we have a clear-cut indication of the end point.

To meet requirement 4, which is that the volume of each solution used must be known, we use volumetric glassware such as volumetric flasks, burets, and pipets.

Data for a typical acid–base titration are given in Example 7.7. Note that the experiment is run in triplicate, a standard procedure for checking the precision of a titration.

EXAMPLE 7.7

Following are data for the titration of 0.108 M H$_2$SO$_4$ with a solution of NaOH of unknown concentration. What is the concentration of the NaOH solution?

	Volume of 0.108 M H$_2$SO$_4$	Volume of NaOH
Trial I	25.0 mL	33.48 mL
Trial II	25.0 mL	33.46 mL
Trial III	25.0 mL	33.50 mL

Solution

From the balanced equation for this acid–base reaction, we know the stoichiometry: Two moles of NaOH react with one mole of H_2SO_4. From the experimental data, we calculate that the average volume of the NaOH required for complete reaction is 33.48 mL. Because the units of molarity are moles/liter, we must convert volumes of reactants from milliliters to liters. We can then use the factor-label method (Section 1.5) to calculate the molarity of the NaOH solution. What we wish to calculate is the number of moles of NaOH per liter of NaOH.

$$\frac{\text{mol NaOH}}{\text{L NaOH}} = \frac{0.108 \text{ mol } H_2SO_4}{1 \text{ L } H_2SO_4} \times \frac{0.0250 \text{ L } H_2SO_4}{0.03348 \text{ L NaOH}} \times \frac{2 \text{ mol NaOH}}{1 \text{ mol } H_2SO_4}$$

$$= \frac{0.161 \text{ mol NaOH}}{\text{L NaOH}} = 0.161 \text{ } M$$

Problem 7.7

Calculate the concentration of an acetic acid solution using the following data. Three 25.0-mL samples of acetic acid were titrated to a phenolphthalein end point with 0.121 M NaOH. The volumes of NaOH were 19.96 mL, 19.73 mL, and 19.79 mL.

It is important to understand that titration is not a method of determining the acidity (or basicity) of a solution. If we want to do that, we measure the sample's pH, which is the only measurement of solution acidity or basicity. Rather, titration is a method for determining the total acid or base concentration of a solution, which is not the same as the acidity. For example, a 0.1 M solution of HCl in water has a pH of 1, but a 0.1 M solution of acetic acid has a pH of 2.9. These two solutions have the same concentration of acid and each neutralizes the same volume of NaOH solution, but they have very different acidities.

7.11 Buffers

A What Is a Buffer and How Does One Work?

As noted earlier, the body must keep the pH of blood between 7.35 and 7.45. Yet we frequently eat acidic foods such as oranges, lemons, sauerkraut, and tomatoes, and doing so eventually adds considerable quantities of H_3O^+ to the blood. Despite these additions of acidic or basic substances, the body manages to keep the pH of blood remarkably constant. The body does it with buffers. A **buffer** is a solution whose pH changes very little when H_3O^+ or OH^- ions are added to it. In a sense, a pH buffer is an acid or base "shock absorber."

The most common buffers consist of approximately equal molar amounts of a weak acid and a salt of the weak acid. Put another way, they consist of approximately equal amounts of a weak acid and its conjugate base. For example, if we dissolve 1.0 mol of acetic acid (a weak acid) and 1.0 mol of its conjugate base (in the form of CH_3COONa, sodium acetate) in 1.0 L of water, we have a good buffer solution. The equilibrium present in this buffer solution is

Buffer A solution that resists change in pH when limited amounts of an acid or base are added to it; an aqueous solution containing a weak acid and its conjugate base

Added as
CH$_3$COOH

Added as
CH$_3$COO$^-$Na$^+$

$$CH_3COOH + H_2O \rightleftharpoons CH_3COO^- + H_3O^+$$

Acetic acid
(A weak acid)

Acetate ion
(Conjugate base
of a weak acid)

A buffer resists any change in pH upon the addition of small quantities of acid or base. To see how, we will use an acetic acid–sodium acetate buffer as an example. If a strong acid such as HCl is added to this buffer solution, the added H$_3$O$^+$ ions react with CH$_3$COO$^-$ ions and are removed from solution.

$$CH_3COO^- + H_3O^+ \longrightarrow CH_3COOH + H_2O$$

Acetate ion
(Conjugate base
of a weak acid)

Acetic acid
(A weak acid)

There is a slight increase in the concentration of CH$_3$COOH as well as a slight decrease in the concentration of CH$_3$COO$^-$, but there is no appreciable change in pH. We say that this solution is buffered because it resists a change in pH on the addition of small quantities of a strong acid.

If NaOH or another strong base is added to the buffer solution, the added OH$^-$ ions react with CH$_3$COOH molecules and are removed from solution.

See the **Interactive General, Organic, and Biochemistry CD-ROM, version 2.0,** for further exploration on this topic.

$$CH_3COOH + OH^- \longrightarrow CH_3COO^- + H_2O$$

Acetic acid
(A weak acid)

Acetate ion
(Conjugate base
of a weak acid)

Here there is a slight decrease in the concentration of CH$_3$COOH as well as a slight increase in the concentration of CH$_3$COO$^-$, but, again, there is no appreciable change in pH.

The important point about this or any other buffer solution is that when the conjugate base of the weak acid removes H$_3$O$^+$, it is converted to the undissociated weak acid. Because a substantial amount of weak acid is already present, there is no appreciable change in its concentration and, because H$_3$O$^+$ ions are removed from solution, there is no appreciable change in pH. By the same token, when the weak acid removes OH$^-$ ions from solution, it is converted to its conjugate base. Because OH$^-$ ions are removed from solution, there is no appreciable change in pH.

The effect of a buffer can be quite powerful. Addition of either dilute HCl or NaOH to pure water, for example, causes a dramatic change in pH (Figure 7.5).

When HCl or NaOH is added to a phosphate buffer, the results are quite different. Suppose we have a phosphate buffer solution of pH 7.21 prepared by dissolving 0.10 mol NaH$_2$PO$_4$ (a weak acid) and 0.10 mol Na$_2$HPO$_4$ (its conjugate base) in enough water to make 1.00 L of solution. If we add 0.010 mol of HCl to 1.0 L of this solution, the pH decreases to only 7.12. If we add 0.1 mol of NaOH, the pH increases to only 7.30.

Phosphate buffer (pH 7.21) + 0.010 mol HCl   pH 7.21 → 7.12

Phosphate buffer (pH 7.21) + 0.010 mol NaOH   pH 7.21 → 7.30

Figure 7.6 shows the effect of adding acid to a buffer solution.

(a) pH 7.00   (b) pH 2.00   (c) pH 12.00

Figure 7.5 The addition of HCl and NaOH to pure water. (a) The pH of pure water is 7.0. (b) The addition of 0.01 mol of HCl to 1 L of pure water causes the pH to decrease to 2. (c) The addition of 0.010 mol of NaOH to 1 L of pure water causes the pH to increase to 12.

B Buffer pH

In the previous example, the pH of the buffer containing equal molar amounts of $H_2PO_4^-$ and HPO_4^{2-} is 7.21. From Table 7.3, we see that 7.21 is the pK_a of the acid $H_2PO_4^-$. This is not a coincidence. If we make a buffer solution by mixing equimolar concentrations of any weak acid and its conjugate base, the pH of the solution will equal the pK_a of the weak acid.

Figure 7.6 Buffer solutions. The solution in the Erlenmeyer flask on the right in both (a) and (b) is a buffer of pH 7.40, the same pH as human blood. The buffer solution also contains bromocresol green, an acid–base indicator that is blue at pH 7.40 (see Figure 7.2). (a) The beaker contains some of the pH 7.40 buffer and the bromocresol green indicator to which has been added 5 mL of 0.1 M HCl. After the addition of the HCl, the pH of the buffer solution drops only 0.65 unit to 6.75. (b) The beaker contains pure water and bromocresol green indicator to which has been added 5 mL of 0.10 M HCl. After the addition of the HCl, the pH of the unbuffered solution drops by 4.38 units to pH 3.02.

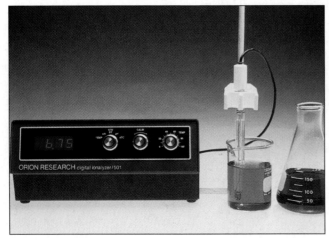 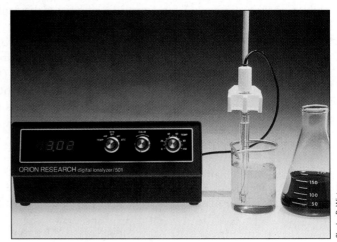

(a)   (b)

This fact allows us to prepare buffer solutions to maintain almost any pH. For example, if we want to maintain a pH of 9.14, we could make a buffer solution from boric acid, H_3BO_3, and sodium dihydrogen borate, NaH_2BO_3, the sodium salt of its conjugate base (see Table 7.3).

See the **Interactive General, Organic, and Biochemistry CD-ROM, version 2.0,** for further exploration on this topic.

EXAMPLE 7.8

What is the pH of a buffer solution containing equimolar quantities of:
(a) H_3PO_4 and NaH_2PO_4?
(b) H_2CO_3 and $NaHCO_3$?

Solution

The pH is equal to the pK_a of the weak acid, which we find in Table 7.3.
(a) pH = 2.12
(b) pH = 6.37

Problem 7.8

What is the pH of a buffer solution containing equimolar quantities of:
(a) NH_4Cl and NH_3?
(b) CH_3COOH and CH_3COONa?

C Buffer Capacity

Buffer capacity is the amount of hydrogen or hydroxide ions that a buffer can absorb without a significant change in its pH. We have already mentioned that a pH buffer is an acid–base "shock absorber." We now ask what makes one solution a better acid–base shock absorber than another solution. The capacity of a pH buffer depends on both its pH and its concentration:

> **Buffer capacity** The extent to which a buffer solution can prevent a significant change in pH of a solution upon addition of an acid or a base

pH:	The closer the pH of the buffer is to the pK_a of the weak acid, the greater the buffer capacity.
Concentration:	The greater the concentration of the weak acid and its conjugate base, the greater the buffer capacity.

An effective buffer has a pH equal to the pK_a of the weak acid ±1. For acetic acid, for example, the pK_a is 4.75. Therefore, a solution of acetic acid and sodium acetate functions as an effective buffer within the pH range of approximately 3.75–5.75. The most effective buffer is one with equal concentrations of the weak acid and its salt—that is, one in which the pH of the buffer solution equals the pK_a of the weak acid.

Buffer capacity also depends on concentration. The greater the concentration of weak acid and its conjugate base, the greater the buffer capacity. We could make a buffer solution by dissolving 1.0 mol each of CH_3COONa and CH_3COOH in 1 L of H_2O, or we could use only 0.10 mol of each. Both solutions have the same pH of 4.75. However, the former has a buffer capacity ten times that of the latter. If we add 0.2 mol of HCl to the former solution, it performs the way we expect—the pH drops to 4.57. If we add 0.2 mol of HCl to the latter solution, however, the pH drops to 1.0 because the buffer has been "swamped out." That is, the amount of H_3O^+ added has exceeded the buffer capacity. The first 0.10 mol of HCl completely neutralizes essentially all the CH_3COO^- present. After that, the solution contains only CH_3COOH and is no longer a buffer, so that the second 0.10 mol of HCl decreases the pH to 1.0.

D Blood Buffers

The average pH of human blood is 7.4. Any change larger than 0.10 pH unit in either direction may cause illness. If the pH goes below 6.8 or above 7.8, death may result. To hold the pH of the blood close to 7.4, the body uses three buffer systems: carbonate, phosphate, and proteins (proteins are discussed in Chapter 13).

The most important of these systems is the carbonate buffer. The weak acid of this buffer is carbonic acid, H_2CO_3; the conjugate base is the bicarbonate ion, HCO_3^-. The pK_a of H_2CO_3 is 6.37 (from Table 7.3). Because the pH of an equal mixture of a weak acid and its salt is equal to the pK_a of the weak acid, a buffer with equal concentrations of H_2CO_3 and HCO_3^- has a pH of 6.37.

Blood, however, has a pH of 7.4. The carbonate buffer can maintain this pH only if $[H_2CO_3]$ and $[HCO_3^-]$ are not equal. In fact, the necessary $[HCO_3^-]/[H_2CO_3]$ ratio is about $10:1$. The normal concentrations of these species in blood are about $0.025\ M\ HCO_3^-$ and $0.0025\ M\ H_2CO_3$. This buffer works because any added H_3O^+ is neutralized by the HCO_3^- and any added OH^- is neutralized by the H_2CO_3.

The fact that the $[HCO_3^-]/[H_2CO_3]$ ratio is $10:1$ means that this system is a better buffer for acids, which lower the ratio and thus improve buffer efficiency, than for bases, which raise the ratio and decrease buffer capacity. This result is in harmony with the actual functioning of the body—under normal conditions, larger amounts of acidic than basic substances enter the blood. The $10:1$ ratio is easily maintained under normal conditions, because the body can very quickly increase or decrease the amount of CO_2 entering the blood.

The second most important buffering system of the blood is a phosphate buffer made up of hydrogen phosphate ion, HPO_4^{2-}, and dihydrogen phosphate ion, $H_2PO_4^-$. In this case, a $1.6:1$ $[HPO_4^{2-}]/[H_2PO_4^-]$ ratio is necessary to maintain a pH of 7.4. This ratio is well within the limits of good buffering action.

7.12 The Henderson-Hasselbalch Equation

Suppose we want to make a phosphate buffer solution of pH 7.00. The weak acid with a pK_a closest to this desired pH is $H_2PO_4^-$; it has a pK_a of 7.21. If we use equal concentrations of NaH_2PO_4 and Na_2HPO_4, however, we will have a buffer of pH 7.21. We want a phosphate buffer that is slightly more acidic than 7.21, so it would seem reasonable to use more of the weak acid, $H_2PO_4^-$, and less of its conjugate base, HPO_4^{2-}. But what proportions of these two salts do we use? Fortunately, we can calculate these proportions using the **Henderson-Hasselbalch equation.**

The Henderson-Hasselbalch equation is a mathematical relationship between pH, the pK_a of a weak acid, and the concentrations of the weak acid and its conjugate base. The equation is derived in the following way. Assume that we are dealing with weak acid, HA, and its conjugate base, A^-.

$$HA + H_2O \rightleftharpoons A^- + H_3O^+$$

$$K_a = \frac{[A^-][H_3O^+]}{[HA]}$$

Taking the logarithm of this equation gives

$$\log K_a = \log [H_3O^+] + \log \frac{[A^-]}{[HA]}$$

CHEMICAL CONNECTIONS 7D

Acidosis and Alkalosis

The pH of blood is normally between 7.35 and 7.45. If the pH goes lower than this level, the condition is called **acidosis.** A blood pH higher than 7.45 produces **alkalosis.** Both are abnormal conditions. Acidosis leads to depression of the nervous system. Mild acidosis can result in fainting; a more severe case can cause coma. Alkalosis leads to overstimulation of the nervous system, muscle cramps, and convulsions. If the acidosis or alkalosis persists for a sufficient period of time, or if the pH gets too far away from 7.35 to 7.45, death may result.

Acidosis has several causes. One type, called respiratory acidosis, results from difficulty in breathing (hypoventilation). An obstruction in the windpipe or diseases such as pneumonia, emphysema, asthma, or congestive heart failure may diminish the amount of CO_2 that leaves the body through the lungs. (You can even produce mild acidosis by holding your breath.) The pH of the blood decreases because the CO_2, unable to escape fast enough, remains in the blood, where it lowers the $[HCO_3^-]/[H_2CO_3]$ ratio.

Respiratory alkalosis involves the reverse process. It arises from rapid or heavy breathing, called hyperventilation. This condition can come about from fever, infection, the action of certain drugs, or even hysteria. Here the excessive loss of CO_2 raises the $[HCO_3^-]/[H_2CO_3]$ ratio and the pH.

Acidosis caused by other factors is called metabolic acidosis. Two causes of this condition are starvation (or fasting) and heavy exercise. When the body doesn't get enough food, it burns its own fat, and the products of this reaction are acidic compounds that enter the blood. This problem sometimes arises in people on fad diets. Heavy exercise causes the muscles to produce excessive amounts of lactate (the anion of lactic acid),

Charles D. Winters

■ **High blood pH brought on by severe anxiety can cause alkalosis. One remedy is to breathe into a paper bag.**

which makes muscles feel tired and sore. Metabolic acidosis is also caused by a number of metabolic irregularities. For example, the disease diabetes mellitus produces acidic compounds called ketone bodies (Section 19.7).

Metabolic alkalosis can result from various metabolic irregularities as well. In addition, it can be caused by excessive vomiting. The contents of the stomach are strongly acidic (pH 1.0–3.0), and loss of substantial amounts of this acidic material raises the pH of the blood.

Rearranging terms gives us a new expression, in which $-\log K_a$ is by definition pK_a, and $-\log [H_3O^+]$ is by definition pH. Making these substitutions gives the Henderson-Hasselbalch equation:

$$-\log [H_3O^+] = -\log K_a + \log \frac{[A^-]}{[HA]}$$

$$pH = pK_a + \log \frac{[A^-]}{[HA]} \quad \text{Henderson-Hasselbalch equation}$$

The Henderson-Hasselbalch equation gives us a convenient way to calculate the pH of a buffer when the concentrations of the weak acid and its conjugate base are not equal.

EXAMPLE 7.9

What is the pH of a phosphate buffer solution containing 1.0 mol/L of sodium dihydrogen phosphate, NaH_2PO_4, and 0.50 mol/L of sodium hydrogen phosphate, Na_2HPO_4?

Solution

The weak acid in this problem is $H_2PO_4^-$; its ionization produces HPO_4^{2-}. The pK_a of this acid is 7.21 (from Table 7.3). Under the weak acid and its conjugate base are shown their concentrations.

$$H_2PO_4^- + H_2O \rightleftharpoons HPO_4^{2-} + H_3O^+ \quad pK_a = 7.21$$

1.0 mol/L   0.50 mol/L

Substituting these values in the Henderson-Hasselbalch equation gives a pH of 6.91.

$$pH = 7.21 + \log \frac{0.50}{1.0}$$

$$= 7.21 - 0.30 = 6.91$$

Problem 7.9

What is the pH of a boric acid buffer solution containing 0.25 mol/L of boric acid, H_3BO_3, and 0.50 mol/L of sodium dihydrogen borate, NaH_2BO_3? See Table 7.3 for the pK_a of boric acid.

Returning to the problem posed at the beginning of this section, how do we calculate the proportions of NaH_2PO_4 and Na_2HPO_4 needed to make up a phosphate buffer of pH 7.00? We know that the pK_a of $H_2PO_4^-$ is 7.21 and that the buffer we wish to prepare has a pH of 7.00. We can substitute these two values in the Henderson-Hasselbalch equation as follows:

$$7.00 = 7.21 + \log \frac{[HPO_4^{2-}]}{[H_2PO_4^-]}$$

Rearranging and solving gives

$$\log \frac{[HPO_4^{2-}]}{[H_2PO_4^-]} = 7.00 - 7.21 = -0.21$$

Taking the antilog of both sides gives

$$\frac{[HPO_4^{2-}]}{[H_2PO_4^-]} = \frac{0.62}{1}$$

Thus, to prepare a phosphate buffer of pH 7.00, we can use 0.62 mol of Na_2HPO_4 and 1.0 mol of NaH_2PO_4. Alternatively, we can use any other amounts of these two salts, as long as their mole ratio is 0.62 : 1.0.

S U M M A R Y

By the **Arrhenius** definitions, acids are substances that produce H_3O^+ ions in aqueous solution, and bases are substances that produce OH^- ions in aqueous solution (Section 7.1). A strong acid reacts completely or almost completely with water to form H_3O^+ ions (Sec-

tion 7.2). A strong base reacts completely or almost completely with water to form OH^- ions.

The **Brønsted-Lowry** definitions expand the definitions of acid and base to beyond water (Section 7.3): An acid is a proton donor, and a base is a proton acceptor (Section 7.3). Every acid has a **conjugate base,** and every base has a **conjugate acid.** The stronger the acid, the weaker its conjugate base. Conversely, the stronger the base, the weaker its conjugate acid. An **amphiprotic substance,** such as water, can act as either an acid or a base (Section 7.3).

In an acid–base reaction, the position of equilibrium favors the reaction of the stronger acid and the stronger base to form the weaker acid and the weaker base (Section 7.4).

The strength of a weak acid is expressed by its **ionization constant, K_a** (Section 7.5). The larger the value of K_a, the stronger the acid $pK_a = -\log [K_a]$.

Acids react with metals, metal hydroxides, and metal oxides to give **salts,** which are ionic compounds made up of cations from the base and anions from the acid (Sections 7.6B, 7.6C, and 7.6D). Acids also react with carbonates, bicarbonates, ammonia, and amines (Sections 7.6E and 7.6F) to give salts.

In pure water, a small percentage of molecules undergo self-ionization:

$$H_2O + H_2O \rightleftharpoons H_3O^+ + OH^-$$

As a result, pure water has a concentration of $10^{-7}\,M$ for H_3O^+ and $10^{-7}\,M$ for OH^- (Section 7.7). The **ion product of water, K_w,** is equal to 1.0×10^{-14}. **$pK_w = 14$.**

Hydronium ion concentrations are generally expressed in **pH** units, with pH $= -\log [H_3O^+]$ (Section 7.8). **pOH** $= -\log [OH^-]$. Solutions with pH less than 7 are acidic; those with pH greater than 7 are basic. A **neutral solution** has a pH of 7. The pH of an aqueous solution is measured with an acid–base indicator or with a pH meter.

Aqueous solutions of salts of strong acids and weak bases are acidic; those of weak acids and strong bases are basic; and those of strong acids and strong bases are neutral (Section 7.9).

We can measure the concentration of aqueous solutions of acids and bases by using a **titration,** (Section 7.10). In an acid–base titration, a base of known concentration is added to an acid of unknown concentration (or vice versa) until an end point is reached, at which point the acid or base being titrated is completely neutralized.

A **buffer** does not change its pH very much when either hydronium ions or hydroxide ions are added to it (Section 7.11A). Buffer solutions consist of approximately equal concentrations of a weak acid and its conjugate base. The **buffer capacity** depends on both its pH and its concentration (Section 7.11C). The most effective buffer solutions have a pH equal to the pK_a of the weak acid (Section 7.11C). The greater the concentration of the weak acid and its conjugate base, the greater the buffer capacity (Section 7.11C). The most important buffers for blood are carbonate and phosphate (Section 7.11D).

The Henderson-Hasselbalch equation is a mathematical relationship between pH, the pK_a of a weak acid, and the concentrations of the weak acid and its conjugate base (Section 7.12):

$$pH = pK_a + \log \frac{[A^-]}{[HA]}$$

P R O B L E M S

Numbers that appear in color indicate difficult problems.
▶ designates problems requiring application of principles.

Arrhenius Acids and Bases

7.10 Define:
  (a) An Arrhenius acid
  (b) An Arrhenius base

7.11 Write an equation for the reaction that takes place when each acid is added to water. For a diprotic or triprotic acid, consider only its first ionization.
  (a) HNO_3 (b) HBr   (c) H_2SO_3
  (d) H_2SO_4 (e) HCO_3^- (f) H_3BO_3

7.12 Which of these acids are monoprotic, which are diprotic, and which are triprotic? Which are amphiprotic?
  (a) $H_2PO_4^-$ (b) HBO_3^{2-} (c) $HClO_4$
  (d) C_2H_5OH (e) HSO_3^- (f) HS^-
  (g) H_2CO_3

7.13 Write an equation for the reaction that takes place when each base is added to water.
  (a) LiOH
  (b) $(CH_3)_2NH$

Brønsted-Lowry Acids and Bases

7.14 Define (a) Brønsted-Lowry acid and (b) Brønsted-Lowry base.

7.15 Write the formula for the conjugate base of each acid.
  (a) H_2SO_4 (b) H_3BO_3 (c) HI
  (d) H_3O^+ (e) NH_4^+ (f) HPO_4^{2-}

7.16 Write the formula for the conjugate base of each acid.
  (a) $H_2PO_4^-$ (b) H_2S (c) HCO_3^-
  (d) CH_3CH_2OH (e) H_2O

7.17 Write the formula for the conjugate acid of each base.
(a) OH^- (b) HS^- (c) NH_3
(d) $C_6H_5O^-$ (e) CO_3^{2-} (f) HCO_3^-

7.18 Write the formula for the conjugate acid of each base.
(a) H_2O (b) HPO_4^{2-} (c) CH_3NH_2
(d) PO_4^{3-} (e) H_2O (f) HBO_3^{2-}

7.19 For each equilibrium, label the stronger acid, the stronger base, the weaker acid, and the weaker base. For which reaction(s) does the position of equilibrium lie toward the right? For which does it lie toward the left?
(a) $H_3PO_4 + OH^- \rightleftharpoons H_2PO_4^- + H_2O$
(b) $H_2O + Cl^- \rightleftharpoons HCl + OH^-$
(c) $HCO_3^- + OH^- \rightleftharpoons CO_3^{2-} + H_2O$

7.20 For each equilibrium, label the stronger acid, the stronger base, the weaker acid, and the weaker base. For which reaction(s) does the position of equilibrium lie toward the right? For which does it lie toward the left?
(a) $C_6H_5OH + C_2H_5O^- \rightleftharpoons C_6H_5O^- + C_2H_5OH$
(b) $HCO_3^- + H_2O \rightleftharpoons H_2CO_3 + OH^-$
(c) $CH_3COOH + H_2PO_4^- \rightleftharpoons CH_3COO^- + H_3PO_4$

7.21 Unless under pressure, carbonic acid in aqueous solution breaks down into carbon dioxide and water, and carbon dioxide is evolved as bubbles of gas. Write an equation for the conversion of carbonic acid to carbon dioxide and water.

7.22 Will carbon dioxide be evolved as a gas when sodium bicarbonate is added to an aqueous solution of each compound? Explain.
(a) Sulfuric acid
(b) Ethanol, C_2H_5OH
(c) Ammonium chloride, NH_4Cl

Acid Ionization Constants

7.23 Which has the larger numerical value?
(a) The pK_a of a strong acid or the pK_a of a weak acid
(b) The K_a of a strong acid or the K_a of a weak acid

7.24 In each pair, select the stronger acid.
(a) Pyruvic acid ($pK_a = 2.49$) or lactic acid ($pK_a = 3.08$)
(b) Citric acid ($pK_a = 3.08$) or phosphoric acid ($pK_a = 2.10$)
(c) Benzoic acid ($K_a = 6.5 \times 10^{-5}$) or lactic acid ($K_a = 8.4 \times 10^{-4}$)
(d) Carbonic acid ($K_a = 4.3 \times 10^{-7}$) or boric acid ($K_a = 7.3 \times 10^{-10}$)

7.25 Which solution will be more acidic—that is, which will have a lower pH?
(a) $0.10\ M$ CH_3COOH or $0.10\ M$ HCl
(b) $0.10\ M$ CH_3COOH or $0.10\ M$ H_3PO_4
(c) $0.010\ M$ H_2CO_3 or $0.010\ M$ $NaHCO_3$
(d) $0.10\ M$ NaH_2PO_4 or $0.10\ M$ Na_2HPO_4
(e) $0.10\ M$ aspirin ($pK_a = 3.47$) or $0.10\ M$ acetic acid

7.26 Which solution will be more acidic—that is, which will have a lower pH?
(a) $0.10\ M$ C_6H_5OH (phenol) or $0.10\ M$ C_2H_5OH (ethanol)
(b) $0.10\ M$ NH_3 or $0.10\ M$ NH_4Cl
(c) $0.10\ M$ NaCl or $0.10\ M$ NH_4Cl
(d) $0.10\ M$ $CH_3CH(OH)COOH$ (lactic acid) or $0.10\ M$ CH_3COOH
(e) $0.10\ M$ ascorbic acid (vitamin C, $pK_a = 4.1$) or $0.10\ M$ acetic acid

Acid and Base Properties

7.27 Write an equation for the reaction of HCl with each compound. Which are acid–base reactions? Which are redox reactions?
(a) Na_2CO_3 (b) Mg (c) NaOH
(d) Fe_2O_3 (e) NH_3 (f) CH_3NH_2
(g) $NaHCO_3$

7.28 When a solution of sodium hydroxide is added to a solution of ammonium carbonate and the solution is heated, ammonia gas, NH_3, is released. Write a net ionic equation for this reaction. Both NaOH and $(NH_4)_2CO_3$ exist as dissociated ions in aqueous solution.

Self-Ionization of Water

7.29 Given the following values of $[H_3O^+]$, calculate the corresponding value of $[OH^-]$ for each solution.
(a) $10^{-11}\ M$ (b) $10^{-4}\ M$
(c) $10^{-7}\ M$ (d) $10\ M$

7.30 Given the following values of $[OH^-]$, calculate the corresponding value of $[H_3O^+]$ for each solution.
(a) $10^{-10}\ M$ (b) $10^{-2}\ M$
(c) $10^{-7}\ M$ (d) $10\ M$

pH and pK_a

7.31 What is the pH of each solution, given the following values of $[H_3O^+]$? Which solutions are acidic, which are basic, and which are neutral?
(a) $10^{-8}\ M$ (b) $10^{-10}\ M$ (c) $10^{-2}\ M$
(d) $10\ M$ (e) $10^{-7}\ M$

7.32 What is the pH and pOH of each solution given the following values of $[OH^-]$? Which solutions are acidic, which are basic, and which are neutral?
(a) $10^{-3}\ M$ (b) $10^{-1}\ M$
(c) $10^{-5}\ M$ (d) $10^{-7}\ M$

7.33 What is the pH of each solution, given the following values of $[H_3O^+]$? Which solutions are acidic, which are basic, and which are neutral?

(a) $3.0 \times 10^{-9} M$
(b) $6.0 \times 10^{-2} M$
(c) $8.0 \times 10^{-12} M$
(d) $5.0 \times 10^{-7} M$

7.34 Which is more acidic, a beer with $[H_3O^+] = 3.16 \times 10^{-5}$ or a wine with $[H_3O^+] = 5.01 \times 10^{-4}$?

7.35 What is the $[OH^-]$ and pOH of each solution?

(a) Potassium hydroxide (0.10 M), pH 13.0
(b) Sodium carbonate (0.10 M), pH 11.6
(c) Trisodium phosphate (0.10 M), pH 12.0
(d) Sodium bicarbonate (0.10 M), pH 8.4

7.36 Following are pH ranges for several human biological materials. From the pH at the midpoint of each range, calculate the corresponding $[H_3O^+]$. Which materials are acidic, which are basic, and which are neutral?

(a) Milk, pH 6.6–7.6
(b) Gastric contents, pH 1.0–3.0
(c) Spinal fluid, pH 7.3–7.5
(d) Saliva, pH 6.5–7.5
(e) Urine, pH 4.8–8.4
(f) Blood plasma, pH 7.35–7.45
(g) Feces, pH 4.6–8.4
(h) Bile, pH 6.8–7.0

The pH of Aqueous Salt Solutions

7.37 Write a balanced equation for the reaction of an acid and a base that gives each salt.

(a) $BaCl_2$
(b) $NaNO_3$
(c) $HCOONH_4$
(d) $CaSO_4$
(e) $MgSO_4$
(f) Na_2HPO_4
(g) NH_4Cl
(h) Li_2CO_3

7.38 Which salts in Problem 7.37 are acidic, which are basic, and which are neutral?

7.39 When sodium carbonate is dissolved in pure water, the resulting solution is basic; that is, its pH is greater than 7. Explain how this happens.

Titration

7.40 What is the purpose of an acid–base titration?

7.41 What is the molarity of a solution made by dissolving 12.7 g of HCl in enough water to make 1.00 L of solution?

7.42 What is the molarity of a solution made by dissolving 3.4 g of $Ba(OH)_2$ in enough water to make 450 mL of solution? Assume that $Ba(OH)_2$ ionizes completely in water to Ba^{2+} and OH^- ions.

7.43 Describe how you would prepare each of the following solutions (in each case assume that you have the solid bases).

(a) 400.0 mL of 0.75 M NaOH
(b) 1.0 L of 0.071 M $Ba(OH)_2$

7.44 If 25.0 mL of an aqueous solution of H_2SO_4 requires 19.7 mL of 0.72 M NaOH to reach the end point, what is the molarity of the H_2SO_4 solution?

7.45 A sample of 27.0 mL of 0.310 M NaOH is titrated with 0.740 M H_2SO_4. How many milliliters of the H_2SO_4 solution are required to reach the end point?

7.46 A 0.300 M solution of H_2SO_4 was used to titrate 10.00 mL of an unknown base; 15.00 mL of acid was required to neutralize the basic solution. What was the molarity of the base?

7.47 A solution of an unknown base was titrated with 0.150 M HCl, and 22.0 mL of acid was needed to reach the end point of the titration. How many moles of the unknown base were in the solution?

7.48 The usual concentration of HCO_3^- ions in blood plasma is approximately 24 millimoles per liter (mmol/L). How would you make up 1.00 L of a solution containing this concentration of HCO_3^- ions?

Buffers

7.49 Write equations to show what happens when, to a buffer solution containing equimolar amounts of CH_3COOH and CH_3COO^-, we add (a) H_3O^+ (b) OH^-.

7.50 Write equations to show what happens when, to a buffer solution containing equimolar amounts of HPO_4^{2-} and $H_2PO_4^-$, we add (a) H_3O^+ (b) OH^-.

7.51 What is the pH of a buffer solution made by dissolving 0.10 mol of formic acid, HCOOH, and 0.10 mol of sodium formate, HCOONa, in 1 L of water?

7.52 We commonly refer to a buffer as consisting of approximately equal molar amounts of a weak acid and its conjugate base—for example, CH_3COOH and CH_3COO^-. Is it also possible to have a buffer consisting of approximately equal molar amounts of a weak base and its conjugate acid? Explain.

7.53 What is meant by buffer capacity?

7.54 How can you change the pH of a buffer? How can you change the capacity of a buffer?

7.55 The pH of a solution made by dissolving 1.0 mol of propanoic acid and 1.0 mol of sodium propanoate in 1.0 L of water is 4.85.

(a) What would the pH be if we used 0.10 mol of each (in 1 L of water) instead of 1.0 mol?

(b) With respect to buffer capacity, how would the two solutions differ?

7.56 What is the connection between buffer action and Le Chatelier's principle?

The Henderson-Hasselbalch Equation

7.57 Show that when the concentration of the weak acid, [HA], in an acid–base buffer equals that of the conjugate base of the weak acid, $[A^-]$, the pH of the buffer solution is equal to the pK_a of the weak acid.

7.58 Calculate the pH of an aqueous solution containing the following:
(a) 0.80 M lactic acid and 0.40 M lactate ion
(b) 0.30 M NH_3 and 1.50 M NH_4^+

7.59 What is the ratio of $HPO_4^{2-}/H_2PO_4^-$ in a phosphate buffer of pH 7.40 (the average pH of human blood plasma)?

7.60 What is the ratio of $HPO_4^{2-}/H_2PO_4^-$ in a phosphate buffer of pH 7.9 (the pH of human pancreatic fluid)?

7.61 The pH of 0.10 M HCl is 1.0. When 0.10 mol of sodium acetate, CH_3COONa, is added to this solution, its pH changes to 2.9. Explain why the pH changes, and why it changes to this particular value.

Chemical Connections

7.62 (Chemical Connections 7A) Which weak base is used as a flame retardant in plastics?

7.63 (Chemical Connections 7B) What is the most important immediate first aid in any chemical burn of the eyes?

7.64 (Chemical Connections 7B) With respect to corneal burns, are strong acids or strong bases more dangerous?

7.65 (Chemical Connections 7C) Name the most common bases used in over-the-counter antacids.

7.66 (Chemical Connections 7D) What causes (a) hypoventilation, (b) respiratory alkalosis, and (c) metabolic alkalosis?

Additional Problems

7.67 4-Methylphenol, $CH_3C_6H_4OH$ (pK_a = 10.26), is only slightly soluble in water, but its sodium salt, $CH_3C_6H_4O^-Na^+$, is quite soluble in water. In which of the following solutions will 4-methylphenol dissolve more readily than in pure water?
(a) Aqueous NaOH
(b) Aqueous $NaHCO_3$
(c) Aqueous NH_3

7.68 Benzoic acid, C_6H_5COOH (pK_a = 4.19), is only slightly soluble in water, but its sodium salt, $C_6H_5COO^-Na^+$, is quite soluble in water. In which of the following solutions will benzoic acid dissolve more readily than in pure water?
(a) Aqueous NaOH
(b) Aqueous $NaHCO_3$
(c) Aqueous Na_2CO_3

7.69 Assume that you have a dilute solution of HCl (0.10 M) and a concentrated solution of acetic acid (5.0 M). Which solution is more acidic? Explain.

7.70 If the $[OH^-]$ of a solution is 1×10^{-14},
(a) What is the pH of the solution?
(b) What is the $[H_3O^+]$?

7.71 What is the molarity of a solution made by dissolving 0.583 g of the diprotic acid oxalic acid, $H_2C_2O_4$, in enough water to make 1.75 L of solution?

7.72 Following are three organic acids and the pK_a of each: butanoic acid, 4.82; barbituric acid, 5.00; and lactic acid, 3.85.
(a) What is the K_a of each acid?
(b) Which of the three is the strongest acid and which is the weakest?

7.73 The pK_a value of barbituric acid is 5.0. If the H_3O^+ and barbiturate ion concentrations are each 0.0030 M, what is the concentration of the undissociated barbituric acid?

7.74 If pure water self-ionizes to give H_3O^+ and OH^- ions, why doesn't pure water conduct an electric current?

7.75 Can an aqueous solution have a pH of zero? Explain your answer using aqueous HCl as your example.

7.76 A scale of K_b values for bases could be set up in a manner similar to that for the K_a scale for acids. However, this step is generally considered unnecessary. Explain.

7.77 Do a 1.0 M CH_3COOH solution and a 1.0 M HCl solution have the same pH? Explain.

7.78 Suppose you wish to make a buffer whose pH is 7.21. You have available 1.00 L of 0.100 M NaH_2PO_4 and solid Na_2HPO_4. How many grams of the solid Na_2HPO_4 must be added to the solution to accomplish this task? (Assume that the volume remains 1.00 L.)

7.79 In the past, boric acid has been used to rinse an inflamed eye. How would you make up one liter of a $H_3BO_3/H_2BO_3^-$ buffer solution that has a pH of 7.40?

7.80 Suppose you want to make a CH_3COOH/CH_3COO^- buffer solution with a pH of 5.60. The acetic acid concentration is to be 0.10 M. What should the acetate ion concentration be?

7.81 For an acid–base reaction, one way to determine the position of equilibrium is to say that the larger of the equilibrium arrow pair points to the acid with the higher value of pK_a. For example,

$$CH_3COOH + HCO_3^- \rightleftharpoons CH_3COO^- + H_2CO_3$$
$$\text{p}K_a = 4.75 \qquad\qquad \text{p}K_a = 6.37$$

Explain why this rule works.

7.82 When a solution prepared by dissolving 4.00 g of an unknown acid in 1.00 L of water is titrated with 0.600 M NaOH, 38.7 mL of the NaOH solution is needed to neutralize the acid. What was the molarity of the acid solution?

7.83 Write equations to show what happens when, to a buffer solution containing equal amounts of HCOOH and $HCOO^-$, we add (a) H_3O^+ (b) OH^-.

7.84 If we add 0.10 mol of NH_3 to 0.50 mol of HCl dissolved in enough water to make 1.0 L of solution, what happens to the NH_3? Will any NH_3 remain? Explain.

7.85 Suppose you have an aqueous solution prepared by dissolving 0.050 mol of NaH_2PO_4 in 1 L of water. This solution is not a buffer, but suppose you want to make it into one. How many moles of solid Na_2HPO_4 must you add to 1.0 L of this aqueous solution to make it into (a) a buffer of pH 7.21, (b) a buffer of pH 6.21, and (c) a buffer of pH 8.21?

7.86 The pH of a 0.10 *M* solution of acetic acid is 2.93. When 0.10 mol of sodium acetate, CH_3COONa, is added to this solution, its pH changes to 4.74. Explain why the pH changes, and why it changes to this particular value.

7.87 Suppose you have a phosphate buffer of pH 7.21. If you add more solid NaH_2PO_4 to this buffer, would you expect the pH of the buffer to increase, decrease, or remain unchanged? Explain.

7.88 Suppose you have a bicarbonate buffer containing carbonic acid, H_2CO_3, and sodium bicarbonate, $NaHCO_3$, and that the pH of the buffer is 6.37. If you add more solid $NaHCO_3$ to this buffer solution, would you expect its pH to increase, decrease, or remain unchanged? Explain.

InfoTrac College Edition

For additional readings, go to InfoTrac College Edition, your online research library, at

http://infotrac.thomsonlearning.com

CHAPTER 8

8.1 Introduction

8.2 Structure and Classification

8.3 Nomenclature

8.4 Physical Properties

8.5 Basicity

8.6 Reaction with Acids

8.7 Epinephrine:
A Prototype for the
Development of New
Bronchodilators

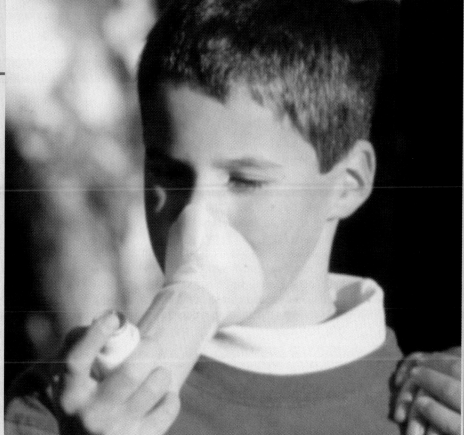

This inhaler delivers puffs of albuterol (Proventil), a potent synthetic bronchodilator whose structure is patterned after that of epinephrine (adrenaline). See Section 8.7.

Jack Ballard/Visuals Unlimited, Inc.

Amines

8.1 Introduction

Carbon, hydrogen, and oxygen are the three most common elements in organic compounds. Because of the wide distribution of amines in the biological world, nitrogen is the fourth most common element of organic compounds. The most important chemical property of amines is their basicity.

8.2 Structure and Classification

Amines are classified as **primary (1°), secondary (2°),** or **tertiary (3°),** depending on the number of carbon groups bonded to nitrogen (Section 1.4B).

$$CH_3-NH_2 \qquad CH_3-\overset{\overset{\textstyle H}{|}}{N}-CH_3 \qquad CH_3-\overset{\overset{\textstyle CH_3}{|}}{N}-CH_3$$

Methylamine (a 1° amine)   Dimethylamine (a 2° amine)   Trimethylamine (a 3° amine)

Amines are further classified as aliphatic or aromatic. In an **aliphatic amine,** all the carbons bonded to nitrogen are derived from alkyl groups. In an **aromatic amine,** one or more of the groups bonded to nitrogen are aryl groups.

Aliphatic amine An amine in which nitrogen is bonded only to alkyl groups

Aromatic amine An amine in which nitrogen is bonded to one or more aromatic rings

---

CHEMICAL CONNECTIONS 8A

Amphetamines (Pep Pills)

Amphetamine, methamphetamine, and phentermine—all synthetic amines—are powerful stimulants of the central nervous system. Like most other amines, they are stored and administered as their salts. The sulfate salt of amphetamine is prescribed as Benzedrine, the hydrochloride salt of the s enantiomer of methamphetamine as Methedrine, and the hydrochloride salt of phentermine as Fastin.

These three amines all have similar physiological effects and are referred to by the general name *amphetamines.* Structurally, they have in common a benzene ring with a three-carbon side chain and an amine nitrogen on the second carbon of the chain. Physiologically, they have in common the ability to reduce fa-

tigue and diminish hunger by raising the glucose level of the blood. Because of these properties, amphetamines are widely prescribed to counter mild depression, reduce hyperactivity in children, and suppress appetite in people who are trying to lose weight. They are also used illegally to reduce fatigue and elevate mood.

Abuse of amphetamines can have severe effects on both the body and the mind. They are addictive, concentrate in the brain and nervous system, and can lead to long periods of sleeplessness, loss of weight, and paranoia.

The action of amphetamines is similar to that of epinephrine (Section 8.7), the hydrochloride salt of which is named Adrenalin.

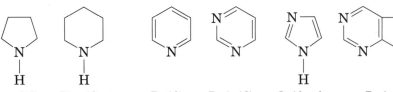

Amphetamine
(Benzedrine)

(s)-Methamphetamine
(Methedrine)

Phentermine
(Fastin)

Aniline
(a 1° aromatic amine)

N-Methylaniline
(a 2° aromatic amine)

Benzyldimethylamine
(a 3° aliphatic amine)

Heterocyclic amine An amine in which nitrogen is one of the atoms of a ring

Heterocyclic aliphatic amine A heterocyclic amine in which nitrogen is bonded only to alkyl groups

Heterocyclic aromatic amine An amine in which nitrogen is one of the atoms of an aromatic ring

An amine in which the nitrogen atom is part of a ring is classified as a **heterocyclic amine.** When the ring is saturated, the amine is classified as a **heterocyclic aliphatic amine.** When the nitrogen is part of an aromatic ring (Section 4.2), the amine is classified as a **heterocyclic aromatic amine.** The most important of these compounds are pyridine, pyrimidine, imidazole, and purine. Pyridine and pyrimidine are analogs of benzene in which first one and then two CH groups are replaced by nitrogen atoms. Pyrimidine and purine serve as the building blocks for the amine bases of DNA and RNA (Chapter 16).

Pyrrolidine   Piperidine
(heterocyclic aliphatic amines)

Pyridine   Pyrimidine   Imidazole   Purine
(heterocyclic aromatic amines)

CHEMICAL CONNECTIONS 8B

Alkaloids

Alkaloids are basic nitrogen-containing compounds found in the roots, bark, leaves, berries, or fruits of plants. In almost all alkaloids, the nitrogen atom is part of a ring. The name "alkaloid" was chosen because these compounds are alkali-like (*alkali* is an older term for a basic substance) and react with strong acids to give water-soluble salts. Thousands of different alkaloids have been extracted from plant sources, many of which are used in modern medicine.

When administered to animals, including humans, alkaloids have pronounced physiological effects. Whatever their individual effects, most alkaloids are toxic in large enough doses. For some, the toxic dose is very small!

© Inga Spence/Visuals Unlimited

■ **Tobacco plants.**

(*s*)-Coniine is the toxic principal of water hemlock (a member of the carrot family). Its ingestion can cause weakness, labored respiration, paralysis, and eventually death. It was the toxic substance in the "poison hemlock" used in the death of Socrates. Water hemlock is easily confused with Queen Anne's lace, a type of wild carrot—a mistake that has killed numerous people.

(*s*)-Nicotine occurs in the tobacco plant. In small doses, it is an addictive stimulant. In larger doses, this substance causes depression, nausea, and vomiting. In still larger doses, it is a deadly poison. Solutions of nicotine in water are used as insecticides.

Cocaine is a central nervous system stimulant obtained from the leaves of the coca plant. In small doses, it decreases fatigue and gives a sense of well-being. Prolonged use of cocaine leads to physical addiction and depression.

EXAMPLE 8.1

How many hydrogens does piperidine have? How many does pyridine have? Write the molecular formula of each amine.

Solution

Recall that hydrogen atoms are not shown in line-angle formulas. Piperidine has 11 hydrogens, and its molecular formula is $C_5H_{11}N$. Pyridine has 5 hydrogens, and its molecular formula is C_5H_5N.

Problem 8.1

How many hydrogens does pyrrolidine have? How many does purine have? Write the molecular formula of each amine.

8.3 Nomenclature

A IUPAC Names

IUPAC names for aliphatic amines are derived just as they are for alcohols. The final **-e** of the parent alkane is dropped and replaced by **-amine.** The location of the amino group on the parent chain is indicated by a number.

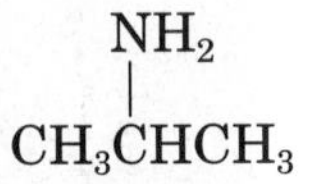

$$NH_2 \qquad NH_2$$

$$CH_3CHCH_3 \qquad \text{(cyclohexane ring)} \qquad H_2N\text{~~~~~~~}NH_2$$

2-Propanamine 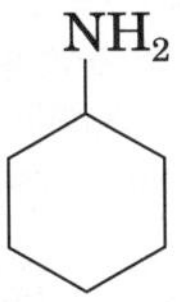   Cyclohexanamine 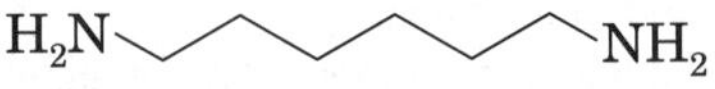   1,6-Hexanediamine

IUPAC nomenclature retains the common name **aniline** for $C_6H_5NH_2$, the simplest aromatic amine. Its simple derivatives are named using numbers to locate substituents or, alternatively, the prefixes ortho (o), meta (m), and para (p). Several derivatives of aniline have common names that remain in wide use. Among them is **toluidine** for a methyl-substituted aniline.

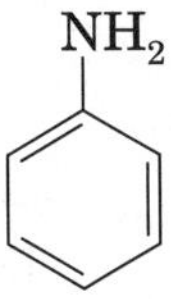

Aniline 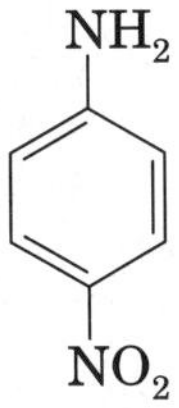   4-Nitroaniline   3-Methylaniline
         (*p*-Nitroaniline)   (*m*-Toluidine)

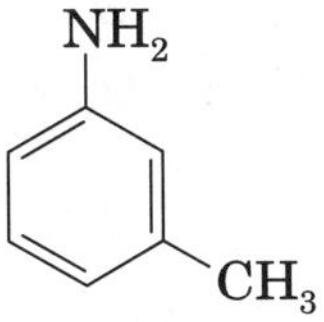

Unsymmetrical secondary and tertiary amines are commonly named as *N*-substituted primary amines. The largest group bonded to nitrogen is taken as the parent amine; the smaller groups bonded to nitrogen are named, and their locations are indicated by the prefix *N* (indicating that they are bonded to nitrogen).

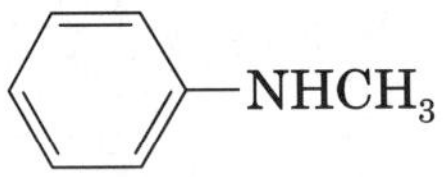

N-Methylaniline 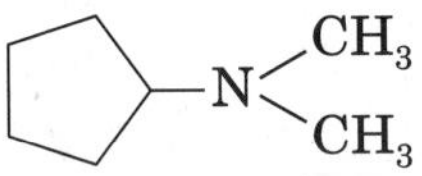   *N,N*-Dimethyl-
         cyclopentanamine

EXAMPLE 8.2

Write the IUPAC name for each amine. Be certain to specify the configuration of the stereocenter in (c).

(a)

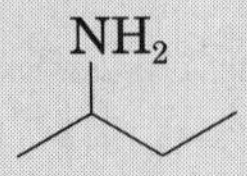

$$NH_2$$

(b) $H_2N(CH_2)_5NH_2$

(c)

$$H \quad NH_2$$

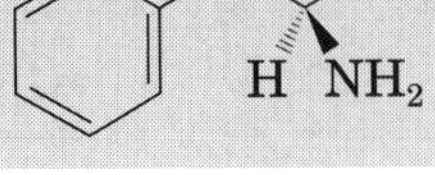

Solution

(a) The parent alkane has four carbon atoms and is therefore butane. The amino group is on carbon 2, giving the IUPAC name 2-butanamine.

(b) The parent chain has five carbon atoms and is therefore pentane. There are amino groups on carbons 1 and 5, giving the IUPAC name 1,5-pentanediamine. The common name of this diamine is cadaverine, which should give you a hint of where it occurs in nature and of its odor. Cadaverine, one of the end products of decaying flesh, is quite poisonous.

(c) The parent chain has three carbon atoms and is therefore propane. To have the lowest numbers possible, we number from the end that places the phenyl group on carbon 1 and the amino group on carbon 2. The priorities for determining R or S configuration are $NH_2 > C_6H_5CH_2 > CH_3 > H$. Its systematic name is (S)-1-phenyl-2-propanamine. Its common name is amphetamine.

Problem 8.2

Write a structural formula for each amine.

(a) 2-Methyl-1-propanamine
(b) Cyclopentanamine
(c) 1,4-Butanediamine

B Common Names

Common names for most aliphatic amines list the groups bonded to nitrogen in alphabetical order in one word ending in the suffix **-amine.**

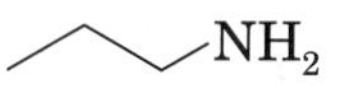

Propylamine 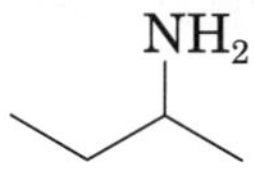   *sec*-Butylamine 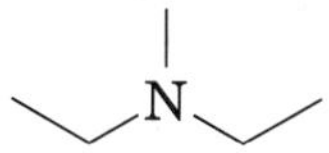   Diethylmethylamine 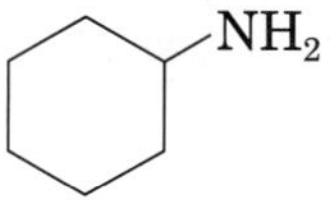   Cyclohexylamine

EXAMPLE 8.3

Write a structural formula for each amine.
(a) Isopropylamine
(b) Cyclohexylmethylamine
(c) Triethylamine

Solution

(a) $(CH_3)_2CHNH_2$ (b) 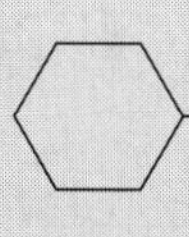 ⬡—$NHCH_3$ (c) $(CH_3CH_2)_3N$

Problem 8.3

Write a structural formula for each amine.
(a) Isobutylamine
(b) Diphenylamine
(c) Diisopropylamine

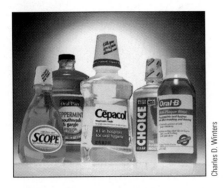

■ **Several over-the-counter mouthwashes contain an *N*-alkylpyridinium chloride as an antibacterial.**

Hydrogen bonding

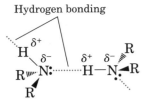

Figure 8.1 Hydrogen bonding between two molecules of a secondary amine.

When four atoms or groups of atoms are bonded to a nitrogen atom—as, for example, in NH_4^+ and $CH_3NH_3^+$—nitrogen bears a positive charge and is associated with an anion as a salt. The compound is named as a salt of the corresponding amine. The ending **-amine** (or aniline or pyridine or the like) is replaced by **-ammonium** (or anilinium, pyridinium, or the like) and the name of the anion (chloride, acetate, and so on) is added.

$$(CH_3CH_2)_3NH^+Cl^-$$

Triethylammonium chloride

8.4 Physical Properties

Like ammonia, low-molecular-weight amines have very sharp, penetrating odors. Trimethylamine, for example, is the pungent principal in the smell of rotting fish. Two other particularly pungent amines are 1,4-butanediamine (putrescine) and 1,5-pentanediamine (cadaverine).

Amines are polar compounds because of the difference in electronegativity between nitrogen and hydrogen ($3.0 - 2.1 = 0.9$). Both primary and secondary amines have N—H bonds, and they can form hydrogen bonds with one another (Figure 8.1). Tertiary amines do not have a hydrogen bonded to nitrogen and, therefore, do not form hydrogen bonds with one another.

An N — H ----- N hydrogen bond is weaker than an O — H ----- O hydrogen bond because the difference in electronegativity between nitrogen and hydrogen ($3.0 - 2.1 = 0.9$) is less than that between oxygen and hydrogen ($3.5 - 2.1 = 1.4$). To see the effect of hydrogen bonding between molecules of comparable molecular weight, we can compare the boiling points of

---

CHEMICAL CONNECTIONS 8C

Tranquilizers

Most people have experienced anxiety and stress at some time in their lives, and each person develops various ways to cope with these factors. Perhaps it is meditation, or exercise, or psychotherapy, or drugs. One modern way of coping is to use tranquilizers.

The first modern tranquilizers were derivatives of a compound called benzodiazepine. The first of these compounds, chlorodiazepoxide, better known as Librium, was introduced in

1960 and was soon followed by more than two dozen related compounds. Diazepam, better known as Valium, became one of the most widely used of these drugs.

Librium, Valium, and other benzodiazepines are central nervous system sedatives/hypnotics. As sedatives, they diminish activity and excitement, thereby exerting a calming effect. As hypnotics, they produce drowsiness and sleep. We will discuss their action on the central nervous system in Section 15.4.

Benzodiazepine

Chlorodiazepoxide
(Librium)

Diazepam
(Valium)

See the **Interactive General, Organic, and Biochemistry CD-ROM, version 2.0,** for further exploration on this topic.

ethane, methanamine, and methanol. Ethane is a nonpolar hydrocarbon, and the only attractive forces between its molecules are weak London dispersion forces. Both methanamine and methanol have polar molecules that interact in the pure liquid by hydrogen bonding. Methanol has the higher boiling point because hydrogen bonding between its molecules is stronger than that between methanamine molecules.

	CH_3CH_3	CH_3NH_2	CH_3OH
Molecular weight (amu)	30.1	31.1	32.0
Boiling point (°C)	−88.6	−6.3	65.0

All classes of amines form hydrogen bonds with water and are more soluble in water than are hydrocarbons of comparable molecular weight. Most low-molecular-weight amines are completely soluble in water, but higher-molecular-weight amines are only moderately soluble in water or are insoluble.

8.5 Basicity

Like ammonia, amines are weak bases, and aqueous solutions of amines are basic. The following acid–base reaction between an amine and water is written using curved arrows to emphasize that, in this proton-transfer reaction (Section 7.1), the unshared pair of electrons on nitrogen forms a new covalent bond with hydrogen and displaces a hydroxide ion.

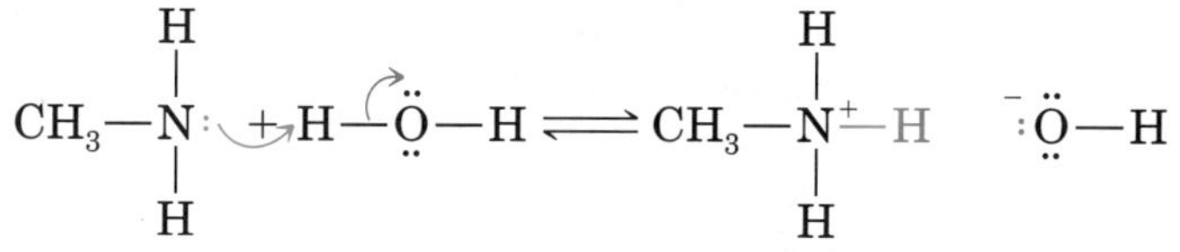

$$CH_3-\overset{\overset{\displaystyle H}{|}}{\underset{\underset{\displaystyle H}{|}}{N}}:\ +H-\overset{..}{\underset{..}{O}}-H \rightleftharpoons CH_3-\overset{\overset{\displaystyle H}{|}}{\underset{\underset{\displaystyle H}{|}}{N}}\overset{+}{-}H \quad :\overset{..}{O}-H$$

Methylamine   Methylammonium
(a base)   hydroxide

The base dissociation constant, K_b, for the reaction of an amine with water has the following form, illustrated here for the reaction of methylamine with water to give methylammonium hydroxide. pK_b is defined as the negative logarithm of K_b.

$$K_b = \frac{[CH_3NH_3^+][OH^-]}{[CH_3NH_2]} = 4.37 \times 10^{-4}$$

$$pK_b = -\log 4.37 \times 10^{-4} = 3.360$$

All aliphatic amines have about the same base strength, pK_b 3.0–4.0, and are slightly stronger bases than ammonia (Table 8.1). Aromatic amines

Table 8.1	Approximate Base Strengths of Amines			
Class	**pK_a**	**Example**	**Name**	
Aliphatic	3.0–4.0	$CH_3CH_2NH_2$	Ethanamine	Stronger base
Ammonia	4.74			↑
Aromatic	8.5–9.5	$C_6H_5NH_2$	Aniline	Weaker base

and heterocyclic aromatic amines (pK_b 8.5–9.5) are considerably weaker bases than aliphatic amines. One additional point about the basicities of amines: While aliphatic amines are weak bases by comparison with inorganic bases such as NaOH, they are strong bases among organic compounds.

Given the basicities of amines, we can determine which form of an amine exists in body fluids—say, blood. In a normal, healthy person, the pH of blood is approximately 7.40, which is slightly basic. If an aliphatic amine is dissolved in blood, it is present predominantly as its protonated or conjugate acid form.

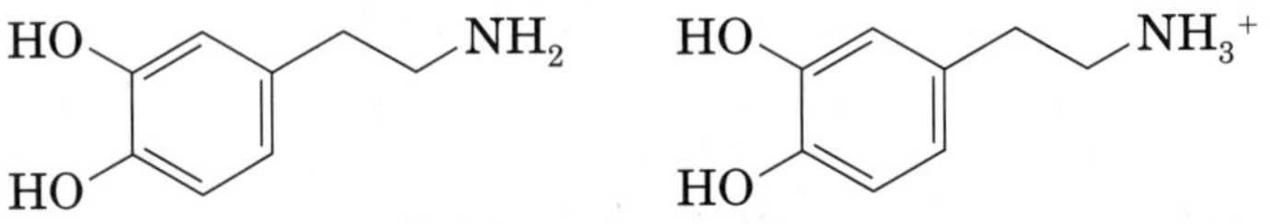

Dopamine

Conjugate acid of dopamine
(the major form present
in blood plasma)

We can show that an aliphatic amine such as dopamine dissolved in blood is present largely as its protonated or conjugate acid form in the following way. Assume that the amine, RNH_2, has a pK_b of 3.50 and that it is dissolved in blood, pH 7.40. We first write the base dissociation constant for the amine and then solve for the ratio of RNH_3^+ to RNH_2.

$$RNH_2 + H_2O \rightleftharpoons RNH_3^+ + OH^-$$

$$K_b = \frac{[RNH_3^+][OH^-]}{[RNH_2]}$$

$$\frac{K_b}{[OH^-]} = \frac{[RNH_3^+]}{[RNH_2]}$$

We now substitute the appropriate values for K_b and $[OH^-]$ in this equation. Taking the antilog of 3.50 gives a K_b of 4.0×10^{-4}. Calculating the concentration of hydroxide requires two steps. First recall from Section 7.8 that pH + pOH = 14. If the pH of blood is 7.40, then its pOH is 6.60 and its $[OH^-]$ is 2.5×10^{-7}. Substituting these values in the appropriate equation gives a ratio of 1600 parts RNH_3^+ to 1 part RNH_2.

$$\frac{4.0 \times 10^{-4}}{2.5 \times 10^{-7}} = \frac{[RNH_3^+]}{[RNH_2]} = 1600$$

As this calculation demonstrates, an aliphatic amine present in blood is more than 99.9% in the protonated form. Thus, even though we may write the structural formula of dopamine as the free amine, it is present as the protonated form. It is important to realize, however, that the amine and ammonium ion forms are always in equilibrium, so some of the unprotonated form is nevertheless present in solution.

Aromatic amines, on the other hand, are considerably weaker bases than aliphatic amines and are present in blood largely in the unprotonated form. Performing the same type of calculation for an aromatic amine, $ArNH_2$, with pK_b of approximately 10, we find that the aromatic amine is more than 99.9% in its unprotonated ($ArNH_2$) form.

EXAMPLE 8.4

Select the stronger base in each pair of amines.

(a) 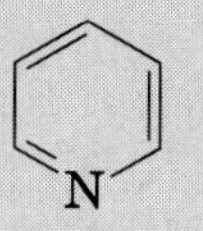 [structure] (A)   or   [structure] (B)

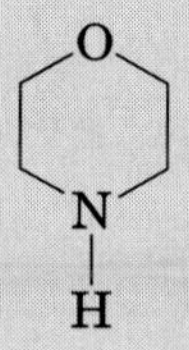

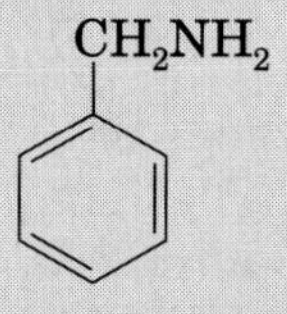

(b) 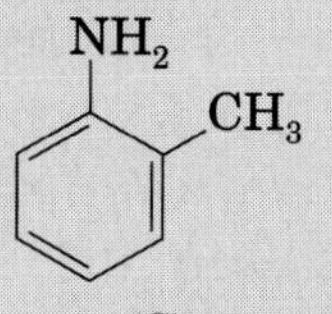 [structure] (C)   or   [structure] (D)

Solution
(a) Morpholine (B), a 2° aliphatic amine, is the stronger base. Pyridine
   (A), a heterocyclic aromatic amine, is the weaker base.
(b) Benzylamine (D), a 1° aliphatic amine, is the stronger base. Even
   though it contains an aromatic ring, it is not an aromatic amine because the amine nitrogen is not bonded to the aromatic ring.
   o-Toluidine (C), a 1° aromatic amine, is the weaker base.

Problem 8.4
Select the stronger base from each pair of amines.

(a) 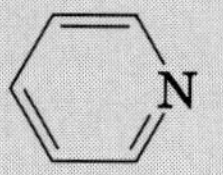 [structure] (A)   or   [structure] (B)

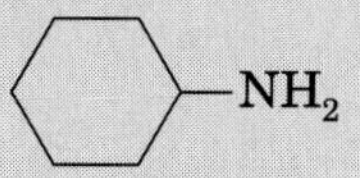

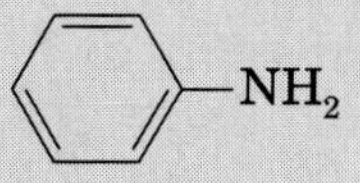

(b) NH₃ (C)   or   [structure] (D)

8.6 Reaction with Acids

The most important chemical property of amines is their basicity. Amines,
whether soluble or insoluble in water, react quantitatively with strong acids
to form water-soluble salts as illustrated by the reaction of (*R*)-norepineph-
rine (noradrenaline) with aqueous HCl to form a hydrochloride salt.

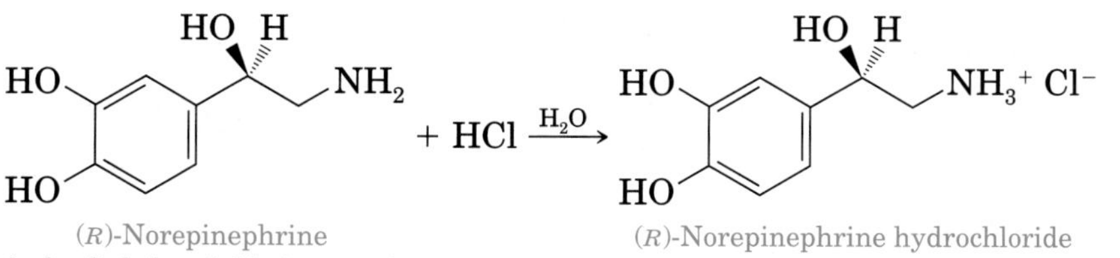

[reaction structures]

(*R*)-Norepinephrine
(only slightly soluble in water)

(*R*)-Norepinephrine hydrochloride
(a water-soluble salt)

The Solubility of Drugs in Body Fluids

Many drugs have "·HCl" or some other acid as part of their chemical formulas and occasionally as part of their generic names. Invariably these drugs are amines that are insoluble in aqueous body fluids such as blood plasma and cerebrospinal fluid. For the administered drug to be absorbed and carried by body fluids, it must be treated with an acid to form a water-soluble ammonium salt. Methadone, a narcotic analgesic, is marketed as its water-soluble hydrochloride salt. Novocain, one of the first local anesthetics, is the hydrochloride salt of procaine.

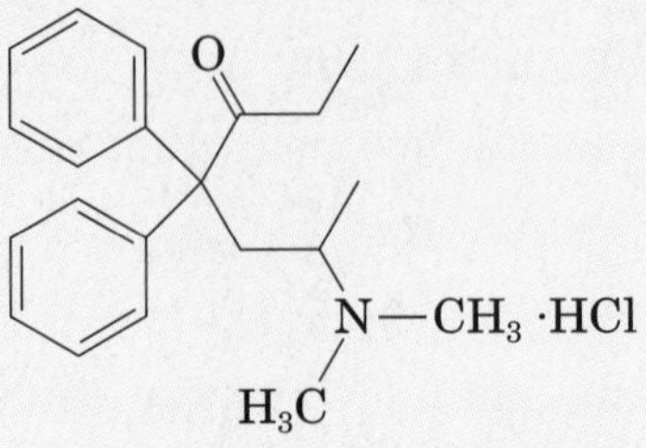

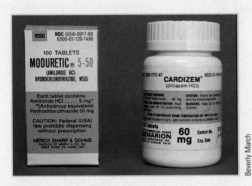

These two drugs are amine salts and are labeled as hydrochlorides.

Methadone ·HCl

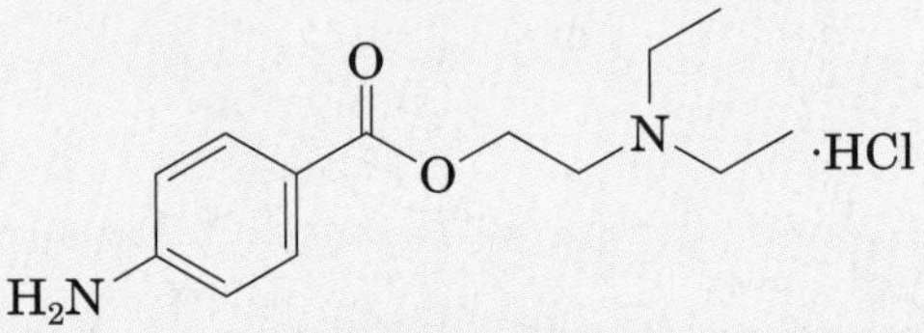

Procaine ·HCl
(Novocain, a local anesthetic)

There is another reason besides increased water solubility for preparing these and other amine drugs as salts. Amines are very susceptible to oxidation and decomposition by atmospheric oxygen, with a corresponding loss of biological activity. By comparison, their amine salts are far less susceptible to oxidation; they retain their effectiveness for a much longer time.

See the **Interactive General, Organic, and Biochemistry CD-ROM, version 2.0,** for further exploration on this topic.

EXAMPLE 8.5

Complete the equation for each acid–base reaction, and name the salt formed.

(a) $(CH_3CH_2)_2NH + HCl \longrightarrow$ (b) 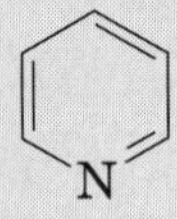 $+ CH_3COOH \longrightarrow$

Solution

(a) $(CH_3CH_2)_2NH_2{}^+Cl^-$ (b)
Diethylammonium chloride

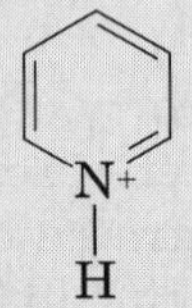

CH_3COO^-

Pyridinium acetate

Problem 8.5

Complete the equation for each acid–base reaction and name the salt formed.

(a) $(CH_3CH_2)_3N + HCl \longrightarrow$

(b) 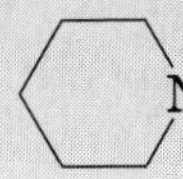 ⬡NH $+ CH_3COOH \longrightarrow$

8.7 Epinephrine: A Prototype for the Development of New Bronchodilators

Epinephrine was first isolated in pure form in 1897 and its structure determined in 1901. It occurs in the adrenal gland (hence the common name adrenalin) as a single enantiomer with the R configuration at its stereocenter. Epinephrine is commonly referred to as a catecholamine: The common name of 1,2-dihydroxybenzene is catechol (Section 4.5A), and amines containing a benzene ring with ortho hydroxyl groups are called catecholamines.

Early on, it was recognized that epinephrine is a vasoconstrictor, a bronchodilator, and a cardiac stimulant. The fact that it has these three major effects has stimulated much research, one line of which has sought to develop compounds that are even more effective bronchodilators than epinephrine but that lack epinephrine's cardiac-stimulating and vasoconstricting effects.

Soon after epinephrine became commercially available, it became an important treatment for asthma and hayfever. It has been marketed for the relief of bronchospasms under several trade names, including Bronkaid Mist and Primatine Mist.

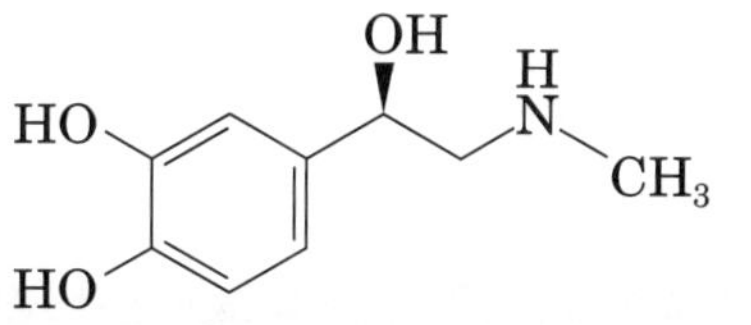

Epinephrine

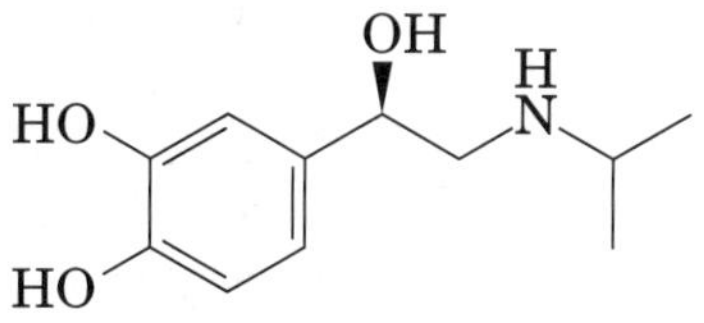

(R)-Isoproterenol

One of the most important of the synthetic catecholamines was isoproterenol, the levorotatory enantiomer of which retains the bronchodilating effects of epinephrine but is free from its cardiac-stimulating effects. (R)-Isoproterenol was introduced into clinical medicine in 1951 and for the next two decades was the drug of choice for the treatment of asthmatic attacks. Interestingly, the hydrochloride salt of (S)-isoproterenol is a nasal decongestant and was marketed under several trade names, including Sudafed.

A problem with the first synthetic catecholamines (and with epinephrine itself) is the fact that they are inactivated by an enzyme-catalyzed reaction that converts one of the —OH groups on the catechol unit to an OCH_3 group. A strategy to circumvent this enzyme-catalyzed inactivation was to replace the catechol unit with one that would allow the drug to bind to the catecholamine receptors in the bronchi but would not be inactivated by this enzyme.

In terbutaline (Brethaire), inactivation is prevented by placing the —OH groups meta to each other on the aromatic ring. In addition, the isopropyl group of isoproterenol is replaced by a *tert*-butyl group. In albuterol (Proventil), the commercially most successful of the antiasthma medications, one —OH group of the catechol unit is replaced by a —CH_2OH group and the isopropyl group is replaced by a *tert*-butyl group. When terbutaline and albuterol were introduced into clinical medicine in the 1960s, they almost immediately replaced isoproterenol as the drugs of choice for the treatment of asthmatic attacks.

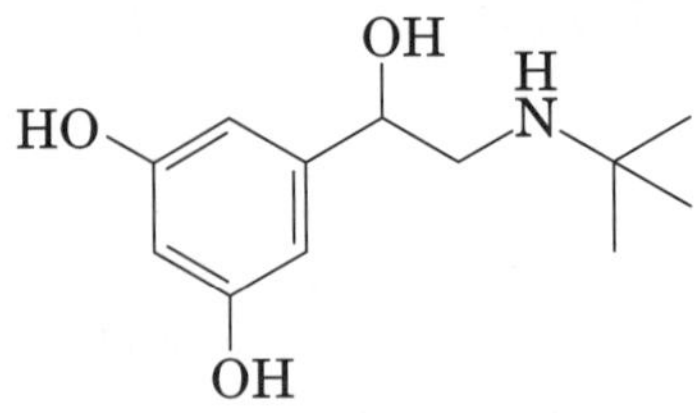

Terbutaline

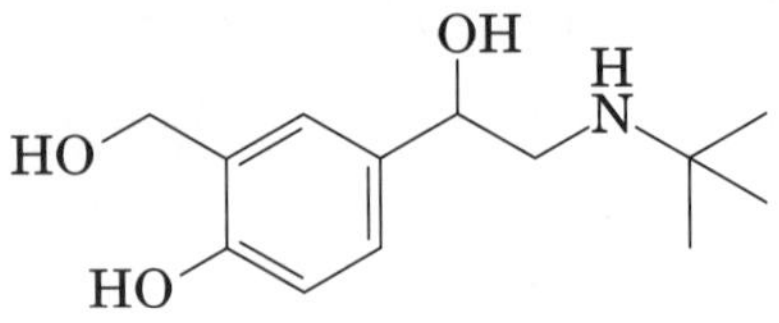

Albuterol

In their search for a longer-acting bronchodilator, scientists reasoned that extending the side chain on nitrogen might strengthen the binding of the drug to the adrenoreceptors in the lungs, thereby increasing the duration of the drug's action. A result of this line of reasoning is salmeterol (Serevent), a bronchodilator that is approximately ten times more potent than albuterol and much longer acting.

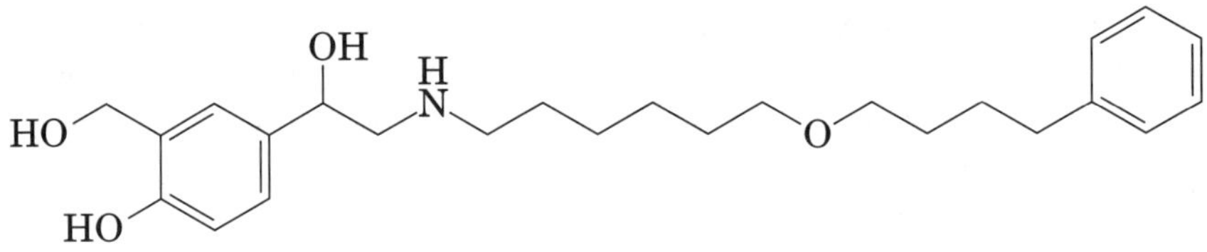

Salmeterol

SUMMARY

Amines are classified as **primary, secondary,** or **tertiary,** depending on the number of carbon atoms bonded to nitrogen (Section 8.2). In an **aliphatic amine,** all carbon atoms bonded to nitrogen are derived from alkyl groups. In an **aromatic amine,** one or more of the groups bonded to nitrogen are aryl groups. In a **heterocyclic amine,** the nitrogen atom is part of a ring.

In IUPAC nomenclature, aliphatic amines are named by changing the final *-e* of the parent alkane to *-amine* and using a number to indicate the location of the amino group on the parent chain (Section 8.3A). In the common system of nomenclature, aliphatic amines are named by listing carbon groups in alphabetical order in one word ending in the suffix *-amine* (Section 8.3B).

Amines are polar compounds, and primary and secondary amines associate by intermolecular hydrogen bonding (Section 8.4). All classes of amines form hydrogen bonds with water and are more soluble in water than are hydrocarbons of comparable molecular weight.

Amines are weak bases, and aqueous solutions of amines are basic. The base ionization constant for an amine in water is denoted by the symbol K_b (Section 8.5). Aliphatic amines are stronger bases than are aromatic amines.

All amines, whether soluble or insoluble in water, react with strong acids to form water-soluble salts (Section 8.6). We can use this chemical property to separate water-insoluble amines from water-insoluble neutral organic compounds.

KEY REACTIONS

1. **Basicity of Aliphatic Amines (Section 8.5)** Most aliphatic amines have about the same basicity (pK_b 3.0–4.0) and are slightly stronger bases than ammonia (pK_b 4.74).

$$CH_3NH_2 + H_2O \rightleftharpoons CH_3NH_3^+ + OH^-$$
$$pK_b = 3.36$$

2. **Basicity of Aromatic Amines (Section 8.5)** Most aromatic amines (pK_b 9.0–10.0) are considerably weaker bases than ammonia and aliphatic amines.

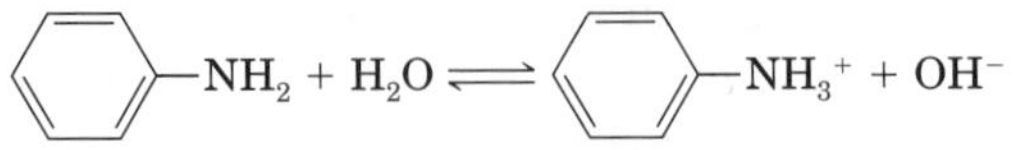

$$pK_b = 9.37$$

3. **Reaction with Acids (Section 8.5)** All amines, whether water-soluble or water-insoluble, react quantitatively with strong acids to form water-soluble salts.

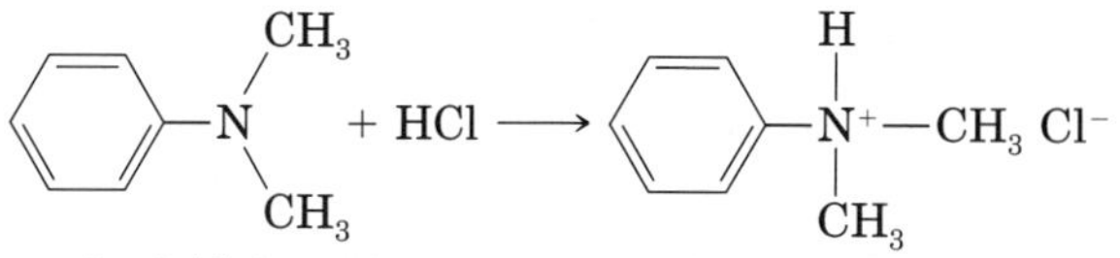

Insoluble in water   A water-soluble salt

P R O B L E M S

Numbers that appear in color indicate difficult problems.
▶ designates problems requiring application of principles.

Structure and Nomenclature

8.6 We use the terms primary, secondary, and tertiary to classify both alcohols and amines. Cyclohexanol, for example, is classified as a secondary alcohol, and cyclohexanamine is classified as a primary amine. In each compound, the functional group is attached to a carbon of the cyclohexane ring. Explain why one compound is classified as secondary whereas the other is primary.

8.7 What is the difference in structure between an aliphatic amine and an aromatic amine?

8.8 In what way are pyridine and pyrimidine related to benzene?

8.9 Draw a structural formula for each amine.
(a) 2-Butanamine
(b) 1-Octanamine
(c) 2,2-Dimethyl-1-propanamine
(d) 1,5-Pentanediamine
(e) 2-Bromoaniline
(f) Tributylamine

8.10 Draw a structural formula for each amine.
(a) 4-Methyl-2-pentanamine
(b) *trans*-2-Aminocyclohexanol
(c) *N,N*-Dimethylaniline
(d) Dicyclohexylamine
(e) *sec*-Butylamine
(f) 2,4-Dimethylaniline

8.11 Classify each amino group as primary, secondary, or
▶ tertiary, and as aliphatic or aromatic.

(a)

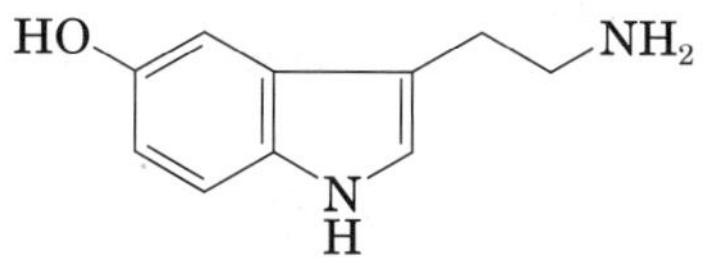

Serotonin
(a neurotransmitter)

(b)

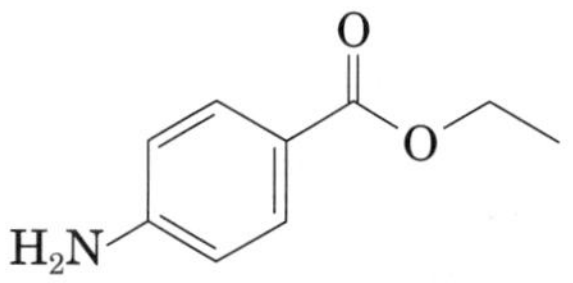

Benzocaine
(a topical anesthetic)

(c)

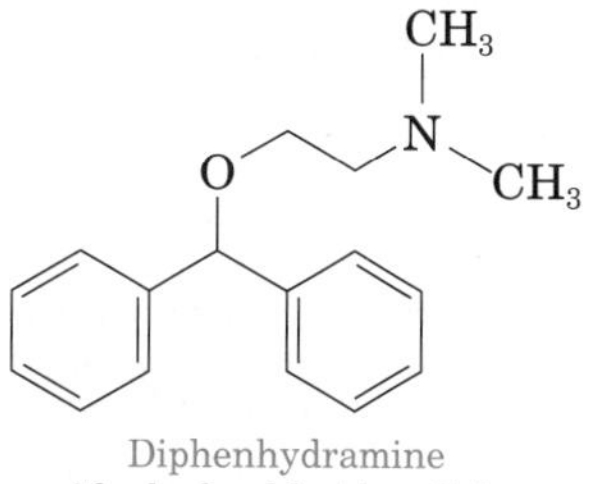

Diphenhydramine
(the hydrochloride salt is
the antihistamine Benadryl)

8.12 Classify each amino group as primary, secondary, or tertiary, and as aliphatic or aromatic.

(a)

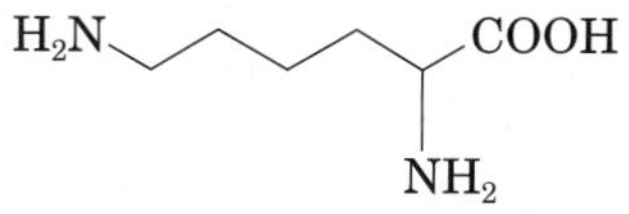

Lysine
(an amino acid)

(b)

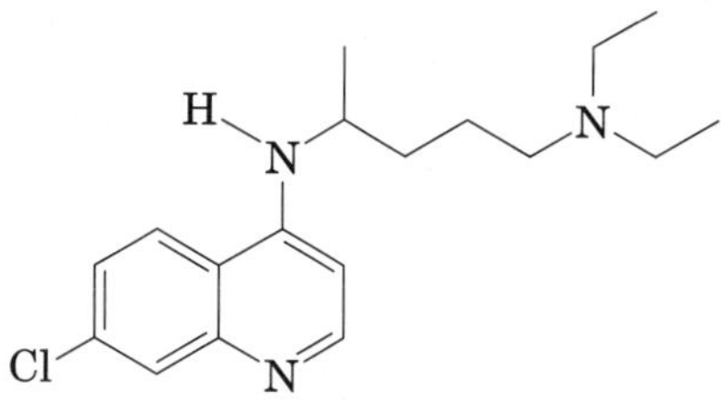

Chloroquine
(an antimalaria drug)

(c) H_2N⌒⌒COOH

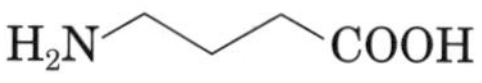

4-Aminobutanoic acid
(a neurotransmitter)

8.13 There are eight constitutional isomers with molecular formula $C_4H_{11}N$.

(a) Name and draw a structural formula for each amine.

(b) Classify each amine as primary, secondary, or tertiary.

(c) Which are chiral?

8.14 There are eight primary amines with molecular formula $C_5H_{13}N$.

(a) Name and draw a structural formula for each amine.

(b) Which are chiral?

Physical Properties

8.15 Propylamine (bp 48°C), ethylmethylamine (bp 37°C), and trimethylamine (bp 3°C) are constitutional isomers with molecular formula C_3H_9N. Account for the fact that trimethylamine has the lowest boiling point of the three.

8.16 Account for the fact that 1-butanamine (bp 78°C) has a lower boiling point than 1-butanol (bp 117°C).

8.17 2-Methylpropane (bp −12°C), 2-propanol (bp 82°C), and 2-propanamine (bp 32°C) all have approximately the same molecular weight, yet their boiling points are quite different. Explain the reason for these differences.

8.18 Account for the fact that most low-molecular-weight amines are very soluble in water whereas low-molecular-weight hydrocarbons are not.

Basicity of Amines

8.19 Compare the basicities of amines with those of alcohols.

8.20 Write structural formulas for these amine salts.

(a) Ethyltrimethylammonium hydroxide

(b) Dimethylammonium iodide

(c) Tetramethylammonium chloride

(d) Anilinium bromide

8.21 Name these amine salts.

(a) $CH_3CH_2NH_3^+Cl^-$

(b) $(CH_3CH_2)_2NH_2^+Cl^-$

(c) 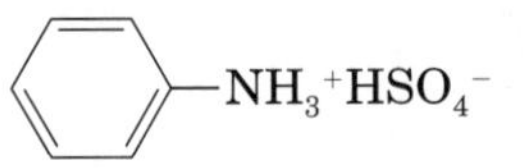 —$NH_3^+HSO_4^-$

8.22 From each pair of compounds, select the stronger base.

(a)

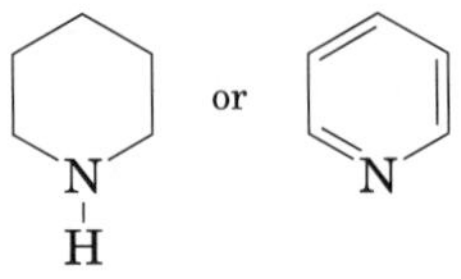

(b)

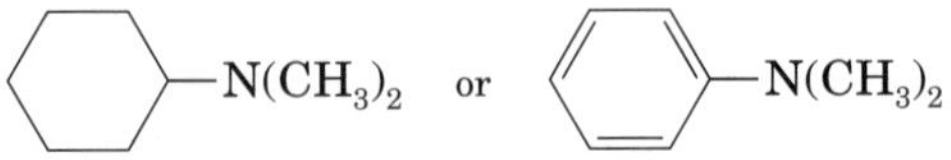

(c)

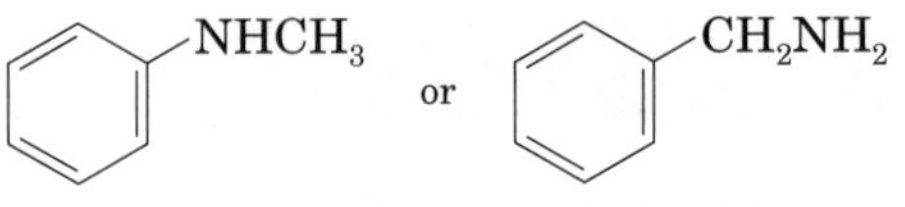

8.23 Following are two structural formulas for 4-aminobutanoic acid, a neurotransmitter. Is this compound better represented by structural formula (A) or (B)? Explain.

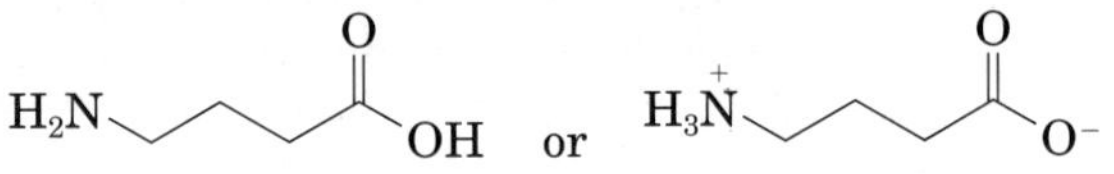

8.24 Suppose you have two test tubes, one containing 2-methylcyclohexanol and the other containing 2-methylcyclohexanamine, both insoluble in water, but you do not know which test tube contains which compound. Describe a simple chemical test by which you could tell which compound is the alcohol and which is the amine.

8.25 Complete the equations for the following acid–base reactions.

(a)

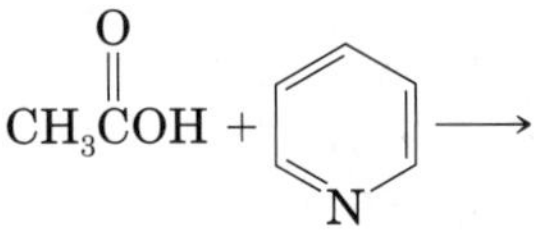

Acetic acid   Pyridine

(b)

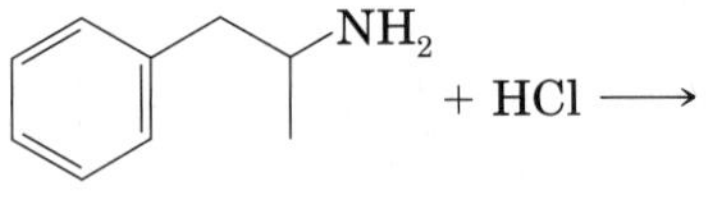

1-Phenyl-2-propanamine
(Amphetamine)

(c)

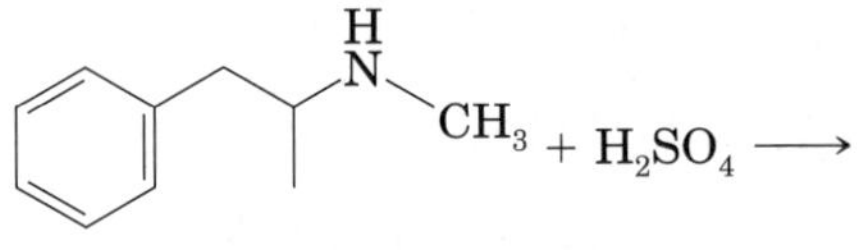

Methamphetamine

8.26 The pK_b of amphetamine [Problem 8.25(b)] is approximately 3.2.

(a) Which form of amphetamine would you expect to be present at pH 1.0, the pH of stomach acid?

(b) Which form of amphetamine would you expect to be present at pH 7.40, the pH of blood plasma?

8.27 Pyridoxamine is one form of vitamin B_6.

Pyridoxamine
(Vitamin B_6)

$$HO \overset{CH_2NH_2}{\underset{H_3C}{\bigcirc}} CH_2OH$$

(a) Which nitrogen atom of pyridoxamine is the stronger base?

(b) Draw the structural formula of the salt formed when pyridoxamine is treated with one mole of HCl.

8.28 Many tumors of the breast are correlated with estrogen levels in the body. Drugs that interfere with estrogen binding have antitumor activity and may even help prevent tumor occurrence. A widely used antiestrogen drug is tamoxifen.

Tamoxifen

(a) Name the functional groups in tamoxifen.

(b) Classify the amino group in tamoxifen as primary, secondary, or tertiary.

(c) How many stereoisomers are possible for tamoxifen?

8.29 Classify each amino group in epinephrine and albuterol (Section 8.7) as primary, secondary, or tertiary. In addition, list the similarities and differences between the structural formulas of these two compounds.

Chemical Connections

8.30 (Chemical Connections 8A) What are the differences in structure between the natural hormone epinephrine (Section 8.7) and the synthetic pep pill amphetamine? Between amphetamine and methamphetamine?

8.31 (Chemical Connections 8A) What are the possible negative effects of illegal use of amphetamines such as methamphetamine?

8.32 (Chemical Connections 8B) What is an alkaloid? Are all alkaloids basic as measured by a litmus test?

8.33 (Chemical Connections 8B) Identify all stereocenters in coniine and nicotine. How many stereoisomers are possible for each?

8.34 (Chemical Connections 8B) Of the two nitrogen atoms in nicotine, which is converted to its salt by reaction with one mole of HCl? Draw the structural formula of this salt.

8.35 (Chemical Connections 8B) Cocaine has four stereocenters. Identify each. Draw the structural formula for the salt formed by treatment of cocaine with one mole of HCl.

8.36 (Chemical Connections 8C) What structural feature is common to all benzodiazepines?

8.37 (Chemical Connections 8C) Is Librium chiral? Is Valium chiral?

8.38 (Chemical Connections 8C) Benzodiazepines affect neural pathways in the central nervous system mediated by GABA, whose IUPAC name is 4-aminobutanoic acid. Draw a structural formula of GABA.

8.39 (Chemical Connections 8D) Suppose you saw this label on a decongestant: phenylephrine ·HCl. Should you worry about being exposed to a strong acid such as HCl? Explain.

8.40 (Chemical Connections 8D) Give two reasons why amine-containing drugs are most commonly administered as their salts.

Additional Problems

8.41 Draw a structural formula for a compound with each molecular formula.

(a) A 2° aromatic amine, C_7H_9N

(b) A 3° aromatic amine, $C_8H_{11}N$

(c) A 1° aliphatic amine, C_7H_9N

(d) A chiral 1° amine, $C_4H_{11}N$

(e) A 3° heterocyclic amine, $C_5H_{11}N$

(f) A trisubstituted 1° aromatic amine, $C_9H_{13}N$

(g) A chiral quaternary ammonium salt, $C_9H_{22}NCl$

8.42 Arrange these three compounds in order of decreasing ability to form intermolecular hydrogen bonds: CH_3OH, CH_3SH, and $(CH_3)_2NH$.

8.43 Consider these three compounds: CH_3OH, CH_3SH, and $(CH_3)_2NH$.

(a) Which is the strongest acid?

(b) Which is the strongest base?

(c) Which has the highest boiling point?

(d) Which forms the strongest hydrogen bonds in the pure state?

8.44 Arrange these compounds in order of increasing boiling point: $CH_3CH_2CH_2CH_3$, $CH_3CH_2CH_2OH$, and $CH_3CH_2CH_2NH_2$.

8.45 Account for the fact that amines have about the same solubility in water as alcohols of similar molecular weight.

8.46 If you dissolved $CH_3CH_2CH_2OH$ and $CH_3CH_2CH_2NH_2$ in the same container of water and lowered the pH of the solution to 2 by adding HCl, would anything happen to the structures of these compounds? If so, write the formula of the species present in solution at pH 2.

8.47 The compound phenylpropanolamine hydrochloride is used as both a decongestant and an anorexic. The chemical name of this compound is 2-amino-1-phenyl-1-propanol.

(a) Draw the structural formula of 2-amino-1-phenyl-1-propanol.

(b) How many stereocenters are present in this molecule? How many stereoisomers are possible for it?

8.48 Procaine was one of the first local anesthetics. Its hydrochloride salt is marketed as Novocain.

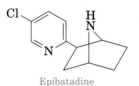

Procaine

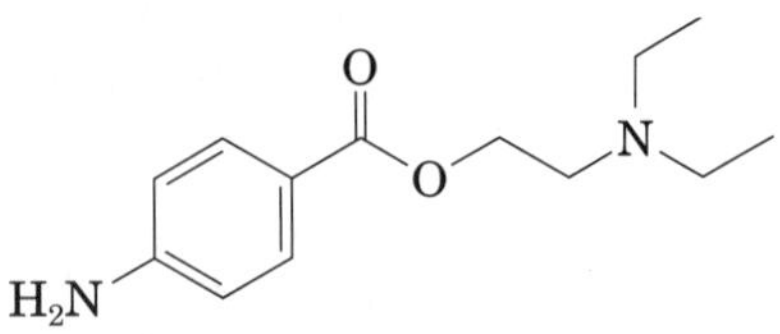

(a) Is procaine chiral? Does it contain a stereocenter?

(b) Which nitrogen atom of procaine is the stronger base?

(c) Draw a structural formula of the salt formed by treating procaine with one mole of HCl, showing which nitrogen is protonated and bears the positive charge.

8.49 Following are two structural formulas for gabapentin, a drug used to treat epilepsy. Is gabapentin better represented by structure (A) or (B)? Explain.

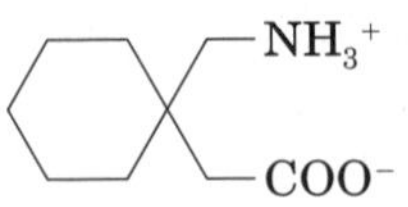

(A)   or   (B)

8.50 Several poisonous plants, including *Atropa belladonna,* contain the alkaloid atropine. The name "belladonna" (which means beautiful lady) probably comes from the fact that Roman women used extracts from this plant to make themselves more attractive. Atropine is widely used by ophthalmologists and optometrists to dilate the pupils for eye examination.

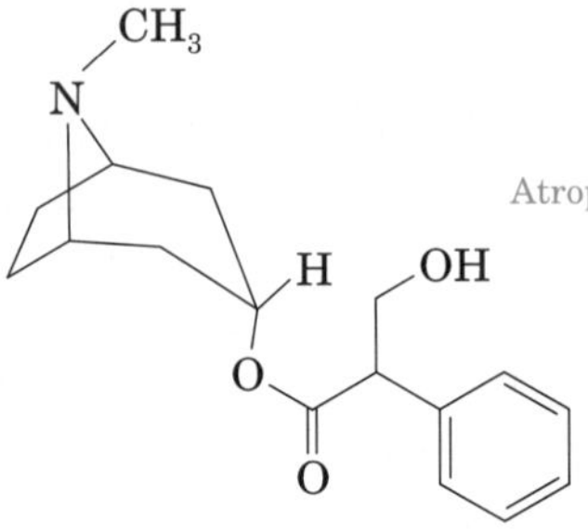

Atropine

(a) Classify the amino group in atropine as primary, secondary, or tertiary.

(b) Locate the single stereocenter in atropine.

(c) Account for the fact that atropine is almost insoluble in water (1 g in 455 mL cold water), but atropine hydrogen sulfate is very soluble (1 g in 5 mL of cold water).

(d) Account for the fact that a dilute aqueous solution of atropine is basic (pH approximately 10.0).

8.51 Epibatadine, a colorless oil isolated from the skin of the Equadorian poison frog *Epipedobates tricolor,* has several times the analgesic potency of morphine. It is the first chlorine-containing, non-opioid (nonmorphine-like in structure) analgesic ever isolated from a natural source.

Epibatadine

■ **Poison arrow frog.**

(a) Which of the two nitrogens in epibatadine is the stronger base?

(b) Mark the three stereocenters in this molecule.

InfoTrac College Edition

For additional readings, go to InfoTrac College Edition, your online research library, at

http://infotrac.thomsonlearning.com

CHAPTER 9

9.1 *Introduction*

9.2 *Structure and Bonding*

9.3 *Nomenclature*

9.4 *Physical Properties*

9.5 *Oxidation*

9.6 *Reduction*

9.7 *Addition of Alcohols*

9.8 *Keto-Enol Tautomerism*

Benzaldehyde is found in the kernels of bitter almonds, and cinnamaldehyde is found in Ceylonese and Chinese cinnamon oils.

Charles D. Winters

Aldehydes and Ketones

9.1 Introduction

In this and the following chapter, we study the physical and chemical properties of compounds containing the **carbonyl group,** $C=O$. Because the carbonyl group is present in aldehydes, ketones, and carboxylic acids and their derivatives, it is one of the most important functional groups in organic chemistry. Its chemical properties are straightforward, and an understanding of its characteristic reaction patterns leads very quickly to an understanding of a wide variety of organic and biochemical reactions.

9.2 Structure and Bonding

The functional group of an **aldehyde** is a carbonyl group bonded to a hydrogen atom (Section 1.4C). In methanal, the simplest aldehyde, the carbonyl group is bonded to two hydrogen atoms. In other aldehydes, it is bonded to one hydrogen atom and one carbon atom. The functional group

183

of a ketone is a carbonyl group bonded to two carbon atoms (Section 1.4C).

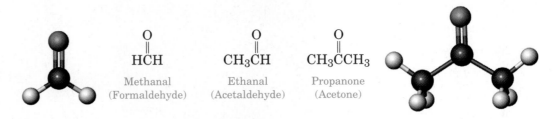

$$\overset{O}{\underset{}{\overset{\parallel}{HCH}}}$$

Methanal
(Formaldehyde)

$$\overset{O}{\underset{}{\overset{\parallel}{CH_3CH}}}$$

Ethanal
(Acetaldehyde)

$$\overset{O}{\underset{}{\overset{\parallel}{CH_3CCH_3}}}$$

Propanone
(Acetone)

Because aldehydes always contain at least one hydrogen bonded to the C=O group, they are often written RCH=O or RCHO. Similarly, ketones are often written RCOR′.

9.3 Nomenclature

See the **Interactive General, Organic, and Biochemistry CD-ROM, version 2.0,** for further exploration on this topic.

A IUPAC Names

The IUPAC names for aldehydes and ketones follow the familiar pattern of selecting as the parent alkane the longest chain of carbon atoms that contains the functional group (Section 2.4A). To name an aldehyde, we change the suffix *-e* of the parent alkane to *-al*. Because the carbonyl group of an aldehyde can appear only at the end of a parent chain and numbering must start with it as carbon 1, there is no need to use a number to locate the aldehyde group.

For **unsaturated aldehydes,** we show the presence of a carbon–carbon double bond and an aldehyde by changing the ending of the parent alkane from *-ane* to *-enal:* "-en-" to show the carbon–carbon double bond, and "-al" to show the aldehyde. We number the carbon chain with the aldehyde as carbon 1. We show the location of the carbon–carbon double bond by the number of its first carbon.

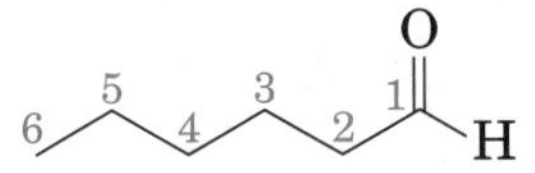

Hexanal

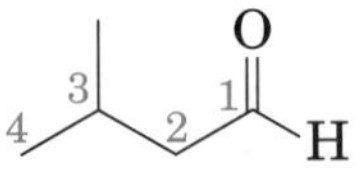

3-Methylbutanal

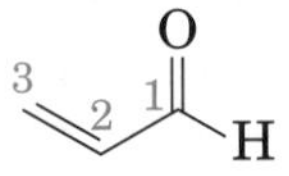

2-Propenal
(Acrolein)

In the IUPAC system, we name ketones by selecting as the parent alkane the longest chain that contains the carbonyl group and then indicating the presence of this group by changing the *-e* of the parent alkane *-one*. The parent chain is numbered from the direction that gives the smaller number to the carbonyl carbon. While the systematic name of the simplest ketone is 2-propanone, the IUPAC retains its common name, acetone.

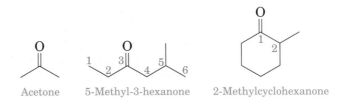

Acetone

5-Methyl-3-hexanone

2-Methylcyclohexanone

EXAMPLE 9.1

Write the IUPAC name for each compound:

(a) 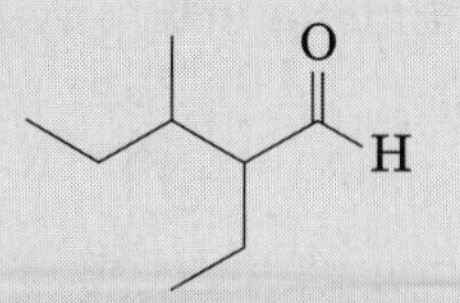 (b) 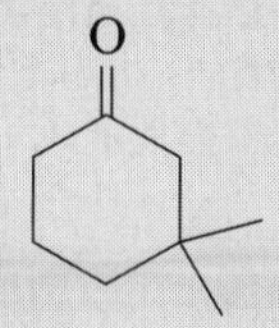 (c)

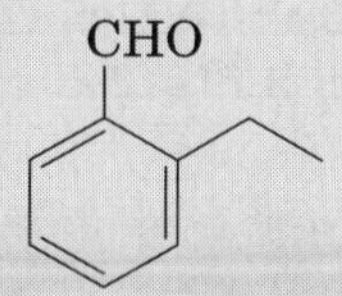

Solution

(a) The longest chain has six carbons, but the longest chain that contains the carbonyl carbon has five carbons. Its IUPAC name is 2-ethyl-3-methylpentanal.
(b) We number the six-membered ring beginning with the carbonyl carbon. Its IUPAC name is 3,3-dimethylcyclohexanone.
(c) This molecule is derived from benzaldehyde. Its IUPAC name is 2-ethylbenzaldehyde.

Problem 9.1

Write the IUPAC name for each compound. Specify the R or S configuration of (c).

(a) 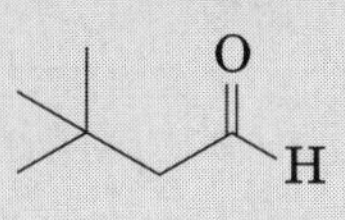 (b) 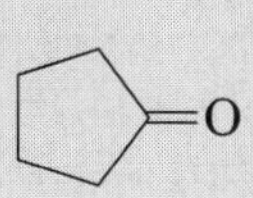 (c) C_6H_5—C

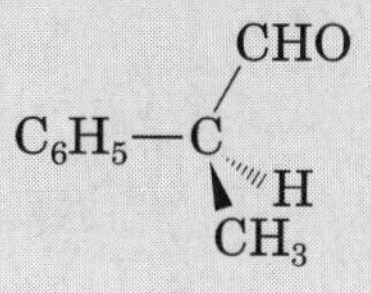

EXAMPLE 9.2

Write structural formulas for all ketones with molecular formula $C_6H_{12}O$ and give the IUPAC name of each. Which of these ketones are chiral?

Solution

There are six ketones with this molecular formula: two with a six-carbon chain, three with a five-carbon chain and a methyl branch, and one with a four-carbon chain and two methyl branches. Only 3-methyl-2-pentanone has a stereocenter and is chiral.

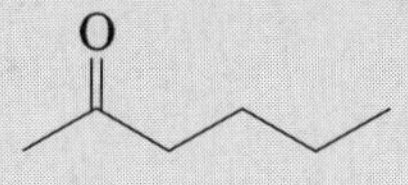

2-Hexanone	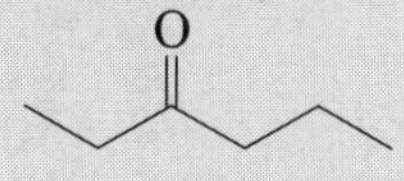 3-Hexanone	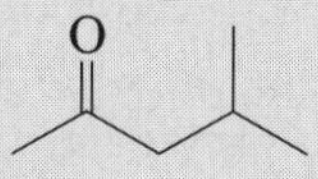 4-Methyl-2-pentanone
3-Methyl-2-pentanone	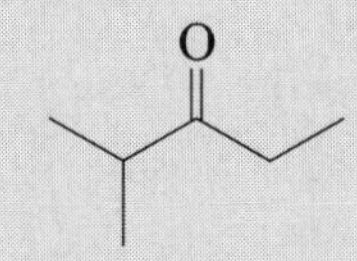 2-Methyl-3-pentanone	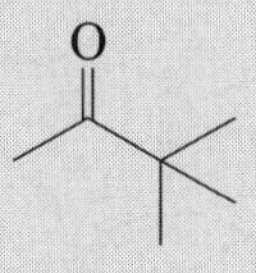 3,3-Dimethyl-2-butanone

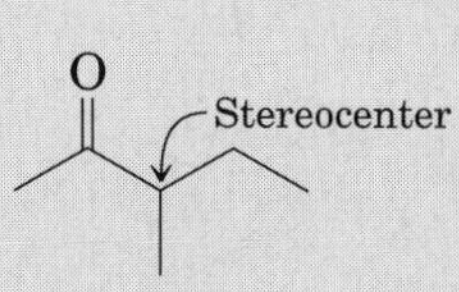

Stereocenter

Problem 9.2

Write structural formulas for all aldehydes with molecular formula $C_6H_{12}O$ and give each its IUPAC name. Which of these aldehydes are chiral?

In naming aldehydes or ketones that also contain an —OH or —NH$_2$ group elsewhere in the molecule, the parent chain is numbered to give the lower number to the carbonyl group. An —OH substituent is indicated by *hydroxy,* and an —NH$_2$ substituent is indicated by *amino-*. Hydroxy and amino substituents are numbered and alphabetized along with any other substituents that might be present.

EXAMPLE 9.3

Write the IUPAC name for each compound.

(a) 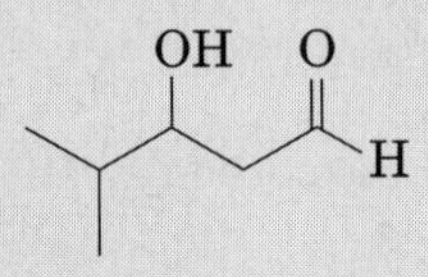 (b)

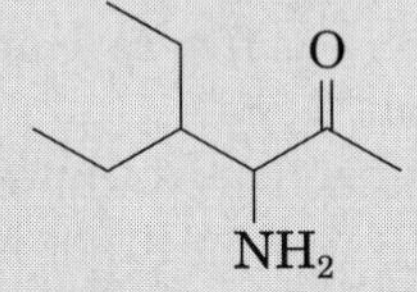

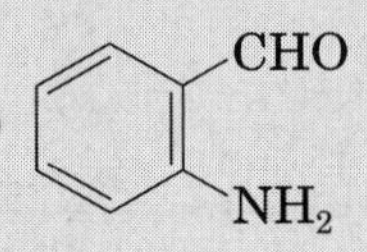

Solution

(a) We number the parent chain beginning with the CHO as carbon 1. There is a hydroxyl group on carbon 3 and a methyl group on carbon 4. The IUPAC name of this compound is 3-hydroxy-4-methylpentanal.

(b) The IUPAC name of this compound is 3-amino-4-ethyl-2-hexanone.

Problem 9.3

Write the IUPAC name for each compound.

(a) 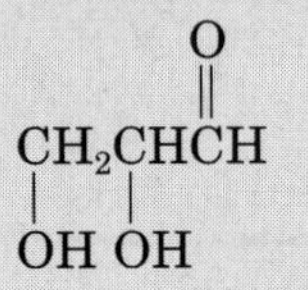 CH$_2$CHCH
    | |
    OH OH

(b)

(c)

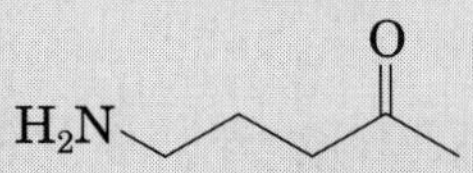

CHEMICAL CONNECTIONS 9A

Some Naturally Occurring Aldehydes and Ketones

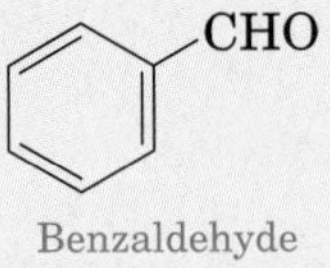

Benzaldehyde
(oil of almonds)

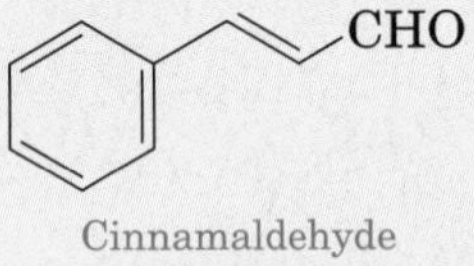

Cinnamaldehyde
(oil of cinnamon)

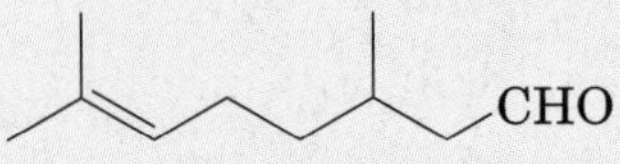

Citronellal
(citronella oils; also in lemon and lemon grass oils)

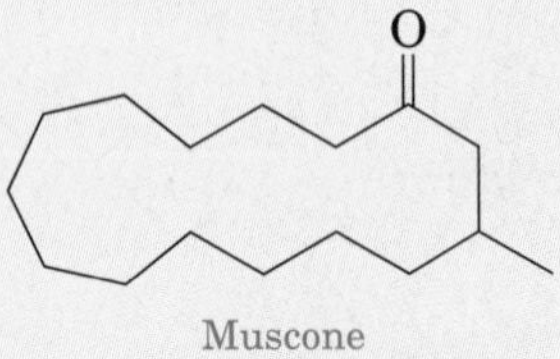

Muscone
(from the musk deer; used in perfumes)

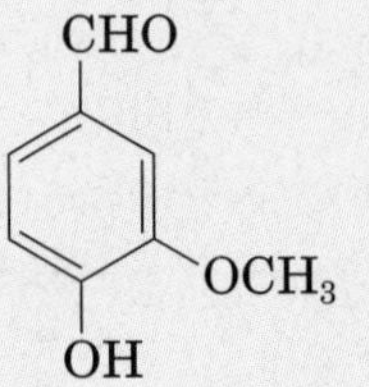

Vanillin
(vanilla bean)

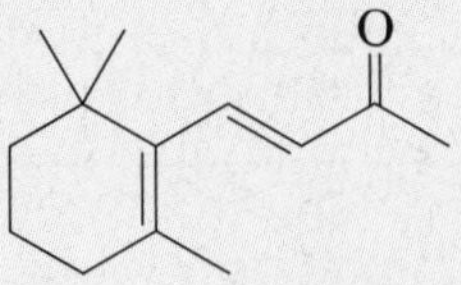

β-Ionone
(from violets)

B Common Names

We derive the common name for an aldehyde from the common name of the corresponding carboxylic acid. The word "acid" is dropped, and the suffix "-ic" or "-oic" is changed to "-*aldehyde*." Because we have not yet studied common names for carboxylic acids, we are not in a position to discuss common names for aldehydes. We can, however, illustrate how they are derived by reference to two common names with which you are familiar. The name formaldehyde is derived from formic acid, and the name acetaldehyde is derived from acetic acid.

$\overset{O}{\overset{\|}{HCH}}$	$\overset{O}{\overset{\|}{HCOH}}$	$\overset{O}{\overset{\|}{CH_3CH}}$	$\overset{O}{\overset{\|}{CH_3COH}}$
Formaldehyde	Formic acid	Acetaldehyde	Acetic acid

We derive common names for ketones by naming each alkyl or aryl group bonded to the carbonyl group as a separate word, followed by the word "ketone." The alkyl or aryl groups are generally listed in order of increasing molecular weight.

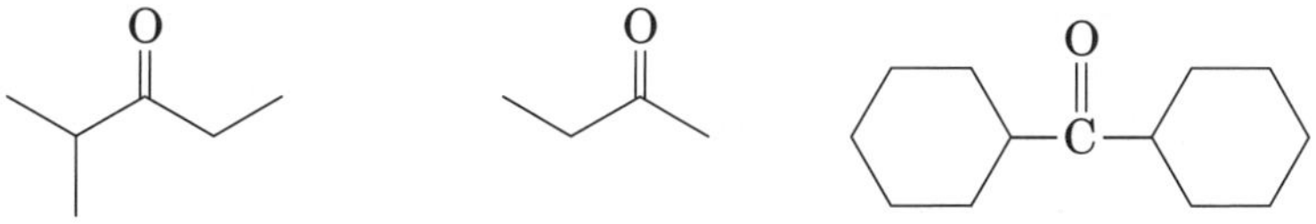

Ethyl isopropyl ketone   Methyl ethyl ketone   Dicyclohexyl ketone

2-Butanone, more commonly called methyl ethyl ketone (MEK), is used as a solvent for paints and varnishes.

9.4 Physical Properties

Oxygen is more electronegative than carbon: Therefore, a carbon–oxygen double bond is polar, with oxygen bearing a partial negative charge and carbon bearing a partial positive charge (Figure 9.1).

In liquid aldehydes and ketones, intermolecular attractions occur between the partial positive charge on the carbonyl carbon of one molecule and the partial negative charge on the carbonyl oxygen of another molecule. There is no possibility for hydrogen bonding between aldehyde or ketone molecules, which explains why these compounds have lower boiling points than alcohols (Section 5.2C) and carboxylic acids (Section 10.2), compounds in which there is hydrogen bonding between molecules.

Table 9.1 lists structural formulas and boiling points of six compounds of similar molecular weight. Of the six, pentane and diethyl ether have the lowest boiling points. The boiling point of 1-butanol, which can associate by

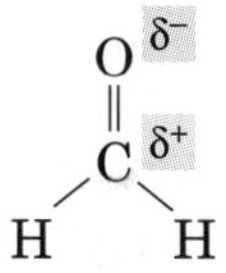

Polarity of a carbonyl group

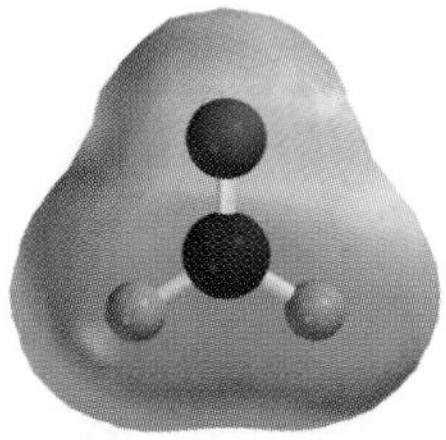

Figure 9.1 The polarity of a carbonyl group. Oxygen bears a partial negative charge and carbon bears a partial positive charge.

 See the **Interactive General, Organic, and Biochemistry CD-ROM, version 2.0,** for further exploration on this topic.

Table 9.1	Boiling Points of Six Compounds of Comparable Molecular Weight		
Name	**Structural Formula**	**Molecular Weight**	**Boiling Point (°C)**
diethyl ether	$CH_3CH_2OCH_2CH_3$	74	34
pentane	$CH_3CH_2CH_2CH_3CH_3$	72	36
butanal	$CH_3CH_2CH_2CHO$	72	76
2-butanone	$CH_3CH_2COCH_3$	72	80
1-butanol	$CH_3CH_2CH_2CH_2OH$	74	117
propanoic acid	CH_3CH_2COOH	72	141

intermolecular hydrogen bonding, is higher than that of either butanal or 2-butanone. Propanoic acid, in which intermolecular association by hydrogen bonding is the strongest, has the highest boiling point.

Because the oxygen atom of their carbonyl groups are hydrogen bond acceptors, the low-molecular-weight aldehydes and ketones are more soluble in water than are nonpolar compounds of comparable molecular weight. Formaldehyde, acetaldehyde, and acetone are infinitely soluble in water. As the hydrocarbon portion of the molecule increases in size, aldehydes and ketones become less soluble in water.

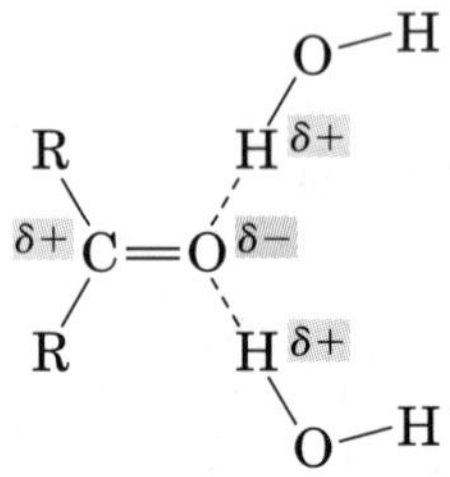

Most aldehydes and ketones have strong odors. The odors of ketones are generally pleasant, and many are used in perfumes and as flavoring agents. The odors of aldehydes vary. You may be familiar with the smell of formaldehyde; if so, you know that it is not pleasant. Many higher aldehydes, however, have pleasant odors and are used in perfumes.

9.5 Oxidation

Aldehydes are oxidized to carboxylic acids by a variety of oxidizing agents, including potassium dichromate (Section 5.3C).

The body uses nicotinamide adenine dinucleotide, NAD^+, for this type of oxidation (Section 26.3).

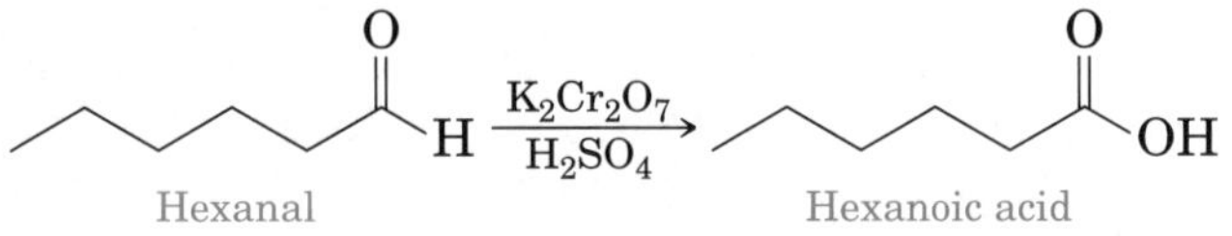

Hexanal   Hexanoic acid

They are also oxidized to carboxylic acids by the oxygen in the air. In fact, aldehydes that are liquid at room temperature are so sensitive to oxidation that they must be protected from contact with air during storage. Often this is done by sealing the aldehyde in a container under an atmosphere of nitrogen.

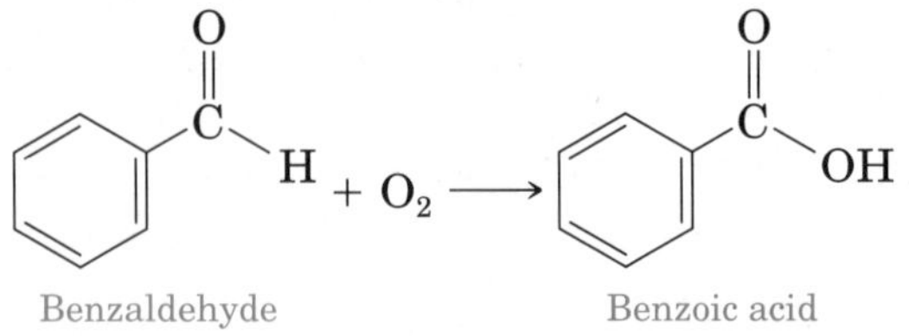

Benzaldehyde   Benzoic acid

*See the **Interactive General, Organic, and Biochemistry CD-ROM, version 2.0,** for further exploration on this topic.*

Ketones, in contrast, resist oxidation by most oxidizing agents, including potassium dichromate and molecular oxygen.

The fact that aldehydes are so easy to oxidize and ketones are not allows us to use simple chemical tests to distinguish between these types of compounds. Suppose that we have a compound we know is either an aldehyde or a ketone. To identify which it is, all we need to do is treat it with a mild oxidizing agent. If it can be oxidized, it is an aldehyde; otherwise, it is a ketone. One reagent that has been used for this purpose is Tollens' reagent.

Tollens' reagent contains silver nitrate and ammonia in water. When these two compounds are mixed, silver ion combines with NH_3 to form the ion $Ag(NH_3)_2^+$. When this solution is added to an aldehyde, the aldehyde acts as a reducing agent and reduces the complexed silver ion to silver metal. If this reaction is carried out properly, the silver metal precipitates as a smooth, mirror-like deposit, leading to the name **silver-mirror test.** If the remaining solution is then acidified with HCl, the carboxylic anion, $RCOO^-$, is converted to the carboxylic acid, RCOOH.

$$\underset{\text{Aldehyde}}{R-\overset{\overset{\displaystyle O}{\|}}{C}-H} + \underset{\substack{\text{Tollens'}\\\text{reagent}}}{2Ag(NH_3)_2^+} + 3OH^- \longrightarrow \underset{\substack{\text{Carboxylic}\\\text{anion}}}{R-\overset{\overset{\displaystyle O}{\|}}{C}-O^-} + \underset{\substack{\text{Silver}\\\text{mirror}}}{2Ag} + 4NH_3 + 2H_2O$$

Today, silver(I) is rarely used for the oxidation of aldehydes because of its high cost and because of the availability of other, more convenient methods for this oxidation. This reaction, however, is still used for making (silvering) mirrors.

■ **A silver mirror has been deposited on the inside of this flask by the reaction between an aldehyhde and Tollens' reagent.**

EXAMPLE 9.4

Draw a structural formula for the product formed by treating each compound with Tollens' reagent followed by acidification with aqueous HCl.
(a) Pentanal
(b) 4-Hydroxybenzaldehyde

Solution

The aldehyde group in each compound is oxidized to a carboxylic anion, $-COO^-$. Acidification with HCl converts the anion to a carboxylic acid, $-COOH$.

(a)

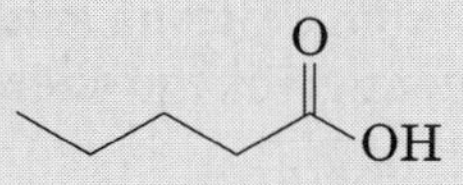

Pentanoic acid

(b)

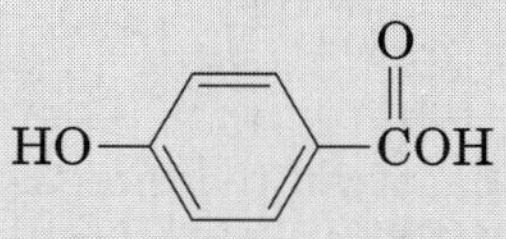

4-Hydroxybenzoic acid

Problem 9.4

Complete the equations for these oxidations.
(a) Hexanedial + $O_2 \longrightarrow$
(b) 3-Phenylpropanal + $Ag(NH_3)_2^+ \longrightarrow$

9.6 Reduction

In Section 3.6D, we saw that the $C=C$ double bond of an alkene can be reduced by hydrogen in the presence of a transition-metal catalyst to a $C-C$ single bond. The same is true of the $C=O$ double bond of an aldehyde or ketone. Aldehydes are reduced to primary alcohols and ketones are reduced to secondary alcohols.

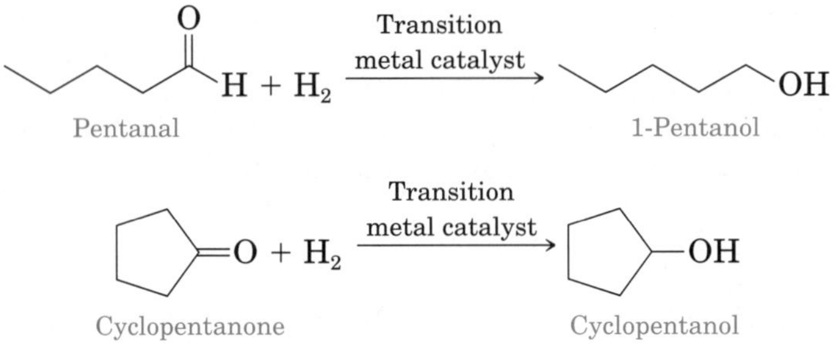

Pentanal → 1-Pentanol (Transition metal catalyst)

Cyclopentanone → Cyclopentanol (Transition metal catalyst)

The reduction of a C=O double bond under these conditions is slower than the reduction of a C=C double bond. Thus, if the same molecule contains both C=C and C=O double bonds, the C=C double bond is reduced first.

The most common reagent used in the laboratory for the reduction of an aldehyde or ketone is sodium borohydride, $NaBH_4$. This reagent contains hydrogen in the form of hydride ion, $H:^-$. In the hydride ion, hydrogen has two valence electrons and bears a negative charge. In a reduction by sodium borohydride, hydride ion is attracted to and then adds to the partially positive carbonyl carbon, which leaves a negative charge on the carbonyl oxygen. Reaction of this intermediate with aqueous acid gives the alcohol.

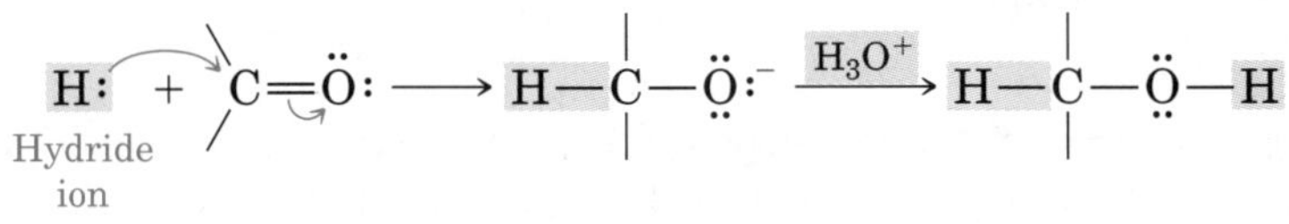

Of the two hydrogens added to the carbonyl group in this reduction, one comes from the reducing agent and the other comes from aqueous acid. Reduction of cyclohexanone, for example, with this reagent gives cyclohexanol:

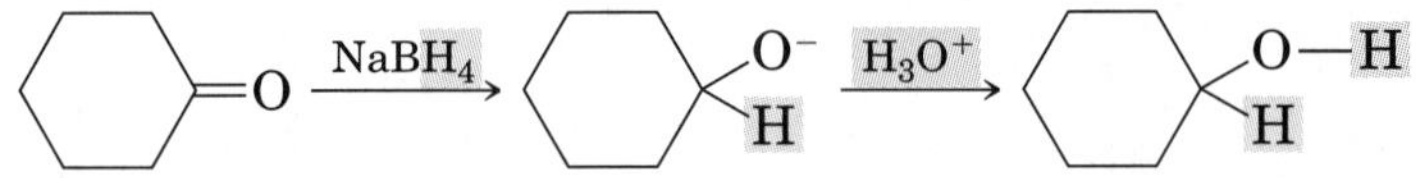

An advantage of $NaBH_4$ over the H_2/metal catalyst reaction is that $NaBH_4$ does not reduce carbon–carbon double bonds. The reason for this selectivity is quite straightforward. There is no polarity (no partial positive or negative charges) on a carbon–carbon double bond. Therefore, it has no site to attract the hydride ion. In the following reaction, $NaBH_4$ selectively reduces the aldehyde to a primary alcohol:

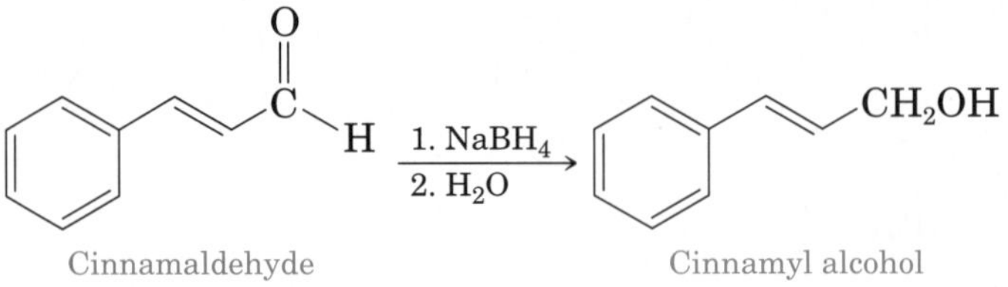

Cinnamaldehyde → Cinnamyl alcohol (1. NaBH₄ 2. H₂O)

In biological systems, the agent for the reduction of aldehydes and ketones is the reduced form of the coenzyme nicotinamide adenine dinucleotide, abbreviated NADH (Section 18.3). This reducing agent, like NaBH$_4$, delivers a hydride ion to the carbonyl carbon of the aldehyde or ketone. Reduction of pyruvate by NADH, for example, gives lactate:

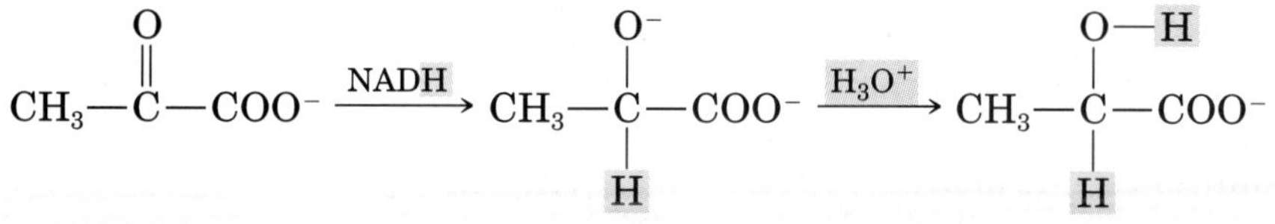

Pyruvate   Lactate

Pyruvate is the end product of glycolysis, a series of enzyme-catalyzed reactions that converts glucose to two molecules of this ketoacid (Section 19.1). Under anaerobic conditions, NADH reduces pyruvate to lactate. The build-up of lactate in the bloodstream leads to acidosis, and in muscle tissue is associated with muscle fatigue. When blood lactate reaches a concentration of about 0.4 mg/100 mL, muscle tissue becomes almost completely exhausted.

EXAMPLE 9.5

Complete the equations for these reductions.

(a)

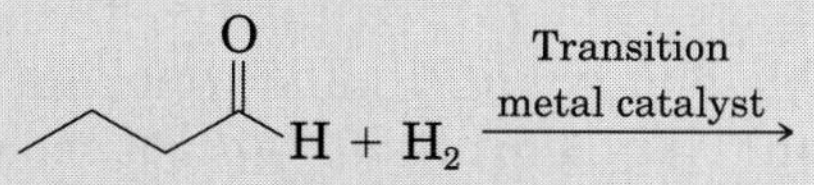

(b)

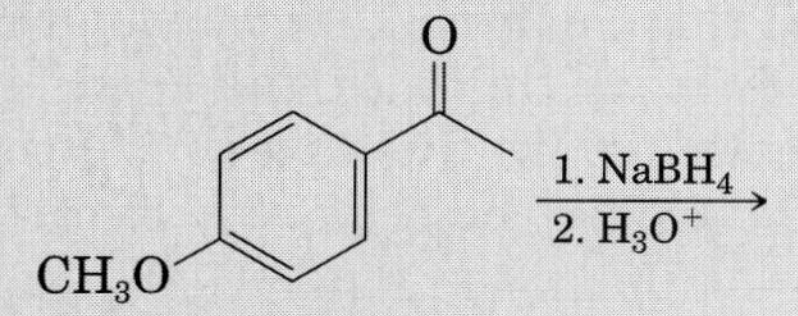

Solution

The carbonyl group of the aldehyde in (a) is reduced to a primary alcohol and that of the ketone in (b) is reduced to a secondary alcohol.

(a) 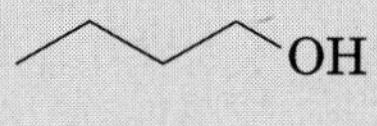   (b)

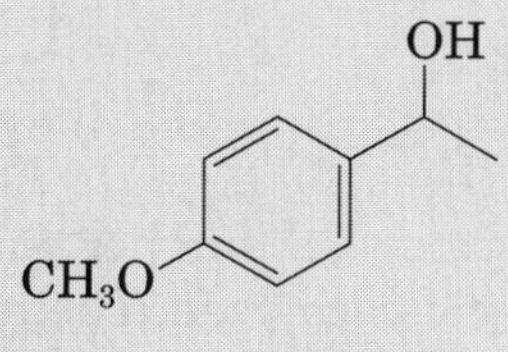

Problem 9.5

Which aldehyde or ketone gives these alcohols upon reduction with H$_2$/ metal catalyst?

(a)

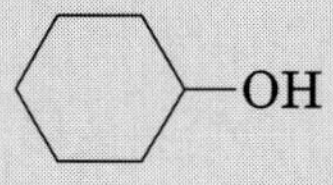

(b) 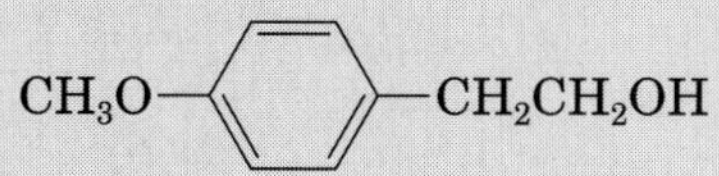 CH$_3$O—⬡—CH$_2$CH$_2$OH

(c)

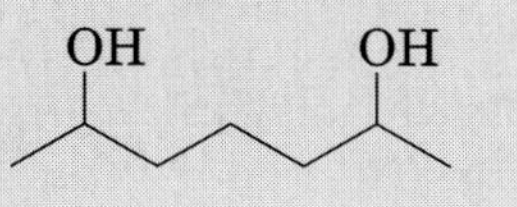

9.7 Addition of Alcohols

> **Hemiacetal** A molecule containing a carbon bonded to one —OH and one —OR group; the product of adding one molecule of alcohol to the carbonyl group of an aldehyde or ketone

Addition of a molecule of alcohol to the carbonyl group of an aldehyde or ketone forms a **hemiacetal** (a half-acetal). The functional group of a hemiacetal is a carbon bonded to one —OH group and one —OR group. In forming a hemiacetal, the H of the alcohol adds to the carbonyl oxygen and the OR adds to the carbonyl carbon. Shown here are the hemiacetals formed by addition of one molecule of ethanol to benzaldehyde and to cyclohexanone:

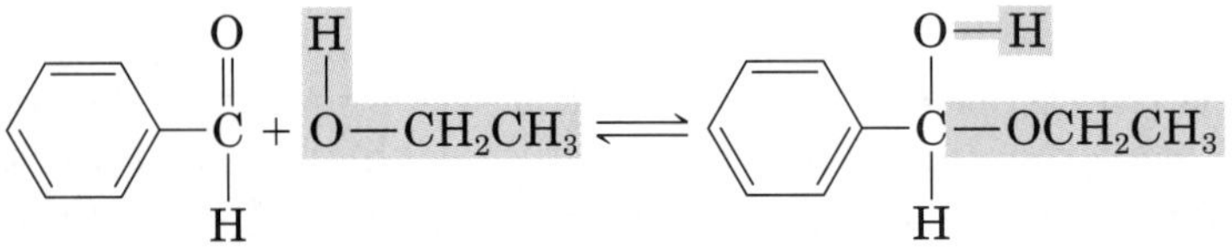

Benzaldehyde   Ethanol   A hemiacetal

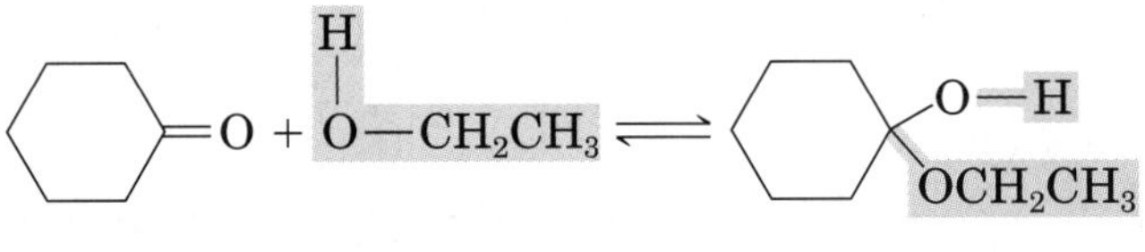

Cyclohexanone   Ethanol   A hemiacetal

Hemiacetals are generally unstable and are only minor components of an equilibrium mixture, except in one very important type of molecule. When a hydroxyl group is part of the same molecule that contains the carbonyl group and a five- or six-membered ring can form, the compound exists almost entirely in a cyclic hemiacetal form. In this case, the —OH group adds to the C=O group of the same molecule. We will have much more to say about cyclic hemiacetals when we consider the chemistry of carbohydrates in Chapter 11.

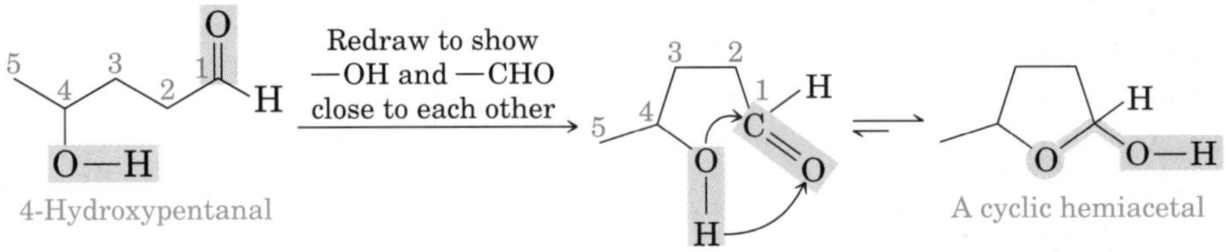

4-Hydroxypentanal   A cyclic hemiacetal

> **Acetal** A molecule containing two —OR groups bonded to the same carbon

Hemiacetals can react further with alcohols to form **acetals** plus water. This reaction is acid-catalyzed. The functional group of an acetal is a carbon bonded to two —OR groups.

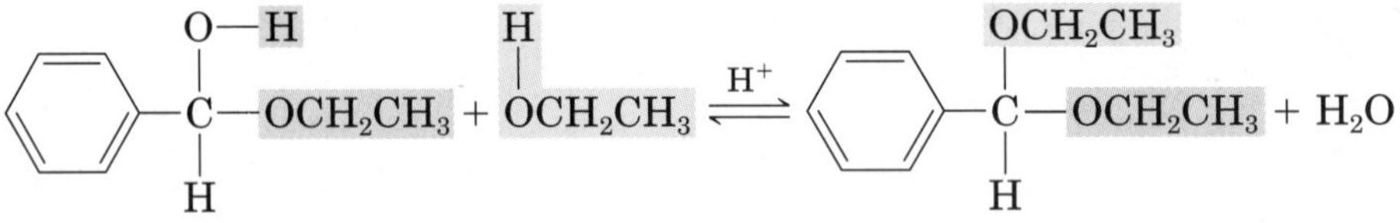

A hemiacetal   Ethanol   An acetal
(from benzaldehyde)

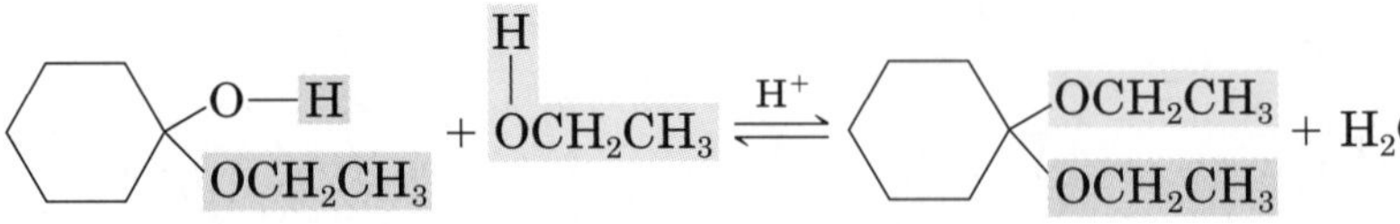

A hemiacetal   Ethanol   An acetal
(from cyclohexanone)

All steps in hemiacetal and acetal formation are reversible. As with any other equilibrium, we can make this one go in either direction by using Le Chatelier's principle. If we want to drive it to the right (formation of the acetal), we either use a large excess of alcohol or remove water from the equilibrium mixture. If we want to drive it to the left (hydrolysis of the acetal to the original aldehyde or ketone and alcohol), we use a large excess of water.

EXAMPLE 9.6

Show the reaction of 2-butanone with one molecule of ethanol to form a hemiacetal and then with a second molecule of ethanol to form an acetal.

Solution

Given are structural formulas of the hemiacetal and then the acetal.

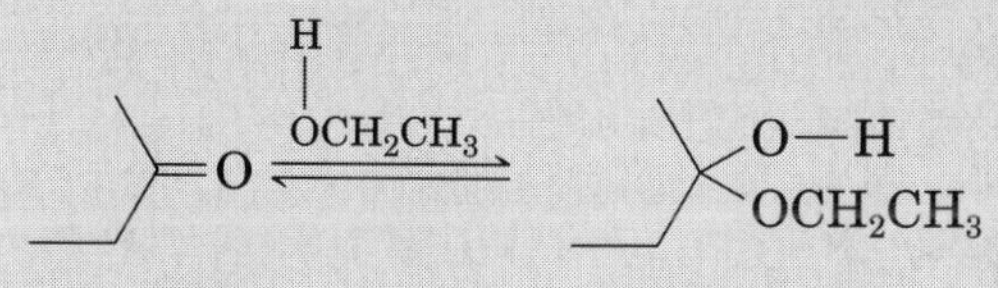

2-Butanone   A hemiacetal

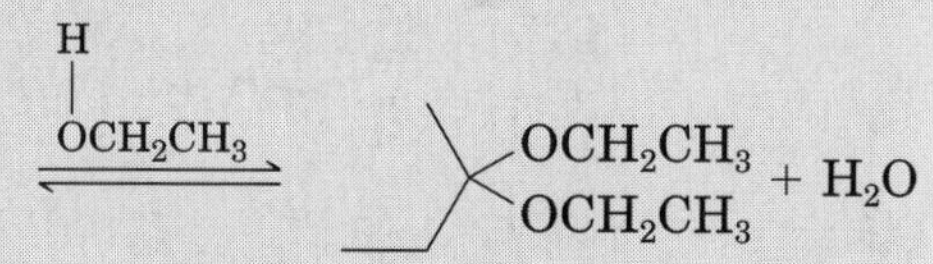

An acetal

Problem 9.6

Show the reaction of benzaldehyde with one molecule of methanol to form a hemiacetal and then with a second molecule of methanol to form an acetal.

EXAMPLE 9.7

Identify all hemiacetals and acetals in the following structures, and tell whether they are formed from aldehydes or ketones.

(a) 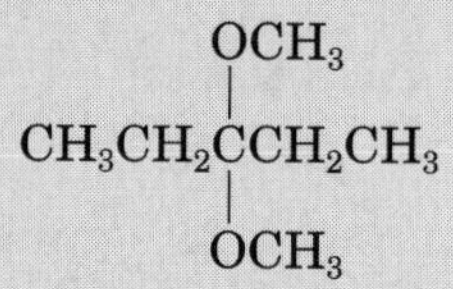 $CH_3CH_2\overset{\overset{\displaystyle OCH_3}{|}}{\underset{\underset{\displaystyle OCH_3}{|}}{C}}CH_2CH_3$

(b) $CH_3CH_2OCH_2CH_2OH$

(c)

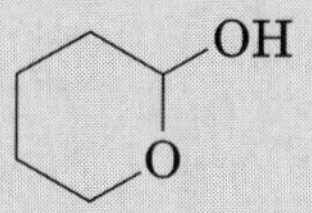

Solution

Compound (a) is an acetal derived from a ketone. Compound (b) is neither a hemiacetal nor an acetal because it does not have a carbon bonded to two oxygens; its functional groups are an ether and a primary alcohol. Compound (c) is a hemiacetal derived from an aldehyde.

(a) 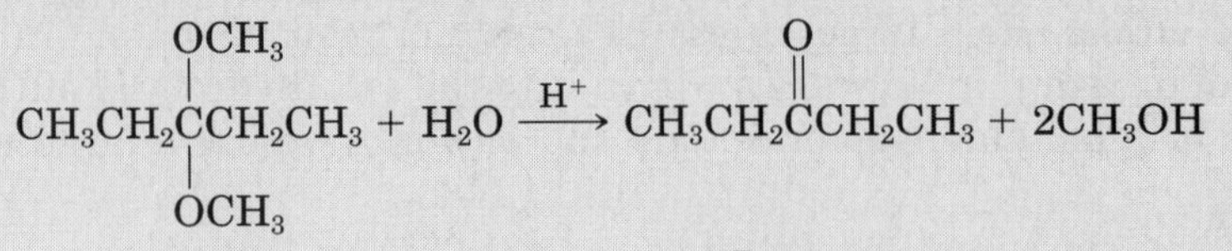
$$\underset{OCH_3}{\overset{OCH_3}{CH_3CH_2CCH_2CH_3}} + H_2O \xrightarrow{H^+} CH_3CH_2\overset{O}{\overset{\|}{C}}CH_2CH_3 + 2CH_3OH$$

An acetal   2-Pentanone

(c) 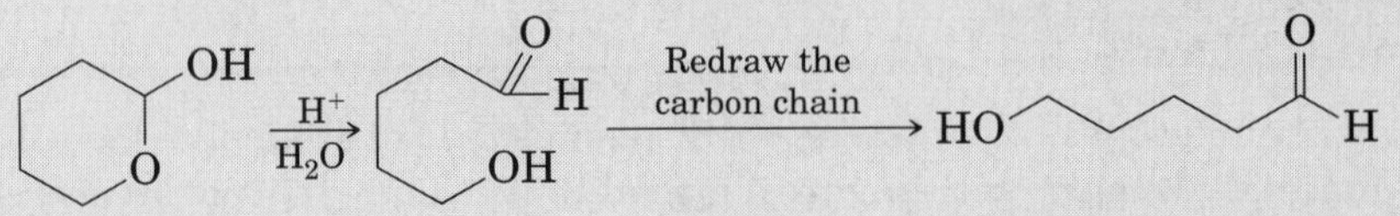 A hemiacetal $\xrightarrow[H_2O]{H^+}$ (aldehyde/OH) $\xrightarrow{\text{Redraw the carbon chain}}$ HO—...—CHO

A hemiacetal   5-Hydroxypentanal

Problem 9.7

Identify all hemiacetals and acetals in the following structures, and tell whether they are formed from aldehydes or ketones.

(a) 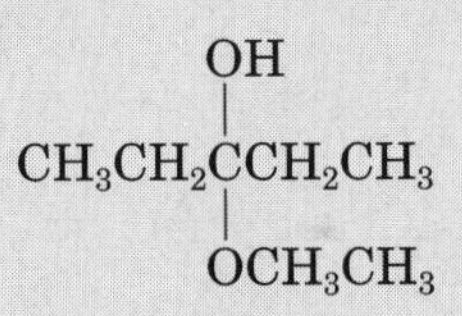
$$\underset{OCH_3CH_3}{\overset{OH}{CH_3CH_2CCH_2CH_3}}$$

(b) $CH_3OCH_2CH_2OCH_3$

(c) 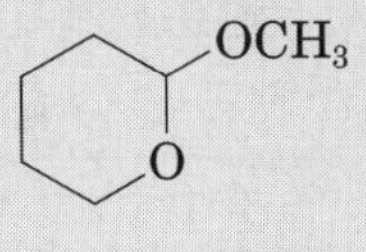 [cyclic structure with OCH_3 and O]

9.8 Keto-Enol Tautomerism

A carbon atom adjacent to a carbonyl group is called an α-carbon, and a hydrogen atom bonded to it is called an α-hydrogen.

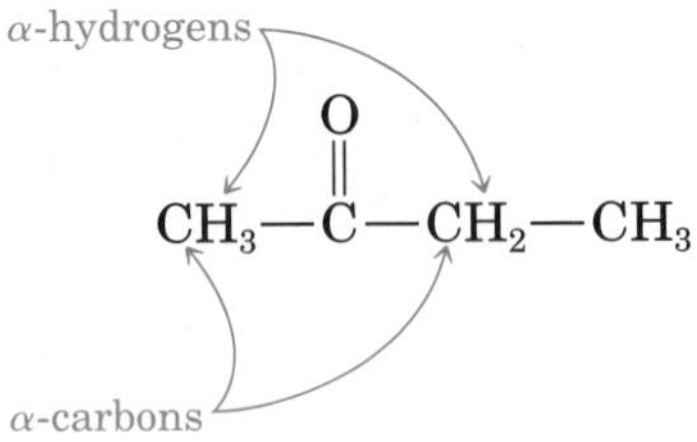

α-hydrogens

$$CH_3-\overset{O}{\overset{\|}{C}}-CH_2-CH_3$$

α-carbons

A carbonyl compound that has a hydrogen on an α-carbon is in equilibrium with a constitutional isomer called an **enol.** The name *enol* is derived from the IUPAC designation of it as both an alkene (*-en-*) and an alcohol (*-ol*).

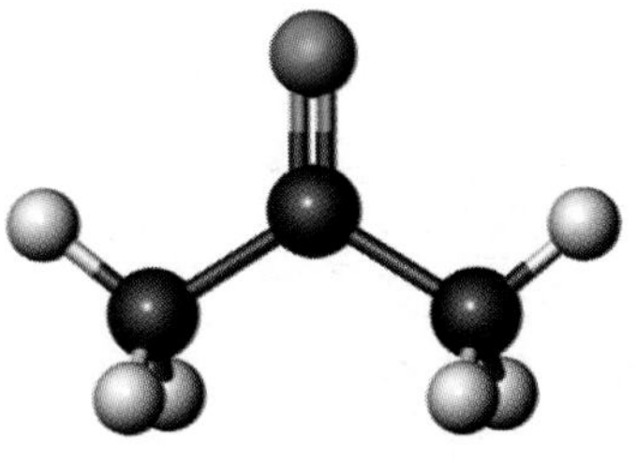

Enol A molecule containing an —OH group bonded to a carbon of a carbon–carbon double bond

$$CH_3-\overset{O}{\overset{\|}{C}}-CH_3 \rightleftharpoons CH_3-\overset{OH}{\overset{\|}{C}}=CH_2$$

Acetone
(keto form)

Acetone
(enol form)

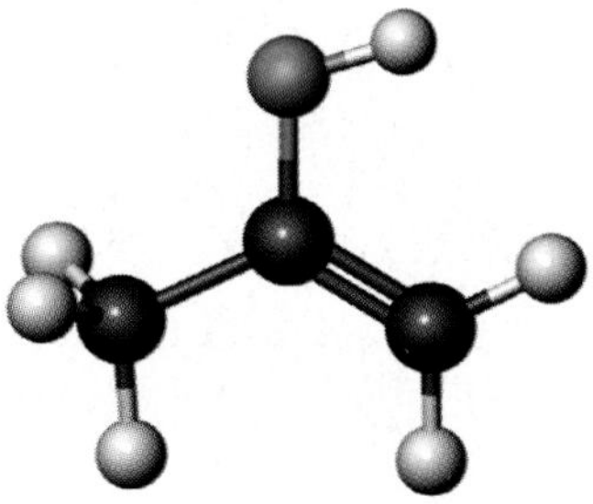

Keto and enol forms are examples of **tautomers,** constitutional isomers in equilibrium with each other that differ in the location of a hydrogen atom relative to an oxygen atom. This type of isomerism is called **keto-enol tautomerism.** For any pair of keto-enol tautomers, the keto form generally predominates at equilibrium.

> **Tautomers** Constitutional isomers that differ in the location of a hydrogen atom and a double bond relative to an oxygen or nitrogen atom

EXAMPLE 9.8

Draw structural formulas for the two enol forms for each ketone.

(a)

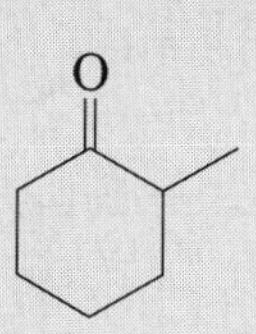

(b)

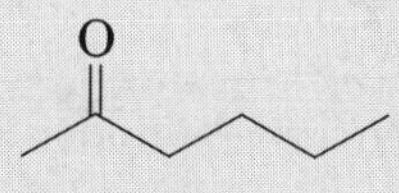

Solution

(a)

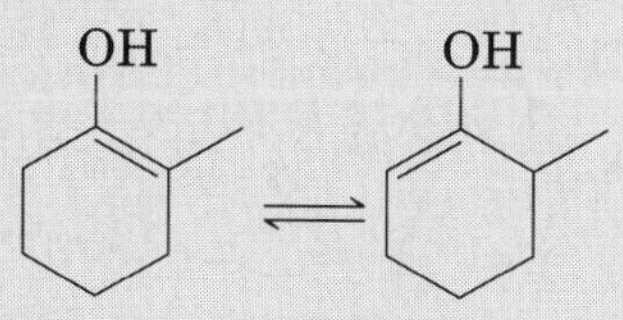

(b)

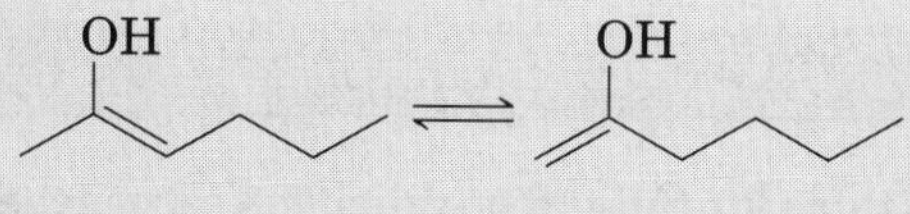

Problem 9.8

Draw a structural formula for the keto form of each enol.

(a) 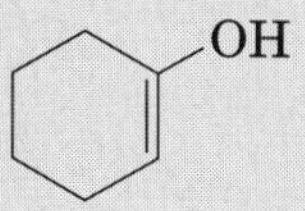   (b) 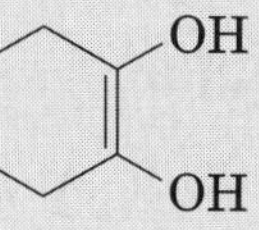   (c)

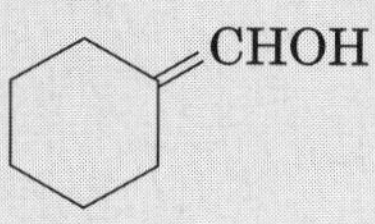

S U M M A R Y

An **aldehyde** contains a carbonyl group bonded to at least one hydrogen atom (Section 9.2). A **ketone** contains a carbonyl group bonded to two carbon atoms. The IUPAC name of an aldehyde is derived by changing -e of the parent alkane to -al (Section 9.3A). The IUPAC name of a ketone is derived by changing -e of the parent alkane to -one and using a number to locate the carbonyl carbon.

Aldehydes and ketones are polar compounds. They have higher boiling points and are more soluble in water than nonpolar compounds of comparable molecular weight (Section 9.4).

K E Y   R E A C T I O N S

1. Oxidation of an Aldehyde to a Carboxylic Acid (Section 9.5) The aldehyde group is among the most easily oxidized functional groups. Oxidizing agents include $K_2Cr_2O_7$, Tollens' reagent, and O_2.

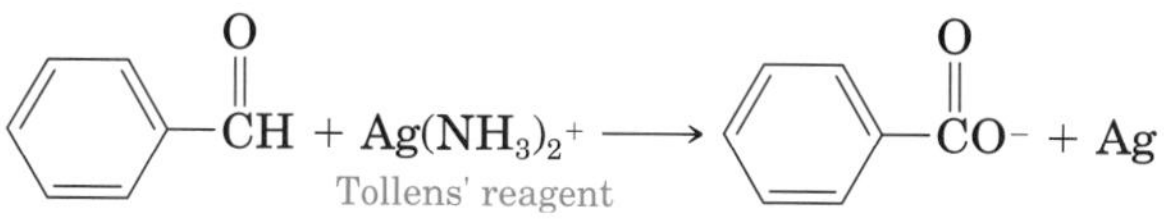

2. **Reduction (Section 9.6)** Aldehydes are reduced to primary alcohols and ketones to secondary alcohols by H_2 in the presence of a transition metal catalyst such as Pt or Ni. They are also reduced to alcohols by sodium borohydride, $NaBH_4$, followed by protonation.

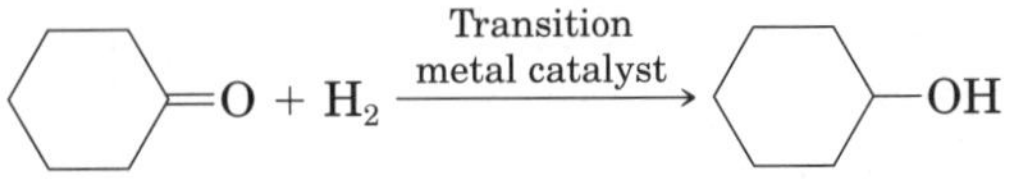

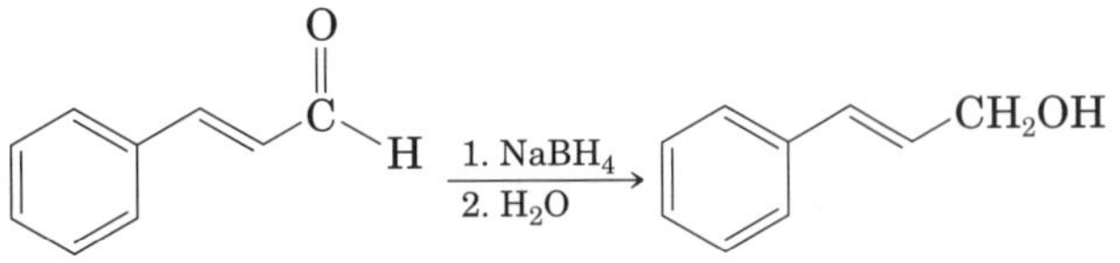

3. **Addition of Alcohols to Form Hemiacetals (Section 9.7)** Hemiacetals are only minor components of an equilibrium mixture of aldehyde or ketone and alcohol, except where the —OH and C=O groups are parts of the same molecule and a five- or six-membered ring can form.

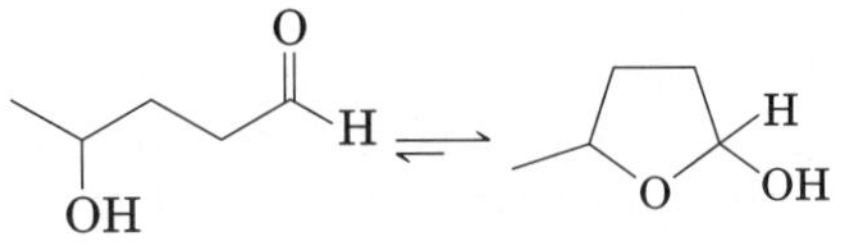

4. **Addition of Alcohols to Form Acetals (Section 9.7)** Formation of acetals is catalyzed by acid. Acetals are hydrolyzed in aqueous acid to an aldehyde or ketone and two molecules of an alcohol.

Cyclohexanone Methanol An acetal

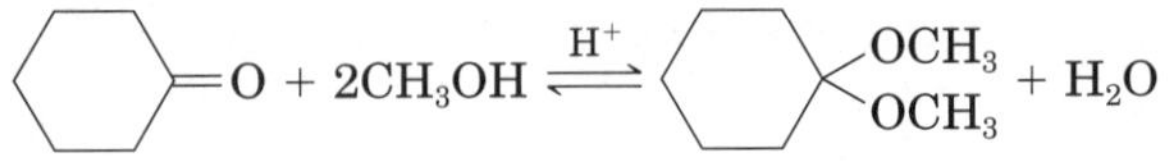

5. **Keto-Enol Tautomerism (Section 9.8)** The keto form generally predominates at equilibrium.

Keto form (approximately 99.9%) Enol form

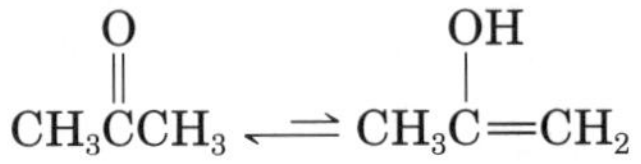

PROBLEMS

Numbers that appear in color indicate difficult problems. ▶ designates problems requiring application of principles.

Preparation of Aldehydes and Ketones

9.9 Complete the equations for these reactions.

(a)

(b)

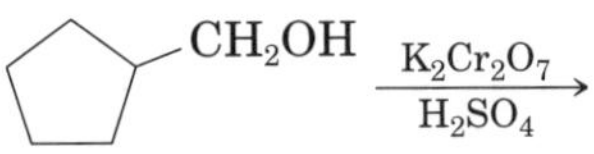

9.10 Write an equation for each conversion.
(a) 1-Pentanol to pentanal
(b) 1-Pentanol to pentanoic acid
(c) 2-Pentanol to 2-pentanone
(d) 2-Propanol to acetone
(e) Cyclohexanol to cyclohexanone

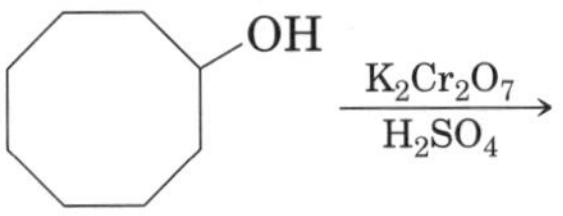

Structure and Nomenclature

9.11 What is the difference in structure between an aldehyde and a ketone?

9.12 What is the difference in structure between an aromatic aldehyde and an aliphatic aldehyde?

9.13 Is it possible for the carbon atom of a carbonyl group to be a stereocenter? Explain.

9.14 Which compounds contain carbonyl groups?

(a)

(b)

(c)

(d)

(e)

(f)

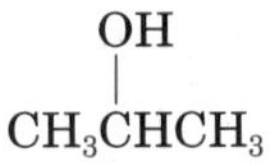

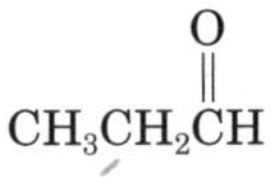

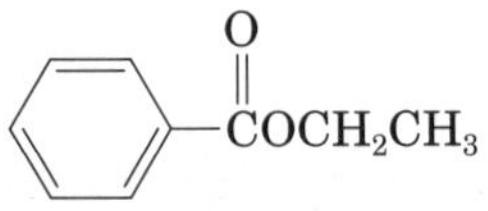

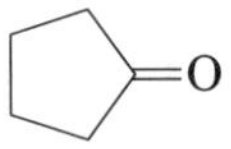

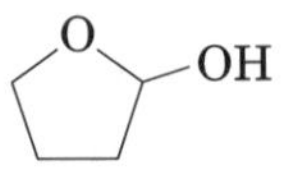

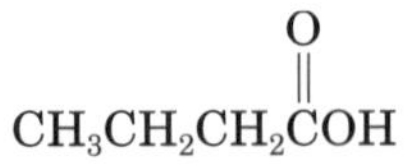

9.15 Following are structural formulas for two steroid hormones. Name the functional groups in each.

(a)

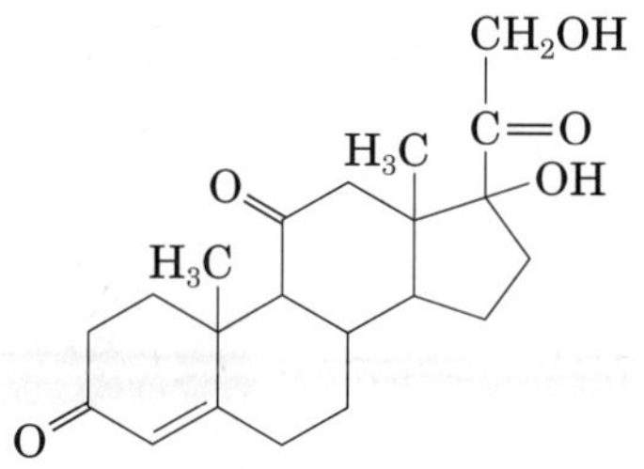

Cortisone

(b)

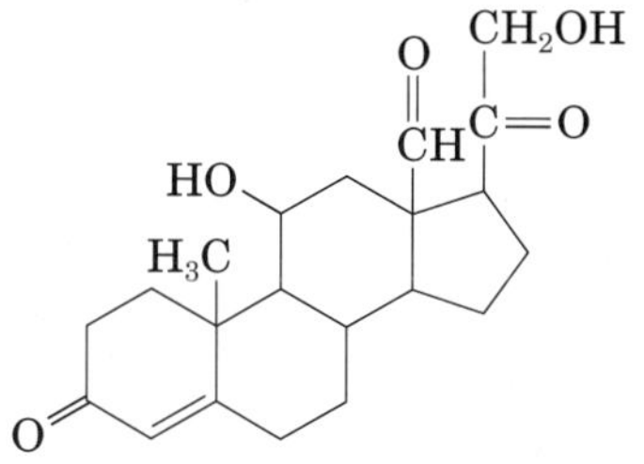

Aldosterone

9.16 Draw structural formulas for the four aldehydes with molecular formula $C_5H_{10}O$. Which of these aldehydes are chiral?

9.17 Draw structural formulas for these aldehydes.
(a) Formaldehyde
(b) Propanal
(c) 3,7-Dimethyloctanal
(d) Decanal
(e) 4-Hydroxybenzaldehyde
(f) 2,3-Dihydroxypropanal

9.18 Draw structural formulas for these ketones.
(a) Ethyl isopropyl ketone
(b) 2-Chlorocyclohexanone
(c) 2,4-Dimethyl-3-pentanone
(d) Diisopropyl ketone
(e) Acetone
(f) 2,5-Dimethylcyclohexanone

9.19 Write IUPAC names for these compounds.
(a) $(CH_3CH_2CH_2)_2C{=}O$

(b)

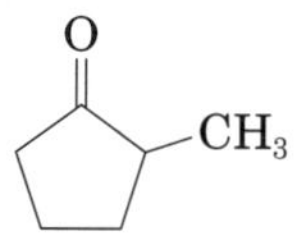

(c)

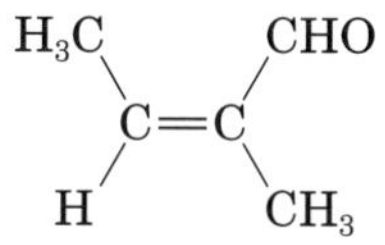

(d)

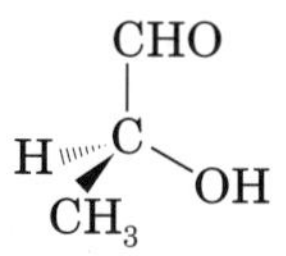

(e)

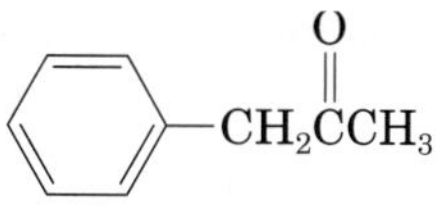

(f) $\overset{O}{\overset{\|}{HC}}(CH_2)_4\overset{O}{\overset{\|}{CH}}$

9.20 Write IUPAC names for these compounds.

(a)

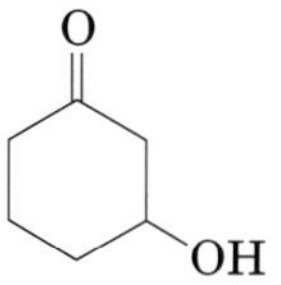

(b)

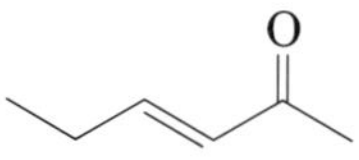

(c)

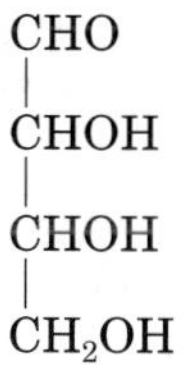

(d)

9.21 Explain why each name is incorrect. Write the correct IUPAC name for the intended compound.
(a) 3-Butanone
(b) 1-Butanone
(c) 4-Methylbutanal
(d) 2,2-Dimethyl-3-butanone

9.22 Explain why each name is incorrect. Write the correct IUPAC name for the intended compound.
(a) 2-Pentanal
(b) Cyclopentanal
(c) 3-Ethyl-2-butanone
(d) 5-Aminobenzaldehyde

Physical Properties

9.23 In each pair of compounds, select the one with the higher boiling point.
(a) Acetaldehyde or ethanol
(b) Acetone or 3-pentanone
(c) Butanal or butane
(d) Butanone or 2-butanol

9.24 Acetone is completely soluble in water, but 4-heptanone is completely insoluble. Explain.

9.25 Why does acetone have a higher boiling point (56°C) than ethyl methyl ether (11°C), even though their molecular weights are almost the same?

9.26 Pentane, 1-butanol, and butanal all have approximately the same molecular weights but different boiling points. Arrange them in order of increasing boiling point. Explain the basis for your ranking.

9.27 Show how acetaldehyde can form hydrogen bonds with water.

9.28 Why can't two molecules of acetone form a hydrogen bond with each other?

Oxidation–Reduction of Aldehydes and Ketones

9.29 Draw the structural formula for the principal organic product formed when each compound is treated with $K_2Cr_2O_7/H_2SO_4$. If no reaction takes place, say so.
(a) Butanal
(b) Benzaldehyde
(c) Cyclohexanone
(d) Cyclohexanol

9.30 Draw the structural formula for the principal organic product formed when each compound in Problem 9.29 is treated with Tollens' reagent. If no reaction takes place, say so.

9.31 What simple chemical test could you use to distinguish between the members of each pair of compounds? Tell what you would do, what you would expect to observe, and how you would interpret your experimental observation.
(a) Pentanal and 2-pentanone
(b) 2-Pentanone and 2-pentanol

9.32 Explain why liquid aldehydes are often stored under an atmosphere of nitrogen rather than air.

9.33 Suppose that you take a bottle of benzaldehyde (a liquid, bp 179°C) from a shelf and find a white solid in the bottom of the bottle. The solid turns litmus red; that is, it is acidic. Yet aldehydes are neutral compounds. How can you explain these observations?

9.34 Explain why the reduction of an aldehyde always gives a primary alcohol and the reduction of a ketone always gives a secondary alcohol.

9.35 Write the structural formula for the principal organic product formed by treating each compound with H_2/transition metal catalyst.

(a) $CH_3\overset{\overset{\displaystyle O}{\|}}{C}CH_2CH_3$

(b) $CH_3(CH_2)_4\overset{\overset{\displaystyle O}{\|}}{C}H$

(c) [cyclopentanone ring with CH₃ substituent]

(d) [benzaldehyde ring with CH and OH substituents]

9.36 Write the structural formula for the principal organic product formed by treating each compound in Problem 9.35 with $NaBH_4$ followed by H_2O.

9.37 1,3-Dihydroxy-2-propanone, more commonly known as dihydroxyacetone, is the active ingredient in artificial tanning agents such as Man-Tan and Magic Tan.
(a) Write a structural formula for this compound.
(b) Would you expect it to be soluble or insoluble in water?
(c) Write the structural formula of the product formed by its reduction with H_2/metal catalyst.

9.38 Draw a structural formula for the product formed by treatment of butanal with each set of reagents.
(a) H_2/metal catalyst
(b) $NaBH_4$, then H_2O
(c) $Ag(NH_3)_2^+$ (Tollens' reagent)
(d) $K_2Cr_2O_7/H_2SO_4$

9.39 Draw a structural formula for the product formed by treatment of acetophenone, $C_6H_5COCH_3$, with each set of reagents given in Problem 9.38.

Addition of Alcohols

9.40 What is the characteristic structural feature of a hemiacetal? Of an acetal?

9.41 Which compounds are hemiacetals, which are acetals, and which are neither?

(a) [benzene ring]$-\overset{\overset{\displaystyle OCH_3}{|}}{C}HOCH_3$

(b) $CH_3CH_2\overset{\overset{\displaystyle OH}{|}}{C}HOCH_3$

(c) $CH_3OCH_2OCH_3$

(d) [cyclic structure with O, O and CH_2CH_3 substituent]

(e) [cyclic structure with O and OCH_2CH_3 substituent]

(f) [cyclic structure with O and OH substituent]

9.42 Which compounds are hemiacetals, which are acetals, and which are neither?

(a)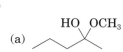

(b) 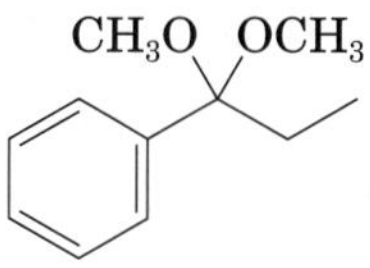 [structure with CH₃O and OCH₃ on benzene-attached carbon]

(c) 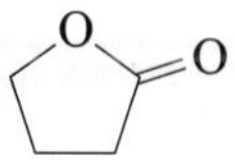 [γ-butyrolactone type structure with O and =O]

(d) 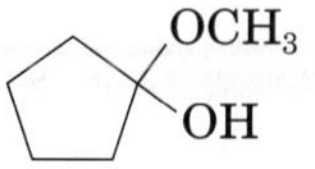 [cyclopentane with OCH₃ and OH]

(e) 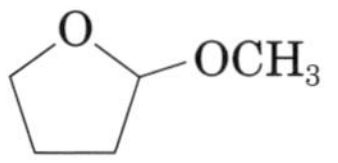 [tetrahydrofuran ring with O and —OCH₃]

(f) 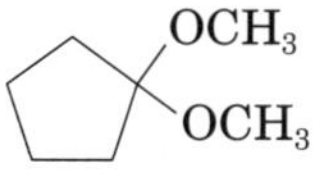 [cyclopentane with OCH₃ and OCH₃]

9.43 Draw the hemiacetal and then the acetal formed in each reaction. In each case, assume an excess of the alcohol.

(a) Propanal + methanol $\longrightarrow$

(b) Cyclopentanone + methanol $\longrightarrow$

9.44 Draw the structures of the aldehydes or ketones and alcohols formed when these acetals are treated with aqueous acid and hydrolyzed.

(a) 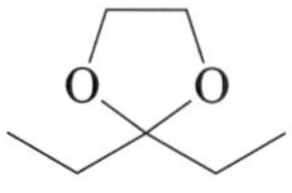 [cyclic acetal with O—O ring and ethyl groups]

(b) 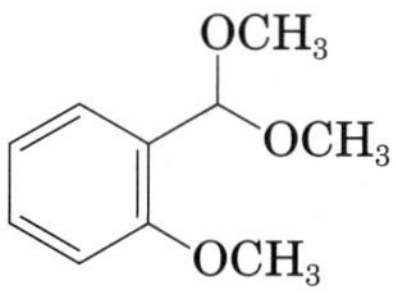 [benzene with CH(OCH₃)₂ and OCH₃]

(c) 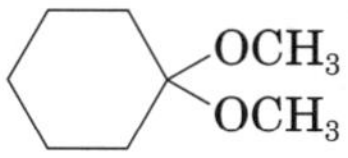 [cyclohexane with OCH₃ and OCH₃]

(d) 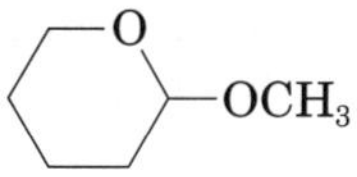 [tetrahydropyran ring with O and —OCH₃]

9.45 What is the difference in meaning between the terms "hydration" and "hydrolysis"? Give an example of each.

9.46 What is the difference in meaning between the terms "hydration" and "dehydration"? Give an example of each.

Keto-Enol Tautomerism

9.47 Which of these compounds undergo keto-enol tautomerism?

(a) 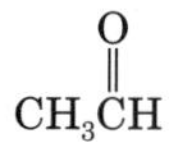 CH_3CH [with =O]

(b) 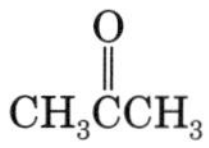 CH_3CCH_3 [with =O]

(c) 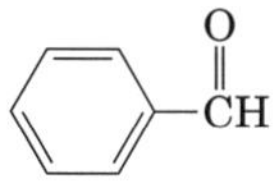 [benzene ring]—CH [with =O]

(d) 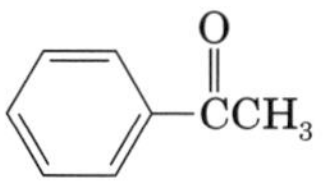 [benzene ring]—CCH₃ [with =O]

(e) 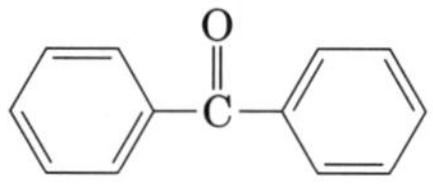 [benzene ring]—C—[benzene ring] [with =O]

(f) 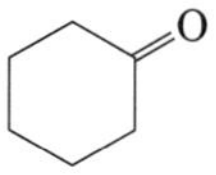 [cyclohexanone]

9.48 Draw all enol forms of each aldehyde and ketone.

(a) 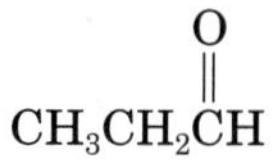 CH_3CH_2CH [with =O]

(b) 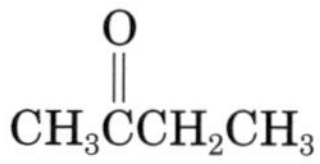 $CH_3CCH_2CH_3$ [with =O]

(c) 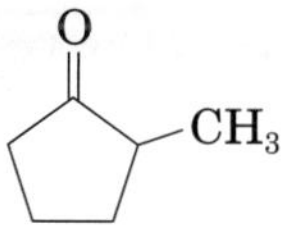 [cyclopentanone with CH₃ substituent and =O]

9.49 Draw the structural formula of the keto form of each enol.

(a) 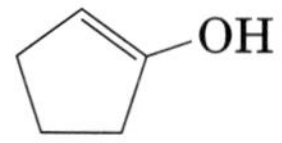 [cyclopentene with OH]

(b)  $CH_3C=CHCH_2CH_2CH_3$ [with OH]

(c) 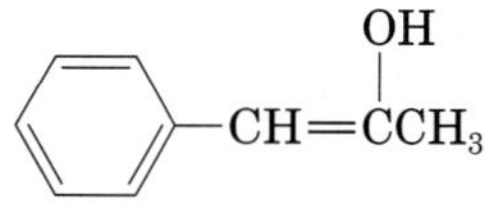 [benzene]—CH=CCH₃ [with OH]

9.50 Starting with cyclohexanone, show how to prepare these compounds. In addition to the given starting material, use any other organic or inorganic reagents as necessary.

(a) Cyclohexanol

(b) Cyclohexene

(c) 1,2-Dibromocyclohexane

(d) Cyclohexane

(e) Chlorocyclohexane

9.51 Draw the structural formula of an aldehyde or ketone that can be reduced to produce each alcohol. If none exists, say so.

(a) 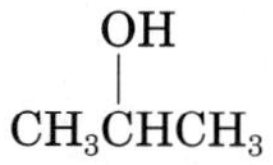 CH_3CHCH_3 [with OH]

(b) 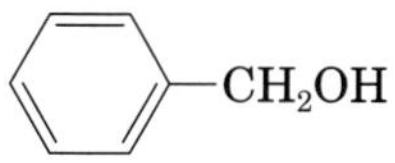 [benzene]—CH₂OH

(c) CH_3OH

(d) 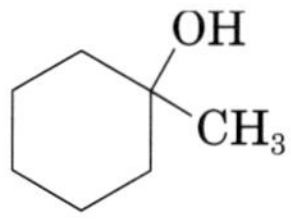 [cyclohexane with OH and CH₃]

9.52 Draw the structural formula of an aldehyde or ke-tone that can be reduced to produce each alcohol. If none exists, say so.

(a)

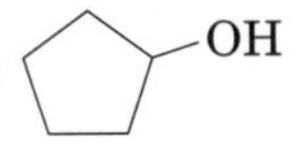

(b)

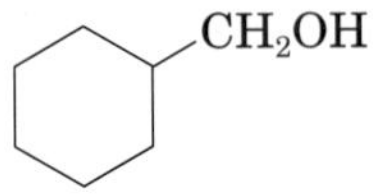

(c) $CH_3\overset{\displaystyle CH_3}{\underset{\displaystyle CH_3}{C}}OH$

(d)  HO‿‿‿‿OH

9.53 1-Propanol can be prepared by the reduction of an aldehyde, but it cannot be prepared by the acid-catalyzed hydration of an alkene. Explain why it can't be prepared from an alkene.

9.54 Show how to bring about these conversions. In addition to the given starting material, use any other organic or inorganic reagents as necessary.

(a) $C_6H_5\overset{\displaystyle O}{\overset{\|}{C}}CH_2CH_3 \longrightarrow C_6H_5\overset{\displaystyle OH}{\overset{|}{C}}HCH_2CH_3$

$\longrightarrow C_6H_5CH{=}CHCH_3$

(b)

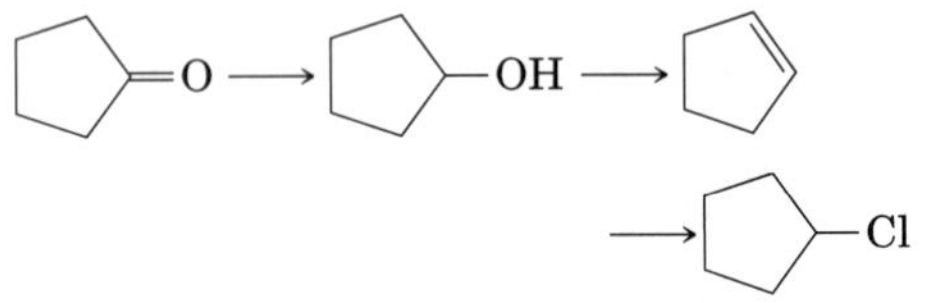

9.55 Show how to bring about these conversions. In addition to the given starting material, use any other organic or inorganic reagents as necessary.
(a) 1-Pentene to 2-pentanone
(b) Cyclohexene to cyclohexanone

9.56 Describe a simple chemical test by which you could distinguish between the members of each pair of compounds.
(a) Cyclohexanone and aniline
(b) Cyclohexene and cyclohexanol
(c) Benzaldehyde and cinnamaldehyde

Additional Problems

9.57 Indicate the aldehyde or ketone group in these compounds.

(a) $H\overset{\displaystyle O}{\overset{\|}{C}}CH_2CH_2CH_2\overset{\displaystyle O}{\overset{\|}{C}}CH_3$

(b)

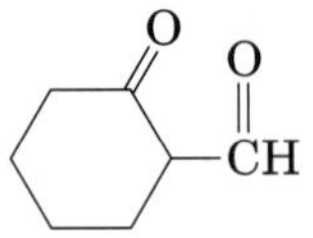

(c) $HOCH_2\overset{\displaystyle HO}{\overset{|}{C}}H\overset{\displaystyle O}{\overset{\|}{C}}H$

(d)

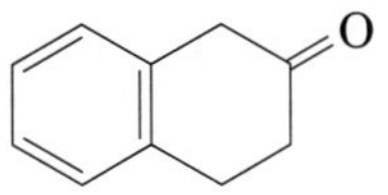

(e) 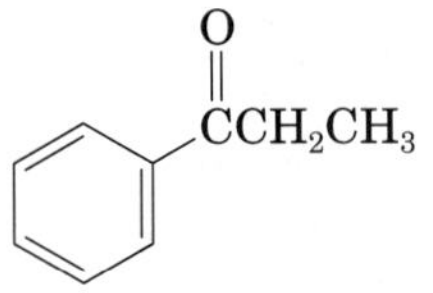 $\overset{\displaystyle O}{\overset{\|}{C}}CH_2CH_3$

(f) 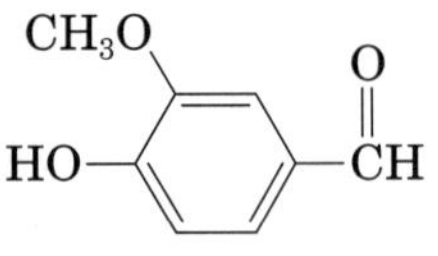 $HO{-}\overset{\displaystyle CH_3O}{}\cdots{-}\overset{\displaystyle O}{\overset{\|}{C}}H$

9.58 Draw the structural formula for the product formed by treatment of each compound in Problem 9.57 with sodium borohydride, $NaBH_4$.

9.59 Sodium borohydride is a laboratory reducing agent. NADH is a biological reducing agent. In what way is the chemistry by which they reduce aldehydes and ketones similar?

9.60 Name the functional groups in each steroid hormone. Methandrostenolone is a synthetic anabolic steroid; that is, it promotes tissue and muscle growth and development. What are the differences in structural formula between testosterone and methandrostenolone?

(a)

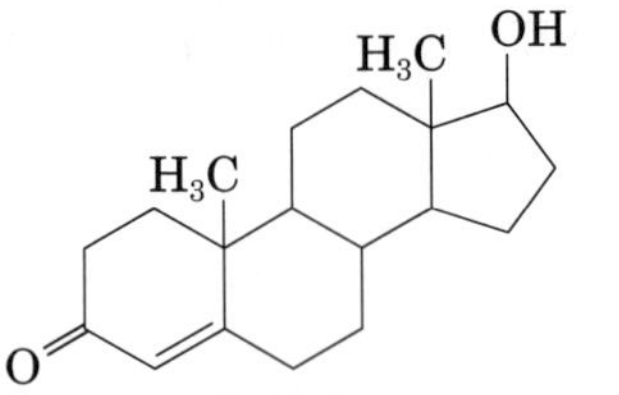

Testosterone

(b)

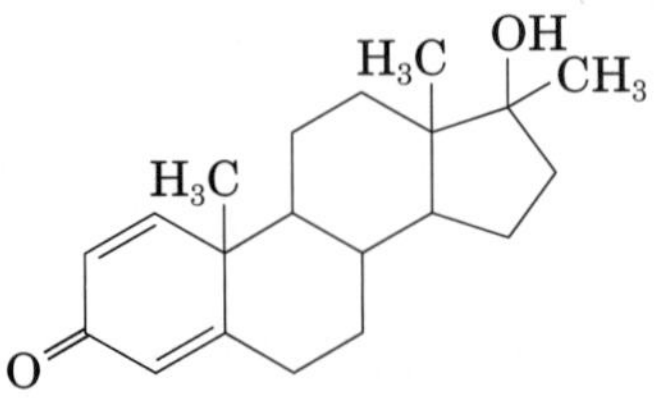

Methandrostenolone

9.61 Draw structural formulas for the (a) one ketone and (b) two aldehydes with molecular formula C_4H_8O.

9.62 Draw structural formulas for these compounds.
(a) 1-Chloro-2-propanone
(b) 3-Hydroxybutanal
(c) 4-Hydroxy-4-methyl-2-pentanone
(d) 3-Methyl-3-phenylbutanal
(e) 1,3-Cyclohexanedione
(f) 5-Hydroxyhexanal

9.63 Why does acetone have a lower boiling point (56°C) than 2-propanol (82°C), even though their molecular weights are almost the same?

9.64 Propanal (bp 49°C) and 1-propanol (bp 97°C) have about the same molecular weight, yet their boiling points differ by almost 50°C. Explain this fact.

9.65 What simple chemical test could you use to distinguish between the members of each pair of compounds? Tell what you would do, what you would expect to observe, and how you would interpret your experimental observation.
(a) Benzaldehyde and cyclohexanone
(b) Acetaldehyde and acetone

9.66 5-Hydroxyhexanal forms a six-membered cyclic hemiacetal, which predominates at equilibrium in aqueous solution.
(a) Draw a structural formula for this cyclic hemiacetal.
(b) How many stereoisomers are possible for 5-hydroxyhexanal?
(c) How many stereoisomers are possible for this cyclic hemiacetal?

9.67 The following molecule is an enediol; each carbon of the double bond carries an —OH group. Draw structural formulas for the α-hydroxyketone and the α-hydroxyaldehyde with which this enediol is in equilibrium.

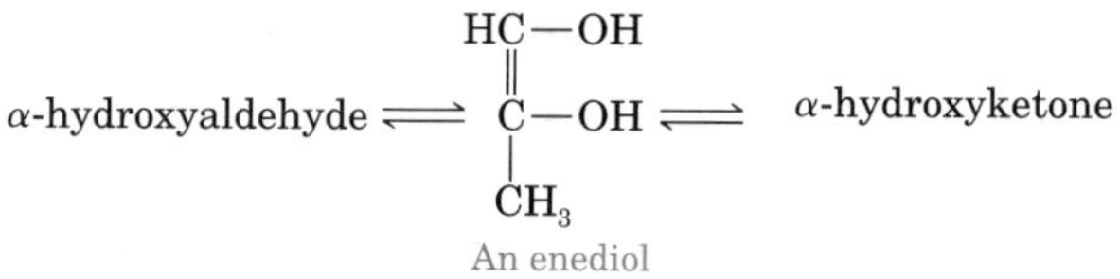

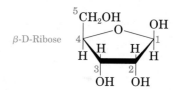

α-hydroxyaldehyde ⇌ ⇌ α-hydroxyketone

An enediol

9.68 Alcohols can be prepared by the acid-catalyzed hydration of alkenes (Section 3.6B) and by the reduction of aldehydes and ketones (Section 9.6). Show how you might prepare each of the following alcohols by (1) acid-catalyzed hydration of an alkene and (2) reduction of an aldehyde or ketone.
(a) Ethanol
(b) Cyclohexanol
(c) 2-Propanol
(d) 1-Phenylethanol

9.69 Glucose, $C_6H_{12}O_6$, contains an aldehyde group but exists predominantly in the form of the cyclic hemiacetal shown here:

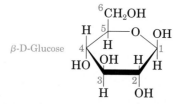

β-D-Glucose

(a) A cyclic hemiacetal is formed when the —OH group of one carbon bonds to the carbonyl group of another carbon. Which carbon in glucose provides the —OH group and which provides the CHO group?
(b) Write the structural formula of the free aldehyde form of glucose.

9.70 Ribose, $C_5H_{10}O_5$, contains an aldehyde group but exists predominantly in the form of the cyclic hemiacetal shown here:

β-D-Ribose

(a) Which carbon of ribose provides the —OH group and which provides the CHO group for formation of this cyclic hemiacetal?
(b) Write the structural formula of the free aldehyde form of ribose.

InfoTrac College Edition

For additional readings, go to InfoTrac College Edition, your online research library, at

http://infotrac.thomsonlearning.com

CHAPTER **10**

10.1 Introduction

10.2 Carboxylic Acids

10.3 Anhydrides, Esters, and Amides

10.4 Preparation of Esters—Fischer Esterification

10.5 Preparation of Amides

10.6 Hydrolysis of Anhydrides, Esters, and Amides

10.7 Phosphoric Anhydrides and Esters

10.8 Step-Growth Polymerizations

Citrus fruits are sources of citric acid, a tricarboxylic acid.

Charles D. Winters

Carboxylic Acids, Anhydrides, Esters, and Amides

10.1 Introduction

In this chapter, we study carboxylic acids, another class of organic compounds containing the carbonyl group. We also study three classes of organic compounds derived from carboxylic acids: anhydrides, esters, and amides. Below, under the general formula of each carboxylic acid derivative, is a drawing to help you see how the functional group of the derivative is formally related to a carboxyl group. The loss of —OH from a carboxyl group and H— from ammonia, for example, gives an amide.

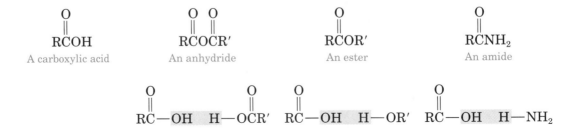

Of these three carboxylic derivatives, anhydrides are so reactive that they are rarely found in nature. Esters and amides, however, are widespread in the biological world.

10.2 Carboxylic Acids

The functional group of a **carboxylic acid** is a **carboxyl group** (Section 1.4D), which can be represented in any one of three ways:

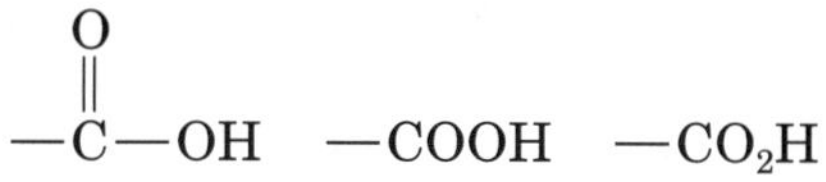

$$
\begin{matrix} & O \\ & \| \\ -C&-OH \end{matrix} \quad -COOH \quad -CO_2H
$$

A IUPAC Names

We derive the IUPAC name of an acyclic carboxylic acid from the name of the longest carbon chain that contains the carboxyl group. The final **-e** from the name of the parent alkane is dropped and replaced by **-oic acid.** The chain is numbered beginning with the carbon of the carboxyl group. Because the carboxyl carbon is understood to be carbon 1, there is no need to give it a number. In the following examples, the common name is given in parentheses.

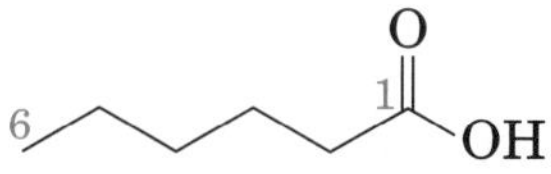

Hexanoic acid
(Caproic acid)

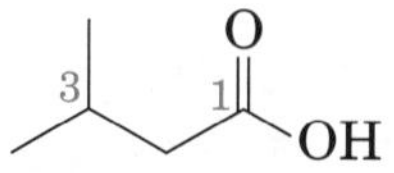

3-Methylbutanoic acid
(Isovaleric acid)

When a carboxylic acid also contains an alcohol, we indicate the presence of the —OH group by adding the prefix **hydroxy-.** When it contains an amine, we indicate the presence of the —NH$_2$ group by **amino-.**

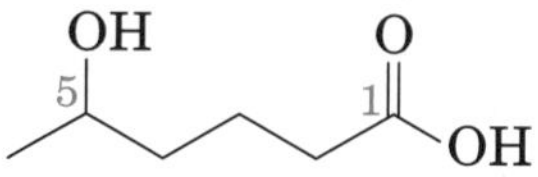

5-Hydroxyhexanoic acid

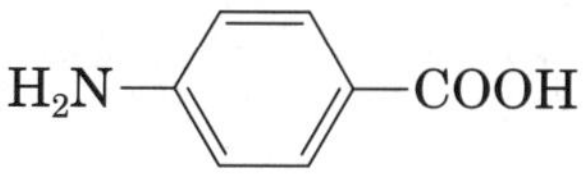

4-Aminobenzoic acid

To name dicarboxylic acids, we add the suffix **-dioic acid** to the name of the parent alkane that contains both carboxyl groups. The numbers of the carboxyl carbons are not indicated because they can be only at the ends of the parent chain.

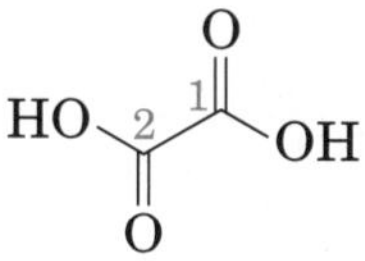

Ethanedioic acid
(Oxalic acid)

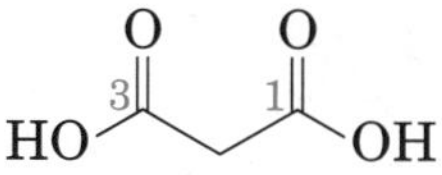

Propanedioic acid
(Malonic acid)

■ Leaves of the rhubarb plant contain the poison oxalic acid as its potassium or sodium salt.

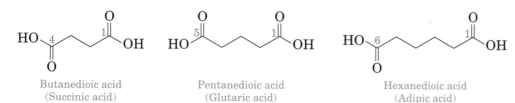

Butanedioic acid
(Succinic acid)

Pentanedioic acid
(Glutaric acid)

Hexanedioic acid
(Adipic acid)

The unbranched carboxylic acids having between 12 and 20 carbon atoms are known as fatty acids. We study them in Chapter 12.

Table 10.1   Several Aliphatic Carboxylic Acids and Their Common Names

Structure	IUPAC Name	Common Name	Derivation
HCOOH	methanoic acid	formic acid	Latin: *formica,* ant
CH_3COOH	ethanoic acid	acetic acid	Latin: *acetum,* vinegar
CH_3CH_2COOH	propanoic acid	propionic acid	Greek: *propion,* first fat
$CH_3(CH_2)_2COOH$	butanoic acid	butyric acid	Latin: *butyrum,* butter
$CH_3(CH_2)_3COOH$	pentanoic acid	valeric acid	Latin: *valere,* to be strong
$CH_3(CH_2)_4COOH$	hexanoic acid	caproic acid	Latin: *caper,* goat
$CH_3(CH_2)_6COOH$	octanoic acid	caprylic acid	Latin: *caper,* goat
$CH_3(CH_2)_8COOH$	decanoic acid	capric acid	Latin: *caper,* goat
$CH_3(CH_2)_{10}COOH$	dodecanoic acid	lauric acid	Latin: *laurus,* laurel
$CH_3(CH_2)_{12}COOH$	tetradecanoic acid	myristic acid	Greek: *myristikos,* fragrant
$CH_3(CH_2)_{14}COOH$	hexadecanoic acid	palmitic acid	Latin: *palma,* palm tree
$CH_3(CH_2)_{16}COOH$	octadecanoic acid	stearic acid	Greek: *stear,* solid fat
$CH_3(CH_2)_{18}COOH$	eicosanoic acid	arachidic acid	Greek: *arachis,* peanut

■ **Formic acid was first obtained in 1670 from the destructive distillation of ants, whose Latin genus is *Formica*. It is one of the components of the venom injected by stinging ants.**

The name *oxalic acid* is derived from one of its sources in the biological world—plants of the genus *Oxalis,* one of which is rhubarb. Oxalic acid also occurs in human and animal urine, and calcium oxalate is a major component of kidney stones. Succinic acid is an intermediate in the citric acid cycle (Section 18.4). Adipic acid is one of the two monomers required for the synthesis of the polymer nylon-66 (Section 10.8A).

B Common Names

Common names for aliphatic carboxylic acids, many of which were known long before the development of IUPAC nomenclature, are often derived from the name of a natural substance from which the acid can be isolated. Table 10.1 lists several of the unbranched aliphatic carboxylic acids found in the biological world along with the common name of each. Those with 16, 18, and 20 carbon atoms are particularly abundant in animal fats and vegetable oils (Section 12.2), and the phospholipid components of biological membranes (Section 12.5).

When common names are used, the Greek letters alpha (α), beta (β), gamma (γ), and so forth, are often added as a prefix to locate substituents.

GABA is a neurotransmitter in the central nervous system.

4-Aminobutanoic acid
(γ-Aminobutyric acid; GABA)

2-Hydroxypropanoic acid
(Lactic acid)

EXAMPLE 10.1

Write the IUPAC name for each carboxylic acid:

(a)

(b) HO—⟨ ⟩—COOH

(c)

Solution
(a) The longest carbon chain that contains the carboxyl group has five carbons and, therefore, the parent alkane is pentane. The IUPAC name is 2-ethylpentanoic acid.
(b) 4-Hydroxybenzoic acid
(c) (*R*)-2-Hydroxypropanoic acid

Problem 10.1

Each of these compounds has a well-recognized common name. A derivative of glyceric acid is an intermediate in glycolysis (Section 19.2). β-Alanine is a building block of pantothenic acid (Section 18.5). Mevalonic acid is an intermediate in the biosynthesis of steroids (Section 28.4). Write the IUPAC name for each compound.

(a)
COOH
|
CHOH
|
CH₂OH
Glyceric acid

(b) H₂NCH₂CH₂COOH
β-Alanine

(c) 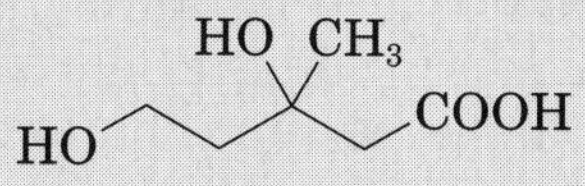
HO CH₃
HO⟋⟍⟋COOH
Mevalonic acid

C Physical Properties

A major feature of carboxylic acids is the polarity of the carboxyl group. This group contains three polar covalent bonds: C=O, C—O, and O—H. It is the polarity of these bonds that determines the major physical properties of carboxylic acids.

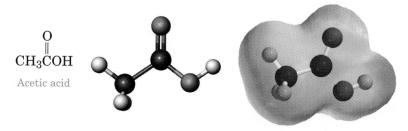

O
‖
CH₃COH
Acetic acid

Carboxylic acids have significantly higher boiling points than other types of organic compounds of comparable molecular weight (Table 10.2). Their higher boiling points result from their polarity and the fact that hydrogen bonding between two carboxyl groups creates a dimer that behaves as a higher-molecular-weight compound.

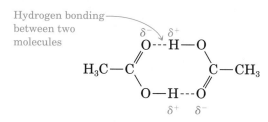

Hydrogen bonding between two molecules

$$H_3C-C \begin{matrix} O \cdots H-O \\ O-H \cdots O \end{matrix} C-CH_3$$

Carboxylic acids are more soluble in water than are alcohols, ethers, aldehydes, and ketones of comparable molecular weight. This increased sol-

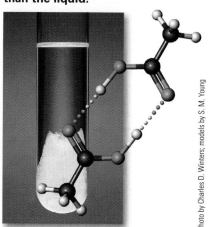

■ Acetic acid molecules can interact through hydrogen bonds. Shown here is partly solid glacial acetic acid. Like most substances (water being a notable exception), the solid is denser than the liquid.

Table 10.2   Boiling Points and Solubilities in Water of Two Groups of Carboxylic Acids, Alcohols, and Aldehydes of Comparable Molecular Weight

Structure	Name	Molecular Weight	Boiling Point (°C)	Solubility (g/100 mL H_2O)
CH_3COOH	acetic acid	60.5	118	infinite
$CH_3CH_2CH_2OH$	1-propanol	60.1	97	infinite
CH_3CH_2CHO	propanal	58.1	48	16
$CH_3(CH_2)_2COOH$	butanoic acid	88.1	163	infinite
$CH_3(CH_2)_3CH_2OH$	1-pentanol	88.1	137	2.3
$CH_3(CH_2)_3CHO$	pentanal	86.1	103	slight

ubility is due to their strong association with water molecules by hydrogen bonding through both their carbonyl and hydroxyl groups. The first four aliphatic carboxylic acids (formic, acetic, propanoic, and butanoic) are infinitely soluble in water. As the size of the hydrocarbon chain increases relative to that of the carboxyl group, however, water solubility decreases. The solubility of hexanoic acid (six carbons) in water is 1.0 g/100 mL water.

We must mention two other properties of carboxylic acids. First, the liquid carboxylic acids from propanoic acid to decanoic acid have sharp, often disagreeable odors. Butanoic acid is found in stale perspiration and is a major component of "locker room odor." Pentanoic acid smells even worse, and goats, which secrete C_6, C_8, and C_{10} acids (Table 10.1), are not famous for their pleasant odors. Second, carboxylic acids have a characteristic sour taste. The sour taste of pickles and sauerkraut is due to the presence of lactic acid. The sour taste of limes (pH 1.9), lemons (pH 2.3), and grapefruit (pH 3.2) is due to the presence of citric and other acids.

D Acidity

Carboxylic acids are weak acids. Values of K_a for most unsubstituted aliphatic and aromatic carboxylic acids fall within the range of 10^{-4} to 10^{-5} ($pK_a = 4.0–5.0$). The value of K_a for acetic acid, for example, is 1.74×10^{-5}; its pK_a is 4.76.

$$CH_3\overset{O}{\overset{\|}{C}}OH + H_2O \rightleftharpoons CH_3\overset{O}{\overset{\|}{C}}O^- + H_3O^+ \qquad K_a = \frac{[CH_3COO^-][H_3O^+]}{[CH_3COOH]} = 1.74 \times 10^{-5}$$

$$pK_a = 4.76$$

Substituents of high electronegativity (especially —OH, —Cl, and —NH_3^+) near the carboxyl group increase the acidity of carboxylic acids, often by several orders of magnitude. Compare, for example, the acidities of acetic acid and the chlorine-substituted acetic acids. Both dichloroacetic acid and trichloroacetic acid are stronger acids than H_3PO_4 (pK_a 2.1).

Dichloroacetic acid is used as a topical astringent and as a treatment for genital warts in males.

Dentists use a 50% aqueous solution of trichloroacetic acid to cauterize gums. This strong acid kills the bleeding and diseased tissue and allows the growth of healthy gum tissue.

Formula:	CH_3COOH	$ClCH_2COOH$	$Cl_2CHCOOH$	Cl_3CCOOH
Name:	Acetic acid	Chloroacetic acid	Dichloroacetic acid	Trichloroacetic acid
pK_a:	4.76	2.86	1.48	0.70

Increasing acid strength →

Electronegative atoms on the carbon adjacent to a carboxyl group increase acidity because they pull electron density away from the O—H bond, thereby facilitating ionization of the carboxyl group and making it a stronger acid.

E Reaction with Bases

All carboxylic acids, whether soluble or insoluble in water, react with NaOH, KOH, and other strong bases to form water-soluble salts.

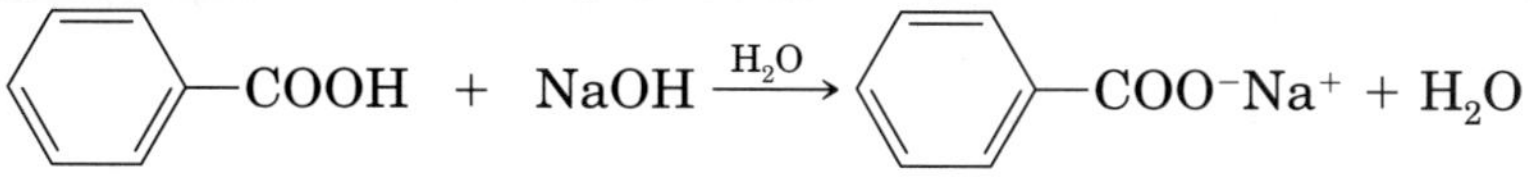

$$\text{Benzoic acid} \quad —COOH \ + \ NaOH \ \xrightarrow{\ H_2O\ } \quad —COO^-Na^+ \ + \ H_2O$$

Benzoic acid
(slightly soluble in water)

Sodium benzoate
(60 g/100 mL water)

Carboxylic acids also form water-soluble salts with ammonia and amines.

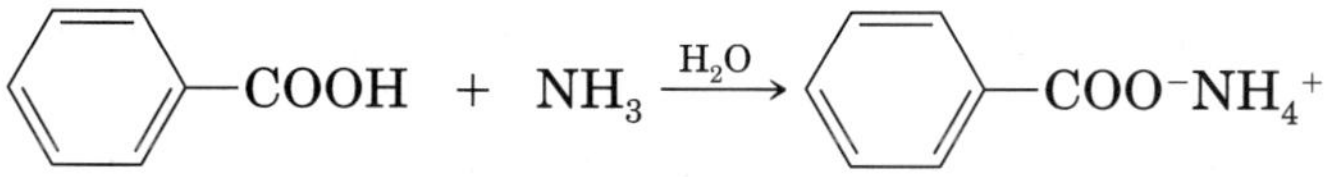

$$—COOH \ + \ NH_3 \ \xrightarrow{\ H_2O\ } \quad —COO^-NH_4^+$$

Benzoic acid
(slightly soluble in water)

Ammonium benzoate
(20 g/100 mL water)

Carboxylic acids, like inorganic acids (Section 7.6E), react with sodium bicarbonate and sodium carbonate to form water-soluble sodium salts and carbonic acid. Carbonic acid then decomposes to give water and carbon dioxide, which evolves as a gas.

$$CH_3COOH \ + \ NaHCO_3 \ \xrightarrow{\ H_2O\ } \ CH_3COO^-Na^+ \ + \ CO_2 \ + \ H_2O$$

Acetic acid   Sodium acetate

Salts of carboxylic acids are named in the same manner as the salts of inorganic acids: the cation is named first and then the anion. We derive the name of the anion from the name of the carboxylic acid by dropping the suffix *-ic acid* and adding the suffix *-ate*. For example, $CH_3CH_2COO^-Na^+$ is named sodium propanoate.

■ **Sodium benzoate and calcium propanoate are used as preservatives in baked goods.**

Charles D. Winters

EXAMPLE 10.2

Complete the equation for each acid–base reaction and name the salt formed.

(a) $CH_3(CH_2)_2COOH + NaOH \longrightarrow$

(b) $CH_3\overset{\displaystyle OH}{\underset{\displaystyle |}{C}}HCOOH + NaHCO_3 \longrightarrow$

Solution
Each carboxylic acid is converted to its sodium salt. In (b), carbonic acid forms, which then decomposes to carbon dioxide and water.

(a) $CH_3(CH_2)_2COOH + NaOH \longrightarrow CH_3(CH_2)_2COO^-Na^+ + H_2O$
      Butanoic acid   Sodium butanoate

OH
|
(b) $CH_3CHCOOH$ + $NaHCO_3$
2-Hydroxypropanoic acid
(Lactic acid)

OH
|
$\longrightarrow CH_3CHCOO^-Na^+$ + H_2O + CO_2
Sodium 2-hydroxypropanoate
(Sodium lactate)

Problem 10.2

Write an equation for the reaction of each acid in Example 10.2 with ammonia and name the salt formed.

One last point about the acidity of carboxylic acids: The form in which they exist in an aqueous solution depends on the solution's pH. If we dissolve a carboxylic acid in water, the solution will be slightly acidic due to ionization of the carboxyl group. If we then add enough HCl to the solution to bring its pH to 2.0 or lower (more acidic), almost all carboxylic acid molecules are present in their un-ionized (—COOH) form. When the pH of the solution is equal to the pK_a of the acid, the un-ionized and ionized forms are present in approximately equal amounts. When the pH of the solution is brought to 8.0 or higher (more basic), almost all carboxylic molecules are present in their ionized (—COO⁻) form.

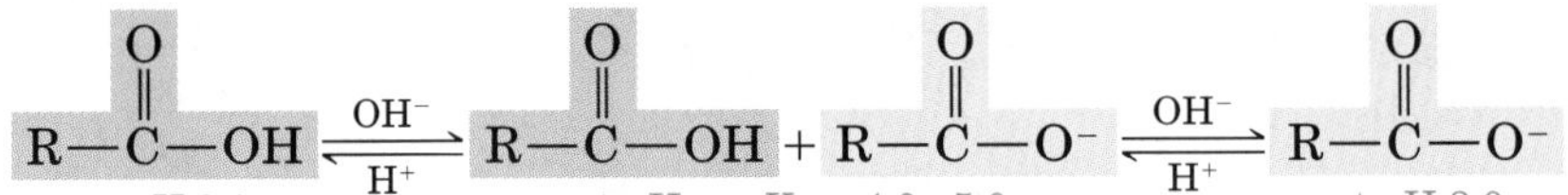

$$R-\overset{O}{\overset{\|}{C}}-OH \underset{H^+}{\overset{OH^-}{\rightleftharpoons}} R-\overset{O}{\overset{\|}{C}}-OH + R-\overset{O}{\overset{\|}{C}}-O^- \underset{H^+}{\overset{OH^-}{\rightleftharpoons}} R-\overset{O}{\overset{\|}{C}}-O^-$$

at pH 2.0   at pH = pK_a = 4.0–5.0   at pH 8.0
or lower   (both forms are present in   or higher
(more acidic)   approximately equal amounts)   (more basic)

10.3 Anhydrides, Esters, and Amides

A Anhydrides

Anhydride A compound in which two carbonyl groups are bonded to the same oxygen atom; RCO—O—COR′

The functional group of an **anhydride** consists of two carbonyl groups bonded to the same oxygen atom. The anhydride may be symmetrical (from two identical **acyl groups**), or mixed (from two different acyl groups). To name anhydrides, we drop the word "acid" from the name of the carboxylic acid from which the anhydride is derived and add the word "anhydride."

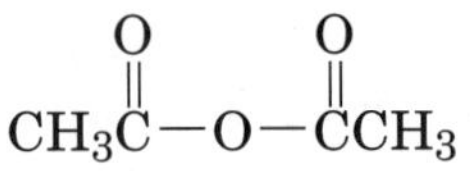

Acyl group An $R\overset{O}{\overset{\|}{C}}$ group

$$CH_3\overset{O}{\overset{\|}{C}}-O-\overset{O}{\overset{\|}{C}}CH_3$$
Acetic anhydride

B Esters

Ester A compound in which the H atom of a carboxyl group, RCOOH, is replaced by an —R′ group; RCOOR′

The functional group of an **ester** is a carbonyl group bonded to an —OR group. Both IUPAC and common names of esters are derived from the names of the parent carboxylic acids. The alkyl group bonded to oxygen is named first, followed by the name of the acid in which the suffix **-ic acid** is replaced by the suffix **-ate.**

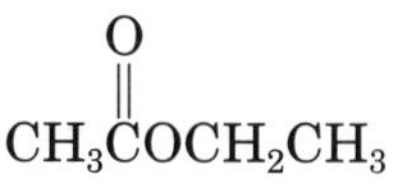

$$CH_3\overset{\displaystyle O}{\overset{\displaystyle \|}{C}}OCH_2CH_3$$

Ethyl ethanoate
(Ethyl acetate)

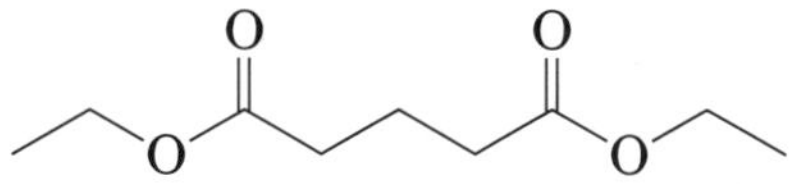

Diethyl pentanedioate
(Diethyl glutarate)

Cyclic esters are called **lactones.** Following are structural formulas for a five-membered and a seven-membered lactone. Note that the oxygen atom counts as one atom of the ring.

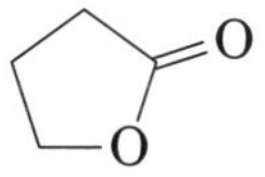

A five-membered
lactone

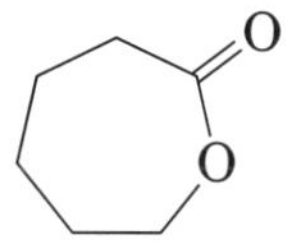

A seven-membered
lactone

> **Lactone** A cyclic ester

Esters are slightly polar compounds because of the presence of the C=O and C—O—C bonds. Because they have no available hydrogens for hydrogen bonding, however, they are generally insoluble in water. Almost all low-molecular-weight esters are liquids.

■ **Some products that contain ethyl acetate, a common carboxylic ester.**

Charles D. Winters

EXAMPLE 10.3

Write the IUPAC name of each ester.

(a)

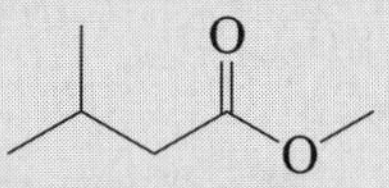

(b)

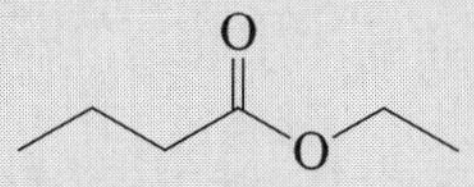

Solution

The IUPAC name is given first and the common name is given in parentheses.

(a) Methyl 3-methylbutanoate (methyl isovalerate, from isovaleric acid)
(b) Ethyl butanoate (ethyl butyrate, from butyric acid)

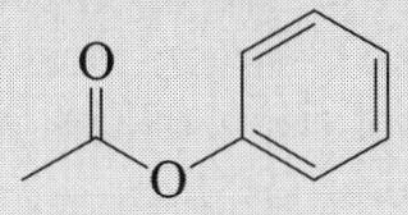

See the **Interactive General, Organic, and Biochemistry CD-ROM, version 2.0,** for further exploration on this topic.

Problem 10.3

Write the IUPAC name for each ester.

(a) $\overset{\displaystyle O}{\overset{\displaystyle \|}{C}}OCH_2CH_3$

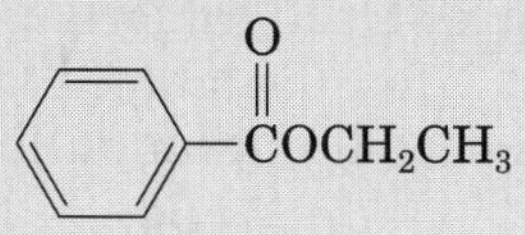

(b)

C Amides

The functional group of an **amide** is a carbonyl group bonded to a nitrogen atom. Amides are named by dropping the suffix **-oic acid** from the IUPAC name of the parent acid, or **-ic acid** from its common name, and adding **-amide.** If the nitrogen atom of the amide is bonded to an alkyl group, the

> **Amide** A compound in which a carbonyl group is bonded to a nitrogen atom, $RCONR'_2$; one or both of the R' groups may be H or alkyl or aryl groups

CHEMICAL CONNECTIONS 10A

The Pyrethrins: Natural Insecticides of Plant Origin

Pyrethrym is a natural insecticide obtained from the powdered flower heads of several species of *Chrysanthemum,* particularly *C. cinerariaefolium.* The active substances in pyrethrum, principally pyrethrins I and II, are contact poisons for insects and cold-blooded vertebrates. Because their concentrations in the pyrethrum powder used in chrysanthemum-based insecticides are nontoxic to plants and higher animals, pyrethrum powder has found wide application in household and livestock sprays as well as in dusts for edible plants. Natural pyrethrins are esters of chrysanthemic acid.

While pyrethrum powders are effective insecticides, the active substances in them are destroyed rapidly in the environment. In an effort to develop synthetic compounds as effective as these natural insecticides but with greater biostability, chemists have prepared a series of esters related in structure to chrysanthemic acid. Permethrin is one of the most commonly used synthetic pyrethrin-like compounds in household and agricultural products.

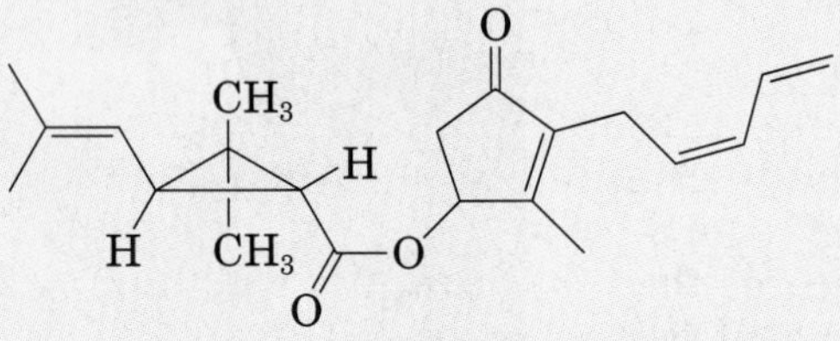

Pyrethrin I

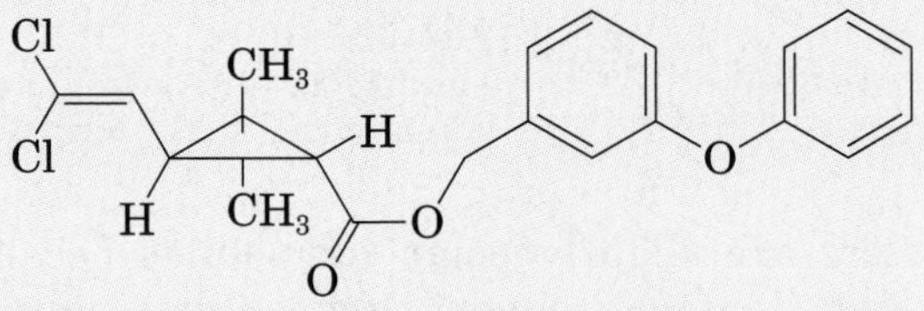

Permethrin

group is named and its location on nitrogen is indicated by *N*-. Two alkyl groups are indicated by *N,N*-di-.

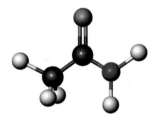

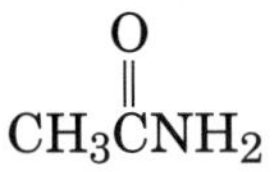

$$CH_3\overset{\displaystyle O}{\overset{\displaystyle \|}{C}}NH_2$$

Acetamide
(a 1° amide)

$$CH_3\overset{\displaystyle O}{\overset{\displaystyle \|}{C}}NHCH_3$$

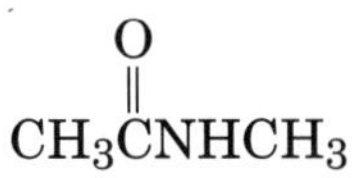

N-Methylacetamide
(a 2° amide)

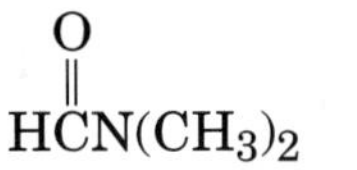

$$H\overset{\displaystyle O}{\overset{\displaystyle \|}{C}}N(CH_3)_2$$

N, N-Dimethylformamide
(a 3° amide)

Lactam A cyclic amide

Cyclic amides are called **lactams.** Following are structural formulas for a four-membered and a seven-membered lactam. A four-membered lactam is essential to the function of the penicillin and cephalosporin antibiotics (Chemical Connections 10B).

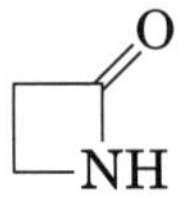

A four-membered
lactam

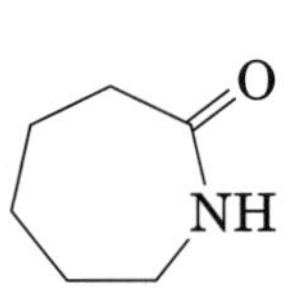

A seven-membered
lactam

CHEMICAL CONNECTIONS 10B

The Penicillins and Cephalosporins: β-Lactam Antibiotics

The **penicillins** were discovered in 1928 by the Scottish bacteriologist Sir Alexander Fleming. Thanks to the brilliant experimental work of Sir Howard Florey, an Australian pathologist, and Ernst Chain, a German chemist who fled Nazi Germany, penicillin G was introduced into the practice of medicine in 1943. For their pioneering work in developing one of the most effective antibiotics of all time, Fleming, Florey, and Chain were awarded the Nobel Prize for medicine or physiology in 1945.

The mold from which Fleming discovered penicillin was *Penicillium notatum,* a strain that gives a relatively low yield of penicillin. It was replaced in commercial production of the antibiotic by *P. chrysogenum,* a strain cultured from a mold found growing on a grapefruit in a market in Peoria, Illinois. The structural feature common to all penicillins is the four-membered lactam ring, called a **β-lactam,** bonded to a five-membered, sulfur-containing ring. The penicillins owe their antibacterial activity to a common mechanism that inhibits the biosynthesis of a vital part of bacterial cell walls.

Andrew McClenaghan/Science Photo Library/Photo Researchers, Inc.

■ **Macrophotograph of the fungus *Penicillium notatum* growing in a petri dish culture of Whickerham's agar. This fungus is the species that was used as an early source of the first penicillin antibiotic.**

The β-lactam ring

The penicillins differ in the group bonded to the carbonyl carbon

Penicillin G

Amoxicillin

Soon after the penicillins were introduced into medical practice, penicillin-resistant strains of bacteria began to appear. They have since proliferated dramatically. One approach to combating resistant strains is to synthesize newer, more effective penicillins, such as ampicillin, methicillin, and amoxicillin.

Another approach is to search for newer, more effective β-lactam antibiotics. The most effective of these agents discovered so far are the **cephalosporins,** the first of which was isolated from the fungus *Cephalosporium acremonium.*

The cephalosporins differ in the group bonded to the carbonyl carbon ...

... and the group bonded to this carbon of the six-membered ring

Cefalexin
(a β-lactam antibiotic)

This class of β-lactam antibiotics has an even broader spectrum of antibacterial activity than the penicillins and is effective against many penicillin-resistant bacterial strains. Cephalexin (Keflex) is presently one of the most widely prescribed of the cephalosporin antibiotics.

EXAMPLE 10.4

Write the IUPAC name for each amide.

(a) $CH_3CH_2CH_2CNH_2$ with =O on the C

(b) H_2N—...—NH_2 (hexanediamide structure)

Solution

Given is the IUPAC name and then, in parentheses, the common name.

(a) Butanamide (butyramide, from butyric acid)

(b) Hexanediamide (adipamide, from adipic acid)

Problem 10.4

Draw a structural formula for each amide.

(a) *N*-Cyclohexylacetamide

(b) Benzamide

10.4 Preparation of Esters—Fischer Esterification

One of the most commonly used preparations of esters involves treating a carboxylic acid with an alcohol in the presence of an acid catalyst, such as concentrated sulfuric acid. In this reaction, two changes take place. First, OH is removed from the carboxyl group and H from the alcohol to form a molecule of H_2O. Second, the carbon of the $C{=}O$ group bonds to the O of the alcohol.

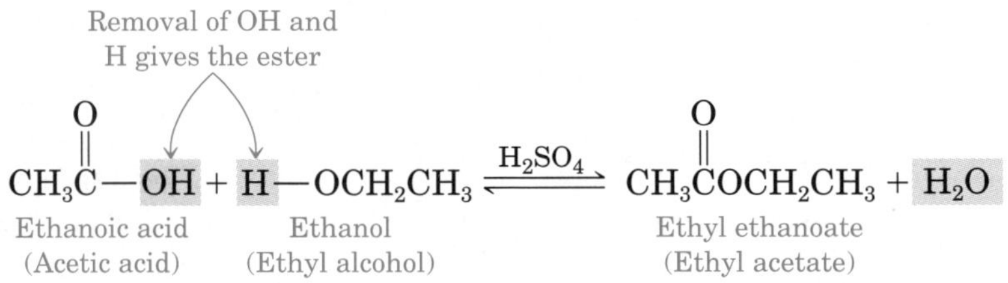

Preparing an ester in this manner is given the special name **Fischer esterification,** after the German chemist Emil Fischer (1852–1919).

In the process of Fischer esterification, the alcohol adds to the carbonyl group of the carboxylic acid to form a tetrahedral carbonyl addition intermediate. Note how closely this step resembles the addition of an alcohol to the carbonyl group of an aldehyde or ketone to form a hemiacetal (Section 9.7). In the case of Fischer esterification, the intermediate then loses a molecule of water to give an ester.

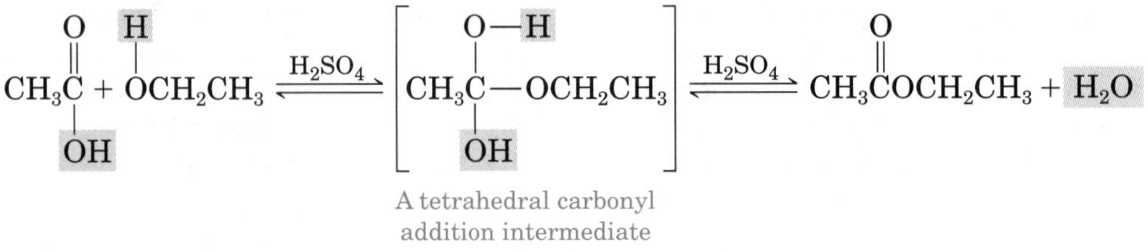

Fischer esterification is reversible, and generally, at equilibrium, the quantities of carboxylic acid and alcohol present are appreciable. By controlling reaction conditions, however, we can use Fischer esterification to prepare esters in high yields. If the alcohol is inexpensive, we can use a large excess of it to drive the equilibrium to the right and increase the yield of ester. Alternatively, we can remove water from the reaction mixture as it forms.

> **Fischer esterification** The process of forming an ester by refluxing a carboxylic acid and an alcohol in the presence of an acid catalyst, commonly sulfuric acid

EXAMPLE 10.5

Complete these Fischer esterification reactions:

(a)

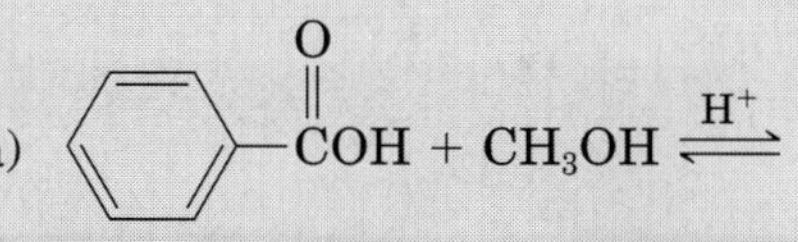

Benzoic acid

(b)

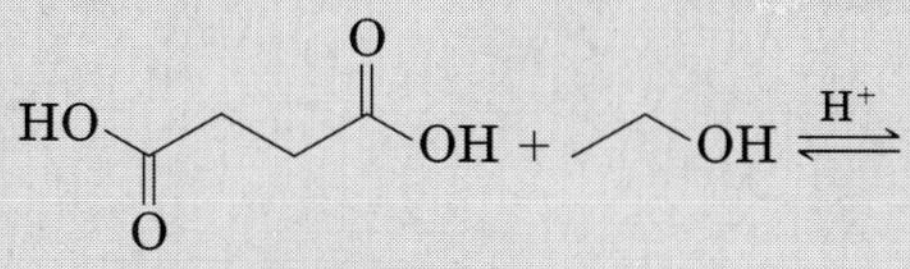

Butanedioic acid   (excess)
(Succinic acid)

Solution

Here is a structural formula for the ester produced in each reaction.

(a)

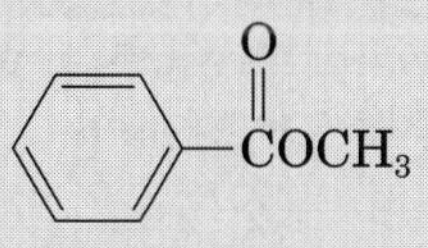

Methyl benzoate

(b)

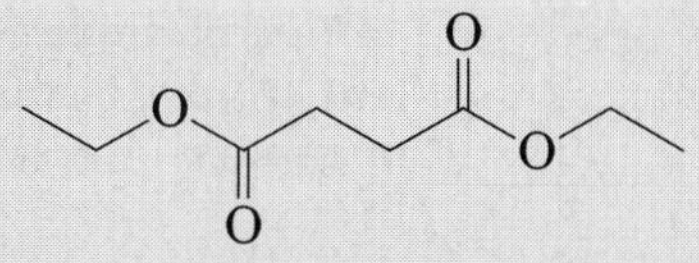

Diethyl butanedioate
(Diethyl succinate)

Problem 10.5

Complete these Fischer esterification reactions:

(a)

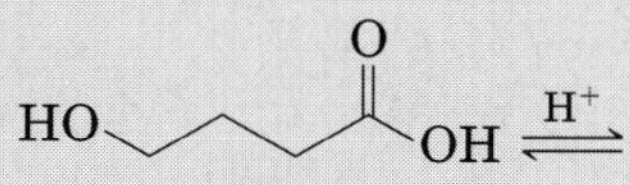

(b)

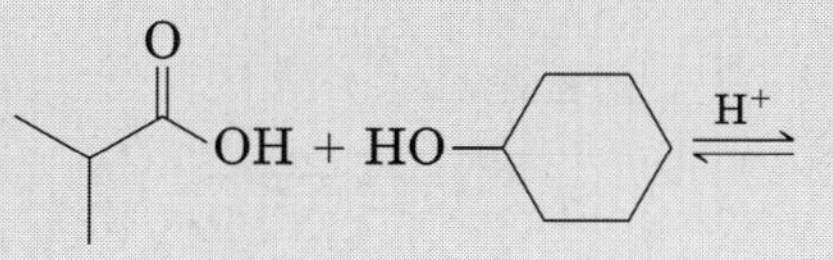

10.5 Preparation of Amides

In principle, we can form an amide by treating a carboxylic acid with an amine and removing —OH from the acid and —H from the amine. In practice, mixing the two leads to an acid–base reaction that forms an ammonium salt. If this salt is heated to a high enough temperature, water splits out and an amide forms.

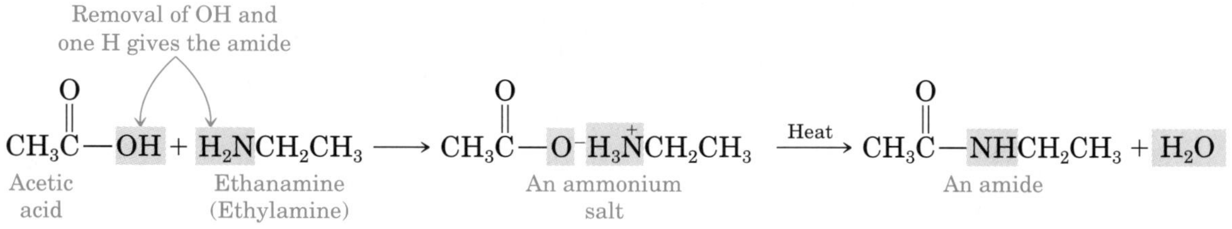

Removal of OH and
one H gives the amide

$$CH_3\overset{O}{\overset{||}{C}}-OH + H_2NCH_2CH_3 \longrightarrow CH_3\overset{O}{\overset{||}{C}}-O^- \overset{+}{H_3}NCH_2CH_3 \xrightarrow{\text{Heat}} CH_3\overset{O}{\overset{||}{C}}-NHCH_2CH_3 + H_2O$$

Acetic   Ethanamine   An ammonium   An amide
acid   (Ethylamine)   salt

CHEMICAL CONNECTIONS 10C

From Willow Bark to Aspirin—and Beyond

The story of the development of this modern pain reliever goes back more than 2000 years. In 400 BCE, the Greek physician Hippocrates recommended chewing the bark of the willow tree to alleviate the pain of childbirth and to treat eye infections.

The active component of willow bark was found to be salicin, a compound composed of salicyl alcohol bonded to a unit of β-D-glucose (Section 11.3). Hydrolysis of salicin in aqueous acid followed by oxidation gave salicylic acid. Salicylic acid proved to be an even more effective reliever of pain, fever, and inflammation than salicin, and without the latter's extremely bitter taste. Unfortunately, patients quickly recognized salicylic acid's major side effect; it causes severe irritation of the mucous membrane lining of the stomach.

CH₂OH — O-glucose
Salicin

COH — OH (O)
Salicylic acid

In the search for less irritating but still effective derivatives of salicylic acid, chemists at the Bayer division of I. G. Farben in Germany in 1883 treated salicylic acid with acetic anhydride and prepared acetylsalicylic acid. They gave this new compound the name *aspirin*.

Acetylsalicylic acid (Aspirin)
COOH — OCCH₃ (O)

Aspirin proved to be less irritating to the stomach than salicylic acid as well as more effective in relieving the pain and inflammation of rheumatoid arthritis. Aspirin, however, remains irritating to the stomach and frequent use of it can cause duodenal ulcers in susceptible persons.

In the 1960s, in a search for even more effective and less irritating analgesics and nonsteroidal anti-inflammatory drugs (NSAIDs), chemists at the Boots Pure Drug Company in England, who were studying compounds structurally related to salicylic acid, discovered an even more potent compound, which they named *ibuprofen.* Soon thereafter, Syntex Corporation in the United States developed naproxen, the active ingredient in

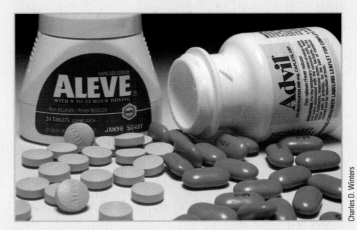

■ **Two nonprescription pain relievers. The active ingredient in Advil is ibuprofen; that in Aleve is sodium naproxen.**

Aleve. Both ibuprofen and naproxen have one stereocenter and can exist as a pair of enantiomers. For each drug, the active form is the s enantiomer. Naproxen is administered as its water-soluble sodium salt.

H CH₃ — COOH
(S)-Ibuprofen

H CH₃ — COOH
CH₃O—
(S)-Naproxen

In the 1960s, researchers discovered that aspirin acts by inhibiting cyclooxygenase (COX), a key enzyme in the conversion of arachidonic acid to prostaglandins (Chemical Connections 12I). With this discovery, it became clear why only one enantiomer of ibuprofen and naproxen is active: Only the s enantiomer of each has the correct handedness to bind to COX and inhibit its activity.

It is much more common, however, to prepare amides by treating an amine with an anhydride.

$$CH_3\overset{O}{\overset{\|}{C}}-O-\overset{O}{\overset{\|}{C}}CH_3 + H_2NCH_2CH_3 \longrightarrow CH_3\overset{O}{\overset{\|}{C}}-NHCH_2CH_3 + CH_3\overset{O}{\overset{\|}{C}}OH$$

Acetic anhydride · · · An amide

10.6 Hydrolysis of Anhydrides, Esters, and Amides

A Anhydrides

Carboxylic anhydrides, particularly the low-molecular-weight ones, react readily with water to give two carboxylic acids.

See the **Interactive General, Organic, and Biochemistry CD-ROM, version 2.0,** for further exploration on this topic.

$$CH_3\overset{O}{\overset{\|}{C}}O\overset{O}{\overset{\|}{C}}CH_3 + H_2O \longrightarrow CH_3\overset{O}{\overset{\|}{C}}OH + HO\overset{O}{\overset{\|}{C}}CH_3$$

Acetic anhydride · · · Acetic acid · · · Acetic acid

B Esters

Esters are hydrolyzed only very slowly, even in boiling water. Hydrolysis becomes considerably more rapid, however, when the ester is heated in aqueous acid or base. When we discussed acid-catalyzed Fischer esterification in Section 10.4, we pointed out that it is an equilibrium reaction. Hydrolysis of esters in aqueous acid, also an equilibrium reaction, is the reverse of Fischer esterification. A large excess of water drives the equilibrium to the right to form the carboxylic acid and alcohol.

$$CH_3\overset{O}{\overset{\|}{C}}OCH_2CH_3 + H_2O \underset{}{\overset{H^+}{\rightleftharpoons}} CH_3\overset{O}{\overset{\|}{C}}OH + CH_3CH_2OH$$

Ethyl acetate · · · Acetic acid · · · Ethanol

Hydrolysis of an ester can also be carried out using a hot aqueous base, such as aqueous NaOH. This reaction is often called **saponification,** a reference to its use in the manufacture of soaps (Chemical Connections 12C). The carboxylic acid formed in the hydrolysis reacts with hydroxide ion to form a carboxylic acid anion. Thus each mole of ester hydrolyzed requires one mole of base, as shown in the following balanced equation:

Saponification Hydrolysis of an ester in aqueous NaOH or KOH to an alcohol and the sodium or potassium salt of a carboxylic acid

$$CH_3\overset{O}{\overset{\|}{C}}OCH_2CH_3 + NaOH \overset{H_2O}{\longrightarrow} CH_3\overset{O}{\overset{\|}{C}}O^-Na^+ + CH_3CH_2OH$$

Ethyl acetate · · · Sodium hydroxide · · · Sodium acetate · · · Ethanol

EXAMPLE 10.6

Complete the equation for each hydrolysis reaction. Show the products as they are ionized under these experimental conditions.

CHEMICAL CONNECTIONS 10D

Ultraviolet Sunscreens and Sunblocks

Ultraviolet (UV) radiation penetrating the earth's ozone layer is arbitrarily divided into two regions: UVB (290–320 nm) and UVA (320–400 nm). UVB, a more energetic form of radiation than UVA, creates more radicals and hence does more oxidative damage (Section 16.7). UVB radiation interacts directly with biomolecules of the skin and eyes, causing skin cancer, skin aging, eye damage leading to cataracts, and delayed sunburn that appears 12–24 hours after exposure. UVA radiation, by contrast, causes tanning. It also damages skin, albeit much less efficiently than UVB. Its role in promoting skin cancer is less well understood.

Commercial sunscreen products are rated according to their sun protection factor (SPF), which is defined as the minimum effective dose of UV radiation that produces a delayed sunburn on protected skin compared to unprotected skin. Two types of active ingredients are found in commercial sunblocks and sunscreens. The most common sunblock agent is titanium dioxide, TiO_2, which reflects and scatters UV radiation. (TiO_2 is also widely used in paints because it covers better than any other inorganic white pigment.) Sunscreens, the second type of active ingredient, absorb UV radiation and then reradiate it as heat. These compounds are most effective in screening UVB radiation, but they do not screen UVA radiation. Thus they allow tanning but prevent the UVB-associated damage. Given here are structural formulas for three common esters used as UVB-screening agents, along with the name by which each is most commonly listed in the Active Ingredients label on commercial products:

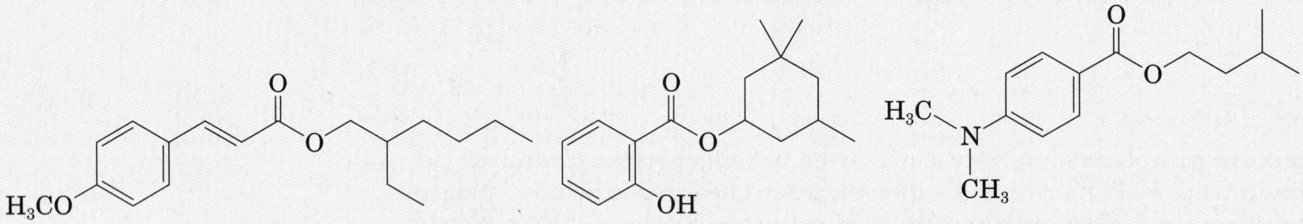

Octyl *p*-methoxycinnamate   Homosalate   Padimate A

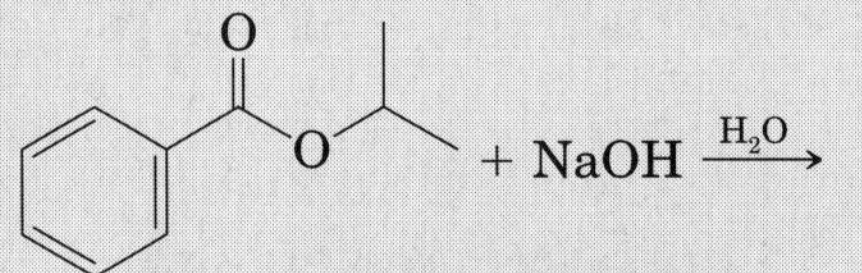

Solution

The products of hydrolysis of compound (a) are benzoic acid and 2-propanol. In aqueous NaOH, benzoic acid is converted to its sodium salt. In this reaction, one mole of NaOH is required for hydrolysis of each mole of this ester. Compound (b) is a diester of ethylene glycol. Two moles of NaOH are required for its hydrolysis.

(a) Sodium benzoate + 2-Propanol (Isopropyl alcohol)

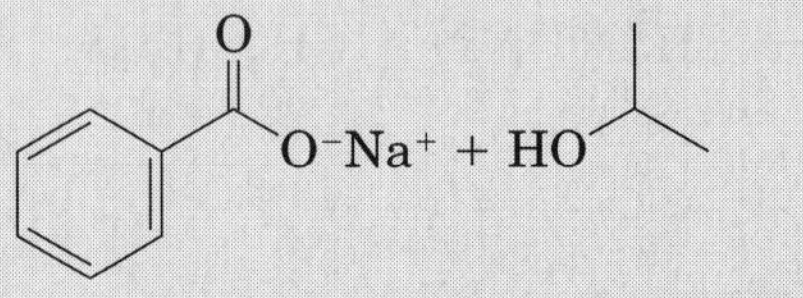

(b) $2CH_3CO^-Na^+$ + $HOCH_2CH_2OH$
Sodium acetate   1,2-Ethanediol (Ethylene glycol)

Problem 10.6

Complete the equation for each hydrolysis reaction. Show all products as they are ionized under these experimental conditions.

(a) 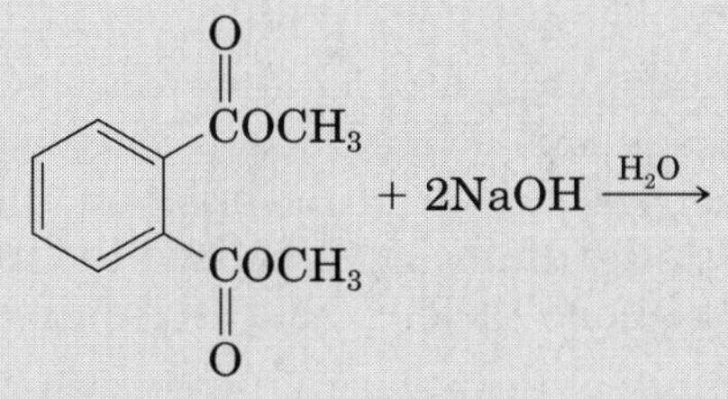 [structure: benzene ring with two COCH₃ ester groups] + 2NaOH $\xrightarrow{H_2O}$

(b) 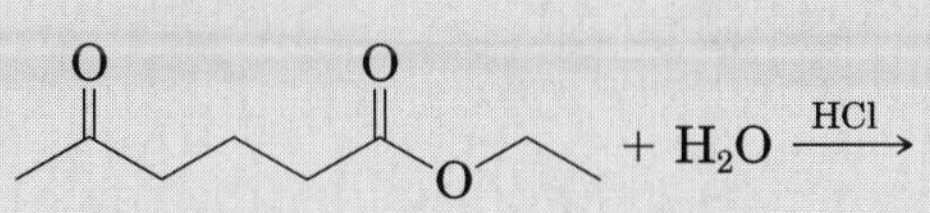 [structure: ketoester] + H_2O $\xrightarrow{HCl}$

CHEMICAL CONNECTIONS 10E

Barbiturates

In 1864, Adolph von Baeyer (1835–1917) discovered that heating the diethyl ester of malonic acid with urea in the presence of sodium ethoxide (like sodium hydroxide, a strong base) gives a cyclic compound that he named *barbituric acid.*

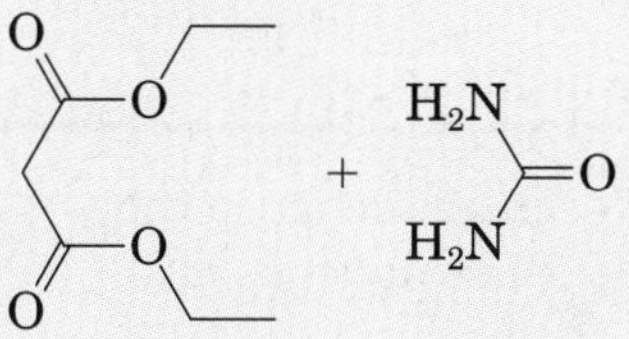

[structure: Diethyl propanedioate + Urea]

Diethyl propanedioate   Urea
(Diethyl malonate)

$\xrightarrow[CH_3CH_2OH]{CH_3CH_2O^-Na^+}$ [structure: Barbituric acid] $+\ 2CH_3CH_2OH$

Barbituric acid

[structure: Pentobarbital]

Pentobarbital

[structure: Sodium pentobarbital]

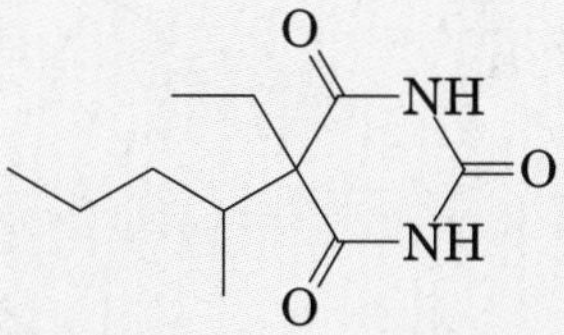

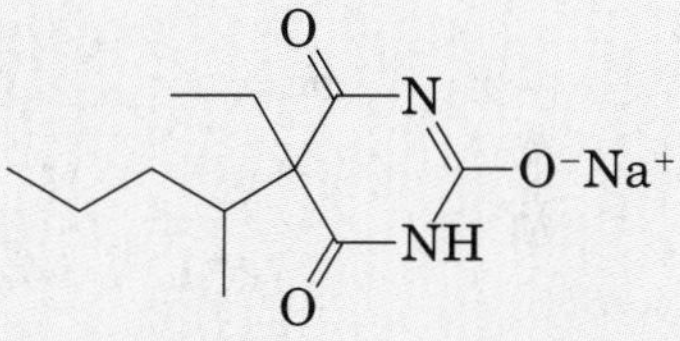

Sodium pentobarbital
(Nembutal)

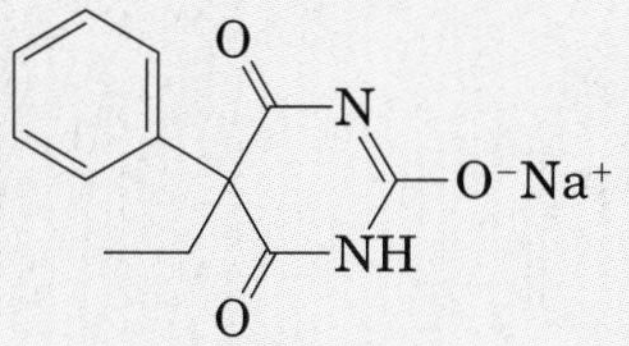

[structure: Phenobarbital]

Phenobarbital

Some say that Baeyer named this compound after a friend named Barbara. Others claim that he named it after St. Barbara, the patron saint of artillerymen.

A number of derivatives of barbituric acid have powerful sedative and hypnotic effects. One such derivative is pentobarbital. Like the other derivatives of barbituric acid, pentobarbital is quite insoluble in water and body fluids. To increase its solubility in these fluids, pentobarbital is converted to its sodium salt, which is given the name Nembutal. Phenobarbital, also administered as its sodium salt, is an anticonvulsant, sedative, and hypnotic.

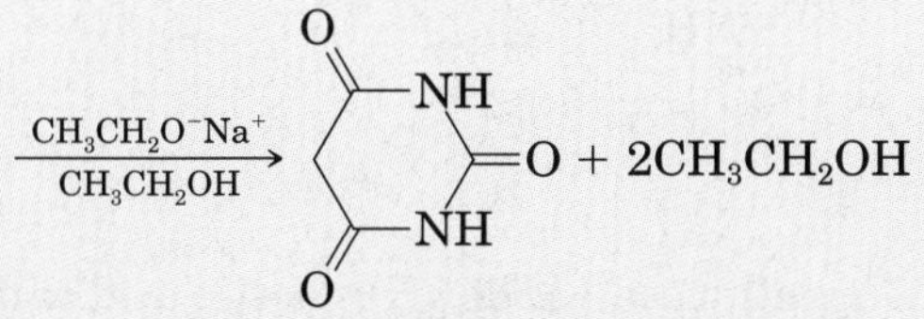

Technically speaking, only the sodium salts of these compounds should be called barbiturates. In practice, however, all derivatives of barbituric acid are called barbiturates, whether they are the un-ionized form or the ionized, water-soluble salt form.

Barbiturates have two principal effects. In small doses, they are sedatives (tranquilizers); in larger doses, they induce sleep. Barbituric acid, in contrast, has neither of these effects. Barbiturates are dangerous because they are addictive, which means that a regular user will suffer withdrawal symptoms when their use is stopped. They are especially dangerous when taken with alcohol because the combined effect (called a *synergistic effect*) is usually greater than the sum of the effects of either drug taken separately.

C Amides

See the **Interactive General, Organic, and Biochemistry CD-ROM, version 2.0,** for further exploration on this topic.

Amides require more vigorous conditions for hydrolysis in both acid and base than do esters. Hydrolysis in hot aqueous acid gives a carboxylic acid and an ammonium ion. This reaction is driven to completion by the acid–base reaction between ammonia or the amine and the acid to form an ammonium ion. For complete hydrolysis, one mole of acid is required per mole of amide.

$$CH_3CH_2CH_2\overset{\displaystyle O}{\overset{\|}{C}}NH_2 + H_2O + HCl \xrightarrow[\text{heat}]{H_2O} CH_3CH_2CH_2\overset{\displaystyle O}{\overset{\|}{C}}OH + NH_4^+\,Cl^-$$

Butanamide $\qquad\qquad\qquad\qquad\qquad$ Butanoic acid

In aqueous base, the products of amide hydrolysis are a carboxylic acid salt and ammonia or an amine. This hydrolysis is driven to completion by the acid–base reaction between the carboxylic acid and base to form a salt. To carry out this process, one mole of base is required per mole of amide.

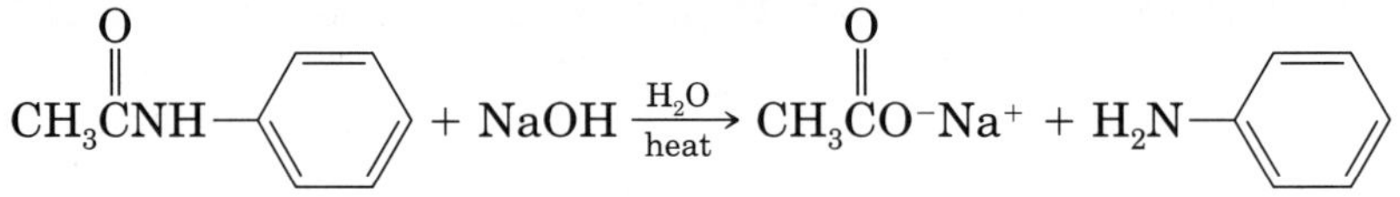

$$CH_3\overset{\displaystyle O}{\overset{\|}{C}}NH\!-\!\bigcirc + NaOH \xrightarrow[\text{heat}]{H_2O} CH_3\overset{\displaystyle O}{\overset{\|}{C}}O^-Na^+ + H_2N\!-\!\bigcirc$$

N-Phenylethanamide $\qquad\qquad$ Sodium acetate $\qquad$ Aniline
(*N*-Phenylacetamide,
Acetanilide)

EXAMPLE 10.7

Write a balanced equation for the hydrolysis of each amide in concentrated aqueous HCl. Show all products as they exist in aqueous HCl.

(a) $CH_3\overset{\displaystyle O}{\overset{\|}{C}}N(CH_3)_2$ $\quad$ (b) 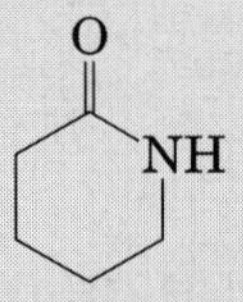 [cyclic lactam with NH and C=O]

Solution

(a) Hydrolysis of *N,N*-dimethylacetamide gives acetic acid and dimethylamine. Dimethylamine, a base, reacts with HCl to form dimethylammonium ion, shown here as dimethylammonium chloride.

$$CH_3\overset{\displaystyle O}{\overset{\|}{C}}N(CH_3)_2 + H_2O + HCl \xrightarrow{\text{Heat}} CH_3\overset{\displaystyle O}{\overset{\|}{C}}OH + (CH_3)_2NH_2^+Cl^-$$

(b) Hydrolysis of this lactam gives the protonated form of 5-aminopentanoic acid.

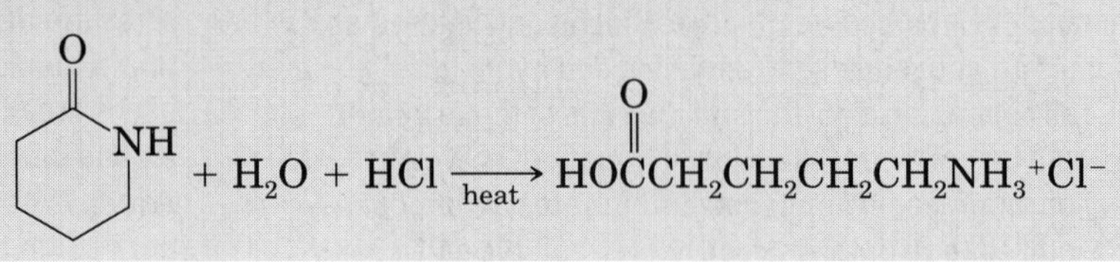

$$[\text{cyclic lactam}] + H_2O + HCl \xrightarrow[\text{heat}]{} HO\overset{\displaystyle O}{\overset{\|}{C}}CH_2CH_2CH_2CH_2NH_3^+Cl^-$$

Problem 10.7

Write a balanced equation for the hydrolysis of the amides in Example 10.7 in concentrated aqueous NaOH. Show all products as they exist in aqueous NaOH.

10.7 Phosphoric Anhydrides and Esters

A Phosphoric Anhydrides

Because of the special importance of phosphoric anhydrides in biochemical systems, we discuss them here to show the similarity between them and the anhydrides of carboxylic acids. The functional group of a **phosphoric anhydride** consists of two phosphoryl (P=O) groups bonded to the same oxygen atom. Shown here are structural formulas for two anhydrides of phosphoric acid and the ions derived by ionization of the acidic hydrogens of each:

At physiological pH (approximately 7.3–7.4), pyrophosphoric acid exists as the pyrophosphate ion and triphosphoric acid exists as the triphosphate ion.

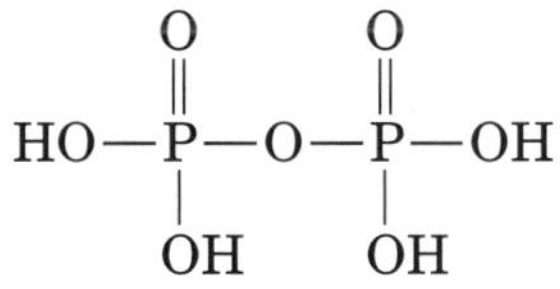

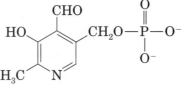

Diphosphoric acid (Pyrophosphoric acid)	Diphosphate ion (Pyrophosphate ion)	Triphosphoric acid	Triphosphate ion

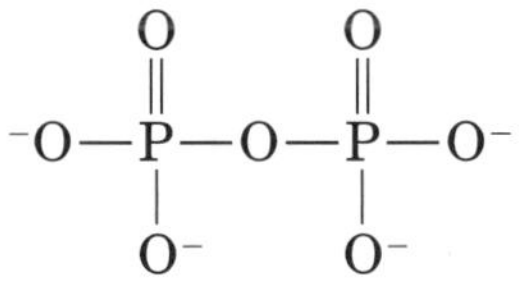

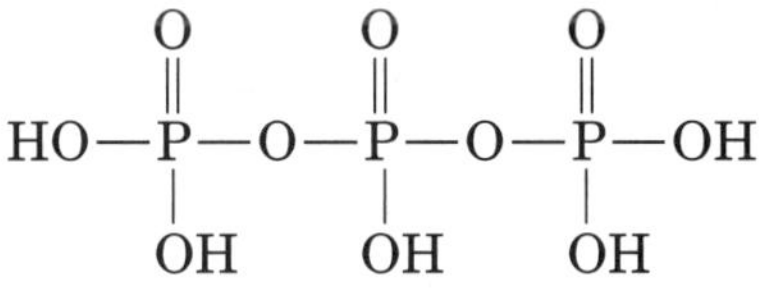

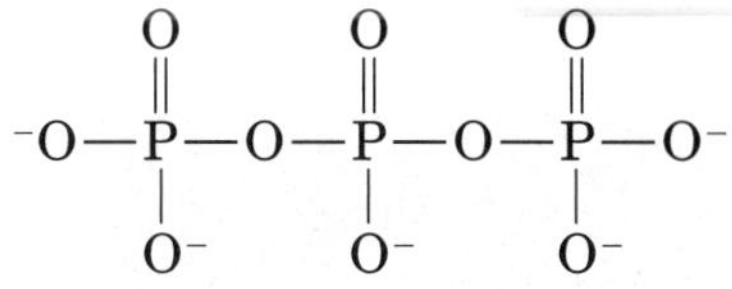

B Phosphoric Esters

Phosphoric acid has three —OH groups. It forms mono-, di-, and triphosphoric esters, which we name by giving the name(s) of the alkyl group(s) bonded to oxygen followed by the word "phosphate" (for example, dimethyl phosphate). In more complex **phosphoric esters,** it is common practice to name the organic molecule and then indicate the presence of the phosphoric ester by including either the word "phosphate" or the prefix *phospho-*. Dihydroxyacetone phosphate, for example, is an intermediate in glycolysis (Section 19.2). Pyridoxal phosphate is one of the metabolically active forms of vitamin B_6. The last two phosphoric esters are shown here as they are ionized at pH 7.4, the pH of blood plasma.

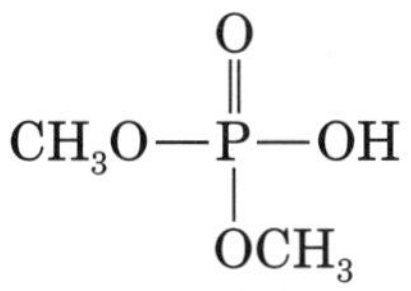

Dimethyl phosphate 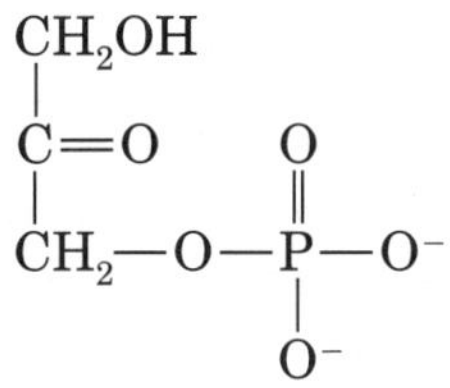   Dihydroxyacetone phosphate   Pyridoxal phosphate

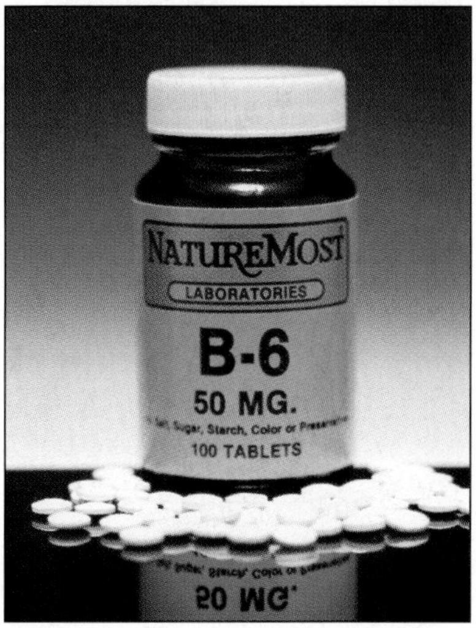

■ **Vitamin B_6, pyridoxal.**

10.8 Step-Growth Polymerizations

Step-growth polymers form from the reaction of molecules containing two functional groups, with each new bond being created in a separate step. In this section, we discuss three types of step-growth polymers: polyamides, polyesters, and polycarbonates.

Step-growth polymerization
A polymerization in which chain growth occurs in a stepwise manner between difunctional monomers, as, for example, between adipic acid and hexamethylenediamine to form nylon-66

A Polyamides

Polyamide A polymer in which each monomer unit is bonded to the next by an amide bond—for example, nylon-66

In the early 1930s, chemists at E. I. DuPont de Nemours & Company began fundamental research into the reactions between dicarboxylic acids and diamines to form **polyamides.** In 1934, they synthesized nylon-66, the first purely synthetic fiber. Nylon-66 is so named because it is synthesized from two different monomers, each containing six carbon atoms.

In the synthesis of nylon-66, hexanedioic acid and 1,6-hexanediamine are dissolved in aqueous ethanol and then heated in an autoclave to 250°C and an internal pressure of 15 atm. Under these conditions, —COOH and —NH$_2$ groups react by loss of H$_2$O to form a polyamide, similar to the formation of amides described in Section 10.5.

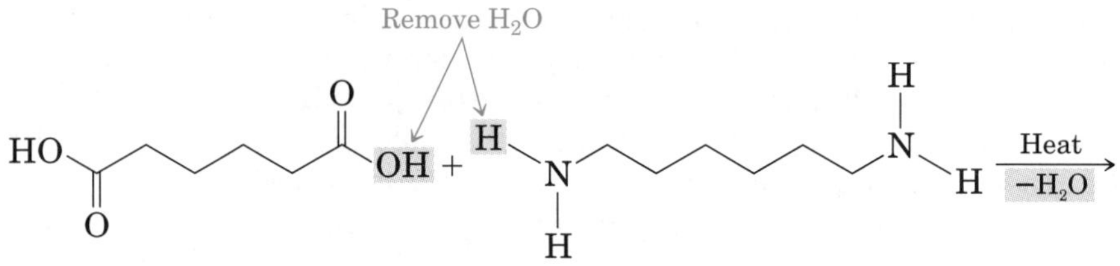

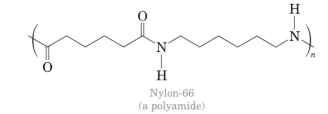

Hexanedioic acid
(Adipic acid)

1,6-Hexanediamine
(Hexamethylenediamine)

Nylon-66
(a polyamide)

Charles D. Winters

■ **Bulletproof vests have a thick layer of Kevlar.**

Based on extensive research into the relationships between molecular structure and bulk physical properties, scientists at DuPont reasoned that a polyamide containing benzene rings would be even stronger than nylon-66. Their line of reasoning eventually produced the following polyamide that DuPont named Kevlar.

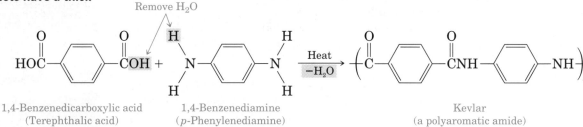

1,4-Benzenedicarboxylic acid
(Terephthalic acid)

1,4-Benzenediamine
(*p*-Phenylenediamine)

Kevlar
(a polyaromatic amide)

One remarkable feature of Kevlar is that it weighs less than other materials of similar strength. For example, a cable woven of Kevlar has a strength equal to that of a similarly woven steel cable. Yet the Kevlar cable has only 20% of the weight of the steel cable! Kevlar now finds use in such articles as anchor cables for offshore drilling rigs and reinforcement fibers for automobile tires. It is also woven into a fabric that is so tough that it can be used for bulletproof vests, jackets, and raincoats.

B Polyesters

Polyester A polymer in which each monomer unit is bonded to the next by an ester bond—for example, poly(ethylene terephthalate)

The first **polyester,** developed in the 1940s, involved polymerization of benzene 1,4-dicarboxylic acid with 1,2-ethanediol to give poly(ethylene terephthalate), abbreviated PET. Virtually all PET is now made from the dimethyl ester of terephthalic acid by the following reaction:

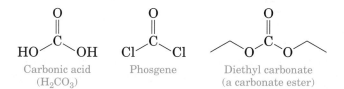

Remove CH₃OH

Dimethyl terephthalate + 1,2-Ethanediol (Ethylene glycol) → [Heat, $-CH_3OH$] → Poly(ethylene terephthalate) (Dacron, Mylar)

The crude polyester can be melted, extruded, and then drawn to form the textile fiber Dacron polyester. Dacron's outstanding features include its stiffness (about four times that of nylon-66), very high tensile strength, and remarkable resistance to creasing and wrinkling. Because the early Dacron polyester fibers were harsh to the touch due to their stiffness, they were usually blended with cotton or wool to make acceptable textile fibers. Newly developed fabrication techniques now produce less harsh Dacron polyester textile fibers. PET is also fabricated into Mylar films and recyclable plastic beverage containers.

C Polycarbonates

A **polycarbonate,** the most familiar of which is Lexan, forms from the reaction of the disodium salt of bisphenol A and phosgene. Phosgene is a derivative of carbonic acid, H_2CO_3, in which both —OH groups have been replaced by chlorine atoms. An ester of carbonic acid is called a carbonate.

Carbonic acid (H_2CO_3)   Phosgene   Diethyl carbonate (a carbonate ester)

■ **Mylar can be made into extremely strong films. Because the film has very tiny pores, it is used for balloons that can be inflated with helium; the helium atoms diffuse only slowly through the pores of the film.**

Polycarbonate A polyester in which the carboxyl groups are derived from carbonic acid

CHEMICAL CONNECTIONS **10F**

Stitches That Dissolve

As the technological capabilities of medicine have grown, the demand for synthetic materials that can be used inside the body has increased as well. Polymers already have many of the characteristics of an ideal biomaterial: They are lightweight and strong, are inert or biodegradable depending on their chemical structure, and have physical properties (softness, rigidity, elasticity) that are easily tailored to match those of natural tissues.

Even though most medical uses of polymeric materials require biostability, some applications require them to be biodegradable. An example is the polymer of glycolic acid and

lactic acid used in absorbable sutures, which are marketed under the trade name of Lactomer.

Traditional suture materials such as catgut must be removed by a health care specialist after they have served their purpose. Stitches of Lactomer, however, are hydrolyzed slowly over a period of approximately two weeks. By the time the torn tissues have healed, the stitches have hydrolyzed, and no suture removal is necessary. The body metabolizes and excretes the glycolic and lactic acids formed during this hydrolysis.

Remove H_2O

Glycolic acid + Lactic acid → [Polymerization, $-nH_2O$] → A polymer of glycolic acid and lactic acid

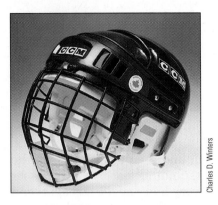

■ **A polycarbonate hockey mask.**

In forming a polycarbonate, each mole of phosgene reacts with two moles of the sodium salt of a phenol called bisphenol A:

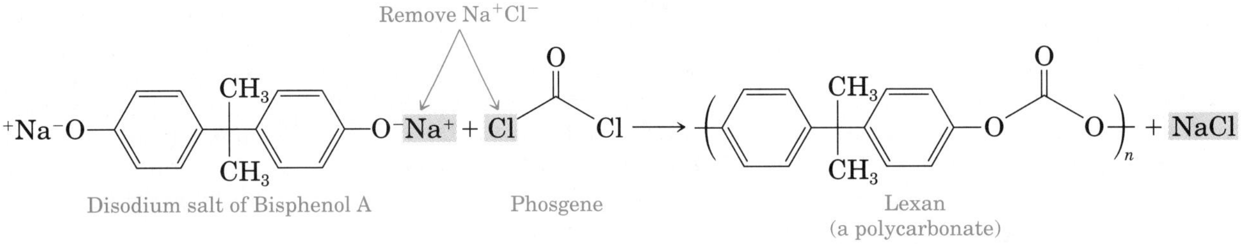

Lexan is a tough, transparent polymer that has high impact and tensile strengths and retains its properties over a wide temperature range. It is used in sporting equipment (helmets and face masks); to make light, impact-resistant housings for household appliances; and in the manufacture of safety glass and unbreakable windows.

S U M M A R Y

The functional group of a **carboxylic acid** is the **carboxyl group, —COOH** (Section 10.1). IUPAC names of carboxylic acids are derived from the name of the parent alkane by dropping the suffix *-e* and adding *-oic acid*. Dicarboxylic acids are named as *-dioic acids* (Section 10.2A). Common names for many carboxylic and dicarboxylic acids remain in wide use.

Carboxylic acids are polar compounds; consequently, they have higher boiling points and are more soluble in water than alcohols, aldehydes, ketones, and ethers of comparable molecular weight (Section 10.2C). Carboxylic acids are weak acids that react with strong bases to form salts.

An **anhydride** contains two carbonyl groups bonded to the same oxygen (Section 10.3A). An **ester** contains a carbonyl group bonded to an —OR group derived from an alcohol or phenol (Section 10.3B). An **amide** contains a carbonyl group bonded to a nitrogen atom derived from an amine (Section 10.3C).

Step-growth polymerizations involve the stepwise reaction of difunctional monomers (Section 10.8). Important commercial polymers synthesized through step-growth processes include polyamides, polyesters, and polycarbonates.

K E Y   R E A C T I O N S

1. Acidity of Carboxylic Acids (Section 10.2D) Values of pK_a for most unsubstituted aliphatic and aromatic carboxylic acids are within the range of 4.0 to 5.0.

$$\underset{O}{\overset{\parallel}{CH_3COH}} + H_2O \rightleftharpoons \underset{O}{\overset{\parallel}{CH_3CO^-}} + H_3O^+ \quad pK_a = 4.76$$

2. Reaction of Carboxylic Acids with Bases (Section 10.2E) Carboxylic acids form water-soluble salts with alkali metal hydroxides, carbonates, bicarbonates, ammonia, and amines.

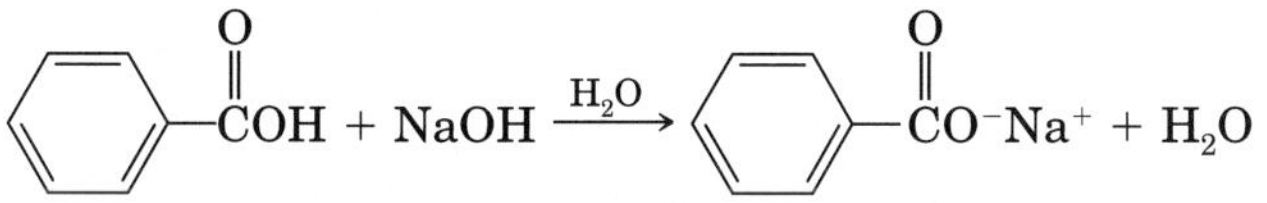

$$\text{(phenyl)}\overset{O}{\overset{\parallel}{COH}} + NaOH \xrightarrow{H_2O} \text{(phenyl)}\overset{O}{\overset{\parallel}{CO^-Na^+}} + H_2O$$

3. Fischer Esterification (Section 10.4) Fischer esterification is reversible. To achieve high yields of ester, it is necessary to force the equilibrium to the right. One way to accomplish this goal is to use an excess of the alcohol. Another way is to remove water as it is formed.

$$\underset{O}{\overset{\parallel}{CH_3COH}} + CH_3CH_2CH_2OH$$

$$\xrightarrow{H_2SO_4} \underset{O}{\overset{\parallel}{CH_3COCH_2CH_2CH_3}} + H_2O$$

4. Preparation of an Amide (Section 10.5) Reaction of an amine with an anhydride gives an amide.

$$\underset{\substack{O \\ \text{Acetic} \\ \text{anhydride}}}{\overset{\parallel}{CH_3C}} - O - \overset{\overset{O}{\parallel}}{CCH_3} + H_2NCH_2CH_3$$

$$\longrightarrow \underset{\substack{O \\ \text{An amide}}}{\overset{\parallel}{CH_3C}} - NHCH_2CH_3 + \overset{O}{\overset{\parallel}{CH_3COH}}$$

5. Hydrolysis of an Ester (Section 10.6B) Esters are hydrolyzed rapidly only in the presence of acid or base. Acid is a catalyst. Base is required in an equimolar amount.

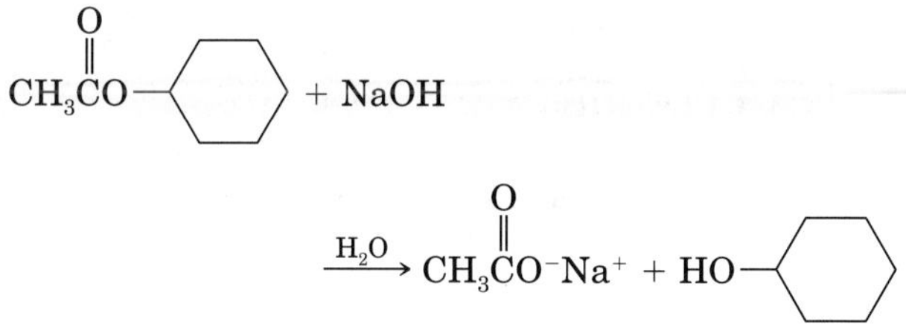

$$\underset{O}{\overset{\parallel}{CH_3CO}}-\text{(cyclohexyl)} + NaOH$$

$$\xrightarrow{H_2O} \underset{O}{\overset{\parallel}{CH_3CO^-Na^+}} + HO-\text{(cyclohexyl)}$$

6. Hydrolysis of an Amide (Section 10.5C) Either acid or base is required in an amount equivalent to that of the amide.

$$\underset{O}{\overset{\parallel}{CH_3CH_2CH_2CNH_2}} + H_2O + HCl$$

$$\xrightarrow[\text{heat}]{H_2O} \underset{O}{\overset{\parallel}{CH_3CH_2CH_2COH}} + NH_4^+ \ Cl^-$$

$$\underset{O}{\overset{\parallel}{CH_3CH_2CH_2CNH_2}} + NaOH$$

$$\xrightarrow[\text{heat}]{H_2O} \underset{O}{\overset{\parallel}{CH_3CH_2CH_2CO^-}} Na^+ + NH_3$$

P R O B L E M S

Numbers that appear in color indicate difficult problems.
◥ designates problems requiring application of principles.

Structure and Nomenclature

10.8 Name and draw structural formulas for the four carboxylic acids with molecular formula $C_5H_{10}O_2$. Which of these carboxylic acids are chiral?

10.9 Write the IUPAC name for each carboxylic acid.

(a) 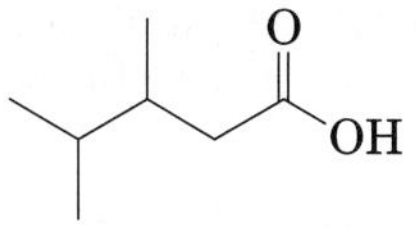 [structure: 3-methylbutyl chain with COOH, O and OH]

(b) 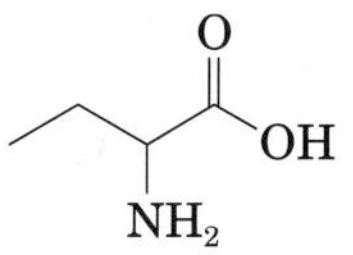 [structure: chain with COOH (O, OH) and NH₂ substituent]

(c) 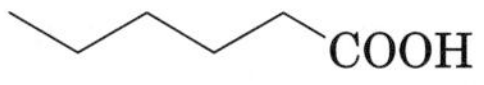 [structure: pentyl chain with COOH]

10.10 Write the IUPAC name for each carboxylic acid.

(a) HOOC—CH(OH)—COOH

(b) (benzene ring with COOH and OH groups)

(c) CCl_3COOH

10.11 Draw a structural formula for each carboxylic acid.
   (a) 4-Nitrophenylacetic acid
   (b) 4-Aminobutanoic acid
   (c) 4-Phenylbutanoic acid
   (d) *cis*-3-Hexenedioic acid

10.12 Draw a structural formula for each carboxylic acid.
   (a) 2-Aminopropanoic acid
   (b) 3,5-Dinitrobenzoic acid
   (c) Dichloroacetic acid
   (d) *o*-Aminobenzoic acid

10.13 Draw a structural formula for each salt.
   (a) Sodium benzoate
   (b) Lithium acetate
   (c) Ammonium acetate
   (d) Disodium adipate
   (e) Sodium salicylate
   (f) Calcium butanoate

10.14 Calcium oxalate is a major component of kidney stones. Draw a structural formula for this compound.

10.15 The monopotassium salt of oxalic acid is present in certain leafy vegetables, including rhubarb. Both oxalic acid and its salts are poisonous in high concentrations. Draw a structural formula for monopotassium oxalate.

Physical Properties

10.16 Draw a structural formula for the dimer formed when two molecules of formic acid form hydrogen bonds with each other.

10.17 Propanedioic (malonic) acid forms an internal hydrogen bond in which the H atom of one COOH group forms a hydrogen bond with an O atom of the other COOH group. Draw a structural formula to show this internal hydrogen bonding. (There are two possible answers.)

10.18 Hexanoic (caproic) acid has a solubility in water of about 1 g/100 mL of water. Which part of the molecule contributes to water solubility, and which part prevents solubility?

10.19 The following compounds all have approximately the same molecular weight: hexanoic acid, heptanal, and 1-heptanol. Arrange them in order of increasing boiling point.

10.20 The following compounds all have approximately the same molecular weight: propanoic acid, 1-butanol, and diethyl ether. Arrange them in order of increasing boiling point.

10.21 Arrange these compounds in order of increasing solubility in water: acetic acid, pentanoic acid, and decanoic acid.

Preparation of Carboxylic Acids

10.22 Complete the equations for these oxidations.

(a) $CH_3(CH_2)_4CH_2OH \xrightarrow[H_2SO_4]{K_2Cr_2O_7}$

(b) (benzene ring with CHO, OCH₃, and OH groups) $+ Ag(NH_3)_2^+ \longrightarrow$

(c) HO——$CH_2OH \xrightarrow[H_2SO_4]{K_2Cr_2O_7}$

10.23 Draw the structural formula for a compound with the given molecular formula that, on oxidation by potassium dichromate, gives the carboxylic acid or dicarboxylic acid shown.

(a) $C_6H_{14}O \xrightarrow{[O]} CH_3(CH_2)_4COH$ (with =O on C)

(b) $C_6H_{12}O \xrightarrow{[O]} CH_3(CH_2)_4COH$ (with =O on C)

(c) $C_6H_{14}O_2 \xrightarrow{[O]} HOC(CH_2)_4COH$ (with =O on each C)

Acidity of Carboxylic Acids

10.24 Alcohols, phenols, and carboxylic acids all contain an —OH group. Which are the strongest acids? Which are the weakest acids?

10.25 Arrange these compounds in order of increasing acidity: benzoic acid, benzyl alcohol, and phenol.

10.26 Complete the equations for these acid–base reactions.

(a) (benzene ring)—$CH_2COOH + NaOH \longrightarrow$

(b) (alkene chain)$COOH + NaHCO_3 \longrightarrow$

(c) (benzene ring with COOH and OCH₃) $+ NaHCO_3 \longrightarrow$

(d) $\overset{\underset{\displaystyle |}{OH}}{CH_3CHCOOH}$ + $H_2NCH_2CH_2OH$ $\longrightarrow$

(e) $\diagdown\diagup\diagdown\diagup COO^-Na^+$ + HCl $\longrightarrow$

10.27 Complete the equations for these acid-base reactions.

(a) [structure: benzene ring with OH and CH₃ substituents] + NaOH $\longrightarrow$

(b) [structure: benzene ring with COO^-Na^+ and OH substituents] + HCl $\longrightarrow$

(c) [structure: benzene ring with COOH and OCH_3 substituents] + $H_2NCH_2CH_2OH$ $\longrightarrow$

(d) [cyclohexane ring]$-$COOH + $NaHCO_3$ $\longrightarrow$

10.28 Formic acid is one of the components responsible for the sting of biting ants and is injected under the skin by bee and wasp stings. The pain can be relieved by rubbing the area of the sting with a paste of baking soda ($NaHCO_3$) and water that neutralizes the acid. Write an equation for this reaction.

10.29 The pK_a of acetic acid is 4.76. What form(s) of acetic acid are present at pH 2.0 or lower? At pH 4.76? At pH 8.0 or higher?

10.30 The normal pH range for blood plasma is 7.35 to 7.45. Under these conditions, would you expect the carboxyl group of lactic acid (pK_a 4.07) to exist primarily as a carboxyl group or as a carboxylic anion? Explain.

10.31 The pK_a of ascorbic acid (Chemical Connections 11B) is 4.10. Would you expect ascorbic acid dissolved in blood plasma, pH 7.35 to 7.45, to exist primarily as ascorbic acid or as ascorbate anion? Explain.

10.32 In Chapter 13, we discuss a class of compounds called amino acids, so named because they contain both an amino group and a carboxyl group. Following is a structural formula for the amino acid alanine.

$$\overset{\underset{\displaystyle |}{\underset{\displaystyle NH_3^+}{}}}{CH_3\overset{\overset{\displaystyle O}{\displaystyle \|}}{C}HCO^-}$$
Alanine

What would you expect to be the major form of alanine present in aqueous solution at (a) pH 2.0, (b) at pH 5–6, and (c) at pH 11.0? Explain.

10.33 Complete the equations for the following acid–base reactions. Assume one mole of NaOH per mole of amino acid.

(a) $\overset{\underset{\displaystyle |}{\underset{\displaystyle NH_3^+}{}}}{CH_3CHCOOH}$ + NaOH $\longrightarrow$

(b) $\overset{\underset{\displaystyle |}{\underset{\displaystyle NH_3^+}{}}}{CH_3CHCOO^-Na^+}$ + NaOH $\longrightarrow$

10.34 Which is the stronger base: $CH_3CH_2NH_2$ or $CH_3CH_2COO^-$? Explain.

10.35 Complete the equations for the following acid–base reactions. Assume one mole of HCl per mole of amino acid.

(a) $\overset{\underset{\displaystyle |}{\underset{\displaystyle NH_2}{}}}{CH_3CHCOO^-Na^+}$ + HCl $\longrightarrow$

(b) $\overset{\underset{\displaystyle |}{\underset{\displaystyle NH_3^+}{}}}{CH_3CHCOO^-Na^+}$ + HCl $\longrightarrow$

Reactions of Carboxylic Acids

10.36 Define and give an example of Fischer esterification.

10.37 Complete these examples of Fischer esterification. In each case, assume an excess of alcohol.

(a) CH_3COOH + HO$\diagup\diagdown\diagup\diagdown$ $\overset{H^+}{\rightleftharpoons}$

(b) CH_3COOH + HO$-$[cyclohexane ring] $\overset{H^+}{\rightleftharpoons}$

(c) [benzene ring with two COOH groups] + CH_3CH_2OH $\overset{H^+}{\rightleftharpoons}$

10.38 Complete these examples of Fischer esterification. In each case, assume an excess of the alcohol.

(a) [benzene ring with COOH and OCH_3 substituents] + $HOCH_2CH_2NH_2$ $\overset{H^+}{\rightleftharpoons}$

(b) [cyclopentane ring]$-$COOH + $(CH_3)_2CHOH$ $\overset{H^+}{\rightleftharpoons}$

(c) HO$\overset{\overset{\displaystyle O}{\displaystyle \|}}{C}\diagup\diagdown\diagup\overset{\overset{\displaystyle O}{\displaystyle \|}}{C}$OH + CH_3OH $\overset{H^+}{\rightleftharpoons}$

10.39 From what carboxylic acid and alcohol is each ester derived?

(a) 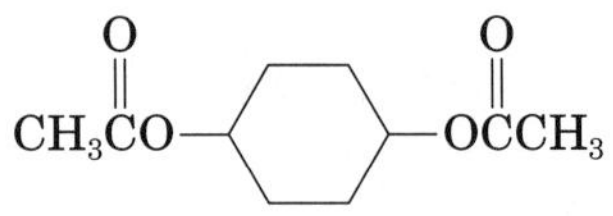 CH_3CO—cyclohexane—$OCCH_3$ (with two C=O groups)

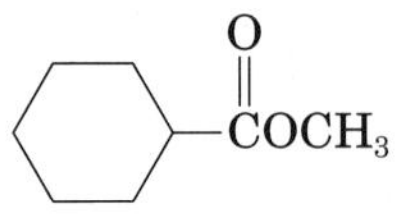

(b) cyclohexane—$COCH_3$ (with C=O)

(c) $CH_3OCCH_2CH_2COCH_3$ (with two C=O)

(d) 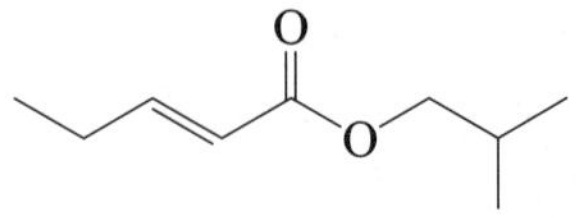 (structure of an ester)

10.40 Methyl 2-hydroxybenzoate (methyl salicylate) has the odor of oil of wintergreen. This compound is prepared by Fischer esterification of 2-hydroxybenzoic acid (salicylic acid) with methanol. Draw a structural formula for methyl 2-hydroxybenzoate.

10.41 Methyl 4-hydroxybenzoate (methylparaben) is used as a preservative in foods, beverages, and cosmetics. This compound is prepared by Fischer esterification of 4-hydroxybenzoic acid with methanol. Draw a structural formula for methylparaben.

10.42 When 5-hydroxypentanoic acid is treated with an acid catalyst, it forms a lactone (a cyclic ester). Draw a structural formula for this lactone.

10.43 From what carboxylic acid and amine or ammonia can each amide be synthesized?

(a) 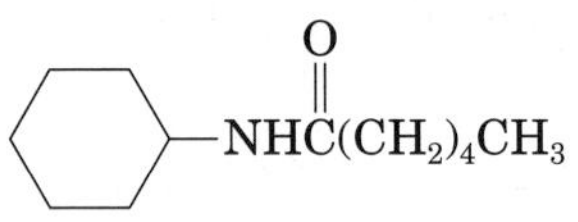 cyclohexane—$NHC(CH_2)_4CH_3$ (with C=O)

(b) $(CH_3)_2CHCN(CH_3)_2$ (with C=O)

(c) $H_2NC(CH_2)_4CNH_2$ (with two C=O)

10.44 *N,N*-Diethyl *m*-toluamide (DEET) is the active ingredient in several common insect repellents. From what acid and amine can DEET by synthesized?

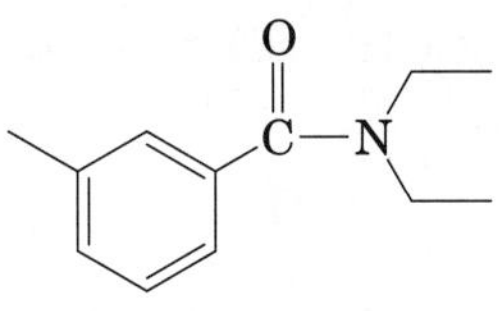

N,N-Diethyl *m*-toluamide
(DEET)

10.45 Following are structural formulas for two local anesthetics. Lidocaine was introduced in 1948 and is now the most widely used local anesthetic for infiltration and regional anesthesia. Its hydrochloride salt is marketed under the name Xylocaine. Mepivacaine is faster acting and somewhat longer in duration than lidocaine. Its hydrochloride salt is marketed under the name Carbocaine.

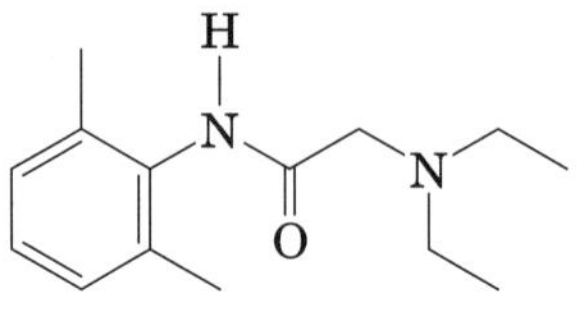

Lidocaine
(Xylocaine)

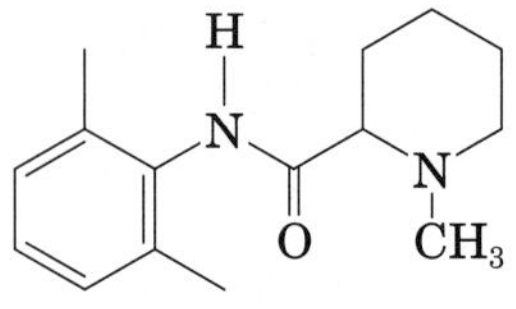

Mepivacaine
(Carbocaine)

(a) Name the functional groups in each anesthetic.

(b) What similarities in structure do you find between these compounds?

Structure and Nomenclature of Anhydrides, Esters, and Amides

10.46 Draw a structural formula for each compound.

(a) Diethyl carbonate

(b) *p*-Nitrobenzamide

(c) Ethyl 3-hydroxybutanoate

(d) Diethyl oxalate

(e) Ethyl *trans*-2-pentenoate

(f) Butanoic anhydride

10.47 Flavoring agents are the largest class of food additives. Each ester listed here is a synthetic flavor additive, with a flavor very close to the natural flavor. While these esters are the major components of the natural flavors, each natural flavor contains many more components, including some in only trace amounts. Draw a structural formula for each ester. (Isopentane is the common name for 2-methylbutane.)

(a) Ethyl formate (rum)

(b) Isopentyl acetate (banana)

(c) Octyl acetate (orange)

(d) Methyl butanoate (apple)

(e) Ethyl butanoate (pineapple)

(f) Methyl 2-aminobenzoate (grape)

10.48 Write the IUPAC name for each compound.

(a) 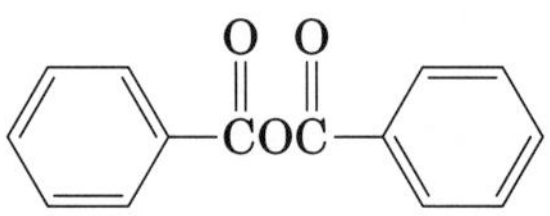 benzene—COC—benzene (with two C=O)

(b) $CH_3(CH_2)_8COCH_3$ (with C=O)

(c) $CH_3(CH_2)_4CNHCH_3$ (with C=O)

(d) H_2N—⬡—$\overset{\overset{\displaystyle O}{\|}}{C}NH_2$

(e) $CH_3\overset{\overset{\displaystyle O}{\|}}{C}O$—⬠

(f) $CH_3\overset{\overset{\displaystyle OH}{|}}{C}HCH_2\overset{\overset{\displaystyle O}{\|}}{C}OCH_2CH_3$

Reactions of Anhydrides, Esters, and Amides

10.49 Define and give an example of saponification.

10.50 What product forms when ethyl benzoate is treated with each reagent?
(a) H_2O, NaOH, heat
(b) H_2O, HCl, heat

10.51 What product forms when benzamide is treated with each reagent?
(a) H_2O, NaOH, heat
(b) H_2O, HCl, heat

10.52 Which of these types of compounds will produce bubbles of CO_2 when added to an aqueous solution of sodium bicarbonate?
(a) A carboxylic acid
(b) A carboxylic ester
(c) The sodium salt of a carboxylic acid

10.53 Complete the equations for these reactions.

(a) CH_3O—⬡—NH_2 + $CH_3\overset{\overset{\displaystyle O}{\|}}{C}O\overset{\overset{\displaystyle O}{\|}}{C}CH_3$ $\longrightarrow$

(b) ⬡NH + $CH_3\overset{\overset{\displaystyle O}{\|}}{C}O\overset{\overset{\displaystyle O}{\|}}{C}CH_3$ $\longrightarrow$

10.54 The analgesic phenacetin is synthesized by treating 4-ethoxyaniline with acetic anhydride. Draw a structural formula for phenacetin.

CH_3CH_2O—⬡—NH_2

4-Ethoxyaniline

10.55 Phenobarbital is a long-acting sedative, hypnotic, and anticonvulsant.

Phenobarbital

(a) Name all functional groups in this compound.
(b) Draw structural formulas for the products from complete hydrolysis of all amide groups in aqueous NaOH.

10.56 Following is a structural formula for aspartame, an artificial sweetener about 180 times as sweet as sucrose (table sugar):

Aspartame

(a) Name all functional groups in aspartame.
(b) Is aspartame chiral? If so, how many stereoisomers are possible for it?
(c) Draw structural formulas for the products formed on hydrolysis of all ester and amide bonds in aspartame.

10.57 Why are nylon-66 and Kevlar referred to as polyamides?

10.58 Draw short sections of two parallel chains of nylon-66 (each chain running in the same direction) and show how it is possible to align them such that there is hydrogen bonding between the N—H groups of one chain and C=O groups of the parallel chain.

10.59 Why are Dacron and Mylar referred to as polyesters?

Anhydrides and Esters of Phosphoric Acid

10.60 What type of structural feature do the anhydrides of phosphoric acid and carboxylic acids have in common?

10.61 Draw structural formulas for the mono-, di-, and tri-ethyl esters of phosphoric acid.

10.62 1,3-Dihydroxy-2-propanone (dihydroxyacetone) and phosphoric acid form a monoester called dihydroxyacetone phosphate, which is an intermediate in glycolysis (Section 19.2). Draw a structural formula for this monophosphate ester.

10.63 Show how triphosphoric acid can form from three molecules of phosphoric acid. How many H_2O molecules are split out?

10.64 Write an equation for the hydrolysis of trimethyl phosphate to dimethyl phosphate and methanol.

Chemical Connections

10.65 (Chemical Connections 10A) Locate the ester group in pyrethrin I and draw a structural formula for chrysanthemic acid, the carboxylic acid from which this ester is derived.

10.66 (Chemical Connections 10A) What structural features do pyrethrin I (a natural insecticide) and permethrin (a synthetic pyrethrenoid) have in common?

10.67 (Chemical Connections 10A) A commercial Clothing & Gear Insect Repellant gives the following information about permethrin, its active ingredient:

Cis/trans ratio: Minimum 35% (+/−) cis and maximum 65% (+/−) trans

(a) To what does the cis/trans ratio refer?

(b) To what does the designation "(+/−)" refer?

10.68 (Chemical Connections 10B) Identify the β-lactam portion of amoxicillin and cephalexin.

10.69 (Chemical Connections 10C) What is the compound in willow bark that is responsible for its ability to relieve pain? How is this compound related to salicylic acid?

10.70 (Chemical Connections 10C) Name the two functional groups in aspirin.

10.71 (Chemical Connections 10C) Once it has been opened, and particularly if it has been left open to the air, a bottle of aspirin may develop a vinegar-like odor. Explain how this might happen.

10.72 (Chemical Connections 10C) What is the structural relationship between aspirin and ibuprofen? Between aspirin and naproxen?

10.73 (Chemical Connections 10D) What is the difference in meaning between *sunblock* and *sunscreen?*

10.74 (Chemical Connections 10D) How do sunscreens prevent UV radiation from reaching the skin?

10.75 (Chemical Connections 10D) What structural features do the three sunscreens given in this box have in common?

10.76 (Chemical Connections 10E) Barbiturates are derived from urea. Identify the portion of the structure of pentobarbital and phenobarbital that is derived from urea.

10.77 (Chemical Connections 10F) Why do Lactomer stitches dissolve within two to three weeks following surgery?

Additional Problems

10.78 Potassium sorbate is added as a preservative to certain foods to prevent bacteria and molds from causing food spoilage and to extend the foods' shelf lives. The IUPAC name of potassium sorbate is potassium (*trans,trans*-2,4-hexadienoate). Draw a structural formula of potassium sorbate.

10.79 Propanoic acid and methyl acetate are constitutional isomers, and both are liquids at room temperature. One of these compounds has a boiling point of 141°C; the other has a boiling point of 57°C. Which compound has which boiling point? Explain.

10.80 Excess ascorbic acid is excreted in the urine, the pH of which is normally in the range 4.8–8.4. What form of ascorbic acid (pK_a 4.10) would you expect to be present in urine of pH 8.4: ascorbic acid or ascorbate anion? Explain.

10.81 The pH of human gastric juice is normally in the range 1.0–3.0. What form of lactic acid (pK_a 4.07), would you expect to be present in the stomach: lactic acid or its anion? Explain.

10.82 Benzocaine, a topical anesthetic, is prepared by treatment of 4-aminobenzoic acid with ethanol in the presence of an acid catalyst followed by neutralization. Draw a structural formula for benzocaine.

10.83 The analgesic acetaminophen is synthesized by treating 4-aminophenol with one equivalent of acetic anhydride. Write an equation for the formation of acetaminophen. (*Hint:* The —NH$_2$ group is more reactive with acetic anhydride than the —OH group.)

10.84 Procaine (its hydrochloride is marketed as Novocain) was one of the first local anesthetics developed for infiltration and regional anesthesia. It is synthesized by the following Fischer esterification. Draw a structural formula for procaine.

$$\underset{\substack{p\text{-Aminobenzoic}\\ \text{acid}}}{\ce{H2N-C6H4-COOH}} + \underset{\text{2-Diethylaminoethanol}}{\ce{HO-CH2CH2-N(C2H5)2}} \xrightarrow[\text{esterification}]{\text{Fischer}} \text{Procaine}$$

10.85 1,3-Diphosphoglycerate, an intermediate in glycolysis (Section 19.2), contains a mixed anhydride (an anhydride of a carboxylic acid and phosphoric acid) and a phosphoric ester. Draw structural formulas for the products formed by hydrolysis of the anhydride and ester bonds. Show each product as it would exist in solution at pH 7.4.

1,3-Diphosphoglycerate

InfoTrac College Edition

For additional readings, go to InfoTrac College Edition, your online research library, at

http://infotrac.thomsonlearning.com

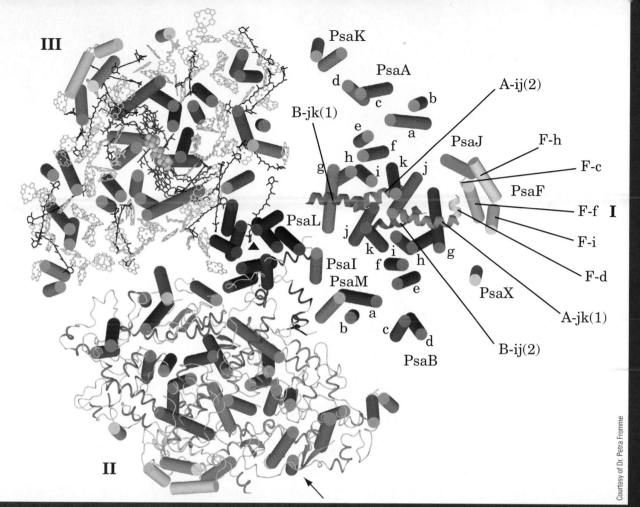

Courtesy of Dr. Petra Fromme

PART 2 BIOCHEMISTRY

Chapter 11
Carbohydrates

Chapter 12
Lipids

Chapter 13
Proteins

Chapter 14
Enzymes

Chapter 15
Chemical Communications:
Neurotransmitters and Hormones

Chapter 16
Nucleotides, Nucleic Acids,
and Heredity

Chapter 17
Gene Expression
and Protein Synthesis

Chapter 18
Bioenergetics: How the Body
Converts Food to Energy

Chapter 19
Specific Catabolic Pathways:
Carbohydrate, Lipid, and
Protein Metabolism

Chapter 20
Biosynthetic Pathways

Chapter 21
Nutrition

Chapter 22
Immunochemistry

Chapter 23
Body Fluids

229

CHAPTER 11

11.1 Introduction

11.2 Monosaccharides

11.3 The Cyclic Structure of Monosaccharides

11.4 Reactions of Monosaccharides

11.5 Disaccharides and Oligosaccharides

11.6 Polysaccharides

11.7 Acidic Polysaccharides

Breads, grains, and pasta are sources of carbohydrates.

Charles D. Winters

Carbohydrates

11.1 Introduction

Carbohydrates are the most abundant organic compounds in the plant world. They act as storehouses of chemical energy (glucose, starch, glycogen); are components of supportive structures in plants (cellulose), crustacean shells (chitin), and connective tissues in animals (acidic polysaccharides); and are essential components of nucleic acids (D-ribose and 2-deoxy-D-ribose). Carbohydrates account for approximately three-fourths of the dry weight of plants. Animals (including humans) get their carbohydrates by eating plants, but they do not store much of what they consume. In fact, less than 1% of the body weight of animals is made up of carbohydrates.

The word *carbohydrate,* means "hydrate of carbon" and derives from the formula $C_n(H_2O)_m$. Two examples of carbohydrates with molecular formulas that can be written alternatively as hydrates of carbon are

- Glucose (blood sugar): $C_6H_{12}O_6$, which can be written as $C_6(H_2O)_6$
- Sucrose (table sugar): $C_{12}H_{22}O_{11}$, which can be written as $C_{12}(H_2O)_{11}$

Not all carbohydrates, however, have this general formula. Some contain too few oxygen atoms to fit it; others contain too many oxygens. Some also contain nitrogen. The term *carbohydrate* has become so firmly rooted in the chemical nomenclature that, although not completely accurate, it persists as the name for this class of compounds.

At the molecular level, most **carbohydrates** are polyhydroxyaldehydes, polyhydroxyketones, or compounds that yield them after hydrolysis. The simpler members of the carbohydrate family are often referred to as **saccharides** because of their sweet taste (Latin: *saccharum,* sugar). Carbohydrates are classified as monosaccharides, oligosaccharides, or polysaccharides depending on the number of simple sugars they contain.

> **Carbohydrate** A polyhydroxyaldehyde or polyhydroxyketone, or a substance that gives these compounds on hydrolysis

11.2 Monosaccharides

A Structure and Nomenclature

Monosaccharides have the general formula $C_nH_{2n}O_n$, with one of the carbons being the carbonyl group of either an aldehyde or a ketone. The most common monosaccharides have three to nine carbon atoms. The suffix **-ose** indicates that a molecule is a carbohydrate, and the prefixes **tri-, tetr-, pent-,** and so forth indicate the number of carbon atoms in the chain. Monosaccharides containing an aldehyde group are classified as **aldoses;** those containing a ketone group are classified as **ketoses.**

There are only two trioses: the aldotriose glyceraldehyde and the ketotriose dihydroxyacetone.

> **Monosaccharide** A carbohydrate that cannot be hydrolyzed to a simpler compound

> **Aldose** A monosaccharide containing an aldehyde group

> **Ketose** A monosaccharide containing a ketone group

$$
\begin{array}{cc}
\text{CHO} & \text{CH}_2\text{OH} \\
| & | \\
\text{CHOH} & \text{C}{=}\text{O} \\
| & | \\
\text{CH}_2\text{OH} & \text{CH}_2\text{OH} \\
\text{Glyceraldehyde} & \text{Dihydroxyacetone} \\
\text{(an aldotriose)} & \text{(a ketotriose)}
\end{array}
$$

Often the designations *aldo-* and *keto-* are omitted, and these molecules are referred to simply as trioses, tetroses, and the like.

See the **Interactive General, Organic, and Biochemistry CD-ROM, version 2.0,** for further exploration on this topic.

B Fischer Projection Formulas

Glyceraldehyde contains a stereocenter and therefore exists as a pair of enantiomers (Figure 11.1).

Chemists commonly use two-dimensional representations called **Fischer projections** to show the configuration of carbohydrates. To draw a Fischer projection, you draw a three-dimensional representation of the molecule oriented so that the vertical bonds from the stereocenter are directed away from you and the horizontal bonds from it are directed toward you.

> **Fischer projection** A two-dimensional representation for showing the configuration of a stereocenter; horizontal lines represent bonds projecting forward from the stereocenter, and vertical lines represent bonds projecting toward the rear

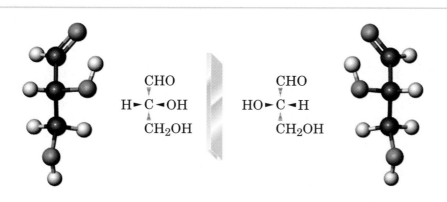

Figure 11.1 The enantiomers of glyceraldehyde.

You then write the molecule as a cross, with the stereocenter indicated by the point at which the bonds cross.

$$
\begin{array}{c}
\text{CHO} \\
| \\
\text{H} \blacktriangleright \text{C} \blacktriangleleft \text{OH} \\
| \\
\text{CH}_2\text{OH}
\end{array}
\xrightarrow[\text{Fischer projection}]{\text{convert to a}}
\begin{array}{c}
\text{CHO} \\
| \\
\text{H} \!-\!\!\!|\!\!\!-\! \text{OH} \\
| \\
\text{CH}_2\text{OH}
\end{array}
$$

(R)-Glyceraldehyde
(three-dimensional representation)

(R)-Glyceraldehyde
(Fischer projection)

The horizontal segments of this Fischer projection represent bonds directed toward you, and the vertical segments represent bonds directed away from you. The only atom in the plane of the paper is the stereocenter.

C D- and L-Monosaccharides

Emil Fischer, who in 1902 became the second Nobel Prize winner in chemistry, made many fundamental discoveries in the chemistry of carbohydrates, proteins, and other areas of organic and biochemistry.

Even though the R,S system is widely accepted today as a standard for designating configuration, the configuration of carbohydrates is commonly designated using the D,L system proposed by Emil Fischer in 1891. At that time, it was known that one enantiomer of glyceraldehyde has a specific rotation (Section 6.5B) of +13.5°; the other has a specific rotation of −13.5°. Fischer proposed that these enantiomers be designated D and L, but he had no experimental way to determine which enantiomer has which specific rotation. Fischer, therefore, did the only possible thing—he made an arbitrary

Table 11.1 Configurational Relationships among the Isomeric D-Aldotetroses, D-Aldopentoses, and D-Aldohexoses

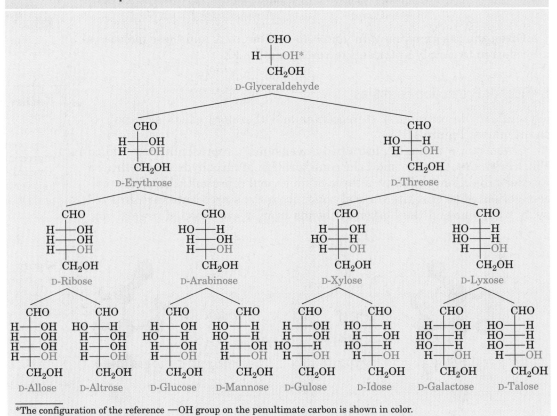

*The configuration of the reference —OH group on the penultimate carbon is shown in color.

assignment. He assigned the dextrorotatory enantiomer the following configuration and named it D-glyceraldehyde. He named its enantiomer L-glyceraldehyde. Fischer could have been wrong, but by a stroke of good fortune, he wasn't. In 1952, x-ray crystallography proved his assignment of configuration to the enantiomers of glyceraldehyde to be correct.

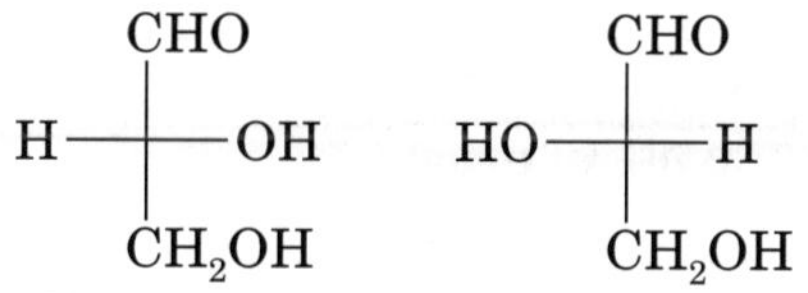

CHO   CHO

H———OH   HO———H

CH₂OH   CH₂OH

D-Glyceraldehyde   L-Glyceraldehyde
$[\alpha]_D^{25} = +13.5°$ $[\alpha]_D^{25} = -13.5°$

D-glyceraldehyde and L-glyceraldehyde serve as reference points for the assignment of relative configurations to all other aldoses and ketoses. The reference point is the penultimate carbon—that is, the next-to-last carbon on the chain. A **D-monosaccharide** has the same configuration at its penultimate carbon as D-glyceraldehyde (its —OH group is on the right); an **L-monosaccharide** has the same configuration at its penultimate carbon as L-glyceraldehyde (its —OH group is on the left).

Tables 11.1 and 11.2 show names and Fischer projections for all D-aldo- and D-2-ketotetroses, pentoses, and hexoses. Each name consists of three parts. The D specifies the configuration at the stereocenter farthest from the

D-Monosaccharide A monosaccharide that, when written as a Fischer projection, has the —OH group on its penultimate carbon to the right

L-Monosaccharide A monosaccharide that, when written as a Fischer projection, has the —OH group on its penultimate carbon to the left

Table 11.2   Configurational Relationships among the D-2-Ketopentoses and D-2-Ketohexoses

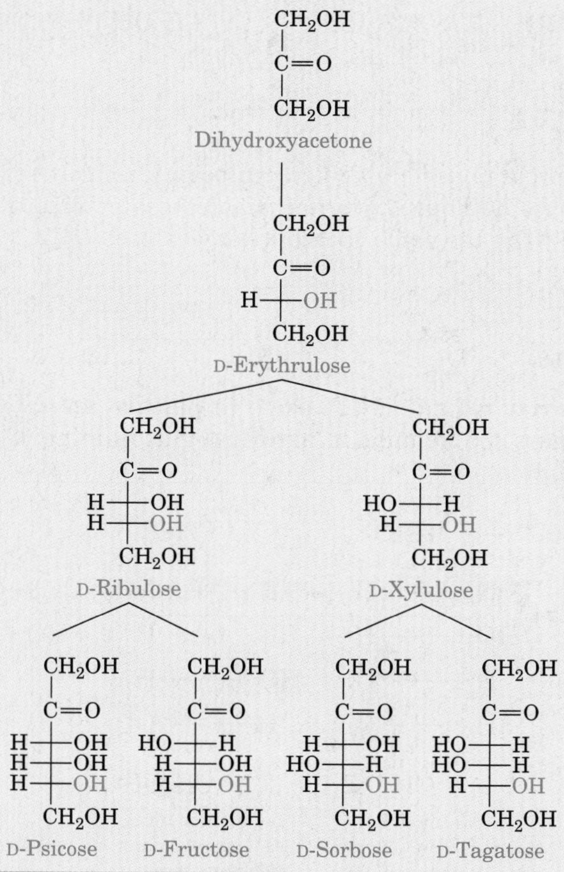

CH₂OH
|
C=O
|
CH₂OH
Dihydroxyacetone

CH₂OH
|
C=O
H——OH
CH₂OH
D-Erythrulose

CH₂OH   CH₂OH
| |
C=O   C=O
H——OH   HO——H
H——OH   H——OH
CH₂OH   CH₂OH
D-Ribulose   D-Xylulose

CH₂OH   CH₂OH   CH₂OH   CH₂OH
| | | |
C=O   C=O   C=O   C=O
H——OH   HO——H   H——OH   HO——H
H——OH   H——OH   HO——H   HO——H
H——OH   H——OH   H——OH   H——OH
CH₂OH   CH₂OH   CH₂OH   CH₂OH
D-Psicose   D-Fructose   D-Sorbose   D-Tagatose

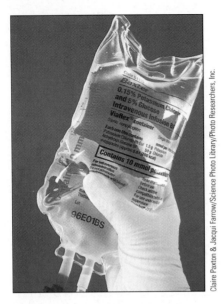

■ **Gloved hand holding an intravenous (i.v.) drip bag containing 0.15% potassium chloride (saline) and 5% glucose.**

carbonyl group. Prefixes such as *rib-, arabin-,* and *gluc-* specify the configuration of all other stereocenters in the monosaccharide relative to one another. The suffix *-ose* indicates that the compound is a carbohydrate.

The three most abundant hexoses in the biological world are D-glucose, D-galactose, and D-fructose. The first two are D-aldohexoses; the third is a D-2-ketohexose. Glucose, by far the most abundant of the three, is also known as dextrose because it is dextrorotatory. Other names for this monosaccharide include grape sugar and blood sugar. Human blood normally contains 65–110 mg of glucose/100 mL of blood.

EXAMPLE 11.1

Draw Fischer projections for the four aldotetroses. Which are D-monosaccharides, which are L-monosaccharides, and which are enantiomers? Refer to Table 11.1 and write the name of each aldotetrose.

Solution

Following are Fischer projections for the four aldotetroses. The D- and L- refer to the configuration of the penultimate carbon, which, in the case of aldotetroses, is carbon 3. In the Fischer projection of a D-aldotetrose, the —OH on carbon 3 is on the right; in an L-aldotetrose, it is on the left.

One pair of enantiomers		A second pair of enantiomers	
CHO	CHO	CHO	CHO
H——OH	HO——H	HO——H	H——OH
H——OH	HO——H	H——OH	HO——H
CH₂OH	CH₂OH	CH₂OH	CH₂OH
D-Erythrose	L-Erythrose	D-Threose	L-Threose

Problem 11.1

Draw Fischer projections for all 2-ketopentoses. Which are D-2-ketopentoses, which are L-2-ketopentoses, and which are enantiomers? Refer to Table 11.2 and write the name of each 2-ketopentose.

D Amino Sugars

Amino sugar A monosaccharide in which an —OH group is replaced by an —NH₂ group

Amino sugars contain an —NH₂ group in place of an —OH group. Only three amino sugars are common in nature: D-glucosamine, D-mannosamine, and D-galactosamine.

CHO	CHO	CHO	CHO
H——NH₂	H₂N——H	H——NH₂	H——NHCCH₃ (O)
HO——H	HO——H	HO——H	HO——H
H——OH	H——OH	HO——H	H——OH
H——OH	H——OH	H——OH	H——OH
CH₂OH	CH₂OH	CH₂OH	CH₂OH
D-Glucosamine	D-Mannosamine (C-2 stereoisomer of D-glucosamine)	D-Galactosamine (C-4 stereoisomer of D-glucosamine)	*N*-Acetyl-D-glucosamine

Galactosemia

One out of every 18,000 infants is born with a genetic defect that renders the child unable to utilize the monosaccharide galactose. Galactose is part of lactose (milk sugar, Section 11.5B). When the body cannot utilize galactose, it accumulates in the blood and in the urine. This buildup in the blood is harmful because it can lead to mental retardation, failure to grow, cataract formation in the eye and, in severe cases, death due to liver damage. When galactose accumulation results from a deficiency of the enzyme galactokinase, the disorder, known as galactosuria, has only mild symptoms. When the enzyme

galactose-1-phosphate uridinyltransferase is deficient, however, the disorder is called galactosemia, and its symptoms are severe.

The deleterious effects of galactosemia can be avoided by giving the infant a milk formula in which sucrose is substituted for lactose. Because sucrose contains no galactose, the infant thus consumes a galactose-free diet. A galactose-free diet is critical only in infancy. With maturation, most children develop another enzyme capable of metabolizing galactose. As a consequence, they are able to tolerate galactose as they mature.

N-Acetyl-D-glucosamine, a derivative of D-glucosamine, is a component of connective tissue such as cartilage. It is also a component of chitin, the hard shell-like exoskeleton of lobsters, crabs, shrimp, and other shellfish. Several other amino sugars are components of naturally occurring antibiotics.

■ **Northern lobster.**

E Physical Properties

Monosaccharides are colorless, crystalline solids. Because hydrogen bonding is possible between their polar —OH groups and water, all monosaccharides are very soluble in water. They are only slightly soluble in ethanol and are insoluble in nonpolar solvents such as diethyl ether, dichloromethane, and benzene.

11.3 The Cyclic Structure of Monosaccharides

In Section 9.7, we saw that aldehydes and ketones react with alcohols to form **hemiacetals.** We also saw that cyclic hemiacetals form very readily when hydroxyl and carbonyl groups are part of the same molecule and that their interaction can produce a ring. For example, 4-hydroxypentanal forms a five-membered cyclic hemiacetal. Note that 4-hydroxypentanal contains one stereocenter and that hemiacetal formation generates a second stereocenter at carbon 1.

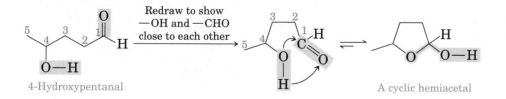

See the **Interactive General, Organic, and Biochemistry CD-ROM, version 2.0,** for further exploration on this topic.

Monosaccharides have hydroxyl and carbonyl groups in the same molecule. As a result, they also exist almost exclusively as five- and six-membered cyclic hemiacetals.

> **Haworth projection** A way to view furanose and pyranose forms of monosaccharides; the ring is drawn flat and viewed through its edge with the anomeric carbon on the right and the oxygen atom to the rear

A Haworth Projections

A common way of representing the cyclic structure of monosaccharides is the **Haworth projection,** named after the English chemist Sir Walter N.

See the **Interactive General, Organic, and Biochemistry CD-ROM, version 2.0,** for further exploration on this topic.

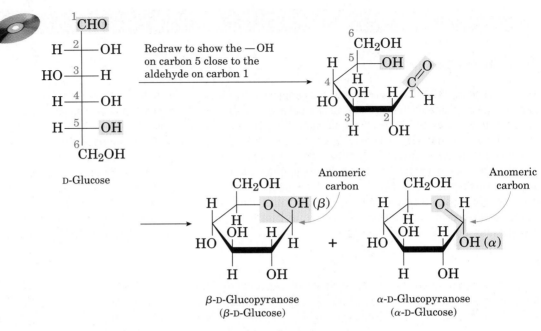

D-Glucose

Figure 11.2   Haworth projections for α-D-glucopyranose; and β-D-glucopyranose.

Haworth (Nobel Prize for chemistry, 1937). In a Haworth projection, a five- or six-membered cyclic hemiacetal is represented as a planar pentagon or hexagon, respectively, lying roughly perpendicular to the plane of the paper. Groups attached to the carbons of the ring then lie either above or below the plane of the ring. The new carbon stereocenter created in forming the cyclic structure is called an **anomeric carbon.** Stereoisomers that differ in configuration only at the anomeric carbon are called **anomers.** The anomeric carbon of an aldose is carbon 1; that of the most common ketoses is carbon 2.

Typically, Haworth projections are drawn with the anomeric carbon to the right and the hemiacetal oxygen to the back (Figure 11.2).

In the terminology of carbohydrate chemistry, the designation β means that the —OH on the anomeric carbon of the cyclic hemiacetal lies on the same side of the ring as the terminal —CH_2OH. Conversely, the designation α means that the —OH on the anomeric carbon of the cyclic hemiacetal lies on the side of the ring opposite from the terminal —CH_2OH.

A six-membered hemiacetal ring is indicated by **-pyran-,** and a five-membered hemiacetal ring is indicated by **-furan-.** The terms **furanose** and **pyranose** are used because monosaccharide five- and six-membered rings correspond to the heterocyclic compounds furan and pyran.

Anomeric carbon   The hemiacetal carbon of the cyclic form of a monosaccharide

Anomers   Monosaccharides that differ in configuration only at their anomeric carbons

Furanose   A five-membered cyclic hemiacetal form of a monosaccharide

Pyranose   A six-membered cyclic hemiacetal form of a monosaccharide

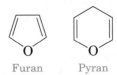

Furan   Pyran

Because the α and β forms of glucose are six-membered cyclic hemiacetals, they are named α-D-glucopyranose and β-D-glucopyranose, respectively. The designations *-furan-* and *-pyran-* are not always used in monosaccharide names, however. Thus, the glucopyranoses, for example, are often named simply α-D-glucose and β-D-glucose.

CHEMICAL CONNECTIONS 11B

L-Ascorbic Acid (Vitamin C)

The structure of L-ascorbic acid (vitamin C) resembles that of a monosaccharide. In fact, this vitamin is synthesized both biochemically by plants and some animals and commercially from D-glucose. Humans do not have the enzymes required to carry out this synthesis. For this reason, we must obtain vitamin C in the food we eat or as a vitamin supplement. Approximately 66 million kg of vitamin C is synthesized annually in the United States.

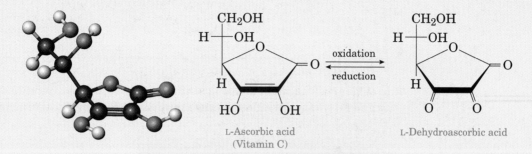

L-Ascorbic acid
(Vitamin C)

L-Dehydroascorbic acid

Charles D. Winters

■ **Oranges are a major source of vitamin C.**

You would do well to remember the configuration of groups on the Haworth projections of α-D-glucopyranose and β-D-glucopyranose as reference structures. Knowing how the open-chain configuration of any other aldohexose differs from that of D-glucose, you can then construct its Haworth projection by referring to the Haworth projection of D-glucose.

EXAMPLE 11.2

Draw Haworth projections for the α and β anomers of D-galactopyranose.

Solution

One way to arrive at these projections is to use the α and β forms of D-glucopyranose as references and to remember (or discover by looking at Table 11.1) that D-galactose differs from D-glucose only in the configuration at carbon 4. Thus, you can begin with the Haworth projections shown in Figure 11.2 and then invert the configuration at carbon 4.

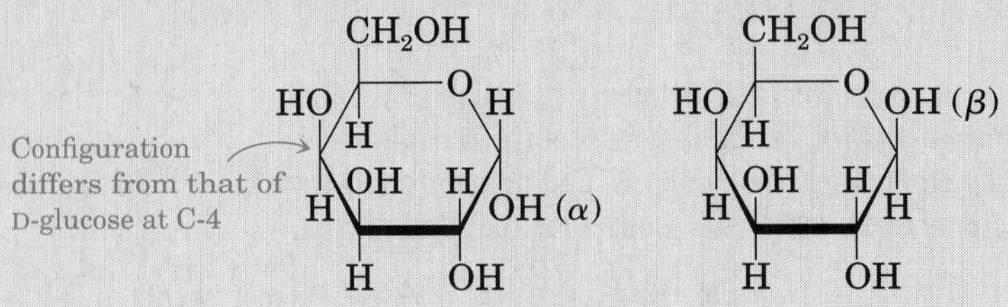

Configuration differs from that of D-glucose at C-4

α-D-Galactopyranose
(α-D-Galactose)

β-D-Galactopyranose
(β-D-Galactose)

Problem 11.2

D-Mannose exists in aqueous solution as a mixture of α-D-mannopyranose and β-D-mannopyranose. Draw Haworth projections for these molecules.

Aldopentoses also form cyclic hemiacetals. The most prevalent forms of D-ribose and other pentoses in the biological world are furanoses. Following are Haworth projections for α-D-ribofuranose (α-D-ribose) and β-2-deoxy-D-ribofuranose (β-2-deoxy-D-ribose). The prefix *2-deoxy* indicates the absence of oxygen at carbon 2. Units of D-ribose and 2-deoxy-D-ribose in nucleic acids and most other biological molecules are found almost exclusively in the β configuration.

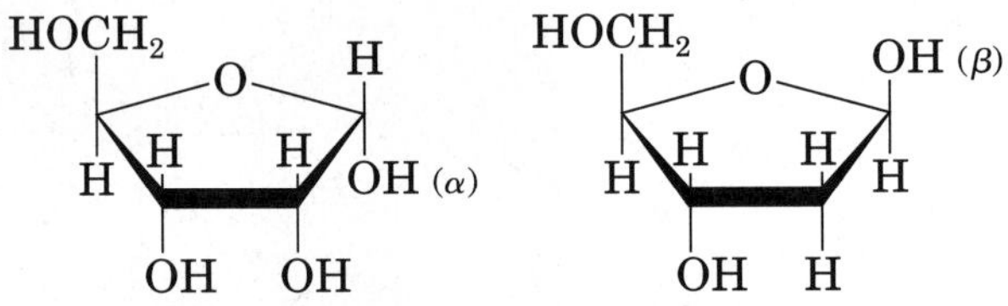

α-D-Ribofuranose
(α-D-Ribose)

β-2-Deoxy-D-ribofuranose
(β-2-Deoxy-D-ribose)

Fructose also forms five-membered cyclic hemiacetals. β-D-Fructofuranose, for example, is found in the disaccharide sucrose (Section 11.5A).

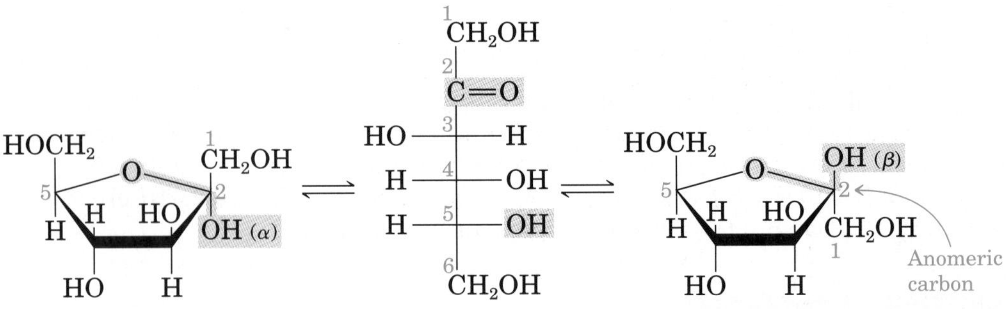

α-D-Fructofuranose
(α-D-Fructose)

D-Fructose

β-D-Fructofuranose
(β-D-Fructose)

Anomeric carbon

B Conformation Representations

A five-membered ring is so close to being planar that Haworth projections provide adequate representations of furanoses. For pyranoses, however, the six-membered ring is more accurately represented as a **chair conformation**

(Section 2.6B). Following are structural formulas for α-D-glucopyranose and β-D-glucopyranose, both drawn as chair conformations. Also shown is the open-chain or free aldehyde form with which the cyclic hemiacetal forms are in equilibrium in aqueous solution. Notice that each group, including the anomeric —OH, on the chair conformation of β-D-glucopyranose is equatorial. Notice also that the —OH group on the anomeric carbon is axial in α-D-glucopyranose. Because the —OH on the anomeric carbon of β-D-glucopyranose is in the more stable equatorial position (Section 2.6B), the β anomer predominates in aqueous solution.

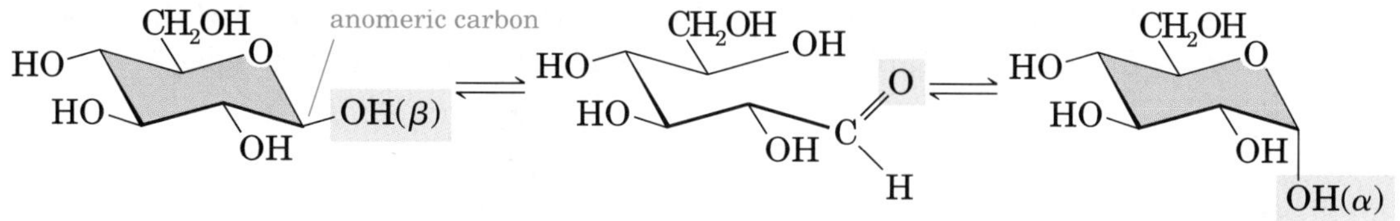

β-D-Glucopyranose
$[\alpha]_D^{25} + 18.7°$

D-Glucose

α-D-Glucopyranose
$[\alpha]_D^{25} + 112°$

At this point, you should compare the relative orientations of groups on the D-glucopyranose ring in the Haworth projection and chair conformation. The orientations of groups on carbons 1 through 5 of β-D-glucopyranose, for example, are up, down, up, down, and up in both representations.

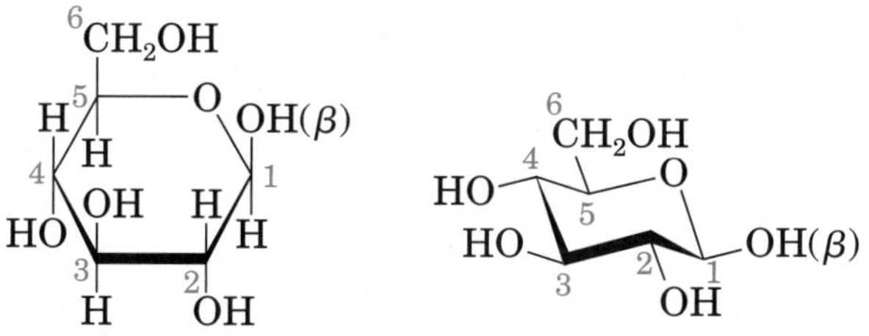

β-D-Glucopyranose
(Haworth projection)

β-D-Glucopyranose
(chair conformation)

We do not show hydrogen atoms bonded to the ring in chair conformations. We often show them, however, in Haworth projections.

EXAMPLE 11.3

Draw chair conformations for α-D-galactopyranose and β-D-galactopyranose. Label the anomeric carbon in each.

Solution
The configuration of D-galactose differs from that of D-glucose only at carbon 4. Therefore, draw the α and β forms of D-glucopyranose and then interchange the positions of the —OH and —H groups on carbon 4.

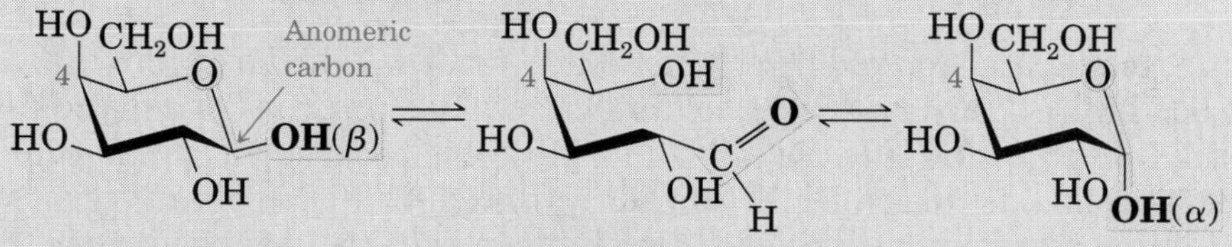

β-D-Galactopyranose
(β-D-Galactose)

D-Galactose

α-D-Galactopyranose
(α-D-Galactose)

Problem 11.3

Draw chair conformations for α-D-mannopyranose and β-D-mannopyranose. Label the anomeric carbon in each.

C Mutarotation

Mutarotation The change in specific rotation that occurs when an α or β form of a carbohydrate is converted to an equilibrium mixture of the two forms

Mutarotation is the change in specific rotation that accompanies the equilibration of α- and β-anomers in aqueous solution. For example, a solution prepared by dissolving crystalline α-D-glucopyranose in water has a specific rotation of +112°, which gradually decreases to an equilibrium value of +52.7° as α-D-glucopyranose reaches equilibrium with β-D-glucopyranose. A solution of β-D-glucopyranose also undergoes mutarotation, during which the specific rotation changes from +18.7° to the same equilibrium value of +52.7°. The equilibrium mixture consists of 64% β-D-glucopyranose and 36% α-D-glucopyranose, with only a trace (0.003 percent) of the open-chain form. Mutarotation is common to all carbohydrates that exist in hemiacetal forms.

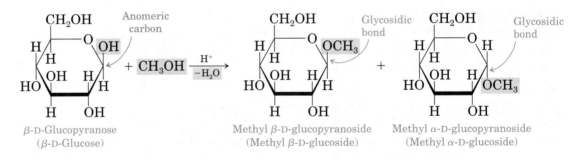

β-D-Glucopyranose
$[\alpha]_D^{25} = +18.7°$

Open-chain form

α-D-Glucopyranose
$[\alpha]_D^{25} = +112°$

11.4 Reactions of Monosaccharides

See the **Interactive General, Organic, and Biochemistry CD-ROM, version 2.0,** for further exploration on this topic.

A Formation of Glycosides (Acetals)

As we saw in Section 9.7, treatment of an aldehyde or ketone with one molecule of alcohol yields a hemiacetal, and treatment of the hemiacetal with a molecule of alcohol yields an acetal. Treatment of a monosaccharide, all forms of which exist almost exclusively as cyclic hemiacetals, with an alcohol also yields an acetal, as illustrated by the reaction of β-D-glucopyranose with methanol.

β-D-Glucopyranose
(β-D-Glucose)

Methyl β-D-glucopyranoside
(Methyl β-D-glucoside)

Methyl α-D-glucopyranoside
(Methyl α-D-glucoside)

Glycoside A carbohydrate in which the —OH group on its anomeric carbon is replaced by an —OR group

A cyclic acetal derived from a monosaccharide is called a **glycoside,** and the bond from the anomeric carbon to the —OR group is called a **glycosidic bond.** Mutarotation is not possible in a glycoside because an acetal—unlike a hemiacetal—is no longer in equilibrium with the open-chain, carbonyl-containing compound. Glycosides are stable in water and aqueous base, like other acetals (Section 9.7), however, they are hydrolyzed in aqueous acid to an alcohol and a monosaccharide.

Glycosidic bond The bond from the anomeric carbon of a glycoside to an —OR group

We name glycosides by listing the alkyl or aryl group bonded to oxygen, followed by the name of the carbohydrate in which the ending **-e** is replaced by **-ide.** For example, the methyl glycoside derived from β-D-glucopyranose is named methyl β-D-glucopyranoside; that derived from β-D-ribofuranose is named methyl β-D-ribofuranoside.

EXAMPLE 11.4

Draw a structural formula for methyl β-D-ribofuranoside (methyl β-D-riboside). Label the anomeric carbon and the glycosidic bond.

Solution

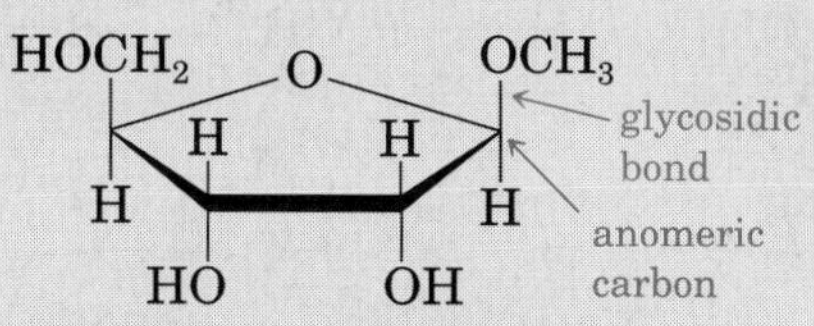

Methyl β-D-ribofuranoside
(Methyl β-D-riboside)

Problem 11.4

Draw a Haworth projection and a chair conformation for methyl α-D-mannopyranoside (methyl α-D-mannoside). Label the anomeric carbon and the glycosidic bond.

B Reduction to Alditols

The carbonyl group of a monosaccharide can be reduced to an hydroxyl group by a variety of reducing agents, including hydrogen in the presence of a transition metal catalyst and sodium borohydride (Section 9.6). The reduction products are known as **alditols.** Reduction of D-glucose gives D-glucitol, more commonly known as D-sorbitol. Here, D-glucose is shown in the open-chain form. Only a small amount of this form is present in solution but, as it is reduced, the equilibrium between cyclic hemiacetal forms (only the β form is shown here) and the open-chain form shifts to replace it.

Alditol The product formed when the CHO group of a monosaccharide is reduced to a CH_2OH group

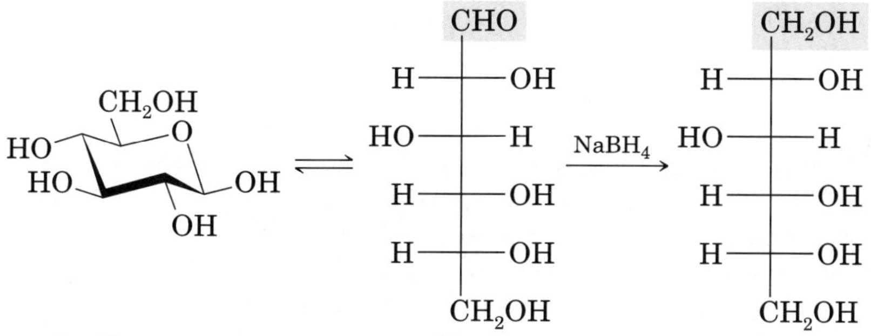

β-D-Glucopyranose   D-Glucose   D-Glucitol
(D-Sorbitol)

■ Many "sugar free" products contain sugar alcohols, such as D-sorbitol and xylitol.

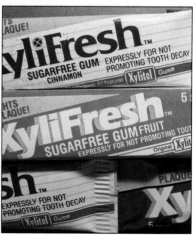

Gregory Smolin

We name alditols by dropping the **-ose** from the name of the monosaccharide and adding **-itol.** Sorbitol is found in the plant world in many berries and in cherries, plums, pears, apples, seaweed, and algae. It is about 60% as sweet as sucrose (table sugar) and is used in the manufacture of candies and as a sugar substitute for diabetics. Other alditols commonly found in the biological world include erythritol, D-mannitol, and xylitol. Xylitol is used as a sweetening agent in "sugarless" gum, candy, and sweet cereals.

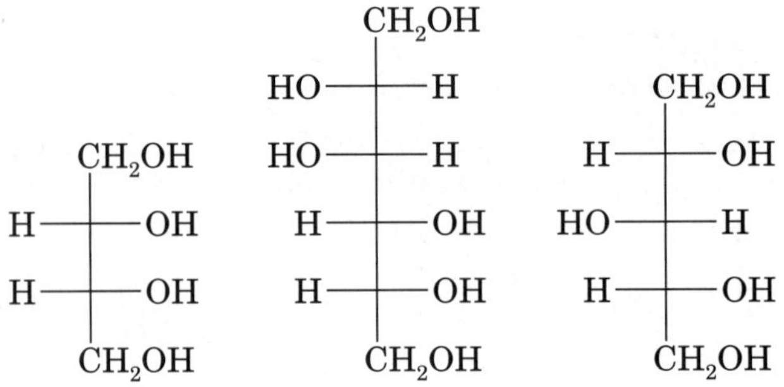

Erythritol   D-Mannitol   Xylitol

C Oxidation to Aldonic Acids (Reducing Sugars)

Oxidations and reductions of monosaccharides in nature are catalyzed by specific enzymes, as for example glucose oxidase.

As we saw in Section 9.5, aldehydes (RCHO) are oxidized to carboxylic acids (RCOOH) by several agents, including oxygen, O_2. Similarly, the aldehyde group of an aldose can be oxidized, under basic conditions, to a carboxylate group. Under these conditions, the cyclic form of the aldose is in equilibrium with the open-chain form, which is then oxidized by the mild oxidizing agent. D-Glucose, for example, is oxidized to D-gluconate (the anion of D-gluconic acid).

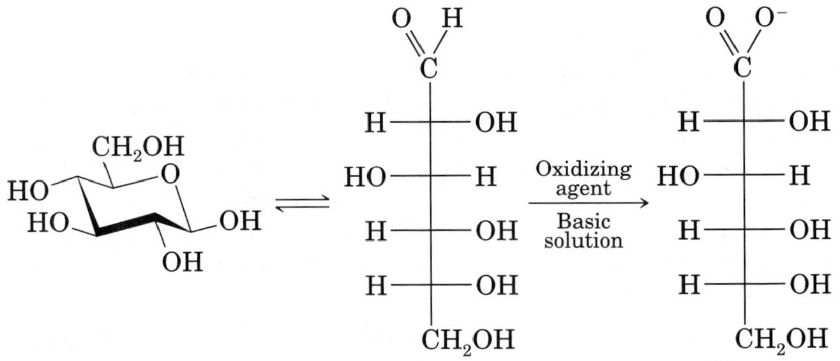

β-D-Glucopyranose   D-Glucose   D-Gluconate
(β-D-Glucose)

> **Reducing sugar** A carbohydrate that reacts with a mild oxidizing agent under basic conditions to give an aldonic acid; the carbohydrate reduces the oxidizing agent

Any carbohydrate that reacts with an oxidizing agent to form an aldonic acid is classified as a **reducing sugar** (it reduces the oxidizing agent).

Surprisingly, 2-ketoses are also reducing sugars. Carbon 1 (a CH_2OH group) of a ketose is not oxidized directly. Rather, under the basic conditions of this oxidation, a 2-ketose exists in equilibrium with an aldose by way of an enediol intermediate. The aldose is then oxidized by the mild oxidizing agent.

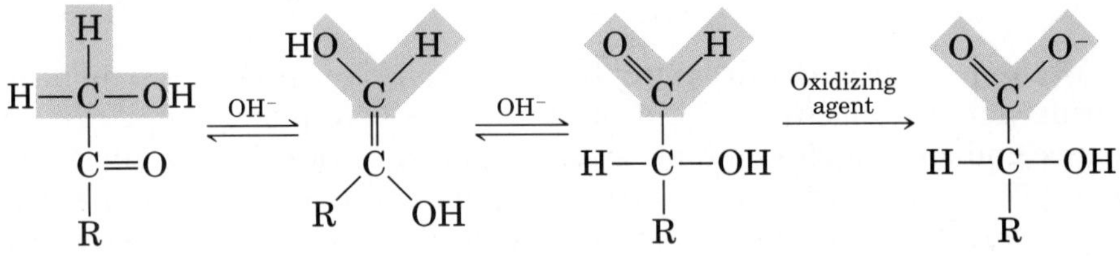

A 2-ketose   An enediol   An aldose   An aldonate

Testing for Glucose

The analytical procedure most often performed in a clinical chemistry laboratory is the determination of glucose in blood, urine, or other biological fluids. The high frequency with which this test is performed reflects the high incidence of diabetes mellitus. Approximately 2 million known diabetics live in the United States, and it is estimated that another 1 million remain undiagnosed.

Diabetes mellitus (Chemical Connections 15G) is characterized by insufficient blood levels of the hormone insulin. If the blood concentration of insulin is too low, muscle and liver cells do not absorb glucose from the blood; this problem, in turn, leads to increased levels of blood glucose (hyperglycemia), impaired metabolism of fats and proteins, ketosis, and possible diabetic coma. A rapid test for blood glucose levels is critical for early diagnosis and effective management of this disease. In addition to giving results quickly, a test must be specific for D-glucose; that is, it must give a positive test for glucose but not react with any other substances normally present in biological fluids.

Today blood glucose levels are measured by an enzyme-based procedure using the enzyme glucose oxidase. This enzyme catalyzes the oxidation of β-D-glucose to D-gluconic acid.

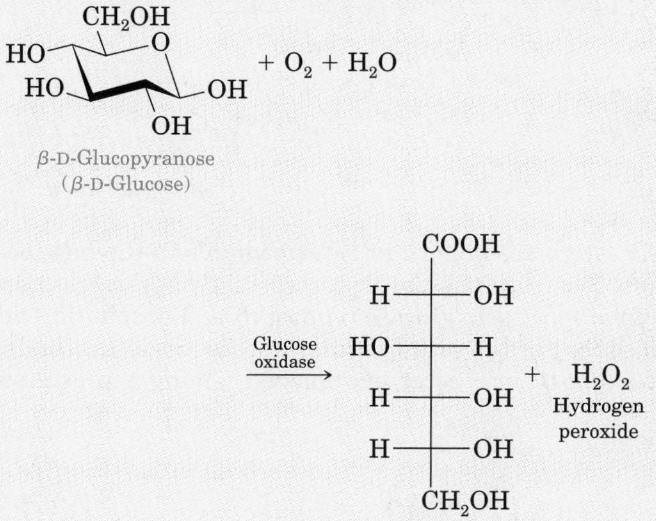

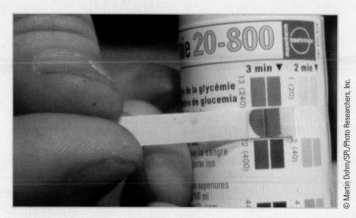

■ **Chemstrip kit for blood glucose.**

Glucose oxidase is specific for β-D-glucose. Therefore, complete oxidation of any sample containing both β-D-glucose and α-D-glucose requires conversion of the α form to the β form. Fortunately, this interconversion is rapid and complete in the short time required for the test.

Molecular oxygen, O_2, is the oxidizing agent in this reaction and is reduced to hydrogen peroxide, H_2O_2. In one procedure, hydrogen peroxide formed in the glucose oxidase-catalyzed reaction oxidizes colorless o-toluidine to a colored product in a reaction catalyzed by the enzyme peroxidase. The concentration of the colored oxidation product is determined spectrophotometrically and is proportional to the concentration of glucose in the test solution.

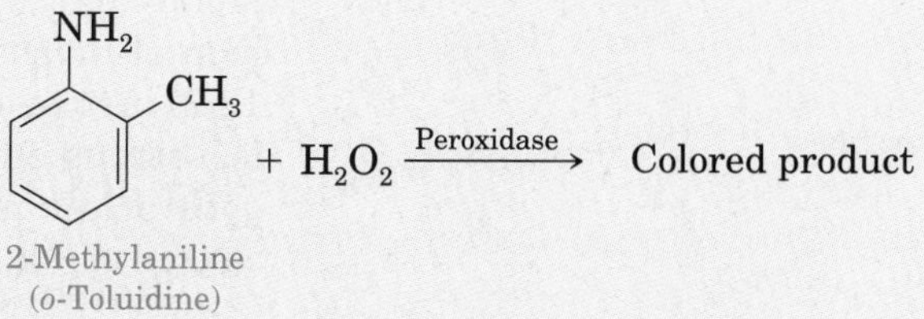

β-D-Glucopyranose
(β-D-Glucose)

$+ \; O_2 \; + \; H_2O$

Glucose oxidase →

D-Gluconic acid

$+ \; H_2O_2$
Hydrogen peroxide

2-Methylaniline
(o-Toluidine)

$+ \; H_2O_2 \; \xrightarrow{\text{Peroxidase}}$ Colored product

Several commercially available test kits use the glucose oxidase reaction for qualitative determination of glucose in urine.

D Oxidation to Uronic Acids

Enzyme-catalyzed oxidation of the primary alcohol at carbon 6 of a hexose yields a uronic acid. Enzyme-catalyzed oxidation of D-glucose, for example, yields D-glucuronic acid, shown here in both its open-chain and cyclic hemiacetal forms.

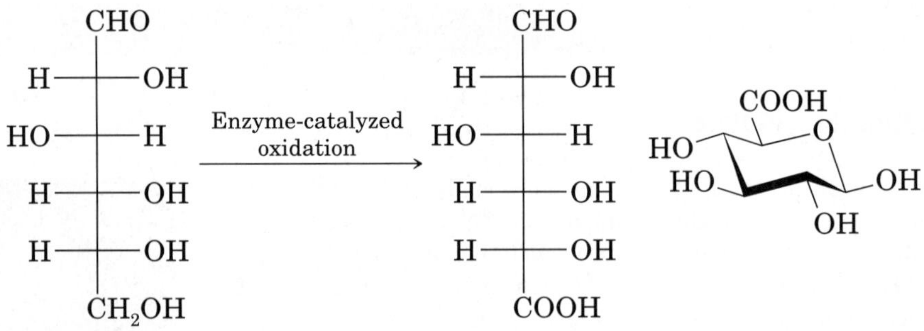

D-Glucose → (Enzyme-catalyzed oxidation) → D-Glucuronic acid (a uronic acid) [Fischer projection | Chair conformation]

D-Glucuronic acid is widely distributed in both the plant and animal worlds. In humans, it serves as an important component of the acidic polysaccharides of connective tissues (Section 11.7). The body also uses it to detoxify foreign phenols and alcohols. In the liver, these compounds are converted to glycosides of glucuronic acid (glucuronides) and then excreted in urine. The intravenous anesthetic propfol, for example, is converted to the following glucuronide and excreted in urine:

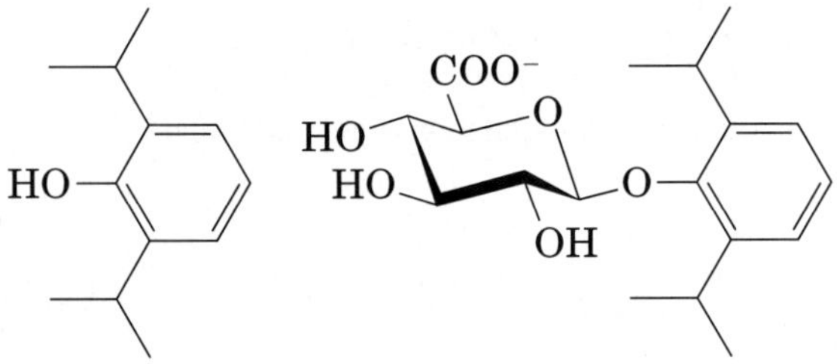

Propofol   A urine-soluble glucuronide

E Phosphoric Esters

Mono- and diphosphoric esters are important intermediates in the metabolism of monosaccharides. For example, the first step in glycolysis (Section 19.2) involves conversion of glucose to glucose 6-phosphate. Phosphoric acid is a strong enough acid so that at the pH of cellular and intercellular fluids, both acidic protons of a phosphoric ester are ionized, giving the ester a charge of -2.

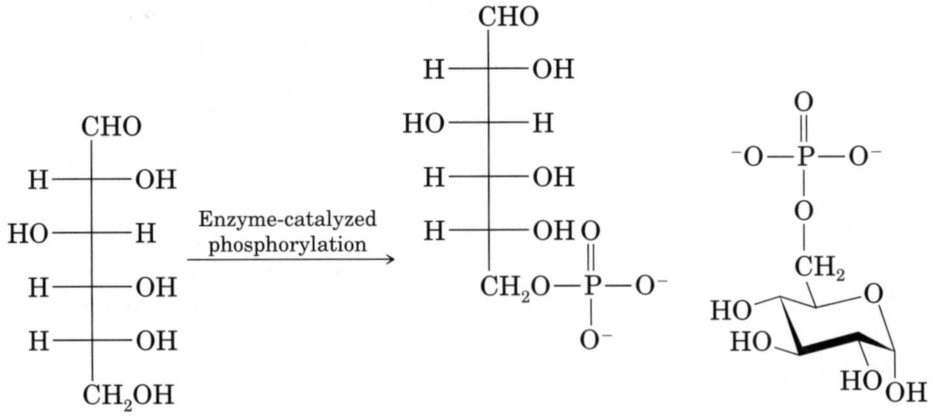

D-Glucose → (Enzyme-catalyzed phosphorylation) → D-Glucose 6-phosphate α-D-Glucose 6-phosphate

11.5 Disaccharides and Oligosaccharides

Most carbohydrates in nature contain more than one monosaccharide unit. Those that contain two units are called **disaccharides,** those that contain three units are called **trisaccharides,** and so forth. We use the general term **oligosaccharide** to describe any of the carbohydrates that contain from six to ten monosaccharide units. Carbohydrates containing larger numbers of monosaccharide units are called **polysaccharides.**

In a disaccharide, two monosaccharide units are joined by a glycosidic bond between the anomeric carbon of one unit and an —OH group of the other unit. Three important disaccharides are sucrose, lactose, and maltose.

> **Disaccharide** A carbohydrate containing two monosaccharide units joined by a glycosidic bond

> **Oligosaccharide** A carbohydrate containing from 6 to 10 monosaccharide units, each joined to the next by a glycosidic bond

> **Polysaccharide** A carbohydrate containing a large number of monosaccharide units, each joined to the next by one or more glycosidic bonds

A Sucrose

Sucrose (table sugar) is the most abundant disaccharide in the biological world. It is obtained principally from the juice of sugar cane and sugar beets. In sucrose, carbon 1 of α-D-glucopyranose bonds to carbon 2 of D-fructofuranose by an α-1,2-glycosidic bond. Because the anomeric carbons of both the glucopyranose and fructofuranose units are involved in formation of the glycosidic bond, neither monosaccharide unit is in equilibrium with its open-chain form. Thus sucrose is a nonreducing sugar.

In the production of sucrose, sugar cane or sugar beet is boiled with water, and the resulting solution is cooled. Sucrose crystals separate and are collected. Subsequent boiling to concentrate the solution followed by cooling yields a dark, thick syrup known as molasses.

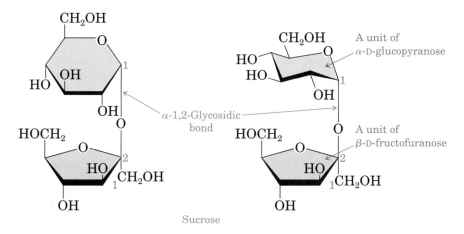

Sucrose

B Lactose

Lactose is the principal sugar present in milk. It accounts for 5 to 8% of human milk and 4 to 6% of cow's milk. This disaccharide consists of D-galactopyranose bonded by a β-1,4-glycosidic bond to carbon 4 of D-glucopyranose. Lactose is a reducing sugar, because the cyclic hemiacetal of the D-glucopyranose unit is in equilibrium with its open-chain form and can be oxidized to a carboxyl group.

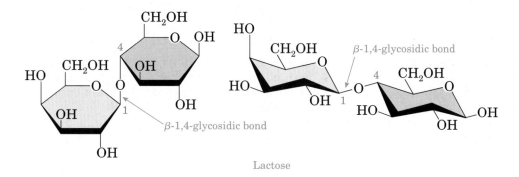

Lactose

CHEMICAL CONNECTIONS 11D

A, B, AB, and O Blood Types

Membranes of animal plasma cells have large numbers of relatively small carbohydrates bound to them. In fact, the outsides of most plasma cell membranes are literally "sugar-coated." These membrane-bound carbohydrates are part of the mechanism by which cell types recognize one another and, in effect, act as biochemical markers. Typically, they contain from 4 to 17 monosaccharide units consisting primarily of relatively few monosaccharides, the most common of which are D-galactose, D-mannose, L-fucose, *N*-acetyl-D-glucosamine, and *N*-acetyl-D-galactosamine. L-Fucose is a 6-deoxyaldohexose.

CHO
HO——H
H——OH
H——OH
HO——H
CH₃

An L-monosaccharide because this —OH group is on the left in the Fischer projection

Carbon 6 is —CH₃ rather than —CH₂OH

L-Fucose

To see the importance of these membrane-bound carbohydrates, consider the ABO blood group system, discovered in 1900 by Karl

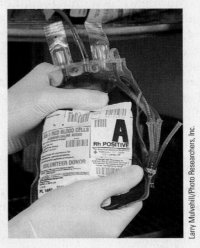

■ **Bag of blood showing blood type.**

Landsteiner (1868–1943). Whether an individual belongs to type A, B, AB, or O is genetically determined and depends on the type of trisaccharide or tetrasaccharide bound to the surface of the red blood cells. These surface-bound carbohydrates, designated as A, B, and O, act as antigens. The type of glycosidic bond joining each monosaccharide is shown in the figure.

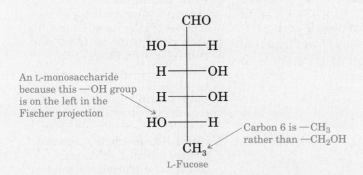

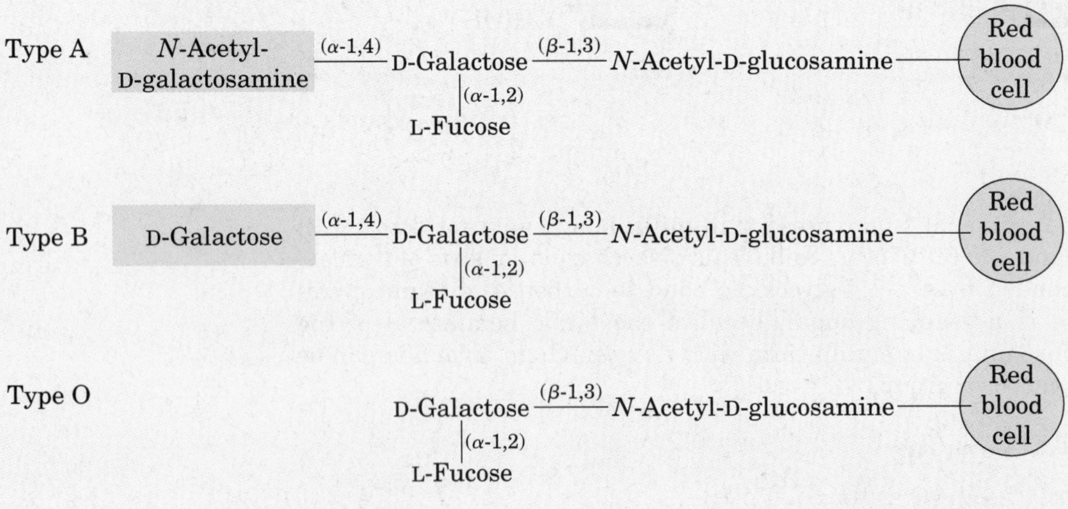

The blood carries antibodies against foreign substances. When a person receives a blood transfusion, the antibodies clump (aggregate) the foreign blood cells. Type A blood, for example, has A antigens (*N*-acetyl-D-galactosamine) on the surfaces of its red blood cells, and carries anti-B antibodies (against B antigen). B-type blood carries B antigen (D-galactose)

and has anti-A antibodies (against A antigens). Transfusion of type A blood into a person with type B blood can be fatal, and vice versa. The relationships between blood type and donor/receiver relationships are summarized in the figure.

(continued on next page)

CHEMICAL CONNECTIONS 11D

A, B, AB, and O Blood Types, *(continued)*

Sugar on cell surface: O
Has antibodies against: A and B
Can receive blood from: O
Can donate blood to: O, A, B, and AB

Type O

Type A   Type B

Sugar on cell surface: A
Has antibodies against: B
Can receive blood from: A or O
Can donate blood to: A and AB

Sugar on cell surface: B
Has antibodies against: A
Can receive blood from: B and O
Can donate blood to: B and AB

Type AB

Sugars on cell surface: A and B
Has antibodies against: None
Can receive blood from: O, A, B, and AB
Can donate blood to: AB

People with type O blood are universal donors, and those with type AB blood are universal acceptors. People with A-type blood can accept from type A or type O donors only. Those with type B blood can accept from type B or type O donors only, and type O donors can accept only from type O donors.

C Maltose

Maltose derives its name from its presence in malt, the juice from sprouted barley and other cereal grains. It consists of two units of D-glucopyranose joined by a glycosidic bond between carbon 1 (the anomeric carbon) of one unit and carbon 4 of the other unit. Because the oxygen atom on the anomeric carbon of the first glucopyranose unit is alpha, the bond joining the two units is called an α-1,4-glycosidic bond. Following are a Haworth projection and a chair conformation for β-maltose, so named because the —OH group on the anomeric carbon of the glucose unit on the right is beta.

Maltose is an ingredient in most syrups.

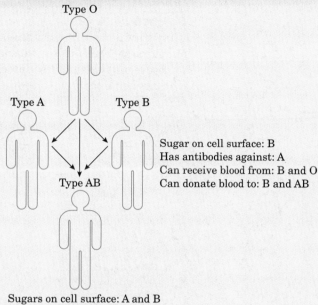

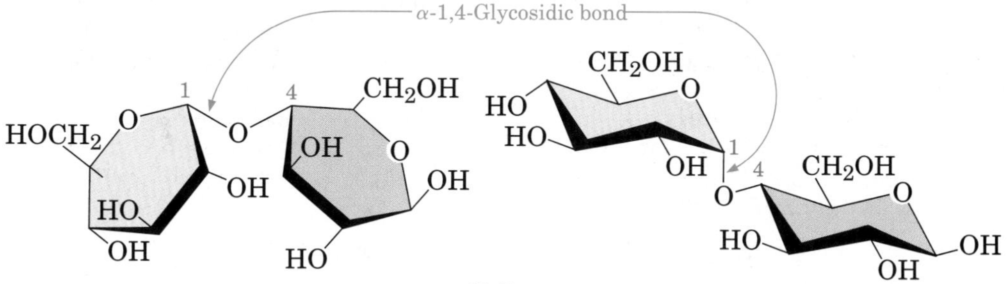

Maltose

Maltose is a reducing sugar: The hemiacetal group on the right unit of D-glucopyranose is in equilibrium with the free aldehyde and can be oxidized to a carboxylic acid.

EXAMPLE 11.5

Draw a chair conformation for the β anomer of a disaccharide in which two units of D-glucopyranose are joined by an α-1,6-glycosidic bond.

Solution

First draw a chair conformation of α-D-glucopyranose. Then connect the anomeric carbon of this monosaccharide to carbon 6 of a second D-glucopyranose unit by an α-1,6 glycosidic bond. The resulting molecule is either α or β depending on the orientation of the —OH group on the reducing end of the disaccharide. The disaccharide shown here is the β form.

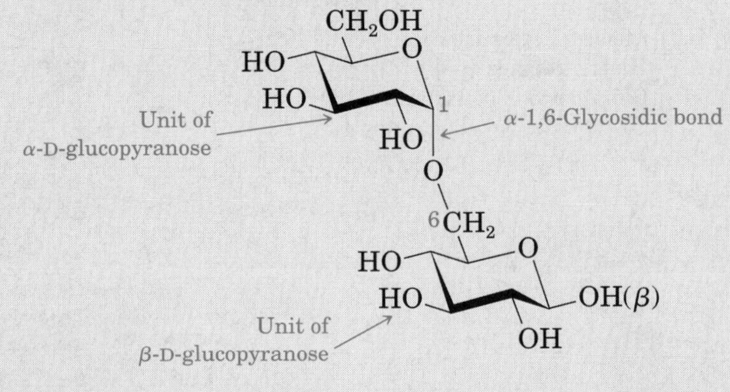

Problem 11.5

Draw a chair conformation for the α form of a disaccharide in which two units of D-glucopyranose are joined by a β-1,3-glycosidic bond.

D Relative Sweetness

Among the saccharide sweetening agents, D-fructose tastes the sweetest—even sweeter than sucrose (Table 11.3). The sweet taste of honey is due largely to D-fructose and D-glucose. Lactose has almost no sweetness and is sometimes added to foods as a filler. Some people cannot tolerate lactose well, however, and should avoid these foods.

We have no mechanical way to measure sweetness. Such testing is done by having a group of people taste solutions of varying sweetness.

Table 11.3 Relative Sweetness of Some Carbohydrate and Artificial Sweetening Agents

Carbohydrate	Sweetness Relative to Sucrose	Artificial Sweetener	Sweetness Relative to Sucrose
fructose	1.74	saccharine	450
sucrose (table sugar)	1.00	acesulfame-K	200
honey	0.97	aspartame	180
glucose	0.74		
maltose	0.33		
galactose	0.32		
lactose (milk sugar)	0.16		

High-Fructose Corn Syrup

If you read the labels of soft drinks and other artificially sweetened food products, you will find that many of them contain high-fructose corn syrup. Looking at Table 11.3, you will see one reason for its use: Fructose is more than 70% sweeter than sucrose (table sugar).

The production of high-fructose corn syrup begins with the partial hydrolysis of corn starch catalyzed by the enzyme α-amylase.

This enzyme catalyzes the hydrolysis of α-glucosidic bonds and breaks corn starch into small polysaccharides called dextrins. The enzyme glucoamylase then catalyzes the hydrolysis of the dextrins to D-glucose. Finally, enzyme-catalyzed isomerization of D-glucose gives D-fructose. Several billion pounds of high-fructose corn syrup are produced each year in this way for use by the food processing industry.

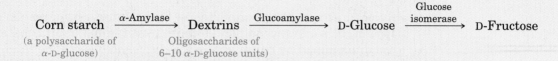

$$\text{Corn starch} \xrightarrow{\alpha\text{-Amylase}} \text{Dextrins} \xrightarrow{\text{Glucoamylase}} \text{D-Glucose} \xrightarrow[\text{isomerase}]{\text{Glucose}} \text{D-Fructose}$$

(a polysaccharide of α-D-glucose)   Oligosaccharides of 6–10 α-D-glucose units)

11.6 Polysaccharides

Polysaccharides consist of large numbers of monosaccharide units bonded together by glycosidic bonds. Three important polysaccharides, all made up of glucose units, are starch, glycogen, and cellulose.

A Starch: Amylose and Amylopectin

Starch is used for energy storage in plants. It is found in all plant seeds and tubers and is the form in which glucose is stored for later use. Starch can be separated into two principal polysaccharides: amylose and amylopectin. Although the starch from each plant is unique, most starches contain 20 to 25% amylose and 75 to 80% amylopectin.

Complete hydrolysis of both amylose and amylopectin yields only D-glucose. Amylose is composed of continuous, unbranched chains of as many as 4000 D-glucose units joined by α-1,4-glycosidic bonds. Amylopectin contains chains of as many as 10,000 D-glucose units, also joined by α-1,4-glycosidic bonds. In addition, there is considerable branching from this linear network. At branch points, new chains of 24 to 30 units are started by α-1,6-glycosidic bonds (Figure 11.3).

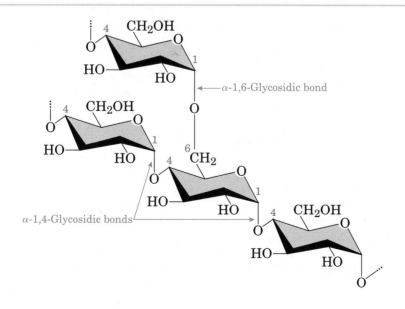

Figure 11.3 Amylopectin is a highly branched polymer of D-glucose. Chains consist of 24–30 units of D-glucose joined by α-1,4-glycosidic bonds and branches created by α-1,6-glycosidic bonds.

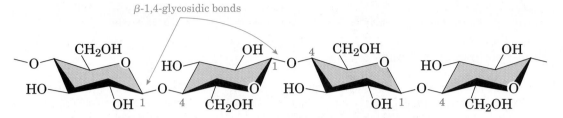

Figure 11.4 Cellulose is a linear polymer containing as many as 3000 units of D-glucose joined by β-1,4-glycosidic bonds.

B Glycogen

Glycogen acts as the energy-reserve carbohydrate for animals. Like amylopectin, it is a branched polysaccharide containing approximately 10^6 glucose units joined by α-1,4- and α-1,6-glycosidic bonds. The total amount of glycogen in the body of a well-nourished adult human is about 350 g, divided almost equally between liver and muscle.

C Cellulose

Cellulose, the most widely distributed plant skeletal polysaccharide, constitutes almost half of the cell-wall material of wood. Cotton is almost pure cellulose.

Cellulose is a linear polysaccharide of D-glucose units joined by β-1,4-glycosidic bonds (Figure 11.4). It has an average molecular weight of 400,000 g/mol, corresponding to approximately 2200 glucose units per molecule.

Cellulose molecules act much like stiff rods, a feature that enables them to align themselves side by side into well-organized, water-insoluble fibers in which the OH groups form numerous intermolecular hydrogen bonds. This arrangement of parallel chains in bundles gives cellulose fibers their high mechanical strength. It also explains why cellulose is insoluble in water. When a piece of cellulose-containing material is placed in water, there are not enough —OH groups on the surface of the fiber to pull individual cellulose molecules away from the strongly hydrogen-bonded fiber.

Humans and other animals cannot use cellulose as food because our digestive systems do not contain β-glucosidases, enzymes that catalyze the hydrolysis of β-glucosidic bonds. Instead, we have only α-glucosidases; hence, we use the polysaccharides starch and glycogen as sources of glucose. On the other hand, many bacteria and microorganisms do contain β-glucosidases and so can digest cellulose. Termites (much to our regret) have such bacteria in their intestines and can use wood as their principal food. Ruminants (cud-chewing animals) and horses can also digest grasses and hay because β-glucosidase-containing microorganisms are present in their alimentary systems.

11.7 Acidic Polysaccharides

Acidic polysaccharides—polysaccharides that contain carboxyl groups and/or sulfuric ester groups—play important roles in the structure and function of connective tissues. Because acidic polysaccharides contain amino sugars, a more current name for these substances is glycosaminoglycans. There is no single general type of connective tissue. Rather, there are a large number of highly specialized forms, such as cartilage, bone, synovial fluid, skin, tendons, blood vessels, intervertebral disks, and cornea. Most connective tissues consist of collagen, a structural protein, combined with a variety of acidic polysaccharides (glycosaminoglycans) that interact with collagen to form tight or loose networks.

A Hyaluronic Acid

Hyaluronic acid is the simplest acidic polysaccharide present in connective tissue. It has a molecular weight of between 10^5 and 10^7 g/mol and contains from 300 to 100,000 repeating units, depending on the organ in which it occurs. It is most abundant in embryonic tissues and in specialized connective tissues such as synovial fluid, the lubricant of joints in the body, and the vitreous of the eye, where it provides a clear, elastic gel that holds the retina in its proper position.

In rheumatoid arthritis, inflammation of the synovial tissue results in swelling of the joints.

Hyaluronic acid is composed of D-glucuronic acid linked by a β-1,3-glycosidic bond to N-acetyl-D-glucosamine, which is in turn linked to D-glucuronic acid by a β-1,4-glycosidic bond.

D-glucuronic acid

N-acetyl-D-glucosamine

The repeating unit of hyaluronic acid

B Heparin

Heparin is a heterogeneous mixture of variably sulfonated polysaccharide chains, ranging in molecular weight from 6000 to 30,000 g/mol. This acidic polysaccharide is synthesized and stored in mast cells of various tissues—particularly the liver, lungs, and gut. Heparin has many biological functions, the best known and most fully understood of which is its anticoagulant activity. It binds strongly to antithrombin III, a plasma protein involved in terminating the clotting process. A heparin preparation with good anticoagulant activity contians a minimum of eight repeating units (Figure 11.5). The larger the molecule, the better its anticoagulant activity.

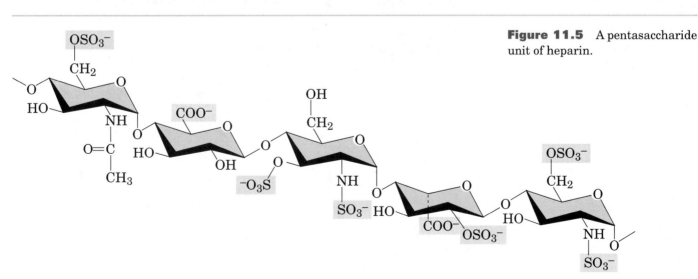

Figure 11.5 A pentasaccharide unit of heparin.

Monosaccharides are polyhydroxyaldehydes or polyhydroxyketones (Section 11.2A). The most common have the general formula $C_nH_{2n}O_n$ where n varies from 3 to 8. Their names contain the suffix *-ose,* and prefixes *tri-, tetr-,* and so on indicate the number of carbon atoms in the chain. The prefix *aldo-* indicates an aldehyde and the prefix *keto-* indicates a ketone.

In a **Fischer projection** of a carbohydrate, we write the carbon chain vertically with the most highly oxidized carbon toward the top (Section 11.2B). Horizontal lines represent groups projecting above the plane of the page; vertical lines represent groups projecting behind the plane of the page.

The penultimate carbon of a monosaccharide is the next-to-last carbon of a Fischer projection (Section 11.2C). A monosaccharide that has the same configuration at the penultimate carbon as D-glyceraldehyde is called a **D-monosaccharide;** one that has the same configuration at the penultimate carbon as L-glyceraldehyde is called an **L-monosaccharide.**

Monosaccharides exist primarily as cyclic hemiacetals (Section 11.3). A six-membered cyclic hemiacetal is a **pyranose;** a five-membered cyclic hemiacetal is a **furanose.** The new stereocenter resulting from hemiacetal formation is called an **anomeric carbon,** and the stereoisomers formed in this way are called **anomers.** The symbol β-indicates that the —OH group on the anomeric carbon lies on the same side of the ring as the terminal —CH$_2$OH. The symbol α-indicates that the —OH group on the anomeric carbon lies on the opposite side from the terminal —CH$_2$OH. Furanoses and pyranoses can be drawn as **Haworth projections** (Section 11.3A). Pyranoses can also be drawn as **chair conformations** (Section 11.3B). **Mutarotation** is the change in specific rotation that accompanies formation of an equilibrium mixture of α and β anomers in aqueous solution (Section 11.3C).

A **glycoside** is a cyclic acetal derived from a monosaccharide (Section 11.4A). An **alditol** is a polyhydroxy compound formed when the carbonyl group of a monosaccharide is reduced to a hydroxyl group (Section 11.4B). An **aldonic acid** is a carboxylic acid formed when the aldehyde group of an aldose is oxidized to a carboxyl group (Section 11.4C). Any carbohydrate that reacts with an oxidizing agent to form an aldonic acid is classified as a **reducing sugar** (it reduces the oxidizing agent).

A **disaccharide** contains two monosaccharide units joined by a glycosidic bond (Section 11.5). Terms applied to carbohydrates containing larger numbers of monosaccharides are **trisaccharide, tetrasaccharide, oligosaccharide,** and **polysaccharide. Sucrose** is a disaccharide consisting of D-glucose joined to D-fructose by an α-1,2-glycosidic bond (Section 11.5A). **Lactose** is a disaccharide consisting of D-galactose joined to D-glucose by a β-1,4-glycosidic bond (Section 11.5B). **Maltose** is a disaccharide of two molecules of D-glucose joined by an α-1,4-glycosidic bond (Section 11.5C).

Starch can be separated into two fractions, amylose and amylopectin (Section 11.6A). **Amylose** is a linear polysaccharide of as many as 4000 units of D-glucopyranose joined by α-1,4-glycosidic bonds. **Amylopectin** is a highly branched polysaccharide of D-glucose joined by α-1,4-glycosidic bonds and, at branch points, by α-1,6-glycosidic bonds. **Glycogen,** the reserve carbohydrate of animals, is a highly branched polysaccharide of D-glucopyranose joined by α-1,4-glycosidic bonds and, at branch points, by α-1,6-glycosidic bonds (Section 11.6B). **Cellulose,** the skeletal polysaccharide of plants, is a linear polysaccharide of D-glucopyranose joined by β-1,4-glycosidic bonds (Section 11.6C).

The carboxyl and sulfate groups of **acidic polysaccharides** are ionized to —COO$^-$ and —SO$_3^-$ at the pH of body fluids, which gives these polysaccharides net negative charges (Section 11.7).

K E Y   R E A C T I O N S

1. Formation of Cyclic Hemiacetals (Section 11.3A) A monosaccharide existing as a five-membered ring is a furanose; one existing as a six-membered ring is a pyranose. A pyranose is most commonly drawn as either a Haworth projection or a chair conformation.

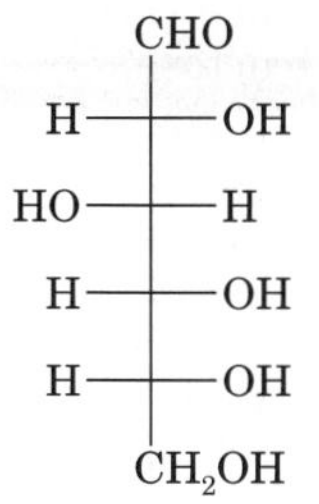

D-Glucose

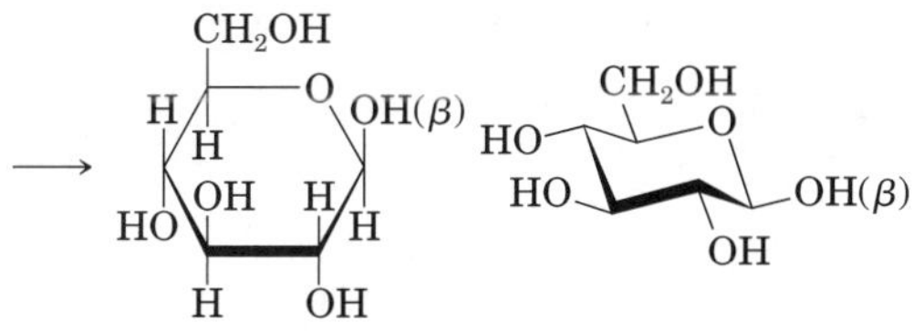

β-D-Glucopyranose
(β-D-Glucose)

2. Mutarotation (Section 11.3C) Anomeric forms of a monosaccharide are in equilibrium in aqueous solution. Mutarotation is the change in specific rotation that accompanies this equilibration.

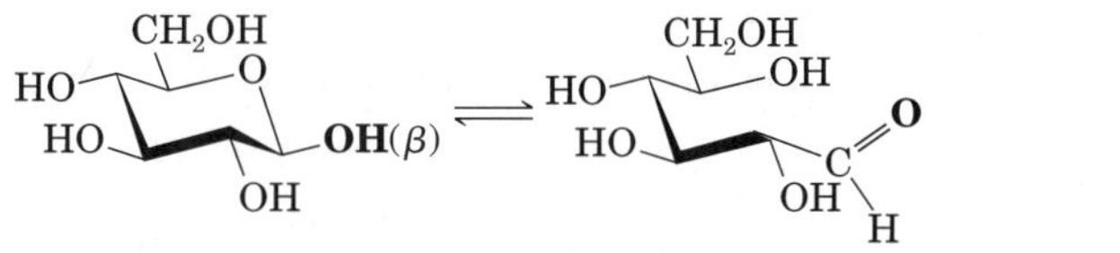

β-D-Glucopyranose
$[\alpha]_{D}^{25} + 18.7°$

Open-chain form

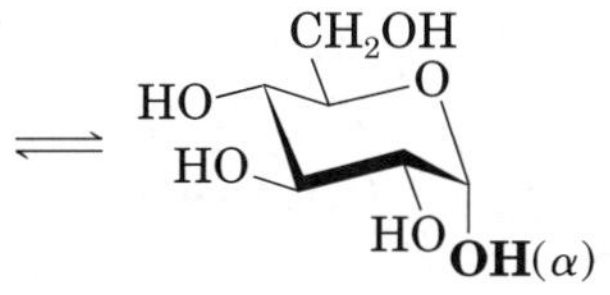

α-D-Glucopyranose
$[\alpha]_{D}^{25} + 112°$

3. Formation of Glycosides (Section 11.4A) Treatment of a monosaccharide with an alcohol in the presence of an acid catalyst forms a cyclic acetal called a glycoside. The bond to the new —OR group is called a glycosidic bond.

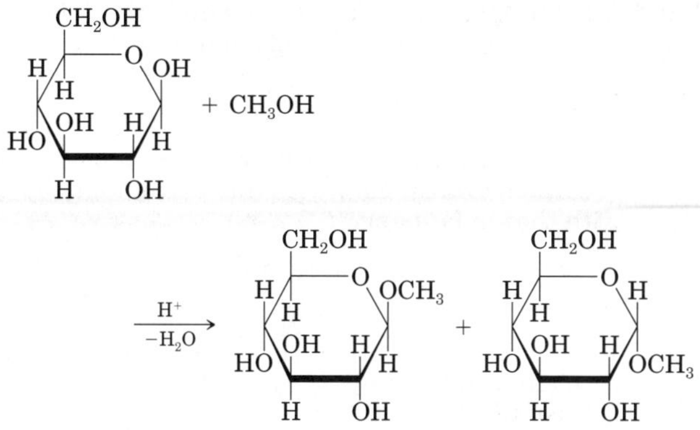

$+ CH_3OH$

$\xrightarrow[-H_2O]{H^+}$

$+$

4. Reduction to Alditols (Section 11.4B) Reduction of the carbonyl group of an aldose or ketose to a hydroxyl group yields a polyhydroxy compound called an alditol.

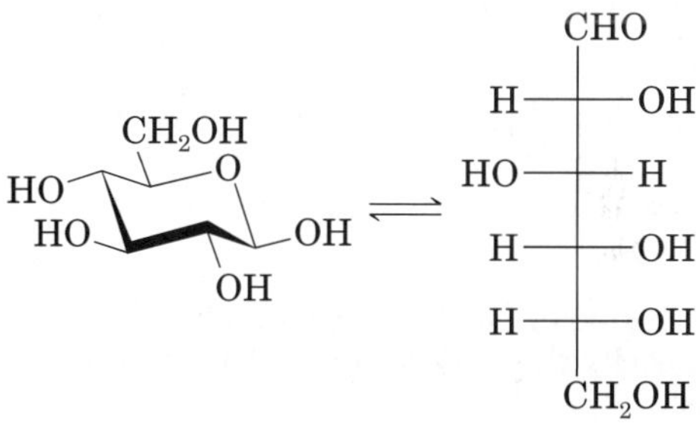

β-D-Glucopyranose   D-Glucose

$\xrightarrow{NaBH_4}$

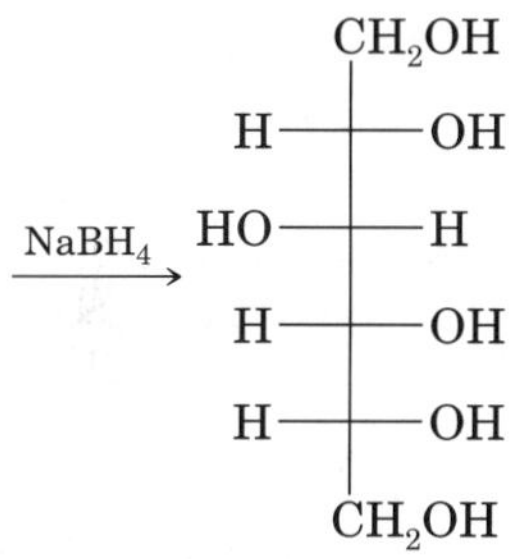

D-Glucitol
(D-Sorbitol)

5. Oxidation to an Aldonic Acid (Section 11.4C) Oxidation of the aldehyde group of an aldose to a carboxyl group by a mild oxidizing agent gives a polyhydroxycarboxylic acid called an aldonic acid.

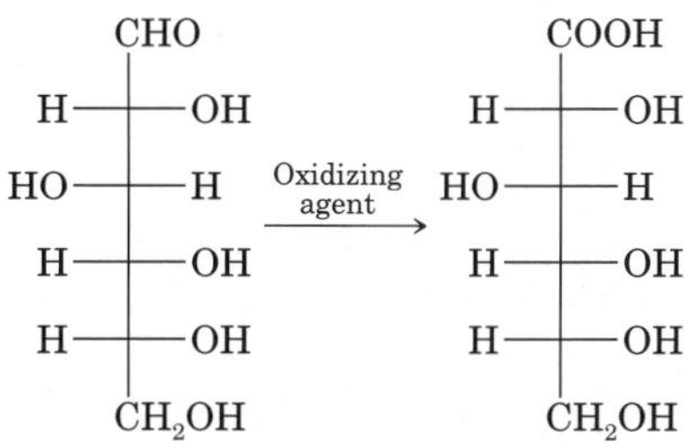

D-Glucose

$\xrightarrow[\text{agent}]{\text{Oxidizing}}$

D-Gluconic acid

P R O B L E M S

Numbers that appear in color indicate difficult problems.
▶ designates problems requiring application of principles.

11.6 Define carbohydrate.

11.7 What percentage of the dry weight of plants consists of carbohydrates? How does this percentage compare with that in humans?

Monosaccharides

11.8 What is the difference in structure between an aldose and a ketose? Between an aldopentose and a ketopentose?

11.9 Of the eight D-aldohexoses, which is the most abundant in the biological world?

11.10 Name the three most abundant hexoses in the biological world. Which are aldohexoses, and which are ketohexoses?

11.11 Which hexose is also known as "dextrose"?

11.12 What does it mean to say that D- and L-glyceraldehyde are enantiomers?

11.13 Explain the meaning of the designations D and L as used to specify the configuration of a monosaccharide.

11.14 Which carbon of an aldopentose determines whether the pentose has a D or L configuration?

11.15 How many stereocenters are present in D-glucose? In D-ribose? How many stereoisomers are possible for each monosaccharide?

11.16 Which of the following compounds are D-monosaccharides, and which are L-monosaccharides?

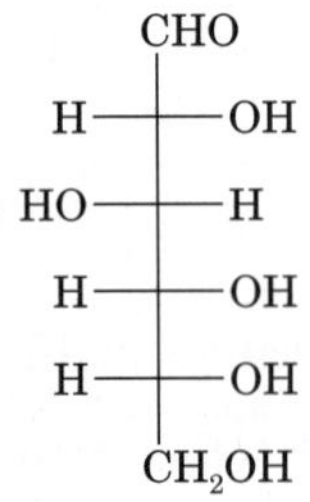

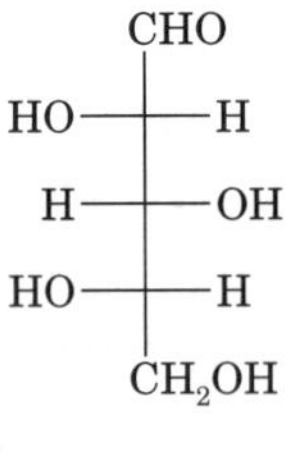

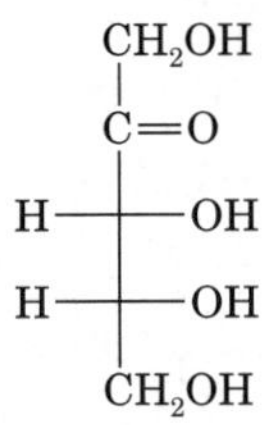

11.17 Draw Fischer projections for L-ribose and L-arabinose.

11.18 Draw a Fischer projection for a D-2-ketoheptose.

11.19 Explain why all mono- and disaccharides are soluble in water.

11.20 What is an amino sugar? Name the three amino sugars most commonly found in nature.

The Cyclic Structure of Monosaccharides

11.21 Define the term *anomeric carbon*. Which carbon is the anomeric carbon in glucose? In fructose?

11.22 Define:
(a) Pyranose
(b) Furanose

11.23 Explain the conventions for using α and β to designate the configurations of the cyclic forms of monosaccharides.

11.24 Are α-D-glucose and β-D-glucose anomers? Explain. Are they enantiomers? Explain.

11.25 Are the hydroxyl groups on carbons 1, 2, 3, and 4 of α-D-glucose all in equatorial positions?

11.26 In what way are chair conformations of aldohexoses a more accurate representation of the molecular shape than Haworth projections?

11.27 Convert each of the following Haworth projections to an open-chain form and to a Fischer projection. Name the monosaccharides you have drawn.

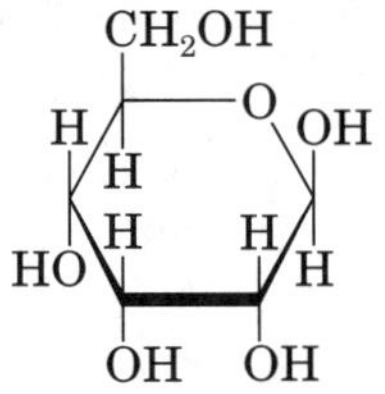

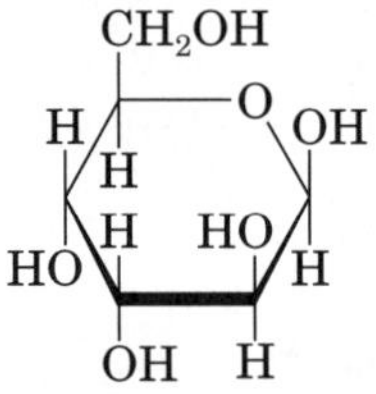

11.28 Convert each of the following chair conformations to an open-chain form and to a Fischer projection. Name the monosaccharides you have drawn.

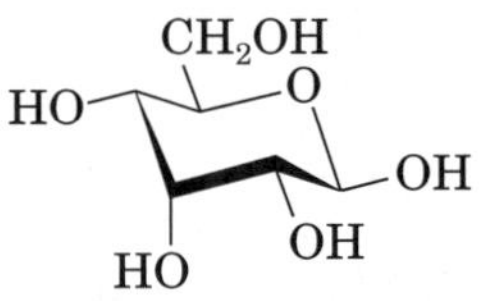

11.29 Explain the phenomenon of mutarotation. How is it detected?

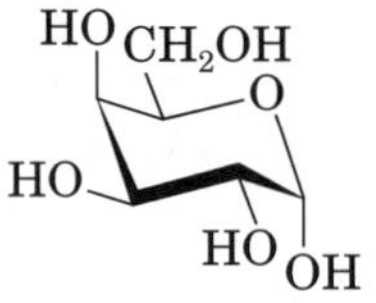

11.30 The specific rotation of α-D-glucose is $+112.2°$. What is the specific rotation of α-L-glucose?

11.31 When α-D-glucose is dissolved in water, the specific rotation of the solution changes from $+112.2°$ to $+52.7°$. Does the specific rotation of α-L-glucose also change when it is dissolved in water? If so, to what value does it change?

Reactions of Monosaccharides

11.32 Define the terms *glycoside* and *glycosidic bond*.

11.33 What is the difference in meaning between the terms *glycosidic bond* and *glucosidic bond*?

11.34 Do glycosides undergo mutarotation?

11.35 Draw Fischer projections for the product formed by treating each of the following monosaccharides with sodium borohydride, $NaBH_4$, in water.
(a) D-Galactose
(b) D-Ribose

11.36 Reduction of D-glucose by $NaBH_4$ gives D-sorbitol, a compound used in the manufacture of sugar-free gums and candies. Draw a structural formula for D-sorbitol.

11.37 Reduction of D-fructose by $NaBH_4$ gives two alditols, one of which is D-sorbitol. Name and draw a structural formula for the other alditol.

11.38 Ribitol and β-D-ribose 1-phosphate are both derivatives of D-ribose. Draw a structural formula for each compound.

11.39 One pathway for the metabolism of D-glucose 6-phosphate is its enzyme-catalyzed conversion to D-fructose 6-phosphate. Show that this transformation can be regarded as two enzyme-catalyzed keto-enol tautomerizations (Section 9.8).

CHO
H——OH
HO——H
H——OH
H——OH
$CH_2OPO_3{}^{2-}$

D-Glucose 6-phosphate

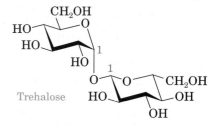

CH_2OH
C=O
HO——H
H——OH
H——OH
$CH_2OPO_3{}^{2-}$

D-Fructose 6-phosphate

11.40 One step in glycolysis, the pathway that converts glucose to pyruvate (Section 19.2), involves an enzyme-catalyzed conversion of dihydroxyacetone phosphate to D-glyceraldehyde 3-phosphate. Show that this transformation can be regarded as two enzyme-catalyzed keto-enol tautomerizations (Section 9.8).

CH_2OH
C=O
$CH_2OPO_3{}^{2-}$

Dihydroxyacetone phosphate

Enzyme catalysis

CHO
H——OH
$CH_2OPO_3{}^{2-}$

D-Glyceraldehyde 3-phosphate

Disaccharides and Oligosaccharides

11.41 Name three important disaccharides. From which monosaccharides is each derived?

11.42 What does it mean to describe a glycosidic bond as β-1,4-? To describe it as α-1,6-?

11.43 Both maltose and lactose are reducing sugars, but sucrose is not. Explain why.

11.44 Following is a structural formula for a disaccharide.

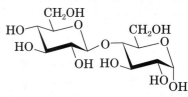

(a) Name the two monosaccharide units in the disaccharide.
(b) Describe the glycosidic bond.
(c) Is this disaccharide a reducing sugar?
(d) Will this disaccharide undergo mutarotation?

11.45 The disaccharide trehalose is found in young mushrooms and is the chief carbohydrate in the blood of certain insects.

CH₂OH
HO
HO
HO 1
1
O O CH₂OH
Trehalose HO OH
OH

(a) Name the two monosaccharide units in trehalose.
(b) Describe the glycosidic bond in trehalose.
(c) Is trehalose a reducing sugar?
(d) Will trehalose undergo mutarotation?

Polysaccharides

11.46 What is the difference in structure between oligosaccharides and polysaccharides?

11.47 Name three polysaccharides that are composed of units of D-glucose. In which are the glucose units joined by α-glycosidic bonds? In which are they joined by β-glycosidic bonds?

11.48 Starch can be separated into two principal polysaccharides, amylose and amylopectin. What is the major difference in structure between the two?

11.49 Where is glycogen stored in the human body?

11.50 Why is cellulose insoluble in water?

11.51 How is it possible that cows can digest grass but humans cannot?

11.52 A Fischer projection of *N*-acetyl-D-glucosamine is given in Section 11.2.

(a) Draw a Haworth projection and a chair conformation for the β-pyranose form of this monosaccharide.

(b) Draw a Haworth projection and a chair conformation for the disaccharide formed by joining two units of the pyranose form of *N*-acetyl-D-glucosamine by a β-1,4-glycosidic bond. If you draw them correctly, you will have a structural formula for the repeating dimer of chitin, the structural polysaccharide component of the shell of lobsters and other crustaceans.

11.53 Propose structural formulas for the repeating disaccharide unit in these polysaccharides.

(a) Alginic acid, isolated from seaweed, is used as a thickening agent in ice cream and other foods. Alginic acid is a polymer of D-mannuronic acid in the pyranose form joined by β-1,4-glycosidic bonds.

(b) Pectic acid is the main component of pectin, which is responsible for the formation of jellies from fruits and berries. Pectic acid is a polymer of D-galacturonic acid in the pyranose form joined by α-1,4-glycosidic bonds.

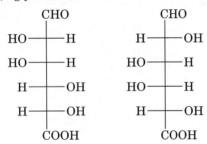

D-Mannuronic acid   D-Galacturonic acid

11.54 Hyaluronic acid acts as a lubricant in the synovial fluid of joints. In rheumatoid arthritis, inflammation breaks hyaluronic acid down to smaller molecules. Under these conditions, what happens to the lubricating power of the synovial fluid?

11.55 The anticlotting property of heparin is partly due to the negative charges it carries.

(a) Identify the functional groups that provide the negative charges.

(b) Which type of heparin is a better anticoagulant, one with a high or a low degree of polymerization?

Chemical Connections

11.56 (Chemical Connections 11A) Why does congenital galactosemia appear only in infants? Why can galactosemia be relieved by feeding an infant a formula containing sucrose as the only carbohydrate?

11.57 (Chemical Connections 11B) What is the difference in structure between L-ascorbic acid and L-dehydroascorbic acid? What does the designation L indicate in these molecules?

11.58 (Chemical Connections 11B) When L-ascorbic acid participates in a redox reaction, it is converted to L-dehydroascorbic acid. In this reaction, is L-ascorbic acid oxidized or reduced? Is L-ascorbic acid a biological oxidizing agent or a biological reducing agent?

11.59 (Chemical Connections 11C) Why is the glucose assay one of the most common analytical tests performed in clinical chemistry laboratories?

11.60 (Chemical Connections 11D) What monosaccharides do types A, B, and O blood have in common? In which monosaccharides do they differ?

11.61 (Chemical Connections 11D) L-Fucose is a monosaccharide unit common to A, B, AB, and O blood types.

(a) Is this monosaccharide an aldose or a ketose? A hexose or a pentose?

(b) What is unusual about this monosaccharide?

(c) If L-fucose were to undergo a reaction in which its terminal —CH₃ group was converted to a —CH₂OH group, which monosaccharide would result?

11.62 (Chemical Connections 11D) Why can't a person with A-type blood donate to a person with B-type blood?

11.63 (Chemical Connections 11E) What is high-fructose corn syrup, and how is it produced?

Additional Problems

11.64 2,6-Dideoxy-D-altrose, also known as D-digitoxose, is a monosaccharide obtained from the hydrolysis of digitoxin, a natural product extracted from purple foxglove (*Digitalis purpurea*). Digitoxin has found wide use in cardiology because it reduces pulse rate, regularizes heart rhythm, and strengthens heart beat. Draw a structural formula for the open-chain form of 2,6-dideoxy-D-altrose.

11.65 In making candy or sugar syrups, sucrose is boiled in water with a little acid, such as lemon juice. Why does the product mixture taste sweeter than the initial sucrose solution?

11.66 Hot-water extracts of ground willow bark are an effective pain reliever (Chemical Connections 10C). Unfortunately, the liquid is so bitter that most persons refuse it. The pain reliever in these infusions is salicin. Name the monosaccharide unit in salicin.

Salicin

11.67 Following is a Haworth projection and a chair conformation of the repeating disaccharide unit in chondroitin 6-sulfate. This biopolymer acts as the flexible connecting matrix between the tough protein filaments in cartilage. It is available as a dietary supplement, often combined with D-glucosamine sulfate. Some believe this combination can strengthen and improve joint flexibility.

(a) From what two monosaccharide units is the repeating disaccharide unit of chondroitin 6-sulfate derived?

(b) Describe the glycosidic bond between the two units.

InfoTrac College Edition

For additional readings, go to InfoTrac College Edition, your online research library, at

http://infotrac.thomsonlearning.com

CHAPTER **12**

12.1 Introduction

12.2 The Structure of Triglycerides

12.3 Properties of Triglycerides

12.4 Complex Lipids

12.5 Membranes

12.6 Glycerophospholipids

12.7 Sphingolipids

12.8 Glycolipids

12.9 Steroids

12.10 Steroid Hormones

12.11 Bile Salts

12.12 Prostaglandins, Thromboxanes, and Leukotrienes

Doug Perrine/TCL/Masterfile

Sea lions are marine mammals that require a heavy layer of fat to survive in cold waters.

Lipids

12.1 Introduction

Unlike in the case of carbohydrates, we define lipids in terms of a property and not in terms of their structure.

Found in living organisms, **lipids** are a family of substances that are insoluble in water but soluble in nonpolar solvents and solvents of low polarity, such as diethyl ether.

A Classification by Function

Lipids play three major roles in human biochemistry: (1) to store energy within fat cells, (2) as parts of membranes to separate compartments of aqueous solutions from each other, and (3) to serve as chemical messengers.

Storage An important use for lipids, especially in animals, is the storage of energy. As we saw in Section 11.6, plants store energy in the form of starch. Animals (including humans) find it more economical to use fats instead. Although our bodies do store some carbohydrates in the form of glycogen for quick energy when we need it, energy stored in the form of fats has much greater importance for us. The reason is simple: The burning of fats produces more than twice as much energy (about 9 kcal/g) as the burning of an equal weight of carbohydrates (about 4 kcal/g).

258

Membrane Components The lack of water solubility of lipids is an important property because our body chemistry is so firmly based on water. Most body constituents, including carbohydrates and proteins, are soluble in water. But the body also needs insoluble compounds for the membranes that separate compartments containing aqueous solutions from each other, whether they are cells or organelles within the cells. Lipids provide these membranes. Their water insolubility derives from the fact that the polar groups they contain are much smaller than their alkane-like (nonpolar) portions. These nonpolar portions provide the water-repellent, or *hydrophobic,* property.

Messengers Lipids also serve as chemical messengers. Primary messengers, such as steroid hormones, deliver signals from one part of the body to another part. Secondary messengers, such as prostaglandins and thromboxanes, mediate the hormonal response.

Classification by Structure

For purposes of study, we can classify lipids into four groups: (1) simple lipids, such as fats and waxes; (2) complex lipids; (3) steroids; and (4) prostaglandins, thromboxanes, and leukotrienes.

12.2 The Structure of Triglycerides

Animal fats and vegetable oils are triglycerides. **Triglycerides** are triesters of glycerol and long-chain carboxylic acids called fatty acids. In Section 10.3, we saw that esters are made up of an alcohol part and an acid part. As the name indicates, the alcohol of triglycerides is always glycerol.

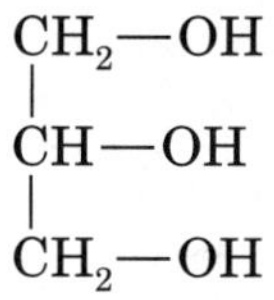

$$CH_2{-}OH$$
$$|$$
$$CH{-}OH$$
$$|$$
$$CH_2{-}OH$$

Glycerol

In contrast to the alcohol part, the acid component of fats may be any number of acids, which do, however, have certain things in common:

1. Fatty acids are practically all unbranched carboxylic acids.

2. They range in size from about 10 to 20 carbons.

3. They contain an even number of carbon atoms.

4. Apart from the —COOH group, they have no functional groups except that some do have double bonds.

5. In most fatty acids that have double bonds, the cis isomers predominate.

Only even-numbered acids are found in fats because the body builds these acids entirely from acetate units and therefore puts the carbons in two at a time (Section 20.3). Table 12.1 shows the most important acids found in fats.

The fatty acids can be divided into two groups: saturated and unsaturated. **Saturated fatty acids** have only single bonds in the hydrocarbon chain. **Unsaturated fatty acids** have at least one C=C double bond in the chain. All the unsaturated fatty acids listed in Table 12.1 are cis isomers. This fact explains their physical properties, as reflected in their melting points. Saturated fatty acids are solids at room temperature because the

Table 12.1   The Most Important Fatty Acids in Triglycerides

Carbon Atoms: Double Bonds*	Structure	Common Name	Melting Point (°C)
Saturated Fatty Acids			
12 : 0	$CH_3(CH_2)_{10}COOH$	lauric acid	44
14 : 0	$CH_3(CH_2)_{12}COOH$	myristic acid	58
16 : 0	$CH_3(CH_2)_{14}COOH$	palmitic acid	63
18 : 0	$CH_3(CH_2)_{16}COOH$	stearic acid	70
20 : 0	$CH_3(CH_2)_{18}COOH$	arachidic acid	77
Unsaturated Fatty Acids			
16 : 1	$CH_3(CH_2)_5CH{=}CH(CH_2)_7COOH$	palmitoleic acid	1
18 : 1	$CH_3(CH_2)_7CH{=}CH(CH_2)_7COOH$	oleic acid	16
18 : 2	$CH_3(CH_2)_4(CH{=}CHCH_2)_2(CH_2)_6COOH$	linoleic acid	−5
18 : 3	$CH_3CH_2(CH{=}CHCH_2)_3(CH_2)_6COOH$	linolenic acid	−11
20 : 4	$CH_3(CH_2)_4(CH{=}CHCH_2)_4(CH_2)_2COOH$	arachidonic acid	−49

*The first number is the number of carbons in the fatty acid; the second number is the number of carbon–carbon double bonds in its hydrocarbon chain.

regular nature of their aliphatic chains allows the molecules to be packed in a close, parallel alignment:

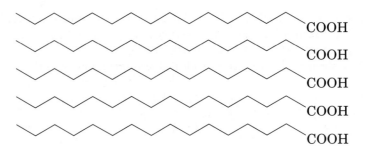

> The longer the aliphatic chain, the higher the melting point.

The interactions (London dispersion forces) between neighboring chains are weak. Nevertheless, the regular packing allows these forces to operate over a large portion of the chain, ensuring that a considerable amount of energy is needed to melt them.

Unsaturated fatty acids, in contrast, are all liquids at room temperature because the cis double bonds interrupt the regular packing of the chains:

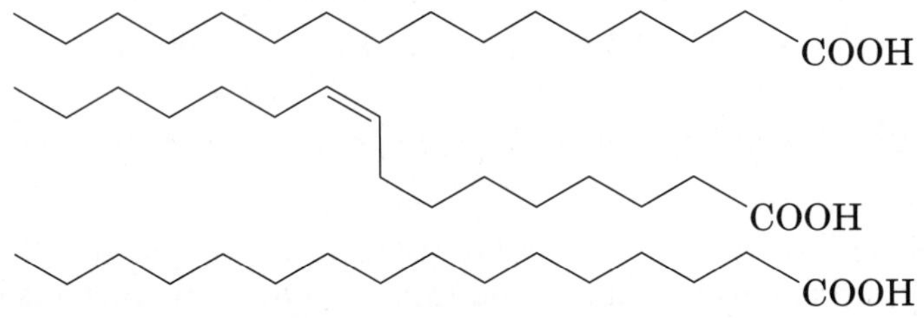

Here the London dispersion forces act over shorter segments of the chain, so much less energy is required to melt them. The greater the degree of unsaturation, the lower the melting point because each double bond introduces more disorder into the packing of the molecules (Table 12.1).

In **triglycerides** (also called **triacylglycerols**) all three —OH groups of glycerol are esterified. Thus a typical fat molecule might be

Oleate (18:1)   Palmitate (16:0)

$$CH_3(CH_2)_7CH=CH(CH_2)_7\overset{\overset{\displaystyle O}{\|}}{C}O\overset{|}{C}H \quad \begin{array}{c} \overset{\overset{\displaystyle O}{\|}}{C}H_2O\overset{\overset{\displaystyle O}{\|}}{C}(CH_2)_{14}CH_3 \\ \\ CH_2O\overset{\overset{\displaystyle O}{\|}}{C}(CH_2)_{16}CH_3 \end{array}$$

Stearate (18:0)

A triglyceride

Triglycerides are the most common lipid materials, although **mono-** and **diglycerides** are not infrequent. In the latter two types, only one or two —OH groups of the glycerol are esterified by fatty acids.

Triglycerides are complex mixtures. Although some of the molecules contain three identical fatty acids, in most cases two or three different acids are present. The hydrophobic character of fats is caused by the long hydrocarbon chains. The ester groups (—$\overset{\overset{\displaystyle O}{\|}}{C}$—O—) although polar themselves, are buried in a nonpolar environment, which makes the fats insoluble in water.

■ **Some vegetable oils.**

Charles D. Winters

12.3 Properties of Triglycerides

A Physical State

With some exceptions, **fats** that come from animals are generally solids at room temperature, and those from plants or fish are usually liquids. Liquid fats are often called **oils,** even though they are esters of glycerol just like solid fats and should not be confused with petroleum, which is mostly alkanes.

What is the structural difference between solid fats and liquid oils? In most cases, it is the degree of unsaturation. The physical properties of the fatty acids are carried over to the physical properties of the triglycerides. Solid animal fats contain mainly saturated fatty acids, whereas vegetable

> **Fat** A mixture of triglycerides containing a high proportion of long-chain, saturated fatty acids

> **Oil** A mixture of triglycerides containing a high proportion of long-chain, unsaturated fatty acids or short-chain, saturated fatty acids

 CHEMICAL CONNECTIONS 12A

Rancidity

The double bonds in fats and oils are subject to oxidation by the air (see Section 4.5C). When a fat or oil is allowed to stand out in the open, this reaction slowly turns some of the molecules into aldehydes and other compounds with bad tastes and odors. We then say that the fat or oil has become rancid and is no longer edible. Vegetable oils, which generally contain more double bonds, are more susceptible to this than solid fats, but even fats contain some double bonds, so rancidity can be a problem here, too.

Another cause of unpleasant taste is hydrolysis. The hydrolysis of triglycerides may produce short-chain fatty acids, such as butanoic acid (butyric acid), which have bad odors. To prevent rancidity, fats and oils should be kept refrigerated (these reactions are slower at low temperatures) and in dark bottles (the oxidation is catalyzed by ultraviolet light). In addition, antioxidants are often added to fats and oils to prevent rancidity.

CHEMICAL CONNECTIONS 12B

Waxes

Plant and animal waxes are simple esters. They are solids because of their high molecular weights. As in fats, the acid portions of the esters consist of a mixture of fatty acids, but the alcohol portions are not glycerol but simple long-chain alcohols. For example, a major component of beeswax is 1-triacontyl palmitate:

Palmitic acid portion   1-Triacontanol portion

$$CH_3(CH_2)_{13}CH_2\overset{O}{\overset{\|}{C}}-OCH_2(CH_2)_{28}CH_3$$

Triacontyl palmitate

Waxes generally have higher melting points than fats (60 to 100° C) and are harder. Animals and plants often use them for protective coatings. For example, the leaves of most plants are coated with wax, which helps to prevent microorganisms from attacking them and allows the plants to conserve water. The feathers of birds and the fur of animals are also coated with wax.

Important waxes include carnauba wax (from a Brazilian palm tree), lanolin (from lamb's wool), beeswax, and spermaceti (from whales). These substances are used to make cosmetics, polishes, candles, and ointments. Paraffin waxes are not esters. They are mixtures of high-molecular-weight alkanes. Neither is ear wax a simple ester. It is a gland secretion and contains a mixture of fats (triglycerides), phospholipids, and esters of cholesterol.

oils contain high amounts of unsaturated fatty acids. Table 12.2 shows the average fatty acid content of some common fats and oils. Note that even solid fats contain some unsaturated acids and that liquid fats contain some saturated acids. Some unsaturated fatty acids (linoleic and linolenic acids) are called *essential fatty acids* because the body cannot synthesize them from precursors; they must therefore be included in the diet.

Although most vegetable oils contain high amounts of unsaturated fatty acids, there are exceptions. Coconut oil, for example, has only a small amount of unsaturated acids. This oil is a liquid not because it contains

Table 12.2 Average Percentage of Fatty Acids of Some Common Fats and Oils

	Saturated				Unsaturated			
	Lauric	Myristic	Palmitic	Stearic	Oleic	Linoleic	Linolenic	Other
Animal Fats								
Beef tallow	—	6.3	27.4	14.1	49.6	2.5	—	0.1
Butter	2.5	11.1	29.0	9.2	26.7	3.6	—	17.9
Human	—	2.7	24.0	8.4	46.9	10.2	—	7.8
Lard	—	1.3	28.3	11.9	47.5	6.0	—	5.0
Vegetable Oils								
Coconut	45.4	18.0	10.5	2.3	7.5	—	—	16.3
Corn	—	1.4	10.2	3.0	49.6	34.3	—	1.5
Cottonseed	—	1.4	23.4	1.1	22.9	47.8	—	3.4
Linseed	—	—	6.3	2.5	19.0	24.1	47.4	0.7
Olive	—	—	6.9	2.3	84.4	4.6	—	1.8
Peanut	—	—	8.3	3.1	56.0	26.0	—	6.6
Safflower	—	—	6.8	—	18.6	70.1	3.4	1.1
Soybean	0.2	0.1	9.8	2.4	28.9	52.3	3.6	2.7
Sunflower	—	—	5.6	2.2	25.1	66.2	—	—

many double bonds, but rather because it is rich in low-molecular-weight fatty acids (chiefly lauric acid).

Oils with an average of more than one double bond per fatty acid chain are called *polyunsaturated*. Their role in the human diet is discussed in Section 21.4.

Pure fats and oils are colorless, odorless, and tasteless. This statement may seem surprising because we all know the tastes and colors of such fats and oils as butter and olive oil. The tastes, odors, and colors are caused by small amounts of other substances dissolved in the fat or oil.

■ **Many common products contain hydrogenated vegetable oils.**

B Hydrogenation

In Section 3.6D, we learned that we can reduce carbon–carbon double bonds to single bonds by treating them with hydrogen (H$_2$) and a catalyst. It is, therefore, not difficult to convert unsaturated liquid oils to solids. For example:

$$
\begin{array}{l}
\underset{}{CH_2-O-\overset{O}{\overset{\|}{C}}-(CH_2)_7-CH=CH-(CH_2)_7-CH_3} \quad \text{Oleic acid}\\[4pt]
\underset{}{CH-O-\overset{O}{\overset{\|}{C}}-(CH_2)_7-CH=CHCH_2CH=CH-(CH_2)_4-CH_3} \quad \text{Linoleic acid} \;+\; 5H_2 \xrightarrow{\ Pt\ }\\[4pt]
\underset{}{CH_2-O-\overset{}{C}-(CH_2)_7-CH=CHCH_2CH=CH-(CH_2)_4-CH_3}\\[2pt]
\quad\quad\quad\quad\;\; O \quad\quad\quad\quad\quad\quad\quad \text{Linoleic acid}
\end{array}
$$

$$
\begin{array}{l}
CH_2-O-\overset{O}{\overset{\|}{C}}-(CH_2)_{16}-CH_3 \quad \text{Stearic acid}\\[4pt]
CH-O-\overset{O}{\overset{\|}{C}}-(CH_2)_{16}-CH_3\\[4pt]
CH_2-O-\overset{}{C}-(CH_2)_{16}-CH_3\\[2pt]
\quad\quad\quad\;\; O
\end{array}
$$

This hydrogenation is carried out on a large scale to produce the solid shortening sold in stores under such brand names as Crisco, Spry, and Dexo. In making such products, manufacturers must be careful not to hydrogenate all the double bonds, because a fat with no double bonds would be *too* solid. Partial, but not complete, hydrogenation results in a product with just the right consistency for cooking.

Margarine is also made by partial hydrogenation of vegetable oils.

Because less hydrogen is used, margarine contains more unsaturation than hydrogenated shortenings.

C Saponification

Glycerides, being esters, are subject to hydrolysis, which can be carried out with either acids or bases. As we saw in Section 10.6, the use of bases is more practical. An example of the saponification of a typical fat is

$$
\underset{\text{A triglyceride}}{\begin{array}{l}
\quad\quad\;\; \overset{O}{\overset{\|}{}}\\[-2pt]
\overset{O}{\overset{\|}{R}}\;\; CH_2OCR\\[-2pt]
RCOCH \quad\; O\\[-2pt]
\quad\;\; CH_2OCR
\end{array}}
\;+\; 3NaOH \xrightarrow{\text{saponification}}
\underset{\substack{\text{1,2,3-Propanetriol}\\\text{(Glycerol; Glycerin)}}}{\begin{array}{l}
CH_2OH\\
CHOH\\
CH_2OH
\end{array}}
\;+\;
\underset{\text{Sodium soaps}}{3R\overset{O}{\overset{\|}{C}}O^-Na^+}
$$

Thus saponification is the base-catalyzed hydrolysis of fats and oils producing glycerol and a mixture of fatty acid salts called soaps. **Soap** has been used for thousands of years, and saponification is one of the oldest known chemical reactions (see Chemical Connections 12C).

CHEMICAL CONNECTIONS 12C

Soaps and Detergents

Soaps clean because each soap molecule has a hydrophilic head and a hydrophobic tail. The —COO⁻ end of the molecule (the hydrophilic head), being ionic, is soluble in water. The other end (the hydrophobic tail) is a long-chain alkane-like portion, which is insoluble in water but soluble in organic compounds. The soap is needed to remove the water-insoluble portion of dirt. The hydrophilic head of the soap molecule dissolves in the water; the hydrophobic tail dissolves in the dirt. This causes an emulsifying action in which the soap molecules surround the dirt particles: The hydrophobic tail interacts with the hydrophobic dirt particle and the hydrophilic head provides the attraction for water molecules (figure, part c). The cluster of soap molecules in which the dirt is embedded is called a **micelle** (figure, part d).

The solid soaps and soap flakes with which we are all familiar are sodium salts of fatty acids. Liquid soaps are generally potassium salts of the same acids. Soap has the disadvantage of forming precipitates in "hard water," which is water that contains relatively high amounts of Ca^{2+} and Mg^{2+} ions. Most community water supplies in the United States are hard water. Calcium and magnesium salts of fatty acids are insoluble in water. Therefore, when soaps are used in hard water, the Ca^{2+} and Mg^{2+} ions precipitate the fatty acid anions, which deposit on clothes, dishes, sinks, and bathtubs (forming the "ring" around the bathtub).

Because of these difficulties, much of the soap used in the United States has been replaced in the last 50 years by **detergents** for both household and industrial cleaning. Two typical detergent molecules are

$$CH_3(CH_2)_{11}—O—\overset{\displaystyle O}{\underset{\displaystyle O}{\overset{\|}{\underset{\|}{S}}}}—O^- \ Na^+$$

Sodium dodecyl sulfate
(Sodium lauryl sulfate)

$$CH_3(CH_2)_{10}CH_2—\langle\text{ring}\rangle—\overset{\displaystyle O}{\underset{\displaystyle O}{\overset{\|}{\underset{\|}{S}}}}—O^- \ Na^+$$

Sodium dodecylbenzenesulfonate

These molecules also have hydrophobic tails and hydrophilic heads and so can function in the same way as soap molecules. However, because they do not precipitate with Ca^{2+} and Mg^{2+}, they work very well in hard water.

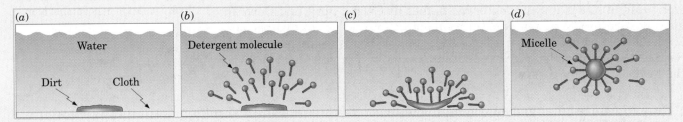

(a) Water — Dirt — Cloth   (b) Detergent molecule   (c)   (d) Micelle

■ The cleaning action of soap. (*a*) **Without soap, water molecules cannot penetrate through molecules of hydrophobic dirt. (*b* and *c*) Soap molecules line up at the interface between dirt and water. (*d*) Soap molecules carry dirt particles away.**

12.4 Complex Lipids

The triglycerides discussed in the previous sections are significant components of fat storage cells. Other kinds of lipids, called complex lipids, are important in a different way. They constitute the main components of membranes (Section 12.5). Complex lipids can be divided into two groups: phospholipids and glycolipids.

Phospholipids contain an alcohol, two fatty acids, and a phosphate group. There are two types: **glycerophospholipids** and **sphingolipids.** In glycerophospholipids, the alcohol is glycerol (Section 12.6). In sphingolipids, the alcohol is sphingosine (Section 12.7).

Glycolipids are complex lipids that contain carbohydrates (Section 12.8). Figure 12.1 shows schematic structures for all of these lipids.

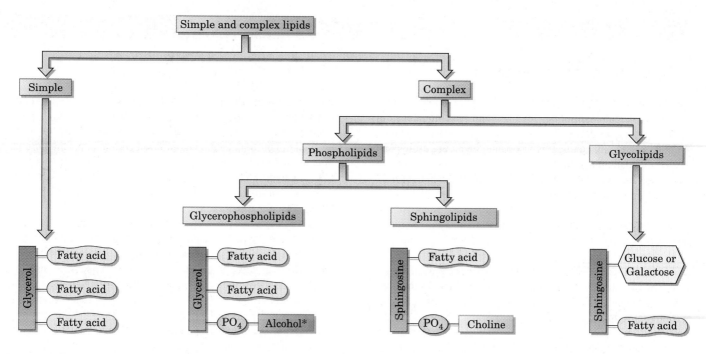

Figure 12.1 Schematic diagram of simple and complex lipids.

*The alcohol can be choline, serine, ethanolamine, inositol, or certain others.

12.5 Membranes

The complex lipids mentioned in Section 12.4 form the **membranes** around body cells and around small structures inside the cells. Unsaturated fatty acids are important components of these lipids. The cell membranes separate cells from the external environment and provide selective transport for nutrients and waste products into and out of cells.

These membranes are made of **lipid bilayers** (Figure 12.2). In a lipid bilayer, two rows (layers) of complex lipid molecules are arranged tail to tail. The hydrophobic tails point toward each other, which enables them to get as far away from the water as possible. This arrangement leaves the hydrophilic heads projecting to the inner and outer surfaces of the membrane. Cholesterol (Section 12.9), another membrane component, also positions the hydrophilic portion of the molecule on the surface of the membranes and the hydrophobic portion inside the bilayer.

The unsaturated fatty acids prevent the tight packing of the hydrophobic chains in the lipid bilayer, thereby providing a liquid-like character to the membranes. This property is of extreme importance because many products of the body's biochemical processes must cross the membrane, and the liquid nature of the lipid bilayer allows such transport.

The lipid part of the membrane serves as a barrier against any movement of ions or polar compounds into and out of the cells. In the lipid bilayer, protein molecules are either suspended on the surface (peripheral proteins) or partly or fully embedded in the bilayer (integral proteins). These proteins stick out either on the inside or on the outside of the membrane; others are thoroughly embedded, going through the bilayer and projecting from both sides. The model shown in Figure 12.2, called the **fluid mosaic model** of membranes, allows the passage of nonpolar compounds by diffusion, as these compounds are soluble in the lipid membranes. In

These small structures inside the cell are called *organelles*.

Most lipid molecules in the bilayer contain at least one unsaturated fatty acid.

This effect is similar to the one that causes unsaturated fatty acids to have lower melting points than saturated fatty acids.

The term *mosaic* refers to the topography of the bilayers: protein molecules dispersed in the lipid. The term *fluid* is used because the free lateral motion in the bilayers makes it liquid-like.

CHEMICAL CONNECTIONS 12D

Transport Across Cell Membranes

Membranes are not just random assemblies of complex lipids that provide a nondescript barrier. In human red blood cells, for example, the outer part of the bilayer is made largely of phosphatidylcholine and sphingomyelin, while the inner part consists mostly of phosphatidylethanolamine and phosphatidyl serine (Sections 12.6 and 12.7). In another example, in the membrane called sarcoplasmic reticulum in the heart muscles, phosphatidylethanolamine is found in the outer part of the membrane, phosphatidylserine is found in the inner part, and phosphatidyl choline is equally distributed in the two layers of the membrane.

Membranes are not static structures, either. In many processes, they fuse with one another; in others, they disintegrate and their building blocks are used elsewhere. When membranes fuse in vacuole fusions inside of cells, for example, certain restrictions prevent incompatible membranes from intermixing.

The protein molecules are not dispersed randomly in the bilayer. Sometimes they cluster in patches, but they also appear in regular geometric patterns. An example of the latter are **gap junctions,** channels made of six proteins that create a central pore. These channels allow neighboring cells to communicate. Gap junctions are an example of **passive transport.** Small polar molecules—which include such essential nutrients as inorganic ions, sugars, amino acids, and nucleotides—can readily pass through gap junctions. Large molecules, such as proteins, polysaccharides, and nucleic acids, cannot.

In **facilitated transport,** a specific interaction takes place between the transporter and the transported molecule. Consider the **anion transporter** of the red blood cells, through which chloride and bicarbonate ions are exchanged in a one-for-one ratio. The transporter is a protein with 14 helical structures that span the membrane. One side of the helices contains the hydrophobic parts of the protein, which can interact with the lipid membrane. The other side of the helices forms a channel. This channel contains the hydrophilic portions of the protein, which can interact with the hydrated ions. In this manner, the anion passes through the erythrocyte membrane.

Active transport involves the passage of ions through a concentration gradient. For example, a higher concentration of K^+ is found inside cells than outside in the surrounding environment. Nevertheless, potassium ions can be transported from the outside into a cell, but at the expense of energy. The transporter, a membrane protein called Na^+, K^+ ATPase, uses the energy from the hydrolysis of ATP molecules to change the conformation of the transporter, which brings in K^+ and exports Na^+. Detailed studies of K^+ ion channels have revealed that K^+ ions enter the channel in pairs. Each of these hydrated ions carries eight water molecules in its solvation layers, with the negative pole of the water molecules (the oxygen atoms) surrounding the positive ion. Deeper in the channel, K^+ ions encounter a constriction, called a selectivity filter. To pass through, the K^+ ions must shed their water molecules. The pairing of the naked K^+ ions generates enough electrostatic repulsion to keep the ions moving through the channel. The channel itself is lined with oxygen atoms to provide a similar attractive environment like that offered by the stable hydrated form before entry.

Polar compounds, in general, are transported through specific **transmembrane channels.**

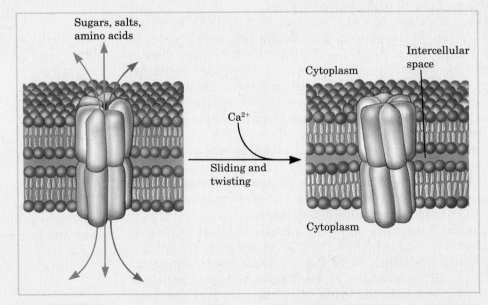

■ **Gap junctions are made of six cylindrical protein subunits. They line up in two plasma membranes parallel to each other, forming a pore. The pores of gap junctions are closed by a sliding and twisting motion of the cylindrical subunits.**

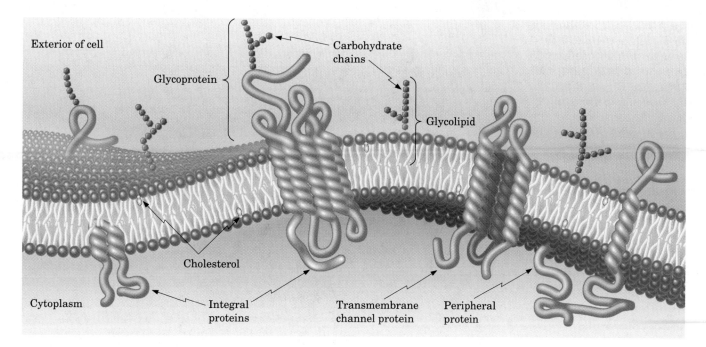

Figure 12.2 The fluid mosaic model of membranes. Note that proteins are embedded in the lipid matrix.

---

contrast, polar compounds are transported either through specific channels through the protein regions or by a mechanism called active transport (See Chemical Connections 12D). For any transport process, the membrane must behave like a nonrigid liquid so that the proteins can move sideways within the membrane.

See the **Interactive General, Organic, and Biochemistry CD-ROM, version 2.0,** for further exploration on this topic.

12.6 Glycerophospholipids

The structure of glycerophospholipids (also called phosphoglycerides) is very similar to that of fats. The alcohol is glycerol. Two of the three —OH groups are esterified by fatty acids. As with the simple fats, these fatty acids may be any long-chain carboxylic acids, with or without double bonds. The third —OH group is esterified not by a fatty acid, but rather by a phosphate group, which is also esterified to another alcohol. If the other alcohol is choline, a quaternary ammonium compound, the glycerophospholipids are called **phosphatidylcholines** (common name **lecithin**):

Glycerophospholipids are membrane components of cells throughout the body.

In all glycerophospholipids, lecithins, cephalins, and phosphatidylinositols, the fatty acid on carbon 2 of glycerol is always unsaturated.

This typical lecithin molecule has stearic acid on one end and linoleic acid in the middle. Other lecithin molecules contain other fatty acids, but the one on the end is always saturated and the one in the middle is always unsaturated. Lecithin is a major component of egg yolk. Having both polar and nonpolar portions within one molecule, it is an excellent emulsifier and is used in mayonnaise.

Figure 12.3 Space-filling molecular models of complex lipids in a bilayer. *(From* Biochemistry *by Lubert Stryer © 1975, 1981, 1988, 1995 by Lubert Stryer; © 2002 by W. H. Freeman and Company. Used with the permission of W. H. Freeman and Company.)*

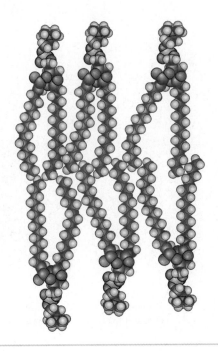

R = fatty acid portions.

Note that lecithin has a negatively charged phosphate group and a positively charged quaternary nitrogen from the choline. These charged parts of the molecule provide a strongly hydrophilic head, whereas the rest of the molecule is hydrophobic. Thus, when a phospholipid such as lecithin is part of a lipid bilayer, the hydrophobic tail points toward the middle of the bilayer, and the hydrophilic heads line both the inner and outer surfaces of the membranes (Figures 12.2 and 12.3).

Lecithins are just one example of glycerophospholipids. Another is the **cephalins,** which are similar to the lecithins in every way except that, instead of choline, they contain other alcohols, such as ethanolamine or serine:

$$
\begin{array}{ll}
CH_2-O-\overset{\displaystyle O}{\overset{\|}{C}}-R & HOCH_2CH_2NH_2 \\
& \text{Ethanolamine} \\
CH-O-\overset{\displaystyle O}{\overset{\|}{C}}-R' & \\
CH_2-O-\overset{\displaystyle O}{\underset{\underset{\displaystyle O^-}{\|}}{P}}-O-CH_2CH_2NH_3{}^+ &
\end{array}
$$

A phosphatidylethanolamine
(a cephalin)

$$
\begin{array}{ll}
CH_2-O-\overset{\displaystyle O}{\overset{\|}{C}}-R & HOCH_2CHCOO^- \\
& \qquad\quad NH_3{}^+ \\
& \qquad\text{Serine} \\
CH-O-\overset{\displaystyle O}{\overset{\|}{C}}-R' & \\
CH_2-O-\overset{\displaystyle O}{\underset{\underset{\displaystyle O^-}{\|}}{P}}-O-CH_2CHCOO^- \\
& \qquad\qquad\quad NH_3{}^+
\end{array}
$$

A phosphatidylserine
(a cephalin)

Another important group of glycerophospholipids is the **phosphatidylinositols (PI).** In PI, the alcohol inositol is linked to the rest of the molecule by a phosphate ester linkage. Such compounds not only are integral structural parts of the biological membranes, but also, in their higher phosphorylated form, such as **phosphatidylinositol 4,5-bisphosphates (PIP2),** serve as signaling molecules in chemical communication (see Chapter 15).

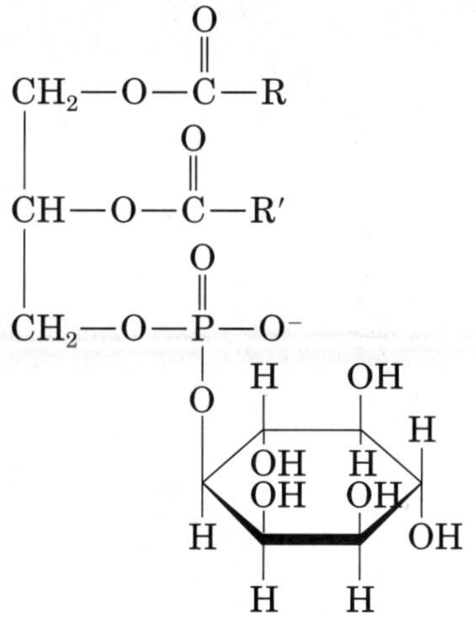

$$CH_2-O-\overset{\overset{O}{\|}}{C}-R$$

$$CH-O-\overset{\overset{O}{\|}}{C}-R'$$

$$CH_2-O-\overset{\overset{\|}{P}}{O}-O^-$$

Phosphatidylinositols, PI

12.7 Sphingolipids

Myelin, the coating of nerve axons, contains a different kind of complex lipid: **sphingolipids.** In sphingolipids, the alcohol portion is sphingosine:

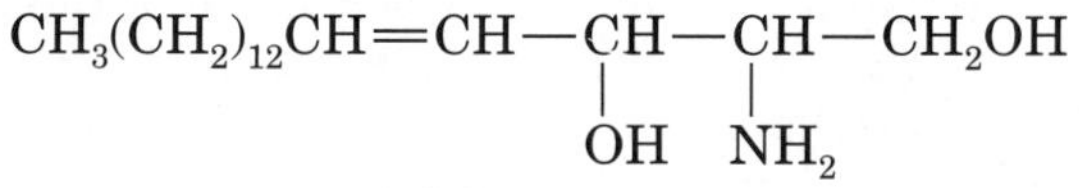

$$CH_3(CH_2)_{12}CH=CH-\underset{\underset{OH}{|}}{CH}-\underset{\underset{NH_2}{|}}{CH}-CH_2OH$$

Sphingosine

A long-chain fatty acid is connected to the —NH$_2$ group by an amide bond, and the —OH group at the end of the chain is esterified by phosphoryl-choline:

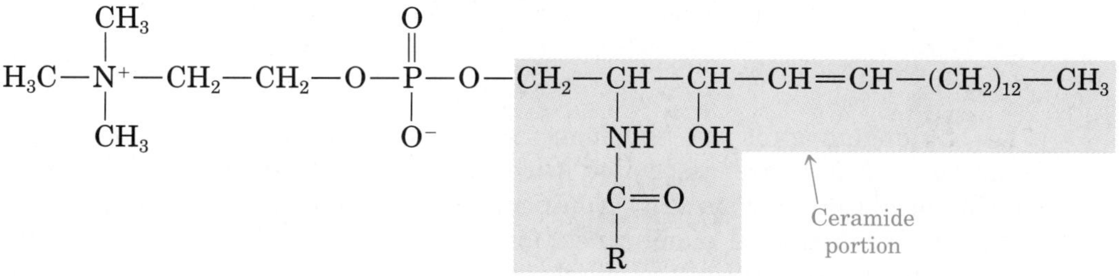

A sphingomyelin
(a sphingolipid)

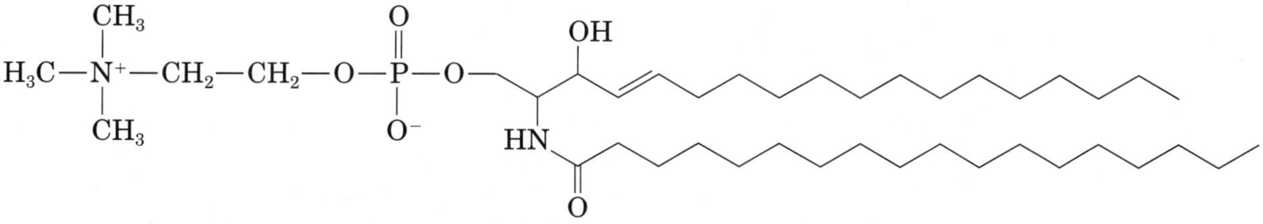

Ceramide portion

Sphingomyelin
(schematic diagram)

The combination of a fatty acid and sphingosine (shown in color) is called the **ceramide** portion of the molecule, because many of these compounds are also found in cerebrosides (Section 12.8). The ceramide part of complex lipids may contain different fatty acids; stearic acid, for example, occurs mainly in sphingomyelin.

Johann Thudichum, who discovered sphingolipids in 1874, named these brain lipids after the monster of Greek mythology, the sphinx. Part woman and part winged lion, the sphinx devoured all who could not provide the correct answer to her riddles. Sphingolipids appeared to Thudichum as part of a dangerous riddle of the brain.

CHEMICAL CONNECTIONS 12E

The Myelin Sheath and Multiple Sclerosis

The human brain and spinal cord can be divided into gray and white regions. Forty % of the human brain is white matter. Microscopic examination reveals that the white matter consists of nerve axons wrapped in a white lipid coating, called the **myelin sheath,** which provides insulation and thereby allows the rapid conduction of electrical signals. The myelin sheath consists of 70% lipids and 30% proteins in the usual lipid bilayer structure.

Specialized cells, called **Schwann cells,** wrap themselves around the peripheral nerve axons to form numerous concentric layers. In the brain, other cells do the wrapping in a similar manner.

Multiple sclerosis affects 250,000 people in the United States. In this disease, the myelin sheath gradually deteriorates. Symptoms, which include muscle weariness, lack of coordination, and loss of vision, may vanish for a time but later return with greater severity. Autopsy of multiple sclerotic brains shows scar-like plaques of white matter, with bare axons not covered by myelin sheaths. The symptoms occur because the demyelinated axons cannot conduct nerve impulses. A secondary effect of the demyelination is damage to the axon itself.

Similar demyelination occurs in the Guillain-Barré syndrome that follows certain viral infections. In 1976, fears of a "swine flu" epidemic prompted a vaccination program that precipitated a number of cases of Guillain-Barré syndrome. This disease can lead to a paralysis that can cause death unless artificial breathing is supplied. In the 1976 cases, the U.S. government assumed

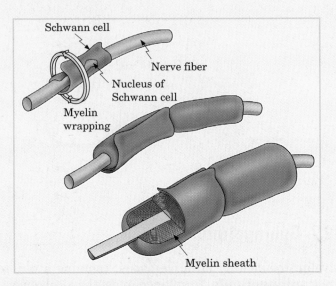

■ **Myelination of a nerve axon outside the brain by a Schwann cell. The myelin sheath is produced by the Schwann cell and is rolled around the nerve axon for insulation.**

legal responsibility for the few bad vaccines and paid compensation to the victims and their families.

Sphingomyelins are the most important lipids of myelin sheaths of nerve cells and are associated with diseases such as multiple sclerosis (Chemical Connections 12E). Sphingolipids are not randomly distributed in membranes. In viral membranes, for example, most of the sphingomyelin appears on the inside of the membrane.

12.8 Glycolipids

Glycolipids are complex lipids that contain carbohydrates and ceramides. One group, the **cerebrosides,** are ceramide mono- or oligosaccharides. Other groups, such as the **gangliosides,** contain a more complex carbohydrate structure (see Chemical Connections 12F). In cerebrosides, the fatty acid of the ceramide part may contain either 18-carbon or 24-carbon chains; the latter is found only in these complex lipids. A glucose or galactose carbohydrate unit forms a beta glycosidic bond with the ceramide portion of the molecule. The cerebrosides occur primarily in the brain (7% of the dry weight) and at nerve synapses.

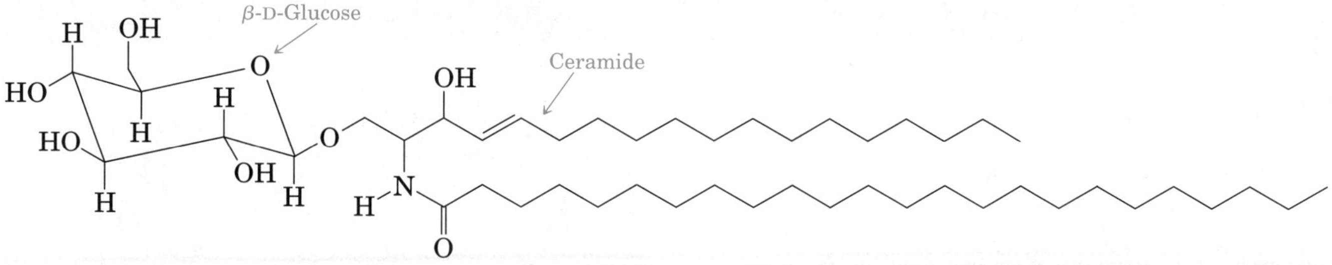

β-D-Glucose

Ceramide

Glucocerebroside

EXAMPLE 12.1

A lipid isolated from the membrane of red blood cells had the following structure:

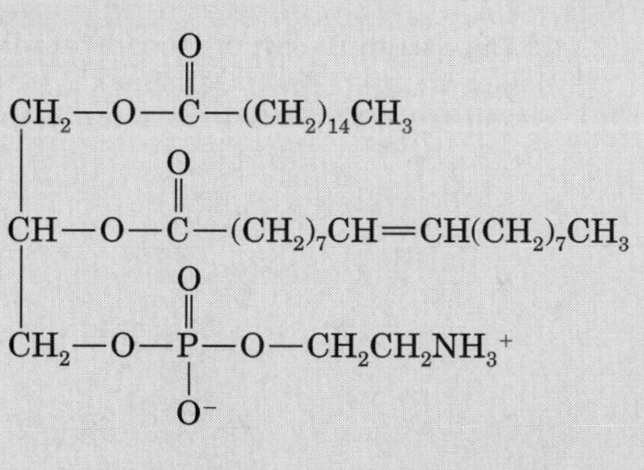

$$CH_2-O-\overset{\displaystyle O}{\overset{\|}{C}}-(CH_2)_{14}CH_3$$

$$CH-O-\overset{\displaystyle O}{\overset{\|}{C}}-(CH_2)_7CH=CH(CH_2)_7CH_3$$

$$CH_2-O-\overset{\displaystyle O}{\overset{\|}{P}}-O-CH_2CH_2NH_3{}^+$$
$$\overset{\displaystyle |}{\underset{O^-}{}}$$

(a) To what group of complex lipids does this compound belong?
(b) What are the components?

Solution

(a) The molecule is a triester of glycerol and contains a phosphate group; therefore, it is a glycerophospholipid.
(b) Besides glycerol and phosphate, it has a palmitic acid and an oleic acid component. The other alcohol is ethanolamine. Therefore, it belongs to the subgroup of cephalins.

Problem 12.1

A complex lipid had the following structure:

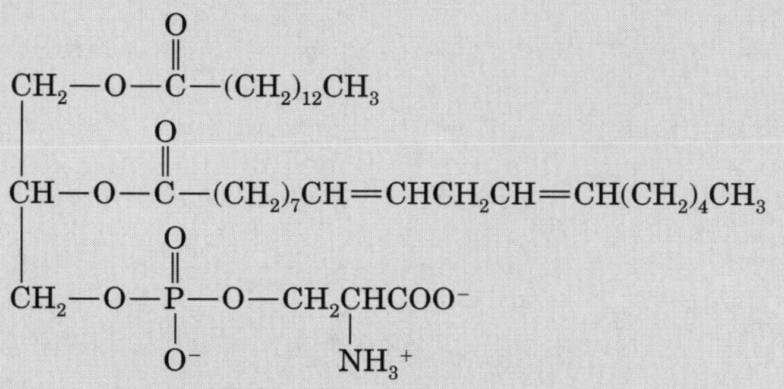

$$CH_2-O-\overset{\displaystyle O}{\overset{\|}{C}}-(CH_2)_{12}CH_3$$

$$CH-O-\overset{\displaystyle O}{\overset{\|}{C}}-(CH_2)_7CH=CHCH_2CH=CH(CH_2)_4CH_3$$

$$CH_2-O-\overset{\displaystyle O}{\overset{\|}{P}}-O-CH_2CHCOO^-$$
$$\overset{\displaystyle |}{\underset{O^-}{}}\qquad\overset{\displaystyle |}{\underset{NH_3{}^+}{}}$$

(a) To what group of complex lipids does this compound belong?
(b) What are the components?

CHEMICAL CONNECTIONS 12F

Lipid Storage Diseases

Complex lipids are constantly being synthesized and decomposed in the body. In several genetic diseases classified as lipid storage diseases, some of the enzymes needed to decompose the complex lipids are defective or missing. As a consequence, the complex lipids accumulate and cause enlarged liver and spleen, mental retardation, blindness, and, in certain cases, early death. Table 12F summarizes some of these diseases and indicates the missing enzyme and the accumulating complex lipid.

At present there is no treatment for these diseases. The best way to prevent them is by genetic counseling. Some of the diseases can be diagnosed during fetal development. For example, Tay-Sachs disease, which affects about 1 in every 30 Jewish Americans (versus 1 in 300 in the non-Jewish population), can be diagnosed from amniotic fluid obtained by amniocentesis.

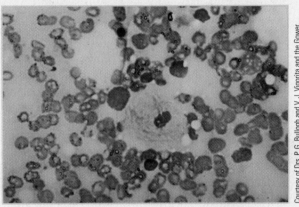

■ **The accumulation of glucocerebrosides in the cell of a patient with Gaucher's disease. These Gaucher cells infiltrate the bone marrow.**

Courtesy of Drs. P. G. Bullogh and V. J. Vigorita and the Gower Medical Publishing Co. NY

Table 12F   Lipid Storage Diseases

Name	Accumulating Lipid	Missing or Defective Enzyme
Gaucher's disease	Glucocerebroside	β-Glucosidase
Krabbe's leukodystrophy	Galactocerebroside	β-Galactosidase
Fabry's disease	Ceramide trihexoside	α-Galactosidase

CHEMICAL CONNECTIONS 12F

Lipid Storage Diseases, *(continued)*

Table 12F Lipid Storage Diseases, *(continued)*

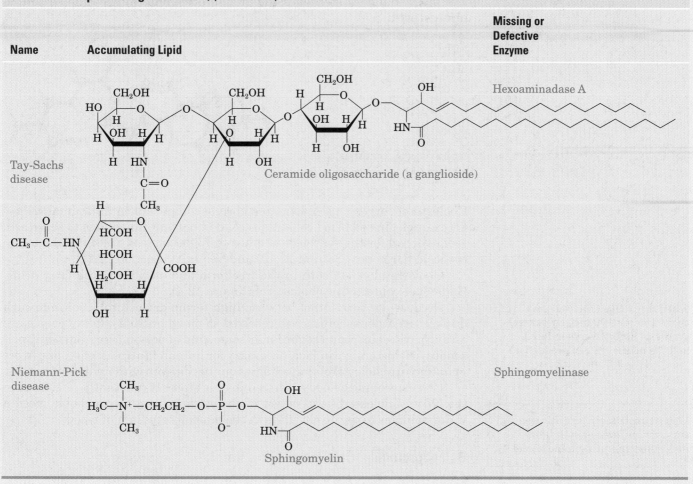

Name	Accumulating Lipid	Missing or Defective Enzyme
Tay-Sachs disease	Ceramide oligosaccharide (a ganglioside)	Hexoaminadase A
Niemann-Pick disease	Sphingomyelin	Sphingomyelinase

12.9 Steroids

The third major class of lipids is the **steroids,** which are compounds containing this ring system:

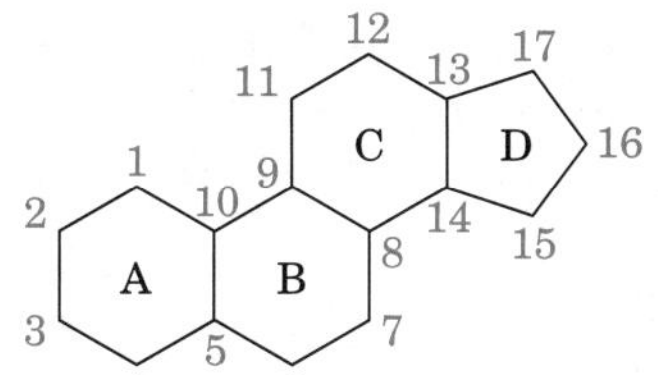

In this structure, three cyclohexane rings (A, B, and C) are connected in the same way as in phenanthrene (Section 4.3D); a fused cyclopentane ring (D)

is also present. Steroids are thus completely different in structure from the lipids already discussed. Note that they are not necessarily esters, although some of them are.

A Cholesterol

The most abundant steroid in the human body, and the most important, is **cholesterol:**

Cholesterol

Cholesterol serves as a plasma membrane component in all animal cells—for example, in red blood cells. Its second important function is to serve as a raw material for the synthesis of other steroids, such as the sex and adrenocorticoid hormones (Section 12.10) and bile salts (Section 12.11).

Cholesterol exists both in the free form and esterified with fatty acids. Gallstones contain free cholesterol (Figure 12.4).

Because the correlation between high serum cholesterol levels and such diseases as atherosclerosis has received so much publicity, many people are afraid of cholesterol and regard it as some kind of poison. Far from being poisonous, cholesterol is, in fact, necessary for human life. In essence, our livers manufacture cholesterol that satisfies our needs even without dietary intake.

Cholesterol in the body is in a dynamic state. It constantly circulates in the blood. Cholesterol and esters of cholesterol, being hydrophobic, need a water-soluble carrier to circulate in the aqueous medium of blood.

When the cholesterol level goes above 150 mg/100 mL, cholesterol synthesis in the liver is reduced to half the normal rate of production.

Lipoproteins Spherically shaped clusters containing both lipid molecules and protein molecules

B Lipoproteins—Carriers of Cholesterol

Cholesterol, along with fat, is transported by **lipoproteins.** Most lipoproteins contain a core of hydrophobic lipid molecules surrounded by a shell of hydrophilic molecules such as proteins and phospholipids (Figure 12.5). As summarized in Table 12.3, there are four kinds of lipoproteins:

● **High-density lipoprotein (HDL),** which has a protein content of about 33% and about 30% cholesterol

● **Low-density lipoprotein (LDL),** in which the protein content is only about 25%, but which contains about 50% cholesterol

● **Very-low-density lipoprotein (VLDL),** which mostly carries triglycerides (fats) synthesized by the liver

● **Chylomicrons,** which carry dietary lipids synthesized in the intestines.

C Transport of Cholesterol in LDL

The transport of cholesterol from the liver starts out as a large VLDL particle (55 nanometers in diameter). The core of VLDL contains triglycerides and cholesteryl esters, mainly cholesteryl linoleate. It is surrounded by a polar coat of phospholipids and proteins (Figure 12.5). The VLDL is carried

Figure 12.4 A human gallstone is almost pure cholesterol. This gallstone measures 5 mm in diameter.

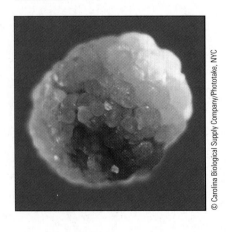

© Carolina Biological Supply Company/Phototake, NYC

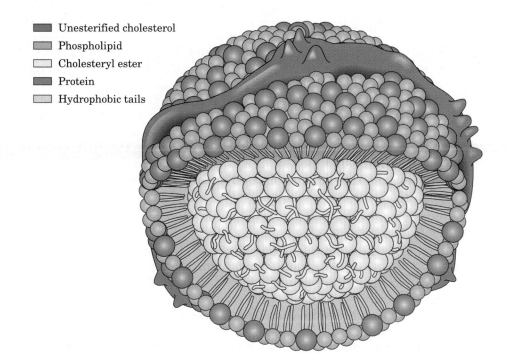

Unesterified cholesterol
Phospholipid
Cholesteryl ester
Protein
Hydrophobic tails

Figure 12.5 Low-density lipoprotein shown schematically.

in the serum. When the capillaries reach muscle or fat tissues, the triglycerides and all proteins except apoB-100 are removed from the VLDL. At this point, the diameter of the lipoprotein shrinks to 22 nanometers and its core contains only cholesteryl esters. Because of the removal of fat, its density increases and it becomes LDL. Low-density lipoprotein stays in the plasma for about 2.5 days.

The LDL carries cholesterol to the cells, where specific LDL-receptor molecules line the cell surface in certain concentrated areas called **coated pits.** The apoB-100 protein on the surface of the LDL binds specifically to the LDL-receptor molecules in the coated pits. After such binding, the LDL is taken inside the cell (endocytosis), where enzymes break down the lipoprotein. In the process, they liberate free cholesterol from cholesteryl esters. In this manner, the cell can, for example, use cholesterol as a component of a membrane. This is the normal fate of LDL and the normal course of cholesterol transport.

Michael Brown and Joseph Goldstein of the University of Texas shared the Nobel Prize in medicine in 1986 for the discovery of the LDL-receptor–mediated pathway.

Table 12.3   Compositions and Properties of Human Lipoproteins

Property	HDL	LDL	VLDL	Chylomicrons
Core				
Cholesterol and Cholesteryl esters (%)	30	50	22	8
Triglycerides (%)	8	4	50	84
Surface				
Phospholipids (%)	29	21	18	7
Proteins (%)	33	25	10	1–2
Density (g/mL)	1.05–1.21	1.02–1.06	0.95–1.00	<0.95
Diameter (nm)	5–15	18–28	30–80	100–500

Percentages are given as % dry weight.

D Transport of Cholesterol in HDL

High-density lipoprotein transports cholesterol from peripheral tissues to the liver and transfers cholesterol to LDL. While in the serum, the free cholesterols in HDL are converted to cholesteryl esters. These esterified cholesterols are delivered to the liver for synthesis of bile acids and steroid hormones. The cholesterol uptake from HDL differs from that noted with LDL. This process does not involve endocytosis and degradation of the lipoprotein particle. Instead, in a selective lipid uptake, the HDL binds to the liver cell surface and transfers its cholesteryl ester to the cell. After that the HDL, depleted from its lipid content, re-enters the circulation.

E Levels of LDL and HDL

Like all lipids, cholesterol is insoluble in water. If its level is elevated in the blood serum, plaque-like deposits may form on the inner surfaces of the arteries. The resulting decrease in the diameter of the blood vessels may lead to a decrease in the flow of blood. This **atherosclerosis,** along with accompanying high blood pressure, may lead to heart attack, stroke, or kidney dysfunction.

Atherosclerosis may exacerbate the blockage of some arteries by a clot at the point where the arteries are constricted by plaque. Furthermore, blockage may deprive cells of oxygen, causing them to cease to function. The death of heart muscles due to lack of oxygen is called *myocardial infarction.*

Most cholesterol is transported by low-density lipoproteins. If a sufficient number of LDL receptors are found on the surface of the cells, LDL is effectively removed from the circulation and its concentration in the blood plasma drops. The number of LDL receptors on the surface of cells is controlled by a feedback mechanism (see Section 14.6). That is, when the concentration of cholesterol molecules inside the cells is high, the synthesis of the LDL receptor is suppressed. Thus less LDL is taken into the cells from the plasma and the LDL concentration in the plasma rises. On the other hand, when the cholesterol level inside the cells is low, the synthesis of the LDL receptor increases. As a consequence, the LDL is taken up more rapidly and its level in the plasma falls.

In certain cases, however, there are not enough LDL receptors. In the disease called *familial hypercholesterolemia,* the cholesterol level in the plasma may be as high as 680 mg/100 mL, as opposed to 175 mg/100 mL in normal subjects. These high levels of cholesterol can lead to premature atherosclerosis and heart attack. The high plasma cholesterol levels in affected patients are caused because the body lacks enough functional LDL receptors or, if enough are produced, they are not concentrated in the coated pits.

In general, high LDL content means high cholesterol content in the plasma because LDL cannot enter the cells and be metabolized. Therefore, a high LDL level combined with a low HDL level is a symptom of faulty cholesterol transport and a warning for possible atherosclerosis.

The serum cholesterol level controls the amount of cholesterol synthesized by the liver. When serum cholesterol is high, the synthesis is at a low level. Conversely, when the serum cholesterol level is low, the synthesis of cholesterol increases.

Diets low in cholesterol and saturated fatty acids usually reduce the serum cholesterol level, and a number of drugs can inhibit the synthesis of cholesterol in the liver. The commonly used statin drugs, such as atorvastatin (Lipicor) or Simvastatin (Zocor) inhibit one of the key enzymes (biocatalyst) in cholesterol synthesis, HMG-CoA reductase (Section 20.4). In this way, they block the synthesis of cholesterol inside the cells and stimulate the synthesis of LDL-receptor proteins. More LDL then enters the cells, diminishing the amount of cholesterol that will be deposited on the inner walls of arteries.

It is generally considered desirable to have high levels of HDL and low levels of LDL in the bloodstream. HDL carries cholesterol from plaques deposited in the arteries to the liver, thereby reducing the risk of atherosclerosis. Premenopausal women have more HDL than men, which is why women have a lower risk of coronary heart disease. HDL levels can be increased by exercise and weight loss.

12.10 Steroid Hormones

Cholesterol is the starting material for the synthesis of steroid hormones. In this process, the aliphatic side chain on the D ring is shortened by the removal of a six-carbon unit, and the secondary alcohol group on carbon 3 is oxidized to a ketone. The resulting molecule, *progesterone,* serves as the starting compound for both the sex hormones and the adrenocorticoid hormones (Figure 12.6).

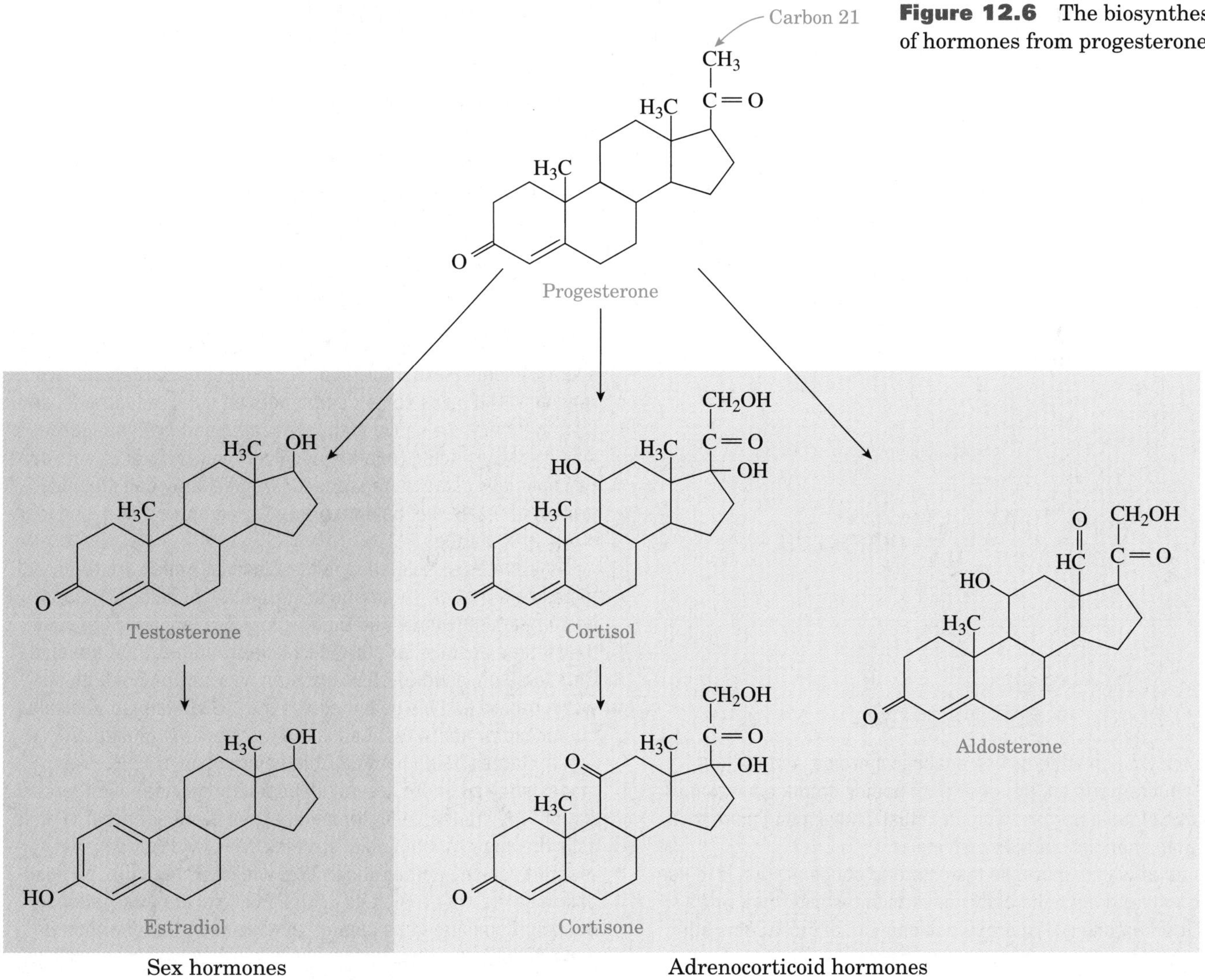

Figure 12.6 The biosynthesis of hormones from progesterone.

Sex hormones

Adrenocorticoid hormones

Anabolic Steroids

Testosterone, the principal male hormone, is responsible for the buildup of muscles in men. Recognizing this fact, many athletes have taken this drug in an effort to increase their muscular development. This practice is especially common among athletes in sports in which strength and muscle mass are important, including weight lifting, shot put, and hammer throw. Participants in other sports, such as running, swimming, and cycling, would also like larger and stronger muscles.

Although used by many athletes, testosterone has two disadvantages:

1. Besides its effect on muscles, it affects secondary sexual characteristics, and too much of it can result in undesired side effects.
2. It is not very effective when taken orally and must be injected to achieve the best results.

For these reasons, a large number of other anabolic steroids, all of them synthetic, have been developed. Examples include the following compounds:

Methandienone

Methenolone

Nandrolone decanoate

Jed Jacobson/AllSport USA/Getty Images

■ **Home run champion Mark McGwire used androstenedione, a muscle-building dietary supplement that is allowed in baseball, but is banned in professional football, college athletics, and the Olympic Games.**

4-Androstene-3,17-dione

Some women athletes use anabolic steroids, just as their male counterparts do. Because their bodies produce only small amounts of testosterone, women actually have much more to gain from anabolic steroids than men.

Another way to increase testosterone concentration is to use prohormones, which the body converts to testosterone. One such prohormone is 4-androstenedione, or "andro." Some athletes have used it to enhance performance.

The use of anabolic steroids is forbidden in many sporting events, especially in international competition, largely for two reasons: (1) It gives some competitors an unfair advantage, and (2) these drugs can have many unwanted and even dangerous side effects, ranging from acne to liver tumors. Side effects can be especially disadvantageous for women; they can include growth of facial hair, baldness, deepening of the voice, and menstrual irregularities.

All athletes participating in the Olympic Games are required to pass a urine test for anabolic steroids. A number of medal-winning athletes have had their victories taken away because they tested positive for steroid use. For example, the Canadian Ben Johnson, a world-class sprinter, was stripped of both his world record and his gold medal in the 1988 Olympiad. A similar failed test for andro resulted in the U.S. shot put champion Randy Barnes being banned from competition in 1998. Prohormones such as andro are not listed under the Anabolic Steroid Act of 1990; hence, their nonmedical use is not a federal offense, as is the case with anabolic steroids. Mark McGwire hit his record-breaking home runs in 1998 while taking andro, as baseball rules did not prohibit its usage. Even so, the International Olympic Committee has banned the use of both prohormones and anabolic steroids.

A Adrenocorticoid Hormones

The adrenocorticoid hormones (Figure 12.6) are products of the adrenal glands. We divide them into two groups according to function: *Mineralocorticoids* regulate the concentrations of ions (mainly Na$^+$ and K$^+$), and *glucocorticoids* control carbohydrate metabolism.

Aldosterone is one of the most important mineralocorticoids. Increased secretion of aldosterone enhances the reabsorption of Na$^+$ and Cl$^-$ ions in the kidney tubules and increases the loss of K$^+$. Because Na$^+$ concentration controls water retention in the tissues, aldosterone also controls tissue swelling.

Cortisol is the major glucocorticoid. Its function is to increase the glucose and glycogen concentrations in the body. This is done at the expense of other nutrients. Fatty acids from fat storage cells and amino acids from body proteins are transported to the liver, which, under the influence of cortisol, manufactures glucose and glycogen from these sources.

Cortisol and its ketone derivative, *cortisone,* have remarkable anti-inflammatory effects. These or similar synthetic derivatives, such as prednisolone, are used to treat inflammatory diseases of many organs, rheumatoid arthritis, and bronchial asthma.

The term *adrenal* means "adjacent to the renal" (which refers to the kidney).

The term *corticoid* indicates that the site of the secretion is the cortex (outer part) of the gland.

B Sex Hormones

The most important male sex hormone is testosterone (Figure 12.6). This hormone, which promotes the normal growth of the male genital organs, is synthesized in the testes from cholesterol. During puberty, increased testosterone production leads to such secondary male sexual characteristics as deep voice and facial and body hair.

Female sex hormones, the most important of which is estradiol (Figure 12.6), are synthesized from the corresponding male hormone (testosterone) by aromatization of the A ring:

Testosterone
(Partial Structure)

Estradiol
(Partial Structure)

Estradiol, together with its precursor progesterone, regulates the cyclic changes occurring in the uterus and ovaries known as the *menstrual cycle.* As the cycle begins, the level of estradiol in the body rises, which in turn causes the lining of the uterus to thicken. Then another hormone, called luteinizing hormone (LH), triggers ovulation. If the ovum is fertilized, increased progesterone levels will inhibit any further ovulation. Both estradiol and progesterone then promote further preparation of the uterine lining to receive the fertilized ovum. If no fertilization takes place, progesterone production stops altogether, and estradiol production decreases. This halt decreases the thickening of the uterine lining, which is then sloughed off with accompanying bleeding during menstruation (Figure 12.7).

Because progesterone is essential for the implantation of the fertilized ovum, blocking its action leads to termination of pregnancy (see Chemical Connections 12H). Progesterone interacts with a receptor (a protein molecule) in the nucleus of cells. The receptor changes its shape when progesterone binds to it (see Section 15.7).

Figure 12.7 Events of the menstrual cycle. (*a*) Levels of sex hormones in the bloodstream during the phases of one menstrual cycle in which pregnancy does not occur. (*b*) Development of an ovarian follicle during the cycle. (*c*) Phases of development of the endometrium, the lining of the uterus. The endometrium thickens during the proliferative phase. In the secretory phase, which follows ovulation, the endometrium continues to thicken and the glands secrete a glycogen-rich nutritive material in preparation to receive an embryo. When no embryo is implanted, the new outer layers of the endometrium disintegrate and the blood vessels rupture, producing the menstrual flow.

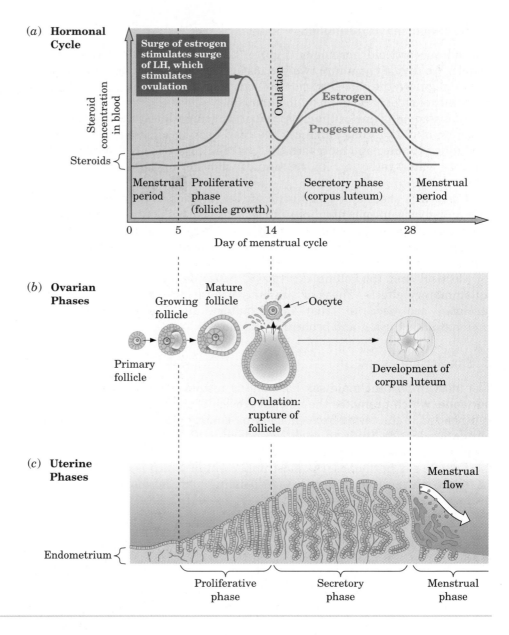

RU486 binds to the receptors of glucocorticoid hormones as well. Its use as an antiglucocorticoid is also recommended to alleviate a disease known as Cushing syndrome, involving the overproduction of cortisone.

A drug, now widely used in France and China, called mifepristone or RU486 acts as a competitor to progesterone:

Mifepristone
(RU486)

CHEMICAL CONNECTIONS 12H

Oral Contraception

Because progesterone prevents ovulation during pregnancy, it occurred to investigators that progesterone-like compounds might be used for birth control. Synthetic analogs of progesterone proved to be more effective than the natural compound. In "the Pill," a synthetic progesterone-like compound is supplied together with an estradiol-like compound (the latter prevents irregular menstrual flow). Triple-bond derivatives of testosterone, such as norethindrone, norethynodrel, and ethynodiol diacetate, are used most often in birth-control pills:

Norethindone

Norethynodrel

Ethynodiol diacetate

Mifepristone binds to the same receptor site as progesterone. Because the progesterone molecule cannot reach the receptor molecule, the uterus is not prepared for the implantation of the fertilized ovum, and the ovum is aborted. Once pregnancy has been established, RU486 can be taken up through 49 days of gestation. This chemical form of abortion has been approved by the Food and Drug Administration and in recent years has been gaining clinical use as a supplement to surgical abortion.

A completely different approach is the "morning after pill" that can be taken orally up to 72 hours after an unprotected intercourse. The "morning after pill" is not an abortion pill because it acts before pregnancy takes place. Actually, the components of the pill are regular contraceptives. Two kinds are on the market as prescription drugs: One is a progesterone-like compound, levonogestrel, the other is a combination of levonogestrel and ethynil estradiol marketed as Preven.

Estradiol and progesterone also regulate secondary female sex characteristics, such as the growth of breasts. Thanks to this property, RU486, as an anti-progesterone, has also been reported to be effective against certain types of breast cancer.

Testosterone and estradiol are not exclusive to either males or females. A small amount of estradiol production occurs in males, and a small amount of testosterone production is normal in females. Only when the proportion of these two hormones (hormonal balance) becomes upset can one observe symptoms of abnormal sexual differentiation.

12.11 Bile Salts

Bile salts are oxidation products of cholesterol. First the cholesterol is converted to the trihydroxy derivative, and the end of the aliphatic chain is oxidized to the carboxylic acid. The latter, in turn, forms an amide linkage with an amino acid, either glycine or taurine:

The dispersion of dietary lipids by bile salts is similar to the action of soap on dirt.

Glycine portion

Taurine portion

Glycocholate

Taurocholate

Bile salts are powerful detergents. One end of the molecule is strongly hydrophilic because of the negative charge, and the rest of the molecule is largely hydrophobic. As a consequence, bile salts can disperse dietary lipids in the small intestine into fine emulsions and thereby help digestion.

Because they are eliminated in the feces, bile salts also remove excess cholesterol in two ways: (1) They are themselves breakdown products of cholesterol (so cholesterol is eliminated via bile salts), and (2) they solubilize deposited cholesterol in the form of bile salt–cholesterol particles.

12.12 Prostaglandins, Thromboxanes, and Leukotrienes

Ulf S. von Euler won the Nobel Prize in physiology and medicine in 1970.

Prostaglandins, a group of fatty-acid-like substances, were discovered by Kurzrok and Leib in the 1930s, when they demonstrated that seminal fluid caused a hysterectomized uterus to contract. Ulf von Euler of Sweden iso-

C H E M I C A L C O N N E C T I O N S 1 2 I

Action of Anti-inflammatory Drugs

Anti-inflammatory steroids (such as cortisone; Section 12.10) exert their function by inhibiting phospholipase A_2, the enzyme that releases unsaturated fatty acids from complex lipids in the membranes. For example, arachidonic acid, one of the components of membranes, is made available to the cell through this process. Because arachidonic acid is the precursor of prostaglandins, thromboxanes, and leukotrienes, inhibiting its release stops the synthesis of these compounds and prevents inflammation.

Steroids such as cortisone have many undesirable side effects (duodenal ulcer and cataract formation, among others). Therefore, their use must be controlled. A variety of nonsteroidal anti-inflammatory agents, including aspirin, ibuprofen, ketoprofen, and indomethacin, are available to serve this function.

Aspirin and other NSAIDs (see Chemical Connections 10C) inhibit the cyclooxygenase enzymes, which synthesize prostaglandins and thromboxanes. Aspirin (acetyl salicylic acid), for example, acetylates the enzymes, thereby blocking the entrance of arachidonic acid to the active site. This inhibition of both COX-1 and COX-2 explains why aspirin and the other anti-inflammatory agents have undesirable side effects. NSAIDs also interfere with the COX-1 isoform of the enzyme, which is needed

for normal physiological function. Their side effects include stomach and duodenal ulceration and renal (kidney) toxicity.

Obviously it would be desirable to have an anti-inflammatory agent without such side effects, and one that inhibits only the COX-2 isoform. To date, the Food and Drug Administration (FDA) has approved two COX-2 inhibitor drugs: Celebrex, which quickly became the most frequently prescribed drug, and VIOXX, a more recent introduction. Despite their selective inhibition of COX-2, however, these drugs also have ulcer-causing side effects. Many other COX-2 inhibitors are in the clinical trial stage.

The use of COX-2 inhibitors is not limited to rheumatoid arthritis and osteoarthiritis. Celebrex has been approved by FDA to treat a type of colon cancer, familial adenomateous polyposis, in an approach called chemoprevention. All anti-inflammatory agents reduce pain and relieve fever and swelling by reducing the prostaglandin production; however, they do not affect the leukotriene production. Thus, asthmatic patients must beware of using these anti-inflammatory agents. Even though they inhibit the prostaglandin synthesis, they may shift the available arachidonic acid to leukotriene production, which could precipitate a severe asthma reaction.

lated these compounds from human semen and, thinking that they had come from the prostate gland, named them **prostaglandin.** Even though the seminal gland secretes 0.1 mg of prostaglandin per day in mature males, small amounts of prostaglandins are present throughout the body in both sexes.

Prostaglandins are synthesized in the body from arachidonic acid by a ring closure at carbons 8 and 12. The enzyme catalyzing this reaction is called **cyclooxygenase** (COX, for short). The product, known as PGG_2, is the common precursor of other prostaglandins, including PGE and PGF. The prostaglandin E group (PGE) has a carbonyl group at carbon 9; the subscript indicates the number of double bonds in the hydrocarbon chain. The prostaglandin F group (PGF) has two hydroxyl groups on the ring at carbons 9 and 11.

Other prostaglandins (PGAs and PGBs) are derived from PGE.

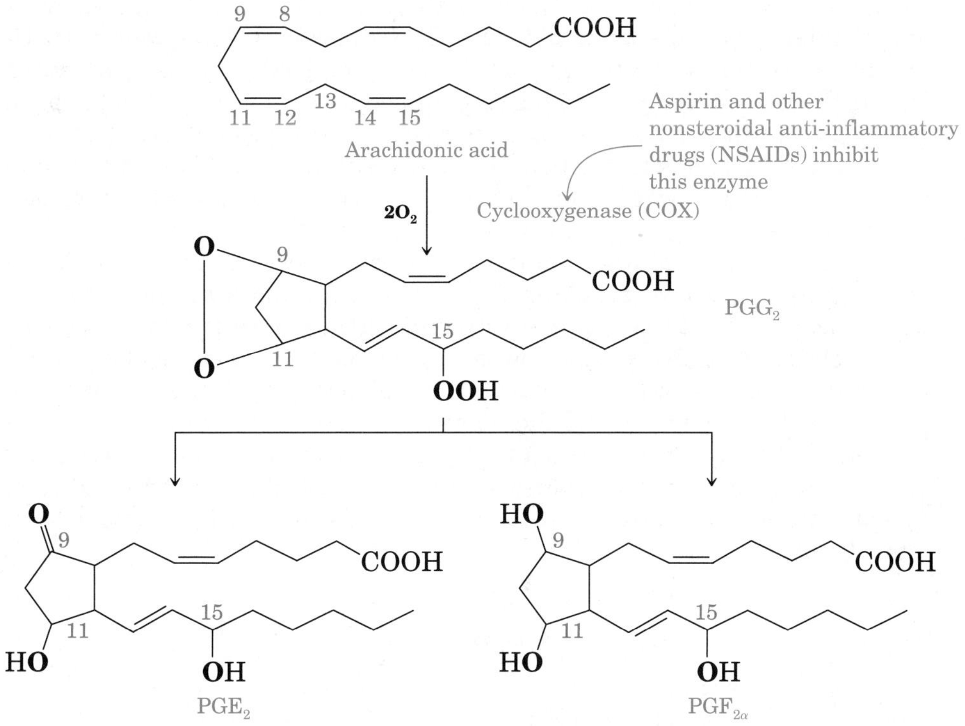

The COX enzyme comes in two forms in the body: COX-1 and COX-2. COX-1 catalyzes the normal physiological production of prostaglandins, which are always present in the body. For example, PGE_2 and $PGF_{2\alpha}$ stimulate uterine contractions and induce labor. PGE_2 lowers blood pressure by relaxing the muscles around blood vessels. In aerosol form, this prostaglandin is used to treat asthma; it opens up the bronchial tubes by relaxing the surrounding muscles. PGE_1 is used as a decongestant; it opens up nasal passages by constricting blood vessels.

COX-2, on the other hand, is responsible for the production of prostaglandins in inflammation. When a tissue is injured or damaged, special inflammatory cells invade the injured tissue and interact with resident cells—for example, smooth muscle cells. This interaction activates the COX-2 enzyme, and prostaglandins are synthesized. Such tissue injury may occur in heart attack (myocardial infarction), rheumatic arthritis, and ulcerative colitis. Nonsteroidal anti-inflammatory drugs (NSAIDs), such as aspirin, inhibit both COX enzymes (see Chemical Connections 12I).

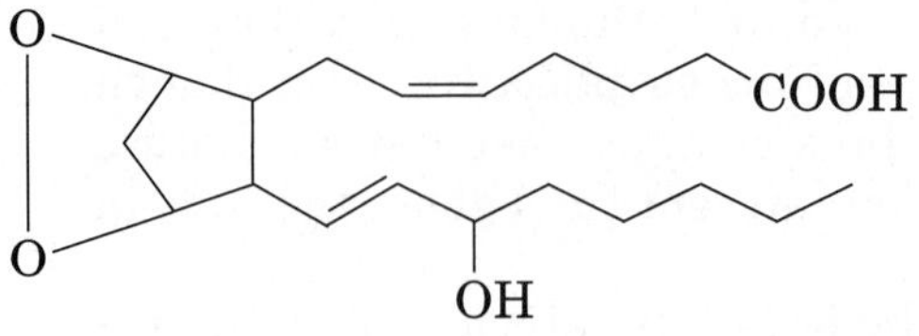

PGH₂

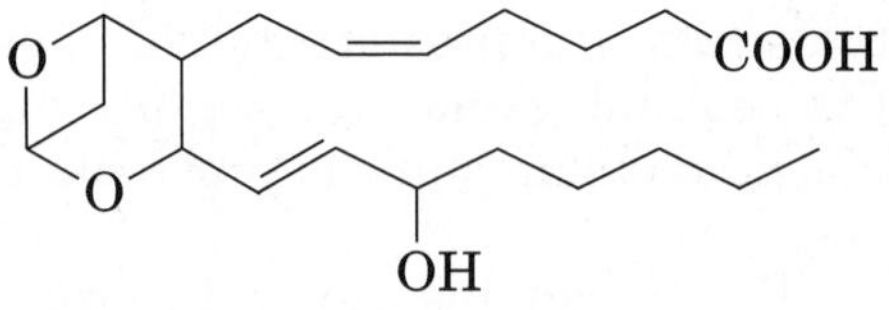

Thromboxane A₂

Another class of arachidonic acid derivatives is the **thromboxanes.** Their synthesis also includes a ring closure. These substances are derived from PGH₂, but their ring is a cyclic acetal. Thromboxane A₂ is known to induce platelet aggregation. When a blood vessel is ruptured, the first line of defense is the platelets circulating in the blood, which form an incipient clot (see Chemical Connections 23B). Thromboxane A₂ causes other platelets to clump, thereby increasing the size of the blood clot. Aspirin and similar anti-inflammatory agents inhibit the COX enzyme. Consequently, PGH₂ and thromboxane A₂ synthesis is inhibited, and blood clotting is impaired. This is the reason that physicians recommend a daily dose of 81 mg aspirin for people at risk for heart attack or stroke. On the other hand, it also explains why physicians forbid patients to use aspirin and other anti-inflammatory agents for a week before a planned surgery because aspirin causes excessive bleeding.

A variety of NSAIDs inhibit COX enzymes. Ibuprofen and indomethacin, both powerful NSAID painkillers, can block the inhibitory effect of aspirin and thus eliminate its anticlotting benefits. Therefore, the use of these NSAIDs together with aspirin is not recommended. Other painkillers, such as acetaminophen and diclofenac, do not interfere with aspirin's anticlotting ability and therefore can be taken together.

Another group of substances that acts to mediate hormonal responses are the **leukotrienes.** Like prostaglandins, leukotrienes are derived from arachidonic acid by an oxidative mechanism. However, in this case, there is no ring closure.

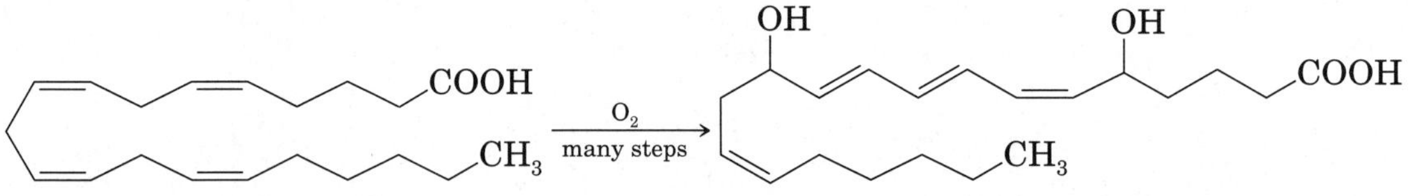

Arachidonic acid   Leukotriene B4

Leukotrienes occur mainly in white blood cells (leukocytes) but are also found in other tissues of the body. They produce long-lasting muscle contractions, especially in the lungs, and can cause asthma-like attacks. In fact, they are 100 times more potent than histamines. Both prostaglandins and leukotrienes cause inflammation and fever, so the inhibition of their production in the body is a major pharmacological concern. One way to counteract the effects of leukotrienes is to inhibit their uptake by leukotriene receptors (LTRs) in the body. A new antagonist of LTRs, zafirlukast, marketed under the brand name Accolate, is used to treat and control chronic asthma. Another antiasthmatic drug, zileuton, inhibits 5-lipoxygenase, which is the initial enzyme in leukotriene biosynthesis from arachidonic acid.

SUMMARY

Lipids are water-insoluble substances (Section 12.1). They are classified into four groups: fats (triglycerides); complex lipids; steroids; and prostaglandins, thromboxanes, and leukotrienes. **Fats** consist of fatty acids and glycerol (Section 12.2). In saturated fatty acids, the hydrocarbon chains have only single bonds; unsaturated fatty acids have hydrocarbon chains with one or more double bonds, all in the cis configuration. Solid fats contain mostly saturated fatty acids, whereas **oils** contain substantial amounts of unsaturated fatty acids (Section 12.3). The alkali salts of fatty acids are called **soaps.**

Complex lipids can be divided into two groups: phospholipids and glycolipids (Section 12.4). **Phospholipids** are made of a central alcohol (glycerol or sphingosine), fatty acids, and a nitrogen-containing phosphate ester, such as phosphorylcholine or inositol phosphate (Section 12.6). **Glycolipids** contain sphingosine and a fatty acid, collectively known as the ceramide portion of the molecule, and a carbohydrate portion (Section 12.8).

Many phospholipids and glycolipids are important constituents of cell **membranes** (Section 12.5). Membranes are made of a **lipid bilayer** in which the hydrophobic parts of phospholipids (fatty acid residues) point toward the middle of the bilayer, and the hydrophilic parts point toward the inner and outer surfaces of the membrane (Section 12.5).

The third major group of lipids comprises the **steroids** (Section 12.9). The characteristic feature of the steroid structure is a fused four-ring nucleus. The most common steroid, **cholesterol,** serves as a start-ing material for the synthesis of other steroids, such as bile salts and sex and other hormones (Section 12.9A). Cholesterol is also an integral part of membranes, occupying the hydrophobic region of the lipid bilayer. Because of its low solubility in water, cholesterol deposits are implicated in the formation of gallstones and the plaque-like deposits of atherosclerosis.

Cholesterol is transported in the blood plasma mainly by two kinds of lipoproteins: **HDL** and **LDL** (Section 12.9B). LDL delivers cholesterol to the cells to be used mostly as a membrane component. HDL delivers cholesteryl esters mainly to the liver to be used in the synthesis of bile acids and steroid hormones (Section 12.9C). High levels of LDL and low levels of HDL are symptoms of faulty cholesterol transport, indicating greater risk of atherosclerosis (Section 12.9E).

An oxidation product of cholesterol is progesterone, a **sex hormone** (Section 12.10). It also gives rise to the synthesis of other sex hormones, such as testosterone and estradiol, as well as to the **adrenocorticoid hormones.** Among the latter, cortisol and cortisone are best known for their anti-inflammatory action. **Bile salts** are also oxidation products of cholesterol (Section 12.11). They emulsify all kinds of lipids, including cholesterol, and are essential in the digestion of fats.

Prostaglandins, thromboxanes, and **leukotrienes** are derived from arachidonic acid (Section 12.12). They have a wide variety of effects on body chemistry; among other things, they can lower or raise blood pressure, cause inflammation and blood clotting, and induce labor. In general, they mediate hormone action.

PROBLEMS

Numbers that appear in color indicate difficult problems. ▶ designates problems requiring application of principles.

Structure and Properties of Fats

12.2 Why are fats a good source of energy for storage in the body?

12.3 What is the meaning of the term *hydrophobic?* Why is the hydrophobic nature of lipids important?

12.4 Draw the structural formula of a fat molecule (triglyceride) made of myristic acid, oleic acid, palmitic acid, and glycerol.

12.5 Oleic acid has a melting point of 16°C. If you converted the cis double bond into a trans double bond, what would happen to the melting point? Explain.

12.6 Draw schematic formulas for all possible 1,3-diglycerides made up of glycerol, oleic acid, or stearic acid. How many are there? Draw the structure of one of the diglycerides.

12.7 For the diglycerides in Problem 12.6, predict which two will have the highest melting points and which two will have the lowest melting points.

12.8 Predict which acid in each pair has the higher melting point and explain why.

(a) Palmitic acid or stearic acid

(b) Arachidonic acid or arachidic acid

12.9 Which has the higher melting point: (a) a triglyceride containing only lauric acid and glycerol or (b) a triglyceride containing only stearic acid and glycerol?

12.10 Explain why the melting points of the saturated fatty acids increase as we move from lauric acid to stearic acid in Table 12.1.

12.11 Predict the order of the melting points of triglycerides containing fatty acids, as follows.

(a) Palmitic, palmitic, stearic

(b) Oleic, stearic, palmitic

(c) Oleic, linoleic, oleic

12.12 Look at Table 12.2. Which animal fat has the highest percentage of unsaturated fatty acids?

12.13 Rank the following in order of increasing solubility in water (assuming that all are made with the same fatty acids): (a) triglycerides, (b) diglycerides, and (c) monoglycerides. Explain your answer.

12.14 How many moles of H_2 are used up in the catalytic hydrogenation of 1 mol of a triglyceride containing glycerol, palmitic acid, oleic acid, and linoleic acid?

12.15 Name the products of the saponification of this triglyceride:

$$CH_2-O-\overset{\displaystyle O}{\overset{\displaystyle \|}{C}}-(CH_2)_{14}CH_3$$
$$CH-O-\overset{\displaystyle O}{\overset{\displaystyle \|}{C}}-(CH_2)_{16}CH_3$$
$$CH_2-O-\overset{\displaystyle O}{\overset{\displaystyle \|}{C}}-(CH_2)_7(CH=CHCH_2)_3CH_3$$

12.16 Using the equation in Section 12.3 as a guideline for stoichiometry, calculate the number of moles of NaOH needed to saponify 5 mol of (a) triglycerides, (b) diglycerides, and (c) monoglycerides.

Membranes

12.17 Which portion of the phosphatidylinositol molecule contributes to (a) the fluidity of the bilayer and (b) the surface polarity of the bilayer?

12.18 How do the unsaturated fatty acids of the complex lipids contribute to the fluidity of a membrane?

12.19 Which type of lipid molecules is most likely to be present in membranes?

12.20 What is the difference between an integral and a peripheral membrane protein?

Complex Lipids

12.21 Which glycerophospholipid has the most polar groups capable of forming hydrogen bonds with water?

12.22 Draw the structure of a phosphatidylinositol that contains oleic and arachidonic acid.

12.23 Among the glycerophospholipids containing palmitic acid and linolenic acid, which will have the greatest

solubility in water: (a) phosphatidylcholine, (b) phosphatidylethanolamine, or (c) phosphatidylserine? Explain.

12.24 Name all the groups of complex lipids that contain ceramides.

12.25 Are the various phospholipids randomly distributed in membranes? Give an example.

12.26 Enumerate the functional groups that contribute to the hydrophilic character of (a) glucocerebroside and (b) sphingomyelin.

Steroids

12.27 Cholesterol has a fused four-ring steroid nucleus and is a part of body membranes. The —OH group on carbon 3 is the polar head, and the rest of the molecule provides the hydrophobic tail that does not fit into the zig-zag packing of the hydrocarbon portion of the saturated fatty acids. Considering this structure, tell whether small amounts of cholesterol well dispersed in the membrane contribute to the stiffening (rigidity) or the fluidity of a membrane. Explain.

12.28 Where can pure cholesterol crystals be found in the body?

12.29 (a) Find all the carbon stereocenters in a cholesterol molecule.

(b) How many total stereoisomers are possible?

(c) How many of these stereoisomers do you think are found in nature?

12.30 Look at the structures of cholesterol and the hormones shown in Figure 12.6. Which ring of the steroid structure undergoes the most substitution?

12.31 What makes LDL soluble in blood plasma?

12.32 How does LDL deliver its cholesterol to the cells?

12.33 How does lovastatin reduce the severity of atherosclerosis?

12.34 How does VLDL become LDL?

12.35 How does HDL deliver its cholesteryl esters to liver cells?

12.36 How does the serum cholesterol level control both cholesterol synthesis in the liver and LDL uptake?

Steroid Hormones and Bile Salts

12.37 What physiological functions are associated with cortisol?

12.38 Estradiol in the body is synthesized starting from progesterone. What chemical modifications occur when estradiol is synthesized?

12.39 Describe the difference in structure between the male hormone testosterone and the female hormone estradiol.

12.40 Considering that RU486 can bind to the receptors of progesterone as well as those of cortisone and cortisol, what can you say regarding the importance of the functional group on carbon 11 of the steroid ring in drug and receptor binding?

12.41 (a) How does the structure of RU486 resemble that of progesterone?

(b) How do the two structures differ?

12.42 What are the common structural features of the oral contraceptive pills, including mifepristone?

12.43 List all of the functional groups that make taurocholate water-soluble.

12.44 Explain how the constant elimination of bile salts through the feces can reduce the danger of plaque formation in atherosclerosis.

Prostaglandins and Leukotrienes

12.45 What is the basic structural difference between:

(a) Arachidonic acid and prostaglandin PGE_2?

(b) PGE_2 and $PGF_{2\alpha}$?

12.46 Find and name all the functional groups in (a) glycocholate, (b) cortisone, (c) prostaglandin PGE_2, and (d) leukotriene B4.

12.47 What are the chemical and physiological functions of the COX-2 enzyme?

12.48 How does aspirin, an anti-inflammatory drug, prevent strokes caused by blood clots in the brain?

Chemical Connections

12.49 (Chemical Connections 12A) What causes rancidity? How can it be prevented?

12.50 (Chemical Connections 12B) What makes waxes harder and more difficult to melt than fats?

12.51 (Chemical Connections 12C) What is the major structural difference between soaps and detergents?

12.52 (Chemical Connections 12D) How do the gap junctions prevent the passage of proteins from cell to cell?

12.53 (Chemical Connections 12D) How does the anion transporter provide a suitable environment for the passage of hydrated chloride ions?

12.54 (Chemical Connections 12D) In what sense is the active transport of K^+ selective? How does K^+ pass through the transporter?

12.55 (Chemical Connections 12E) (a) What role does sphingomyelin play in the conductance of nerve signals?

(b) What happens to this process in multiple sclerosis?

12.56 (Chemical Connections 12F) Compare the complex lipid structures listed for the lipid storage diseases (Table 12F) with the missing or defective enzymes. Explain why the missing enzyme in Fabry's disease is α-galactosidase and not β-galactosidase.

12.57 (Chemical Connections 12F) Identify the monosaccharides in the accumulating glycolipid of Fabry's disease.

12.58 (Chemical Connections 12G) How does the oral anabolic steroid methenolone differ structurally from testosterone?

12.59 (Chemical Connections 12H) What is the role of progesterone and similar compounds in contraceptive pills?

12.60 (Chemical Connections 12I) How does cortisone prevent inflammation?

12.61 (Chemical Connections 12I) How does indomethacin act in the body to reduce inflammation?

12.62 (Chemical Connections 12I) What kind of prostaglandins are synthesized by COX-1 and COX-2 enzymes?

12.63 (Chemical Connections 12I) Steroids prevent asthma-causing leukotriene synthesis as well as inflammation-causing prostaglandin synthesis. Nonsteroidal anti-inflammatory agents (NSAIDs) such as aspirin reduce only prostaglandin production. Why do NSAIDs not affect leukotriene production?

Additional Problems

12.64 What is the role of taurine in lipid digestion?

12.65 Draw a schematic diagram of a lipid bilayer. Show how the bilayer prevents the passage by diffusion of a polar molecule such as glucose. Show why nonpolar molecules, such as CH_3CH_2—O—CH_2CH_3, can diffuse through the membrane.

12.66 How many different triglycerides can you create using three different fatty acids (A, B, and C) in each case?

12.67 Prostaglandins have a five-membered ring closure; thromboxanes have a six-membered ring closure. The synthesis of both groups of compounds are prevented by COX inhibitors; the COX enzymes catalyze ring closure. Explain this fact.

12.68 Which lipoprotein is instrumental in removing the cholesterol deposited in the plaques on arteries?

12.69 What are coated pits? What is their function?

12.70 What are the constituents of sphingomyelin?

12.71 What structural feature do detergents (Chemical Connections 12C) and the bile salt, taurocholate, share?

12.72 (Chemical Connections 12D) What is the difference between a facilitated transporter and an active transporter?

12.73 What part of LDL interacts with the LDL receptor?

12.74 What is the major difference between aldosterone and all the other hormones listed in Figure 12.6?

12.75 (Chemical Connections 12I) The new anti-inflammatory drug Celebrex does not have the usual side effect of stomach upset or ulceration commonly observed with the other NSAIDs. Why?

12.76 How many grams of H_2 are needed to saturate 100.0 g of a triglyceride made of glycerol and one unit each of lauric, oleic, and linoleic acids?

12.77 Prednisolone is the synthetic glucocorticoid medicine most frequently prescribed to combat autoimmune diseases. Compare its structure to the natural glucocorticoid hormone, cortisone. What are the similarities and differences in structure?

InfoTrac College Edition

For additional readings, go to InfoTrac College Edition, your online research library, at

http://infotrac.thomsonlearning.com

CHAPTER **13**

13.1   Introduction

13.2   Amino Acids

13.3   Zwitterions

13.4   Cysteine: A Special
        Amino Acid

13.5   Peptides and Proteins

13.6   Some Properties of
        Peptides and Proteins

13.7   The Primary Structure of
        Proteins

13.8   The Secondary Structure
        of Proteins

13.9   The Tertiary and
        Quaternary Structures of
        Proteins

13.10   Glycoproteins

13.11   Denaturation

Spider silk is a fibrous protein that exhibits unmatched strength and toughness.

Hans Strand/Stone/Getty Images

Proteins

13.1 Introduction

Protein A large biological molecule made of numerous amino acids linked together by amide bonds

Proteins are by far the most important of all biological compounds. The very word "protein" is derived from the Greek *proteios,* meaning "of first importance," and the scientists who named these compounds more than 100 years ago chose an appropriate term. There are many types of proteins, and they perform a variety of functions, including the following roles:

1. **Structure** In Section 11.6, we saw that the main structural material for plants is cellulose. For animals, it is structural proteins, which are the chief constituents of skin, bones, hair, and nails. Two important structural proteins are collagen and keratin.

2. **Catalysis** Virtually all the reactions that take place in living organisms are catalyzed by proteins called enzymes. Without enzymes, the reactions would take place so slowly as to be useless. We will discuss enzymes in depth in Chapter 14.

The heart itself is a muscle, expanding and contracting about 70 to 80 times per minute.

3. **Movement** Every time we crook a finger, climb stairs, or blink an eye, we use our muscles. Muscle expansion and contraction are involved in every movement we make. Muscles are made up of protein molecules called myosin and actin.

289

4. **Transport** A large number of proteins perform transportation duties. For example, hemoglobin, a protein in the blood, carries oxygen from the lungs to the cells in which it is used and carbon dioxide from the cells to the lungs. Other proteins transport molecules across cell membranes.

5. **Hormones** Many hormones are proteins, including insulin, oxytocin, and human growth hormone.

Antibodies are present in the gamma globulin fraction of blood plasma.

6. **Protection** When a protein from an outside source or some other foreign substance (called an antigen) enters the body, the body makes its own proteins (called antibodies) to counteract the foreign protein. This antibody production is the major mechanism that the body uses to fight disease. Blood clotting is another protective function carried out by a protein, called fibrinogen. Without blood clotting, we would bleed to death from any small wound.

7. **Storage** Some proteins store materials in the way that starch and glycogen store energy. For example, casein in milk and ovalbumin in eggs store nutrients for newborn mammals and birds. Ferritin, a protein in the liver, stores iron.

8. **Regulation** Some proteins not only control the expression of genes, thereby regulating the kind of proteins synthesized in a particular cell, but also dictate when such manufacture takes place.

These are not the only functions of proteins, but they are among the most important. Clearly, any individual needs a great many proteins to carry out all these varied functions. A typical cell contains about 9000 different proteins; the entire human body has about 100,000 different proteins.

Collagen, actin, and keratin are fibrous proteins. Albumin, hemoglobin, and immunoglobulins are globular proteins.

We can classify proteins into two major types: **fibrous proteins,** which are insoluble in water and are used mainly for structural purposes, and **globular proteins,** which are more or less soluble in water and are used mainly for nonstructural purposes.

13.2 Amino Acids

See the **Interactive General, Organic, and Biochemistry CD-ROM, version 2.0,** for further exploration on this topic.

Although a wide variety of proteins exist, they all have basically the same structure: They are chains of amino acids. As its name implies, an **amino acid** is an organic compound containing an amino group and a carboxyl group. Organic chemists can synthesize many thousands of amino acids, but nature is much more restrictive and uses only 20 different amino acids to make up proteins. Furthermore, all but one of the 20 fit the formula

$$\begin{array}{c} \text{H} \\ | \\ \text{R}-\text{C}-\text{COO}^- \\ | \\ \text{NH}_3^+ \end{array}$$

Alpha (α-) amino acid An amino acid in which the amino group is linked to the carbon atom next to the —COOH carbon

Even the one that doesn't fit this formula (proline) comes fairly close: It differs only in that it has a bond between the R and the N. The 20 amino acids found in proteins are called **alpha amino acids.** They are listed in Table 13.1, which also shows the one- and three-letter abbreviations that chemists and biochemists use for them.

The one-letter abbreviations are more recent inventions, but the three-letter abbreviations are still frequently used.

The most important aspect of the R groups is their polarity. On that basis we can classify amino acids into the four groups shown in Table 13.1: nonpolar, polar but neutral, acidic, and basic. Note that the nonpolar side

Table 13.1 **The 20 Amino Acids Commonly Found in Proteins (both the three-letter and one-letter abbreviations are shown) and Their Isoelectric Points (pI)**

Nonpolar Side Chains

alanine (Ala, A)
pI = 6.01

$$CH_3\underset{\underset{NH_3^+}{|}}{CH}COO^-$$

glycine (Gly, G)
pI = 5.97

$$H\underset{\underset{NH_3^+}{|}}{CH}COO^-$$

isoleucine (Ile, I)
pI = 6.02

$$CH_3CH_2\underset{\underset{NH_3^+}{|}}{\overset{\overset{CH_3}{|}}{CH}}CHCOO^-$$

leucine (Leu, L)
pI = 5.98

$$(CH_3)_2CHCH_2\underset{\underset{NH_3^+}{|}}{CH}COO^-$$

methionine (Met, M)
pI = 5.74

$$CH_3SCH_2CH_2\underset{\underset{NH_3^+}{|}}{CH}COO^-$$

phenylalanine (Phe, F)
pI = 5.48

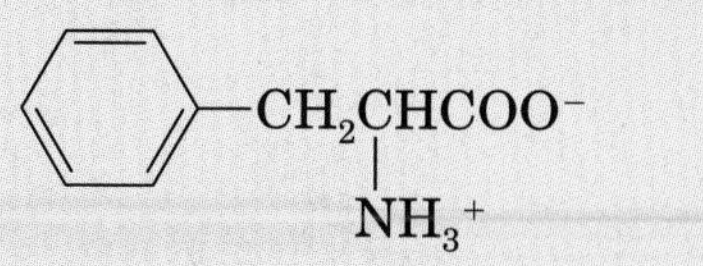

proline (Pro, P)
pI = 6.48

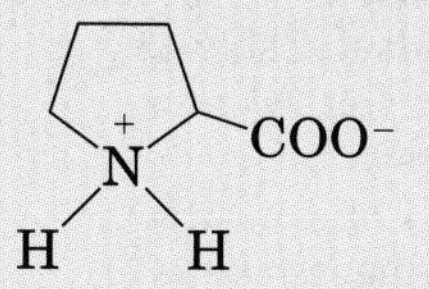

tryptophan (Trp, W)
pI = 5.88

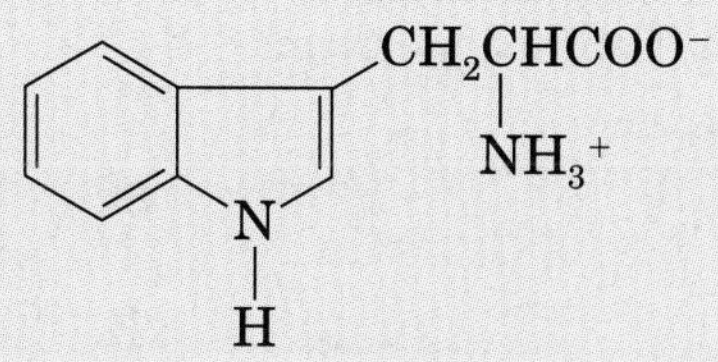

valine (Val, V)
pI = 5.97

$$(CH_3)_2\underset{\underset{NH_3^+}{|}}{CH}CHCOO^-$$

Polar but Neutral Side Chains

asparagine (Asn, N)
pI = 5.41

$$H_2N\overset{\overset{O}{\|}}{C}CH_2\underset{\underset{NH_3^+}{|}}{CH}COO^-$$

glutamine (Gln, Q)
pI = 5.65

$$H_2N\overset{\overset{O}{\|}}{C}CH_2CH_2\underset{\underset{NH_3^+}{|}}{CH}COO^-$$

cysteine (Cys, C)
pI = 5.07

$$HSCH_2\underset{\underset{NH_3^+}{|}}{CH}COO^-$$

tyrosine (Tyr, Y)
pI = 5.66

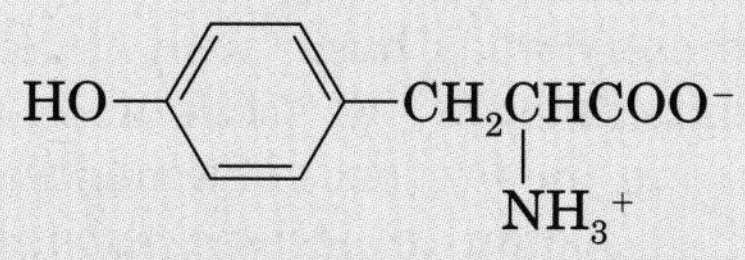

serine (Ser, S)
pI = 5.68

$$HOCH_2\underset{\underset{NH_3^+}{|}}{CH}COO^-$$

threonine (Thr, T)
pI = 5.87

$$CH_3\underset{\underset{NH_3^+}{|}}{\overset{\overset{OH}{|}}{CH}}CHCOO^-$$

Acidic Side Chains

aspartic acid (Asp, D)
pI = 2.77

$$^-OOCCH_2\underset{\underset{NH_3^+}{|}}{CH}COO^-$$

glutamic acid (Glu, E)
pI = 3.22

$$^-OOCCH_2CH_2\underset{\underset{NH_3^+}{|}}{CH}COO^-$$

Basic Side Chains

arginine (Arg, R)
pI = 10.76

$$H_2N\overset{\overset{NH_2^+}{\|}}{C}NHCH_2CH_2CH_2\underset{\underset{NH_3^+}{|}}{CH}COO^-$$

histidine (His, H)
pI = 7.59

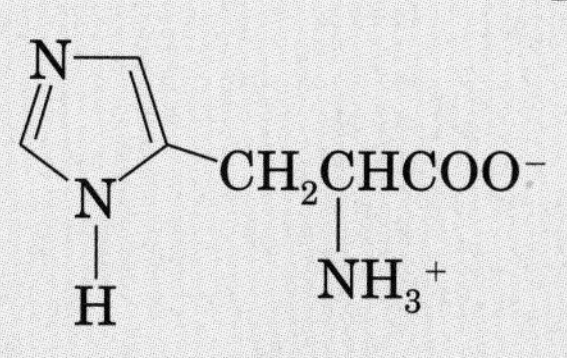

lysine (Lys, K)
pI = 9.74

$$\overset{+}{H_3}NCH_2CH_2CH_2CH_2\underset{\underset{NH_3^+}{|}}{CH}COO^-$$

*Each ionizable group is shown in the form present in highest concentration at pH 7.0.

chains are *hydrophobic* (they repel water), whereas polar but neutral, acidic, and basic side chains are *hydrophilic* (attracted to water). This aspect of the R groups is very important in determining both the structure and the function of each protein molecule.

When we look at the general formula for the 20 amino acids, we see at once that all of them (except glycine, in which R = H) are chiral with (carbon) stereocenters, since R, H, COOH, and NH₂ are four different groups. This means that each of the amino acids with one stereocenter exists as two enantiomers. As is the case for most examples of this kind, nature makes only one of the two possible enantiomers for each amino acid, and it is virtually always the L form. Except for glycine, which exists in only one form, all the amino acids in all the proteins in your body are the L form. D amino acids are extremely rare in nature; some are found, for example, in the cell walls of a few types of bacteria.

In Section 11.2, we learned about the systematic use of the D,L system. There we used glyceraldehyde as a reference point for the assignment of relative configuration. Here again, we can use glyceraldehyde as a reference point with amino acids.

Two of the amino acids in Table 13.1 have a second carbon stereocenter and exist as two pairs of enantiomers, but only one of these four possible stereoisomers exists in proteins. Can you find them?

$$
\begin{array}{cc}
\overset{\displaystyle O}{\underset{\displaystyle }{\diagdown}} & \\
\text{C—H} & \\
\text{HO—C—H} & \\
\text{CH}_2\text{OH} & \\
\text{L-Glyceraldehyde} &
\end{array}
\qquad
\begin{array}{c}
\overset{\displaystyle O}{\diagdown}\\
\text{C—O}^- \\
\text{H}_3\text{N}^+\text{—C—H} \\
\text{CH}_3 \\
\text{L-Alanine}
\end{array}
$$

The spatial relationship of the functional groups around the carbon stereocenter in L-amino acids, as in L-alanine, can be compared to that of L-glyceraldehyde. When we put the carbonyl groups of both compounds in the same position (top), the —OH of L-glyceraldehyde and the NH₃⁺ of L-alanine lie to the left of the carbon stereocenter.

13.3 Zwitterions

See the **Interactive General, Organic, and Biochemistry CD-ROM, version 2.0,** for further exploration on this topic.

In Section 10.2 we learned that carboxylic acids, RCOOH, cannot exist in the presence of a moderately weak base (such as NH_3). They donate a proton to become carboxylate ions, RCOO⁻. Likewise, amines, RNH_2 (Section 8.6), cannot exist as such in the presence of a moderately weak acid (such as acetic acid). They gain a proton to become substituted ammonium ions, RNH_3^+.

An amino acid has —COOH and —NH₂ groups in the same molecule. Therefore, in water solution, the —COOH donates a proton to the —NH₂, so that an amino acid actually has the structure

$$
\begin{array}{c}
\text{H} \\
| \\
\text{R—C—COO}^- \\
| \\
\text{NH}_3^+
\end{array}
$$

Zwitterion comes from the German *zwitter,* meaning "hybrid."

Compounds that have a positive charge on one atom and a negative charge on another are called **zwitterions.** Amino acids are zwitterions, not only in water solution but also in the solid state. They are therefore ionic compounds—that is, internal salts. *Un-ionized RCH(NH₂)COOH molecules do not actually exist, in any form.*

The fact that amino acids are zwitterions explains their physical properties. All of them are solids with high melting points (for example, glycine melts at 262°C), just as we would expect for ionic compounds. The 20 amino acids are also fairly soluble in water, as ionic compounds generally are; if they had no charges, we would expect only the smaller ones to be soluble.

If we add an amino acid to water, it dissolves and then has the same zwitterionic structure that it has in the solid state. Let us see what happens if we change the pH of the solution, as we can easily do by adding a source of H_3O^+, such as HCl solution (to lower the pH), or a strong base, such as NaOH (to raise the pH). Because H_3O^+ is a stronger acid than a typical carboxylic acid (Section 10.2), it donates a proton to the —COO^- group, turning the zwitterion into a positive ion. This happens to all amino acids if the pH is sufficiently lowered—let's say, to pH 2.

$$R-\underset{\underset{NH_3^+}{|}}{\overset{\overset{H}{|}}{C}}-COO^- + H_3O^+ \longrightarrow R-\underset{\underset{NH_3^+}{|}}{\overset{\overset{H}{|}}{C}}-COOH + H_2O$$

Addition of OH^- to the zwitterion causes the —NH_3^+ to donate its proton to OH^-, turning the zwitterion into a negative ion. This happens to all amino acids if the pH is sufficiently raised—let's say, to pH 10.

$$R-\underset{\underset{NH_3^+}{|}}{\overset{\overset{H}{|}}{C}}-COO^- + OH^- \longrightarrow R-\underset{\underset{NH_2}{|}}{\overset{\overset{H}{|}}{C}}-COO^- + H_2O$$

In both cases, the amino acid is still an ion so it is still soluble in water. There is no pH at which an amino acid has no ionic character at all. If the amino acid is a positive ion at low pH and a negative ion at high pH, there must be some pH at which all the molecules have equal positive and negative charges. This pH is called the **isoelectric point (pI).**

Every amino acid has a different isoelectric point, although most of them are not very far apart (see the values in Table 13.1). Fifteen of the 20 have isoelectric points near 6. However, the three basic amino acids have higher isoelectric points, and the two acidic amino acids have lower values.

At or near the isoelectric point, amino acids exist in aqueous solution largely or entirely as zwitterions. As we have seen, they react with either a strong acid, by taking a proton (the —COO^- becomes —$COOH$), or a strong base, by giving a proton (the —NH_3^+ becomes —NH_2). To summarize:

$$R-\underset{\underset{NH_3^+}{|}}{\overset{\overset{H}{|}}{C}}-COOH \underset{H_3O^+}{\overset{OH^-}{\rightleftharpoons}} R-\underset{\underset{NH_3^+}{|}}{\overset{\overset{H}{|}}{C}}-COO^- \underset{H_3O^+}{\overset{OH^-}{\rightleftharpoons}} R-\underset{\underset{NH_2}{|}}{\overset{\overset{H}{|}}{C}}-COO^-$$

In Section 7.3, we learned that a compound that is both an acid and a base is called amphiprotic. We also learned, in Section 7.11, that a solution that neutralizes both acid and base is a buffer solution. Amino acids are therefore *amphiprotic* compounds, and aqueous solutions of them are *buffers*.

Proteins also have isoelectric points and act as buffers. This property will be discussed in Section 13.6.

Isoelectric point (pI) A pH at which a sample of amino acids or protein has an equal number of positive and negative charges

Acidic amino acids have two —COO^- groups, one on the α-carbon and one on the side chain. At the isoelectric point, one of these must be in the un-ionized form (—$COOH$). To achieve this, H_3O^+ must be added to lower the pH. Similar considerations explain why the isoelectric points of basic amino acids are higher than pH 6.

13.4 Cysteine: A Special Amino Acid

One of the 20 amino acids in Table 13.1 has a chemical property not shared by any of the others. This amino acid, cysteine, can easily be dimerized by many mild oxidizing agents:

We met these reactions in Section 5.5D.

[O] symbolizes oxidation, and [H] stands for reduction.

A disulfide bond

$$2\,HS-CH_2-CH-COO^- \underset{[H]}{\overset{[O]}{\rightleftharpoons}}\ ^-OOC-CH-CH_2-S-S-CH_2-CH-COO^-$$

$$\underset{\substack{| \\ NH_3^+}}{} \qquad\qquad \underset{\substack{| \\ NH_3^+}}{} \qquad\qquad \underset{\substack{| \\ NH_3^+}}{}$$

Cysteine · Cystine

A dimer is a molecule made up of two units.

The dimer of cysteine, which is called **cystine,** can in turn be fairly easily reduced to give two molecules of cysteine. As we shall see, the presence of cystine has important consequences for the chemical structure and shape of protein molecules of which it is part. The **S—S bond** (shown in color) is also called a **disulfide bond** (Section 5.5D).

13.5 Peptides and Proteins

Each amino acid has a carboxyl group and an amino group. In Section 18.5, we saw that a carboxylic acid and an amine could be combined to form an amide:

$$R-\overset{\overset{\textstyle O}{\|}}{C}-O^- + R'-NH_3^+ \longrightarrow R-\overset{\overset{\textstyle O}{\|}}{C}-NH-R' + H_2O$$

In the same way, the —COO⁻ group of one amino acid molecule, say glycine, can combine with the —NH₃⁺ group of a second molecule, say alanine:

$$H_3\overset{+}{N}-CH_2-\overset{\overset{\textstyle O}{\|}}{C}-O^- + H_3\overset{+}{N}-\underset{\substack{| \\ CH_3}}{CH}-\overset{\overset{\textstyle O}{\|}}{C}-O^- \longrightarrow H_3\overset{+}{N}-CH_2-\overset{\overset{\textstyle O}{\|}}{C}-NH-\underset{\substack{| \\ CH_3}}{CH}-\overset{\overset{\textstyle O}{\|}}{C}-O^- + H_2O$$

Glycine · · · · · · · · · · · · Alanine · · · · · · · · · · · · Glycylalanine
(Gly—Ala)

Peptide bond An amide bond that links two amino acids

This reaction takes place in the cells by a mechanism that we will examine in Section 17.5. The product is an amide. The two amino acids are joined together by a **peptide bond** (also called a **peptide linkage**). The product is a **dipeptide.**

The naming of the peptides begins with the N-terminal amino acids.

It is important to realize that glycine and alanine could also be linked the other way:

$$H_3\overset{+}{N}-\underset{\substack{| \\ CH_3}}{CH}-\overset{\overset{\textstyle O}{\|}}{C}-O^- + H_3\overset{+}{N}-CH_2-\overset{\overset{\textstyle O}{\|}}{C}-O^- \longrightarrow H_3\overset{+}{N}-\underset{\substack{| \\ CH_3}}{CH}-\overset{\overset{\textstyle O}{\|}}{C}-NH-CH_2-\overset{\overset{\textstyle O}{\|}}{C}-O^- + H_2O$$

Alanine · · · · · · · · · · · · Glycine · · · · · · · · · · · · Alanylglycine
(Ala—Gly)

In Ala—Gly, the —NH₃⁺ is connected to a —CHCH₃; in Gly—Ala, the —NH₃⁺ is connected to a —CH₂.

In this case we get a *different* dipeptide. The two dipeptides are constitutional isomers, of course; they are different compounds in all respects, with different properties.

EXAMPLE 13.1

Show how to form the dipeptide aspartylserine (Asp—Ser).

Solution

The name implies that this dipeptide is made of two amino acids, aspartic acid (Asp) and serine (Ser), with the amide being formed between the α-carboxyl group of aspartic acid and the amino group of serine. Therefore, we write the formula of aspartic acid with its amino group on the left side. Next we place the formula of serine to the right, with its amino group facing the α-carboxyl group of aspartic acid. Finally, we eliminate a water molecule between the —COO⁻ and —NH₃⁺ groups next to each other, forming the peptide bond:

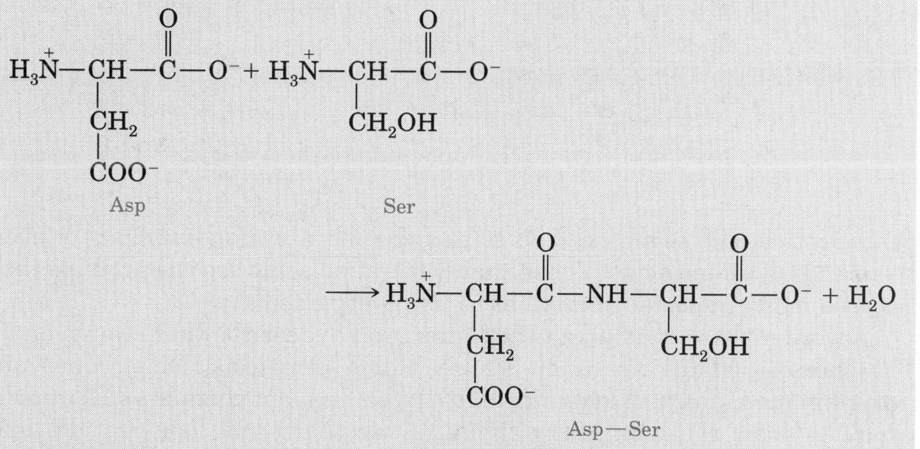

$$
\underset{\text{Asp}}{\overset{\displaystyle H_3\overset{+}{N}-CH-\overset{O}{\overset{\|}{C}}-O^-}{\underset{\overset{|}{CH_2}}{\underset{\overset{|}{COO^-}}{}}}} + \underset{\text{Ser}}{\overset{\displaystyle H_3\overset{+}{N}-CH-\overset{O}{\overset{\|}{C}}-O^-}{\underset{CH_2OH}{}}}
$$

$$
\longrightarrow \underset{\text{Asp—Ser}}{\overset{\displaystyle H_3\overset{+}{N}-CH-\overset{O}{\overset{\|}{C}}-NH-CH-\overset{O}{\overset{\|}{C}}-O^-}{\underset{\overset{|}{CH_2}\qquad\quad CH_2OH}{\underset{\overset{|}{COO^-}}{}}}} + H_2O
$$

Problem 13.1

Show how to form the dipeptide valylphenylalanine (Val—Phe).

Any two amino acids, whether the same or different, can be linked together to form dipeptides in a similar manner. But the possibilities do not end there. Each dipeptide still contains a —COO⁻ and an —NH₃⁺ group. We can therefore add a third amino acid to alanylglycine, say lysine:

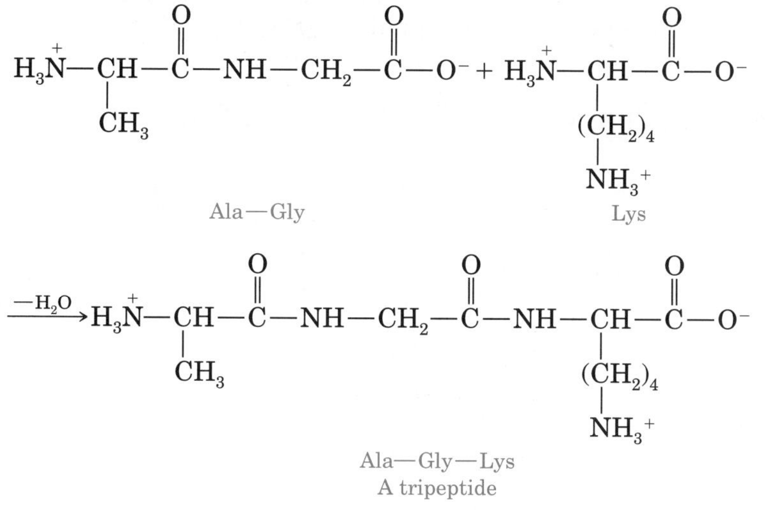

$$
\underset{\text{Ala—Gly}}{\overset{\displaystyle H_3\overset{+}{N}-CH-\overset{O}{\overset{\|}{C}}-NH-CH_2-\overset{O}{\overset{\|}{C}}-O^-}{\underset{CH_3}{}}} + \underset{\text{Lys}}{\overset{\displaystyle H_3\overset{+}{N}-CH-\overset{O}{\overset{\|}{C}}-O^-}{\underset{\overset{|}{(CH_2)_4}}{\underset{\overset{|}{NH_3^+}}{}}}}
$$

$$
\xrightarrow{-H_2O} \underset{\substack{\text{Ala—Gly—Lys}\\\text{A tripeptide}}}{\overset{\displaystyle H_3\overset{+}{N}-CH-\overset{O}{\overset{\|}{C}}-NH-CH_2-\overset{O}{\overset{\|}{C}}-NH-CH-\overset{O}{\overset{\|}{C}}-O^-}{\underset{CH_3\qquad\qquad\qquad (CH_2)_4}{\underset{NH_3^+}{}}}}
$$

The product is a **tripeptide.** Because it also contains a —COO⁻ and an —NH₃⁺ group, we can continue the process to get a tetrapeptide, a

CHEMICAL CONNECTIONS 13A

Glutathione

A very important tripeptide present in high concentrations in all tissues is glutathione. Glutathione contains L-glutamic acid, L-cysteine, and glycine. Its structure is unusual because the glutamic acid is linked to the cysteine by its γ-carboxyl group rather than by the α-carboxyl group, as is usual in most peptides and proteins:

Glutathione functions in the cells as a general protective agent. Oxidizing agents that would damage the cells, such as peroxides, will oxidize glutathione instead (the cysteine portion; see also Section 13.4), thereby protecting the proteins and nucleic acids. Many foreign chemicals also get attached to glutathione, so in a sense it acts as a detoxifying agent.

$$H_3\overset{+}{N}-CH-CH_2-CH_2-\underset{O}{\overset{}{C}}-NH-CH-\underset{O}{\overset{}{C}}-NH-CH_2-COO^-$$

$$\underset{COO^-}{} \qquad \qquad \underset{\underset{SH}{CH_2}}{}$$

Glutathione
(Glu—Cys—Gly)

pentapeptide, and so on, until we have a chain containing hundreds or even thousands of amino acids. These chains of amino acids are the proteins that serve so many important functions in living organisms.

A word must be said about the terms used to describe these compounds. The shortest chains are often simply called **peptides,** longer ones are **polypeptides,** and still longer ones are **proteins,** but chemists differ about where to draw the line. Many chemists use the terms "polypeptide" and "protein" almost interchangeably. In this book, we will consider a protein to be a polypeptide chain that contains a minimum of 30 to 50 amino acids.

The amino acids in a chain are often called **residues.** It is customary to use either the one-letter or the three-letter abbreviations shown in Table 13.1 to represent peptides and proteins. For example, the tripeptide shown on the previous page, alanylglycyllysine, is AGK or Ala—Gly—Lys. **C-terminal amino acid** is the residue with the free COO⁻ group (lysine in Ala—Gly—Lys), and **N-terminal amino acid** is the residue with the free NH₃⁺ group (alanine in Ala—Gly—Lys). It is the universal custom to write peptide and protein chains with the N-terminal residue on the left. No matter how long a protein chain gets—hundreds or thousands of units—it always has just two ends: one C-terminal and one N-terminal.

13.6 Some Properties of Peptides and Proteins

Figure 13.1 Six atoms of the peptide backbone lie in an imaginary (shaded) plane.

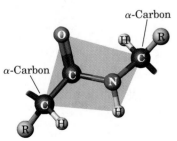

The continuing pattern of peptide bonds is the backbone of the peptide or protein molecule; the R groups are called the **side chains.** The six atoms of the peptide backbone lie in the same plane (Figure 13.1), and two adjacent peptide bonds can rotate relative to one another about the C—N and C—C bonds.

$$\underset{R_1}{\overset{H}{\underset{|}{N}}}-CH-\overset{O}{\overset{\|}{C}}-\underset{R_2}{\overset{H}{\underset{|}{N}}}-CH-\overset{O}{\overset{\|}{C}}$$

The 20 different amino acid side chains supply variety and determine the physical and chemical properties of proteins. Among these properties,

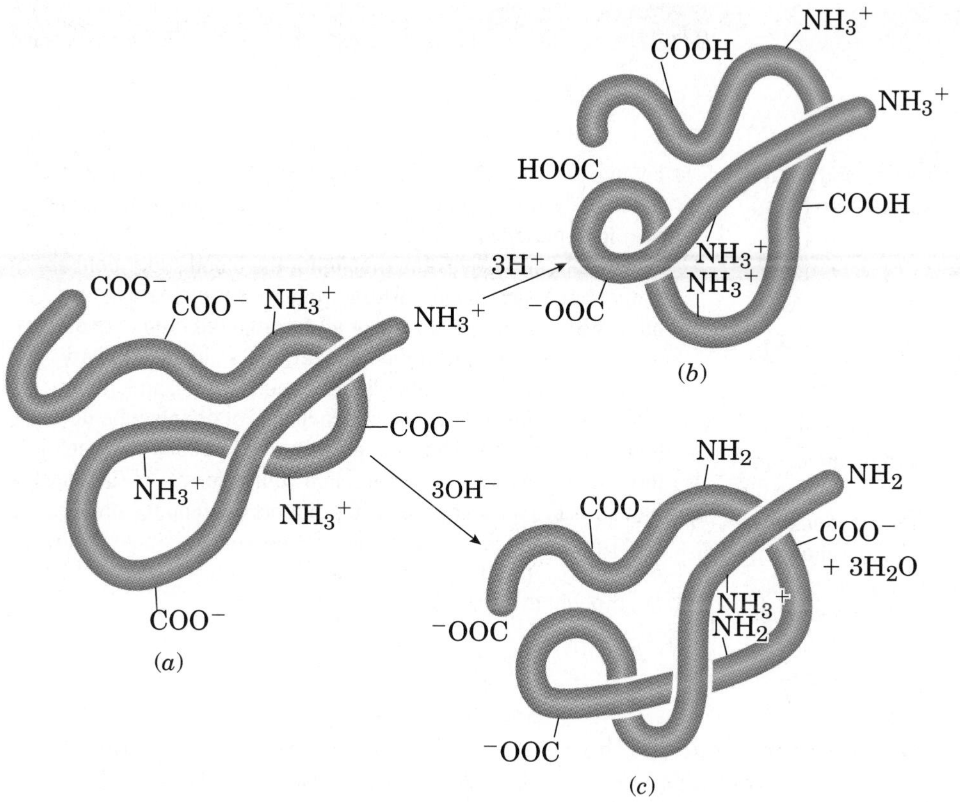

Figure 13.2 Schematic diagram of a protein (*a*) at its isoelectric point and its buffering action when (*b*) H$^+$ or (*c*) OH$^-$ ions are added.

acid–base behavior is one of the most important. Like amino acids (Section 13.3), proteins behave as zwitterions. The side chains of glutamic and aspartic acids provide acidic groups, whereas lysine and arginine provide basic groups (histidine does, too, but this side chain is less basic than the other two). (See the structures of these amino acids in Table 13.1.)

The isoelectric point of a protein occurs at the pH at which there are an equal number of positive and negative charges (the protein has no *net* charge). At any pH above the isoelectric point, the protein molecules have a net negative charge; at any pH below the isoelectric point, they have a net positive charge. Some proteins, such as hemoglobin, have an almost equal number of acidic and basic groups; the isoelectric point of hemoglobin is at pH 6.8. Others, such as serum albumin, have more acidic groups than basic groups; the isoelectric point of this protein is 4.9. In each case, however, because proteins behave like zwitterions, they act as buffers, for example, in the blood (Figure 13.2).

The water solubility of large molecules such as proteins often depends on the repulsive forces between like charges on their surfaces. When protein molecules are at a pH at which they have a net positive or negative charge, the presence of these like charges causes the protein molecules to repel each other. These repulsive forces are smallest at the isoelectric point, when the net charges are close to zero. When there are no repulsive forces, the protein molecules tend to clump together to form aggregates of two or more molecules, reducing their solubility. Therefore, *proteins are least soluble in water at their isoelectric points and can be precipitated from their solutions.*

As we pointed out in Section 13.1, proteins have many functions. To understand these functions, we must look at four levels of organization in their structures. The *primary structure* describes the linear sequence of amino acids in the polypeptide chain. The *secondary structure* refers to certain re-

The terminal —COOH and —NH$_2$ groups also ionize, but they are only 2 out of 50 or more residues.

Carbonates and phosphates are the other blood buffers (see Section 7.11).

CHEMICAL CONNECTIONS 13B

AGE and Aging

A reaction can take place between a primary amine and an aldehyde or a ketone, linking the two molecules (shown here for an aldehyde):

$$R{-}\overset{\overset{\displaystyle O}{\|}}{C}{-}H + H_2N{-}R' \longrightarrow R{-}CH{=}N{-}R' + H_2O$$

An imine

Because proteins have NH_2 groups and carbohydrates have aldehyde or keto groups, they can undergo this reaction, establishing a link between a sugar and a protein molecule. When this reaction is not catalyzed by enzymes, it is called *glycation* of proteins. The process, however, does not stop there. When these linked products are heated in a test tube, high-molecular-weight, water-insoluble, brownish complexes are formed. These complexes are called ***advanced glycation end-products (AGE).*** In the body, they cannot be heated, but the same result happens over long periods of time.

The longer we live, and the higher the blood sugar concentration becomes, the more AGE products accumulate in the body. AGE products show up in all the afflicted organs of diabetic patients: in the lens of the eye (cataracts), in the capillary blood vessels of the retina (diabetic retinopathy), and in the glomeruli of the kidneys (kidney failure). For people who do not have diabetes, these harmful protein modifications become disturbing only in advanced years. In a young person, metabolism functions properly and the AGE products decompose and are eliminated from the body. In an older person, metabolism slows and the AGE products accumulate. AGE products themselves are thought to enhance oxidative damages.

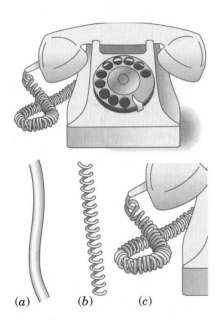

Figure 13.3 The "structure" of a telephone cord: (*a*) primary, (*b*) secondary, (*c*) tertiary.

See the **Interactive General, Organic, and Biochemistry CD-ROM, version 2.0,** for further exploration on this topic.

peating patterns, such as the α-helix conformation or the pleated sheet (Section 13.8), or the absence of a repeating pattern, as with the random coil (Section 13.8). The *tertiary structure* describes the overall conformation of the polypeptide chain.

A good analogy for these three types of structure is a coiled telephone cord (Figure 13.3). The primary structure is the stretched-out cord. The secondary structure is the coil in the form of a helix. We can also take the entire coil and twist it into various shapes, thereby creating a tertiary structure. As we shall see, protein molecules twist and curl in a very similar manner.

The quaternary structure (Section 13.9) applies mainly to proteins containing more than one polypeptide chain (subunit) and deals with how the different chains are spatially related to each other. We also call a structure quaternary when a single polypeptide chain interacts with lipid bilayers and that interaction determines how the different parts of the polypeptide chains are related to each other.

13.7 The Primary Structure of Proteins

Very simply, the primary structure of a protein consists of the sequence of amino acids that makes up the chain. Each of the very large number of peptide and protein molecules in biological organisms has a different sequence of amino acids—and that sequence allows the protein to carry out its function, whatever it may be.

How can so many different proteins arise from different sequences of only 20 amino acids? Let us look at a little arithmetic, starting with a dipeptide. How many different dipeptides can be made from 20 amino acids? There are 20 possibilities for the N-terminal amino acid, and for each of these 20 there are 20 possibilities for the C-terminal amino acid. This means that there are $20 \times 20 = 400$ different dipeptides possible from the 20 amino acids. What about tripeptides? We can form a tripeptide by taking any of the 400 dipeptides and adding any of the 20 amino acids. Thus, there are $20 \times 20 \times 20 = 8000$ tripeptides, all different. It is easy to see that we can calculate the total number of possible peptides or proteins for a chain of n amino acids simply by raising 20 to the nth power (20^n).

Taking a typical small protein to be one with 60 amino acid residues, the number of proteins that can be made from the 20 amino acids is $20^{60} = 10^{78}$. This is an enormous number, possibly greater than the total number of atoms in the universe. Clearly, only a tiny fraction of all possible protein molecules has ever been made by biological organisms.

Each peptide or protein in the body has its own unique sequence of amino acids. As with naming of peptides, the *assignment of positions of the amino acids in the sequence starts at the N-terminal end.* Thus in Figure 13.4 glycine is in the number 1 position on the A chain and phenylalanine is number 1 on the B chain. We mentioned that proteins also have secondary, tertiary, and, in some cases, quaternary structures. We will deal with these in Sections 13.8 and 13.9, but here we can say that *the primary structure of a protein determines to a large extent the native* (most frequently occurring) *secondary and tertiary structures.* That is, the particular sequence of amino acids on the chain enables the whole chain to fold and curl in such a way as to assume its final shape. As we will see in Section 13.11, without its particular three-dimensional shape, a protein cannot function.

Just how important is the exact amino acid sequence to the function of a protein? Can a protein perform the same function if its sequence is a little different? The answer to this question is that a change in amino acid sequence may or may not matter, depending on what kind of a change it is. Consider, as an example, cytochrome c, which is a protein of terrestrial vertebrates. Its chain consists of 104 amino acid residues. It performs the same function (electron transport) in humans, chimpanzees, sheep, and other animals. While humans and chimpanzees have exactly the same amino acid sequence of this protein, sheep cytochrome c differs in 10 positions out of the 104. (You can find more about biochemical evolution in Chemical Connections 17E).

Another example is the hormone insulin. Human insulin consists of two chains having a total of 51 amino acids. The two chains are connected by disulfide bonds. Figure 13.4 shows the sequence of amino acids. Insulin is necessary for proper utilization of carbohydrates (Section 19.2), and people with severe diabetes (Chemical Connections 15G) must take insulin injections. The amount of human insulin available is far too small to meet the need for it, so bovine insulin (from cattle) or insulin from hogs or sheep is used instead. Insulin from these sources is similar, but not identical, to human insulin. The differences are entirely in the 8, 9, and 10 positions of the A chain and the C-terminal position (30) of the B chain:

	A Chain			B Chain
	8	9	10	30
Human	—Thr—	Ser—	Ile—	—Thr
Bovine	—Ala—	Ser—	Val—	—Ala
Hog	—Thr—	Ser—	Ile—	—Ala
Sheep	—Ala—	Gly—	Val—	—Ala

The remainder of the molecule is the same in all four varieties of insulin. Despite the slight differences in structure, all these insulins perform the same function and even can be used by humans. However, none of the other three is quite as effective in humans as human insulin.

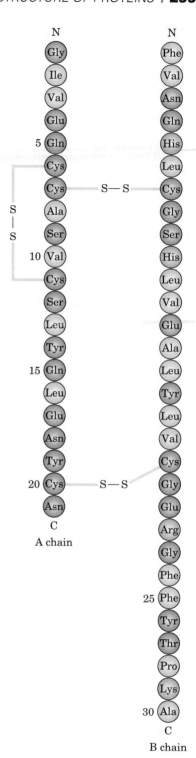

Figure 13.4 The hormone insulin consists of two polypeptide chains, A and B, held together by two disulfide cross-bridges (S—S). The sequence shown is for bovine insulin.

CHEMICAL CONNECTIONS 13C

The Use of Human Insulin

Although human insulin, as manufactured by recombinant DNA techniques (see Section 17.8), is on the market, many diabetic patients continue to use hog or sheep insulin because it is cheaper. Changing from animal to human insulin creates an occasional problem for diabetics. All diabetics experience an insulin reaction (hypoglycemia) when the insulin level in the blood is too high relative to the blood sugar level. Hypoglycemia is preceded by symptoms of hunger, sweating, and poor coordination. These symptoms, called *hypoglycemic awareness,* signal the

patient that hypoglycemia is coming and that it must be reversed, which the patient can do by eating sugar.

Some diabetics who changed from animal to human insulin reported that the hypoglycemic awareness from recombinant DNA human insulin is not as strong as that from animal insulin. This lack of recognition can create some hazards, and the effect is probably due to different rates of absorption in the body. The literature supplied with human insulin now incorporates a warning that hypoglycemic awareness may be altered.

The first sequence of an important protein, insulin, was obtained by Frederick Sanger (1918–) in England, for which he received the Nobel Prize in 1958.

Both oxytocin and vasopressin are secreted by the pituitary gland.

Both hormones are used as drugs—vasopressin to combat loss of blood pressure after surgery and oxytocin to induce labor.

Another factor showing the effect of substituting one amino acid for another is that sometimes patients become allergic to, say, bovine insulin but can switch to hog or sheep insulin without experiencing an allergic reaction.

In contrast to the previous examples, there are small changes in amino acid sequence that make a great deal of difference. Consider two peptide hormones, oxytocin and vasopressin (Figure 13.5). These nonapeptides have identical structures, including a disulfide bond, except for different amino acids in positions 2 and 7. Yet their biological functions are quite different. Vasopressin is an antidiuretic hormone. It increases the amount of water reabsorbed by the kidneys and raises blood pressure. Oxytocin has no effect on water in the kidneys and slightly lowers blood pressure. It affects contractions of the uterus in childbirth and the muscles in the breast that aid in the secretion of milk. Vasopressin also stimulates uterine contractions, but to a much lesser extent than oxytocin.

Another instance where a minor change makes a major difference is in the blood protein hemoglobin. A change in only one amino acid in a chain of 146 is enough to cause a fatal disease—sickle cell anemia (Chemical Connections 13D).

In some cases, slight changes in amino acid sequence make little or no difference to the functioning of peptides and proteins, but most times, the sequence is highly important. The sequences of 10,000 protein and peptide molecules have now been determined. The methods for doing so are complicated and will not be discussed in this book.

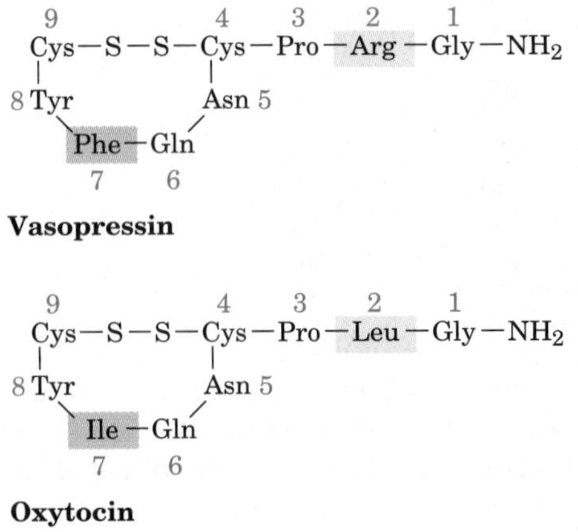

Figure 13.5 The structures of vasopressin and oxytocin. Differences are shown in color.

```
 9              4    3    2    1
Cys—S—S—Cys—Pro—Arg—Gly—NH₂
 |              |
8 Tyr          Asn 5
      Phe—Gln
       7    6
```
Vasopressin

```
 9              4    3    2    1
Cys—S—S—Cys—Pro—Leu—Gly—NH₂
 |              |
8 Tyr          Asn 5
       Ile—Gln
        7    6
```
Oxytocin

CHEMICAL CONNECTIONS 13D

Sickle Cell Anemia

Normal adult human hemoglobin has two alpha chains and two beta chains (see Figure 13.13). Some people, however, have a slightly different kind of hemoglobin in their blood. This hemoglobin (called HbS) differs from the normal type only in the beta chains and only in one position on these two chains: The glutamic acid in the sixth position of normal Hb is replaced by a valine residue in HbS.

	4	5	6	7	8	9
Normal Hb—	Thr—	Pro—	Glu—	Glu—	Lys—	Ala—
Sickle cell Hb—	Thr—	Pro—	Val—	Glu—	Lys—	Ala—

This change affects only two positions in a molecule containing 574 amino acid residues. Yet, it is enough to produce a very serious disease, *sickle cell anemia.*

Red blood cells carrying HbS behave normally when there is an ample oxygen supply. When the oxygen pressure decreases, however, the red blood cells become sickle-shaped. This malformation occurs in the capillaries. As a result of this change in shape, the cells may clog the capillaries. The body's defenses destroy the clogging cells, and the loss of the blood cells causes anemia.

This change at only a single position of a chain consisting of 146 amino acids is severe enough to cause a high death rate. A child who inherits two genes programmed to produce sickle cell hemoglobin (a homozygote) has an 80% smaller chance of surviving to adulthood than a child with only one such gene (a heterozygote) or a child with two normal genes. Despite the high mortality of homozygotes, the genetic trait survives. In central Africa, 40% of the population in malaria-ridden areas carry the sickle cell gene, and 4% are homozygotes. It seems that the sickle cell genes help to acquire immunity against malaria in early childhood so that in malaria-ridden areas the transmission of these genes is advantageous.

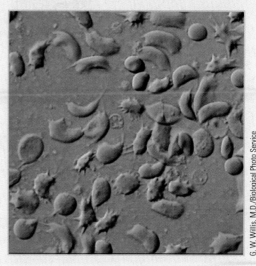

Figure 13D Blood cells from a patient with sickle cell anemia. Both normal cells (round) and sickle cells (shriveled) are visible.

There is no known cure for sickle cell anemia. Recently the Food and Drug Administration approved hydroxyurea, (sold under the name Droxia) to treat and control the symptoms of the disease. Hydroxyurea prompts the bone marrow to manufacture fetal hemoglobin (HbF), which does not have beta chains where the mutation occurs. Thus red blood cells containing HbF do not sickle and do not clog the capillaries. With hydroxyurea therapy, the bone marrow still manufactures mutated HbS, but the presence of cells with fetal hemoglobin dilutes the concentration of the sickling cells, thereby relieving the symptoms of the disease.

13.8 The Secondary Structure of Proteins

Proteins can fold or align themselves in such a manner that certain patterns repeat themselves. These repeating patterns are referred to as **secondary structures.** The two most common secondary structures encountered in proteins are the α-helix and the β-pleated sheet (Figure 13.6), which were proposed by Linus Pauling and Robert Corey in the 1940s. In contrast, those protein conformations that do not exhibit a repeated pattern are called random coils (Figure 13.7).

In the α-helix form, a single protein chain twists in such a manner that its shape resembles a right-handed coiled spring—that is, a helix. The shape of the helix is maintained by numerous **intramolecular hydrogen bonds** that exist between the backbone —C=O and H—N— groups. As shown in Figure 13.6, there is a hydrogen bond between the —C=O oxygen atom of each peptide linkage and the —N—H hydrogen atom of another peptide linkage four amino acid residues farther along the chain.

Linus Pauling (1901–1994), who won the 1954 Nobel Prize in chemistry, identified these structures. He also discovered or contributed much to our understanding of certain fundamental concepts, including chemical bonds and electronegativity.

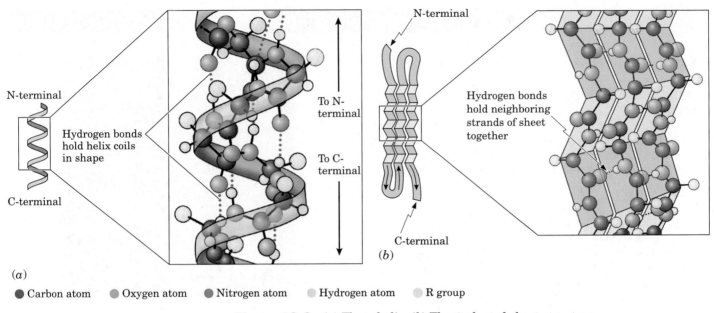

● Carbon atom   ● Oxygen atom   ● Nitrogen atom   ● Hydrogen atom   ● R group

Figure 13.6 (*a*) The α-helix. (*b*) The β-pleated sheet structure.

These hydrogen bonds are in just the right position to cause the molecule (or a portion of it) to maintain a helical shape. Each —N—H points upward and each C=O points downward, roughly parallel to the axis of the helix. All the R— groups (the amino acid side chains) point outward from the helix.

The other important orderly structure in proteins is the **β-pleated sheet.** In this case, the orderly alignment of protein chains is maintained by **intermolecular or intramolecular hydrogen bonds.** The β-pleated sheet structure can occur between molecules when polypeptide chains run parallel (all the N-terminal ends on one side) or antiparallel (neighboring N-terminal ends on opposite sides). β-Pleated sheets can also occur intramolecularly, when the polypeptide chain makes a U-turn, forming a hairpin structure, and the pleated sheet is antiparallel (Figure 13.6).

In all secondary structures, the hydrogen bonding is between backbone —C=O and H—N— groups. This is the distinction between secondary and tertiary structures. In the latter, as we shall see, the hydrogen bonding takes place between R groups on the side chains.

Few proteins have predominantly α-helix or β-pleated sheet structures. Most proteins, especially globular ones, have only certain portions of their molecules in these conformations. The rest of the molecule consists of **random**

> **Beta (β)-pleated sheet** A secondary protein structure in which the backbone of two protein chains in the same or different molecules is held together by hydrogen bonds

An intramolecular hydrogen bond goes from a hydrogen atom in a molecule to an O, N, or F atom in the same molecule.

Figure 13.7 A random coil.

Figure 13.8 Schematic structure of the enzyme carboxypeptidase. The β-pleated sheet portions are shown in blue, the green structures are the α-helix portions, and the orange strings are the random coil areas.

CHEMICAL CONNECTIONS 13E

Protein/Peptide Conformation–Dependent Diseases

In a number of diseases, a normal protein or peptide becomes pathological when its conformation changes. A common feature of these proteins is the property to self-assemble into β-sheet–forming amyloid (starch-like) plaques. These amyloid structures appear in several diseases.

One example of this process is the prion protein, the discovery of which brought Stanley Prusiner of the University of California, San Francisco, the Nobel Prize in 1997. When prion undergoes conformational change, it can cause the so-called mad cow disease. During the conformational change, the α-helical content of the normal prion protein unfolds and reassembles in the β-pleated sheet form. This new form supposedly has the potential to cause more normal prion proteins to undergo conformational change. In humans, it causes spongiform encephalitis; Creutzfeld-Jakob disease is one variant that

mainly afflicts the elderly. Although the transmission of this infection from diseased cows to humans is rare, fear of it caused the wholesale slaughter of British cattle in 1998 and, for a while, an embargo was placed on the importation of such meat in most of Europe and America.

Another such conformational change is found in the protein transthyretin. This protein normally forms a tetramer that carries vitamin A and thyroid hormone in the serum. When an individual inherits a mutant form of this protein from one of his or her parents, the tetramer becomes unstable and dissociates. These monomers, in turn, partially unfold and aggregate to form amyloid fibers that interfere with nerve and muscle function. The resulting disease is called senile systematic amyloidoses. β-amyloid plaques also appear in Alzheimer's disease brains (see Chemical Connections 15D).

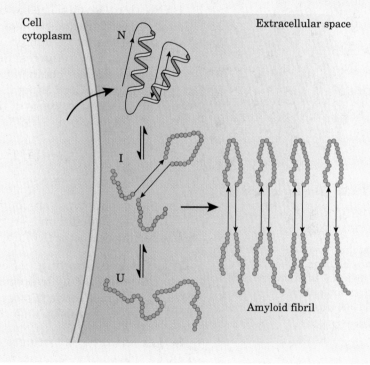

■ **Figure 13E** Schematic representation of a possible mechanism of amyloid fibril formation. After synthesis, the protein is assumed to fold in native (N) secondary structure aided by chaperones. Under certain conditions, the native structure can partially unfold (I) and form β-pleated sheets of amyloid fibrils or even completely unfold (U) as a random coil.

coil. Many globular proteins contain all three kinds of secondary structures in different parts of their molecules: α-helix, β-pleated sheet, and random coil. Figure 13.8 shows a schematic representation of such a structure.

Keratin, a fibrous protein of hair, fingernails, horns, and wool, is one protein that does have a predominantly α-helix structure. Silk is made of fibroin, another fibrous protein, which exists mainly in the β-pleated sheet form. Silkworm silk and especially spider silk exhibit a combination of strength and toughness unmatched by high-performance synthetic fibers. In its primary structure, silk contains sections that consist of only alanine (25%) and glycine (42%). The formation of β-pleated sheets, largely by the alanine sections, allows microcrystals to orient themselves along the fiber axis, which accounts for the material's superior mechanical strength.

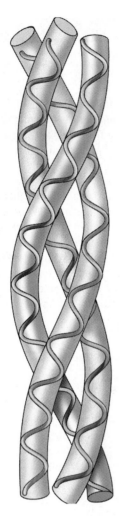

Figure 13.9 The triple helix of collagen.

Hydroxyproline is not one of the 20 amino acids listed in Table 13.1, but the body makes it from proline and uses it in certain proteins, of which collagen is but one example.

Another repeating pattern classified as a secondary structure is the **extended helix** of collagen (Figure 13.9). It is quite different from the α-helix. Collagen is the structural protein of connective tissues (bone, cartilage, tendon, blood vessels, skin), where it provides strength and elasticity. The most abundant protein in humans, it makes up about 30% by weight of all the body's protein. The extended helix structure is made possible by the primary structure of collagen. Each strand of collagen consists of repetitive units that can be symbolized as Gly—X—Y; that is, every third amino acid in the chain is glycine. Glycine, of course, has the shortest side chain (—H) of all amino acids. About one-third of the X amino acid is proline, and the Y is often hydroxyproline.

13.9 The Tertiary and Quaternary Structures of Proteins

A Tertiary Structure

In general, tertiary structures are stabilized four ways:

1. Covalent Bonds The covalent bond most often involved in stabilization of the tertiary structure of proteins is the disulfide bond. In Section 13.4, we noted that the amino acid cysteine is easily converted to the dimer cystine. When a cysteine residue is in one chain and another cysteine residue is in another chain (or in another part of the same chain), formation of a disulfide bond provides a covalent linkage that binds together the two chains or the two parts of the same chain:

$$\text{—SH HS—} \xrightarrow{[O]} \text{—S—S—}$$

Examples of both types are found in the structure of insulin (Figure 13.4).

Besides covalent bonds, three other interactions (Figure 13.10) can stabilize tertiary structures.

2. Hydrogen Bonding In Section 13.8, we saw that secondary structures are stabilized by hydrogen bonding between backbone —C=O and H—N— groups. Tertiary structures are stabilized by hydrogen bonding between polar groups on side chains.

3. Salt Bridges Salt bridges occur only between two amino acids with ionized side chains—that is, between an acidic amino acid (—COO⁻) and a basic amino acid, (—NH₃⁺ or =NH₂⁺) side chain. The two are held together by simple ion–ion attraction.

4. Hydrophobic Interactions In aqueous solution, globular proteins usually turn their polar groups outward, toward the aqueous solvent, and their nonpolar groups inward, away from the water molecules. The nonpolar groups prefer to interact with each other, excluding water from these regions. The result is hydrophobic interactions (see Section 12.1). Although this type of interaction is weaker than hydrogen bonding or salt bridges, it usually acts over large surface areas so that cooperatively the interactions are strong enough to stabilize a loop or some other tertiary structure formation.

Figure 13.11 illustrates the four types of interactions that stabilize the tertiary structures of proteins.

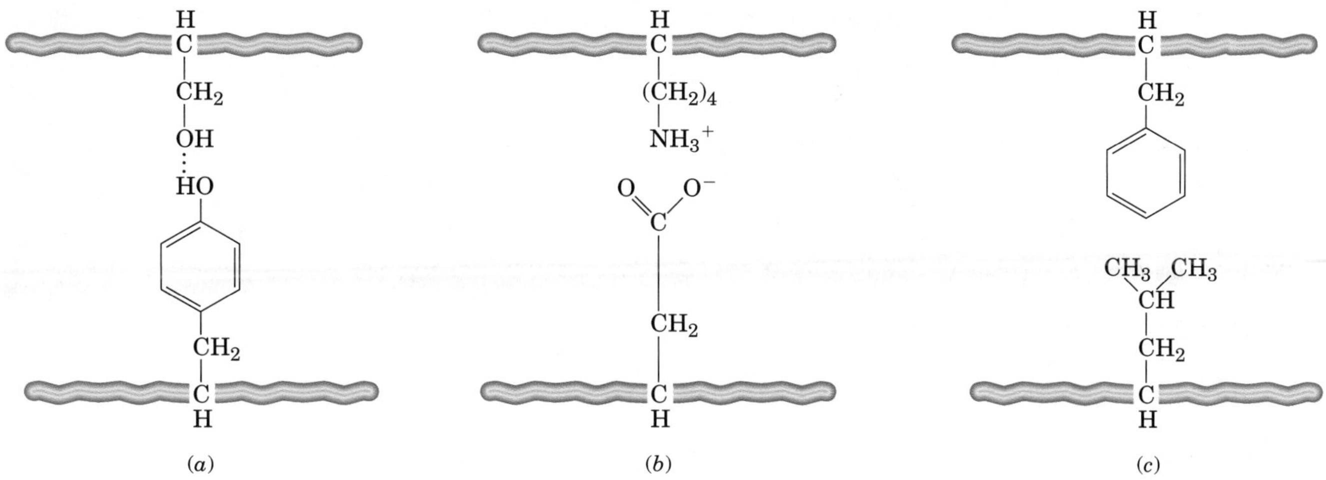

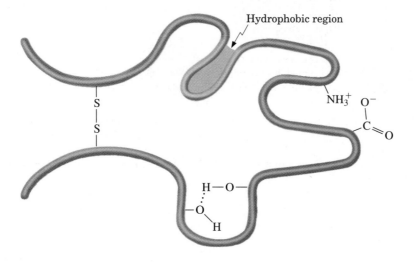

(a)   (b)   (c)

Figure 13.10 Noncovalent interactions that stabilize the tertiary and quaternary structures of proteins: (a) hydrogen bonding, (b) salt bridge, (c) hydrophobic interaction.

EXAMPLE 13.2

What kind of noncovalent interaction occurs between the side chains of serine and glutamine?

Solution
The side chain of serine ends in an —OH group; that of glutamine ends in an amide, CO—NH$_2$ group. The two groups can form hydrogen bonds.

Problem 13.2

What kind of noncovalent interaction occurs between the side chains of arginine and glutamic acid?

In Section 13.7, we pointed out that the primary structure of a protein largely determines its secondary and tertiary structures. We can now see the reason for this. When the particular R groups are in the proper positions, all the hydrogen bonds, salt bridges, disulfide linkages, and hydrophobic interactions that stabilize the three-dimensional structure of that molecule can form.

Figure 13.11 The four types of interactions that stabilize the tertiary structure of proteins, shown schematically.

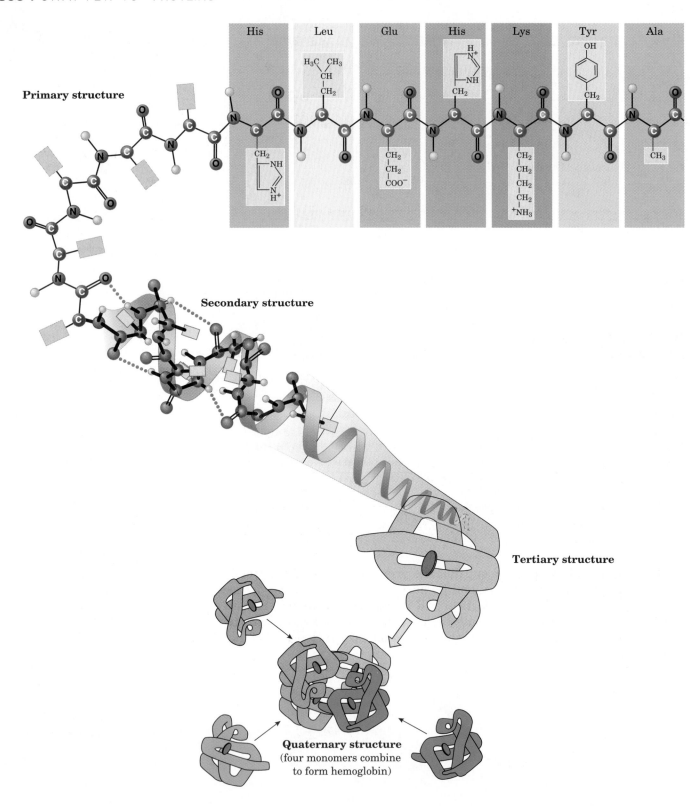

Figure 13.12 Primary, secondary, tertiary, and quaternary structures of protein.

CHEMICAL CONNECTIONS 13F

Proteomics, Ahoy!

Proteins in the body are in a state of dynamic flux. Their multiple functions necessitate that they change constantly: Some are rapidly synthesized, others have their synthesis inhibited; some are degraded, others are modified. There is a new concerted effort being made to catalogue all the proteins in their various forms in a particular cell or a tissue. The name of this venture is *proteomics,* a term coined as an analogy to *genomics* (see Chemical Conenctions 16D), in which all the genes of an organism and their locations in the chromosomes were determined. In proteomics, all the proteins and peptides of a cell or tissue are separated, and their masses are then characterized and compared with a computerized database to obtain and identify a comprehensive protein profile. In this manner, the dynamic states of a large number of proteins are obtained simultaneously and the status of a cell or tissue can be ascertained whether it is healthy or pathological.

The side chains of some proteins allow them to fold (form a tertiary structure) in only one way; other proteins, especially those with long polypeptide chains, can fold in a number of possible ways. Certain proteins in living cells, called **chaperones,** help a newly synthesized polypeptide chain assume the proper secondary and tertiary structures that are necessary for the functioning of that molecule and prevent foldings that would yield biologically inactive molecules.

Chaperone A protein molecule that helps other proteins to fold into the biologically active conformation and enables partially denatured proteins to regain their biologically active conformation

When a protein consists of more than one polypeptide chain, each is called a *subunit.*

B Quaternary Structure

The highest level of protein organization is the quaternary structure, which applies mainly to proteins with more than one polypeptide chain. Figure 13.12 summarizes schematically the four levels of protein structure. Quaternary structure determines how the different subunits of the protein fit into an organized whole. The subunits are packed and held together by hydrogen bonds, salt bridges, and hydrophobic interactions—the same forces that operate within tertiary structures. The quaternary structure also determines how the different segments of a single polypeptide chain fit into an organized whole when they traverse and interact with lipid bilayers of membranes.

1. Hemoglobin Hemoglobin in adult humans is made of four chains (called globins): two identical chains (alpha) of 141 amino acid residues each and two identical chains (beta) of 146 residues each. Figure 13.13 shows how the four chains fit together.

In hemoglobin, each globin chain surrounds an iron-containing heme unit, the structure of which is shown in Figure 13.14. Proteins that contain

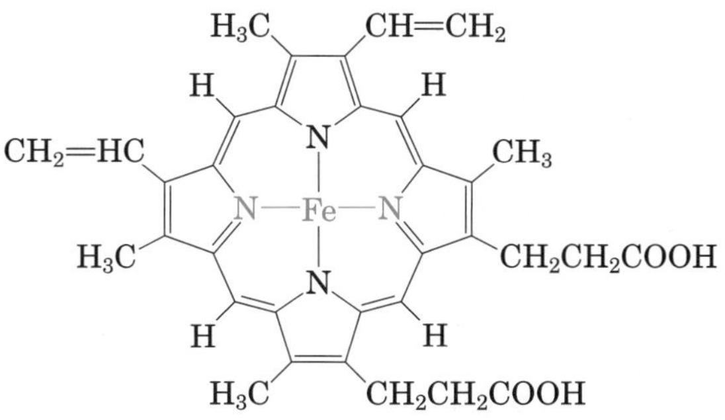

Heme molecules with iron ion

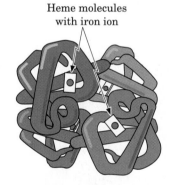

Hemoglobin

Figure 13.13 The quaternary structure of hemoglobin.

Figure 13.14 The structure of heme.

CHEMICAL CONNECTIONS 13G

The Role of Quaternary Structures in Mechanical Stress and Strain

Collagen is the main component of the extracellular matrix that exists between cells of higher organisms. The triple-helix units of collagen (Figure 13.9) are further organized into a pattern called the quarter-stagger arrangement, forming fibrils (Figure 13G.1). Fibers made of these fibrils provide elasticity to such tissues as skin and cornea of the eye.

In the **quarter-stagger arrangement,** the units along a row are not spaced end to end. Rather, there is a gap between the end of one unit and the beginning of another unit. These gaps play an important role in bone formation. Collagen is one of the main constituents of bones and teeth, and the gaps in the quarter-stagger arrangement are essential for the deposition of inorganic crystals of calcium hydroxyapatite, $Ca_5(PO_4)_3OH$. The gaps serve as nucleation sites for the growth of these crystals. The combination of hydroxyapatite crystals and collagen creates a hard material that still has some springiness, owing to the presence of collagen. Dentine, the main constituent of the internal part of a tooth, contains a higher percentage (about 75%) of inorganic crystals than does bone and is therefore harder. Enamel, the outer part of the tooth, has a still higher mineral content (about 95%) and is even harder.

The quaternary structures of other protein molecules residing inside cells provide both shape and strength. These fibrous filaments, commonly called cytoskeleton, come in different sizes.

The microfilaments are 7 nm in diameter and are made of globular G-actin molecules (Figure 13G.2), which under physiological conditions form fibrous F-actin. In F-actin, the single globes are polymerized to form a double-helical structure, which provides the elasticity required when cells move, change their shape, or divide. Some of these actin filaments become anchored to cell membranes by other proteins. Through these proteins, they are connected to the collagen in the extracellular matrix. In this fashion, stresses and strains inside and outside the cells are coordinated. This same F-actin, in the specialized muscle cells, interacts with myosin (see Figure 18.11) to provide the contraction of muscles.

The cytoskeleton with the largest diameter is called a microtubule. These are 30-nm-wide cylindrical structures are formed by assembling two similar proteins, α- and β-tubulin. In the presence of Ca^{2+} ions, the dimer is quite stable. Some of these dimers form a helical structure in which 13 units form a turn. The microtubules are not static, however. At the expense of energy, they constantly shed dimeric units at the "minus end" and add new dimeric units at the "plus end" at about the same rate (Figure 13G.3). Microtubules provide the internal scaffolding of cells in much the same way that bones are the scaffolding in the human body.

(continued on next page)

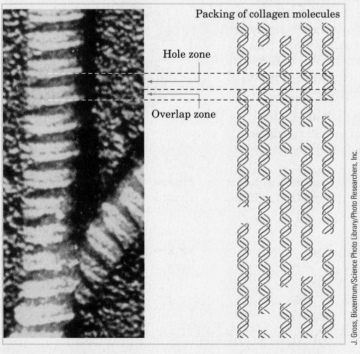

J. Gross, Biozentrum/Science Photo Library/Photo Researchers, Inc.

■ **Figure 13G.1** Under the electron microscope, collagen fibers exhibit alternating light and dark bands. The dark bands correspond to the 40-nm gaps or "holes" between pairs of aligned collagen triple helices.

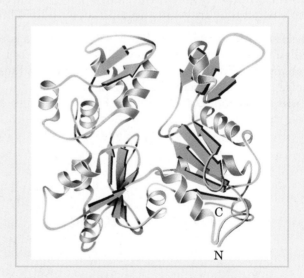

■ **Figure 13G.2** The three-dimensional structure of an actin monomer from skeletal muscle. This view shows the two domains *(left and right)* of actin.

CHEMICAL CONNECTIONS 13G

The Role of Quaternary Structures in Mechanical Stress and Strain *(continued)*

Microtubules also act as "railroads" along which molecules and organelles are transported from one part of the cell to another. For example, in nerve cells organelles and vesicles move at a fast rate, 2–5 μm/s along microtubules. This movement is unidirectional from the minus end to the plus end of microtubules, resulting in the transport from the cell body to the peripheral axons. A 360-kD protein, called kinesin, provides the walking leg of this motion. Kinesin molecules attach themselves with their tails to the organelle or vesicle to be transported. In this way, the heads of kinesin molecules "walk" along the microtubule in 8 nm small steps. At each step, the kinesin hydrolyzes an ATP molecule to supply the energy of the motion.

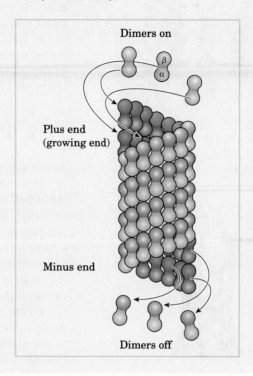

■ **Figure 13G.3**
A model treadmilling of microtubules.

---

non-amino acid portions are called **conjugated proteins.** The non-amino acid portion of a conjugated protein is called a **prosthetic group.** In hemoglobin, the globins are the amino acid portions and the heme units are the prosthetic groups.

Hemoglobin containing two alpha and two beta chains is not the only kind existing in the human body. In the early development stage of the fetus, the hemoglobin contains two alpha and two gamma chains. The fetal hemoglobin has greater affinity for oxygen than does the adult hemoglobin. In this way, the mother's red blood cells carrying oxygen can pass it to the fetus for its own use. Fetal hemoglobin also alleviates some of the symptoms of sickle cell anemia (see Chemical Connections 13D).

The oxygen- and carbon dioxide-carrying functions of hemoglobin are discussed in Sections 23.3 and 23.4.

2. Collagen Another example of quaternary structure and higher organizations of subunits can be seen in collagen. The triple helix units, called *tropocollagen,* constitute the soluble form of collagen; they are stabilized by hydrogen bonding between the backbones of the three chains. Collagen is made of many tropocollagen units. Tropocollagen is found only in fetal or young connective tissues. With aging, the triple helixes (Figure 13.9) that organize themselves into fibrils cross-link and form insoluble collagen. In collagen, the **cross-linking** consists of covalent bonds that link together two lysine residues on adjacent chains of the helix. This cross-linking of collagen is an example of the tertiary structures that stabilize the three-dimensional conformations of protein molecules. The quarter-stagger alignment of tropocollagen (Figure 13G.1) forms collagen fibrils, and the twisting of these fibrils into a five- or six-strand helix builds the collagen fibers. These higher organizations provide the superior strength that these fibers exhibit in tissues (Chemical Connections 13G).

Figure 13.15 Integral membrane protein of rhodopsin made of α-helices.

Figure 13.16 A β-barrel of integral membrane protein of the outer membrane of mitochondrion made of eight β-pleated sheets.

O-linked saccharide.

α-N-Acetylgalactosyl-serine

N-linked saccharide.

β-N-Acetylglucosyl-aspargine

3. Integral Membrane Proteins Integral membrane proteins traverse partly or completely a membrane bilayer. (See Figure 12.2.) It is estimated that about one-third of all proteins are integral membrane proteins. To keep the protein stable in the nonpolar environment of a lipid bilayer, it must form quaternary structures in which the outer surface is largely nonpolar and interacts with the lipid bilayer. Thus most of the polar groups of the protein must turn inward. Two such quaternary structures exist in integral membrane proteins: (1) often 6 to 10 α-helices crossing the membrane and (2) β-barrels made of 8, 12, 16, or 18 anti-parallel β-sheets (Figures 13.15 and 13.16).

13.10 Glycoproteins

Although many proteins, such as serum albumin, consist exclusively of amino acids, others also contain covalently linked carbohydrates and are therefore classified as **glycoproteins.** These include most of the plasma proteins (for example, fibrinogen), enzymes such as ribonuclease, hormones such as thyroglobulin, storage proteins such as casein and ovalbumins, and protective proteins such as immunoglobulins and interferon. The carbohydrate content of these proteins may vary from a few percent (immunoglobulins) up to 85% (blood group substances, see Chemical Connections 11D). Most of the proteins in membranes (lipid bilayers; see Section 12.5) are glycoproteins.

Glycoproteins have predominantly two kinds of linkages between the protein and the carbohydrate parts. The **O-linked saccharides** are bonded to an —OH group on the serine or threonine side chain of the protein. The linkage itself is a glycosidic bond. For example, mucins, which coat and protect the mucous membranes, are O-linked saccharides of glycoproteins. They line the respiratory and gastrointestinal tracts as well as the cervix. These highly viscous glycoproteins help to protect the underlying cells from harmful environmental agents.

The second kind of linkage between the carbohydrate and the protein is the N-glycosidic bond (—C—N—). This bond exists between the N of an asparagine residue of the protein chain and the C-1 (anomeric carbon) of N-acetylglucosamine (Section 11.2). This kind of glycoprotein is called an **N-linked saccharide.** N-Acetylglucosamine is the first unit in the fork-like branched oligosaccharidic chain. Immunoglobulins are an example of N-linked saccharides. In many cases, the function involved in adding N-linked saccharides to the protein core is to make them more water-soluble and to facilitate their transport within the cell.

A special class of glycoproteins are the proteoglycans. These compounds have a protein core, and the side chains are long polysaccharide chains made of acidic polysaccharides (also called glycosaminoglycans), as discussed in Section 11.7. Proteoglycans are essential parts of the extracellular matrix that provide strength, flexibility, and elasticity in tissues such as skin, cartilage, and the cornea of the eye.

13.11 Denaturation

Protein conformations are stabilized in their native states by secondary and tertiary structures and through the aggregation of subunits in quaternary structure. Any physical or chemical agent that destroys these stabilizing structures changes the conformation of the protein (Table 13.2). We call this process **denaturation.**

Table 13.2	Modes of Protein Denaturation (Destruction of secondary and higher structures)
Denaturing Agent	**Affected Regions**
Heat	H bonds
6 *M* urea	H bonds
Detergents	Hydrophobic regions
Acids, bases	Salt bridges, H bonds
Salts	Salt bridges
Reducing agents	Disulfide bonds
Heavy metals	Disulfide bonds
Alcohol	Hydration layers

For example, heat cleaves hydrogen bonds, so boiling a protein solution destroys the α-helical and β-pleated sheet structure (compare Figures 13.6 and 13.7). In collagen, the triple helixes disappear upon boiling, and the molecules have a largely random-coil conformation in the denatured state, which is gelatin. In other proteins, especially globular proteins, heat causes the unfolding of the polypeptide chains; because of subsequent intermolecular protein–protein interactions, precipitation or coagulation then takes place. That is what happens when we boil an egg.

Similar conformational changes can be brought about by the addition of denaturing chemicals. Solutions such as 6 *M* aqueous urea, $H_2N—CO—NH_2$, break hydrogen bonds and cause the unfolding of globular proteins. Surface-active agents (detergents) change protein conformation by opening up the hydrophobic regions, whereas acids, bases, and salts affect both salt bridges and hydrogen bonds.

Reducing agents, such as 2-mercaptoethanol ($HOCH_2CH_2SH$), can break the —S—S— disulfide bonds, reducing them to —SH groups. The processes of permanent waving and straightening of curly hair are examples of the latter (Figure 13.17). The protein keratin, which makes up human hair, contains a high percentage of disulfide bonds. These bonds are primarily responsible for the shape of the hair, whether straight or curly. In either permanent waving or straightening, the hair is first treated with a reducing agent that cleaves some of the —S—S— bonds. This treatment allows the molecules to lose their rigid orientations and become more flexi-

Denaturation The loss of the secondary, tertiary, and quaternary structures of a protein by a chemical or physical agent that leaves the primary structure intact

■ **Reduction of disulfide bond.**

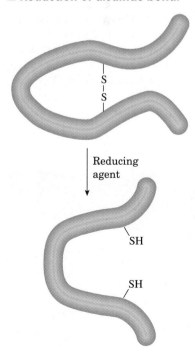

Charles D. Winters

Figure 13.17 Permanent wave alters the shape of hair through reduction and oxidation of disulfides and thiols.

C H E M I C A L   C O N N E C T I O N S   1 3 H

Laser Surgery and Protein Denaturation

Proteins can be denatured by physical means, most notably by heat. For instance, bacteria are killed and surgical instruments are sterilized by heat. A special method of heat denaturation that is seeing increasing use in medicine relies on lasers. A laser beam (a highly coherent light beam of a single wavelength) is absorbed by tissues, and its energy is converted to heat energy. This process can be used to cauterize incisions so that a minimal amount of blood is lost during the operation.

Laser beams can be delivered by an instrument called a **fiberscope.** The laser beam is guided through tiny fibers, thousands of which are fitted into a tube only 1 mm in diameter. In this way, the laser delivers the energy for denaturation only where it is needed. It can, for example, seal wounds or join blood vessels without the necessity of cutting through healthy tissues. Fiberscopes have been used successfully to diagnose and treat many bleeding ulcers in the stomach, intestines, and colon.

A novel use of the laser fiberscope is in treating tumors that cannot be reached for surgical removal. A drug called Photofrin, which is activated by light, is given to patients intravenously. The drug in this form is inactive and harmless. The patient then waits 24 to 48 h, during which time the drug accumulates in the tumor but is removed and excreted from healthy tissues. A laser fiberscope with 630 nm of red light is then directed toward the tumor. An exposure between 10 and 30 min is applied. The energy of the laser beam activates the Photofrin, which destroys the tumor.

This technique does not offer a complete cure because the tumor may grow back, or it may have spread before the treatment. The treatment has only one side effect: The patient remains sensitive to exposure to strong light for approximately 30 days (so sunlight must be avoided). But this inconvenience is minor compared to the pain, nausea, hair loss, and other side effects that accompany radiation or chemotherapy of tumors.

In the United States, Photofrin is approved only to treat esophageal cancer. In Europe, Japan, and Canada, it is also

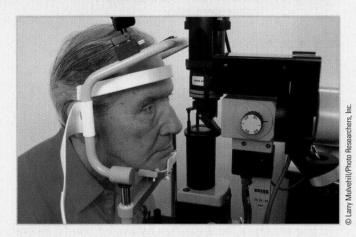

■ **Figure 13H**   Argon krypton laser surgery.

used to treat lung, bladder, gastric, and cervical cancers. The light that activates Photofrin penetrates only a few millimeters, but the new drugs under development may use radiation in the near infrared spectrum that can penetrate tumors up to a few centimeters.

The most common use of laser technology in surgery is its application to correct near-sightedness and astigmatism. In a computer-assisted laser surgery process, the curvature of the cornea is changed. Using the energy of the laser beam, physicians remove part of the cornea. In the procedure called photorefractive keratectomy (PRK), the outer layers of the cornea are denatured—that is, burned off. In the LASIK (*laser in situ keratomileusis*) procedure, the surgeon creates a flap or a hinge of the outer layers of the cornea and then with the laser beam burns off a computer-programmed amount under the flap to change the shape of the cornea. After the 5- to 10-minute procedure is complete, the flap is put back, and it heals without stitches. In successful surgeries, the patient regains good vision one day after the surgery and no longer needs prescription lenses.

■ **Raw egg whites are an antidote to heavy metal poisoning.**

ble. The hair is then set into the desired shape, using curlers or rollers, and an oxidizing agent is applied. The oxidizing agent reverses the preceding reaction, forming new disulfide bonds, which now hold the molecules together in the desired positions.

Heavy metal ions (for example, Pb^{2+}, Hg^{2+}, and Cd^{2+}) also denature protein by attacking the —SH groups. They form salt bridges, as in —$S^-Hg^{2+-}S$—. This very feature is taken advantage of in the antidote for heavy metal poisoning: raw egg whites and milk. The egg and milk proteins are denatured by the metal ions, forming insoluble precipitates in the stomach. These must be pumped out or removed by inducing vomiting. In this way, the poisonous metal ions are removed from the body. If the antidote is not pumped out of the stomach, the digestive enzymes would degrade the proteins and release the poisonous heavy metal ions, which would then be absorbed in the bloodstream.

Other chemical agents such as alcohol also denature proteins, coagulating them. This process is used in sterilizing the skin before injections.

At a concentration of 70%, ethanol penetrates bacteria and kills them by coagulating their proteins, whereas 95% alcohol denatures only surface proteins.

Denaturation changes secondary, tertiary, and quaternary structures. It does not affect primary structures (that is, the sequence of amino acids that make up the chain). If these changes occur to a small extent, denaturation can be reversed. For example, when we remove a denatured protein from a urea solution and put it back into water, it often reassumes its secondary and tertiary structure. This is reversible denaturation. In living cells, some denaturation caused by heat can be reversed by **chaperones.** These proteins help a partially heat-denatured protein to regain its native secondary, tertiary, and quaternary structures. Some denaturation, however, is irreversible. We cannot unboil a hard-boiled egg, for example.

S U M M A R Y

Proteins are giant molecules made of amino acids linked together by **peptide bonds.** They have many functions (Section 13.1): structural (collagen), enzymatic (ribonuclease), carrier (hemoglobin), storage (casein), protective (immunoglobulin), and hormonal (insulin). **Amino acids** are organic compounds containing an amino ($-NH_2$) and a carboxylic ($-COOH$) acid group (Section 13.2). The 20 amino acids found in proteins are classified by their side chains: nonpolar, polar but neutral, acidic, and basic. All amino acids in human tissues are L amino acids. Amino acids in the solid state, as well as in water, carry both positive and negative charges; they are called **zwitterions** (Section 13.3). The pH at which the number of positive charges equals the number of negative charges is the **isoelectric point** of an amino acid or protein.

When the amino group of one amino acid condenses with the carboxyl group of another amino acid, an amide (peptide) linkage is formed, with the elimination of water. The two amino acids form a dipeptide (Section 13.5). Three amino acids form a tripeptide. Many amino acids form a **polypeptide chain.** Proteins are made of one or more polypeptide chains.

The linear sequence of amino acids is the **primary structure** of proteins (Section 13.7). The repeating short-range conformations (**α-helix, β-pleated sheet, extended helix of collagen,** and **random coil**) are the **secondary structures** (Section 13.8). The **tertiary structure** is the three dimensional conformation of the protein molecule (Section 13.9). Tertiary structures are maintained by covalent cross-links such as **disulfide bonds** and by **salt bridges, hydrogen bonds,** and **hydrophobic interactions** between the side chains. The precise fit of polypeptide subunits into an aggregated whole is called the **quaternary structure** (Section 13.9).

Many proteins are classified as **glycoproteins** because they contain carbohydrate units (Section 13.10).

Secondary and tertiary structures stabilize the native conformation of proteins; physical and chemical agents, such as heat or urea, destroy these structures and **denature** the protein (Section 13.11). Protein functions depend on native conformation; when a protein is denatured, it can no longer carry out its function. Some (but not all) denaturation is reversible; in some cases, **chaperone** molecules reverse denaturation.

P R O B L E M S

Numbers that appear in color indicate difficult problems. ► designates problems requiring application of principles.

13.3 What are the functions of (a) ovalbumin and (b) myosin?

13.4 The members of which class of proteins are insoluble in water and can serve as structural materials?

Amino Acids

13.5 What is the difference in structure between tyrosine and phenylalanine?

13.6 Classify the following amino acids as nonpolar, polar but neutral, acidic, or basic.

(a) Arginine   (b) Leucine
(c) Glutamic acid   (d) Asparagine
(e) Tyrosine   (f) Phenylalanine
(g) Glycine

13.7 Which amino acid has the highest percent nitrogen (g N/100 g amino acid)?

13.8 Why does glycine have no D or L form?

13.9 Draw the structure of proline. To which class of heterocyclic compounds does this molecule belong? (See Section 8.2.)

13.10 Which amino acid is also a thiol?

13.11 Why is it necessary to have proteins in our diets?

13.12 Which amino acids in Table 13.1 have more than one stereocenter?

13.13 What are the similarities and differences in the structures of alanine and phenylalanine?

13.14 Two of the 20 amino acids can be obtained by hydroxylation of other amino acids listed in Table 13.1. What are those two, and what are the precursor amino acids?

13.15 Draw the structures of L- and D-valine.

Zwitterions

13.16 Why are all amino acids solids at room temperature?

13.17 Show how alanine, in solution at its isoelectric point, acts as a buffer (write equations to show why the pH does not change much if we add an acid or a base).

13.18 Explain why an amino acid cannot exist in an un-ionized form [$RCH(NH_2)COOH$] at any pH.

13.19 Draw the structure of valine at pH 1 and at pH 12.

Peptides and Proteins

13.20 A tetrapeptide is abbreviated as DPKH. Which amino acid is at the N-terminal and which is at the C-terminal?

13.21 Draw the structure of a tripeptide made of threonine, arginine, and methionine.

13.22 (a) Use the three-letter abbreviations to write the representation of the following tripeptide.

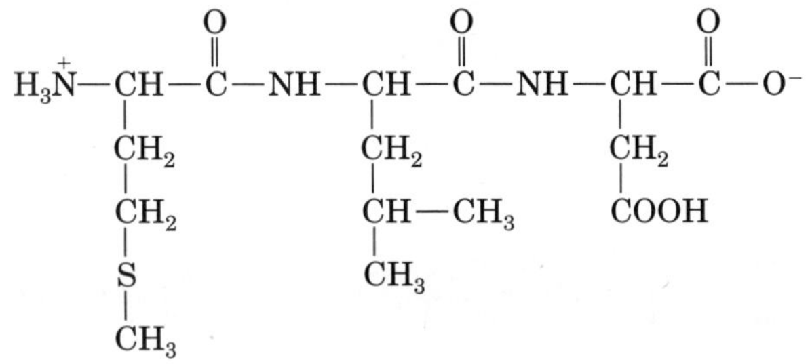

$$
\overset{+}{H_3}N-CH-\overset{\overset{\displaystyle O}{\|}}{C}-NH-CH-\overset{\overset{\displaystyle O}{\|}}{C}-NH-CH-\overset{\overset{\displaystyle O}{\|}}{C}-O^-
$$

with side chains:
- CH_2 — CH_2 — CH_2
- CH_2 — $CH-CH_3$ — $COOH$
- S — CH_3
- CH_3

(b) Which amino acid is the C-terminal end, and which is the N-terminal end?

13.23 Draw the structure of the dipeptide leucylproline.

13.24 A polypeptide chain is made of alternating valine and phenylalanine. Which part of the polypeptide is polar (hydrophilic)?

13.25 Show by chemical equations how alanine and glutamine can be combined to give two different dipeptides.

Properties of Peptides and Proteins

13.26 (a) How many atoms of the peptide bond lie in the same plane?
(b) Which atoms are they?

13.25 (a) Draw the structural formula of the tripeptide Met—Ser—Cys.
(b) Draw the different ionic structures of this tripeptide at pH 2.0, 7.0, and 10.0.

13.28 How can a protein act as a buffer?

13.29 Proteins are least soluble at their isoelectric points. What would happen to a protein precipitated at its isoelectric point if a few drops of dilute HCl were added?

Primary Structure of Proteins

13.30 How many different tripeptides can be made (a) using one, two, or three residues each of leucine, threonine, and valine and (b) using all 20 amino acids?

13.31 How many different tetrapeptides can be made (a) if the peptides contain the residues of asparagine, proline, serine, and methionine and (b) if all 20 amino acids can be used?

13.32 How many amino acid residues in the A chain of insulin are the same in insulin from humans, cattle (bovine), hogs, and sheep?

13.33 Based on your knowledge of the chemical properties of amino acid side chains, suggest a substitution for leucine in the primary structure of a protein that would probably not change the character of the protein very much.

Secondary Structure of Proteins

13.34 Is a random coil a (a) primary, (b) secondary, (c) tertiary, or (d) quaternary structure? Explain.

13.35 Decide whether the following structures that exist in collagen are primary, secondary, tertiary, or quaternary.
(a) Tropocollagen
(b) Collagen fibril
(c) Collagen fiber
(d) The proline-hydroxyproline-glycine repeating sequence

13.36 Proline is often called an α-helix terminator; that is, it is usually in the random-coil secondary structure following an α-helix portion of a protein chain. Why does proline not fit easily into an α-helix structure?

Tertiary and Quaternary Structures of Proteins

13.37 Polyglutamic acid (a polypeptide chain made only of glutamic acid residues) has an α-helix conformation below pH 6.0 and a random-coil conformation above pH 6.0. What is the reason for this conformational change?

13.38 Distinguish between intermolecular and intramolecular hydrogen bonding between backbone groups. Where in protein structures do you find one and where do you find the other?

13.45 *(continued)*
two entities form? To what group of proteins does cytochrome c belong?

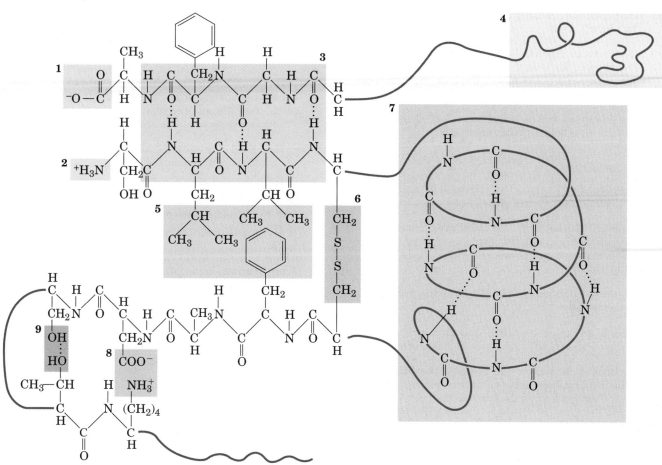

13.39 Identify the different primary, secondary, and tertiary structures in the numbered boxes:

13.40 If both cysteine residues on the B chain of insulin were changed to alanine residues, how would it affect the quaternary structure of insulin?

13.41 In a 6 *M* urea solution, a protein that contained mostly anti-parallel β-pleated sheet became a random coil. Which groups and bonds were affected by urea?

13.42 (a) What is the difference in the quaternary structure between fetal hemoglobin and adult hemoglobin?
(b) Which can carry more oxygen?

13.43 Where are the nonpolar side chains of proteins located in the β-barrel of an integral membrane protein?

13.44 What kind of changes are necessary to transform a protein having a predominantly α-helical structure into one having a β-pleated sheet structure?

13.45 The cytochrome c protein is important in producing energy from food. It contains a heme surrounded by a polypeptide chain. What kind of structure do these

Glycoproteins

13.46 In many glycoproteins, the protein carbohydrate linkage occurs on a serine side chain between the —OH group of the protein and the hemiacetal,

$$\overset{-O\quad H}{\underset{R\quad OH}{C}}$$, of the monosaccharide—for example,

D-glucose. Give the structure of such a protein–carbohydrate linkage in which a β-glycosidic bond has been established (see Section 11.5).

13.47 Draw the structure of an N-linked saccharide where the amide nitrogen group on the side chain of aspargine is linked to C-1 of N-acetylglucosamine by a β-N-glycosidic bond.

13.48 What is the difference between a glycoprotein and a proteoglycan?

13.49 Collagen is the most abundant protein in the human body. This fibrous protein contains short, mainly disaccharidic side chains, attached to hydroxylysyl residues. How would you classify collagen?

Protein Denaturation

13.50 Which amino acid side chain is most frequently involved in denaturation by reduction?

13.51 What does the reducing agent do in straightening curly hair?

13.52 Silver nitrate, $AgNO_3$, is sometimes put into the eyes of newborn infants as a disinfectant against gonorrhea. Silver is a heavy metal. Explain how this treatment may work against bacteria.

13.53 Why do nurses and physicians use 70% alcohol to wipe the skin before giving injections?

Chemical Connections

13.54 (Chemical Connections 13A) Write the structure of the oxidation product of glutathione.

13.55 (Chemical Connections 13B) AGE products become disturbing only in elderly people, even though they also form in younger people. Why don't they harm younger people?

13.56 (Chemical Connections 13C) Define *hypoglycemic awareness*.

13.57 (Chemical Connections 13D) How does hydroxyurea therapy alleviate the symptoms of sickle cell anemia?

13.58 (Chemical Connections 13E) What is the difference in the conformation between normal prion protein and the amyloid prion that causes mad cow disease?

13.59 (Chemical Connections 13F) What is the aim of proteomics?

13.60 (Chemical Connections 13G) What constituent of bone determines its hardness?

13.61 (Chemical Connections 13H) How does the fiberscope help to heal bleeding ulcers?

Additional Problems

13.62 Which diseases are associated with amyloid plaques?

13.63 How many different dipeptides can be made (a) using only alanine, tryptophan, glutamic acid, and arginine and (b) using all 20 amino acids?

13.64 Denaturation is usually associated with transitions from helical structures to random coils. If an imaginary process were to transform the keratin in your hair from an α-helix to a β-pleated sheet structure, would you call the process denaturation? Explain.

13.65 What are β-barrels of integral membrane proteins made of?

13.66 Draw the structure of lysine (a) above, (b) below, and (c) at its isoelectric point.

13.67 In collagen, some of the chains of the triple helices in tropocollagen are cross-linked by covalent bonds between two lysyl residues, $-(CH_2)_4-NH-(CH_2)_4-$. What kind of structure is formed by these cross-links? Explain.

13.68 Considering the vast number of animal and plant species on Earth (including those now extinct) and the large variety of protein molecules in each organism, have all possible protein molecules been used already by some species or other? Explain.

13.69 What kind of noncovalent interaction occurs between the following amino acids?

(a) Valine and isoleucine

(b) Glutamic acid and lysine

(c) Tyrosine and threonine

(d) Alanine and alanine

13.70 How many different decapeptides (peptides containing 10 amino acids each) can be made from the 20 amino acids?

13.71 Which amino acid does not rotate the plane of polarized light?

13.72 Write the expected products of the acid hydrolysis of the following tetrapeptide:

$$H_3\overset{+}{N}-CH-\overset{O}{\overset{\|}{C}}-NH-CH-\overset{O}{\overset{\|}{C}}-NH-CH-\overset{O}{\overset{\|}{C}}-NH-CH-\overset{O}{\overset{\|}{C}}-O^-$$

13.73 What charges are on aspartic acid at pH 2.0?

13.74 How many ways can you link the two amino acids in the dipeptide Val—Lys? Which of these peptide bonds will you find in proteins?

13.75 Carbohydrates are integral parts of glycoproteins. Does the addition of carbohydrate to the protein core make the protein more or less soluble in water? Explain.

InfoTrac College Edition

For additional reading, go to InfoTrac College Edition, your online research library, at

http://infotrac.thomsonlearning.com

CHAPTER 14

14.1 Introduction

14.2 Naming and Classifying Enzymes

14.3 Common Terms in Enzyme Chemistry

14.4 Factors Affecting Enzyme Activity

14.5 Mechanism of Enzyme Action

14.6 Enzyme Regulation

14.7 Enzymes in Medical Diagnosis and Treatment

A hydrothermal vent on the ocean floor along with marine life, including tube worms.

© NSF Oasis Project/Norbert Wu Photography

Enzymes

14.1 Introduction

The cells in your body are chemical factories. Only a few of the thousands of compounds necessary for the operation of the human organism are obtained from the diet. Instead, most of them are synthesized within the cells, which means that hundreds of chemical reactions take place in your cells every second of your life.

Nearly all of these reactions are catalyzed by **enzymes,** which are large protein molecules that increase the rates of chemical reactions without themselves undergoing any change. Without enzymes, life as we know it would not be possible.

Proteins are not the only biological catalysts, however. **Ribozymes** are enzymes made of ribonucleic acids. They catalyze the self-cleavage of certain portions of their own molecules. Many biochemists believe that during evolution RNA catalysts emerged first, with protein enzymes arriving on the scene later. (We will learn more about RNA catalysts in Section 16.4.)

Like all catalysts, enzymes do not change the position of equilibrium. That is, enzymes cannot make a reaction take place that would not occur without them. What they do is increase the rate; they cause reactions to

For the discovery of ribozymes, Sidney Altman of Yale University and Thomas R. Cech of the University of Colorado were awarded the 1989 Nobel Prize in chemistry.

317

take place faster by lowering the activation energy. As catalysts, enzymes are remarkable in two respects:

1. They are extremely effective, increasing reaction rates by anywhere from 10^9 to 10^{20} times.

2. Most of them are extremely **specific.**

> **Enzyme specificity** The limitation of an enzyme to catalyze one specific reaction with one specific substrate

As an example of their effectiveness, consider the oxidation of glucose. A lump of glucose or even a glucose solution exposed to oxygen under sterile conditions would show no appreciable change for months. In the human body, however, the same glucose is oxidized within seconds.

Every organism has many enzymes—many more than 3000 in a single cell. Most enzymes are very specific, each of them speeding up only one particular reaction or class of reactions. For example, the enzyme urease catalyzes only the hydrolysis of urea and not that of other amides, even closely related ones.

$$(NH_2)_2C{=}O + H_2O \xrightarrow{\text{urease}} 2\,NH_3 + CO_2$$
$$\text{Urea}$$

Another type of specificity can be seen with trypsin, an enzyme that cleaves the peptide bonds of protein molecules—but not every peptide bond, only those on the carboxyl side of lysine and arginine residues (Figure 14.1). The enzyme carboxypeptidase specifically catalyzes the hydrolysis on only the last amino acid on a protein chain—the one at the C-terminal end. Lipases are less specific: They cleave any triglyceride, but they still don't affect carbohydrates or proteins.

> The enzyme glucose oxidase that oxidizes D-glucose does not work on L-glucose. It is specific for β-D-glucose.

The specificity of enzymes also extends to stereospecificity. The enzyme arginase hydrolyzes the amino acid L-arginine (the naturally occurring form) to a compound called L-ornithine and urea (Section 19.8) but has no effect on its mirror image, D-arginine.

Enzymes are distributed according to the body's need to catalyze specific reactions. A large number of protein-splitting enzymes are in the blood, ready to promote clotting. Digestive enzymes, which also catalyze the hydrolysis of proteins, are located in the secretions of the stomach and pancreas. Even within the cells themselves, some enzymes are localized according to the need for specific reactions. The enzymes that catalyze the oxidation of compounds that are part of the citric acid cycle (Section 18.4) are located in the mitochondria, and special organelles such as lysosomes contain an enzyme (lysozyme) that catalyzes the dissolution of bacterial cell walls.

Figure 14.1 A typical amino acid sequence. The enzyme trypsin catalyzes the hydrolysis of this chain only at the points marked with an arrow (the —COOH side of lysine and arginine).

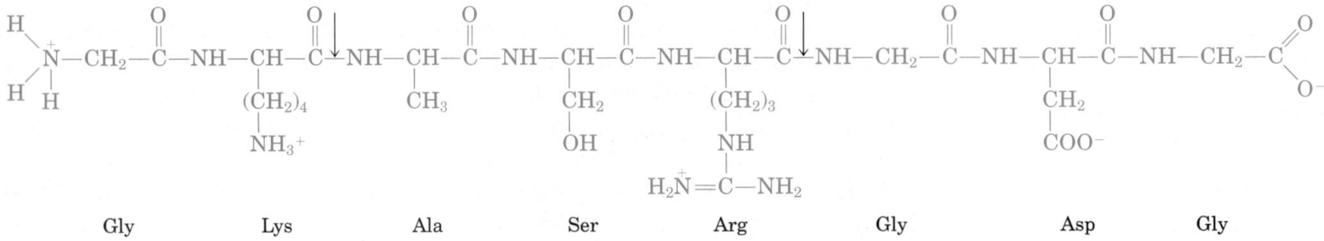

C H E M I C A L   C O N N E C T I O N S   1 4 A

Muscle Relaxants and Enzyme Specificity

In the body, nerves transmit signals to the muscles. Acetylcholine is a neurotransmitter (Section 15.1) that operates between the nerve endings and muscles. It attaches itself to a specific receptor in the muscle end plate. This attachment transmits a signal to the muscle to contract; shortly thereafter, the muscles relax. A specific enzyme, acetylcholinesterase, then catalyzes the hydrolysis of the acetylcholine, removing it from the receptor site and thus preparing it for the next signal transmission—that is, the next contraction.

Succinylcholine is sufficiently similar to acetylcholine so that it, too, attaches itself to the receptor of the muscle end plate.

However, acetylcholinesterase can hydrolyze succinylcholine only very slowly. While it remains attached to the receptor, no new signal can reach the muscle to allow it to contract again. Thus the muscle stays relaxed for a long time.

This feature makes succinylcholine a good muscle relaxant during minor surgery, especially when a tube must be inserted into the bronchus (bronchoscopy). For example, after intravenous administration of 50 mg of succinylcholine, paralysis and respiratory arrest are observed within 30 s. While respiration is carried on artificially, the bronchoscopy can be performed within minutes.

$$CH_3-\overset{\overset{\displaystyle O}{\|}}{C}-O-CH_2-CH_2-\overset{\overset{\displaystyle CH_3}{|}}{\underset{\underset{\displaystyle CH_3}{|}}{\overset{+}{N}}}-CH_3 + H_2O \xrightarrow{\text{acetylcholin-esterase}} CH_3-\overset{\overset{\displaystyle O}{\|}}{C}-OH + HO-CH_2-CH_2-\overset{\overset{\displaystyle CH_3}{|}}{\underset{\underset{\displaystyle CH_3}{|}}{\overset{+}{N}}}-CH_3$$

Acetylcholine   Acetic acid   Choline

$$CH_3-\overset{\overset{\displaystyle CH_3}{|}}{\underset{\underset{\displaystyle CH_3}{|}}{\overset{+}{N}}}-CH_2-CH_2-O-\overset{\overset{\displaystyle O}{\|}}{C}-CH_2-CH_2-\overset{\overset{\displaystyle O}{\|}}{C}-O-CH_2-CH_2-\overset{\overset{\displaystyle CH_3}{|}}{\underset{\underset{\displaystyle CH_3}{|}}{\overset{+}{N}}}-CH_3$$

Succinylcholine

14.2 Naming and Classifying Enzymes

Enzymes are commonly given names derived from the reaction that they catalyze and/or the compound or type of compound on which they act. For example, lactate dehydrogenase speeds up the removal of hydrogen from lactate (a redox reaction). Acid phosphatase catalyzes the hydrolysis of phosphate ester bonds under acidic conditions. As can be seen from these examples, the names of most enzymes end in "-ase." Some enzymes, however, have older names, which were assigned before their actions were clearly understood. Among these are pepsin, trypsin, and chymotrypsin—all enzymes of the digestive tract.

Enzymes can be classified into six major groups according to the type of reaction they catalyze (see also Table 14.1):

1. **Oxidoreductases** catalyze oxidations and reductions.
2. **Transferases** catalyze the transfer of a group of atoms, such as CH_3, CH_3CO, or NH_2, from one molecule to another.
3. **Hydrolases** catalyze hydrolysis reactions.
4. **Lyases** catalyze the addition of a group to a double bond or the removal of two groups from adjacent atoms to create a double bond.
5. **Isomerases** catalyze isomerization reactions.
6. **Ligases,** or synthetases, catalyze the joining of two molecules.

See the **Interactive General, Organic, and Biochemistry CD-ROM, version 2.0,** for further exploration on this topic.

Table 14.1 Classification of Enzymes

Class	Typical Example	Reaction Catalyzed	Section Number in This Book
1. Oxidoreductases	Lactate dehydrogenase	$CH_3-C-COO^- \longrightarrow CH_3-CH-COO^-$ $\quad\quad\parallel O \quad\quad\quad\quad\quad\quad\quad\quad OH$ Pyruvate $\quad\quad\quad\quad$ L-(+)-Lactate	19.2
2. Transferases	Aspartate amino transferase or Aspartate transaminase	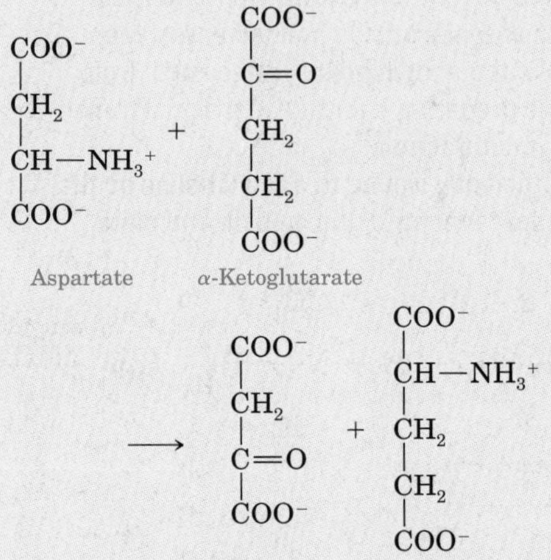 COO^- CH_2 $CH-NH_3^+$ $\quad + \quad$ $C=O$ COO^- $\quad\quad\quad\quad$ CH_2 $\quad\quad\quad\quad\quad\quad$ CH_2 $\quad\quad\quad\quad\quad\quad$ COO^- Aspartate $\quad$ α-Ketoglutarate $\longrightarrow$ COO^- $\quad + \quad$ $CH-NH_3^+$ $\quad\quad CH_2 \quad\quad\quad\quad CH_2$ $\quad\quad C=O \quad\quad\quad\quad CH_2$ $\quad\quad COO^- \quad\quad\quad COO^-$ Oxaloacetate $\quad$ Glutamate	19.8
3. Hydrolases	Acetylcholinesterase	$CH_3-C-OCH_2CH_2\overset{+}{N}(CH_3)_3 + H_2O$ $\quad\quad\parallel O$ Acetylcholine $\longrightarrow CH_3COOH + HOCH_2CH_2\overset{+}{N}(CH_3)_3$ Acetic acid $\quad\quad\quad$ Choline	14.3
4. Lyases	Aconitase	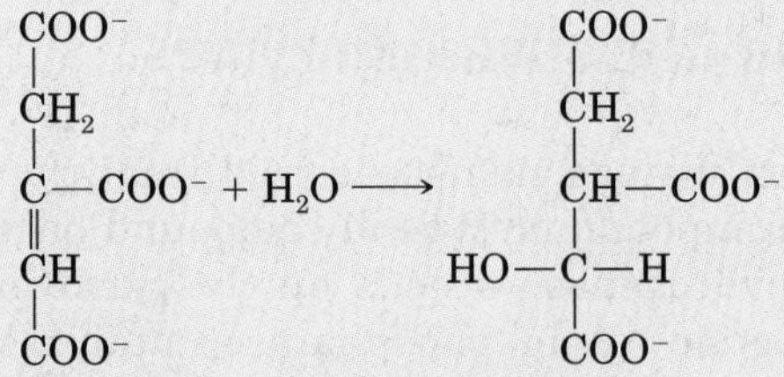 $COO^- \quad\quad\quad\quad\quad COO^-$ $CH_2 \quad\quad\quad\quad\quad\quad CH_2$ $C-COO^- + H_2O \longrightarrow CH-COO^-$ $\parallel \quad\quad\quad\quad\quad\quad\quad\quad HO-C-H$ $CH \quad\quad\quad\quad\quad\quad\quad\quad COO^-$ COO^- *cis*-Aconitate $\quad\quad\quad$ Isocitrate	18.4
5. Isomerases	Phosphohexose isomerase	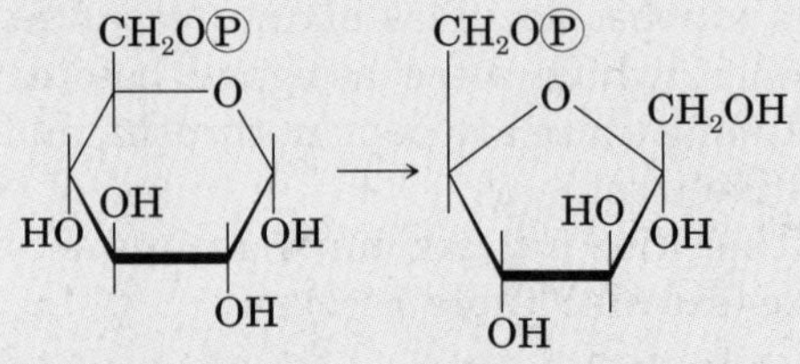 Glucose 6-phosphate $\longrightarrow$ Fructose 6-phosphate	19.2
6. Ligases	Tyrosine-tRNA synthetase	ATP + L-tyrosine + tRNA $\longrightarrow$ L-tyrosyltRNA + AMP + PP_i	17.6

14.3 Common Terms in Enzyme Chemistry

Cofactor The nonprotein part of an enzyme necessary for its catalytic function

Some enzymes, such as pepsin and trypsin, consist of polypeptide chains only. Other enzymes contain nonprotein portions called **cofactors.** The protein (polypeptide) portion of the enzyme is called an **apoenzyme.**

The cofactors may be metallic ions, such as Zn^{2+} or Mg^{2+}, or organic compounds. Organic cofactors are called **coenzymes.** An important group of coenzymes are the B vitamins, which are essential to the activity of many enzymes (Section 18.3). Another important coenzyme is heme (Figure 13.14), which is part of several oxidoreductases as well as of hemoglobin. In any case, an apoenzyme cannot catalyze a reaction without its cofactor, nor can the cofactor function without the apoenzyme. When a metal ion is a cofactor, it can bind directly to the protein or to the coenzyme, if the enzyme contains one.

The compound on which the enzyme works, and whose reaction it speeds up, is called the **substrate.** The substrate usually binds to the enzyme surface while it undergoes the reaction. The substrate binds to a specific portion of the enzyme during the reaction, called the **active site.** If the enzyme has coenzymes, they are located at the active site. Therefore, the substrate is simultaneously surrounded by parts of the apoenzyme, coenzyme, and metal ion cofactor (if any), as shown in Figure 14.2.

Activation is any process that initiates or increases the action of an enzyme. It can be the simple addition of a cofactor to an apoenzyme or the cleavage of a polypeptide chain of a proenzyme (Section 14.6).

Inhibition is the opposite—any process that makes an active enzyme less active or inactive (Section 14.5). Inhibitors are compounds that accomplish this. **Competitive inhibitors** bind to the active site of the enzyme surface, thereby preventing the binding of substrate. **Noncompetitive inhibitors,** which bind to some other portion of the enzyme surface, may sufficiently alter the tertiary structure of the enzyme so that its catalytic effectiveness is slowed down. Both competitive and noncompetitive inhibition are *reversible,* but some compounds alter the structure of the enzyme *permanently* and thus make it *irreversibly* inactive.

> **Coenzyme** A nonprotein organic molecule, frequently a B vitamin, that acts as a cofactor

Some enzymes have two kinds of cofactors: a coenzyme and a metallic ion.

> **Active site** A three-dimensional cavity of the enzyme with specific chemical properties that enable it to accommodate the substrate

> **Competitive inhibition** An enzyme regulation in which an inhibitor competes with the substrate for the active site

> **Noncompetitive inhibition** An enzyme regulation in which an inhibitor binds to the enzyme outside of the active site, thereby changing the shape of the active site and reducing its catalytic activity

14.4 Factors Affecting Enzyme Activity

Enzyme activity is a measure of how much reaction rates are increased. In this section, we examine the effects of concentration, temperature, and pH on enzyme activity.

A Enzyme and Substrate Concentration

If we keep the concentration of substrate constant and increase the concentration of enzyme, the rate increases linearly (Figure 14.3). That is, if the enzyme concentration doubles, the rate doubles as well; if the enzyme concentration triples, the rate also triples. This is the case in practically all enzyme reactions, because the molar concentration of enzyme is almost always much lower than that of substrate (that is, many more molecules of substrate are typically present than molecules of enzyme).

Conversely, if we keep the concentration of enzyme constant and increase the concentration of substrate, we get an entirely different type of curve, called a saturation curve (Figure 14.4). In this case, the rate does not increase continuously. Instead, a point is reached after which the rate stays the same even if we increase the substrate concentration further. This happens because, at the saturation point, substrate molecules are bound to all available active sites of the enzymes. Because the reactions take place at the active sites, once they are all occupied the reaction is proceeding at its maximum rate. Increasing the substrate concentration can no longer increase the rate because the excess substrate cannot find any active sites to which to attach.

Figure 14.2 Schematic diagram of the active site of an enzyme and the participating components.

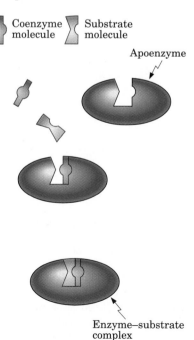

Coenzyme molecule   Substrate molecule

Apoenzyme

Enzyme–substrate complex

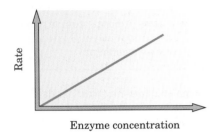

Figure 14.3 The effect of enzyme concentration on the rate of an enzyme-catalyzed reaction. Substrate concentration, temperature, and pH are constant.

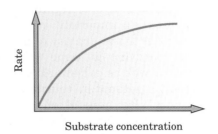

Figure 14.4 The effect of substrate concentration on the rate of an enzyme-catalyzed reaction. Enzyme concentration, temperature, and pH are constant.

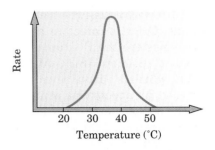

Figure 14.5 The effect of temperature on the rate of an enzyme-catalyzed reaction. Substrate and enzyme concentrations and pH are constant.

B. Murton/Southampton Oceanography Centre/Science Photo Library/Photo Researchers, Inc.

■ **A "black smoker" vent pours out a sulfurous (as indicated by the yellow) mineral-rich fluid from its mound or chimney.**

B Temperature

Temperature affects enzyme activity because it changes the three-dimensional structure of the enzyme. In uncatalyzed reactions, the rate usually increases as the temperature increases. The effect of temperature on enzyme-catalyzed reactions is different. When we start at a low temperature (Figure 14.5), an increase in temperature first causes an increase in rate. However, protein conformations are very sensitive to temperature changes. Once the optimal temperature is reached, any further increase in temperature causes changes in enzyme conformation. The substrate may then not fit properly onto the changed enzyme surface, so the rate of reaction actually *decreases*.

After a *small* temperature increase above the optimum, the decreased rate could still be increased again by lowering the temperature because, over a narrow temperature range, changes in conformation are reversible. However, at some higher temperature above the optimum, we reach a point where the protein denatures (Section 13.11); the conformation is then altered irreversibly, and the polypeptide chain cannot refold. At this point, the enzyme is completely inactivated. The inactivation of enzymes at low temperatures is used in the preservation of food by refrigeration.

Most enzymes from bacteria and higher organisms have an optimal temperature around 37°C. However, the enzymes of organisms that live at the ocean floor at 2°C have an optimal temperature in that range. Other organisms live in ocean vents under extreme conditions, and their enzymes have optimal conditions at ranges from 90 to 105°C. The enzymes of these hyperthermophile organisms also have other extreme requirements, such as pressures up to 100 atm, and some of them have an optimal pH in the range of 1 to 4. Enzymes from these hyperthermophiles, especially polymerases that catalyze the polymerization of DNA (Section 16.6), have gained commercial importance.

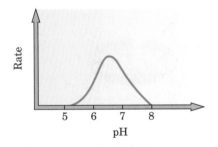

Figure 14.6 The effect of pH on the rate of an enzyme-catalyzed reaction. Substrate and enzyme concentrations and temperature are constant.

C pH

As the pH of its environment changes the conformation of a protein (Section 13.11), we would expect pH-related effects to resemble those observed when the temperature changes. Each enzyme operates best at a certain pH (Figure 14.6). Once again, within a narrow pH range, changes in enzyme activity are reversible. However, at extreme pH values (either acidic or basic), enzymes are denatured irreversibly, and enzyme activity cannot be restored by changing back to the optimal pH.

C H E M I C A L   C O N N E C T I O N S   1 4 B

Acidic Environment and *Helicobacter*

Certain organisms can live in an extreme pH environment. An example is the bacterium *Helicobacter pylori,* which survives happily in the acidic medium of the stomach wall. These bacteria live under the mucus coating of the stomach and release a toxin that causes ulcers. The role of *H. pylori* in ulceration was discovered only in the last decade, because people did not believe that enzymes of a bacterium could possibly function in such a harsh acidic environment with pH 1.4. In reality, the enzymes in the cytoplasm of *H. pylori* are not really exposed to low pH. The bacteria contain a special enzyme, **urease,** which converts urea to the basic ammonia; therefore, despite the acidic environment, *H. py-*

lori lives in a protective, neutralizing basic cloud. The urease itself, which is found on the surface of the bacteria, is protected against acid denaturation by its quaternary structure. The subunits of urease form a central cavity that contains the active sites. This cavity is filled with a buffer that protects the active sites before the actual urea hydrolysis starts.

After the destructive role played by *H. pylori* in ulceration was established, new treatments for gastric disorders were designed. Antibiotics such as amoxicillin or tetracyclin can eradicate the bacteria; the ulcers can then heal, and their recurrence is prevented.

14.5 Mechanism of Enzyme Action

We have seen that the action of enzymes is highly specific for a substrate. What kind of mechanism can account for such specificity? About 100 years ago, Arrhenius suggested that catalysts speed up reactions by combining with the substrate to form some kind of intermediate compound. In an enzyme-catalyzed reaction, the intermediate is called the **enzyme–substrate complex.**

A Lock-and-Key Model

To account for the high specificity of most enzyme-catalyzed reactions, a number of models have been proposed. The simplest and most frequently quoted is the **lock-and-key model** (Figure 14.7). This model assumes that the enzyme is a rigid, three-dimensional body. The surface that contains the active site has a restricted opening into which only one kind of substrate can fit, just as only the proper key can fit exactly into a lock and turn it open.

According to the lock-and-key mechanism, an enzyme molecule has its particular shape because that shape is necessary to maintain the active site in exactly the geometric alignment required for that particular reaction. An enzyme molecule is very large (typically consisting of 100 to 200 amino acid residues), but the active site is usually composed of only two or a few amino acid residues, which may well be located at different places in the chain. The other amino acids—those not part of the active site—are located in the sequence in which we find them because that sequence causes the whole molecule to fold up in exactly the required way. This arrangement emphasizes that the shape and the functional groups on the surface of the active site are of utmost importance in recognizing a substrate.

The lock-and-key model was the first to explain the action of enzymes. For most enzymes, there is evidence that this model is too restrictive. Enzyme molecules are in a dynamic state, not a static one. There are constant motions within them, so that the active site has some flexibility.

B Induced-Fit Model

From X-ray diffraction, we know that the size and shape of the active site cavity change when the substrate enters. To explain this phenomenon, the American biochemist Daniel Koshland introduced the **induced-fit model**

Lock-and-key model A model explaining the specificity of enzyme action by comparing the active site to a lock and the substrate to a key

See the **Interactive General, Organic, and Biochemistry CD-ROM, version 2.0,** for further exploration on this topic.

 Substrate

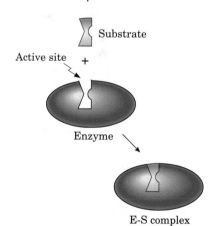

Figure 14.7 The lock-and-key model of the enzyme mechanism.

Induced-fit model A model explaining the specificity of enzyme action by comparing the active site to a glove and the substrate to a hand

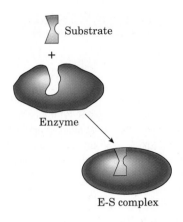

Figure 14.8 The induced-fit model of the enzyme mechanism.

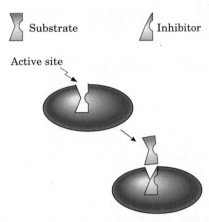

Figure 14.9 The mechanism of competitive inhibition. When a competitive inhibitor enters the active site, the substrate cannot enter.

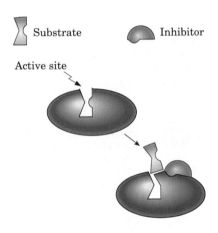

Figure 14.10 Mechanism of noncompetitive inhibition. The inhibitor attaches itself to a site other than the active site (allosterism), thereby changing the conformation of the active site.

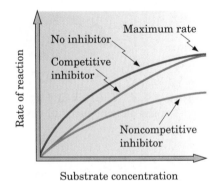

Figure 14.11 Enzyme kinetics in the presence and the absence of inhibitors.

(Figure 14.8), in which he compared the changes occurring in the shape of the cavity upon substrate binding to the changes in the shape of a glove when a hand is inserted. That is, the enzyme modifies the shape of the active site to accommodate the substrate. Recent experiments during actual catalysis have demonstrated that not only does the shape of the active site change with the binding of substrate, but even in the bound state both the backbone and the side chains of the enzyme are in constant motion. Furthermore, as a key residue—for example, the arginyl residue—pulsates, its motion correlates with the catalytic activity.

Both the lock-and-key and the induced-fit model explain the phenomenon of competitive inhibition (Section 14.3). The inhibitor molecule fits into the active site cavity in the same way the substrate does (Figure 14.9), thereby preventing the substrate from entering. The result is that whatever reaction is supposed to take place on the substrate does not occur.

Both the lock-and-key model and the induced-fit model emphasize the shape of the active site. However, the chemistry at the active site is actually the most important factor. A survey of known active sites of enzymes shows that five amino acids participate in the active sites in more than 65% of all cases. They are, in order of their dominance, His > Cys > Asp > Arg > Glu. A quick glance at Table 13.1 reveals that most of these amino acids have either acidic or basic side chains. Thus acid–base chemistry often underlies the mode of catalysis. The example given in Chemical Connections 14C confirms this relationship. Out of the eight amino acids in the catalytic site, two are Arg, one is Glu, one is Asp, and two are the related Asn.

Many cases of noncompetitive inhibition can also be explained by the induced-fit model. In this case, the inhibitor does not bind to the active site but rather to a different part of the enzyme. Nevertheless, the binding causes a change in the three-dimensional shape of the enzyme molecule, which so alters the shape of the active site that the substrate can no longer fit (Figure 14.10).

If we compare enzyme activity in the presence and the absence of an inhibitor, we can tell whether competitive or noncompetitive inhibition is taking place (Figure 14.11). The maximum reaction rate is the same without an inhibitor and in the presence of a competitive inhibitor. The only difference

CHEMICAL CONNECTIONS 14C

Active Sites

The perception of the active site as either a rigid cavity (lock-and-key model) or a partly flexible template (induced-fit model) is an oversimplification. Not only is the geometry of the active site important, but so are the specific interactions that take place between the enzyme surface and the sub-strate. To illustrate, we take a closer look at the active site of the enzyme pyruvate kinase. This enzyme catalyzes the transfer of the phosphate group from phosphoenol pyruvate (PEP) to ADP, an important step in glycolysis (Section 19.2).

$$CH_2=C-COO^- + R-O-P-O-P-O^- \longrightarrow CH_3-C-COO^- + R-O-P-O-P-O-P-O^-$$

| Phosphoenol pyruvate | ADP | Pyruvate | ATP |

The active site of the enzyme binds both substrates, PEP and ADP (Figure 14C.1). The rabbit muscle pyruvate kinase has two cofactors, K^+ and either Mn^{2+} or Mg^{2+}. The divalent cation is coordinated to the carbonyl and carboxylate oxygen of the pyruvate substrate and to a glutamyl 271 and aspartyl 295 residue of the enzyme. (The numbers indicate the position of the amino acid in the sequence.) The nonpolar $=CH_2$ group lies in a hydrophobic pocket formed by an alanyl 292, glycyl 294, and threonyl 327 residue. The K^+ on the other side of the active site is coordinated with the phosphate of the substrate and the seryl 76 and aspargyl 74 residue of the enzyme. Lysyl 269 and arginyl 72 are also part of the catalytic apparatus anchoring the ADP. This arrangement of the active site illustrates that specific folding into secondary and tertiary structures is required to bring important functional groups together. The residues of amino acids participating in the active sites are sometimes close in the sequence (aspargyl 74 and seryl 76), but mostly remain far apart (glutamyl 271 and aspartyl 295). Figure 14C.2 illustrates the secondary and tertiary structures providing such a stable active site.

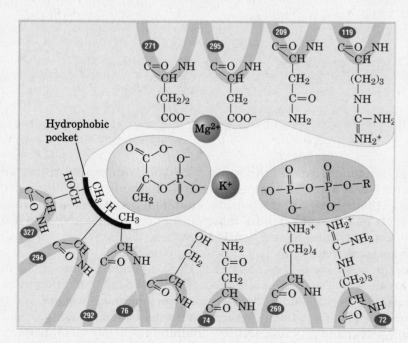

■ **Figure 14C.1** The active site and substrates of pyruvate kinase.

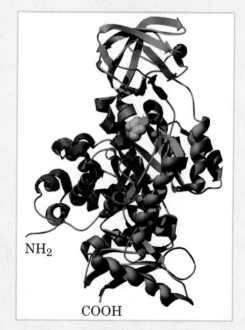

■ **Figure 14C.2** Ribbon cartoon of pyruvate kinase. Pyruvate, Mg^{2+}, K^+ are depicted as space-filling models.

Sulfa Drugs as Competitive Inhibitors

The vitamin folic acid (see Table 21.2), in its reduced form as tetrahydrofolate, acts as a coenzyme in a number of biosynthetic processes, including the synthesis of amino acids and of nucleotides. Humans obtain folic acid from the diet or from microorganisms in the intestinal tract. These microorganisms can synthesize folic acid if *para*-aminobenzoic acid is available to them.

In the 1930s, researchers discovered that the compound sulfanilamide could kill many types of harmful bacteria, thereby curing several diseases. Because sulfanilamide itself has an unacceptably high toxicity to humans, a number of derivatives of this compound, such as sulfapyridine and sulfathiazole, which act in a similar way, are used instead. These drugs work by "tricking" the bacteria, which normally use *para*-aminobenzoic acid as a raw material in the synthesis of folic acid. When the bacteria get a sulfa drug instead, they cannot tell the difference and use it to make a molecule that also has a folic acid type of structure but is not exactly the same. When they try to use this fake folic acid as a coenzyme, not only doesn't it work, but it is also now a competitive inhibitor of the enzyme's action. Consequently, many of the bacteria's amino acids and nucleotides cannot be synthesized, and the bacteria die.

Tetrahydrofolate

p-Aminobenzoic acid

Sulfapyridine

Sulfanilamide

Sulfathiazole

is that this maximum rate is achieved at a low substrate concentration with no inhibitor but at a high substrate concentration when an inhibitor is present. This is the true sign of competitive inhibition because in this scenario the substrate and the inhibitor are competing for the same active site. If the substrate concentration is sufficiently increased, the inhibitor will be displaced from the active site by Le Chatelier's principle.

If, on the other hand, the inhibitor is noncompetitive, it cannot be displaced by addition of excess substrate because it is bound to a different site. In this case, the enzyme cannot be restored to its maximum activity, and the maximum rate of the reaction is lower than it would be in the absence of the inhibitor. An example of noncompetitive inhibition is heavy metal poisoning (Chemical Connections 14E).

Enzymes can also be inhibited irreversibly if a compound is bound covalently and permanently to or near the active site. Such inhibition occurs with penicillin, which inhibits the enzyme transpeptidase; this enzyme is necessary for cross-linking bacterial cell walls. Without cross-linking, the bacterial cytoplasm spills out, and the bacteria die (Chemical Connections 10B).

C H E M I C A L C O N N E C T I O N S 1 4 E

Noncompetitive Inhibition and Heavy Metal Poisoning

Many enzymes contain a number of —SH groups (cysteine residues) that are easily cross-linked by heavy metal ions (Section 13.11):

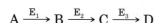

$$\begin{array}{c} \text{SH} \\ \\ \text{SH} \end{array} + Pb^{2+} \longrightarrow \begin{array}{c} S^- \\ \\ S^- \end{array} Pb^{2+} + 2H^+$$

If these cross-linked cysteine residues are found at or near active sites, the cross-linking can noncompetitively inhibit enzyme activity. This inhibition is the basis of lead and mercury poisoning.

In the seventeenth century, felt hats were made of rabbit hair. The workers treated the hair with mercury salt to enhance the felting process. This step contaminated the air with mercury vapors, which the hat makers inhaled. As a consequence they developed twitches, a symptom of nervous disorders. The term "mad hatter" (as in *Alice in Wonderland*) originates from this industrial disease.

Mercury poisoning has become an acute problem in recent years because the mercury and mercury compounds that factories dumped into streams and lakes for decades have entered the food chain. Mercury is also released into the environment following its use in mining. Certain microorganisms convert mercury metal to organic mercury compounds, mainly dimethylmercury, $(CH_3)_2Hg$. This compound then enters the food chain when it is absorbed by algae. Later, it becomes concentrated in fish, and people who eat the contaminated fish can get mercury poisoning.

In Japan, where mercury was dumped in the Bay of Minamata in the 1950s, more than 100 people suffered mercury poisoning from a diet high in locally caught seafood, and 44 people died. Once such contaminated food was removed from the markets, the number of mercury-poisoning cases reported dropped drastically.

A current campaign focuses on another potential source of contamination: mercury-containing medical thermometers in the home. Advocates are now seeking to replace them with non-mercury thermometers.

14.6 Enzyme Regulation

A Feedback Control

Enzymes are often regulated by environmental conditions. **Feedback control** is an enzyme regulation process in which formation of a product inhibits an earlier reaction in the sequence. The reaction product of one enzyme may control the activity of another, especially in a complex system in which enzymes work cooperatively. For example, in such a system, each step is catalyzed by a different enzyme:

$$A \xrightarrow{E_1} B \xrightarrow{E_2} C \xrightarrow{E_3} D$$

The last product in the chain, D, may inhibit the activity of enzyme E_1 (by competitive or noncompetitive inhibition). When the concentration of D is low, all three reactions proceed rapidly. As the concentration of D increases, however, the action of E_1 becomes inhibited and eventually stops. In this manner, the accumulation of D serves as a message that tells enzyme E_1 to shut down because the cell has enough D for its present needs. Shutting down E_1 stops the entire process.

B Proenzymes

Some enzymes are manufactured by the body in an inactive form. To make them active, a small part of their polypeptide chain must be removed. These inactive forms of enzymes are called **proenzymes** or **zymogens.** After the excess polypeptide chain is removed, the enzyme becomes active. For example, trypsin is manufactured as the inactive molecule trypsinogen (a

See the **Interactive General, Organic, and Biochemistry CD-ROM, version 2.0,** for further exploration on this topic.

Proenzyme (zymogen) A protein that becomes an active enzyme after undergoing a chemical change

CHEMICAL CONNECTIONS 14 F

A Zymogen and Hypertension

Many high blood pressure (hypertension) medications come under the heading of ACE inhibitors. ACE stands for *angiotensin-converting enzyme*, a key player in controlling blood pressure. It catalyzes the activation of angiotensin, a vasoconstrictive hormone, from its inactive zymogen, angiotensinogen.

$$\text{Angiotensinogen} \xrightarrow{\text{ACE}} \text{Angiotensin}$$

This process involves the cleaving of the last two amino acids from the carboxyl terminal of the decapeptide: Asp—Arg—Val—Tyr—Ile—His—Pro—Phe—His—Leu. The octapeptide, angiotensin, is a potent constrictor of the blood vessels and, therefore, increases blood pressure. When a competitive inhibitor occupies the active site of ACE, however, the hormone stays in its inactive zymogen state. As a consequence, blood pressure is lowered.

A large number of ACE inhibitors are currently on the market. Some are sulfur-containing compounds, such as Captopril; others are non–sulfur-containing compounds, such as Accupril and Altace.

The active site of the enzyme, ACE, contains several key chemical entities: arginyl, aspartyl residues, and a Zn^{2+} ion. Each of these interacts with a group on the inhibitor. The arginyl residue attracts the —COO^- group; the aspartyl amide attracts the —C=O group; and the Zn^{2+} forms a salt with the negatively charged sulfide ion or —COO^- group. The key element is the proper separation of these functional groups on the inhibitor molecule.

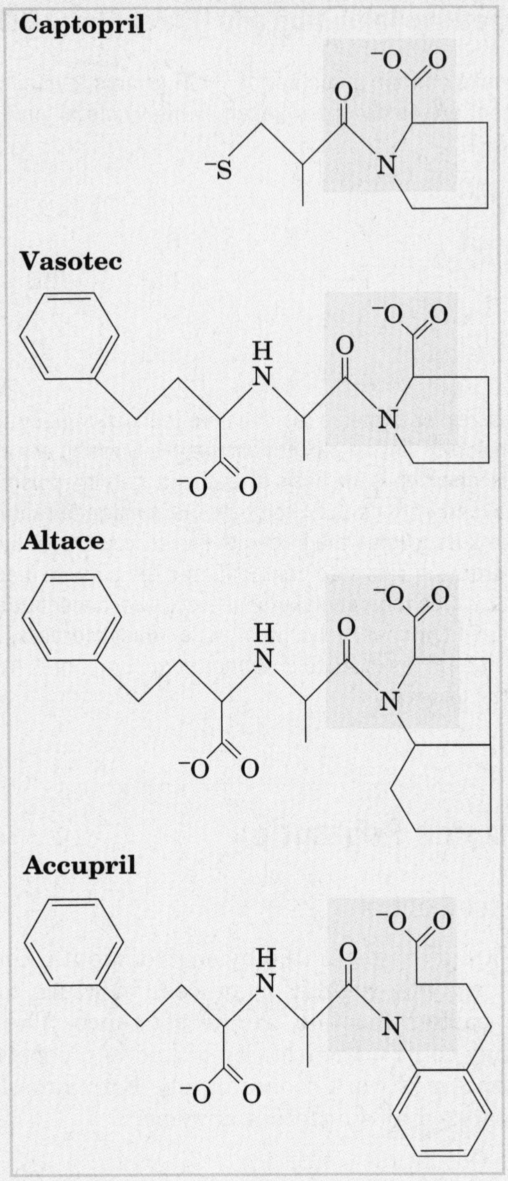

Captopril

Vasotec

Altace

Accupril

■ **Figure 14F** Structures of the active forms of four ACE inhibitors.

The removal of the six-amino-acid fragment is, of course, also catalyzed by an enzyme.

zymogen). When a fragment containing six-amino-acid residues is removed from the N-terminal end, the molecule becomes a fully active trypsin molecule. Removal of the fragment not only shortens the chain but also changes the three-dimensional structure (the tertiary structure), thereby allowing the molecule to achieve its active form.

Why does the body go to so much trouble? Why not just make the fully active trypsin to begin with? The reason is very simple. Trypsin is a protease—it catalyzes the hydrolysis of proteins (Figure 14.1)—and is, therefore, an important catalyst for the digestion of the proteins we eat. But it would not be good if it cleaved the proteins of which our own bodies are made! Therefore, the body makes trypsin in an inactive form; only after it enters the digestive tract is it allowed to become active.

C Allosterism

Sometimes regulation takes place by means of an event that occurs at a site other than the active site but that eventually affects the active site. This type of interaction is called **allosterism,** and any enzyme regulated by this mechanism is called an **allosteric enzyme.** If a substance binds noncovalently and reversibly to a site *other than the active site,* it may affect the enzyme in either of two ways: It may inhibit enzyme action (**negative modulation**) or it may stimulate enzyme action (**positive modulation**).

The substance that binds to the allosteric enzyme is called a **regulator,** and the site to which it attaches is called a **regulatory site.** In most cases, allosteric enzymes contain more than one polypeptide chain (subunits); the regulatory site is on one polypeptide chain and the active site on another.

Specific regulators can bind reversibly to the regulatory sites. For example, a protein kinase is an allosteric enzyme. In this case, the enzyme has only one polypeptide chain, so it carries both the active site and the regulatory site at different parts of this chain (Figure 14.12). The regulator—another protein molecule—binds reversibly to the regulatory site. As long as the regulator remains bound to the regulatory site, the total enzyme–regulator complex will be inactive. When the regulator is removed from the regulatory site, the protein kinase becomes active. In this way, the regulator controls the allosteric enzyme action.

Allosteric regulation may occur with proteins other than enzymes. Section 23.3 explains how modulators affect the oxygen-carrying ability of hemoglobin, an allosteric protein.

D Protein Modification

The activity of an enzyme may also be controlled by **protein modification.** The modification is usually a change in the primary structure, typically by addition of a functional group covalently bound to the apoenzyme.

The best-known example of protein modification is the activation/inhibition of enzymes by phosphorylation. A phosphate group is often bonded to a serine or tyrosine residue. In some enzymes, such as glycogen phosphorylase (Section 20.1), the phosphorylated form is the active form of the enzyme. Without it, the enzyme would be less active or inactive.

The opposite example is the enzyme pyruvate kinase, PK (discussed in Chemical Connections 14C). Pyruvate kinase from the liver is inactive when it is phosphorylated. Enzymes that catalyze such phosphorylation go under the common name of *kinases.* When the activity of PK is not needed, it is phosphorylated (to PKP) by a protein kinase using ATP as a substrate as well as a source of energy (Section 18.3). When the system wants to turn on PK activity, the phosphate group, P_i, is removed by another enzyme, phosphatase, which renders PK active.

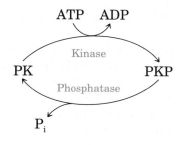

> **Allosterism** An enzyme regulation in which the binding of a regulator on one site on the enzyme modifies the enzyme's ability to bind the substrate in the active site

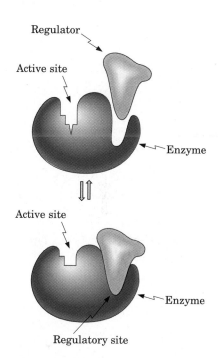

Figure 14.12 The allosteric effect. Binding of a regulator to a site other than the active site changes the shape of the active site.

Phosphoserine

$$H_3N^+—CH—COO^-$$
$$|$$
$$CH_2$$
$$|$$
$$O—PO_3{}^{2-}$$

Phosphoserine

One Enzyme, Two Functions

The enzyme called prostaglandin enderoperoxide synthase (PGHS) catalyzes the conversion of arachidonic acid to prostaglandin PGH_2 (Section 12.12) in two steps:

Arachidonic acid

Cyclooxygenase

PGG_2

Peroxidase

PGH_2

In the process, PGHS inserts two molecules of oxygen into arachidonic acid.

The enzyme itself is a single protein molecule. It is associated with the cell membrane and has a heme coenzyme. It has two distinct functions. The enzyme's first activity is to close a substituted cyclopentane ring; this is the *cyclooxygenase activity*. The *peroxidase activity* of the same enzyme yields the 15-hydroxy derivative prostaglandin, PGH_2.

Commercial painkillers act two ways inhibiting the formation of prostaglandin (PGH_2: (1) PGHS's cyclooxygenase activity is inhibited by aspirin and related NSAIDs (Chemical Connections 12I). Aspirin inhibits PGHS by acetylating serine in the active site; as a result, the active site is no longer able to accommodate arachidonic acid. Other NSAIDs also inhibit the cyclooxygenase activity by competitive inhibition, but they do not inhibit the peroxidase activity. (2) Another class of inhibitor has antioxidant activity. For example, acetaminophen (Tylenol) acts as a pain killer by inhibiting the peroxidase activity of PGHS.

E Isoenzymes

Another form of regulation of enzyme activity occurs when the same enzyme appears in different forms in different tissues. Lactate dehydrogenase catalyzes the oxidation of lactate to pyruvate, and vice versa (Figure 19.3, step 11). The enzyme has four subunits (tetramer). Two kinds of subunits, called H and M, exist. The enzyme that dominates in the heart is an H_4 enzyme, meaning that all four subunits are of the H type, although some M-type subunits are present as well. In the liver and skeletal muscles, the M type dominates. Other types of tetramer combinations exist in different tissues: H_3M, H_2M_2, and HM_3. These different forms of the same enzyme are called **isozymes** or **isoenzymes.**

The H_4 enzyme is allosterically inhibited by high levels of pyruvate, while the M_4 enzyme is not. In diagnosing the severity of heart attacks (Section 14.7), the release of H_4 isoenzyme is monitored in the serum.

> **Isozymes** Enzymes that perform the same function but have different combinations of subunits and thus different quaternary structures

14.7 Enzymes in Medical Diagnosis and Treatment

Most enzymes are confined within the cells of the body. However, small amounts of them can also be found in body fluids such as blood, urine, and cerebrospinal fluid. The level of enzyme activity in these fluids can easily be monitored. This information can prove extremely useful: Abnormal activity (either high or low) of particular enzymes in various body fluids signals either the onset of certain diseases or their progression. Table 14.2 lists some enzymes used in medical diagnosis and their activities in normal body fluids.

Table 14.2 Enzyme Assays Useful in Medical Diagnosis

Enzyme	Normal Activity	Body Fluid	Disease Diagnosed
Alanine amino-transferase (ALT)	3–17 U/L*	Serum	Hepatitis
Acid phosphatase	2.5–12 U/L	Serum	Prostate cancer
Alkaline phosphatase (ALP)	13–38 U/L	Serum	Liver or bone disease
Amylase	19–80 U/L	Serum	Pancreatic disease or mumps
Aspartate amino-transferase (AST)	7–19 U/L	Serum	Heart attack or hepatitis
	7–49 U/L	Cerebrospinal fluid	
Lactate dehydro-genase (LDH)	100–350 WU/mL	Serum	Heart attack
Creatine phospho-kinase (CPK)	7–60 U/L	Serum	
Phosphohexose isomerase (PHI)	15–75 U/L	Serum	

*U/L = International units per liter; WU/mL = Wrobleski units per milliliter.

For example, a number of enzymes are assayed (measured) during myocardial infarction to diagnose the severity of the heart attack. Dead heart muscle cells spill their enzyme contents into the serum. As a consequence, the level of creatine phosphokinase (CPK) in the serum rises rapidly, reaching a maximum within two days. This increase is followed by a rise in aspartate aminotransferase (AST; formerly called glutamate-oxaloacetate transaminase, or GOT). This second enzyme reaches a maximum two to three days after the heart attack. In addition to CPK and AST, lactate dehydrogenase (LDH) levels are monitored; they peak after five to six days. In infectious hepatitis, the alanine aminotransferase (ALT; formerly called glutamate-pyruvate transaminase, or GPT) level in the serum can rise to ten times normal. There is also a concurrent increase in AST activity in the serum.

In some cases, administration of an enzyme is part of therapy. After duodenal or stomach ulcer operations, for instance, patients are advised to take tablets containing digestive enzymes that are in short supply in the stomach after surgery. Such enzyme preparations contain lipases, either alone or combined with proteolytic enzymes. They are sold under such names as Pancreatin, Acro-lase, and Ku-zyme.

S U M M A R Y

Enzymes are proteins that catalyze chemical reactions in the body. Most enzymes are very specific—they catalyze only one particular reaction (Section 14.1). The compound whose reaction is catalyzed by an enzyme is called the **substrate.** Enzymes are classified into six major groups according to the type of reaction they catalyze (Section 14.2). They are typically named after the substrate and the type of reaction they catalyze, by adding the ending "-ase."

Some enzymes are made of polypeptide chains only. Others have, besides the polypeptide chain (the **apoenzyme**), nonprotein **cofactors**, which are either organic compounds (**coenzymes**) or inorganic ions (Section 14.3). Only a small part of the enzyme surface, called the **active site,** participates in the actual catalysis of chemical reactions. Cofactors, if any, are part of the active site.

Compounds that slow enzyme action are called **inhibitors. A competitive inhibitor** attaches itself to the active site; a **noncompetitive inhibitor** binds to other parts of the enzyme surface (Section 14.3). The higher the enzyme and substrate concentrations, the higher the enzyme activity, except that, at sufficiently high substrate concentrations, a saturation point is reached. After this point, increasing substrate concentration no longer increases the rate (Section 14.4). Each enzyme has an optimal temperature and pH at which it has its greatest activity (Section 14.4).

Two closely related mechanisms by which enzyme activity and specificity are explained are the **lock-and-key model** and the **induced-fit model** (Section 14.5).

Enzyme activity is regulated by five mechanisms. In **feedback control,** the concentration of products influences the rate of the reaction (Section 14.6A). In **allosterism,** an interaction takes place at a position other than the active site but affects the active site, either positively or negatively (Section 14.6C). Enzymes can be activated or inhibited by **protein modification** (Section 14.6D). Some enzymes, called **proenzymes** or **zymogens,** must be activated by removing a small portion of the polypeptide chain (Section 14.6B). Finally, enzyme activity is also regulated by **isozymes,** which are different forms of the same enzyme (Section 14.6E).

Abnormal enzyme activity can be used to diagnose certain diseases (Section 14.7).

PROBLEMS

Numbers that appear in color indicate difficult problems.
◢ designates problems requiring application of principles.

Enzymes

14.1 What is the difference between the terms *catalyst* and *enzyme?*

14.2 What are ribozymes made of?

14.3 Would a lipase hydrolyze two triglycerides, one containing only oleic acid and the other containing only palmitic acid, with equal ease?

14.4 Compare the activation energy in uncatalyzed reactions and in enzyme-catalyzed reactions.

14.5 Why does the body need so many different enzymes?

14.6 Trypsin cleaves polypeptide chains at the carboxyl side of a lysine or arginine residue (Figure 14.1). Chymotrypsin cleaves polypeptide chains on the carboxyl side of an aromatic amino acid residue or any other nonpolar, bulky side chain. Which enzyme is more specific? Explain.

Naming and Classifying Enzymes

14.7 Both lyases and hydrolases catalyze reactions involving water molecules. What is the difference in the types of reactions that these two enzymes catalyze?

14.8 Monoamine oxidases are important enzymes in brain chemistry. Judging from the name, which of the following would be a suitable substrate for this class of enzymes:

(a) $HO-\!\!\bigcirc\!\!-\overset{\displaystyle OH}{\underset{\displaystyle |}{CH}}-CH_2NH_2$

(b) $CH_3-\overset{\displaystyle O}{\overset{\displaystyle \|}{C}}-N(CH_3)_2$

(c) $\bigcirc\!\!-NO_2$

14.9 On the basis of the classification given in Section 14.2, decide to which group each of the following enzymes belongs:

(a) Phosphoglyceromutase

$$^-OOC-\underset{\underset{\displaystyle OH}{\displaystyle |}}{CH}-CH_2-OPO_3{}^{2-}$$
3-Phosphoglycerate

$$\rightleftharpoons {}^-OOC-\underset{\underset{\displaystyle OPO_3{}^{2-}}{\displaystyle |}}{CH}-CH_2-OH$$
2-Phosphoglycerate

(b) Urease

$$H_2N-\overset{\displaystyle O}{\overset{\displaystyle \|}{C}}-NH_2 + H_2O \rightleftharpoons 2NH_3 + CO_2$$
Urea

(c) Succinate dehydrogenase

$$^-OOC-CH_2-CH_2-COO^- \quad + \quad FAD$$
Succinate — Coenzyme (oxidized form)

$$\rightleftharpoons \underset{^-OOC}{\overset{H}{}}C=C\underset{H}{\overset{COO^-}{}} \quad + \quad FADH_2$$
Fumarate — Coenzyme (reduced form)

(d) Aspartase

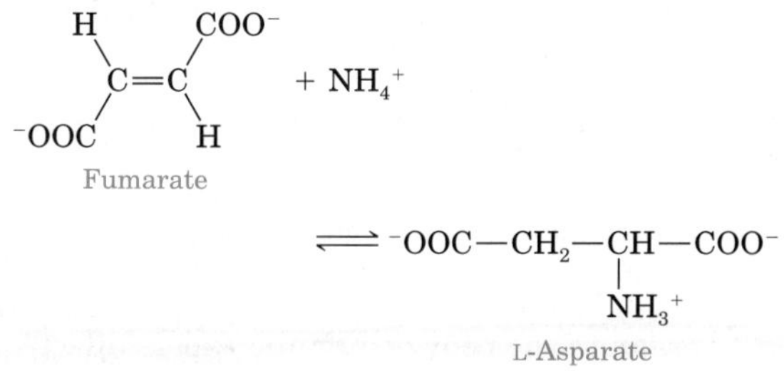

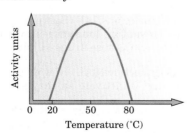

Fumarate

$$\rightleftharpoons {}^-OOC-CH_2-CH-COO^-$$
$$| $$
$$NH_3^+$$

L-Asparate

14.10 What kind of reaction does each of the following enzymes catalyze?

   (a) Deaminases   (b) Hydrolases

   (c) Dehydrogenases   (d) Isomerases

Common Terms in Enzyme Chemistry

14.11 What is the difference between a *coenzyme* and a *cofactor?*

14.12 In the citric acid cycle, an enzyme converts succinate to fumarate (see the reaction in Problem 14.9c). The enzyme consists of a protein portion and an organic molecule portion called FAD. What term do we use to refer to (a) the protein portion and (b) the organic molecule portion?

14.13 What is the difference between reversible and irreversible noncompetitive inhibition?

Factors Affecting Enzyme Activity

14.14 In most enzyme-catalyzed reactions, the rate of reaction reaches a constant value with increasing substrate concentration. This relationship is described in a saturation curve diagram (Figure 14.4). If the enzyme concentration, on a molar basis, is twice the maximum substrate concentration, would you obtain a saturation curve?

14.15 At a very low concentration of a certain substrate, we find that, when the substrate concentration doubles, the rate of the enzyme-catalyzed reaction also doubles. Would you expect the same finding at a very high substrate concentration? Explain.

14.16 If we wish to double the rate of an enzyme-catalyzed reaction, can we do so by increasing the temperature 10°C? Explain.

14.17 A bacterial enzyme has the following temperature-dependent activity.

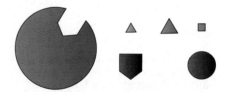

   (a) Is this enzyme more or less active at normal body temperature than when a person has a fever?

   (b) What happens to the enzyme activity if the patient's temperature is lowered to 35°C?

14.18 The optimal temperature for the action of lactate dehydrogenase is 36°C. It is irreversibly inactivated at 85°C, but a yeast containing this enzyme can survive for months at −10°C. Explain how this can happen.

14.19 The activity of pepsin was measured at various pH values. When the temperature and the concentrations of pepsin and substrate were held constant, the following activities were obtained:

pH	Activity
1.0	0.5
1.5	2.6
2.0	4.8
3.0	2.0
4.0	0.4
5.0	0.0

   (a) Plot the pH dependence of pepsin activity.

   (b) What is the optimal pH?

   (c) Predict the activity of pepsin in the blood at pH 7.4.

Mechanism of Enzyme Action

14.20 Urease can catalyze the hydrolysis of urea but not the hydrolysis of diethylurea. Explain why diethylurea is not hydrolyzed.

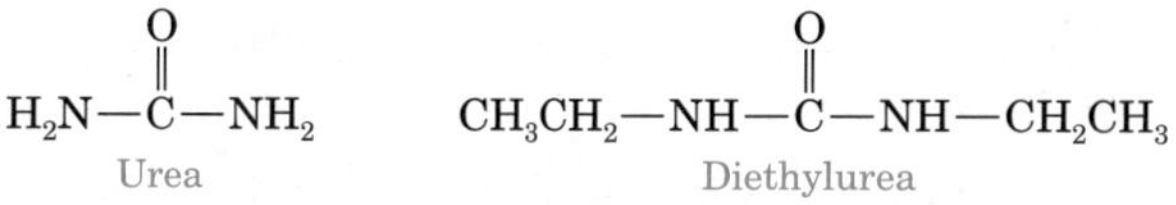

$$\underset{\text{Urea}}{H_2N-\overset{\overset{\displaystyle O}{\|}}{C}-NH_2} \qquad \underset{\text{Diethylurea}}{CH_3CH_2-NH-\overset{\overset{\displaystyle O}{\|}}{C}-NH-CH_2CH_3}$$

14.21 The following reaction may be represented by the cartoon figures:

Glucose + ATP $\rightleftharpoons$ glucose 6-phosphate + ADP

In this enzyme-catalyzed reaction, Mg^{2+} is a cofactor, fluoroglucose is a competitive inhibitor, and Cd^{2+} is a noncompetitive inhibitor. Identify each component of the reaction by a cartoon figure and assemble them to show (a) the normal enzyme reaction, (b) a competitive inhibition, and (c) a noncompetitive inhibition.

14.22 Which amino acids appear most frequently in the active sites of enzymes?

14.23 What kind of chemical reaction occurs most frequently at the active site?

14.24 Which of the following is a correct statement describing the induced-fit model of enzyme action? Substrates fit into the active site

  (a) Because both are exactly the same size and shape.

  (b) By changing their size and shape to match those of the active site.

  (c) By changing the size and shape of the active site upon binding.

14.25 What is the maximum rate that can be achieved in competitive inhibition compared with noncompetitive inhibition?

14.26 Enzymes are long protein chains, usually containing more than 100 amino acid residues. Yet the active site contains only a few amino acids. Explain why the other amino acids of the chain are present and what would happen to the enzyme activity if significant changes were made in the enzyme's structure.

Enzyme Regulation

14.27 The hydrolysis of glycogen to yield glucose is catalyzed by the enzyme phosphorylase. Caffeine, which is not a carbohydrate and not a substrate for the enzyme, inhibits phosphorylase. What kind of regulatory mechanism is at work?

14.28 Can the product of a reaction that is part of a sequence act as an inhibitor for another reaction in the sequence? Explain.

14.29 What is the difference between a zymogen and a proenzyme?

14.30 The enzyme trypsin is synthesized by the body in the form of a long polypeptide chain containing 235 amino acids (trypsinogen) from which a piece must be cut before the trypsin can be active. Why does the body not synthesize trypsin directly?

14.31 Give the structure of a tyrosyl residue of an enzyme modified by a protein kinase.

14.32 What is an isozyme?

14.33 The enzyme glycogen phosphorylase initiates the hydrolysis of glycogen to glucose 1-phosphate. It comes in two forms: Phosphorylase b is less active, and phosphorylase a is more active. The difference between b and a is the modification of the apoenzyme. Phosphorylase a has two phosphate groups added to the polypeptide chain. In analogy with the pyruvate kinase discussed in the text, give a scheme indicating the transition between b and a. What enzymes and what cofactors control this reaction?

Enzymes in Medical Diagnosis and Treatment

14.34 After a heart attack, the levels of certain enzymes rise in the serum. Which enzyme would you monitor within 24 hours following a suspected heart attack?

14.35 The enzyme formerly known as GPT (glutamate-pyruvate transaminase) has a new name: ALT (alanine aminotransferase). Looking at the equation in Section 19.9, which is catalyzed by this enzyme, what prompted this change of name?

14.36 If an examination of a patient indicated elevated levels of AST but normal levels of ALT, what would be your tentative diagnosis?

14.37 Which LDH isozyme is monitored in the case of a heart attack?

14.38 Chemists who have been exposed for years to organic vapors usually show higher-than-normal activity when given the alkaline phosphatase test. What organ in the body do organic vapors affect?

14.39 What enzyme preparation is given to patients after duodenal ulcer surgery?

14.40 Chymotrypsin is secreted by the pancreas and passed into the intestine. The optimal pH for this enzyme is 7.8. If a patient's pancreas cannot manufacture chymotrypsin, would it be possible to supply it orally? What happens to chymotrypsin's activity during its passage through the gastrointestinal tract?

Chemical Connections

14.41 (Chemical Connections 14A) Acetylcholine causes muscles to contract. Succinylcholine, a close relative, is a muscle relaxant. Explain the different effects of these related compounds.

14.42 (Chemical Connections 14A) An operating team usually administers succinylcholine before bronchoscopy. What is achieved by this procedure?

14.43 (Chemical Connections 14B) How does *Helicobacter pylori* protect its enzymes against stomach acid?

14.44 (Chemical Connections 14B) How does the quaternary structure of urease of *H. pylori* protect this enzyme against acid denaturation?

14.45 (Chemical Connections 14B) How do we treat ulcers caused by infection with *H. pylori?*

14.46 (Chemical Connections 14C) What role does Mn^{2+} play in anchoring the substrate in the active site of protein kinase?

14.47 (Chemical Connections 14C) Which amino acids of the active site interact with the $=CH_2$ group of the phosphoenol pyruvate? Do both provide the same surface environment? What is the nature of the interaction?

14.48 (Chemical Connections 14D) What part of folic acid is the key structure mimicked by sulfa drugs?

14.49 (Chemical Connections 14D) Sulfa drugs kill certain bacteria by preventing the synthesis of folic acid, a vitamin. Why don't these drugs kill people?

14.50 (Chemical Connections 14E) How does a heavy metal ion such as Hg^{2+} inactivate an enzyme?

14.51 (Chemical Connections 14F) Identify the functional groups that appear in all ACE inhibitors presented in Chemical Connections 14F.

14.52 (Chemical Connections 14G) Which activity of prostaglandin enderoperoxide synthase (PGHS) is inhibited by Tylenol, and which is inhibited by aspirin?

Additional Problems

14.53 Where can one find enzymes that are both stable and active at 90°C?

14.54 What kind of bonds does ACE (angiotensin-converting enzyme) hydrolyze?

14.55 Food can be preserved by inactivation of enzymes that would cause spoilage—for example, by refrigeration. Give an example of food preservation in which the enzymes are inactivated (a) by heat and (b) by lowering the pH.

14.56 Why is enzyme activity during myocardial infarction measured in the patients' serum rather than in their urine?

14.57 In activating angiotensinogen, which two amino acids are cleaved from the decapeptide?

14.58 What is the common characteristic of the amino acids of which the carboxyl groups of the peptide bonds can be hydrolyzed by trypsin?

14.59 Many enzymes are active only in the presence of Zn^{2+}. What common term is used for ions such as Zn^{2+} when discussing enzyme activity?

14.60 An enzyme has the following pH dependence:

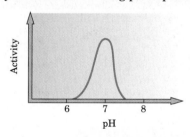

At what pH do you think this enzyme works best?

14.61 What enzyme is monitored in the diagnosis of infectious hepatitis?

14.62 The enzyme chymotrypsin catalyzes the following type of reaction:

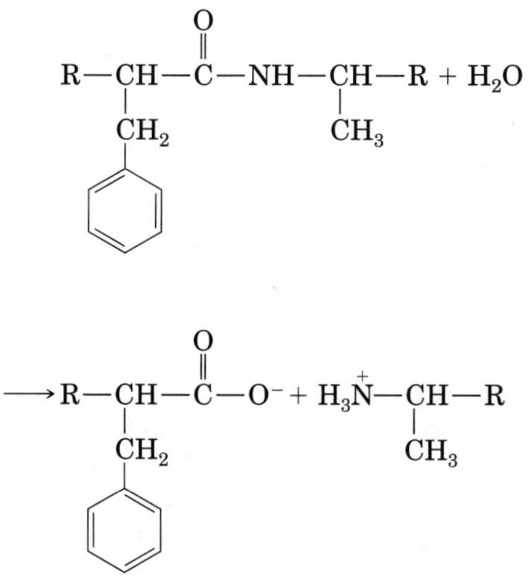

$$R-CH-\overset{\overset{\displaystyle O}{\|}}{C}-NH-CH-R + H_2O$$
with CH_2 below the first CH (bearing a phenyl ring) and CH_3 below the second CH

$$\longrightarrow R-CH-\overset{\overset{\displaystyle O}{\|}}{C}-O^- + H_3\overset{+}{N}-CH-R$$
with CH_2 below the first CH (bearing a phenyl ring) and CH_3 below the second CH

On the basis of the classification given in Section 14.2, to which group of enzymes does chymotrypsin belong?

14.63 Nerve gases operate by forming covalent bonds at the active site of cholinesterase. Is this an example of competitive inhibition? Can the nerve gas molecules be removed by simply adding more substrate (acetylcholine) to the enzyme?

14.64 What would be the appropriate name for an enzyme that catalyzes each of the following reactions:

(a) $CH_3CH_2OH \longrightarrow CH_3\overset{\overset{\displaystyle O}{\|}}{C}-H$

(b) $CH_3\overset{\overset{\displaystyle O}{\|}}{C}-O-CH_2CH_3 + H_2O$

$\longrightarrow CH_3\overset{\overset{\displaystyle O}{\|}}{C}-OH + CH_3CH_2OH$

14.65 In Section 20.5, a reaction between pyruvate and glutamate, to form alanine and α-ketoglutarate, is given. How would you classify the enzyme that catalyzes this reaction?

14.66 A liver enzyme is made of four subunits: 2A and 2B. The same enzyme, when isolated from the brain, has the following subunits: 3A and 1B. What would you call these two enzymes?

14.67 What is the function of a ribozyme?

InfoTrac College Edition

For additional reading, go to InfoTrac College Edition, your online research library, at

http://infotrac.thomsonlearning.com

15.1   Introduction

15.2   Chemical Messengers: Neurotransmitters and Hormones

15.3   Cholinergic Messenger: Acetylcholine

15.4   Amino Acids as Messengers: Neurotransmitters

15.5   Adrenergic Messengers: Neurotransmitters / Hormones

15.6   Peptidergic Messengers: Neurotransmitters / Hormones

15.7   Steroid Messengers: Hormones

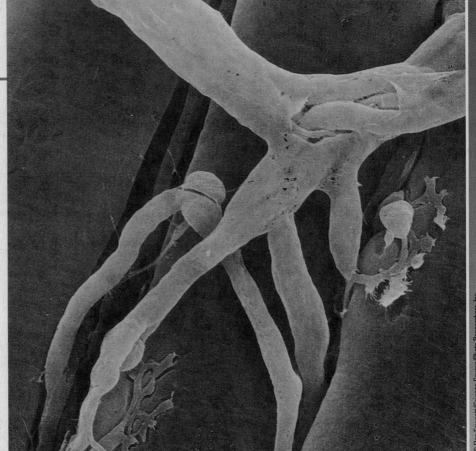

© Don Fawcett/Science Source/Photo Researchers, Inc.

Scanning electron micrograph of neuron touching a muscle cell.

Chemical Communications: Neurotransmitters and Hormones

15.1 Introduction

The importance of chemical communications in health can be ascertained by a quick look at Table 15.1. The list contains only a small sample of drugs pertaining to this chapter and it is of utmost interest to health care providers. In fact, a vast number of drugs in the pharmacopoeia act, in one way or another, to influence chemical communications.

Each cell in the body is an isolated entity enclosed in its own membrane. Furthermore, within each cell of higher organisms, organelles, such as the nucleus or the mitochondrion, are also enclosed by membranes separating them from the rest of the cell. If cells could not communicate with one another, the thousands of reactions in each cell would be uncoordinated. The same is true for organelles within a cell. Such communication allows the activity of a cell in one part of the body to be coordinated with the activity of cells in a different part of the body. There are three principal types of molecules for communications:

- **Receptors** are protein molecules on the surface of cells embedded in the membrane.

- **Chemical messengers** also called ligands, interact with the receptors.

- **Secondary messengers** in many cases carry the message from the receptor to the inside of the cell and amplify the message.

If your house is on fire and the fire threatens your life, external signals—light, smoke, and heat—register alarm at specific receptors in your eyes, nose, and skin. From there the signals are transmitted by specific compounds to nerve cells, or **neurons.** In the neurons, the signals travel as electric impulses along the axons (Figure 15.1). When they reach the end of the neuron, the signals are transmitted to adjacent neurons by specific compounds called **neurotransmitters.** Communication between the eyes and the brain, for example, is by neural transmission.

As soon as the danger signals are processed in the brain, other neurons carry messages to the muscles and to the endocrine glands. The message to the muscles is to run away or to take some other action in response to the fire (save the baby or run to the fire extinguisher, for example). To do so, the muscles must be activated. Again, neurotransmitters carry the necessary messages from the neurons to the muscle cells and the endocrine glands. The endocrine glands are stimulated, and a different chemical signal, called a **hormone,** is secreted into your bloodstream. "The adrenaline begins to flow." The danger signal carried by adrenaline makes quick energy available so that the muscles can contract and relax rapidly, allowing your body to take rapid action to avoid the danger.

Without these chemical communicators, the whole organism—you—would not survive because there is a constant need for coordinated efforts to face a complex outside world. The chemical communications between different cells and different organs play a role in the proper functioning of our bodies. Its significance is illustrated by the fact that *a large percentage of the drugs we encounter in medical practice try to influence this communication.* The scope of these drugs covers all fields—from prescriptions against hypertension, heart disease, through antidepressants, to pain killers, just to mention a few. Principally, there are four ways these drugs act in the body. A drug may affect either the **receptor** or the **messenger.**

1. An **antagonist** drug blocks the receptor and prevents its stimulation.

2. An **agonist** drug competes with the natural messenger for the receptor site. Once there, it stimulates the receptor.

3. Other drugs may decrease the concentration of the messenger by controlling the release of messengers from their storage.

4. Still other drugs may increase the concentration of the messenger by inhibiting its removal from the receptors.

Table 15.1 presents selected drugs and their modes of action that affect neurotransmission.

Chemical messengers fit into the receptor sites in a manner reminiscent of the lock-and-key model mentioned in Section 14.5.

Nerve cells are present throughout the body and, together with the brain, constitute the nervous system.

Neurotransmitter A chemical messenger between a neuron and another target cell: neuron, muscle cell, or cell of a gland

Hormone A chemical messenger released by an endocrine gland into the bloodstream and transported there to reach its target cell

Figure 15.1 Neuron and synapse.

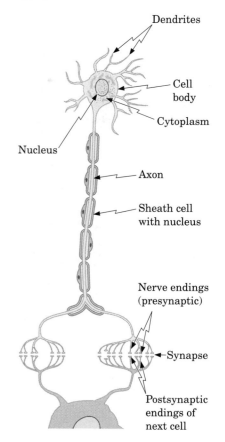

Dendrites

Cell body

Cytoplasm

Nucleus

Axon

Sheath cell with nucleus

Nerve endings (presynaptic)

Synapse

Postsynaptic endings of next cell

Table 15.1 Drugs That Affect Nerve Transmission

Messenger	Drugs That Affect Receptor Sites		Drugs That Affect Available Concentration of Neurotransmitter or Removal from Receptor Sites	
	Agonists (Activate Receptor Sites)	Antagonists (Block Receptor Sites)	Increase Concentration	Decrease Concentration
Acetylcholine (cholinergic)	Nicotine Pilocarpine Carbachol Succinylcholine Tacrine (Cognex)	Curare Atropine Propantheline (Pro-Banthine)	Malathion Nerve gases Succinylcholine Donepezil (Aricept)	*Clostridium botulinum* toxin
Calcium ion	Nifedipine (Adalat) Diltiazem (Cardizem)		Digitoxin (Lanoxicaps)	
Epinephrine (α-adrenergic)	Terazosin (Hytrin) Prazosin (Minipress)	Clonidine (Catapres)		
Norepinephrine (β-adrenergic)	Phenylephrine (Neo-Synephrine) Epinephrine (Adrenalin) Sotalol (Betaplace)	Propranolol (Inderal) Metoprolol (Lopressor) Selegiline (Deprenyl)	Amphetamines Iproniazide Imirapin (Tofranil) Amitriptyline (Elavil)	Reserpine Methyldopa (Aldomet) Metyrosine (Demser)
Dopamine (adrenergic)	Methylphenidate (Ritalin)	Clozapine (Clorazil) Metoclopramide (Reglan) Promethazine (Phenergan)	(Deprenyl) Entacapon (Comtan)	
Serotonin (adrenergic)		Ondansetron (Zofran)	Antidepressant Fluoxetine (Prozac)	
Histamine (adrenergic)	2-Methylhistamine Betazole Pentagastin	Fexofenadine (Allegra) Promethazine (Phenergan) Diphenhydramine (Benadryl) Ranitidine (Zantac) Cimetidine (Tagamet)	Histidine	Hydrazino-histidine
Glutamic acid (amino acid)	N-Methyl-D-Aspartate	Phencyclidine		
Enkephalin (peptidergic)	Opiate	Morphine Heroin Meperidine (Demerol)	Naloxone (Narcan)	

15.2 Chemical Messengers: Neurotransmitters and Hormones

As mentioned earlier, **neurotransmitters** are compounds that communicate between two nerve cells or between a nerve cell and another cell (such as a muscle cell). A nerve cell (Figure 15.1) consists of a main cell body from which projects a long, fiber-like part called an **axon.** Coming off the other side of the main body are hair-like structures called **dendrites.**

Table 15.2 The Principal Hormones and Their Actions

Gland	Hormone	Action	Structures Shown in
Parathyroid	Parathyroid hormone	Increases blood calcium Excretion of phosphate by kidney	
Thyroid	Thyroxine (T_4)	Growth, maturation, and metabolic rate	Chemical Connections 4C
	Triiodothyronine (T_3)	Metamorphosis	
Pancreatic islets			
Beta cells	Insulin	Hypoglycemic factor Regulation of carbohydrates, fats, and proteins	Section 13.7 Chemical Connections 15G
Alpha cells	Glucagon	Liver glycogenolysis	
Adrenal medulla	Epinephrine Norepinephrine	Liver and muscle glycogenolysis	Section 15.5
Adrenal cortex	Cortisol Aldosterone Adrenal androgens	Carbohydrate metabolism Mineral metabolism Androgenic activity (especially females)	Section 12.10 Section 12.10
Kidney	Renin	Hydrolysis of blood precursor protein to yield angiotensin	
Anterior pituitary	Luteinizing hormone Interstitial cell-stimulating hormone Prolactin Mammotropin	Causes ovulation Formation of testosterone and progesterone in interstitial cells Growth of mammary gland Lactation Corpus luteum function	
Posterior pituitary	Vasopressin	Contraction of blood vessels Kidney reabsorption of water	Section 13.7
	Oxytocin	Stimulates uterine contraction and milk ejection	Section 13.7
Ovaries	Estradiol Progesterone	Estrous cycle Female sex characteristics	Section 12.10 Section 12.10
Testes	Testosterone Androgens	Male sex characteristics Spermatogenesis	Section 12.10

Typically, neurons do not touch each other. Between the axon end of one neuron and the cell body or dendrite end of the next neuron is a space filled with an aqueous fluid, called a **synapse.** If the chemical signal travels, say, from axon to dendrite, we call the nerve ends on the axon the **presynaptic** site. The neurotransmitters are stored at the presynaptic site in **vesicles,** which are small, membrane-enclosed packages. Receptors are located on the **postsynaptic** site of the cell body or the dendrite.

Hormones are diverse compounds secreted by specific tissues (the endocrine glands), released into the bloodstream, and then adsorbed onto specific receptor sites, usually relatively far from their source. (This is the physiological definition of a hormone.) Table 15.2 lists some of the principal hormones. Figure 15.2 shows the target organs of hormones secreted by the pituitary gland.

The distinction between hormones and neurotransmitters is physiological, not chemical. Whether a certain compound is considered to be a neurotransmitter or a hormone depends on whether it acts over a short distance across a synapse (2×10^{-6} cm), in which case it is a neurotransmitter, or over a long distance (20 cm) from the secretory gland through the blood-

Synapse An aqueous small space between the tip of a neuron and its target cell

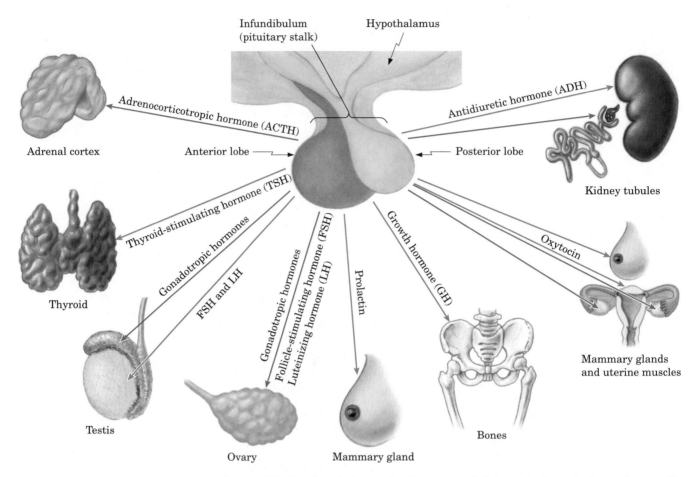

Figure 15.2 The pituitary gland is suspended from the hypothalamus by a stalk of neural tissue. The hormones secreted by the anterior and posterior lobes of the pituitary gland and the target tissues they act upon are shown. *(Modified from P. W. Davis and E. P. Solomon,* The World of Biology. *Philadelphia: Saunders College Publishing, 1986)*

stream to the target cell, in which case it is a hormone. For example, epinephrine and norepinephrine are both neurotransmitters and hormones.

There are, broadly speaking, five classes of chemical messengers: *cholinergic, amino acid, adrenergic, peptidergic,* and *steroid* messengers. This classification is based on the chemical nature of the important messenger (ligand) in each group. Neurotransmitters can belong to all five classes, and hormones can belong to the last three classes.

Messengers can also be classified according to how they work. Some of them—epinephrine, for example—*activate enzymes.* Others affect the *synthesis of enzymes and proteins* by working on the transcription of genes (Section 17.2); steroid hormones (Section 12.10) work in this manner. Finally, some affect the *permeability of membranes;* acetylcholine, insulin, and glucagon belong to this class.

Still another way of classifying messengers is according to their potential to *act directly* or through a *secondary messenger.* The steroid hormones act directly. They can penetrate the cell membrane and also pass through the membrane of the nucleus. For example, estradiol stimulates uterine growth.

Other chemical messengers act through secondary messengers. For example, epinephrine, glucagon, luteinizing hormone, norepinephrine, and vasopressin use cAMP as a secondary messenger (see details in Section 15.5C).

In the following sections, we will sample the mode of communication within each of the five chemical categories of messengers.

15.3 Cholinergic Messenger: Acetylcholine

The main **cholinergic neurotransmitter** is acetylcholine:

$$CH_3-\overset{\overset{\textstyle O}{\|}}{C}-O-CH_2-CH_2-\overset{\overset{\textstyle CH_3}{|}}{\underset{\underset{\textstyle CH_3}{|}}{N^+}}-CH_3$$

Acetylcholine

A Cholinergic Receptors

There are two kinds of receptors for this messenger. We will look at one that exists on the motor end plates of skeletal muscles or in the sympathetic ganglia. The nerve cells that bring messages contain stored acetylcholine in the vesicles in their axons. The receptor on the muscle cells or neurons is also known as nicotinic receptor because nicotine (see Chemical Connections 8B) inhibits the neurotransmission of these nerves. The receptor itself is a *transmembrane protein* (Figure 12.2) made of five different subunits. The central core of the receptor is an ion channel through which, when open, Na^+ and K^+ ions can pass (Figure 15.3). When the ion channels are closed, the K^+ ion concentration is higher inside the cell than outside; the reverse is true for the Na^+ ion concentration.

B Storage of Messengers

Events begin when a message is transmitted from one neuron to the next by neurotransmitters. The message is initiated by calcium ions (see Chemical Connections 15A). When the Ca^{2+} concentration in a neuron reaches a certain level (more than 0.1 μM), the vesicles containing acetylcholine fuse

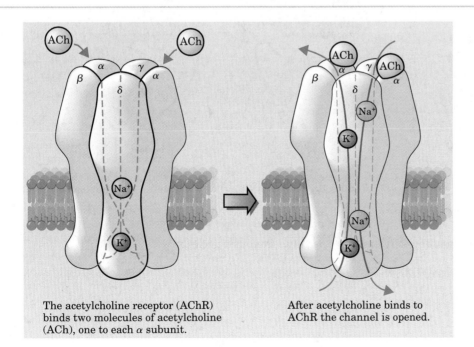

The acetylcholine receptor (AChR) binds two molecules of acetylcholine (ACh), one to each α subunit.

After acetylcholine binds to AChR the channel is opened.

Figure 15.3 Acetylcholine in action. The receptor protein has five subunits. When two molecules of acetylcholine bind to the two α subunits, a channel opens to allow the passage of Na^+ and K^+ ions. *(Courtesy of Anthony Tu, Colorado State University)*

CHEMICAL CONNECTIONS 15A

Calcium as a Signaling Agent (Secondary Messenger)

The message delivered to the receptors on the cell membranes by neurotransmitters or hormones must be delivered intracellularly to the various locations within the cell. Figure 15.4 depicts one mode of intracellular signaling, in which cAMP acts as a second messenger or signaling agent. The most universal and, at the same time, most versatile signaling agent is the cation Ca^{2+}.

Calcium ions in the cells come from either extracellular sources or intracellular stores, such as the endoplasmic reticulum. If the ions come from the outside, they enter the cells through specific calcium channels. Calcium ions control our heart beats, our movements through the action of skeletal muscles, and through the release of neurotransmitters in our neurons, learning and memory. They are also involved in signaling the beginning of life at fertilization and its end at death. Calcium ion signaling controls these different functions by two mechanisms: (1) increased concentration (forming sparks and puffs) and (2) duration of the signals.

In the resting state of the neuron, the Ca^{2+} concentration is about 0.1 μM. When neurons are stimulated, this level may increase to 0.5 μM. However, to elicit a fusion between the synaptic vesicles and the plasma membrane of the neuron, much higher concentrations are needed (10–25 μM).

An increase in calcium ion concentration may take the form of sparks or puffs. The source of calcium ions may be external (calcium influx caused by the electric signal of nerve transmission) or internal (calcium released from the stores of the endoplasmic reticulum). Upon receiving the signal of calcium puffs, the vesicles storing acetylcholine travel to the membrane of the presynaptic cleft. There they fuse with the membrane and empty their contents into the synapse.

Calcium ions can also control signaling by the duration of the signal. The signal in arterial smooth muscle lasts for 0.1 to 0.5 s. The wave of Ca^{2+} in the liver lasts 10 to 60 s. The calcium wave in the human egg lasts 1 to 35 min after fertilization. Thus, combining the concentration, localization, and duration of the signal, calcium ions can deliver messages to perform a variety of functions.

The effects of Ca^{2+} are modulated through specific calcium-binding proteins. In all nonmuscle cells and in smooth muscles, *calmodulin* is the calcium-binding protein. Calmodulin-bound calcium activates an enzyme, protein kinase II, which then phosphorylates an appropriate protein substrate. In this way, the signal is translated into metabolic activity.

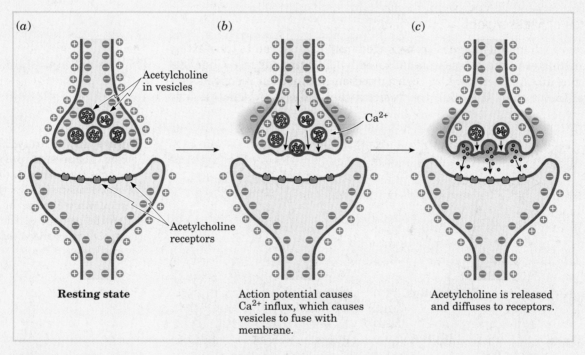

■ **Figure 15A** Calcium signaling to release acetylcholine from its vesicles.

Nerve Gases and Antidotes

Most nerve gases in the military arsenal exert their lethal effect by binding to acetylcholinesterase. Under normal conditions, this enzyme hydrolyzes the synaptic neurotransmitter acetylcholine within a few milliseconds after it is released at the nerve end-

Sarin
(Agent GB)

Soman
(Agent GD)

Parathion

ings. Nerve gases such as Sarin (agent GB, also called Tabun), Soman (agent GD), and agent VX are organic phosphonates related to such pesticides as parathion (the latter being much less lethal, of course):

Agent VX

Phosphonates, $X{-}\overset{O}{\underset{R}{P}}{-}O{-}$, are related to

phosphates, $-O{-}\overset{O}{\underset{O}{P}}{-}O{-}$, where R is an alkyl group.

If any of these phosphonates binds to acetylcholinesterase, the enzyme is irreversibly inactivated, and the transmission of nerve signals stops. The result is a cascade of symptoms: sweating, bronchial constriction due to mucus buildup, dimming of vision, vomiting, choking, convulsions, paralysis, and respiratory failure. Direct inhalation of as little as 0.5 mg can cause death within a few minutes. If the dosage is less or if the nerve gas is absorbed through the skin, the lethal effect may take several hours.

In warfare, protective clothing and gas masks are effective countermeasures. An enzyme of the bacteria *Pseudomonas diminuta*, organophosphorous hydroxylase, is capable of detoxify-

ing nerve gases. Their inclusion in cotton towelettes may help to remove nerve gases from contaminated skin. Also, first-aid kits containing injectable antidotes are available. These antidotes contain the alkaloid atropine and pralidoxime chloride. The two substances must be used in tandem—and very quickly—after exposure to nerve gases.

The nations of the world, with the exception of a few Arab countries, signed a treaty in 1993 that obliges all signatory governments to destroy their stockpiles of chemical weapons. In addition, the signees pledged to dismantle the plants that manufacture them within 10 years.

with the presynaptic membrane of the nerve cells. Then they empty the neurotransmitters into the synapse. The messenger molecules travel across the synapse and are adsorbed onto specific receptor sites.

C The Action of Messengers

The presence of acetylcholine molecules at the postsynaptic receptor site triggers a conformational change (Section 13.9A) in the receptor protein. This opens the *ion channel* and allows ions to cross membranes freely. Na^+ ions have higher concentration outside the neuron than do K^+ ions; thus, more Na^+ enters the cell than K^+ leaves. Because it involves ions, which carry electric charges, this process is translated into an electric signal. After a few milliseconds, the channel closes again. The acetylcholine still occupies the receptor. In order that the channel should reopen and transmit a new signal, the acetylcholine must be removed, and the neuron must be reactivated.

CHEMICAL CONNECTIONS 15C

Botulism and Acetylcholine Release

When meat or fish is improperly cooked or preserved, a deadly food poisoning, called botulism, may result. The culprit is the bacterium *Clostridium botulinum,* whose toxin prevents the release of acetylcholine from the presynaptic vesicles. Therefore, no neurotransmitter reaches the receptors on the surface of muscle cells, and the muscles no longer contract. Left untreated, the affected person may die.

Surprisingly, the botulin toxin has a valuable medical use. It is used in the treatment of involuntary muscle spasms—for example, in facial tics. These tics are caused by the uncontrolled release of acetylcholine. Controlled administration of the toxin, when applied locally to the facial muscles, stops the uncontrolled contractions and relieves the facial distortions.

"Facial distortion" has multiple meanings in the cosmetics industry. "Frown lines" and other wrinkles can also be removed by temporarily paralyzing facial muscles. The Food and Drug Administration has recently approved Botox (botulin toxin) for cosmetic use, meaning that its manufacturer can now advertise its rejuvenating appeal. Its use is fast spreading. Botox has been used on an off-label basis for a number of years, especially in Hollywood. Indeed, several movie directors have complained that some actors have used so much Botox that they can no longer show a variety of facial expressions.

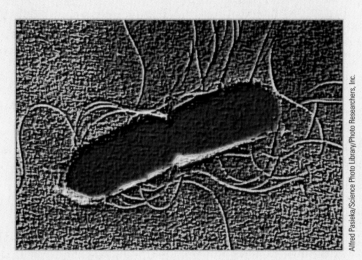

■ **Figure 15C** *Clostridium botulinum* is a food-poisoning bacterium.

D The Removal of Messengers

Acetylcholine is removed rapidly from the receptor site by the enzyme *acetylcholinesterase,* which hydrolyzes it.

$$CH_3-\overset{\overset{\textstyle O}{\|}}{C}-O-CH_2-CH_2-\overset{\overset{\textstyle CH_3}{|}}{\underset{\underset{\textstyle CH_3}{|}}{N^+}}-CH_3 + H_2O$$

Acetylcholine

$$\xrightarrow[\text{esterase}]{\text{Acetylcholin-}} CH_3-\overset{\overset{\textstyle O}{\|}}{C}-O^- + HO-CH_2-CH_2-\overset{\overset{\textstyle CH_3}{|}}{\underset{\underset{\textstyle CH_3}{|}}{N^+}}-CH_3$$

Acetate   Choline

This rapid removal enables the nerves to transmit more than 100 signals per second. By this means, the message moves from neuron to neuron until finally it gets transmitted, again by acetylcholine molecules, to the muscles or endocrine glands that are the ultimate target of the message.

Obviously, the action of the acetylcholinesterase enzyme is essential to the entire process. When this enzyme is inhibited, the removal of acetylcholine is incomplete, and nerve transmission ceases.

Alzheimer's Disease and Acetylcholine Transferase

Alzheimer's disease is the name given to the symptoms of severe memory loss and other senile behavior that afflict about 1.5 million people in the United States. Postmortem identification of this disease focuses on two pathological hallmarks in the brain: (1) buildup of protein deposits known as β-amyloid plaques outside the nerve cells and (2) neurofibrillar tangles composed of tau proteins. Controversy exists as to which of the two is the primary cause of the neurodegeneration in Alzheimer's disease. Each has its advocates. Those who favor the primacy of the plaque are nicknamed β-aptists, and the other camp sports the name of tauists.

Tau protein binds to microtubules, one of the major cytoskeletons (see Chemical Connections 13G and Figure 13.G3). Genetic mutation of tau protein or environmental factors such as hyperphosphorylation may alter the ability of tau to bind to microtubules. These altered tau proteins form tangles in the cytoplasm of neurons. Neurofibrillar tangles have been found in Alzheimer's brains in the absence of plaque as well, which suggests that tau abnormality may be sufficient to cause neurodegeneration.

In most Alzheimer's brains, the dominant feature is plaques. The plaques consist of protein fibers, some 7–10 nm thick, that are mixed with small peptides called β-amyloid peptides. These peptides originate from a water-soluble precursor, amyloid precursor protein (APP); this transmembrane protein has an unknown function. Certain enzymes, called secretases, cut peptides containing 38, 40, and 42 amino acids from the transmembrane region of APP. Mutations of APP proteins in Alzheimer's disease patients cause preferential accumulation of the 42 amino acid peptide forming β-pleated sheets, which precipitates creating plaques.

In Alzheimer's disease, nerve cells in the cerebral cortex die, the brain becomes smaller, and part of the cortex atrophies. The depression of the folds on the brain surface become deeper.

People with Alzheimer's disease are forgetful, especially about recent events. As the disease advances, they become confused and, in severe cases, lose their ability to speak; at that point, they need total care. As yet, there is no cure for this disease.

Patients with Alzheimer's disease have significantly diminished acetylcholine transferase activity in their brains. This enzyme synthesizes acetylcholine by transferring the acetyl group from acetyl-CoA to choline:

$$\underset{\text{Acetyl-CoA}}{CH_3\overset{\overset{\displaystyle O}{\|}}{C}-S-CoA} + \underset{\text{Choline}}{HO-CH_2CH_2\overset{\overset{\displaystyle CH_3}{|}}{\underset{\underset{\displaystyle CH_3}{|}}{\overset{+}{N}}}CH_3}$$

$$\longrightarrow \underset{\text{Acetylcholine}}{CH_3\overset{\overset{\displaystyle O}{\|}}{C}-O-CH_2CH_2\overset{\overset{\displaystyle CH_3}{|}}{\underset{\underset{\displaystyle CH_3}{|}}{\overset{+}{N}}}CH_3} + \underset{\text{Coenzyme A}}{CoA-SH}$$

The diminished concentration of acetylcholine can be partially compensated for by inhibiting the enzyme acetylcholinesterase, which decomposes acetylcholine. Certain drugs, which act as acetylcholinesterase inhibitors, have been shown to improve memory and other cognitive functions in some people with the disease. Drugs such as tacrine (Cognex), donepezil (Aricept), rivastigmine (Exelon), and galantamine (Reminyl) belong to this category; they all moderate the symptoms of Alzheimer's disease. The alkaloid huperzine A, an active ingredient of Chinese herb tea that has been used for centuries to improve memory, is also a potent inhibitor of acetylcholinesterase.

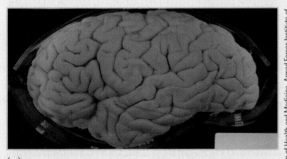

(a)

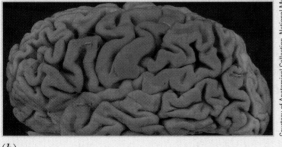

(b)

Courtesy of Anatomical Collection, National Museum of Health and Medicine, Armed Forces Institute of Pathology, Washington, D.C.

(c)

Science VU/Visuals Unlimited

■ **Figure 15D** (a) Normal brain. (b) Brain of a person with Alzheimer's disease. (c) PET scan comparing a normal brain with the brain of a person with Alzheimer's disease.

E Control of Neurotransmission

Acetylcholinesterase is inhibited irreversibly by phosphonates in nerve gases and pesticides (Chemical Connections 15B) or reversibly by succinylcholine (Chemical Connections 14A) and decamethonium bromide.

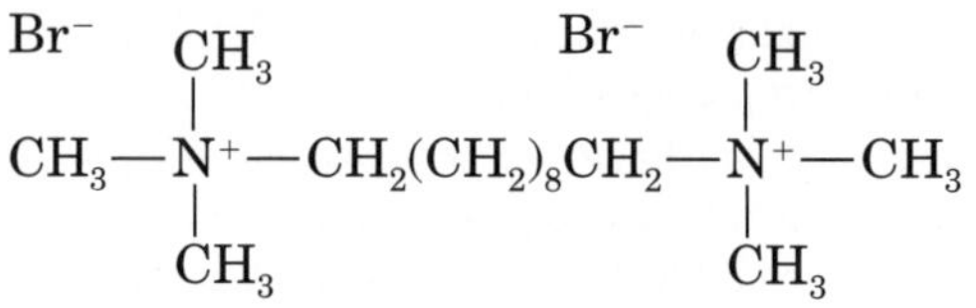

$$\text{Br}^-\quad \underset{\underset{\text{CH}_3}{|}}{\overset{\overset{\text{CH}_3}{|}}{\text{CH}_3-\text{N}^+}}-\text{CH}_2(\text{CH}_2)_8\text{CH}_2-\underset{\underset{\text{CH}_3}{|}}{\overset{\overset{\text{Br}^-\quad\text{CH}_3}{|}}{\text{N}^+}}-\text{CH}_3$$

Decamethonium bromide

Succinylcholine and decamethonium bromide resemble the choline end of acetylcholine and, therefore, act as competitive inhibitors of acetylcholinesterase. In small doses, these reversible inhibitors relax the muscles temporarily, making them useful as muscle relaxants in surgery. In large doses, they are just as deadly as the irreversible inhibitors.

The inhibition of acetylcholinesterase is but one way in which cholinergic neurotransmission is controlled. Another way is to modulate the action of the receptor. Because acetylcholine enables the ion channels to open and propagate signals, this mode of action is called *ligand-gated ion channels*. The attachment of the ligand to the receptor is critical in signaling. Nicotine in low doses is a stimulant; it is an agonist because it prolongs the receptor's biochemical response. In large doses, however, nicotine becomes an antagonist and blocks the action on the receptor. As such, it may cause convulsions and respiratory paralysis. Succinylcholine, besides being a reversible inhibitor of acetylcholinesterase, also has this concentration-dependent agonist/antagonist effect on the receptor. A strong antagonist, which blocks the receptor completely, can interrupt the communication between neuron and muscle cell. The venom of a number of snakes, especially that of the cobra, cobratoxin, exerts its deadly influence in this manner. The plant extract curare, which was used in poisoned arrows by the Amazon Indians, works in the same way. In small doses, curare is also used as a muscle relaxant.

Finally, the supply of the acetylcholine messenger can influence the proper nerve transmission. If the acetylcholine messenger is not released from its storage as in botulism (Chemical Connections 15C) or if its synthesis is impaired as in Alzheimer's disease (Chemical Connections 15D), the concentration of acetylcholine is reduced, and nerve transmission is impaired.

15.4 Amino Acids as Messengers: Neurotransmitters

A Messengers

Amino acids are distributed throughout the neurons individually or as parts of peptides and proteins. They can also act as neurotransmitters. Some of them, such as glutamic acid, aspartic acid, and cysteine, act as **excitatory neurotransmitters** similar to acetylcholine and norepinephrine. Others, such as glycine, β-alanine, taurine, and mainly γ-aminobutyric acid (GABA), are **inhibitory neurotransmitters;** they reduce neurotransmission. Note that some of these neurotransmitter amino acids are not found in proteins.

$$^+H_3NCH_2CH_2SO_3^-$$
Taurine

$$^+H_3NCH_2CH_2COO^-$$
β-Alanine

$$^+H_3NCH_2CH_2CH_2COO^-$$
γ-Aminobutyric acid
(GABA)
(IUPAC name:
4-aminobutanoic acid)

B Receptors

Each of these amino acids has its own receptor; in fact, glutamic acid has at least five subclasses of receptors. The best known among them is the *N*-Methyl-D-aspartate (NMDA) receptor. It is a ligand-gated ion channel similar to the nicotinic cholinergic receptor discussed in Section 15.3:

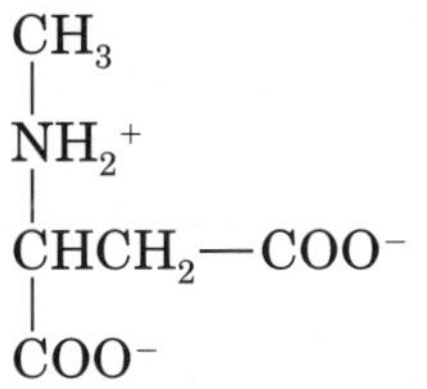

$$
\begin{array}{l}
CH_3 \\
| \\
NH_2^+ \\
| \\
CHCH_2-COO^- \\
| \\
COO^-
\end{array}
$$
N-methyl-D-aspartate

When glutamic acid binds to this receptor, the ion channel opens, Na^+ and Ca^{2+} flow into the neuron, and K^+ flows out of the neuron. The same thing happens when NMDA, being an agonist, stimulates the receptor. The gate of this channel is closed by a Mg^{2+} ion.

Phencyclidine (PCP), an antagonist of this receptor, induces hallucination. PCP, known by the street name of "angel dust," is a controlled substance; it causes bizarre psychotic behavior and long-term psychological problems.

C Removal of Messengers

In contrast to acetylcholine, there is no enzyme that would degrade glutamic acid and thereby remove it from its receptor once the signaling has been done. Glutamic acid is removed by **transporter** molecules, which bring it back through the presynaptic membrane into the neuron. This process is called **reuptake.**

> **Transporter** A protein molecule that carries small molecules, such as glucose or glutamic acid, across a membrane

15.5 Adrenergic Messengers: Neurotransmitters/Hormones

A Monoamine Messengers

The third class of neurotransmitters/hormones, the adrenergic messengers, includes such monoamines as epinephrine, serotonin, dopamine, and histamine. (Structures of these compounds can be found later in this section and in Chemical Connections 15E.) These monoamines transmit signals by a mechanism whose beginning is similar to the action of acetylcholine. That is, they are adsorbed on a receptor.

Martin Rodbell (1925–1998) of the National Institutes of Health was one of the pioneers in discovering and naming signal transduction. For this discovery, he was awarded the Nobel Prize in physiology and medicine in 1994.

Signal transduction A cascade of events through which the signal of a neurotransmitter or hormone delivered to its receptor is carried inside the target cell and amplified into many signals that can cause protein modifications, enzyme activation, and the opening of membrane channels

G-protein also participates in another signal transduction cascade, which involves inositol-based compounds (Section 12.6) as signaling molecules. Phospho-inositoldiphosphate (PIP2) mediates the action of hormones and neurotransmitters. These messengers can stimulate the phosphorylation of enzymes, in a manner similar to the cAMP cascade. They also play an important role in the release of calcium ions from their storage areas in the endoplasmic reticulum (ER) or sarcoplasmic reticulum (SR).

B Action of Messengers

Once the monoamine neurotransmitter/hormone (for example, norepinephrine) is adsorbed onto the receptor site, the signal will be amplified inside the cell. In the example shown in Figure 15.4, the receptor has an associated protein called G-protein. This protein is the key to the cascade that produces many signals inside the cell (amplification). The entire process that occurs after the messenger binds to the receptor is called **signal transduction.** The active G-protein has an associated nucleotide, guanosine triphosphate (GTP). It is an analog of adenosine triphosphate (ATP), in which the aromatic base adenine is substituted by guanine (Section 16.2). The G-protein becomes inactive when its associated nucleotide is hydrolyzed to guanosine diphosphate (GDP). Signal transduction starts with the active G-protein, which activates the enzyme adenylate cyclase.

C Secondary Messengers

Adenylate cyclase produces a secondary messenger inside the cell, called cyclic AMP (cAMP). The manufacture of cAMP activates processes that result in the transmission of an electrical signal. The cAMP is manufactured by adenylate cyclase from ATP:

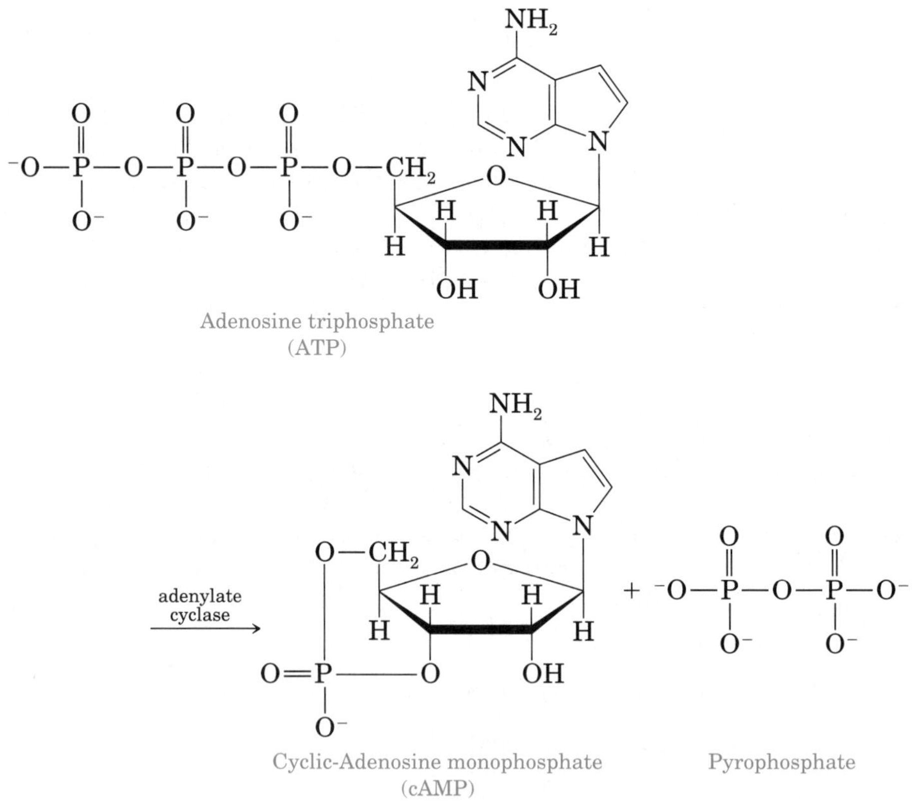

Adenosine triphosphate
(ATP)

Cyclic-Adenosine monophosphate   Pyrophosphate
(cAMP)

The activation of adenylate cyclase accomplishes two important goals:

1. It converts an event occurring at the outer surface of the target cell (adsorption onto receptor site) to a change inside the target cell (formation of cAMP). Thus the primary messenger (neurotransmitter or hormone) does not have to cross the membrane.

2. It amplifies the signal. One molecule adsorbed on the receptor triggers the adenylate cyclase to make many cAMP molecules. Thus the signal is amplified many thousands of times.

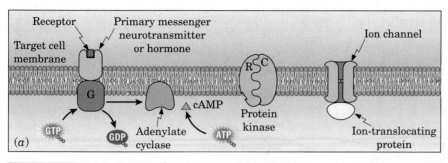

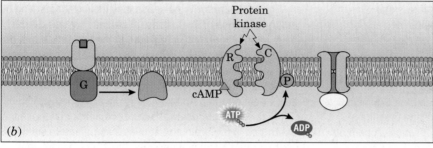

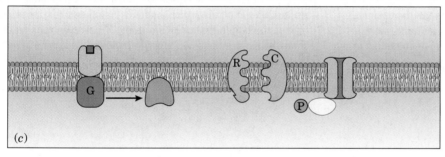

Figure 15.4 The sequence of events in the postsynaptic membrane when norepinephrine is absorbed onto the receptor site. (*a*) The active G-protein hydrolyzes GTP. The energy of hydrolysis of GTP to GDP activates adenylate cyclase. A molecule of cAMP forms when adenylate cyclase cleaves ATP into cAMP and pyrophosphate. (*b*) Cyclic AMP activates a protein kinase by dissociating the regulatory (R) unit from the catalytic unit (C). A second molecule of ATP, shown in (*b*), has phosphorylated the catalytic unit and been converted to ADP. (*c*) The catalytic unit phosphorylates the ion-translocating protein that blocked the channel for ion flow. The phosphorylated ion-translocating protein changes its shape and position and opens the ion gates.

D Removal of Signal

How does this signal amplification stop? When the neurotransmitter or hormone dissociates from the receptor, the adenylate cyclase halts the manufacture of cAMP. The cAMP already produced is destroyed by the enzyme phosphodiesterase, which catalyzes the hydrolysis of the phosphoric ester bond, yielding AMP.

The amplification through the secondary messenger (cAMP) is a relatively slow process. It may take from 0.1 s to a few minutes. Therefore, in cases where the transmission of signals must be fast (ms or μs), a neurotransmitter such as acetylcholine acts on membrane permeability directly, without the mediation of a secondary messenger.

E Control of Neurotransmission

The G-protein–adenylate cyclase cascade in transduction signaling is not limited to monoamine messengers. A wide variety of peptide hormones and neurotransmitters (Section 15.6) use this signaling pathway. Included among them are glucagon, vasopressin, luteinizing hormone, enkephalins, and P-protein. Neither is the opening of ion channels, depicted in Figure 15.4, the only target of this signaling. A number of enzymes can be phosphorylated by protein kinases, and the phosphorylation controls whether these enzymes will be active or inactive (Section 14.6).

The fine control of the G-protein–adenylate cyclase cascade is essential for health. Consider the toxin of the bacteria *Vibrio cholerae,* which perma-

CHEMICAL CONNECTIONS 15E

Parkinson's Disease: Depletion of Dopamine

Parkinson's disease is characterized by spastic motion of the eyelids as well as rhythmic tremors of the hands and other parts of the body, often when the patient is at rest. The posture of the patient changes to a forward, bent-over position; walking becomes slow, with shuffling footsteps. The cause of this degenerative nerve disease is unknown, but genetic factors and environmental effects, such as exposure to pesticides and/or high concentrations of metals such as Mn^{2+} ion, have been implicated.

The neurons affected contain, under normal conditions, mostly dopamine as a neurotransmitter. People with Parkinson's disease have depleted amounts of dopamine in their brains, but the dopamine receptors are not affected. Thus the first line of remedy is to *increase the concentration of dopamine.* Dopamine cannot be administered directly because it cannot penetrate the blood–brain barrier and therefore does not reach the tissue where its action is needed. L-Dopa, on the other hand, is transported through the arterial wall and converted to dopamine in the brain:

(S)-3,4-Dihydroxyphenylalanine
(L-Dopa)

enzyme-catalyzed decarboxylation →

Dopamine $+ CO_2$

When L-dopa is administered, many Parkinson's disease patients are able to synthesize dopamine and resume normal nerve transmission. In these individuals, L-dopa reverses the symptoms of Parkinson's disease, although the respite is only temporary. In other patients, the L-dopa regimen provides little benefit.

Another way to increase dopamine concentration is to *prevent its metabolic elimination.* The drug entacapon (Comtan) inhibits an enzyme that is instrumental in clearing dopamine from the brain. The enzyme (catechol-O-methyl transferase, COMT) converts dopamine to 3-methoxy-4-hydroxy-L-phenylalanine, which is then eliminated. Entacapon is usually administered together with L-dopa. Another drug, (R)-selegiline (L-Deprenyl), is a monoamine oxidase (MAO) inhibitor. L-Deprenyl, which is also given together with L-dopa, can reduce the symptoms of Parkinson's disease and even increase the life span of patients. It increases the level of dopamine by *preventing its oxidation by MAOs.*

Other drugs may treat only the symptoms of Parkinson's disease: the spastic motions and the tremors. These drugs, such as benzotropin (Cogentin), are similar to atropine (Problem 8.45) and act on cholinergic receptors, thereby preventing muscle spasms.

The real cure for the disease may lie in transplanting human embryonic dopamine neurons. In preliminary work, such grafts have been functionally integrated in the patient's brain, producing dopamine. In the most successful cases, patients have been able to resume normal, independent life following the transplant.

Certain drugs designed to affect one neurotransmitter may also affect another. An example is the drug methylphenidate (Ritalin). In higher doses, this drug promotes the dopamine concentration in the brain and acts as a stimulant. In small doses, it is prescribed to calm hyperactive children or to minimize ADD (attention deficit disorder). It seems that in smaller doses Ritalin raises the concentration of serotonin. This neurotransmitter decreases hyperactivity without affecting the dopamine levels of the brain.

Serotonin

The close connection between two monoamine neurotransmitters, dopamine and serotonin, is also evident in their roles in controlling the nausea and vomiting that often follow general anesthesia and chemotherapy. Blockers of dopamine receptors in the brain, such as metoclopramide (Reglan) or promethazine (Phenergan), can alleviate the symptoms after anesthesia. However, a blocker of serotonin receptors in the brain as well as on the terminals of the vagus nerve in the stomach, such as ondansetrone (Zofran), is the drug of choice for preventing chemotherapy-induced vomiting.

Synthesis and degradation of dopamine are not the only way that the brain keeps its concentration at a steady state. The concentration is also controlled by specific proteins, called *transporters,* that ferry the used dopamine from the receptor back across the synapse into the original neuron for reuptake. Cocaine addiction works through such a transporter. Cocaine binds to the dopamine transporter, like a reversible inhibitor, thereby preventing the reuptake of dopamine. As a consequence, dopamine is not transported back to the original neuron and stays in the synapse, increasing the continuous firing of signals, which is the psychostimulatory effect associated with a cocaine "high."

nently activates G-protein. The result is the symptoms of cholera—namely, severe dehydration as a result of diarrhea. This problem arises because the activated G-proteins overproduce cAMP. This excess, in turn, opens the ion channels, which produces a large outflow of Na^+ ions and accompanying water from the epithelial cells to the intestines. Therefore, the first measure taken in treating cholera victims is to replace the lost water and salt.

F Removal of Neurotransmitters

The inactivation of the adrenergic neurotransmitters differs somewhat from the invactivation of the cholinergic transmitters. While acetylcholine is decomposed by acetylcholinesterase, most of the adrenergic neurotransmitters are inactivated in a different way. *The body inactivates monoamines by oxidizing them to aldehydes.* Enzymes that catalyze these reactions, called monoamine oxidases (MAOs), are very common in the body. For example, one MAO converts both epinephrine and norepinephrine to the corresponding aldehyde:

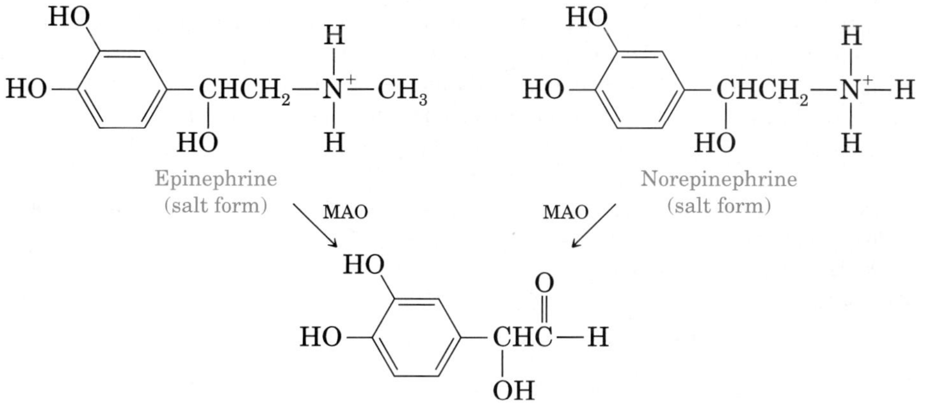

Epinephrine
(salt form)   MAO   MAO   Norepinephrine
(salt form)

Many drugs that are used as antidepressants or antihypertensive agents are inhibitors of monoamine oxidases—for example, Tofranil and Elavil. These inhibitors prevent MAOs from converting monoamines to aldehydes, thereby increasing the concentration of the active adrenergic neurotransmitters.

There is also an alternative way to remove adrenergic neurotransmitters. Shortly after adsorption onto the postsynaptic membrane, the neurotransmitter comes off the receptor site and is reabsorbed through the presynaptic membrane and stored again in the vesicles.

G Histamines

The neurotransmitter histamine is present in mammalian brains. It is synthesized from the amino acid, histidine, by decarboxylation:

Histamine cannot readily pass through the blood–brain barrier (Section 23.1) and must be synthesized in the brain neurons.

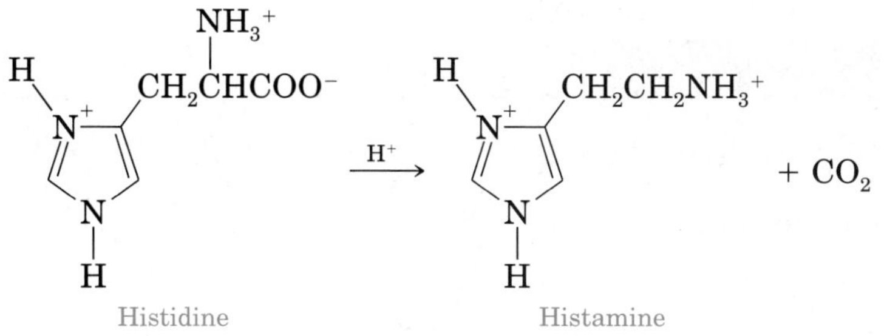

Histidine   Histamine

CHEMICAL CONNECTIONS 15F

Nitric Oxide as a Secondary Messenger

The toxic effects of the simple gaseous molecule NO have long been recognized. Therefore, it came as a surprise to find that this compound plays a major role in chemical communications. This simple molecule is synthesized in the cells when arginine is converted to citrulline. (These two compounds appear in the urea cycle; see Section 19.8). Nitric oxide is a relatively nonpolar molecule, and shortly after it has been produced in the nerve cell, it quickly diffuses across the lipid bilayer membrane. During its short half-life (4–6 s), it can reach a neighboring cell. Because NO passes through membranes, it does not need extracellular receptors to deliver its message. NO is very unstable, so there is no need for a special mechanism for its destruction.

NO acts as an intercellular messenger between the endothelial cells surrounding the blood vessels and the smooth muscles encompassing these cells. It relaxes the muscle cells, thereby dilating the blood vessels. The outcome is less-restricted blood flow and a drop in blood pressure. This reason also explains why nitroglycerin (Chemical Connections 4D) is effective against angina: It produces NO in the body.

Another role of NO in dilating blood vessels lies in remedying impotence. The new impotence-relieving drug, Viagra, enhances the activity of NO by inhibiting an enzyme (phosphodiesterase) that otherwise would reduce NO's effect on smooth muscles.

When the NO concentration is sufficiently high, the blood vessels dilate, allowing enough blood to flow to provide an erection. In most cases, this happens within an hour after taking the pill.

Sometimes the dilation of blood vessels is not so beneficial. Headaches are caused by dilated arteries in the head. NO-producing compounds in food—nitrites in smoked and cured meats and sodium glutamate in seasoning (Chemical Connections 19D)—can cause such headaches. Nitroglycerin itself often induces headaches.

Nitric oxide is toxic. This toxicity is used by our immune system (Section 22.2B) to fight infections by viruses.

The toxic effect of NO is also evident in strokes. A blocked artery restricts the blood flow to certain parts of the brain; the oxygen-starved neurons then die. Next, neurons in the surrounding area, 10 times larger than the place of the initial attack, release glutamic acid, which stimulates other cells. They, in turn, release NO, which kills all cells in the area. Thus the damage to the brain is spread tenfold. A concentrated effort is under way to find inhibitors of the NO-producing enzyme, nitric oxide synthase, that can be used as antistroke drugs. For the discovery of NO and its role in blood pressure control, three pharmacologists—Robert Furchgott, Louis Ignarro, and Ferid Murad—received the 1998 Nobel Prize in physiology.

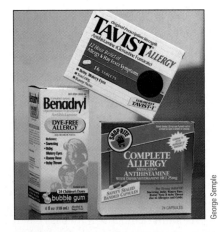

■ **Antihistamines block the H$_1$ receptor for histamine.**

Sir James W. Black of England received the 1988 Nobel Prize in medicine for the invention of cimetidine and such other drugs as propranolol (Table 15.1).

The action of histamine as a neurotransmitter is very similar to that of other monoamines. There are two kinds of receptors for histamine. One receptor, H$_1$, can be blocked by antihistamines such as dimenhydrinate (Dramamine) and diphenhydramine (Benadryl). The other receptor, H$_2$, can be blocked by ranitidine (Zantac) and cimetidine (Tagamet).

H$_1$ receptors are found in the respiratory tract. They affect the vascular, muscular, and secretory changes associated with hay fever and asthma. Therefore, antihistamines that block H$_1$ receptors relieve these symptoms. The H$_2$ receptors are found mainly in the stomach and affect the secretion of HCl. Cimetidine and ranitidine, both H$_2$ blockers, reduce acid secretion and, therefore, are effective drugs for ulcer patients. However, the real cure for ulcers is to kill the bacteria *Helicobacter pylori* (Chemical Connections 14B) by administering either antibiotics or a combination of antibiotics and H$_2$ blockers.

EXAMPLE 15.1

Three enzymes in the adrenergic neurotransmission pathway affect the transduction of the signals. Identify them and describe how they affect the neurotransmission.

Solution

The enzyme adenylate cyclase amplifies the signal by producing cAMP secondary messengers. The enzyme phosphatase terminates the signal by hydrolyzing cAMP. The enzyme monoamine oxidase (MAO) reduces the frequency of signals by oxidizing the monoamine neurotransmitters to the corresponding aldehydes.

Problem 15.1

What is the functional difference between G-protein and GTP?

15.6 Peptidergic Messengers:
Neurotransmitters/Hormones

A Messengers

Many of the most important hormones affecting metabolism belong to the peptidergic messengers group. Among them are insulin (Section 13.7 and Chemical Connections 15G) and glucagons, hormones of the pancreatic islets, and vasopressin and oxytocin (Section 13.7) products of the posterior pituitary gland.

In the last few years, scientists have isolated a number of brain peptides that have affinity for certain receptors and, therefore, act as if they were neurotransmitters. Some 25 or 30 such peptides are now known.

The first brain peptides isolated were the **enkephalins.** These pentapeptides are present in certain nerve cell terminals. They bind to specific pain receptors and seem to control pain perception. Because they bind to the receptor site that also binds the pain-killing alkaloid morphine, it is assumed that the N-terminal end of the pentapeptide fits the receptor (Figure 15.5).

Even though morphine remains the most effective agent in reducing pain, its clinical use is limited because of its side effects. These include respiratory depression, constipation, and, most significantly, addiction. The

Figure 15.5 Similarities between the structure of morphine and that of the brain's own pain regulators, the enkephalins.

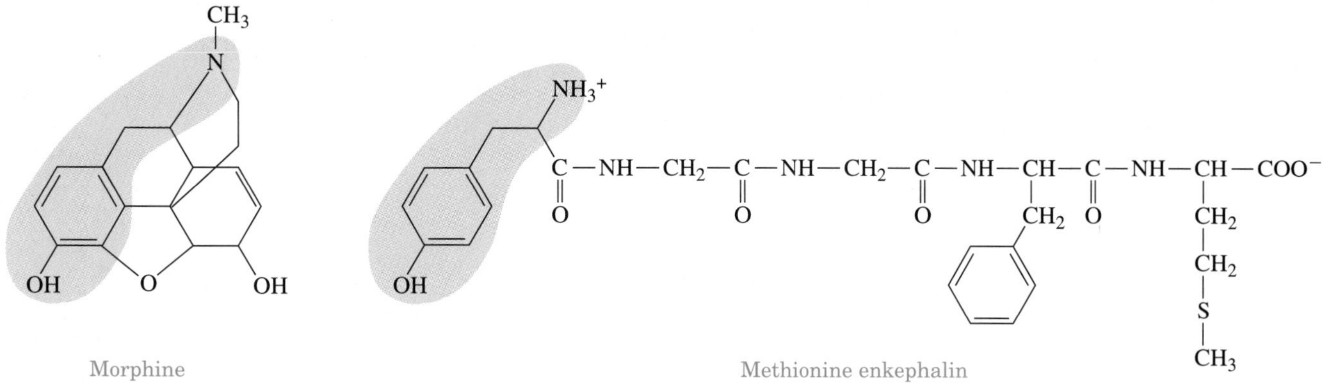

Morphine

Methionine enkephalin

Diabetes

The disease diabetes mellitus affects approximately 16 million people in the United States. In a normal person, the pancreas, a large gland behind the stomach, secretes insulin as well as several other hormones. Diabetes usually results from low insulin secretion. Insulin is necessary for glucose molecules to penetrate such cells as brain, muscle, and fat cells, where they can be used. It accomplishes this task by being adsorbed onto the receptors in the target cells. This adsorption triggers the manufacture of cyclic GMP (not cAMP), and this secondary messenger in turn increases the transport of glucose molecules into the target cells.

In the resulting cascade of events, the first step is the self-(auto)phosphorylation of the receptor molecule itself, on the cytoplasmic side. The phosphorylated insulin receptor activates enzymes and regulatory proteins by phosphorylating them. As a consequence, glucose transporter molecules (GLUT4) that are stored inside the cells migrate to the plasma membrane. Once there, they facilitate the movement of glucose across the membrane. This transport relieves the accumulation of glucose in blood serum and makes it available for metabolic activity inside the cells. Thus glucose can be used as an energy source, stored as glycogen, or even diverted to enter fat and other molecular biosynthetic pathways.

In diabetic patients, the glucose level rises to 600 mg/100 mL of blood or higher (normal is 80 to 100 mg/100 mL). There are two kinds of diabetes. In insulin-dependent diabetes, patients do not manufacture enough of this hormone in the pancreas. This Type I disease develops early, before the age of 20, and must be treated with daily injections of insulin. Even with daily injections of insulin, the blood sugar level fluctuates, which may cause other disorders, such as cataracts, retinal dystrophy leading to blindness, kidney disease, heart attack, and nervous disorders.

One way to counteract these fluctuations is to monitor the blood sugar and, as the glucose level rises, to administer insulin. Such monitoring requires pricking the finger six times per day, an invasive regimen that few diabetics follow faithfully. Recently, noninvasive techniques have been developed. One of the most promising employs contact lenses. The blood sugar's fall and rise is mimicked by the glucose content of tears. A fluorescent sensor in the contact lens monitors these glucose fluctuations; its data can be read via a hand-held fluorimeter. Thus the patterns of glucose fluctuation can be obtained noninvasively and, if it rises to the danger zone, it can be counteracted by insulin intake.

The delivery of insulin also has undergone a revolution. The tried-and-true methods of injections and insulin pumps are still widely used, but new delivery is available orally or by nasal sniffing.

In Type II (non–insulin-dependent) diabetes, the patient has enough insulin in the blood but cannot utilize it properly because the target cells have an insufficient number of receptors. Such patients typically develop the disease after age 40 and are likely to be obese. Overweight people usually have a lower-than-normal number of insulin receptors in their adipose (fat) cells.

Oral drugs can help this second type of diabetic patient in several ways. For example, sulfonyl urea compounds, such as tolbutamide (Orinase), increase insulin secretion. In addition, insulin concentration in the blood can be increased by enhancing its release from the β-cells of pancreatic islets. The drug repaglinide (Prandin) blocks the K^+-ATP channels of the β-cells, allowing Ca^{2+} influx, which induces the release of insulin from the cells.

These drugs seem to control the symptoms of diabetes, but fluctuations in insulin levels may turn high blood sugar (hyperglycemia) into low blood sugar (hypoglycemia) which is just as dangerous. Other drugs for Type II diabetes that do not elicit hypoglycemia attempt to control the glucose level at its source. Migitol (Glyset), an anti-α-glucosidase drug, inhibits the enzyme that converts glycogen or dietary starch into glucose. The drug metformin (Glucophage) decreases glucose production in the liver, carbohydrate absorption in the intestines, and glucose uptake by fat cells.

Tolbutamide
(Orinase)

Metformin
(Glucophage)

clinical use of enkephalins yielded only modest relief. The challenge is to develop analgesic drugs that do not involve the opiate receptors in the brain.

Another brain peptide, **neuropeptide Y,** affects the hypothalamus, a region that integrates the body's hormonal and nervous systems. Neuropeptide Y is a potent orexic (appetite-stimulating) agent. When its receptors are blocked (for example, by leptin, the "thin" protein) appetite is suppressed. Leptin is an anorexic agent.

Still another peptidergic neurotransmitter is **substance P** (*P* for "pain"). This 11-amino-acid peptide is involved in transmission of pain signals. In injury or inflammation, sensory nerve fibers transmit signals from the peripheral nervous system (where the injury occurred) to the spinal cord, which processes the pain. The peripheral neurons synthesize and release substance P, which bonds to receptors on the surface of the spinal cord. The substance P, in its turn, removes the magnesium block at the *N*-methyl-D-aspartate (NMDA) receptor. Glutamic acid, an excitatory amino acid, can then bind to this receptor. In doing so, it amplifies the pain signal going to the brain.

B Secondary Messengers

All peptidergic messengers, hormones, and neurotransmitters act through secondary messengers. Glucagon, luteinizing hormone, antidiuretic hormone, angiotensin, enkaphalin, and substance P use the G-protein–adenylate cyclase cascade. Others, such as vasopressin, use membrane-derived phosphatidylinositol (PI) derivates (Section 12.6). As with the cAMP cascade, control of PI secondary messengers is done by phosphorylation. Activation may be attained via phosphorylation, and inactivation by removing a phosphate. The transition between phosphatidylinositol and its various phosphates is exercised by a kinase/phosphatase mechanism:

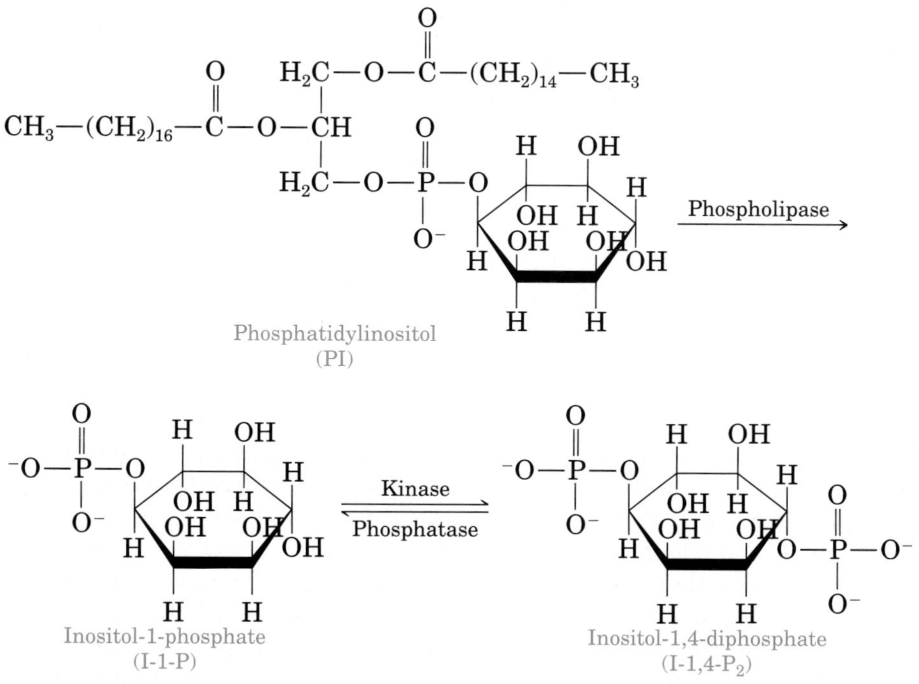

Phosphatidylinositol
(PI)

Inositol-1-phosphate
(I-1-P)

Inositol-1,4-diphosphate
(I-1,4-P$_2$)

Still other substances use calcium (Chemical Connections 15A) as secondary messengers.

CHEMICAL CONNECTIONS 15H

Tamoxifen and Breast Cancer

Large-scale clinical studies have shown that tamoxifen (Nolvadex) can reduce the recurrence of breast cancer in women who had been successfully treated for the disease. Even more importantly, a study involving 13,000 women showed that tamoxifen reduced the occurrence of breast cancer by 45% in women who were at high risk because of family history, previous breast abnormality, or age. Tamoxifen is a nonsteroidal compound that binds to the estradiol receptors situated on the nuclei of cells. By binding to the receptor, it causes conformational changes. This then prevents estradiol's binding to its receptor and thus hampers the stimulation of the growth of breast cancer. In addition, tamoxifen has shown promise in treating refractory ovarian cancer.

Although it reduces the risk of breast cancer, tamoxifen increases the risk of both uterine cancer and clot formation. The latter condition can be life-threatening if clots migrate to the lungs, causing pulmonary embolism. However, these last two diseases are much rarer than breast cancer.

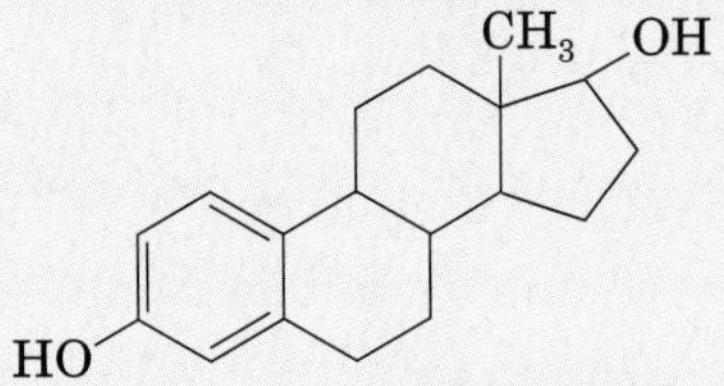

Estradiol

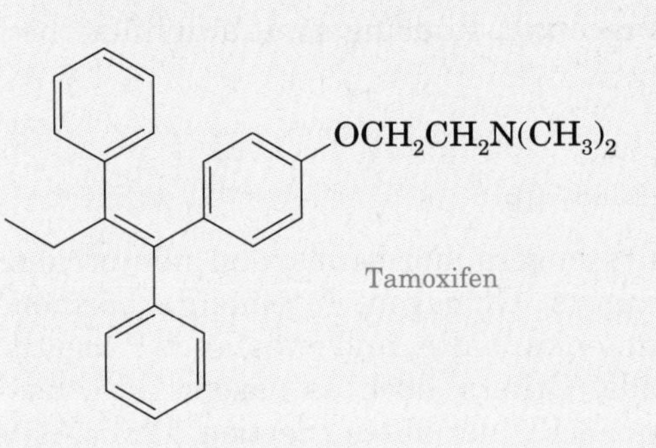

Tamoxifen

15.7 Steroid Messengers: Hormones

In Section 12.10, we saw that a large number of hormones possess steroid ring structures. These hormones, the sex hormones among them, are hydrophobic; therefore, they can cross plasma membranes of the cell by diffusion.

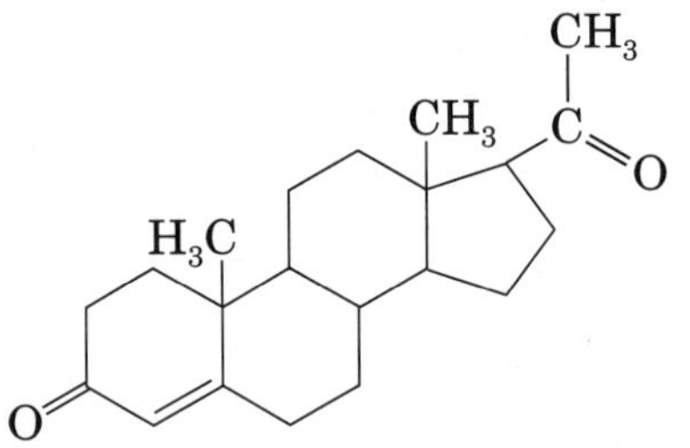

Progesterone

There is no need for special receptors embedded in the membrane for these hormones. It has been shown, however, that **steroid hormones** interact inside the cell with protein receptors. Most of these receptors are localized in the nucleus of the cell, but small amounts also exist in the cytoplasm. When they interact with steroids, they facilitate their migration through the aqueous cytoplasm; the proteins themselves are hydrophilic.

Once inside the nucleus, the steroid–receptor complex can either bind directly to the DNA or combine with a transcription factor (Section 17.2), influencing the synthesis of a certain key protein. Thyroid hormones, which also have large hydrophobic domains, have protein receptors as well, which facilitates their transport through aqueous media.

The steroid hormonal response through protein synthesis is not fast. In fact, it takes hours to occur. Steroids can also act at the cell membrane, influencing ligand-gate ion channels. Such a response would take only seconds. An example of such a fast response occurs in fertilization. The sperm head contains proteolytic enzymes, which act on the egg to facilitate its penetration. These enzymes are stored in acrosomes, organelles found on the sperm head. During fertilization, progesterone originating from the follicle cells surrounding the egg acts on the acrosome outer membrane, which disintegrates within seconds and releases the proteolytic enzymes.

The same steroid hormones depicted in Figure 12.6 act as neurotransmitters, too. These neurosteroids are synthesized in the brain cells in both neurons and glia, and they affect receptors, mainly the NMDA and GABA receptors (Section 15.4). Progesterone and progesterone metabolites in brain cells can induce sleep, have analgesic and anticonvulsive effects, and can even serve as natural anesthetics.

S U M M A R Y

Cell-to-cell communications are carried out by three kinds of molecules (Section 15.1). **Receptors** are protein molecules embedded in the membranes of cells. **Chemical messengers,** or ligands, interact with receptors. **Secondary messengers** carry and amplify the signals from the receptor to inside the cell.

Neurotransmitters send chemical messengers across a short distance—the **synapse** between two neurons or between a neuron and a muscle or endocrine gland cell (Section 15.2). This communication occurs in milliseconds. **Hormones** transmit their signals more slowly and over a longer distance, from the source of their secretion (endocrine gland), through the bloodstream, into target cells (Section 15.2). Many drugs affect these chemical communications: **Antagonists** block receptors; **agonists** stimulate receptors.

Five kinds of chemical messengers exist: **cholinergic, amino acid, adrenergic, peptidergic,** and **steroid.** Neurotransmitters may belong to the first four classes, hormones to the last three classes. Acetylcholine is cholinergic, glutamic acid is an amino acid, epinephrine (adrenaline) and norepinephrine are adrenergic, enkephalins are peptidergic and progesterone is a steroid.

Nerve transmission starts with the neurotransmitters packaged in **vesicles** in the **presynaptic end** of neurons. When these neurotransmitters are re-leased, they cross the membrane and the synapse and are adsorbed onto receptor sites on the **postsynaptic** membranes. This adsorption triggers an electrical response. Some neurotransmitters act directly, whereas others act through a secondary messenger, **cyclic AMP.** After the electrical signal is triggered, the neurotransmitter molecules must be removed from the postsynaptic end. In the case of acetylcholine (Section 15.3), this removal is done by an enzyme called acetylcholinesterase; in the case of monoamines (Section 15.5), enzymes (MAOs) oxidize them to aldehydes.

Amino acids (Section 15.4) bind to their receptors, which are ligand-gated ion channels. Peptides and proteins (Sections 15.6) bind to receptors on the target cell membrane and use secondary messengers to exert their influence. **Signal transduction** is the process that occurs after a ligand binds to its receptor: The signal is carried inside the cell and is amplified. Steroids (Section 15.7) penetrate the cell membrane, and their receptors are found in the cytoplasm. Together with their receptors, they penetrate the cell nucleus. These hormones can act in three ways: (1) They activate enzymes, (2) they affect the gene transcription of an enzyme or protein, and (3) they change membrane permeability. The same steroids can also act as neurotransmitters, when synthesized in neurons.

P R O B L E M S

Numbers that appear in color indicate difficult problems.
⯈ designates problems requiring application of principles.

Chemical Communications

15.2 What kind of signal travels along the axon of a neuron?

15.3 What is the difference between a *chemical messenger* and a *secondary messenger*?

Neurotransmitters and Hormones

15.4 Define the following:
(a) Synapse   (b) Receptor
(c) Presynaptic   (d) Postsynaptic
(e) Vesicle

15.5 What is the role of Ca^{2+} in releasing neurotransmitters into the synapse?

15.6   Which signal takes longer: (a) neurotransmitter or (b) hormone? Explain.

15.7   Which gland controls lactation?

15.8   To which of the three groups of chemical messengers do these hormones belong?
   (a) Norepinephrine
   (b) Thyroxine
   (c) Oxytocin
   (d) Progesterone

Cholinergic Neurotransmitters

15.9   How does acetylcholine transmit an electric signal from neuron to neuron?

15.10 Which end of the acetylcholine molecule fits into the receptor site?

15.11 Cobra venom and botulin are both deadly toxins, but they affect cholinergic neurotransmission differently. How does each cause paralysis?

15.12 Different ion concentrations across a membrane generate a potential (voltage). We call such a membrane *polarized*. What happens when acetylcholine is adsorbed on its receptor?

Amino Acid Neurotransmitters

15.13 List two features by which taurine differs from the amino acids found in proteins.

15.14 How is glutamic acid removed from its receptor?

15.15 What is unique in the structure of GABA that distinguishes it from all the amino acids that are present in proteins?

15.16 What is the structural difference between NMDA, an agonist of a glutamic acid receptor, and L-aspartic acid?

Adrenergic Neurotransmitters

15.17 (a) Identify two monoamine neurotransmitters in Table 15.1.
   (b) Explain how they act.
   (c) What medication controls the particular diseases caused by the lack of monoamine neurotransmitters?

15.18 What bond is hydrolyzed and what bond is formed in the synthesis of cAMP?

15.19 How is the catalytic unit of protein kinase activated in adrenergic neurotransmission?

15.20 The formation of cyclic AMP is described in Section 15.5. Show by analogy how cyclic GMP is formed from GTP.

15.21 By analogy to the action of MAO on epinephrine, write the structural formula of the product of the corresponding oxidation of dopamine.

15.22 The action of protein kinase is the next-to-last step in the G-protein–adenylate cyclase cascade signal transduction. What kind of effects can elicit the phosphorylation by this enzyme?

15.23 Explain how adrenergic neurotransmission is affected by (a) amphetamines and (b) reserpine. (See Table 15.1.)

15.24 Which step in the events depicted in Figure 15.3 provides an electrical signal?

15.25 What kind of product is the MAO-catalyzed oxidation of epinephrine?

15.26 How is histamine removed from the receptor site?

15.27 Cyclic AMP affects the permeability of membranes for ion flow.
   (a) What blocks the ion channel?
   (b) How is this blockage removed?
   (c) What is the direct role of cAMP in this process?

15.28 Dramamine and cimetidine are both antihistamines. Would you expect Dramamine to cure ulcers and cimetidine to relieve the symptoms of asthma? Explain.

Peptidergic Neurotransmitters

15.29 What is the chemical nature of enkephalins?

15.30 What is the mode of action of Demerol as a pain killer? (See Table 15.1.)

15.31 What enzyme catalyzes the formation of inositol-1,4,5-triphosphate from inositol-1,4-diphosphate? Give the structures of the reactant and the product.

15.32 How is the secondary messenger inositol-1,4,5-triphosphate inactivated?

Chemical Connections

15.33 (Chemical Connections 15A) What is the difference between calcium sparks (puffs) and calcium waves?

15.34 (Chemical Connections 15A) What is the role of calmodulin in signaling by Ca^{2+} ions?

15.35 (Chemical Connections 15A) To enable a fusion between the synaptic vessel and the plasma membrane, the calcium concentration is increased. How many-fold of an increase in Ca^{2+} concentration is needed?

15.36 (Chemical Connections 15B) (a) What is the effect of nerve gases?
   (b) What molecular substance do they affect?

15.37 (Chemical Connections 15C) What is the mode of action of botulinum toxin?

15.38 (Chemical Connections 15C) How can a deadly botulinum toxin contribute to the facial beauty of Hollywood actresses?

15.39 (Chemical Connections 15D) What are the neurofibrillar tangles in the brains of Alzheimer's disease patients made of? How do they affect the cell structure?

15.40 (Chemical Connections 15D) What are the plaques in the brains of Alzheimer's disease patients made of?

15.41 (Chemical Connections 15D) Alzheimer's disease causes loss of memory. What kind of drugs may provide some relief for, if not cure, this disease? How do they act?

15.42 (Chemical Connections 15E) Why would a dopamine pill be ineffective in treating Parkinson's disease?

15.43 (Chemical Connections 15E) What is the mechanism by which cocaine stimulates the continuous firing of signals between neurons?

15.44 (Chemical Connections 15E) Parkinson's disease is due to a paucity of dopamine neurons, yet its symptoms are relieved by drugs that block cholinergic receptors. Explain.

15.45 (Chemical Connections 15E) In certain cases, embryonic dopamine neurons transplanted into the brains of patients with advanced Parkinson's disease resulted in complete remission. How was this result possible?

15.46 (Chemical Connections 15F) How can NO cause headaches?

15.47 (Chemical Connections 15F) How is nitric oxide synthesized in the cells?

15.48 (Chemical Connections 15F) How is the toxicity of NO detrimental in strokes?

15.49 (Chemical Connections 15G) Tolbutamide is called a sulfonyl urea compound. Identify the sulfonyl urea moiety in the structure of the drug.

15.50 (Chemical Connections 15G) What is the difference between insulin-dependent and non–insulin-dependent diabetes?

15.51 (Chemical Connections 15G) How does insulin facilitate the absorption of glucose from blood serum into adipocytes (fat cells)?

15.52 (Chemical Connections 15G) Diabetic patients must frequently monitor the fluctuation of glucose levels in their blood. What is the advantage of the latest technique of monitoring the glucose content in tears over the older technique of obtaining frequent blood samples?

15.53 (Chemical Connections 15H) Tamoxifen is an antagonist. How does it work in reducing the reoccurrence of breast cancer?

Additional Problems

15.54 Considering its chemical nature, how does aldosterone (Section 12.10) affect mineral metabolism (Table 15.2)?

15.55 What is the function of the ion-translocating protein in adrenergic neurotransmission?

15.56 Decamethonium acts as a muscle relaxant. If an overdose of decamethonium occurs, can paralysis be prevented by administering large doses of acetylcholine? Explain.

15.57 Endorphin, a potent pain killer, is a peptide containing 22 amino acids; among them are the same five N-terminal amino acids found in the enkephalins. Does this explain endorphin's pain-killing action?

15.58 How do alanine and beta alanine differ in structure?

15.59 Where is a G-protein located in adrenergic neurotransmission?

15.60 (Chemical Connections 15F) List a number of effects that are caused when NO, acting as a secondary messenger, relaxes smooth muscles.

15.61 (a) In terms of their action, what do the hormone vasopressin and the neurotransmitter dopamine have in common?

(b) What is the difference in their mode's of action?

15.62 What is the difference in the models of action between acetylcholinesterase and acetylcholine transferase?

15.63 How does cholera toxin exert its effect?

15.64 Give the formulas for the following reaction:
$$GTP + H_2O \rightleftharpoons GDP + P_i.$$

15.65 Insulin is a hormone that, when it binds to a receptor, enables glucose molecules to enter the cell and be metabolized. If you have a drug that is an agonist, how would the glucose level in the serum change upon administering the drug?

15.66 (Chemical Connections 15E) Ritalin is used to alleviate hyperactivity in attention deficit disorder of children. How does this drug work?

15.67 The pituitary gland releases luteinizing hormone (LH), which enhances the production of progesterone in the uterus. Classify these two messengers and discuss how each delivers its message.

InfoTrac College Edition

For additional reading, go to InfoTrac College Edition, your online research library, at

http://infotrac.thomsonlearning.com

CHAPTER 16

16.1 Introduction

16.2 Components of Nucleic Acids

16.3 Structure of DNA and RNA

16.4 RNA

16.5 Genes, Exons, and Introns

16.6 DNA Replication

16.7 DNA Repair

16.8 Cloning

The world's first sheep clones. These Welsh Mountain sheep are the product of research by Dr. Ian Wilmut and colleagues at the Roslin Institute in Edinburgh, Scotland.

James King-Holmes/Science Photo Library/Photo Researchers, Inc.

Nucleotides, Nucleic Acids, and Heredity

16.1 Introduction

Each cell of our bodies contains thousands of different protein molecules. Recall from Chapter 13 that all of these molecules are made up of the same 20 amino acids, just arranged in different sequences. That is, the hormone insulin has a different amino acid sequence than the globin of the red blood cell. Even the same protein—for example, insulin—has different sequence in different species (Section 13.7). Within the same species, individuals may have some differences in their proteins, although the differences are much less dramatic than those seen between species. This variation is most obvious in cases where people have such conditions as hemophilia, albinism, or color-blindness because they lack certain proteins that normal people have or because the sequence of their amino acids is somewhat different (see Chemical Connections 13D).

After scientists became aware of the differences in amino acid sequences, their next question was, how do the cells know which proteins to synthesize out of the extremely large number of possible amino acid se-

■ **Human chromosomes magnified about 8000 times.**

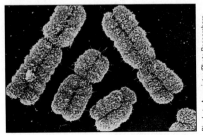

Biophoto Associates/Photo Researchers

quences? The answer to this question is that an individual gets the information from its parents through *heredity*. Heredity is the transfer of characteristics, anatomical as well as biochemical, from generation to generation. We all know that a pig gives birth to a pig and a mouse gives birth to a mouse.

It was easy to determine that the information is obtained from the parent or parents, but what form does this information take? During the last 60 years, revolutionary developments have enabled us to answer this question—the transmission of heredity on the molecular level.

From about the end of the nineteenth century, biologists suspected that the transmission of hereditary information from one generation to another took place in the nucleus of the cell. More precisely, they believed that structures within the nucleus, called **chromosomes,** have something to do with heredity. Different species have different numbers of chromosomes in the nucleus. The information that determines external characteristics (red hair, blue eyes) and internal characteristics (blood group, hereditary diseases) was thought to reside in **genes** located inside the chromosomes.

Chemical analysis of nuclei showed that they are largely made up of special basic proteins called *histones* and a type of compound called *nucleic acids*. By 1940, it became clear through the work of Oswald Avery (1877–1955) that, of all the material in the nucleus, only a nucleic acid called deoxyribonucleic acid (DNA) carries the hereditary information. That is, the genes are located in the DNA. Other work in the 1940s by George Beadle (1903–1989) and Edward Tatum (1909–1975) demonstrated that each gene controls the manufacture of one protein, and that external and internal characteristics are expressed through this gene. Thus the expression of the gene (DNA) in terms of an enzyme (protein) led to the study of protein synthesis and its control. *The information that tells the cell which proteins to manufacture is carried in the molecules of DNA.*

16.2 Components of Nucleic Acids

Two kinds of nucleic acids are found in cells: **ribonucleic acid (RNA)** and **deoxyribonucleic acid (DNA).** Each has its own role in the transmission of hereditary information. As we just saw, DNA is present in the chromosomes of the nucleus. RNA is not found in the chromosomes, but rather is located elsewhere in the nucleus and even outside the nucleus, in the cytoplasm. As we will see in Section 16.4, there are four types of RNA, all with similar structures.

Both DNA and RNA are polymers. Just as proteins consist of chains of amino acids, and polysaccharides consist of chains of monosaccharides, so nucleic acids are also chains. The building blocks (monomers) of nucleic acid chains are *nucleotides*. Nucleotides themselves, however, are composed of three simpler units: a base, a monosaccharide, and a phosphate. We will look at each of these components in turn.

A Bases

The **bases** found in DNA and RNA are chiefly those shown in Figure 16.1. All of them are basic because they are heterocyclic aromatic amines (Section 8.2). Two of these bases—adenine (A) and guanine (G)—are purines, and the other three—cytosine (C), thymine (T), and uracil (U)—are pyrimidines. The two purines (A and G) and one of the pyrimidines (C) are found in both DNA and RNA, but uracil (U) is found only in RNA, and thymine (T) is found only in DNA. Note that thymine differs from uracil only in the

There are two parents if reproduction is sexual but one parent if it is asexual.

Some simple organisms, such as bacteria, do not have a nucleus. Their chromosomes are condensed in a central region.

Gene The unit of heredity; a DNA segment that codes for one protein

Beadle and Tatum shared the 1958 Nobel Prize in physiology with Joshua Lederberg.

The early hypothesis of "one gene, one protein" has undergone modifications. We now know that one gene can participate in the manufacturing of many proteins through processes such as recombination.

Bases Purines and pyrimidines, which are components of nucleotides, DNA, and RNA

The three pyrimidines and guanine are in their keto forms rather than their enol forms (see Section 9.8).

The initial letter of each base is used as an abbreviation for that base.

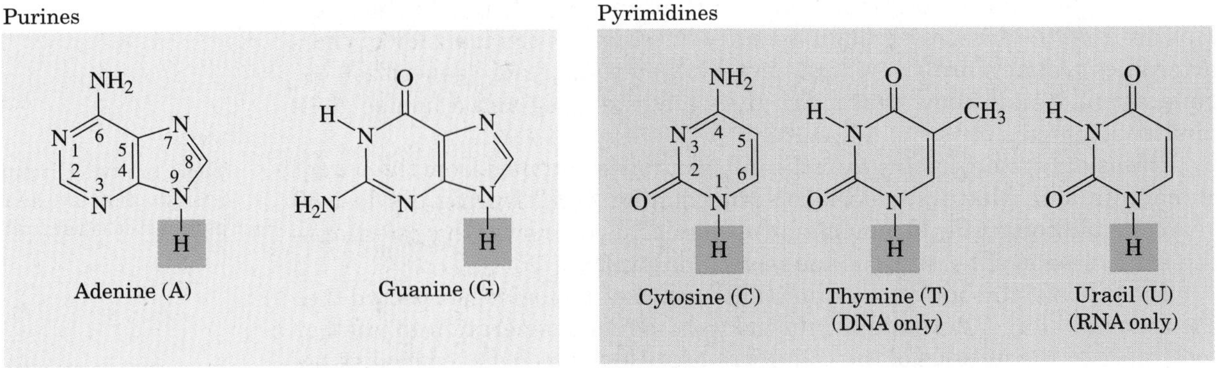

Purines

Pyrimidines

Figure 16.1 The five principal bases of DNA and RNA. Note how the rings are numbered. The hydrogens shown in blue are lost when the bases are bonded to monosaccharides.

methyl group in the 5 position. Thus both DNA and RNA contain four bases: two pyrimidines and two purines. For DNA, the bases are A, G, C, and T; for RNA, the bases are A, G, C, and U.

B Sugars

The sugar component of RNA is D-ribose (Section 11.2). In DNA, it is 2-deoxy-D-ribose (hence the name deoxyribonucleic acid):

The only difference between these molecules is that ribose has an OH group in the 2 position not found in deoxyribose.

The full name of β-D-ribose is β-D-ribofuranose and that of β-2-deoxy-D-ribose is β-2-deoxy-D-ribofuranose (see Section 11.3).

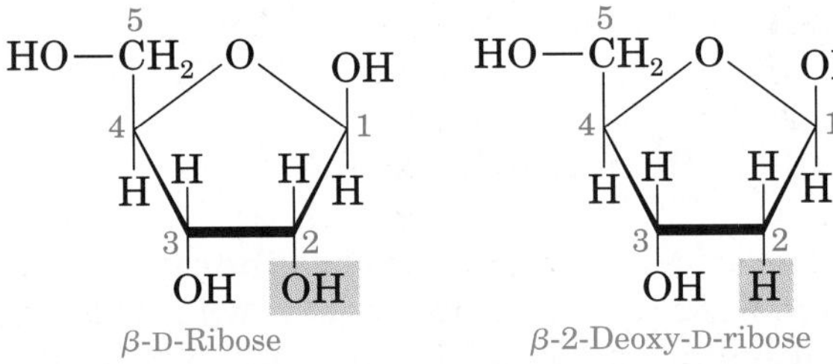

β-D-Ribose

β-2-Deoxy-D-ribose

Nucleoside A compound composed of ribose or deoxyribose and a base

The combination of sugar and base is known as a **nucleoside.** The purine bases are linked to C-1 of the monosaccharide through N-9 (the nitrogen at position 9 of the five-membered ring) by a β-*N*-glycosidic bond:

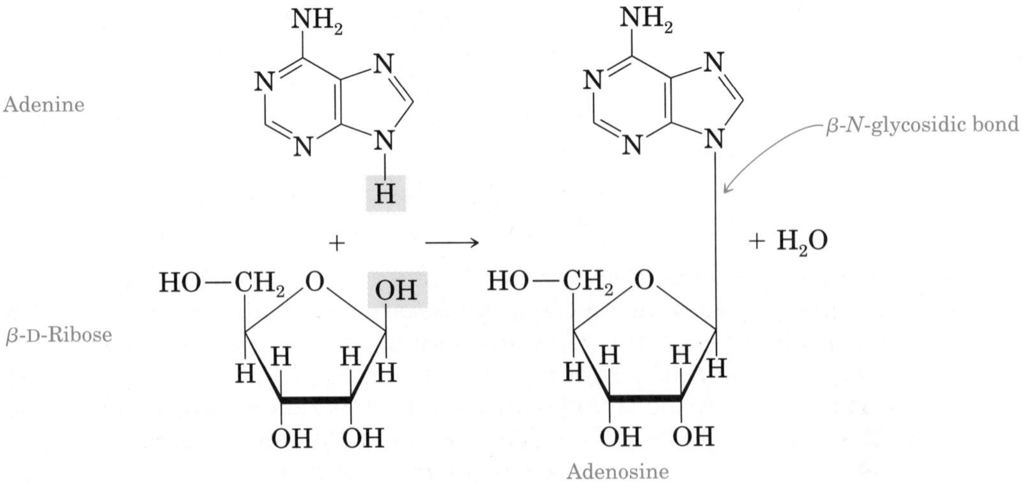

Adenine

β-D-Ribose

β-*N*-glycosidic bond

+ H_2O

Adenosine

The nucleoside made of guanine and ribose is called **guanosine.** Table 16.1 gives the names of the other nucleosides.

The pyrimidine bases are linked to C-1 of the monosaccharide through their N-1 by a *β-N*-glycosidic bond.

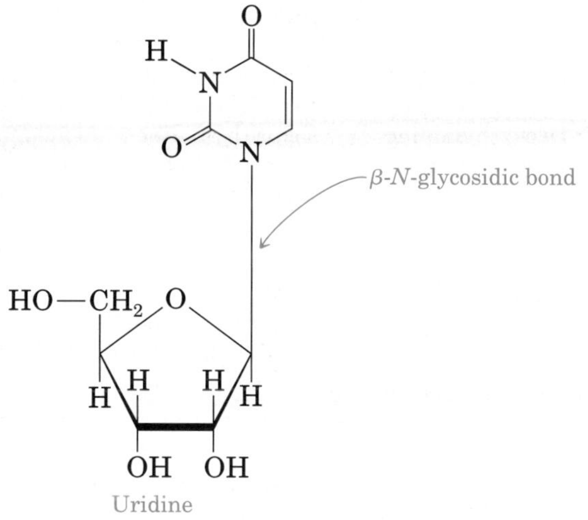

Uridine

C Phosphate

The third component of nucleic acids is phosphoric acid. When this group forms a phosphate ester (Section 10.7) bond with the —CH$_2$OH group of a nucleoside, the result is a compound known as a **nucleotide.** For example, adenosine combines with phosphate to form the nucleotide adenosine 5'-monophosphate (AMP):

> **Nucleotide** A nucleoside bonded to one, two, or three phosphate groups

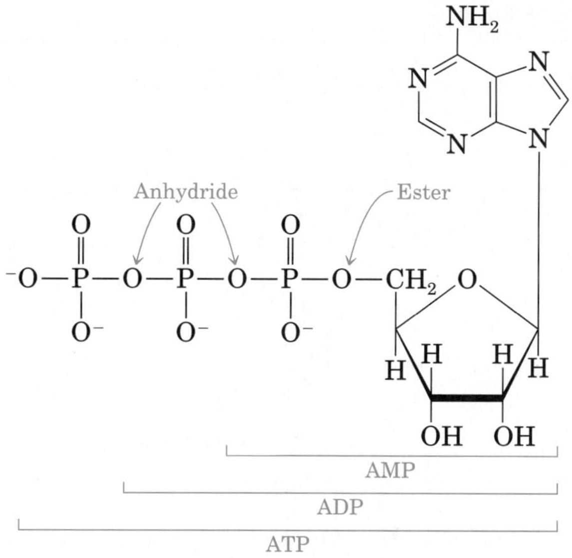

Table 16.1 gives the names of the other nucleotides. Some of these nucleotides play an important role in metabolism. They are part of the structure of key coenzymes, cofactors, and activators (Section 18.3 and 20.2). Most notably, adenosine 5'-triphosphate (ATP) serves as a common currency into which the energy gained from food is converted and stored. In ATP, two more phosphate groups are joined to AMP in phosphate anhydride bonds (Section 10.7). In adenosine 5'-diphosphate (ADP), only one phosphate group is bonded to the AMP. All other nucleotides have important

Primed numbers are used for the ribose and deoxyribose portions of nucleosides, nucleotides, and nucleic acids. Unprimed numbers are used for the bases.

Table 16.1 The Eight Nucleosides and Eight Nucleotides in DNA and RNA

Base	Nucleoside	Nucleotide
		DNA
Adenine (A)	Deoxyadenosine	Deoxyadenosine 5′-monophosphate (dAMP)*
Guanine (G)	Deoxyguanosine	Deoxyguanosine 5′-monophosphate (dGMP)*
Thymine (T)	Deoxythymidine	Deoxythymidine 5′-monophosphate (dTMP)*
Cytosine (C)	Deoxycytidine	Deoxycytidine 5′-monophosphate (dCMP)*
		RNA
Adenine (A)	Adenosine	Adenosine 5′-monophosphate (AMP)
Guanine (G)	Guanosine	Guanosine 5′-monophosphate (GMP)
Uracil (U)	Uridine	Uridine 5′-monophosphate (UMP)
Cytosine (C)	Cytidine	Cytidine 5′-monophosphate (CMP)

*The *d* indicates that the sugar is deoxyribose.

multiphosphorylated forms. For example, guanosine exists as GMP, GDP, and GTP.

In Section 16.3, we will see how DNA and RNA are chains of nucleotides. In summary:

A nucleoside = Base + Sugar

A nucleotide = Base + Sugar + Phosphoric acid

A nucleic acid = A chain of nucleotides

EXAMPLE 16.1

GTP is an important store of energy. Draw the structure of guanosine triphosphate.

Solution
The base guanine is linked to a ribose unit by β-*N*-glycosidic linkage. The triphosphate is linked to C-5′ of the ribose by an ester bond.

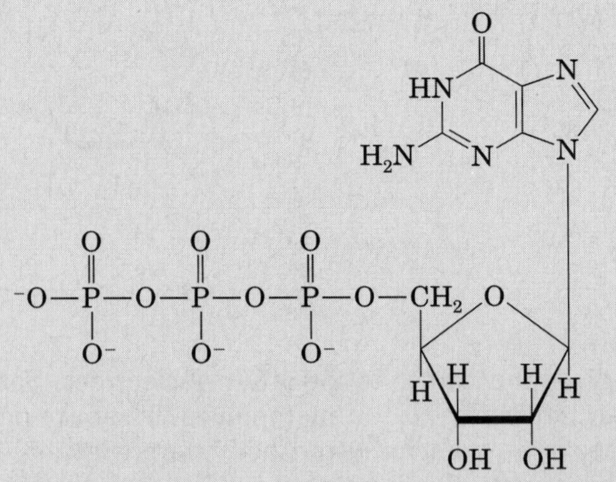

Problem 16.1
Draw the structure of UMP.

CHEMICAL CONNECTIONS 16A

Anticancer Drugs

A major difference between cancer cells and most normal cells is that the cancer cells divide much more rapidly. Rapidly dividing cells require a constant new supply of DNA. One component of DNA is the nucleoside deoxythymidine, which is synthesized in the cell by the methylation of the uracil base.

Fluorouracil

If fluorouracil is administered to a cancer patient as part of chemotherapy, the body converts it to fluorouridine, a compound that irreversibly inhibits the enzyme that manufactures thymidine from uridine, greatly decreasing DNA synthesis. Because this inhibition affects the rapidly dividing cancer cells more than the healthy cells, the growth of the tumor and the spread of the cancer are arrested. Unfortunately, chemotherapy with fluorouracil or other anticancer drugs also weakens the body, because it interferes with DNA synthesis in normal cells.

Chemotherapy is used intermittently to give the body time to recover from the side effects of the drug. During the period after chemotherapy, special precautions must be taken so that bacterial infections do not debilitate the already-weakened body.

16.3 Structure of DNA and RNA

In Chapter 13, we saw that proteins have primary, secondary, and higher structures. Nucleic acids, which are chains of monomers, also have primary, secondary, and higher structures.

A Primary Structure

Nucleic acids are polymers of nucleotides, as shown schematically in Figure 16.2. Primary structure is the sequence of nucleotides. Note that it can be divided into two parts: (1) the backbone of the molecule and (2) the bases that are the side-chain groups. The backbone in DNA consists of alternating deoxyribose and phosphate groups. Each phosphate group is linked to the 3′ carbon of one deoxyribose unit and simultaneously to the 5′ carbon of the next deoxyribose unit (Figure 16.3). Similarly, each monosaccharide unit forms a phosphate ester at the 3′ position and another at the 5′ position. The primary structure of RNA is the same except that each sugar is ribose (so there is an —OH group in the 2′ position) rather than deoxyribose and U is present instead of T.

Thus the backbone of the DNA and RNA chains has two ends: a 3′—OH end and a 5′—OH end. These two ends have roles similar to those of the C-terminal and N-terminal ends in proteins. The backbone provides structural stability for the DNA and RNA molecules.

The bases that are linked, one to each sugar unit, are the side chains, and they carry all the information necessary for protein synthesis. Analysis of the base composition of DNA molecules from many different species was done by Erwin Chargaff (1905–), who showed that in DNA taken from many different species, the quantity of adenine (in moles) is always approximately equal to that of thymine, and the quantity of guanine is always approximately equal to that of cytosine, although the adenine/guanine ratio varies widely from species to species (see Table 16.2). This important information helped to establish the secondary structure of DNA, as we will soon see.

Just as the order of the amino acid residues of protein side chains determines the primary structure of the protein (for example,

See the **Interactive General, Organic, and Biochemistry CD-ROM, version 2.0,** for further exploration on this topic.

Nucleic acid A polymer composed of nucleotides

Figure 16.2 Schematic diagram of a nucleic acid molecule. The four bases of each nucleic acid are arranged in various specific sequences.

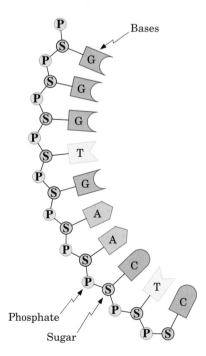

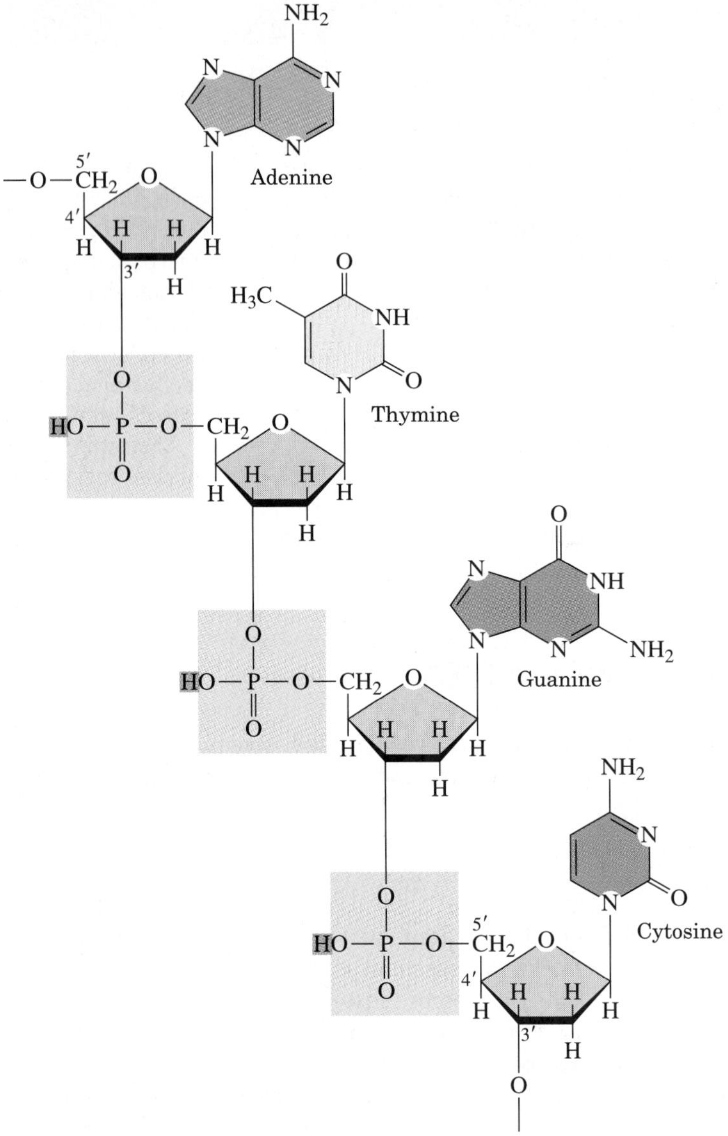

Figure 16.3 Primary structure of the DNA backbone. The hydrogens shown in blue cause the acidity of nucleic acids. In the body, at neutral pH, the phosphate groups carry a charge of −1, and the hydrogens are replaced by Na^+ and K^+.

The sequence TGA is not the same as AGT, but rather its opposite.

—Ala—Gly—Glu—Met—), the order of the bases (for example, —ATTGAC—) provides the primary structure of DNA. As with proteins, we need a convention to tell us which end to start with when we write the sequence of bases. For nucleic acids, the convention is to begin the sequence with the nucleotide that has the free 5′ terminal. Thus the sequence AGT means that adenine is the base at the 5′ terminal and thymine is the base at the 3′ terminal.

Table 16.2 Base Composition and Base Ratio in Two Species

Organism	Base Composition (mol %)				Base Ratio	
	A	G	C	T	A/T	G/C
Human	30.9	19.9	19.8	29.4	1.05	1.00
Wheat germ	27.3	22.7	22.8	27.1	1.01	1.00

B Secondary Structure of DNA

In 1953, James Watson (1928–) and Francis Crick (1916–) established the three-dimensional structure of DNA. Their work is a cornerstone in the history of biochemistry. The model of DNA established by Watson and Crick was based on two important pieces of information obtained by other workers: (1) the Chargaff rule that (A and T) and (G and C) are present in equimolar quantities and (2) X-ray diffraction photographs obtained by Rosalind Franklin (1920–1958) and Maurice Wilkins (1916–). By the clever use of these facts, Watson and Crick concluded that DNA is composed of two strands entwined around each other in a **double helix,** as shown in Figure 16.4.

In the DNA double helix, the two polynucleotide chains run in opposite directions. Thus, at each end of the double helix, there is one 5′—OH and one 3′—OH terminal. The sugar–phosphate backbone is on the outside, exposed to the aqueous environment, and the bases point inward. The bases are hydrophobic, so they try to avoid contact with water. Through their hydrophobic interactions, they stabilize the double helix. The bases are paired according to Chargaff's rule: For each adenine on one chain, a thymine is aligned opposite it on the other chain; each guanine on one chain has a cytosine aligned with it on the other chain. *The bases so paired form hydrogen bonds with each other, two for A—T and three for G—C, thereby stabilizing the double helix* (Figure 16.5). A—T and G—C are **complementary base pairs.**

Watson, Crick, and Wilkins were awarded the 1962 Nobel Prize in medicine for their discovery. Franklin died in 1958. The Nobel Committee does not award the Nobel Prize posthumously.

> **Double helix** The arrangement in which two strands of DNA are coiled around each other in a screw-like fashion

Recall from Section 13.8 that a helix has a shape like a coiled spring or a spiral staircase. It was Pauling's discovery that human hair protein is helical that led Watson and Crick to look for helixes in DNA.

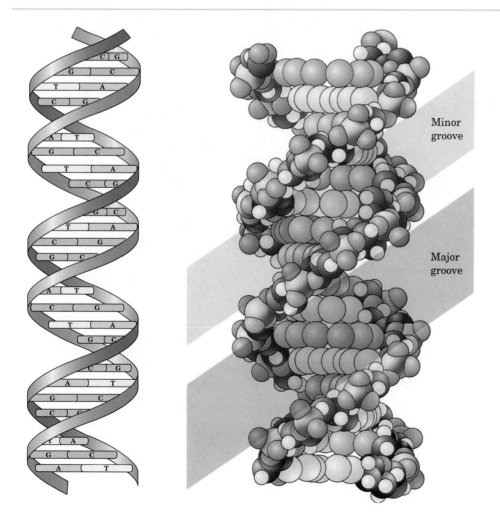

Figure 16.4 Three-dimensional structure of the DNA double helix.

Minor groove

Major groove

■ **Rosalind Franklin (1920–1958).**

■ **Watson and Crick with their model of the DNA molecule.**

The important thing, as Watson and Crick realized, is that only adenine could fit with thymine and only guanine could fit with cytosine. Let us consider the other possibilities. Can two purines (AA, GG, or AG) fit opposite each other? Figure 16.6 shows that they would overlap. How about two pyrimidines (TT, CC, or CT)? As shown in Figure 16.6, they would be too far

Figure 16.5 A and T pair up by forming two hydrogen bonds; G and C pair up by forming three hydrogen bonds.

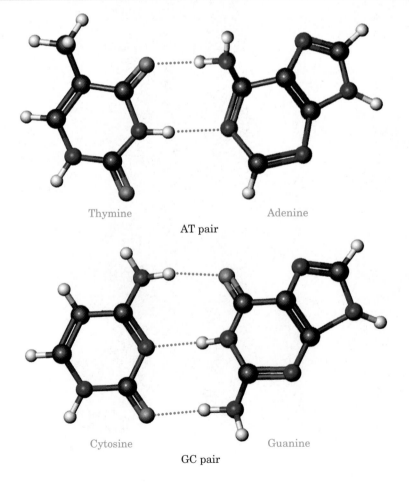

Thymine   Adenine

AT pair

Cytosine   Guanine

GC pair

apart. *There must be a pyrimidine opposite a purine.* But could A fit opposite C, or G opposite T? Figure 16.7 shows that the hydrogen bonding would be much weaker.

The entire action of DNA—and of the heredity mechanism—depends on the fact that, *wherever there is an adenine on one strand of the helix, there must be a thymine on the other strand because that is the only base that fits and forms strong hydrogen bonds, and similarly for G and C.* The entire heredity mechanism rests on these slender hydrogen bonds (Figure 16.5), as we will see in Section 16.6.

C Higher Structures of DNA

If a human DNA molecule were fully stretched out, its length would be perhaps 1 m. However, the DNA molecules in the nuclei are not stretched out, but rather coiled around basic protein molecules called **histones.** The acidic DNA and the basic histones attract each other by electrostatic (ionic) forces, combining to form units called **nucleosomes.** In a nucleosome, eight histone molecules form a core, around which a 147-base-pair DNA double helix is wound. Nucleosomes are further condensed into chromatin when a 30-nm-wide fiber forms in which nucleosomes are wound in a **solenoid** fashion, with six nucleosomes forming a repeating unit (Figure 16.8). Chromatin fibers are organized still further into loops, and loops are arranged into bands to provide the superstructure of chromosomes.

The beauty of establishing the three-dimensional structure of the DNA molecule was that the knowledge of this structure immediately led to the

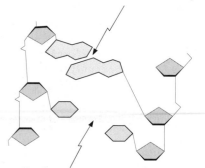

Two purines overlap

Gap between two pyrimidines

Figure 16.6 The bases of DNA cannot stack properly in the double helix if a purine is opposite a purine or a pyrimidine is opposite a pyrimidine.

Solenoid A coil wound in the form of a helix

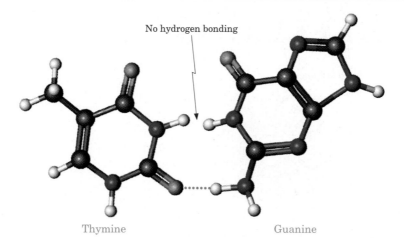

No hydrogen bonding

Thymine   Guanine

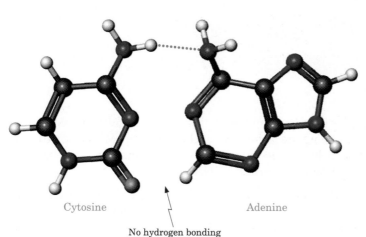

Cytosine   Adenine

No hydrogen bonding

Figure 16.7 Only one hydrogen bond is possible for GT or CA. These combinations are not found in DNA. Compare this figure with Figure 16.5.

Figure 16.8 Superstructure of chromosomes. In nucleosomes, the bandlike DNA double helix winds around cores consisting of eight histones. Solenoids of nucleosomes form 30 nm filament. Loops and minibands are other substructures.

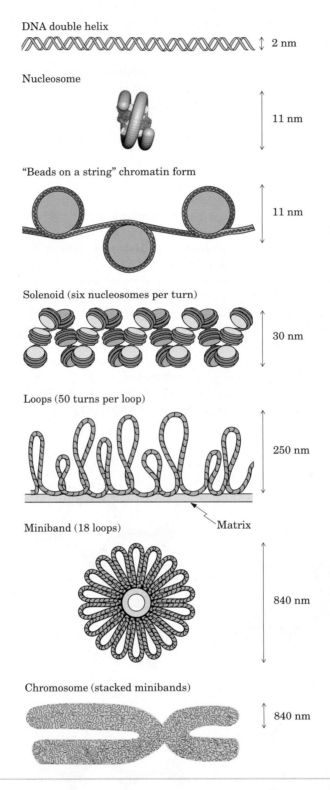

DNA double helix

2 nm

Nucleosome

11 nm

"Beads on a string" chromatin form

11 nm

Solenoid (six nucleosomes per turn)

30 nm

Loops (50 turns per loop)

250 nm

Matrix

Miniband (18 loops)

840 nm

Chromosome (stacked minibands)

840 nm

explanation for the transmission of heredity—how the genes transmit traits from one generation to another. Before we look at the mechanism of DNA replication (in Section 16.6), let us summarize the three differences in structure between DNA and RNA:

1. DNA has four bases: A, G, C, and T. RNA has three of these bases—A, G, and C—but its fourth base is U, not T.
2. In DNA, the sugar is 2-deoxy-D-ribose. In RNA, it is D-ribose.

3. DNA is almost always double-stranded, with the helical structure shown in Figure 16.4.

There are several kinds of RNA (as we will see in Section 16.4); all of them are single-stranded, although base-pairing can occur within a chain (see, for example, Figure 16.9). When it does, adenine pairs with uracil because thymine is not present.

In DNA–RNA interactions, the complementary bases are

DNA	RNA
A	U
G	C
C	G
T	A

16.4 RNA

We previously noted that there are four types of RNA.

1. **Messenger RNA (mRNA)**   mRNA molecules carry the genetic information from the DNA in the nucleus directly to the cytoplasm, where the protein is synthesized. Messenger RNA consists of a chain of nucleotides whose sequence is exactly complementary to that of one of the strands of the DNA. This type of RNA is not long-lived, however. It is synthesized as needed and then degraded, so its concentration at any given time is rather low. The size of mRNA varies widely, with the average unit containing perhaps 750 nucleotides.

> **Messenger RNA (mRNA)** The RNA that carries genetic information from DNA to the ribosome and acts as a template for protein synthesis

2. **Transfer RNA (tRNA)**   Containing from 73 to 93 nucleotides per chain, tRNAs are relatively small molecules. There is at least one different tRNA molecule for each of the 20 amino acids from which the body makes its proteins. The three-dimensional tRNA molecules are L-shaped, but they are conventionally represented as a cloverleaf in two dimensions. Figure 16.9 shows a typical structure. Transfer RNA mole-

> **Transfer RNA (tRNA)** The RNA that transports amino acids to the site of protein synthesis in ribosomes

Figure 16.9   Structure of tRNA. (*a*) Two-dimensional simplified cloverleaf structure. (*b*) Three-dimensional structure. (*From* Biochemistry *by Lubert Stryer* © 1975, 1981, 1988, 1995 by Lubert Stryer; © 2002 by W. H. Freeman and Company. Used with the permission of W. H. Freeman and Company.)

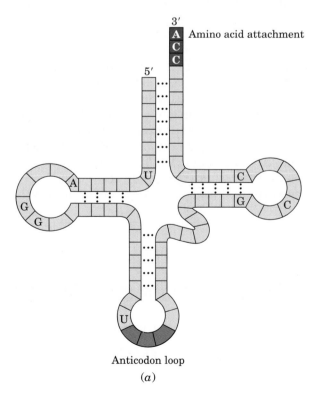

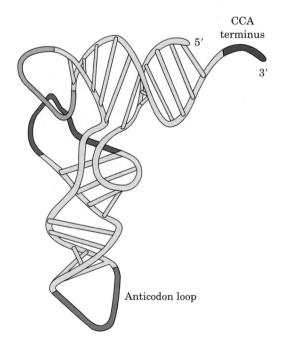

(*a*)

(*b*)

1-Methylguanosine

Ribosomal RNA (rRNA) The RNA complexed with proteins in ribosomes

Ribosome Small spherical bodies in the cell made of protein and RNA; the site of protein synthesis

Splicing The removal of an internal RNA segment and the joining of the remaining ends of the RNA molecule

Exon Nucleotide sequence in mRNA that codes for a protein

Intron A nucleotide sequence in mRNA that does not code for a protein

Philip A. Sharp (1944–) and Richard J. Roberts (1943–) were awarded the 1993 Nobel Prize in medicine for their discovery of the noncoding nature of introns.

cules contain not only cytosine, guanine, adenine, and uracil, but also several other modified nucleotides, such as 1-methylguanosine.

3. **Ribosomal RNA (rRNA)** **Ribosomes,** which are small spherical bodies located in the cells but outside the nuclei, contain rRNA. They consist of about 35% protein and 65% ribosomal RNA (rRNA). These large molecules have molecular weights up to 1 million. As discussed in Section 25.5, protein synthesis takes place on the ribosomes.

4. **Ribozymes** Ribozymes (catalytic RNA) are RNAs with special enzyme functions. In messenger RNA, the precursor molecule is shortened by **splicing** to mature mRNA. The splicing is catalyzed by ribozymes, which are segments of the mRNAs themselves. Similar splicing is catalyzed by ribozymes in forming mature tRNA from its precursor. This ribozyme, however, has both RNA and protein subunits. Both are necessary for the catalytic activity in vivo, although the RNA subunit contains the active site. In higher organisms, ribozymes reside in the ribosome (Section 17.5); hence, they are made of rRNA.

Even though ribozymes have simpler primary structures than proteins (only 4 building blocks in ribozymes versus 20 in proteins), they form complicated secondary structures by hydrogen bonding. These structures are, in turn, twisted into different tertiary structures. Ribozymes also have an active site just like protein enzymes where the substrate binds.

16.5 Genes, Exons, and Introns

A gene is a stretch of DNA, containing a few hundred nucleotides, that carries one particular message—for example, "make a globin molecule." One DNA molecule may have between 1 million and 100 million bases. Therefore, there are many genes in one DNA molecule. In bacteria, this message is continuous; in higher organisms, it is not. That is, stretches of DNA that spell out (code for) the amino acid sequence to be assembled are interrupted by long stretches that seemingly do not code for anything. The coding sequences are called **exons,** short for "expressed sequences," and the noncoding sequences are called **introns,** short for "intervening sequences."

For example, the globin gene has three exons broken up by two introns. Because DNA contains exons and introns, the mRNA transcribed from it also contains exons and introns. The introns are cut out by ribozymes, and the exons are spliced together before the mRNA is used to synthesize a protein (Figure 16.10). In other words, the introns function as spacers and, in rare instances, as enzymes, catalyzing the splicing of exons into mature mRNA.

In humans, only 3% of the DNA codes for proteins or RNA with clear functions. Introns are not the only noncoding DNA sequences, however. **Satellites** are DNA molecules in which short nucleotide sequences are repeated hundreds or thousands of times. Large satellite stretches appear at

Figure 16.10 Introns are cut out of mRNA before the protein is synthesized.

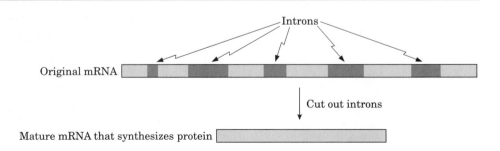

the ends and centers of chromosomes and provide stability for the chromosomes. Smaller repetitive sequences, called **mini-** or **microsatellites,** are associated with cancer when they mutate.

16.6 DNA Replication

The DNA in the chromosomes carries out two functions: (1) It reproduces itself, and (2) it supplies the information necessary to make all the proteins in the body, including enzymes. The second function is covered in Chapter 17. Here we are concerned with the first, **replication.**

Each gene is a section of a DNA molecule that contains a specific sequence of the four bases A, G, T, and C, typically comprising about 1000 to 2000 nucleotides. The base sequence of the gene carries the information necessary to produce one protein molecule. If the sequence is changed (for example, if one A is replaced by a G, or if an extra T is inserted), a different protein is produced, which might have an impaired function as in sickle cell anemia (Chemical Connections 13D).

But consider the task that must be accomplished by the organism. When an individual is conceived, the egg and sperm cells unite to form the zygote. This cell contains only a small amount of DNA, but it nevertheless provides all the genetic information that the individual will ever have. In a human cell, some 6 billion base pairs must be duplicated at each cell cycle, and a fully grown human being may contain more than 1 trillion cells. Each cell contains the same amount of DNA as the original single cell. Furthermore, cells are constantly dying and being replaced. Thus, there must be a mechanism by which DNA molecules can be copied over and over again without error. In Section 17.7, we will see that such errors sometimes do happen and can have serious consequences. Here, however, we want to examine this remarkable mechanism that takes place every day in billions of organisms, from microbes to whales, and has been taking place for billions of years—with only a tiny percentage of errors.

In human cells, the average chromosome has several hundred origins of replication where the copying occurs simultaneously. The DNA double helix has two strands running in opposite directions. The point on the DNA where replication takes place is called the **replication fork.** (See Figure 16.11.)

 See the **Interactive General, Organic, and Biochemistry CD-ROM, version 2.0,** for further exploration on this topic.

Replication The process by which copies of DNA are made during cell division

The pigments responsible for the blue or brown color of eyes are synthesized with the help of specific enzymes. If one of these enzymes is lacking, the eye color may be different.

Figure 16.11 General features of the replication of DNA. The two strands of the DNA double helix are shown separating at the replication fork.

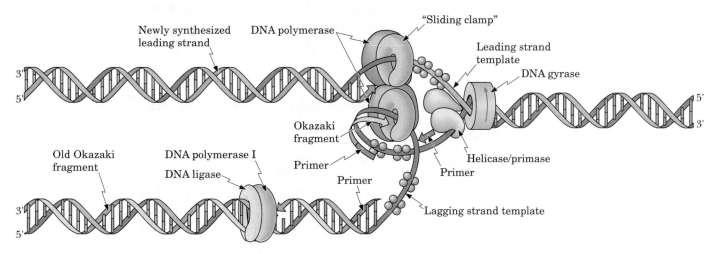

Telomers, Telomerase, and Immortality

Every person has a genetic makeup consisting of about 3 billion pairs of nucleotides, distributed over 46 chromosomes. Telomers are specialized structures at the ends of chromosomes. In vertebrates, telomers are TTAGGG sequences that are repeated hundreds to thousands of times. In normal **somatic cells** that divide in a cyclic fashion throughout the life of the organism (via mitosis), chromosomes lose about 50 to 200 nucleotides from their telomers at each cell division.

DNA polymerase, the enzyme that links the fragments, does not work at the end of linear DNA. This fact results in the short-

ening of the telomers at each replication. The telomer shortening acts as a clock by which the cells count the number of times they have divided. After a certain number of divisions, the cells stop dividing, having reached the limit of the aging process.

In contrast to somatic cells, all immortal cells (germ cells in proliferative stem cells, normal fetal cells, and cancer cells) possess an enzyme, telomerase, that can extend the shortened telomers by synthesizing new chromosomal ends. Telomerase is a ribonucleoprotein; that is, it is made of RNA and protein. The activity of this enzyme seems to confer immortality to the cells.

The newly synthesized strands are called daughter strands.

The base sequence of each newly synthesized DNA chain is complementary to the chain already there.

If the unwinding begins in the middle, the synthesis of new DNA molecules on the old templates continues in both directions until the whole molecule is duplicated. Alternatively, the unwinding can start at one end and proceed in one direction until the whole double helix is unwound.

Replication is bidirectional and it is done with the same speed in both directions. An interesting detail of DNA replication is that the two daughter strands are synthesized in different ways. One of the syntheses is continuous in the $5' \longrightarrow 3'$ direction (see Section 16.3). It is called the **leading strand.** Along the other strand that runs $3' \longrightarrow 5'$ direction, the synthesis is discontinuous. It is called the **lagging strand.**

The replication process is called **semiconservative** because each daughter molecule has one parental strand (conserved) and one newly synthesized one.

Replication is a very complex process involving a number of enzymes and binding proteins. A growing body of evidence indicates that these enzymes assemble in "factories" through which the DNA moves. Such factories may be bound to membranes in bacteria. In higher organisms, the replication factories are not permanent structures. Instead, they may be disassembled and their parts reassembled in ever-larger factories. These assemblies of enzyme "factories" go by the name of **replisomes,** and they contain key enzymes such as polymerases, helicases, and primases (Table 16.3). The primases can shuttle in and out of the replisomes. Other proteins, such as clamp loaders and clamp proteins, through which the newly synthesized primer is threaded, are also parts of the replisomes.

Table 16.3   Components of Replisomes and Their Functions

Component	Function
Helicase	Unwinds the DNA double helix
Primase	Synthesizes short oligonucleotides (primers)
Clamp protein	Allows the leading strand to be threaded through
DNA polymerase	Joins the assembled nucleotides
Ligase	Joins Okazaki fragments in the lagging strand

DNA Fingerprinting

The base sequence in the nucleus of every one of our billions of cells is identical. However, except for people who have an identical twin, the base sequence in the total DNA of one person is different from that of every other person. This uniqueness makes it possible to identify suspects in criminal cases from a bit of skin or a trace of blood left at the scene of the crime and to prove the identity of a child's father in paternity cases.

To do so, the nuclei of the cells of the criminal evidence are extracted. Their DNA is amplified by PCR techniques (Section 16.8). With the aid of restriction enzymes, the DNA molecules are cut at specific points. The resulting DNA fragments are put on a gel and subjected to **electrophoresis.** In this process, the DNA fragments move with different velocities; the smaller fragments move faster and the larger fragments move slower. After a sufficient amount of time, the fragments separate. When they are made visible in the form of an autoradiogram, one can discern bands in a lane. This is called a **DNA fingerprint.**

When the DNA fingerprint made from a sample taken from a suspect matches that from a sample obtained at the scene of the crime, the police have a positive identification. The accompanying figure shows DNA fingerprints derived by using one particular restriction enzyme. Here, a total of nine lanes can be seen. Three (numbers 1, 5, and 9) are control lanes. They contain the DNA fingerprint of a virus, using one particular restriction enzyme.

Three other lanes (2, 3, and 4) were used in a paternity suit: They contain the DNA fingerprints of the mother, the child, and the alleged father. The child's DNA fingerprint (lane 3) contains six bands. The mother's DNA fingerprint (lane 4) has five bands, all of which match those of the child. The alleged father's DNA fingerprint (lane 2) also contains six bands, of which three match those of the child. This is a positive identification. In such cases, one cannot expect a perfect match even if the man is really the father because the child has inherited only half of its genes from the father. Thus, of six bands, only three are expected to match.

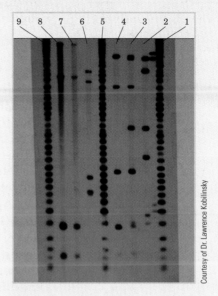

■ **Figure 16C** DNA fingerprint.

In the case just described, the paternity suit was won on the basis of the DNA fingerprint matching.

In the left area of the radiogram are three more lanes (6, 7, and 8). These DNA fingerprints were used in an attempt to identify a rapist. In lanes 7 and 8 are the DNA fingerprints of semen obtained from the rape victim. In lane 6 is the DNA fingerprint of the suspect. The DNA fingerprints of the semen do not match those of the suspect. This is a negative identification and excluded the suspect from the case. When positive identification occurs, the probability that a positive match is due to chance is 1 in 100 billion.

DNA fingerprints are now routinely accepted in court cases. Many convictions are based on such evidence and, just as importantly, many jailed suspects have been released when DNA fingerprinting proved them innocent.

The replication of DNA occurs in a number of distinct steps. A few of the salient features are enumerated below:

1. **Opening up the superstructure.** During replication, the very condensed superstructure of chromosomes must be opened up to be accessible to enzymes and other proteins. A complicated signal transduction mechanism accomplishes this feat. One notable step of the signal transduction is the acetylation and deacetylation of key lysine residues of histones. When histone acetylase, an enzyme, puts acetyl groups on key lysyl residues, some positive charges are eliminated and the strength of the DNA–histone interaction is weakened:

$$\text{Histone}-(\text{CH}_2)_4-\text{NH}_3^+ + \text{CH}_3-\text{COO}^- \underset{\text{deacetylation}}{\overset{\text{acetylation}}{\rightleftharpoons}} \text{Histone}-(\text{CH}_2)_4-\text{NH}-\overset{\overset{\text{O}}{\|}}{\text{C}}-\text{CH}_3$$

Chromatin The DNA complexed with histone and nonhistone proteins that exists in eukaryotic cells between cell divisions

This process allows the opening up of key regions on the DNA molecule. When another enzyme, histone deacetylase, removes these acetyl groups, the positive charges are reestablished. That, in turn, facilitates regaining the highly condensed structure of **chromatin.**

2. **Relaxation of higher structures of DNA.**   Topoisomerases (also called gyrases) are enzymes that facilitate the relaxation of supercoiling in DNA. They do so during replication by temporarily introducing either single or double strand breaks in the DNA. The transient break forms a phosphodiester linkage between a tyrosyl residue of the enzyme and either the 5′ or 3′ end of a phosphate on the DNA. Once the supercoiling is relaxed, the broken strands are joined together, and the topoisomerase diffuses from the location of the replicating fork. Topoisomerases are also involved in the untangling of the replicated chromosomes, before cell division can occur.

3. **Unwinding the DNA double helix.**   The replication of DNA molecules starts with the unwinding of the double helix, which can occur at either end or in the middle. Special unwinding protein molecules, called **helicases,** attach themselves to one DNA strand (Figure 16.11) and cause the separation of the double helix. Helicases of eukaryotes are made of six different protein subunits. The subunits form a ring with a hollow core, where the single-stranded DNA sits. The helicases hydrolyze ATP as the DNA strand moves through. The energy of the hydrolysis (Section 16.2C) promotes this movement.

4. **Primer/Primases.**   Primers are short—4 to 15 nucleotides long—RNA oligonucleotides synthesized from ribonucleoside triphosphates. They are needed to start the synthesis of both daughter strands. The enzyme catalyzing this synthesis is called primase. Primases form complexes with DNA polymerase in eukaryotes. Primers are placed about every 50 nucleotides in the lagging-strand synthesis.

5. **DNA Polymerase.**   The key enzymes in replication are the DNA polymerases. Once the two strands are separated at the replication fork, the DNA nucleotides must be lined up. All four kinds of free DNA nucleotide molecules are present in the vicinity of the replication fork. These nucleotides constantly move into the area and try to fit themselves into new chains. The key to the process is that, as we saw in Section 16.3, *only thymine can fit opposite adenine, and only cytosine can fit opposite guanine.* Wherever a cytosine, for example, is present on one of the strands of an unwound portion of the helix, all four nucleotides may approach, but three of them will be turned away because they do not fit. Only the nucleotide of guanine fits.

In the absence of an enzyme, this alignment is extremely slow. The speed and specificity are provided by DNA polymerase. The active site of this enzyme is quite snug. It surrounds the end of the DNA template–primer complex, creating a specifically shaped pocket for the incoming nucleotide. With such a close contact, the activation energy is lowered and the polymerase enables complementary base pairing with high specificity at a rate of 100 times per second. While the bases of the newly arrived nucleotides are being hydrogen-bonded to their partners, polymerases join the nucleotide backbones.

Along the lagging strand 3′ ⟶ 5′, the enzymes can synthesize only short fragments because the only way they can work is from 5′ to 3′. These short fragments consist of about 200 nucleotides each, named **Okazaki fragments** after their discoverer.

Okazaki fragment A short DNA segment made of about 200 nucleotides in higher organisms (eukaryotes) and 2000 nucleotides in prokaryotes

6. **Ligation.**   The Okazaki fragments and any nicks remaining are eventually joined together by another enzyme, DNA ligase. At the end of the
process, there are two double-stranded DNA molecules, each exactly the
same as the original molecule because only T fits opposite A and only G
fits against C in the active site of the polymerase.

Tuneko and Reiji Okazaki provided
biochemical verification for the
semi-continuous replication of DNA
in 1968.

16.7 DNA Repair

The viability of cells depends on DNA repair enzymes that can detect, recognize, and remove mutations from DNA. Such mutations may arise from external or internal sources. Externally, UV radiation or highly reactive
oxidizing agents, such as superoxide, O_2^-, may damage a base. Errors in copying or internal chemical reactions—for example, deamidation of a base—can
create damage internally. Deamidation of the base cytosine turns it into
uracil (Figure 16.1), which creates a mismatch. The former C-G pair becomes
a U-G mispair that must be removed.

The same deamidation reaction that
can initiate BER plays an important
role in providing the enormous
variability of immunoglobulins (see
Section 22.4).

CHEMICAL CONNECTIONS 16D

Pharmacogenomics: Tailoring Medication to an Individual's Predisposition

A complete DNA sequence of an organism is called a **genome.**
The genome of an average human contains approximately 3 billion base pairs, distributed among 22 pairs of chromosomes plus
2 sex chromosomes. Each chromosome consists of a single DNA
molecule. Among the 3 billion base pairs are some 90 million that
represent 30,000 human genes. It was the task of the Human
Genome Project to determine the complete sequence of the
genome and, in the process, to identify the sequence and location of the genes. The Human Genome Project was completed in
the year 2000. Many other genomes have also been established,
ranging from the lowly bacteria *Escherischia coli* (3 million base
pairs) to the mouse (3 billion base pairs).

It has always been known that an individual's genetic inheri-
tance plays a role in disease and drug effectiveness. As early as
510 B.C., Phytagoras wrote that some individuals developed hemolytic anemia when eating fava beans, while others thrived on
it. Adverse drug reactions are the sixth leading cause of death
in the United States (more than 100,000 deaths per year). Problems caused by ineffective drugs are even more numerous. With
the new knowledge of an individual's genetic makeup it is now
within our power to prescribe drugs in the dosage that best
suits the individual and that minimizes adverse reactions or ineffectiveness. **Pharmacogenomics** is the study of how genetic
variation influences individual responses to a given drug or
class of drugs.

A case in point is CYP2D6, the gene of a member of the cytochrome P-450 enzyme group. This enzyme detoxifies drugs by
adding an —OH group, making them more water-soluble and

thus able to be excreted in the urine. The normal or wild type
of this enzyme is correlated with *extensive metabolism* (EM) of
drugs. A mutation, in which a guanine (G) becomes adenine (A)
in the CYP2D6 gene, exists approximately in 25% of the population. Its presence makes the individual a *poor metabolizer*
(PM). Thus a prescribed drug dosage will stay much longer
than normal in the body of the individual having this genetic
makeup, possibly leding to toxic effects. Another mutation, in
which an adenine (A) base was deleted on the CYP2D6 gene,
exists in approximately 3% of the population. It is associated
with very fast, *ultra-extensive metabolism* (UEM). In these individuals the drug may be cleared from the body before it can be
effective.

These three classes (EM, PM, and UEM) have been known
for some time and could be monitored by frequent blood sample analyses during a six-week course of therapy with a drug.
With the advances of the Human Genome Project, an individual
can now be tested *before* and not during the administration of
a drug therapy. This strategy means that the dosage can be ad-
justed to the individual's particular needs. Drug companies
have developed DNA chips that can read a patient's predisposition to a drug using the DNA of the individual taken in one
simple blood test. The genetic predisposition is merely one factor among many determining the body's total response to a
drug, of course. Nevertheless, eliminating errors based on
known genetic makeup can only minimize adverse effects or
ineffectiveness.

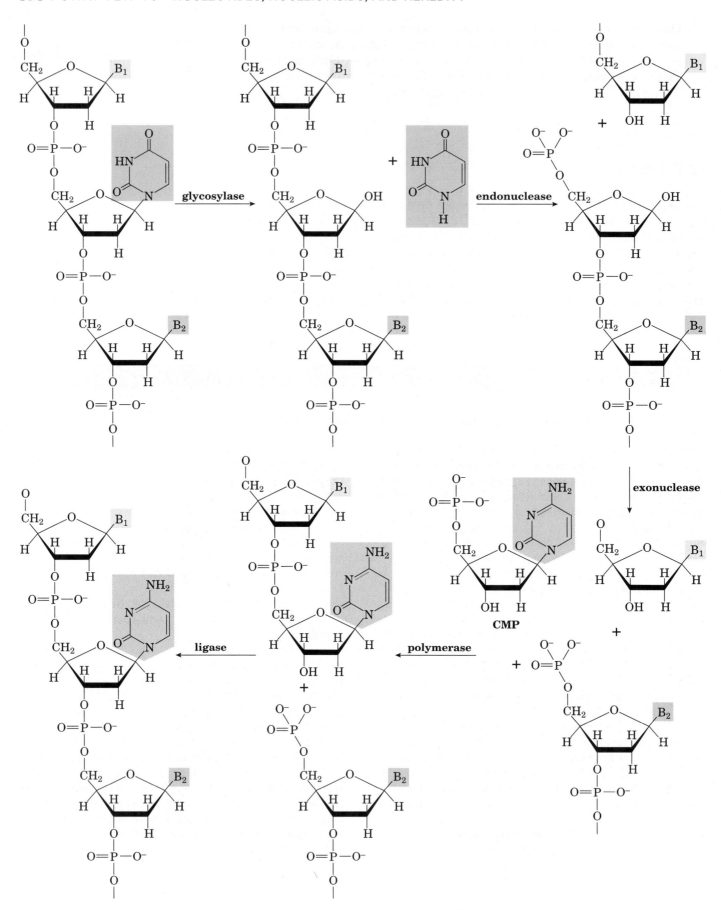

Figure 16.12 Base excision repair (BER) pathway.

The repair can be effected in a number of ways. One of the most common is called BER, *b*ase *e*xcision *r*epair (Figure 16.12). This pathway contains two steps:

1. A specific DNA glycosylase recognizes the damaged base. It hydrolyzes the N—C′ β-glycosidic bond between the uracil base and the deoxyribose, then flips the damaged base, completing the excision. The sugar–phosphate backbone is still intact. At the **AP site** (*a*purinic or *a*pyrimidinic site) created in this way, the backbone is cleaved by a second enzyme, endonuclease. A third enzyme, exonuclease, liberates the sugar–phosphate unit of the damaged site.

2. In the synthesis step, the enzyme DNA polymerase inserts the correct nucleotide, cytidine, and the enzyme DNA ligase seals the backbone to complete the repair.

CHEMICAL CONNECTIONS 16E

Apoptosis: Programmed Cell Death

In necrosis, cells of an organism die in large clusters when exposed to toxins or deprived of oxygen. Such a typical necrosis occurs in heart attack and stroke. *Apoptosis*, or programmed cell death, is clearly distinguishable from necrosis. Only scattered cells die at a time, and no inflammation or scar results. In apoptosis, the cells cleave into apoptopic bodies containing intact organelles and large nuclear fragments. These bodies are then swallowed up by roving scavenger cells.

At the heart of this self-immolation process is a family of protein-cleaving enzymes, called **caspases.** The name is an acronym for "cysteine-containing aspartate-specific proteases," which describes the mode of action: Cysteine is in the active site of the enzyme, and the target protein is cleaved by hydrolyzing the amide bond formed by the α-carboxyl group of the amino acid **aspartate.** Caspases are zymogens, and they are activated

by self-clipping. Once active, they attack targeted proteins. For example, they cleave gelsolin, a protein that normally binds to actin filaments and that maintains the shape of the cell. As a result of actin degradation, the apoptotic cells acquire one of their signature features: They lose their normal shape, become more rounded, and develop blister-like bumps on their surfaces.

Caspases also attack another protein, a companion of the enzyme, **endonuclease.** When this enzyme is paired with its companion, it is unable to enter the cell nucleus. After it is freed from the companion, however, endonuclease can enter the nucleus and cleave DNA. The result is the second distinguishing fingerprint of death by apoptosis: the appearance of DNA fragments in multiples of 180 base pairs (180, 360, or 540 base pairs). These appear in the membrane-enclosed apoptotic bodies, the remnants of the dying cells.

(*a*)

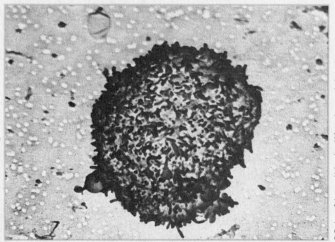

(*b*)

© Dr. Larry Schwartz, University of Massachusetts, Amherst

■ **Figure 16E**   Signs of apoptosis. (*a*) Blisters. (*b*) Apoptotic bodies.

A second repair mechanism removes not a single mismatched base but rather whole nucleotides—as many as 24 to 32 residue oligonucleotides. Known as NER, for *n*ucleotide *e*xcision *r*epair, it similarly involves a number of repair enzymes.

Any defect in the repair mechanisms may lead to harmful or even lethal mutations. For example, individuals with inherited XP (xeroderma pigmentosa), a condition in which one or another enzyme in the NER repair pathway is missing or defective, have 1000 times greater risk of skin cancer than normal individuals.

16.8 Cloning

The term **clone** refers to a genetically identical population, be that a sheep, a cell, a virus, or a DNA molecule. Molecular **cloning** is the exact copying of segments of DNA. Millions of copies of selected DNA fragments can be made within a few hours with high precision by a technique called **polymerase chain reaction (PCR),** discovered by Kary B. Mullis (1945–), who shared the 1993 Nobel Prize in chemistry for this achievement.

PCR techniques can be used only if the sequence of a gene to be copied is known. In such a case, one can synthesize two primers that are complementary to the 3′ ends of the gene. The primers are polynucleotides consisting of 12 to 16 nucleotides. When added to a target DNA segment, they hybridize with the 3′ end of each strand of the gene.

5′CATAGGACAGC—OH   Primer

3′TACGTATCCTGTCGTAGG—   Gene

Figure 16.13   Schematic diagram of PCR. *(Adapted from* The Unusual Origin of the Polymerase Chain Reaction, *by Kary B. Mullis, illustrated by Michael Goodman. Scientific American, April 1990. Reprinted with the permission of Michael Goodman.)*

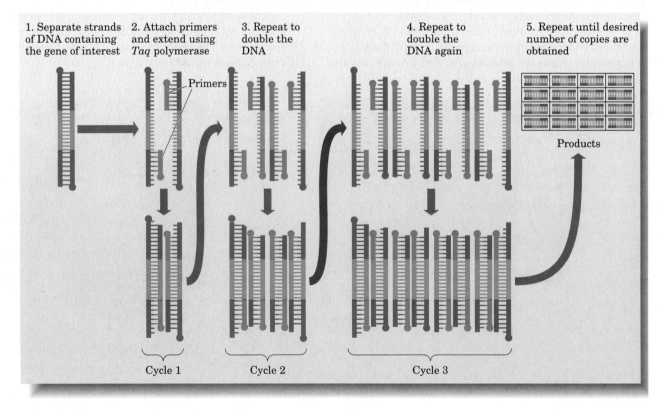

1. Separate strands of DNA containing the gene of interest
2. Attach primers and extend using *Taq* polymerase
3. Repeat to double the DNA
4. Repeat to double the DNA again
5. Repeat until desired number of copies are obtained

Primers

Products

Cycle 1   Cycle 2   Cycle 3

In cycle 1 (Figure 16.13), the polymerase extends the primers in each direction as individual nucleotides are assembled and connected on the template DNA. In this way, two new copies are created. The two-step process is repeated (cycle 2) when the primers are **hybridized** with the new strands, and the primers are extended again. At that point, four new copies have been created. The process continues, and in 25 cycles, 2^{25} or some 33 million copies can be made. In practice, only a few million are produced, which is sufficient for the isolation of a gene.

This fast process is practical because of the discovery of heat-resistant polymerases isolated from bacteria that live in hot thermal vents on the sea floor (Section 14.4). A temperature of 95°C is needed because the double helix must be unwound to hybridize the primer to the target DNA. Once single strands of DNA have been exposed, the mixture is cooled to 70°C. The primers are hybridized and subsequent extensions take place. The 95°C and 70°C cycles are repeated over and over. No new enzyme is required because the polymerase is stable at both temperatures.

PCR techniques are routinely used when a gene or a segment of DNA must be amplified from a few molecules. It is used in genome study (Chemical Connections 16D), in obtaining evidence from a crime scene (Chemical Connections 16C), and even in obtaining the genes of long-extinct species found fossilized in amber.

> **Hybridization** The process in which two strands of nucleic acids or segments thereof form a double-stranded structure through hydrogen-bonding of complementary base pairs

The temperature at which the hydrogen bonds of double-stranded DNA break and the two strands separate is referred to as the *annealing* temperature.

S U M M A R Y

Nucleic acids are composed of sugars, phosphates, and organic bases (Section 16.2). Two kinds exist: **ribonucleic acid (RNA)** and **deoxyribonucleic acid (DNA)**. In DNA, the sugar is the monosaccharide 2-deoxy-D-ribose; in RNA, it is D-ribose. In DNA, the heterocyclic amine bases are adenine (A), guanine (G), cytosine (C), and thymine (T). In RNA, they are A, G, C, and uracil (U). Nucleic acids are giant molecules with backbones made of alternating units of sugar and phosphate. The bases are side chains joined by β-*N*-glycosidic bonds to the sugar units (Section 16.2).

DNA is made of two strands that form a double helix (Section 16.3). The sugar–phosphate backbone runs on the outside of the double helix, and the hydrophobic bases point inward. **Complementary pairing** of the bases occurs in the double helix, such that each A on one strand is hydrogen-bonded to a T on the other, and each G is hydrogen-bonded to a C. No other pairs fit. DNA is coiled around basic protein molecules called **histones.** Together they form **nucleosomes,** which are further condensed into the chromatins of chromosome.

The DNA molecule carries, in the sequence of its bases, all the information necessary to maintain life. When cell division occurs and this information is passed from parent cell to daughter cells, the sequence of the parent DNA is copied.

A **gene** is a segment of a DNA molecule that carries the sequence of bases that directs the synthesis of one particular protein (Section 16.5). There are four kinds of RNA (Section 16.4): **messenger RNA (mRNA), transfer RNA (tRNA), ribosomal RNA (rRNA),** and **ribozyme (catalytic RNA).** DNA in higher organisms contains sequences, called **introns,** that do not code for proteins (Section 16.5). The sequences that do code for proteins are called **exons.**

DNA replication occurs in several distinct steps (Section 16.6). First, the superstructures of chromosomes are loosened by acetylation of histones. **Topoisomerases** relax the higher structures. **Helicases** at the replication fork separate the two strands of DNA. RNA primers and primerase are needed to start the synthesis of daughter strands. The **leading strand** is synthesized continuously by **DNA polymerase.** The **lagging strand** is synthesized discontinuously as **Okazaki fragments. DNA ligase** seals the nicks and the Okazaki fragments.

An important **DNA repair** mechanism is BER, or single base excision repair (Section 16.7).

Cloning is copying genes or segments of DNA (Section 16.8). The **polymerase chain reaction (PCR)** technique can make millions of copies with high precision in a few hours.

P R O B L E M S

Numbers that appear in color indicate difficult problems.
▶ designates problems requiring application of principles.

Nucleic Acids and Heredity

16.2 What structures of the cell, visible in a microscope, contain hereditary information?

16.3 Name one hereditary disease.

Components of Nucleic Acids

16.4 (a) Where in a cell is the DNA located?
(b) Where in a cell is the RNA located?

16.5 What are the components of (a) a nucleotide and (b) a nucleoside?

16.6 Draw the structures of ADP and GDP. Are these structures parts of nucleic acids?

16.7 What is the difference in structure between thymine and uracil?

16.8 Which DNA and RNA bases contain a carbonyl group?

16.9 Draw the structures of (a) cytidine and (b) deoxycytidine.

16.10 Which DNA and RNA bases are primary amines?

16.11 What is the difference in structure between D-ribose and 2-deoxy-D-ribose?

16.12 What is the difference between a nucleoside and a nucleotide?

Structure of DNA and RNA

16.13 In RNA, which carbons of the ribose are linked to the phosphate group and which are linked to the base?

16.14 What constitutes the backbone of DNA?

16.15 Draw the structures of (a) UDP and (b) dAMP.

16.16 In DNA, which carbon atoms of 2-deoxy-D-ribose are bonded to the phosphate groups?

16.17 The sequence of a short DNA segment is ATGGCAATAC.
(a) What name do we give to the two ends (terminals) of a DNA molecule?
(b) In this segment, which end is which?

16.18 Chargaff showed that, in samples of DNA taken from many different species, the molar quantity of A was always approximately equal to the molar quantity of T; the same is true for C and G. How did this information help to establish the structure of DNA?

16.19 How many hydrogen bonds can be formed between uracil and adenine?

16.20 How many histones are in a nucleosome?

16.21 What is the nature of the interaction between histones and DNA in nucleosomes?

16.22 What are chromatin fibers made of?

16.23 What constitutes the superstructure of chromosomes?

RNA

16.24 Which type of RNA has enzyme activity? Where does it function mostly?

16.25 Which has the longest chains: tRNA, mRNA, or rRNA?

16.26 Which type of RNA contains modified nucleotides?

16.27 Which type of RNA has a sequence exactly complementary to that of DNA?

16.28 Where is rRNA located in the cell?

16.29 What kind of functions do ribozymes, in general, perform?

Genes, Exons, and Introns

16.30 Define:
(a) Intron (b) Exon

16.31 Does mRNA also have introns and exons? Explain.

16.32 (a) What percentage of human DNA codes for proteins?
(b) What is the function of the rest of the DNA?

16.33 Do satellites code for a particular protein?

DNA Replication

16.34 A DNA molecule normally replicates itself millions of times, with almost no errors. What single fact about the structure is most responsible for this fidelity of replication?

16.35 What functional groups on the bases form hydrogen bonds in the DNA double helix?

16.36 Draw the structures of adenine and thymine, and show with a diagram the two hydrogen bonds that stabilize A-T pairing in DNA.

16.37 Draw the structures of cytosine and guanine, and show with a diagram the three hydrogen bonds that stabilize C-G pairing in nucleic acids.

16.38 How many different bases are in a DNA double helix?

16.39 What is a replication fork? How many replication forks may exist simultaneously on an average human chromosome?

16.40 Why is the replication called semiconservative?

16.41 How does the removal of some positive charges from histones enable the opening of the chromosomal superstructure?

16.42 Write the chemical reaction for the deacetylation of acetyl-histone.

16.43 What is the quaternary structure of helicases in eukaryotes?

16.44 What are helicases? What is their function?

16.45 Can dATP serve as a source for a primer?

16.46 What are the side products of the action of primase in forming primers?

16.47 What do we call the enzymes that join nucleotides into a DNA strand?

16.48 In which direction is the DNA molecule synthesized continuously?

16.49 What kind of bond formation do polymerases catalyze?

16.50 Which enzyme catalyzes the joining of Okazaki fragments?

DNA Repair

16.51 As a result of damage, a few of the G residues in a gene are methylated. What kind of mechanism could the cell use to repair the damage?

16.52 What is the function of endonuclease in the BER repair mechanism?

16.53 When cytosine is deaminated, uracil is formed. Uracil is a naturally occurring base. Why would the cell use base excision repair to remove it?

16.54 What bonds are cleaved by glycosylase?

16.55 What are AP sites? What enzyme creates them?

16.56 Why are xeroderma pigmentosa patients 1000 times more prone to develop skin cancer than normal individuals are?

Cloning

16.57 What is the advantage of using polymerase from thermophile bacteria that live in hot thermal vents in cloning?

16.58 What 12-nucleotide primer would you use in the PCR technique when you want to clone a gene whose 3′ end is as follows: 3′TACCGTCATCCGGTG—?

Chemical Connections

16.59 (Chemical Connections 16A) Draw the structure of the fluorouridine nucleoside that inhibits DNA synthesis.

16.60 (Chemical Connections 16A) Give an example of how anticancer drugs work in chemotherapy.

16.61 (Chemical Connections 16B) What sequence of nucleotides is repeated many times in telomers?

16.62 (Chemical Connections 16B) Why are up to 200 nucleotides lost at each replication?

16.63 (Chemical Connections 16B) How does telomerase make a cancer cell immortal?

16.64 (Chemical Connections 16C) After having been cut by restriction enzymes, how are DNA fragments separated from each other?

16.65 (Chemical Connections 16C) How is DNA fingerprinting used in paternity suits?

16.66 (Chemical Connections 16D) What is the function of cytochrome P-450?

16.67 (Chemical Connections 16D) How does knowledge of the human genome enable patients to be screened for individual drug tolerance?

16.68 (Chemical Connections 16E) How do caspase enzymes, which operate in the cytoplasm of the cell, cause the cleavage of DNA in the nucleus?

16.69 (Chemical Connections 16E) What are the distinguishing fingerprints of cell death by apoptosis?

Additional Problems

16.70 What is the active site of a ribozyme?

16.71 Why is it important that a DNA molecule be able to replicate itself millions of times without error?

16.72 Draw the structures of (a) uracil and (b) uridine.

16.73 How would you classify the functional groups that bond together the three different components of a nucleotide?

16.74 Why do we call DNA or RNA *nucleic acids?*

16.75 Which nucleic acid molecule is the largest?

16.76 What kind of bonds are broken during replication? Does the primary structure of DNA change during replication?

16.77 In sheep DNA, the mol % of adenine (A) was found to be 29.3. Based on Chargaff's rule, what would be the approximate mol % of G , C, and T?

InfoTrac College Edition

For additional reading, go to InfoTrac College Edition, your online research library, at

http://infotrac.thomsonlearning.com

CHAPTER **17**

17.1 Introduction

17.2 Transcription

17.3 The Role of RNA in Translation

17.4 The Genetic Code

17.5 Translation and Protein Synthesis

17.6 Gene Regulation

17.7 Mutations, Mutagens, and Genetic Diseases

17.8 Recombinant DNA

The diagonal that divides the panel is the decoded sequence of chromosome X; hence it represents half of a woman. The upper-left panel is assembled from the molecular fragments of life. Its dominant motif is genomics, written in the language of nucleotides. The lower-right panel depicts cellular components. Its dominant motif is proteomics, written in the language of amino acids.

"New Literature", Collage by Frederick A. Bettelheim

Gene Expression and Protein Synthesis

17.1 Introduction

We have seen that the DNA molecule is a storehouse of information. We can compare it to a loose-leaf cookbook, each page of which contains one recipe. The pages are the genes. To prepare a meal, we use a number of recipes. Similarly, to provide a certain inheritable trait, a number of genes—segments of DNA—are needed.

Of course, the recipe itself is not the meal. The information in the recipe must be expressed in the proper combination of food ingredients. Similarly, the information stored in DNA must be expressed in the proper combination of amino acids representing a particular protein. The way this expression works is now so well established that it is called the **central dogma of molecular biology.** The dogma states that *the information contained in DNA molecules is transferred to RNA molecules, and then from the RNA molecules the information is expressed in the structure of proteins.* **Gene expression** is the turning on or activation of a gene. Transmission of information occurs in two steps: transcription and translation.

Although this statement is correct in the vast majority of cases, in certain viruses the flow of information goes from RNA to DNA.

Gene expression The activation of a gene to produce a specific protein; it involves both transcription and translation

384

Transcription Because the information (that is, the DNA) is in the nucleus of the cell and the amino acids are assembled outside the nucleus, the information must first be carried out of the nucleus. This step is analogous to copying a recipe from a cookbook. All the necessary information is copied, albeit in a slightly different format, as if we were converting the printed page into handwriting. On the molecular level, this task is accomplished by transcribing the information from the DNA molecule onto a molecule of messenger RNA (Section 16.4), so named because it carries the message from the nucleus to the site of protein synthesis. Other RNAs are similarly transcribed. rRNA is needed to form ribosomes, and tRNA is required to carry out the translation into protein language. The transcribed information on the different RNA molecules is then carried out of the nucleus.

> **Transcription** The process in which information encoded in a DNA molecule is copied into an mRNA molecule

Translation The mRNA serves as a template on which the amino acids are assembled in the proper sequence. To complete the assembly, the information that is written in the language of nucleotides must be translated into the language of amino acids. The translation is done by the second type of RNA, transfer RNA (Section 16.4). An exact, word-to-word translation occurs. Each amino acid in the protein language has a corresponding word in the RNA language. Each word in the RNA language is a sequence of three bases. This correspondence between three bases and one amino acid is called the genetic code (we will discuss the code in Section 17.4).

> **Translation** The process in which information encoded in an mRNA molecule is used to assemble a specific protein

A summary of the process follows:

$$\text{DNA} \xrightarrow{\text{replication}} \text{DNA} \xrightarrow{\text{transcription}} \text{mRNA} \xrightarrow{\text{translation}} \text{protein}$$

In higher organisms (eukaryotes), transcription and translation occur sequentially. The transcription takes place in the nucleus. After RNA leaves the nucleus and enters the cytoplasm, the translation takes place there. In lower organisms (prokaryotes), there is no nucleus and thus transcription and translation occur simultaneously in the cytoplasm. This extended form of the "central dogma" was challenged in 2001, when it was found that even in eukaryotes about 15% of the proteins are produced in the nucleus itself. Thus some simultaneous transcription and translation do occur even in higher organisms.

We know more about bacterial transcription and translation because they are simpler than the processes operating in higher organisms. Nevertheless, we will concentrate on studying gene expression and protein synthesis in the eukaryotic systems because they are more relevant to human health care.

17.2 Transcription

Transcription starts when the DNA double helix begins to unwind at a point near the gene that is to be transcribed (Figure 17.1). As we saw in Section 16.3C, nucleosomes form chromatin and higher condensed structures in the chromosomes. To make the DNA available for transcription, these superstructures change constantly. There are specific **binding proteins** that bind to the nucleosomes, making the DNA become less dense and more accessible. Only then can the enzyme **helicase,** a ring-shaped complex of six protein subunits, unwind the double helix.

 See the **Interactive General, Organic, and Biochemistry CD-ROM, version 2.0,** for further exploration on this topic.

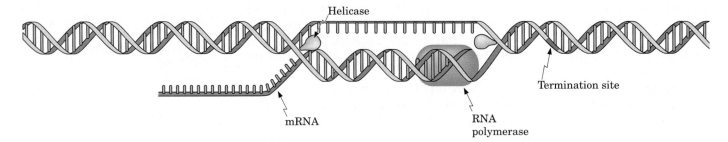

Figure 17.1 Transcription of a gene. The information in one DNA strand is transcribed to a strand of RNA. The termination site is the locus of termination of transcription.

Note again that RNA contains no thymine but has uracil instead.

Only one strand of the DNA molecule is transcribed. Ribonucleotides assemble along the unwound DNA strand in the complementary sequence. Opposite each C on the DNA is a G on the growing mRNA; the other complementary bases follow the patterns G $\longrightarrow$ C, A $\longrightarrow$ U, and T $\longrightarrow$ A. The ribonucleotides, thus aligned, are linked to form the appropriate RNA.

Three kinds of **polymerases** catalyze the transcription. RNA polymerase I (poly I) catalyzes the formation of rRNA, poly II catalyzes mRNA formation, and poly III catalyzes tRNA formation. Each enzyme is a complex of 10 or more subunits. Some subunits are unique to each kind of polymerase, whereas other subunits appear in all three polymerases.

The eukaryotic gene has two major parts: the **structural gene** that is transcribed into RNA and the **regulatory gene** that controls the transcription. The structural gene is made of exons and introns (Figure 17.2). The regulatory gene is not transcribed, but rather has control elements. One such control element is the **promoter.** On the DNA strand, there is always a sequence of bases that the polymerase recognizes as an **initiation signal,** saying in essence, "Start here." A promoter is unique to each gene. Besides unique nucleotide sequences, promoters contain **consensus sequences,** such as the TATA box. In TATA boxes the two nucleotides, T and A, are repeated many times. A TATA box lies approximately 25 base pairs upstream—

Figure 17.2 Organization and transcription of a split eukaryote gene.

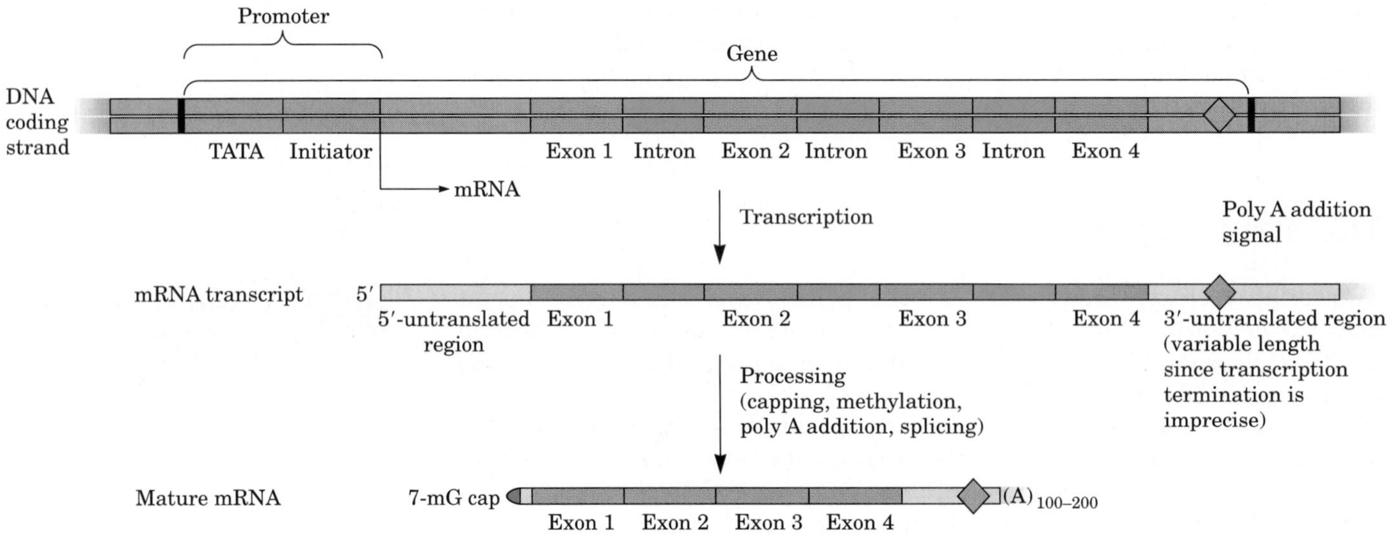

CHEMICAL CONNECTIONS 17A

"Antisense" Makes Sense

Many diseases are caused by faulty or excessive production of certain proteins. As we have seen, the message to make proteins is in the mRNA. Antisense drugs are chemically modified strands of short stretches of DNA that are exactly opposite—hence "anti"—the coding, or "sense" of a mRNA. Thus, if a stretch of mRNA is AACGUU, its antisense will be TTGCAA. When such an antisense drug is delivered to a cell, it strongly binds to the mRNA and prevents the translation of the unwanted protein.

The trick is to prevent the degradation of the antisense drug by the cell's enzymes before or after it attaches to the mRNA. One way to protect the antisense drug is to modify the backbone by substituting a sulfur atom for an oxygen atom in the phosphate group of the backbone, thereby forming phosphorothioate moities.

There is only one antisense drug on the market today. Vitravene is used in treatment of cytomegalovirus-induced retinitis in AIDS patients (see Chemical Connections 17B). Vitravene is a stretch of 21 nucleotides. The structure of the first four of these nucleotides is shown in Figure 17A , spelling out GCGT. Currently, many other antisense oligonucleotide drugs are in clinical trials as potential treatments for various cancers, such as lymphoma.

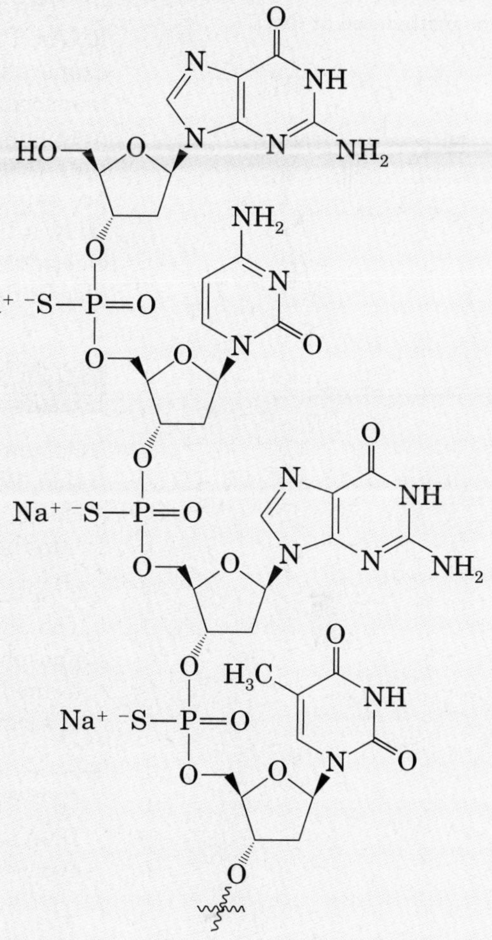

■ **Figure 17A** **First four of the 21 segments of oligonucleotide drug Vitravene.**

that is, before the start of the transcription process (see Figure 17.2). TATA boxes are common to all eukaryotes. All three RNA polymerases interact with their promoter regions via **transcription factors** that are binding proteins.

After initiation, the RNA polymerase zips up the complementary bases by forming a phosphate ester bond (Section 10.7) between each ribose and the next phosphate group. This process is called **elongation.**

At the end of the gene is a **termination sequence** that tells the enzyme, "Stop the transcription."

The enzyme poly II has two different forms. At the C-terminal domain, poly II has serine and threonine repeats that can be phosphorylated. When poly II starts the initiation, the enzyme is in its unphosphorylated form. Upon phosphorylation, it performs the elongation process. After termination of the transcription, poly II is dephosphorylated by a phosphatase. In this manner, poly II is constantly recycled between its initiation and elongation roles.

The enzyme synthesizes the mRNA molecule from the 5′ to the 3′ end (the zipper can move in only one direction). Because the complementary chains (RNA and DNA) run in opposite directions, however, the enzyme must move along the DNA template in the 3′ ⟶ 5′ direction of the DNA

After RNA molecules are synthesized, they move out of the nucleus and into the cytoplasm.

Guanine methylated at the 7′ position

The S, or Svedberg unit, is a measure of the size of these bodies.

Codon The sequence of three nucleotides in messenger RNA that codes for a specific amino acid

The 20 amino acids are always available in the cytoplasm, near the site of protein synthesis.

(Figure 17.1). As the mRNA is synthesized, it moves away from the DNA template, which then rewinds to the original double-helix form. Transfer RNA and ribosomal RNA are also synthesized on DNA templates in this manner.

The RNA products of transcription are not necessarily the functional RNAs. Previously, we have seen that in higher organisms mRNA contains exons and introns (Section 16.5). To ensure that mRNA is functional, the transcribed product is capped at both ends. The 5′ end acquires a methylated guanine (7-mG cap) and a polyA tail that may contain 100 to 200 adenine residue caps at the 3′ end. Once the two ends are capped, the introns are spliced out. This is a **post-transcription process** (Figure 17.2). Similarly, a transcribed tRNA must be trimmed and capped, and some of its nucleotides methylated before becoming functional tRNA. Functional rRNA also undergoes post-transcriptional methylation.

EXAMPLE 17.1

Polymerase II both initiates the transcription and performs the elongation. What are the two forms of the enzyme in these processes? What chemical bond formation occurs in the conversion of these two forms one into the other?

Solution

The phosphorylated form of poly II performs the elongation and the unphosphorylated form initiates the transcription. The chemical bond formed in the phosphorylation is the phosphoric ester between the —OH of serine and threonine residues of the enzyme and phosphoric acid.

Problem 17.1

DNA is highly condensed in the chromosomes. What is the sequence of events that enables the transcription of a gene to begin?

17.3 The Role of RNA in Translation

Translation is the process by which the genetic information preserved in the DNA and transcribed into the mRNA is converted to the language of proteins—that is, the amino acid sequence. Three of the four types of RNA (mRNA, rRNA, and tRNA) participate in the process.

The synthesis of proteins takes place on the ribosomes (Section 16.4). These spheres dissociate into two parts—a larger and a smaller body. Each of these bodies contains rRNA and some polypeptide chains that act as enzymes, speeding up the synthesis. In higher organisms, including humans, the larger ribosomal body is called the 60S ribosome, and the smaller one is called the 40S ribosome. The mature mRNA is transported into the cytoplasm, where its 5′ end is attached to the smaller ribosomal body and later joined by the larger body. Together they form a unit on which the mRNA is stretched out. Triplets of bases on the mRNA are called **codons.** After the mRNA is attached to the ribosome in this way, the 20 amino acids are brought to the site, each carried by its own particular tRNA molecule.

The most important segments of the tRNA molecule are (1) the site to which enzymes attach the amino acids and (2) the recognition site. Figure 17.3 shows that the 3′ terminal of the tRNA molecule is single-stranded; this end carries the amino acid.

As we have said, each tRNA is specific for one amino acid only. How does the body make sure that alanine, for example, attaches only to the one tRNA molecule that is specific for alanine? The answer is that each cell carries at least 20 specific enzymes for this purpose. Each of these enzymes recognizes only one amino acid and only one tRNA. The enzyme attaches the activated amino acid to the 3′ terminal —OH group of the tRNA, forming an ester bond. The second important segment of the tRNA molecule carries the **codon recognition site,** which is a sequence of three bases called an **anticodon** located at the opposite end of the molecule in the three-dimensional structure of tRNA (see Figure 17.3). This triplet of bases can align itself in a complementary fashion to the codon triplet on mRNA.

17.4 The Genetic Code

By 1961, it was apparent that the order of bases in a DNA molecule corresponds to the order of amino acids in a particular protein. But the code was unknown. Obviously, it could not be a one-to-one code. There are only four bases, so if A coded for glycine, G for alanine, C for valine, and T for serine, there would be 16 amino acids that could not be coded.

In 1961, Marshall Nirenberg (1927–) and his co-workers attempted to break the code in a very ingenious way. They made a synthetic molecule of mRNA consisting of uracil bases only. They put this molecule into a cell-free system that synthesized proteins and then supplied the system with all 20 amino acids. The only polypeptide produced was a chain consisting solely of the amino acid phenylalanine. This experiment showed that the code for phenylalanine must be UUU or some other multiple of U.

A series of similar experiments by Nirenberg and other workers followed, and by 1967 the entire genetic code had been broken. *Each amino acid is coded for by a sequence of three bases,* called a *codon.* Table 17.1 shows the complete code.

The first important aspect of the **genetic code** is that it is almost universal. In virtually every organism, from a bacterium to an elephant to a human, the same sequence of three bases codes for the same amino acid.

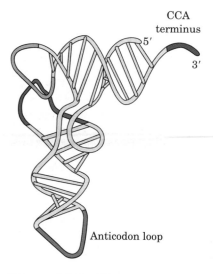

CCA terminus

Anticodon loop

Figure 17.3 Three-dimensional structure of tRNA.

Anticodon A sequence of three nucleotides on tRNA complementary to the codon in mRNA

Genetic code The sequence of triplets of nucleotides (codons) that determines the sequence of amino acids in a protein

Some exceptions to the genetic code in Table 17.1 occur in mitochondrial RNA. Because of that and other evidence, it is thought that the mitochondrion may have been an ancient entity. During evolution it developed a symbiotic relationship with eukaryotic cells. For example, some of the respiratory enzymes located on the cristae of the mitochondrion (see Section 18.2) are encoded in the mitochondrial DNA, and other members of the same respiratory chain are encoded in the nucleus of the eukaryotic cell.

Table 17.1 The Genetic Code

First Position (5′-end)	Second Position								Third Position (3′-end)
	U		**C**		**A**		**G**		
U	UUU	Phe	UCU	Ser	UAU	Tyr	UGU	Cys	U
	UCC	Phe	UCC	Ser	UAC	Tyr	UGC	Cys	C
	UUA	Leu	UCA	Ser	UAA	Stop	UGA	Stop	A
	UUG	Leu	UCG	Ser	UAG	Stop	UGG	Trp	G
C	CUU	Leu	CCU	Pro	CAU	His	CGU	Arg	U
	CUC	Leu	CCC	Pro	CAC	His	CGC	Arg	C
	CUA	Leu	CCA	Pro	CAA	Gln	CGA	Arg	A
	CUG	Leu	CCG	Pro	CAG	Gln	CGG	Arg	G
A	AUU	Ile	ACU	Thr	AAU	Asn	AGU	Ser	U
	AUC	Ile	ACC	Thr	AAC	Asn	AGC	Ser	C
	AUA	Ile	ACA	Thr	AAA	Lys	AGA	Arg	A
	AUG*	Met	ACG	Thr	AAG	Lys	AGG	Arg	G
G	GUU	Val	GCU	Ala	GAU	Asp	GGU	Gly	U
	GUC	Val	GCC	Ala	GAC	Asp	GGC	Gly	C
	GUA	Val	GCA	Ala	GAA	Glu	GGA	Gly	A
	GUG	Val	GCG	Ala	GAG	Glu	GGG	Gly	G

*AUG also serves as the principal initiation codon.

The universality of the genetic code implies that all living matter on earth arose from the same primordial organisms. This finding is perhaps the strongest evidence for Darwin's theory of evolution.

There are 20 amino acids in proteins, but 64 possible combinations of four bases into triplets. All 64 codons (triplets) have been deciphered. Three of them—UAA, UAG, and UGA—are "stop signs." They terminate protein synthesis. The remaining 61 codons code for amino acids. Since there are only 20 amino acids, there must be more than one codon for each amino acid. Indeed, some amino acids have as many as six codons. Leucine, for example, is coded by UUA, UUG, CUU, CUC, CUA, and CUG.

Because of this multiple coding, the genetic code is called a *multiple code* or a *degenerate code.*

Just as there are three stop signs in the code, there is also an initiation sign. The initiation sign is AUG, which is also the codon for the amino acid methionine. This means that, in all protein synthesis, the first amino acid is always methionine. Methionine can also be put into the middle of the chain because there are two kinds of tRNA for it.

Although all protein synthesis starts with methionine, most proteins in the body do not have a methionine residue at the N-terminal of the chain. In most cases, the initial methionine is removed by an enzyme before the polypeptide chain is completed. The code on the mRNA is always read in the $5' \longrightarrow 3'$ direction, and the first amino acid to be linked to the initial methionine is the N-terminal end of the translated polypeptide chain.

EXAMPLE 17.2

Which amino acid is represented by the codon CGU? What is its anticodon?

Solution
Looking at Table 17.1, we find that CGU corresponds to arginine; the anticodon is GCA.

Problem 17.2
What are the codons for histidine? What are the anticodons?

See the **Interactive General, Organic, and Biochemistry CD-ROM, version 2.0,** for further exploration on this topic.

All protein synthesis takes place outside the nucleus, in the cytosol.

17.5 Translation and Protein Synthesis

So far we have met the molecules that participate in protein synthesis (Section 17.3) and the dictionary of the translation, the genetic code. Now let us look at the actual mechanism by which the polypeptide chain is assembled.

There are four major stages in protein synthesis: activation, initiation, elongation, and termination. At each stage, a number of molecular entities participate in the process (Table 17.2).

Table 17.2   Molecular Components of Reactions at Four Stages of Protein Synthesis

Stage	Molecular Components
Activation	Amino acid, ATP, tRNA, aminoacyl-tRNA synthetase
Initiation	tRNA$_i^{Met}$, 40S ribosome, initiation factor proteins, mRNA, 60S ribosome
Elongation	40S and 60S ribosomes, tRNA, elongation factor proteins, mRNA, peptidyl transferase
Termination	Releasing factor proteins, tRNA, mRNA, 40S and 60S ribosomes

A Activation

Each amino acid is first activated by reacting with a molecule of ATP:

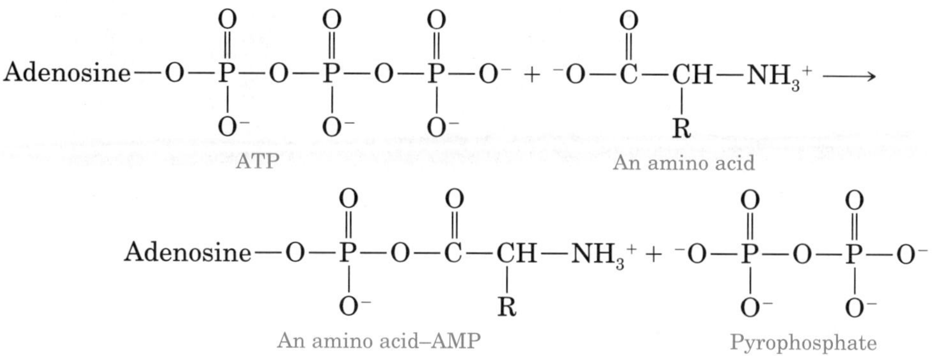

$$
\underset{\text{ATP}}{\text{Adenosine}-O-\overset{\overset{\displaystyle O}{\|}}{\underset{\underset{\displaystyle O^-}{|}}{P}}-O-\overset{\overset{\displaystyle O}{\|}}{\underset{\underset{\displaystyle O^-}{|}}{P}}-O-\overset{\overset{\displaystyle O}{\|}}{\underset{\underset{\displaystyle O^-}{|}}{P}}-O^-} + \underset{\text{An amino acid}}{{}^-O-\overset{\overset{\displaystyle O}{\|}}{C}-\underset{\underset{\displaystyle R}{|}}{C}H-NH_3{}^+} \longrightarrow
$$

$$
\underset{\text{An amino acid–AMP}}{\text{Adenosine}-O-\overset{\overset{\displaystyle O}{\|}}{\underset{\underset{\displaystyle O^-}{|}}{P}}-O-\overset{\overset{\displaystyle O}{\|}}{C}-\underset{\underset{\displaystyle R}{|}}{C}H-NH_3{}^+} + \underset{\text{Pyrophosphate}}{{}^-O-\overset{\overset{\displaystyle O}{\|}}{\underset{\underset{\displaystyle O^-}{|}}{P}}-O-\overset{\overset{\displaystyle O}{\|}}{\underset{\underset{\displaystyle O^-}{|}}{P}}-O^-}
$$

The activated amino acid is then bonded to its own particular tRNA molecule with the aid of an enzyme (a synthetase) that is specific for that particular amino acid and that particular tRNA molecule:

In the activation step, energy is used up. Two high-energy phosphate bonds are broken for each amino acid added to the chain:
ATP $\longrightarrow$ AMP + PPi and
PPi $\longrightarrow$ 2Pi

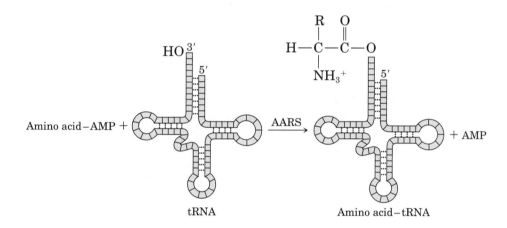

The specific recognition of an enzyme, aminoacyl-tRNA synthetase (AARS), *of its own tRNA and amino acid is often referred to as the second genetic code.* The 20 different synthetases recognize their substrates by stretches of nucleotide sequences on the tRNA.

B Initiation

The initiation stage in eukaryotes consists of three steps:

1. **Forming the pre-initiation complex** To initiate the protein synthesis a unique tRNA is used, designated as **tRNA$_i$**^{Met}**.** This unique tRNA carries a methionine residue, but it is used solely for the initiation step. It is attached to the 40S ribosomal body and forms the pre-initiation complex (Figure 17.4a). Just as in transcription, each step in translation is aided by a number of factors; these proteins are called **initiation factors.**

2. **Migration to mRNA** Next, the pre-initiation complex binds to the mRNA near the 5′ end that has been capped by a methyl GTP. The complex then migrates to the initiation codon of mRNA, AUG. The anticodon of the tRNA, UAC, lines up against the codon (Figure 17.4b). A number of factors aid this step, too.

In prokaryotes, the corresponding ribosomes are smaller, 30S and 50S.

Figure 17.4 The initiation of protein synthesis. (*a*) A unique tRNA molecule, methionine tRNA, attaches itself to the 40S ribosome. (*b*) The 40S ribosomal body is attached to the mRNA. It stays there during the entire translational process. (See Figure 17.5.) (*c*) The 60S ribosomal body joins the unit.

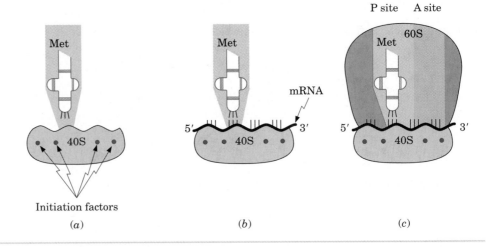

(*a*)   (*b*)   (*c*)

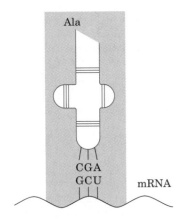

Figure 17.5 Alanine tRNA aligning on the ribosome with its complementary codon.

This enzyme, which is part of the 60S ribosome unit, is not a protein but a ribozyme.

3. Forming the full ribosomal complex The 60S ribosomal body joins the complex formed in step 2. Both 40S and 60S ribosomes carry two binding sites. The one shown on the left in Figure 17.4(c) is called the **P site,** because it is where the growing peptide chain will bind. The one next to it is called the **A (acceptor) site,** because that site accepts the incoming tRNA bringing the next amino acid.

C Elongation

1. Binding to the A site At this point, the A site is vacant, and each of the 20 tRNA molecules can come in and try to fit itself in. Only one of the 20 tRNAs carries exactly the right anticodon that corresponds to the next codon on the mRNA. (This is an alanine tRNA in Figures 17.5 and 17.6a). The binding of this tRNA to the A site takes place with the aid of proteins called **elongation factors** (Figure 17.6b).

Figure 17.6 Phases of elongation in protein synthesis. (*a*) After initiation, the tRNA of the second amino acid (alanine in this case) approaches the ribosome. (*b*) The tRNA binds to the A site. (*c*) The enzyme peptidyl transferase connects methionine to alanine, forming a peptide bond. (*d*) The peptide tRNA is moved (translocated) from the A to the P site while the ribosome moves to the right and simultaneously releases the empty tRNA. In this diagram, the alanine tRNA is shown in blue.

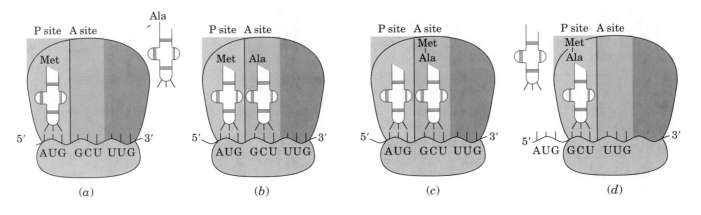

(*a*)   (*b*)   (*c*)   (*d*)

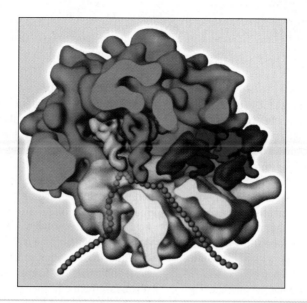

Figure 17.7 Ribosome in action. The lower yellow half represents the 30S ribosome; the upper blue half represents the 50S ribosome. The yellow and green twisted cones are tRNAs, and the chain of beads stand for mRNA. The elongation factors are in dark blue. *(Courtesy of Dr. J. Frank, Wadsworth Center, Albany, New York)*

2. **Forming the first peptide bond** At the A site, the new amino acid, alanine (Ala), is linked to the Met in a peptide bond by the enzyme **peptidyl transferase.** The empty tRNA remains on the P site (Figure 17.6c).

3. **Translocation** In the next phase of elongation, the whole ribosome moves one codon along the mRNA. Simultaneously with this move, the dipeptide is **translocated** from the A site to the P site, as shown in Figure 17.6(d); the empty tRNA dissociates and goes back to the tRNA pool to pick up another amino acid.

4. **Forming the second peptide bond** After the translocation, the A site is associated with the next codon on the mRNA, which is UUG in Figure 17.6(d). Once again, each tRNA can try to fit itself in, but only the one whose anticodon is AAC can align itself with UUG. This tRNA, which carries leucine (Leu), now comes in. The transferase establishes a new peptide bond between Leu and Ala, moving the dipeptide from the P site to the A site and forming a tripeptide.

These elongation steps are repeated until the last amino acid is attached.

Figure 17.7 shows a three-dimensional model of the translational process, which has been constructed on the basis of recent cryoelectron microscopy and X ray diffraction studies. This model clearly shows how the elongation factor proteins (in dark blue) are fitting in a cleft between the 50S (blue) and the 30S (pale yellow) bodies of prokaryotic ribosomes. The tRNAs on the P site (green) and on the A site (yellow) occupy a central cavity in the ribosomal complex. The orange beads represent the mRNA.

A similar arrangement exists in eukaryotes, which have 60S and 40S ribosomes.

More recently, a third site was identified on the ribosome. The site from which the tRNA exits when it is no longer needed, is called the E site.

D Termination

After the last translocation, the next codon reads "stop" (UAA, UGA, or UAG). At this point, no more amino acids can be added. Releasing factors then cleave the polypeptide chain from the last tRNA via a mechanism not yet fully understood. The tRNA itself is released from the P site. At the end, the whole mRNA is released from the ribosome. While the mRNA is attached to the ribosomes, many polypeptide chains are synthesized on it simultaneously.

Their work done, the two parts of the ribosome separate.

CHEMICAL CONNECTIONS 17B

Viruses

Nucleic acids are essential for life as we know it. No living thing can exist without them because they carry the information necessary to make protein molecules. The smallest forms of life, the viruses, consist only of a molecule of nucleic acid surrounded by a "coat" of protein molecules. In some viruses, the nucleic acid is DNA; in others, it is RNA. No virus has both. Whether viruses can be considered a true form of life became a topic of debate recently. In the summer of 2002, a group of scientists from the State University of New York, Stony Brook, reported that they had synthesized the polymyelitis virus in the laboratory from fragments of DNA. This new "synthetic" virus caused the same polio symptoms and death as the wild virus.

The shapes and sizes of viruses vary greatly, as shown in Figure 17B.1. Because their structures are so simple, viruses cannot reproduce themselves in the absence of other organisms. They carry DNA or RNA but do not have the nucleotides, enzymes, amino acids, and other molecules necessary to replicate their nucleic acid (Section 16.6) or to synthesize proteins (Section 17.5). Instead, viruses invade the cells of other organisms and cause those cells (the hosts) to do these tasks for them. Typically, the protein coat of a virus remains outside the host cell, attached to the cell wall, while the DNA or RNA is pushed inside.

Once the viral nucleic acid is inside the cell, the cell stops replicating its own DNA and making its own proteins. Instead, it replicates the viral nucleic acid and synthesizes the viral protein, according to the instructions on the viral nucleic acid. One host cell can make many copies of the virus.

In many cases, the cell bursts when a large number of new viruses have been synthesized, sending the new viruses out into the intercellular material, where they can infect other cells. This kind of process causes the host organism to get sick, and perhaps to die. Among the many human diseases caused by viruses are measles, hepatitis, mumps, influenza, the common cold, rabies, and smallpox. There is no cure for most viral diseases. Antibiotics, which can kill bacteria, have no effect on viruses. So far, the best defense against these diseases has been immunization (Chemical Connections 22B), which under the proper circumstances can work spectacularly well. Smallpox, once one of the most dreaded diseases, has been eradicated from this planet by many years of vaccination, and comprehensive programs of vaccination against such diseases as polio and measles have greatly reduced the incidence of these diseases.

Lately, a number of antiviral agents have been developed. They completely stop the reproduction of viral nucleic acids (DNA or RNA) inside infected cells without preventing the DNA of normal cells from replicating. One such drug is called vidarabine, or Ara-A, and is sold under the trade name Vira-A.

Antiviral agents often act like anticancer drugs (Chemical Connections 16A), in that they have structures similar to one of the nucleotides necessary for the synthesis of nucleic acids. Vidarabine is the same as adenosine, except that the sugar is arabinose instead of ribose (Figure 17B.2). Vidarabine is used to fight a life-threatening viral illness, herpes encephalitis. It is also effective in neonatal herpes infection and chickenpox. However, like many other anticancer and antiviral drugs, vidarabine is toxic, causing nausea and diarrhea. In some cases, it has caused chromosomal damage.

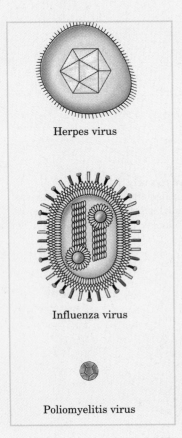

■ **Figure 17B.1** The shape of three viruses. *(Illustration by Irving Geis, from* Scientific American, *Jan. 1963. Rights owned by Howard Hughes Medical Institute. Not to be reproduced without permission.)*

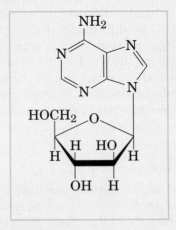

■ **Figure 17B.2** Vidarabine.

CHEMICAL CONNECTIONS 17C

AIDS

In recent years a deadly virus, commonly called the AIDS virus (for acquired immune deficiency syndrome) or HIV (for human immunodeficiency virus), has spread with alarming speed around the world. Perhaps 100 million people are infected, and this number is increasing daily, especially in underdeveloped countries. The AIDS virus invades the human immune system, and especially enters the T (lymphocyte) cells and kills them (Figure 17C.1). This RNA-containing virus, or **retrovirus,** decreases the population of T cells (blood cells that fight invading foreign cells; see Section 22.2) and thus allows other opportunistic invaders, such as the protozoan *Pneumocystis carinii,* to proliferate, causing pneumonia and eventual death.

At the time of the writing, there is no cure for AIDS. Drug treatment can slow down and even seemingly arrest the development of the disease, however. There are four different types of drugs.

The first type is a **nucleoside analog** that competitively inhibits the enzyme **reverse transcriptase,** which is necessary for the multiplication of the virus. Among the most frequently used agents are azidothymidine (AZT), dideoxycytidine (ddCyd), dideoxyinosine (ddI), and (-)-2-deoxy-3'-thiacytidine (3TC). As illustrated with ddCyd in Figure 17C.2, these drugs resemble nucleosides—in this case, cytidine (Section 16.2), but the two —OH groups on the ribose have been removed. Thus they can enter the active site of the enzyme transcriptase and competitively inhibit its action.

A second type of anti-AIDS drugs is the **non-nucleoside inhibitors** of the enzyme reverse transcriptase. Among these agents are efavirenze and nevirapine. They inhibit the enzyme by noncompetitive binding.

A third group of drugs aims to inhibit a different enzyme, **protease,** that is also necessary for the propagation of the virus. Abacavir, indinavir (Figure 17C.2), nelfinavir, ritonavir, and saquinavir are the most frequently prescribed drugs in this category.

Because the HIV virus quickly mutates and thus becomes resistant to an inhibitor, the present practice is to prescribe a cocktail of drugs called HAART, or *highly active antiretroviral treatment.* HAART may contain, for example, ritonavir, ddCyd, and 3TC, or any other combination of the antitranscriptase and antiprotease drugs. Such a two- or three-drug cocktail greatly reduces the presence of HIV in the patient's blood and decreases the death rate. It even allows, if only temporarily, more or less normal functions for patients.

A fourth group of anti-AIDS drugs, now in development, aims to inhibit the fusion of the virus to the T-cell membrane, thereby preventing its entrance into the cell. HIV particles carry densely arranged spikes on their surfaces. These spikes are made of two glycoproteins: **gp120,** a surface glycoprotein, and **gp41,** a transmembrane glycoprotein. The latter mediates the entry of the

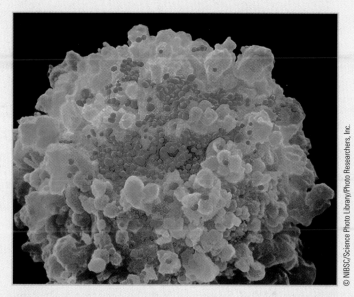

© NIBSC/Science Photo Library/Photo Researchers, Inc.

■ **Figure 17C.1** HIV particles (pink) on the surface of a lymphocyte (green).

virus into the target cell. Drugs such as T20 and T1249 are synthetic polypeptides made of 36 and 39 amino acids, respectively, that interact with gp41 (Figure 17C.3).

A **vaccine** against AIDS would be the ideal remedy. People who are ill with AIDS or who have been exposed to the virus carry antibodies against the virus in their blood. That is, the immune system fights back against the infection. If the immune system could be bolstered either by vaccination with "dead" virus or by cytokines such as interleukines (see Section 22.6), AIDS would be curable. Efforts in the direction of vaccination have been hampered by the deadly nature of the virus, however. Even if only one out of a million "dead" viruses survived, it could still kill the vaccinated person.

In the beginning of the AIDS epidemic, some people were infected by blood transfusion. About two to three of every 10,000 blood samples in the United States carry antibodies against the virus. All such blood samples are now rejected; patented screening tests are performed on every blood donor, thereby making blood transfusion a safe procedure.

The strategy of treating a patient with a battery of drugs developed in fighting AIDS paid unexpected dividends in treating other diseases. Hepatitis viruses B and C are present in 1.5% of the U.S. population. A combination of interferon, which bolsters the immune system, and ribavirin, another viral protease inhibitor, is now the cocktail treatment of choice for hepatitis C.

(continued on next page)

C H E M I C A L   C O N N E C T I O N S   17C

AIDS, *(continued)*

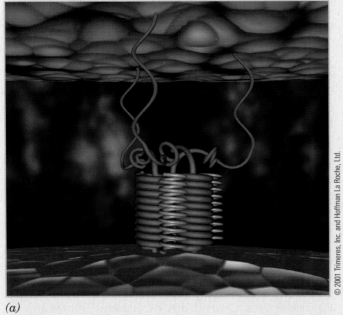

2′,3′-Dideoxycytidine (ddCyd)
(Zalcitabine)

Indinavir

■ **Figure 17C.2** Structural formulas for some AIDS drugs.

> **Reverse transcriptase** The enzyme that enables a virus to copy the information encoded in an RNA molecule into a DNA molecule

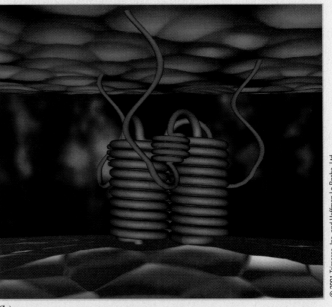

(a)   *(b)*

■ **Figure 17C.3** AIDS drug preventing the fusion of HIV virus to the T-cell membrane. (*a*) T-20 (yellow coils), a fusion inhibitor, binds to the matching coil segments that are part of HIV's gp41 (blue threads and red coils), preventing the viral fusion with the cell membrane of the T-cell above. (*b*) Without the fusion inhibitor, gp41 binds tightly to receptors on the T-cell membrane.

EXAMPLE 17.3

A tRNA has an anticodon, GAA. What amino acid will this tRNA carry? What are the steps necessary for the amino acid to bind to the tRNA?

Solution

The amino acid is leucine, as the codon is CUU. Leucine has to be activated by ATP. A specific enzyme, leucine-tRNA synthetase, catalyzes the carboxyl ester bond formation between the carboxyl group of leucine and the —OH group of tRNA.

Problem 17.3

What are the reactants in the reaction forming valine-tRNA?

17.6 Gene Regulation

Every embryo that is formed by sexual reproduction inherits its genes from both the parent sperm and the egg cells. But the genes in its chromosomal DNA are not active all the time. Rather, they are switched on and off during development and growth of the organism. Soon after formation of the embryo, the cells begin to differentiate. Some cells become neurons, some become muscle cells, some become liver cells, and so on. Each cell is a specialized unit that uses only some of the many genes it carries in its DNA. Thus each cell must switch some of its genes on and off—either permanently or temporarily. How this is done is the subject of **gene regulation.**

We know less about gene regulation in eukaryotes than in the simpler prokaryotes. Even with our limited knowledge, however, we can state that organisms do not have a single, unique way of controlling genes. Many gene regulations occur at the **transcriptional level** (DNA $\longrightarrow$ RNA). Others operate at the **translational level** (mRNA $\longrightarrow$ protein). A few of these processes are listed here as examples.

> **Gene regulation** The control process by which the expression of a gene is turned on or off. Since RNA synthesis proceeds in one direction ($5' \longrightarrow 3'$) (see Figure 17.2), the gene (DNA) to be transcribed runs from 3' to 5' Thus the control sites are in front of, or upstream of, the 3' end of the structural gene.

A Control on the Transcriptional Level

In eukaryotes, transcription is regulated by three entities: promoters, enhancers, and response elements. A fourth way of regulation is the phosphorylation of transcription factors.

1. Promoters of a gene are located adjacent to the transcription site. They are defined by an initiator and conserved sequences, such as a TATA box (see also Section 17.2 and Figure 17.2) or one or more copies of a GGGCGG sequence called a GC box. In eukaryotes, the enzyme RNA-polymerase has little affinity for binding to DNA. Instead, different transcription factors, or binding proteins, bind to the different modules of the promoter. With the aid of these transcription factors, the promoter functions to control the transcription on a steady, normal level. The binding, one after another, of these factors to the promoter site controls the rate of transcription initiation.

Transcription factors may allow the synthesis of mRNA (and from there the target protein) to vary by a factor of 1 million. This wide variation in eukaryotic cells is exemplified by the α-A-crystallin gene, which can be expressed in the lens of the eye, at a rate a-millionfold higher than that in the liver cell of the same organism.

CHEMICAL CONNECTIONS 17D

The Promoter: A Case of Targeted Expression

Although the major part of a gene contains the region that codes for a particular protein, a portion of the gene does not code for any part of the protein. Instead, it acts as a switch that turns the transcription process on or off. This regulatory part of the gene is called the **promoter.** By using the gene insertion technique discussed in Section 17.8, one can combine the coding sequence of one gene with the promoter of another gene, thereby targeting the expression of a protein to a certain organ.

A recent development in AIDS research (Chemical Connections 17C) involved an enzyme called HIV-1 protease. This enzyme is essential if the virus that causes AIDS is to multiply inside the host cell. It has been argued that if suitable drugs that can inhibit HIV-1 protease could be found, the development of AIDS could be arrested, and perhaps the disease cured. But how can we test the hundreds of organic compounds that show some promise of being HIV-1 protease inhibitors? Here is where the concept of the promoter comes in. For example, if the coding region of the HIV-1 protease is linked to the promoter of α-A-crystallin, a protein molecule that appears mostly in the lens of the eye, the expression of the HIV-1 protease could be targeted to the eye only. True to the design, a DNA sequence containing the gene for HIV-1 protease and the promoter of α-A-crystallin were transfected into a mouse. The result was transgenic mice that became blind 24 days after birth. Otherwise, the mice were not affected at all—they ate, ran, and bred as normal mice do. The blindness was the result of the action of HIV-1 protease in the lens, causing cataract formation.

These transgenic mice could be used to test drugs to determine whether they would be good inhibitors of HIV-1 protease. If injection of a potential drug delayed cataract formation, that would be proof that the drug truly acts as an inhibitor of HIV-1

protease in a living organism and not just in test tubes. Many of the new drugs now used in treatment of AIDS patients (see Chemical Connections 17C), such as indinavir (Figure 17C.2), have first been tested in this manner. This strategy is clearly more efficient than testing all such potential inhibitors directly in humans.

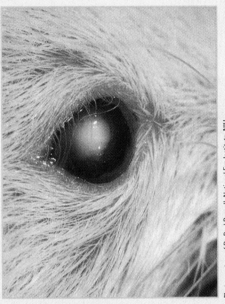

Photo courtesy of Dr. Paul Russell, National Eye Institute, NIH

■ **Eye with cataractous lens of a transgenic mouse containing the HIV-1 protease linked to α-A-crystallin promoter, 27 days after birth.**

How do these transcription factors find the specific gene control sequences into which they fit, and how do they bind to them? The interaction between the protein and DNA involves nonspecific electrostatic interactions (positive ions attracting negative ions and repelling other positive ions) as well as more specific hydrogen bonding. The transcription factors find their targeted sites by twisting their protein chains so that a certain amino acid sequence is present at the surface. One such conformational twist is provided by **metal-binding fingers** (Figure 17.8). These finger shapes are created by Zn^{2+} ions, which form covalent bonds with the amino acid side chains of the protein.

The zinc fingers interact with specific DNA (or sometimes RNA) sequences. The recognition comes by hydrogen bonding between a nucleotide (for example, guanine) and the side chain of a specific amino acid (for example, arginine). Besides metal-binding fingers, at least two other prominent transcription factors exist: **helix-turn-helix** and **leucine zipper.** Transcription factors can also be repressors, reducing the rate of transcription. Some can serve as both activators and repressors.

2. A second group of transcription factors, known as **enhancers,** also speeds up the transcription process. These factors may be located several thousand nucleotides away from the transcription site. To stimulate transcription, an enhancer must be brought in the vicinity of the promoter. With the aid of special proteins, called **sensors,** the DNA forms a loop. The enhancer with the sensor interacts with a trancription factor. This complex then allows the RNA polymerase II to speed up the transcription when higher-than-normal production of proteins is needed (Figure 17.9).

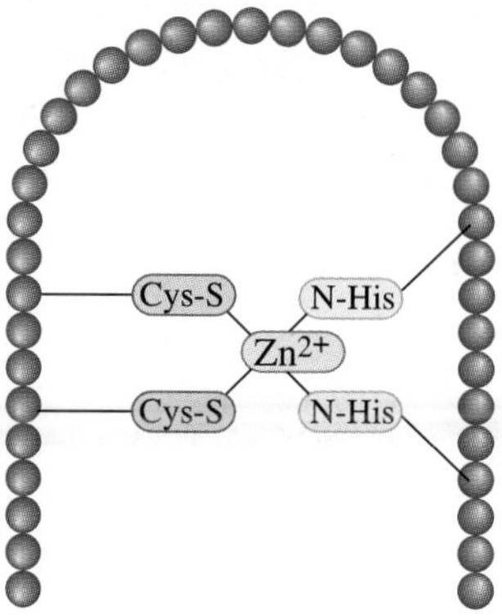

3. The third group of transcription factors, **response elements,** are activated by their transcription factors in response to an outside stimulus. The stimulus may be heat shock, heavy metal toxicity, or simply a hormonal signal, such as binding a steroid hormone to its receptor. The response element of steroids is in front of and 250 base pairs upstream from the starting point of transcription. Only the receptor with the bound steroid hormone can interact with its response element, thereby initiating transcription.

4. Transcription does not occur at the same rate throughout the cell life cycle. It is accelerated or slowed down as the need arises. The signal to speed up transcription may originate from outside of the cell. One such signal, in the GTP-adenylate cyclase-cAMP pathway (Section 15.5), produces **phosphorylated protein kinase.** This enzyme enters the nucleus, where it phosphorylates transcription factors, which aid in the transcription cascade.

Figure 17.8 Schematic view of a metal-binding finger (a common motif in many transcription factors). The zinc ion forms covalent bonds to two cysteines and two histidines.

B Control on the Translational Level

During translation, there are also a number of mechanisms that ensure quality control.

1. The Specificity of a tRNA for Its Unique Amino Acid First, the attachment of the proper amino acid to the proper tRNA must be achieved. The enzyme that catalyzes this reaction, aminoacyl-tRNA synthetase (AARS), is specific for each tRNA and amino acid combination. The AARS enzymes recognize their tRNA by specific nucleotide sequences. Furthermore, the active site of the enzyme has two **sieving portions.** For example, in isoleucyl-tRNA synthetase the first sieve excludes any amino acids that are larger than isoleucine. If a similar amino acid such as valine, which is smaller than isoleucine, arrives at the active site, the second sieve eliminates it. The second sieving site is thus a "proofreading" site.

2. Recognition of the Stop Codon Another quality-control measure occurs at the termination. The stop codons must be recognized by release factors, leading to the release of the polypeptide chain and allowing the recycling of the ribosomes. Otherwise, a longer polypeptide chain may be toxic. The release factor combines with GTP and binds to the ribosomal A

Figure 17.9 The enhancer process. A loop in DNA brings the enhancer to the initiation site.

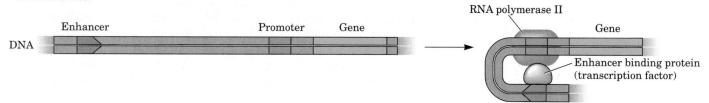

site when that site is occupied by the termination codon. Both the GTP and the peptidyl-tRNA ester bond are hydrolyzed. This hydrolysis releases the polypeptide chain and the deacylated tRNA. Finally, the ribosome dissociates from the mRNA.

3. Post-translational Controls

(a) *Removal of methionine.* In most proteins, the methionine residue at the N-terminal, which was added in the initiation step, is removed. A special enzyme, metaminopeptidase, cleaves the peptide bond.

(b) *Chaperoning.* The tertiary structure of a protein is largely determined by the amino acid sequence (primary structure). Proteins begin to fold even as they are synthesized on ribosomes. Nevertheless, misfolding may occur due to mutation in a gene, lack of fidelity in transcription, or translational errors. All of these errors may lead to aggregation of misfolded proteins that can be detrimental to the cell—for example, causing amyloid diseases. Certain proteins in living cells, called **chaperones,** help the newly synthesized polypeptide chains to fold properly. They recognize hydrophobic regions exposed on unfolded proteins and bind to them. Chaperones then shepherd the proteins to the biologically desirable folding as well as to their local destinations within the cell.

(c) *Degradation of misfolded proteins.* A third post-translational control exists in the form of **proteasomes.** These are cylindrical assemblies of a number of protein subunits with proteolytic activity in the core of the cylinder. If the rescue by chaperones fails, proteases may degrade the misfolded protein first by targeting it with ubiquitination (see Chemical Connections 19E) and finally by proteolysis.

17.7 Mutations, Mutagens, and Genetic Diseases

In Section 16.6, we saw that the base-pairing mechanism provides an almost perfect way to copy a DNA molecule during replication. The key word here is "almost." No machine, not even the copying mechanism of DNA replication, is totally without error. It has been estimated that, on average, one error occurs for every 10^{10} bases (that is, one in 10 billion). An error in the copying of a sequence of bases is called a **mutation.** Mutations can occur during replication. Base errors can also occur during transcription in protein synthesis (a nonheritable error).

These errors may have widely varying consequences. For example, the codon for valine in mRNA can be GUA, GUG, GUC, or GUU. In DNA, these codons correspond to CAT, CAC, CAG, and CAA, respectively. Assume that the original codon in the DNA is CAT. If a mistake is made during replication and CAT is spelled as CAG in the copy, there will be no harmful mutation. Instead, when a protein is synthesized, the CAG will be transcribed onto the mRNA as GUC, which also codes for valine. Therefore, although a mutation has occurred, the same protein is manufactured.

Now assume that the original sequence in the gene's DNA is CTT, which transcribes onto mRNA as GAA and codes for glutamic acid. If a mutation occurs during replication and CTT becomes ATT, the new cells will probably die. The reason is that ATT transcribes to UAA, which does not code for any amino acid but rather is a stop signal. Thus, instead of continuing to build a protein chain with glutamic acid, the synthesis will stop altogether. An important protein will not be manufactured, and the organism may die. In this way, very harmful mutations are not carried over from one generation to the next.

Sickle cell anemia (Chemical Connections 13D) is caused by a single amino acid change: valine for glutamic acid. In this case, CTT is mutated to CAT.

CHEMICAL CONNECTIONS 17E

Mutations and Biochemical Evolution

We can trace the genetic relationship of different species through the variability of their amino acid sequences in different proteins. For example, the blood of all mammals contains hemoglobin, but the amino acid sequences of the hemoglobins are not identical. In the table, we see that the first 10 amino acids in the β-globin of humans and gorillas are exactly the same. As a matter of fact, there is only one amino acid difference, at position 104, between us and apes. The β-globin of the pig differs from ours at 10 positions, of which 2 are in the N-terminal decapeptide. That of the horse differs from ours in 26 positions, of which 4 are in this decapeptide. β-Globin seems to have gone through many mutations during the evolutionary process, because only 26 of the 146 sites are invariant—that is, exactly the same in all species studied so far.

The relationship between different species can also be established by finding similarities in their mRNA primary structures. As the mutations actually occurred on the original DNA molecule, and then were perpetuated in the progeny by the mutant DNA, it is instructive to learn how a point mutation may occur in different species. Looking at position 4 of the β-globin

molecule, we see a change from serine to threonine. The code for serine is AGU or AGC, whereas that for threonine is ACU or ACC (Table 17.1). Thus a change from G to C in the second position of the codon brought the divergence between the β-globins of humans and horses. The genes of closely related species, such as humans and apes, have very similar primary structures, presumably because these two species diverged on the evolutionary tree only recently. On the other hand, species far removed from each other diverged long ago and have undergone more mutations, which show up in differences in the primary structures of their DNA, mRNA, and consequently proteins.

The *number* of amino acid substitutions is significant in the evolutionary process caused by mutation, but the *kind* of substitution is even more important. If the substitution involves an amino acid with physicochemical properties similar to those of the amino acid in the ancestor protein, the mutation is most probably viable. For example, in human and gorilla β-globin, position 4 is occupied by threonine, but it is occupied by serine in the pig and horse. Both amino acids provide an —OH-carrying side chain.

Amino Acid Sequence of the N-Terminal Decapeptides of β-Globin in Different Species

Species	Position									
	1	2	3	4	5	6	7	8	9	10
Human	Val	His	Leu	Thr	Pro	Glu	Glu	Lys	Ser	Ala
Gorilla	Val	His	Leu	Thr	Pro	Glu	Glu	Lys	Ser	Ala
Pig	Val	His	Leu	Ser	Ala	Glu	Glu	Lys	Ser	Ala
Horse	Val	Glu	Leu	Ser	Gly	Glu	Glu	Lys	Ala	Ala

Ionizing radiation (X rays, ultraviolet light, gamma rays) can cause mutations. Furthermore, a large number of organic compounds can induce mutations by reacting with DNA. Such compounds are called **mutagens.** Many changes caused by radiation and mutagens do not become mutations because the cell has its own repair mechanism, called nucleotide excision repair (NER), which can prevent mutations by cutting out damaged areas and resynthesizing them (Section 16.7). In spite of this defense mechanism, certain errors in copying that result in mutations do slip by. Many compounds (both synthetic and natural) are mutagens, and some can cause cancer when introduced into the body. These substances are called **carcinogens** (Chemical Connections 4B). One of the main tasks of the U.S. Food and Drug Administration and the Environmental Protection Agency is to identify carcinogens and eliminate them from our food, drugs, and environment. Even though most carcinogens are mutagens, the reverse is not true.

Not all mutations are harmful. Some are beneficial because they enhance the survival rate of the species. For example, mutation is used to develop new strains of plants that can withstand pests.

Examples of mutagens are benzene, carbon tetrachloride, and vinyl chloride.

CHEMICAL CONNECTIONS 17F

Oncogenes

An **oncogene** is a gene that in some way or other participates in the development of cancer. Cancer cells differ from normal cells in a variety of structural and metabolic ways; however, the most important difference is their uncontrolled proliferation. Most ordinary cells are quiescent. When they lose their controls, they give rise to tumors—benign or malignant. The uncontrolled proliferation allows these cells to spread, invade other tissues, and colonize them, in a process called *metastasis.*

Malignant tumors can be caused by a variety of agents: chemical carcinogens such as benzo(a)pyrene (Chemical Connections 4B), ionizing radiation such as X rays, and viruses. However, for a normal cell to become a cancer cell, certain transformations must occur. Many of these transformations take place on the level of the genes. Certain genes in normal cells have counterparts in viruses, especially in retroviruses (which carry their genes in the form of RNA rather than DNA). Thus normal cells in a human and in a virus both have genes that code for similar proteins. Usually these genes code for proteins that control cell growth.

One such protein is epidermal growth factor (EGF). When a retrovirus gets into the cell, it takes control of the cell's own EGF gene. At that point, the EGF gene becomes an oncogene, a cancer-causing gene, because it produces large amounts of the EGF protein. This excess production, in turn, allows the epidermal cells to grow in an uncontrolled manner.

Because the normal EGF gene in the normal cell was turned into an oncogene, we call it a **proto-oncogene.** Such conversions of a proto-oncogene to an oncogene can occur not just by viral invasion, but also by a mutation of the gene. Although most cancers are caused by multiple mutations, one form of cancer, called human bladder carcinoma, is caused by a single mutation of a gene—a change from guanine (G) to thymine (T). The resulting protein has a valine in place of a glycine. This change is sufficient to transform the cell into a cancerous state because the protein for which the gene codes amplifies signals (Section 15.5). Thus, because of the mutation that caused the proto-oncogene to become an oncogene, the cell stays "on"—its metabolic process is not shut off—and it is now a cancer cell that proliferates without control.

There are a number of other mechanisms by which a proto-oncogene is transformed into an oncogene, but in each case the gene product has something to do with growth, hormonal action, or some other kind of cell regulation. About 100 viral and cellular oncogenes have been identified to date.

CHEMICAL CONNECTIONS 17G

p53: A Central Tumor Suppressor Protein

Not all cancer-causing gene mutations have their origin in oncogenes. There are some 36 known **tumor suppressor genes,** the products of which are proteins controlling cell growth. None of them is more important than a protein with a molar mass of 53,000 simply named **p53.** This protein responds to a variety of cellular stresses including DNA damage, lack of oxygen (hypoxia) and aberrantly activated oncogenes (Chemical Connections 17F). In about 40% of all cancer cases, the tumor contains p53 that underwent mutation. Mutated p53 protein can be found in 55% of lung cancers, about half of all colon and rectal cancers, and some 40% of lymphomas, stomach cancers, and pancreatic cancers. In addition, in one-third of all soft tissue sarcomas, p53 is inactive, even though it did not undergo mutation.

These statistics indicate that the normal function of p53 protein is to suppress tumor growth. When it is mutated or otherwise not present in sufficient or active form, it is unable to perform this protective function and cancer spreads. p53 protein binds to specific sequences of double-stranded DNA. When X rays or γ-rays damage DNA, an increase in p53 protein concentration is observed. The increased binding of p53 controls the cell cycle; it holds it between cell division and DNA replication. The time gained in this arrested cell cycle allows the DNA to repair its damage. If that fails, the p53 protein triggers apoptosis, the programmed cell death of the injured cell (see also Chemical Connections 16E).

Recently, it was reported that p53 has a finely tuned function in cells. It suppresses tumor growth, but if p53 is overly expressed (that is, its concentration is high), it contributes to the premature aging of the organism. In those conditions, p53 arrests the cell cycle not only of damaged cells but also of stem cells. These cells normally differentiate into various types (muscle cells, nerve cells, and so forth) and replace those cells that die with aging. Excess p53 slows down this differentiation. In mice, it was found that the abundance of p53 protein made the animal cancer-free but at a cost: The mice lost weight and muscle, their bones became brittle, and their wounds took longer to heal. All in all, their average life span was 20% shorter than that of normal mice.

If a mutation is harmful, it results in an inborn genetic disease. This condition may be carried as a recessive gene from generation to generation with no individual demonstrating the symptoms of the disease. When both parents carry recessive genes, however, an offspring has a 25% chance of inheriting the disease. If the defective gene is dominant, on the other hand, every carrier will develop symptoms.

Specific diseases that result from defective or missing genes are discussed in Chemical Connections 12F, 13D, 19B, and 19F.

A recessive gene expresses its message only if both parents transmit it to the offspring.

17.8 Recombinant DNA

There are no cures for the inborn genetic diseases that we discussed in the preceding section. The best we can do is detect the carriers and, through genetic counseling of prospective parents, try not to perpetuate the defective genes. However, **recombinant DNA techniques** give us some hope for the future. At this time, recombinant DNA techniques are used mostly in bacteria, plants, and test animals (such as mice), but they could someday be successfully extended to human cells as well.

One example of recombinant DNA techniques begins with certain circular DNA molecules found in the cells of the bacteria *Escherichia coli*. These molecules, called **plasmids** (Figure 17.10), consist of double-stranded DNA arranged in a ring. Certain highly specific enzymes called **restriction endonucleases** cleave DNA molecules at specific locations (a different location for each enzyme). For example, one of these enzymes may split a double-stranded DNA as follows:

$$
\text{B} \sim \text{GAATTC} \sim \text{B} \xrightarrow{\text{enzyme}} \text{B} \sim \text{G} \qquad \text{AATTC} \sim \text{B}
$$
$$
\text{B} \sim \text{CTTAAG} \sim \text{B} \qquad\qquad \text{B} \sim \text{CTTAA} \quad + \quad \text{G} \sim \text{B}
$$

We use "B" to indicate a bacterial plasmid.

The enzyme is so programmed that whenever it finds this specific sequence of bases in a DNA molecule, it cleaves it as shown. Since a plasmid is circular, cleaving it in this way produces a double-stranded chain with two ends (Figure 17.11). These are called "sticky ends" because on one strand each has several free bases that are ready to pair up with a complementary section if they can find one.

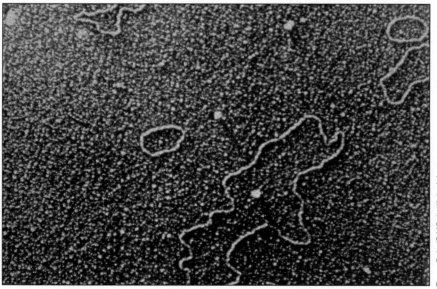

Figure 17.10 Plasmids from a bacterium used in the recombinant DNA technique.

The next step is to give them such a section. This is done by adding a gene from some other species. The gene is a strip of double-stranded DNA that has the necessary base sequence. For example, we can put in the human gene that manufactures insulin, which we can get in two ways:

1. It can be made in a laboratory by chemical synthesis; that is, chemists can combine the nucleotides in the proper sequence to make the gene.
2. We can cut a human chromosome with the same restriction enzyme.

Since it is the same enzyme, it cuts the human gene so as to leave the same sticky ends:

$$\text{H—GAATTC—H} \xrightarrow{\text{enzyme}} \text{H—G} \quad \text{CTTAA} \; + \; \text{AATTC—H} \quad \text{G—H}$$
$$\text{H—CTTAAG—H}$$

The human gene must be cut at two places so that a piece of DNA that carries two sticky ends is freed. To splice the human gene into the plasmid, the two are mixed in the presence of DNA ligase, and the sticky ends come together:

$$\text{H—G} \quad \text{CTTAA} \; + \; \text{AATTC—B} \quad \text{G—B} \xrightarrow{\text{enzyme}} \text{H—GAATTC—B}$$
$$\text{H—CTTAAG—B}$$

This reaction takes place at both ends of the human gene, turning the plasmid into a circle once again (Figure 17.11).

The modified plasmid is then put back into a bacterial cell, where it replicates naturally every time the cell divides. Bacteria multiply quickly, so soon we have a large number of bacteria, all containing the modified plasmid. All these cells now manufacture human insulin by transcription and translation. Thus, we can use bacteria as a factory to manufacture specific proteins. This new industry has tremendous potential for lowering the price of drugs that are now manufactured by isolation from human or animal tissues (for example, human interferon, a molecule that fights infection). Not only bacteria but also plant cells can be used (Figure 17.12).

Ultimately, if recombinant DNA techniques can be applied to humans and not just to bacteria, genetic diseases might someday be cured by this powerful technology. For instance, an infant or fetus who is missing a gene might be given that gene. Once in the cells, the gene would reproduce itself for the individual's entire lifetime.

We use "H" to indicate a human gene.

Figure 17.11 The recombinant DNA technique can be used to turn a bacterium into an insulin "factory."

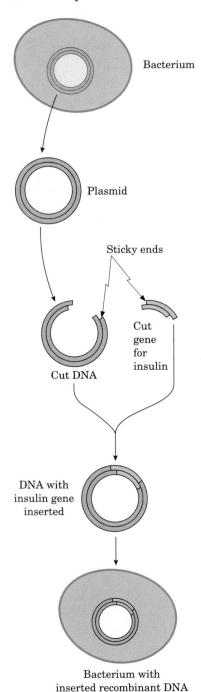

Bacterium

Plasmid

Sticky ends

Cut DNA

Cut gene for insulin

DNA with insulin gene inserted

Bacterium with inserted recombinant DNA

EXAMPLE 17.4

Two different restriction endonucleases act on the following sequence of a double-stranded DNA:

$$\text{\~\~\~AATGAATTCGAGGC\~\~\~}$$
$$\text{\~\~\~TTACTTAAGCTCCG\~\~\~}$$

The endonuclease EcoRI recognizes the sequence GAATTC and cuts the sequence between G and A. The other endonuclease, TaqI, recognizes the sequence TCGA and cuts the sequence between T and C. What are the sticky ends that each of these endonucleases will create?

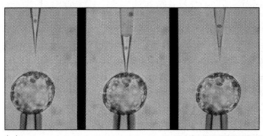

(a)

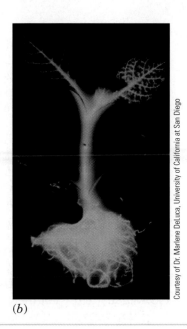

(b)

Courtesy of Dr. Anne Crossway, Calgene, Inc., Davis, California, and Biotechniques, Eaton Publishing, Natick, Massachusetts

Courtesy of Dr. Marlene DeLuca, University of California at San Diego

Figure 17.12 (*a*) Injection of an aqueous DNA solution into the nucleus of a protoplast. (*b*) A luminescent tobacco plant. The gene of the enzyme luciferase (from a firefly) has been incorporated into the genetic material of the tobacco plant.

Solution

EcoRI	∿∿∿AATG	AATTCGAGGC∿∿∿
	∿∿∿TTACTTAA	GCTCCG∿∿∿
TaqI	∿∿∿AATGAATT	CGAGGC∿∿∿
	∿∿∿TTACTTAAGC	TCCG∿∿∿

Problem 17.4

Show the sticky end of the following double-stranded DNA sequence that is cut by TaqI:

$$\text{∿∿∿CCTCGATTG∿∿∿}$$
$$\text{∿∿∿GGAGCTAAC∿∿∿}$$

Human insulin is now marketed by the Lilly Corporation in two forms, called Humulin R and Humulin N (Chemical Connections 13C). The R form has a faster onset of action. Both versions are manufactured by the recombinant DNA technique.

As of mid-2002, no such gene therapy had been successfully achieved in humans. This failure was not for lack of trying. A tremendous amount of research has been devoted to not only trying to cure genetic diseases but also to delivering a fatal bullet to cancer cells by recombinant DNA techniques. The difficulties are numerous, but finding the proper vector (a harmless virus, for example) is one of the bigger challenges. The virus must be proficient enough to deliver the selected gene without eliciting an immune reaction from the organism itself. The hope is that, in the future, genes can be inserted in human organs and they will express themselves in sufficient quantity. Even then the product of the healthy gene will compete with the product of the faulty gene, ideally alleviating the symptoms of the disease or curing it altogether.

By far, the greatest potential of **genetic engineering**—that is, inserting new genes into cells—is in the field of agriculture. Genetic engineering has produced tomatoes that can be picked in their naturally ripened states yet do not spoil quickly on the shelves of supermarkets. Other genetically engineered plants—corn and cotton, for example—can resist damage by insects, fungi, or herbicides. Most of the soybeans grown in the United States today are genetically engineered. In addition to creating herbicide- and insect-resistant plants, genetic engineering offers opportunities to improve crop yields and resist freezing temperatures. All these factors will contribute to the increase of the food supply as is necessary to feed the ever-growing world population.

Some people believe that genetically engineered organisms may create havoc with the ecology of the planet by introducing new, dominant species. Others express concern that genetically engineered organisms may enter the human food chain in ways for which the product has not been approved. This concern is legitimate. For example, genetically engineered corn that has been approved only for animal feed has already found its way in trace amounts into corn chips intended for human consumption. This happened at a single manufacturer. No adverse medical effect has been reported, although some people may prove to be allergic to the introduced gene. Such concerns have restricted the export to European markets of beef that was fed on genetically engineered corn.

Plant breeding has been going on for centuries, with the goal of achieving exactly the same goals as genetic engineering does. Obviously, care must be taken and controls exercised so as not to release harmful newly engineered life-forms into the environment.

SUMMARY

A **gene** is a segment of a DNA molecule that carries the sequence of bases that directs the synthesis of one particular protein (Section 17.1). The information stored in the DNA is transcribed onto RNA and then expressed in the synthesis of a protein molecule. This is done in two steps: **transcription** and **translation.** In transcription (Section 17.2), the information is copied from DNA onto mRNA by complementary base pairing (A $\longrightarrow$ U, T $\longrightarrow$ A, G $\longrightarrow$ C, C $\longrightarrow$ G). There are also start and stop signals.

The mRNA is strung out along the ribosomes (Section 17.3). Transfer RNA carries the individual amino acids, with each tRNA going to a specific site on the mRNA. A sequence of three bases (a triplet) on mRNA constitutes a **codon.** It spells out the particular amino acid that the tRNA brings to this site. Each tRNA has a recognition site, the **anticodon,** that pairs up with the codon. When two tRNA molecules are aligned at adjacent sites, the amino acids that they carry are linked by an enzyme forming a peptide bond. The translation process continues until the whole protein is synthesized (Section 17.5). The **genetic code** (Section 17.4) provides the correspondence between a codon and an amino acid. In most cases, there is more than one codon for each amino acid. Most of human DNA (96 to 98%) does not code for proteins.

There are a number of mechanisms for gene regulation on both the transcriptional level and the translational level (Section 17.6). **Promoters** have initiator and conserved sequences. **Transcription factors** bind to the promoter, thereby regulating the rate of transcription. **Enhancers** are nucleotide sequences far removed from the transcription sites. **Sensors** are proteins that bring the enhancers in the vicinity of the promoter and in this way control the rate of transcription. Some translational controls, such as **release factors,** act during translation, others, such as **chaperones,** act after translation is completed.

A change in the sequence of bases is called a **mutation** (Section 17.7). Mutations can be caused by an internal mistake or induced by chemicals or radiation. In fact, a change in just one base can cause a mutation. This may be harmful or beneficial, or it may cause no change whatsoever in the amino acid sequence. If a mutation is very harmful, the organism may die. Chemicals that cause mutations are called **mutagens** (Section 17.7). Chemicals that cause cancer are called **carcinogens.** Many carcinogens are mutagens, but the reverse is not true.

With the discovery of restriction enzymes that can cut DNA molecules at specific points, scientists have found ways to splice DNA segments together. In this manner, a human gene (for example, the one that codes for insulin) can be spliced into a bacterial plasmid. The bacteria, when multiplied, can then transmit this new information to the daughter cells so that the ensuing generations of bacteria can manufacture human insulin. This powerful method is called the **recombinant DNA technique** (Section 17.8). Genetic engineering is the process by which genes are inserted into cells.

P R O B L E M S

Numbers that appear in color indicate difficult problems.
▶ designates problems requiring application of principles.

17.5 Does the term *gene expression* refer to (a) transcription, (b) translation, or (c) transcription plus translation?

17.6 In what part of the eukaryote cell does transcription occur?

17.7 Where does most of the translation occur in a eukaryote cell?

Transcription

17.8 What is the function of RNA polymerase?

17.9 What is the role of helicase in transcription?

17.10 Where is an initiation signal located?

17.11 Which end of the DNA contains the termination signal?

17.12 What would happen to the transcription process if
▶ a drug added to a eukaryote cell inhibits the phosphatase?

17.13 Where is the methyl group located in the guanine cap?

17.14 How are the adenine nucleotides linked together in
▶ the polyA tails?

The Role of RNA in Translation

17.15 Where are the codons located?

17.16 What are the two most important sites on tRNA molecules?

The Genetic Code

17.17 (a) If a codon is GCU, what is the anticodon?
(b) For what amino acid does this codon code?

17.18 If a segment of DNA is 981 units long, how many amino acids appear in the protein for which this DNA segment codes? (Assume that the entire segment is used to code for the protein and that there is no methionine at the N-terminal end of the protein.)

17.19 In what sense does the universality of the genetic code support the theory of evolution?

Translation and Protein Synthesis

17.20 To which end of the tRNA is the amino acid bonded? Where does the energy come from to form the tRNA–amino acid bond?

17.21 There are three sites on the ribosome, each partici-
▶ pating in the translation. Identify them and describe what is happening at each site.

17.22 What is the main role of (a) the 40S ribosome and (b) the 60S ribosome?

17.23 What is the function of elongation proteins?

Gene Regulation

17.24 Which molecules are involved in gene regulation at the transcriptional level?

17.25 Where are enhancers located? How do they work?

17.26 Where are the sieving portions of AARS enzymes located? What is their function?

17.27 What is the function of proteosomes in quality control?

17.28 What kind of interactions exist between metal-binding fingers and DNA?

Mutations

17.29 Using Table 17.1, give an example of a mutation that
▶ (a) does not change anything in a protein molecule and (b) might cause fatal changes in a protein.

17.30 How do cells repair mutations caused by X rays?
▶

17.31 Can a harmful mutation-causing genetic disease exist from generation to generation without exhibiting the symptoms of the disease? Explain.

17.32 Are all mutagens also carcinogens?

Recombinant DNA

17.33 How do restriction endonucleases operate?

17.34 What are sticky ends?

17.35 A new genetically engineered corn has been approved
▶ by the Food and Drug Administration. This new corn shows increased resistance to a destructive insect called a corn borer. What is the difference, in principle, between this genetically engineered corn and one that developed insect resistance by mutation (natural selection)?

17.36 EcoRI restriction endonuclease recognizes the se-
▶ quence GAATTC and cuts it between G and A. What will be the sticky ends of the following double-helical sequence when EcoRI acts on it?

C A A A G A A T T C G
G T T T C T T A A G C

Chemical Connections

17.37 (Chemical Connections 17A) What is unique in the backbone of the antisense drug Vitravene?

17.38 (Chemical Connections 17B) What is a viral "coat"? Where do the ingredients—amino acids, enzymes, and so forth—necessary to synthesize the coat come from?

17.39 (Chemical Connections 17C) What are the target cells for the AIDS virus?

17.40 (Chemical Connections 17C) What HIV-1 viral enzymes are inhibited by (a) AZT and (b) indinavir?

17.41 (Chemical Connections 17D) What was the role of the α-A-crystallin promoter in the development of protease-inhibitor drugs as a treatment for AIDS?

17.42 (Chemical Connections 17E) What is an invariant site?

17.43 (Chemical Connections 17F) What kind of mutation transforms the EGF proto-oncogene into an oncogene in human bladder carcinoma?

17.44 (Chemical Connections 17G) What is p53? Why is its mutated form associated with cancer?

17.45 (Chemical Connections 17G) How does p53 promote DNA repair?

Additional Problems

17.46 In both the transcription and the translation steps of protein synthesis, a number of different molecules come together to act as a factor unit. What are these units of (a) transcription and (b) translation?

17.47 In the tRNA structure, there are stretches where complementary base pairing is necessary and other areas where it is absent. Describe two functionally critical areas (a) where base pairing is mandatory and (b) where it is absent.

17.48 Is there any way to prevent a hereditary disease? Explain.

17.49 How does the cell make sure that a specific amino acid (say, valine) attaches itself only to the one tRNA molecule that is specific for valine?

17.50 (a) What is a plasmid?

(b) How does it differ from a gene?

17.51 Why do we call the genetic code *degenerate?*

17.52 Glycine, alanine, and valine are classified as nonpolar amino acids. Compare their codons. What similarities do you find? What differences do you find?

17.53 Looking at the multiplicity (degeneracy) of the genetic code, you may get the impression that the third base of the codon is irrelevant. Point out where this is not the case. Out of the 16 possible combinations of the first and second bases, in how many cases is the third base irrelevant?

17.54 What polypeptide is coded for by the mRNA sequence 5′-GCU-GAA-GUC-GAG-GUG-UGG-3′?

17.55 A new endonuclease is found. It cleaves double-stranded DNA at every location where C and G are paired on opposite strands. Could this enzyme be used in producing human insulin by the recombinant DNA technique? Explain.

InfoTrac College Edition

For additional readings, go to InfoTrac College Edition, your online research library, at

http://infotrac.thomsonlearning.com

CHAPTER 18

18.1 Introduction

18.2 Cells and Mitochondria

18.3 The Principal Compounds of the Common Catabolic Pathway

18.4 The Citric Acid Cycle

18.5 Electron and H^+ Transport

18.6 Phosphorylation and the Chemiosmotic Pump

18.7 The Energy Yield

18.8 Conversion of Chemical Energy to Other Forms of Energy

Wailua Falls, Hawaii, is a natural demonstration of two pathways ending in a common pool.

Bioenergetics: How the Body Converts Food to Energy

18.1 Introduction

The same compounds may be synthesized in one part of a cell and broken down in a different part of the cell.

Living cells are in a dynamic state, which means that compounds are constantly being synthesized and then broken down into smaller fragments. Thousands of different reactions take place at the same time. **Metabolism** is the sum total of all the chemical reactions involved in maintaining the dynamic state of the cell.

In general, we can divide metabolic reactions into two broad groups: (1) those in which molecules are broken down to provide the energy needed by cells and (2) those that synthesize the compounds needed by cells—both simple and complex. **Catabolism** is the process of breaking down molecules to supply energy. The process of synthesizing (building up) molecules is **anabolism.**

In spite of the large number of chemical reactions, a mere handful dominate cell metabolism. In this chapter and Chapter 19, we focus our attention on the catabolic pathways that yield energy. A **biochemical pathway** is a series of consecutive biochemical reactions. We will see the actual reactions that enable the chemical energy stored in our food to be converted to

409

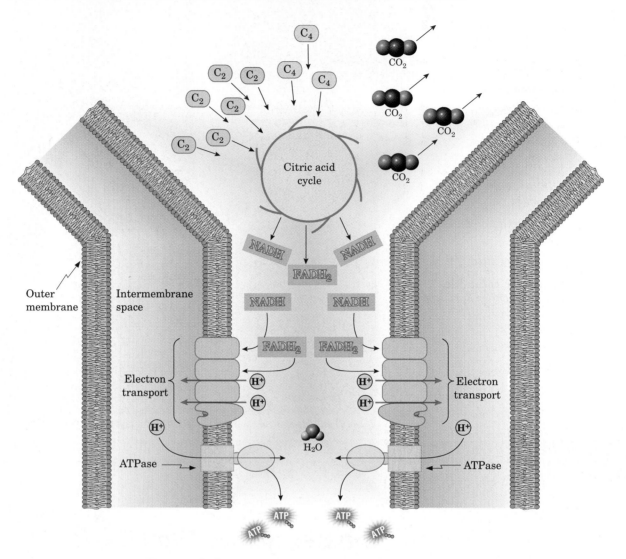

Figure 18.1 In this simplified schematic diagram of the common catabolic pathway, an imaginary funnel represents what happens in the cell. The diverse catabolic pathways drop their products into the funnel of the common catabolic pathway, mostly in the form of C_2 fragments (Section 18.4). (The source of the C_4 fragments will be shown in Section 19.9.) The spinning wheel of the citric acid cycle breaks these molecules down further. The carbon atoms are released in the form of CO_2, and the hydrogen atoms and electrons are picked up by special compounds such as NAD^+ and FAD. Then the reduced NADH and $FADH_2$ cascade down into the stem of the funnel, where the electrons are transported inside the walls of the stem and the H^+ ions are expelled to the outside. In their drive to get back, the H^+ ions form the energy carrier ATP. Once back inside, they combine with the oxygen that picked up the electrons and produce water.

Common catabolic pathway
A series of chemical reactions through which most degraded food goes to yield energy in the form of ATP; the common catabolic pathway consists of (1) the citric acid cycle (Section 18.4) and (2) oxidative phosphorylation (Sections 18.5 and 18.6)

the energy we use every minute of our lives—to think, to breathe, and to use our muscles to walk, write, eat, and everything else. In Chapter 20, we will look at some synthetic (anabolic) pathways.

The food we eat consists of many types of compounds, largely the ones we discussed in earlier chapters: carbohydrates, lipids, and proteins. All of them can serve as fuel, and we derive our energy from them. To convert those compounds to energy, the body uses a different pathway for each type of compound. *All these diverse pathways converge to one* **common catabolic pathway,** which is illustrated in Figure 18.1. In the figure, the diverse pathways are shown as different food streams. The small C_2 and C_4 molecules produced from the original large molecules in food drop into an imaginary collecting funnel that represents the common catabolic pathway. At the end of the funnel appears the energy carrier molecule adenosine triphosphate (ATP).

The purpose of catabolic pathways is to convert the chemical energy in foods to molecules of ATP. In the process, foods also yield metabolic intermediates, which the body can use for synthesis. In this chapter, we deal with the common catabolic pathway. In Chapter 19, we discuss the ways in which the different types of food (carbohydrates, lipids, and proteins) feed molecules into the common catabolic pathway.

18.2 Cells and Mitochondria

A typical animal cell has many components, as shown in Figure 18.2. Each serves a different function. For example, the replication of DNA (Section 16.6) takes place in the **nucleus; lysosomes** remove damaged cellular components and some unwanted foreign materials; and **Golgi bodies** package and process proteins for secretion and delivery to other cellular compartments. The specialized structures within cells are called **organelles.**

An animal cell

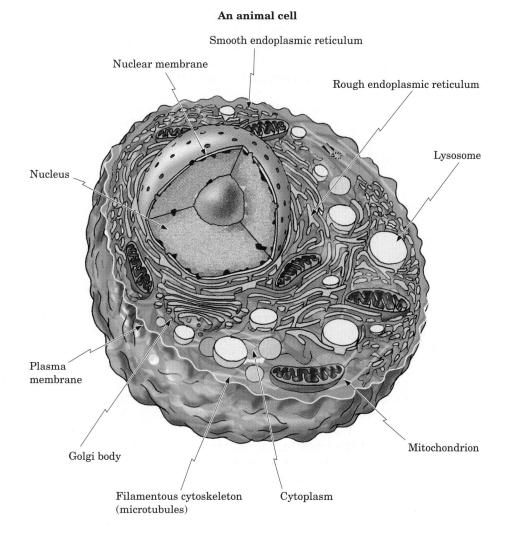

Figure 18.2 Diagram of a rat liver cell, a typical higher animal cell.

Smooth endoplasmic reticulum

Nuclear membrane

Rough endoplasmic reticulum

Lysosome

Nucleus

Plasma membrane

Golgi body

Filamentous cytoskeleton (microtubules)

Cytoplasm

Mitochondrion

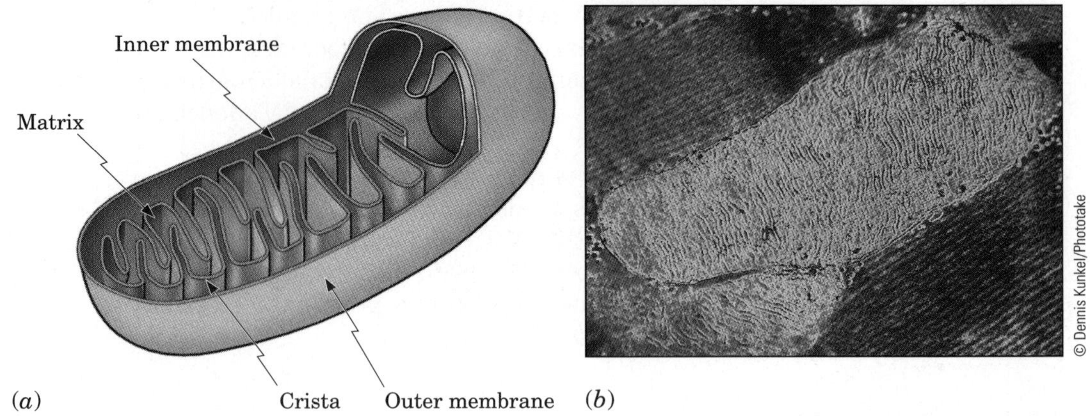

Inner membrane

Matrix

(a) Crista Outer membrane (b)

Figure 18.3 (*a*) Schematic of a mitochondrion cut to reveal the internal organization. (*b*) Colorized transmission electron micrograph of a mitochondrion in a heart muscle cell.

Mitochondria is the plural form; *mitochondrion* is the singular form.

The **mitochondria,** which possess two membranes (Figure 18.3), are the organelles in which the common catabolic pathway takes place in higher organisms. The enzymes that catalyze the common pathway are all located in these organelles. Because these enzymes are synthesized in the cytosol they must be imported through the two membranes. They cross the outer membrane through *t*ranslocator *o*uter *m*embrane (TOM) channels and are accepted in the intermembrane space by chaperone-like *t*ranslocator *i*nner *m*embrane (TIM) complexes, which also insert them into the inner membrane.

Because the enzymes are located inside the inner membrane of mitochondria, the starting materials of the reactions in the common pathway must pass through the two membranes to enter the mitochondria. Products must leave the same way.

The inner membrane of a mitochondrion is quite resistant to the penetration of any ions and of most uncharged molecules. However, ions and molecules can still get through the membrane—they are transported across it by the numerous protein molecules embedded in it (Figure 12.2). The outer membrane, on the other hand, is quite permeable to small molecules and ions and does not have transporting membrane proteins.

The **matrix** is the inner nonmembranous portion of a mitochondrion (Figure 18.3). The inner membrane is highly corrugated and folded. On the basis of electron microscopic studies, Paladin proposed his baffle model of the mitochondrion in 1952. The baffles, which are called **cristae,** project into the matrix like the bellows of an accordion. The enzymes of the oxidative phosphorylation cycle are localized on the cristae. The space between the inner and outer membranes is the **intermembrane space.**

This classical baffle model of mitochondria underwent some changes in the late 1990s after three-dimensional pictures were obtained by the new technique electron-microscope tomography. The 3-D images indicate that the cristae have narrow tubular connections to the inner membrane. These tubular connections may control the diffusion of metabolites from the inside to the intermembrane space. Furthermore, the space between the outer and the inner membrane varies during metabolism, possibly controlling the rate of reactions.

The enzymes of the citric acid cycle are located in the matrix. We will soon see in detail how the specific sequence of these enzymes causes the chain of events in the common catabolic pathway. Furthermore, we will discuss the ways in which nutrients and reaction products move into and out of the mitochondria.

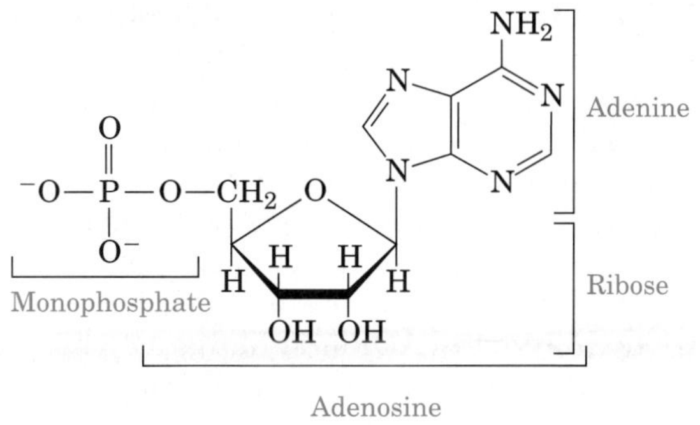

Figure 18.4 Adenosine 5′-monophosphate (AMP).

18.3 The Principal Compounds of the Common Catabolic Pathway

The common catabolic pathway has two parts: the **citric acid cycle** (also called the tricarboxylic acid cycle or the Krebs cycle) and the **electron transport chain** and **phosphorylation,** together called the **oxidative phosphorylation pathway.** To understand what actually happens in these reactions, we must first introduce the principal compounds participating in the common catabolic pathway.

A Agents for Storage of Energy and Transfer of Phosphate Groups

The most important of these agents are three rather complex compounds: **adenosine monophosphate (AMP), adenosine diphosphate (ADP),** and **adenosine triphosphate (ATP)** (Figures 18.4 and 18.5). All three of these molecules contain the heterocyclic amine adenine (Section 16.2) and the sugar D-ribose (Section 11.3A) joined together by a β-N-glycosidic bond, forming adenosine (Section 16.2).

AMP, ADP, and ATP all contain adenosine connected to phosphate groups. The only difference between the three molecules is the number of phosphate groups. As you can see from Figure 18.5, each phosphate is attached to the next by an anhydride bond (Section 10.7A). ATP contains three phosphates—one phosphoric ester and two phosphoric anhydride bonds. In all three molecules, the first phosphate is attached to the ribose by a phosphoric ester bond (Section 10.7B).

A phosphoric anhydride bond, P—O—P, contains more chemical energy (7.3 kcal/mol) than a phosphoric–ester linkage, C—O—P (3.4 kcal/mol). Thus, when ATP and ADP are hydrolyzed to yield phosphate (Figure 18.5), they release more energy per phosphate group than does AMP.

Figure 18.5 Hydrolysis of ATP produces ADP plus dihydrogen phosphate ion plus energy.

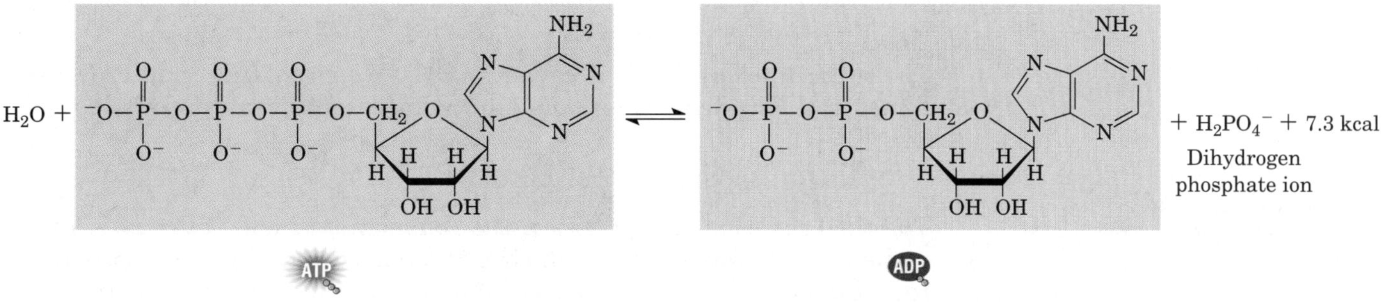

The PO_4^{3-} ion is generally called inorganic phosphate.

When one phosphate group is hydrolyzed from each, the following energy yields are obtained:
ATP = 7.3 kcal/mol
ADP = 7.3 kcal/mol
AMP = 3.4 kcal/mol

The body is able to extract only 40 to 60% of the total caloric content of food.

Conversely, when inorganic phosphate bonds to AMP or ADP, greater amounts of energy are added to the chemical bond than when it bonds to adenosine. ADP and ATP contain *high-energy* phosphoric anhydride bonds.

ATP releases the most energy and AMP releases the least energy when each gives up one phosphate group. This property makes ATP a very useful compound for energy storage and release. The energy gained in the oxidation of food is stored in the form of ATP, albeit only for a short while. ATP molecules in the cells normally do not last longer than about 1 min. They are hydrolyzed to ADP and inorganic phosphate to yield energy that drives other processes, such as muscle contraction, nerve signal conduction, and biosynthesis. As a consequence, ATP is constantly being formed and decomposed. Its turnover rate is very high. Estimates suggest that the human body manufactures and degrades as much as 40 kg (approximately 88 lb) of ATP every day.

B Agents for Transfer of Electrons in Biological Oxidation–Reduction Reactions

The $^+$ in NAD^+ refers to the positive charge on the nitrogen.

Nicotinamide and riboflavin are both members of the vitamin B group (see Table 21.1).

Two other actors in this drama are the coenzymes (Section 14.3), NAD^+ (nicotinamide adenine dinucleotide) and FAD (flavin adenine dinucleotide), both of which contain an ADP core (Figure 18.6). In NAD^+, the operative part of the coenzyme is the nicotinamide part. In FAD, the operative part is the flavin portion. In both molecules, the ADP is the handle by which the apoenzyme holds onto the coenzyme; the other end of the molecule carries out the actual chemical reaction. For example, when NAD^+ is reduced, the nicotinamide part of the molecule gets reduced:

The R stands for the rest of the NAD^+ molecule; R' stands for the rest of the FAD molecule.

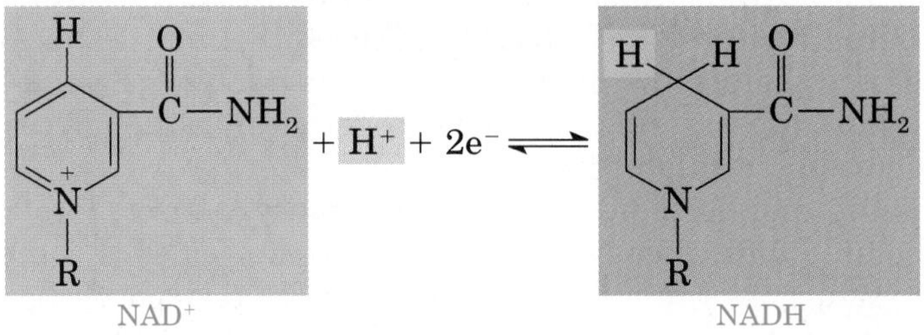

The reduced form of NAD^+ is called NADH. The same reduction happens on two nitrogens of the flavin portion of FAD:

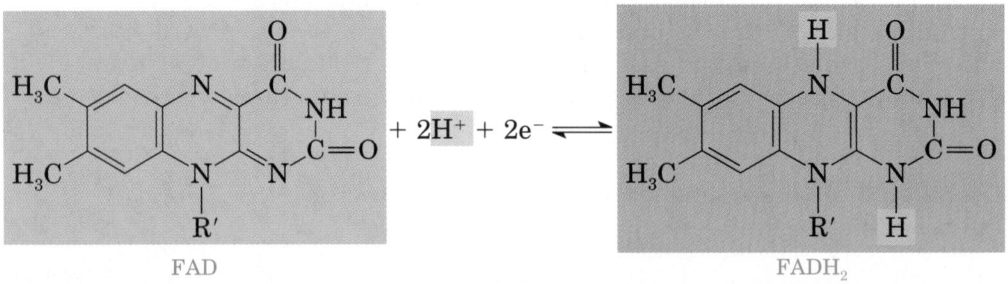

The reduced form of FAD is called $FADH_2$. We view NAD^+ and FAD coenzymes as the **hydrogen ion** and **electron-transporting molecules.**

C Agent for Transfer of Acetyl Groups

Acetyl group The CH_3CO- group

The final principal compound in the common catabolic pathway is **coenzyme A** (CoA; Figure 18.7), which is the **acetyl (CCH_3CO-)-transporting molecule.** Coenzyme A also contains ADP, but here the next structural

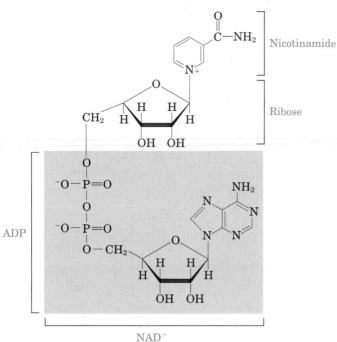

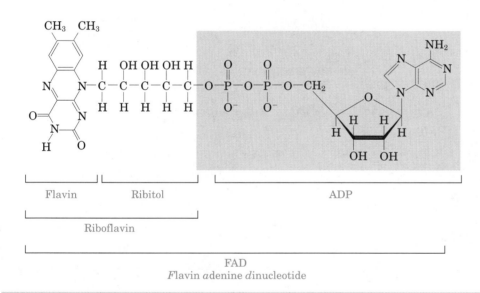

Figure 18.6 The structures of NAD$^+$ and FAD.

Figure 18.7 The structure of coenzyme A.

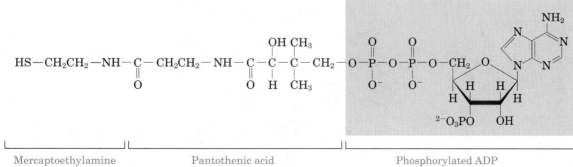

The hydrolysis of acetyl coenzyme A yields 7.51 kcal/mol.

Hans Krebs (1900–1981), Nobel laureate in 1953, established the relationships among the different components of the cycle.

unit is pantothenic acid, another B vitamin. Just as ATP can be viewed as an ADP molecule to which a $-PO_3^{2-}$ group is attached by a high-energy bond, so **acetyl coenzyme A** can be considered a CoA molecule linked to an acetyl group by a high-energy thioester bond (7.51 kcal/mol). The active part of coenzyme A is the mercaptoethylamine. The acetyl group of acetyl coenzyme A is attached to the SH group:

$$CH_3-\overset{\overset{\displaystyle O}{\|}}{C}-S-CoA$$

Acetyl coenzyme A

Figure 18.8 The citric acid (Krebs) cycle. The numbered steps are explained in detail in the text.

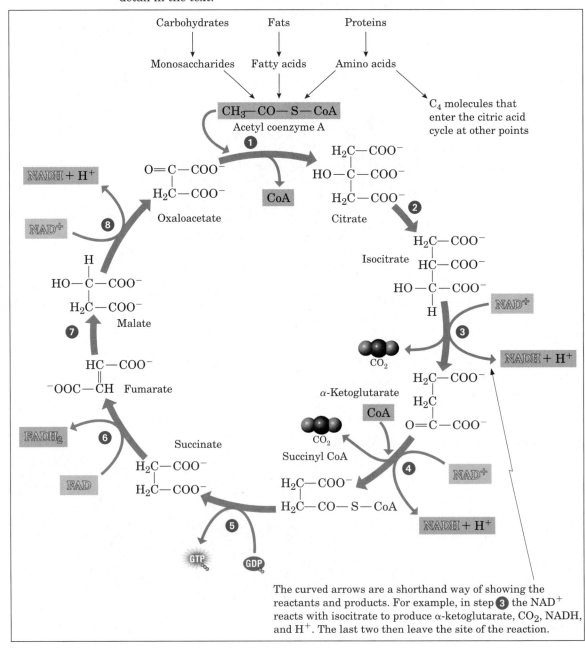

The curved arrows are a shorthand way of showing the reactants and products. For example, in step ❸ the NAD^+ reacts with isocitrate to produce α-ketoglutarate, CO_2, NADH, and H^+. The last two then leave the site of the reaction.

18.4 The Citric Acid Cycle

The common catabolism of carbohydrates and lipids begins when they have been broken down into pieces of two-carbon atoms each. The two-carbon fragments are the acetyl portions of acetyl coenzyme A. The acetyl is now fragmented further in the citric acid cycle.

Figure 18.8 gives the details of the citric acid cycle. A good way to gain an insight into this cycle is to use Figure 18.8 in conjunction with the simplified schematic diagram shown in Figure 18.9, which shows only the carbon balance.

We will now follow the two carbons of the acetyl group (C_2) through each step in the citric acid cycle. The circled numbers correspond to those in Figure 18.8.

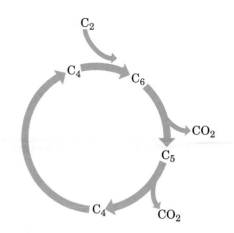

Figure 18.9 A simplified view of the citric acid cycle, showing only the carbon balance.

Step 1 Acetyl coenzyme A enters the cycle by combining with a C_4 compound called oxaloacetate:

$$O=C-COO^-$$
$$H_2C-COO^-$$
Oxaloacetate

$$+ \quad CH_3-\overset{\overset{\displaystyle O}{\|}}{C}-S-CoA$$
Acetyl CoA

$$\xrightarrow{\text{Citrate synthase}}$$

$$H_2C-\overset{\overset{\displaystyle O}{\|}}{C}-S-CoA$$
$$HO-C-COO^-$$
$$H_2C-COO^-$$
Citryl CoA

$$\xrightarrow{H_2O}$$

$$H_2C-COO^-$$
$$HO-C-COO^-$$
$$H_2C-COO^- \quad + \quad HS-CoA$$
Citrate

The first thing that happens is the addition of the $-CH_3$ group of the acetyl CoA to the C=O of the oxaloacetate, catalyzed by the enzyme citrate synthase. This event is followed by hydrolysis of the thioester to produce the C_6 compound citrate ion and CoA. Therefore, Step ① is a building-up rather than a breaking-down process.

In Step ⑧ we will see where the oxaloacetate comes from.

Step 2 The citrate ion is dehydrated to *cis*-aconitate, after which the *cis*-aconitate is hydrated, but this time to isocitrate instead of citrate:

$$H_2C-COO^-$$
$$HO-C-COO^-$$
$$H_2C-COO^-$$
Citrate

$$\xrightarrow{-H_2O}$$

$$H_2C-COO^-$$
$$HC-COO^-$$
$$C-COO^-$$
$$H$$
cis-Aconitate

$$\xrightarrow[\text{aconitase}]{H_2O}$$

$$H_2C-COO^-$$
$$HC-COO^-$$
$$HO-C-COO^-$$
$$H$$
Isocitrate

See the **Interactive General, Organic, and Biochemistry CD-ROM, version 2.0,** for further exploration on this topic.

In citrate, the alcohol is a tertiary alcohol. We learned in Section 5.3C that tertiary alcohols cannot be oxidized. The alcohol in the isocitrate is a secondary alcohol, which upon oxidation yields a ketone.

Step 3 The isocitrate is now oxidized and **decarboxylated** at the same time:

Decarboxylation The process that leads to the loss of CO_2 from a —COOH group

Oxidation:

$$H_2C-COO^-$$
$$HC-COO^-$$
$$HO-C-COO^-$$
$$H$$
Isocitrate

$$+ \quad \textbf{NAD}^+ \quad \xrightarrow{\text{Isocitrate dehydrogenase}}$$

$$H_2C-COO^-$$
$$HC-COO^-$$
$$O=C-COO^-$$
Oxalosuccinate

$$+ \quad \textbf{NADH} + H^+$$

Decarboxylation:

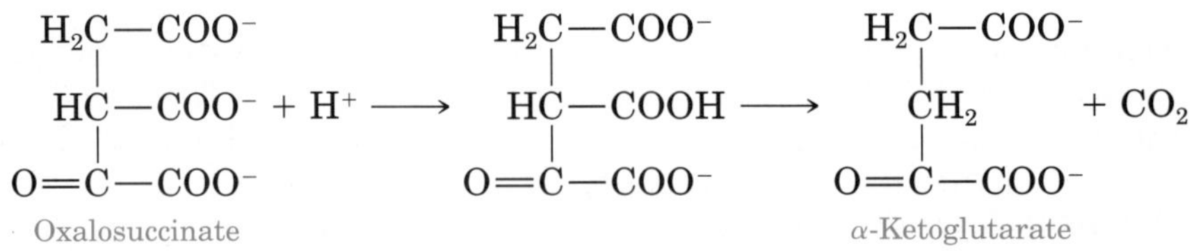

$$\underset{\text{Oxalosuccinate}}{\begin{array}{c} H_2C-COO^- \\ | \\ HC-COO^- \\ | \\ O=C-COO^- \end{array}} + H^+ \longrightarrow \begin{array}{c} H_2C-COO^- \\ | \\ HC-COOH \\ | \\ O=C-COO^- \end{array} \longrightarrow \underset{\alpha\text{-Ketoglutarate}}{\begin{array}{c} H_2C-COO^- \\ | \\ CH_2 \\ | \\ O=C-COO^- \end{array}} + CO_2$$

In oxidizing the secondary alcohol to a ketone, the oxidizing agent NAD^+ removes two hydrogens. One of them is used to reduce NAD^+ to NADH. The other replaces the COO^- that goes into making CO_2. Note that the CO_2 given off comes from the original oxaloacetate and not from the two carbons of the acetyl CoA. Both of these carbons are still present in the α-ketoglutarate. Also note that we are now down to a C_5 compound, α-ketoglutarate.

The CO_2 molecules given off in Steps ③ and ④ are the ones we exhale.

The P_i is inorganic phosphate.

Steps 4 and 5 Next, a complex system removes another CO_2, once again from the original oxaloacetate portion rather than from the acetyl CoA portion:

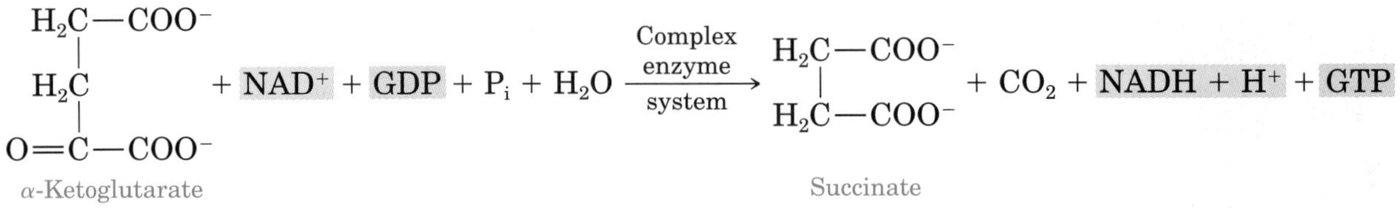

$$\underset{\alpha\text{-Ketoglutarate}}{\begin{array}{c} H_2C-COO^- \\ | \\ H_2C \\ | \\ O=C-COO^- \end{array}} + NAD^+ + GDP + P_i + H_2O \xrightarrow[\text{system}]{\underset{\text{enzyme}}{\text{Complex}}} \underset{\text{Succinate}}{\begin{array}{c} H_2C-COO^- \\ | \\ H_2C-COO^- \end{array}} + CO_2 + NADH + H^+ + GTP$$

We are now down to a C_4 compound, succinate. This oxidative decarboxylation is more complex than the first. It occurs in many steps and requires a number of cofactors. For our purpose, it is sufficient to know that, during this second oxidative decarboxylation, a high-energy compound called **guanosine triphosphate (GTP)** is also formed.

GTP is similar to ATP, except that guanine replaces adenine. Otherwise, the linkages of the base to ribose and the phosphates are exactly the same as in ATP. The function of GTP is also similar to that of ATP—namely, it stores energy in the form of high-energy phosphoric anhydride bonds (chemical energy). The energy from the hydrolysis of GTP drives many important biochemical reactions—for example, the signal transduction in neurotransmission (Section 15.5).

Step 6 In this step, the succinate is oxidized by FAD, which removes two hydrogens to give fumarate (the double bond in this molecule is trans):

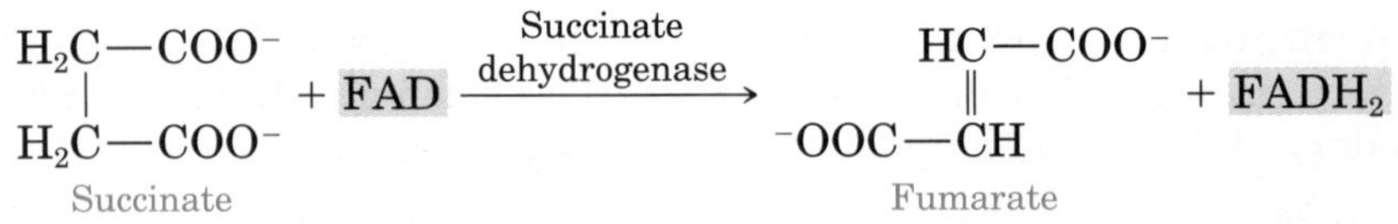

$$\underset{\text{Succinate}}{\begin{array}{c} H_2C-COO^- \\ | \\ H_2C-COO^- \end{array}} + FAD \xrightarrow[\text{dehydrogenase}]{\text{Succinate}} \underset{\text{Fumarate}}{\begin{array}{c} HC-COO^- \\ \| \\ {}^-OOC-CH \end{array}} + FADH_2$$

This reaction cannot be carried out in the laboratory, but with the aid of an enzyme catalyst, the body does it easily.

Step 7 The fumarate is now hydrated to give the malate ion:

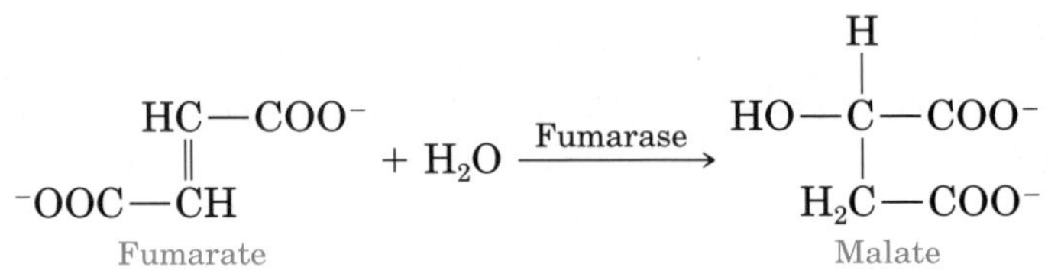

$$\underset{\text{Fumarate}}{\begin{array}{c} HC-COO^- \\ \| \\ {}^-OOC-CH \end{array}} + H_2O \xrightarrow{\text{Fumarase}} \underset{\text{Malate}}{\begin{array}{c} H \\ | \\ HO-C-COO^- \\ | \\ H_2C-COO^- \end{array}}$$

Step 8 In the final step of the cycle, malate is oxidized by NAD^+ to give oxaloacetate:

$$\underset{\text{Malate}}{\overset{\displaystyle H}{\underset{\displaystyle H_2C-COO^-}{HO-\overset{|}{\underset{|}{C}}-COO^-}}} + \boxed{NAD^+} \xrightarrow[\text{dehydrogenase}]{\text{Malate}} \underset{\text{Oxaloacetate}}{\overset{\displaystyle O=C-COO^-}{\underset{\displaystyle H_2C-COO^-}{|}}} + \boxed{NADH + H^+}$$

Thus, the final product of the Krebs cycle is oxaloacetate, which is the compound with which we started in Step ①.

What has happened in the entire process is that the original two acetyl carbons of acetyl CoA were added to the C_4 oxaloacetate to produce a C_6 unit, which then lost two carbons in the form of CO_2 to produce, at the end of the process, the C_4 unit oxaloacetate. The net effect is that one two-carbon acetyl group enters the cycle and two carbon dioxides are given off.

How does the citric acid cycle produce energy? We have already learned that one step in the process produces a high-energy molecule of GTP. However, most of the energy is produced in the other steps that convert NAD^+ to NADH and FAD to $FADH_2$. These reduced coenzymes carry the H^+ and electrons that eventually will provide the energy for the synthesis of ATP (discussed in detail in Sections 18.5 and 18.6).

This stepwise degradation and oxidation of acetate in the citric acid cycle results in the most efficient extraction of energy. Rather than in one burst, the energy is released in small packets carried away step by step in the form of NADH and $FADH_2$.

The cyclic nature of this acetate degradation has other advantages besides maximizing energy yield:

1. The citric acid cycle components provide raw materials for amino acid synthesis as the need arises (Chapter 20).

2. The many-component cycle provides an excellent method for regulating the speed of catabolic reactions.

For example, α-ketoglutaric acid is used to synthesize glutamic acid.

The regulation can occur at many different parts of the cycle, so that feedback information can be used at many points to speed up or slow down the process as necessary.

The following equation represents the overall reactions in the citric acid cycle:

$$GDP + P_i + \boxed{CH_3-CO-S-CoA} + 2H_2O + \boxed{3NAD^+ + FAD}$$
$$\longrightarrow CoA + GTP + 2CO_2 + \boxed{3NADH + FADH_2 + 3H^+} \quad \text{(Eq. 18.1)}$$

The citric acid cycle is controlled by feedback mechanisms. When the essential product of this cycle, $NADH + H^+$, and the end product of the common catabolic pathway, ATP, accumulate, they inhibit some of the enzymes in the cycle. Citrate synthase (Step ①), isocitrate dehydrogenase (Step ③), and α-ketoglutarate dehydrogenase (part of the complex enzyme system in Step ④) are inhibited by ATP and/or by $NADH + H^+$. This inhibition slows down or shuts off the cycle. On the other hand, when the feed material, acetyl CoA, is in abundance, the cycle accelerates. The enzyme isocitrate dehydrogenase (Step ③) is stimulated by ADP and NAD^+, which are the essential reactants from which the end products of the cycle are derived.

18.5 Electron and H⁺ Transport

The reduced coenzymes NADH and FADH$_2$ are end products of the citric acid cycle. They carry hydrogen ions and electrons and, therefore, have the potential to yield energy when these combine with oxygen to form water:

The oxygen in this reaction is the oxygen we breathe.

$$4H^+ + 4e^- + O_2 \longrightarrow 2H_2O + \text{energy}$$

This simple exothermic reaction is carried out in many steps.

A number of enzymes are involved in this reaction, all of which are embedded in the inner membrane of the mitochondria. These enzymes are situated in a particular *sequence* in the membrane so that the product from one enzyme can be passed on to the next enzyme, in a kind of assembly line. The enzymes are arranged in order of increasing affinity for electrons, so electrons flow through the enzyme system (Figure 18.10).

The sequence of the electron-carrying enzyme systems starts with complex I. The largest complex, it contains some 40 subunits, among them a flavoprotein and several FeS clusters. **Coenzyme Q** (CoQ; also called ubiquinone) is associated with complex I, which oxidizes the NADH produced in the citric acid cycle and reduces the CoQ:

$$NADH + H^+ + CoQ \longrightarrow NAD^+ + CoQH_2$$

Some of the energy released in this reaction is used to move 2H⁺ across the membrane, from the matrix to the intermembrane space. The CoQ is soluble in lipids and can move laterally within the membrane.

See the **Interactive General, Organic, and Biochemistry CD-ROM, version 2.0,** for further exploration on this topic.

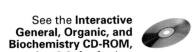

Figure 18.10 Schematic diagram of the electron and H⁺ transport chain and subsequent phosphorylation. The combined processes are also known as oxidative phosphorylation.

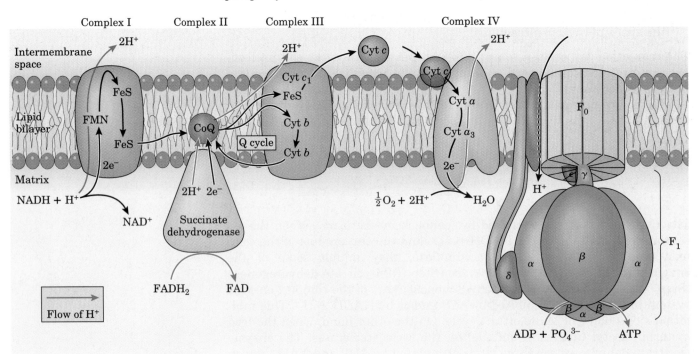

The Importance of Iron–Sulfur Cluster Assembly

As we have seen, iron–sulfur clusters play an important role in the electron transport system. We have also learned that metals play a critical role in redox reactions. A less appreciated fact is that sulfur can play an even more active role.

Sulfur has many oxidation states and thus can attract or donate electrons quickly with ease. The careful packaging of the resulting sulfur–iron clusters in surrounding proteins, as in complexes I and II, is necessary because the naked iron and sulfur ions are even more toxic to a cell than is cyanide. A large number of proteins, enzymes, and chaperones facilitate these assemblies. The iron–sulfur clusters themselves are formed inside

the mitochondrion. For their role inside the mitochondrion, they are inserted into protein molecules. To perform their role on the inner membrane of mitochondrion, they first must be transported through the membrane by special iron–sulfur cluster transporters. Only then can they be packaged into complexes of the electron transport chain.

In an inherited disease called sideroblastic anemia, the iron–sulfur cluster transporter is inactive in cells that should eventually become red blood cells (RBC). As a consequence, iron accumulates inside the mitochondrion of these RBC precursor cells, causing anemia.

Complex II also catalyzes the transfer of electrons to CoQ. The source of the electrons is the oxidation of succinate in the citric acid cycle, producing $FADH_2$. The final reaction is

$$FADH_2 + CoQ \longrightarrow FAD + CoQH_2$$

The energy of this reaction is not sufficient to pump two protons across the membrane, nor is there an appropriate channel for such a transfer.

Complex III delivers the electrons from $CoQH_2$ to **cytochrome c.** This integral membrane complex contains 11 subunits, including cytochrome b, cytochrome c_1, and FeS clusters. Each cytochrome has an iron-ion-containing heme center (Section 13.9) embedded in its own protein. Complex III has two channels through which two H^+ ions are pumped from the $CoQH_2$ into the intermembrane space. The process is very complicated. To simplify matters, we can imagine that it occurs in two distinct steps, as does the electron transfer. Since each cytochrome c can pick up only one electron, two cytochrome c's are needed:

CoQH₂ + 2 cytochrome c (reduced)

$$\longrightarrow CoQ + 2H^+ + 2 \text{ cytochrome c (oxidized)}$$

Cytochrome c is also a mobile carrier of electrons—it can move laterally in the intermembrane space.

Complex IV, known as cytochrome oxidase, contains 13 subunits—most importantly, cytochrome a_3, a heme that has an associated copper center. Complex IV is an integral membrane protein complex. The electron movement flows from cytochrome c to cytochrome a to cytochrome a_3. There, the electrons are transferred to the oxygen molecule, and the O—O bond is cleaved. The oxidized form of the enzyme takes up two H^+ ions from the matrix for each oxygen atom. The water molecule formed in this way is released into the matrix:

$$\tfrac{1}{2}O_2 + 2H^+ + 2e^- \longrightarrow H_2O$$

The letters used to designate the cytochromes were given in order of their discovery.

CHEMICAL CONNECTIONS 18B

2,4-Dinitrophenol as an Uncoupling Agent

Polynitrated organic compounds are highly explosive (see Chemical Connections 4D). Trinitrotoluene (TNT) is the best known of these agents. During World War I, many ammunition workers were exposed to 2,4-dinitrophenol (DNP), a compound used to prepare the explosive picric acid. After it was observed that these workers lost weight, DNP was used as a weight-reducing drug during the 1920s. Unfortunately, DNP eliminated not only the fat but sometimes also the patient, and its use as a diet pill was discontinued after 1929.

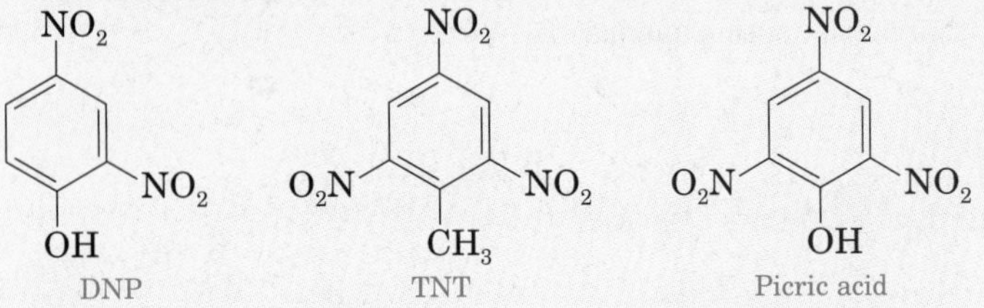

DNP   TNT   Picric acid

Today we know why DNP works as a weight-reducing drug. It is an effective protonophore, which is a compound that transports H$^+$ ions through cell membranes passively, without the expenditure of energy. As noted earlier, H$^+$ ions accumulate in the intermembrane space of mitochondria and, under normal condi-

tions, drive the synthesis of ATP while they are going back to the inside. This is Mitchell's chemiosmotic principle in action. When DNP is ingested, it transfers the H$^+$ back to the mitochondrion easily, and no ATP is manufactured. The energy of the electron separation is dissipated as heat and is not built in as chemical energy in ATP. The loss of this energy-storing compound makes the utilization of food much less efficient, resulting in weight loss.

The medical literature includes a case history of a woman whose muscles contained mitochondria in which electron transport was not coupled to oxidative phosphorylation. This unfortunate woman could not utilize electron transport to generate ATP. As a consequence, she was severely incapacitated and bedridden. When the uncoupling worsened, the heat generated by the uncontrolled oxidation (no ATP production) was so severe that the patient required continuous cooling.

A similar mechanism provides heat in hibernating bears. The bears have brown fat; its color is derived from the numerous mitochondria in the tissue. The brown fat also contains an uncoupling protein, a protonophore that allows the H$^+$ ions to stream back into the mitochondrial matrix without manufacturing ATP. The heat generated in this manner keeps the animal alive during cold winter days.

During this process, two more H$^+$ ions are pumped out of the matrix and into the intermembrane space. Although the mechanism of pumping out protons from the matrix is not known, the energy driving this process is derived from the energy of water formation. This final pumping into the intermembrane space makes a total of six H$^+$ ions per NADH + H$^+$ and four H$^+$ ions per FADH$_2$ molecules.

18.6 Phosphorylation and the Chemiosmotic Pump

How do the electron and H$^+$ transports produce the chemical energy of ATP? In 1961, Peter Mitchell (1920–1992), an English chemist, proposed the **chemiosmotic theory** to answer this question: The energy in the electron transfer chain creates a proton gradient. A **proton gradient** is a continuous variation in the H$^+$ concentration along a given region. In this case, there is a higher concentration of H$^+$ in the intermembrane space than inside the mitochondrion. The driving force, which is the result of the spontaneous flow of ions from a region of high concentration to a region of low concentration, propels the protons back to the mitochondrion through a complex known as **proton translocating ATPase.** This compound is located on the inner membrane of the mitochondrion (Figure 18.10) and is the active enzyme that catalyzes the conversion of ADP and inorganic phosphate to ATP (the reverse of the reaction shown in Figure 18.5):

Chemiosmotic theory Mitchell proposed that the electron transport is accompanied by an accumulation of protons in the intermembrane space of the mitochondrion, which in turn creates an osmotic pressure; the protons driven back to the mitochondrion under this pressure generate ATP

Mitchell received the Nobel Prize in chemistry in 1978.

$$\text{ADP} + \text{P}_\text{i} \xrightleftharpoons{\text{ATPase}} \text{ATP} + \text{H}_2\text{O}$$

CHEMICAL CONNECTIONS 18C

Protection Against Oxidative Damage

As we have seen, most of the oxygen we breathe in is used in the final step of the common pathway—namely, in the conversion of our food supply to ATP. This reaction occurs in the mitochondria of cells. However, some 10% of the oxygen we consume is used elsewhere, in specialized oxidation reactions. One such reaction is the hydroxylation of compounds (for example, steroids). It is catalyzed by the enzyme cytochrome P-450, which usually resides outside the mitochondria in the endoplasmic reticulum of cells. P-450 catalyzes the following reaction, where RH represents a steroid or other molecule:

$$RH + O_2 + \boxed{NADPH} + H^+$$
$$\longrightarrow ROH + H_2O + \boxed{NADP^+}$$

Our bodies use this reaction to detoxify foreign substances (for example, barbiturates). The action of P-450 is not always beneficial, however. Some of the most powerful carcinogens are converted to their active forms by P-450 (see Chemical Connections 4B).

Another use of oxygen for detoxification is in the fight against infection by bacteria. Specialized white blood cells called leukocytes (Section 23.2B) engulf bacteria and kill them by generating superoxide, O_2^-, a highly reactive form of oxygen. But superoxide and other highly reactive forms of oxygen, such as ·OH radicals and hydrogen peroxide (H_2O_2), not only destroy foreign cells but also damage our own cells. They especially attack unsaturated fatty acids, thereby damaging cell membranes (Section 12.5). Our bodies have their own defenses against these very reactive compounds. One example is glutathione (Chemical Connections 13A), which protects us against oxidative damage. Furthermore, our cells contain two powerful enzymes that decompose these highly reactive oxidizing agents:

$$2O_2^- + 2H^+ \xrightarrow{\text{superoxide dismutase}} H_2O_2 + O_2$$

$$2H_2O_2 \xrightarrow{\text{catalase}} 2H_2O + O_2$$

These enzymes also play an important defensive role in heart attacks and strokes. When oxygen deprivation of the heart muscles or the brain occurs because arteries are blocked, a condition known as ischemia reperfusion may arise. This means that when these tissues suddenly become reoxygenated, superoxide (O_2^-) can form. This molecule, and the ·OH radicals derived from it, can further damage the tissues. Superoxide dismutase and catalase prevent reperfusion injuries.

Furthermore, the presence of superoxide activates uncoupling proteins (Chemical Connections 18B) and thus allows the leakage of H^+ ions across the mitochondrial membrane. These H^+ ions then serve as substrates for superoxide dismutase in forming hydrogen peroxide from superoxide.

One theory that has been advanced to explain why humans and animals show symptoms of old age suggests that these oxidizing agents damage healthy cells over the years. All the enzymes involved in reactions with highly reactive forms of oxygen contain a heavy metal: iron in the case of P-450 and catalase, and copper and/or zinc in the case of superoxide dismutase. If the theory is correct, then these protecting agents keep us alive for many years, but the oxidizing processes do so much damage that we eventually exhibit symptoms of aging and die.

The proton translocating ATPase is a complex "rotor engine" made of 16 different proteins. The F_0 sector, which is embedded in the membrane, contains the **proton channel** (Figure 18.10). The 12 subunits that form this channel rotate every time a proton passes from the cytoplasmic side (intermembrane) to the matrix side of the mitochondrion. This rotation is transmitted to a "rotor" in the F_1 sector. F_1 contains five kinds of polypeptides. The rotor (γ and ϵ subunits) is surrounded by the catalytic unit (made of α and β subunits) that synthesizes the ATP. The F_1 catalytic unit converts the mechanical energy of the rotor into chemical energy of the ATP molecule. The last unit, the "stator," containing the δ subunit, stabilizes the whole complex.

The proton translocating ATPase can catalyze the reaction in both directions. When protons that have accumulated on the outer surface of the mitochondrion stream inward, the enzyme manufactures ATP and stores the electrical energy (due to the flow of charges) in the form of chemical energy. In the reverse reaction, the enzyme hydrolyzes ATP and, as a consequence, pumps out H^+ from the mitochondrion. Each pair of protons that is translocated gives rise to the formation of one ATP molecule. Only when the two parts of the proton translocating ATPase F_1 and F_0 are linked is energy production possible. When the interaction between F_1 and F_0 is disrupted, the energy transduction is lost.

The protons that enter a mitochondrion combine with the electrons transported through the electron transport chain and with oxygen to form water. The net result of the two processes (electron/H^+ transport and ATP formation) is that each oxygen molecule we breathe in combines with four H^+ ions and four electrons to give two water molecules. The four H^+ ions and four electrons come from the NADH and $FADH_2$ molecules produced in the citric acid cycle. The oxygen, therefore, has two functions:

● To oxidize NADH to NAD^+ and $FADH_2$ to FAD so that these molecules can go back and participate in the citric acid cycle

● To provide energy for the conversion of ADP to ATP

The electron and H^+ transport chain and subsequent phosphorylation process are collectively known as oxidative phosphorylation.

The latter function is accomplished indirectly, not through the reduction of O_2 to H_2O. The entrance of the H^+ ions into the mitochondrion drives the ATP formation, but the H^+ ions enter the mitochondrion because the O_2 depleted the H^+ ion concentration when water was formed. This rather complex process involves the transport of electrons along a whole series of enzyme molecules (which catalyze all these reactions); however, the cell cannot utilize the O_2 molecules without it and eventually will die. The following equations represent the overall reactions in oxidative phosphorylation:

$$NADH + 3\ ADP + \tfrac{1}{2}O_2 + 3P_i + H^+ \longrightarrow NAD^+ + 3\ ATP + H_2O \qquad \text{(Eq. 18.2)}$$

$$FADH_2 + 2\ ADP + \tfrac{1}{2}O_2 + 2P_i \longrightarrow FAD + 2\ ATP + H_2O \qquad \text{(Eq. 18.3)}$$

18.7 The Energy Yield

The energy released during electron transport is now finally built into the ATP molecule. Therefore, it is instructive to look at the energy yield in the universal biochemical currency: the number of ATP molecules.

Each pair of protons entering a mitochondrion results in the production of one ATP molecule. For each NADH molecule, three pairs of protons are pumped into the intermembrane space in the electron transport process. Therefore, for each NADH molecule, we get three ATP molecules, as can be seen in Equation 18.2. For each $FADH_2$ molecule, only four protons are pumped out of the mitochondrion. Therefore, only two ATP molecules are produced for each $FADH_2$, as seen in Equation 18.3.

Now we can produce the energy balance for the entire common catabolic pathway (citric acid cycle and oxidative phosphorylation combined). For each C_2 fragment entering the citric acid cycle, we obtain three NADH and one $FADH_2$ (Equation 18.1) plus one GTP, which is equivalent in energy to one ATP. Thus, the total number of ATP molecules produced per C_2 fragment is

$$3\ NADH \times 3\ ATP/NADH = \quad 9\ ATP$$
$$1\ FADH_2 \times 2\ ATP/FADH_2 = \quad 2\ ATP$$
$$1\ GTP = \quad \underline{1\ ATP}$$
$$= 12\ ATP$$

Each C_2 fragment that enters the cycle produces 12 ATP molecules and uses up two O_2 molecules. The total effect of the energy-production chain of reactions discussed in this chapter (the common catabolic pathway) is to oxidize one C_2 fragment with two molecules of O_2 to produce two molecules of CO_2 and 12 molecules of ATP:

$$C_2 + 2O_2 + 12\,ADP + 12P_i \longrightarrow 12\,ATP + 2CO_2$$

The important thing is not the waste product, CO_2, but rather the 12 ATP molecules. These molecules will now release their energy when they are converted to ADP.

18.8 Conversion of Chemical Energy to Other Forms of Energy

As mentioned in Section 18.3, the storage of chemical energy in the form of ATP lasts only a short time. Usually, within a minute, the ATP is hydrolyzed (an exothermic reaction) and releases its chemical energy. How does the body utilize this chemical energy? To answer this question, let us look at the different forms in which energy is needed in the body.

A Conversion to Other Forms of Chemical Energy

The activity of many enzymes is controlled and regulated by phosphorylation. For example, the enzyme phosphorylase, which catalyzes the breakdown of glycogen (Chemical Connections 19B), occurs in an inactive form, phosphorylase b. When ATP transfers a phosphate group to a serine residue, the enzyme becomes active. Thus, the chemical energy of ATP is used in the form of chemical energy to activate phosphorylase b so that glycogen can be utilized. We will see several other examples of this energy conversion in Chapters 19 and 20.

B Electrical Energy

The body maintains a high concentration of K^+ ions inside the cells despite the fact that the K^+ concentration is low outside the cells. The reverse is true for Na^+. So that K^+ does not diffuse out of the cells and Na^+ does not enter them, special transport proteins in the cell membranes constantly pump K^+ into and Na^+ out of the cells. This pumping requires energy, which is supplied by the hydrolysis of ATP to ADP. With this pumping, the charges inside and outside the cell are unequal, which generates an electric potential. Thus, the chemical energy of ATP is transformed into electrical energy, which operates in neurotransmission (Section 15.2).

C Mechanical Energy

ATP is the immediate source of energy in muscle contraction. In essence, muscle contraction takes place when thick and thin filaments slide past each other (Figure 18.11). The thick filament is myosin, an ATPase enzyme (that is, one that hydrolyzes ATP). The thin filament, actin, binds strongly to myosin in the contracted state. However, when ATP binds to myosin, the actin–myosin complex dissociates, and the muscle relaxes. When myosin hydrolyzes ATP, it interacts with actin once more, and a new contraction occurs. In this way, the hydrolysis of ATP drives the alternating association and dissociation of actin and myosin and, consequently, the contraction and relaxation of the muscle.

Figure 18.11 Schematic diagram of muscle contraction.

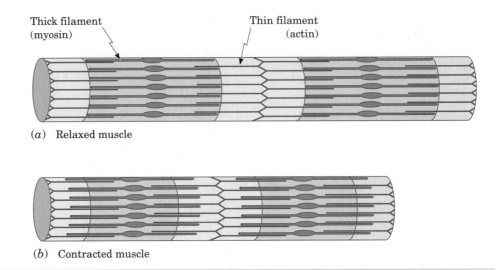

Thick filament (myosin)

Thin filament (actin)

(*a*)   Relaxed muscle

(*b*)   Contracted muscle

D Heat Energy

One molecule of ATP upon hydrolysis to ADP yields 7.3 kcal/mol. Some of this energy is released as heat and used to maintain body temperature. If we estimate that the specific heat of the body is about the same as that of water, a person weighing 60 kg would need to hydrolyze approximately 99 moles (approximately 50 kg) of ATP to raise the temperature of the body from room temperature, 25°C, to 37°C. Not all body heat is derived from ATP hydrolysis; some other exothermic reactions in the body also make heat contributions.

S U M M A R Y

The sum total of all the chemical reactions involved in maintaining the dynamic state of cells is called **metabolism** (Section 18.1). The breaking down of molecules is **catabolism;** the building up of molecules is **anabolism.** The **common metabolic pathway** uses a two-carbon C_2 fragment (acetyl) from different foods. Through the **citric acid cycle** and **oxidative phosphorylation (electron transport chain),** the C_2 fragment is oxidized. The products formed are water and carbon dioxide. The energy from oxidation is built into the high-chemical-energy-storing molecule ATP.

Both the citric acid cycle and oxidative phosphorylation take place in the **mitochondria** (Section 18.2). The enzymes of the citric acid cycle are located in the matrix, whereas the enzymes of the oxidative phosphorylation chain are on the inner mitochondrial membrane. Some of them project into the intermembrane space.

The principal carriers in the common catabolic pathway (Section 18.3) are as follows: **ATP** is the phosphate carrier; **CoA** is the C_2 fragment carrier; and **NAD$^+$** and **FAD** carry the hydrogen ions and electrons. The unit common to all these carriers is **ADP.**

The nonactive end of the carriers acts as a handle that fits into the active sites of the enzymes.

In the citric acid cycle (Section 18.4), the C_2 fragment first combines with a C_4 fragment (oxaloacetate) to yield a C_6 fragment (citrate). An oxidative decarboxylation yields a C_5 fragment. One CO_2 is released, and one NADH + H$^+$ is passed to the electron transport chain. Another oxidative decarboxylation provides a C_4 fragment. Once again, a CO_2 is released, and another NADH + H$^+$ is passed to the electron transport chain. Subsequently, two dehydrogenation (oxidation) steps yield one FADH$_2$ and one additional NADH + H$^+$, along with an analog of ATP called GTP. This cycle is controlled by a feedback mechanism.

The electrons of NADH enter the electron transport chain (Section 18.5) at the complex I stage. The coenzyme Q (CoQ) of this complex picks up the electrons and the H$^+$ and becomes CoQH$_2$. The energy of this reduction reaction is used to expel two H$^+$ ions from the matrix into the intermembrane space. Complex II also has CoQ. Electrons and H$^+$ are passed to this complex, and complex II catalyzes the transfer of electrons from FADH$_2$. However, no H$^+$ ions are pumped into the intermembrane space at this point. The electrons are passed along by $CoQH_2$ to complex III of the electron transport chain. At complex III, the two H^+ ions from the $CoQH_2$ are expelled into the intermembrane space. Cytochrome c of complex III transfers electrons to complex IV through redox reactions. As the electrons are transported from cytochrome c to complex IV, an additional two H^+ ions are expelled from the matrix of the mitochondrion to the intermembrane space. For each NADH, six H^+ ions are expelled. For each $FADH_2$, four H^+ ions are expelled. Finally, the electrons passed to complex IV return to the matrix, where they combine with oxygen and H^+ to form water.

When the expelled H^+ ions stream back into the mitochondrion, they drive a complex enzyme called **proton translocating ATPase,** which makes one ATP molecule for each two H^+ ions that enter the mitochondrion (Section 18.6). Therefore, for each $NADH + H^+$ coming from the citric acid cycle, three ATP molecules are formed. For each $FADH_2$, two ATP molecules are formed. The overall result: For each C_2 fragment that enters the citric acid cycle, 12 ATP molecules are produced (Section 18.7).

The proton translocating ATPase is a complex "rotor engine" (Section 18.6). The proton channel part (F_0) is embedded in the membrane, and the catalytic unit (F_1) converts mechanical energy to chemical energy of the ATP molecule. The chemical energy is stored in ATP only for a short time—ATP is quickly hydrolyzed, usually within a minute. This chemical energy is used to do chemical, mechanical, and electrical work in the body and to maintain body temperature (Section 18.8).

P R O B L E M S

Numbers that appear in color indicate difficult problems.
▶ designates problems requiring application of principles.

Metabolism

18.1   To what end product is the energy of foods converted in the catabolic pathways?

18.2   (a) How many sequences are in the common catabolic pathway?
(b) Name these sequences.

Cells and Mitochondria

18.3   (a) How many membranes do mitochondria have?
(b) Which membrane is permeable to ions and small molecules?

18.4   How do the enzymes of the common pathway find their way into the mitochondria?

18.5   What are cristae, and how are they related to the inner membrane of mitochondria?

18.6   (a) Where are the enzymes of the citric acid cycle located?
(b) Where are the enzymes of oxidative phosphorylation located?

Principal Compounds of the Common Catabolic Pathway

18.7   How many high-energy phosphate bonds are in the ATP molecule?

18.8   What are the products of the following reaction?

$$AMP + H_2O \xrightarrow{H^+}$$

18.9   Which yields more energy, the hydrolysis (a) of ATP to ADP or (b) of ADP to AMP?

18.10   How much ATP is needed for normal daily activity in humans?

18.11   What kind of chemical bond exists between the ribitol and the flavin in FAD?

18.12   When NAD^+ is reduced, two electrons enter the molecule, together with one H^+ ion. Where in the product will the two electrons be located?

18.13   Which atoms in the flavin portion of FAD are reduced to yield $FADH_2$?

18.14 NAD^+ has two ribose units in its structure; FAD has a ribose and ribitol. What is the relationship between these molecules?

18.15   In the common catabolic pathway, a number of important molecules act as carriers (transfer agents).
(a) Which is the carrier of phosphate groups?
(b) Which are the coenzymes transferring hydrogen ions and electrons?
(c) What kind of groups does coenzyme A carry?

18.16   The ribitol in FAD is bound to phosphate. What is nature of this bond? On the basis of the energies of the different bonds in ATP, estimate how much energy (in kcal/mol) would be obtained from the hydrolysis of this bond.

18.17   What kind of chemical bond exists between the pantothenic acid and mercaptoethylamine in the structure of CoA?

18.18   Name the vitamin B molecules that are a part of the structure of (a) NAD^+, (b) FAD, and (c) coenzyme A.

18.19 In both NAD^+ and FAD, the vitamin B portion of the molecule is the active part. Is this also true for CoA?

18.20 What type of compound is formed when coenzyme A reacts with acetate?

18.21 The fats and carbohydrates metabolized by our bodies are eventually converted to a single compound. What is it?

The Citric Acid Cycle

18.22 The first step in the citric acid cycle is abbreviated as

$$C_2 + C_4 = C_6$$

(a) What do these symbols stand for?

(b) What are the common names of the three compounds involved in this reaction?

18.23 What is the only C_5 compound in the citric acid cycle?

18.24 Identify by number those steps of the citric acid cycle that are not redox reactions.

18.25 Which substrate in the citric acid cycle is oxidized by FAD? What is the oxidation product?

18.26 In Steps ③ and ⑤ of the citric acid cycle, the compounds are shortened by one carbon each time. What is the form of this one-carbon compound? What happens to it in the body?

18.27 According to Table 14.1, to what class of enzymes does fumarase belong?

18.28 List all the enzymes or enzyme systems of the citric acid cycle that could be classified as oxidoreductases.

18.29 Is ATP directly produced during any step of the citric acid cycle? Explain.

18.30 There are four dicarboxylic acid compounds, each containing four carbons, in the citric acid cycle. Which is (a) the least oxidized and (b) the most oxidized?

18.31 Why is a many-step cyclic process more efficient in utilizing energy from food than a single-step combustion?

18.32 Did the two CO_2 molecules given off in one turn of the citric acid cycle originate from the entering acetyl group?

18.33 Which intermediates of the citric acid cycle contain $C=C$ double bonds?

18.34 The citric acid cycle can be regulated by the body. It can be slowed down or speeded up. What mechanism controls this process?

18.35 Oxidation is defined as loss of electrons. When oxidative decarboxylation occurs, as in Step ④, where do the electrons of the α-ketoglutarate go?

Oxidative Phosphorylation

18.36 What is the main function of oxidative phosphorylation (the electron transport chain)?

18.37 What are the mobile electron carriers of the oxidative phosphorylation?

18.38 In each complex of the electron transport system, the redox reaction occurs around Fe ions.

(a) Identify the compounds that contain such Fe centers.

(b) Identify the compounds that contain ion centers other than iron.

18.39 What kind of motion is set up in the proton translocating ATPase by the passage of H^+ from the intermembrane space into the matrix?

18.40 The following reaction is a reversible reaction:

$$NADH \rightleftharpoons NAD^+ + H^+ + 2e^-$$

(a) Where does the forward reaction occur in the common catabolic pathway?

(b) Where does the reverse reaction occur?

18.41 In oxidative phosphorylation, water is formed from H^+, e^-, and O_2. Where does this take place?

18.42 At what points in oxidative phosphorylation are the H^+ ions and the electrons separated from each other?

18.43 How many ATP molecules are generated (a) for each H^+ translocated through the ATPase complex and (b) for each C_2 fragment that goes through the complete common catabolic pathway?

18.44 When H^+ is pumped out into the intermembrane space, is the pH there increased, decreased, or unchanged compared with that in the matrix?

The Chemiosmotic Pump

18.45 What is the channel through which H^+ ions reenter the matrix of mitochondria?

18.46 The proton gradient accumulated in the intermembrane area of a mitochondrion drives the ATP manufacturing enzyme, ATPase. Why do you think Mitchell called this the "chemiosmotic theory"?

18.47 Which part of the proton translocating ATPase machinery is the catalytic unit? What chemical reaction does it catalyze?

18.48 When the interaction between the two parts of proton translocating ATPase, F_0 and F_1, are disrupted, no energy production is possible. Which subunits maintain connections between F_0 and F_1, and what names are designated for these subunits?

The Energy Yield

18.49 If each mole of ATP yields 7.3 kcal of energy upon hydrolysis, how many kilocalories of energy would you get from 1 g of CH_3COO^- (C_2) entering the citric acid cycle?

18.50 A hexose (C_6) enters the common metabolic pathway in the form of two C_2 fragments.
   (a) How many molecules of ATP are produced from one hexose molecule?
   (b) How many O_2 molecules are used up in the process?

Conversion of Chemical Energy to Other Forms

18.51 (a) How do muscles contract?
   (b) Where does the energy in muscle contraction come from?

18.52 Give an example of the conversion of the chemical energy of ATP to electrical energy.

18.53 How is the enzyme phosphorylase activated?

Chemical Connections

18.54 (Chemical Connections 18A) How can sulfur act as a redox center?

18.55 (Chemical Connections 18A) Sulfur usually is poisonous to cells. How can Fe-S clusters exist and not kill the cells?

18.56 (Chemical Connections 18B) What is a protonophore?

18.57 (Chemical Connections 18B) Oligomycin is an antibiotic that allows electron transport to continue but stops phosphorylation in both bacteria and humans. Would you use it as an antibacterial drug for people? Explain.

18.58 (Chemical Connections 18C) How does superoxide dismutase prevent damage to the brain during a stroke?

18.59 (Chemical Connections 18C) (a) What kind of reaction is catalyzed by cytochrome P-450?
   (b) Where does the oxygen reactant come from?

Additional Problems

18.60 (a) What is the difference in structure between ATP and GTP?
   (b) Compared with ATP, would you expect GTP to carry more, less, or about the same amount of energy?

18.61 How many grams of CH_3COOH (from acetyl CoA) molecules must be metabolized in the common metabolic pathway to yield 87.6 kcal of energy?

18.62 What is the basic difference in the functional groups between citrate and isocitrate?

18.63 The passage of H^+ ions from the cytoplasmic side into the matrix generates mechanical energy. Where is this energy of motion exhibited first?

18.64 What kind of reaction occurs in the citric acid cycle when a C_6 compound is converted to a C_5 compound?

18.65 What structural characteristics do citric acid and malic acid have in common?

18.66 Two keto acids are important in the citric acid cycle. Identify them, and tell how they are manufactured.

18.67 Which filament of muscles is an enzyme, catalyzing the reaction that converts ATP to ADP?

18.68 One of the end products of food metabolism is water. How many molecules of H_2O are formed from the entry of each molecule of (a) NADH + H^+ and (b) $FADH_2$? (*Hint:* Use Figure 18.10.)

18.69 How many stereocenters are in isocitrate?

18.70 Acetyl CoA is labeled with radioactive carbon as shown: CH_3^*CO—S—CoA. This compound enters the citric acid cycle. If the cycle is allowed to progress to only the α-ketoglutarate level, will the CO_2 expelled by the cell be radioactive?

18.71 Where is the H^+ ion channel located in the proton translocating ATPase complex?

18.72 Is the passage of H^+ ion through the channel converted directly into chemical energy?

18.73 Does all the energy used in ATP synthesis come from the mechanical energy of rotation?

18.74 (a) In the citric acid cycle, how many steps can be classified as decarboxylation reactions?
   (b) In each case, what is the concurrent oxidizing agent? (*Hint:* See Table 14.1.)

18.75 What is the role of succinate dehydrogenase in the citric acid cycle?

18.76 How many stereocenters are in malate?

InfoTrac College Edition

For additional reading, go to InfoTrac College Edition, your online research library, at

http://infotrac.thomsonlearning.com

CHAPTER 19

19.1 Introduction

19.2 Glycolysis

19.3 The Energy Yield from Glucose

19.4 Glycerol Catabolism

19.5 β-Oxidation of Fatty Acids

19.6 The Energy Yield from Stearic Acid

19.7 Ketone Bodies

19.8 Catabolism of the Nitrogen of Amino Acids

19.9 Catabolism of the Carbon Skeleton of Amino Acids

19.10 Catabolism of Heme

Ballet dancer leaping.

© Dennis Degnan/Corbis

Specific Catabolic Pathways: Carbohydrate, Lipid, and Protein Metabolism

19.1 Introduction

The food we eat serves two main purposes: (1) It fulfills our energy needs, and (2) it provides the raw materials to build the compounds our bodies need. Before either of these processes can take place, food—carbohydrates, fats, and proteins—must be broken down into small molecules that can be absorbed through the intestinal walls.

A Carbohydrates

Complex carbohydrates (di- and polysaccharides) in the diet are broken down by enzymes and stomach acid to produce monosaccharides, the most important of which is glucose (Section 21.4). Glucose also comes from the enzymatic breakdown of glycogen that is stored in the liver and muscles until needed. Once monosaccharides are produced, they can be used either to build new oligo- and polysaccharides or to provide energy. The specific pathway by which energy is extracted from monosaccharides is called glycolysis (Sections 19.2 and 19.3).

B Lipids

Ingested fats are broken down by lipases to glycerol and fatty acids or to monoglycerides, which are absorbed through the intestine (Section 21.5). In a similar way, complex lipids are hydrolyzed to smaller units before their absorption. As with carbohydrates, these smaller molecules (fatty acids, glycerol, and so on) can be used to build the complex molecules needed in membranes, or they can be oxidized to provide energy, or they can be stored in **fat storage depots** (Figure 19.1). The stored fats can later be hydrolyzed to glycerol and fatty acids whenever they are needed as fuel.

The specific pathway by which energy is extracted from glycerol involves the same glycolysis pathway as that used for carbohydrates (Section 19.4). The specific pathway used by the cells to obtain energy from fatty acids is called β-oxidation (Section 19.5).

C Proteins

As you might expect from a knowledge of their structures, proteins are broken down by HCl in the stomach and by digestive enzymes in the stomach (pepsin) and intestines (trypsin, chymotrypsin, and carboxypeptidases) to produce their constituent amino acids. The amino acids absorbed through the intestinal wall enter the **amino acid pool.** They serve as building blocks for proteins as needed and, to a smaller extent (especially during starvation), as a fuel for energy. In the latter case, the nitrogen of the amino acids is catabolized through oxidative deamination and the urea cycle and is expelled from the body as urea in the urine (Section 19.8). The carbon skeletons of the amino acids enter the common catabolic pathway (Chapter 18) as either α-ketoacids (pyruvic, oxaloacetic, and α-ketoglutaric acids) or acetyl coenzyme A (Section 19.9).

In all cases, *the specific pathways of carbohydrate, triglyceride (fat), and protein catabolism converge to the common catabolic pathway* (Figure 19.2, next page). In this way, the body needs fewer enzymes to get energy from diverse food materials. Efficiency is achieved because a minimal number of chemical steps are required and because the energy-producing factories of the body are localized in the mitochondria.

19.2 Glycolysis

Glycolysis is the specific pathway by which the body gets energy from monosaccharides. The detailed steps in glycolysis are shown in Figure 19.3, and the most important features are shown schematically in Figure 19.4.

A Glycolysis of glucose

In the first steps of glucose metabolism, energy is consumed rather than released. At the expense of two molecules of ATP (which are converted to ADP), glucose (C_6) is phosphorylated. First, glucose 6-phosphate is formed in Step ①, then, after isomerization to fructose 6-phosphate in Step ②, a

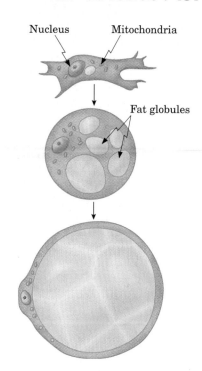

Figure 19.1 Storage of fat in a fat cell. As more and more fat droplets accumulate in the cytoplasm, they coalesce to form a very large globule of fat. Such a fat globule may occupy most of the cell, pushing the cytoplasm and the organelles to the periphery. *(Modified from C. A. Villee, E. P. Solomon, and P.W. Davis,* Biology, *Philadelphia, Saunders College Publishing, 1985.)*

Amino acid pool The free amino acids found both inside and outside cells throughout the body

Glycolysis The biochemical pathway that breaks down glucose to pyruvate, which yields chemical energy in the form of ATP and reduced coenzymes

See the **Interactive General, Organic, and Biochemistry CD-ROM, version 2.0,** for further exploration on this topic.

Figure 19.2 The convergence of the specific pathways of carbohydrate, fat, and protein catabolism into the common catabolic pathway, which is made up of the citric acid cycle and oxidative phosphorylation.

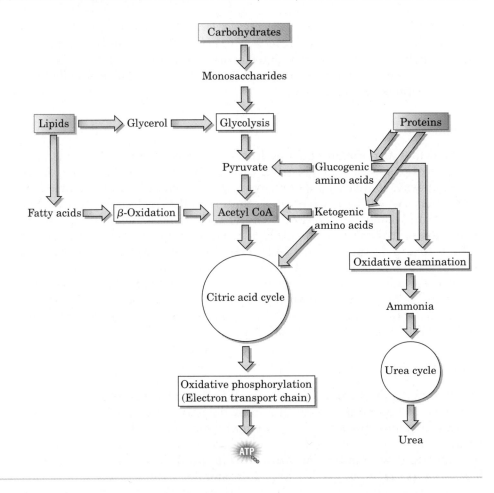

In Step ⑨, after hydrolysis of the phosphate, the resulting enol of pyruvic acid tautomerizes to the more stable keto form (Section 9.8).

second phosphate group is attached to yield fructose 1,6-bisphosphate in Step ③. We can consider these steps to be the activation process.

In the second stage, the C_6 compound, fructose 1,6-bisphosphate, is broken into two C_3 fragments in Step ④. The two C_3 fragments, glyceraldehyde 3-phosphate and dihydroxyacetone phosphate, are in equilibrium (they can be converted to each other). Only glyceraldehyde 3-phosphate is oxidized in glycolysis, but as this species is removed from the equilibrium mixture, the equilibrium shifts and dihydroxyacetone phosphate is converted to glyceraldehyde 3-phosphate.

In the third stage, glyceraldehyde 3-phosphate is oxidized to 1,3-bisphosphoglycerate in Step ⑤. The hydrogen of the aldehyde group is removed by the NAD^+ coenzyme. In Step ⑥, the phosphate from the carboxyl group is transferred to ADP, yielding ATP and 3-phosphoglycerate. The latter compound, after isomerization in Step ⑦ and dehydration in Step ⑧, is converted to phosphoenolpyruvate, which loses its remaining phosphate in Step ⑨ and yields pyruvate and another ATP molecule. This last Step (⑨) is also the "payoff" step, as the two ATP molecules produced here (one for each C_3 fragment) represent the net yield of ATPs in glycolysis. Step ⑨ is catalyzed by an enzyme, pyruvate kinase, whose active site was depicted in Chemical Connections 14C. This enzyme plays a key role in the regulation of glycolysis. For example, pyruvate kinase is inhibited by ATP and activated by AMP. Thus, when plenty of ATP is available, glycolysis is shut down; when ATP is scarce and AMP levels are high, the glycolytic pathway is speeded up.

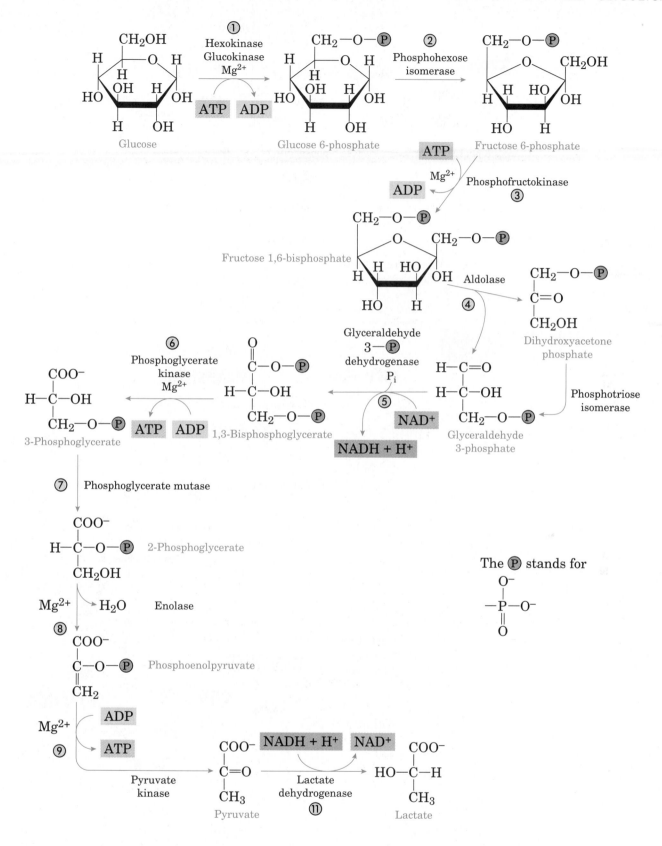

Figure 19.3 Glycolysis, the pathway of glucose metabolism. (Steps ⑩, ⑫, and ⑬ are shown in Figure 19.4.) Some of the steps are reversible, but equilibrium arrows are not shown (they appear in Figure 19.4).

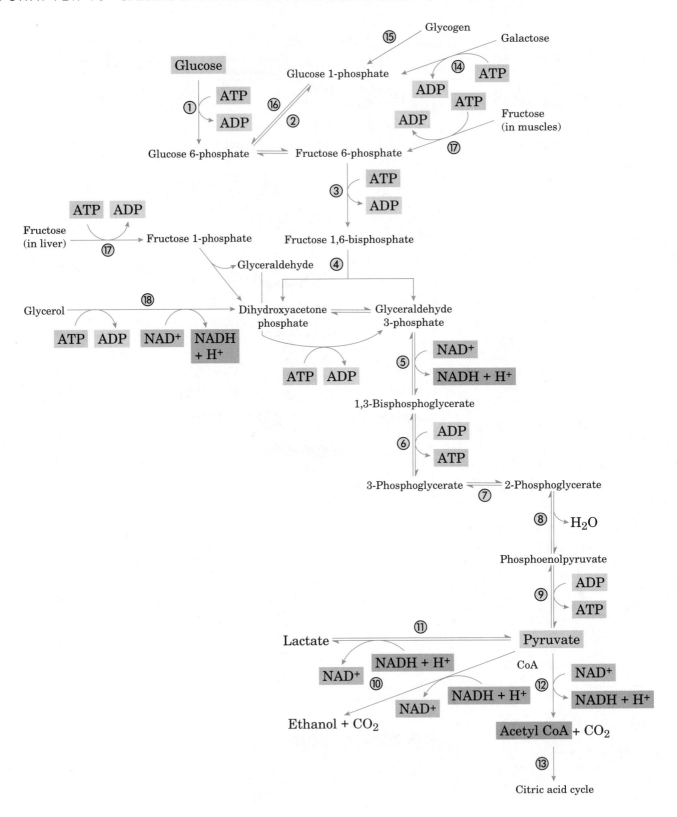

Figure 19.4 An overview of glycolysis and the entries to it and exits from it.

Lactate Accumulation

Many athletes suffer muscle cramps when they engage in strenuous exercise (see Chapter 7 opener). This problem results from a shift from normal glucose catabolism (glycolysis : citric acid cycle : oxidative phosphorylation) to that of lactate production (see Step ⑪ in Figure 19.4). During exercise, oxygen is used up rapidly, which slows down the rate of the common catabolic pathway. The demand for energy makes anaerobic glycolysis proceed at a high rate, but because the aerobic (oxygen-demanding) pathways are slowed down, not all the pyruvate produced in glycolysis can enter the citric acid cycle. The excess pyruvate ends up as lactate, which causes painful muscle contractions.

The same shift in catabolism also occurs in heart muscle when coronary thrombosis leads to cardiac arrest. The blockage of the artery to the heart muscles cuts off the oxygen supply. The common catabolic pathway and its ATP production are consequently shut off. Glycolysis proceeds at an accelerated rate, causing lactate to accumulate. The heart muscle contracts, producing a cramp. Just as in skeletal muscle, massage of heart muscles can relieve the cramp and start the heart beating. Even if heartbeat is restored within 3 min (the amount of time the brain can survive without being damaged), acidosis may develop as a result of the cardiac arrest. Therefore, at the same time that efforts are under way to start the heart beating by chemical, physical, or electrical means, an intravenous infusion of 8.4% bicarbonate solution is given to combat acidosis.

All of these glycolysis reactions occur in the cytoplasm outside the mitochondria. Because they occur in the absence of O_2, they are also called reactions of the **anaerobic pathway.** As indicated in Figure 19.4, the end product of glycolysis, pyruvate, does not accumulate in the body. In certain bacteria and yeast, pyruvate undergoes reductive decarboxylation in Step ⑩ to produce ethanol. In some bacteria, and in mammals in the absence of oxygen, pyruvate is reduced to lactate in Step ⑪.

B Entrance to the Citric Acid Cycle

Pyruvate is not the end product of glucose metabolism. Most importantly, pyruvate goes through an oxidative decarboxylation in the presence of coenzyme A in Step ⑫ to produce acetyl CoA:

$$NAD^+ + CH_3-\overset{\overset{\displaystyle O}{\|}}{C}-COO^- + CoA-SH \longrightarrow CH_3-\overset{\overset{\displaystyle O}{\|}}{C}-S-CoA + CO_2 + NADH + H^+$$

Pyruvate   Acetyl coenzyme A

This reaction is catalyzed by a complex enzyme system, pyruvate dehydrogenase, that sits on the inner membrane of the mitochondrion. The reaction produces acetyl CoA, CO_2, and NADH + H^+. The acetyl CoA then enters the citric acid cycle in Step ⑬ and goes through the common catabolic pathway.

In summary, after converting complex carbohydrates to glucose, the body gets energy from glucose by converting it to acetyl CoA (by way of pyruvate) and then using the acetyl CoA as a starting material for the common catabolic pathway.

C Pentose Phosphate Pathway

As we saw in Figure 19.4, glucose 6-phosphate plays a central role in different entries into the glycolytic pathway. However, glucose 6-phosphate can also be utilized by the body for other purposes, not just for the production of

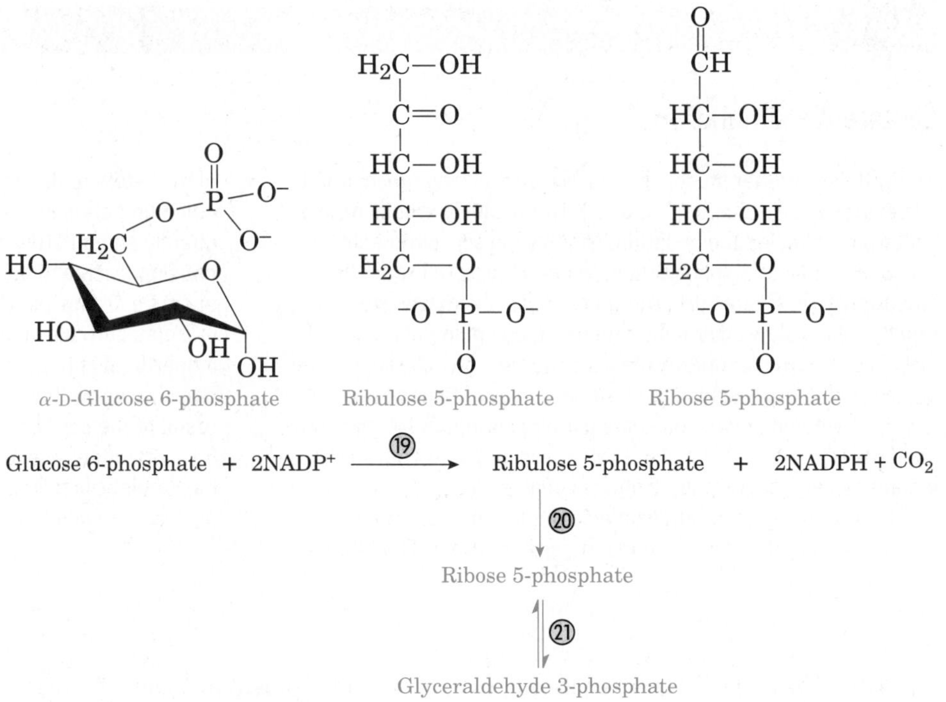

Glucose 6-phosphate + 2NADP⁺ $\xrightarrow{\text{⑲}}$ Ribulose 5-phosphate + 2NADPH + CO_2

Figure 19.5 Simplified schematic representation of the pentose phosphate pathway, also called a shunt.

Pentose phosphate pathway
The biochemical pathway that produces ribose and NADPH from glucose 6-phosphate or, alternatively, energy

NADPH is badly needed in red blood cells (RBC) for defense against oxidative damages. Glutathione (Chemical Connections 13A) is the main agent used to keep the hemoglobin in its reduced form. It is regenerated by NADPH, so an insufficient supply of NADPH leads to the destruction of RBC, causing severe anemia.

energy in the form of ATP. Most importantly, glucose 6-phosphate can be shunted to the **pentose phosphate pathway** in Step ⑲ (Figure 19.5). This pathway has the capacity to produce NADPH and ribose in Step ⑳ as well as energy.

NADPH is needed in many biosynthetic processes, including synthesis of unsaturated fatty acids (Section 20.3), cholesterol, and amino acids as well as photosynthesis (Chemical Connections 20A) and the reduction of ribose to deoxyribose for DNA. Ribose is needed for the synthesis of RNA (Section 16.3). Therefore, when the body needs these synthetic ingredients more than energy, glucose 6-phosphate is utilized through the pentose phosphate pathway. When energy is needed, glucose 6-phosphate remains in the glycolytic pathway and even ribose 5-phosphate can be channeled back to glycolysis through glyceraldehyde 3-phosphate. Through this reversible reaction, the cells can also obtain ribose directly from the glycolytic intermediates.

19.3 The Energy Yield from Glucose

Using Figure 19.4, let us sum up the energy derived from glucose catabolism in terms of ATP production. First, however, we must take into account the fact that glycolysis takes place in the cytoplasm, whereas oxidative phosphorylation occurs in the mitochondria. Therefore, the NADH + H⁺ produced in glycolysis must penetrate the mitochondrial membrane to be utilized in oxidative phosphorylation.

NADH is too large to cross the mitochondrial membrane. Two routes are available to get the electrons of NADH + H⁺ into the mitochondria; they have different efficiencies. In glycerol 3-phosphate transport, which operates in muscle and nerve cells, only two ATP molecules are produced for each $NADH + H^+$. In the other transport route, which operates in the heart and the liver, three ATP molecules are produced for each $NADH + H^+$ produced in the cytoplasm, just as is the case in the mitochondria (Section 18.7). Since most energy production takes place in skeletal muscle cells, when we construct the energy balance sheet, we use two ATP molecules for each $NADH + H^+$ produced in the cytoplasm.

Armed with this knowledge, we are ready to calculate the energy yield of glucose in terms of ATP molecules produced in skeletal muscles. Table 19.1 shows this calculation. In the first stage of glycolysis (Steps ①, ②, and ③), two ATP molecules are used up, but this loss is more than compensated for by the production of 14 ATP molecules in Steps ⑤, ⑥, ⑨ and ⑫, and in the conversion of pyruvate to acetyl CoA. The net yield of these steps is 12 ATP molecules. As we saw in Section 18.7, the oxidation of one acetyl CoA produces 12 ATP molecules, and one glucose molecule provides two acetyl CoA molecules. Therefore, the total net yield from metabolism of one glucose molecule in skeletal muscle is 36 molecules of ATP, or 6 ATP molecules per carbon atom.

$$C_6H_{12}O_6 + 6O_2 \longrightarrow 6CO_2 + 6H_2O$$

If the same glucose is metabolized in the heart or liver, the electrons of the two NADH produced in the glycolysis are transported into the mitochondrion by the malate-aspartate shuttle. Through this shuttle, the two NADH yield a total of 6 ATP molecules, so that, in this case, 38 ATP molecules are produced for each glucose molecule. It is instructive to note that most of the energy (in the form of ATP) from the glucose is produced in the common metabolic pathway.

Glucose is not the only monosaccharide that can be used as an energy source. Other hexoses, such as galactose (Step ⑭) and fructose (Step ⑰), enter the glycolysis pathway at the stages indicated in Figure 19.4. They also yield 36 molecules of ATP per hexose molecule. Furthermore, the glycogen stored in the liver and muscle cells and elsewhere can be converted by

Nicotinamide adenine dinucleotide phosphate ($NADP^+$)

Muscles attached to bones are called skeletal muscles.

Recent investigations suggest that only about 30 to 32 ATP molecules are produced per glucose molecule (2.5 ATP/NADH and 1.5 ATP/$FADH_2$). Futher research into the complexity of the oxidative phosphorylation pathway is needed to verify these numbers.

Table 19.1   ATP Yield from Complete Glucose Metabolism

Step Numbers in Figure 19.4	Chemical Steps	Number of ATP Molecules Produced
① ② ③	Activation (glucose $\longrightarrow$ 1,6-fructose bisphosphate)	−2
⑤	Oxidative phosphorylation 2(glyceraldehyde 3-phosphate $\longrightarrow$ 1,3-bisphosphoglycerate), producing 2($NADH + H^+$) in cytosol	4
⑥ ⑨	Dephosphorylation 2(1,3-bisphosphoglycerate $\longrightarrow$ pyruvate)	4
⑫	Oxidative decarboxylation 2(pyruvate $\longrightarrow$ acetyl CoA), producing 2($NADH + H^+$) in mitochondrion	6
⑬	Oxidation of two C_2 fragments in citric acid cycle and oxidative phosphorylation common pathways, producing 12 ATP for each C_2 fragment	24
	Total	36

Effects of Signal Transduction on Metabolism

The ligand-binding/G-protein/adenylate cyclase cascade, which activates proteins by phosphorylation, has a wide range of effects other than opening or closing ion-gated channels (Figure 15.4). An example of such a target is the glycogen phosphorylase enzyme that participates in the breakdown of glycogen stored in muscles. The enzyme cleaves units of glucose 1-phosphate from glycogen, which enters into the glycolytic pathway and yields quick energy (Section 19.3). The active form of the enzyme, phosphorylase a, is phosphorylated. When it is dephosphorylated (phosphorylase b), it is inactive. When epinephrine signals danger to a muscle cell, phosphorylase is activated through the cascade, and quick energy is produced. Thus, the signal is converted to a metabolic event, allowing the muscles to contract rapidly and enabling the person in danger to fight or run away.

Not all phosphorylation of enzymes results in activation. Consider glycogen synthase. In this case, the phosphorylated form of the enzyme is inactive, and the dephosphorylated form is active. This enzyme participates in glycogenesis, the conversion of glucose to, and storage in the form of, glycogen. The action of glycogen synthase is the opposite of phosphorylase. Nature provides a beautiful balance, inasmuch as the danger signal of the epinephrine hormone has a dual target: It activates phosphorylase to get quick energy, but simultaneously it inactivates glycogen synthase so that the available glucose will be used solely for energy and will not be stored away.

Glycogenolysis The biochemical pathway for the breakdown of glycogen to glucose

enzymatic breakdown and phosphorylation to glucose 1-phosphate (Step ⑮). This compound in turn isomerizes to glucose 6-phosphate, providing an entry to the glycolytic pathway (Step ⑯). The pathway in which glycogen breaks down to glucose is called **glycogenolysis.**

19.4 Glycerol Catabolism

Glycerol 1-phosphate is the same as glycerol 3-phosphate.

The glycerol hydrolyzed from fats or complex lipids (Chapter 12) can also be a rich energy source. The first step in glycerol utilization is an activation step. The body uses one ATP molecule to form glycerol 1-phosphate:

$$
\begin{array}{ccccc}
CH_2OH & & CH_2O\!-\!\text{P} & & CH_2O\!-\!\text{P} \\
| & \xrightarrow[\text{ATP}\quad\text{ADP}]{} & | & \xrightarrow[\text{NAD}^+\quad\text{NADH + H}^+]{} & | \\
CHOH & & CHOH & & C\!=\!O \\
| & & | & & | \\
CH_2OH & & CH_2OH & & CH_2OH \\
\text{Glycerol} & & \text{Glycerol 1-phosphate} & & \text{Dihydroxyacetone phosphate}
\end{array}
$$

The glycerol phosphate is oxidized by NAD^+ to dihydroxyacetone phosphate, yielding $NADH + H^+$ in the process. Dihydroxyacetone phosphate then enters the glycolysis pathway (Step ⑱ in Figure 19.4) and is isomerized to glyceraldehyde 3-phosphate. A net yield of 20 ATP molecules is produced from each glycerol molecule, or 6.7 ATP molecules per carbon atom.

19.5 β-Oxidation of Fatty Acids

The β-carbon is the second carbon from the COOH group.

As early as 1904, Franz Knoop, working in Germany, proposed that the body utilizes fatty acids as an energy source by breaking them down into C_2 fragments. Prior to fragmentation, the β-carbon is oxidized:

$$
-C-C-C-\overset{\beta}{C}-\overset{\alpha}{C}-COOH
$$

The name **β-oxidation** has its origin in Knoop's prediction. It took about 50 years to establish the mechanism by which fatty acids are utilized as an energy source.

Figure 19.6 (next page) depicts the overall process of fatty acid metabolism. As is the case with the other foods we have seen, the first step involves activation. This activation occurs in the cytosol, where the fat was previously hydrolyzed to glycerol and fatty acids. It converts ATP to AMP and inorganic phosphate (Step ①), which is equivalent to the cleavage of two high-energy phosphate bonds. The chemical energy derived from the splitting of ATP is built into the compound acyl CoA, which is formed when the fatty acid combines with coenzyme A. The fatty acid oxidation occurs inside the mitochondrion, so the acyl group of acyl CoA must pass through the mitochondrial membrane. Carnitine is the acyl group transporter.

Once the fatty acid in the form of acyl CoA is inside the mitochondrion, the β-oxidation starts. In the first oxidation (dehydrogenation; Step ②), two hydrogens are removed, creating a trans double bond between the alpha and beta carbons of the acyl chain. The hydrogens and electrons are picked up by FAD.

In Step ③, the double bond is hydrated. An enzyme specifically places the hydroxy group on C-3, the β-carbon. The second oxidation (dehydrogenation; Step ④) requires NAD^+ as a coenzyme. The two hydrogens and electrons removed are transferred to the NAD^+ to form $NADH + H^+$. In the process, a secondary alcohol group is oxidized to a keto group at the beta carbon. In Step ⑤, the enzyme thiolase cleaves the terminal C_2 fragment (an acetyl CoA) from the chain, and the rest of the molecule is attached to a new molecule of coenzyme A.

The cycle then starts again with the remaining acyl CoA, which is now two carbon atoms shorter. At each turn of the cycle, one acetyl CoA is produced. Most fatty acids contain an even number of carbon atoms. The cyclic spiral continues until we reach the last four carbon atoms. When this fragment enters the cycle, two acetyl CoA molecules are produced in the fragmentation step.

The β-oxidation of unsaturated fatty acids proceeds in the same way. An extra step is involved, in which the cis double bond is isomerized to a trans bond, but otherwise the spiral is the same.

> **β-oxidation** The biochemical pathway that degrades fatty acids to acetyl CoA by removing two carbons at a time and yielding energy

Transfer into the mitochondrion is accomplished by an enzyme system called carnitine acyltransferase.

See the **Interactive General, Organic, and Biochemistry CD-ROM, version 2.0,** for further exploration on this topic.

19.6 The Energy Yield from Stearic Acid

To compare the energy yield from fatty acids with that of other foods, let us select a typical and quite abundant fatty acid—stearic acid, the C_{18} saturated fatty acid.

We start with the initial step, in which energy is used up rather than produced. The reaction breaks two high-energy phosphoric anhydride bonds:

$$ATP \longrightarrow AMP + 2\,P_i$$

This reaction is equivalent to hydrolyzing two molecules of ATP to ADP. In each cycle of the spiral, we obtain one $FADH_2$, one $NADH^+ + H^+$, and one acetyl CoA. Stearic acid (C_{18}) goes through seven cycles in the spiral before it reaches the final C_4 stage. In the last (eighth) cycle, one $FADH_2$, one $NADH^+ + H^+$, and two acetyl CoA molecules are produced. Now we can add up the energy. Table 19.2 shows that, for a C_{18} compound, we obtain a total of 146 ATP molecules.

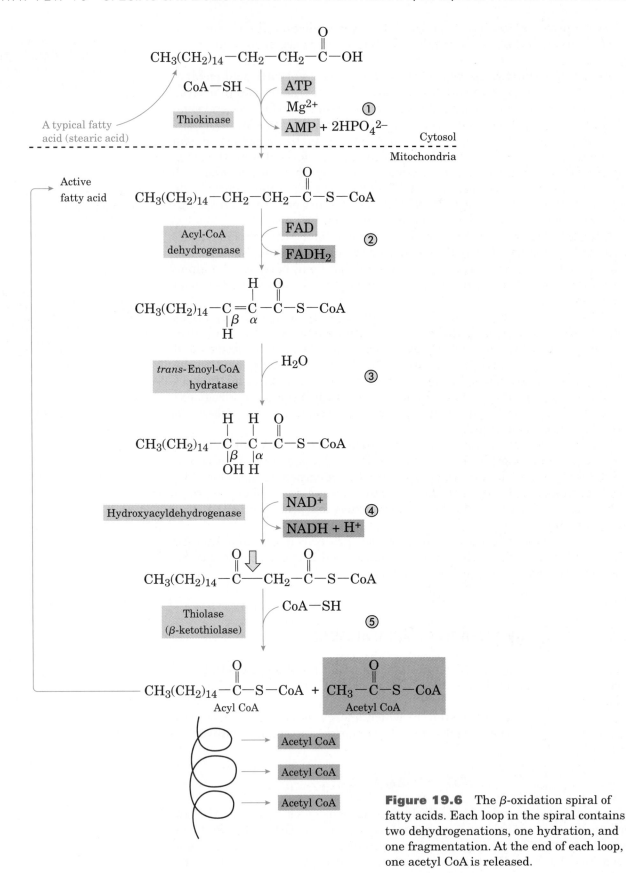

Figure 19.6 The β-oxidation spiral of fatty acids. Each loop in the spiral contains two dehydrogenations, one hydration, and one fragmentation. At the end of each loop, one acetyl CoA is released.

Table 19.2 ATP Yield from Complete Stearic Acid Metabolism

Step Number in Figure 19.6	Chemical Steps	Happens	Number of ATP Molecules Produced
①	Activation (stearic acid $\longrightarrow$ stearyl CoA)	Once	-2
②	Dehydrogenation (acyl CoA $\longrightarrow$ *trans*-enoyl CoA), producing $FADH_2$	8 times	16
④	Dehydrogenation (hydroxyacyl CoA $\longrightarrow$ ketoacyl CoA), producing $NADH + H^+$	8 times	24
	C_2 fragment (acetyl CoA $\longrightarrow$ common catabolic pathway), producing 12 ATP for each C_2 fragment	9 times	108
		Total	$\overline{146}$

It is instructive to compare the energy yield from fats with that from carbohydrates, since both are important constituents of the diet. In Section 19.3 we saw that glucose, $C_6H_{12}O_6$, produces 36 ATP molecules—that is, 6 ATPs for each carbon atom. For stearic acid, there are 146 ATP molecules and 18 carbons, or 146/18 = 8.1 ATP molecules per carbon atom. Since additional ATP molecules are produced from the glycerol portion of fats (Section 19.4), fats have a higher caloric value than carbohydrates.

EXAMPLE 19.1

Ketone bodies are a source of energy, especially during dieting and starvation. If acetoacetate (see Section 19.7) is metabolized through β-oxidation, how many ATPs would be produced?

Solution

In Step ①, activation of acetoacetate to acetoacetyl-CoA requires 2 ATPs. Step ⑤ yields two acetyl CoA, which enter the common catabolic pathway, yielding 12 ATPs for each acetyl CoA, for a total of 24 ATPs. The net yield, therefore, is 22 ATP molecules.

Problem 19.1

Which fatty acid yields more ATP molecules per carbon atom: (a) stearic acid or (b) lauric acid?

19.7 Ketone Bodies

In spite of the high caloric value of fats, the body preferentially uses glucose as an energy supply. When an animal is well fed (plenty of sugar intake), fatty acid oxidation is inhibited, and fatty acids are stored in the form of neutral fat in fat depots. When physical exercise demands energy, when the glucose supply dwindles (as in fasting or starvation), or when glucose cannot be utilized (as in the case of diabetes), the β-oxidation pathway of fatty acid metabolism is mobilized.

In some pathological conditions, glucose may not be available at all.

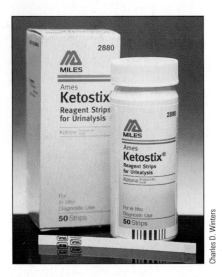

■ **Test kit for the presence of ketone bodies in the urine.**

Unfortunately, low glucose supply also slows down the citric acid cycle. This lag happens because some oxaloacetate is essential for the continuous operation of the citric acid cycle (Figure 18.8). Oxaloacetate is produced from malate, but it is also produced by the carboxylation of phospho-enolpyruvate (PEP):

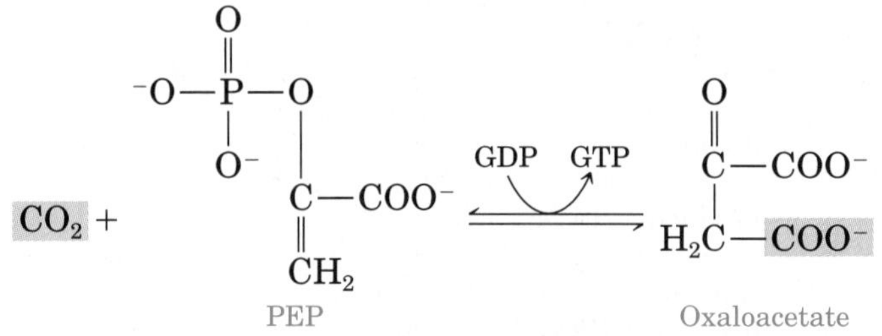

$$CO_2 + \text{PEP} \xrightleftharpoons[\text{GDP}]{\text{GTP}} \text{Oxaloacetate}$$

If there is no glucose, there will be no glycolysis, no PEP formation, and, therefore, greatly reduced oxaloacetate production.

Thus, even though the fatty acids are oxidized, not all of the resulting C_2 fragments (acetyl CoA) can enter the citric acid cycle because not enough oxaloacetate is present. As a result, acetyl CoA builds up in the body, with the following consequences.

The liver is able to condense two acetyl CoA molecules to produce ace-toacetyl CoA:

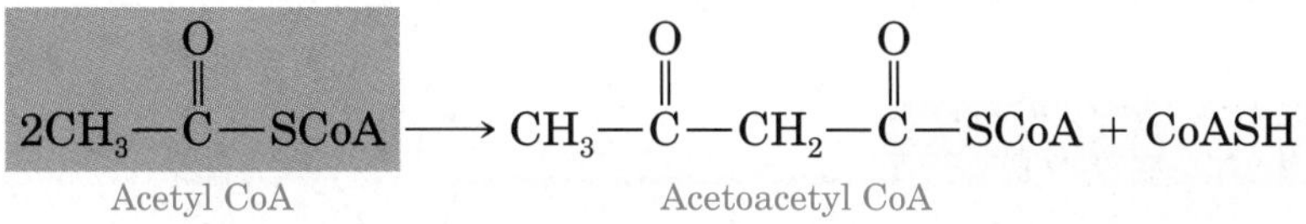

$$2CH_3\text{—}\overset{O}{\overset{\|}{C}}\text{—SCoA} \longrightarrow CH_3\text{—}\overset{O}{\overset{\|}{C}}\text{—}CH_2\text{—}\overset{O}{\overset{\|}{C}}\text{—SCoA} + \text{CoASH}$$

Acetyl CoA   Acetoacetyl CoA

When the acetoacetyl CoA is hydrolyzed, it yields acetoacetate, which can be reduced to form β-hydroxybutyrate:

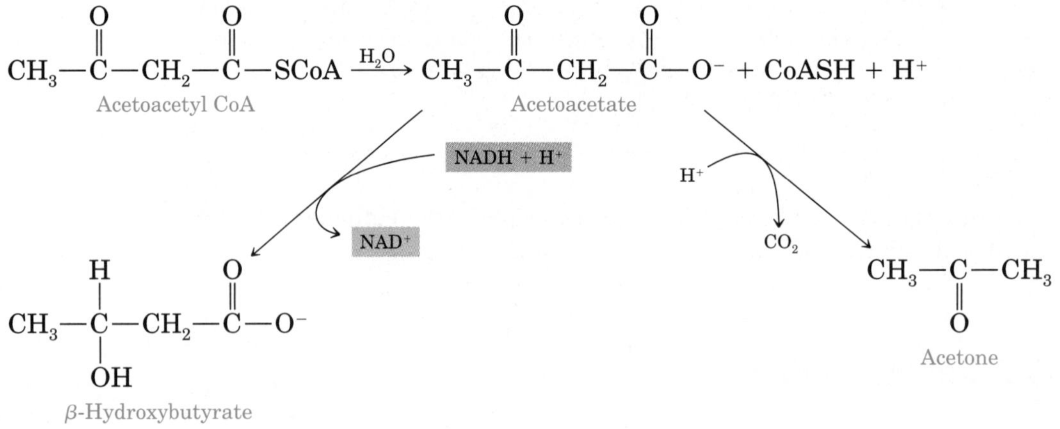

$$CH_3\text{—}\overset{O}{\overset{\|}{C}}\text{—}CH_2\text{—}\overset{O}{\overset{\|}{C}}\text{—SCoA} \xrightarrow{H_2O} CH_3\text{—}\overset{O}{\overset{\|}{C}}\text{—}CH_2\text{—}\overset{O}{\overset{\|}{C}}\text{—O}^- + \text{CoASH} + H^+$$

Acetoacetyl CoA   Acetoacetate

$$CH_3\text{—}\overset{H}{\underset{OH}{\overset{|}{C}}}\text{—}CH_2\text{—}\overset{O}{\overset{\|}{C}}\text{—O}^-$$

β-Hydroxybutyrate

$$CH_3\text{—}\overset{O}{\underset{O}{\overset{\|}{C}}}\text{—}CH_3$$

Acetone

(NADH + H⁺ → NAD⁺) (H⁺ → CO₂)

Ketone bodies A collective name for acetone, acetoacetate and β-hydroxybutyrate; compounds produced from acetyl CoA in the liver that are used as a fuel for energy production by muscle cells and neurons

These two compounds, together with smaller amounts of acetone, are collec-tively called **ketone bodies.** Under normal conditions, the liver sends these compounds into the bloodstream to be carried to tissues and utilized there as a source of energy via the common catabolic pathway. The brain, for ex-ample, normally uses glucose as an energy source. During periods of starva-tion, however, ketone bodies may be the major energy source for the brain. Normally, the concentration of ketone bodies in the blood is low. But during

Ketoacidosis in Diabetes

In untreated diabetes, the glucose concentration in the blood is high because the lack of insulin prevents utilization of glucose by the cells. Regular injections of insulin can remedy this situation. However, in some stressful conditions, **ketoacidosis** can still develop.

A typical case was a diabetic patient admitted to the hospital in a semicomatous state. He showed signs of dehydration, his skin was inelastic and wrinkled, his urine showed high concentrations of glucose and ketone bodies, and his blood contained excess glucose and had a pH of 7.0, a drop of 0.4 pH unit from normal, which is an indication of severe acidosis. The patient's urine also contained the bacterium *Escherichia coli.* This indication of urinary tract infection explained why the normal doses of insulin were insufficient to prevent ketoacidosis.

The stress of infection can upset the normal control of diabetes by changing the balance between administered insulin and other hormones produced in the body. This happened during the patient's infection, and his body started to produce ketone bodies in large quantities. Both glucose and ketone bodies appear in the blood before they show up in the urine.

The acidic nature of ketone bodies (acetoacetic acid and β-hydroxybutyric acid) lowers the blood pH. A large drop in pH is prevented by the bicarbonate/carbonic acid buffer (Section 7.11), but even a drop of 0.3 to 0.5 pH unit is sufficient to decrease the Na^+ concentration. Such a decrease of Na^+ ions in the interstitial fluids draws out K^+ ions from the cells. This, in turn, impairs brain function and leads to coma. During the secretion of ketone bodies and glucose in the urine, a lot of water is lost, the body becomes dehydrated, and the blood volume shrinks. As a consequence, the blood pressure drops, and the pulse rate increases to compensate for it. Smaller quantities of nutrients reach the brain cells, which can also cause coma.

The patient mentioned here was infused with physiological saline solution to remedy his dehydration. Extra doses of insulin restored his glucose level to normal, and antibiotics cured the urinary infection.

starvation and in untreated diabetes mellitus, ketone bodies accumulate in the blood and can reach high concentrations. When this occurs, the excess is secreted in the urine. A check of urine for ketone bodies is used in the diagnosis of diabetes.

19.8 Catabolism of the Nitrogen of Amino Acids

The proteins of our foods are hydrolyzed to amino acids in digestion. These amino acids are primarily used to synthesize new proteins. However, unlike carbohydrates and fats, they cannot be stored, so excess amino acids are catabolized for energy production. Section 19.9 explains what happens to the carbon skeleton of the amino acids. Here we discuss the catabolic fate of the nitrogen. Figure 19.7 (next page) gives an overview of the whole process of protein catabolism.

In the tissues, amino groups ($-NH_2$) freely move from one amino acid to another. The enzymes that catalyze these reactions are the transaminases. In essence, there are three stages in nitrogen catabolism in the liver: transamination, oxidative deamination, and the urea cycle.

Lipomics

Lipomics is the comprehensive analyses of lipid profiles in a cell or a tissue. It is a new approach modeled on the success of genomics (Chemical Connections 16D). It can characterize a sample as small as 50 mg by separating out all the lipid components it contains. By comparing the data with a computer database, lipomics can provide information on the metabolic status of as many as 300 components in one analysis.

Figure 19.7 The overview of pathways in protein catabolism.

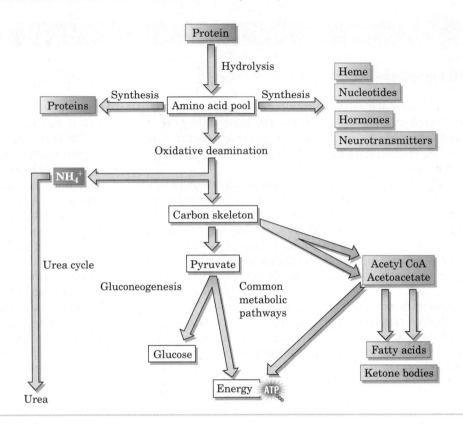

A Transamination

In the first stage, **transamination,** amino acids transfer their amino groups to α-ketoglutarate:

> **Transamination** The exchange of the amino group of an amino acid and a keto group of an α-ketoacid

$$R-\underset{\underset{NH_3^+}{|}}{CH}-COO^- + \underset{\underset{COO^-}{|}}{\overset{COO^-}{\underset{|}{\underset{CH_2}{\underset{|}{\underset{CH_2}{|}}}}}}{\overset{|}{C=O}} \xrightarrow{\text{transaminase}} R-\underset{\underset{O}{\|}}{C}-COO^- + \underset{\underset{COO^-}{|}}{\overset{COO^-}{\underset{|}{\underset{CH_2}{\underset{|}{\underset{CH_2}{|}}}}}}{\overset{|}{CH-NH_3^+}}$$

α-Amino acid   α-Ketoglutarate   α-Ketoacid   Glutamate
(zwitterion form)

The catabolism of the α-ketoacid is discussed in the next section.

The carbon skeleton of the amino acid remains behind as an α-ketoacid.

B Oxidative Deamination

The second stage of the nitrogen catabolism is the **oxidative deamination** of glutamate, which occurs in the mitochondrion:

> **Oxidative deamination** The reaction in which the amino group of an amino acid is removed and an α-ketoacid is formed

$$\underset{\underset{COO^-}{|}}{\overset{COO^-}{\underset{|}{H-\underset{\underset{CH_2}{\underset{|}{\underset{CH_2}{|}}}}{\overset{|}{C}}-NH_3^+}}} + NAD^+ + H_2O \rightleftharpoons NH_4^+ + \underset{\underset{COO^-}{|}}{\overset{COO^-}{\underset{|}{\underset{CH_2}{\underset{|}{\underset{CH_2}{|}}}}{\overset{|}{C=O}}}} + NADH + H^+$$

Glutamate   α-Ketoglutarate

The oxidative deamination yields NH_4^+ and regenerates α-ketoglutarate, which can again participate in the first stage (transamination). The $NADH + H^+$ produced in the second stage enters the oxidative phosphorylation pathway and eventually produces three ATP molecules.

The body must get rid of NH_4^+ because both it and NH_3 are toxic.

C Urea Cycle

In the third stage, the NH_4^+ is converted to urea through the **urea cycle** (Figure 19.8, next page). In Step ①, NH_4^+ is condensed with CO_2 in the mitochondrion to form an unstable compound, carbamoyl phosphate. This condensation occurs at the expense of two ATP molecules. In Step ②, carbamoyl phosphate is condensed with ornithine, a basic amino acid similar in structure to lysine:

> **Urea cycle** A cyclic pathway that produces urea from ammonia and carbon dioxide

Not all organisms dispose of the metabolic nitrogen in the form of urea. Bacteria and fish release ammonia in the surrounding water as such. Birds and reptiles secrete nitrogen in the form of uric acid, the concentrated white solid so familiar in bird droppings.

Ornithine does not occur in proteins.

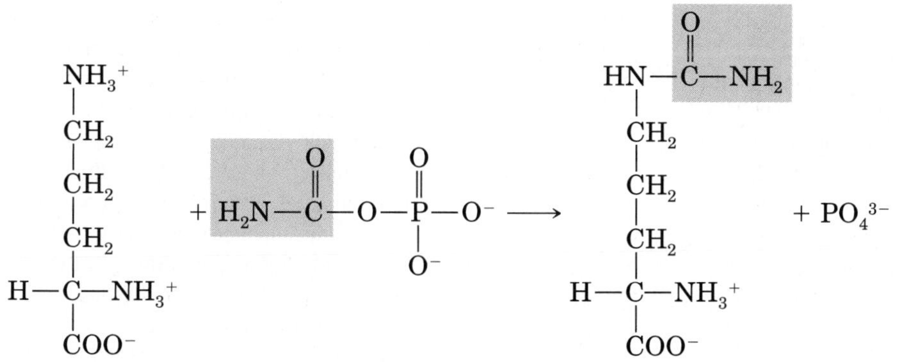

Ornithine   Carbamoyl phosphate   Citrulline

The resulting citrulline diffuses out of the mitochondrion into the cytoplasm.

A second condensation reaction in the cytoplasm takes place between citrulline and aspartate, forming argininosuccinate (Step ③):

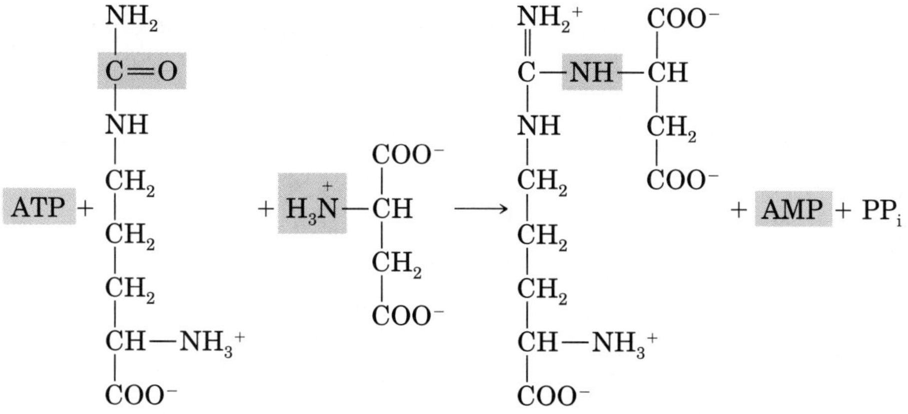

Citrulline   Aspartate   Argininosuccinate

The energy for this reaction comes from the hydrolysis of ATP to AMP and pyrophosphate (PP_i).

PP_i stands for pyrophosphate.

Ubiquitin and Protein Targeting

In previous sections, we dealt with the catabolism of dietary proteins, especially with their use as an energy source. The body's own cellular proteins are also broken down and degraded. This occurs sometimes in response to a stress such as starvation, but more often it is done to maintain a steady-state level of regulatory proteins or to eliminate damaged proteins. But how does the cell know which proteins to degrade and which to leave alone?

One pathway that targets proteins for destruction depends on an ancient protein, called **ubiquitin.** (As the name implies, ubiquitin is present in all cells belonging to higher organisms.) Its sequence of 76 amino acids is essentially identical in yeast and in humans. The C-terminal amino acid of ubiquitin is glycine, the carboxyl group of which forms an amide linkage with the side chain of a lysine residue in the protein that is targeted for destruction.

This process, known as **ubiquitinylation,** requires three enzymes and costs energy by using up one ATP molecule per ubiquitin molecule. In many cases, a polymer of ubiquitin molecules (polyubiquitin) is attached to the target protein molecule. In attaching one ubiquitin molecule to the next, the same linkage is used as with the targeted protein: The glycine at the C-terminal of one ubiquitin forms an amide linkage with the amino group on the side chain of a lysyl residue of a second ubiquitin. Once a protein molecule becomes "flagged" by several ubiquitin molecules, it is delivered to an organelle containing the machinery for proteolysis. Part of the proteolytic system is proteasome, a large, water-soluble complex assembly of proteases and regulatory proteins. In this assembly, the polyubiquitin is removed, and the protein is fully or partially degraded. Cleavage of a protein is called **proteolysis.**

One role of the ubiquitin targeting is that of "garbage collector." Damaged proteins that chaperones (Section 13.9) cannot

salvage by proper refolding must be removed. For example, oxidation of hemoglobin causes unfolding, and rapid degradation by ubiquitin targeting follows. A second role is the control of the concentration of some regulatory proteins. The protein products of some oncogenes (Chemical Connections 17F) are also removed by ubiquitin targeting.

Both of these functions result in complete degradation of the targeted proteins. A third role involves only a partial degradation of the target protein. As an example, suppose a virus invades a cell. The infected cell recognizes a foreign protein from the virus and targets it for destruction through the ubiquitin system. The targeted protein is partially degraded. A peptide segment of the viral protein is incorporated into a system that rises to the surface of the infected cell and displays the viral peptide as an **antigen.** An antigen is a foreign body recognized by the immune system (Section 22.3). The T cells (Section 22.3) can now recognize the infected cell as a foreign cell; they attack and kill it.

The excessive zeal of ubiquitinylation as a garbage collector lies at the heart of cystic fibrosis, a genetic disease. Cystic fibrosis causes severe bronchopulmonary disorders and pancreatic insufficiency. The molecular culprit is a protein in the membranes that serves as a chloride channel, providing for the transport of chloride ions in and out of cells. In cystic fibrosis, this protein has a mutation that does not allow it to fold properly. The system perceives these chloride channel proteins as faulty and targets them for destruction by ubiquitinylation. The result is an insufficient number of functioning chloride channels. This shortage produces obstructions in the respiratory and intestinal tracts by clogging up the secretion pathways with highly viscous mucins.

In Step ④, the argininosuccinate is split into arginine and fumarate:

$$
\underset{\text{Argininosuccinate}}{
\begin{array}{c}
\overset{+}{NH_2} \\
\parallel \\
C-NH-CH \\
| \qquad\qquad | \\
NH \qquad\quad CH_2 \\
| \qquad\qquad | \\
CH_2 \qquad\; COO^- \\
| \\
CH_2 \\
| \\
CH_2 \\
| \\
CH-\overset{+}{NH_3} \\
| \\
COO^-
\end{array}}
\;\longrightarrow\;
\underset{\text{Arginine}}{
\begin{array}{c}
\overset{+}{NH_2} \\
\parallel \\
C-NH_2 \\
| \\
NH \\
| \\
CH_2 \\
| \\
CH_2 \\
| \\
CH_2 \\
| \\
CH-\overset{+}{NH_3} \\
| \\
COO^-
\end{array}}
\;+\;
\underset{\text{Fumarate}}{
\begin{array}{c}
H \qquad COO^- \\
\;\diagdown\; C \diagup \\
\parallel \\
\diagup C \diagdown \\
\;^-OOC \qquad H
\end{array}}
$$

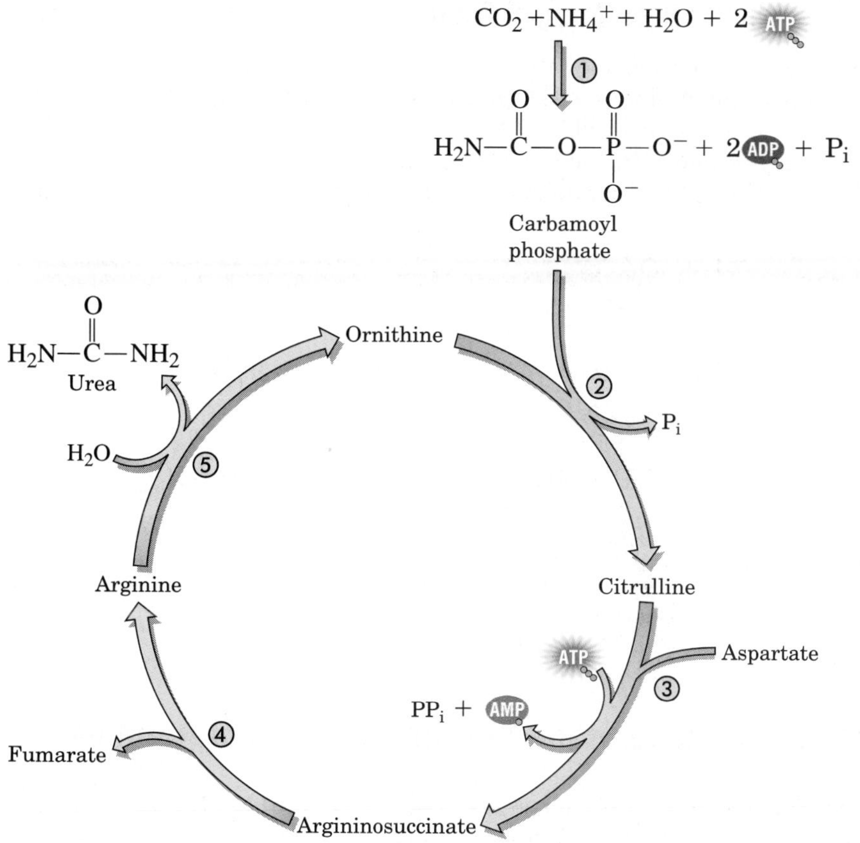

$$CO_2 + NH_4^+ + H_2O + 2 \; \text{ATP}$$

①

$$H_2N-\overset{\overset{\displaystyle O}{\|}}{C}-O-\overset{\overset{\displaystyle O}{\|}}{\underset{\underset{\displaystyle O^-}{|}}{P}}-O^- + 2\,\text{ADP} + P_i$$

Carbamoyl
phosphate

Figure 19.8 The urea cycle.

See the **Interactive General, Organic, and Biochemistry CD-ROM, version 2.0,** for further exploration on this topic.

$$H_2N-\overset{\overset{\displaystyle O}{\|}}{C}-NH_2$$
Urea

H_2O

⑤

Ornithine

②

P_i

Arginine

Citrulline

ATP

③

Aspartate

Fumarate

④

$PP_i +$ AMP

Argininosuccinate

In Step ⑤, the final step, arginine is hydrolyzed to urea and ornithine:

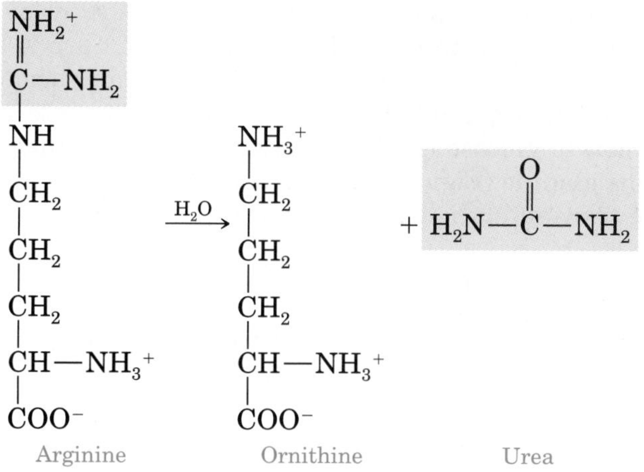

$$
\begin{array}{ccc}
\underset{\displaystyle \text{C}-\text{NH}_2}{\overset{\displaystyle \overset{+}{N}H_2}{\|}} & & \\
\text{NH} & \text{NH}_3^+ & \\
\text{CH}_2 & \text{CH}_2 & \overset{\displaystyle O}{\|} \\
\text{CH}_2 \xrightarrow{\;H_2O\;} & \text{CH}_2 & + \; H_2N-\text{C}-NH_2 \\
\text{CH}_2 & \text{CH}_2 & \\
\text{CH}-\text{NH}_3^+ & \text{CH}-\text{NH}_3^+ & \\
\text{COO}^- & \text{COO}^- & \\
\text{Arginine} & \text{Ornithine} & \text{Urea}
\end{array}
$$

The final product of the three stages is urea, which is excreted in the urine of mammals. The ornithine reenters the mitochondrion, completing the cycle. It is then ready to pick up another carbamoyl phosphate.

An important aspect of carbamoyl phosphate's role as an intermediate is that it can be used for synthesis of nucleotide bases (Chapter 16). Furthermore, the urea cycle is linked to the citric acid cycle in that both involve fumarate.

Hans Krebs, who elucidated the citric acid cycle, was also instrumental in establishing the urea cycle.

D Other Pathways of Nitrogen Catabolism

The urea cycle is not the only way that the body can dispose of the toxic NH_4^+ ions. The oxidative deamination process, which produced the NH_4^+ in the first place, is reversible. Therefore, the buildup of glutamate from α-ketoglutarate and NH_4^+ is always possible. A third possibility for disposing of NH_4^+ is the ATP-dependent amidation of glutamate to yield glutamine:

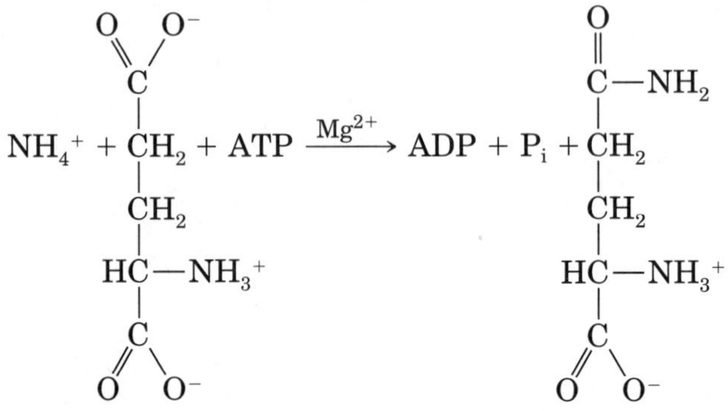

Glutamate   Glutamine

19.9 Catabolism of the Carbon Skeleton of Amino Acids

After transamination of amino acids (Section 19.8) to glutamate, the alpha amino group is removed from glutamate by oxidative deamination (Section 19.8). The remaining carbon skeleton is used as an energy source (Figure 19.9). We will not study the pathways involved except to point out the eventual fate of the skeleton. Not all the carbon skeletons of amino acids are used as fuel. Some may be degraded up to a certain point, and the resulting intermediate then used as a building block to construct another needed molecule.

For example, if the carbon skeleton of an amino acid is catabolized to pyruvate, the body has two possible choices: (1) use the pyruvate as an energy supply via the common catabolic pathway or (2) use it as a building block to synthesize glucose (Section 20.2). Those amino acids that yield a carbon skeleton that is degraded to pyruvate or another intermediate capable of conversion to glucose (such as oxaloacetate) are called **glucogenic.** One example is alanine (Figure 19.9). When alanine reacts with α-ketoglutaric acid, the transamination produces pyruvate directly:

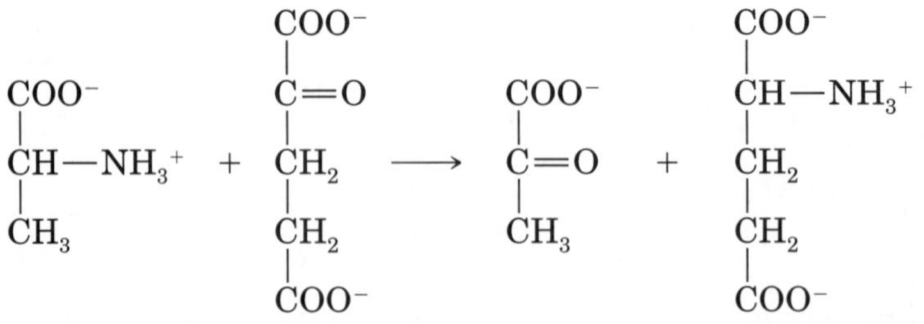

Alanine α-Ketoglutarate   Pyruvate   Glutamate

On the other hand, many amino acids are degraded to acetyl CoA and acetoacetic acid. These compounds cannot form glucose but are capable of yielding ketone bodies; they are called **ketogenic.** Leucine is an example of a ketogenic amino acid. Some amino acids are both glucogenic and ketogenic—for example, phenylalanine.

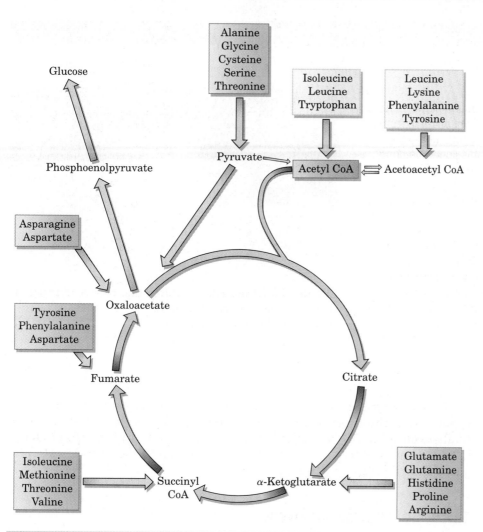

Figure 19.9 Catabolism of the carbon skeletons of amino acids. The glucogenic amino acids are in the purple boxes; the ketogenic ones in the gold boxes.

CHEMICAL CONNECTIONS 19F

Hereditary Defects in Amino Acid Catabolism: PKU

Many hereditary diseases involve missing or malfunctioning enzymes that catalyze the breakdown of amino acids. The oldest known of such diseases is cystinuria, which was described as early as 1810. In this disease, cystine shows up as flat hexagonal crystals in the urine. Stones form because of cystine's low solubility in water. This problem leads to blockage in the kidneys or the ureters and requires surgery. One way to reduce the amount of cystine secreted is to remove as much methionine as possible from the diet. Beyond that, an increased fluid intake increases the volume of the urine, reducing the solubility problem. In addition, penicillamine can prevent cystinuria.

An even more important genetic defect is the absence of the enzyme phenylalanine hydroxylase, which causes a disease called phenylketonuria (PKU). In normal catabolism, this enzyme helps degrade phenylalanine by converting it to tyrosine. If the enzyme is defective, phenylalanine is converted to phenylpyruvate (see the discussion of the conversion of alanine to pyruvate in Section 19.9). Phenylpyruvate (a ketoacid) accumulates in the body and inhibits the conversion of pyruvate to acetyl CoA, thus depriving the cells of energy via the common catabolic pathway. This effect is most important in the brain, which gets its energy from the utilization of glucose. PKU results in mental retardation. This genetic defect can be detected early because phenylpyruvic acid appears in the urine. A federal regulation requires that all infants be tested for this disease. When PKU is detected, mental retardation can be prevented by restricting the intake of phenylalanine in the diet. In particular, patients with PKU should avoid the artificial sweetener aspartame (Chemical Connections 21C) because it yields phenylalanine when hydrolyzed in the stomach.

CHEMICAL CONNECTIONS 19G

Jaundice

In normal individuals, there is a balance between the destruction of heme in the spleen and the removal of bilirubin from the blood by the liver. When this balance is upset, the bilirubin concentration in the blood increases. If this condition continues, the skin and whites of the eyes both become yellow—a condition known as **jaundice**. Jaundice can signal a malfunctioning of the liver, the spleen, or the gallbladder, where bilirubin is stored before excretion.

Hemolytic jaundice is the acceleration of the destruction of red blood cells in the spleen to such a level that the liver cannot cope with the excess bilirubin production. When gallstones obstruct the bile ducts and the bilirubin cannot be excreted in the feces, the backup of bilirubin causes jaundice. In infectious hepatitis, the liver is incapacitated, and bilirubin is not removed from the blood, causing jaundice.

Science Photo Library/Photo Researchers, Inc.

■ **Yellowing of the sclera (white outer coat of eyeball) in jaundice. The yellowing is caused by an excess of bilirubin, a bile pigment, in the blood.**

Both glucogenic and ketogenic amino acids, when used as an energy supply, enter the citric acid cycle at some point (Figure 19.9) and are eventually oxidized to CO_2 and H_2O. The oxaloacetate (a C_4 compound) produced in this manner enters the citric acid cycle, adding to the oxaloacetate produced from PEP and in the cycle itself.

19.10 Catabolism of Heme

Red blood cells are continuously being manufactured in the bone marrow. Their life span is relatively short—about four months. Aged red blood cells are destroyed in the phagocytic cells.

When a red blood cell is destroyed, its hemoglobin is metabolized: The globin (Section 13.9) is hydrolyzed to amino acids, and the heme is first oxidized to biliverdin and finally reduced to bilirubin (Figure 19.10). The color

Phagocytes are specialized blood cells that destroy foreign bodies.

Figure 19.10 Heme degradation from heme to biliverdine to bilirubin.

Heme   Biliverdine   Bilirubin

changes observed in bruises signal the redox reactions occuring in heme catabolism: black and blue are due to the congealed blood, green to the biliverdin, and yellow to the bilirubin. The iron is preserved in ferritin, an iron-carrying protein, and reused. The bilirubin enters the liver via the blood and is then transferred to the gallbladder, where it is stored in the bile and finally excreted via the small intestine. The color of feces is provided by urobilin, an oxidation product of bilirubin.

S U M M A R Y

The foods we eat are broken down into small molecules in the stomach and intestines before being absorbed. These small molecules serve two purposes: (1) They can be the building blocks of new materials that the body needs to synthesize (anabolism) or (2) they can be used for energy supply (catabolism). Each group of compounds—carbohydrates, fats, and proteins—has its own catabolic pathway. All the different catabolic pathways converge to the common catabolic pathway (Section 19.1).

The specific pathway of carbohydrate catabolism is **glycolysis** (Section 19.2A). In this process, hexose monosaccharides are activated by ATP and eventually converted to two C_3 fragments, dihydroxyacetone phosphate and glyceraldehyde phosphate. The glyceraldehyde phosphate is further oxidized and eventually ends up as pyruvate. All of these reactions occur in the cytosol. Pyruvate is converted to acetyl CoA, which is further catabolized in the common pathway. When completely metabolized, a hexose yields the energy of 36 ATP molecules (Section 19.3).

When the body needs intermediates for synthesis rather than energy, the glycolytic pathway can be shunted to the **pentose phosphate pathway** (Section 19.2C). In this way, NADPH is obtained, which is necessary for reduction. The pentose phosphate pathway also yields ribose, which is necessary for synthesis of RNA.

Fats are broken down to glycerol and fatty acids. Glycerol is catabolized in the glycolysis pathway and yields 20 ATP molecules (Section 19.4).

Fatty acids are broken down into C_2 fragments in the **β-oxidation** spiral (Section 19.5). At each turn of the spiral, one acetyl CoA is released along with one $FADH_2$ and one $NADH^+ + H^+$. These products go through the common catabolic pathway. Stearic acid, a C_{18} compound, yields 146 molecules of ATP (Section 19.6). In starvation and under certain pathological conditions, not all the acetyl CoA produced in the β-oxidation of fatty acids enters the common catabolic pathway. Instead, some of it forms acetoacetate, β-hydroxybutyrate, and acetone, commonly called **ketone bodies** (Section 19.7). Excess ketone bodies in the blood are secreted in the urine.

Proteins are broken down to amino acids. The nitrogen of the amino acids is first transferred to glutamate (Section 19.8A). This, in turn, is **oxidatively deaminated** to yield ammonia (Section 19.8B). Mammals get rid of the toxic ammonia by converting it to urea in the **urea cycle** (Section 19.8C); urea is secreted in the urine. The carbon skeletons of amino acids are catabolized via the citric acid cycle (Section 19.9). Some of them enter as pyruvate or other intermediates of the citric acid cycle; these are **glucogenic amino acids.** Others are incorporated into acetyl CoA or ketone bodies and are called **ketogenic amino acids.** Heme is catabolized to bilirubin, which is excreted in the feces (Section 19.10).

P R O B L E M S

Numbers that appear in color indicate difficult problems.
▶ designates problems requiring application of principles.

Specific Pathways

19.2 What are the products of the lipase-catalyzed hydrolysis of fats?

19.3 What is the main use of amino acids in the body?

Glycolysis

19.4 Although catabolism of a glucose molecule eventually produces a lot of energy, the first step uses up energy. Explain why this step is necessary.

19.5   In one step of the glycolysis pathway, a C_6 chain is broken into two C_3 fragments, only one of which can be further degraded in the glycolysis pathway. What happens to the other C_3 fragment?

19.6   Kinases are enzymes that catalyze the addition (or removal) of a phosphate group to (or from) a substance. ATP is also involved. How many kinases are in glycolysis? Name them.

19.7   (a) Which steps in glycolysis of glucose need ATP?
       (b) Which steps in glycolysis yield ATP directly?

19.8   At which intermediate of the glycolytic pathway does the oxidation—and hence the energy production—begin? In what form is the energy produced?

19.9   At what point in glycolysis can ATP act as an inhibitor? What kind of enzyme regulation occurs in this inhibition?

19.10 The end product of glycolysis, pyruvate, cannot enter as such into the citric acid cycle. What is the name of the process that converts this C_3 compound to a C_2 compound?

19.11 What essential compound is produced in the pentose phosphate pathway that is needed for synthesis as well as for defense against oxidative damages?

19.12 Which of the following steps yields energy and which consumes energy?
       (a) Pyruvate $\longrightarrow$ lactate
       (b) Pyruvate $\longrightarrow$ acetyl CoA + CO_2

19.13 How many moles of lactate are produced from 3 moles of glucose?

19.14 How many moles of net $NADH^+ + H^+$ are produced from 1 mol of glucose going to
       (a) Acetyl CoA?   (b) Lactate?

Energy Yield from Glucose

19.15 Of the 36 molecules of ATP produced by the complete metabolism of glucose, how many are produced in glycolysis alone—that is, before the common pathway?

19.16 How many net ATP molecules are produced in the skeletal muscles for each glucose molecule
       (a) In glycolysis alone (up to pyruvate)?
       (b) In converting pyruvate to acetyl CoA?
       (c) In the total oxidation of glucose to CO_2 and H_2O?

19.17 (a) If fructose is metabolized in the liver, how many moles of net ATP are produced from each mole during glycolysis?
       (b) How many moles are produced if the same thing occurs in a muscle cell?

19.18 In Figure 19.3, Step ⑤ yields one NADH. Yet in Table 19.1, the same step indicates the yield of 2 NADH + H^+. Is there a discrepancy between these two statements? Explain.

Glycerol Catabolism

19.19 Based on the names of the enzymes participating in glycolysis, what would be the name of the enzyme catalyzing the activation of glycerol?

19.20 Which yields more energy upon hydrolysis, ATP or glycerol 1-phosphate? Why?

β-Oxidation of Fatty Acids

19.21 Two enzymes participating in β-oxidation have the word "thio" in their names.
       (a) Name the two enzymes.
       (b) To what chemical group does this name refer?
       (c) What is the common feature in the action of these two enzymes?

19.22 (a) Which part of the cells contains the enzymes needed for the β-oxidation of fatty acids?
       (b) How does the activated fatty acid get there?

19.23 Assume that lauric acid (C_{12}) is metabolized through β-oxidation. What are the products of the reaction after three turns of the spiral?

19.24 Is the β-oxidation of fatty acid (without the subsequent metabolism of C_2 fragments via the common metabolic pathway) more efficient with a short-chain fatty acid than with a long-chain fatty acid? Is more ATP produced per C atom in a short-chain fatty acid than in a long-chain fatty acid during β-oxidation?

Energy Yield from Fatty Acids

19.25 Calculate the number of ATP molecules obtained in the β-oxidation of myristic acid, $CH_3(CH_2)_{12}COOH$.

19.26 Assume that the cis-trans isomerization in the β-oxidation of unsaturated fatty acids does not require energy. Which C_{18} fatty acid yields the greater amount of energy, saturated (stearic acid) or monounsaturated (oleic acid)? Explain.

19.27 Assuming that both fats and carbohydrates are available, which does the body preferentially use as an energy source?

19.28 If equal weights of fats and carbohydrates are eaten, which will give more calories? Explain.

Ketone Bodies

19.29 Acetoacetate is the common source of acetone and β-hydroxybutyrate. Name the type of reactions that yield these ketone bodies from acetoacetate.

19.30 Do ketone bodies have nutritional value?

19.31 What happens to the oxaloacetate produced from carboxylation of phosphoenolpyruvate?

$$\begin{array}{c}\text{H}_3\text{C} \\ \quad \text{CH}-\overset{\text{H}^-}{\underset{\text{NH}_3{}^+}{\text{C}}}-\text{COO}^- \\ \text{H}_3\text{C}\end{array} + \begin{array}{c}\text{COO}^- \\ \text{C}=\text{O} \\ \text{CH}_2 \\ \text{CH}_2 \\ \text{COO}^-\end{array} \longrightarrow \begin{array}{c}\text{H}_3\text{C} \\ \quad \text{CH}-\overset{}{\underset{\text{O}}{\text{C}}}-\text{COO}^- \\ \text{H}_3\text{C}\end{array} + \begin{array}{c}\text{COO}^- \\ \text{H}-\text{C}-\text{NH}_3{}^+ \\ \text{CH}_2 \\ \text{CH}_2 \\ \text{COO}^-\end{array}$$

Catabolism of Amino Acids

19.32 What kind of reaction is the above, and what is its function in the body?

19.33 Write an equation for the oxidative deamination of alanine.

19.34 Ammonia, NH_3, and ammonium ion, $NH_4{}^+$, are both soluble in water and could easily be excreted in the urine. Why does the body convert them to urea rather than excreting them directly?

19.35 The metabolism of the carbon skeleton of tyrosine yields pyruvate. Why is tyrosine a glucogenic amino acid?

19.36 What are the sources of the nitrogen in urea?

19.37 What compound is common to both the urea and citric acid cycles?

19.38 (a) What is the toxic product of the oxidative deamination of glutamate?
(b) How does the body get rid of it?

19.39 If the urea cycle is inhibited, in what other ways can the body get rid of $NH_4{}^+$ ions?

Catabolism of Heme

19.40 Why is a high bilirubin content in the blood an indication of liver disease?

19.41 When hemoglobin is fully metabolized, what happens to the iron in it?

19.42 Describe which groups on the biliverdin are the oxidation products and which are the reduction products in the degradation of heme.

Chemical Connections

19.43 (Chemical Connections 19A) What causes cramps of the muscles when a person is fatigued?

19.44 (Chemical Connections 19B) How does the signal of epinephrine result in depletion of glycogen in the muscle?

19.45 (Chemical Connections 19C) What system counteracts the acidic effect of ketone bodies in the blood?

19.46 (Chemical Connections 19C) The patient whose condition is described in Chemical Connections 19C was transferred to a hospital in an ambulance. Could a nurse in the ambulance tentatively diagnose his diabetic condition without running blood and urine tests? Explain.

19.47 (Chemical Connections 19D) Assume that you have a 100-mg sample of a diethyl ether extract of a gallstone. What kind of information can you obtain through lipomics?

19.48 (Chemical Connections 19E) What is meant by the following expression: "the amino acid sequence of ubiquitin is highly conserved"?

19.49 (Chemical Connections 19E) How does ubiquitin attach itself to a target protein?

19.50 (Chemical Connections 19E) How does the body eliminate damaged (oxidized) hemoglobin from the blood?

19.51 (Chemical Connections 19F) Draw structural formulas for each reaction component, and complete the following equation:

Phenylalanine $\longrightarrow$ Phenylpyruvate + ?

19.52 (Chemical Connections 19G) What compound causes the yellow coloration of jaundice?

Additional Problems

19.53 If you receive a laboratory report showing the presence of a high concentration of ketone bodies in the urine of a patient, what disease would you suspect?

19.54 What compounds give the sequence in coloration of bruises from black and blue to green and yellow?

19.55 (a) At which step of the glycolysis pathway does NAD^+ participate (see Figures 19.3 and 19.4)?
(b) At which step does $NADH + H^+$ participate?
(c) As a result of the overall pathway, is there a net increase of NAD^+, of $NADH + H^+$, or of neither?

19.56 What is the net energy yield in moles of ATP produced when yeast converts one mole of glucose to ethanol?

19.57 Can the intake of alanine, glycine, and serine relieve hypoglycemia caused by starvation? Explain.

19.58 How can glucose be utilized to produce ribose for RNA synthesis?

19.59 Write the products of the transamination reaction between alanine and oxaloacetate:

$$
\begin{array}{c}
\text{COO}^- \\
| \\
\text{CH}-\text{NH}_3^+ \\
| \\
\text{CH}_3
\end{array}
\;+\;
\begin{array}{c}
\text{COO}^- \\
| \\
\text{C}=\text{O} \\
| \\
\text{CH}_2 \\
| \\
\text{COO}^-
\end{array}
\;\longrightarrow
$$

19.60 Phosphoenolpyruvate (PEP) has a high-energy phosphate bond that has more energy than the anhydride bonds in ATP. What step in glycolysis suggests that this is so?

19.61 Suppose that a fatty acid labeled with radioactive carbon-14 is fed to an experimental animal. Where would you look for the radioactivity?

19.62 What functional groups are in carbamoyl phosphate?

19.63 Is the urea cycle an energy-producing or energy-consuming pathway?

19.64 What intermediate of the glycolytic pathway can replenish oxaloacetate in the citric acid cycle?

19.65 How many turns of the spiral are there in the β-oxidation of (a) lauric acid and (b) palmitic acid?

InfoTrac College Edition

For additional readings, go to InfoTrac College Edition, your online research library, at

http://infotrac.thomsonlearning.com

CHAPTER **20**

20.1 *Introduction*

20.2 *Biosynthesis of Carbohydrates*

20.3 *Biosynthesis of Fatty Acids*

20.4 *Biosynthesis of Membrane Lipids*

20.5 *Biosynthesis of Amino Acids*

Algae on mudflats.

© Bruce M. Herman/Photo Researchers, Inc.

Biosynthetic Pathways

20.1 Introduction

Anabolic pathways are also called biosynthetic pathways.

In the human body, and in most other living tissues, the pathways by which a compound is synthesized (anabolism) are usually different from the pathways by which it is degraded (catabolism). There are several reasons why it is biologically advantageous for anabolic and catabolic pathways to be different. We will give two of them here:

1. **Flexibility** If the normal **biosynthetic pathway** is blocked, the body can often use the reverse of the degradation pathway instead (recall that most steps in degradation are reversible), thus providing another way to make the necessary compounds.

2. **Overcoming the effect of Le Chatelier's principle** This can be illustrated by the cleavage of a glucose unit from a glycogen molecule, an equilibrium process:

$$\underset{\text{Glycogen}}{(\text{Glucose})_n} + \text{P}_\text{i} \underset{\text{phosphorylase}}{\rightleftharpoons} \underset{\substack{\text{Glycogen} \\ \text{(one unit smaller)}}}{(\text{Glucose})_{n-1}} + \text{Glucose 1-phosphate} \quad (20.1)$$

455

Phosphorylase catalyzes not only glycogen degradation (the forward reaction), but also glycogen synthesis (the reverse reaction). However, the body contains a large excess of inorganic phosphate, P_i. This excess would drive the reaction, on the basis of Le Chatelier's principle, to the right, which represents glycogen degradation. To provide a method for the synthesis of glycogen even in the presence of excess inorganic phosphate, a different pathway is needed in which P_i is not a reactant. Thus, the body uses the following synthetic pathway:

$$\underset{\text{Glycogen}}{(\text{Glucose})_{n-1}} + \text{UDP-glucose} \longrightarrow \underset{\substack{\text{Glycogen} \\ \text{(one unit larger)}}}{(\text{Glucose})_n} + \text{UDP} \quad (20.2)$$

Not only are the synthetic pathways different from the catabolic pathways, but the energy requirements are also different, as are the pathways' locations. Most catabolic reactions occur in the mitochondria, whereas anabolic reactions generally take place in the cytoplasm. We will not describe the energy balances of the biosynthetic processes in detail as we did for catabolism. However, keep in mind that, while energy (in the form of ATP) is *obtained* in the degradative processes, biosynthetic processes *consume* energy.

20.2 Biosynthesis of Carbohydrates

We discuss the biosynthesis of carbohydrates by looking at three examples:

- Conversion of atmospheric CO_2 to glucose in plants
- Synthesis of glucose in animals and humans
- Conversion of glucose to other carbohydrate molecules in animals and humans

A Conversion of Atmospheric CO_2 to Glucose in Plants

The most important biosynthesis of carbohydrates takes place in plants, green algae, and cyanobacteria, with the last two representing an important part of the marine food web. This is **photosynthesis.** In this process, the energy of the sun is built into chemical bonds of carbohydrates; the overall reaction is

Photosynthesis The process in which plants synthesize carbohydrates from CO_2 and H_2O with the help of sunlight and chlorophyll

$$6H_2O + 6CO_2 \xrightarrow[\text{chlorophyll}]{\substack{\text{energy in} \\ \text{the form of} \\ \text{sunlight}}} \underset{\text{Glucose}}{C_6H_{12}O_6} + 6O_2 \quad (20.3)$$

Although the primary product of photosynthesis is glucose, it is largely converted to other carbohydrates, mainly cellulose and starch.

This very complicated process takes place in large protein-cofactor complexes (Chemical Connections 20A). We will not discuss it further here except to note that the carbohydrates of plants—starch, cellulose, and other mono- and polysaccharides—serve as the basic carbohydrate supply of all animals, including humans.

B Synthesis of Glucose in Animals

In Chapter 19, we saw that when the body needs energy, carbohydrates are broken down via glycolysis. When energy is not needed, glucose can be syn-

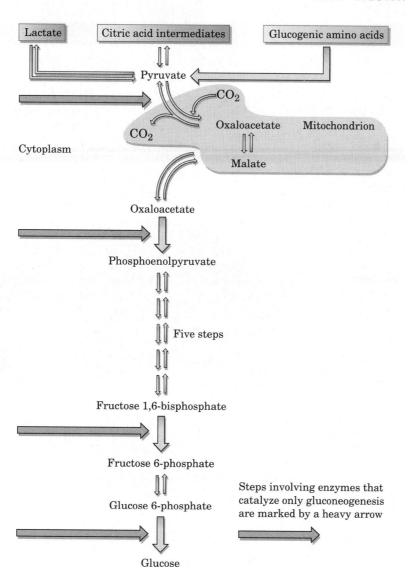

Figure 20.1 Gluconeogenesis. All reactions take place in cytosol, except for those shown in the mitochondria.

thesized from the intermediates of the glycolytic and citric acid pathways. This process is called **gluconeogenesis.** As shown in Figure 20.1, a large number of intermediates—pyruvate, lactate, oxaloacetate, malate, and several amino acids (the glucogenic amino acids we met in Section 19.9)—can serve as starting compounds. Gluconeogenesis proceeds in the reverse order from glycolysis, and many of the enzymes of glycolysis also catalyze gluconeogenesis. At four points, however, unique enzymes (marked in Figure 20.1) catalyze only gluconeogenesis and not the breakdown reactions. These four enzymes make *gluconeogenesis a pathway that is distinct from glycolysis.*

C Conversion of Glucose to Other Carbohydrates in Animals

The third important biosynthetic pathway for carbohydrates is the conversion of glucose to other hexoses and hexose derivatives and the synthesis of di-, oligo-, and polysaccharides. The common step in all of these processes is the activation of glucose by uridine triphosphate (UTP) to form UDP-glucose:

> **Gluconeogenesis** The process by which glucose is synthesized in the body.

ATP is used up in gluconeogenesis and produced in glycolysis.

CHEMICAL CONNECTIONS 20A

Photosynthesis

Photosynthesis requires sunlight, water, CO_2, and pigments found in plants, mainly chlorophyll. The overall reaction shown in Equation 20.3 actually occurs in two distinct steps. First, the light interacts with the pigments that are located in highly membranous organelles of plants, called **chloroplasts** (Figure 20A.1). Chloroplasts resemble mitochondria (Section 18.2) in many respects: They contain a whole chain of oxidation–reduction enzymes similar to the cytochrome and iron–sulfur complexes of mitochondrial membranes, and they contain a proton translocating ATPase. In a manner similar to mitochondria, the proton gradient accumulated in the intermembrane region drives the synthesis of ATP in chloroplasts (see the discussion of the chemiosmotic pump in Section 18.6).

The chlorophyll is the central part of a complex machinery called photosystem I and II. The detailed structure of photosystem I was elucidated in 2001. The photosystem consists of three monomeric entities. These monomeric entities are designated by I, II and III (see Biochemistry Part Opener on page 229). Each monomer contains 12 different proteins, 96 chlorophyll molecules, and 30 cofactors that include iron clusters, lipids, and Ca^{2+} ions. Its most important feature shows that the central Mg^{2+} is bound to the sulfur of a methionine residue of a surrounding protein. This Mg-S linkage makes this whole assembly a powerful oxidizing agent, meaning that it can readily accept electrons.

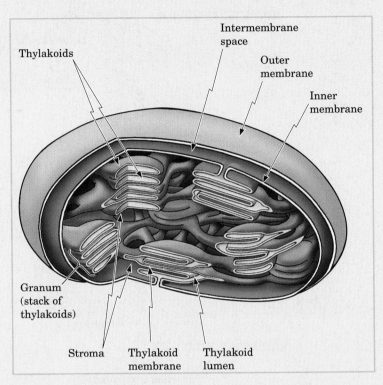

■ **Figure 20.A1** Inside of a chloroplast.

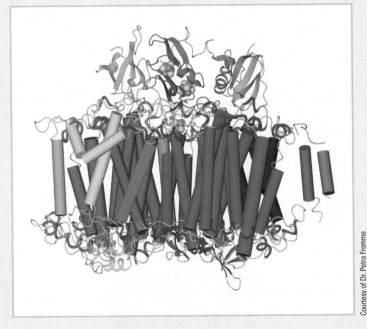

■ **Figure 20.A2** Side view of monomer III in photosystem I.

CHEMICAL CONNECTIONS 20A

Photosynthesis, *(continued)*

The chlorophyll itself, buried in a complex protein that traverses the chloroplast membranes, is a molecule similar to the heme we have already encountered in hemoglobin (Figure 13.12). In contrast to heme, however, chlorophyll contains Mg^{2+} instead of Fe^{2+}:

Chlorophyll a

The first reaction that takes place in photosynthesis is called the light reaction because chlorophyll captures the energy of sunlight and, with its aid, strips the electrons and protons from water to form oxygen, ATP, and NADPH + H^+ (see Section 19.2C):

$$H_2O + ADP + P_i + NADP^+ + sunlight$$
$$\longrightarrow O_2 + ATP + NADPH + H^+$$

The second reaction, called the dark reaction because it does not need light, in essence converts CO_2 to carbohydrates:

$$CO_2 + ATP + NADPH + H^+$$
$$\longrightarrow (CH_2O)_n + ADP + P_i + NADP^+$$
Carbohydrates

In this step, the energy, now in the form of ATP, is used to help NADPH + H^+ reduce carbon dioxide to carbohydrates. Thus, the protons and electrons stripped in the light reaction are added to the carbon dioxide in the dark reaction. This reduction takes place in a multistep cyclic process called the **Calvin cycle,** named after its discoverer, Melvin Calvin (1911–1997), who was awarded the 1961 Nobel Prize in chemistry for his work. In this cycle, CO_2 is first attached to a C_5 fragment and breaks down to two C_3 fragments (triose phosphates). Through a complex series of steps, these fragments are converted to a C_6 compound and, eventually, to glucose.

$$CO_2 + C_5 = 2C_3 = C_6$$

The critical step in the dark reaction (Calvin cycle) is the attachment of CO_2 to ribulose 1,5-bisphosphate, a compound derived from ribulose (Table 11.2). The enzyme that catalyzes this reaction, ribulose-1,5-bisphosphate carboxylase-oxygenase, nicknamed RuBisCO, is one of the slowest in nature. As in traffic, the slowest-moving vehicle determines the overall flow, so RuBisCO is the main factor in the low efficiency of the Calvin cycle. Because of this enzyme, most plants convert less than 1% of the absorbed radiant energy into carbohydrates. To overcome its inefficiency, plants must synthesize large quantities of this enzyme. More than half of the soluble proteins in plant leaves are RuBisCO enzymes, whose synthesis requires a lot of energy expenditure.

Uridine diphosphate (UDP)
Uridine diphosphate glucose (UDP-glucose)

UDP is similar to ADP except that the base is uracil instead of adenine. UTP, an analog of ATP, contains two high-energy phosphate anhydride

Glycogenesis The conversion of glucose to glycogen

bonds. For example, when the body has excess glucose and wants to store it as glycogen (a process called **glycogenesis**), the glucose is first converted to glucose 1-phosphate, but then a special enzyme catalyzes the reaction:

$$\text{glucose 1-phosphate} + \boxed{\text{UTP}} \longrightarrow \text{UDP-glucose} + {}^{-}\text{O}-\overset{\overset{\displaystyle O}{\|}}{\underset{\underset{\displaystyle O^-}{|}}{P}}-\text{O}-\overset{\overset{\displaystyle O}{\|}}{\underset{\underset{\displaystyle O^-}{|}}{P}}-\text{O}^-$$

$$\underset{\text{Glycogen}}{\text{UDP-glucose} + (\text{glucose})_n} \longrightarrow \boxed{\text{UDP}} + \underset{\substack{\text{Glycogen} \\ \text{(one unit larger)}}}{(\text{glucose})_{n+1}}$$

The biosynthesis of many other di- and polysaccharides and their derivatives also uses the common activation step: forming the appropriate UDP compound.

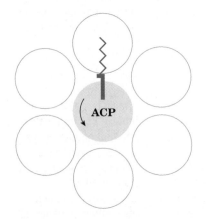

Figure 20.2 The biosynthesis of fatty acids. The ACP (central blue sphere) has a long side chain (⏄) that carries the growing fatty acid (⌇). The ACP rotates counterclockwise, and its side chain sweeps over a multienzyme system (empty spheres). As each cycle is completed, a C_2 fragment is added to the growing fatty acid chain.

20.3 Biosynthesis of Fatty Acids

The body can synthesize all the fatty acids it needs except for linoleic and linolenic acids (essential fatty acids; see Section 12.2). The source of carbon in this synthesis is acetyl CoA. Since acetyl CoA is also a degradation product of the β-oxidation spiral of fatty acids (Section 19.5), we might expect the synthesis to be the reverse of the degradation. This is not the case. For one thing, the majority of fatty acid synthesis occurs in the cytoplasm, while degradation takes place in the mitochondria. Fatty acid synthesis is catalyzed by a multienzyme system.

However, one aspect of fatty acid synthesis is the same as in fatty acid degradation: Both processes involve acetyl CoA, so both proceed in units of two carbons. Fatty acids are built up two carbons at a time, just as they are broken down two carbons at a time (Section 19.5).

Most of the time, fatty acids are synthesized when excess food is available. That is, when we eat more food than we need for energy, our bodies turn the excess acetyl CoA (produced by catabolism of carbohydrates; see Section 19.2) into fatty acids and then to fats. The fats are stored in the fat depots, which are specialized fat-carrying cells (see Figure 19.1).

The key to fatty acid synthesis is an **acyl carrier protein (ACP).** It can be looked upon as a merry-go-round—a rotating protein molecule to which the growing chain of fatty acids is bonded. As the growing chain rotates with the ACP, it sweeps over the multienzyme complex; at each enzyme, one reaction of the chain is catalyzed (Figure 20.2).

At the beginning of this cycle, the ACP picks up an acetyl group from acetyl CoA and delivers it to the first enzyme, fatty acid synthase, here called synthase for short:

$$\underset{}{\overset{\overset{\displaystyle O}{\|}}{\text{CH}_3\text{C}}-\text{S}-\text{CoA}} + \text{HS}-\text{ACP} \longrightarrow \text{HS}-\text{CoA} + \overset{\overset{\displaystyle O}{\|}}{\text{CH}_3\text{C}}-\text{S}-\text{ACP}$$

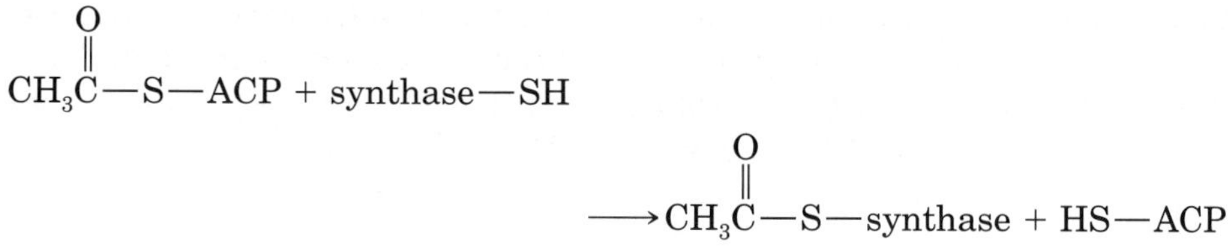

$$\underset{\displaystyle CH_3\overset{\displaystyle O}{\overset{\|}{C}}-S-ACP}{} + synthase-SH$$

$$\longrightarrow CH_3\overset{\displaystyle O}{\overset{\|}{C}}-S-synthase + HS-ACP$$

The C_2 fragment on the synthase is condensed with a C_3 fragment attached to the ACP in a process in which CO_2 is given off:

The —SH group is the site of acyl binding as a thioester.

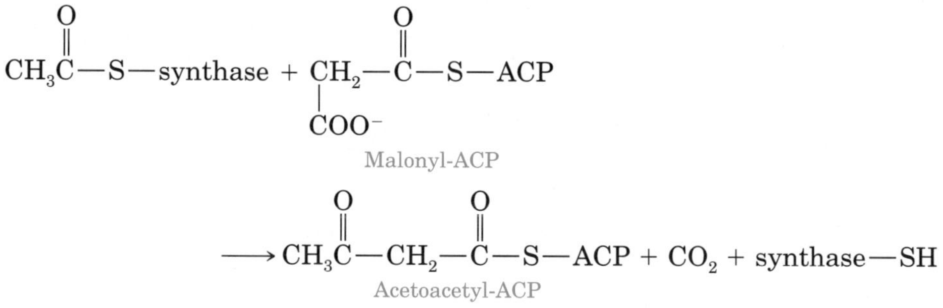

$$CH_3\overset{\displaystyle O}{\overset{\|}{C}}-S-synthase + \underset{\displaystyle COO^-}{CH_2}-\overset{\displaystyle O}{\overset{\|}{C}}-S-ACP$$

Malonyl-ACP

$$\longrightarrow CH_3\overset{\displaystyle O}{\overset{\|}{C}}-CH_2-\overset{\displaystyle O}{\overset{\|}{C}}-S-ACP + CO_2 + synthase-SH$$

Acetoacetyl-ACP

The result is a C_4 fragment, which is reduced twice and dehydrated before it becomes a fully saturated C_4 group. This event marks the end of one cycle of the merry-go-round.

The three steps are the reverse of what we saw in the β-oxidation of fatty acids in Section 19.5.

In the next cycle, the C_4 fragment is transferred to the synthase, and another malonyl-ACP (C_3 fragment) is added. When CO_2 splits out, a C_6 fragment is obtained. The merry-go-round continues to turn. At each turn, another C_2 fragment is added to the growing chain. Chains up to C_{16} (palmitic acid) can be obtained in this process. If the body needs longer fatty acids—for example, stearic (C_{18})—another C_2 fragment is added to palmitic acid by a different enzyme system.

Unsaturated fatty acids are obtained from saturated fatty acids by an oxidation step in which hydrogen is removed and combined with O_2 to form water:

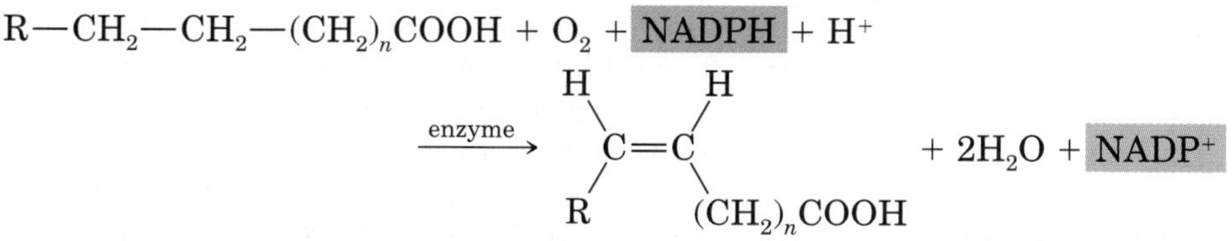

$$R-CH_2-CH_2-(CH_2)_n COOH + O_2 + \boxed{NADPH} + H^+$$

$$\xrightarrow{enzyme} \underset{\displaystyle R \qquad (CH_2)_n COOH}{\overset{\displaystyle H \qquad H}{C=C}} + 2H_2O + \boxed{NADP^+}$$

An example of such elongation and unsaturation is the fatty acid, docosahexenoic acid that as its name indicates is a 22-carbon fatty acid with 6 cis double bonds (22:6). Docosahexenoic acid is part of the glycerophospholipids prevalent in the membranes where the visual pigment rhodopsin resides. Its presence is necessary to provide fluidity to the membrane to process the light signals reaching our retina.

20.4 Biosynthesis of Membrane Lipids

The various membrane lipids (Sections 12.6 to 12.8) are assembled from their constituents. We just saw how fatty acids are synthesized in the body. These fatty acids are activated by CoA, forming acyl CoA. Glycerol 3-

We encountered glycerol 3-phosphate as a vehicle for transporting electrons in and out of mitochondria (Section 19.3).

phosphate, which is obtained from the reduction of dihydroxyacetone phosphate (a C_3 fragment of glycolysis; see Figure 19.4), is the second building block of glycerophospholipids. This compound combines with two acyl CoA molecules, which can be the same or different:

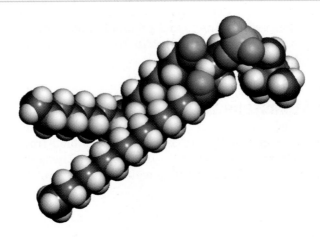

To complete the molecule, an activated serine or choline or ethanolamine is added to the $-OPO_3^{2-}$ group, forming phosphate esters (see the structures in Section 12.6; Figure 20.3 shows a model of phosphatidylcholine). Choline is activated by cytidine triphosphate (CTP), yielding CDP-choline. This process is similar to the activation of glucose by UTP (Section 20.2) except that the base is cytosine rather than uracil (Section 16.2).

Sphingolipids (Section 12.7) are similarly built up from smaller molecules. An activated phosphocholine is added to the sphingosine part of ceramide (Section 12.7) to make sphingomyelin.

The glycolipids are constructed in a similar fashion. Ceramide is assembled as described above, and the carbohydrate is added one unit at a time in the form of activated monosaccharides (UDP glucose and so on).

Cholesterol, the molecule that controls the fluidity of membranes and is a precursor of all steroid hormones and bile salts, is also synthesized by the human body. It is assembled in the liver from C_2 fragments that come from the acetyl group of acetyl CoA. All the carbon atoms of cholesterol come from the carbons of acetyl CoA molecules (Figure 20.4). Cholesterol in the brain is synthesized in nerve cells themselves; its presence is necessary to form synapses. Cholesterol from our diets and cholesterol that is synthesized in the liver and circulates in the plasma as LDL (see Section 12.9) are not available for synapse formation because LDL cannot cross the blood–brain barrier.

Figure 20.3 A model of phosphatidylcholine, commonly called lecithin.

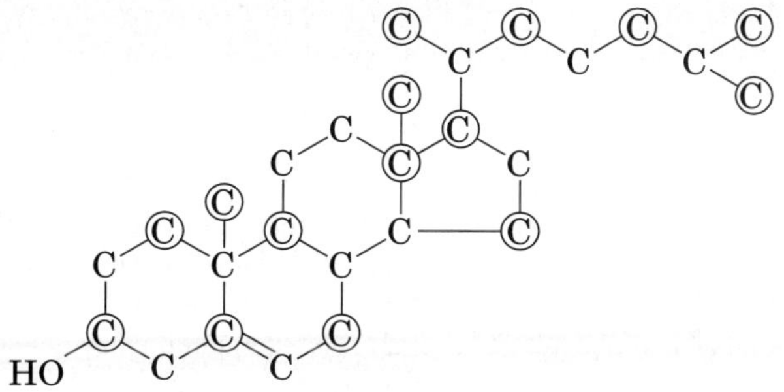

Figure 20.4 Biosynthesis of cholesterol. The circled carbon atoms come from the —CH$_3$ group, and the others come from the —CO— group of the acetyl group of acetyl CoA.

Cholesterol synthesis starts with the sequential condensation of three acetyl CoA molecules to form a C$_6$ compound, hydroxymethylglutaryl CoA (HMGCoA):

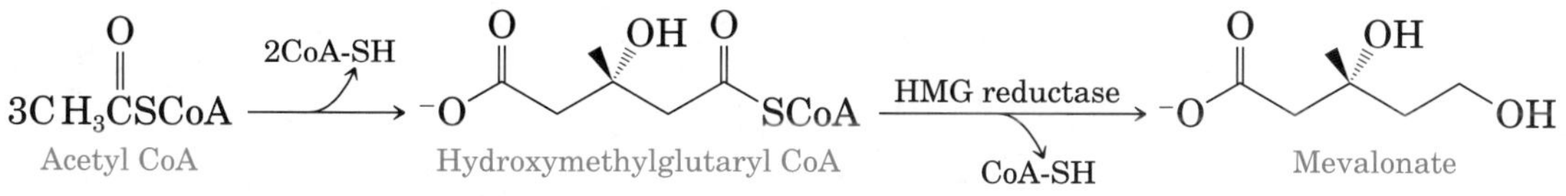

A key enzyme, HMG reductase, controls the rate of cholesterol synthesis. It reduces HMGCoA to a primary alcohol, yielding CoA in the process. The resulting compound, mevalonate, undergoes phosphorylation and decarboxylation to yield a C$_5$ compound, isopentenyl pyrophosphate:

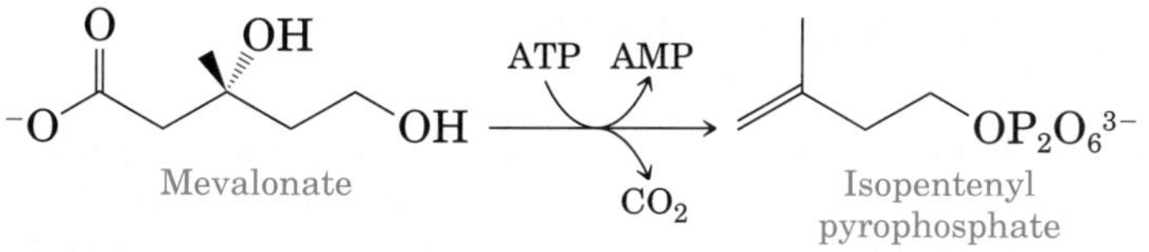

From this basic C$_5$ unit, the other multiple C$_5$ compounds are formed that eventually lead to cholesterol synthesis. These intermediates are geranyl, C$_{10}$, and farnesyl, C$_{15}$, pyrophosphates:

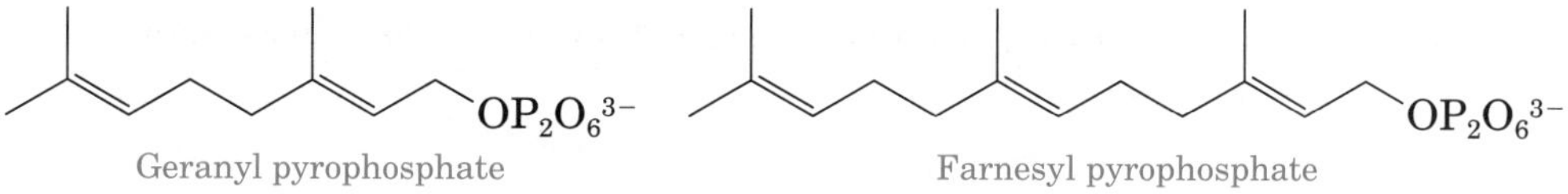

Finally, cholesterol is synthesized from the condensation of two farnesyl pyrophosphate molecules.

The statin drugs, such as lovastatin, competitively inhibit the key enzyme HMG reductase and thereby the biosynthesis of cholesterol. They are frequently prescribed to control the cholesterol level in the blood so as to prevent atherosclerosis (Section 12.9E).

Note that the intermediates in cholesterol synthesis, the geranyl and farnesyl pyrophosphates, are made of isoprene units; we discussed these C$_5$ compounds in Section 3.5. These C$_{10}$ and C$_{15}$ compounds are also used to enable protein molecules to be dispersed in the lipid bilayers of membranes. When these multiple isoprene units are attached to a protein in a process called **prenylation,** the protein becomes more hydrophobic and is able to move laterally within the bilayer with greater ease. Prenylation marks proteins to be associated with membranes and to perform other cellular functions, such as signal transduction of G protein (Chemical Connections 20B).

The name *prenylation* originates from isoprene, the five-carbon unit from which the cholesterol intermediates C$_{10}$, C$_{15}$, and C$_{30}$ are made.

CHEMICAL CONNECTIONS 20B

Prenylation of Ras Protein and Cancer

Prenylation, or the attachment of multiple isoprene units to a protein, enables the protein to be part of a membrane, an integral protein. The C_{10} and C_{15} units are attached to the protein via a thioether linkage, —C—S—C—. For example, the farnesyl pyrophosphate interacts with a cysteine residue that is located near the carboxyl terminal of the target protein. As a thioether linkage is formed, the pyrophosphate is liberated simultaneously. The resultant modified protein acquires a hydrophobic tail that makes its association with membranes feasible.

A protein named Ras is a GTP-binding protein and plays a key role in the signal transduction (Section 15.5) that regulates cell growth. It is an oncogene (Chemical Connections 17E) in that a Ras-related mutation may cause cancer. The signaling activity of Ras depends on its prenylation. Mutations that inhibit prenylation of Ras cause the cell to die. Other mutations of prenylated Ras may cause uncontrolled growth; indeed, one-third of all human cancers have Ras mutations.

$$S—CH_2—CH—CO—\sim\!\!\sim\!\!\sim$$

Farnesyl (or geranyl)   Cysteine

20.5 Biosynthesis of Amino Acids

The human body needs 20 different amino acids to make its protein chains—all 20 are found in a normal diet. Some of the amino acids can be synthesized from other compounds; these are the nonessential amino acids. Others cannot be synthesized by the human body and must be supplied in the diet; these are the **essential amino acids** (see Section 21.6). Most nonessential amino acids are synthesized from some intermediate of either glycolysis (Section 19.2) or the citric acid cycle (Section 18.4). Glutamate plays a central role in the synthesis of five nonessential amino acids. Glutamate itself is synthesized from α-ketoglutarate, one of the intermediates in the citric acid cycle:

Table 13.1 shows the essential and nonessential amino acids.

$$NADH + H^+ + NH_4^+ + \begin{array}{c} COO^- \\ | \\ C\!=\!O \\ | \\ CH_2 \\ | \\ CH_2 \\ | \\ COO^- \end{array} \rightleftharpoons \begin{array}{c} COO^- \\ | \\ CH\!-\!NH_3^+ \\ | \\ CH_2 \\ | \\ CH_2 \\ | \\ COO^- \end{array} + NAD^+ + H_2O$$

α-Ketoglutarate   Glutamate

The forward reaction is the synthesis, and the reverse reaction is the oxidative deamination (degradation) reaction we encountered in the catabolism of amino acids (Section 19.8). In this case, the synthetic and degradative pathways are exactly the reverse of each other.

Glutamate can serve as an intermediate in the synthesis of alanine, serine, aspartate, asparagine, and glutamine. For example, the transamination reaction we saw in Section 19.8 leads to alanine formation:

CHEMICAL CONNECTIONS 20C

Amino Acid Transport and Blue Diaper Syndrome

Amino acids, both essential and nonessential, are usually more concentrated inside the cells than in the cells' surroundings. In the body, special transport mechanisms are available to carry the amino acids into the cells. Sometimes these transport mechanisms are faulty, however, and the cells lack an amino acid despite the fact that the diet provides the particular amino acid.

An interesting example of this phenomenon is blue diaper syndrome, in which infants excrete blue urine. Indigo blue, a dye, is an oxidation product of the amino acid tryptophan. But how does this oxidation product get into the system to be ex-

creted in the urine? The patient's tryptophan transport mechanism is faulty. Although enough tryptophan is being supplied by the diet, most of it is not absorbed through the intestine. Instead, it accumulates there and is oxidized by bacteria in the gut. The oxidation product is moved into the cells, but because the cells cannot use it, it is excreted in the urine.

Much of the blue diaper syndrome can be eliminated with antibiotics. This treatment kills many of the gut's bacteria, and the nonabsorbed tryptophan is then excreted in the feces because the bacteria cannot now oxidize it to indigo blue.

L-Tryptophan   Indigo

■ **Structures of tryptophan and indigo**

Pyruvate   Glutamate   Alanine   α-Ketoglutarate

Besides being the building blocks of proteins, amino acids serve as intermediates for a large number of biological molecules. We have already seen that serine is needed in the synthesis of membrane lipids (Section 20.4). Certain amino acids are also intermediates in the synthesis of heme and of the purines and pyrimidines that are the raw materials for DNA and RNA (Chapter 16).

SUMMARY

For most biochemical compounds, the biosynthetic pathways are different from the degradation pathways. In **photosynthesis,** carbohydrates are synthesized in plants from CO_2 and H_2O, using sunlight as an energy source (Section 20.2). Glucose can be synthesized by animals from the intermediates of glycolysis, from those of the citric acid cycle, and from glucogenic amino acids. This process is called **gluconeogenesis** (Section 20.2). When glucose or other monosaccharides are built into di-, oligo-, and polysac-

charides, each monosaccharide unit in its activated form is added to a growing chain.

Fatty acid biosynthesis is accomplished by a multienzyme system (Section 20.3). The key to this process is the **acyl carrier protein, ACP,** which acts as a merry-go-round transport system: It carries the growing fatty acid chain over a number of enzymes, each of which catalyzes a specific reaction. With each complete turn of the merry-go-round, a C_2 fragment is added to the growing fatty acid chain. The source of the C_2 fragment is malonyl ACP, a C_3 compound attached to the ACP. It becomes C_2 with the loss of CO_2.

Glycerophospholipids (Section 20.4) are synthesized from glycerol 3-phosphate, fatty acids that are activated by conversion to acyl CoA, and activated alcohols such as choline. Cholesterol (Section 20.4) is synthesized from acetyl CoA. Three C_2 fragments are condensed to form C_6, hydroxymethylglutaryl CoA. After reduction and decarboxylation, C_5 isoprene

units are formed that condense to the C_{10} and C_{15} intermediates from which cholesterol is built.

Many nonessential amino acids are synthesized in the body from the intermediates of glycolysis or of the citric acid cycle (Section 20.5). In half of these cases, glutamate is the donor of the amino group in transamination. Amino acids serve as building blocks for proteins.

P R O B L E M S

Numbers that appear in color indicate difficult problems.
◀ designates problems requiring applications of principles.

Biosynthesis

20.1 Why are the pathways that the body uses for anabolism and catabolism mostly different?

20.2 How does the large excess of inorganic phosphate in a cell affect the amount of glycogen present? Explain.

20.3 Glycogen can be synthesized in the body by the same enzymes that degrade it. Why is this process utilized in glycogen synthesis only to a small extent, while most glycogen biosynthesis occurs via a different synthetic pathway?

20.4 Do most anabolic and catabolic reactions take place in the same location?

Carbohydrate Biosynthesis

20.5 What is the difference in the overall chemical equations between photosynthesis and respiration?

20.6 In photosynthesis, what are the sources of (a) carbon, (b) hydrogen, and (c) energy?

20.7 Name a compound that can serve as a raw material for gluconeogenesis and is (a) from the glycolytic pathway, (b) from the citric acid cycle, and (c) an amino acid.

20.8 How is glucose activated for glycogen synthesis?

20.9 Glucose is the only carbohydrate compound that the brain can use for energy. Which pathway is mobilized to supply the need of the brain during starvation: (a) glycolysis, (b) gluconeogenesis, or (c) glycogenesis? Explain.

20.10 Are the enzymes that combine two C_3 compounds into a C_6 compound in gluconeogenesis the same as or different from those that cleave the C_6 compound into two C_3 compounds in glycolysis?

20.11 Devise a scheme in which maltose is formed starting with UDP-glucose.

20.12 Glycogen is written as (glucose)$_n$.
(a) What does the n stand for?
(b) What is the approximate value of n?

20.13 What are the constituents of UTP?

Fatty Acid Biosynthesis

20.14 What is the source of carbon in fatty acid synthesis?

20.15 (a) Where in the body does fatty acid synthesis occur?
(b) Does fatty acid degradation occur in the same location?

20.16 Is ACP an enzyme?

20.17 In fatty acid biosynthesis, which compound is added repeatedly to the synthase?

20.18 (a) What is the name of the first enzyme in fatty acid synthesis?
(b) What does it do?

20.19 From which compound is CO_2 released in fatty acid synthesis?

20.20 What are the common functional groups in CoA, ACP, and synthase ?

20.21 In the synthesis of unsaturated fatty acids, $NADPH + H^+$ is converted to $NADP^+$. Yet this is an oxidation step and not a reduction step. Explain.

20.22 Which of these fatty acids can be synthesized by the multienzyme fatty acid synthesis complex alone?
(a) Oleic   (b) Stearic
(c) Myristic   (d) Arachidonic
(e) Lauric

20.23 Some enzymes can use NADH as well as NADPH as a coenzyme. Other enzymes use one or the other exclusively. Which features would prevent NADPH from fitting into the active site of an enzyme that otherwise can accommodate NADH?

20.24 Are fatty acids for energy, in the form of fat, synthesized in the same way as fatty acids for the lipid bilayer of membrane?

20.25 Linoleic and linolenic acids cannot be synthesized in the human body. Does this mean that the human body cannot make an unsaturated fatty acid from a saturated one?

Biosynthesis of Membrane Lipids

20.26 When the body synthesizes the following membrane lipid, from which building blocks is it assembled?

$$CH_2-O-\overset{\overset{\displaystyle O}{\|}}{C}-(CH_2)_{14}CH_3$$

$$CH-O-\overset{\overset{\displaystyle O}{\|}}{C}-(CH_2)_{10}CH_3$$

$$CH_2-O-\overset{\overset{\displaystyle O}{\|}}{\underset{\underset{\displaystyle O^-}{|}}{P}}-O-CH_2-\underset{\underset{\displaystyle NH_3^+}{|}}{CH}-COO^-$$

20.27 Name the activated constituents necessary to form the glycolipid glucoceramide.

20.28 Why is HMG reductase a key enzyme in cholesterol synthesis?

20.29 Describe by carbon skeleton designation how a C_2 compound ends up as a C_5 compound.

Biosynthesis of Amino Acids

20.30 Which reaction is the reverse of the synthesis of glutamate from α-ketoglutarate, ammonia, and $NADH + H^+$?

20.31 Which amino acid will be synthesized by the following process?

$$\begin{array}{l} COO^- \\ | \\ C=O \\ | \quad\quad + NADH + H^+ + NH_4^+ \longrightarrow \\ CH_2 \\ | \\ COO^- \end{array}$$

20.32 Draw the structure of the compound needed to synthesize asparagine from glutamate by transamination.

20.33 Name the products of the following transamination reaction:

$$(CH_3)_2CH-\overset{\overset{\displaystyle O}{\|}}{C}-COO^-$$

$$+ \ ^-OOC-CH_2-CH_2-\underset{\underset{\displaystyle NH_3^+}{|}}{CH}-COO^- \longrightarrow$$

Chemical Connections

20.34 (Chemical Connections 20A) Photosystem I and II are complex factories of proteins, chlorophyll, and many cofactors. Where are these photosystems located in the plant and in which reaction of photosynthesis do they participate?

20.35 (Chemical Connections 20A) Which coenzyme reduces CO_2 in the Calvin cycle?

20.36 (Chemical Connections 20B) Why would an unprenylated Ras protein fail to transduct signals across a membrane?

20.37 (Chemical Connections 20C) (a) What compound produces the colored urine of blue diaper syndrome?

(b) What is the source of this compound?

Additional Problems

20.38 In the structure of $NADP^+$ what are the bonds that connect nicotinamide and adenine to the ribose units?

20.39 Which C_3 fragment carried by ACP is used in fatty acid synthesis?

20.40 When glutamate transaminates phenylpyruvate, what amino acid is produced?

$$C_6H_5-CH_2-\overset{\overset{\displaystyle O}{\|}}{C}-COO^-$$

$$+ \ ^-OOC-CH_2-CH_2-\underset{\underset{\displaystyle NH_3^+}{|}}{CH}-COO^- \longrightarrow$$

20.41 Does the prenylation of Ras protein itself cause cancer?

20.42 Each activation step in the synthesis of complex lipids occurs at the expense of one ATP molecule. How many ATP molecules are used in the synthesis of one molecule of lecithin?

20.43 Consider the fact that the deamination of glutamic acid and its synthesis from α-ketoglutaric acid are equilibrium reactions. Which way will the equilibrium shift when the human body is exposed to cold temperature?

20.44 Which compound reacts with glutamate in a transamination process to yield serine?

20.45 What are the names of the C_{10} and C_{15} intermediates in cholesterol biosynthesis?

20.46 Which is carbon number 1 in HMGCoA (3-hydroxy-3-methylglutaryl CoA)?

20.47 In most biosynthetic processes, the reactant is reduced to obtain the desired product. Verify whether this statement holds for the overall reaction of photosynthesis.

20.48 What is the major difference in structure between chlorophyll and heme?

20.49 Can the complex enzyme system participating in every fatty acid synthesis manufacture fatty acids of any length?

InfoTrac College Edition

For additional reading, go to InfoTrac College Edition, your online research library, at

http://infotrac.thomsonlearning.com

CHAPTER **21**

21.1   Introduction

21.2   Nutrition

21.3   Calories

21.4   Carbohydrates in the Diet and Their Digestion

21.5   Fats in the Diet and Their Digestion

21.6   Proteins in the Diet and Their Digestion

21.7   Vitamins, Minerals, and Water

Charles D. Winters

Foods high in fiber include whole grains, legumes, fruits, and vegetables.

Nutrition

21.1 Introduction

In Chapters 18 and 19, we saw what happens to the food that we eat in its final stages—after the proteins, lipids, and carbohydrates have been broken down into their components. In this chapter, we discuss the earlier stages—nutrition and diet—and then the digestive processes that break down the large molecules to the small ones that undergo metabolism. The purpose of food is to provide energy and new molecules to replace those that the body uses. This synthesis of new molecules is especially important for the period during which a child becomes an adult.

21.2 Nutrition

The components of food and drink that provide growth, replacement, and energy are called **nutrients.** Not all components of food are nutrients. Some components of food and drinks, such as those that provide flavor, color, or aroma, enhance our pleasure in the food but are not themselves nutrients.

In the days of sailing ships, many sailors contracted scurvy because they had no source of fresh fruit or vegetables on voyages that could last many weeks away from land. To counter this problem, British ships carried limes, which could be stored for long periods. Limes are an excellent source of vitamin C. British sailors acquired the nickname "limey" from this practice, a name that was often extended to any Englishman.

468

CHEMICAL CONNECTIONS 21A

Parenteral Nutrition

In clinical practice, it occasionally happens that a patient cannot be fed through the gastrointestinal tract. Under these conditions, intravenous feeding or parenteral nutrition is required.

In the past, patients could be maintained on parenteral nutrition for only short periods. Calories were provided in the form of glucose, and amino acids were supplied in the form of protein hydrolysates. Over long periods, however, the use of parenteral nutrition resulted in deterioration and progressive emaciation of the patient.

More recently, the solutions used in parenteral nutrition have been manufactured to contain both essential and nonessential amino acids in fixed proportions. Calories now do not come only from glucose. Glucose supplies only about 20% of the caloric needs, fat emulsions provide the rest of the caloric requirements and the essential fatty acids. The parenteral solution also supplies the full vitamin and mineral requirements.

Nutritionists classify nutrients into six groups:

1. Carbohydrates
2. Lipids
3. Proteins
4. Vitamins
5. Minerals
6. Water

For food to be used in our bodies, it must be absorbed through the intestinal walls into the bloodstream or lymph system. Some nutrients, such as vitamins, minerals, glucose, and amino acids, can be absorbed directly. Others, such as starch, fats, and proteins, must first be broken down into smaller components before they can be absorbed. This breakdown process is called **digestion.**

A healthy body needs the proper intake of all nutrients. However, nutrient requirements vary from one person to another. For example, more energy is needed to maintain the body temperature of an adult than that of a child. For this reason, nutritional requirements are usually given per kilogram of body weight. Furthermore, the energy requirements of a physically active body are greater than those of a person in a sedentary occupation. Therefore, when average values are given, as in **Dietary Reference Intakes (DRI)** and in the former guidelines of **Recommended Daily Allowances (RDA),** one should be aware of the wide range that these average values represent.

The public interest in nutrition and diet changes with time and geography. Seventy or eighty years ago, the main nutritional interest of most Americans was getting enough food to eat and avoiding diseases caused by vitamin deficiencies, such as scurvy or beriberi. This issue is still the main concern of the large majority of the world's population. In affluent societies such as ours, however, today's nutritional message is no longer "eat more" but rather "eat less and discriminate more in your selection of food." Dieting to reduce body weight is a constant effort in a sizable percentage of the American population. Many people discriminate in their selection of food to avoid cholesterol (Section 12.9E) and saturated fatty acids to reduce the risk of heart attacks.

Along with such **discriminatory curtailment diets** came many different faddish diets. **Diet faddism** is an exaggerated belief in the effects of

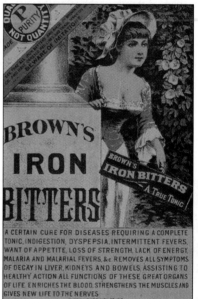

BROWN'S IRON BITTERS

A CERTAIN CURE FOR DISEASES REQUIRING A COMPLETE TONIC, INDIGESTION, DYSPEPSIA, INTERMITTENT FEVERS, WANT OF APPETITE, LOSS OF STRENGTH, LACK OF ENERGY, MALARIA AND MALARIAL FEVERS, &c. REMOVES ALL SYMPTOMS OF DECAY IN LIVER, KIDNEYS AND BOWELS, ASSISTING TO HEALTHY ACTION ALL FUNCTIONS OF THESE GREAT ORGANS OF LIFE. ENRICHES THE BLOOD, STRENGTHENS THE MUSCLES AND GIVES NEW LIFE TO THE NERVES.

■ A late-nineteenth-century advertisement for a cure-all tonic.

Digestion The process in which the body breaks down large molecules into smaller ones that can then be absorbed and metabolized

Discriminatory curtailment diet A diet that avoids certain food ingredients that are considered harmful to the health of an individual—for example, low-sodium diets for people with high blood pressure

Nutrition Facts

Serving size 1 Bar (28g)
Servings per Container 6

Amount Per Serving

Calories 120   Calories from Fat 35

% Daily Value*

Total Fat 4g	6%
Saturated Fat 2g	10%
Cholesterol 0mg	0%
Sodium 45mg	2%
Potassium 100mg	3%
Total Carbohydrate 19g	6%
Dietary fiber 2g	8%
Sugars 13g	
Protein 2g	

Vitamin A 15%	•	Vitamin C 15%
Calcium 15%	•	Iron 15%
Vitamin D 15%	•	Vitamin E 15%
Thiamin 15%	•	Riboflavin 15%
Niacin 15%	•	Vitamin B6 15%
Folate 15%	•	Vitamin B12 15%
Biotin 10%	•	Pantothenic Acid 10%
Phosphorus 15%	•	Iodine 2%
Magnesium 4%	•	Zinc 4%

*Percent Daily Values are based on a 2,000 calorie diet. Your daily values may be higher or lower depending on your calorie needs.

	Calories:	2,000	2,500
Total Fat	Less than	65g	80g
Sat Fat	Less than	20g	25g
Cholesterol	Less than	300mg	300mg
Sodium	Less than	2,400mg	2,400mg
Potassium		3,500mg	3,500mg
Total Carbohydrate		300g	375g
Dietary Fiber		25g	30g

Figure 21.1 A food label for a peanut butter crunch bar. The portion at the bottom (following the asterisk) is the same on all labels that carry it.

Fiber The cellulosic non-nutrient component in our food

nutrition upon health and disease. This phenomenon is not new; it has been prevalent for many years. Many times it is driven by visionary beliefs, but backed by little science. In the nineteenth century, Dr. Kellogg (of the cornflakes fame) recommended a largely vegetarian diet based on his belief that meat produces sexual excess. Eventually his religious fervor withered and his brother made a commercial success of his inventions of grain-based food. The latest fad is raw food, which bans any application of heat higher than 118°F to food. Heat, these faddists maintain, depletes the nutritional value of proteins and vitamins and concentrates pesticides in food. Obviously raw food diet is vegetarian, as it excludes meat and meat products.

A recommended food is rarely as good, and a condemned food is never as bad, as faddists claim. Scientific studies have proved, for instance, that food grown with chemical fertilizers is just as healthy as food grown with organic (natural) fertilizers.

Each food contains a large variety of nutrients. For example, a typical breakfast cereal lists the following items as its ingredients: milled corn, sugar, salt, malt flavoring, and vitamins A, B, C, and D, plus flavorings and preservatives. U.S. consumer laws require that most packaged food be labeled in a uniform manner to show the nutritional values of the food. Figure 21.1 shows a typical label of the type found on almost every can, bottle, or box of food that we buy.

Such labels must list the percentages of **Daily Values** for four key vitamins and minerals: vitamins A and C, calcium, and iron. If other vitamins or minerals have been added, or if the product makes a nutritional claim about other nutrients, their values must be shown as well. The percent daily values on the labels are based on a daily intake of 2000 Cal. For anyone who eats more than that amount, the actual percentage figures would be lower (and higher for those who eat less). Note that each label specifies the serving size; the percentages are based on that portion, not on the entire contents of the package. The section at the bottom of the label is exactly the same on all labels, no matter what the food, and shows the daily amounts of nutrients recommended by the government, based on a consumption of either 2000 or 2500 Cal. Some food packages are allowed to carry shorter labels, either because they have only a few nutrients or because the package has limited label space. The uniform labels make it much easier for the consumer to know exactly what he or she is eating.

The U.S. Department of Agriculture issues occasional guidelines as to what constitutes a healthy diet, depicted in the form of a pyramid (Figure 21.2). These guidelines consider the basis of a healthy diet to be the foods richest in starch (bread, rice, and so on), plus lots of fruits and vegetables (which are rich in vitamins and minerals). Protein-rich foods (meat, fish, dairy products) are to be consumed more sparingly, and fats, oils, and sweets are not considered necessary at all.

An important non-nutrient in some foods is **fiber,** which generally consists of the indigestible portion of vegetables and grains. Lettuce, cabbage, celery, whole wheat, brown rice, peas, and beans are all high in fiber. Chemically, fiber is made of cellulose, which, we saw in Section 11.7C, cannot be digested by humans. But though we cannot digest it, fiber is necessary for proper operation of the digestive system; without it constipation may result. In more serious cases, a diet lacking sufficient fiber may lead to colon cancer. The DRI recommendation is to ingest 35 g/day for men 50 and younger, and 25 g/day for women of the same age.

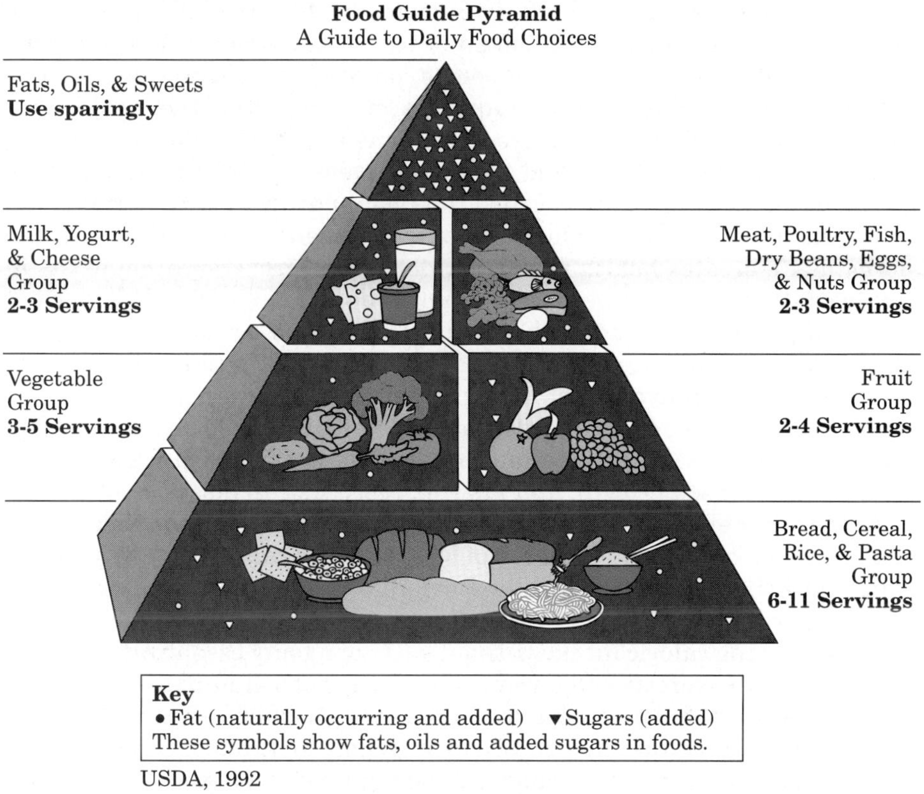

Food Guide Pyramid
A Guide to Daily Food Choices

Figure 21.2 The Food Guide Pyramid developed by the U.S. Department of Agriculture as a general guide to a healthful diet.

Fats, Oils, & Sweets
Use sparingly

Milk, Yogurt, & Cheese Group
2-3 Servings

Meat, Poultry, Fish, Dry Beans, Eggs, & Nuts Group
2-3 Servings

Vegetable Group
3-5 Servings

Fruit Group
2-4 Servings

Bread, Cereal, Rice, & Pasta Group
6-11 Servings

Key
• Fat (naturally occurring and added)   ▼ Sugars (added)
These symbols show fats, oils and added sugars in foods.

USDA, 1992

FOOD GROUP SERVING SIZES

Bread, Cereal, Rice, and Pasta
1/2 cup cooked cereal, rice, pasta
1 ounce dry cereal
1 slice bread
2 cookies
1/2 medium doughnut

Vegetables
1/2 cup cooked or raw chopped
  vegetables
1 cup raw leafy vegetables
3/4 cup vegetable juice
10 french fries

Fruit
1 medium apple, banana, or orange
1/2 cup chopped, cooked, or canned
  fruit
3/4 cup fruit juice
1/4 cup dried fruit

Milk, Yogurt, and Cheese
1 cup milk or yogurt
$1\frac{1}{2}$ ounces natural cheese
2 ounces process cheese
2 cups cottage cheese
$1\frac{1}{2}$ cups ice cream
1 cup frozen yogurt

Meat, Poultry, Fish, Dry Beans, Eggs, and Nuts
2–3 ounces cooked lean meat, fish,
  or poultry
2–3 eggs
4–6 tablespoons peanut butter
$1\frac{1}{2}$ cups cooked dry beans
1 cup nuts

Fats, Oils, and Sweets
Butter, mayonnaise, salad dressing,
  cream cheese, sour cream, jam, jelly

21.3 Calories

The largest part of our food supply goes to provide energy for our bodies. As we saw in Chapters 18 and 19, this energy comes from the oxidation of carbohydrates, fats, and proteins. The energy derived from food is usually measured in calories. One nutritional calorie (Cal) equals 1000 cal or 1 kcal. Thus, when we say that the average daily nutritional requirement for a young adult male is 2900 Cal, we mean the same amount of energy needed

One calorie is the amount of heat necessary to raise the temperature of 1 g of liquid water by 1°C.

■ Different activity levels are associated with different caloric needs.

Basal caloric requirement The caloric requirement for a resting body

Excessive caloric intake results from all sources of food—carbohydrates and proteins as well as fats.

Obesity is defined by the National Institutes of Health as being a person who has a body mass index (BMI) of 30 or greater. The BMI is a measure of body fat based on height and weight that applies to both adult men and women. For example, a person 70 inches high is *normal* (BMI less than 25) if he weighs 174 lb or less. A person of the same height is *overweight* if he weighs more than 174 lb but less than 209 lb; he is *obese* if he weighs more than 209 lb. More than 97 million Americans are overweight or obese.

to raise the temperature of 2900 kg of water by 1°C or of 29 kg (64 lb) of water by 100°C. A young adult female needs 2100 Cal/day. These are peak requirements—children and older people, on average, require less energy. Keep in mind that these energy requirements apply to active people. For bodies completely at rest, the corresponding energy requirement for young adult males is 1800 Cal/day, and that for females is 1300 Cal/day. The requirement for a resting body is called the **basal caloric requirement.**

An imbalance between the caloric requirement of the body and the caloric intake creates health problems. Chronic caloric starvation exists in many parts of the world where people simply do not have enough food to eat because of prolonged drought, the devastation of war, natural disasters, or overpopulation. Famine particularly affects infants and children. Chronic starvation, called **marasmus,** increases infant mortality by as much as 50%. It results in arrested growth, muscle wasting, anemia, and general weakness. Even if starvation is later alleviated, it leaves permanent damage, insufficient body growth, and lowered resistance to disease.

At the other end of the caloric spectrum is excessive caloric intake. It results in obesity, or the accumulation of body fat. **Obesity** is becoming an epidemic in the U.S. population, with important consequences: It increases the risk of hypertension, cardiovascular disease, and diabetes. Reducing diets aim at decreasing caloric intake without sacrificing any essential nutrients. A combination of exercise and lower caloric intake can eliminate obesity, but usually these diets must achieve their goal over an extended period.

Crash diets give the illusion of quick weight loss, but most of this decrease is due to loss of water, which can be regained very quickly. To reduce obesity, we must lose body fat and not water; achieving this goal takes a lot of effort, because fats contain so much energy. A pound of body fat is equivalent to 3500 Cal. Thus, to lose 10 pounds, it is necessary either to consume 35,000 fewer Cal, which can be achieved if one reduces caloric intake by 350 Cal every day for 100 days (or by 700 Cal daily for 50 days), or to use up, through exercise, the same number of food calories.

21.4 Carbohydrates in the Diet and Their Digestion

Carbohydrates are the major source of energy in the diet. They also furnish important compounds for the synthesis of cell components (Chapter 20). The main dietary carbohydrates are the polysaccharide starch, the disaccharides lactose and sucrose, and the monosaccharides glucose and fructose. Before the body can absorb carbohydrates, it must break down di-, oligo-, and polysaccharides into monosaccharides, because only monosaccharides can pass into the bloodstream.

The monosaccharide units are connected to each other by glycosidic bonds. Glycosidic bonds are cleaved by hydrolysis. In the body, this hydrolysis is catalyzed by acids and by enzymes. When a metabolic need arises, storage polysaccharides—amylose, amylopectin, and glycogen—are hydrolyzed to yield glucose and maltose.

This hydrolysis is aided by a number of enzymes:

● **α-Amylase** attacks all three storage polysaccharides at random, hydrolyzing the α-1,4 glycosidic bonds,

● **β-Amylase** also hydrolyzes the α-1,4 glycosidic bonds but in an orderly fashion, cutting disaccharidic maltose units one by one from the nonreducing end of a chain.

● The **debranching enzyme** hydrolyzes the α-1,6 glycosidic bonds (Figure 21.3).

In acid-catalyzed hydrolysis, storage polysaccharides are cut at random points. At body temperature, acid catalysis is slower than the enzyme-catalyzed hydrolysis.

The digestion (hydrolysis) of starch and glycogen in our food supply starts in the mouth, where α-amylase is one of the main components of saliva. Hydrochloric acid in the stomach and other hydrolytic enzymes in the intestinal tract decompose starch and glycogen to produce mono- and disaccharides (D-glucose and maltose).

D-glucose enters the bloodstream and is carried to the cells to be utilized (Section 19.2). For this reason, D-glucose is often called blood sugar. In healthy people, little or none of this sugar ends up in the urine except for short periods of time (binge eating).

In diabetes mellitus, however, glucose is not completely metabolized and does appear in the urine (Section 23.5B). Because of this, it is necessary to test the urine of diabetic patients for the presence of glucose (Chemical Connections 11C).

The latest DRI guideline, issued by the National Academy of Sciences in 2002, recommends a minimum carbohydrate intake of 130 g/day. Most people exceed this value. Artificial sweeteners (Chemical Connections 21C) can be used to reduce mono- and disaccharide intake.

Not everyone considers this recommendation to be the final word. Certain diets, such as the Atkins diet, recommend reducing carbohydrate intake to force the body to burn stored fat for energy supply. For example, during the introductory two-week period in the Atkins diet, only 20 g of carbohydrates per day is recommended in the form of salad and vegetables; no fruits or starchy vegetables are allowed. For the longer term, an additional 5 g of carbohydrates per day is added in the form of fruits. This restriction induces ketosis, the production of ketone bodies (Section 19.7) that may create muscle weakness and kidney problems.

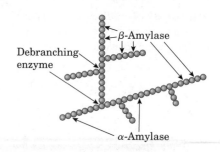

Figure 21.3 The action of different enzymes on glycogen and starch.

21.5 Fats in the Diet and Their Digestion

Fats are the most concentrated source of energy. About 98% of the lipids in our diet are fats and oils (triglycerides); the remaining 2% consist of complex lipids and cholesterol.

The lipids in the food we eat must be hydrolyzed into smaller components before they can be absorbed into the blood or lymph system through the intestinal walls. The enzymes that promote this hydrolysis are located in the small intestine and are called *lipases*. However, since lipids are insoluble in the aqueous environment of the gastrointestinal tract, they must be dispersed into fine colloidal particles before the enzymes can act on them.

Bile salts (Section 12.11) perform this important function. Bile salts are manufactured in the liver from cholesterol and stored in the gallbladder. From there, they are secreted through the bile ducts into the intestine. Lipases act on the emulsion produced by bile salts and dietary fats, breaking the fats into glycerol and fatty acids and complex lipids into fatty acids, alcohols (glycerol, choline, ethanolamine, sphingosine), and carbohydrates. These hydrolysis products are then absorbed through the intestinal walls.

Only two fatty acids are essential in higher animals, including humans: linolenic and linoleic acids (Section 12.3). Nutritionists occasionally list arachidonic acid as an **essential fatty acid.** However, our bodies can synthesize arachidonic acid from linoleic acid.

Figure 21.4 Different enzymes hydrolyze peptide chains in different but specific ways.

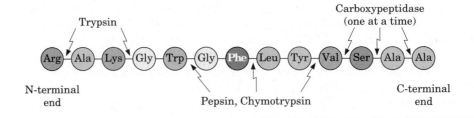

21.6 Proteins in the Diet and Their Digestion

Although the proteins in our diet can be used for energy (Section 19.9), their main use is to furnish amino acids from which the body synthesizes its own proteins (Section 17.5).

The digestion of dietary proteins begins with cooking, which denatures proteins. (Denatured proteins are hydrolyzed more easily by hydrochloric acid in the stomach and by digestive enzymes than are native proteins.) *Stomach acid* contains about 0.5% HCl. This HCl both denatures the proteins and hydrolyzes the peptide bonds randomly. *Pepsin,* the proteolytic enzyme of stomach juice, breaks peptide bonds on the C=O side of three amino acids only: tryptophan, phenylalanine, and tyrosine (Figure 21.4).

Most protein digestion occurs in the small intestine. There, the enzyme *chymotrypsin* breaks internal peptide bonds at the same positions as does pepsin. Another enzyme, *trypsin,* hydrolyzes them only on the C=O side of arginine and lysine. Other enzymes, such as *carboxypeptidase,* hydrolyze amino acids one by one from the C-terminal end of the protein. The amino acids and small peptides are then absorbed through the intestinal walls.

The human body is incapable of synthesizing ten of the amino acids needed to make proteins. These ten (the **essential amino acids**) must be obtained from our food; they are shown in Table 13.1. The body hydrolyzes food proteins into their amino acid constituents and then puts the amino acids together again to make body proteins. For proper nutrition, the human diet should contain about 20% protein.

A dietary protein that contains all of the essential amino acids is called a **complete protein.** Casein, the protein of milk, is a complete protein, as are most other animal proteins—those found in meat, fish, and eggs. People who eat adequate quantities of meat, fish, eggs, and dairy products get all the amino acids they need to keep healthy. About 50 g/day of complete proteins constitutes an adequate quantity.

An important animal protein that is not complete is gelatin, which is made by denaturing collagen (Section 13.11). Gelatin lacks tryptophan and is low in several other amino acids, including isoleucine and methionine. Many people on quick-reducing diets consume "liquid protein." This substance is simply denatured and partially hydrolyzed collagen (gelatin). Therefore, if it is the only protein source in the diet, some essential amino acids will be lacking.

Most plant proteins are incomplete. For example, corn protein lacks lysine and tryptophan; rice protein lacks lysine and threonine; wheat protein lacks lysine; and legumes are low in methionine and cystein. Even soy protein, one of the best plant proteins, is very low in methionine. Adequate amino acid nutrition is possible with a vegetarian diet, but only if a wide range of vegetables is eaten. **Protein complementation** is one such diet. In protein complementation, two or more foods complement the others' deficiencies. For example, grains and legumes complement each other, with grains being low in lysine but high in methionine. Over time, such protein

Essential amino acid An amino acid that the body cannot synthesize in the required amounts and so must be obtained in the diet

Table 21.1   Vitamins and Trace Elements as Coenzymes and Cofactors

Vitamin/Trace Element	Form of Coenzyme	Representative Enzyme	Reference
B_1, thiamine	Thiamine pyrophosphate, TPP	Pyruvate dehydrogenase	Step 12, Section 19.2
B_2, riboflavin	Flavin adenine dinucleotide, FAD	Succinate dehydrogenase	Step 6, Section 18.4
Niacin	Nicotinamide adenine dinucleotide, NAD^+	D-Glyceraldehyde-3-phosphate dehydrogenase	Step 5, Section 19.2
Panthotenic acid	Coenzyme A, CoA	Fatty acid synthase	Step 1, Section 20.3
B_6, pyridoxal	Pyridoxal phosphate, PLP	Aspartate amino transferase	Class 2, Section 14.2
B_{12}, cobalamine		Ribose reductase	Step 1, Section 16.2
Biotin	N-carboxybiotin	Acetyl CoA carboxylase	Malonyl CoA, Section 20.3
Folic acid		Purin biosynthesis	Section 16.2
Mg		Pyruvate kinase	Chemical Connections 14C
Fe		Cytochrome oxidase	Section 18.5
Cu		Cytochrome oxidase	Section 18.5
Zn		DNA polymerase	Section 16.6
Mn		Arginase	Step 5, Section 19.8
K		Pyruvate kinase	Chemical Connections 14C
Ni		Urease	Section 14.1
Mo		Nitrate reductase	
Se		Glutathione peroxidase	Chemical Connections 13A

complementation in vegetarian diets became the staple in many parts of the world—corn tortillas and beans in Central and South America, rice and lentils in India, and rice and tofu in China and Japan.

In many developing countries, protein deficiency diseases are widespread because the people get their protein mostly from plants. Among these diseases is **kwashiorkor,** whose symptoms include a swollen stomach, skin discoloration, and retarded growth.

21.7 Vitamins, Minerals, and Water

Vitamins and **minerals** are essential for good nutrition. Animals maintained on diets that contain sufficient carbohydrates, fats, and proteins and provided with an ample water supply cannot survive on these alone; they also need the essential organic components called vitamins and the inorganic ions called minerals. Many vitamins, especially those in the B group, act as coenzymes and inorganic ions as cofactors in enzyme-catalyzed reactions (Table 21.1). Table 21.2 lists the structures, dietary sources and functions of the vitamins and minerals. Deficiencies in vitamins and minerals lead to many nutritionally controllable diseases (one example is shown in Figure 21.5); these conditions are also listed in Table 21.2.

The recent trend in vitamin appreciation is connected to their general role rather than any specific action against a particular disease. For example, today the role of vitamin C in prevention of scurvy is barely mentioned but it is hailed as an important antioxidant. Similarly, other antioxidant vitamins or vitamin precursors dominate the medical literature. As an example, it has been shown that consumption of carotenoids (other than β-carotene), as well as vitamin E and C contributes significantly to respiratory health. The most important of the three is vitamin E. Furthermore, the loss of vitamin C during hemodialysis contributes significantly to oxidative damage in patients, leading to accelerated atherosclerosis.

Some vitamins have unusual effects besides their normal participation as coenzymes in many metabolic pathways or action as antioxidants in the body. Among these are the well-known effect of riboflavin, vitamin B_2, as a

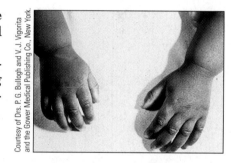

Courtesy of Drs. P. G. Bullogh and V. J. Vigorita and the Gower Medical Publishing Co., New York.

Figure 21.5   Symptoms of ricketse, a vitamin D deficiency in children. The nonmineralization of the bones of the radius and the ulna results in prominence of the wrist.

The name "vitamin" comes from a mistaken generalization. Casimir Funk (1884–1967) discovered that certain diseases such as beriberi, scurvy, and pellagra are caused by lack of certain nutrients. He found that the "antiberiberi factor" compound is an amine and mistakenly thought that all such nutrients are amines. Hence, Funk coined the name "Vitamine."

Table 21.2 Vitamins and Minerals: Sources, Functions, Deficiency Diseases, and Daily Requirements

Name	Structure	Best Food Source	Function	Deficiency Symptoms and Diseases	Recommended Dietary Allowance[a]
Fat-Soluble Vitamins					
A	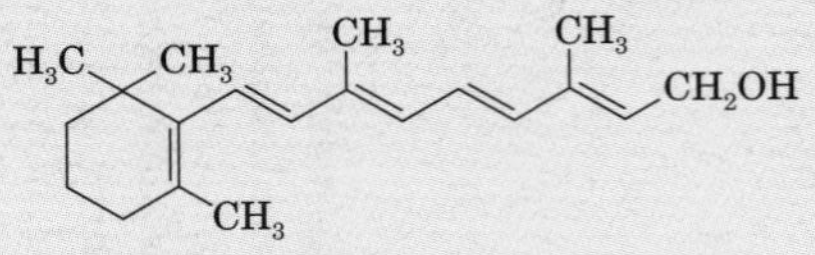	Liver, butter, egg yolk, carrots, spinach, sweet potatoes	Vision; to heal eye and skin injuries	Night blindness; blindness; keratinization of epithelium and cornea	800 μg (1500 μg)[b]
D	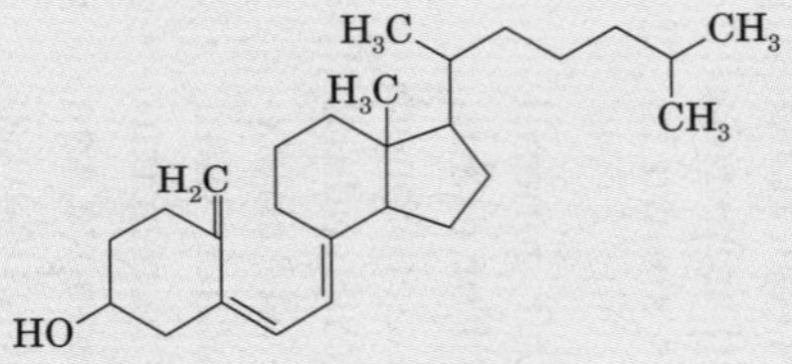	Salmon, sardines, cod liver oil, cheese, eggs, milk	Promotes calcium and phosphate absorption and mobilization	Rickets (in children): pliable bones; osteomalacia (in adults); fragile bones	5–10 μg; exposure to sunlight
E	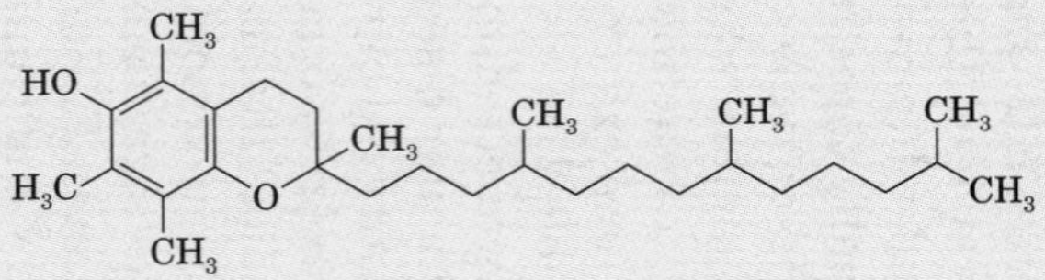	Vegetable oils, nuts, potato chips, spinach	Antioxidant	In cases of malabsorption such as in cystic fibrosis: anemia. In premature infants: anemia	8–10 mg
K	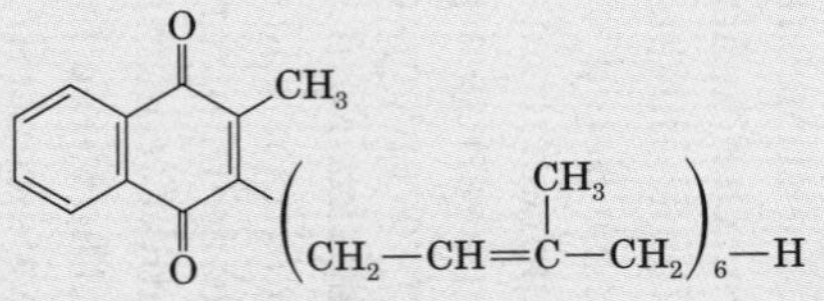	Spinach, potatoes, cauliflower, beef liver	Blood clotting	Uncontrolled bleeding (mostly in newborn infants)	65–80 μg
Water-Soluble Vitamins					
B$_1$ (thiamine)	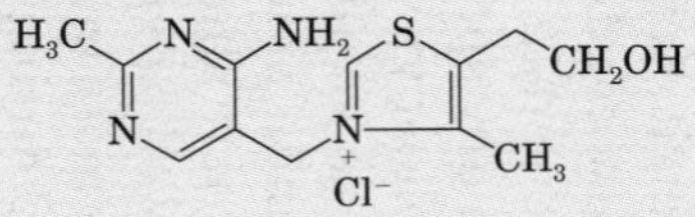	Beans, soybeans, cereals, ham, liver	Coenzyme in oxidative decarboxylation and in pentose phosphate shunt	Beriberi. In alcoholics: heart failure; pulmonary congestion	1.1 mg
B$_2$ (riboflavin)	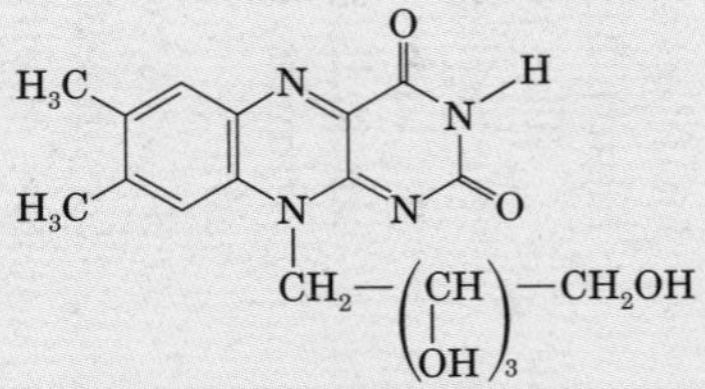	Kidney, liver, yeast, almonds, mushrooms, beans	Coenzyme of oxidative processes	Invasion of cornea by capillaries; cheilosis; dermatitis	1.4 mg

Vitamin	Source	Function	Deficiency symptoms	Recommended daily allowance
Nicotinic acid (niacin) 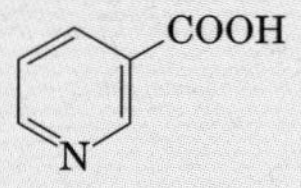	Chickpeas, lentils, prunes, peaches, avocados, figs, fish, meat, mushrooms, peanuts, bread, rice, beans, berries	Coenzyme of oxidative processes	Pellagra	15–18 mg
B$_6$ (pyridoxal) 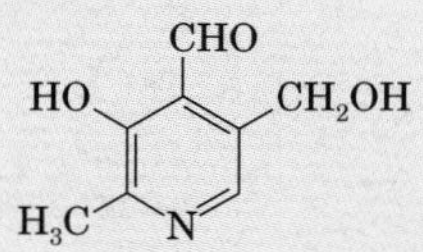	Meat, fish, nuts, oats, wheat germ, potato chips	Coenzyme in transamination; heme synthesis	Convulsions; chronic anemia; peripheral neuropathy	1.6–2.2 mg
Folic acid 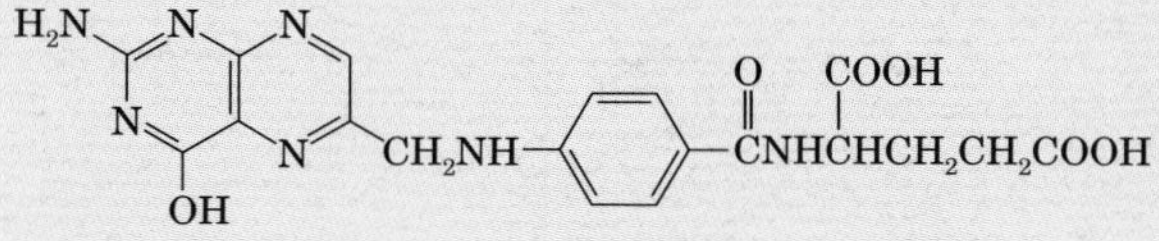	Liver, kidney, eggs, spinach, beets, orange juice, avocados, cantaloupe	Coenzyme in methylation and in DNA synthesis	Anemia	400 μg
B$_{12}$ 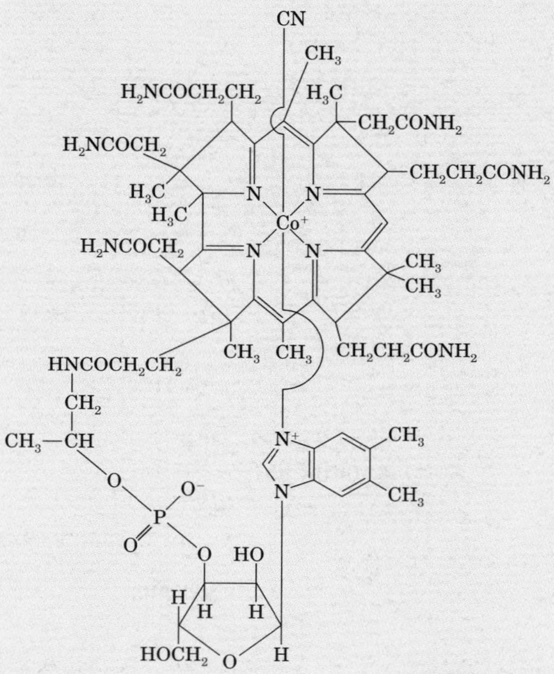	Oysters, salmon, liver, kidney	Part of methyl-removing enzyme in folate metabolism	Patchy demyelination; degradation of nerves, spinal cord, and brain	1–3 μg

Table 21.2   Vitamins and Minerals: Sources, Functions, Deficiency Diseases, and Daily Requirements *(continued)*

Name	Structure	Best Food Source	Function	Deficiency Symptoms and Diseases	Recommended Dietary Allowance[a]
Pantothenic acid	CH₃ O | || HOCH₂C—CHNHCH₂CH₂COOH | CH₃ OH	Peanuts, buckwheat, soybeans, broccoli, lima beans, liver, kidney, brain, heart	Part of CoA; fat and carbohydrate metabolism	Gastrointestinal disturbances; depression	4–7 mg
Biotin	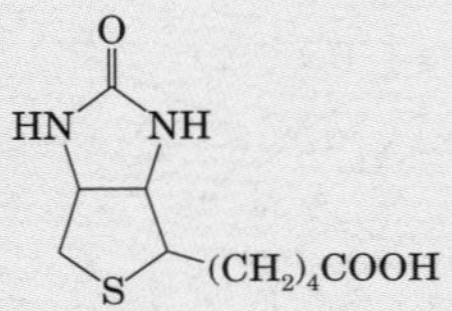 (structure shown)	Yeast, liver, kidney, nuts, egg yolk	Synthesis of fatty acids	Dermatitis; nausea; depression	30–100 µg
C (ascorbic acid)	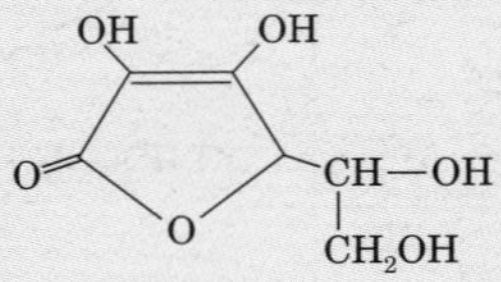 (structure shown)	Citrus fruit, berries, broccoli, cabbage, peppers, tomatoes	Hydroxylation of collagen, wound healing; bond formation; antioxidant	Scurvy; capillary fragility	60 mg
Minerals					
Potassium		Apricots, bananas, dates, figs, nuts, raisins, beans, chickpeas, cress, lentils	Provides membrane potential	Muscle weakness	3500 mg
Sodium		Meat, cheese, cold cuts, smoked fish, table salt	Osmotic pressure	None	2000–2400 mg
Calcium		Milk, cheese, sardines, caviar	Bone formation; hormonal function; blood coagulation; muscle contraction	Muscle cramps; osteoporosis; fragile bones	800–1200 mg
Chloride		Meat, cheese, cold cuts, smoked fish, table salt	Osmotic pressure	None	1700–5100 mg
Phosphorus		Lentils, nuts, oats, grain flours, cocoa, egg yolk, cheese, meat (brain, sweetbreads)	Balancing calcium in diet	Excess causes structural weakness in bones	800–1200 mg

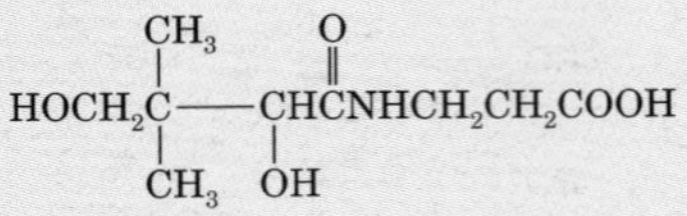

Element	Sources	Function	Deficiency Symptom	RDA
Magnesium	Cheeses, cocoa, chocolate, nuts, soybeans, beans	Cofactor in enzymes	Hypocalcemia	280–350 mg
Iron	Raisins, beans, chickpeas, parsley, smoked fish, liver, kidney, spleen, heart, clams, oysters	Oxidative phosphorylation; hemoglobin	Anemia	15 mg
Zinc	Yeast, soybeans, nuts, corn, cheese, meat, poultry	Cofactor in enzymes, insulin	Retarded growth; enlarged liver	12–15 mg
Copper	Oysters, sardines, lamb, liver	Oxidative enzymes cofactor	Loss of hair pigmentation, anemia	1.5–3 mg
Manganese	Nuts, fruits, vegetables, whole-grain cereals	Bone formation	Low serum cholesterol levels; retarded growth of hair and nails	2.0–5.0 mg
Chromium	Meat, beer, whole wheat and rye flours	Glucose metabolism	Glucose not available to cells	0.05–0.2 mg
Molybdenum	Liver, kidney, spinach, beans, peas	Protein synthesis	Retarded growth	0.075–0.250 mg
Cobalt	Meat, dairy products	Component of vitamin B_{12}	Pernicious anemia	0.05 mg (20–30 mg)[b]
Selenium	Meat, seafood	Fat metabolism	Muscular disorders	0.05–0.07 mg (2.4–3.0 mg)[b]
Iodine	Seafood, vegetables, meat	Thyroid glands	Goiter	150–170 μg (1000 μg)[b]
Fluorine	Fluoridated water; fluoridated toothpaste	Enamel formation	Tooth decay	1.5–4.0 mg (8–20 mg)[b]

[a]The U.S. RDAs are set by the Food and Nutrition Board of the National Research Council. The numbers given here are based on the latest recommendations (*National Research Council Recommended Dietary Allowances*, 10th ed., 1989, National Academy Press, Washington, D.C.). The RDA varies with age, sex, and level of activity; the numbers given are average values for both sexes between the ages of 18 and 54.

[b]Toxic if doses above the level shown in parentheses are taken.

Too Much Vitamin Intake?

One might suppose that, since vitamins are essential nutrients, the more you take, the better off you are. Using this reasoning, some people consider that the RDAs listed for each vitamin and mineral in Table 21.2 are merely a minimum, and that larger quantities will be even more beneficial. It has become fashionable to take large doses of vitamins, especially those with antioxidant properties—namely, vitamins A, C, and E. No scientific proof exists for such claims. Larger doses are usually harmless because the body excretes the excess, but not in all cases.

High dosages can do unexpected harm in certain instances, and complying with the RDA can prevent harm not usually associated with the vitamin. The case of folic acid illustrates this point. Table 21.2 shows an RDA of 400 μg for this vitamin, as the amount necessary to prevent anemia. It has also been shown

that this amount of folic acid is especially important during pregnancy. Consumption of this vitamin can prevent birth defects that result in spina bifida, the birth of a child with an open spine, and anencephaly, the birth of a child without most of the brain. Since B vitamins are not stored in the body, it is important that a pregnant woman take the RDA every day—but not more. An intake of 1000 μg or more of folic acid daily is harmful because it can hide the symptoms of certain anemias without actually preventing these diseases. The U.S. Food and Drug Administration now requires that food labeled as "enriched" include certain amounts of folic acid. The amount for a slice of enriched bread is 40 to 70 μg, and that for enriched cereals is 100 μg per serving.

Similarly, high dosages for niacin and of vitamins A and D have been shown to cause health problems.

Dieting and Artificial Sweeteners

Obesity is a serious problem in the United States. Many people would like to lose weight but are unable to control their appetites enough to do so, which is why artificial sweeteners are popular. These substances have a sweet taste but do not add calories. Many people restrict their sugar intake. Some are forced to do so by diseases such as diabetes; others, by the desire to lose weight. Since most of us like to eat sweet foods, artificial sweeteners are added to many foods and drinks for those who must (or want to) restrict their sugar intake.

Four noncaloric artificial sweeteners are now approved by the U.S. Food and Drug Administration (FDA). The oldest of these, saccharin, is 450 times sweeter than sucrose. Saccharin has been in use for about 100 years. Unfortunately, some tests have shown that this sweetener, when fed in massive quantities to rats, caused some of these rats to develop cancer. Other tests

have given negative results. Recent research has shown that this cancer in rats is not relevant to human consumption. Saccharin continues to be sold on the market.

A newer artificial sweetener, aspartame, does not have the slight aftertaste that saccharin does. Aspartame is the methyl ester of a dipeptide, Asp-Phe (aspartylphenylalanine). Its sweetness was discovered in 1969. After extensive biological testing, it was approved by the FDA in 1981 for use in cold cereals, drink mixes, and gelatins and as tablets or powder to be used as a sugar substitute. Aspartame is 100 to 150 times sweeter than sucrose. It is made from natural amino acids, so both the aspartic acid and the phenylalanine have the L configuration. Other possibilities have also been synthesized: the L-D, D-L and D-D configurations. However, they are all bitter rather than sweet.

(continued on next page)

Saccharin

Aspartame

Acesulfame-K

CHEMICAL CONNECTIONS 21C

Dieting and Artificial Sweeteners *(continued)*

Aspartame is sold under such brand names as Equal and NutraSweet. A newer version of aspartame, neotame, was approved by the FDA in 2002. Twenty-five times sweeter than aspartame, it is essentially the same compound but with one —H substituted by a —CH_2—CH_2—$C(CH_3)_3$ group at the amino terminal.

A third artificial sweetener, acesulfame-K, is 200 times sweeter than sucrose and is used (under the name Sunette) mostly in dry mixes. The fourth option, Sucralose, is used in sodas, baked goods, and tabletop packets. This trichloro derivative of a disaccharide is 600 times sweeter than sucrose and has no aftertaste. Neither Sucralose nor acesulfame-K is metabolized in the body; that is, they pass through unchanged.

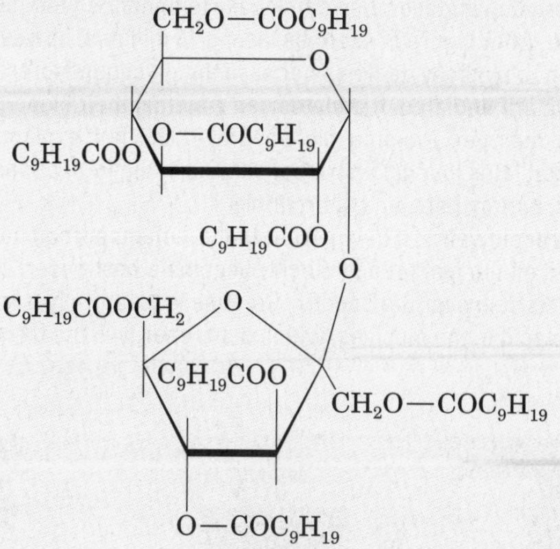

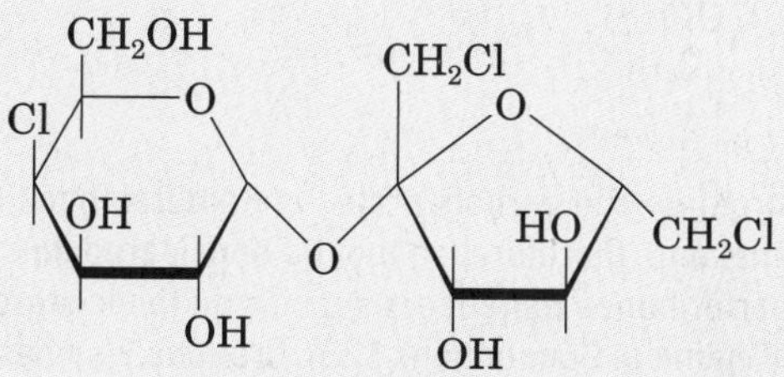

Sucralose

Of course, calories in the diet do not come only from sugar. Dietary fat is an even more important source (Section 21.5). It has long been hoped that some kind of artificial fat, which would taste the same but have no (or only a few) calories, would help in losing weight. Procter & Gamble Company has developed such a product, called Olestra. Like natural fats, this molecule is a carboxylic ester; but instead of glycerol, the alcohol component is sucrose (Section 11.5). All eight of sucrose's OH groups are converted to ester groups; the carboxylic acids are the fatty acids containing eight to ten carbons. The one illustrated here is made with the C_{10} decanoic (capric) acid.

Olestra

Although Olestra has a chemical structure similar to a fat, the human body cannot digest it because the enzymes that digest ordinary fats are not designed for the particular size and shape of this molecule. Therefore, it passes through the digestive system unchanged, and we derive no calories from it. Olestra can be used instead of an ordinary fat in cooking such items as cookies and potato chips. People who eat these will then be consuming fewer calories.

On the downside, Olestra can cause diarrhea, cramps, and nausea in some people, effects that increase with the amount consumed. Individuals who are susceptible to such side effects will have to decide if the potential to lose weight is worth the discomfort involved. Furthermore, Olestra dissolves and sweeps away some vitamins and nutrients in other foods being digested at the same time. To counter this effect, food manufacturers must add some of these nutrients (vitamins A, D, E, and K) to the product. FDA requires that all packages of Olestra-containing foods carry a warning label explaining these side effects.

---

photosensitizer. Niacinamide, the amide form of vitamin B (niacin), is used in megadoses (2 g/day) to treat an autoimmune disease called bullus pemphigoid, which causes blisters on the skin. Conversely, the same megadoses in healthy patients may cause harm (Chemical Connections 21B).

Although the concept of RDA has been used since 1940 and periodically updated as new knowledge dictated, a new concept is being developed in the field of nutrition. The DRI, currently in the process of completion, is designed to replace the RDA and is tailor-made to different ages and genders. It gives a set of two to four values for a particular nutrient in RDI:

● The estimated average requirement

● The recommended dietary allowance

PABA or *p*-aminobenzoic acid, a precursor of folic acid (Chemical Connections 14D), actually protects the skin against sunburn. As such, it is an integral part of many sunscreen lotions.

CHEMICAL CONNECTIONS 21D

Food for Performance Enhancement

Athletes do whatever they can legally to enhance their performance. After vigorous exercise lasting 15 minutes, however, performance typically declines because the glycogen stores in the muscle are depleted. If athletes can maintain their blood glucose levels, their performance might improve. This fact explains why many athletes load up on carbohydrates in the form of pasta meals, energy bars, and sport drinks.

In recent years, a new performance-enhancer food has appeared on the market and quickly become a best-seller: creatine. It is sold over the counter. Creatine is a naturally occurring amino acid in the muscles, which store energy in the form of

high-energy phosphocreatine. During a short and strenuous bout of exercise, such as a 100-meter sprint, the muscles first use up the ATP obtainable from the reaction of phosphocreatine with ADP; only then do they rely on the glycogen stores.

Phosphocreatine   Creatine

Both creatine and carbohydrates are natural food and body components and, therefore, cannot be considered equivalent to banned performance enhancers such as anabolic steroids or "andro" (Chemical Connections 12G). Creatine also has no known hazards, even in long-term use. It is beneficial in improving performance in sports where short bursts of energy are needed, such as weight lifting, jumping, and sprinting. Lately, it has been used experimentally to preserve muscle neurons in degenerative diseases such as Parkinson's disease, Huntington's disease, and muscular dystrophy.

■ Figure 21D Creatine is sold over the counter.

- The adequate intake level
- The tolerable upper intake level

For example, the RDA for vitamin D is 5-10 μg, the adequate intake listed in the DRI for vitamin D for a person between age 9 and 50 years is 5 μg, and the tolerable upper intake level for the same age group is 50 μg.

A third set of standards appears on food labels in the United States, the Daily Values discussed in Section 21.2. Each give a single value for each nutrient and reflects the need of an average healthy person eating 2000 to 2500 calories per day. The Daily Value for vitamin D, as appears on vitamin bottles, is 400 International Units, which is 10 μg, the same as the RDA.

Water makes up 60% of our body weight. Most of the compounds in our body are dissolved in water, which also serves as a transporting medium to carry nutrients and waste materials. These functions are discussed in Chapter 23. We must maintain a proper balance between water intake and water excretion via urine, feces, sweat, and exhalation of breath. A normal diet requires about 1200 to 1500 mL of water per day, in addition to the water consumed as part of our foods.

Public drinking water systems are regulated by the Environmental Protection Agency (EPA) which sets minimum standards for protection of public health. Public water systems are treated by disinfectants (typically chlorine) to kill microorganisms. Chlorinated water may have both an aftertaste and an odor. Private wells are not regulated by EPA standards.

Bottled water may come from springs, streams, or public water sources. Since bottled water is classified as a food, it comes under the FDA supervision, which requires it to meet the same standards for purity and sanitation as tap water. Most bottled water is disinfected by ozone, which leaves no flavor or odor in water.

S U M M A R Y

Nutrients are components of foods that provide growth, replacement, and energy (Section 21.2). Nutrients are classified into six groups: carbohydrates, lipids, proteins, vitamins, minerals, and water. Each food contains a variety of nutrients. The largest part of our food intake is used to provide energy for our bodies. A typical young adult needs 2900 Cal (male) or 2100 Cal (female) as an average daily caloric intake (Section 21.3). **Basal caloric requirements,** the energy needs when the body is completely at rest, are less than the normal requirements. An Imbalance between need and caloric intake may create health problems, for example, chronic starvation increases infant mortality, whereas obesity leads to hypertension, cardiovascular disease, and diabetes.

Carbohydrates are the major source of energy in our diet (Section 21.4). Monosaccharides are directly absorbed in the intestines, while oligo- and polysaccharides, such as starch, are digested with the aid of stomach acid, α- and β-amylases, and debranching enzymes.

Fats are the most concentrated source of energy (Section 21.5). They are emulsified by bile salts and digested by lipase before being absorbed as fatty acids and glycerol in the intestines. Essential fatty and amino acids are needed as building blocks because our bodies cannot synthesize them.

Proteins are hydrolyzed by stomach acid and further digested by enzymes such as pepsin and trypsin before being absorbed as amino acids (Section 21.6).

Vitamins and minerals are essential constituents of the diet that are needed in small quantities (Section 21.7). The fat-soluble vitamins are A, D, E, and K. Vitamins C and the B group are water-soluble vitamins. Most of the B vitamins are essential coenzymes. The most important dietary minerals are Na^+, Cl^-, K^+, PO_4^{3-}, Ca^{2+}, Fe^{2+}, and Mg^{2+}, but trace minerals are also necessary.

Water makes up 60% of body weight.

P R O B L E M S

Numbers that appear in color indicate difficult problems. ◀ designates problems requiring application of principles.

Nutrition

21.1 Are nutrient requirements uniform for everyone?

21.2 Is banana flavoring, isopentyl acetate, a nutrient?

21.3 If sodium benzoate, a food preservative, is excreted as such, and if calcium propionate, another food preservative, is metabolized to CO_2 and H_2O, would you consider either of these preservatives to be a nutrient? If so, why?

21.4 Is corn grown solely with organic fertilizers more nutritious than that grown with chemical fertilizers?

21.5 Which part of the Nutrition Facts label found on food packages is the same for all labels carrying it?

21.6 Of which kinds of food does the government recommend that we have the most servings each day?

21.7 What is the importance of fiber in the diet?

21.8 Can a chemical that, in essence, goes through the body unchanged be an essential nutrient? Explain.

Calories

21.9 A young adult female needs a caloric intake of 2100 Cal/day. Her basal caloric requirement is only 1300 Cal/day. Why is the extra 800 Cal needed?

21.10 What ill effects may obesity bring?

21.11 Assume that you want to lose 20 lb of body fat in 60 days. Your present dietary intake is 3000 Cal/day. What should your caloric intake be, in Cal/day, to achieve this goal, assuming no change in exercise habits?

21.12 What is marasmus?

21.13 Diuretics help to secrete water from the body. Would diuretic pills be a good way to reduce body weight?

Carbohydrates in the Diet and Their Digestion

21.14 Humans cannot digest wood; termites do so with the aid of bacteria in their digestive tract. Is there a basic difference in the digestive enzymes present in humans and termites?

21.15 What is the product of the reaction when α-amylase acts on amylose?

21.16 Does HCl in the stomach hydrolyze both the 1,4- and 1,6-glycosidic bonds?

21.17 Beer contains maltose. Can beer consumption be detected by analyzing the maltose content of a blood sample?

Fats in the Diet and Their Digestion

21.18 Which nutrient provides energy in its most concentrated form?

21.19 What is the precursor of arachidonic acid in the body?

21.20 How many (a) essential fatty acids and (b) essential amino acids do humans need in their diets?

21.21 Do lipases degrade (a) cholesterol or (b) fatty acids?

21.22 What is the function of bile salts in digestion of fats?

Proteins in the Diet and Their Digestion

21.23 Is it possible to get a sufficient supply of nutritionally adequate proteins by eating only vegetables?

21.24 Suggest a way to cure kwashiorkor.

21.25 What is the difference between protein digestion by trypsin and by HCl?

21.26 Which one will be digested faster: (a) a raw egg or (b) a hard-boiled egg? Explain.

Vitamins, Minerals, and Water

21.27 In a prison camp during a war, the prisoners are fed plenty of rice and water but nothing else. What would be the result of such a diet in the long run?

21.28 (a) How many milliliters of water per day does a normal diet require?
(b) How many calories does this amount of water contribute?

21.29 Why did British sailing ships carry a supply of limes?

21.30 What are the symptoms of vitamin A deficiency?

21.31 What is the function of vitamin K?

21.32 (a) Which vitamin contains cobalt?
(b) What is the function of this vitamin?

21.33 Vitamin C is recommended in megadoses by some people for prevention of all kinds of diseases, ranging from colds to cancer. What disease has been scientifically proven to be prevented when sufficient daily doses of vitamin C are in the diet?

21.34 Why is the Recommended Dietary Allowances (RDA) being phased out in preference of Daily Reference Intake (DRI)?

21.35 What are the nonspecific effects of vitamin E, C, and carotenoids?

21.36 What are the best dietary sources of calcium, phosphorus, and cobalt?

21.37 Which vitamin contains a sulfur atom?

21.38 What are the symptoms of vitamin B_{12} deficiency?

Chemical Connections

21.39 (Chemical Connections 21A) Why cannot glucose supply all the caloric needs in parenteral nutrition over a long period?

21.40 (Chemical Connections 21B) The RDA for niacin is 18 mg. Would it be still better to take a 100-mg tablet of this vitamin every day?

21.41 (Chemical Connections 21C) Describe the difference between the structure of aspartame and the methyl ester of phenylalanylaspartic acid.

21.42 (Chemical Connections 21C) (a) Which artificial sweeteners are not metabolized in the body at all?
(b) What could be products of digested aspartame?

21.43 (Chemical Connections 21C) What is common in the structures of Olestra and Sucralose?

21.44 (Chemical Connections 21D) Looking in Table 13.1, which lists the common amino acids found in proteins, which amino acid most resembles creatine?

Additional Problems

21.45 Which two chemicals are used most frequently to disinfect public water supply?

21.46 Which vitamin is part of coenzyme A (CoA)? List a step (or the enzyme) that has CoA as a coenzyme in (a) glycolysis and (b) fatty acid synthesis.

21.47 What vitamin in megadoses is prescribed to combat autoimmune blisters?

21.48 Why is it necessary to have proteins in our diets?

21.49 What chemical processes take place during digestion?

21.50 According to the government's food pyramid, are there any foods that we can completely omit from our diets and still be healthy?

21.51 Does the debranching enzyme help in digesting amylose?

21.52 As an employee of a company that markets walnuts, you are asked to provide information for an ad that would stress the nutritional value of walnuts. What information would you provide?

21.53 In diabetes, insulin is administered intravenously. Explain why this hormone protein cannot be taken orally.

21.54 Egg yolk contains a lot of lecithin (a phosphoglyceride). After ingesting a hard-boiled egg, would you find an increase in the lecithin level of your blood? Explain

21.55 What would you call a diet that scrupulously avoids phenylalanine-containing compounds? Could aspartame be used in such a diet?

21.56 What kind of supplemental enzyme would you recommend for a patient after a peptic ulcer operation?

21.57 In a trial, a woman was accused of poisoning her husband by adding arsenic to his meals. Her attorney stated that this was done to promote her husband's health, as arsenic is an essential nutrient. Would you accept this argument?

InfoTrac College Edition

For additional reading, go to InfoTrac College Edition, your online research library, at

http://infotrac.thomsonlearning.com

CHAPTER **22**

22.1 Introduction

22.2 Organs and Cells of the Immune System

22.3 Antigens and Their Presentation by Major Histocompatibility Complex

22.4 Immunoglobulin

22.5 T Cells and Their Receptors

22.6 Control of Immune Response: Cytokines

22.7 How to Recognize "Self"

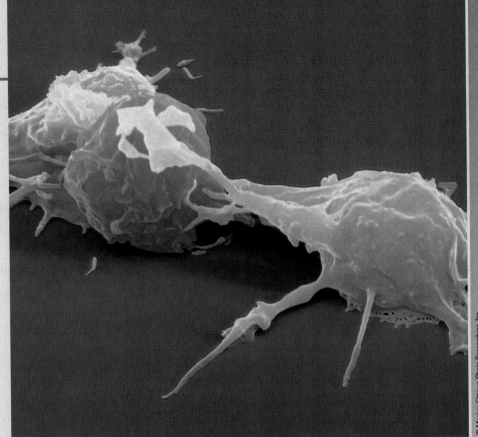

Two natural killer (NK) cells, shown in yellow-orange, attacking a leukemia cell, shown in red.

© Meckes/Ottawa/Photo Researchers, Inc.

Immunochemistry

22.1 Introduction

When you were in elementary school, you may have had chickenpox along with many other children. The viral disease passed from one person to the next and ran its course, but after the children recovered, they never had chickenpox again. Those who were infected became *immune* to this disease. Humans and other vertebrates possess a highly developed, complex immune system that defends the body against foreign invaders. Figure 22.1, which presents an overview of this complex system, could serve as a road map as you read the different sections in this chapter. Referring back to Figure 22.1, you will find the precise location of the topic under discussion and its relationship to the immune system as a whole.

A Innate Immunity

The first line of defense is the natural resistance of the body: the **innate immunity.** It comes in two forms, external and internal. The salient characteristics of innate immunity are that it is nonspecific and without memory.

One example of **external innate immunity** is the skin, which provides a barrier against penetration of pathogens. The skin also secretes lactic acid

Innate immunity The natural nonspecific resistance of the body against foreign invaders, which has no memory

486

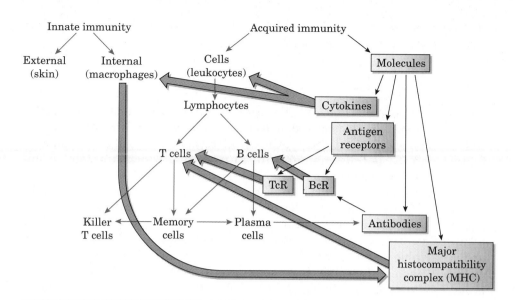

Figure 22.1 Overview of the immune system: its components and their interactions.

and fatty acids, both of which create a low pH, uncomfortable environment for bacteria. Tears in the eyes and mucus in the respiratory and gastrointestinal tracts perform the same function.

After a pathogen penetrates a tissue, **internal innate immunity** takes over. This system creates physiological barriers such as high oxygen pressures that kill anaerobic bacteria. Natural killer cells (NK) function as policemen of internal innate immunity. When they encounter cancerous cells, cells infected by a virus, or any other suspicious cells, they attach themselves to those abnormal cells (see the chapter opening photo). Other nonspecific responses include the proliferation of macrophages, which engulf and digest bacteria and reduce inflammation. In the inflammatory response to injury or infection, the capillaries dilate to allow greater flow of blood to the site of the injury, enabling agents of the internal innate immunity system to congregate there.

Acute inflammation and its symptoms were described two millennia ago by one of the founding fathers of systematic medicine, the Roman physician Celsus. These symptoms are redness, swelling, heat, and pain.

B Adaptive Immunity

Vertebrates have a second line of defense, called the **adaptive or acquired immunity.** We refer to this type of immunity when we colloquially talk about the **immune system.** The key features of the immune system are *specificity* and *memory*. The immune system uses antibodies designed specifically for each type of invader. In a second encounter with the same invader, the response is more rapid, more vigorous, and more prolonged than it was in the first case, because the immune system remembers the nature of the invader from the first encounter.

This is why, for example, if we have measles once, we will not get this disease a second time.

The invaders may be bacteria, viruses, molds, or pollen grains. A body with no defense against such invaders could not survive. In a rare genetic disease, a person is born without a functioning immune system. Attempts have been made to bring up such children in an enclosure totally sealed from the environment. While in this environment, they can survive, but when the environment is removed, such people always die within a short time. The disease AIDS (Chemical Connections 17C) slowly destroys the immune system, leaving its victims to die from some invading organism that a person without AIDS would easily be able to fight off.

As we shall see, the beauty of the body's immune system is that it is flexible. The system is capable of making millions of potential defenders, so that it can almost always find just the right one to counter the invader, even when it has never seen that particular organism before.

C Components of the Immune System

Foreign substances that invade the body are called **antigens.** The immune system is made of both cells and molecules. Two types of **white blood cells,** called lymphocytes, fight against the invaders: (1) T cells kill the invader by contact and (2) B cells manufacture **antibodies,** which are soluble immunoglobulin molecules that immobilize antigens.

The basic molecules of the immune system belong to the **immunoglobulin superfamily.** All molecules of this class have a certain portion of the molecule that can interact with antigens, and all are glycoproteins. The polypeptide chains in this superfamily have two domains: a constant region and a variable region. The constant region has the same amino acid sequence in each of the same class molecules. In contrast, the variable region is antigen-specific, which means that the amino acid sequence in this region is unique for each antigen. The variable regions are designed to recognize only one specific antigen.

There are three representatives of the immunoglobulin superfamily in the immune system:

1. Antibodies are soluble immunoglobulins secreted by the plasma cells.
2. Receptors on the surface of T cells (TcR) recognize and bind antigens presented to them.
3. Molecules that present antigens also belong to this superfamily. They reside inside the cells. These protein molecules are known as major histocompatibility complex (MHC).

When a cell is infected by an antigen, MHC molecules interact with it and bring a characteristic portion of the antigen to the surface of the cell. Such a surface presentation then marks the diseased cell for destruction. This can happen in a cell that was infected by a virus, and it can happen in macrophages that engulf and digest bacteria and virus.

D The Speed of the Immune Response

The process of finding and making just the right immunoglobulin to fight a particular invader is relatively slow, compared to the actions of the chemical messengers discussed in Chapter 15. While neurotransmitters act within a millisecond and hormones within seconds, minutes, or hours, immunoglobulins respond to an antigen over a longer span of time—weeks and months.

Although the immune system can be considered another form of chemical communication (Chapter 15), it is much more complex than neurotransmission because it involves molecular signals and interplay between various cells. In these constant interactions, the major elements are as follows: (1) the cells of the immune system, (2) the antigens and their perception by the immune system, (3) the antibodies (immunoglobulin molecules) that are designed to immobilize antigens, (4) the receptor molecules on the surface of the cells that recognize antigens, and (5) the cytokine molecules that control these interactions. Because the immune system is the cornerstone of the body's defenses, its importance for students in health-related sciences is undeniable.

Antigen A substance foreign to the body that triggers an immune response

Antibody A glycoprotein molecule that interacts with an antigen

Immunoglobulin superfamily Glycoproteins that are composed of constant and variable protein segments having significant homologies to suggest that they evolved from a common ancestry

Other glycoproteins belonging to the immunoglobulin superfamily are the cell-adhesion molecules that operate among cells and cause loose cells to adhere to each other to form layers and then tissues such as epithelium. They also play a significant part in the development of embryos.

Blood flow

Arterial end
of capillary

Lymph

Venous end
of capillary

Lymph
capillary

Intracellular
fluid

Interstitial
fluid

Erythrocyte

Cells forming wall
of capillary

Figure 22.2 Exchange of compounds among three body fluids: blood, interstitial fluid, and lymph. (*After J. R. Holum,* Fundamentals of General, Organic and Biological Chemistry. *New York: John Wiley & Sons, 1978, p. 569*)

22.2 Organs and Cells of the Immune System

The blood plasma circulates in the body and comes in contact with the other body fluids through the semipermeable membranes of the blood vessels. Therefore, blood can exchange chemical compounds with other body fluids and, through them, with the cells and organs of the body (Figure 22.2).

A Lymphoid Organs

The lymphatic capillary vessels drain the fluids that bathe the cells of the body. The fluid within these vessels is the **lymph.** Lymphatic vessels circulate throughout the body and enter certain organs, called **lymphoid organs,** such as the thymus, spleen, tonsils, and lymph nodes (Figure 22.3). The cells primarily responsible for the functioning of the immune system are the specialized white blood cells called **lymphocytes.** As their name implies, these cells are mostly found in the lymphoid organs. Lymphocytes may be either specific for a given antigen or nonspecific.

   T cells are lymphocytes that originate in the bone marrow but mature in the thymus gland. **B cells** are lymphocytes that originate and mature in the bone marrow. Both B and T cells are found mostly in the lymph, where they circulate looking for invaders. Small numbers of lymphocytes are also found in the blood. To get there, they must squeeze through tiny openings between the endothelial cells. This process is aided by signaling molecules called cytokines (Section 22.6). The sequence of the response of the body to a foreign invader is depicted schematically in Table 22.1.

B Cells of the Internal Innate Immunity

The two most important cells of innate immunity are macrophages and natural killer cells. **Macrophages** are the first cells in the blood that encounter an antigen. They belong to the internal innate immune system; inasmuch as they are nonspecific, macrophages attack virtually anything that is not recognized as part of the body, including pathogens, cancer cells, and damaged tissues. Macrophages engulf an invading bacterium or virus and kill it. The "magic bullet" in this case is the NO molecule, which we have seen is both toxic and can act as a secondary messenger (Chemical Connections 15F).

■ **Macrophage-ingesting bacteria (the rod-shaped structures). The bacteria will be pulled inside the cell within a membrane-bound vesicle and quickly killed.**

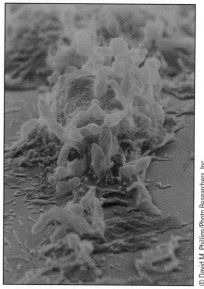

© David M. Phillips/Photo Researchers, Inc.

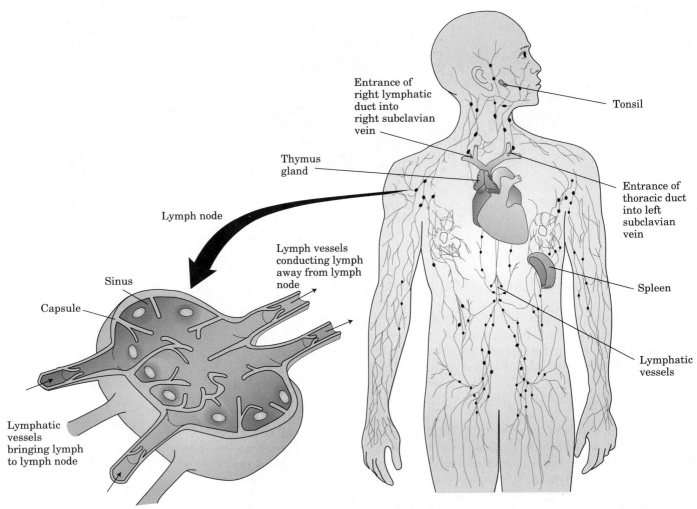

Figure 22.3   The lymphatic system is a web of lymphatic vessels containing a clear fluid, called lymph, and various lymphatic tissues and organs located throughout the body. Lymph nodes are masses of lymphatic tissue covered with a fibrous capsule. Lymph nodes filter the lymph. In addition, they are packed with macrophages and lymphocytes.

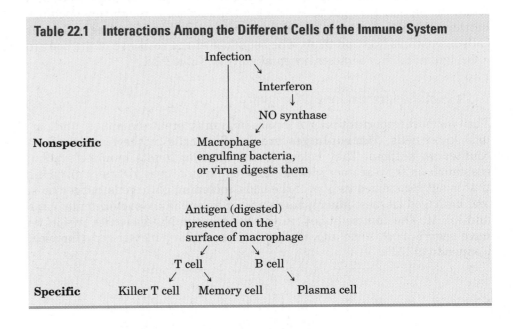

Table 22.1   Interactions Among the Different Cells of the Immune System

The NO molecule is short-lived and must be constantly manufactured anew. When an infection begins, the immune system manufactures the protein interferon. It, in turn, activates a gene that produces an enzyme, nitric oxide synthase. With the aid of this enzyme, the macrophages, endowed with NO, kill the invading organisms. Macrophages then digest the engulfed antigen and display a small portion of it on their surface.

Natural killer (NK) cells target abnormal cells. Once in physical contact with such cells, NK cells release proteins, aptly called perforins, that perforate the target cell membranes and create pores. The membrane of the target cell becomes leaky, allowing hypotonic liquid from the surroundings to enter the cell, which swells and eventually bursts.

The interferon-induced, enzyme-manufactured NO is critical for the immune system. This enzyme is also implicated in the overproduction of NO, however. Such an overproduction may result in pathologies, such as septic shock, multiple sclerosis, rheumatoid arthritis, and cancer formation.

C Cells of Adaptive Immunity: T and B Cells

T cells interact with the antigen presented by the macrophage and produce other T cells that are highly specific to the antigen. When these T cells differentiate, some of them become **killer T cells,** which kill the invading foreign cells by cell-to-cell contact. Killer T cells, like the natural killer cells, act through perforins, which attach themselves to the target cell, in effect punching holes in its membranes. Through these holes water rushes into the target cell; it swells and eventually bursts.

Other T cells will become **memory cells.** They will remain in the bloodstream, so that if the same antigen enters the body again, even years after the primary infection, the body will not need to build up its defenses anew but is ready to kill the invader instantly.

The production of antibodies is the task of **plasma cells.** These cells are derived from B cells after the B cells have been exposed to an antigen.

The lymphatic vessels, in which most of the antigen attacking takes place, flow through a number of lymph nodes (Figure 22.3). These nodes are essentially filters. Most plasma cells reside in lymph nodes, so most antibodies are produced there. Each lymph node is also packed with millions of other lymphocytes. More than 99% of all invading bacteria and foreign particles are filtered out in the lymph nodes. As a consequence, the outflowing lymph is almost free of invaders and is packed with antibodies produced by the plasma cells.

Both the natural killer cells and the T killer cells act the same way, through perforin. The T killer cells attack specific targets; the natural killer cells attack all suspicious targets.

22.3 Antigens and Their Presentation by Major Histocompatibility Complex

A Antigens

Antigens are any foreign substances that elicit an immune response; for this reason, they are also called immunogens. Three features characterize an antigen. The first condition is foreignness—molecules of your own body will not elicit an immune response. The second condition is that the antigen must be of molecular weight greater than 6000. The third condition is that the molecule must have sufficient complexity. A polypeptide made of lysine only, for example, is not immunogenic.

Antigens can be proteins, polysaccharides, or nucleic acids, as all of these substances are large molecules. Antigens may be soluble in the cytoplasm, or they may be found at the surface of cells, either embedded in the membrane or just absorbed on it. An example of polysaccharidic antigenicity is ABO blood groups (Chemical Connections 11D).

In protein antigens, only part of the primary structure is needed to cause an immune response. Between 5 and 7 amino acids are needed to interact with an antibody, and between 10 and 15 amino acids are necessary

There are rare exceptions: In an autoimmune disease, the body mistakes its own protein as foreign (see Chemical Connections 22D).

Epitope The smallest number of amino acids on an antigen that elicits an immune response

---

to bind to a receptor on a T cell. The smallest unit of an antigen capable of binding with an antibody is called the **epitope.** The amino acids in an epitope do not have to be in sequence in the primary structure, as folding and secondary structures may bring amino acids that are not in sequence into each other's proximity. For example, the amino acids in positions 20 and 28 may form part of an epitope. *Antibodies can recognize all types of antigens, but T cell receptors recognize only peptide antigens.*

As noted earlier, antigens may be in the interior of an infected cell or on the surface of a virus or bacteria that penetrated the cell. To elicit an immune reaction, the antigen or its epitope must be brought to the surface of the infected cell. Similarly, after a macrophage swallows up and partially digests an antigen, the macrophage must bring the epitope back to its surface to elicit an immune response from T cells (Table 22.1).

B Major Histocompatibility Complexes

The task of bringing an antigen's epitope to the infected cell's surface is performed by a protein complex called **major histocompatibility complex (MHC).** As the name implies, its role in immune response was first discovered in organ transplants. MHC molecules are transmembrane proteins belonging to the immunoglobulin superfamily. The two classes of MHC molecules (Figure 22.4) both have peptide-binding variable domains. Class I MHC is made of a single polypeptide chain, whereas class II MHC is a dimer. Class I MHC seeks out antigen molecules that have been *synthesized*

Figure 22.4 Differential processing of antigens in the MHC class II pathway (left) or MHC class I pathway (right). Cluster determinants (CD) are parts of the T cell receptor complex (see Section 22.5B).

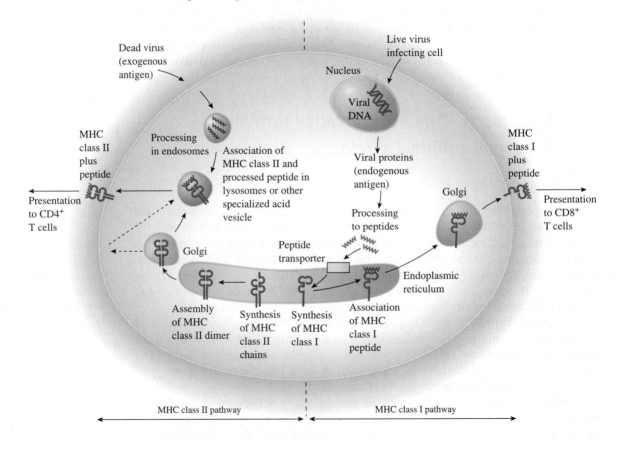

Table 22.2 Immunoglobulin Classes

Class	Molecular Weight (MW)	Carbohydrate Content (%)	Concentration in Serum (mg/100 mL)
IgA	200 000–700 000	7–12	90–420
IgD	160 000	<1	1–40
IgE	190 000	10–12	0.01–0.1
IgG	150 000	2–3	600–1800
IgM	950 000	10–12	50–190

inside a virus-infected cell. Class II MHC molecules pick up exogeneous *"dead" antigens.* In each case, the epitope attached to the MHC is brought to the cell surface to be presented to T cells.

For example, if a macrophage engulfed and digested a virus, the result would be dead antigens. The digestion occurs in several steps. First, the antigen is processed in lysosomes, special organelles of cells that contain proteolytic enzymes. A special enzyme called GILT (gamma-interferon inducible lysosomal thiol reductase) breaks the disulfide bridges of the antigen by reduction. The reduced peptide antigen unfolds and is exposed to proteolytic enzymes that hydrolyze it to smaller peptides. These peptides serve as epitopes that are recognized by class II MHC.

22.4 Immunoglobulin

A Classes of Immunoglobulins

Immunoglobulins are glycoproteins—that is, carbohydrate-carrying protein molecules. Not only do the different classes of immunoglobulins vary in molecular weight and carbohydrate content, but their concentration in the blood differs dramatically as well (Table 22.2). The IgG and IgM antibodies are the most important antibodies in the blood. They interact with antigens and trigger the swallowing up (phagocytosis) of these cells by such specialized cells as macrophages. The IgA molecules are found mostly in secretions: tears, milk, and mucus. Therefore, these immunoglobulins attack the invading material before it gets into the bloodstream. IgE plays a part in such allergic reactions as asthma and hay fever.

In immunoglobulin IgG, the light chains have a molecular weight of 25,000 and the heavy chains have a molecular weight of 50,000.

B Structure of Immunoglobulins

Each immunoglobulin molecule is made of four polypeptide chains: two identical light chains and two identical heavy chains. The four polypeptide chains are arranged symmetrically, forming a Y shape (Figure 22.5). Four disulfide bridges link the four chains into a unit. Both light and heavy chains have constant and variable regions. The constant regions have the same amino acid sequences in different antibodies, and the variable regions have different amino acid sequences in different antibodies.

The variable regions of the antibody recognize the foreign substance (the antigen) and bind to it (Figure 22.6). Since each antibody contains two variable regions, it can bind two antigens, thereby forming a large aggregate.

The binding of the antigen to the variable region of the antibody is not by covalent bonds but by much weaker intermolecular forces such as London dispersion forces, dipole–dipole interactions, and hydrogen bonds. This

■ **Antibody binding to the epitope of an antigen.**

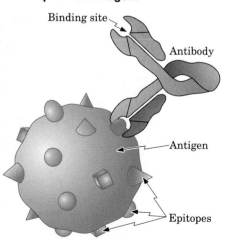

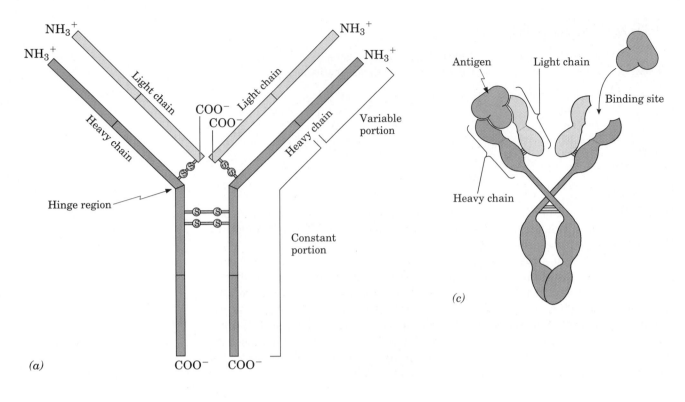

(a)

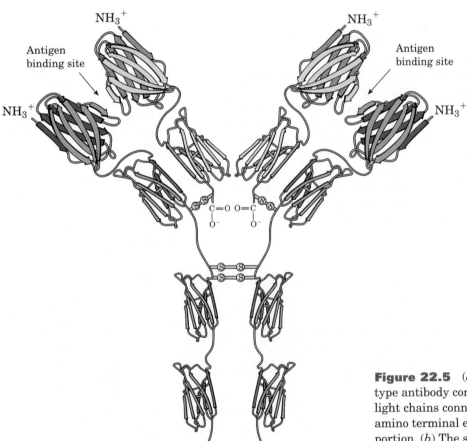

(b)

Figure 22.5 (*a*) Schematic diagram of an IgG-type antibody consisting of two heavy chains and two light chains connected by disulfide linkages. The amino terminal end of each chain has the variable portion. (*b*) The same in the ribbon model. (*c*) Model showing how an antibody bonds to an antigen.

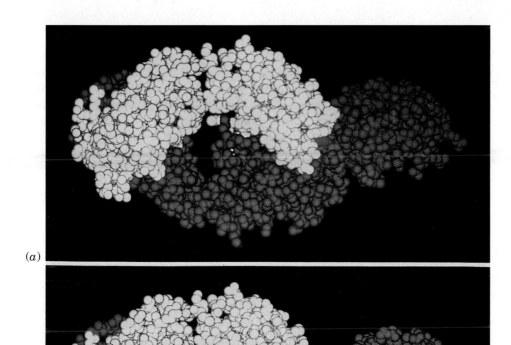

(a)

(b)

Figure 22.6 (*a*) An antigen–antibody complex. The antigen (shown in green) is lysozyme. The heavy chain of the antibody is shown in blue; the light chain in yellow. The most important amino acid residue (glutamine in the 121 position) on the antigen is the one that fits into the antibody groove (shown in red). (*b*) The antigen–antibody complex has been pulled apart. Note how they fit each other. (*Courtesy of Dr. R. J. Poijak, Pasteur Institute, Paris, France*)

binding is similar to the way in which substrates bind to enzymes or hormones and neurotransmitters bind to a receptor site. That is, the antigen must fit into the antibody surface. Since there are a large variety of antigens against which the human body must fight, our systems contain more than 10,000 different antibodies.

Studies have shown that many other molecules that function as communication signals between cells also have the same basic structural design as the immunoglobulins.

C B Cells and Antibodies

Each B cell synthesizes only one unique immunoglobulin antibody, and that antibody contains a unique antigen-binding site to one epitope. Before encountering an antigen, these antibodies are inserted in the plasma membrane of the B cells, where they serve as receptors. When an antigen interacts with its receptor, it stimulates the B cell to divide and differentiate into plasma cells. These daughter cells secrete soluble antibodies that have the same antigen-binding sites as the original antibody/receptor. The soluble secreted antibodies appear in the serum (the noncellular part of blood) and can react with the antigen. Thus, an immunoglobulin produced in B cells acts both as a receptor to be stimulated by the antigen and as a secreted messenger ready to neutralize and eventually destroy the antigen.

D How Does the Body Acquire the Diversity Needed to React to Different Antigens?

During B cell development, the variable regions of the H chains are assembled by a process called V(J)D recombination. There are a number of exons in each of three different areas—V, J, and D—of the immunoglobin gene. Combining one exon from each area creates a *new V(J)D gene*. This process creates a great diversity because of the large number of ways that this combination can be performed (Figure 22.7). But that is merely the first step. When the body encounters a new antigen, mutations are induced in the new V(J)D genes; these mutations then create even greater diversity.

There are three ways to create mutations, two of which affect the variable regions and one that alters the constant region:

1. Somatic hypermutation (SHM) creates a point mutation (only one nucleotide). The resultant protein of the mutation is able to bind more strongly with the antigen.

2. In a gene conversion (GC) mutation, stretches of nucleotide sequences are copied from a pseudogene V and entered into the V(J)D. This also enables the proteins synthesized from the GC mutation to make stronger contact with the antigen than without the mutation.

3. Mutations on the constant region of the chain can be achieved by class switch recombination (CSR). In this case, exons of the constant regions are swapped between highly repetitive regions.

The diversity of antibodies created by the V(J)D recombination is greatly amplified and finely tuned by the mutations on these genes. Since the response to the antigen occurs on the gene level, it is easily preserved and transmitted from one generation of cells to the next.

Figure 22.7 Diversification of immunoglobulins by V(J)D recombination. Exons of three genes—the V(ariable) (A, B, C, D), J(oin) (1, 2, 3, 4), and D(iverse) (a, b, c, d) genes—combine to form new V(J)D genes that are transcribed to the corresponding mRNAs. The expressions of these new genes result in a wide variety of immunoglobulins having different variable regions on their heavy chains.

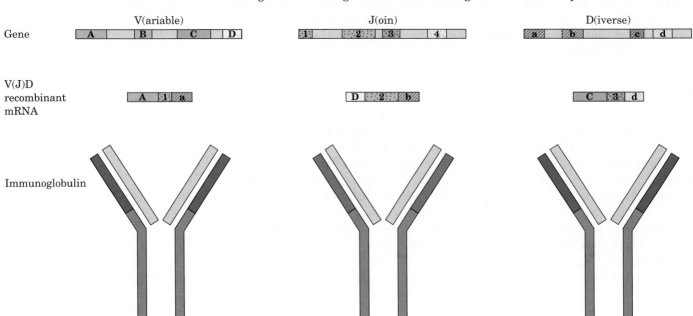

CHEMICAL CONNECTIONS 22A

Antibodies and Cancer Therapy

Antibodies come in a variety of forms. As a response to a specific antigen, the body may develop a number of closely related antibodies, the so-called **multiclonal antibodies.** Monoclonal antibodies, in contrast, are identical copies of a single antibody molecule that bind to a specific antigen. Georg Kohler and Cesar Milstein (1927–2002) developed techniques for producing monoclonal antibodies in large quantities. For their pioneering work, they received the Nobel Prize in 1984.

Monoclonal antibody therapy is the latest advance in fighting cancer. For instance, the drug Rituxan is a monoclonal antibody against the CD20 antigen, which appears in high concentration on the surface of B cells in patients suffering from non-Hodgkins lymphoma. Rituxan kills tumor cells in two ways. First, by binding to the diseased B cells, it enlists the whole immune system to attack the marked cells and to destroy them. Second, the monoclonal antibody causes the tumor cells to stop growing and die through apoptosis, programmed cell death (Chemical Connections 16E). Another monoclonal antibody is Herceptin, which is recommended as a treatment for metastatic breast cancers that have a specific antigen, c-erb-B2, on the surface of cancer cells.

One advantage that monoclonal antibody treatment has over chemotherapy is its relatively minimal toxicity. Even when the monoclonal antibody is employed to deliver damaging radiation, as in the case of the antibody against the prostate-specific antigen to which actinium-225 is attached, the antibody delivers the radiation to only the specific cancer cells, killing them without significantly harming healthy tissues.

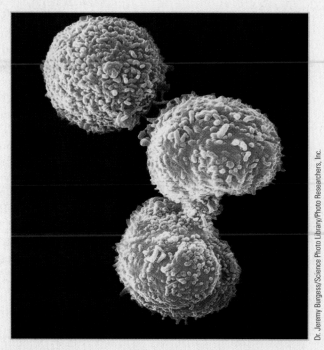

Dr. Jeremy Burgess/Science Photo Library/Photo Researchers, Inc.

■ **False-color scanning electron micrograph of hybridoma cells producing a monoclonal antibody to cytoskeleton protein.**

E Monoclonal Antibodies

When an antigen is injected into an organism (for example, human lysozyme into a rabbit), the initial response is quite slow. It may take from one to two weeks before the antilysozyme antibody shows up in the rabbit's serum. Those antibodies, however, are not uniform. The antigen may have many epitopes, and the antisera contains a mixture of immunoglobulins with varying specificity for all the epitopes. Even antibodies to a single epitope usually have a variety of specificities. **Monoclonal antibodies,** in contrast, are the products of cells cloned from a single B cell and have a single specificity.

F Roles of Antibodies That Are Not Antigen-Specific

Recent discoveries indicate that antibodies are versatile and may play other roles besides binding to an antigen and marking it for destruction. Antibodies are able to form hydrogen peroxide, H_2O_2, from water and singlet oxygen. The latter is a free radical and very toxic to cells. Thus, when singlet oxygen is formed in a process, antibodies are able to detoxify it and convert it to the less toxic H_2O_2. Furthermore, the hydrogen peroxide formed in this way may be used to kill pathogens directly.

22.5 T Cells and Their Receptors

A T Cell Receptors

Like B cells, T cells carry on their surface unique receptors that interact with antigens. We noted earlier that T cells respond only to protein antigens. An individual has millions of different T cells, each of which carries on its surface a unique T cell receptor (TcR) that is specific for one antigen only. The TcR is a glycoprotein made of two subunits cross-linked by disulfide bridges. Like immunoglobulins, TcRs have constant (C) and variable (V) regions. The antigen binding occurs on the variable region. The similarity in amino acid sequence between immunoglobulins (Ig) and TcR, as well as the organization of the polypeptide chains, makes TcR molecules members of the immunoglobulin superfamily.

There are, however, basic differences between immunoglobulins and TcRs. For instance, immunoglobulins have four polypeptide chains, whereas TcRs contain only two subunits. Immunoglobulins can interact directly with antigens, but TcRs can interact with them only when the epitope of an antigen is presented by an MHC molecule. Lastly, immunoglobulins can undergo somatic mutation. This kind of mutation can occur in all body cells except the ones involved in sexual reproduction. Thus, Ig molecules can increase their diversity by somatic mutation; TcRs cannot.

B T Cell Receptor Complex

A TcR is anchored in the membrane by hydrophobic transmembrane segments (Figure 22.8). TcR alone, however, is not sufficient for antigen binding. Also needed are other protein molecules that act as coreceptors and/or

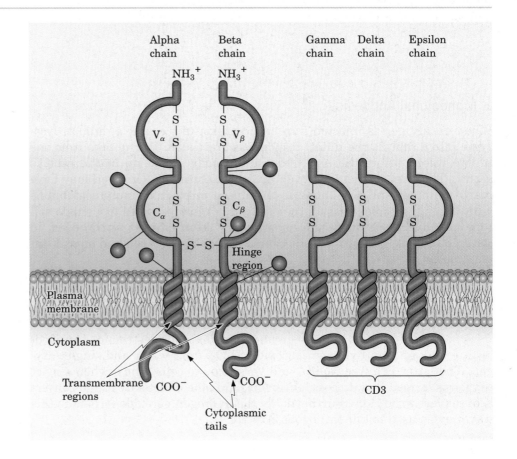

Figure 22.8 Schematic structure of TcR complex. TcR consists of two chains, α and β. Each has two extracellular domains, an amino-terminal V domain and a carboxyl-terminal C domain. Domains are stabilized by intrachain disulfide bonds between cysteine residues. The α and β chains are linked by an interchain disulfide bridge near the cell membrane (hinge region). Each chain is anchored on the membrane by a hydrophobic transmembrane segment and ends in the cytoplasm with a carboxyl-terminal segment rich in cationic residues. Both chains are glycosylated (red spheres). The cluster determinant (CD) coreceptor consists of three chains, γ, δ, and ϵ. Each is anchored in the plasma membrane by a hydrophobic transmembrane segment. Each is also cross-linked by a disulfide bridge, and the carboxylic terminal is located in the cytoplasm.

Immunization

Smallpox was a scourge over the centuries, each outbreak leaving many dead and others maimed by deep pits on the face and body. A form of immunization was practiced in ancient China and the Middle East by intentionally exposing people to scabs and fluids of lesions of victims of smallpox. This practice was known as variolation in the Western world, where the disease was called variola. Variolation was introduced to England and the American Colonies in 1721.

Edward Jenner, an English physician, noted that milkmaids who had contracted cowpox from infected cows seemed to be immune to smallpox. Cowpox was a mild disease, whereas smallpox could be lethal. In 1796, Jenner performed a daring experiment: He dipped a needle into the pus of a cowpox-infected milkmaid and then scratched a boy's hand with the needle. Two months later, Jenner injected the boy with a lethal dose of smallpox-carrying agent. The boy survived, and did not develop any symptoms of the disease. The word spread, and Jenner was soon established in the immunization business. When the news reached France, the skeptics there coined a derogatory term, *vaccination,* which literally means "encowment." The derision did not last long, and the practice was soon adopted worldwide.

A century later, in 1879, Louis Pasteur found that tissue infected with rabies has much weakened virus in it. When injected into patients, it elicits an immune response that protects against rabies. Pasteur named these attenuated, protective antigens *vaccines* in honor of Jenner's work. Today, immunization and vaccination are synonymous.

Vaccines are available for several diseases, including polio, measles, and smallpox, to name a few. A vaccine may be made of either dead viruses or bacteria or weakened ones. For example, the Salk polio vaccine is a polio virus that has been made harmless by treatment with formaldehyde; it is given by intramuscular injection. In contrast, the Sabin polio vaccine is a mu-

tated form of the wild virus; the mutation makes the virus sensitive to temperature. The mutated live virus is taken orally. The body temperature and the gastric juices render it harmless before it penetrates the bloodstream.

Many cancers have specific carbohydrate cell surface tumor markers. If such antigens could be introduced by injection without endangering the individual, they could provide an ideal vaccine. Obviously, one cannot use live or even attenuated cancer cells for vaccination. However, the expectation is that a synthetic analog of a natural tumor marker will elicit the same immune reaction that the original cancer surface makers do. Thus injecting such innocuous synthetic compounds may prompt the body to produce immunoglobulins that can provide a cure for the cancer—or at least prevent the occurrence of it. A compound called 12:13 dEpoB, a derivative of the macrolid epothilon B, is currently being explored for its potential to become the first anticancer vaccine.

Vaccines change lymphocytes into plasma cells that produce large quantities of antibodies to fight any invading antigens. This is only the immediate, short-term response, however. Some lymphocytes become memory cells rather than plasma cells. These memory cells do not secrete antibodies, but instead store them to serve as a detecting device for future invasion of the same foreign cells. In this way, long-term immunity is conferred. If a second invasion occurs, these memory cells divide directly into both antibody-secreting plasma cells and more memory cells. This time the response is faster because it does not have to go through the process of activation and differentiation into plasma cells, which usually takes two weeks.

Smallpox, which was once one of humanity's worst scourges, has been totally wiped out, and smallpox vaccination is no longer required. Because smallpox is a potential threat of bioterrorism, the U.S. government has recently begun gearing up its smallpox vaccine production.

---

signal transducers. These molecules go under the name of CD3, CD4, and CD8, with "CD" standing for **cluster determinant.** TcR and CD together form the **T cell receptor complex.**

The CD3 molecule adheres to the TcR in the complex not through covalent bonds but by intermolecular forces. It is a signal transducer because, upon antigen binding, CD3 becomes phosphorylated. This event sets off a signaling cascade inside the cell, which is carried out by different kinases. We saw a similar signaling cascade in neurotransmission (Section 15.5).

The CD4 and CD8 molecules act as **adhesion molecules** as well as signal transducers. A T cell has either a CD4 or a CD8 molecule to help bind the antigen to the receptor (Figure 22.4). A unique characteristic of the CD4 molecule is that it binds strongly to a special glycoprotein that has a molecular weight of 120,000 (gp120). This glycoprotein exists on the surface of the human immunodeficiency virus (HIV). Through this binding to CD4, the HIV virus can enter and infect T cells and cause AIDS (Chemical Connec-

> **Adhesion molecules** Various protein molecules that help to bind an antigen to the T cell receptor

tions 17C). T cells die as a result of HIV infection, which depletes the T cell population so drastically that the immune system can no longer function. As a consequence, the body succumbs to opportunistic pathogen infections.

22.6 Control of Immune Response: Cytokines

A Nature of Cytokines

> **Cytokine** A glycoprotein that traffics between cells and alters the function of a target cell

Cytokines are glycoprotein molecules that are produced by one cell but alter the function of another cell. They have no antigen specificity. Cytokines transmit intercellular communications between different types of cells at diverse sites in the body. They are short-lived and are not stored in cells.

Cytokines facilitate a coordinated and appropriate inflammatory response by controlling many aspects of the immune reaction. They are released in bursts, in response to all manner of insult or injury (real or perceived). They travel and bind to specific cytokine receptors on the surface of macrophages and B and T cells and induce cell proliferation.

One set of cytokines are called **interleukins** (ILs) because they communicate between and coordinate the actions of leukocytes (all kinds of white blood cells). Macrophages secrete IL-1 upon a bacterial infection. The presence of IL-1 then induces shivering and fever. The elevated body temperature both reduces the bacterial growth and speeds up the mobilization of the immune system. One leukocyte may make many different cytokines, and one cell may be the target of many cytokines.

B Classes of Cytokines

Cytokines can be classified according to their mode of action, origin, or target. The best way to classify them is by their structure—namely, the secondary structure of their polypeptide chains.

1. One class of cytokines is made of four α-helical segments. A typical example is interleukin-2 (IL-2), which is a 15,000 MW polypeptide chain. A prominent source of IL-2 is T cells. IL-2 activates other B and T cells

CHEMICAL CONNECTIONS 22C

Mobilization of Leukocytes: Tight Squeeze

When injury occurs, leukocytes in the bloodstream must migrate into the lymph or inflammatory sites to provide protection. A number of adhesion and signaling molecules guide this migration, and two major protein molecules on the surface of leukocytes facilitate their movement.

Selectins are protein molecules belonging to the CD (cluster determinant) class. When activated, they interact with the carbohydrate portions of glycoproteins on the surface of endothelial cells. These endothelial cells line the inner wall of blood vessels and, with tight junctions, seal the vessels against leakiness. Endothelial cells have their own selectin molecules that interact with the carbohydrates of the leukocytes. These alternate interactions create a rolling motion of leukocytes along the endothelial wall. When a leukocyte arrives at a point near the injury or infection, another surface molecule is activated. This molecule, called integrin, exists in almost every cell type of the animal

kingdom. Activated integrin on endothelial cells strongly interacts with surface adhesion molecules of the leukocyte, which stops rolling. This attachment and the force of the streaming blood make the leukocyte flatten out.

Chemokines, such as IL-8, also play a role in this leukocyte trafficking. When a chemokine is bound to its receptor on the leukocyte, the cell undergoes a remarkable transformation. Inside the cell, the cytoskeleton—actin—is first broken down and then repolymerized, resulting in armlike and leglike projections (lamellipodia) that enable the cells to migrate. In addition, chemokines signal the endothelial cells to change their shape. This results in the opening of gaps between the cells. The flattened cells then manage to squeeze through the gaps between the tight junctions and enter the tissue or lymph to migrate to the point of injury. After accumulating in the inflamed tissue, the leukocytes may kill the invading bacteria by engulfing and digesting them.

and macrophages. Its action is to enhance proliferation and differentiation of the target cells.

2. Another class of cytokines has only β-pleated sheets in its secondary structure. Tumor necrosis factor (TNF), for example, is produced mostly by T cells and macrophages. Its name comes from its ability to destroy susceptible tumor cells through lysis, after it binds to receptors on the tumor cell.

3. A third class of cytokines has both α-helix and β-sheet secondary structures. A representative of this class is epidermal growth factor (EGF), which is a cysteine-rich protein. As its name indicates, EGF stimulates the growth of epidermal cells and its main function is in wound healing.

4. A subgroup of cytokines, the chemotactic cytokines, are also called **chemokines.** In humans, there are some 40 chemokines, all low-molecular-weight proteins with distinct structural characteristics. They attract leukocytes to the site of infection or inflammation. All chemokines have four cysteine residues, which form two disulfide bridges: Cys1-Cys3 and Cys2-Cys4. Chemokines have a variety of names, such as interleukin-8 (IL-8) and monocyte chemotactic proteins (MCP-1 to MCP-4). Chemokines interact with specific receptors, which consist of seven helical segments coupled to GTP-binding proteins.

C Mode of Action of Cytokines

When a tissue is injured, leukocytes are rushed to the inflamed area. Chemokines help leukocytes migrate out of the blood vessels to the site of injury (Chemical Connections 22C). There leukocytes, in all their forms—neutrophils, monocytes, lymphocytes—accumulate and attack the invaders by engulfing them (phagocytosis) and later killing them. Other phagocytic cells, the macrophages, which reside in the tissues and thus do not have to migrate, do the same. These activated phagocytic cells destroy their prey by releasing endotoxins that kill bacteria and/or by producing such highly reactive oxygen intermediates as superoxide, singlet oxygen, hydrogen peroxide, and hydroxyl radicals (Chemical Connections 18C).

Chemokines are also major players in chronic inflammation, autoimmune diseases, asthma, and other forms of allergic inflammation, and even in transplant rejection.

22.7 How to Recognize "Self"

One major problem facing body defenses is how to recognize the foreign body as "not self" and thereby avoid attacking the "self"—that is, the healthy cells of the organism.

A Selection of T and B Cells

The members of the adaptive immunity system, B cells and T cells, are all specific and have memory, so they target only truly foreign invaders. The T cells mature in the thymus gland. During the maturation process, those T cells that fail to interact with MHC and thus cannot respond to foreign antigens are eliminated through a selection process. T cells that express receptors (TcR) that are prone to interact with normal self antigens are also eliminated through the selection process (Figure 22.9). Thus the activated T cells leaving the thymus gland carry TcRs that can respond to foreign antigens. Even if some T cells prone to react with self antigen escape the selection detection, they can be deactivated through the signal transduction

TNF targets other cells as well. For example, it plays a role in rheumatoid arthritis. When the immune system goes awry, the TNF attacks the patient's own cartilage. It binds to the receptors of the cartilage cells, which lyse. In this manner, it eats away even the bones. A new drug, called Enebrel, interacts with TNF in the blood. It immobilizes the cytokine and prevents its migration to the joints, where it would cause damage.

Chemokine A low-molecular-weight polypeptide that interacts with special receptors on target cells and alters their functions

---

Two subgroups or chemokines exist in which the first two Cys residues (1) are adjacent (CC) or (2) are separated by one amino acid (CXC).

Special receptors for the CC and CXC chemokines are found on the surface of leukocytes and lymphocytes.

The crucial parts of the chemokines in these interactions are the N-terminal domain and the loop between Cys2 and Cys4.

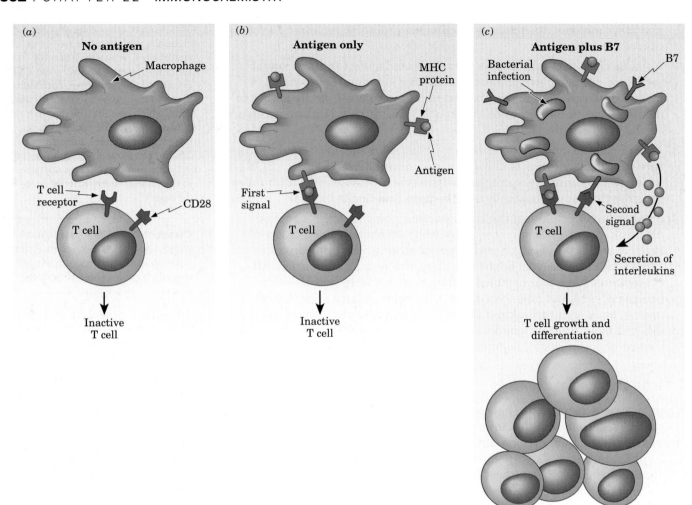

Figure 22.9 Maturation and activation of T cells. (*a*) T cells in the absence of an antigen do not proliferate. (*b*) In the presence of an antigen, the TcR binds to the antigen presented by MHC. No proliferation occurs because a second signal is needed. In this manner the body can avoid responding to a "self" antigen. (*c*) A second signal is provided when the infected cell produces a stress signal—for example, a "heat shock" protein that normally has a chaperoning function (Section 13.11). The primary MHC-antigen-TcR signal and the second signal are enhanced by the synthesis of cytokines, such as interleukins. Under such triple stimulation, T cells mature and proliferate.

system that, among other functions, performs tyrosine kinase activation and phosphatase deactivation similar to those processes seen in adrenergic neurotransmitter signaling (Section 15.5).

Similarly, the maturation of B cells in the bone marrow depends on the engagement of their receptors, BcR, with antigen. Those B cells that are prone to interact with self antigen are also eliminated before they leave the bone marrow. Again, as with the T cells, many signaling pathways control the proliferation of B cells. Among them, the activation by tyrosine kinase and the deactivation by phosphatase provide a secondary control.

B Discrimination of the Cells of the Innate Immunity System

The first line of defense is the innate immunity system, in which cells such as natural killer cells or macrophages have no specific targets and no memory of which epitope signals danger. Still, these cells must somehow dis-

Myasthenia Gravis: An Autoimmune Disease

Myasthenia gravis manifests itself as muscular weakness and excessive fatigue. In its severe form, the patient has difficulty chewing, swallowing, and even breathing. Death follows from respiratory failure. Patients with myasthenia gravis develop antibodies against their own acetylcholine receptors (Figure 15.3). This condition is an example of an autoimmune disease, in which the body mistakes its own protein for a foreign body. The anti-bodies produced interact with the acetylcholine receptor in the neuromuscular junctions. This interaction blocks both the reception of acetylcholine molecules and the signaling nerve impulses. As a result, the muscles do not respond, and paralysis sets in. In addition, the interaction leads to the degradation and lysis of the blocked receptors.

criminate between normal and abnormal cells to identify targets. The mechanism by which this identification is accomplished has only recently been explored and is not yet fully understood. The main point is that the cells of the innate immunity have two kinds of receptors on their surface: an **activating receptor** and an **inhibitory receptor.** When a healthy cell of the body encounters a macrophage or a natural killer cell, the inhibitory receptor on the latter's surface recognizes the epitope of the normal cell, binds to it, and prevents the activation of the killer cell or macrophage. On the other hand, when a macrophage encounters a bacterium with a foreign antigen on its surface, the antigen binds to the activating receptor of the macrophage. This ligand binding prompts the macrophage to engulf the bacterium by phagocytosis. Such foreign antigens may be the lipopolysaccharides of gram-negative bacteria or the peptidoglycans (Chemical Connections 10B) of gram-positive bacteria.

When a cell is infected, damaged, or transformed into a malignant cell, the epitopes that signaled healthy cells diminish greatly, and unusual epitopes are presented on the surface of these cancer cells. The effect is that fewer inhibitory receptors of macrophages or natural killer cells can bind to the surface of the target cell and more activating receptors find inviting ligands. As a consequence, the balance shifts in favor of activation and the macrophages or killer cells will do their job.

C Autoimmune Diseases

In spite of the safeguards in the body intended to prevent acting against "self," in the form of healthy cells, many diseases exist in which some part of a pathway in the immune system goes awry. The skin disease psoriasis is thought to be a T cell-mediated disease in which cytokines and chemokines play an essential role. Other autoimmune diseases, such as myasthenia gravis (Chemical Connections 22D), rheumatoid arthritis, multiple sclerosis (Chemical Connections 12E), and insulin-dependent diabetes (Chemical Connections 15G) also involve cytokines and chemokines. Allergies are another example of malfunctioning of the immune system. Pollens and animal furs are allergens that can provoke asthma attacks. Some people are so sensitive to certain food allergens that even a knife used in smearing peanut butter may prove fatal in a person known to be allergic to peanuts.

The major drug treatment for autoimmune diseases involves the glucocorticoids, the most important of which is cortisol (Section 12.10). It is a standard therapy for rheumatoid arthritis, asthma, inflammatory bone diseases (Crohn's disease), psoriasis, and eczema. The wide beneficial effects of glucocorticoids are overshadowed, however, by their undesirable side effects, which include osteoporosis, skin atrophy, and diabetes. Glucocorti-

Glycomics

You may have noticed that in this chapter we frequently called the participant molecules *glycoproteins*. Immunoglobulins, cytokines, and T cell receptors are all glycoproteins. As we saw in Section 13.10, a glycoprotein is a complex molecule with a protein core and carbohydrate side chains. In essence, the carbohydrate portions of glycoproteins as well as those of glycolipids constitute a third language of chemical communication.

The language of DNA was deciphered through genomics (Chemical Connections 16D), which in essence elucidated all the information coded in the human chromosomes as well as in the chromosomes of other species. In the wake of the success of genomics, a similar effort was launched under the name of proteomics (Chemical Connections 13F) to find and describe all the proteins of an organism and their functions. Besides the language of the nucleotides and the language written in amino acids, there is a third language of communication written in monosaccharides. This language is more variable and specific to individuals. For example, the ABO blood group to which an individual belongs is written in the carbohydrate language (Chemical Connections 11D). Another example of information written in the carbohydrate language is the coat around the *Pneumococcus* bacterium. Different carbohydrates make the same bacterium ei-

ther highly infectious or dormant. Cancer cells often display abnormal sugars on their surface, marking their malignancy.

Because of the importance of deciphering this information, an effort was launched recently to elucidate the structures of all the carbohydrates (including those that are parts of the glycoproteins and glycolipids) that an organism possesses and, if possible, to describe their functions. This effort is called **glycomics** in analogy with genomics and proteomics.

But consider the complexity of this third language. Genomics has an alphabet of 4 letters (the nucleotides); the alphabet of proteomics consists of 20 letters (amino acids). If we include only the D-sugars of pentoses and hexoses in Table 11.1, we have 12 original letters in the glycomics alphabet. While nucleotides and amino acids can form only one kind of bond, each of these sugars can form a minimum of four kinds of bonds (1,2; 1,4; and so on) and in the α and β configurations. Thus the carbohydrate alphabet has $12 \times 4 \times 2 = 96$ letters at minimum. Now consider just the two-letter words. There are $4^2 = 16$ such words in the nucleotide language, $20^2 = 400$ words in the amino acid language, and $96^2 = 9216$ such words in the monosaccharide language. Clearly, the task of glycomics is much more daunting than that of the genomics and even that of the proteomics.

coids regulate the synthesis of cytokines either directly, by interacting with their genes, or indirectly, through transcription factors.

Macrolide drugs are used to suppress the immune system during tissue transplantation or in the case of certain autoimmune diseases. Drugs such as cyclosporin A or rapamycin bind to receptors in the cytosol and, through secondary messengers, inhibit the entrance of nuclear factors into the nucleus. Normally those nuclear factors signal a need for transcription, so their absence prevents the transcription of cytokines—for example, interleukin-2.

S U M M A R Y

The human immune system protects us against foreign invaders. It consists of two parts: (1) the natural resistance of the body, called innate immunity, and (2) adaptive or acquired immunity (Section 22.1). **Innate immunity** is nonspecific. **Macrophages** and **natural killer (NK) cells** are cells of innate immunity that function as policemen. **Adaptive or acquired immunity** is highly specific, being directed against one particular invader. Acquired immunity (known as the immune system) also has memory, unlike innate immunity.

The principal cellular components of the immune system are the white blood cells, or **leukocytes** (Section 22.2). The specialized leukocytes in the lymph system are called **lymphocytes** (Section 22.2). They circulate mostly in the **lymphoid organs.** The **lymph** is a collection of vessels extending throughout the body and connected to the interstitial fluid on the one hand, and to the blood vessels on the other hand. Lymphocytes that mature in the bone marrow and produce soluble immunoglobulins are **B cells.** Lymphocytes that mature in the thymus gland are **T cells.**

Antigens are large, complex molecules of foreign origin (Section 22.3). An antigen may be a bacterium, a virus, or a toxin. Such an invader may interact with antibodies, T cell receptors (TcR), or major histocompatibility complex (MHC) molecules. All three types of molecules belong to the **immunoglobulin superfamily.** An **epitope** is the smallest part of an antigen that binds to antibodies, TcRs, and MHCs. Antibodies are **immunoglobulins** (Section 22.4). These water-soluble glycoproteins are made of two heavy chains and two light chains. All four chains are linked together by disulfide bridges. Immunoglobulins contain variable regions in which the amino acid composition of each antibody is different. These regions interact with antigens to form insoluble large aggregates.

A large variety of antibodies are synthesized by a number of processes in the body. During B cell development, the **variable regions** of the H chains are assembled by a process called V(J)D recombination. Mutations on these new genes then create even greater diversity. Somatic hypermutation (SHM) that creates a point mutation (only one nucleotide) is one way; patches of mutation introduced into the V(J)D gene are the gene conversion (GC) mutation. Immunoglobulins respond to an antigen over a long span of time, lasting for weeks and even months.

All antigens—whether proteins, polysaccharides, or nucleic acids—interact with soluble immunoglobulins produced by B cells. Protein antigens also interact with T cells. The binding of the epitope to the TcR is facilitated by MHC, which carries the epitope to the T cell surface where it is presented to the receptor (Section 22.5).

Upon epitope binding to the receptor, the T cell is stimulated. It proliferates and can differentiate into (1) killer T cells or (2) memory cells. The T cell receptor has a number of helper molecules, such as CD4 or CD8, that enable it to bind the epitope tightly. CD **(cluster determinant)** molecules also belong to the immunoglobulin superfamily. Antibodies can recognize all antigens, but T cell receptors recognize only peptide antigens.

The control and coordination of the immune response are handled by **cytokines,** which are small protein molecules (Section 22.6). Chemotactic cytokines, the **chemokines,** such as interleukin-8, facilitate the migration of leukocytes from blood vessels into the site of injury or inflammation. Other cytokines activate B and T cells and macrophages, enabling them to engulf the foreign body, digest it, or destroy it by releasing special toxins. Some cytokines, such as tumor necrosis factor (TNF), can lyse tumor cells.

A number of mechanisms in the body ensure that the body recognizes "self" (Section 22.8). In adaptive immunity, the selection of T and B cells that are prone to interact with "self" antigens are eliminated. In innate immunity, two kinds of receptors exist on their surface: an **activating receptor** and an **inhibitory receptor.** The inhibitory receptor on their surface recognizes the epitope of a normal cell, binds to it, and prevents the activation of the killer cell or macrophage. Many autoimmune diseases are T cell-mediated diseases, in which cytokines and chemokines play an essential role. The standard drug treatment for autoimmune diseases involves glucocorticoids, which prevent the transcription and hence the synthesis of cytokines.

P R O B L E M S

Numbers that appear in color indicate difficult problems.
▶ designates problems requiring application of principles.

22.1 Give two examples of external innate immunity in humans.

22.2 Which form of immunity is characteristic of vertebrates only?

22.3 How does the skin fight bacterial invasion?

22.4 ▶ T cell receptors and MHC molecules both interact with an antigen. What is the difference in the mode of interaction between the two?

22.5 What differentiates innate immunity from adaptive (acquired) immunity?

Cells and Organs of the Immune System

22.6 Where in the body do you find the largest concentration of antibodies as well as T cells?

22.7 Where do T and B cells mature and differentiate?

22.8 What are memory cells? What is their function?

22.9 What are the favorite targets of macrophages? How do they kill the target cells?

Antigens and Their Presentation

22.10 Would a foreign substance, such as aspirin (MW 180), be considered an antigen by the body?

22.11 What kind of antigen does a T cell recognize?

22.12 What is the smallest unit of an antigen that is capable of binding to an antibody?

22.13 ▶ How does the body process antigens to be recognized by class II MHC?

22.14 What role does the MHC play in the immune response of the ABO blood groups?

22.15 To what class of compounds do MHCs belong? Where would you find them?

Immunoglobulins

22.16 When a foreign substance is injected in a rabbit, how long does it take to find antibodies against the foreign substance in the rabbit serum?

22.17 Distinguish among the roles of the IgA, IgE, and IgG immunoglobulins.

22.18 (a) Which immunoglobulin has the highest carbohydrate content and the lowest concentration in the serum?

(b) What is its main function?

22.19 Chemical Connections 11D states that the antigen in the red blood cells of a person with B-type blood is a galactose unit. Show schematically how the antibody of a person with A-type blood would aggregate the red blood cells of a B-type person if such a transfusion were made by mistake.

22.20 In the immunoglobulin structure, the "hinge region" joins the stem of the Y to the arms. The hinge region can be cleaved by a specific enzyme to yield one Fc fragment (the stem of the Y) and two Fab fragments (the two arms). Which of these two kinds of fragments can interact with an antigen? Explain.

22.21 How are the light and heavy chains of an antibody held together?

22.22 What do we mean by the term *immunoglobulin superfamily?*

22.23 If you could isolate two monoclonal antibodies from a certain population of lymphocytes, in what sense would they be similar to each other and in what sense would they differ?

22.24 What kind of interaction takes place between an antigen and an antibody?

22.25 How is a new protein created on the variable H chain by V(J)D recombination?

T Cells and Their Receptors

22.26 T cell receptor molecules are made of two polypeptide chains. Which part of the chain acts as a binding site and what binds to it?

22.27 What is the difference between a T cell receptor (TcR) and a TcR complex?

22.28 What kind of tertiary structure characterizes the TcR?

22.29 What are the components of the T cell receptor complex?

22.30 By what chemical process does CD3 transduct signals inside the cell?

22.31 Which adhesion molecule in the TcR complex helps the HIV virus infect a leukocyte?

22.32 Three kinds of molecules in the T cell belong to the immunoglobulin superfamily. List them and indicate briefly their function.

Control of Immune Response: Cytokines

22.33 What kind of molecules are cytokines?

22.34 With what do cytokines interact? Do they bind to antigens?

22.35 As in most biochemistry literature, one encounters a veritable alphabet soup when reading about cytokines. Identify by their full name (a) TNF, (b) IL, and (c) EGF.

22.36 What are chemokines? How do they deliver their message?

22.37 What is the characteristic chemical signature in the structure of chemokines?

22.38 What are the chemical characteristics of cytokines that allow their classification?

22.39 What specific amino acid appears in all chemokines?

How to Recognize "Self"

22.40 How does the body prevent T cells from being active against a "self" antigen?

22.41 What makes a tumor cell different from a normal cell?

22.42 Name a signaling pathway that controls the maturation of B cells and prevents those with affinity toward self antigen from becoming active.

22.43 How does the inhibitory receptor on a macrophage prevent an attack on normal cells?

22.44 Which components of the immune system are principally involved in autoimmune diseases?

22.45 How do glucocorticoids make autoimmune disease sufferers feel more comfortable?

Chemical Connections

22.46 (Chemical Connections 22A) How does Rituxan, a monoclonal antibody, find its way to the surface of non-Hodgkins lymphoma cells?

22.47 (Chemical Connections 22A) Why is monoclonal antibody treatment better than chemotherapy?

22.48 (Chemical Connections 22B) What made Edward Jenner the father of immunization? In your opinion, could one do such an experiment today legally?

22.49 (Chemical Connections 22B) What is the difference between memory cells and plasma cells?

22.50 (Chemical Connections 22C) In leukocyte trafficking, what keeps the leukocytes rolling along the endothelial wall of the blood vessels?

22.51 (Chemical Connections 22C) What is the role of a chemokine in facilitating the passage of leukocytes from a blood vessel to the site of injury?

22.52 (Chemical Connections 22D) What causes the weakening and eventual paralysis of the muscles in myasthenia gravis?

22.53 (Chemical Connections 22D) One finds elevated antibodies in the serum of myasthenia gravis patients. What is the source of the antibodies? Which cells of the immune system manufacture them?

Additional Problems

22.54 Which immunoglobulins form the first line of defense against invading bacteria?

22.55 Where do all the lymphatic vessels come together?

22.56 (Chemical Connections 22A) Why is a monoclonal antibody treatment preferable to a multiclonal antibody treatment?

22.57 Which compound or complex of compounds of the immune system is mostly responsible for the proliferation of leukocytes?

22.58 Name a process beside the V(J)D recombination that can enhance immunoglobulin diversity in the variable region.

22.59 Name a tumor cell marker, a synthetic analogue of which may be the first anticancer vaccine.

22.60 Is the light chain of an immunoglobulin the same as the V region?

22.61 Where are TNF receptors located?

22.62 The variable regions of immunoglobulins bind the antigens. How many polypeptide chains carry variable regions in one immunoglobulin molecule?

InfoTrac College Edition

For additional readings, go to InfoTrac College Edition, your online research library, at

http://infotrac.thomsonlearning.com

CHAPTER 23

23.1   Introduction

23.2   Functions and Composition of Blood

23.3   Blood as a Carrier of Oxygen

23.4   Transport of Carbon Dioxide in the Blood

23.5   Blood Cleansing: A Kidney Function

23.6   Buffer Production: Another Kidney Function

23.7   Water and Salt Balance in Blood and Kidneys

23.8   Blood Pressure

Human circulatory cells: red blood cells, platelets, and white blood cells.

© Ken Eward/BioGrafx/Photo Researchers, Inc.

Body Fluids

23.1   Introduction

Single-cell organisms receive their nutrients directly from the environment and discard waste products directly into it. In multicellular organisms, the situation is not so simple. There, too, each cell needs nutrients and produces wastes, but most of the cells are not directly in contact with the environment. Body fluids serve as a medium for carrying nutrients to and waste products from the cells as well as a means for carrying the chemical communicators (Chapter 15) that coordinate activities among cells.

    The body fluids that are not inside the cells are collectively known as **extracellular fluids.** These fluids make up about one-fourth of a person's body weight. The most abundant is **interstitial fluid,** which directly surrounds most cells and fills the spaces between them. It makes up about 17% of body weight. Another body fluid is **blood plasma,** which flows in the arteries and veins. It makes up about 5% of body weight. Other body fluids that occur in lesser amounts are urine, lymph, cerebrospinal fluid, aqueous humor, and synovial fluid. All body fluids are aqueous solutions—water is the only solvent in the body.

    The blood plasma circulates in the body and comes in contact with the other body fluids through the semipermeable membranes of the blood ves-

The fluid inside the cells is called **intracellular fluid.**

Interstitial fluid The fluid that surrounds the cells and fills the spaces between them

Blood plasma The noncellular portion of blood

508

CHEMICAL CONNECTIONS 23A

Using the Blood–Brain Barrier to Eliminate Undesirable Side Effects of Drugs

Many drugs have undesirable side effects. For example, many antihistamines, such as Dramamine and Benadryl (Section 15.5), cause drowsiness. These antihistamines are supposed to act on the peripheral H_1 histamine receptors to relieve seasickness, hay fever, or asthma. Because they penetrate the blood–brain barrier, they also act as antagonists to the H_1 receptors in the brain, causing sleepiness.

One drug that also acts on the peripheral H_1 receptors, fexofenadine (sold under the trade name Allegra), cannot penetrate the blood–brain barrier. This antihistamine alleviates seasickness and asthma in the same way as the old antihistamines, but it does not cause drowsiness as a side effect.

sels (Figure 22.2). In this way, blood can exchange chemical compounds with other body fluids, such as lymph and interstitial fluid, and, through them, with the cells and organs of the body.

There is only a limited exchange between blood and cerebrospinal fluid, on the one hand, and between blood and the interstitial fluid of the brain, on the other hand. The limited exchange between blood and interstitial fluid of the brain is referred to as the **blood–brain barrier.** This barrier is permeable to water, oxygen, carbon dioxide, glucose, alcohols, and most anesthetics, but only slightly permeable to electrolytes such as Na^+, K^+, and Cl^- ions. Many higher-molecular-weight compounds are also excluded.

The blood–brain barrier protects the cerebral tissue from detrimental substances in the blood and allows it to maintain a low K^+ concentration, which is needed to generate the high electrical potential essential for neurotransmission. It is vital that the body maintain a proper balance of levels of salts, proteins, and all other components in the blood. **Homeostasis** is the process of maintaining the levels of nutrients in the blood as well as the body's temperature.

Body fluids have special importance for the health care professions. Samples of body fluid can be taken with relative ease. The chemical analysis of blood plasma, blood serum, urine, and occasionally cerebrospinal fluid is of major importance in diagnosing disease.

> **Blood–brain barrier** A barrier limiting the exchange of blood components between blood and the cerebrospinal and brain interstitial fluids to water, carbon dioxide, glucose, and other small molecules but excluding electrolytes and large molecules

23.2 Functions and Composition of Blood

It has been known for centuries that "life's blood" is essential to human life. Blood has many functions, including the following:

1. It carries O_2 from the lungs to the tissues.
2. It carries CO_2 from the tissues to the lungs.
3. It carries nutrients from the digestive system to the tissues.
4. It carries waste products from the tissues to the excretory organs.
5. With its buffer systems, it maintains the pH of the body (with the help of the kidneys).
6. It maintains a constant body temperature.
7. It carries hormones from the endocrine glands to wherever they are needed.
8. It transports infection-fighting blood cells (leukocytes) and antibodies.

Figure 23.1 Nutrients, oxygen, and other materials diffuse out of the blood and through the tissue fluid that bathes the cells. Carbon dioxide and other waste products diffuse out of the cells and enter the blood through the capillary wall.

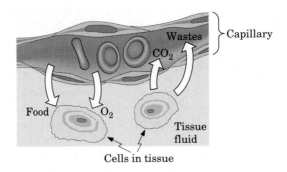

Figure 23.1 shows some of these functions. The rest of this chapter describes how the blood carries out some of these functions.

Whole blood is a complicated mixture. It contains several types of cells (Figure 23.2) plus a liquid, noncellular portion called **plasma,** in which many substances are dissolved (Table 23.1). There are three main types of cellular elements of blood: erythrocytes, leukocytes, and platelets.

A Erythrocytes

The most prevalent cells of blood are the red blood cells, which are called **erythrocytes.** There are about 5 million red blood cells in every cubic millimeter of blood, or roughly 100 million in every drop. Erythrocytes are very specialized cells. They have no nuclei and, hence, no DNA. Their main function is to carry oxygen to the cells and carbon dioxide away from the cells.

Erythrocytes are formed in the bone marrow, and they stay in the bloodstream for about 120 days. Old erythrocytes are removed by the liver and

A 150-lb (68-kg) man has about 6 L of whole blood, 50 to 60% of which is plasma.

Erythrocytes Red blood cells; transporters of blood gases

In an adult male, there are approximately 30 trillion erythrocytes.

Figure 23.2 Some cellular elements of blood. The dimensions are shown in microns (μm; also called micrometers).

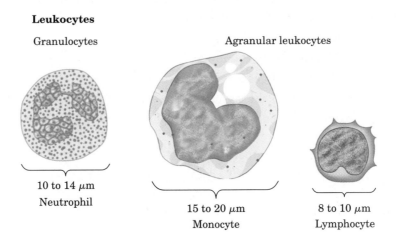

Erythrocytes

7 μm

Thrombocytes (platelets)

1 to 2 μm

Leukocytes

Granulocytes

10 to 14 μm
Neutrophil

Agranular leukocytes

15 to 20 μm
Monocyte

8 to 10 μm
Lymphocyte

Table 23.1   Blood Components and Some Diseases Associated with Their Abnormal Presence in the Blood

Whole Blood → *Plasma*

Cellular Elements   *Fibrinogen*   *Serum*

Cellular Elements	Serum
Erythrocytes (high: polycythemia; low: anemia)	Water (high: edema; low: dehydration)
Leukocytes (high: leukemia; low: typhoid fever)	Albumin (low: edema)
Platelets (low: thrombocytopenia)	Globulins (high: transplant rejection; low: infection)
	Clotting factors (low: hemophilia)
	Glucose (high: diabetes; low: hypoglycemia)
	Cholesterol (high: gallstones, atherosclerosis)
	Urea
	Inorganic salts
	Gases (N_2, O_2, CO_2)
	Enzymes, hormones, vitamins

spleen and destroyed. The constant formation and destruction of red blood cells maintain a steady number of erythrocytes in the body.

B Leukocytes

Leukocytes (white blood cells) are relatively minor cellular components (minor in numbers, not in function). For each 1000 red blood cells, there are only one or two white blood cells. Most of the different leukocytes destroy invading bacteria or other foreign substances by devouring them (**phagocytosis).**

Like erythrocytes, leukocytes are manufactured in the bone marrow. Specialized white blood cells formed in the lymph nodes and in the spleen are called lymphocytes. They synthesize immunoglobulins (antibodies; see Section 22.4) and store them.

Leukocytes White blood cells; part of the immune system

C Platelets

When a blood vessel is cut or injured, the bleeding is controlled by a third type of cellular element: the **platelets** (also called **thrombocytes**). They are formed in the bone marrow and spleen and are more numerous than leukocytes but less numerous than erythrocytes.

Platelets Cellular element of blood; essential to clot formation

There are about 300,000 platelets in each mm^3 of blood, or one for every 10 or 20 red blood cells.

D Plasma

If all the cellular elements from whole blood are removed by centrifugation, the resulting liquid is the plasma. The cellular elements—mainly red blood cells, which settle at the bottom of the centrifuge tube—account for between 40 and 50% of the blood volume.

Blood plasma is 92% water. The dissolved solids in the plasma are mainly proteins (7%). The remaining 1% contains glucose, lipids, enzymes, vitamins, hormones, and waste products such as urea and CO_2. Of the plasma proteins, 55% is albumin, 38.5% is globulin, and 6.5% is fibrinogen. If plasma is allowed to stand, it forms a clot, a gel-like substance. We

A dried clot becomes a scab.

CHEMICAL CONNECTIONS 23B

Blood Clotting

When body tissues are damaged, the blood flow must be stopped or else enough of it will pour out to cause death. The mechanism used by the body to stem leaks in the blood vessels is clotting. This complicated process involves many factors. Here we mention only a few important steps.

When a blood vessel is injured, the first line of defense is the platelets, which circulate constantly in the blood. They rush to the site of injury, adhere to the collagen molecules in the capillary wall that are exposed by the cut, and form a gel-like plug. At the signal of thromboxane A_2, more platelets enlarge the size of the clot (Section 12.12). This plug is porous, however, and a firmer gel (a clot) is needed to seal the site. A clot is a three-dimensional network of fibrin molecules that also contains platelets. The fibrin network is formed from the blood fibrinogen by the enzyme thrombin. Together with the embedded platelets, it constitutes the blood clot.

Why doesn't the blood clot in the blood vessels under normal conditions (with no injury or disease)? The reason is that the enzyme that starts clot formation, thrombin, exists in the blood only in its inactive form, called prothrombin. Prothrombin itself is manufactured in the liver, and vitamin K is needed for its production. Even when prothrombin is in sufficient supply, a number of proteins are needed to change it to thrombin. These proteins are given the collective name thromboplastin. Any thromboplastic substance can activate prothrombin in the presence of Ca^{2+} ions. Thromboplastic substances exist in the platelets, in the plasma, and in the injured tissue itself.

Clotting is nature's way of protecting us from loss of blood. However, we don't want the blood to clot during blood transfusions because this would stop the flow. To prevent this problem, we add sodium citrate to the blood. Sodium citrate interacts with Ca^{2+} ions and removes them from the solution, so the thromboplastic substances cannot activate prothrombin and no clot forms.

A clot is not dangerous if it stays near the injury because, once the body repairs the tissue, the clot is digested and removed. However, a clot formed in one part of the body may break loose and travel to other parts, where it may lodge in an artery. This condition is called **thrombosis.** If the clot then blocks the oxygen and nutrient supply to the heart and brain, it can result in paralysis and death. After surgery, **anticoagulant drugs** are occasionally administered to prevent clot formation.

The most common anticoagulants are heparin and dicumarol. Heparin enhances the inhibition of thrombin by antithrombin (Section 11.7B), and dicumarol blocks the transport of vitamin K to the liver, preventing prothrombin formation.

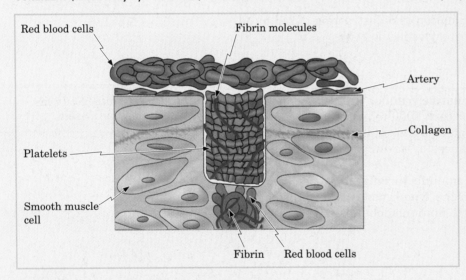

Red blood cells

Fibrin molecules

Artery

Collagen

Platelets

Smooth muscle cell

Fibrin   Red blood cells

■ **Components of a blood clot.** (*From* The Functioning of Blood Platelets *by M. B. Zucker. Copyright 1980 by Scientific American, Inc. All rights reserved.*)

Serum Blood plasma from which fibrinogen has been removed

can squeeze out a clear liquid from the clot, called **serum.** It contains all the components of the plasma except the fibrinogen. This protein is involved in the complicated process of clot formation (Chemical Connections 23B).

As for the other plasma proteins, most of the globulins take part in the immune reactions (Section 22.4), and the albumin provides proper osmotic pressure. If the albumin concentration drops (from malnutrition or from kidney disease, for example), the water from the blood oozes into the interstitial fluid and creates the swelling of tissues called **edema.**

23.3 Blood as a Carrier of Oxygen

One of the most important functions of blood is to carry oxygen from the lungs to the tissues. This task is accomplished by hemoglobin molecules located inside the erythrocytes. As we saw in Section 13.9, hemoglobin is made up of two alpha and two beta protein chains, each attached to a molecule of heme.

The active sites are the hemes, and at the center of each heme is an iron(II) ion. The heme with its central Fe^{2+} ion forms a plane. Because each hemoglobin molecule has four hemes, it can hold a total of four O_2 molecules. In reality, the ability of the hemoglobin molecule to hold O_2 depends on how much oxygen is in the environment. Consider Figure 23.3, which shows how the oxygen-carrying ability of hemoglobin depends on oxygen pressure. When oxygen enters the lungs, the pressure is high (100 mm Hg). At this pressure, all the Fe^{2+} ions of the hemes bind oxygen molecules; they are fully saturated. By the time the blood reaches the muscles through the capillaries, the oxygen pressure in the muscle is only 20 mm Hg (Chemical Connections 23C). At this pressure, only 30% of the binding sites carry oxygen. Thus, 70% of the oxygen carried will be released to the tissues. The S shape of the binding (dissociation) curve (Figure 23.3) implies that the reaction is not a simple equilibrium reaction.

$$HbO_2 \rightleftharpoons Hb + O_2$$

In hemoglobin, each heme has a cooperative effect on the other hemes. This cooperative action allows hemoglobin to deliver twice as much oxygen to the tissues as it would if each heme acted independently. The reason for this is as follows: When a hemoglobin molecule is carrying no oxygen, the four globin units coil into a certain shape (Section 13.9). When the first oxygen molecule attaches to one of the heme subunits, it changes the shape not only of that subunit but also of a second subunit, making it easier for the second subunit to bind an oxygen. When the second heme binds, the shape

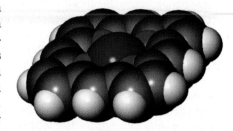

■ **A model of heme with a central Fe^{2+} ion.**

The O_2 picked up by the hemoglobin binds to the Fe^{2+}.

This allosteric effect (Section 14.6) explains the S-shaped curve shown in Figure 23.3.

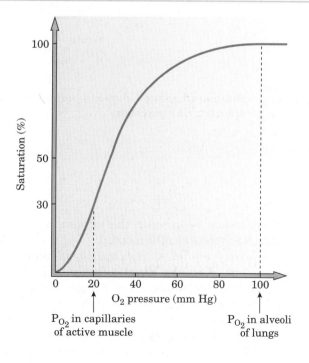

Figure 23.3 An oxygen dissociation curve. Saturation (%) means the percentage of Fe^{2+} ions that carry O_2 molecules.

CHEMICAL CONNECTIONS 23C

Breathing and Dalton's Law

A gas naturally flows from an area of higher pressure to one of lower pressure. This fact is responsible for respiration, though we must look not at the total pressure of air but only at the partial pressure of O_2 and CO_2. Respiration is the process by which blood carries O_2 from the lungs to tissues, collects the CO_2 manufactured by cells, and brings it to the lungs, where it is breathed out.

The air we breathe contains about 21% O_2 at a partial pressure of approximately 159 mm Hg. The partial pressure of O_2 in the alveoli (tiny air sacs) in the lungs is about 100 mm Hg. Because the partial pressure is higher in the inhaled air, O_2 flows to the alveoli. At this point, venous blood flows past the alveoli. The partial pressure of O_2 in the venous blood is only 40 mm Hg, so O_2 flows from the alveoli to the venous blood. This new supply of O_2 raises the partial pressure to about 100 mm Hg, and the blood becomes arterial blood and flows to the body tissues. There, the partial pressure of O_2 is 30 mm Hg or less (because the metabolic activity of the tissue cells has used up most of their oxygen to provide energy). Oxygen now flows from the arterial blood to the body tissues. The partial pressure of O_2 in the blood decreases (to about 40 mm Hg), and the blood becomes venous blood again and returns to the lungs for a fresh supply of O_2. At each step of the cycle, O_2 flows from a region of higher partial pressure to one of lower partial pressure.

Meanwhile, CO_2 goes the opposite way for the same reason. The partial pressure of CO_2 as a result of metabolic activity is about 60 mm Hg in tissues. CO_2 flows from tissues to arterial blood, changing it to venous blood and reaching a CO_2 pressure of 46 mm Hg. The blood then flows to the lungs. The CO_2 pressure in the alveoli is 40 mm Hg, and so the CO_2 flows to the alveoli. Because the partial pressure of CO_2 in air is only 0.3 mm Hg, CO_2 flows from the alveoli and is exhaled.

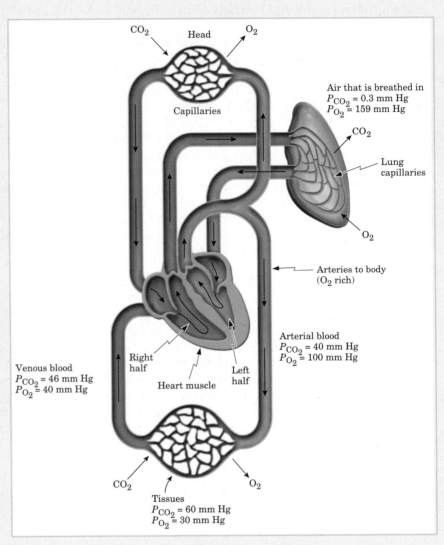

■ **Blood circulation showing partial pressures of O_2 and CO_2 in the different parts of the system.**

Lowering the pH decreases the oxygen-binding capacity of hemoglobin.

Bohr effect The effect caused by a change in the pH on the oxygen-carrying capacity of hemoglobin

This is how the body delivers more oxygen to those tissues that need it.

changes once again, and the oxygen-binding ability of the two remaining subunits increases still more.

The oxygen-delivering capacity of hemoglobin is also affected by its environment. A slight change in the pH of the environment changes the oxygen-binding capacity, a phenomenon called the **Bohr effect.** An increase in CO_2 pressure also decreases the oxygen-binding capacity of hemoglobin. When a muscle contracts, H^+ ions and CO_2 are produced. The newly produced CO_2 enhances the release of oxygen. At the same time, H^+ ions lower the pH of the muscle. Thus, at the same pressure (20 mm Hg), more oxygen is released for an active muscle than for a muscle at rest.

23.4 Transport of Carbon Dioxide in the Blood

What happens to the CO_2 and H^+ produced in metabolically active cells? They bind to the hemoglobin of erythrocytes, and carbaminohemoglobin is formed. This is an equilibrium reaction.

$$Hb-NH_2 + CO_2 \rightleftharpoons Hb-NH-COO^- + H^+$$
$$\text{Carbaminohemoglobin}$$

The CO_2 is bound to the terminal α-NH_2 groups of the four polypeptide chains, so each hemoglobin can carry a maximum of four CO_2 molecules, one for each chain. How much CO_2 each hemoglobin actually carries depends on the pressure of CO_2. The higher the CO_2 pressure, the more carbaminohemoglobin formed.

Only 25% of the total CO_2 produced by the cells is transported to the lungs in the form of carbaminohemoglobin. Another 70% is converted in the red blood cells to carbonic acid by the enzyme carbonic anhydrase. A large part of this H_2CO_3 is carried as such to the lungs, where it is converted back to CO_2 by carbonic anhydrase and released. The remaining 5% of the total CO_2 is carried in the plasma as dissolved gas.

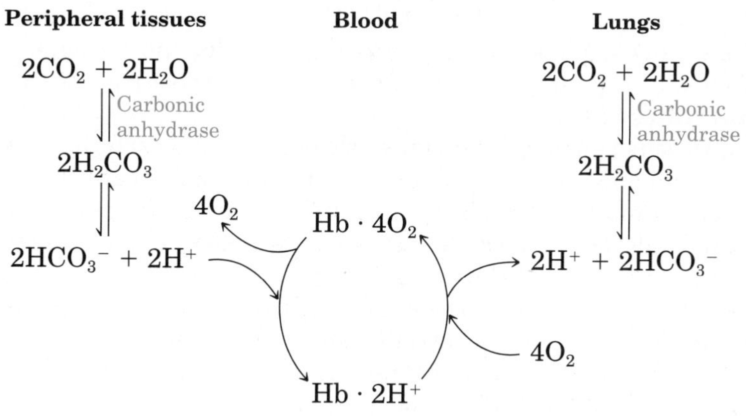

The reaction proceeds to the CO_2 production (top) because loss of CO_2 from the lungs causes this equilibrium to shift to replace it (Le Chatelier's principle).

23.5 Blood Cleansing: A Kidney Function

We have seen that one of blood's functions is to maintain the pH at 7.4 (Section 7.11). Another is to carry waste products away from the cells. CO_2, one of the principal waste products of respiration, is carried by the blood to the lungs and exhaled. The other waste products are filtered out by the kidneys and eliminated in the urine.

The kidney is a superfiltration machine. In the event of kidney failure, filtration can be done by hemodialysis (Figure 23.4). About 100 L of blood passes through a normal human kidney daily. Of this amount, only about 1.5 L of liquid is excreted as urine. Obviously, we do not have here just a simple filtration system in which small molecules are lost and large ones are retained. The kidneys also reabsorb from the urine those small molecules that are not waste products.

Hemodialysis is discussed in Chemical Connections 6F.

The balance between filtration and reabsorption is controlled by a number of hormones.

A Kidney

The biological units inside the kidneys that perform these functions are called **nephrons,** and each kidney contains about a million of them. A nephron is made up of a filtration head called a **Bowman's capsule** con-

Figure 23.4 A patient undergoing hemodialysis for kidney disease.

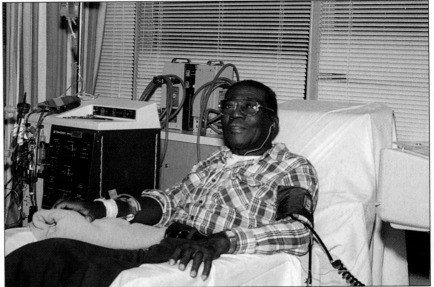

Beverly March; courtesy of Long Island Jewish Hospital

Arteries are blood vessels that carry oxygenated blood away from the heart.

Glomeruli Part of the kidney filtration cell, or nephron; a tangle of capillaries surrounded by a fluid-filled space

The singular is *glomerulus*.

nected to a tiny, U-shaped tube called a tubule. The part of the tubule close to the Bowman's capsule is the **proximal tubule,** the U-shaped twist is called **Henle's loop,** and the part of the tubule farthest from the capsule is the **distal tubule** (Figure 23.5).

Blood vessels penetrate the kidney throughout. The arteries branch into capillaries, and one tiny capillary enters each Bowman's capsule. Inside the capsule, the capillary first branches into even smaller vessels, called **glomeruli,** and then leaves the capsule. The blood enters the glomeruli with every heartbeat, and the pressure forces the water, ions, and all small molecules (urea, sugars, salts, amino acids) through the walls of the glomeruli and the Bowman's capsule. These molecules and ions enter the proximal tubule. Blood cells and large molecules (proteins) are retained in the capillary and exit with the blood.

As shown in Figure 23.5, the tubules and Henle's loop are surrounded by blood vessels; these blood vessels reabsorb vital nutrients. Eighty percent of the water is reabsorbed in the proximal tubule. Almost all the glucose and amino acids are reabsorbed here. When excess sugar is present in the blood (diabetes; see Chemical Connections 15G), some of it passes into the urine. As a consequence, the measurement of glucose concentration in the urine is used in diagnosing diabetes.

By the time the glomerular filtrate reaches Henle's loop, the solids and most of the water have been reabsorbed. Only wastes (such as urea, creatinine, uric acid, ammonia, and some salts) pass into the collecting tubules that bring the urine to the ureter, from which it passes to the bladder, as shown in Figure 23.5(a).

B Urine

Normal urine contains about 4% dissolved waste products; the rest is water. Although the daily amount of urine varies greatly, it averages about 1.5 L per day. The pH varies from 5.5 to 7.5. The main solute is urea, the end product of protein metabolism (Section 19.8). Other nitrogenous waste

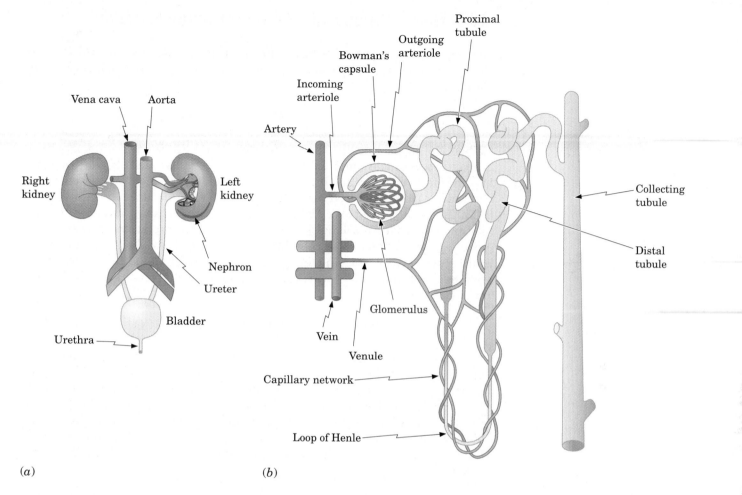

Figure 23.5 Excretion through the kidneys. (*a*) The human urinary system. (*b*) A kidney nephron and its components, with the surrounding circulatory system.

products, such as creatine, creatinine, ammonia, and hippuric acid, are also present, albeit in much smaller amounts.

See the **Interactive General, Organic, and Biochemistry CD-ROM, version 2.0,** for further exploration on this topic.

Creatine   Creatinine   Hippuric acid

In addition, normal urine contains inorganic ions such as Na^+, Ca^{2+}, Mg^{2+}, Cl^-, PO_4^{3-}, SO_4^{2-}, and HCO_3^-.

Under certain pathological conditions, other substances can appear in the urine. This possibility makes urine analysis a highly important part of diagnostic medicine. We have already noted that the presence of glucose and ketone bodies is an indication of diabetes. Other abnormal constituents may include proteins, which can indicate kidney diseases such as nephritis.

23.6 Buffer Production: Another Kidney Function

Among the waste products sent into the blood by the tissues are H^+ ions. They are neutralized by the HCO_3^- ions that are a part of the blood's buffer system:

$$H^+ + HCO_3^- \rightleftharpoons H_2CO_3$$

When the blood reaches the lungs, as we saw in Section 23.4, H_2CO_3 is decomposed by carbonic anhydrase, and CO_2 is exhaled. If the body had no mechanism for replacing HCO_3^-, we would lose most of the bicarbonate buffer from the blood, and the blood pH would eventually drop (acidosis). In reality, the lost HCO_3^- ions are continuously replaced by the kidneys—another principal kidney function. This replacement takes place in the distal tubules. The cells lining the walls of the distal tubules reabsorb the CO_2 that was lost in the glomeruli. With the aid of carbonic anhydrase, carbonic acid forms quickly and then dissociates to HCO_3^- and H^+ ions:

$$CO_2 + H_2O \rightleftharpoons H_2CO_3 \rightleftharpoons H^+ + HCO_3^-$$

The H^+ ions move from the cells into the urine in the tubule, where they are partially neutralized by a phosphate buffer. To compensate for the lost positive ions, Na^+ ions from the tubule enter the cells. When this happens, Na^+ and HCO_3^- ions move from the cells into the capillaries. In this way, the H^+ ions picked up at the tissues and temporarily neutralized in the blood by HCO_3^- are finally pumped out into the urine. At the same time, the HCO_3^- ions lost in the lungs are regained by the blood in the distal tubules.

23.7 Water and Salt Balance in Blood and Kidneys

In Section 23.5, we mentioned that there is a balance in the kidneys between filtration and reabsorption. This balance is under hormonal control. The reabsorption of water is promoted by vasopressin, a small peptide hormone manufactured in the pituitary gland (Figure 15.2). In the absence of this hormone, only the proximal tubules—not the distal tubules or the collecting tubules—reabsorb water. As a consequence, too much water passes into the urine without vasopressin. In the presence of vasopressin, water is reabsorbed in these parts of the nephrons; thus, vasopressin causes the blood to retain more water and produces a more concentrated urine. The production of urine is called **diuresis.** Any agent that reduces the volume of urine is called an **antidiuretic.**

Usually, the vasopressin level in the body is sufficient to maintain the proper amount of water in tissues under various levels of water intake. However, when severe dehydration occurs (as a result of diarrhea, excessive sweating, or insufficient water intake), another hormone helps to maintain proper fluid level. This hormone, aldosterone (Section 12.10), controls the Na^+ ion concentration in the blood. In the presence of aldosterone, the reabsorption of Na^+ ions increases. When more Na^+ ions enter the blood, more Cl^- ions follow (to maintain electroneutrality) as well as more water to solvate these ions. Thus, increased aldosterone production enables the body to retain more water. When the Na^+ ion and water levels in the blood return to normal, aldosterone production stops.

A high concentration of solids in the urine is a symptom of diabetes insipidus.

Diuresis Enhanced excretion of water in urine

For this reason, vasopressin is also called antidiuretic hormone (ADH).

Sex Hormones and Old Age

The same sex hormones that provide a healthy reproductive life in young people can and do cause problems in the aging body. For male fertility, the prostate gland and testis provide semen. The prostate gland is situated at the exit of the bladder, surrounding the ureter. Its growth is promoted by the conversion of the male sex hormone, testosterone, to 5-α-dihydrotestosterone by the enzyme 5-α-reductase. The majority of men over the age of 50 have enlarged prostate glands, a condition known as benign prostatic hyperplasia. In fact, this condition affects more than 90% of men older than age 70. The enlarged prostate gland constricts the ureter and causes a restricted flow of urine.

This symptom can be partially relieved by an oral drug called terazosin (trade name Hytrin). This drug was originally approved for treatment of high blood pressure, where its action results from its ability to block the α-1-adrenoreceptors, thereby reducing the constriction around peripheral blood vessels. In a similar way, it reduces constriction around the ureter (the duct that conveys urine from the kidney to the bladder), allowing an unrestricted flow of urine. Terazosin, if taken as the sole antihypertensive, antiprostate drug, has come under scrutiny lately, however. A study released in 2000 concluded that this drug may contribute to the occurrence of heart attacks.

One measure, the concentration of prostate-specific antigen (PSA) in the serum, may indicate the possibility of developing prostate cancer. A PSA of 4.0 ng/mL or less is considered to be normal or an indicator of benign prostatic hyperplasia. A PSA concentration of more than 10 ng/mL is an indicator for prostate cancer. The area between 4.0 and 10 ng/mL is a gray area that deserves to be closely monitored.

A more accurate measure of prostate cancer is the percentage of free (not bound) PSA. When this level is higher than 15%, the likelihood of cancer diminishes with increasing percentage. When it is 9% or lower, the prostate cancer has probably already spread. The ultimate confirmation of cancer is microscopic examination of tissue taken by biopsy.

Estradiol and progesterone control sexual characteristics in the female body (Section 12.10). With the onset of menopause, smaller amounts of these and other sex hormones are secreted. This drop-off not only stops fertility but occasionally creates medical problems. The most serious complication is osteoporosis, which is the loss of bone tissue through absorption. It results in brittle and occasionally deformed bone structure. Estrogens (a general name for female sex hormones other than progesterone) in the proper amount can help to absorb calcium from the diet and thus prevent osteoporosis. The lack of enough female sex hormones can also result in atrophic vaginitis and atrophic urethritis.

A large number of postmenopausal women take pills that contain a synthetic progesterone analog and natural estrogens. However, continuous medication with these pills—for example, Premarin (mixed estrogens) and Provera (a progesterone analog)—is not without danger. A slight but statistically significant increase in cancer of the uterus has been reported in postmenopausal women who use such medication. In the latest study released in 2002, a slight risk of heart problems was also noticed in women taking such drugs. A physician's advice is needed in weighing the benefits against the risks.

23.8 Blood Pressure

Blood pressure is produced by the pumping action of the heart. The blood pressure at the arterial capillary end is about 32 mm Hg, a level higher than the osmotic pressure of the blood (18 mm Hg). The osmotic pressure of the blood is caused by the fact that more solutes are dissolved in the blood than in the interstitial fluid. At the capillary end, therefore, nutrient solutes flow from the capillaries into the interstitial fluid and from there into the cells (see Figure 22.2). On the other hand, the blood pressure in the venous capillaries is about 12 mm Hg. This level is less than the osmotic pressure, so solutes (waste products) from the interstitial fluid flow into the capillaries.

The blood pressure is maintained by the total volume of blood, by the pumping of the heart, and by the muscles that surround the blood vessels and provide the proper resistance to blood flow. Blood pressure is controlled by several very complex systems—some that act within seconds and some that take days to respond after a change in blood pressure occurs. For example, if a patient hemorrhages, three nervous control systems begin to function within seconds. First, the **baroreceptors** in the neck detect the consequent drop in pressure and send appropriate signals to the heart to pump harder and to the muscles surrounding the blood vessels to contract

> **Baroreceptors** A feedback mechanism in the neck that detects changes in the blood pressure and sends signals to the heart to counteract those changes by pumping harder or slower as the need arises

Hypertension and Its Control

Nearly one-third of the U.S. population has high blood pressure **(hypertension)**. Epidemiological studies have shown that even mild hypertension (diastolic pressure between 90 and 104 mm Hg) brings the risk of cardiovascular disease. It can lead to heart attack, stroke, or kidney failure. Hypertension can be managed effectively with diet and drugs. As noted in the text, blood pressure is under the control of a complex system, so hypertension must be managed on more than one level. Recommended dietary practices are low Na^+ ion intake and abstention from caffeine and alcohol.

The most commonly used drugs for lowering blood pressure are **diuretics**. A number of synthetic organic compounds (mostly thiazides such as Diuril and Enduron) increase urine excretion. By doing so, they decrease the blood volume as well as the Na^+ ion concentration, thereby lowering blood pressure.

Other antihypertensive drugs affect the nerve control of blood pressure. Propranolol hydrochloride blocks the adrenoreceptor sites of beta-adrenergic neurotransmitters (Section 15.5). By doing so, it reduces the flow of nerve signals as well as the blood output of the heart; it affects the responses of baroreceptors and chemoreceptors in the central nervous system and the adrenergic receptors in the smooth muscles surrounding the blood vessels. Propranolol and other beta blockers (such as metoprolol) not only lower blood pressure, but also reduce the risk of myo-

cardial infarction. Similar lowering of blood pressure is achieved by drugs that block calcium channels (for example, nifedipine/Adalat or diltiazem hydrochloride/Cardizem). For proper muscle contraction, calcium ions must pass through muscle cell membranes, an activity facilitated by calcium channels. If a drug blocks these channels, the heart muscles will contract less frequently, pumping less blood and reducing blood pressure.

Other drugs that reduce hypertension include the vasodilators that relax the vascular smooth muscles. However, many of these agents have side effects such as causing fast heartbeat (tachycardia) and promoting Na^+ and water retention. Some drugs on the market act on one specific site rather than on many sites as the blockers do, which obviously reduces the possible side effects. An example is an enzyme inhibitor that prevents the production of angiotensin. Captopril (Chemical Connections 14F) and Accupril are both ACE (angiotensin-converting enzyme) inhibitors that lower blood pressure only if the sole cause of hypertension is angiotensin production. Prostaglandins PGE and PGA (Section 12.12) lower the blood pressure and, when taken orally, have prolonged duration of action. Since hypertension is such a common problem, and it is manageable by drugs, there is great activity in drug research to come up with even more effective and safer antihypertension drugs.

and thus restore pressure. Next, chemical receptors on the cells that detect less O_2 delivery or CO_2 removal send nerve signals. Finally, the central nervous system reacts to oxygen deficiency with a feedback mechanism.

Hormonal controls act somewhat more slowly, taking minutes or even days. The kidneys secrete an enzyme called renin, which acts on an inactive blood protein called angiotensinogen by converting it to **angiotensin,** a potent vasoconstrictor. The action of this peptide increases blood pressure. Aldosterone (Section 23.7) also increases blood pressure by increasing Na^+ ion levels and water reabsorption in the kidneys.

Finally, there is a long-term renal blood control for volume and pressure. When blood pressure falls, the kidneys retain more water and salt, thereby increasing blood volume and pressure.

S U M M A R Y

The most important body fluid is **blood plasma,** or whole blood from which cellular elements have been removed (Section 23.1). Other important body fluids are urine and the **interstitial fluids** that directly surround the cells of the tissues. The cellular elements of the blood are the red blood cells **(erythrocytes),** which carry O_2 to the tissues and CO_2 away from the tissues; the white blood cells **(leukocytes),** which fight infection; and the **platelets,** which control bleeding (Section 23.2). The plasma contains fibrino-

gen, which is necessary for blood clot formation. When fibrinogen is removed from the plasma, what remains is the **serum.** The serum contains albumins, globulins, nutrients, waste products, inorganic salts, enzymes, hormones, and vitamins dissolved in water. The body uses blood to maintain **homeostasis.**

The red blood cells carry O_2 (Section 23.3). The Fe(II) in the heme portion of hemoglobin binds the oxygen, which is then released at the tissues by a combination of factors: low O_2 pressure in the tissue cells, high H^+ ion concentration **(Bohr effect),** and high CO_2 concentration. Of the CO_2 in the venous blood, 25% binds to the terminal $-NH_2$ of the polypeptide chains of hemoglobin to form carbaminohemoglobin, 70% is carried in the plasma as H_2CO_3, and 5% is carried as dissolved gas (Section 23.4).

Waste products are removed from the blood in the kidneys (Section 23.5). The filtration units—**nephrons**—contain an entering blood vessel that branches out into fine vessels called **glomeruli.** Water and all the small molecules in the blood diffuse out of the glomeruli, enter the **Bowman's capsules** of the nephrons, and from there go into the **tubules.** The blood cells and large molecules remain in the blood

vessels. The water, inorganic salts, and organic nutrients are then reabsorbed into the blood vessels. The waste products plus water form the urine, which goes from the tubule into the ureter and finally into the bladder.

The kidneys not only filter out waste products, but also produce HCO_3^- ions for buffering, replacing those lost through the lungs (Section 23.6). The water and salt balance of the blood and urine are under hormonal control (Section 23.7). Vasopressin helps to retain water, and aldosterone increases the Na^+ ion concentration in the blood by reabsorption.

Blood pressure is controlled by a large number of factors (Section 23.8).

P R O B L E M S

Numbers that appear in color indicate difficult problems.
◤ designates problems requiring application of principles.

Body Fluids

23.1 What is the blood–brain barrier?

23.2 What percentage of our body weight is blood plasma?

23.3 (a) Is the blood plasma in contact with the other body fluids?
  (b) If so, in what manner?

Functions and Composition of Blood

23.4 Which blood cell is the most abundant?

23.5 What is the function of erythrocytes?

23.6 Which blood cells are the largest?

23.7 Define:
  (a) Interstitial fluid   (b) Plasma
  (c) Serum   (d) Cellular elements

23.8 Where are the following manufactured?
  (a) Erythrocytes   (b) Leukocytes
  (c) Lymphocytes   (d) Platelets

23.9 How is plasma prepared from whole blood?

23.10 State the function of the following:
  (a) Albumin   (b) Globulin
  (c) Fibrinogen

23.11 Which blood plasma component protects against infections?

23.12 How is serum prepared from blood plasma?

Blood as a Carrier of Oxygen

23.13 How many molecules of O_2 are bound to a hemoglobin molecule at full saturation?

23.14 A chemist analyzes contaminated air from an industrial accident and finds that the partial pressure of oxygen is only 38 mm Hg. How many O_2 molecules, on average, will be transported by each hemoglobin molecule when someone breathes such air? (Consult Figure 23.3.)

23.15 Which inorganic ion is involved in the definition of percent saturation? Explain.

23.16 From Figure 23.3, predict the O_2 pressure at which the hemoglobin molecule binds four, three, two, one, and zero O_2 molecules to its heme.

23.17 Explain how the absorption of O_2 on one binding site in the hemoglobin molecule increases the absorption capacity at other sites.

23.18 (a) What happens to the oxygen-carrying capacity of hemoglobin when the pH is lowered?
  (b) What is the name of this effect?

Transport of CO_2 in the Blood

23.19 (a) Where does CO_2 bind to hemoglobin?
  (b) What is the name of the complex formed?

23.20 In carbaminohemoglobin, the N-terminal ends of globin chains carry a carbamate group, $R-NH-COO^-$. The anionic site can form a salt bridge with a cationic amino group on a side chain of the globin molecule and stabilize it. Draw the formula of the salt bridge.

23.21 If the plasma carries 70% of its carbon dioxide in the form of H_2CO_3 and 5% in the form of dissolved CO_2 gas, calculate the equilibrium constant of the reaction catalyzed by carbonic anhydrase:

$$H_2O + CO_2 \rightleftharpoons H_2CO_3$$

(Assume that the H_2O concentration is included in the equilibrium constant.)

23.22 How is the equilibrium reaction of carbaminohemoglobin formation influenced by the CO_2 concentration of the surroundings? Apply Le Chatelier's principle.

Kidney Functions and Urine

23.23 Where is most of the water reabsorbed in the kidney?

23.24 What nitrogenous waste products are eliminated in the urine?

23.25 Besides nitrogenous waste products, what other components are normal constituents of urine?

23.26 What is the difference between glomeruli and nephrons?

Buffer Production

23.27 What happens to the H^+ ions sent into the blood as waste products by the cells of the tissues?

23.28 The pH of blood is maintained at 7.4. What would be the pH of an active muscle cell relative to that of blood? What component and process generates this pH?

Water and Salt Balance in the Blood and Kidneys

23.29 In which part of the kidneys does vasopressin change the permeability of membranes?

23.30 How does aldosterone production counteract the excessive sweating that usually accompanies a high fever?

23.31 Caffeine is a diuretic. What effect will drinking a lot of coffee have on the specific gravity of your urine?

Blood Pressure

23.32 What is the function of baroreceptors?

23.33 What would happen to the glucose in an arterial capillary if the blood pressure at the end of the capillary were 12 mm Hg rather than the normal 32 mm Hg?

23.34 Some drugs used as antihypertensive agents—for example, Captopril (Chemical Connections 23E)—inhibit the enzyme that activates angiotensinogen. What is the mode of action of such drugs in lowering blood pressure?

23.35 A number of drugs combatting high blood pressure go under the name ACE (angiotensin-converting enzyme) inhibitors. A patient takes diuretics and ACE inhibitors. How do these drugs lower blood pressure?

Chemical Connections

23.36 (Chemical Connections 23A) What side effect of antihistamines is avoided by using the drug fexofenadine?

23.37 (Chemical Connections 23B) Which molecules help to anchor the platelets when a blood vessel is cut?

23.38 (Chemical Connections 23B) What are the functions of (a) thrombin, (b) vitamin K, and (c) thromboplastin in clot formation?

23.39 (Chemical Connections 23B) What is the first line of defense when a blood vessel is cut?

23.40 (Chemical Connections 23C) In circulating blood, where is the partial pressure of CO_2 the highest? The lowest?

23.41 (Chemical Connections 23D) What are the symptoms of an enlarged prostate gland (benign prostatic hyperplasia)?

23.42 (Chemical Connections 23D) What causes osteoporosis in postmenopausal women?

23.43 (Chemical Connections 23E) How do calcium-channel blockers act in lowering blood pressure?

23.44 (Chemical Connections 23E) People with high blood pressure are advised to follow a low-sodium diet. How does such a diet lower blood pressure?

Additional Problems

23.45 Albumin makes up 55% of the protein content of plasma. Is the albumin concentration of the serum higher, lower, or the same as that of the plasma? Explain.

23.46 Which nitrogenous waste product of the body can be classified as an amino acid?

23.47 What happens to the oxygen-carrying ability of hemoglobin when acidosis occurs in the blood?

23.48 Does the heme of hemoglobin participate in CO_2 transport as well as in O_2 transport?

23.49 Is the blood–brain barrier permeable to insulin? Explain.

23.50 When kidneys fail, the body swells, a condition called edema. Why and where does the water accumulate?

23.51 Which hormone controls water retention in the body?

23.52 Describe the action of renin.

23.53 The oxygen dissociation curve of fetal hemoglobin is higher than that of adult hemoglobin. How does the fetus obtain its oxygen supply from the mother's blood when the two circulations meet?

InfoTrac College Edition

For additional readings, go to InfoTrac College Edition, your online research library, at

http://infotrac.thomsonlearning.com

Exponential Notation

The **exponential notation** system is based on powers of 10 (see table). For example, if we multiply $10 \times 10 \times 10 = 1000$, we express this as 10^3. The 3 in this expression is called the **exponent** or the **power,** and it indicates how many times we multiplied 10 by itself and how many zeros follow the 1.

There are also negative powers of 10. For example, 10^{-3} means 1 divided by 10^3:

$$10^{-3} = \frac{1}{10^3} = \frac{1}{1000} = 0.001$$

Numbers are frequently expressed like this: 6.4×10^3. In a number of this type, 6.4 is the **coefficient** and 3 is the exponent, or power of 10. This number means exactly what it says:

$$6.4 \times 10^3 = 6.4 \times 1000 = 6400$$

Similarly, we can have coefficients with negative exponents:

$$2.7 \times 10^{-5} = 2.7 \times \frac{1}{10^5} = 2.7 \times 0.00001 = 0.000027$$

For numbers greater than 10 in exponential notation, we proceed as follows: *Move the decimal point to the left,* to just after the first digit. The (positive) exponent is equal to the number of places we moved the decimal point.

Exponential notation is also called scientific notation.

For example, 10^6 means a one followed by six zeros, or 1,000,000, and 10^2 means 100.

APP. I.1 Examples of Exponential Notation

$10{,}000 = 10^4$
$1000 = 10^3$
$100 = 10^2$
$10 = 10^1$
$1 = 10^0$
$0.1 = 10^{-1}$
$0.01 = 10^{-2}$
$0.001 = 10^{-3}$

EXAMPLE

$$3\,7\,5\,0\,0 = 3.75 \times 10^4$$
4 because we went four places to the left

Four places to the left   Coefficient

$$628 = 6.28 \times 10^2$$

Two places to the left   Coefficient

$$859{,}600{,}000{,}000 = 8.596 \times 10^{11}$$

Eleven places to the left   Coefficient

We don't really have to place the decimal point after the first digit, but by doing so we get a coefficient between 1 and 10, and that is the custom.

Using exponential notation, we can say that there are 2.95×10^{22} copper atoms in a copper penny. For large numbers, the exponent is always *positive.* Note that we do not usually write out the zeros at the end of the number.

For small numbers (less than 1), we move the decimal point *to the right,* to just after the first nonzero digit, and use a *negative exponent.*

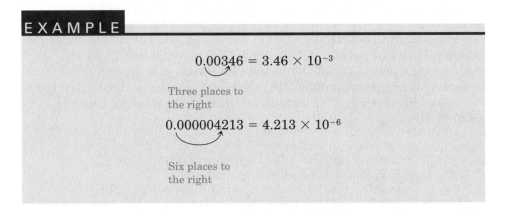

> **EXAMPLE**
>
> $$0.00346 = 3.46 \times 10^{-3}$$
>
> Three places to
> the right
>
> $$0.000004213 = 4.213 \times 10^{-6}$$
>
> Six places to
> the right

In exponential notation, a copper atom weighs 2.3×10^{-25} pounds.

To convert exponential notation into fully written-out numbers, we do the same thing backward.

> **EXAMPLE**
>
> Write out in full: (a) 8.16×10^7 (b) 3.44×10^{-4}.
>
> **Solution**
>
> (a) $8.16 \times 10^7 = 81{,}600{,}000$
>
>   Seven places to the right
>   (add enough zeros)
>
> (b)
>
>   (b) $3.44 \times 10^{-4} = 0.000344$
>
>   Four places to the left

When scientists add, subtract, multiply, and divide, they are always careful to express their answers with the proper number of digits, called significant figures. This method is described in Appendix II.

Adding and Subtracting Numbers in Exponential Notation

We are allowed to add or subtract numbers expressed in exponential notation *only if they have the same exponent.* All we do is add or subtract the coefficients and leave the exponent as it is.

> **EXAMPLE**
>
> Add 3.6×10^{-3} and 9.1×10^{-3}.
>
> **Solution**
>
> $$\begin{array}{r} 3.6 \times 10^{-3} \\ + \ 9.1 \times 10^{-3} \\ \hline 12.7 \times 10^{-3} \end{array}$$

The answer could also be written in other, equally valid ways:

$$12.7 \times 10^{-3} = 0.0127 = 1.27 \times 10^{-2}$$

When it is necessary to add or subtract two numbers that have different exponents, we first must change them so that the exponents are the same.

A calculator with exponential notation changes the exponent automatically.

EXAMPLE

Add 1.95×10^{-2} and 2.8×10^{-3}.

Solution

To add these two numbers, we make both exponents -2. Thus, $2.8 \times 10^{-3} = 0.28 \times 10^{-2}$. Now we can add:

$$
\begin{array}{r}
1.95 \times 10^{-2} \\
+ \; 0.28 \times 10^{-2} \\
\hline
2.23 \times 10^{-2}
\end{array}
$$

Multiplying and Dividing Numbers in Exponential Notation

To multiply numbers in exponential notation, we first multiply the coefficients in the usual way and then algebraically *add* the exponents.

EXAMPLE

Multiply 7.40×10^5 by 3.12×10^9.

Solution

$$7.40 \times 3.12 = 23.1$$

Add exponents:

$$10^5 \times 10^9 = 10^{5+9} = 10^{14}$$

Answer:

$$23.1 \times 10^{14} = 2.31 \times 10^{15}$$

EXAMPLE

Multiply 4.6×10^{-7} by 9.2×10^4

Solution

$$4.6 \times 9.2 = 42$$

Add exponents:

$$10^{-7} \times 10^4 = 10^{-7+4} = 10^{-3}$$

Answer:

$$42 \times 10^{-3} = 4.2 \times 10^{-2}$$

3.3 (a) 1-isopropyl-4-methylcyclohexene   (b) cyclooctene
(c) 4-*tert*-butylcyclohexene

3.4   Line-angle formulas for the other two dienes are:

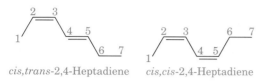

cis,trans-2,4-Heptadiene   *cis,cis*-2,4-Heptadiene

3.5   Four cis-trans isomers are possible.
3.6

(a) CH₃CHCH₃   (b)

3.7   Propose a two-step mechanism similar to that proposed for the addition of HCl to propene.
Step 1: Reaction of H⁺ with the carbon–carbon double bond
gives a 3° carbocation intermediate.

A 3° carbocation
intermediate

Step 2: Reaction of the 3° carbocation intermediate with
bromide ion completes the valence shell of carbon and gives
the product.

3.8   The product from each acid-catalyzed hydration is the
same alcohol.

$$CH_3CCH_2CH_3$$

3.9   Propose a three-step mechanism similar to that for
the acid-catalyzed hydration of propene.
Step 1: Reaction of the carbon–carbon double bond with H⁺
gives a 3° carbocation intermediate.

A 3° carbocation
intermediate

Step 2: Reaction of the 3° carbocation intermediate with
water completes the valence shell of carbon and gives an
oxonium ion.

An oxonium ion

Step 3: Loss of H⁺ from the oxonium ion completes the
reaction and generates a new H⁺ catalyst.

+ H⁺

3.10 (a) $CH_3—C—CH—CH_2$

(b)

3.11 A saturated hydrocarbon contains only carbon–
carbon single bonds. An unsaturated hydrocarbon contains
one or more carbon–carbon double or triple bonds.
3.13

(a) 109.5°   120°

(b) 120°   CH₂OH

(c) HC≡C—CH=CH₂   180° 120°   (d) 120°

3.15 Line-angle formulas for each compound are

(a)   (b)

(c)   (d)

(e)   (f)

3.17 The solutions are:
(a) 1-heptene   (b) 1,4,4-trimethylcyclopentene
(c) 1,3-dimethylcyclohexene   (d) 2,4-dimethyl-2-pentene
(e) 1-octyne   (f) 2,2-dimethyl-3-hexyne
3.19 (a) The longest chain is four carbons. The correct
name is 2-butene.
(b) The chain is not numbered correctly. The correct name
is 2-pentene.
(c) The ring is not numbered correctly. The correct name is
1-methylcyclohexene.
(d) The double bond must be located. The correct name is
3,3-dimethyl-1-pentene.
(e) The chain is not numbered correctly. The correct name
is 2-hexyne.
(f) The longest chain is five carbon atoms. The correct
name is 3,4-dimethyl-2-pentene.

The answer could also be written in other, equally valid ways:

$$12.7 \times 10^{-3} = 0.0127 = 1.27 \times 10^{-2}$$

When it is necessary to add or subtract two numbers that have different exponents, we first must change them so that the exponents are the same.

A calculator with exponential notation changes the exponent automatically.

EXAMPLE

Add 1.95×10^{-2} and 2.8×10^{-3}.

Solution
To add these two numbers, we make both exponents -2. Thus, $2.8 \times 10^{-3} = 0.28 \times 10^{-2}$. Now we can add:

$$\begin{array}{r} 1.95 \times 10^{-2} \\ + \ 0.28 \times 10^{-2} \\ \hline 2.23 \times 10^{-2} \end{array}$$

Multiplying and Dividing Numbers in Exponential Notation

To multiply numbers in exponential notation, we first multiply the coefficients in the usual way and then algebraically *add* the exponents.

EXAMPLE

Multiply 7.40×10^5 by 3.12×10^9.

Solution
$$7.40 \times 3.12 = 23.1$$

Add exponents:
$$10^5 \times 10^9 = 10^{5+9} = 10^{14}$$

Answer:
$$23.1 \times 10^{14} = 2.31 \times 10^{15}$$

EXAMPLE

Multiply 4.6×10^{-7} by 9.2×10^4

Solution
$$4.6 \times 9.2 = 42$$

Add exponents:
$$10^{-7} \times 10^4 = 10^{-7+4} = 10^{-3}$$

Answer:
$$42 \times 10^{-3} = 4.2 \times 10^{-2}$$

To divide numbers expressed in exponential notation, the process is reversed. We first divide the coefficients and then algebraically *subtract* the exponents.

EXAMPLE

Divide: $\dfrac{6.4 \times 10^8}{2.57 \times 10^{10}}$

Solution

$$6.4 \div 2.57 = 2.5$$

Subtract exponents:

$$10^8 \div 10^{10} = 10^{8-10} = 10^{-2}$$

Answer:

$$2.5 \times 10^{-2}$$

EXAMPLE

Divide: $\dfrac{1.62 \times 10^{-4}}{7.94 \times 10^7}$

Solution

$$1.62 \div 7.94 = 0.204$$

Subtract exponents:

$$10^{-4} \div 10^7 = 10^{-4-7} = 10^{-11}$$

Answer:

$$0.204 \times 10^{-11} = 2.04 \times 10^{-12}$$

Scientific calculators do these calculations automatically. All that is necessary is to enter the first number, press $+$, $-$, $\times$, or $\div$, enter the second number, and press $=$. (The method for entering numbers of this form varies; consult the instructions that come with the calculator.) Many scientific calculators also have a key that will automatically convert a number such as 0.00047 to its scientific notation form (4.7×10^{-4}), and vice versa.

Significant Figures

If you measure the volume of a liquid in a graduated cylinder, you might find that it is 36 mL, to the nearest milliliter, but you cannot tell if it is 36.2, or 35.6, or 36.0 mL because this measuring instrument does not give the last digit with any certainty. A buret gives more digits, and if you use one you should be able to say, for instance, that the volume is 36.3 mL and not 36.4 mL. But even with a buret, you could not say whether the volume is 36.32 or 36.33 mL. For that, you would need an instrument that gives still more digits. This example should show you that *no measured number can ever be known exactly*. No matter how good the measuring instrument, there is always a limit to the number of digits it can measure with certainty.

We define the number of **significant figures** as the number of digits of a measured number that have uncertainty only in the last digit.

What do we mean by this definition? Assume that you are weighing a small object on a laboratory balance that can weigh to the nearest 0.1 g, and you find that the object weighs 16 g. Because the balance weighs to the nearest 0.1 g, you can be sure that the object does not weigh 16.1 g or 15.9 g. In this case, you would write the weight as 16.0 g. To a scientist, there is a difference between 16 g and 16.0 g. Writing 16 g says that you don't know the digit after the 6. Writing 16.0 g says that you do know it: It is 0. However, you don't know the digit after that. Several rules govern the use of significant figures in reporting measured numbers.

A. Determining the Number of Significant Figures

1. Nonzero digits are always significant.

For example, 233.1 m has four significant figures; 2.3 g has two significant figures.

2. Zeros at the beginning of a number are never significant.

For example, 0.0055 L has two significant figures; 0.3456 g has four significant figures.

3. Zeros between nonzero digits are always significant.

For example, 2.045 kcal has four significant figures; 8.0506 g has five significant figures.

4. Zeros at the end of a number that contains a decimal point are always significant.

For example, 3.00 L has three significant figures; 0.0450 mm has three significant figures.

5. Zeros at the end of a number that contains no decimal point may or may not be significant.

We cannot tell whether they are significant without knowing something about the number. This is the ambiguous case. If you know that a certain small business made a profit of $36,000 last year, you can be sure that the 3 and 6 are significant, but what about the rest? The profit might have been $36,126 or $35,786.53, or maybe even exactly $36,000. We just don't know because it is customary to round off such numbers. On the other hand, if the profit were reported as $36,000.00, then all seven digits would be significant.

In science, to get around the ambiguous case we use exponential notation (Section 1.3). Suppose a measurement comes out to be 2500 g. If we made the measurement, we of course know whether the two zeros are significant, but we need to tell others. If these digits are *not* significant, we write our number as 2.5×10^3. If one zero is significant, we write 2.50×10^3. If both zeros are significant, we write 2.500×10^3. Since we now have a decimal point, all the digits shown are significant.

B. Multiplying and Dividing

The rule in multiplication and division is that the final answer should have the *same* number of significant figures as there are in the number with the *fewest* significant figures.

EXAMPLE

Do the following multiplications and divisions:
(a) 3.6×4.27
(b) 0.004×217.38
(c) $\dfrac{42.1}{3.695}$
(d) $\dfrac{0.30652 \times 138}{2.1}$

Solution
(a) 15 (3.6 has two significant figures)
(b) 0.9 (0.004 has one significant figure)
(c) 11.4 (42.1 has three significant figures)
(d) 2.0×10^1 (2.1 has two significant figures)

C. Adding and Subtracting

In addition and subtraction, the rule is completely different. The number of significant figures in each number doesn't matter. The answer is given to the *same number of decimal places* as the term with the fewest decimal places.

EXAMPLE

Add or subtract:

(a)	(b)	(c)
320.084	61.4532	
80.47	13.7	
200.23	22	14.26
20.0	0.003	−1.05041
620.8	97	13.21

Solution
In each case, we add or subtract in the normal way but then round off so that the only digits that appear in the answer are those in the columns in which every digit is significant.

D. Rounding Off

When we have too many significant figures in our answer, it is necessary to round off. In this book we have used the rule that if *the first digit dropped* is 5, 6, 7, 8, or 9, we raise *the last digit kept* to the next number; otherwise, we do not.

EXAMPLE

In each case, drop the last two digits:
(a) 33.679   (b) 2.4715   (c) 1.1145   (d) 0.001309   (e) 3.52

Solution
(a) $33.679 = 33.7$
(b) $2.4715 = 2.47$
(c) $1.1145 = 1.11$
(d) $0.001309 = 0.0013$
(e) $3.52 = 4$

E. Counted or Defined Numbers

All of the preceding rules apply to *measured* numbers and **not** to any numbers that are *counted* or *defined*. Counted and defined numbers are known exactly. For example, a triangle is defined as having 3 sides, not 3.1 or 2.9. Here, we treat the number 3 as if it has an infinite number of zeros following the decimal point.

EXAMPLE

Multiply 53.692 (a measured number) $\times$ 6 (a counted number).

Solution

$$322.15$$

Because 6 is a counted number, we know it exactly, and 53.692 is the number with the fewest significant figures. All we really are doing is adding 53.692 six times.

Chapter 1   Organic Chemistry

1.1   Following are Lewis structures showing all bond angles.

(a)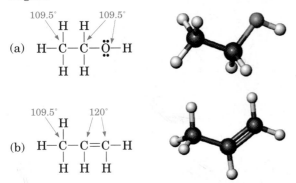

(b)

1.2   Of the four alcohols with the molecular formula $C_4H_{10}O$, two are 1°, one is 2°, and one is 3°. For the Lewis structures the 3° alcohol and two of the 1° alcohols, three $C—CH_3$ bonds are drawn longer to avoid crowding in the formulas.

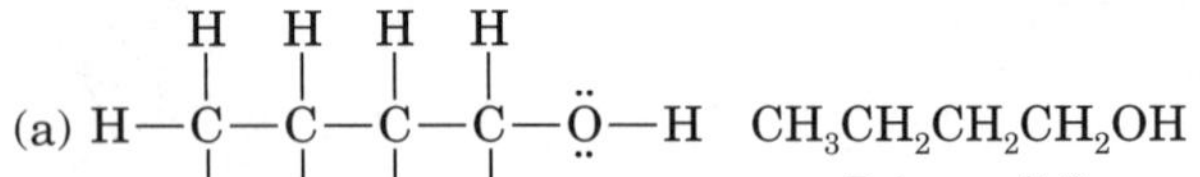

(a) $H—C—C—C—C—\overset{..}{\underset{..}{O}}—H$ $CH_3CH_2CH_2CH_2OH$
Primary (1°)

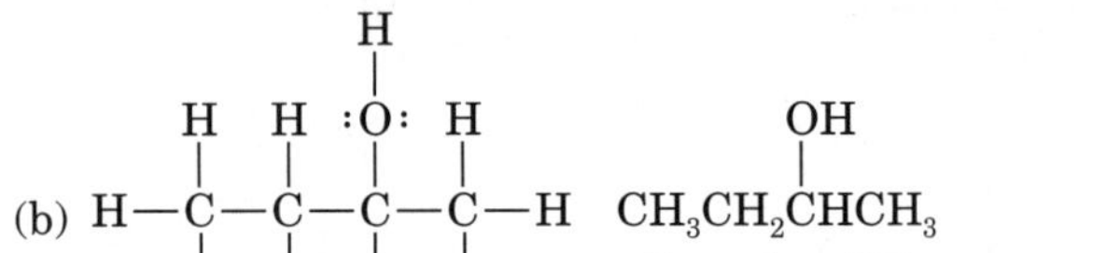

(b) $H—C—C—C—C—H$ $CH_3CH_2CHCH_3$
Secondary (2°)

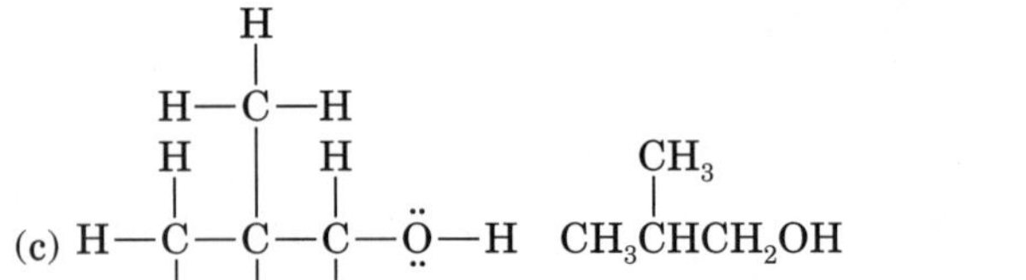

(c) $H—C—C—C—\overset{..}{\underset{..}{O}}—H$ CH_3CHCH_2OH
CH_3
Primary (1°)

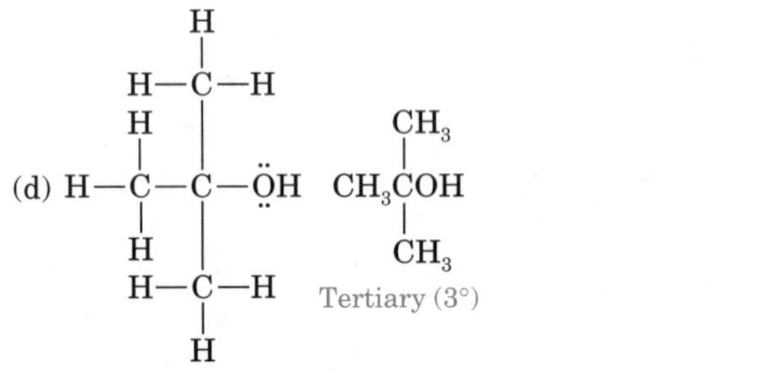

(d) $H—C—C—\overset{..}{O}H$ CH_3COH
CH_3
CH_3
Tertiary (3°)

1.3   The three secondary (2°) amines with molecular formula $C_4H_{11}N$ are

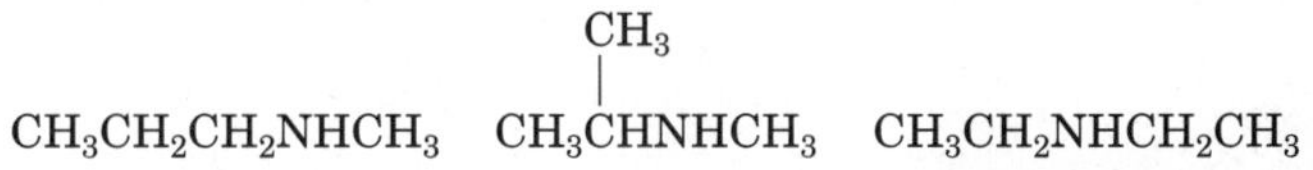

$CH_3CH_2CH_2NHCH_3$ $CH_3\overset{CH_3}{\underset{|}{CH}}NHCH_3$ $CH_3CH_2NHCH_2CH_3$

1.4   The three ketones with the molecular formula $C_5H_{10}O$ are

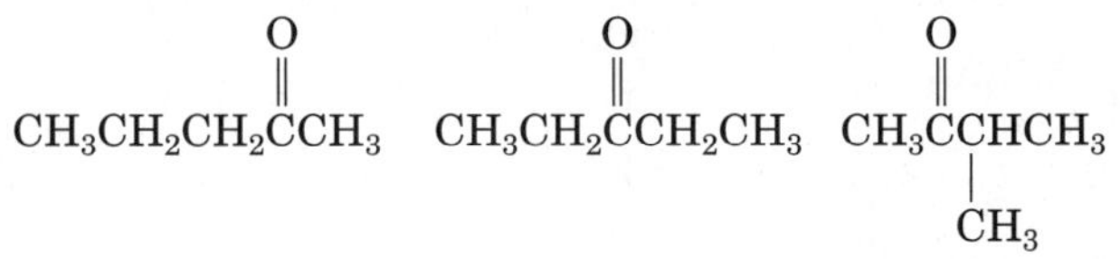

$CH_3CH_2CH_2\overset{O}{\overset{||}{C}}CH_3$ $CH_3CH_2\overset{O}{\overset{||}{C}}CH_2CH_3$ $CH_3\overset{O}{\overset{||}{C}}CHCH_3$
CH_3

1.5   The two carboxylic acids with the molecular formula $C_4H_8O_2$ are

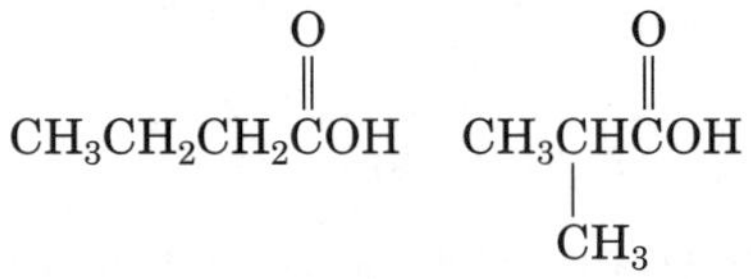

$CH_3CH_2CH_2\overset{O}{\overset{||}{C}}OH$ $CH_3\overset{O}{\overset{||}{CH}}COH$
CH_3

1.7   For hydrogen, the number of valence electrons plus the number of bonds equals 2. For carbon, nitrogen, and oxygen, the number of valence electrons plus the number of bonds equals 8. Carbon, with four valence electrons, forms four bonds. Nitrogen, with five valence electrons, forms three bonds. Oxygen, with six valence electrons, forms two bonds.

1.9   Following are the Lewis structures for each compound:

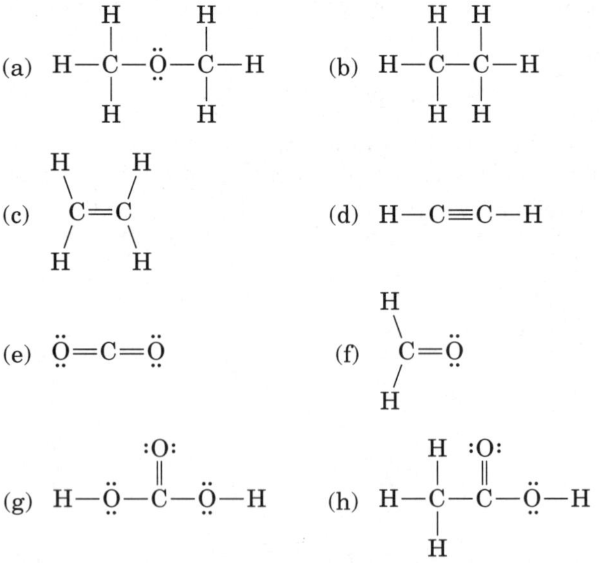

(a) $H—C—\overset{..}{\underset{..}{O}}—C—H$

(b) $H—C—C—H$

(c) $\overset{H}{\underset{H}{C}}=\overset{H}{\underset{H}{C}}$

(d) $H—C≡C—H$

(e) $\overset{..}{\underset{..}{O}}=C=\overset{..}{\underset{..}{O}}$

(f) $\overset{H}{\underset{H}{}}C=\overset{..}{\underset{..}{O}}$

(g) $H—\overset{..}{\underset{..}{O}}—\overset{:O:}{\overset{||}{C}}—\overset{..}{\underset{..}{O}}—H$

(h) $H—C—\overset{:O:}{\overset{||}{C}}—\overset{..}{\underset{..}{O}}—H$

1.11   In stable organic compounds, carbon must have four covalent bonds to other atoms. In (a), carbon has five bonds to other atoms. In (b), one carbon has four bonds to other atoms, but the second carbon has five bonds to other atoms.

1.13   You would predict 120° for the $H—N—H$ bond angles.

1.15   (a) 109.5° about C and O   (b) 120° about C
(c) 180° about C

*Answers to in-text and odd-numbered end-of-chapter problems.

1.17 (a) Predict 120° for all H—C—C and C—C—C bond angles. (b) planar

1.19 Living organisms use the substances they take in from their environment to produce organic compounds. They also use them to produce inorganic compounds—for example, the calcium-containing compounds of bones, teeth, and shells.

1.21 Carbon, which forms four bonds; hydrogen, which forms one bond; nitrogen, which forms three bonds and has one unshared pair of electrons; and oxygen, which forms two bonds and has two unshared pairs of electrons

1.23 A functional group is a part of an organic molecule that undergoes chemical reactions.

1.25

(a) $-\overset{\overset{\displaystyle \cdot\cdot}{O}}{\underset{}{\overset{\|}{C}}}-$ (b) $-\overset{\overset{\displaystyle \cdot\cdot}{O}}{\underset{}{\overset{\|}{C}}}-\ddot{O}-H$

(c) $-\ddot{O}-H$ (d) $-\overset{}{\underset{\overset{|}{H}}{N}}-H$

1.27 (a) Incorrect. The carbon on the left has five bonds. (b) Incorrect. The carbon bearing the Cl has five bonds. (c) Correct. (d) Incorrect. Oxygen has three bonds and one carbon has five bonds. (e) Correct. (f) Incorrect. Carbon on the right has five bonds.

1.29 The one tertiary alcohol with molecular formula $C_4H_{10}O$ is

$$CH_3\underset{\overset{|}{CH_3}}{\overset{\overset{|}{CH_3}}{C}}OH$$

1.31 The one tertiary amine with molecular formula $C_4H_{11}N$ is

$$CH_3CH_2\underset{\overset{|}{CH_3}}{N}CH_3$$

1.33 (a) a ketone (b) two carboxylic acids (c) a primary amine, a secondary amine, and a carboxylic acid (d) two primary alcohols and one ketone

1.35 (a) The four primary alcohols with the molecular formula $C_5H_{12}O$ are

$$CH_3CH_2CH_2CH_2CH_2OH \quad CH_3\underset{\overset{|}{CH_3}}{CH}CH_2CH_2OH$$

$$CH_3CH_2\underset{\overset{|}{CH_3}}{CH}CH_2OH \quad CH_3\underset{\overset{|}{CH_3}}{\overset{\overset{|}{CH_3}}{C}}CH_2OH$$

(b) The three secondary alcohols with the molecular formula $C_5H_{12}O$ are

$$CH_3\underset{\overset{|}{OH}}{CH}CH_2CH_2CH_3 \quad CH_3CH_2\underset{\overset{|}{OH}}{CH}CH_2CH_3 \quad CH_3\underset{\overset{|}{OH}}{CH}\underset{\overset{|}{CH_3}}{CH}CH_3$$

(c) The tertiary alcohol with the molecular formula $C_5H_{12}O$ is

$$CH_3CH_2\underset{\overset{|}{CH_3}}{\overset{\overset{|}{CH_3}}{C}}OH$$

1.37 The eight carboxylic acids with the molecular formula $C_6H_{12}O_2$ are

$$CH_3CH_2CH_2CH_2CH_2COOH \quad CH_3\underset{\overset{|}{CH_3}}{CH}CH_2CH_2COOH \quad CH_3CH_2\underset{\overset{|}{CH_3}}{CH}CH_2COOH$$

$$CH_3CH_2CH_2\underset{\overset{|}{CH_3}}{CH}COOH \quad CH_3\underset{\overset{|}{CH_3}}{\overset{\overset{|}{CH_3}}{C}}CH_2COOH \quad CH_3\underset{\overset{|}{CH_3}}{CH}\underset{\overset{|}{CH_3}}{CH}COOH$$

$$CH_3CH_2\underset{\overset{|}{CH_3}}{\overset{\overset{|}{CH_3}}{C}}COOH \quad CH_3CH_2\underset{\overset{|}{CH_2CH_3}}{CH}COOH$$

1.39 Taxol was discovered by a survey of indigenous plants sponsored by the National Cancer Institute with the goal of discovering new chemicals for fighting cancer.

1.41 The goal of combinatorial chemistry is to synthesize large numbers of closely related compounds in one reaction mixture.

1.43 Predict 109.5° for each C—Si—C bond angle.

1.45

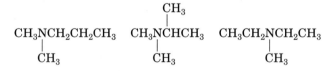

(a) CH_3CH_2OH (b) $CH_3CH_2\overset{\overset{\displaystyle O}{\|}}{C}H$

(c) $CH_3\overset{\overset{\displaystyle O}{\|}}{C}CH_3$ (d) $CH_3CH_2\overset{\overset{\displaystyle O}{\|}}{C}OH$

1.47 The three tertiary amines with the molecular formula $C_5H_{13}N$ are

$$CH_3\underset{\overset{|}{CH_3}}{N}CH_2CH_2CH_3 \quad CH_3\underset{\overset{|}{CH_3}}{\overset{\overset{|}{CH_3}}{N}}CHCH_3 \quad CH_3CH_2\underset{\overset{|}{CH_3}}{N}CH_2CH_3$$

1.49 (a) O—H is the most polar bond. (b) C—C and C=C are the least polar bonds.

Chapter 2 Alkanes

2.1 The alkane is octane, and its molecular formula is C_8H_{18}.

2.2 (a) constitutional isomers (b) the same compound

2.3 The three constitutional isomers with the molecular formula C_5H_{12} are

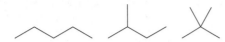

2.4 (a) 5-isopropyl-2-methyloctane, $C_{12}H_{26}$
(b) 4-isopropyl-4-propyloctane, $C_{14}H_{30}$
2.5 (a) isobutylcyclopentane, C_9H_{18}
(b) *sec*-butylcycloheptane, $C_{11}H_{22}$
(c) 1-ethyl-1-methylcyclopropane, C_6H_{12}
2.6 The structure with the three methyl groups equatorial is

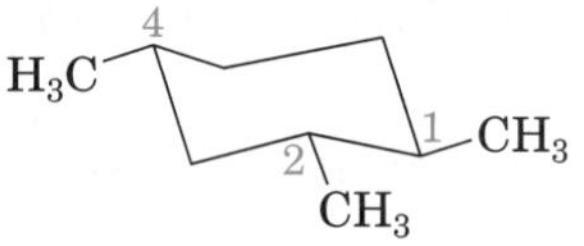

2.7 Cycloalkanes (a) and (c) show cis-trans isomerism.

(a)

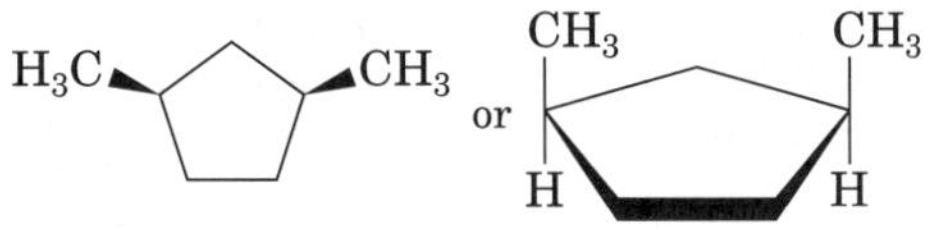

cis-1,3-Dimethylcyclopentane

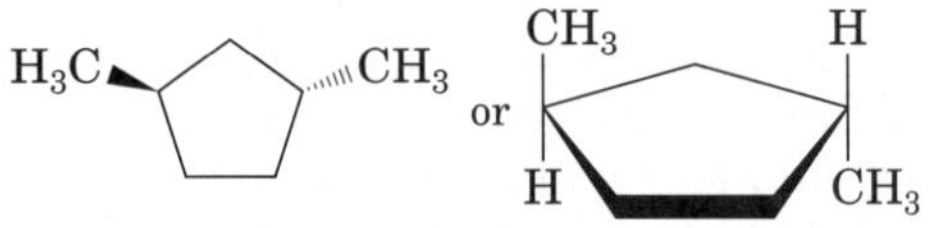

trans-1,3-Dimethylcyclopentane

(c)

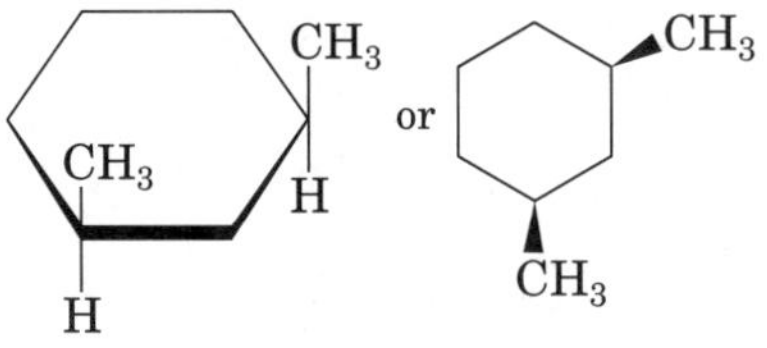

cis-1,3-Dimethylcyclohexane

trans-1,3-Dimethylcyclohexane

2.8 In order of increasing boiling point, they are:
(a) 2,2-dimethylpropane (9.5°C), 2-methylbutane (27.8°C), pentane (36.1°C)
(b) 2,2,4-trimethylhexane, 3,3-dimethylheptane, nonane
2.9 The two products with the formula C_3H_7Cl are

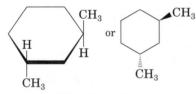

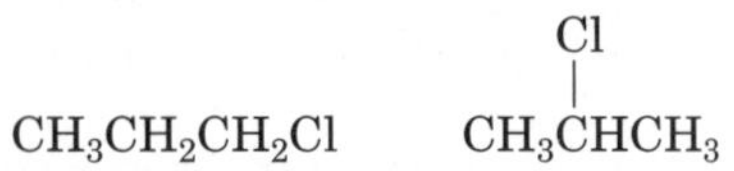

1-Chloropropane   2-Chloropropane
(Propyl chloride)   (Isopropyl chloride)

2.11 The carbon chain of an alkane is not straight; it is bent with all C—C—C bond angles approximately 109.5°.
2.13 Line-angle formulas are

(a)   (b)

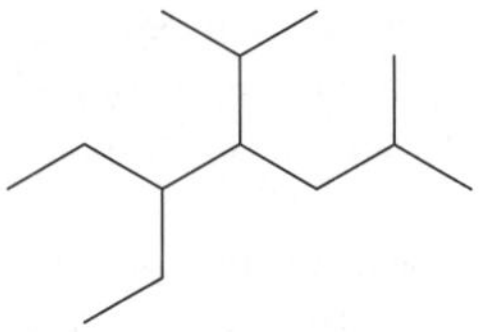

(c) 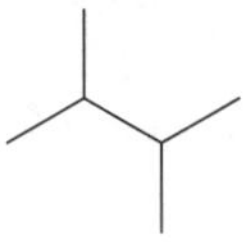   (d)

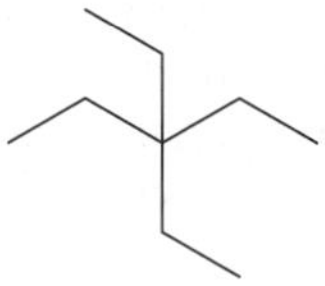

(e) 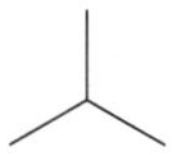   (f)

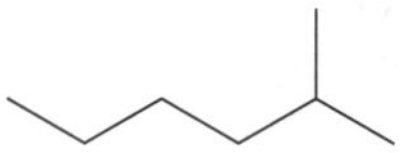

2.15 Statements (a) and (b) are true. Statements (c) and (d) are false.
2.17 Structures (a) and (g) represent the same compound. Structures (a, g), (c), (d), (e), and (f) represent constitutional isomers.
2.19 The structural formulas in parts (b), (c), (e), and (f) represent constitutional isomers.
2.21 (a) ethyl   (b) isopropyl   (c) isobutyl   (d) *tert*-butyl
2.23 (a) 2-methylpentane   (b) 2,5-dimethylhexane
(c) 3-ethyloctane   (d) 1-isopropyl-2-methylcyclohexane
(e) isobutylcyclopentane   (f) 1-ethyl-2,4-dimethyl-cyclohexane
2.25 A conformation is any three-dimensional arrangement of the atoms in a molecule that results from rotation about a single bond.
2.27 Here are representations of two conformations of ethane. In the staggered conformation, the hydrogen atoms on one carbon are as far apart as possible from the hydrogens on the other carbon. In the eclipsed conformation, they are as close as possible.

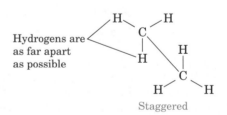

Hydrogens are as far apart as possible

Staggered

Hydrogens are as close as possible

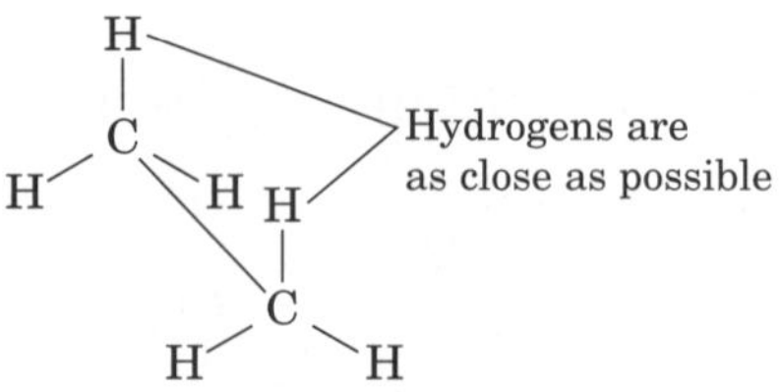

Eclipsed

2.29 no

2.31 Structural formulas for the six cycloalkanes with the molecular formula C_5H_{10} are

Cyclopentane   Methylcyclobutane   1,1-Dimethyl-cyclopropane

trans-1,2-Dimethyl-cyclopropane   cis-1,2-Dimethyl-cyclopropane   Ethylcyclopropane

2.33 The first two representations of menthol show the cyclohexane ring as a planar hexagon. The third shows it as a chair conformation. Notice that in the chair conformation, all three substituents are equatorial.

2.35 The alternative representation is

2.37 Heptane, C_7H_{16}, has a boiling point of 98°C and a molecular weight of 100. Its molecular weight is approximately 5.5 times that of water. Although considerably smaller, water molecules associate in the liquid by relatively strong hydrogen bonding, whereas the much larger heptane molecules associate only by relatively weak London dispersion forces.

2.39 Alkanes are insoluble in water.

2.41 Boiling points of unbranched alkanes are related to their surface area; the larger the surface area, the higher the boiling point. The relative increase in size per CH_2 group is greatest between CH_4 and CH_3CH_3 and becomes progressively smaller as the molecular weight increases. Therefore, the increase in boiling point per added CH_2 group is greatest between CH_4 and CH_3CH_3, and becomes progressively smaller for higher alkanes.

2.43 Balanced equations for the complete combustion of each are

(a) $2CH_3(CH_2)_4CH_3 + 19O_2 \longrightarrow 12CO_2 + 14H_2O$

     Hexane

(b) $+ 9O_2 \longrightarrow 6CO_2 + 6H_2O$

   Cyclohexane

(c) $2CH_3CHCH_2CH_2CH_3 + 19O_2 \longrightarrow 12CO_2 + 14H_2O$

     2-Methylpentane

2.45 Structural formulas for each part are

(a) CH_3Br

(b) —Cl

(c) $BrCH_2CH_2Br$

(d) CH_3CCH_3 (with CH_3 up and Cl down)

(e) CCl_2F_2

2.47 (a) Only one ring contains just carbon atoms. (b) One ring contains two nitrogen atoms. (c) One ring contains two oxygen atoms.

2.49 It has been demonstrated that their presence in the stratosphere results in destruction of the stratospheric ozone layer.

2.51 An octane rating indicates the relative smoothness with which a gasoline blend burns in an automobile engine. The higher the octane rating, the less the engine knocks. The reference hydrocarbons are 2,2,4-trimethylpentane (isooctane), which is assigned an octane rating of 100, and heptane, which is assigned an octane rating of 0.

2.53 2,2,4-Trimethylpentane has a higher heat of combustion both per mole and per gram.

2.55 (a) The longest chain is pentane. Its IUPAC name is 2-methylpentane.

(b) The pentane chain is numbered incorrectly. Its IUPAC name is 2-methylpentane.

(c) The longest chain is pentane. Its IUPAC name is 3-ethyl-3-methylpentane.

(d) The longest chain is hexane. Its IUPAC name is 3,4-dimethylhexane.

(e) The longest chain is heptane. Its IUPAC name is 4-methylheptane.

(f) The longest chain is octane. Its IUPAC name is 3-ethyl-3-methyloctane.

(g) The ring is numbered incorrectly. Its IUPAC name is 1,1-dimethylcyclopropane.

(h) The ring is numbered incorrectly. Its IUPAC name is 1-ethyl-3-methylcyclohexane.

2.57 Tetradecane is a liquid at room temperature.

2.59 Water, a polar molecule, cannot penetrate the surface layer created by this nonpolar hydrocarbon.

Chapter 3   Alkenes

3.1 (a) 3,3-dimethyl-1-pentene   (b) 2,3-dimethyl-2-butene

(c) 3,3-dimethyl-1-butyne

3.2 (a) trans-3,4-dimethyl-2-pentene

(b) cis-4-ethyl-3-heptene

3.3 (a) 1-isopropyl-4-methylcyclohexene (b) cyclooctene
(c) 4-*tert*-butylcyclohexene

3.4 Line-angle formulas for the other two dienes are:

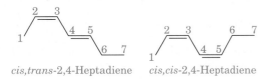

cis,trans-2,4-Heptadiene   *cis,cis*-2,4-Heptadiene

3.5 Four cis-trans isomers are possible.

3.6

(a) CH₃CHCH₃ with Br above   (b) cyclohexane with Br and CH₃

3.7 Propose a two-step mechanism similar to that proposed for the addition of HCl to propene.

Step 1: Reaction of H⁺ with the carbon–carbon double bond gives a 3° carbocation intermediate.

—CH₃ + H⁺ ⟶ ⁺—CH₃

A 3° carbocation
intermediate

Step 2: Reaction of the 3° carbocation intermediate with bromide ion completes the valence shell of carbon and gives the product.

⁺—CH₃ + :Br:⁻ ⟶ Br:—CH₃

3.8 The product from each acid-catalyzed hydration is the same alcohol.

$$CH_3 \quad CH_3CCH_2CH_3 \quad OH$$

3.9 Propose a three-step mechanism similar to that for the acid-catalyzed hydration of propene.

Step 1: Reaction of the carbon–carbon double bond with H⁺ gives a 3° carbocation intermediate.

—CH₃ + H⁺ ⟶ ⁺—CH₃

A 3° carbocation
intermediate

Step 2: Reaction of the 3° carbocation intermediate with water completes the valence shell of carbon and gives an oxonium ion.

⁺—CH₃ + :O—H ⟶ ⁺O—H, CH₃

An oxonium ion

Step 3: Loss of H⁺ from the oxonium ion completes the reaction and generates a new H⁺ catalyst.

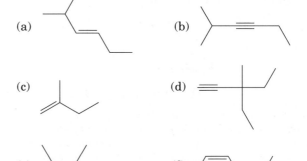

3.10 (a) $CH_3-\underset{\underset{H_3C}{|}}{C}-\underset{\underset{Br}{|}}{CH}-\underset{\underset{Br}{|}}{CH_2}$

(b) cyclohexane with Cl and CH₂Cl

3.11 A saturated hydrocarbon contains only carbon–carbon single bonds. An unsaturated hydrocarbon contains one or more carbon–carbon double or triple bonds.

3.13

(a) 109.5°, 120°, 120°

(b) 120° CH₂OH

(c) HC≡C—CH=CH₂ 180°, 120°

(d) 120°

3.15 Line-angle formulas for each compound are

(a)   (b)

(c)   (d)

(e)   (f)

3.17 The solutions are:
(a) 1-heptene (b) 1,4,4-trimethylcyclopentene
(c) 1,3-dimethylcyclohexene (d) 2,4-dimethyl-2-pentene
(e) 1-octyne (f) 2,2-dimethyl-3-hexyne

3.19 (a) The longest chain is four carbons. The correct name is 2-butene.
(b) The chain is not numbered correctly. The correct name is 2-pentene.
(c) The ring is not numbered correctly. The correct name is 1-methylcyclohexene.
(d) The double bond must be located. The correct name is 3,3-dimethyl-1-pentene.
(e) The chain is not numbered correctly. The correct name is 2-hexyne.
(f) The longest chain is five carbon atoms. The correct name is 3,4-dimethyl-2-pentene.

3.21 Parts (b), (c), and (e) show cis-trans isomerism. Following are line-angle formulas for the cis and trans isomers of each.

(b)

cis-2-Hexene   *trans*-2-Hexene

(c)

cis-3-Hexene   *trans*-3-Hexene

(e)

cis-3-Methyl-   *trans*-3-Methyl-
2-hexene   2-hexene

3.23 There are six alkenes of molecular formula C_5H_{10}.

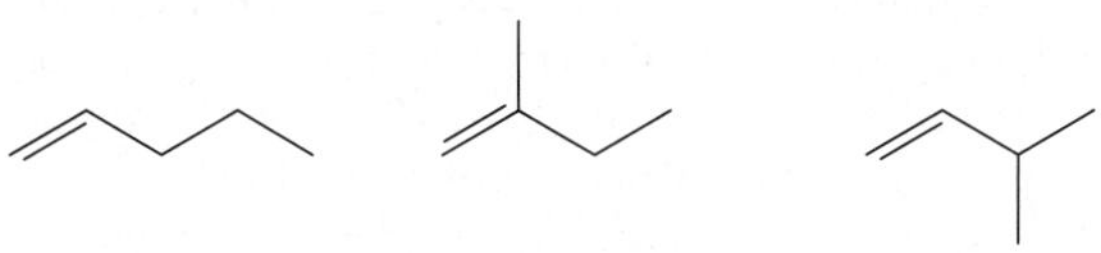

1-Pentene   2-Methyl-1-butene   3-Methyl-1-butene

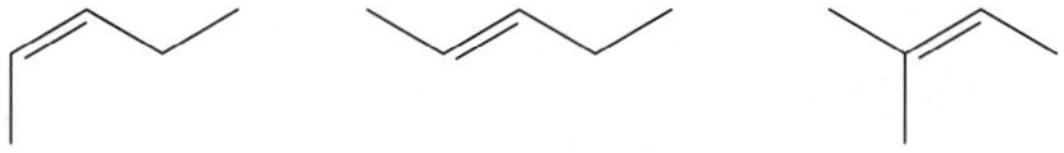

cis-2-Pentene   *trans*-2-Pentene   2-Methyl-2-butene

3.25 Line-angle formulas for these three unsaturated fatty acids are:

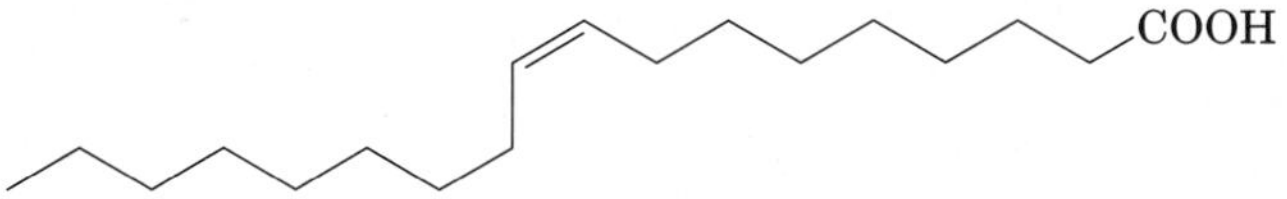

Oleic acid

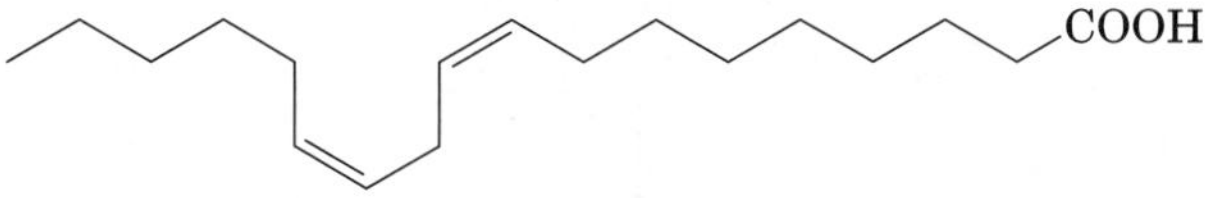

Linoleic acid

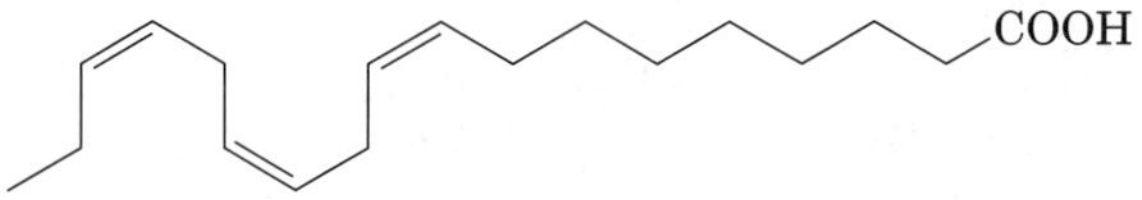

Linolenic acid

3.27

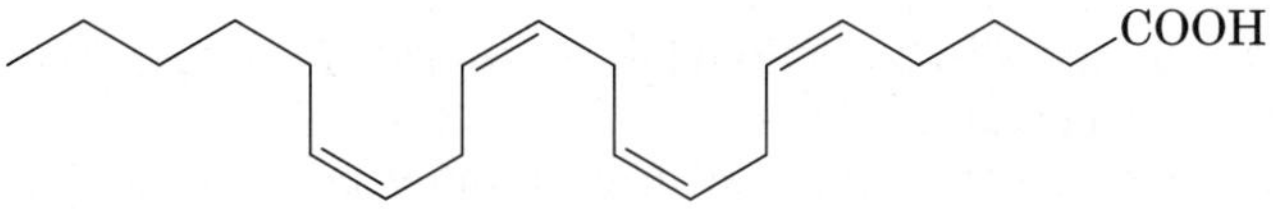

Arachidonic acid

3.29 Only compounds (b) and (d) show cis-trans isomerism.

(b)   (d)

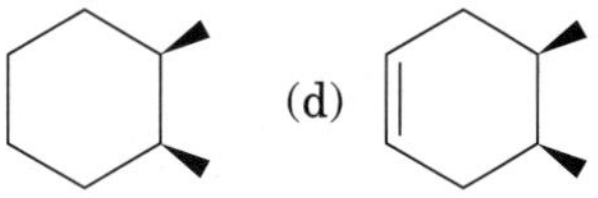

3.31 The structural formula of β-ocimene is drawn on the left showing all atoms and on the right as a line-angle formula.

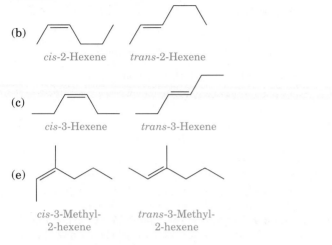

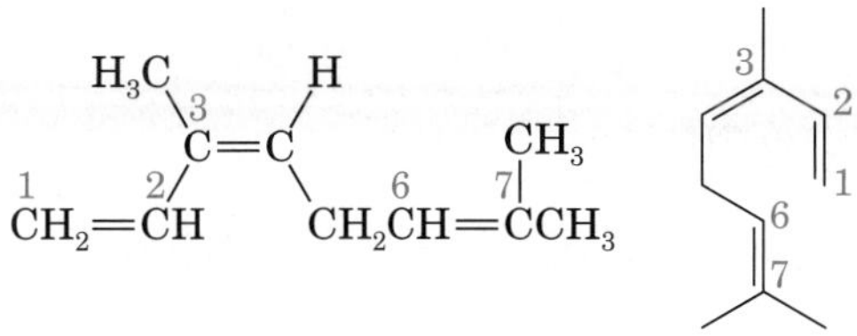

3.33 The four isoprene units in vitamin A are shown in bold.

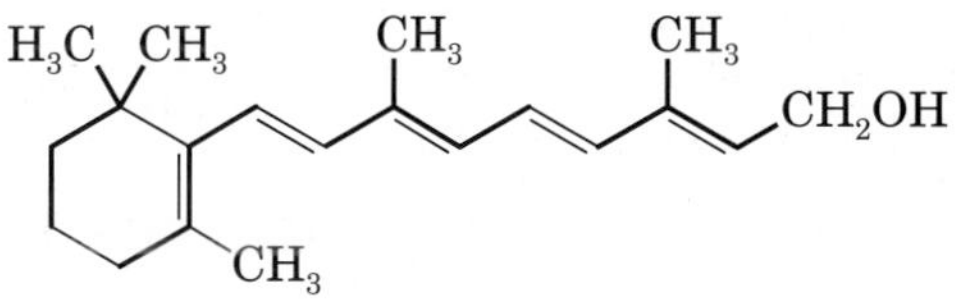

3.35 (a) HBr (b) H_2O/H_2SO_4 (c) HI (d) Br_2
3.37

(a) $CH_3CH_2\overset{+}{\underset{\displaystyle CH_3}{C}}-CH_2CH_3$ and $CH_3CH_2\underset{\displaystyle CH_3}{CH}-\overset{+}{C}HCH_3$

       Tertiary   Secondary

(b) $CH_3CH_2\overset{+}{C}H-CH_2CH_3$ and $CH_3CH_2CH_2-\overset{+}{C}HCH_3$

       Secondary   Secondary

(c) 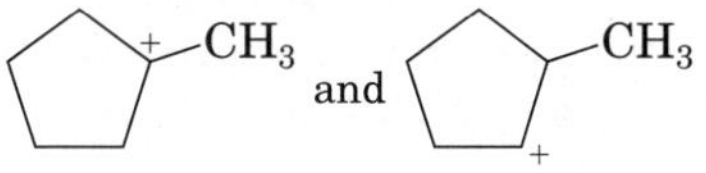 $\overset{+}{}-CH_3$ and $-CH_3$

      Tertiary   Secondary

(d) 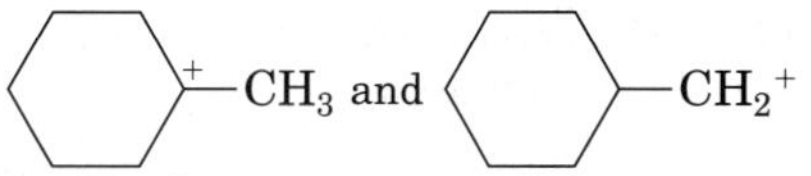 $\overset{+}{}-CH_3$ and $-CH_2^+$

      Tertiary   Primary

3.39

(a) 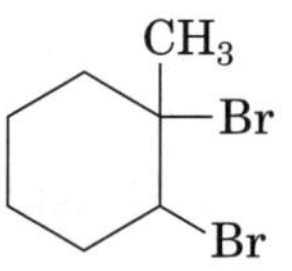 CH_3 / Br / Br

(b) 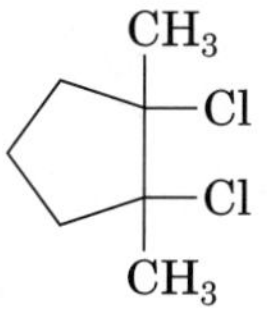 CH_3 / Cl / Cl / CH_3

3.41

(a) $CH_3\underset{\displaystyle CH_3}{C}=CHCH_3$

(b) $CH_2=\underset{\displaystyle CH_3}{C}CH_2CH_3$

(c) $CH_2=CHCH_2CH_2CH_3$

3.43
(a) $CH_3CH_2CH=CHCH_2CH_3$

(b)    or

(c) $CH_2=\overset{\overset{\displaystyle CH_3}{|}}{C}CH_2CH_3$ or $CH_3\overset{\overset{\displaystyle CH_3}{|}}{C}=CHCH_3$

(d) $CH_3CH=CH_2$

3.45
(a) $CH_3CH_2CH_2CH_2CH_3$ (b) $CH_3CH_2CH_2CH_2CH_3$

(c) ⬠   (b) ⬠ with CH_3

3.47 Reagents are shown over each arrow.

$CH_3—CH_3$ ← H₂/M   HBr → $CH_3—CH_2—Br$

$CH_2=CH_2$ Br₂ → $Br—CH_2—CH_2—Br$

H_2O / H_2SO_4 ↙ $CH_3—CH_2—OH$

HCl ↘ $CH_3—CH_2—Cl$

3.49 Ethylene is a natural ripening agent for fruits.

3.51 Its molecular formula is $C_{16}H_{32}O_2$, its molecular weight is 254 amu, and its molar mass is 254 g/mol.

3.53 Rods are primarily responsible for black-and-white vision. Cones are responsible for color vision.

3.55 The most common consumer items made of high-density polyethylene (HDPE) are milk and water jugs, grocery bags, and squeeze bottles. The most common consumer items made of low-density polyethylene (LDPE) are packaging for baked goods, vegetables, and other produce, as well as trash bags. Currently only HDPE materials are recyclable.

3.57 There are five compounds with the molecular formula C_4H_8. All are constitutional isomers. The only cis-trans isomers are cis-2-butene and trans-2-butene.

| Cyclobutane | Methyl-cyclopropane | 1-Butene | cis-2-Butene | trans-2-Butene |

3.59 (a) There are five alkenes with this carbon skeleton.

2-Methyl-1-pentene   2-Methyl-2-pentene   trans-4-Methyl-2-pentene

cis-4-Methyl-2-pentene   4-Methyl-1-pentene

(b) There are two alkenes with this carbon skeleton.

2,3-Dimethyl-1-butene   2,3-Dimethyl-2-butene

(c) There is one alkene with this carbon skeleton.

3,3-Dimethyl-1-butene

3.61 The carbon skeletons of lycopene and β-carotene are almost identical. The difference is that, at each end, the second carbon of β-carotene is bonded to the seventh carbon to form the two six-membered rings of lycopene.

3.63 Each alkene hydration reaction follows Markovnikov's rule. In (a), —H adds preferentially to carbon-1 and —OH to carbon-2 to give 2-hexanol. In (b), each carbon of the double bond has the same pattern of substitution, so 2-hexanol and 3-hexanol are formed in approximately equal amounts. In (c), each carbon of the double bond again has the same pattern of substitution, but no matter which way H—OH adds, the only product possible is 3-hexanol.

(a) 2-Hexanol

(b) 2-Hexanol   +   3-Hexanol

(c) 3-Hexanol

3.65 (a) ⬡

(b) CH_3 cyclohexene   or CH_2 methylenecyclohexane

(c) CH_3 / CH_3 cyclopentene   (d) or

Chapter 4   Benzene and Its Derivatives

4.1 (a) 2,4,6-tri-*tert*-butylphenol   (b) 2,6-dichloroaniline
(c) 3-nitrobenzoic acid

4.2 A saturated compound contains only single covalent bonds. An unsaturated compound contains one or more double or triple bonds. The most common double bonds are $C=C$, $C=O$, and $C=N$. The only triple bond is $C≡C$.

4.3 An aromatic compound is one that contains one or more benzene rings.

4.5 Yes, they have double bonds, at least in the contributing structures we normally use to represent them. Yes, they are unsaturated, because they have fewer hydrogens than a cycloalkane with the same number of carbons.

4.7 (a) An alkene of six carbons has the molecular formula C_6H_{12} and contains one carbon–carbon double bond. Three examples are

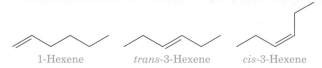

1-Hexene   *trans*-3-Hexene   *cis*-3-Hexene

(b) A cycloalkene of six carbons has the molecular formula C_6H_{10} and contains one ring and one carbon–carbon double bond. Three examples are

Cyclohexene   4-Methylcyclopentene   1-Methylcyclopentene

(c) An alkyne of six carbons has the molecular formula C_6H_{10} and contains one carbon–carbon triple bond. Three examples are

1-Hexyne   2-Hexyne   4-Methyl-2-pentyne

(d) An aromatic hydrocarbon of eight carbons has the molecular formula C_8H_{10} and contains one benzene ring. Three examples are

Ethylbenzene   1,3-Dimethylbenzene   1,4-Dimethylbenzene
               (*m*-Xylene)   (*p*-Xylene)

4.9 Benzene consists of carbons, each surrounded by three regions of electron density, which gives 120° for all bond angles. Bond angles of 120° in benzene can be maintained only if the molecule is planar. Cyclohexane, on the other hand, consists of carbons, each surrounded by four regions of electron density, which gives 109.5° for all bond angles. Angles of 109.5° in cyclohexane can be maintained only if the molecule is nonplanar.

4.11 Neither a unicorn, which has a horn like a rhinoceros, nor a dragon, which has a tough, leathery hide like a rhinoceros, exists. If they did and you made a hybrid of them, you would have a rhinoceros. Furthermore, a rhinoceros is not a dragon part of the time and a unicorn the rest of the time; a rhinoceros is a rhinoceros all of the time. To carry this analogy to aromatic compounds, resonance-contributing structures for them do not exist; they are imaginary. If they did exist and you made a hybrid of them, you would have the real aromatic compound.

4.13

4.15 Only cyclohexene will react with a solution of bromine in dichloromethane. A solution of Br_2/CH_2Cl_2 is red, whereas a dibromocycloalkane is colorless. To tell which bottle contains which compound, place a small quantity of each compound in a test tube and to each add a few drops of Br_2/CH_2Cl_2 solution. If the red color disappears, the compound is cyclohexene. If it remains, the compound is benzene.

4.17

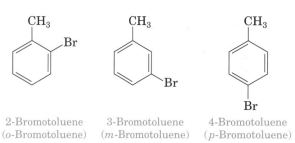

2-Bromotoluene   3-Bromotoluene   4-Bromotoluene
(*o*-Bromotoluene)   (*m*-Bromotoluene)   (*p*-Bromotoluene)

4.19 (a) Nitration using HNO_3/H_2SO_4, followed by sulfonation using H_2SO_4. The order of the steps may also be reversed.

(b) Bromination using $Br_2/FeCl_3$, followed by chlorination using $Cl_2/FeCl_3$. The order of the steps may also be reversed.

4.21 Phenol is a sufficiently strong acid that reacts with strong bases such as sodium hydroxide to form sodium phenoxide, a water-soluble salt. Cyclohexanol has no comparable acidity and does not react with sodium hydroxide.

4.23 *Radical* indicates a molecule or ion with an unpaired electron. *Chain* means a cycle of two or more steps, called propagation steps, that repeat over and over. The net effect of a radical chain reaction is the conversion of starting materials to products. *Chain length* refers to the number of times the cycle of chain propagation steps repeats.

4.25 Vitamin E participates in one or the other of the chain propagation steps and forms a stable radical, which breaks the cycle of propagation steps.

4.27 The abbreviation DDT is derived from **D**ichloro**D**iphenyl**T**richloroethane.

4.29 Insoluble in water. It is entirely a hydrophobic molecule.

4.31 Biodegradable means that a substance can be broken down into one or more harmless compounds by living organisms in the environment.

4.33 Iodine is an element that is found primarily in sea water and, therefore, seafoods are rich sources of it. Individuals in inland areas where seafood is only a limited part of the diet are the most susceptible to developing goiter.

4.35 One of the aromatic rings in Allura Red has a $-CH_3$ group and an $-OCH_3$ group. These groups are not present in Sunset Yellow.

4.37 orange

4.39 Capsaicin is isolated from the fruit of various species of *Capsicum*, otherwise known as chili peppers.

4.41 Two cis-trans isomers are possible for capsaicin.

4.43 Following are the three contributing structures for naphthalene.

(1)   (2)   (3)

4.45 BHT participates in one of the chain propagation steps of autoxidation, forms a stable radical, and thus terminates autoxidation.

4.47

Chapter 5   Alcohols, Ethers, and Thiols

5.1 (a) 2-heptanol (b) 2,2-dimethyl-1-propanol
(c) *trans*-3-isopropylcyclohexanol

5.2 (a) primary (b) secondary (c) primary
(d) tertiary

5.3 In each case, the major product (circled) contains the more substituted double bond.

5.4

| 2-Methyl cyclohexanol | 1-Methyl cyclohexene (C) | 1-Methyl cyclohexanol (D) |

5.5 Each secondary alcohol is oxidized to a ketone.

5.6 (a) ethyl isobutyl ether (b) cyclopentyl methyl ether

5.7 (a) 3-methyl-1-butanethiol (b) 3-methyl-2-butanethiol

5.9 Only (c) and (d) are secondary alcohols.

5.11 (a) 1-pentanol (b) 1,3-propanediol
(c) 1,2-butanediol (d) 3-methyl-1-butanol
(e) *cis*-1,2-cyclohexanediol (f) 2,6-dimethylcyclohexanol

5.13

5.15 (a) ethyl isopropyl ether (b) dibutyl ether
(c) diphenyl ether

5.17 (a) *sec*-butyl mercaptan (b) butyl mercaptan
(c) cyclohexyl mercaptan

5.19 Prednisone contains three ketones, one primary alcohol, one tertiary alcohol, one disubstituted carbon–carbon double bond, and one trisubstituted carbon–carbon double bond. Estradiol contains one secondary alcohol and one disubstituted phenol.

5.21 Low-molecular-weight alcohols form hydrogen bonds with water molecules through both the oxygen and hydrogen atoms of their $-OH$ groups. Low-molecular-weight ethers form hydrogen bonds with water molecules only through the oxygen atom of their $-O-$ group. The greater extent of hydrogen bonding between alcohol and water molecules makes the low-molecular-weight alcohols more soluble in water than the low-molecular-weight ethers.

5.23 Both types of hydrogen bonding are shown on the same illustration.

5.25 In order of increasing boiling point, they are

$$CH_3CH_2CH_3 \quad CH_3CH_2OH$$
$$-42°C \qquad\qquad 78°C$$

$$CH_3CH_2CH_2CH_2OH \quad HOCH_2CH_2OH$$
$$117°C \qquad\qquad 198°C$$

ANSWERS | **A-17**

5.27 Because 1-butanol molecules associate in the liquid state by hydrogen bonding, it has the higher boiling point (117°C). There is very little polarity to an S—H bond. The only interactions among 1-butanethiol molecules in the liquid state are the considerably weaker London dispersion forces. For this reason, 1-butanethiol has the lower boiling point (98°C).

5.29 Evaporation of a liquid from the surface of the skin cools because heat is absorbed from the skin in converting molecules from the liquid state to the gaseous state. 2-Propanol (isopropyl alcohol), which has a boiling point of 82°C, absorbs heat from the skin, evaporates rapidly, and has a cooling effect. 2-Hexanol, which has a boiling point of 140°C, also absorbs heat from the surface of the skin but, because of its higher boiling point, evaporates much more slowly and does not have the same cooling effect as 2-propanol.

5.31 The more water-soluble compound is circled.

(a) $\boxed{CH_3OH}$ or CH_3OCH_3

(b) $\boxed{\underset{\underset{OH}{|}}{CH_3CHCH_3}}$ or $\underset{\underset{CH_2}{\|}}{CH_3CCH_3}$

(c) $CH_3CH_2CH_2SH$ or $\boxed{CH_3CH_2CH_2OH}$

5.33 For three parts, two constitutional isomers will give the desired alcohol. For two parts, only one alkene will give the desired alcohol.

(a) $CH_2{=}CHCH_2CH_3$ or $CH_3CH{=}CHCH_3$

(b) [cyclohexene with CH₃] or [cyclohexane with =CH₂]

(c) $CH_3CH_2CH{=}CHCH_2CH_3$

(d) $CH_2{=}\underset{\underset{CH_3}{|}}{C}CH_2CH_2CH_3$ or $CH_3\underset{\underset{CH_3}{|}}{C}{=}CHCH_2CH_3$

(e) [cyclopentene]

5.35 Phenols are weak acids, with pK_a values approximately equal to 10. Alcohols, considerably weaker acids, have about the same acidity as water.

5.37 The first reaction is an acid-catalyzed dehydration; the second is an oxidation.

(a) $CH_3CH_2CH_2CH_2OH \xrightarrow[\text{heat}]{H_2SO_4} CH_3CH_2CH{=}CH_2 + H_2O$

(b) $CH_3CH_2CH_2CH_2OH \xrightarrow[H_2SO_4]{K_2Cr_2O_7} CH_3CH_2CH_2\overset{\overset{O}{\|}}{C}OH$

5.39 Oxidation of a primary alcohol by $K_2Cr_2O_7/H_2SO_4$ gives a carboxylic acid.

(a) $CH_3(CH_2)_6\overset{\overset{O}{\|}}{C}OH$ (b) $HO\overset{\overset{O}{\|}}{C}CH_2CH_2\overset{\overset{O}{\|}}{C}OH$

5.41 Each can be prepared from 1-propanol (circled) as shown in this flow chart.

5.43 Ethanol and ethylene glycol are derived from ethylene. Ethanol is a solvent and is the starting material for the synthesis of diethyl ether, also an important solvent. Ethylene glycol is used in automotive antifreezes and is one of the two starting materials required for the synthesis of poly(ethylene terephthalate), better known as PET (Section 10.8B).

5.45 (a) The three functional groups are a thiol, a primary amine, and a carboxylic acid. (b) Oxidation of the thiol group gives a disulfide (—S—S—).

(a) $HO\overset{\overset{O}{\|}}{C}\underset{\underset{NH_2}{|}}{C}HCH_2S{-}SCH_2\underset{\underset{NH_2}{|}}{C}H\overset{\overset{O}{\|}}{C}OH$

5.47 Nitroglycerin was discovered in 1847. It is a pale yellow, oily liquid.

5.49 One of the products the body derives from metabolism of nitroglycerin is nitric oxide, which causes the coronary artery to dilate, thus relieving angina.

5.51 The relationship is that 2100 mL of breath contains the same amount of ethanol as 1.00 mL of blood.

5.53 Diethyl ether is easy to use and causes excellent muscle relaxation. Blood pressure, pulse rate, and respiration are usually only slightly affected. Diethyl ether's chief drawbacks are its irritating effect on the respiratory passages and its after-effect of nausea.

5.55 Enflurane and isoflurane are insoluble in water but soluble in hexane.

5.57 $2CH_3OH + 3O_2 \longrightarrow 2CO_2 + 4H_2O$

5.59 The eight isomeric alcohols with the molecular formula C_5H_{12} are

1-Pentanol   2-Pentanol   3-Pentanol   2-Methyl-1-butanol

2-Methyl-2-butanol   3-Methyl-2-butanol   3-Methyl-1-butanol   2,2-Dimethyl-1-propanol

5.61 Ethylene glycol has two —OH groups by which each molecule participates in hydrogen bonding, whereas 1-propanol has only one. The stronger intermolecular forces of attraction between molecules of ethylene glycol give it the higher boiling point.

5.63 Arranged in order of increasing solubility in water, they are

$CH_3CH_2CH_2CH_2CH_2CH_3$ $CH_3CH_2CH_2CH_2CH_2OH$
Hexane   1-Pentanol
(Insoluble)   (2.3 g/mL water)

$HOCH_2CH_2CH_2CH_2OH$
1,4-Butanediol
(Infinitely soluble)

5.65 Each can be prepared from 2-methyl-1-propanol (circled) as shown in this flow chart.

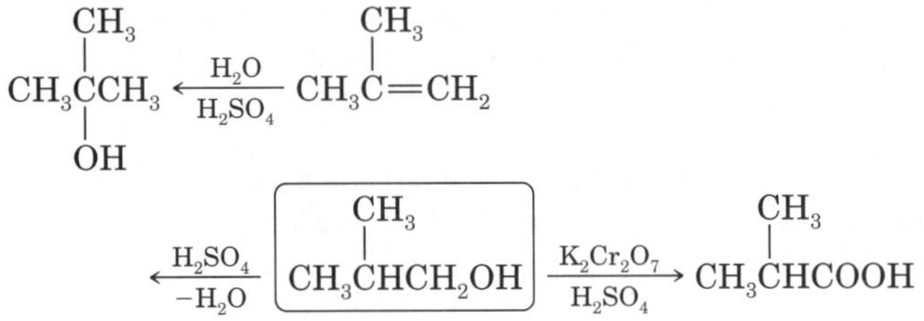

Chapter 6   Chirality

6.1 The enantiomers of each part are drawn with two groups in the plane of the paper, a third group toward you in front of the plane, and the fourth group away from you behind the plane.

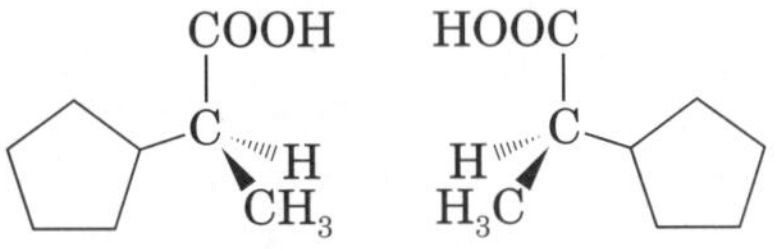

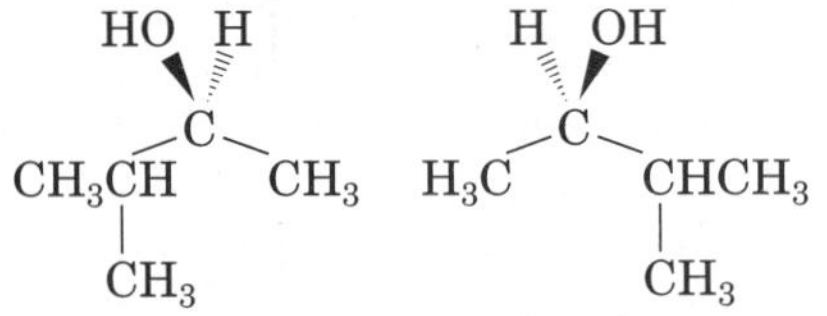

6.2 The group of higher priority in each set is circled.

(a) $\boxed{-CH_2OH}$ and $-CH_2CH_2COOH$

(b) $\boxed{-CH_2NH_2}$ and $-CH_2CH_2COOH$

6.3 The order of priorities and the configuration of each is shown in the drawings.

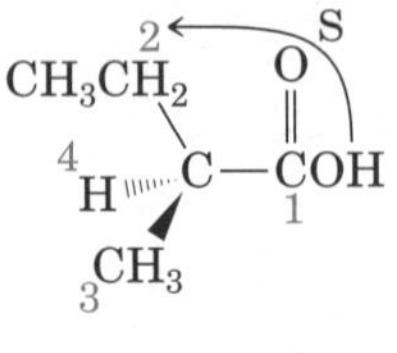

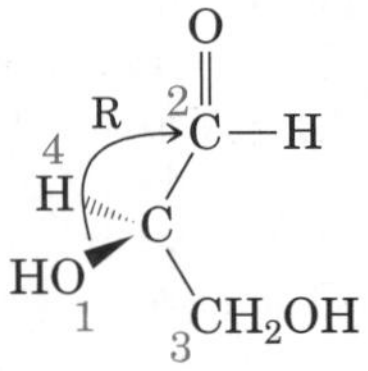

6.4 (a) Compounds 1 and 3 are one pair of enantiomers. Compounds 2 and 4 are a second pair of enantiomers. (b) Compounds 1 and 2, 1 and 4, 2 and 3, and 3 and 4 are diastereomers.

6.5 Four stereoisomers are possible for 3-methylcyclohexanol. The cis isomer is one pair of enantiomers; the trans isomer is a second pair of enantiomers.

6.6 Stereocenters are marked by an asterisk, and the number of stereoisomers possible is shown under the structural formula.

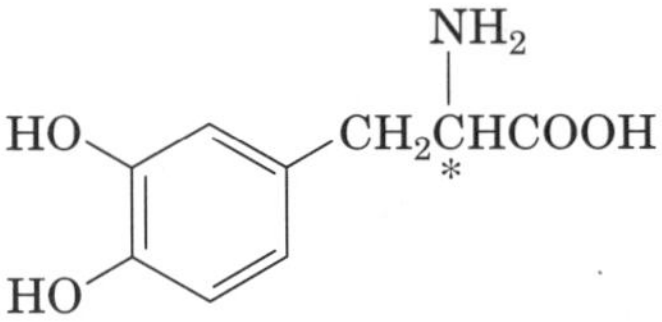

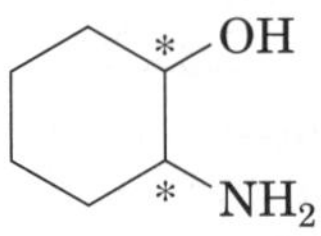

6.7 Chirality is a property of an object. The object is not superposable on its mirror image; that is, the object has handedness. 2-Butanol is a chiral molecule.

6.9 Stereoisomers are isomers that have the same molecular formula and the same connectivity, but a different orientation of their atoms in space. Three examples are cis-trans isomers (in cycloalkanes and alkenes), enantiomers, and diastereomers.

6.11 (a) Chiral. (b) Achiral. (c) Achiral. (d) Achiral. (e) Chiral, unless you are on the equator, in which case it goes straight down and has no chirality.

6.13 Neither *cis*-2-butene nor *trans*-2-butene is chiral. Each is superposable on its mirror image.

6.15 Compounds (a), (c), and (d) contain stereocenters, here marked by asterisks.

6.17 The stereocenter in each chiral compound is marked by an asterisk.

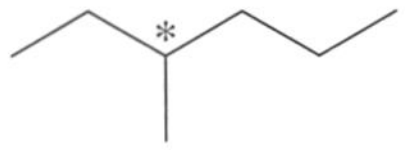

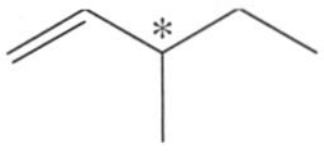

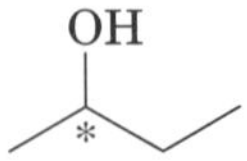

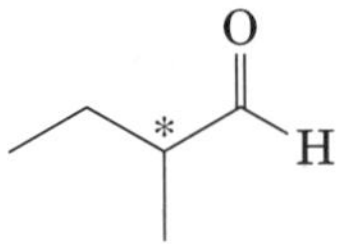

(e) 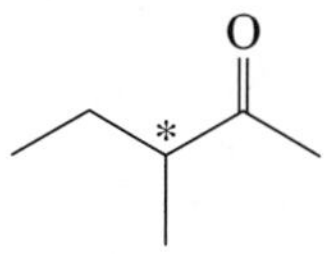 [structure] (f) 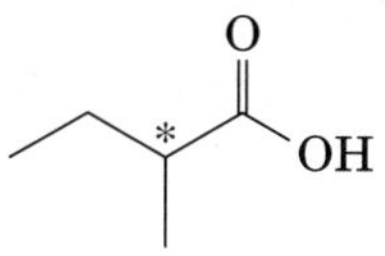 [structure]

6.19 Following are mirror images of each.

(a) 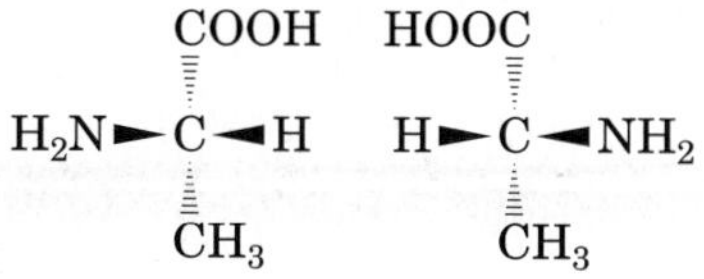 [structures with COOH/HOOC, H₂N—C—H, CH₃]

(b) 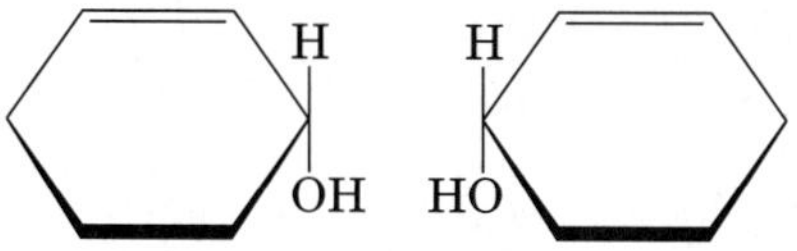 [cyclohexene structures with H, OH]

(c) [sugar ring structures with CH₂OH, OH, HO]

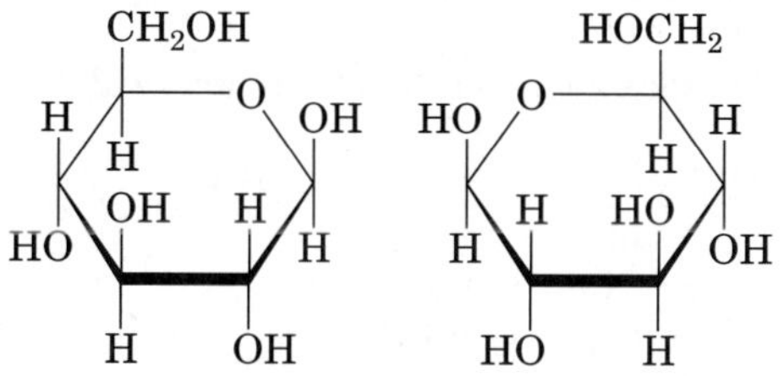

(d) 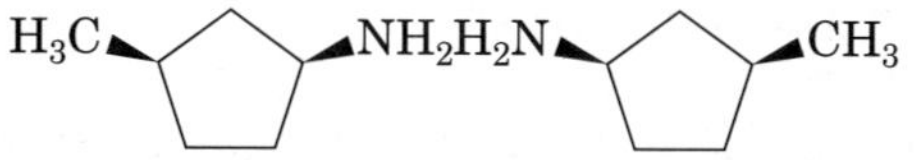 [cyclopentane structures with H₃C, NH₂H₂N, CH₃]

6.21 Parts (b) and (c) contain stereocenters.

(b) 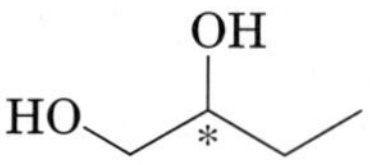 HO [structure with OH] *   (c) 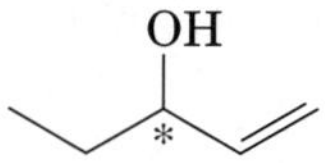 [structure with OH] *

6.23 Each stereocenter is marked by an asterisk. Under each structural formula is the number of stereoisomers possible. Compound (b) has no stereocenter.

(a) 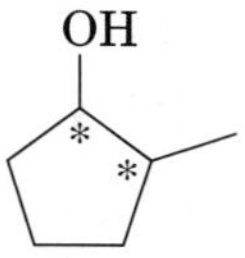 [cyclopentane with OH]
$(2^2 = 4)$

(c) 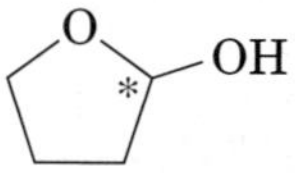 [furan ring with OH]
$(2^1 = 2)$

(d) 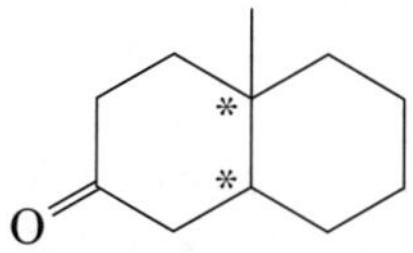 [decalin structure with O]
$(2^2 = 4)$

6.25 Molecules (b) and (d) have R configurations. Molecules (a) and (c) have S configurations.

(a) 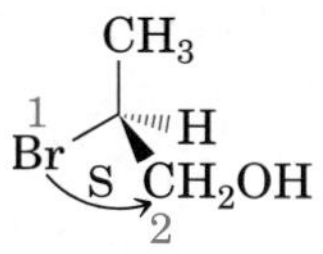 [structure CH₃, Br, S, CH₂OH, H]

(b) 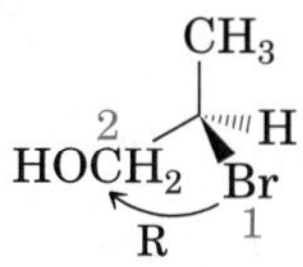 [structure CH₃, HOCH₂, Br, H, R]

(c) 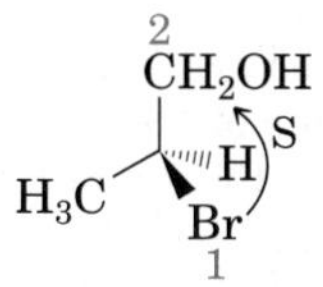 [structure CH₂OH, H₃C, Br, H, S]

(d) 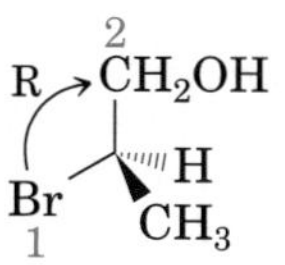 [structure CH₂OH, R, Br, H, CH₃]

6.27 The optical rotation of its enantiomer is +41°.

6.29 To say that a drug is *chiral* means that it has handedness—that it has one or more stereocenters and the possibility for two or more stereoisomers. Just because a compound is chiral does not mean that it will be optically active. It may be chiral and present as a racemic mixture, in which case it will have no effect on the plane of polarized light. If, however, it is present as a single enantiomer, it will rotate the plane of polarized light.

6.31 The two stereocenters are marked by asterisks.

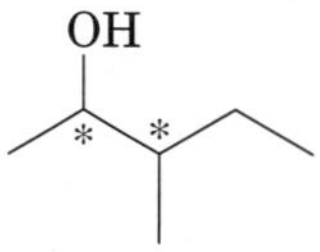

[structure with OH, * *]

3-Methyl-2-pentanol

6.33 Each stereocenter is marked by an asterisk. Under the name of each compound is the number of stereoisomers possible for it.

(a) 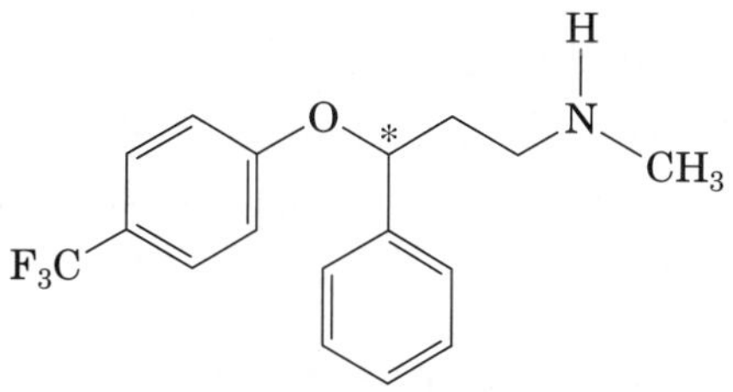 [Fluoxetine structure with F₃C, O, *, N, CH₃, H]

Fluoxetine
(Prozac)
$(2^1 = 2)$

(b) 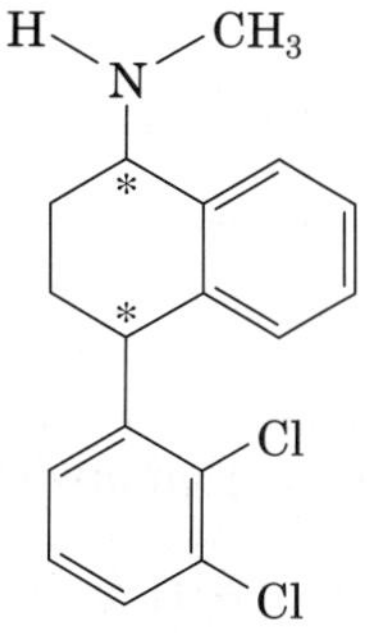 [Sertraline structure with H, N, CH₃, *, *, Cl, Cl]

Sertraline
(Zoloft)
$(2^2 = 4)$

(c) 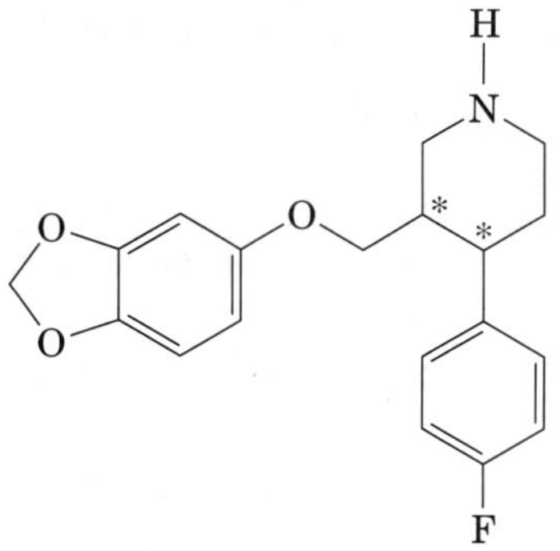 [Paroxetine structure with N, H, O, O, O, *, *, F]

Paroxetine
(Paxil)
$(2^2 = 4)$

6.35 (a) In this chair conformation of glucose, carbons 1, 2, 3, 4, and 5 are stereocenters.   (b) There are $2^5 = 32$ stereoisomers possible.   (c) Because enantiomers always occur in pairs, there are 16 pairs of enantiomers possible. **6.37** (a) The eight stereocenters are marked by asterisks. (b) There are are $2^8 = 256$ stereoisomers possible.

Triamcinolone acetonide

6.39 The majority have a right-handed twist because the machines that make them all impart the same twist.

Chapter 7   Acids and Bases

7.1   (a) toward the left   (b) toward the left

(a)
$$H_3O^+ + I^- \rightleftharpoons H_2O + HI$$
Weaker acid   Weaker base   Stronger base   Stronger acid

(b) $CH_3COO^- + H_2S \rightleftharpoons CH_3COOH + HS^-$
Weaker acid   Weaker acid   Stronger acid   Stronger base

7.2 $pK_a = 9.31$
7.3   (a) ascorbic acid   (b) aspirin
7.4 $[H_3O^+] = 1.0 \times 10^{-2}$
7.5   (a) pH = 2.46   (b) $[H_3O^+] = 7.9 \times 10^{-5}$
7.6   pOH = 4 and pH = 10
7.7   0.0960 M
7.8   (a) pH = 9.25   (b) pH = 4.75
7.9   pH = 9.44
7.10 (a) An acid is a substance that produces H_3O^+ in aqueous solution.   (b) A base is a substance that produces OH^- in aqueous solution.
7.11 (a) $HNO_3(aq) + H_2O(\ell) \longrightarrow NO_3^-(aq) + H_3O^+(aq)$
   (b) $HBr(aq) + H_2O(\ell) \longrightarrow Br^-(aq) + H_3O^+(aq)$
   (c) $H_2SO_3(aq) + H_2O(\ell) \longrightarrow HSO_3^-(aq) + H_3O^+(aq)$
   (d) $H_2SO_4(aq) + H_2O(\ell) \longrightarrow HSO_4^-(aq) + H_3O^+(aq)$
   (e) $HCO_3^-(aq) + H_2O(\ell) \longrightarrow CO_3^{2-}(aq) + H_3O^+(aq)$
   (f) $H_3BO_3(aq) + H_2O(\ell) \longrightarrow H_2BO_3^-(aq) + H_3O^+(aq)$

7.13 (a) $LiOH(s) \xrightarrow{H_2O} Li^+(aq) + OH^-(aq)$

(b)
$(CH_3)_2NH(aq) + H_2O(\ell) \longrightarrow (CH_3)_3NH_2^+(aq) + OH^-(aq)$
7.15 (a) HSO_4^- (b) $H_2BO_3^-$ (c) I^- (d) H_2O (e) NH_3
(f) PO_4^{3-}
7.17 (a) H_2O (b) H_2S (c) NH_4^+ (d) C_6H_5OH
(e) HCO_3^- (f) H_2CO_3

7.19
(a) $H_3PO_4 + OH^- \rightleftharpoons H_2PO_4^- + H_2O$
Stronger acid   Stronger base   Weaker base   Weaker acid

(b) $H_2O + Cl^- \rightleftharpoons HCl + OH^-$
Weaker acid   Weaker base   Stronger acid   Stronger base

(c) $HCO_3^- + OH^- \rightleftharpoons CO_3^{2-} + H_2O$
Stronger acid   Stronger base   Weaker base   Weaker acid

7.21 $H_2CO_3(aq) \longrightarrow CO_2(g) + H_2O(\ell)$
7.23 (a) the pK_a of the weak acid   (b) the K_a of the strong acid
7.25 (a) 0.10 M HCl   (b) 0.10 M H_3PO_4
(c) 0.010 M H_2CO_3 (d) 0.10 M NaH_2PO_4
(e) 0.10 M aspirin
7.27 Only (b) is a redox reaction. All others are acid–base reactions.
(a) $Na_2CO_3 + 2HCl \longrightarrow 2NaCl + CO_2 + H_2O$
(b) $Mg + 2HCl \longrightarrow MgCl_2 + H_2$
(c) $NaOH + HCl \longrightarrow NaCl + H_2O$
(d) $Fe_2O_3 + 6HCl \longrightarrow 2FeCl_3 + 3H_2O$
(e) $NH_3 + HCl \longrightarrow NH_4Cl$
(f) $CH_3NH_2 + HCl \longrightarrow CH_3NH_3Cl$
(g) $NaHCO_3 + HCl \longrightarrow NaCl + H_2O + CO_2$
7.29 (a) $10^{-3} M$ (b) $10^{-10} M$ (c) $10^{-7} M$ (d) $10^{-15} M$
7.31 (a) pH 8; basic   (b) pH = 10; basic   (c) pH = 2; acidic
(d) pH 0; acidic   (e) pH = 7; neutral
7.33 pH = 8.5; basic   (b) pH = 1.2; acidic
(c) pH = 11.1; basic   (d) pH = 6.3; acidic
7.35 (a) $[OH^-] = 0.10 M$; pOH = 1
(b) pOH = 2.4; $[OH^-] = 4.0 \times 10^{-3}$
(c) pOH = 2.0; $[OH^-] = 10^{-2}$
(d) pOH = 5.6; $[OH^-] = 2.5 \times 10^{-6}$
7.37 (a) $Ba(OH)_2 + 2HCl \longrightarrow BaCl_2 + 2H_2O$
(b) $NaOH + HNO_3 \longrightarrow NaNO_3 + H_2O$
(c) $HCOOH + NH_3 \longrightarrow HCOONH_4$
(d) $Ca(OH)_2 + H_2SO_4 \longrightarrow CaSO_4 + 2H_2O$
(e) $Mg(OH)_2 + H_2SO_4 \longrightarrow MgSO_4 + 2H_2O$
(f) $2NaOH + H_3PO_4 \longrightarrow Na_2HPO_4 + 2H_2O$
or $NaOH + NaH_2PO_4 \longrightarrow Na_2HPO_4 + H_2O$
(g) $NH_3 + HCl \longrightarrow NH_4Cl$
7.39 Sodium carbonate is a basic salt. The carbonate ion reacts with water to produce bicarbonate ion and hydroxide ion. Even though the position of this acid–base equilibrium lies far to the left, there is a sufficient concentration of hydroxide ion to make the solution basic.
7.41 0.348 M
7.43 (a) To 12 g NaOH, add enough water to make a volume of 400 mL.
(b) To 12 g of $Ba(OH)_2$, add enough water to make a volume of 1.0 L.
7.45 5.66 mL of 0.740 M H_2SO_4
7.47 3.30×10^{-3} mol of unknown base
7.49 (a) $CH_3COO^- + H_3O^+ \longrightarrow CH_3COOH + H_2O$
(b) $CH_3COOH + OH^- \longrightarrow CH_3COO^- + H_2O$
7.51 The pH of the buffer is 3.75.

7.53 Buffer capacity is the ability of a buffer to neutralize added amounts of H_3O^+ or OH^-. The greater its ability to neutralize these species, the greater its buffer capacity.

7.55 (a) The pH of the buffer would be the same. (b) The buffer containing 1.0 mol of the weak acid and its conjugate base has the greater buffer capacity.

7.57 When [HA] = [A$^-$], then log [A$^-$]/[HA] = 0, and pH = pK_a.

7.59 The HPO_4^{2-}/$H_2PO_4^-$ ratio is 1.55/1.

7.61 The 0.10 mol of sodium acetate reacts completely with the 0.10 mol of HCl to form 0.10 mol of acetic acid. The pH of 0.10 M acetic acid is 2.9.

7.63 Rinse the eyes immediately and thoroughly with a stream of water.

7.65 The bases that do not contain hydroxide ion are $CaCO_3$, $MgCO_3$, and $NaHCO_3$.

7.67 (a) Will dissolve (b) Will not dissolve (c) Will not dissolve

7.69 The 0.10 M HCl is more acidic. Because HCl is completely ionized in aqueous solution, the concentration of H_3O^+ is 0.10 M. Acetic acid is a weak acid and the extent of its ionization in aqueous solution is very small.

7.71 3.70×10^{-3} M

7.73 0.90 M

7.75 Yes. If the concentration of HCl is 1.0 M, $-\log(1.0) = 0$.

7.77 No. The pH of 1.0 M HCl is smaller (it is a more acidic solution) than the pH of 1.0 M acetic acid.

7.79 Dissolve 0.182 mol NaH_2BO_3 and 1.00 mol of H_3BO_3 in enough water to make 1 L of solution.

7.81 The acid with the higher pK_a is the weaker acid. The equilibrium for an acid–base reaction will always lie on the side of the weaker acid and the weaker base.

7.83 (a) $HCOO^- + H_3O^+ \longrightarrow HCOOH + H_2O$
(b) $HCOOH + OH^- \longrightarrow HCOO^- + H_2O$

7.85 (a) 0.050 mol Na_2HPO_4 (b) 0.0050 mol Na_2HPO_4
(c) 0.500 mol Na_2HPO_4

7.87 Decrease (become more acidic). You are adding more of the weak acid to the solution, so it becomes more acidic and its pH decreases.

Chapter 8 Amines

8.1 Pyrollidine has nine hydrogens; its molecular formula is C_4H_9N. Purine has four hydrogens; its molecular formula is $C_5H_4N_4$.

8.2 Following is a line-angle formula for each compound.

(a) 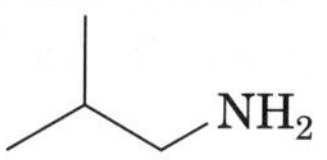 (b)

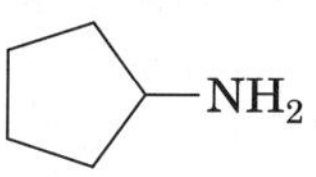

(c)

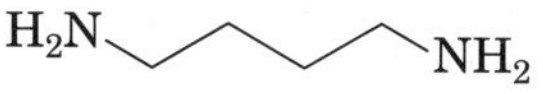

8.3 Following is a line-angle formula for each compound.

(a) 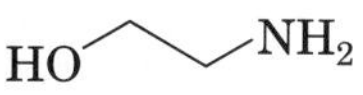 (b)

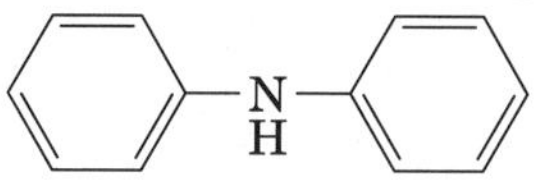

(c)

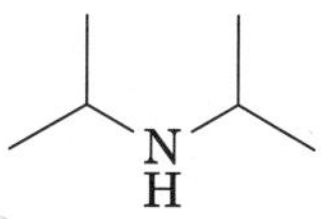

8.4 The stronger base is circled.

(a)

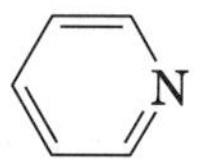

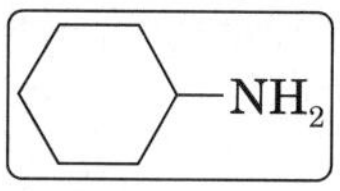

(b)

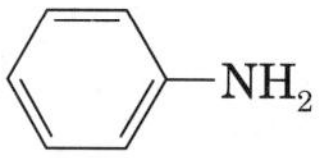

8.5 The product of each reaction is an amine salt.

(a) $(CH_3CH_2)_3\overset{+}{N}H\ Cl^-$

(b)

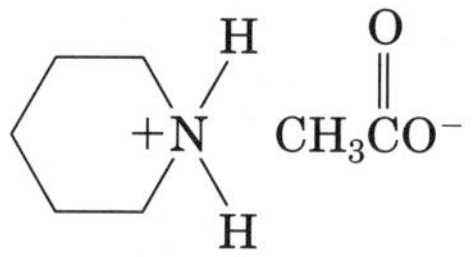

8.7 In an aliphatic amine, all carbon groups bonded to nitrogen are alkyl groups. In an aromatic amine, one or more of the carbon groups bonded to nitrogen are aryl (aromatic) groups.

8.9 Following is a structural formula for each amine.

(a)

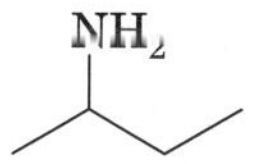

(b) $CH_3(CH_2)_6CH_2NH_2$

(c)

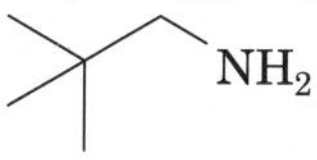

(d) $H_2N(CH_2)_5NH_2$

(e)

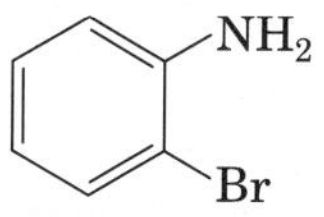

(f) $(CH_3CH_2CH_2CH_2)_3N$

8.11 Each amine is classified by type.

(a)

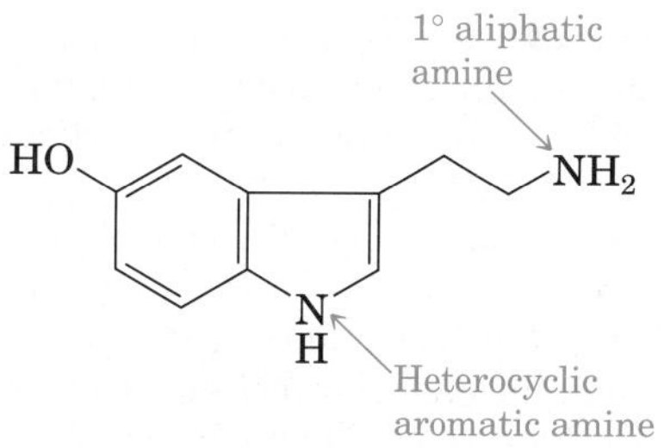

(b)

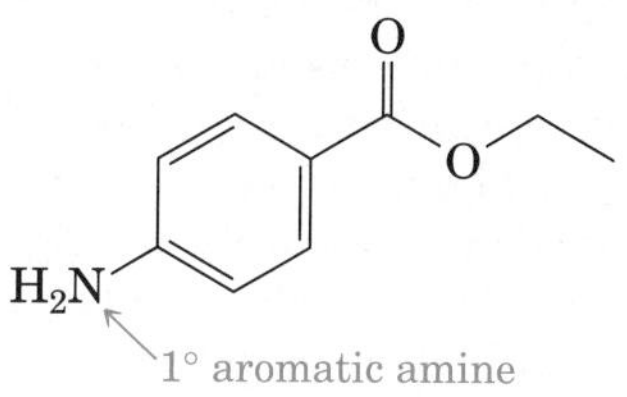

(c)

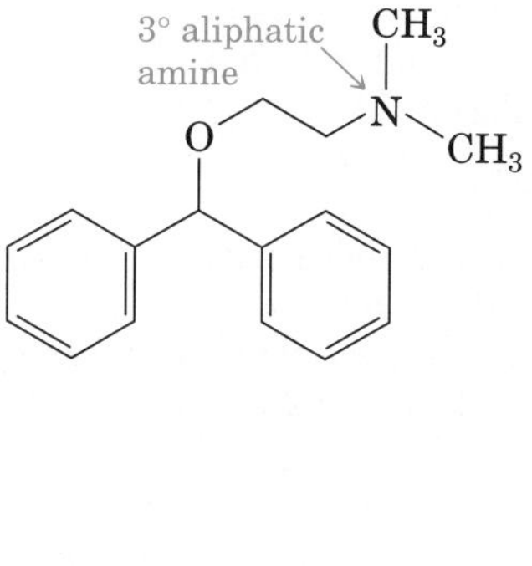

8.13 There are four primary amines of this molecular formula, three secondary amines, and one tertiary amine. Only 2-butanamine is chiral.

1° amines:

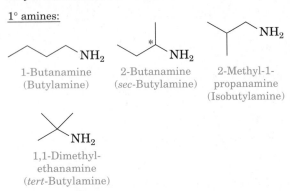

1-Butanamine
(Butylamine)

2-Butanamine
(*sec*-Butylamine)

2-Methyl-1-
propanamine
(Isobutylamine)

1,1-Dimethyl-
ethanamine
(*tert*-Butylamine)

2° amines:

Methylpropyl-
amine

Methylisopropyl-
amine

Diethylamine

3° amines:

Dimethylethylamine

8.15 Both propylamine (a 1° amine) and ethylmethylamine (a 2° amine) have an N—H group and show hydrogen bonding between their molecules in the liquid state. Because of this intermolecular force of attraction, these two amines have higher boiling points than trimethylamine, which has no N—H bond and, therefore, cannot participate in intermolecular hydrogen bonding.

8.17 2-Methylpropane is a nonpolar hydrocarbon and the only attractive forces between its molecules in the liquid state are the very weak London dispersion forces. Both 2-propanol and 2-propanamine are polar molecules and associate in the liquid state by hydrogen bonding. Hydrogen bonding is stronger between alcohol molecules than between amine molecules because of the greater strength of an O—H----O hydrogen bond compared to an N—H----N hydrogen bond. It takes more energy (a higher temperature) to separate an alcohol molecule in the liquid state from its neighbors than to separate an amine molecule from its neighbors and, therefore, the alcohol has the higher boiling point.

8.19 Nitrogen is less electronegative than oxygen and, therefore, more willing to donate its unshared pair of electrons to H^+ in an acid–base reaction to form a salt.

8.21 (a) ethylammonium chloride
(b) diethylammonium chloride
(c) anilinium hydrogen sulfate

8.23 Structural formula A contains both an acid (the carboxyl group) and a base (the 1° amino group). The acid–base reaction between them gives structural formula B, which is the better representation of alanine.

8.25

(a) $CH_3\overset{O}{\overset{\|}{C}}OH$ + [pyridine] ⟶ [pyridinium] CH_3CO^-

(b) [structure] + HCl ⟶ [structure] $NH_3^+Cl^-$

(c) [structure] + H_2SO_4 ⟶ [structure] HSO_4^-

8.27 (a) The primary aliphatic amine is the stronger base and forms the salt with HCl. The salt is named pyridoxamine hydrochloride.

8.29 Albuterol contains one 1° alcohol, one 2° alcohol, one phenol, and one 2° amine. Albuterol differs from epinephrine in that one phenolic —OH group of epinephrine is converted to a —CH₂OH group, and the N-methyl group of epinephrine is converted to a *tert*-butyl group.

8.31 Possible negative effects are long periods of sleeplessness, loss of weight, and paranoia.

8.33 Both coniine and nicotine have one stereocenter; two stereoisomers (one pair of enantiomers) are possible for each.

8.35 The four stereocenters of cocaine are marked with asterisks. Following is the structural formula of the salt formed by the reaction of cocaine with HCl.

8.37 Neither Librium nor Valium is chiral.
8.39 No. No unreacted HCl is present.

8.41

(a) [benzene ring]—NHCH$_3$

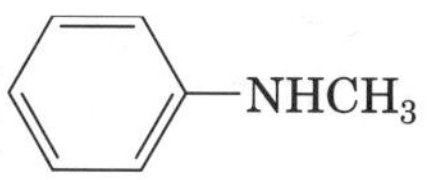

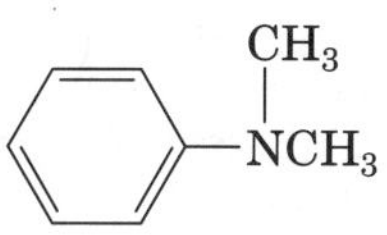

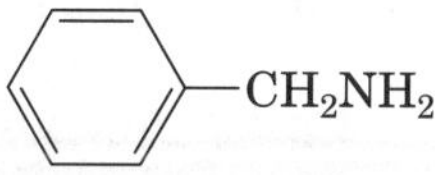

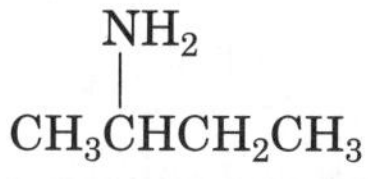

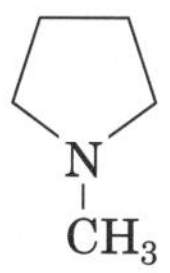

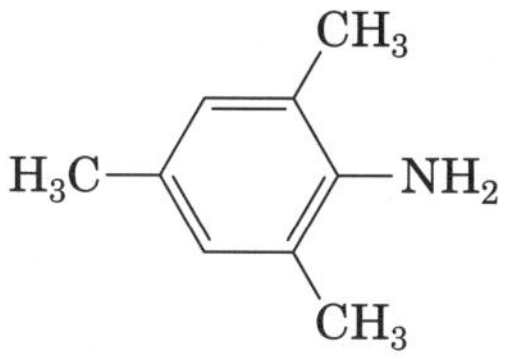

(b) [benzene ring]—N(CH$_3$) with CH$_3$ above

(c) [benzene ring]—CH$_2$NH$_2$

(d) CH$_3$CHCH$_2$CH$_3$ with NH$_2$ above

(e) [pyrrolidine ring with N—CH$_3$]

(f) H$_3$C—[benzene ring with CH$_3$ top and CH$_3$ bottom]—NH$_2$

(g) CH$_3$NCH$_2$CH$_2$CH$_3$ with Cl$^-$, CH$_2$CH$_3$ above (N$^+$) and CH$_3$CHCH$_3$ below

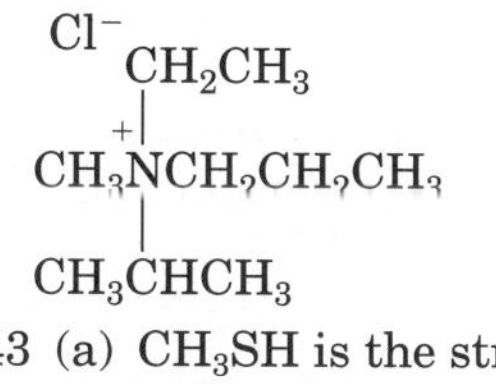

8.43 (a) CH$_3$SH is the stronger acid.
(b) (CH$_3$)$_2$NH is the stronger base.
(c) CH$_3$OH has the highest boiling point.
(d) Molecules of CH$_3$OH form the strongest hydrogen bonds.
8.45 Both alcohols and amines can interact with water by hydrogen bonding. Because amines and alcohols of the same molecular weight have about the same solubility in water, the strength of hydrogen-bonding between their molecules must be comparable.
8.47 (a) Following is its structural formula. (b) It has two stereocenters, marked with asterisks, and $2^2 = 4$ stereoisomers are possible.

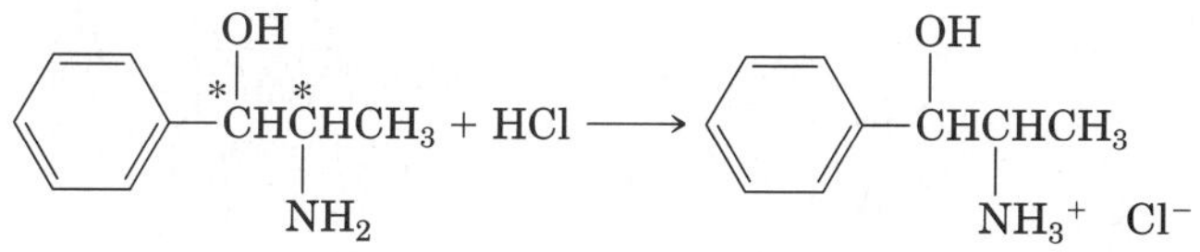

[benzene]—*CHCH*CH$_3$ with OH above, NH$_2$ below + HCl ⟶ [benzene]—CHCHCH$_3$ with OH above, NH$_3$$^+$ Cl$^-$ below

8.49 Gabapentin is better represented by structural formula A, the internal salt. Structural formula B contains both an acid (—COOH) and a base (—NH$_2$) which will react by an internal proton-transfer reaction to form A.
8.51 The 2° aliphatic amine is more basic than the heterocyclic aromatic amine. The three stereocenters are marked by asterisks.

Epibatidine [structure with Cl, pyridine ring N, bicyclic amine, asterisks] — This nitrogen is more basic

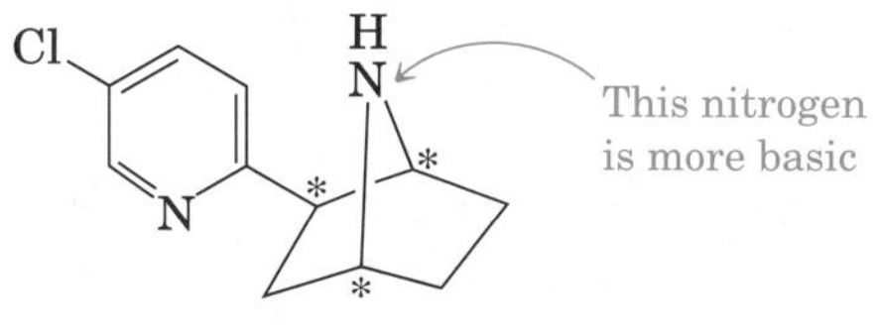

Chapter 9 Aldehydes and Ketones

9.1 (a) 3,3-dimethylbutanal (b) cyclopentanone
(c) (S)-2-phenylpropanal
9.2 Following are line-angle formulas for each aldehyde with the molecular formula C$_6$H$_{12}$O. In the three that are chiral, the stereocenter is marked by an asterisk.

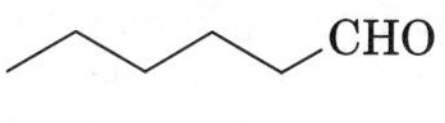

[structures]
Hexanal   4-Methylpentanal

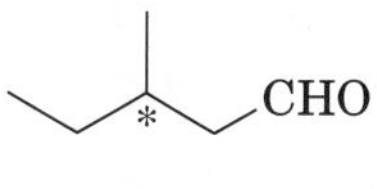

[structures]
3-Methylpentanal*   2-Methylpentanal*

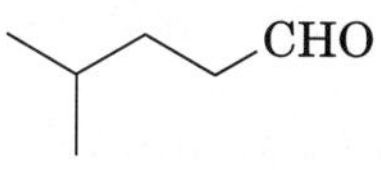

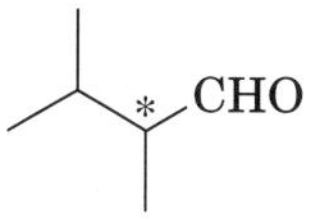

[structures]
2,3-Dimethylbutanal*   3,3-Dimethylbutanal

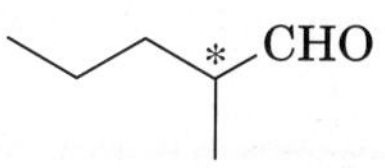

[structure]
2,2-Dimethylbutanal

9.3 (a) 2,3-dihydroxypropanal (b) 2-aminobenzaldehyde
(c) 5-amino-2-pentanone
9.4 Each aldehyde is oxidized to a carboxylic acid.

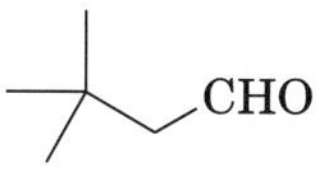

(a) [structure] HO—C(=O)...C(=O)—OH
Hexanedioic acid
(Adipic acid)

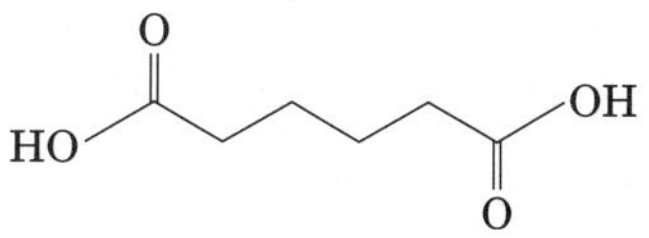

(b) [structure] [benzene]—CH$_2$CH$_2$—C(=O)—OH
3-Phenylpropanoic acid

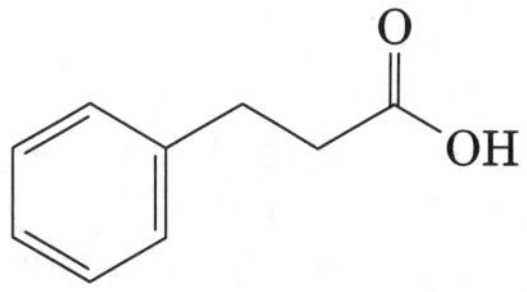

9.5 Each primary alcohol comes from reduction of an aldehyde. Each secondary alcohol comes from reduction of a ketone.

(a) [cyclohexanone structure] =O

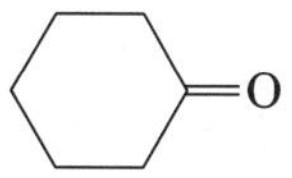

(b) CH$_3$O—[benzene]—CH$_2$CH with O double bond

(c) [structure with two C=O groups]

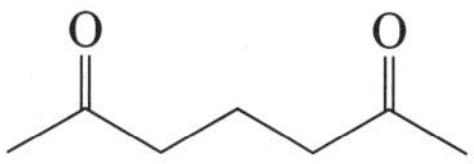

9.6 Shown first is the hemiacetal and then the acetal.

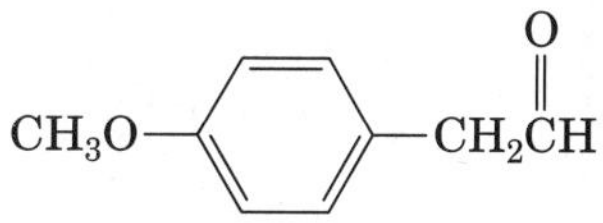

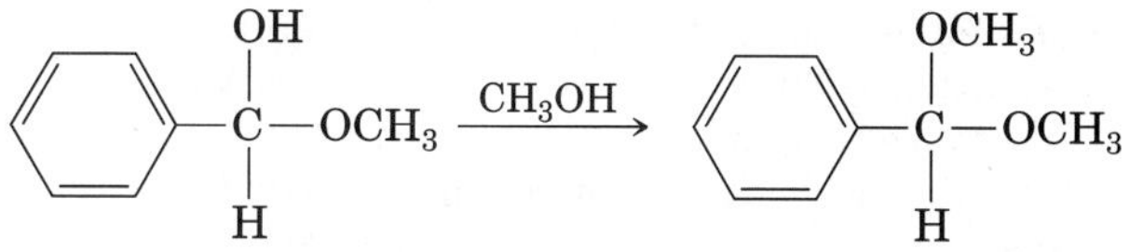

[benzene]—C(—OCH$_3$)(—H) with OH above $\xrightarrow{\text{CH}_3\text{OH}}$ [benzene]—C(—OCH$_3$)(—H) with OCH$_3$ above
Hemiacetal   Acetal

9.7 (a) A hemiacetal formed from 3-pentanone, a ketone, and ethanol.
(b) Neither a hemiacetal nor an acetal. This compound is the dimethyl ether of ethylene glycol.
(c) An acetal derived from 5-hydroxypentanal and methanol.
9.8 Following is the keto form of each enol.

(a) (b)

(c)

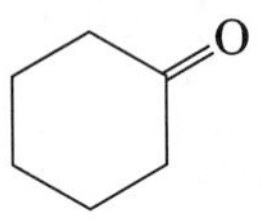

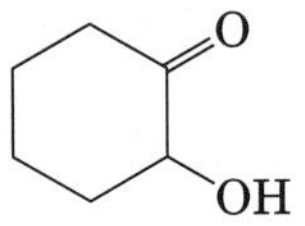

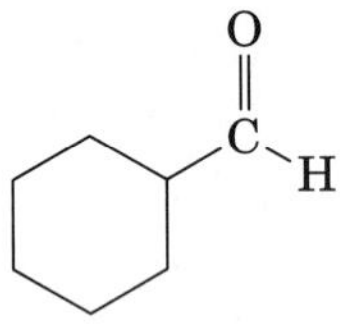

9.9 Oxidation of a primary alcohol gives either an aldehyde or a carboxylic acid, depending on the experimental conditions. Oxidation of a secondary alcohol gives a ketone.

(a)

(b) or

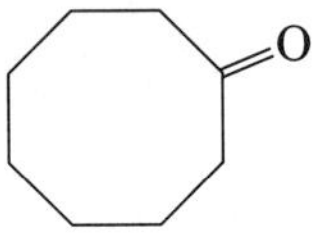

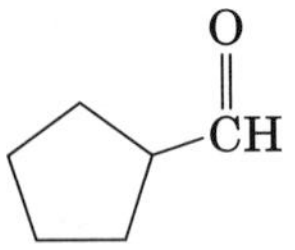

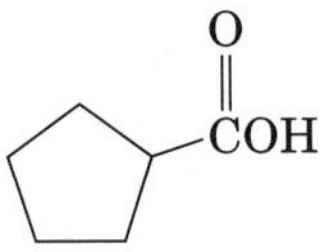

9.11 The carbonyl carbon of an aldehyde is bonded to at least one hydrogen. The carbonyl carbon of a ketone is bonded to two carbon groups.
9.13 No. To be a carbon stereocenter, the carbon atom must have four different groups bonded to it. The carbon atom of a carbonyl group has only three groups bonded to it.
9.15 (a) Cortisone contains three ketones, one 3° alcohol, one 1° alcohol, and one carbon–carbon double bond.
(b) Aldosterone contains two ketones, one aldehyde, one 1° alcohol, one 2° alcohol, and one carbon–carbon double bond.
9.17 Following are structural formulas for each aldehyde.

(a) $\mathrm{H\overset{\overset{O}{\|}}{C}H}$ (b)

(c) (d) $\mathrm{CH_3(CH_2)_8\overset{\overset{O}{\|}}{C}H}$

(e) (f)

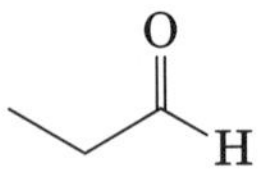

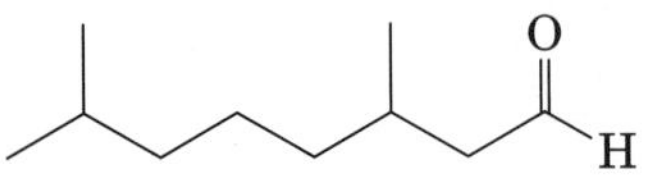

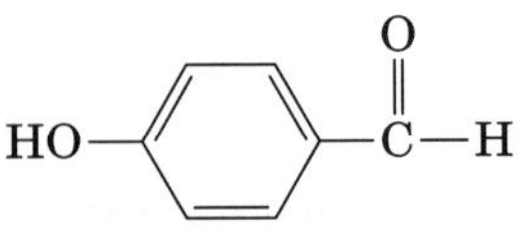

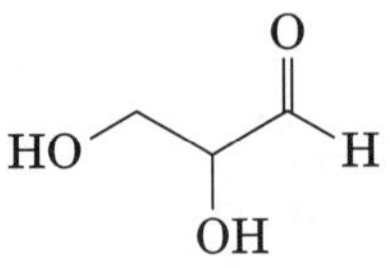

9.19 (a) 4-heptanone (b) 2-methylcyclopentanone
(c) *cis*-2-methyl-2-butenal (d) (*S*)-2-hydroxypropanal
(e) 1-phenyl-2-propanone (f) hexanedial
9.21 (a) The chain is not numbered correctly. Its name is 2-butanone.
(b) The compound is an aldehyde. Its name is butanal.
(c) The longest chain is five carbons. Its name is pentanal.
(d) The location of the ketone takes precedence. Its name is 3,3-dimethyl-2-butanone.
9.23 (a) ethanol (b) 3-pentanone (c) butanal
(d) 2-butanol
9.25 Acetone has the higher boiling point because of the intermolecular attraction between the carbonyl groups of acetone molecules.
9.27 Acetaldehyde forms hydrogen bonds with water primarily through its carbonyl oxygen.

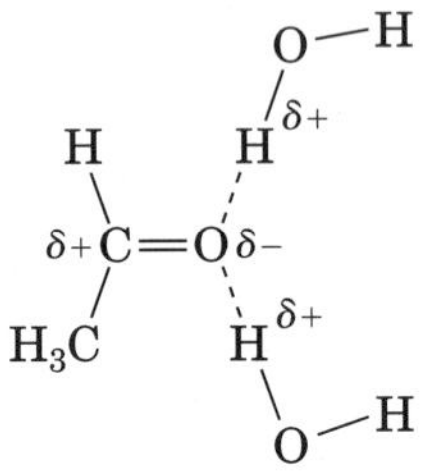

9.29 Aldehydes are oxidized by this reagent to carboxylic acids. Ketones are not oxidized under these conditions. Secondary alcohols are oxidized to ketones.

(a) $\mathrm{CH_3CH_2CH_2\overset{\overset{O}{\|}}{C}OH}$ (b)

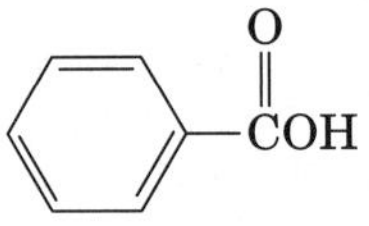

(c) no reaction (d)

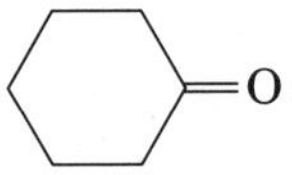

9.31 (a) Treat each with Tollens' reagent. Only pentanal will give a silver mirror.
(b) Treat each with $K_2Cr_2O_7/H_2SO_4$. Only 2-pentanol is oxidized (to 2-pentanone), which causes the red color of $Cr_2O_7^{2-}$ ion to disappear and be replaced by the green color of Cr^{3+} ion.
9.33 The white solid is benzoic acid, formed by air oxidation of benzaldehyde.
9.35 These experimental conditions reduce an aldehyde to a primary alcohol and a ketone to a secondary alcohol.

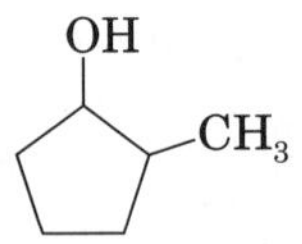

(a) $\mathrm{CH_3\overset{\overset{OH}{|}}{C}HCH_2CH_3}$ (b) $\mathrm{CH_3(CH_2)_4CH_2OH}$

(c) (d)

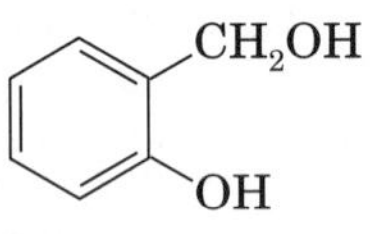

9.37 (a) Following is its structural formula.

$$\underset{\substack{\text{1,3-Dihydroxy-2-propanone}\\ \text{(Dihydroxyacetone)}}}{\mathrm{HOCH_2\overset{\overset{O}{\|}}{C}CH_2OH}} + H_2 \xrightarrow[\text{catalyst}]{\text{Metal}} \underset{\substack{\text{1,2,3-Propanetriol}\\ \text{(Glycerol, glycerin)}}}{\mathrm{HOCH_2\overset{\overset{OH}{|}}{C}HCH_2OH}}$$

(b) Because it has two hydroxyl groups and one carbonyl group, all of which can interact with water molecules by hydrogen bonding, predict that it is soluble in water.
(c) Its reduction gives 1,2,3-propanetriol, better known as glycerol or glycerin.

9.39 The first two reactions are reduction. There is no reaction in (c) or (d); ketones are not oxidized by these reagents.

(a, b) [structure: benzene ring with —CHCH$_3$ bearing OH]

9.41 (a) an acetal (b) a hemiacetal (c) an acetal
(d) an acetal (e) an acetal (f) neither

9.43 Following are equations for the formation of each hemiacetal and acetal.

(a) $CH_3CH_2\overset{O}{\overset{\|}{C}}-H \xrightarrow{CH_3OH} CH_3CH_2\overset{OH}{\underset{OCH_3}{\overset{|}{\underset{|}{C}}}}-H$

$\xrightarrow{CH_3OH} CH_3CH_2\overset{OCH_3}{\underset{OCH_3}{\overset{|}{\underset{|}{C}}}}-H + H_2O$

(b) [cyclopentanone] $=O \xrightarrow{CH_3OH}$ [cyclopentane ring with OH and OCH$_3$]

$\xrightarrow{CH_3OH}$ [cyclopentane ring with OCH$_3$ and OCH$_3$] $+ H_2O$

9.45 *Hydration* refers to the addition of one or more molecules of water to a substance. An example of hydration is the acid-catalyzed addition of water to propene to give 2-propanol. *Hydrolysis* refers to the reaction of a substance with water, generally with the breaking (lysis) of one or more bonds in the substance. An example of hydrolysis is the acid-catalyzed reaction of an acetal with a molecule of water to give an aldehyde or ketone and two molecules of alcohol.

9.47 Compounds (a), (b), (d), and (f) undergo keto-enol tautomerism.

9.49 Following are the keto forms of each enol.

(a) [cyclopentanone structure with =O]

(b) $CH_3\overset{O}{\overset{\|}{C}}CH_2CH_2CH_2CH_3$

(c) [benzene ring]—$CH_2\overset{O}{\overset{\|}{C}}CH_3$

9.51 Compounds (a), (b), and (c) can be formed by reduction of the aldehyde or ketone shown. Compound (d) is a 3° alcohol and cannot be formed in this manner.

(a) $CH_3\overset{O}{\overset{\|}{C}}CH_3$

(b) [benzene ring]—$\overset{O}{\overset{\|}{C}}{\underset{H}{}}$

(c) $H-\overset{O}{\overset{\|}{C}}-H$

9.53 The alkene of three carbons is propene. Acid-catalyzed hydration of this alkene gives 2-propanol as the major product, not 1-propanol.

9.55 Each conversion can be brought about by acid-catalyzed hydration of the alkene to a secondary alcohol, followed by oxidation of the secondary alcohol to a ketone.

(a) $CH_2{=}CHCH_2CH_2CH_3 \xrightarrow[H_2SO_4]{H_2O}$

$CH_3\overset{OH}{\overset{|}{C}}HCH_2CH_2CH_3 \xrightarrow[H_2SO_4]{K_2Cr_2O_7} CH_3\overset{O}{\overset{\|}{C}}CH_2CH_2CH_3$

(b) [cyclohexene] $\xrightarrow[H_2SO_4]{H_2O}$ [cyclohexanol with OH] $\xrightarrow[H_2SO_4]{K_2Cr_2O_7}$ [cyclohexanone with =O]

9.57 The aldehyde and ketone functional groups are circled.

(a) $H\overset{O}{\overset{\|}{C}}CH_2CH_2CH_2\overset{O}{\overset{\|}{C}}CH_3$

(b) [cyclohexane ring with =O and —CH with =O]

(c) $HOCH_2\overset{HO}{\overset{|}{C}}H\overset{O}{\overset{\|}{C}}H$

(d) [bicyclic ring structure with =O]

(e) [benzene ring]—$\overset{O}{\overset{\|}{C}}CH_2CH_3$

(f) HO—[benzene ring with CH$_3$O]—$\overset{O}{\overset{\|}{C}}H$

9.59 Each compound reduces a carbonyl group of an aldehyde or ketone to an alcohol by delivering a hydride ion, $H{:}^-$, to the carbonyl carbon.

9.61 Formulas for this one ketone and two aldehydes of molecular formula C_4H_8O are

(a) $CH_3\overset{O}{\overset{\|}{C}}CH_2CH_3$

(b) $CH_3CH_2CH_2\overset{O}{\overset{\|}{C}}H$ and $CH_3\overset{O}{\overset{\|}{\underset{\underset{CH_3}{|}}{C}}}HCH$

9.63 2-Propanol has the higher boiling point because of the greater attraction between its molecules due to hydrogen bonding through its hydroxyl group.

9.65 (a) Treat each compound with Tollens' reagent. Only benzaldehyde reduces Ag^+ to give a precipitate of silver metal as a silver mirror.
(b) Treat each compound with Tollens' reagent as in part (a). Only acetaldehyde reduces Ag^+ to give a precipitate of silver metal as a silver mirror.

9.67 Shown in the equation are structural formulas for the equilibrium products.

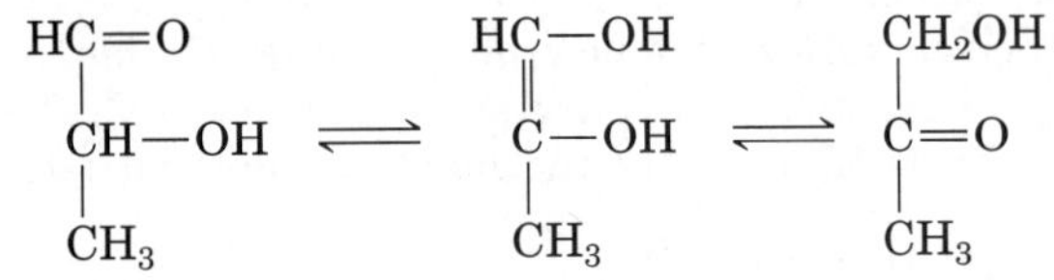

An α-hydroxyaldehyde   An enediol   An α-hydroxyketone

9.69 (a) Carbon 5 provides the —OH group, and carbon 1 provides the —CHO group.   (b) Following is a structural formula for the free aldehyde.

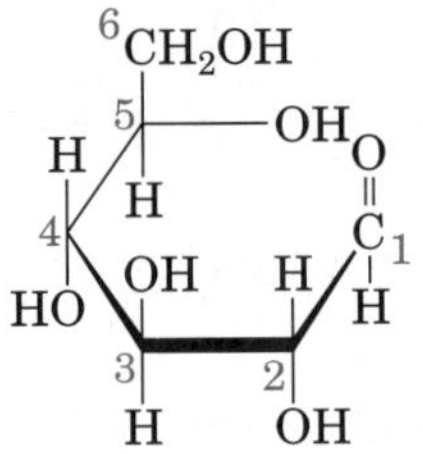

Chapter 10   Carboxylic Acids, Anhydrides, Esters, and Amides

10.1 (a) 2,3-dihydroxypropanoic acid
(b) 3-aminopropanoic acid
(c) 3,5-dihydroxy-3-methylpentanoic acid
10.2 Each acid is converted to its ammonium salt.

(a) $CH_3(CH_2)_2COOH + NH_3 \longrightarrow CH_3(CH_2)_2COO^- NH_4^+$

Butanoic acid   Ammonium butanoate
(Butyric acid)   (Ammonium butyrate)

(b) $CH_3\overset{\text{OH}}{\underset{}{CH}}COOH + NH_3 \longrightarrow CH_3\overset{\text{OH}}{\underset{}{CH}}COO^- NH_4^+$

2-Hydroxypropanoic   Ammonium 2-hydroxypropanoate
acid   (Ammonium lactate)
(Lactic acid)

10.3 (a) ethyl benzoate   (b) phenyl acetate
10.4 Following is a structural formula for each amide.

(a)

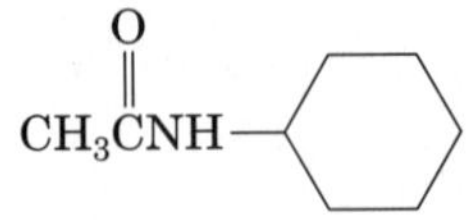

(b)

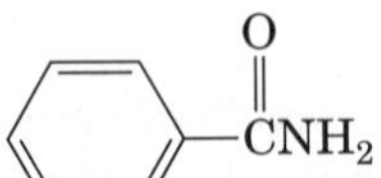

10.5 Following is the structural formula of each ester.

(a)

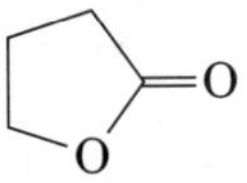

(b)

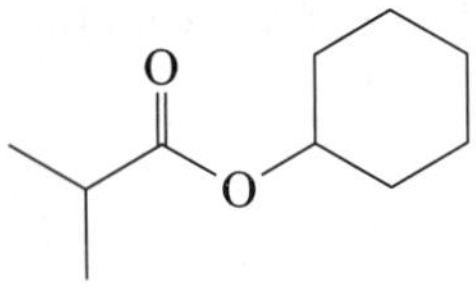

10.6 Under basic conditions, as in part (a), each carboxyl group is present as a carboxylic anion. Under acidic conditions, as in part (b), each carboxyl group is present in its un-ionized form.

(a) 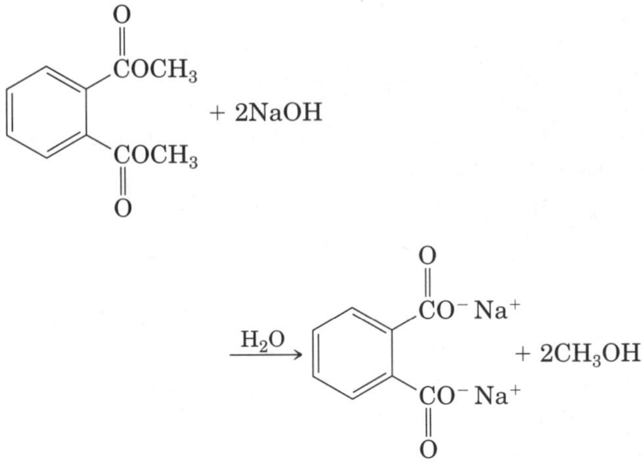   + 2NaOH

$\xrightarrow{H_2O}$ + 2CH$_3$OH

(b) 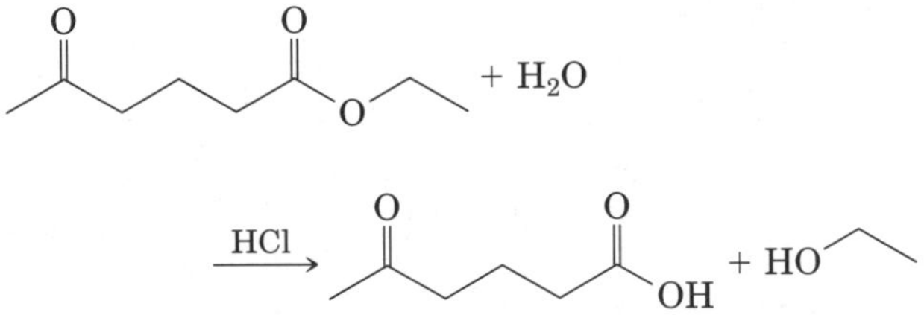   + H$_2$O

$\xrightarrow{HCl}$ + HO$\diagup$

10.7 In aqueous NaOH, each carboxyl group is present as a carboxylic anion, and each amine is present in its unprotonated form.

(a) $CH_3\overset{O}{\overset{\|}{C}}N(CH_3)_2 + NaOH \xrightarrow[\text{heat}]{H_2O} CH_3\overset{O}{\overset{\|}{C}}O^-Na^+ + (CH_3)_2NH$

(b) 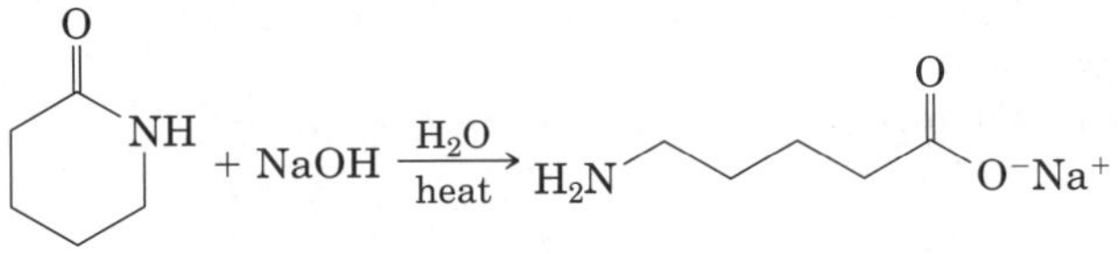   + NaOH $\xrightarrow[\text{heat}]{H_2O}$

10.9 (a) 3,4-dimethylpentanoic acid
(b) 2-aminobutanoic acid   (c) hexanoic acid
10.11 Following are structural formulas for each carboxylic acid.

(a)

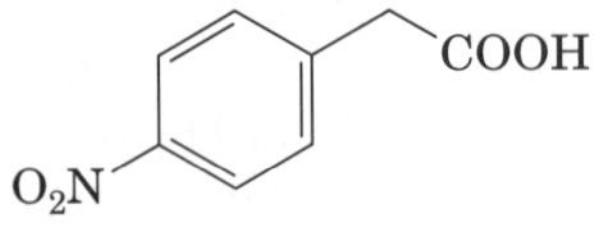

(b) 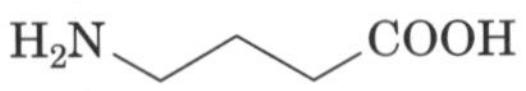 H$_2$N$\diagdown\diagup\diagdown$COOH

(c) 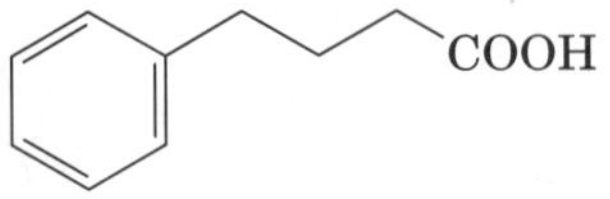   (d)

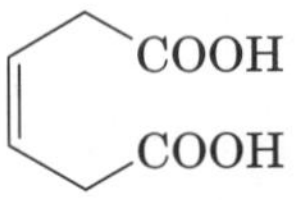

10.13 Following are structural formulas for each salt.

(a) 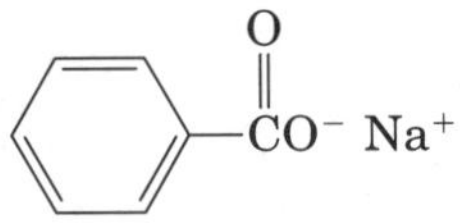   (b) $CH_3\overset{O}{\overset{\|}{C}}O^-$ Li$^+$

(c) $CH_3\overset{O}{\underset{\|}{C}}O^-\ NH_4^+$

(d) $Na^+\ {}^-O\overset{O}{\underset{\|}{C}}(CH_2)_4\overset{O}{\underset{\|}{C}}O^-\ Na^+$

(c)

(e)

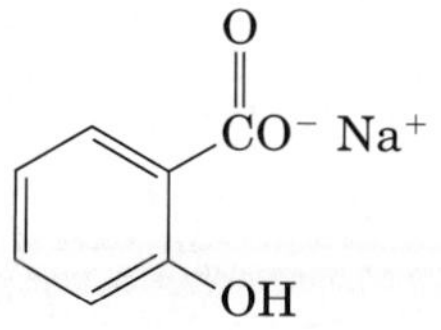

(f) $(CH_3CH_2CH_2\overset{O}{\underset{\|}{C}}O^-)_2Ca^{2+}$

10.15 One of the carboxyl groups in this salt is present as —COO⁻, the other as —COOH.

$HO\overset{O}{\underset{\|}{C}}-\overset{O}{\underset{\|}{C}}O^-\ K^+$

10.17 If you draw this molecule correctly to show the internal hydrogen bonding, you will see that the hydrogen-bonded part of the molecule forms a six-membered ring.

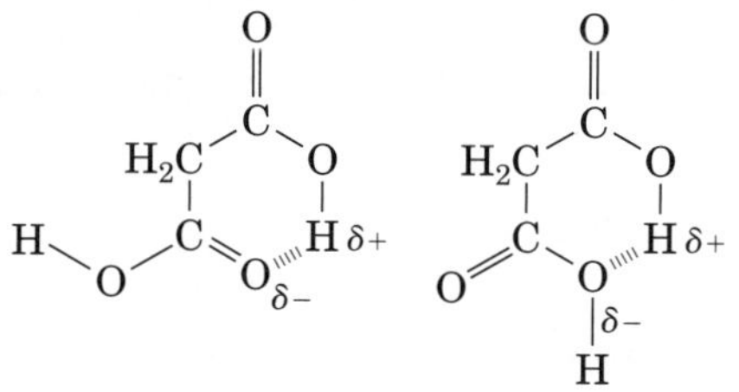

10.19 In order of increasing boiling point, they are heptanal (bp 153°C), 1-heptanol (bp 176°C), and heptanoic acid (bp 223°C).

10.21 In order of increasing solubility in water, they are decanoic acid, pentanoic acid, and acetic acid.

10.23 Following are structural formulas for the indicated starting material.

(a) $CH_3(CH_2)_4CH_2OH$
$C_6H_{14}O$

(b) $CH_3(CH_2)_4\overset{O}{\underset{\|}{C}}H$
$C_6H_{12}O$

(c) $HOCH_2(CH_2)_4CH_2OH$
$C_6H_{14}O_2$

10.25 In order of increasing acidity, they are benzyl alcohol, phenol, and benzoic acid.

10.27 Following are completed equations for these acid–base reactions.

(a)

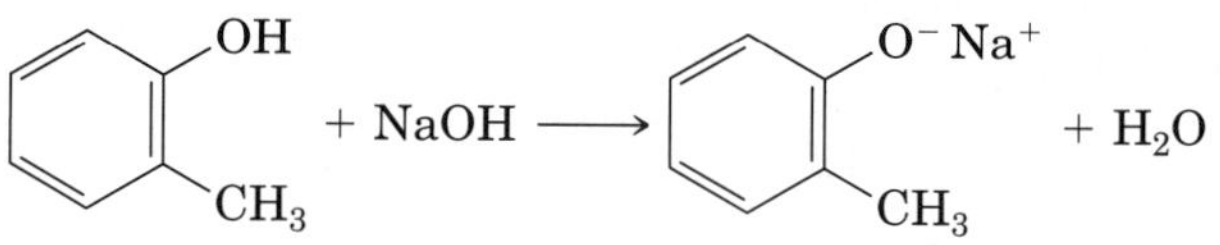

(b)

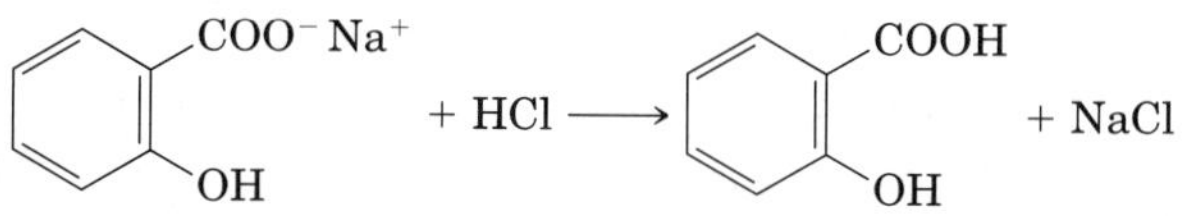

(c)

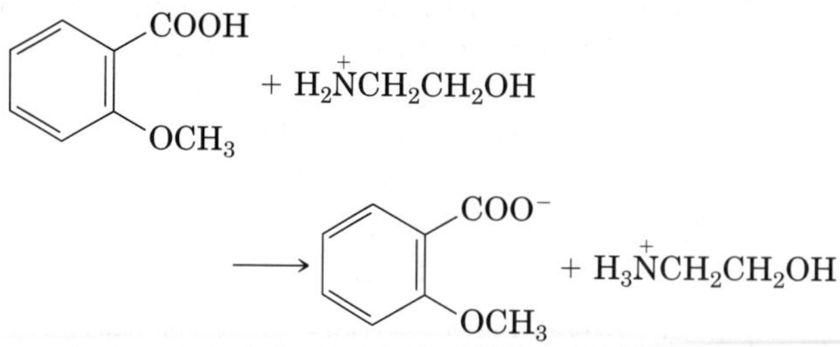

(d)

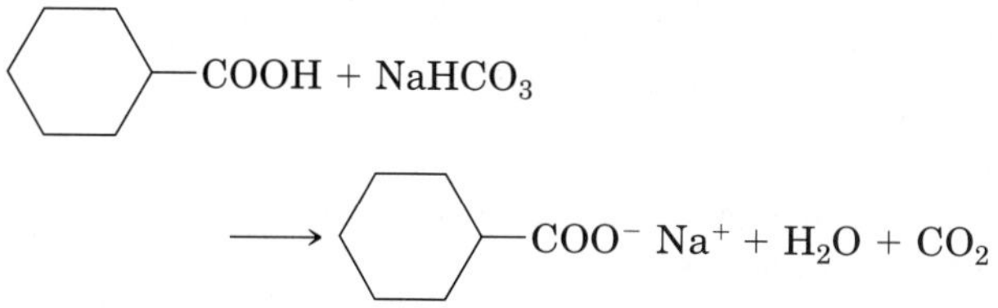

10.29 CH_3COOH at pH 2.0, equal amounts of CH_3COOH and CH_3COO^- at pH 4.76, and CH_3COO^- at pH 8.0 or higher

10.31 The pK_a of ascorbic acid is 4.10. At this pH, ascorbic acid is present 50 percent as ascorbic acid and 50 percent as ascorbate ion. At pH 7.35 to 7.45, which is more basic than pH 4.10, ascorbic acid would be present as ascorbate ion.

10.33 In part (a), the —COOH group is a stronger acid than the —NH₃⁺ group.

(a) $CH_3\underset{\underset{NH_3^+}{|}}{C}HCOOH + NaOH \longrightarrow CH_3\underset{\underset{NH_3^+}{|}}{C}HCOO^- + H_2O$

(b) $CH_3\underset{\underset{NH_3^+}{|}}{C}HCOO^-Na^+ + NaOH$
$\longrightarrow CH_3\underset{\underset{NH_2}{|}}{C}HCOO^-Na^+ + H_2O$

10.35 In part (a), the amine is the stronger base.

(a) $CH_3\underset{\underset{NH_2}{|}}{C}HCOO^-Na^+ + HCl \longrightarrow CH_3\underset{\underset{NH_3^+}{|}}{C}HCOO^-Na^+ + NaCl$

(b) $CH_3\underset{\underset{NH_3^+}{|}}{C}HCOO^-Na^+ + HCl \longrightarrow CH_3\underset{\underset{NH_3^+}{|}}{C}HCOOH + NaCl$

10.37 Following is a structural formula for the ester formed in each reaction.

(a) 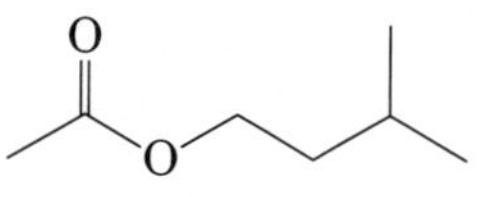 (b)

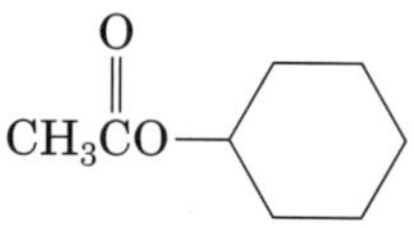

(c)

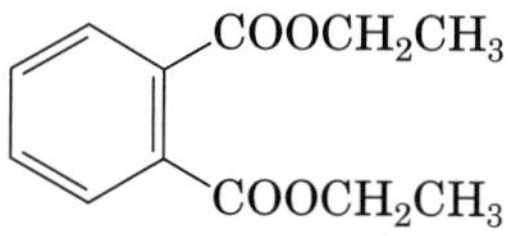

10.39 Following is a structural formula for acid and alcohol from which each ester is derived.

(a) 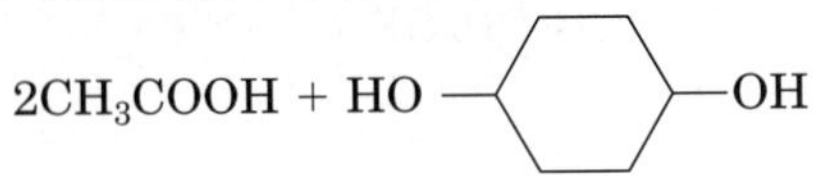 $2CH_3COOH + HO$—⬡—OH

(b)  ⬡—$COOH + CH_3OH$

(c) $2CH_3OH + HOOCCH_2CH_2COOH$

(d) 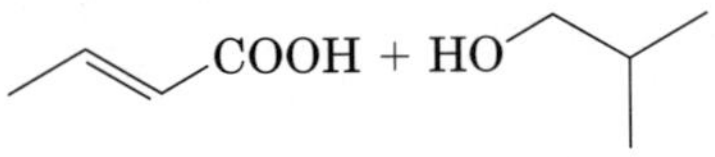 ⟋⟍$COOH + HO$⟋⟍⟨

10.41 Following is a structural formula for methyl 4-hydroxybenzoate.

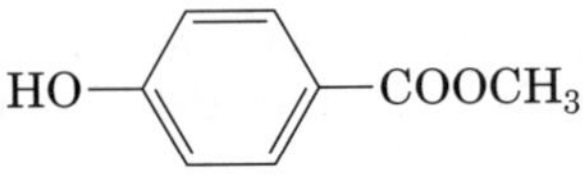

HO—⬡—$COOCH_3$

10.43 Following are structural formulas for the reagents to synthesize each amide.

(a) 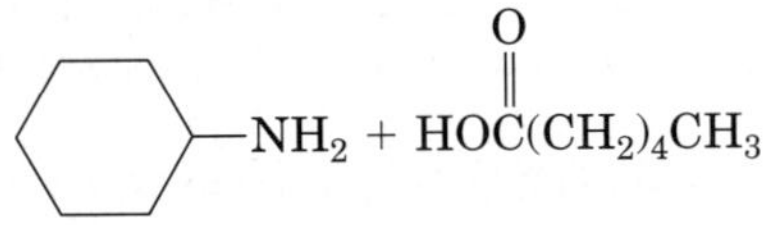 ⬡—$NH_2 + HO\overset{\displaystyle O}{\overset{\|}{C}}(CH_2)_4CH_3$

(b) $(CH_3)_2CH\overset{\displaystyle O}{\overset{\|}{C}}OH + HN(CH_3)_2$

(c) $2NH_3 + HO\overset{\displaystyle O}{\overset{\|}{C}}(CH_2)_4\overset{\displaystyle O}{\overset{\|}{C}}OH$

10.45 (a) Both lidocaine and mepivacaine contain an amide and a 3° amine. (b) Both are derived from 2,6-dimethylaniline. In addition, the aliphatic amine nitrogen in each is separated by one carbon from the carbonyl group of the amide.

10.47 Following is a structural formula for each synthetic flavoring agent.

(a) 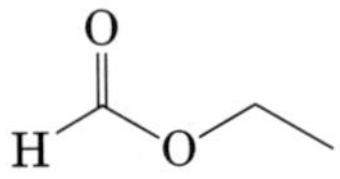 [structure]

(b) 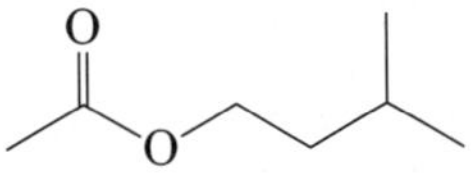 [structure]

(c) 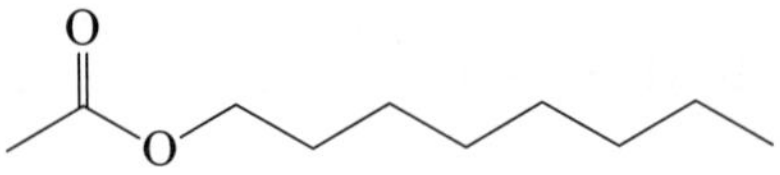 [structure]

(d) 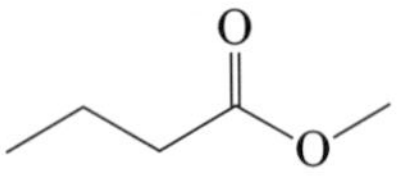 [structure]

(e) 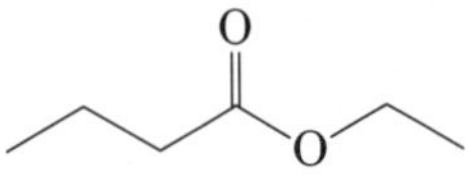 [structure]

(f) 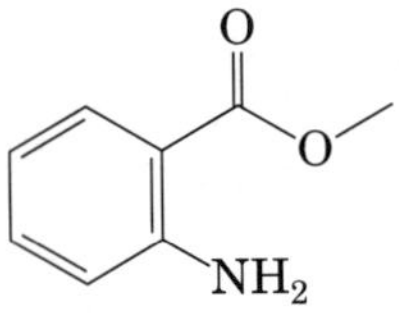 [structure]

10.49 Saponification is the hydrolysis of an ester using aqueous NaOH or KOH to give the sodium or potassium salt of the carboxylic acid and an alcohol. Following is a balanced equation for the saponification of methyl acetate.

$$CH_3\overset{\displaystyle O}{\overset{\|}{C}}OCH_3 + NaOH \xrightarrow{H_2O} CH_3\overset{\displaystyle O}{\overset{\|}{C}}O^-Na^+ + CH_3OH$$

Methyl acetate   Sodium acetate   Methanol

10.51 Each reaction brings about hydrolysis of the amide bond. Each product is shown as it would exist under the specified reaction conditions.

(a) 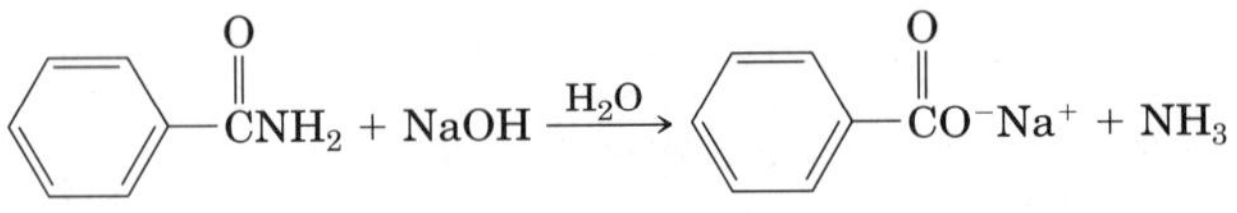 ⬡—$\overset{\displaystyle O}{\overset{\|}{C}}NH_2 + NaOH \xrightarrow{H_2O}$ ⬡—$\overset{\displaystyle O}{\overset{\|}{C}}O^-Na^+ + NH_3$

(b) 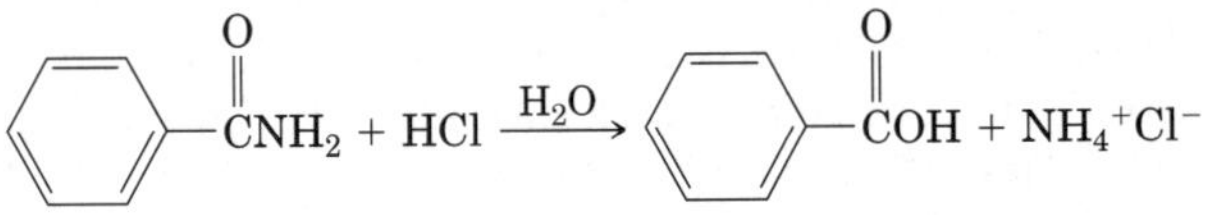 ⬡—$\overset{\displaystyle O}{\overset{\|}{C}}NH_2 + HCl \xrightarrow{H_2O}$ ⬡—$\overset{\displaystyle O}{\overset{\|}{C}}OH + NH_4^+Cl^-$

10.53 The products in each part are an amide and acetic acid.

(a) 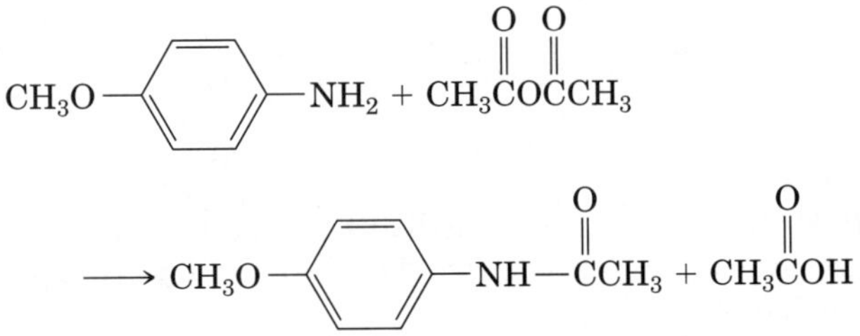 CH_3O—⬡—$NH_2 + CH_3\overset{\displaystyle O}{\overset{\|}{C}}O\overset{\displaystyle O}{\overset{\|}{C}}CH_3$

$\longrightarrow CH_3O$—⬡—$NH-\overset{\displaystyle O}{\overset{\|}{C}}CH_3 + CH_3\overset{\displaystyle O}{\overset{\|}{C}}OH$

(b) 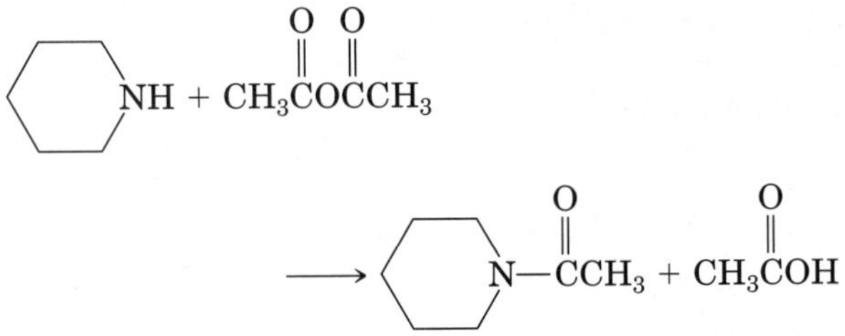 ⬡$NH + CH_3\overset{\displaystyle O}{\overset{\|}{C}}O\overset{\displaystyle O}{\overset{\|}{C}}CH_3$

$\longrightarrow$ ⬡$N-\overset{\displaystyle O}{\overset{\|}{C}}CH_3 + CH_3\overset{\displaystyle O}{\overset{\|}{C}}OH$

10.55 (a) Phenobarbital contains four amide groups. (b) Complete hydrolysis of all amide bonds gives a dicarboxylic acid dianion, two moles of ammonia, and one mole of sodium carbonate.

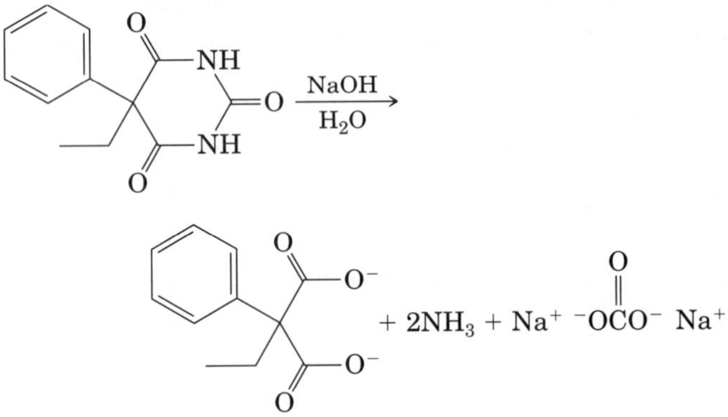

[structure] $\xrightarrow[H_2O]{NaOH}$

[structure] $+ 2NH_3 + Na^+ \, {}^-O\overset{\displaystyle O}{\overset{\|}{C}}O^- \, Na^+$

10.57 In nylon-66 and Kevlar, the monomer units are joined by amide bonds.

10.59 In Dacron and Mylar, the monomer units are joined by ester bonds.

10.61 Following are structural formulas for the mono-, di-, and triethyl esters.

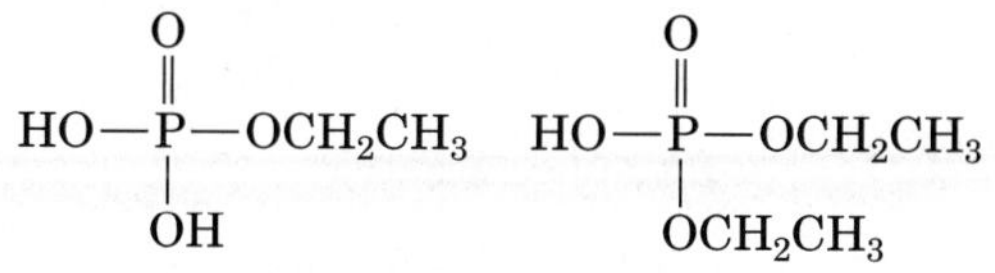

Ethyl phosphate   Diethyl phosphate

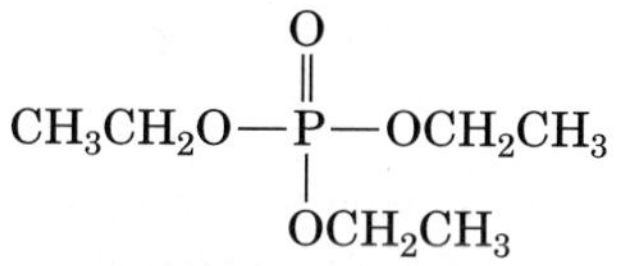

Triethyl phosphate

10.63 Two molecules of water are split out.

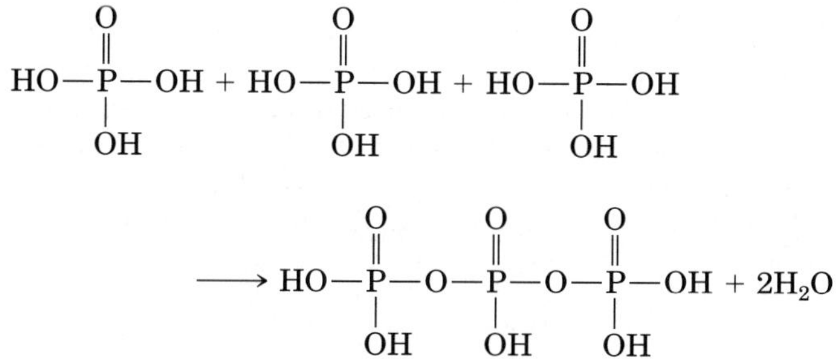

10.65 The arrow points to the ester group. Below is chrysanthemic acid.

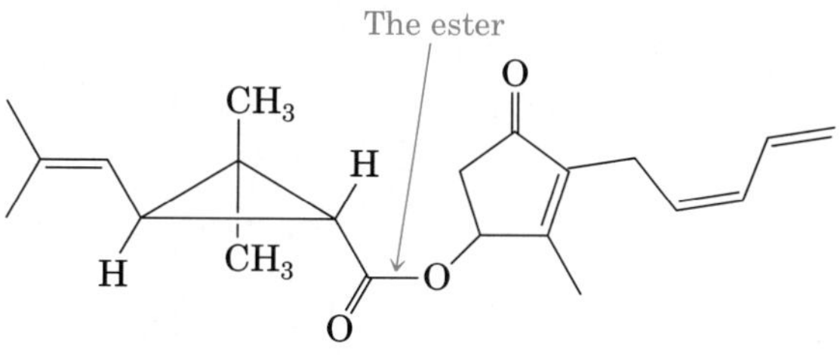

Pyrethrin I

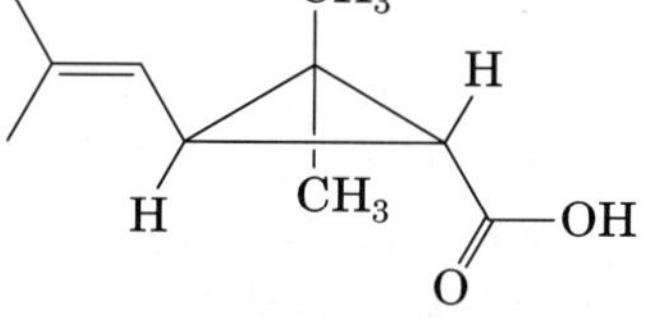

Chrysanthemic acid

10.67 (a) The *cis/trans ratio* refers to the cis-trans relationship between the ester group and the carbon–carbon double bond in the three-membered ring. (b) Permethrin has three stereocenters, and eight stereoisomers (four pairs of enantiomers) are possible for it. The designation "(+/−)" refers to the fact that the members of each pair of possible enantiomers are present in equal amounts; that is, each pair of enantiomers is present as a racemic mixture.

10.69 The compound is salicin. Removal of the glucose unit and oxidation of the primary alcohol to a carboxylic acid gives salicylic acid.

10.71 The moisture present in humid air may be sufficient to bring about hydrolysis of the ester to yield salicylic acid and acetic acid. The vinegar-like odor is due to the presence of acetic acid.

10.73 A *sunblock* prevents all ultraviolet radiation from reaching protected skin by reflecting it away from the skin. A *sunscreen* prevents a portion of the ultraviolet radiation from reaching protected skin. Its effectiveness is related to its skin protection factor (SPF).

10.75 They all contain an ester bonded to an alkyl group as well as a benzene ring. The benzene has either a nitrogen atom or an oxygen atom on it.

10.77 Lactomer stitches dissolve as the ester groups in the polymer chain are hydrolyzed until only glycolic acid and lactic acid remain. These small molecules are metabolized and excreted by existing biochemical pathways.

10.79 Propanoic acid has a boiling point of 141°C, and methyl acetate has a boiling point of 57°C. The boiling point of propanoic acid is higher because of the association of adjacent molecules by hydrogen bonding.

10.81 In a solution of pH 4.07, lactic acid is 50 percent in the un-ionized (lactic acid) form and 50 percent in the ionized (lactate anion) form. In gastric juice of pH 1.0 to 3.0, which is more acidic than pH 4.10, lactic acid is present primarily as un-ionized lactic acid molecules.

10.83 Following is an equation for this synthesis of acetaminophen.

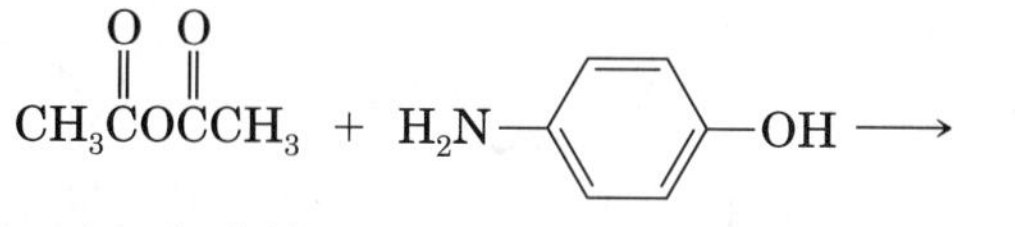

Acetic anhydride   4-Aminophenol

Acetaminophen

10.85 Hydrolysis gives one mole of 2,3-dihydroxypropanoic acid and two moles of phosphoric acid. At pH 7.35–7.45, the carboxyl group is present as its anion, and phosphoric acid is present as its dianion.

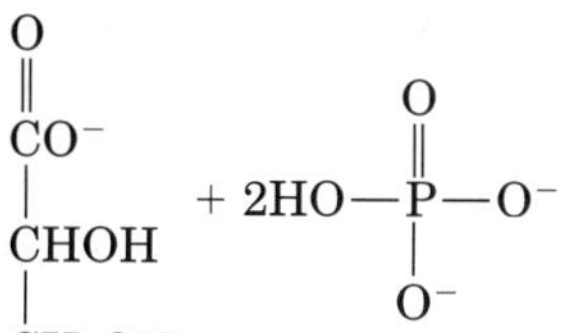

Chapter 11   Carbohydrates

11.1 Following are Fischer projections for the four 2-ketopentoses. They consist of two pairs of enantiomers.

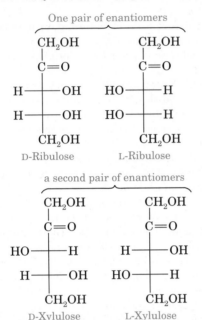

One pair of enantiomers

D-Ribulose   L-Ribulose

a second pair of enantiomers

D-Xylulose   L-Xylulose

11.2 D-Mannose differs in configuration from D-glucose only at carbon 2. One way to arrive at the structures of the α and β forms of D-mannopyranose is to draw the corresponding α and β forms of D-glucopyranose, and then invert the configuration in each at carbon 2.

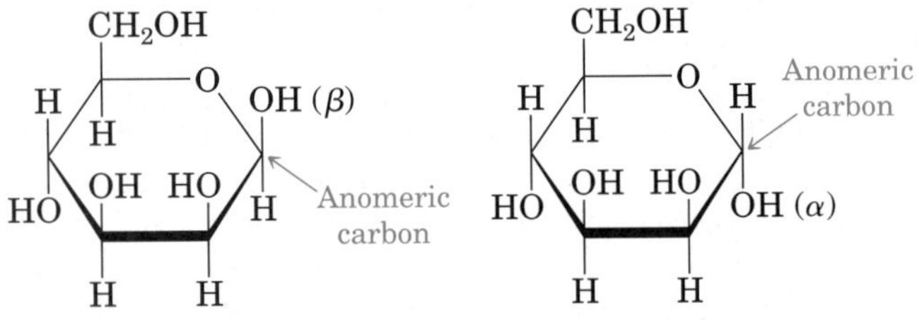

β-D-Mannopyranose
(β-D-Mannose)

α-D-Mannopyranose
(α-D-Mannose)

11.3 D-Mannose differs in configuration from D-glucose only at carbon 2.

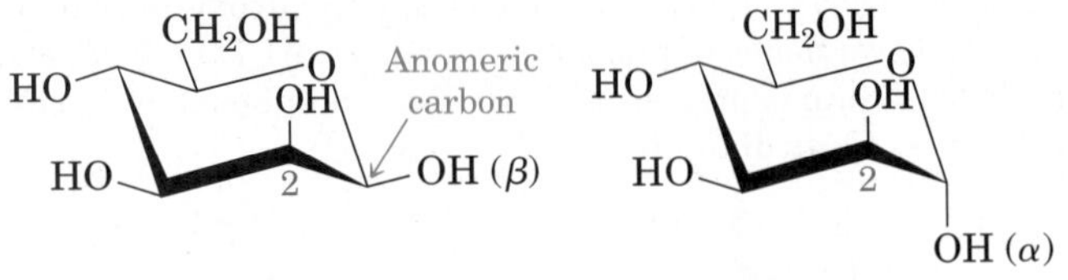

β-D-Mannopyranose
(β-D-Mannose)

α-D-Mannopyranose
(α-D-Mannose)

11.4 Following is a Haworth projection and a chair conformation for this glycoside.

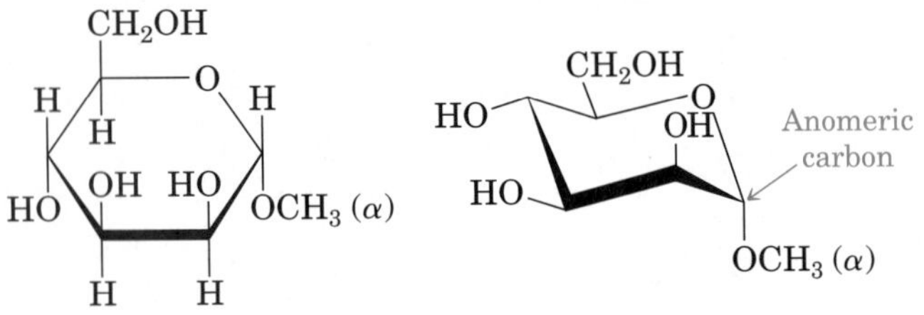

$OCH_3 (\alpha)$

$OCH_3 (\alpha)$

11.5 The β-glycosidic bond is between carbon 1 of the left unit and carbon 3 of the right unit.

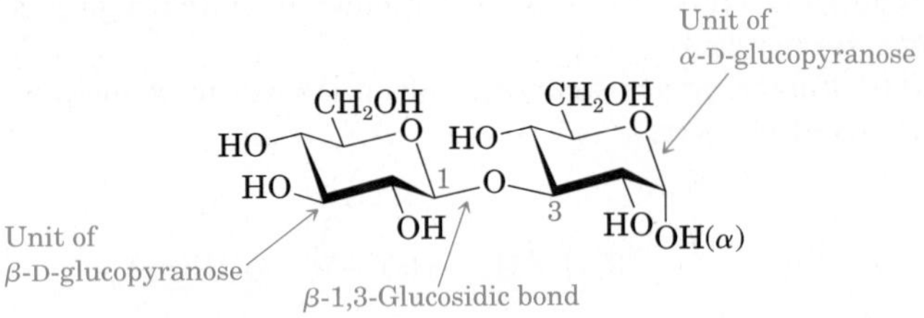

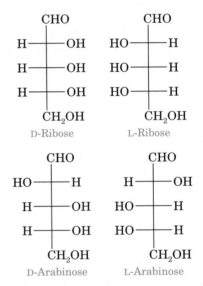

Unit of
α-D-glucopyranose

Unit of
β-D-glucopyranose

β-1,3-Glucosidic bond

11.7 Carbohydrates make up about three-fourths of the dry weight of plants. Less than 1 percent of the body weight of humans is made up of carbohydrates.

11.9 The most abundant D-aldohexose in the biological world is glucose.

11.11 D-glucose

11.13 The designations D and L refer to the configuration of the stereocenter farthest from the aldehyde or ketone group on a monosaccharide chain. When the monosaccharide chain is drawn as a Fischer projection, a D-monosaccharide has the —OH on this carbon on the right; an L monosaccharide has it on the left.

11.15 In D-glucose, carbons 2, 3, 4, and 5 are stereocenters. In D-ribose, carbons 2, 3, and 4 are stereocenters. D-Glucose is one of 16 possible stereoisomers; D-ribose is one of eight possible stereoisomers.

11.17 First draw the D form of each, and then its mirror image.

D-Ribose   L-Ribose

D-Arabinose   L-Arabinose

11.19 Each carbon of a monosaccharide and a disaccharide has an oxygen that is able to participate in hydrogen bonding with water molecules.

11.21 An anomeric carbon is the hemiacetal or acetal carbon in the cyclic form of a monosaccharide. Said another way, it is the carbon that was the carbonyl carbon in the open-chain form of the monosaccharide.

11.23 The designation β means that the —OH group on the anomeric carbon of a cyclic hemiacetal is on the same side of the ring as the terminal —CH₂OH group. The designation α means that it is on the opposite side of the ring from the terminal —CH₂OH group.

11.25 No. The hydroxyl groups on carbon 2, 3, and 4 of α-D-glucose are equatorial, but the hydroxyl group on carbon 1 is axial.

11.27 First compare each Haworth projection with that of β-D-glucose. Compound (a) differs in configuration at carbon 3. Compound (b) differs in configuration at carbons 2 and 3.

(a)

D-Allose

(b)

D-Altrose

11.29 During mutarotation, an α or β form of a carbohydrate is converted to an equilibrium mixture of the two. Mutarotation can be detected by observing the change in optical activity over time as the two forms equilibrate.

11.31 The specific rotation of α-L-glucose changes to $-52.7°$.

11.33 A glycosidic bond is the bond to the —OR group of the cyclic acetal form of any monosaccharide. A glucosidic bond is the bond to the —OR group of a cyclic acetal form of glucose.

11.35 Each aldehyde group is reduced by $NaBH_4$ to a primary alcohol.

(a)

(b)

11.37 In reduction of the ketone group of D-fructose, the —OH group at carbon 2 may be either on the right or the left in the Fischer projection.

D-Fructose → $NaBH_4$ → D-Mannitol + D-Sorbitol

11.39 Review Section 9.8 on keto-enol tautomerism and your answer to Problem 9.67. The intermediate in this conversion is an enediol; that is, it contains a carbon–carbon double bond with two OH groups on it.

D-Glucose 6-phosphate ⇌ An enediol intermediate

⇌ D-Fructose 6-phosphate

11.41 Sucrose contains one unit of D-glucose and one unit of D-fructose. Lactose contains one unit of D-galactose and one unit of D-glucose. Maltose contains two units of D-glucose.

11.43 The anomeric carbons of both monosaccharide units in sucrose are involved in forming the glycosidic bond. In maltose and lactose, one of the monosaccharide units is in the cyclic hemiacetal form, which is in equilibrium with the open-chain form and, therefore, undergoes oxidation.

11.45 (a) Both monosaccharides are units of D-glucose. (b) The glycosidic bond can be described as α-1,1- or β-1,1- depending on which D-glucose unit you take as the reference. (c) Trehalose is not a reducing sugar, because both anomeric carbons participate in forming the glycosidic bond. (d) Trehalose will not undergo mutarotation for the reason given in (c).

11.47 Cellulose, starch, and glycogen are all composed of units of D-glucose. The glycosidic bonds are alpha in starch and glycogen, and beta in cellulose.

11.49 Glycogen is stored in roughly equal amounts in the liver and muscle tissue.

11.51 Within their digestive systems, cows have enzymes that catalyze the hydrolysis of the glycosidic bonds of cellulose. We do not have these enzymes in our digestive systems.

11.53 One way to construct each repeating disaccharide is to draw a disaccharide of two units of D-glucose, and then modify each glucose unit to make it the appropriate unit in alginic acid or pectic acid.

(a)

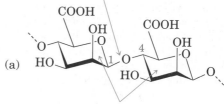

β-1,4-Glycosidic bond

Units of D-mannuronic acid

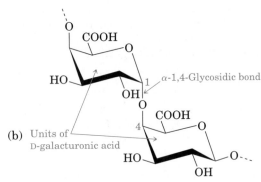

(b) Units of
D-galacturonic acid

11.55 (a) The negative charges are provided by —OSO₃⁻ and —COO⁻ groups. (b) The higher the degree of polymerization, the better the anticoagulant activity.

11.57 The difference is two hydrogens. L-Ascorbic acid is the reduced form of this vitamin; L-dehydroascorbic acid is its oxidized form. The L indicates that each has the L configuration at carbon 6.

11.59 It is used to monitor blood glucose levels in diabetics and pre-diabetics.

11.61 (a) L-Fucose is an L-aldohexose. (b) What is unusual about it in human biochemistry is that it belongs to the L series of monosaccharides, and it has no oxygen atom at carbon 6. (c) If its terminal —CH₃ group were converted to a —CH₂OH group, the monosaccharide formed would be L-galactose.

11.63 High-fructose corn syrup, as its name suggests, is produced from corn syrup by partial enzyme-catalyzed hydrolysis of cornstarch to D-glucose and then partial enzyme-catalyzed isomerization of D-glucose to D-fructose.

11.65 During boiling in water with an acid catalyst, some of the sucrose is hydrolyzed to glucose and fructose, and fructose is sweeter than sucrose.

11.67 (a) The left unit is D-glucuronic acid, which is derived from D-glucose. The right unit is a sulfate ester derived from N-acetyl-D-galactosamine, which is in turn derived from D-galactosamine. (b) The two units are joined by a β-1,3-glycosidic bond.

Chapter 12   Lipids

12.1 It is an ester of glycerol and contains a phosphate group; therefore it is a glycerophospholipid. Besides glycerol and phosphate it has a myristic acid and a linoleic acid component. The other alcohol is serine. Therefore, it belongs to the subgroup of cephalins.

12.3 *Hydrophobic* means water hating. It is important because if the body did not have such molecules there could be no structure since the water would dissolve everything.

12.5 The melting point would increase. This happens because trans double bonds would fit more in the packing of the long hydrophobic tails, creating more order and therefore more interaction between chains. This would require more energy to disrupt, and hence a higher melting point.

12.7 highest: A and B (no double bonds); lowest C and D (two double bonds in each diglyceride)

12.9 (b) because its molecular weight is higher

12.11 lowest (c); then (b); highest (a)

12.13 the more long chain groups, the lower the solubility; lowest (a); then (b); highest (c)

12.15 glycerol, sodium palmitate, sodium stearate, and sodium linolenate

12.17 (a) the unsaturated fatty acid (b) inositol and the phosphate group

12.19 complex lipids and cholesterol

12.21 Phosphatidyl inositol; the inositol has five —OH groups, which can form H-bonds with water.

12.23 (c) because it has three charges, while the others have only two

12.25 No. For example, in red blood cells lecithins are on the outside, facing the plasma; cephalins are on the inside.

12.27 Cholesterol contributes to the fluidity of the membrane because it interrupts the alignments of the saturated fatty acids similar to the cis isomers of the unsaturated fatty acids. Thus there is less London force interaction between the fatty acid tails, contributing to more liquid-like behavior.

12.29 (a) Carbon stereocenters are circled.

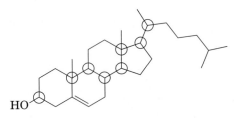

(b) 2⁸ = 256   (c) only 1

12.31 It contains polar groups of phospholipids, cholesterol, and proteins.

12.33 It blocks cholesterol synthesis, thus helping to remove cholesterol from the blood.

12.35 HDL delivers its content by selective lipid uptake that does not involve endocytosis and degradation of the lipoprotein. The lipoprotein never enters the cell but returns to the circulation.

12.37 It increases the glucose and glycogen concentrations in the body. It is also an anti-inflammatory agent.

12.39 Loss of a methyl group at the junction of rings A and B, together with a hydrogen at position 1, produces a double bond. This causes tautomerization of the keto group at C-3, resulting in an aromatic ring A, with a phenolic group at the C-3 position.

12.41 (a) They both have a steroid ring structure.
(b) RU486 has a *para*-aminophenyl group on ring C and a triple-bond group on ring D.

12.43 —OH groups on the A, B, C rings of steroid and the sulfate on taurine.

12.45 (a) PGE₂ has a ring, a ketone group, and two OH groups. (b) The C=O group on PGE₂ is an OH group on PGF₂ₐ.

12.47 COX-2 catalyzes the ring closure between C-8 and C-12 in prostaglandin synthesis. COX-2 produces prostaglandin that take part in inflammation.

12.49 (a) oxidation of the double bond to aldehydes and other compounds
(b) exclude oxygen and sunlight; keep refrigerated

12.51 Soaps are salts of carboxylic acids; detergents are salts of sulfonic acids.

12.53 The transporter is a helical transmembrane protein. The hydrophobic groups on the helices are turned outward and interact with the membrane. The hydrophilic groups of the helices are on the inside and interact with the hydrated chloride ions.

12.55 (a) Sphingomyelin acts as an insulator.
(b) The insulator is degraded, impairing nerve conduction.

12.57 α-D-galactose, β-D-glucose, β-D-glucose

12.59 They prevent ovulation.

12.61 It inhibits prostaglandin formation by preventing ring closure.

12.63 NSAIDS inhibit cyclooxygenases (COX enzymes) that are needed for ring closure. Leukotrienes have no ring in their structure; therefore they are not affected by COX inhibitors.

12.65 (See Figure 12.2.) Polar molecules cannot penetrate the bilayer. They are insoluble in lipids. Nonpolar molecules can interact with the interior of the bilayer (like dissolves like).

12.67 Both groups are derived from a common precursor, PGH_2, which is catalyzed by the COX enzymes.

12.69 Coated pits are concentrations of LDL receptors on the surface of cells. They bind LDL and by endocytosis transfer it inside the cell.

12.71 They are both salts of sulfonic acids.

12.73 apoB-100 protein

12.75 Celebrex inhibits only COX-2 enzymes that cause inflammation, but not COX-1, which is needed to manufacture prostaglandins needed for normal physiological functions.

12.77 Both have the steroid ring structure and the same side chain. Ring A contains one more double bond in prednisolone than in cortisone. In ring C, the carbonyl group of cortisone is reduced to —OH.

Chapter 13   Proteins

13.1

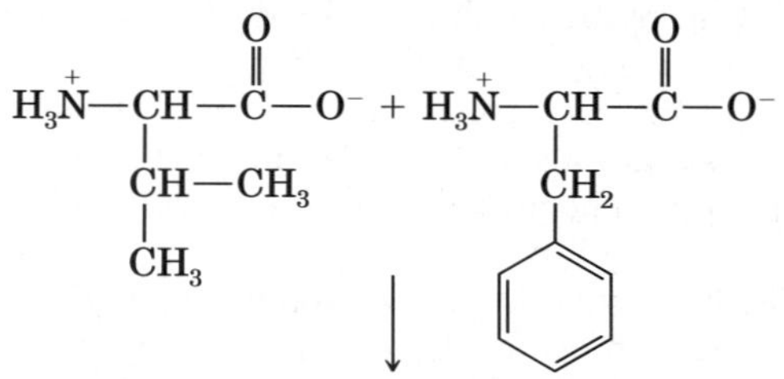

Valine
(Val)

Phenylalanine
(Phe)

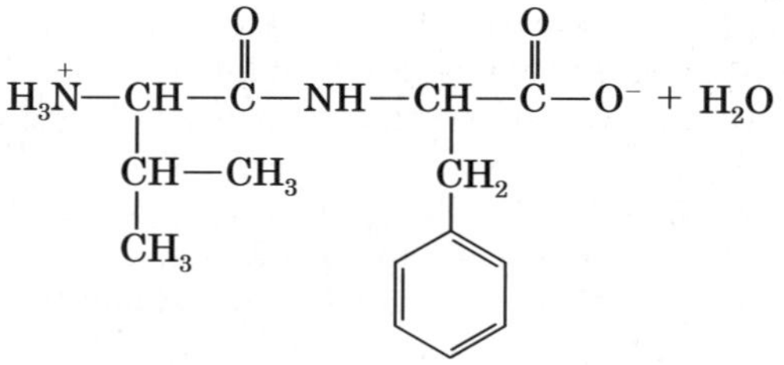

Valylphenylalanine
(Val-Phe)

13.2 salt bridge

13.3 (a) storage   (b) movement

13.5 Tyrosine has an extra —OH group in the side chain. The terminal group of the side chain of tyrosine is a phenol; in phenylalanine, it is a benzene.

13.7 arginine

13.9

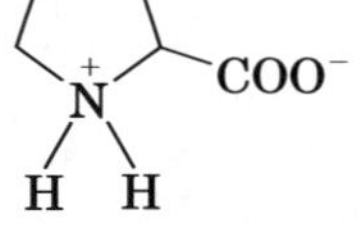

pyrrolidines (saturated pyrroles)

13.11 They supply most of the amino acids we need in our bodies.

13.13 Their structures are the same, except that a hydrogen of alanine is replaced by a phenyl group in phenylalanine.

13.15 These Fischer projections show the relationships between an L-amino acid and an L-monosaccharide and between a D-amino acid and a D-monosaccharide.

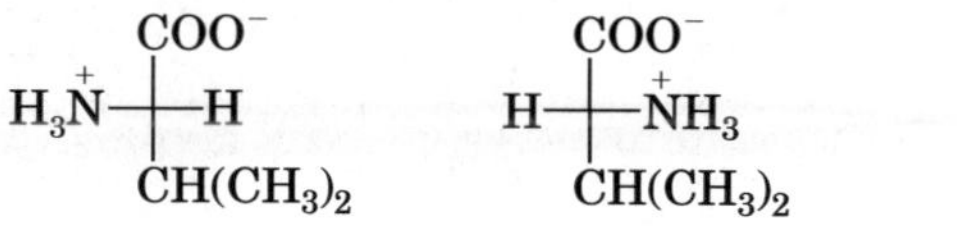

L-Valine   D-Valine

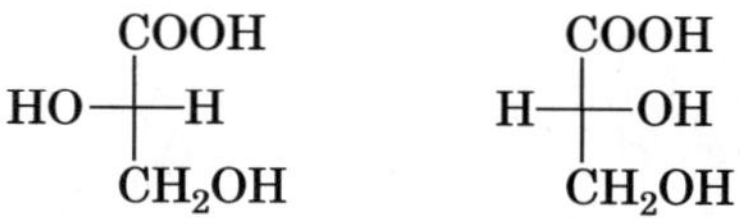

L-Glyceraldehyde   D-Glyceraldehyde

13.17

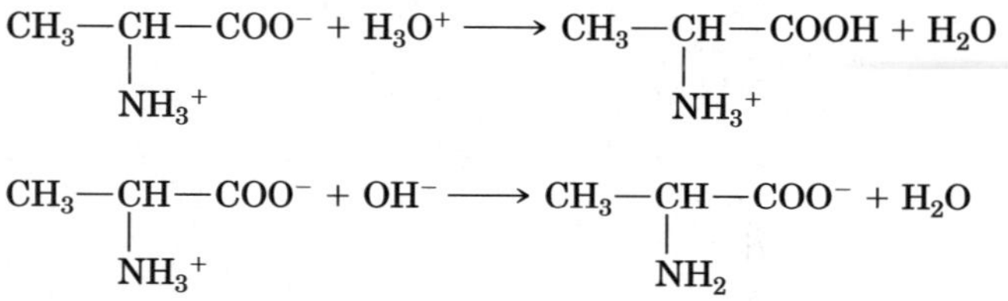

$$CH_3-CH-COO^- + H_3O^+ \longrightarrow CH_3-CH-COOH + H_2O$$
$$\qquad | \qquad\qquad\qquad\qquad\qquad\qquad | $$
$$\quad NH_3^+ \qquad\qquad\qquad\qquad\qquad NH_3^+$$

$$CH_3-CH-COO^- + OH^- \longrightarrow CH_3-CH-COO^- + H_2O$$
$$\qquad | \qquad\qquad\qquad\qquad\qquad\qquad | $$
$$\quad NH_3^+ \qquad\qquad\qquad\qquad\qquad NH_2$$

13.19

at pH1 $CH_3-CH-CH-COOH$ with CH_3 and NH_3^+ groups

at pH12 $CH_3-CH-CH-COO^-$ with CH_3 and NH_2 groups

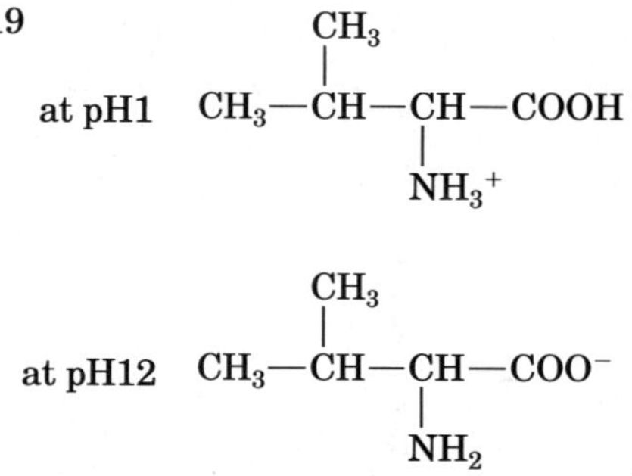

13.21 There are six. One of them is

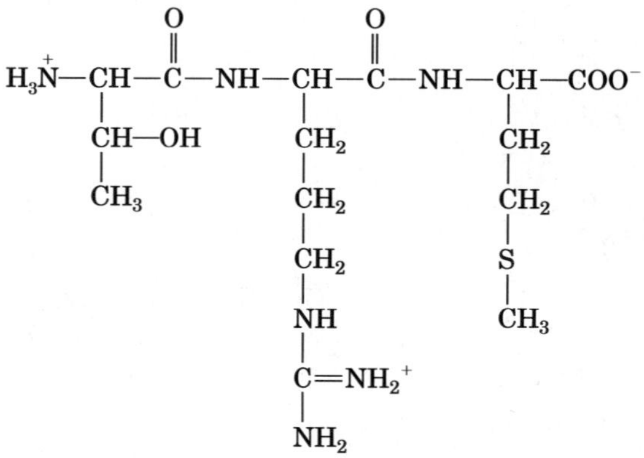

13.23

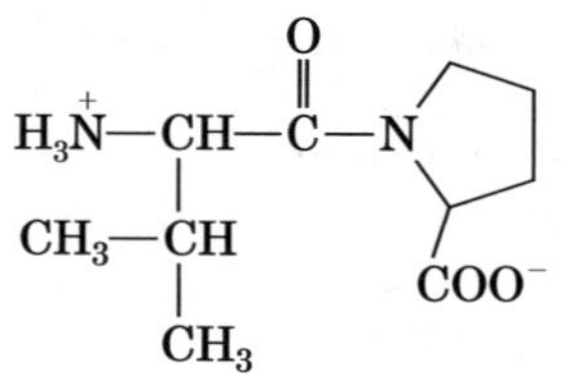

Leu-Pro

13.25

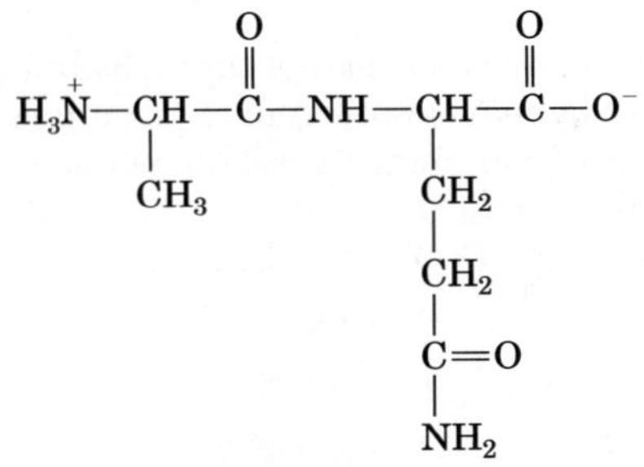

Alanylglutamine
(Ala-Gln)

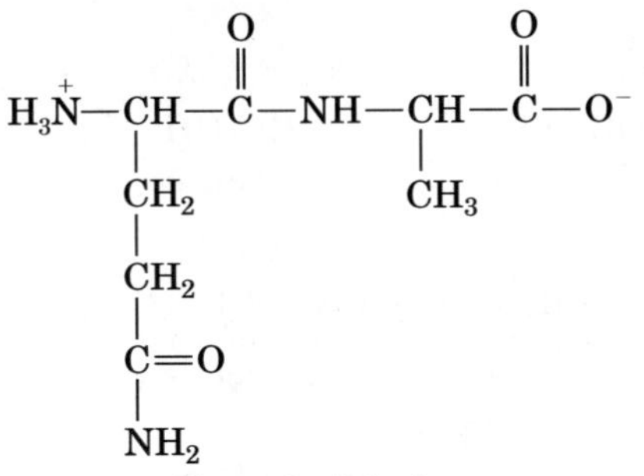

Glutaminylalanine
(Gln-Ala)

13.27 at pH 2.0

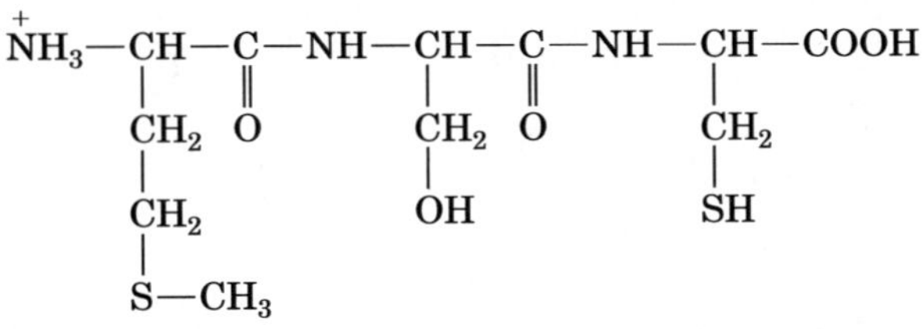

at pH 7.0

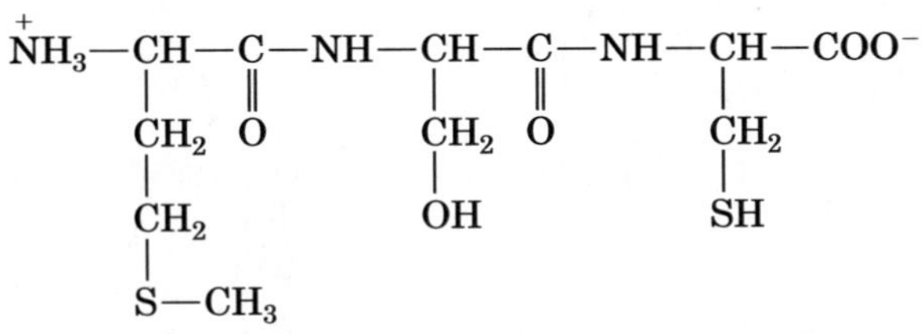

at pH 10.0

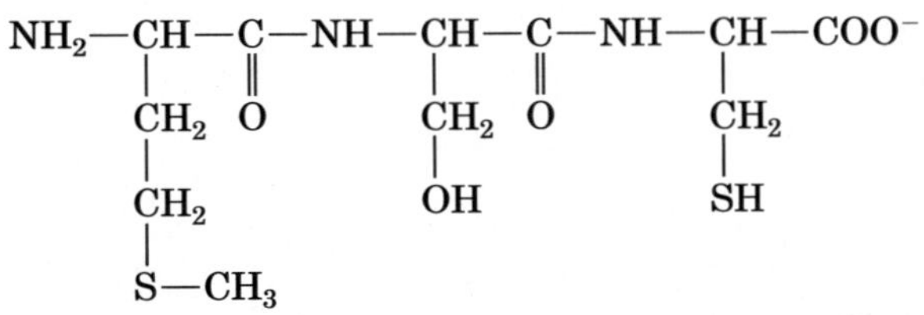

13.29 It would acquire a net positive charge and become more water-soluble.

13.31 (a) 256 (b) $20^4 = 160\,000$

13.33 valine or isoleucine

13.35 (a) quaternary (b) quaternary
(c) quaternary (d) primary

13.37 Above pH 6.0 the COOH groups are converted to COO^- groups. The negative charges repel each other, disrupting the compact α-helix and converting it to a random coil.

13.39 (1) C-terminal end   (2) N-terminal end
(3) pleated sheet   (4) random coil
(5) hydrophobic interaction
(6) disulfide bridge
(7) α-helix   (8) salt bridge
(9) hydrogen bonds

13.41 the intramolecular hydrogen bonds between C=O and —N—H

13.43 on the outside interacting with the lipid bilayer

13.45 The heme and the polypeptide chain form the quaternary structure of cytochrome c. This is a conjugated protein.

13.47

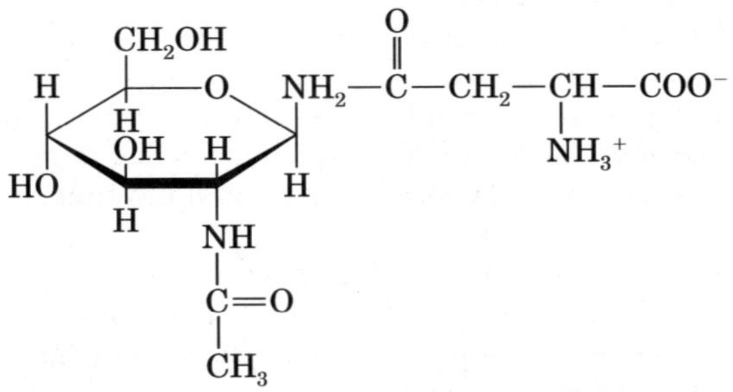

13.49 Collagen is a glycoprotein with O-linked saccharide side chains.

13.51 It breaks the disulfide bridges of keratin and allows the hair to be set into a desired shape.

13.53 to destroy bacteria on the surface of the skin

13.55 In younger people, the AGE products are degraded; in older people, the metabolism slows down and the AGE products accumulate.

13.57 Hydroxyurea therapy promotes the manufacture of fetal hemoglobin, which does not carry beta chains and, therefore, the sickle cell mutations. Cells with fetal hemoglobin do not sickle.

13.59 calcium hydroxyapatite

13.61 The fiberscope carries the laser light in optical fibers in a tiny tube that can be inserted into the body and led to organs of interest. It can repair wounds and seal blood vessels in bleeding ulcers.

13.63 (a) $4^2 = 16$ (b) $20^2 = 400$

13.65 They are made of 8 to 18 β-sheets.

13.67 a quaternary structure, because subunits are cross-linked

13.69 (a) hydrophobic (b) salt bridge
(c) hydrogen bond (d) hydrophobic

13.71 glycine

13.73 one positive charge: NH_3^+

13.75 more soluble because carbohydrates have many OH groups that form hydrogen bonds with water

Chapter 14   Enzymes

14.1 A catalyst is a substance that speeds up a reaction without being used up itself. An enzyme is a catalyst that is a protein or a nucleic acid.

14.3 Yes; lipases are not very specific.

14.5 because enzymes are highly specific, and there are a great many different reactions that must be catalyzed

14.7 In a lyase-catalyzed reaction, the water is either added to a double bond or, if removed from a compound, generates a double bond. Hydrolases, on the other hand, add water to an ester or an amide bond, thereby splitting the molecule into two constituent parts.

14.9 (a) isomerase (b) hydrolase (c) oxidoreductase (d) lyase

14.11 A cofactor is a nonprotein portion of an enzyme. A coenzyme is an organic cofactor.

14.13 See Section 14.3.

14.15 No, at high substrate concentration the enzyme surface is saturated, and doubling of the substrate concentration will produce only a slight increase in the rate of the reaction or no increase at all.

14.17 (a) less active at normal body temperature (b) The activity decreases.

14.19 (a)

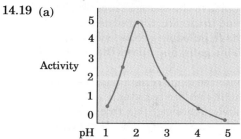

(b) 2 (c) zero activity

14.21 (A) large circle = enzyme
(B) hexagon = active site
(C) triangle = glucose
(D) larger triangle = fluoroglucose
(E) small square = Mg^{2+}
(F) half hexagon = ATP
(G) black circle = Cd^{2+}
(a) assemble ABCEF
(b) ABDEF
(c) ABCEFH

14.23 acid–base reactions

14.25 In competitive inhibition the same maximum rate can be obtained as in the reaction without the inhibitor, although at a higher substrate concentration. In noncompetitive inhibition the maximum rate is always lower than that without the inhibitor.

14.27 an allosteric mechanism

14.29 There is no difference. They are the same.

14.31

$$-NH-CH-C-$$

with side chain CH_2 attached to a para-substituted benzene ring bearing an $O-P(=O)(O^-)(O^-)$ group, and the carbonyl carbon bearing $=O$.

14.33

Phosphorylase b $\rightleftharpoons$ Phosphorylase a

with 2ATP and kinase (forward), ADP product, phosphatase and $2P_i$ (reverse)

14.35 The enzyme is now named for the compound (alanine) that donates the amino group. The former name refers to the products.

14.37 H_4

14.39 digestive enzymes such as Pancreatin or Acro-lase

14.41 Succinylcholine is a competitive inhibitor of acetylcholinesterase and prevents acetylcholine from triggering muscle contractions.

14.43 *Helicobacter pylori* has an enzyme, urease, that catalyzes the hydolysis of urea to ammonia and carbon dioxide. The ammonia is a base and neutralizes the acid in the stomach. In this way the enzymes of the bacteria experience no acidic environment.

14.45 by administering antibiotics

14.47 Alanyl and threonyl residues both provide the same surface environment. In both, the terminal group of the side chain is a nonpolar methyl group. The nature of the interaction is hydrophobic interaction. Glycyl residue also contributes to the hydrophobic interaction.

14.49 Humans do not synthesize their folic acid; they get it in their diets.

14.51

$$-NH-C-COO^-$$

14.53 in hyperthermophile organisms that live in ocean vents

14.55 (a) pasteurized milk, canned tomatoes (b) yogurt, pickles

14.57 His-Leu

14.59 cofactors

14.61 ALT

14.63 No, the nerve gas binds irreversibly by a covalent bond, so it cannot be removed simply by adding substrate.

14.65 a transferase

14.67 They catalyze the self-splicing of introns.

Chapter 15 Chemical Communications: Neurotransmitters and Hormones

15.1 G-protein is an enzyme; it catalyzes the hydrolysis of GTP to GDP. GTP, therefore, is a substrate.

15.3 A chemical messenger operates between cells; secondary messengers signal inside a cell in the cytoplasm.

15.5 The concentration of Ca^{2+} in neurons controls the process. When it reaches $10^{-4}\,M$ the vesicles release the neurotransmitters into the synapse.

15.7 anterior pituitary gland

15.9 Upon binding of acetylcholine, the conformation of the proteins in the receptor changes and the central core of ion channel opens.

15.11 When acetyl choline binds to its receptor, it opens up the ion channels. Thus the concentrations of ions on both sides of the membrane become equalized. The potential (voltage) drops, and the membrane becomes depolarized.

15.13 Taurine is a β-amino acid; its acidic group is —SO$_2$OH instead of —COOH.

15.15 The amino group in GABA is in the gamma position; proteins contain only alpha amino acids.

15.17 (a) norepinephrine and histamine
(b) They activate a secondary messenger, cAMP, inside the cell.
(c) amphetamines and histidine

15.19 It is phosphorylated by an ATP molecule.

15.21

15.23 (a) Amphetamines increase and (b) reserpine decreases the concentration of the adrenergic neuro-transmitter.

15.25 the corresponding aldehyde

15.27 (a) the ion-translocating protein
(b) It gets phosphorylated and changes its shape.
(c) It activates the protein kinase that does the phosphorylation of the ion-translocating protein.

15.29 They are pentapeptides.

15.31 The enzyme is a kinase. The reaction is

I-1, 4-P$_2$

I-1, 4, 5-P$_3$

15.33 Calcium sparks are localized at specific parts of the cell; global calcium waves travel to different parts of the cell.

15.35 100 to 250-fold

15.37 Botulinum toxin prevents the release of acetylcholine from the presynaptic vesticles, preventing neurotransmission.

15.39 They are made of tau proteins. A mutation on tau proteins prevents their normal interaction with micro-tubules, a cytoskeletal component.

15.41 Acetylcholinesterase inhibitors, such as Cognex, inhibit the enzyme that decomposes the neurotransmitter.

15.43 by binding to the dopamine transporter

15.45 The transplanted neurons became integrated into the brain and produce dopamine.

15.47 in the conversion of arginine to citrulline

15.49 —O$_2$S—NH—CO—NH—

15.51 When insulin is adsorbed on its receptor, it elicits the autophosphorylation of the receptor on the cytosolic site. This starts a phosphorylation cascade of a number of compounds, the net result of which is the mobilization of glucose transporters and their movement to the surface of the cell, where they carry the glucose across the membrane.

15.53 Tamoxifen binds to the estrogen receptor. Therefore, estrogen cannot bind to its receptor and cannot reach the DNA in the nucleus to influence protein synthesis.

15.55 It opens the ion gates and allows the translocation of ions.

15.57 Yes. It fits into the same receptor site as enkephalins.

15.59 It is associated with the receptor.

15.61 (a) Both deliver a message to a receptor.
(b) Neurotransmitters act rapidly over short distances. Hormones act more slowly and over longer distances. They are carried by the blood from the source to the target cell.

15.63 Cholera toxin affects G-protein.

15.65 An agonist stimulates the receptor; therefore, more glucose would enter the cells, and its concentration in the serum would decrease.

15.67 The luteinizing hormone is a peptide. It acts through the G-protein-adenylate cyclase cascade, producing secondary messengers. Progesterone is a steroid hormone; it acts in the nucleus of cells, influencing protein synthesis.

Chapter 16   Nucleotides, Nucleic Acids, and Heredity

16.1 Structure of UMP

16.3 hemophilia, sickle-cell anemia, etc.

16.5 (a) base, sugar, phosphate
(b) base, sugar

16.7 Thymine has a methyl group in the 5 position; uracil has a hydrogen there.

16.9 (a)

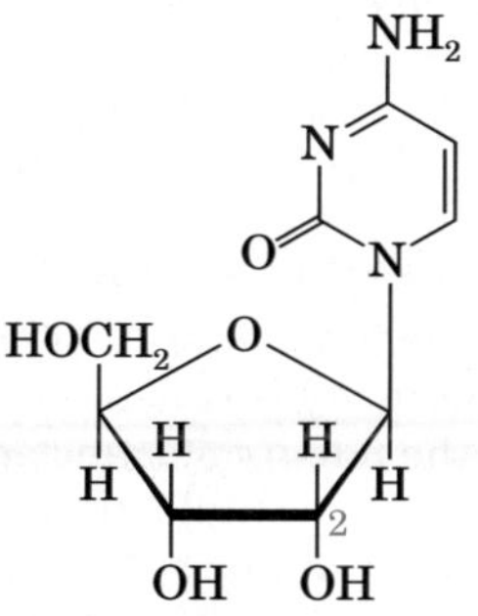

(b) the same, but without the —OH in the 2 position.

16.11 Ribose has an —OH in the 2 position; deoxyribose has an H there.

16.13 C-3 and C-5 to the phosphate group; C-1 to the base

16.15 (a)

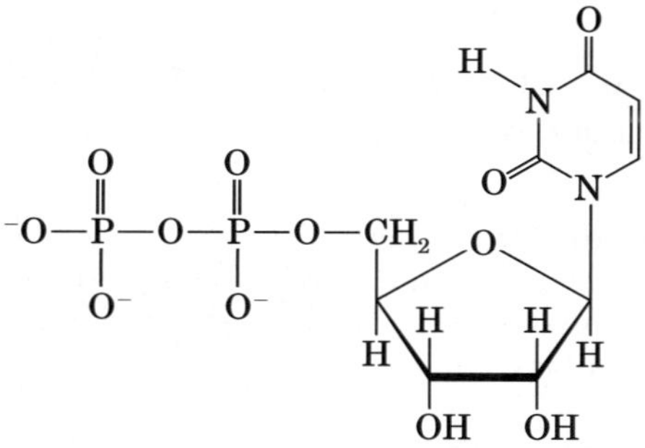

(b)

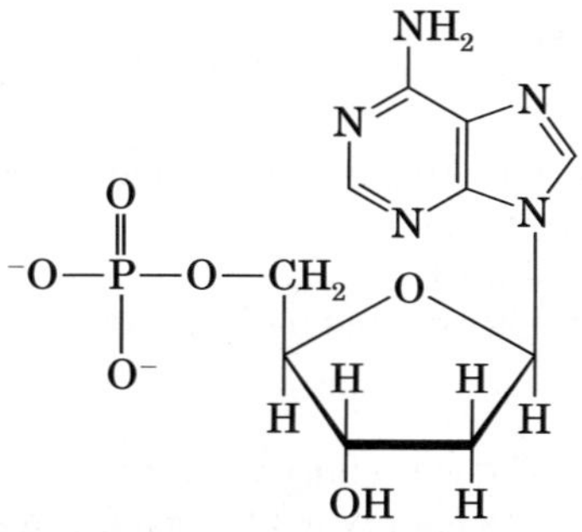

4.17 (a) the 3′—OH end and the 5′—OH end

(b) A is the 5′—OH end. C is the 3′—OH end.

16.19 two

16.21 It is an electrostatic attraction.

16.23 The superstructure of chromosome is made of chromatin fibers, in which the fibers are folded into loops and the loops are further folded into bands.

16.25 rRNA

16.27 mRNA

16.29 Ribozymes in general shorten preexisting RNA molecules: from mRNA to mature RNA, from precursor tRNA to mature tRNA.

16.31 mRNA contains exons and introns when it is first synthesized, but before it actually functions the introns are cut out.

16.33 No; satellites are noncoding regions, repeated nucleotide sequences.

16.35 amino and carbonyl groups

16.37

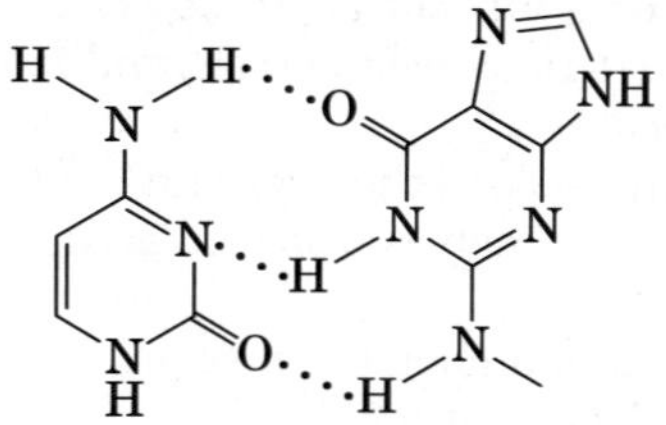

16.39 The replication fork is where the replication starts. There may be hundreds of replication forks on a human chromosome at any time.

16.41 The DNA is wrapped around the histones in the nucleosome. The negative charges on the DNA interact strongly with the positive charges of the histones, making for tight packing. The removal of some of the positive charges loosens the electrostatic interactions.

16.43 Six subunits form a ring with a hollow core through which a single strand of DNA can move.

16.45 No. The sugar in the primer must be ribose. Only ATP, GTP, CTP, and UTP can serve as source for a primer.

16.47 polymerases

16.49 phosphodiester

16.51 If only a few G residues are damaged, the likely mechanism is nucleotide excision repair (NER).

16.53 Uracil is a naturally occurring base only when its sugar is ribose. At the damaged DNA site, the uracil is linked to a deoxyribose.

16.55 AP sites are apurinic or apyrimidinic sites—that is, the sites from which the purine or pyrimidine bases have been excised. The enzyme glycosylase generates AP sites.

16.57 The polymerases from thermophile bacteria are heat resistant, so they can work at the high temperatures that are needed for the unwinding of the DNA double helix.

16.59

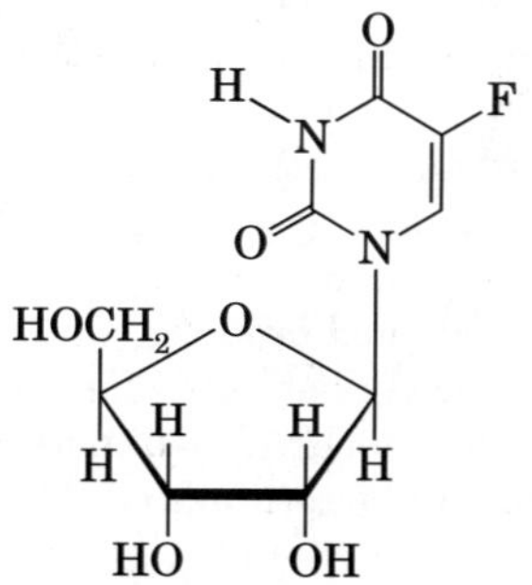

16.61 TTAGGG

16.63 Telomerase resynthesizes the shortened DNA end in telomers; thus, the cell has no signal telling it when to stop dividing and therefore never dies.

16.65 The DNA fragments of the child, mother, and alleged father are run on the same gel, and their patterns are compared.

16.67 One can screen patients for their individual genetic make-up with respect to cytochrome P-450. In this way it can be established whether the patient is an extensive, ultraextensive, or poor metabolizer of a drug even before the drug is administered.

16.69 the DNA fragments in multiple of 180 base pairs

16.71 Every time an error is introduced, it may cause a mutation that can be harmful or even detrimental to the propagation of the species.
16.73 the glycosidic bond between the sugar and the base, and the phosphate ester bond between the sugar and the phosphate
16.75 DNA; it contains up to millions of base pairs.
16.77 T = 29.3; G = 20.7; C = 20.7

Chapter 17   Gene Expression and Protein Synthesis

17.1 First, binding proteins must make the portion of the chromosome where the gene is less condensed and more accessible. Second, the helicase enzyme must unwind the double helix near the gene. Third, the polymerase must recognize the initiation signal on the gene.
17.2 (a) CAU and CAC   (b) GCA and GTG
17.3 valine + ATP + tRNA$_{val}$
17.4 CCT   CGATTG
      GGAGC   TAAC
17.5 (c)
17.7 in the cytoplasm
17.9 to unwind the DNA double helix
17.11 5′ end
17.13 on the N atom of guanine
17.15 on the mRNA
17.17 (a) CGA   (b) alanine
17.19 Because all the diverse life forms on Earth share the same genetic code, they must have evolved from a common ancestry.
17.21 The A site is where the incoming tRNA carrying the amino acid binds. The P site is where the growing peptide chain binds. The E site is where the exiting tRNA binds before being recycled.
17.23 They enable the tRNA to bind to the A site of the rRNA.
17.25 Thousands of nucleotides upstream from the transcription site. By interacting with the promoter, they speed up the transcription process.
17.27 to degrade misfolded proteins
17.29 (a) UAU and UAC; both code for tyrosine.
(b) If UAU is changed to UAA, then instead of producing tyrosine, the chain will be terminated. This might well be fatal.
17.31 Yes; if the mutation occurred on a recessive gene.
17.33 Restriction endonucleases cleave DNA at specific sites.
17.35 In genetic engineering, a new gene is inserted into the genome. However, in the corn that acquired insect resistance by natural selection, mutations occurred in the native genes.
17.37 the —S—P—O— bonding
17.39 T (lymphocyte) cells
17.41 The promoter enabled the HIV protease to be expressed in the lens of the mouse. Thus expressed, the lens became cataractous. In this manner, it could serve as a bioassay to prove the efficacy of new HIV protease inhibitor drugs.
17.43 A change from guanine to thymine in the gene results in a valine in place of a glycine.

17.45 When p53 binds to DNA, it arrests the cell cycle; thus it allows more time for the cell defenses to repair damaged DNA.
17.47 (a) The 5′ end and the section just before the 3′ end must be base paired.   (b) There must be no base pairing in the anticodon loop.
17.49 The aminoacyl-tRNA synthtase is specific for one amino acid and all the possible tRNA molecules carrying anticodons. This specificity is called the second genetic code because it ensures the fidelity of translation.
17.51 *Degenerate* means multiple; that is, more than one codon specifies the same amino acid. There is a certain amount of redundancy in the genetic code.
17.53 The third base is not irrelevant for phe, ile, met, tyr, his, gln, asn, lys, asp, glu, cys, trp, ser, and arg. In eight cases, the third base is irrelevant.
17.55 No. It would produce too many small fragments and no sticky ends.

Chapter 18   Bioenergetics: How the Body Converts Food to Energy

18.1 ATP
18.3 (a) 2   (b) the outer membrane
18.5 Christae are folded membranes originating from the inner membrane. They are connected to the inner membrane by tubular channels.
18.7 two
18.9 Neither; they yield the same energy.
18.11 phosphate ester
18.13 the 2 N atoms that are part of C=N bonds
18.15 (a) adenosine   (b) NAD$^+$ and FAD   (c) acetyl groups
18.17 amide
18.19 No. The pantothenic acid portion is not the active part.
18.21 acetyl coenzyme A
18.23 α-ketoglutarate
18.25 Succinate is oxidized by FAD, and the oxidation product is fumarate.
18.27 lyase
18.29 No, but GTP is produced in step **5**.
18.31 It allows the energy to be released in small packets.
18.33 *cis*-aconitate and fumarate
18.35 to NAD$^+$, which becomes NADH + H$^+$
18.37 cytochrome C and CoQ
18.39 When H$^+$ passes through the ion channel, the proteins of the channel rotate. The kinetic energy of this rotatory motion is converted to and stored as the chemical energy in ATP.
18.41 in the inner membranes of the mitochondria
18.43 (a) 0.5   (b) 12
18.45 the proton-translocating ATPase
18.47 F$_1$; it catalyzes the conversion of ADP to ATP.
18.49 1.5 kcal
18.51 (a) by sliding the thick filaments (myosin) and the thin filaments (actin) past each other   (b) from the hydrolysis of ATP
18.53 ATP transfers a phosphate group to the serine residue at the active site of the enzyme.
18.55 Sulfur, in the Fe-S clusters, is embedded in a protective protein coat.

18.57 No. It would harm humans because they would not synthesize enough ATP molecules.

18.59 (a) hydroxylation   (b) the air we breathe in

18.61 60 g or 1 mol of CH_3COOH

18.63 The energy of motion appears first in the ion channel where, upon passage of H^+, the proteins making this channel rotate.

18.65 They are both hydroxy acids.

18.67 myosin

18.69 two

18.71 F_0; it is made of 12 subunits.

18.73 No, it largely comes from the chemical energy as a result of the breaking of bonds in the O_2 molecule.

18.75 It removes two hydrogens from succinate to produce fumarate.

Chapter 19   Specific Catabolic Pathways: Carbohydrate, Lipid, and Protein Metabolism

19.1 According to Table 19.2 the ATP yield from stearic acid (C_{18}) is 146 ATP. This makes 146/18 = 8.1 ATP/carbon atom. For lauric acid (C_{12}):

Step *1* Activation	−2 ATP
Step *2* Dehydrogenation five times	10 ATP
Step *3* Dehydrogenation five times	15 ATP
Six C_2 fragments in common pathway	72 ATP
Total	95 ATP

95/12 = 7.9 ATP per carbon atom for lauric acid. Thus stearic acid yields more ATP/C atom.

19.3 They serve as building blocks for the synthesis of proteins.

19.5 The two C_3 fragments are in equilibrium. As the glyceraldehyde phosphate is used up, the equilibrium shifts and converts the other C_3 fragment (dihydroxyacetone phosphate) to glyceraldehyde phosphate.

19.7 (a) steps *1* and *3*   (b) steps *6* and *9*

19.9 At step *9*; the pyruvate kinase; it is a feedback regulation

19.11 NADPH

19.13 6 moles

19.15 12

19.17 Two net ATP molecules are produced in both cases.

19.19 glycerol kinase

19.21 (a) thiokinase and thiolase   (b) —SH   (c) Both enzymes insert a CoA—SH into a compound.

19.23 $CH_3(CH_2)_4CO$—CoA, three acetyl CoA, three $FADH_2$, three NADH + H^+

19.25 112

19.27 carbohydrates

19.29 (a) reduction   (b) decarboxylation

19.31 It enters the citric acid cycle.

19.33

(a) CH_3—CH—COO^- + NAD^+ + H_2O $\longrightarrow$
               |
              NH_3^+

                    CH_3—C—COO^- + NADH + H^+ + NH_4^+
                            ‖
                            O

19.35 because pyruvate can be converted to glucose when the body needs it

19.37 fumarate

19.39 (a) by reductive amidation, which is the reverse of oxidative deamidation   (b) by converting glutamate to glutamine

19.41 It is stored in ferritin and reused.

19.43 lactic acid accumulation

19.45 the bicarbonate/carbonic acid buffer

19.47 You can find out what kind of lipids are in the gallstone extract and what their metabolic products are in a single analysis.

19.49 Glycine, the C-terminal amino acid of ubiquitin, forms an amide linkage with a lysine residue in the target protein.

19.51

Phenylalanine   α-Ketoglutarate

Phenylpyruvate   Glutamate

19.53 diabetes

19.55 (a) in steps *5* and *12*
(b) in steps *10* and *11*
(c) If enough O_2 is present in muscles, a net increase of NADH + H^+ occurs. If not enough O_2 is present (as in yeast), there is no net increase of either NAD^+ or NADH + H^+.

19.57 Yes, they are glucogenic and can be converted to glucose.

19.59 CH_3—C—COO^- COO^-
              ‖   + |
              O   CH—NH_3^+
                            |
                            CH_2
                            |
                            COO^-

19.61 in the carbon dioxide exhaled by the animal

19.63 It is energy consuming. Three ATP molecules are hydrolyzed per cycle.

19.65 (a) 5   (b) 7

Chapter 20   Biosynthetic Pathways

20.1 for flexibility and to overcome unfavorable equilibria

20.3 because the presence of a large inorganic phosphate pool would shift the reaction to the degradation process such that no substantial amount of glycogen would be synthesized

20.5 Photosynthesis is the reverse of respiration.

20.7 (a) pyruvate (b) oxaloacetate (c) alanine

20.9 Gluconeogenesis, because the other pathways metabolize glucose, and only gluconeogenesis manufactures it.

20.11 UDP-glucose + glucose = maltose

20.13 uracil, ribose, and three phosphates

20.15 (a) the cytoplasm (b) no

20.17 malonyl ACP

20.19 malonyl ACP

20.21 It is an oxidation step because hydrogen is removed from the substrate. The oxidizing agent is O_2. NADPH is also oxidized during this step.

20.23 NADPH is bulkier than NADH; it also has two more negative charges.

20.25 No, the body makes other unsaturated fatty acids such as oleic acid and arachidonic acid.

20.27 palmitoyl CoA, serine, acyl CoA, UDP-glucose

20.29 $3C_2 \longrightarrow C_6 \longrightarrow C_5 + C$

20.31 aspartic acid

20.33 valine and α-ketoglutarate

20.35 NADPH

20.37 Indigo blue, an oxidation product of tryptophan. Indigo blue cannot enter the cells and ends up in the urine.

20.39 malonyl ACP

20.41 No, the cancer is caused when Ras is mutated, but prenylation is a precondition for the activity of RAS.

20.43 When we are exposed to cold, energy is needed in the form of heat for survival. α-ketoglutarate is part of the citric acid cycle, which produces energy. Thus α-ketoglutarate is used up and the equilibrium shifts to the left, deamidating glutamic acid. This produces α-ketoglutaric acid, replacing that used up in the citric acid cycle.

20.45 geranyl- and farnesylpyrophosphate

20.47 The statement is true. Carbohydrate is synthesized by reducing CO_2.

20.49 No, it can add C_2 fragments to a growing chain up to C_{16}. After that, another enzyme takes over.

Chapter 21 Nutrition and Digestion

21.1 No, they differ according to body weight, age, occupation, and sex.

21.3 Only the calcium propionate. When it is metabolized, it yields energy.

21.5 the section at the bottom that gives daily caloric needs

21.7 It is necessary for proper operation of the digestive system, and a lack of it may lead to colon cancer.

21.9 for daily activities involving motion

21.11 1830 Cal/day

21.13 No, the loss of water is only temporary and the weight is regained very quickly.

21.15 different oligosaccharides of α-D-glucose

21.17 No; maltose cannot be absorbed into the blood; it is hydrolyzed to glucose before being absorbed.

21.19 linoleic acid

21.21 neither

21.23 Yes, but one must eat a wide range of vegetables and cereals.

21.25 Both catalyze the hydrolysis of amide bonds. Acid catalyzes random hydrolysis, and trypsin catalyzes the hydrolysis on the C=O side of arginine and lysine.

21.27 dietary deficiency diseases, since rice cannot supply the essential amino acids lysine or threonine or the essential fatty acids

21.29 to prevent scurvy

21.31 It assists in blood clotting.

21.33 scurvy

21.35 They act as antioxidants in the body.

21.37 biotin, thiamine

21.39 If glucose supplied all caloric needs, it would cause deterioration of the patient.

21.41 No, these are two different dipeptides. In aspartame, aspartic acid is at the N-terminal; in phenylalanylaspartic acid, it is at the C-terminal.

21.43 Both are derivatives of disaccharides.

21.45 chlorine and ozone

21.47 niacinamide

21.49 hydrolysis of acetal, ester, and amide linkages

21.51 No, amylase does not have branches with 1,6-glycosidic linkage.

21.53 The body would treat it as a food protein and digest it (hydrolyze it) before it could get into the blood serum.

21.55 Discriminatory curtailment diet; no, aspartame contains phenylalanine.

21.57 No. Arsenic is not an essential nutrient.

Chapter 22 Immunochemistry

22.1 skin and mucin (or tear)

22.3 The skin fights bacteria by secreting lactic and fatty acids, thereby lowering the pH.

22.5 Acquired immunity is selective, and it has memory.

22.7 T cells mature and differentiate in the thymus gland, B cells in the bone marrow.

22.9 Cancer cells and virus-infected cells; NO molecules kill them.

22.11 only peptide antigens

22.13 The polypeptide antigen is digested by proteolytic enzymes and becomes small peptides. These peptides can serve as targets for the class II MHC bringing them to the surface of the cell.

22.15 They belong to the immunoglobulin superfamily. They are found in the membranes of cells.

22.17 IgA are in secretions. They are the largest, with molecular weights of 200,000 to 700,000. IgE are involved in allergic reactions; their molecular weight is 190,000. IgG are the most important antibodies in blood; their molecular weight is 150,000.

22.19

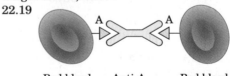

Red blood cell   Anti-A antibody   Red blood cell

22.21 by disulfide bridges

22.23 Two monoclonal antibodies isolated from the same population of lymphocytes would differ in their V region. They would interact with different antigens. They would probably have the same C (constant) region.

22.25 Exons from three regions, V, J, and D, of the variable H gene are selected and combine to form a new gene. The transcription of this new gene yields mRNA that is translated to a new protein.
22.27 TcR receptor complex contains at least two more proteins besides the TcR receptor—for example, the CD3 and CD4 molecules, which are also members of the immunoglobulin superfamily.
22.29 TcR and coreceptors such as CD3 and CD4 or CD8
22.31 CD4 adhesion molecule
22.33 Cytokines are small glycoprotein molecules.
22.35 (a) tumor necrosis factor   (b) interleukin
(c) epidermal growth factor
22.37 They have four cysteine residues. Cysteine residue number 1 is linked to residue number 3 by disulfide linkage. Cysteine number 2 is linked to cysteine number 4.
22.39 cysteine
22.41 In malignant tumor cells, the epitopes that signal healthy cells diminish greatly, and unusual epitopes are presented on the surface of these cells.
22.43 The inhibitory receptor on their surface recognizes the epitope of a normal cell, binds to it, and prevents the activation of the killer cell or macrophage.
22.45 Glucocorticoids regulate the synthesis of TcR-cytokines directly by interacting with their genes or indirectly through transcription factors.
22.47 Monoclonal antibody treatment is only mildly toxic. It specifically attacks the lymphoma cells. Chemotherapy kills many other cells besides the quickly reproducing cancer cells. Therefore, it is highly toxic.
22.49 Plasma cells are differentiated B cells, and memory cells are differentiated T cells.
22.51 The chemokine, when it interacts with the surface of a leukocyte, leads to alternative polymerization and breakdown of actin inside the cells. This allows the cell to change its shape. The leukocyte can then extend leg-like extrusions for moving along the endothelium and can flatten out. The endothelial cells can change their shape and develop gaps among the tight junctions. This process enables the flattened cells to pass through the gaps.
22.53 Antibody is manufactured and secreted by B cells. They are the only cells that manufacture soluble immunoglobulins.
22.55 in the thoracic duct
22.57 Cytokines carry signals from cell to cell to induce proliferation of leukocytes.
22.59 Carbohydrate cell surface marker; epothilon is a synthetic analogue of such marker.
22.61 on many cells, not just tumor cells

Chapter 23   Body Fluids

23.1 the limited exchange of components between blood and the interstitial fluid of brain

23.3 (a) yes   (b) through semipermeable membranes
23.5 They carry oxygen to the cells and carbon dioxide away from the cells.
23.7 See Sections 23.1 and 23.2.
23.9 By centrifugation; the cellular components settle and the supernatant is the plasma.
23.11 globulins
23.13 four
23.15 Fe^{2+}; percent of saturation is the percentage of oxygen carrying Fe^{2+} out of the total iron.
23.17 Each heme has a cooperative effect on the other hemes.
23.19 (a) at the terminal NH_2 groups of the four polypeptide chains   (b) carbaminohemoglobin
23.21 $K = 10$ (significant figures!)
23.23 in the proximal tubule
23.25 inorganic ions
23.27 The H^+ ion from cell metabolism goes to the blood, where it reacts with bicarbonate ion to form H_2CO_3. In the lungs, carbonic anhydrase catalyzes its conversion to CO_2 and water. CO_2 is exhaled. To avoid the loss of bicarbonate ion from blood, CO_2 is reabsorbed into the cells of the distal tubule. There it reacts with water to form carbonic acid, which dissociates and gives bicarbonate and H^+ ion. The latter is filtered out into the urine and gives an acidic pH. In this manner, the H^+ of metabolism ends up in the urine.
23.29 the distal tubule and the collecting tubule
23.31 The specific gravity of urine becomes lower because the urine is more diluted with water.
23.33 No glucose reaches the tissues or the cells. Because 12 mm Hg is lower than the osmotic pressure, nutrients flow from the interstitial fluid into the blood.
23.35 The inhibition of ACE stops the production of the vasoconstriction.
23.37 collagen
23.39 platelets
23.41 restricted flow of urine
23.43 By blocking calcium channels, heart muscles contract less frequently, thereby reducing blood pressure.
23.45 The concentration is higher. Because fibrinogen is removed when plasma becomes serum, the albumin is dissolved in a smaller amount of material, and its concentration is higher.
23.47 Less oxygen is carried; more is released to the tissues.
23.49 No; it is too large of a molecule.
23.51 vasopressin
23.53 Where the two circulations meet, the oxygen pressure is the same. However, the saturation of fetal hemoglobin is higher than that of adult hemoglobin. Thus, at the same oxygen pressure, the fetal hemoglobin can carry more oxygen than the adult hemoglobin. Therefore, some of the oxygen will be transferred from the adult hemoglobin to the fetal hemoglobin when the two circulations meet.

A site *(Section 17.5)* The site on the large ribosomal subunit where the incoming tRNA molecule binds.

Acetal *(Section 9.7)* A molecule containing two —OR groups bonded to the same carbon.

Acetyl coenzyme A *(Section 18.3)* A biomolecule in which an acetyl group is bonded to the —SH group of coenzyme A by a thioester bond.

Achiral *(Section 6.2)* An object that lacks chirality; an object that is superposable on its mirror image.

Acid *(Section 7.1)* An Arrhenius acid is a substance that ionizes in aqueous solution to give a hydronium ion, H_3O^+, and an anion *(Section 7.3)*. A Brønsted-Lowry acid is a proton donor.

Acid ionization constant (K_a) *(Section 7.5)* An equilibrium constant for the ionization of an acid in aqueous solution to H_3O^+ and its conjugate base. K_a is also called an acid dissociation constant.

Acidic salt *(Section 7.9)* The salt of a strong acid and a weak base; an aqueous solution of an acidic salt is basic.

Acidosis *(Chemical Connections 7D)* A condition in which the pH of blood is lower than 7.35.

ACP *(Section 20.3)* Acyl carrier protein, a large molecule that activates a fatty acid for fatty acid synthesis. The fatty acid is bonded to ACP as a thioester.

Acquired immunity *(Section 22.1)* The second line of defense that vertebrates have against invading organisms.

Activation, of an amino acid *(Section 17.5)* The process by which an amino acid is bonded to an AMP molecule and then bonded to the 3'—OH of a tRNA molecule.

Activation, of enzymes *(Section 14.3)* Any process by which an inactive enzyme is transformed into an active enzyme.

Active site *(Section 14.5; Chemical Connections 22C)* A three-dimensional cavity of the enzyme with specific chemical properties to accommodate the substrate.

Acyl group *(Section 10.3A)* An RCO— group.

Adaptive immunity *(Section 22.1)* Acquired immunity with specificity and memory.

Adhesion molecules *(Section 22.5)* Various protein molecules that help to bind an antigen to the T-cell receptor.

Adrenergic neurotransmitter *(Section 15.4)* A monoamine neurotransmitter/hormone, the most common of which are epinephrine (adrenalin), serotonin, histamine, and dopamine.

Adrenocorticoid hormone *(Section 12.10)* A hormone produced by the adrenal glands. Adrenocorticoid hormones regulate the concentrations of ions and carbohydrates.

Agonist *(Section 15.1)* A molecule that mimics the structure of a natural neurotransmitter/hormone, binds to the same receptor, and elicits the same response.

AIDS *(Section 22.5)* Acquired immune deficiency syndrome. The disease caused by the human immunodeficiency virus, which attacks and depletes T cells.

Alcohol *(Section 1.4A)* A compound containing an —OH (hydroxyl) group bonded to a tetrahedral carbon atom.

Aldehyde *(Section 1.4C)* A compound containing a carbonyl group bonded to a hydrogen; a —CHO group.

Alditol *(Section 11.4B)* The product formed when the CHO group of a monosaccharide is reduced to a CH_2OH group.

Aldose *(Section 11.2A)* A monosaccharide containing an aldehyde group.

Aliphatic amine *(Section 8.2)* An amine in which nitrogen is bonded only to alkyl groups.

Aliphatic hydrocarbon *(Section 2.2)* An alkane.

Alkali *(Chemical Connections 7B)* An older term for a base.

Alkaloid *(Chemical Connections 8B)* A basic nitrogen-containing compound of plant origin, many of which have physiological activity when administered to humans.

Alkalosis *(Chemical Connections 7D)* A condition in which the pH of blood is greater than 7.45.

Alkane *(Section 2.2)* A saturated hydrocarbon whose carbon atoms are arranged in an open chain.

Alkene *(Section 3.2)* An unsaturated hydrocarbon that contains a carbon–carbon double bond.

Alkyl group *(Section 14.4A)* A group derived by removing a hydrogen from an alkane; given the symbol R—.

Alkyne *(Section 3.2)* An unsaturated hydrocarbon that contains a carbon–carbon triple bond.

Allosterism *(Section 14.6)* An enzyme regulation in which the binding of a regulator on one site on the enzyme modifies the enzyme's ability to bind the substrate in the active site. Allosteric enzymes often have multiple polypeptide chains with the possibility of chemical communication between the chains.

Alpha (α-) amino acid *(Section 13.2)* An amino acid in which the amino group is linked to the carbon atom next to the —COOH carbon.

Alpha helix *(Section 13.8)* A repeating secondary structure of a protein where the chain adopts a helical conformation and the structure is held together by hydrogen bonds from the peptide backbone N—H to the backbone C=O four amino acids farther up the chain.

Amide *(Section 10.3C)* A compound in which a carbonyl group is bonded to a nitrogen atom, $RCONR'_2$; one or both of the R' groups may be H, alkyl, or aryl groups.

Amino acid *(Section 13.2)* An organic compound containing an amino group and a carboxyl group.

Amino acid neurotransmitter *(Section 15.5)* A class of neurotransmitter/hormones that are amino acids.

Amino acid pool *(Section 19.1)* The name used for the free amino acids found both inside and outside cells throughout the body.

Amino group *(Section 1.4B)* An —NH_2 group.

Amino sugar *(Section 11.2D)* A monosaccharide in which an —OH group is replaced by an —NH_2 group.

Amphiprotic *(Section 7.3)* A substance that can act as either an acid or a base.

Amphoteric *(Section 7.3)* An alternative for amphiprotic.

Anabolism *(Section 18.1)* The pathways by which biomolecules are synthesized.

Anaerobic pathway *(Section 19.2)* The pathway that generates lactic acid from pyruvate when the oxygen supply is too low for the pyruvate to be oxidized aerobically through the common pathway.

Angiotensin *(Section 23.8)* A hormone that is a vasoconstrictor and raises the blood pressure.

Anhydride *(Section 10.3A)* A compound in which two carbonyl groups are bonded to the same oxygen atom; RCO—O—COR′.

Anomeric carbon *(Section 11.3A)* The hemiacetal carbon of the cyclic form of a monosaccharide.

Anomers *(Section 11.3A)* Monosaccharides that differ in configuration only at their anomeric carbons.

Antacid *(Chemical Connections 7C)* A substance taken orally to neutralize stomach acidity.

Antagonist *(Section 15.1)* A molecule that binds to a neurotransmitter receptor but does not elicit the natural response.

Antibody *(Section 22.4)* A defense glycoprotein synthesized by the immune system of vertebrates that interacts with an antigen. It is also called an immunoglobulin.

Anticoagulant *(Chemical Connections 23B)* A drug given to prevent blood clot formation.

Anticodon *(Section 17.3)* A sequence of three nucleotides on tRNA complementary to the codon in mRNA.

Antidiuretic *(Section 23.7)* Any agent that reduces the volume of urine produced.

Antigen *(Section 22.1)* A substance foreign to the body that triggers an immune response.

Apoenzyme *(Section 14.3)* The protein portion of an enzyme that has cofactors or prosthetic groups.

Apoptosis *(Chemical Connections 16E)* Planned cell death characterized by small numbers of cells dying at a time and being cleaved into apoptopic bodies containing organelles and large nuclear fragments.

Ar— *(Section 4.1)* The symbol used for an aryl group.

Arene *(Section 4.1)* A compound containing one or more benzene rings.

Aromatic amine *(Section 8.2)* An amine in which nitrogen is bonded to one or more aromatic rings.

Aromatic compound *(Section 4.1)* A term used to classify benzene and its derivatives.

Aryl group *(Section 4.1)* A group derived from an arene by removal of a hydrogen.

ATP *(Section 18.1)* Adenosine triphosphate.

Autoxidation *(Section 4.5C)* The reaction of a C—H group with oxygen, O_2, to form a hydroperoxide, R—OOH.

Axial position *(Section 2.6B)* A position on a chair conformation of a cyclohexane ring that extends from the ring parallel to the imaginary axis of the ring.

Axon *(Section 15.2)* The long part of a nerve cell that comes out of the main cell body and eventually connects with another nerve cell or a tissue cell.

B cell *(Section 22.2)* A type of lymphocyte that is produced in and matures in the bone marrow. B cells produce antibody molecules.

Baroreceptors *(Section 23.8)* The receptors in the neck that detect a drop in blood pressure and send a signal to the heart to pump harder and to the muscles around the blood vessels to contract to raise blood pressure.

Basal caloric requirement *(Section 21.3)* The caloric requirement for an individual at rest, usually given in Cal/day.

Base *(Section 7.1)* An Arrhenius base is a substance that ionizes in aqueous solution to give hydroxide (OH^-) ions. *(Section 7.3)* A Brønsted-Lowry base is a proton acceptor.

Bases *(Section 16.2)* Purines and pyrimidines, which are components of nucleotides, DNA, and RNA.

Basic salt *(Section 7.9)* The salt of a weak acid and a strong base. An aqueous solution of a basic salt is acidic.

Beta (β) pleated sheet *(Section 13.8)* A secondary protein structure in which the backbone of two protein chains in the same or different molecules is held together by hydrogen bonds.

Beta (β) oxidation *(Section 19.5)* The biochemical pathway that degrades fatty acids to acetyl CoA by removing two carbons at a time and yielding energy.

Bile salts *(Section 12.11)* Oxidation products of cholesterol that are emulsifying agents for fatty acids. They are produced in the liver and stored in the gallbladder.

Biochemical pathway *(Section 18.1)* A series of consecutive biochemical reactions that converts one molecule into another molecule; the pathway may be anabolic or catabolic.

Biosynthetic pathway *(Section 20.1)* A pathway that leads to the production of a compound. These pathways often build larger molecules from smaller ones and consume energy.

Blood cellular elements *(Section 23.2)* The cell components that are suspended in blood—namely, the erythrocytes, leukocytes, and platelets.

Blood plasma *(Section 23.1)* The noncellular portion of blood.

Blood–brain barrier *(Section 23.1)* A barrier limiting the exchange of blood components between blood and the cerebrospinal and brain interstitial fluids to water, carbon dioxide, glucose, and other small molecules but excluding electrolytes and large molecules.

Bohr effect *(Section 23.3)* The effect caused by a change in the pH on the oxygen-carrying capacity of hemoglobin.

Bowman's capsule *(Section 23.5)* A cup-shaped receptacle in the kidney that is the initial, expanded segment of the nephron where filtrate enters from the blood.

Buffer capacity *(Section 7.11C)* The extent to which a buffer solution can prevent a significant change in pH of a solution upon addition of an acid or a base.

Buffer solution *(Section 7.11A)* A solution that resists change in pH when limited amounts of an acid or base are added to it; a aqueous solution containing a weak acid and its conjugate base.

Calvin cycle *(Chemical Connections 20A)* The dark reactions of photosynthesis that produce glucose from CO_2 and H_2O by a cyclic process.

Carbocation *(Section 3.6A)* A species containing a carbon atom with only three bonds to it and bearing a positive charge.

Carbohydrate *(Section 11.1)* A polyhydroxyaldehyde or polyhydroxyketone, or a substance that gives these compounds on hydrolysis.

Carbonyl group *(Section 1.4C)* A C=O group.

Carboxyl group *(Section 1.4D)* A —COOH group.

Carboxylic acid *(Section 1.4D)* A compound containing a —COOH group.

Carcinogen *(Section 17.7)* A chemical mutagen that can cause cancer.

Catabolism *(Section 18.1)* The biochemical pathways that are involved in generating energy by breaking down large nutrient molecules into smaller molecules with the concurrent production of ATP.

Central dogma *(Section 17.1)* Doctrine stating the basic directionality of heredity where DNA leads to RNA, which leads to protein. This is true in almost all life forms, except certain viruses.

Cephalin *(Section 12.6)* A glycerophospholipid similar to phosphatidylcholine with the exception that another alcohol is attached to the phosphate instead of choline.

Ceramide *(Section 12.7)* The combination of a fatty acid linked to sphingosine by an amide bond.

Cerebroside *(Section 12.8)* A glycolipid based on the ceramide backbone where one or more carbohydrates are attached to the ceramide.

Chair conformation *(Section 2.6B)* The most stable conformation of a cyclohexane ring; all bond angles are approximately 109.5°.

Chaperone *(Section 13.11)* Protein molecules that help other proteins to fold into the biologically active conformation and enable partially denatured proteins to regain their biologically active conformation.

Chemical messenger *(Section 15.1)* Any chemical that is released from one location and travels to another location before acting. It may be a hormone, neurotransmitter, or simple ion.

Chemiosmotic hypothesis *(Section 18.6)* The hypothesis proposed by Peter Mitchell to explain how movement of electrons down the electron transport chain creates a proton gradient that, in turn, leads to the production of ATP.

Chemiosmotic theory *(Section 18.6)* Peter Mitchell proposed that electron transport is accompanied by an accumulation of protons in the intermembrane space of the mitochondrion, which in turn creates an osmotic pressure; the protons driven back to the mitochondrion under this pressure generate ATP.

Chemokine *(Section 22.6)* A chemotactic cytokine that facilitates the migration of leukocytes from the blood vessels to the site of injury or inflammation.

Chiral *(Section 6.2)* From the Greek *cheir*, meaning "hand"; an object that is not superposable on its mirror image.

Chloroplast *(Chemical Connections 14A)* The subcellular organelle in plants that contains chlorophyll and is responsible for photosynthesis.

Cholinergic neurotransmitter *(Section 15.3)* A neurotransmitter/ hormone based on acetylcholine.

Chromatin *(Section 16.6)* The DNA complexed with histone and nonhistone proteins that exists in eukaryotic cells between cell divisions.

Chromosomes *(Section 16.1)* Structures within the nucleus of eukaryotes that contain DNA and protein and that are replicated as units during mitosis. Each chromosome is made up of one long DNA molecule that contains many heritable genes.

cis *(Section 2.7)* A prefix meaning "on the same side."

Cis-trans isomers *(Sections 2.7 and 3.2B)* Isomers that have the same order of attachment of their atoms but a different arrangement of their atoms in space due to the presence of either a ring or a carbon–carbon double bond.

Citric acid cycle *(Section 18.4)* The central biochemical pathway in metabolism whereby an acetyl-CoA combines with oxaloacetate to form citrate, which then undergoes a series of reactions to regenerate oxaloacetate while producing energy in the form of GTP, NADH, and $FADH_2$.

Cloning *(Section 16.6)* A process whereby DNA is amplified by inserting it into a host and having the host replicate it along with the host's own DNA.

Cluster determinant *(Section 22.6)* A set of membrane proteins on T cells that help the binding of antigens to the T-cell receptors.

Coated pit *(Section 12.9)* An area of a cell membrane that contains LDL receptors and is involved in bringing LDL into the cell.

Codon *(Section 17.3)* The sequence of three nucleotides in messenger RNA that codes for a specific amino acid.

Coenzyme *(Section 14.3)* An organic molecule, frequently a B vitamin, that acts as a cofactor.

Coenzyme A *(Section 18.3)* An important activating group often attached to a fatty acid or an acetate group. Coenzyme A is based on a structure composed of an adenine-based nucleotide, pantothenic acid, and mercaptoethylamine.

Coenzyme Q *(Section 18.5)* An electron carrier found in the inner mitochondrial membrane that acts as an electron receptor from a variety of sources during electron transport; also called ubiquinone.

Cofactor *(Section 14.3)* The nonprotein part of an enzyme necessary for its catalytic function.

Common catabolic pathway *(Section 18.1)* A series of chemical reactions that most degraded food goes through to yield energy in the form of ATP. The common catabolic pathway consists of (1) the citric acid cycle *(Section 18.4)* and (2) oxidative phosphorylation *(Sections 18.5 and 18.6)*.

Competitive inhibition *(Section 14.5)* An enzyme regulation in which an inhibitor competes with the substrate for the active site.

Complete protein *(Section 21.4)* A protein source that contains sufficient quantities of all amino acids required for normal growth and development.

Complex lipids *(Section 12.4)* Lipids found in membranes. The two main classes are glycolipids and phospholipids.

Conformation *(Section 2.7)* Any three-dimensional arrangement of atoms in a molecule that results from rotation about a single bond.

Conjugate acid *(Section 7.3)* In the Brønsted-Lowry theory, a substance formed when a base accepts a proton.

Conjugate base *(Section 7.3)* In the Brønsted-Lowry theory, a substance formed when an acid donates a proton to another molecule or ion.

Conjugated acid–base pair *(Section 7.3)* A pair of molecules or ions that are related to one another by the gain or loss of a proton.

Conjugated protein *(Section 13.9)* A protein that contains a nonprotein part, such as hemoglobin contains a hemo.

Constitutional isomers *(Section 2.3)* Compounds with the same molecular formula but a different order of attachment of their atoms.

Control site *(Section 17.6)* A DNA sequence that is part of a prokaryotic operon. This sequence is upstream of the structural gene DNA and plays a role in controlling whether the structural gene is transcribed.

Cristae *(Section 18.2)* The inner mitochondrial membrane.

C-terminus *(Section 13.5)* The amino acid at the end of a peptide chain that has a free carboxyl group.

Cyclic ether *(Section 5.4A)* An ether in which the ether oxygen is one of the atoms of a ring.

Cycloalkane *(Section 2.5)* A saturated hydrocarbon that contains carbon atoms bonded to form a ring.

Cyclooxygenase *(Section 12.12)* The enzyme that catalyzes the first step in the conversion of arachidonic acid to prostaglandins and thromboxanes.

Cystine *(Section 13.4)* A dimer of cysteine in which the two amino acids are covalently bonded by disulfide bond between their side-chain —SH groups.

Cytochrome *(Section 18.5)* A class of compounds that act as electron carriers in the inner mitochondrial membrane during electron transport. They are based on an iron-containing porphyrin ring similar to the heme group found in hemoglobin.

Cytokine *(Section 22.6)* A glycoprotein that traffics between cells and alters the function of a target cell.

Cytosol *(Section 19.2)* The soluble portion of a cell that is inside the cell membrane but outside any of the subcellular organelles.

Debranching enzyme *(Section 21.7)* The enzyme that catalyzes the hydrolysis of the 1,6-glycosidic bonds in starch and glycogen.

Decarboxylation *(Section 18.4)* The process that leads to the loss of CO_2 from a —COOH group.

Degenerate code *(Section 17.4)* The genetic code is said to be degenerate because more than one codon can code for the same amino acid.

Dehydration *(Section 5.3B)* Elimination of a molecule of water from an alcohol. An OH is removed from one carbon, and an H is removed from an adjacent carbon.

Dehydrogenase *(Section 14.6)* A class of enzymes that catalyze oxidation–reduction reactions, often using NAD^+.

Denaturation *(Section 13.11)* The loss of the secondary, tertiary, and quaternary structure of a protein by a chemical or physical agent that leaves the primary structure intact.

Dendrite *(Section 15.2)* The hair-like projections that extend from the cell body of a nerve cell on the opposite side from the axon.

Deoxyribonucleic acid *(Section 16.2)* The macromolecule of heredity in eukaryotes and prokaryotes. It is composed of chains of nucleotide monomers of a nitrogenous base, deoxyribose, and phosphate.

Dextrorotatory *(Section 6.5B)* Clockwise (to the right) rotation of the plane of polarized light in a polarimeter.

Diastereomers *(Section 6.4A)* Stereoisomers that are not mirror images of each other.

Digestion *(Section 21.4)* The process in which the body breaks down large molecules into smaller ones that can then be absorbed and metabolized.

Digestion *(Section 21.6)* The process of hydrolysis of starches, fats, and proteins into smaller units.

Diglyceride *(Section 12.2)* An ester of glycerol and two fatty acids.

Diol *(Section 5.2B)* A compound containing two —OH (hydroxyl) groups.

Dipeptide *(Section 13.5)* A peptide with two amino acids.

Diprotic acid *(Section 7.3)* An acid that can give up two protons.

Disaccharide *(Section 11.5)* A carbohydrate containing two monosaccharide units joined by a glycosidic bond.

Discriminatory curtailment diet *(Section 21.2)* A diet that avoids certain food ingredients that are considered harmful to the health of an individual—for example, low-sodium diets for people with high blood pressure.

Distal tubule *(Section 23.5)* The part of the kidney tubule farthest from the Bowman's capsule.

Disulfide *(Section 5.5D)* A compound containing an —S—S— group.

Diuresis *(Section 23.7)* Enhanced excretion of water in urine.

Diuretic *(Chemical Connections 23E)* A compound that increases the flow of urine from the kidney.

DNA *(Section 16.2)* Deoxyribonucleic acid.

Double helix *(Section 16.3)* The arrangement of which two strands of DNA are coiled around each other in a screw-like fashion.

Edema *(Section 23.2)* A swelling of tissues due to water seeping from the blood into the interstitial fluids.

EGF *(Section 22.6)* Epidermal growth factor, a cytokine that stimulates epidermal cells during healing of wounds.

Electron transport chain *(Section 18.6)* A series of electron carriers imbedded or attached to the inner mitochondrial membrane that are alternately reduced and oxidized as electrons pass from an initial electron carrier (NADH or $FADH_2$) to oxygen.

Elongation *(Section 17.5)* The phase of protein synthesis during which activated tRNA molecules deliver new amino acids to ribosomes where they are joined by peptide bonds to form a polypeptide.

Elongation factor *(Section 17.5)* Small protein molecules that are involved in the process of tRNA binding and movement of the ribosome on the mRNA during elongation.

Enantiomers *(Section 6.2)* Stereoisomers that are nonsuperposable mirror images; refers to a relationship between pairs of objects.

End point *(Section 7.10)* The point at which there is an equal amount of acid and base in a neutralization reaction.

Endocrine gland *(Section 15.2)* A gland, such as the pancreas, pituitary, and hypothalamus, that produces hormones involved in the control of the chemical reactions and metabolism.

Energy yield *(Section 18.7)* The calculation of how many ATP molecules are formed from electrons passing down the electron transport chain. It is usually calculated as a number of ATP per NADH used, $FADH_2$ used, or H^+.

Enkephalins *(Section 15.6)* Pentapeptides found in nerve cells of the brain that act to control pain perception.

Enol *(Section 9.8)* A molecule containing an —OH group bonded to a carbon of a carbon–carbon double bond.

Enzyme *(Section 14.1)* A biological catalyst that increases the rate of a chemical reaction by providing an alternative pathway with a lower activation energy.

Enzyme activity *(Section 14.4)* The rate at which an enzyme-catalyzed reaction proceeds; commonly measured as the amount of product produced per minute.

Enzyme specificity *(Section 14.1)* The limitation of an enzyme to catalyze one specific reaction with one specific substrate.

Enzyme–substrate complex *(Section 14.5)* A part of an enzyme reaction mechanism where the enzyme is bound to the substrate.

Epitope *(Section 22.3)* The smallest number of amino acids on an antigen that elicits an immune response.

Equatorial position *(Section 2.6B)* A position on a chair conformation of a cyclohexane ring that extends from the ring roughly perpendicular to the imaginary axis of the ring.

Erythrocytes *(Section 23.2)* Red blood cells; transporters of blood gases.

Essential amino acid *(Section 21.4)* An amino acid that the body cannot synthesize in the required amounts and so must be obtained in the diet.

Essential fatty acid *(Section 12.3)* A polyunsaturated fatty acid that cannot be synthesized in the human body in sufficient quantities for normal growth and development.

Ester *(Section 10.3B)* A compound in which the OH of a carboxyl group, RCOOH, is replaced by an —OR′ group; RCOOR′.

Ether *(Section 5.4A)* A compound containing an oxygen atom bonded to two carbon atoms.

Eukaryote *(Section 17.6)* An organism that has a true nucleus and organelles. Eukaryotes include animals, plants, and fungi.

Excitatory neurotransmitter *(Section 15.4)* A neurotransmitter that increases the transmission of nerve impulses.

Exon *(Section 16.5)* A nucleotide sequence in mRNA that codes for a protein.

External innate immunity *(Section 22.1)* The innate protection against foreign invaders characteristic of the skin barrier, tears, and mucus.

Extracellular fluids *(Section 23.1)* All the body fluids that are not inside the cell.

FAD/FADH₂ *(Section 18.3)* Flavin adenine dinucleotide, a coenzyme involved in metabolic oxidation–reduction reactions.

Fat *(Section 12.2)* A mixture of triglycerides containing a high proportion of long-chain, saturated fatty acids.

Fat depot *(Section 19.1)* The supply of stored fat in the form of triacylglycerols, usually found in fat cells or distributed in other tissues.

Feedback control *(Section 14.6)* A type of enzyme regulation where the product of a series of reactions inhibits the enzyme that catalyzes the first reaction in the series.

Fiber *(Section 21.2)* The cellulosic, non-nutrient component in our food.

Fibrous protein *(Section 13.1)* A protein used for structural purposes. Fibrous proteins are insoluble in water and have a high percentage of secondary structures, such as alpha helices and/or beta-pleated sheets.

Fischer esterification *(Section 10.4)* The process of forming an ester by refluxing a carboxylic acid and an alcohol in the presence of an acid catalyst, commonly sulfuric acid.

Fischer projection *(Section 11.2B)* A two-dimensional representation for showing the configuration of a stereocenter; horizontal lines represent bonds projecting forward from the stereocenter, and vertical lines represent bonds projecting toward the rear.

Fluid mosaic model *(Section 12.5)* A model for the structure and function of biological membranes.

Functional group *(Section 1.4)* An atom or group of atoms within a molecule that shows a characteristic set of physical and chemical properties.

Furanose *(Section 11.3A)* A five-membered cyclic hemiacetal form of a monosaccharide.

Gene *(Section 16.1)* The unit of heredity; a DNA segment that codes for one protein.

Gene expression *(Section 17.1)* The activation of a gene to produce a specific protein. It involves both transcription and translation.

Gene regulation *(Section 17.6)* The various methods used by organisms to control which genes will be expressed and when.

Genetic code *(Section 17.4)* The sequence of triplets of nucleotides (codons) that determines the sequence of amino acids in a protein.

Genetic engineering *(Section 17.8)* The process by which genes are inserted into cells.

Genome *(Chemical Connections 16D)* A complete DNA sequence of an organism.

Globular protein *(Section 13.1)* Protein that is used mainly for nonstructural purposes and is largely soluble in water.

Glomeruli *(Section 23.5)* Part of the kidney filtration cell, nephron; a tangle of capillaries surrounded by a fluid-filled space.

Glucogenic amino acid *(Sections 19.9, 20.2)* An amino acid whose carbon skeleton can be used for the synthesis of glucose.

Gluconeogenesis *(Section 20.2)* The biosynthetic pathway that converts pyruvate, citric acid cycle intermediates, or glucogenic amino acids into glucose.

Glycerophospholipid *(Section 12.4)* A diester of glycerol and two fatty acids. The third —OH group of glycerol is esterified with phosphoric acid, which is, in turn, esterified with a low-molecular-weight alcohol such as choline or ethanolamine.

Glycogenesis *(Section 20.2)* The synthesis of glycogen from glucose.

Glycogenolysis *(Section 19.3)* The biochemical pathway for the breakdown of glycogen to glucose.

Glycol *(Section 5.2B)* A compound with two hydroxyl (—OH) groups on adjacent carbons.

Glycolipid *(Section 12.4)* A complex lipid that contains at least one monosaccharide.

Glycolysis *(Section 19.2)* The biochemical pathway that breaks down glucose to pyruvate, which yields chemical energy in the form of ATP and reduced coenzymes.

Glycomics *(Chemical Connections 22E)* The collective knowledge of all the carbohydrates, including glycoproteins and glycolipids, that a cell or a tissue contains and the determination of their functions.

Glycoprotein *(Section 13.10)* A protein to which one or more carbohydrate molecules are bonded.

Glycoside *(Section 11.4A)* A carbohydrate in which the —OH group on its anomeric carbon is replaced by an —OR group.

Glycosidic bond *(Section 11.4A)* The bond from the anomeric carbon of a glycoside to an —OR group.

Gp120 *(Section 22.5)* A 120,000-molecular-weight glycoprotein on the surface of the human immunodeficiency virus that binds strongly to the CD4 molecules on T-cells.

G-protein *(Section 15.5)* A protein that is either stimulated or inhibited when a hormone binds to a receptor and that subsequently alters the activity of another protein, such as adenylate cyclase.

Haworth projection *(Section 11.3A)* A way to view furanose and pyranose forms of monosaccharides; the ring is drawn flat and viewed through its edge, with the anomeric carbon on the right and the oxygen atom to the rear.

HDL *(Section 12.9)* High-density lipoprotein; a lipoprotein with high protein content (approximately 50 percent).

Helicase *(Section 16.4)* An unwinding protein that acts at a replication fork to unwind DNA so that DNA polymerase can synthesize a new DNA strand.

Hemiacetal *(Section 9.7)* A molecule containing a carbon bonded to one —OH and one —OR group; the product of adding one molecule of alcohol to the carbonyl group of an aldehyde or ketone.

Henderson-Hasselbalch equation *(Section 7.12)* A mathematical relationship between pH, the pK_a of a weak acid, and the concentrations of the weak acid and its conjugate base.

$$pH = pK_a + \log \frac{[A^-]}{[HA]}$$

Henle's loop *(Section 23.5)* A U-shaped twist in a kidney tubule between the proximal tubule and the distal tubule.

Heterocyclic aliphatic amine *(Section 8.2)* A heterocyclic amine in which nitrogen is bonded only to alkyl groups.

Heterocyclic amine *(Section 8.2)* An amine in which nitrogen is one of the atoms of a ring.

Heterocyclic aromatic amine *(Section 8.2)* An amine in which nitrogen is one of the atoms of an aromatic ring.

Histone *(Section 16.1)* A basic protein that is found in complexes with DNA in eukaryotes.

HIV *(Section 22.5)* Human immunodeficiency virus.

Homeostasis *(Section 23.1)* The process of maintaining the proper levels of nutrients and temperature of the blood.

Hormone *(Section 15.2)* A chemical messenger released by an endocrine gland into the bloodstream and transported there to reach its target cell.

Hybridization *(Section 16.6)* A process whereby two strands of nucleic acids or segments thereof form a double-stranded structure through H-bonding of complementary base pairs.

Hydration *(Section 3.6B)* Addition of water.

Hydrocarbon *(Section 2.2)* A compound that contains only carbon and hydrogen atoms.

Hydrolase *(Section 14.2)* An enzyme that catalyzes a hydrolysis reaction.

Hydrolysis *(Section 9.7)* A reaction of a compound with water (*hydro-*) in which one or more bonds are broken (*-lysis*) and the —H and —OH of water add to the ends of the bond or bonds broken.

Hydronium ion *(Section 7.1)* The H_3O^+ ion.

Hydrophobic interaction *(Section 13.9)* Interaction by London dispersion forces between hydrophobic groups.

Hydroxyl group *(Section 1.4A)* An —OH group bonded to a tetrahedral carbon atom.

Hyperglycemia *(Chemical Connections 21D)* A condition whereby the blood glucose level reaches 140 mg/dL or higher.

Hypertension *(Chemical Connections 23E)* The clinical term for high blood pressure.

Hyperthermophile *(Section 14.4)* An organism that lives at extremely high temperatures.

Hypoglycemia *(Chemical Connections 21D)* A condition whereby the blood glucose level drops below the normal 65–100 mg/dL.

Immunogen *(Section 22.3)* Another term for antigen.

Immunoglobulin *(Section 22.4)* An antibody protein generated against and capable of binding specifically to an antigen.

Immunoglobulin superfamily *(Section 22.4)* A family of molecules based on a similar structure that includes the immunoglobulins, T-cell receptors, and other membrane proteins that are involved in cell communications. All molecules in this class have a certain portion that can react with antigens.

Indicator, acid–base *(Section 7.8)* A substance that changes color within a given pH range.

Induced-fit model *(Section 14.5)* A model explaining the specificity of enzyme action by comparing the active site to a glove and the substrate to a hand.

Inhibition, of enzyme activity *(Section 14.3)* Any reversible or irreversible process that makes an enzyme less active.

Inhibitory neurotransmitters *(Section 15.4)* Neurotransmitters that decrease the transmission of nerve impulses.

Initiation, of protein synthesis *(Section 17.5)* The first step in the process whereby the base sequence of an mRNA is translated into the primary structure of a polypeptide.

Initiation signal *(Section 17.2)* A sequence on DNA that identifies the location where transcription is to begin.

Innate immunity *(Section 22.1)* The first line of defense against foreign invaders, which includes skin resistance to penetration, tears, mucus, and nonspecific macrophages that engulf bacteria.

Inorganic phosphate *(Section 18.3)* PO_4^{3-}.

Interleukin *(Section 22.6)* A cytokine that controls and coordinates the action of leukocytes.

Internal innate immunity *(Section 22.1)* The type of innate immunity that is used once a pathogen has already penetrated a tissue.

Interstitial fluid *(Section 22.2)* The fluid that surrounds and fills in spaces between cells.

Intron *(Section 16.5)* A nucleotide sequence in mRNA that does not code for a protein.

Ion product of water, K_w *(Section 7.7)* The concentration of H_3O^+ multiplied by the concentration of OH^-; $[H_3O^+][OH^-] = 1 \times 10^{-14}$.

Isoelectric point *(Section 13.3)* The pH at which a molecule has no net charge; abbreviated pI.

Isoenzyme *(Section 14.6)* An enzyme that can be found in multiple forms but with each form catalyzing the same reaction; also called an isozyme.

Isomerase *(Section 14.2)* An enzyme that catalyzes an isomerization reaction.

Isomers *(Section 2.3)* Different compounds that have the same molecular formula.

Isozymes *(Section 14.6)* Enzymes that perform the same function but have different combinations of subunits—that is, different quaternary structures.

Ketoacidosis *(Chemical Connections 19C)* A biochemical condition caused by starvation or diabetes where ketone bodies build up in the blood and cause the pH to decrease.

Ketogenic amino acid *(Section 19.9)* An amino acid that, when catabolized, can only form acetyl-CoA or acetoacetyl-CoA and therefore cannot lead to glucose production.

Ketone *(Section 1.4C)* A compound containing a carbonyl group bonded to two carbons.

Ketone bodies *(Section 19.7)* A collective name for acetone, acetoacetate and β-hydroxybutyrate; compounds produced from acetyl-CoA in the liver that are used as a fuel for energy production by muscle cells and neurons.

Ketose *(Section 11.2A)* A monosaccharide containing a ketone group.

Killer T cell *(Section 22.2)* A T cell that kills invading foreign cells by cell-to-cell contact.

Kinase *(Section 14.6)* An enzyme that covalently modifies a protein with a phosphate group, usually through the —OH group on the side chain of a serine, threonine, or tyrosine.

Krebs cycle *(Section 18.4)* Another name for the citric acid cycle, used to honor Hans Krebs, the discoverer of the pathway.

Kwashiorkor *(Section 21.4)* A disease caused by insufficient protein intake and characterized by a swollen stomach, skin discoloration, and retarded growth.

Lactam *(Section 10.3C)* A cyclic amide.

Lactone *(Section 10.3B)* A cyclic ester.

Lagging strand *(Section 16.4)* A discontinuously synthesized DNA that elongates in a direction away from the replication fork.

LDL *(Section 12.9)* Low-density lipoprotein; a circulating lipoprotein where the cholesterol content is very high (45 percent) and the protein content is 25 percent or less.

Leading strand *(Section 16.4)* The continuously synthesized DNA strand that elongates toward the replication fork.

Lecithin *(Section 12.6)* The common name for a phosphatidylcholine.

Leukocytes *(Sections 22.2, 23.2)* White blood cells, which are the principal parts of the acquired immunity system and act via phagocytosis or antibody production.

Leukotrienes *(Section 12.12)* A class of acyclic arachidonic acid derivatives that act as mediators of hormonal responses.

Levorotatory *(Section 6.5B)* Counterclockwise rotation of the plane of polarized light in a polarimeter.

Ligases *(Section 14.2)* A class of enzymes that catalyze a reaction joining two molecules. They are often called synthetases or synthases.

Line-angle formula *(Section 2.2)* An abbreviated way to draw structural formulas in which each angle and line terminus represents a carbon atom and each line represents a bond.

Lipase *(Section 21.8)* An enzyme that catalyzes the hydrolysis of an ester bond between a fatty acid and glycerol.

Lipid bilayer *(Section 12.5)* A back-to-back arrangement of phospholipids, often forming a closed vesicle or membrane.

Lipids *(Section 12.1)* A class of compounds found in living organisms that are insoluble in water but soluble in nonpolar solvents.

Lipoproteins *(Section 12.9)* Spherically shaped clusters containing both lipid molecules and protein molecules.

Lock-and-key model *(Section 14.5)* A model for enzyme substrate interaction based on the postulate that the active site of an enzyme is a perfect fit for the substrate.

Lyase *(Section 14.2)* An enzyme that catalyzes the addition of two atoms or groups of atoms to a double bond or their removal to form a double bond.

Lymphocyte *(Section 22.2)* A white blood cell that spends most of its time in the lymphatic tissues. Those that mature in the bone marrow are B cells. Those that mature in the thymus are T cells.

Lymphoid organs *(Section 22.2)* The main organs of the immune system, such as the lymph nodes, spleen, and thymus, which are connected together by lymphatic capillary vessels.

Lysosome *(Section 18.2)* A subcellular organelle that cleanses a cell of damaged cellular components and some foreign material.

Macrophage *(Section 22.2)* An amoeboid white blood cell that moves through tissue fibers, engulfing dead cells and bacteria by phagocytosis, and then displays some of the engulfed antigens on its surface.

Major histocompatibility complex (MHC) *(Section 22.3)* A transmembrane protein complex that brings the epitope of an antigen to the surface of the infected cell to be presented to the T cells.

Marasmus *(Section 21.3)* Another term for chronic starvation whereby the individual does not have adequate caloric intake. It is characterized by arrested growth, muscle wasting, anemia, and general weakness.

Markovnikov's rule *(Section 3.6A)* In the addition of HX or H_2O to an alkene, hydrogen adds to the carbon of the double bond having the greater number of hydrogens.

Membrane *(Section 12.5)* A structure composed of a lipid bilayer, protein, and often carbohydrate that separates one cell compartment from another or one cell from another.

Memory *(Section 22.1)* The characteristic of the acquired immunity system seen as a more rapid and vigorous response to a pathogen the second time it is encountered.

Memory cell *(Section 22.2)* A type of T cell that stays in the blood after an infection is over and acts as a quick line of defense if the same antigen is encountered again.

Mercaptan *(Section 5.5A)* A common name for any molecule containing an —SH group.

Messenger RNA (mRNA) *(Section 16.4)* The RNA that carries genetic information from DNA to the ribosome and acts as a template for protein synthesis.

Meta (m) *(Section 4.3B)* Refers to groups occupying the 1 and 3 positions on a benzene ring.

Metabolism *(Section 18.1)* The sum of all the chemical reactions involved in maintaining the dynamic state of a cell or organism.

Metal-binding finger *(Section 17.6)* A type of transcription factor containing heavy metal ions, such as Zn^{2+}, that is involved in helping RNA polymerase bind to the DNA to be transcribed.

Metastasis *(Chemical Connections 17E)* The process whereby cancerous cells proliferate and spread to other tissues.

Mineral in diet *(Section 21.5)* An inorganic ion that is required for structure and metabolism, such as Ca^{2+}, Fe^{2+}, and Mg^{2+}.

Mirror image *(Section 6.1)* The reflection of an object in a mirror.

Mitochondrion *(Section 18.2)* A subcellular organelle characterized by an outer membrane and a highly folded inner membrane. Mitochondria are responsible for the generation of most of the energy for cells.

Monoclonal antibody *(Section 22.5)* An antibody produced by clones of a single B cell specific to a single epitope.

Monoglyceride *(Section 12.2)* An ester of glycerol and a single fatty acid.

Monomer *(Section 3.7A)* From the Greek *mono*, "single," and *meros*, "part"; the simplest nonredundant unit from which a polymer is synthesized.

Monoprotic acid *(Section 7.3)* An acid that can give up only one proton.

Monosaccharide *(Section 11.2)* A carbohydrate that cannot be hydrolyzed to a simpler compound.

D-Monosaccharide *(Section 11.2C)* A monosaccharide that, when written as a Fischer projection, has the —OH group on its penultimate carbon to the right.

L-Monosaccharide *(Section 11.2C)* A monosaccharide that, when written as a Fischer projection, has the —OH group on its penultimate carbon to the left.

Mutagen *(Section 17.7)* A chemical substance that induces a base change, or mutation, in DNA.

Mutarotation *(Section 11.3C)* The change in specific rotation that occurs when an α or β form of a carbohydrate is converted to a equilibrium mixture of the two forms.

NAD^+/NADH *(Section 18.3)* Nicotinamide adenine dinucleotide, a coenzyme participating in many metabolic oxidation–reduction reactions.

Native structure *(Section 13.10)* The natural or normal conformation of a protein in the biological setting.

Negative modulation *(Section 14.6)* The process whereby an allosteric regulator inhibits enzyme action.

Nephron *(Section 23.5)* The tubular excretory unit of the kidney that filters the blood to remove waste products and to reabsorb nutrients.

Neuron *(Section 15.1)* Another name for a nerve cell.

Neuropeptide Y *(Section 15.6)* A brain peptide that affects the hypothalamus and is an appetite-stimulating agent.

Neurotransmitter *(Section 15.2)* Chemical messengers between a neuron and another target cell: neuron, muscle cell, or cell of a gland.

Neutral salt *(Section 7.9)* A salt whose aqueous solution is neither acidic nor basic.

Neutral solution *(Section 7.7)* An aqueous solution whose pH is 7.0; an aqueous solution that is neither acidic nor basic.

N-linked saccharide *(Section 13.10)* A carbohydrate that is bonded by a N-glycosidic bond to a protein, most commonly to the side-chain nitrogen of asparagine.

Noncompetitive inhibition *(Section 14.5)* An enzyme regulation in which an inhibitor binds to the enzyme outside of the active site, thereby changing the shape of the active site and reducing its catalytic activity.

N-terminus *(Section 13.5)* The amino acid at the end of a peptide chain that has a free amino group.

Nucleic acid *(Section 16.3)* A polymer composed of nucleotides.

Nucleoside *(Section 16.2)* The combination of a heterocyclic aromatic amine attached via a β-glycosidic bond to either D-ribose or 2-deoxy-D-ribose.

Nucleosome *(Section 16.6)* Combinations of DNA and histone proteins.

Nucleotide *(Section 16.2)* A phosphoric ester of a nucleoside.

Nucleus *(Section 18.2)* A subcellular organelle found in eukaryotes where the DNA is stored and replicated.

Nutrient *(Section 21.2)* Components of food and drink that provide energy, replacement, and growth.

Obesity *(Section 21.3)* Accumulation of body fat beyond the norm.

Oil *(Section 12.2)* A mixture of triglycerides containing a high proportion of long-chain unsaturated fatty acids or short-chain saturated fatty acids.

Okazaki fragment *(Section 16.6)* A short DNA segment made of about 200 nucleotides in higher organisms (eukaryotes) and 2000 nucleotides in prokaryotes.

Oligosaccharide *(Section 11.5)* A carbohydrate containing from 4 to 10 monosaccharide units, each joined to the next by a glycosidic bond.

O-linked saccharide *(Section 13.10)* A carbohydrate that is bonded by a glycosidic bond to a protein, most commonly to the side-chain OH of a serine, threonine, or hydroxylysine.

Oncogene *(Chemical Connections 17E)* A gene that in some way participates in the development of cancer.

Optically active *(Section 6.5B)* Showing that a compound rotates the plane of polarized light.

Organelle *(Section 18.2)* A special structure within a cell that carries out a particular part or parts of metabolism.

Organic chemistry *(Section 1.1)* The study of the compounds of carbon.

Ortho (o) *(Section 4.3B)* Refers to groups occupying the 1 and 2 positions on a benzene ring.

Oxidative deamination *(Section 19.8)* A reaction in which the amino group of an amino acid is removed in the form of NH_v^+ and an α-ketoacid is formed.

Oxidative phosphorylation *(Section 18.5)* The process of generating ATPs by producing a hydrogen ion gradient during electron transport and then harnessing the energy of the gradient by letting the ions flow through the ATPase.

Oxidoreductase *(Section 14.2)* An enzyme that catalyzes an oxidation–reduction reaction.

Oxonium ion *(Section 3.6B)* An ion in which oxygen is bonded to three other atoms and bears a positive charge.

P site *(Section 17.5)* The site on the large ribosomal subunit where the current peptide is bound before peptidyl transferase links it to the amino acid attached at the A site during elongation.

Para (p) *(Section 4.3B)* Refers to groups occupying the 1 and 4 positions on a benzene ring.

Parenteral nutrition *(Chemical Connections 21A)* The technical term for intravenous feeding.

Pentose phosphate pathway *(Section 19.2)* A pathway, often called the hexose monophosphate shunt, in which glucose 6-phosphate is decarboxylated to yield ribulose 5-phosphate. When necessary, this pathway can produce ribose for nucleotide synthesis and NADPH, which is required for many synthetic pathways.

Penultimate carbon *(Section 11.2B)* The stereocenter of a monosaccharide farthest from the carbonyl group—for example, carbon 5 of glucose.

Peptide *(Section 13.5)* A short chain of amino acids linked via peptide bonds.

Peptide backbone *(Section 13.6)* The repeating pattern of peptide bonds in a polypeptide or protein.

Peptide bond *(Section 13.5)* An amide bond that links two amino acids.

Peptidergic neurotransmitter *(Section 15.6)* A type of neurotransmitter/hormone that is based on a peptide, such as glucagon, insulin, and the enkephalins.

Perforin *(Section 22.2)* A protein produced by killer T cells that punches holes in the membrane of target cells.

pH *(Section 7.8)* The negative logarithm of the hydronium ion concentration; $pH = -\log[H_3O^+]$.

Phagocytosis *(Section 22.6)* The process by which large particulates, including bacteria, are pulled inside a white cell called a phagocyte.

Phenol *(Section 4.5)* A compound that contains an —OH group bonded to a benzene ring.

Phenyl group *(Section 4.3)* C_6H_5—, the aryl group derived by removing a hydrogen from benzene.

Pheromone *(Chemical Connections 3B)* A chemical secreted by an organism to influence the behavior of another member of the same species.

Phosphatidylcholine *(Section 12.6)* A glycerophospholipid where the substituent attached via a phosphoric ester bond is choline.

Phosphatidylinositol *(Section 12.6)* A glycerophospholipid where the substituent attached via a phosphoric ester bond is inositol.

Phospholipid *(Section 12.4)* A complex lipid comprising an alcohol, such as glycerol or sphingosine, a phosphate group, and fatty acids.

Photosynthesis *(Section 20.2)* The process in which plants synthesize carbohydrates from CO_2 and H_2O with the help of sunlight and chlorophyll.

PIP$_2$ *(Section 12.6)* Phosphatidylinositol 4,5-bisphosphate.

Plane-polarized light *(Section 6.5A)* Light vibrating in only parallel planes.

Plasma cell *(Section 22.2)* A cell derived from a B cell that has been exposed to an antigen.

Plasmid *(Section 17.8)* A small, circular, double-stranded DNA molecule of bacterial origin. Plasmids were found to naturally confer antibiotic resistance for certain bacteria. Now they are used for genetic engineering purposes.

Plastic *(Chemical Connections 3D)* A polymer that can be molded when hot and that retains its shape when cooled.

Platelets *(Section 23.2)* Special cells in the blood that control bleeding when there is an injury; also called thrombocytes.

pOH *(Section 7.8)* The negative logarithm of the hydroxide ion concentration; $pOH = -log[OH^-]$.

Polarimeter *(Section 6.5B)* An instrument for measuring the ability of a compound to rotate the plane of polarized light.

Polyamide *(Section 10.8A)* A polymer in which each monomer unit is bonded to the next by an amide bond—or example, nylon-66.

Polycarbonate *(Section 10.8C)* A polyester in which the carboxyl groups are derived from carbonic acid.

Polyester *(Section 10.8B)* A polymer in which each monomer unit is bonded to the next by an ester bond—for example, poly(ethylene terephthalate).

Polymer *(Section 3.7A)* From the Greek *poly*, "many," and *meros*, "parts"; any long-chain molecule synthesized by bonding together many single parts called monomers.

Polymerase chain reaction (PCR) *(Sections 14.4, 16.6)* An automated technique for amplifying DNA using a heat-stable DNA polymerase from thermophilic bacteria.

Polynuclear aromatic hydrocarbon *(Section 4.3D)* A hydrocarbon containing two or more benzene rings, each of which shares two carbon atoms with another benzene ring.

Polypeptide *(Section 13.5)* A long chain of amino acids bonded via peptide bonds.

Polysaccharide *(Section 11.6)* A carbohydrate containing a large number of monosaccharide units, each joined to the next by one or more glycosidic bonds.

Positive modulation *(Section 14.6)* The process whereby an allosteric regulator increases enzyme action.

Postsynaptic membrane *(Section 15.2)* The membrane on the side of the synapse nearest the dendrite of the neuron receiving the transmission.

Presynaptic membrane *(Section 15.2)* The membrane on the side of the synapse nearest the axon of the neuron transmitting the signal.

Primary (1°) alcohol *(Section 1.4A)* An alcohol in which the carbon atom bearing the —OH group is bonded to only one other carbon.

Primary (1°) amine *(Section 1.4B)* An amine in which nitrogen is bonded to one carbon and two hydrogens.

Primary structure, of DNA *(Section 16.3)* The order of the bases in DNA.

Primary structure, of proteins *(Section 13.7)* The order of amino acids in a peptide, polypeptide, or protein.

Proenzyme *(Section 14.6)* An inactive form of an enzyme that must have part of its polypeptide chain cleaved before it becomes active.

Prokaryote *(Section 17.6)* An organism that has no true nucleus or organelles.

Promoter *(Chemical Connections 17C)* An upstream DNA sequence that is used for RNA polymerase recognition and binding to DNA.

Prostaglandin *(Section 12.12)* A fatty acid-like substance derived from arachidonic acid that is involved in inflammation and smooth muscle metabolism.

Prosthetic group *(Section 13.9)* The non-amino-acid part of a conjugated protein.

Protein *(Section 13.1)* A long chain of amino acids linked via peptide bonds. There are usually 30 to 50 amino acids in a chain before it is considered a protein.

Protein modification *(Section 14.7)* The process of affecting the enzyme activity by covalently modifying the enzyme, such as phosphorylating a particular amino acid.

Proteoglycans *(Section 13.10)* A special class of glycoproteins where the protein is the core and the carbohydrate part is a long side chains of acidic polysaccharides also called glycosaminoglycans.

Proteomics *(Chemical Connections 13G)* The collective knowledge of all the proteins and peptides of a cell or a tissue and their functions.

Proton channel *(Section 18.6)* An opening in a membrane that will let hydrogen ions pass through it. The ATPase has such a channel that will allow hydrogen ions to pass back into the mitochondria while the energy is used to make ATP.

Proton gradient *(Section 18.6)* A difference in proton (hydrogen ion) concentration across a membrane. During the electron transport process, hydrogen ions are pumped out of the mitochondria, causing a gradient to form, with the outside having a higher $[H^+]$.

Proton-translocating ATPase *(Section 18.6)* The ATPase of the inner mitochondrial membrane allows protons to pass through it and uses the energy to phosphorylate ADP to ATP. The enzyme could also hydrolyze ATP and use this energy to move hydrogen ions out of the matrix.

Proximal tubule *(Section 23.5)* The part of the kidney tubule close to the Bowman's capsule.

Pyranose *(Section 11.3A)* A six-membered cyclic hemiacetal form of a monosaccharide.

Quaternary structure *(Section 13.9)* The organization of a protein that has multiple polypeptide chains, or subunits; Refers principally to the way the multiple chains interact.

R *(Section 6.3)* From the Latin *rectus*, meaning "straight, correct"; used in the *R,S* system to show that, when the lowest-priority group is away from you, the order of priority of groups on a stereocenter is clockwise.

R,S system *(Section 6.3)* A set of rules for specifying configuration about a stereocenter.

R— *(Section 2.4)* A symbol used to represent an alkyl group.

Racemic mixture *(Section 6.2)* A mixture of equal amounts of two enantiomers.

Random coils *(Section 13.8)* Proteins that do not exhibit any repeated pattern.

RDA *(Section 21.2)* Recommended Dietary Allowance; an average daily requirement for nutrients published by the U.S. Food and Drug Administration.

Reaction mechanism *(Section 3.6)* A step-by-step description of how a chemical reaction occurs.

Receptor *(Section 15.1)* A membrane protein that can bind a chemical messenger and then perform a function, such as synthesizing a second messenger or opening an ion channel.

Recognition site (*Section 17.3*) The area of the tRNA molecule that recognizes the mRNA codon.

Recombinant DNA (*Section 17.8*) DNAs from two sources that have been combined into one molecule.

Reducing sugar (*Section 11.4C*) A carbohydrate that reacts with a mild oxidizing agent under basic conditions to give an aldonic acid; the carbohydrate reduces the oxidizing agent.

Regioselective reaction (*Section 3.6*) A reaction in which one direction of bond forming or bond breaking occurs in preference to all other directions.

Regulator (*Section 14.6*) A molecule that binds to an allosteric enzyme and changes its activity. This change could be positive or negative.

Regulatory gene (*Section 17.6*) A gene that produces a protein that controls the transcription of a structural gene.

Regulatory site (*Section 14.6*) A site, other than the active site, where a regulator binds to an allosteric enzyme and affects the rate of reaction.

Rem Roentgen equivalent man: a biological measure of radiation.

Replication (*Section 16.4*) The process whereby DNA is duplicated to form two exact replicas from the original DNA molecule.

Residue (*Section 13.5*) Another term for an amino acid in a peptide chain.

Respiratory chain (*Section 18.3*) Another term for the electron transport chain.

Response element (*Section 17.6*) A sequence of DNA upstream from a promoter that interacts with a transcription factor to initiate transcription in eukaryotes.

Restriction endonuclease (*Section 17.8*) An enzyme, usually purified from bacteria, that cuts DNA at a specific base sequence. Restriction endonucleases are the tools that allow the creation of recombinant DNA.

Retrovirus (*Chemical Connections 17B*) A virus, such as HIV, that has an RNA genome.

Reuptake (*Section 15.4*) The transport of a neurotransmitter from its receptor back through the presynaptic membrane into the neuron.

Reverse transcriptase (*Chemical Connections 17C*) The enzyme that enables a virus to copy the information encoded in an RNA molecule into a DNA molecule.

Ribonucleic acid (*Section 16.2*) A type of nucleic acid consisting of nucleotide monomers of a nitrogenous base, ribose, and phosphate.

Ribosomal RNA (rRNA) (*Section 16.5*) The type of RNA that is complexed with proteins and makes up the ribosomes used in translation of mRNA into protein.

Ribosome (*Section 16.4*) Small spherical bodies in the cell made of protein and RNA; the site of protein synthesis.

Ribozyme (*Sections 14.1, 16.5*) An enzyme that is made up of ribonucleic acid. The currently recognized ribozymes catalyze cleavage of part of their own sequences in mRNA and tRNA.

RNA (*Section 16.2*) Ribonucleic acid.

s (*Section 6.3*) From the Latin *sinister,* meaning "left"; used in the *R,S* system to show that, when the lowest-priority group is away from you, the order of priority of groups on a stereocenter is counterclockwise.

Saccharide (*Section 11.1*) A simpler member of the carbohydrate family, such as glucose.

Saponification (*Section 10.6B*) Hydrolysis of an ester in aqueous NaOH or KOH to an alcohol and the sodium or potassium salt of a carboxylic acid.

Satellites (*Section 16.6*) Short sequences of DNA that are repeated hundreds or thousands of times but that do not code for any protein or RNA.

Saturated fatty acid (*Section 12.2*) A fatty acid that has no carbon–carbon double bonds.

Saturated hydrocarbon (*Section 2.2*) A hydrocarbon that contains only carbon–carbon single bonds.

Saturation curve (*Section 14.4*) A graph of enzyme rate versus substrate concentration. At high levels of substrate, the enzyme becomes saturated, and the velocity does not increase linearly with increasing substrate.

Secondary (2°) alcohol (*Section 1.4A*) An alcohol in which the carbon atom bearing the —OH group is bonded to two other carbons.

Secondary (2°) amine (*Section 1.4B*) An amine in which nitrogen is bonded to two carbons and one hydrogen.

Secondary messenger (*Section 15.1*) A molecule that is created/ released due to the binding of a hormone or neurotransmitter, which then proceeds to carry and amplify the signal inside the cell.

Secondary structure, of DNA (*Section 16.3*) Specific forms taken by DNA due to pairing of complementary bases.

Secondary structure, of proteins (*Section 13.8*) Repeating structures within polypeptides that are based solely on interactions of the peptide backbone. Examples are the alpha helix and the beta-pleated sheet.

Semiconservative replication (*Section 16.4*) Replication of DNA strands whereby each daughter molecule has one parental strand and one newly synthesized strand.

Serum (*Section 23.2*) A clear liquid that can be extracted from blood plasma. It has all the soluble components of plasma but lacks the fibrinogen that makes blood clots.

Side chains (*Section 13.6*) The part of an amino acid that varies one from the other. The side chain is attached to the alpha carbon and the nature of the side chain determines the characteristics of the amino acid.

Signal transduction (*Section 15.5*) A cascade of events through which the signal of a neurotransmitter or hormone delivered to its receptor is carried inside the target cell and amplified into many signals that can cause protein modifications, enzyme activation, and the opening of membrane channels.

Significant figures (*Appendix II*) The total number of digits in a number with the exception of placeholders; the digits that express the certainty of a number.

Solenoid (*Section 16.3*) A coil wound in the form of a helix.

Soluble immunoglobulins (*Section 22.1*) The antibodies secreted by B cells, which surround and immobilize antigens.

Specificity (*Section 22.1*) A characteristic of acquired immunity based on the fact that cells make specific antibodies to a wide range of pathogens.

Sphingolipid (*Section 12.4*) A phospholipid with sphingosine as the alcohol backbone.

Splicing *(Section 16.4)* The removal of an internal RNA segment and the joining of the remaining ends of the RNA molecule.

Step-growth polymerization *(Section 10.8)* A polymerization in which chain growth occurs in a stepwise manner between difunctional monomers, as, for example, between adipic acid and hexamethylenediamine to form nylon-66.

Stereocenter *(Section 6.2)* A tetrahedral carbon atom that has four different groups bonded to it.

Stereoisomers *(Section 2.7)* Isomers that have the same connectivity (the same order of attachment of their atoms) but different orientations of their atoms in space.

Steroid *(Section 12.9)* A molecule that contains the characteristic four-ring steroid nucleus.

Strong acid *(Section 7.2)* An acid that ionizes completely in aqueous solution.

Strong base *(Section 7.2)* A base that ionizes completely in aqueous solution.

Structural genes *(Section 17.6)* Genes that code for the product proteins.

Substance P *(Section 15.6)* An 11-amino-acid peptidergic neurotransmitter involved in the transmission of pain signals.

Substrate *(Section 14.3)* The compound or compounds whose reaction an enzyme catalyzes.

Subunit *(Section 14.6)* An individual polypeptide chain of an enzyme that has multiple chains.

Surface presentation *(Section 22.1)* The process whereby a portion of an antigen from a foreign pathogen that infected a cell is brought to the surface of the cell.

Synapse *(Section 15.2)* An aqueous small space between the tip of a neuron and its target cell.

T cell *(Section 22.2)* A type of lymphoid cell that matures in the thymus and that reacts with antigens via bound receptors on its cell surface. T cells can differentiate into memory T cells or killer T cells.

Tautomers *(Section 9.8)* Constitutional isomers that differ in the location of an H atom and a double bond relative to an O or N atom.

T-cell receptor *(Section 22.5)* Glycoprotein of the immunoglobulin superfamily on the surface of T cells that interacts with the epitope presented by MHC.

T-cell receptor complex *(Section 22.5)* The combination of T-cell receptors, antigens, and cluster determinants (CD) that are all involved in the T cell's ability to bind antigen.

Termination *(Section 17.5)* The final stage of translation during which a termination sequence on mRNA tells the ribosomes to dissociate and release the newly synthesized peptide.

Termination sequence *(Section 17.2)* A sequence of DNA that tells the RNA polymerase to terminate synthesis.

Terpene *(Section 3.5)* A compound whose carbon skeleton can be divided into two or more units identical to the carbon skeleton of isoprene.

Tertiary (3°) alcohol *(Section 1.4A)* An alcohol in which the carbon atom bearing the —OH group is bonded to three other carbons.

Tertiary (3°) amine *(Section 1.4B)* An amine in which nitrogen is bonded to three carbons.

Tertiary structure *(Section 13.9)* The overall conformation of a polypeptide chain, including the interactions of the side chains and the position of every atom in the polypeptide.

Thiol *(Section 5.5A)* A compound containing an —SH (sulfhydryl) group bonded to a tetrahedral carbon atom.

Thrombocyte *(Section 23.2)* A cell whose function is to control bleeding after an injury; also called a platelet.

Thrombosis *(Chemical Connections 23B)* The process whereby a blood clot breaks loose from one part of the body and lodges in an artery in another part of the body.

Thromboxane *(Section 12.12)* An arachidonic acid derivative involved in blood clotting.

Titration *(Section 7.10)* An analytical procedure whereby we react a known volume of a solution of known concentration with a known volume of a solution of unknown concentration.

TNF *(Section 22.6)* Tumor necrosis factor; a type of cytokine produced by T cells and macrophages that has the ability to lyse susceptible tumor cells.

trans *(Section 2.7)* A prefix meaning "across from."

Transamination *(Section 19.8)* A reaction that involves the interchange of the alpha amino group from the amino acid and a keto group from an alpha ketoacid.

Transcription *(Section 17.1)* The process in which information encoded in a DNA molecule is copied into an mRNA molecule.

Transcription factors *(Section 17.6)* Binding proteins that facilitate the binding of RNA polymerase to the DNA to be transcribed.

Transfection *(Chemical Connections 17C)* The process wherein a virus infects another organism and the viral genome is then expressed in the host cells.

Transfer RNA (tRNA) *(Section 16.4)* The RNA that transports amino acids to the site of protein synthesis in ribosomes.

Transferases *(Section 14.2)* A class of enzymes that catalyze a reaction where a group of atoms, such as an acetyl group or amino group, is transferred from one molecule to another. *(Section 17.5)* Peptidyl transferase; the enzyme activity of the ribosomal complex that is responsible for formation of peptide bonds between the amino acids of the growing peptide.

Transgenic *(Chemical Connections 17C)* An organism that expresses foreign proteins due to transfection.

Translation *(Section 17.1)* The process in which information encoded in an mRNA molecule is used to assemble a specific protein.

Translocation *(Section 17.5)* The part of translation where the ribosome moves down the mRNA a distance of three bases, so that the new codon is on the A site.

Transporter *(Section 15.5)* A protein molecule carrying small molecules, such as glucose or glutamic acid, across a membrane.

Triacylglycerol *(Section 12.2)* Triglyceride.

Tricarboxylic acid cycle *(Section 18.4)* The citric acid cycle.

Triglyceride *(Section 12.2)* A lipid composed of glycerol esterified to a fatty acid at each of its three hydroxyls.

Triple helix *(Section 13.8)* The collagen triple helix is composed of three peptide chains. Each chain is itself a left-handed helix. These chains are twisted around each other in a right-handed helix.

Triprotic acid *(Section 7.3)* An acid that can give up three protons.

Trisaccharide *(Section 11.7)* A combination of three monosaccharides linked by glycosidic bonds.

Tumor suppression factor *(Chemical Connections 17F)* A protein that controls replication of DNA so that cells do not divide constantly. Many cancers are caused by mutated tumor suppression factors.

Ubiquitinylation *(Chemical Connections 19E)* The process of attaching one or more units of the protein called ubiquitin to an aging protein that is ready for degradation.

Unsaturated fatty acid *(Section 12.2)* A fatty acid containing one or more carbon–carbon double bonds.

Unwinding proteins *(Section 16.4)* Special proteins that help unwind DNA so that it can be replicated.

Urea cycle *(Section 19.8)* A cyclic pathway that produces urea from ammonia and carbon dioxide.

Vesicle, synaptic *(Section 15.2)* A compartment containing a neurotransmitter that fuses with a presynaptic membrane and releases its contents when a nerve impulse arrives.

Vitamin *(Section 21.5)* An organic substance required in small quantities in the diet of most species, which generally functions as a cofactor in important metabolic reactions.

Weak acid *(Section 7.2)* An acid that is only partially ionized in aqueous solution.

Weak base *(Section 7.2)* A base that is only partially ionized in aqueous solution.

Zwitterion *(Section 13.3)* A molecule that has equal numbers of positive and negative charges, giving it a net charge of zero.

Zymogen *(Section 14.6)* An inactive form of an enzyme that must have part of its polypeptide chain cleaved before it becomes active; a proenzyme.

CREDITS

Text and Illustrations

Chapter 12 **Fig. 12.3:** From *Biochemistry* by Lubert Stryer © 1975, 1981, 1988, 1995 by Lubert Stryer; © 2002 by W. H. Freeman and Company. Used with permission of W. H. Freeman and Company.
Chapter 15 **Fig. 15.3:** Courtesy of Anthony Tu, Colorado State University.
Chapter 16 **Fig. 16.9a:** From *Biochemistry* by Lubert Stryer © 1975, 1981, 1988, 1995 by Lubert Stryer; © 2002 by W. H. Freeman and Company. Used with permission of W. H. Freeman and Company. **Fig. 16.9b:** Courtesy of Dr. Sung-Hou Kim. **Fig. 16.13:** Adapted from "The Unusual Origin of the Polymerase Reaction," by Kary B. Mullis, illustrated by Michael Goodman. *Scientific American,* April 1990. Reprinted with permission of Michael Goodman.
Chapter 17 **Figure, p. 391:** Illustration by Irving Geis, from *Scientific American,* Jan. 1963. Rights owned by Howard Hughes Medical Institute. Not to be reproduced without permission. **Fig. 17.7:** Courtesy of Dr. J. Frank, Wadsworth Center, Albany, New York.
Chapter 21 **Fig. 21.3:** Courtesy of Drs. P. G. Bullough and V. J. Vigorita and the Gower Medical Publishing Co., New York.
Chapter 23 **Figure, p. 512:** From *The Functioning of Blood Platelets* by M. B. Zucker. Copyright 1980 by Scientific American, Inc. All rights reserved. Reprinted by permission of Patricia J. Wynne.

Photographs

Pages x, xiv, xv, xvi, 10, 13, 20, 34, 40, 46, 66, 68, 77, 84 (top), 94 (left), 95, 105, 116, 117, 126, 172, 183, 189, 202, 205, 207, 209, 214, 219, 220, 221, 222, 230, 237, 261, 263, 311, 312, 442, 468, 472: Charles D. Winters.
Pages 63, 66, 176, 516: Beverly March.
Contents **p. v:** Tom and Pat Leeson/Photo Researchers, Inc. **p. vii:** Lester Lefkowitz/Tony Stone Worldwide. **p. viii:** Douglas Brown. **p. ix:** NSF Oasis Project/Norbett Wu Photography.
Preface **p. xi:** Christy Carter/Grant Heilman Photography, Inc. **p. xii:** Doug Pernine/TCL/Masterfile. **p. xiii:** NASA.
Revision Summary **p. xvii:** Ken Eward/Biografx/Photo Researchers, Inc. **p. xviii:** Bruce M. Herman/Photo Researchers, Inc. **p. xix:** Meckes/Ottawa/Photo Researchers, Inc.
Chapter 1 **p. 1:** Christi Carter/Grant Heilman Photography, Inc. **p. 2:** Pat and Tom Leeson/Photo Researchers, Inc. **p. 4:** D.E. Cox/Stone/Getty Images. **p. 5:** Pete K. Ziminski/Visuals Unlimited. **p. 6:** George Semple.
Chapter 2 **p. 18:** J.L. Bohin/Photo Researchers, Inc. **p. 31:** Tim Rock/Animals/Animals. **p. 39:** Ashland Oil. Co.
Chapter 3 **p. 55:** David Sieren/Visuals Unlimited, Inc. **p. 56:** DonSuzio. **p. 66 (bottom left):** The Stock Market.
Chapter 4 **p. 75:** Douglas Brown. **p. 84 (center):** John D. Cunningham/Visuals Unlimited, Inc. **p. 85:** Chuck Pefley/Stone/Getty Images.
Chapter 5 **p. 91:** Earl Robber/Photo Researchers, Inc. **p. 94:** The Bettmann Archive/Corbis **p. 100** Kathy Merrifield/Photo Researchers, Inc. **p. 104:** Steven J. Krasemann/Photo Researchers, Inc. **p. 105 (top):** Boston Medical Library in the Francis A. Countway Library. **p. 107:** D. Young/Tom Stack & Associates.
Chapter 6 **p. 114:** Lester Lefkowitz/Stone/Getty Images. **p. 129:** William Brown.

Chapter 7 **p. 134:** RB-GM-J.O. Atlanta/Lisson Agency/Getty Images.
Chapter 8 **p. 167:** Jack Ballard/Visuals Unlimited. **p. 169:** Inga Spence/Visuals Unlimited. **p. 182:** Tom McHugh/Photo Researchers, Inc.
Chapter 9 **p. 198:** George Semple.
Chapter 10 **p. 203:** Hans Reinhard/OKAPIA/Photo Researchers, Inc. **p. 204:** Ted Nelson/Dembinsky Photo Associates. **p. 211:** Andrew McClenaghan/Science Photo Library/Photo Researchers, Inc.
Chapter 11 **p. 229:** Courtesy of Dr. Petra Fromme. **p. 234:** Claire Paxton and Jacqui Farrow/Science Photo Library/Photo Researchers, Inc. **p. 235:** Andrew Martinez 1993/Photo Researchers, Inc. **p. 241:** Gregory Smolin. **p. 243:** Martin Dohrn/SPL/Photo Researchers, Inc. **p. 246:** Larry Mulvehill/Photo Researchers, Inc.
Chapter 12 **p. 258:** Doug Perrine/TCL/Masterfile. **p. 272:** Drs. P. G. Bullogh and V. J. Vigorita and the Gower Medical Publishing Co. Ltd. **p. 274:** Carolina Biological Supply/Phototake, NYC. **p. 278:** Jed Jacobson/AllSport USA/Getty Images.
Chapter 13 **p. 289:** Hans Strand/Stone/Getty Images. **p. 301:** G. W. Willis/Biological Photo Service. **p. 308:** J. Gross/Biozentrum/Science Photo Library/Photo Researchers, Inc. **p. 312 (bottom):** Larry Mulvehill/Photo Researchers, Inc.
Chapter 14 **p. 317:** NSF Oasis Project/Norbert Wu Photography. **p. 322:** B. Murton/Southampton Oceanography Centre/SPL/Photo Researchers, Inc.
Chapter 15 **p. 336:** Don Fawcett/Science Source/Photo Researchers, Inc. **p. 334:** Alfred Pasieka/SPL/Photo Researchers, Inc. **p. 345:** Courtesy of Anatomical Collection, National Museum of Health and Medicine. **p. 352:** George Semple.
Chapter 16 **p. 360:** James King-Holmes/SPL/Photo Researchers, Inc. **p. 360 (bottom):** Biophoto Associates/Photo Researchers, Inc. **p. 368 (left):** Chemical Heritage Foundation. **p. 368 (right):** Barrington Brown/Science Source/Photo Researchers, Inc. **p. 375:** Dr. Lawrence Koblinsky. **p. 379:** Larry Schwartz, University of Massachusetts, Amherst.
Chapter 17 **p. 384:** Fredrick A. Bettelheim. **p. 393:** Dr. J. Frank, Wadsworth Center, Albany, NY. **p. 395:** NIBSC/Science Photo Library/Photo Researchers, Inc. **p. 396:** 2001 Trimeris, Inc. and Hoffman-Laroche, Ltd. **p. 398:** Dr. Paul Russell, National Eye Institute, NIH. **p. 403:** Thomas Broker/Cold Spring Harbor Laboratory. **p. 405: (left)** Courtesy of Dr. Anne Crossway, Calgene, Inc., Biotechniques, and Eaton Publishing. **p. 405 (right):** Dr. Marlene DeLuca, University of CA, San Diego.
Chapter 18 **p. 409:** A & L Sinibaldi/Stone/Getty Images. **p. 412:** Dennis Kunkel/Phototake, NYC.
Chapter 19 **p. 430:** Dennis Degnan/Corbis. **p. 450:** Science Photo Library/Photo Researchers, Inc.
Chapter 20 **p. 455:** Bruce M. Herman/Photo Researchers, Inc. **p. 458:** Dr. Petra Fromme.
Chapter 21 **p. 469:** National Library of Medicine. **p. 475:** Courtesy of Drs. P. G. Bullogh and V. J. Vigorita and the Gower Medical Publishing Co. Ltd. **p. 482:** George Semple.
Chapter 22 **p. 486:** Meckes/Ottawa/Photo Researchers, Inc. **p. 489:** David M. Phillips/Photo Researchers, Inc. **p. 495:** Courtesy of Dr. R. J. Poijak, Pasteur Institute, Paris, France. **p. 497:** Dr. Jeremy Burgess/Science Photo Library/Photo Researchers, Inc.
Chapter 23 **p. 508:** Ken Eward/Biografx/Photo Researchers, Inc.

Key: **Boldface** indicates glossary entry; *italics* indicates figure; 30*t* indicates table.

A (acceptor) site, *392–393*

Abacavir, in AIDS treatment, 395

Accolate, 284

Accupril, 328, 520

ACE (angiotensin-converting enzyme), 328, 520

Acesulfame-K, 248*t*, 480, 481

Acetaldehyde, 12, 184, 187

Acetals, **192–194**, 240–241. *See also* Hemiacetals

Acetamide, 210

Acetaminophen, 284, 330

Acetanilide, hydrolysis of, 218

Acetate ion, 136–137, 138, 140–141, 151, 154–155, 157, 259, 344

Acetic acid, 7–8, 13, 138, 140–141, 151, 154–155, 157, 187, 204*t*, 205–206, 206*t*, 207, 212, 213, 215, 218, 319, 320*t*

Acetic anhydride, 208, 215

Acetoacetate, 441, 442–443, *444*

Acetoacetyl-ACP, in fatty acid biosynthesis, 461

Acetoacyl-CoA, 441, 442

Acetone, 12, 184, *194*, 442–443

Acetylation, during DNA replication, 375–376

Acetylcholine, 338*t*, 319, 320*t*, 340, 341–346

Acetylcholine receptor (AChR), *341*, *342*, 503

Acetylcholinesterase, 319, 320*t*, 343, 344, 346

Acetylcholine transferase, 345

Acetyl-CoA, 345, *416*, 417–419, 431, *432*, 435, 437*t*, 439–441, *440*, 442, 443, *444*, *449*, 460, 463

Acetyl-CoA carboxylase, as cofactor, 475*t*

Acetylene, 7*t*, 19, 46, 50

N-Acetyl-D-galactosamine, 246

N-Acetyl-D-glucosamine, 234, 246, 251, 310

Acetyl group, **414**–416, 417–419

Acetylsalicylic acid, 282

Acetyl-transporting molecules, 414–416

Achiral objects, **118**

Acid–base indicators, 149–150, 150*t*, *151*, 152–*153*, *156*

Acid–base reactions, 138–140, 140–142

Acid–base titration, **152–154**, *153*

Acid dissociation constant, 143

Acidic amino acids, 290, 291*t*, 293

Acidic polysaccharides, 230, 250–251, 310

Acidic salts, 150–152, **151**

Acidic solutions, 148–152, 154–160

Acid ionization constants (K_a), 142–144, **143**

Acidity, 96–97, 106–107. *See also* Acids

Acidosis, 159, 435, 518

Acid phosphatase, 319, 331*t*

Acids, 84, 134–160, *137*, 175–176, 206, 240, 322, 323. *See also* Acidity; Carboxylic acids; Dicarboxylic acids; DNA; RNA
  amphiprotic, monoprotic, diprotic, and triprotic, **139**, 140, 293
  strong and weak, **136**–137, 140–142, **143**

Aconitase, 320*t*, 417

cis-Aconitate, 320*t*, 417

Acquired immunity, *487–488*

Acrolein, molecular structure of, 184

Acrylonitrile, polymers of, 65*t*

ACTH. *See* Adrenocorticotropic hormone (ACTH)

Actin, 289, 290, 308, 425, *426*, 500

Actinium-225, 497

Activation
  of enzymes, 321
  of glycolysis, *433–434*, 437*t*, 441*t*
  of protein synthesis, 390, 390*t*, 391

Active sites, **321**, 323–326, 329

Active transport, across cell membranes, 266, *267*

Acyl carrier protein (ACP), *460*–461

Acyl-CoA, 439, *440*, 441*t*, 461–462

Acyl-CoA dehydrogenase, *440*, 441*t*

Acyl group, **208**, 439

Adalat, 338*t*, 520

Adaptive immunity, 487–488, 491, 501–502

Addition reactions, of alkenes, 57–64

Adenine, 361–362, 364*t*, 365–*366*, *367–369*, *368*, 413

Adenosine, 362, 364*t*, 413–414

Adenosine diphosphate (ADP), 325, *349*, 413–414, *415*, 422–424, 425–426, 431–432, *433–434*, *446*, 459

Adenosine monophosphate (AMP), 320*t*, 391, 413–414, 439–441, *440*, *446*. *See also* Cyclic AMP (cAMP)

Adenosine triphosphate (ATP), 266, 325, 348, *349*, 391, *410*, 411, 413–414, 418, 422–425, 431–*432*, *433–434*, 437, 437*t*, 439–441, *440*, *444*, 445, *446*, 457, 459

Adenylate cyclase, 348, *349*–351, 352–353

Adipamide, 211–212

Adipic acid, 203, 220

Adipocytes, 431

ADP. *See* Adenosine diphosphate (ADP)

Adrenal cortex, 339*t*, *340*

Adrenal glands, 177, 279

Adrenaline, *167*, 177, 337, 338*t*. *See also* Epinephrine

Adrenal medulla, hormone from, 339*t*

Adrenergic neurotransmitters, 338*t*, 340, 347–353

Adrenocorticoid hormones, 277, 279

Adrenocorticotropic hormone (ACTH), action of, *340*

Advanced glycation end-products (AGE), 298

Advil, 214

Aerobic pathways, lactate accumulation and, 435

Agents GB, GD, and VX, 343

Aging, 298, 402, 519

Agonist drugs, 337, 346

AIDS (acquired immunodeficiency syndrome), 387, 395–396, 398, 487, 499–500. *See also* Human immunodeficiency virus (HIV)

Alanine, 291*t*, 292, 294, *306*, 389*t*, 448, *449*, 464–465

β-Alanine, 205, 346–347

Alanine aminotransferase (ALT), 331, 331*t*

Albinism, 360

Albumin, 290, 511, 511*t*, 512

Albuterol, *167*, 177, 178

Alcohols, **9**–11, 55, 91–101, 96*t*, 107–108, 189–191, 192–194, 212–213, 217, *265*, 269, 312–313. *See also* Breath-alcohol screening

Aldehyde group, 9*t*, 231

Aldehydes, **12–13**, 99–101, 183–195, *189*, 298, 351

Alditols, **241**–242

Aldohexoses, **236**–238, 238–239, 240

Aldolase, in glycolysis, *433*

Aldonic acids, 242, 243

Aldopentoses, furanose form of, **236**, 238

Aldoses, **231**, 233–234, 242

Aldosterone, 277, 279, 339*t*, 518

Aleve, 214

Aliphatic amines, *167*–169, 170–171, 173–175

Aliphatic hydrocarbons, **18**

Alkadienes, 53

Alkaline phosphatase (ALP), 331*t*

Alkalis. *See* Bases

Alkaloids, 169, 345

Alkalosis, of blood, 159

Alkanes, **18**–40, 35*t*, 64, 92, 96*t*,
  cis-trans isomerism in, **32–34**
  cyclic, **27–34**
  as hydrocarbons, **18**–19

Alkenes, 18, 19, **46**–68, 57*t*, 76
  cis-trans isomerism in, **48**, 50–52
  converting alcohols to, 97–99
  hydration of, 57*t*, **60**–62, 98–99

Alkylbenzenes, 77–80

Alkylbenzesulfonate detergents, 82–83

Alkyl groups, **24**, 25*t*, 187, 208–210

N-Alkylpyridinium chloride, *172*

Alkyl substitutions, 77–80

Alkynes, 18, 19, **46**, 49, 50, 55

Allegra, 338*t*, 509

Allergens, 503

Allergies, 300, 503. *See also* Side effects

D-Allose, 232*t*

Allosteric effect, 513

Allosteric enzymes, 329

Allosterism, enzyme regulation via, **329**

α-1-Adrenoreceptors, 519

Alpha amino acids, **290**, 291*t*

Altace, molecular structure of, 328

Altman, Sidney, 317

Aluminum (Al), 3

Alzheimer's disease, 303, 345, 346

Amidation, of glutamate, 448

Amides, **209**–212, 213–215, 218–219

Amine group, 9*t*, **11–12**, 186, 234–235

Amines, **11–12**, 145–146, **167–178**, 213–215
  aliphatic and aromatic, *167*–169, 173–175
  heterocyclic, **168**, 174, 175

Amino acid-AMP molecule, 391

Amino acid pool, 431, *444*

Amino acids, **290–292**, 291*t*, 292–304, 318, 323, 325, 338*t*, 345, 360–361, 371, 391–393, 399–400, 401, *416*, 431, *432*, 436, 443–450, *444*, *449*, 469, 493, *494*
  antigenicity of, 491–492
  biosynthesis of, 464–465
  chemical communication via, 340, 346–347, 504
  essential and nonessential, 464, **474**
  genetic code for, 385, 388–389, 389*t*, *389*–390
  peptide bonds between, **294–296**, 318

Amino acid substitutions, in biochemical evolution, 401

Amino acid transport, 465

Amino acid-tRNA molecule, *391*, 397

Amino acid-tRNA synthetase (AARS), 391, 397, 399

p-Aminobenzoic acid, (PABA), 82, 326, 482

4-Aminobutanoic acid, 204, 346–347
γ-Aminobutyric acid (GABA), 204, 346–347
Amino group, 82
Amino sugars, **234**–235
Ammonia, 135–136, 137, 138, 140–141, 145–146, 173, 189, 202, 207, 323, *432*, 445, 448, 517
Ammonium benzoate, 207
Ammonium chloride, 3
Ammonium hydroxide , 135–136
Ammonium ion, 138, 140–141, 145–146, 218, *444*, 445, 448
Ammonium salts, 172, 213
Amniocentesis, Tay-Sachs disease and, 272
Amoxicillin, 211, 323
AMP. *See* Adenosine monophosphate (AMP)
Amphetamine, 168, 170–171
Amphetamines, 168, 338*t*
Amphiprotic acids, **139**, 293
Ampicillin, 211
Amylase, 331*t*
α-Amylase, 249, 472, 473
β-Amylase, in carbohydrate digestion, 472, *473*
Amyloid fibril formation, 303
Amyloid plaques, 303
Amyloid precursor protein (APP), in Alzheimer's disease, 345
Amylopectin, 249, 472
Amylose, 249, 472
Anabolic steroids, 278, 482
Anabolism, 409, 455–456
Anaerobic pathway, reactions of, 435
Analgesics, 214
Androgens, action of, 339*t*
4-Androstene-3,17-dione ("Andro"), 278, 482
Anemia, 300, 301, 373, 377, 400, 421, 435, 511*t*
Anesthetics, ethers as, 91, *105*
"Angel dust," 347
Angiotensin, 355, 520
Angiotensinogen, 328, 520
Anhydrides, 202–203, **208**, 215, 219, 363, 413–414
Aniline, 77, 168, 170, 218
Anilinium, 172
Anion transporter, of red blood cells, 266
Anisole, 77
Anomeric carbon, 235, **236**, 238, 239, 240, 241, 245, 247
Anomers, of monosaccharides, **236**, 237–238, 239, 240, 241
Anorexic agents, 355
Antacids, treating heartburn with, 145
Antagonist drugs, 337, 346
Anterior pituitary gland, hormones secreted by, 339*t*, *340*
Anthracene, 80
Antibiotics, 126, 210, 211, 235, 394, 465
Antibodies, 246–247, 290, 395, 487, **488**, 491, 492, *493*–497, 510
Anticancer drugs, 365, 394
Anticoagulants, 251, 512
Anticodon loops, in tRNA, *371*
Anticodons, **389**, 390, 391, 392–393, 397
Antidepressants, 338*t*
Antidiuretic hormone (ADH), 300, *340*, 355, 518
Antidiuretics, 518
Antifreeze, 95
Antigen–antibody complex, 493–*495*
Antigen receptors, in immune system, *487*, 488
Antigens, 246–247, 290, 447, **488**, 490*t*, 491–*493*, *492, 494, 496*, 498
Antihistamines, 352, 509
Anti-inflammatory drugs, 279, 282
Antiknock ratings, of gasoline, 40

Antioxidants, 84–87, 261, 475
Antisense drugs, 387
Antiseptics, 137
Antithrombin, 512
Antiviral agents, 394
AP (apurinic/apyramidinic) site
apoB-100 protein, in cholesterol transport, 275
Apoenzyme, 320, *321*
Apoptosis, 379, 402
APP. *See* Amyloid precursor protein (APP)
Appetite-stimulating agents, 355
Aqueous humor, 508
Aqueous solution, 95–97, 148, 150–152, 173, 218, 239, 240, 292–293
Ara-A, molecular structure of, 394
D-Arabinose, 232*t*
Arachidic acid, 204*t*, 260*t*
Arachidonic acid, 214, 260*t*, 282, 283, 284, 330, 473
Arenes, 18, 19, **75**–87
Arginase, 318, 475*t*
Arginine (Arg, R), 291*t*, 318, 352, 389*t*, 446–447, *449*
Argininosuccinate, 445–*446*
Argon krypton laser surgery, 312
Aricept, 338*t*, 345
Aromatic amines, **167**–169, 170–171, 173–175
 heterocyclic, **168**, 174, 175
Aromatic compounds, **75**–87
Aromatic sextet, 77
Aromatic substitutions, 81–83
Arrhenius, Svante, 134, 323
Arterial blood, *514*
Arteries, 276, 277, *514*, 516, *517*
Arthritis, 282, 501, 503
Artificial sweeteners, 248*t*, 480–481
Aryl groups, 75, 187, 209–210
Ascorbic acid, 6, 478*t*. *See also* Vitamin C
Asparagine (Asn, N), 291*t*, 310, 389*t*, *449*, 464
Aspartame, 248*t*, 449, 480–481
Aspartate, 320*t*, 445, *446*, 464. *See also* Aspartic acid (Asp, D)
Aspartate aminotransferase (AST), 320*t*, 331, 331*t*, 475*t*
Aspartic acid (Asp, D), 291*t*, 295, 346, 389*t*, *449*. *See also* Aspartate
Aspirin, 6, 214, 282, 284, 330
Asthma, 177, 282, 284, 509
Astigmatism, laser surgery to correct, 312
Atherosclerosis, 274, 276, 463, 475, 511*t*
Atkins diet, 473
Atomic number, 120–123, 121*t*
ATP. *See* Adenosine triphosphate (ATP)
ATPase, *410*, 422–424, 458
Atropine, 338*t*, 343
Attention deficit disorder (ADD), treatment of, 350
Attractive forces, of immunoglobulins, 493–495
Autoimmune diseases, 503–504. *See also* AIDS
 vitamin treatment of, 481
Autoxidation, 84–87
Avery, Oswald, 361
Axial bonds, 30–32
Axons, 338–339
 electrical signal transmission by, *337*
 myelin sheaths on, 269–270
Azidothymidine (AZT), 395

Bacteria, 82, 250, 322, 323, 326, 327, 361, 381, 385, 389–390, 394, *403–404*, 423, 445, 465, 487, *489*, 490*t*, 492, 500, 501, 503
Baking soda, 145. *See also* Sodium bicarbonate
Barbiturates, 217, 423
Barium hydroxide, 136*t*

Barnes, Randy, 278
Baroreceptors, 519–**520**
Basal caloric requirement, **472**
Base excision repair (BER), 377, *378*, 379
Base pairs, in DNA molecule, *367–369*
Bases, 134–160, 137, 106, 167, 173–175, 207–208, 218, 322
 strong, **136**–137, 140–142
 weak, **136**–137, 140–142
Bases (of nucleic acids), **361**–362, 365–366, *367–369*, 370, 373, 376, 377
Basic amino acids, 290, 291*t*, 293
Basic salts, 150–152, **151**
Basic solutions, 148–150, 150–152, 154–160
Bay of Minamata, mercury poisoning in, 327
B cell receptors (BcR), in immune system, *487*, 502
B cells, *487*, 488, 489, 490*t*, 491, 495, *496*, 500, 501–502
Beadle, George, 361
Beeswax, 262
Benadryl, 338*t*, 352, 509
Benzaldehyde, 77, *183*, 185, 186, 188, 192
Benzene, 19, 75, 76–77, 81–83, 401
1-4-Benzenediamine, 220
1,4-Benzenedicarboxylic acid, 220–221
Benzenediol, 84
Benzene rings, 18, 19, 76–77, 273–274
Benzenesulfonic acid, 82
Benzodiazepine, 172
Benzoic acid, 77, 143–144, 188, 207, 213, 215–216
Benzo[a]pyrene, as carcinogen, 80, 402
Benzylamine, 175
Benzyldimethylamine, 168
Beriberi, 469, 475
Beta blockers, in hypertension treatment, 520
Betazole, 338*t*
Bicarbonate ion, 139, *140*, 158, 518
Bicarbonates, reactions with acids, 145–146
Bile, heme catabolism and, 451
Bile salts, 281–282, 473
Bilirubin, 450, 451
Biliverdin, 450, 451
Binding curve, 513
Binding proteins, in transcription, 385
Binding sites, 129–130, 493–495, *494*
Biochemical evolution, mutations and, 401
Biochemical pathways, 409–411
Biodegradable sutures, 221
Bioenergetics, 409–426
Biosynthesis, 277, 455–465. *See also* Anabolism
Bioterrorism, vaccines and, 499
Biotin, 475*t*, 478*t*
Birth-control pills, 281
Bisphenol A, 221–222
1,3-Bisphosphoglycerate, 432, *433–434*, 437*t*
Black, James W., 352
Blood, 174, 509–515
 bilirubin in, 451
 buffers in, 158, 518
 checking alcohol level in, 101
 chemical balance of, 518
 cholesterol in, 274, 276
 clotting of, 251, 284, 511–512
 enzymes in, 318, 330–331
 glucose in, 234, 243
 hormones in, 339–340
 immune system and, *489*
 ketone bodies in, 443
 kidneys and, 510, 515–517, 518
 lactate in, 191
 nitric oxide and, 352
 pH of, 149, 158, 159, 510

salt balance in, 518
thromboxane A$_2$ and, 284
Blood–brain barrier, 462, **509**
Blood circulation, *514*
Blood plasma, **508**–509
Blood pressure, 300, 519–520
Blood sugar, 230, 234, 243, 354, 473
Blood transfusion, 395, 512
Blood types, 246–247, 491, 504
Blood vessels, 352, 516, *517*
Blood volume, in blood pressure regulation, 520
Blow-molding techniques, 66, *67*
Blue diaper syndrome, 465
Blue litmus paper, *137*, 149
Body fluids, 174, 176, 330–331, *489*, 508–520
Body mass index (BMI), 472
Body temperature, 510
Bohr effect, **514**
Boiling point
    of alcohols, 95, 96*t*, 206*t*
    of aldehydes, 187–188, 187*t*, 206*t*
    of alkanes, 34–35, 36
    of amines, 172–173
    of carboxylic acids, 205, 206*t*
    of ethers, 103–104
    of ketones, 187–188, 187*t*
    of thiols, 106, 107*t*
Bond angles, 6–8, 19–20
    in alcohols, *92*
    in alkenes, 47–48
    in ethers, *102*
    in thiols, 104
Bone marrow, 301, 502, 510, 511
Boric acid, 137
Botox (botulin toxin), 344
Botulism, 344, 346
Bowman's capsule, 515–516, *517*
Brain peptides, as neurotransmitters, 353–355
Bread dough, rising of, 145
Breast cancer, 281, 356, 497
Breath-alcohol screening, 101
Breathing, 514, 515
Brethaire, 177
Bromination, 57*t*, 62–63, 81
Bromine (Br$_2$), 2, 7, 37–38, 62–*63*, 76
Bromobenzene, 81
p-Bromobenzoic acid, 78
Bromocresol green, as acid–base indicator, 150*t*, *156*
Bronchodilators, *167*, 177–178
Bronchoscopy, muscle relaxants during, 319
Brønsted, Johannes, 138
Brønsted-Lowry acids and bases, 138–140
Brown, Michael, 275
Brown fat, hibernation and, 422
Bruises, heme catabolism and, 451
Buffer capacity, **157**
Buffers, **154**–160, 293, 518
Bullus pemphigoid, vitamin treatment of, 481
Butanal, 187*t*, 188
Butanamide, 211–212, 218
2-Butanamine, 170–171
Butane, *20*, 21, 27, *29*, 35*t*, 36, 96*t*
Butanedioic acid 203, 213
Butanethiol, 105–106
Butanoic acid, 204*t*, 206, 206*t*, 207, 218, 261
1-Butanol, 92, 96*t*, 107*t*, 187–188, 187*t*
2-Butanol, 92, 97, 114–118
(*R*)-2-Butanol, 122
2-Butanone, 187*t*, 188, 193
2-Butene, 59, 62, 97
Butyl alcohol, 92
sec-Butyl alcohol, 92
tert-Butyl alcohol, 92, 97

sec-Butylamine, 171
Butylated hydroxyanisole (BHA), 87
Butylated hydroxytoluene (BHT), 87
tert-Butyl ethyl ether , 102
Butyl group, 25*t*
sec-Butyl group, 25*t*
tert-Butyl group, 25*t*
sec-Butyl mercaptan, 105–106
2-Butyne, 50
Butyramide, 211–212
Butyric acid, 204*t*, 261

Cadaverine, 170–171
Calcium (Ca), *3*, 478*t*
Calcium channel blockers, 520
Calcium channels, 342
Calcium hydroxyapatite, 308
Calcium ions, 264, 338*t*, 341–343, 458, 517, 520
Calcium oxide 145
Calcium propanoate, *207*
Calmodulin, 342
Calories, 469, 471–472
Calvin, Melvin, 459
Calvin cycle, in photosynthesis, 459
Cancer, 5, 80, 312, 365, 373, 464, 499, 519
Cancer cells, 374, 489, 503, 504
Capillaries, 301, 487, *489*, *509*, *514*, *517*, 519
Capping, in transcription, *386*
Capric acid, 204*t*
Caproic acid, 203, 204*t*
Caprylic acid, 204*t*
Capsaicin, 85
Captopril, 126, 328, 520
Carbachol, 338*t*
Carbaminohemoglobin, 515
Carbamoyl phosphate, 445, *446*, 447–448
Carbocations, *59*, 61, 62
Carbohydrate intake, minimum daily, 473
Carbohydrates, 107, 123–124, 230–251, **231**, 298, 310, 410, *416*, 417–419, 430–431, *432*, 456–460, 469, 472–473, 504
    disaccharides as, **245**–248
    monosaccharides as, **231**–244
    oligosaccharides as, **245**–248
    polysaccharides as, **245**, 249–251
Carbonate buffer system, 158
Carbon dioxide, 145–146, *410*, *416*, 418, 424–425, 445, 456, *457*, 458–459, *509*, 510, 511*t*, 514, 515, 520
Carbonic acid, 139, 145–146, 148, 158, 221–222, 515
Carbonic anhydrase, 515, 518
Carbon radical, in autoxidation, 85–86
Carbon tetrachloride, 37, 38–39, 62–*63*, 401
α-Carbon, 194–195, **290**, 292, *296*, 439
Carbonyl group, *12*–13, 183, 184, 187–188, 194–195, 202–203, 209–210, 235. *See also* Ketone group
N-Carboxybiotin, as coenzyme, 475*t*
Carboxylation, of PEP, 442
Carboxyl group, **13**–14, 202–203, 208–209, 250
Carboxylic acids, 9, 9*t*, **13**–14, 99–101, 187, 188, 189, 202–213, 215, 218. *See also* Fatty acids
Carboxypeptidases, *302*, 318, 431, 474
Carcinogens, 80, 401, 402
Cardiac arrest, lactate accumulation and, 435
Cardiovascular disease, obesity and, 472
Carnauba wax, 262
Carnitine, in fatty acid β-oxidation, 439
Carnitine acyltransferase, 439
Carotene, *46*
Carotenoids, 475
Carson, Rachel, 79

Cartilage, 235, 250–251, 501
Casein, 290, 474
Caspases, in apoptosis, 379
Catabolism, 409, 438, 443–448, 450–451, 455–456
Catalase, in detoxification, 423
Catalysts, 64, 97–99, 190–191, 289, 317–318
Cataracts, 298, *398*
Catechol, 84, 177
Catecholamines, 177
Catechol-*o*-methyl transferase (COMT), 350
CDP-choline, in lipid biosynthesis, 462
CDs. *See* Cluster determinants (CDs)
Cech, Thomas R., 317
Cefalexin, 211
Celebrex, 282
Cell-adhesion molecules. *See* Adhesion molecules
Cell membranes, 246–247, 258, 259, 265–*267*, 282, 336–337, *349*, 352, 356–357, *411*, 447, *458*, 461–464, *498*–499, 508–509. *See also* Thylakoid membrane
Cell nucleus, 336, *337*, 361, 379, 385, 388, 390, *411*, 431, 510
Cells, 308–309, 336–337, 338, 409–426, *411*
Cellulose, 230, 250, 456, 470
Central dogma of molecular biology, 384–385
Cephalins, 268
Cephalosporin, 210, 211
Ceramide, 269, 270–271, 462
Ceramide oligosaccharide, in Tay-Sachs disease, 273
Cerebrosides, 270–271
Cerebrospinal fluid, 330–331, 331*t*, 508, **509**
cGMP. *See* Cyclic GMP (cGMP)
Chain, Ernst, 211
Chain-growth polymers, 64
Chain initiation, in autoxidation, 85
Chain length, in autoxidation, 86
Chain propagation, in autoxidation, 85–86
Chair conformation, **29**–32, 238–239, 240, 245, 247, 248
Chaperones, 303, **307**, 313, 400, 412, 447
Chargaff, Erwin, 365
Chargaff's rule, 365, 367
Chemical bonds, 301
    in acid–base reactions, 138, 140–141
    angles between, 6–8, 19–20, 47–48
Chemical communication, 336–357, 338*t*, **348**, 488, 495, 504
Chemical energy, 425–426
Chemical messengers, 258, 259, 337, 340–341
Chemical reactions, rates of, 321–323
Chemical weapons, 343
Chemiosmotic pump, 422–424, 458
Chemiosmotic theory, **422**
Chemokines, 500, **501**, 503
Chemotactic cytokines, 501
Chemotherapy, 350, 365. *See also* Drugs; Medicine; Therapy
Chemstrip blood sugar test kit, 243
Chickenpox, 394, 486
Chinese herb tea, for Alzheimer's disease, 345
Chiral molecules, 114–130, **117**, **127**, 292
Chitin, 230, 235
Chloride channels, cystic fibrosis and, 447
Chloride ions, 135, 138, 140, 478*t*, 517, 518
Chlorinated water, 483
Chlorination, of benzene, 81
Chlorine , 2, 7, 37–38, 57*t*, 58, 59, 62–63, 221–222
Chloroacetic acid, 206
Chlorobenzene, 81
2-Chlorobutane, 59

Chlorodiazepoxide, 172
Chlorofluorocarbons (CFCs), 38, 39
Chloroform, 37
Chloromethane, 37
Chlorophyll, 456, 458–459
Chloroplasts, *458*
Cholera, 351
Cholesterol, 127, *267*, **274**–277, *275*, 281–282, 436, 462–463, 469, 473, 511*t*
Cholesteryl linoleate, 274–275
Choline, *265*, 267, 319, 320*t*, 344, 345, 462, 473
Cholinergic neurotransmitters, 338*t*, 340, 341–346
Cholinergic receptors, 341
Chromatin, 369, *370*, **376**, 385
Chromium (Cr), 479*t*
Chromosomes, 361, 369, *370*, 375–376, 377, 385
    human, *360*, 373, 374
Chylomicrons, 274
Chymotrypsin, 319, 431, 474
Cinnamaldehyde, *183*, 186, 190
Cinnamyl alcohol, 190
Circulation, *514*
Cis-trans isomers, **32**–34, **48**, 50–52, 53–55, *115*
Citrate synthase, 417, 419
Citric acid, *202*, 206
Citric acid cycle, *410*, 412, 413, *416*, *417*–419, *432*, *434*, 435, 437*t*, 442, *449*, 450, *457*, 464
Citronellal, 186
Citrulline, 352, 445, *446*
Citryl-CoA, in citric acid cycle, 417
Clamp loaders, 374
Clamp protein, *373*, 374*t*
Class I MHC, *492*
Class II MHC, *492*–493
Class switch recombination (CSR) mutation of V(J)D genes, 496
Clones, *360*, 380
Clonidine, 338*t*
Cloning, of DNA, 380–381
Clorazil, 338*t*
Clotting, 251, 284, 290, 318, 511–512
Cloverleaf structure, of tRNA, *371*
Clozapine, 338*t*
Cluster determinants (CDs), *492*, 498, 499
CMP. *See* Cytidine 5′-monophosphate (CMP)
Coated pits, in cholesterol transport, 275
Cobalamine, 475*t*. *See also* Vitamin B₁₂
Cobalt (Co), 479*t*
Cobra venom, 346
Cocaine, 169
Codon recognition site, 389
Codons, **388**, 389–390, 389*t*, 391, 392–393, 400
Coenzyme A, 414–*416*, *415*, 439, *440*, 441*t*, 461–462, 475*t*
Coenzyme Q (CoQ), 420, 421
Coenzymes, **321**, 326, 475*t*
Cofactors, **320**, 325, 458–459, 475*t*
Cold virus, 394
Collagen, 251, 289, 290, 304, 308, 309, 311, 474, 512
Collecting tubule, in kidneys, *517*
Collins, Robert John, *105*
Colon cancer, 282, 470
Color-blindness, 360
Combinatorial chemistry, 6
Combustion, of alkanes, 37
Common catabolic pathway, **410**–411, 413–416, 417–419, 420–422, 431, *432*
Common cold, 394
Competitive inhibition of enzymes, **321**, 324–326

Complementary base pairs, in DNA molecule, *367–369*
Complete protein, 474
Complexes I–IV, in electron and hydrogen ion transport, 420, 421–422
Complex lipids, 259, 264–*265*, *268*, 271, 272–273, 282. *See also* Glycerophospholipids; Glycolipids; Sphingolipids
Compounds
    aromatic, **75**–87
    inorganic, 3–4, 4*t*
    organic, 3–4, 4*t*
    polar, 103–104
Condensed structural formulas, 10, 11, 20*t*
Conformations, **29**–32
(*S*)-Coniine, 169
Conjugate acids and bases, **138**–140, 139*t*, 158–160
Conjugated proteins, 309
Connective tissue, acidic polysaccharides in, 250–251
Consensus sequences, in transcription, 386–387
Constitutional isomers, **21**–23, 23*t*, 23–26, *115*, 294
Contraceptive drugs, 280–281
Contributing structures, 76–77
Cooking oils, autoxidation of, 84
Copper (Cu), 475*t*, 479*t*
Corn, incomplete proteins in, 474, 475
Corn syrup, high-fructose, 249
Cornea, 141, 312
Coronary thrombosis, lactate accumulation and, 435
Cortisol, 277, 279, 339*t*
Cortisone, 277, 279, 280, 282
Cotton, cellulose in, 250
Covalent bonds, 4, 6–8, 304. *See also* Polar covalent bonds
Cowpox, in smallpox immunization, 499
COX-1 and COX-2 enzymes and inhibitors, 282, 283
Cracking, 47
Cramps, lactate accumulation and, 435
Crash diets, 472
Creatine, 482, 517
Creatine phosphokinase (CPK), after heart attack, 331, 331*t*
Creatinine, in urine, 517
Cresol, 84
Creutzfeld-Jakob disease, 303
Crick, Francis, 367, *368*
Cristae, of mitochondria, 412
Crohn's disease, treatment of, 503
Cross-linking, 309, 326, 327
Crutzen, Paul, 39
Cryoelectron microscopy, of ribosomes, 393
C-terminal amino acid in peptides, 296, 298, *302*
Curare, 338*t*, 346
Cure-all tonics, *469*
Cushing syndrome, 280
Cyanobacteria, 456
Cyclic amides, 210, 211
Cyclic AMP (cAMP), 340, 342, 348, *349*, 353
Cyclic esters, 209, 210
Cyclic ethers, **103**
Cyclic GMP (cGMP), diabetes and, 354
Cyclic hemiacetals, creation of, 192
Cyclic hydrocarbons, 27
Cycloalkanes, **27**–32, **32**–34
Cycloalkenes, 52–53, 63
Cyclohexane, 27, **29**–32
Cyclonite (RDX), 82

Cyclooxygenase (COX), 214, 282, 283, 284, 330
    See also COX-1 and COX-2 enzymes and inhibitors
Cyclopentane, 27, *29*
Cyclosporin A, for autoimmune diseases, 504
CYP2D6 gene, mutations of, 377
Cysteine, 291*t*, 294, 304, 305, 327, 346, 389*t*, *449*, 464, 474
Cystic fibrosis, ubiquitin and, 447
Cystine, 294, 449
Cystinuria, 449
Cytidine, 364*t*, 395
Cytidine diphosphate (CDP), 462
Cytidine 5′-monophosphate (CMP), 364*t*
Cytochrome *c*, 299, 421
Cytochrome oxidase, 421, 475*t*
Cytochrome P-450 enzyme, 377, 423
Cytochromes, 421, 458
Cytokines, 395, **500**–501, 503–504
Cytoplasm, *266*, *267*, *337*, 385, 388, *411*, 435, 436, 457, 460
Cytosine, 361–362, 364*t*, 365–*366*, *367*–369, *368*, 462
Cytoskeleton, 308, *411*
Cytosol, 390, 439, *440*

Dacron, 78, 221
Daily Values, for nutrients, 470, 482
Dalton's law of partial pressures, 514
Dark reaction, in photosynthesis, 459
Daughter strands, in DNA replication, 374
Davy, Humphry, 105
ddCyd. *See* Dideoxycytidine (ddCyd)
ddI. *See* Dideoxyinosine (ddI)
DDT (dichlorodiphenyltrichloroethane), 79
Deacetylation, during DNA replication, 375–376
Deamidation, DNA mutations from, 377
Debranching enzyme, in carbohydrate digestion, 472, *473*
Decamethonium bromide, 346
Decarboxylation, 350, 351–352, **417**–418, 463
Deficiency diseases, 469
Dehydration, **97**–99, 351, 443, 511*t*
L-Dehydroascorbic acid, 237
Dehydrogenation, of fatty acids, 439, *440*, 441*t*
Demyelination, in multiple sclerosis, 270
Denaturation, 310–313, **311**
Denaturing agents, of proteins, 311*t*
Dendrites, *337*, 338–339
Density, 35*t*, 36
Dentine, composition of, 308
Deoxyadenosine, 364*t*
Deoxyadenosine 5′-monophosphate (dAMP), 364*t*
Deoxycytidine, 364*t*
Deoxycytidine 5′-monophosphate (dCMP), 364*t*
Deoxyguanosine, 364*t*
Deoxyguanosine 5′-monophosphate (dGMP), 364*t*
β-2-Deoxy-D-ribose, 238, 362–363, 370
(−)-2-Deoxy-3′-thiacytidine (3TC), 395, *396*
Deoxythymidine, 364*t*
Deoxythymidine 5′-monophosphate (dTMP), 364*t*
Dephosphorylation, ATP yield from, 437*t*
Deprenyl, 338*t*, 350
Detergents, 82–83, 264, 281–282
Detoxification, 423
Dextrins, 249
Dextrorotary compounds, **128**, 232–234
Dextrose, 234
dGMP. *See* Deoxyguanosine 5′-monophosphate (dGMP)
Diabetes, 85, 298, 300, 354, 443, 472, 503, 511*t*, 517

Diabetes insipidus, 518
Diabetes mellitus, 243, 354, 443, 473
Diabetic retinopathy, 298
Diastereomers, *115*, 123–**124**
Diatomaceous earth, in dynamite, 94
Diazepam, molecular structure of, 172
Dicarboxylic acids, 203
Dichloroacetic acid, 206
Dichlorodifluoromethane, 38
1,1-Dichloroethylene, 65*t*, *66*
Dichloromethane, 37, 39, 62
Dichromate ion, in breath-alcohol
     screening, 101
Diclofenac, 284
Dicumarol, 512
Dicyclohexyl ether, 102, 187
Dideoxycytidine (ddCyd), 395
Dideoxyinosine (ddI), 395
Dienes, 53
Diet faddism, 469–470
Dietary fat substitutes, 481
Dietary Reference Intakes (DRI), 469, 470,
     476–479*t*, 481–482
Diethylammonium chloride, 176
Diethyl butanedioate, 213
Diethyl carbonate, 221–222
Diethyl ether, 102, 108, 187*t*
Diethyl glutarate, 209
Diethyl malonate, 217
Diethyl pentanedioate, 209
Diethyl propanedioate, 217
Diethyl succinate, 213
Dieting, 469–470, 480–481
Digestion, 3, 250, 443–444, **469**, 472–474
Digestive enzymes, 318, 431
Digitoxin, 338*t*
Diglycerides, 261
5-α-Dihydroxytestosterone, aging and, 519
Dihydroxyacetone, 233*t*
Dihydroxyacetone phosphate, 219, 432,
     433–434, 438
1,2-Dihydroxybenzene, 177
(S)-3,4-Dihydroxyphenylalanine, 350
Dilation of blood vessels, nitric oxide and, 352
Diltiazem, 338*t*, 520
Dimenhydrinate, as antihistamine, 352
Dimerization, 205, 294
N,N-Dimethylacetamide, 218
Dimethylacetylene, 50
Dimethylamine, 11, 167, 218
1,3-Dimethylbenzene, 78
3,3-Dimethyl-2-butanone, 185
cis-1,4-Dimethylcyclohexane, *33*
trans-1,4-Dimethylcyclohexane, *33*
N,N-Dimethylcyclopentanamine, 170
cis-1,2-Dimethylcyclopentane, *33*
trans-1,2-Dimethylcyclopentane, *33*
Dimethyl ether, 103–104
N,N-Dimethylformamide, 210
Dimethyl phosphate, 219
Dimethyl sulfide, 106
Dimethyl terephthalate, 220–221
2,4-Dinitrophenol (DNP), 422
Diols, 95
Dipeptides, amino acids in, 294–295, 298
Diphenhydramine, 338*t*, 352
Diphosphate ion, 219
Diphosphoric acid, 219
Diphosphoric esters, of monosaccharides, 244
Dipole–dipole interactions, *103*, 493–495
Diprotic acids, **139**
Diradicals, in autoxidation, 85
Disaccharides, **245**–248, 430–431, 456,
     457–460, 472–473

Discriminatory curtailment diets, **469**–470
Disease
     autoimmune, 481, 503–504
     of brain and nervous system, 270
     deficiency, 360, 469, 475, 476–479*t*
     diets to prevent, 469–470
     excessive protein production in, 387
     immunity to, 486–488
     kidney, *516*, 517
     lipid storage, 272–273
     protein/peptide-conformation-dependent, 303
     sulfa drugs and, 326
Distal tubules, of kidneys, 516, *517*, 518
Disubstituted alkylbenzenes, 77–78
Disulfide bond, 294, 299, 300, 304, 305,
     311–312, 493, *494*, *498*, 501
Disulfides, **106**–107
Diuresis, **518**, 520
Diuril, 520
D,L system, 123, 232–235, 292
DNA (deoxyribonucleic acid), 80, 168, 361–364,
     364*t*, 365–371, *366*, 366*t*, *367*, *368*, *369*,
     370, 373, 380–381, 394, 400–403, 504
     genes and, 372, 384–385, 389, 389*t*, 397, 398,
     399
     NADPH and, 436
     recombinant, 403–406
     repair of, 377–380
     replication of, *373*–377
     transcription of, **385**–388
DNA chips, pharmacogenomics and, 377
DNA fingerprinting, 375
DNA fragments, in apoptosis, 379
DNA gyrase, *373*
DNA ligase, *373*, 377, *378*, 379
DNA polymerase, *373*, 374, 374*t*, 376, *378*, 379,
     475*t*
DNA polymerase I, *373*
DNA replication, 377–380, 397–400
DNA viruses, 394
DNP. *See* 2,4-Dinitrophenol
Docosahexanoic acid, 461
Dodecanoic acid, 204*t*
Dodecylbenzene, sulfonation of, 82–83
Donepezil, 338*t*, 345
L-Dopa, 350
Dopamine, 174, 338*t*, 347, 350
Double bond, 6, 7, 8, 18, 19, 84, 189–191,
     194–195, 259, 260, 260*t*, 261, 263
     in alkenes, 46, 47, 48, 53–55, 57–64
Double helix, DNA molecule as, **367**–369, *367*,
     370, 371, 376, 385–*386*
Dramamine, 352, 509
DRI. *See* Dietary Reference Intakes (DRI)
Drinks, 468
Drip bag, *234*
Drugs, 177–178, 217, 336–337, 338*t*, 423
     active stereoisomers of, 119–120, 126, 130
     for AIDS, 395–*396*, 398
     for Alzheimer's disease, 345
     anabolic steroids as, 278
     anti-inflammatory, 279, 282
     anticancer, 365
     antihistamine, 352
     antisense, 387
     antistroke, 352
     antiviral, 394
     for autoimmune diseases, 503–504
     blood–brain barrier and, *509*
     cholesterol and, 276–277, 463
     contraceptive, 280–281
     eliminating side effects of, 509
     extensive metabolism of, 377
     for diabetes, 354

     as hydrochlorides, 176
     for hypertension, 520
     monoamine-oxidase-inhibiting, 351
     in monoclonal antibody therapy, 497
     pain-relieving, 214
     for Parkinson's disease, 350
     pharmacogenomics and, 377
     for prostate treatment, 519
     recombinant DNA techniques to produce,
          404–405
     solubility in body fluids, 176
     statin, 463
     sulfa, 326
     synthetic, 6
     weight-loss, 422
dTMP. *See* Deoxythymidine 5′-monophosphate
     (dTMP)
Dynamite, nitroglycerin in, 94
Earth, 2–3
Ear wax, 262
EcoRI restriction endonuclease, 404–405
Edema, 511*t*, 512
E (exit) site, in protein elongation, 393
Efavirenze, in AIDS treatment, 395
EGF. *See* Epidermal growth factor (EGF)
Egg, zygote from, 373
Egg whites, as heavy metal poisoning antidote,
     312
Eicosanoic acid, 204*t*
Elavil, 338*t*
Electrical energy, metabolic conversion of
     chemical energy to, 425
Electrolytes, 137, 147
Electronegativity, 206–207
Electron-microscope tomography, of
     mitochondria, 412
Electron pairs, in acid–base reactions, 138,
     140–141
Electrons. *See* Valence-shell electron-pair
     repulsion (VSEPR) model
Electron transport chain, 413, 414, 420–425,
     432
Electrophoresis, in DNA fingerprinting, 375
Elements, 2–3
Elongation, 387, 388, 390, 390*t*, *392*–393, 461
Elongation factors, 392, *393*
Enamel, composition of, 308
Enantiomers, 33, *115*, **117**, 119–120, 122–123,
     124, 126, 130, 214, *231*, 234, 292
     in racemic mixtures, **118**
     R,S system for naming, 119–123, **120**
Encephalitis, spongiform, 303
Endocrine glands, 337, 339–*340*, 339*t*, 510
Endocytosis, in cholesterol transport, 275, 276
Endometrium, in menstrual cycle, *280*
Endonuclease, *378*, 379
Endoplasmic reticulum (ER), 342, 348, *411*, *492*
Endothelial cells, in leukocyte mobilization, 500
End point, of titration, **152**
Enduron, 520
Enebrel, 501
Enediols, in sugar oxidation, 242
Energy, 413–414, 419, 436–438, 441–442, *444*,
     456, 459, 468, 469, 471–472, 473. *See also*
     Bioenergetics
     from electron transport, 424–425
Energy storage, 258, 413–414
Enflurane, 105
Enhancers, in gene regulation, 399
Enkephalins, 338*t*, 349, 353–355
Enolase, in glycolysis, *433*
Enols, **194**–195
trans-Enoyl-CoA hydratase, *440*
Entacapon, 338*t*

Envelope conformation, of cyclopentane, *29*
Environmental Protection Agency (EPA), public
    drinking water standards by, 482–483
Enzyme assays, 331*t*
Enzymes, 242, 243, 272–273, 289, 317–331,
    320*t*, *321*, 345, 374, 412, 420–422,
    460–461, 472–473, 493. *See also* Proteins;
    Ribozymes
  active sites of, **321**, 323–326
  in amino acid catabolism, 443–448
  in base excision repair, *378*, *379*
  chemical messengers affecting, 340
  chirality of, 129–130
  in cholesterol synthesis, 275, 276–277
  in clotting, 512
  cofactors of, **320**–321
  in DNA fingerprinting, 375
  for eye color, 373
  in gluconeogenesis, *457*
  in immune system, 491
  inhibitors of, **321**, 324–326
  nitric oxide and, 352
  in nucleotide excision repair, 380
  phosphorylation of, 348, 349
  in photosynthesis, 458, 459
  in protein digestion, 474
  in protein metabolism, 431
  in recombinant DNA techniques, 403–404
  in serum, 511*t*
  signal transduction and, 438
  in transcription, 386–388
  for tRNA, 388–389, 391, 397
  ubiquitin and, 447
Enzyme specificity, **318**, 319
Enzyme–substrate complexes, 321, 323–327
EPA. *See* Environmental Protection
    Agency (EPA)
Epichlorohydrin, 108
Epidermal growth factor (EGF), 402, 501
Epinephrine, *167*, 177–178, 338*t*, 339*t*, 340,
    347, 351, 438. *See also* Adrenaline
Epitope, 492–493, 497, 503
  antigenicity of, **492**, *493*, 495
Epothilon B, 499
Epoxides, as carcinogens, 80
Epoxy glues, production of, 108
Epoxy resins, production of, 108
Equal sweetener, 481
Equatorial bonds, **30**–32
Equilibrium
  in acid–base reactions, 140–142
  enzyme catalysis and, 317–318
  of glucose anomers in solution, 239
  between keto and enol tautomers, 194–195
  in self-ionization of water, 147
Equilibrium constant (*K*), 142–144
Equilibrium expressions, 142–144
ER. *See* Endoplasmic reticulum (ER)
Errors, in DNA replication, 377–380, 397–400,
    400–403
Erythritol, 241–242
Erythrocytes, *489*, **510**–511, *510*, 511*t*
Erythromycin, stereoisomers of, 126
Erythrose, 123–124
D-Erythrulose, 233*t*
Escherichia coli, 377, 403, 443
Essential amino acids, 464, **474**
Essential fatty acids, 262, 473
Essential oils, terpenes in, 55–56
Esterification, 212–213, 363
Esters, 202–203, **208**–209, 210, 212–213,
    215–217, 219, 221, 244, 250, 259–261,
    262, 263, 274, 462
Estradiol, 277, 279, 281, 339*t*, 356, 519

Estrogen, *280*, 519
Ethanal, 184
Ethane, 7*t*, 19, 35*t*, 96*t*, 173*t*
Ethanedioic acid, 203
1,2-Ethanediol, 95, 215–216, 220–221
Ethanethiol, 105, 106, 107*t*
Ethanoic acid, 204*t*, 212
Ethanol, 8, 10, 40, 92, 96*t*, 97, 101, 103–104,
    107–108, 107*t*
  in esterification, 212
  from fermentation, 145
  from hydrolysis, 215
  monosaccharide solubility in, 235
  natural versus synthetic, 5
  protein denaturation by, 312–313
  reactions with aldehydes and ketones,
    192–193
Ethanolamine, 268, 462, 473
Ethene, 50
Ethephon, 47
Ethers, 91, **102**–104, 105
Ethrane, 105
Ethyl acetate, 209, 212, 215
Ethyl acrylate, 65*t*
Ethyl alcohol. *See* Ethanol
Ethylamine, 7–8, 213
Ethylbenzene, 77
Ethyl butanoate, 209
Ethyl butyrate, 209
Ethylene, 7*t*, 19, 46, 47, *48*, 50, 57, 58, 60,
    97, 108
  polymerization of, 47, 64–68, 65*t*
Ethylene glycol, *93*, 95, 108, 215–216, 220–221
Ethylene oxide, 103, 108
Ethyl ethanoate, 209
Ethyl group, 25*t*
Ethyl isopropyl ketone, 187
Ethyl mercaptan, 105
2-Ethylpentanoic acid, 204–205
Ethyndiol diacetate, 281
Ethyne, 49, 50
Ethynil estradiol, as contraceptive, 281
Eukaryotes, 385, 386–388, 391–392
  genes of, 375–377, 389–390, 397–400
Evolution, 390, 401
Excitatory neurotransmitters, 346–347
Excretion, 517, 520
Exercise, 442, 472
Exons, *372*, *386*, 496
Exonuclease, in base excision repair, *378*, *379*
Exothermic reactions, 57
Explosives, nitro group in, 82
Extended helix, *304*
External innate immunity, 486–*487*
Extracellular fluids, 508
Eyes, 141, 312, 373, 451, 487

Fabry's disease, characteristics of, 272
Facial tics, botulin treatment of, 344
Facilitated transport, across cell
    membranes, 266
F-actin, quaternary structure of, 308
FAD⁺. *See* Flavin adenine dinucleotide (FAD⁺)
FADH₂. *See* Reduced FAD⁺ (FADH₂)
Familial adenomatous polyposis,
    treatment of, 282
Familial hypercholesterolemia, 276
Famine, 472
Faraday, Michael, 76
Farnesol, 56
Farnesyl pyrophosphate, 463
Fastin, 168
Fasting, fat metabolism and, 442
Fat cells, lipids in, 258

Fat depots, *431*, 441, 460
Fat globules, *431*
Fats, 259, **261**–264, 262*t*, *416*, 422, 430, 431,
    432, 441, 460, 469, *471*, 472, 473, 481. *See
    also* Fat; Lipids
Fat-soluble vitamins, 476*t*
Fatty acids, 85–86, 204*t*, 259–261, 260*t*, 262,
    262*t*, 264–**265**, *416*, 431, *432*, 441–442,
    460–461, 469, 473, 487
  β-oxidation of, 438–439, 439–441, *440*
  cholesterol esterified with, 274
  in cooking oils, 84
  in glycerophospholipids, 267–269
  NADPH in synthesis of, 436
  from protein catabolism, *444*
  saturated, 469
  in sphingolipids, 269
  unsaturated, 265–267, 439, 461
Fatty acid synthase, 460–461, 475*t*
Fava beans, hereditary disease involving, 377
FDA. *See* Food and Drug Administration (FDA)
FD & C. *See* Food, drugs, and cosmetics
    (FD & C)
Feces, 282, 451
Feedback control, 327, 419. *See also* Regulation
Fermentation, 107, 145
Fermentation vats, *91*
Ferredoxin, in photosynthesis, *458*
Ferritin, 290, 451
Fertility, aging and, 519
Fertilization, 279, *280*, 357
Fertilizers, 470
Fetal cells, 374
Fetal hemoglobin (HbF), 301, 309
Fexofenadine, 338*t*, 509
Fiber, dietary, **470**
Fiberscope, for laser surgery, 312
Fibrils, of collagen, 308, 309
Fibrinogen, 290, 511–512, 511*t*
Fibrous proteins, 290, 303
50S ribosome, 391, *393*
Filamentous cytoskeleton, *411*
Filtration, as kidney function, 515–517
Fischer, Emil, 212, 232–233
Fischer esterification, **212**–213
Fischer projections, 231–235
Flavin, in FAD molecular structure, *415*
Flavin adenine dinucleotide (FAD⁺), *410*, 414,
    415, *416*, 418–419, *420*, 421, 422, 424,
    439, *440*, 475*t*
Flavodoxin, in photosynthesis, *458*
Fleming, Alexander, 211
Florey, Howard, 211
Fluid mosaic model, 265–267
Fluorine (F, F₂), 2, 7, 479*t*
Fluorouracil, 365
Fluorouridine, 365
Fluoxetine, 338*t*
Folic acid, 326, 475*t*, 477*t*, 480
Follicle-stimulating hormone (FSH), *340*
Food allergens, 503
Food colorants, synthetic, 83
Food, drugs, and cosmetics (FD & C), 83
Food guide pyramid, 470, *471*
Forane, 105
Formaldehyde, 7*t*, 12, 107, 183–184, 187
Formica, 204
Formic acid, 187, *204*, 204*t*
Formulas, of alkanes, 20*t*. *See also* Structural
    formulas
40S ribosome, 388, 390*t*, 391, *392*
Fractional distillation, of petroleum, 39–40
Franklin, Rosalind, 367, *368*
Free cholesterol, 274

Freons, 38, 39
α-D-Fructofuranose, 238
β-D-Fructofuranose, 238, 245
D-Fructose, 233t, 234, 238, 248, 249
Fructose 1,6-biphosphate, 432, 433–434, 437t, 457
Fructose 1-phosphate, 434, 438
Fructose 6-phosphate, 320t, 431–432, 433–434, 457
Fruits, 468, 471
L-Fucose, molecular structure of, 246
Fugu fish, 31
Fumarase, in citric acid cycle, 418
Fumarate, 416, 418, 446, 448, 449
Fumigants, ethylene oxide as, 103
Functional groups, 8–14, 9t, 202–203
Funk, Casimir, 475
Fur, as allergen, 503
Furan, 236
Furanose form, of monosaccharides, 235, 236–240, 241
Furchgott, Robert, 352

GABA. See γ-Aminobutyric acid (GABA)
G-actin, 308
Galactocerebroside, in Krabbe's leukodystrophy, 272
Galactokinase deficiency, 235
D-Galactopyranose, 237–238, 239, 245
D-Galactosamine, 234
Galactose, 235, 248t, 265, 270–271, 434, 437–438
D-Galactose, 232t, 234, 237–238, 239, 246
Galactosemia, 235
Galactose-1-phosphate uridinyltransferase deficiency, 235
α-Galactosidase, in Fabry's disease, 272
β-Galactosidase, in Krabbe's leukodystrophy, 272
Galactosuria, 235
Galantimine, for Alzheimer's disease, 345
Gallbladder, 451, 473
Gallstones, 274, 511t
Gamma rays, 401, 402
Gangliosides, 270, 273
Gap junctions, 266
Garlic, thiols in, 105
Gases, 511t, 514, 515. See also Noble gases
Gasohol, 91, 107–108
Gasoline, 40, 107–108
Gastric disorders, 323
Gastric fluid, hydrochloric acid in, 137
Gaucher's disease, characteristics of, 272
GC box, 397
GDP. See Guanosine diphosphate (GDP)
Gelatin, as incomplete protein, 474
Gelsolin, in apoptosis, 379
Gene conversion (GC) mutation, of V(J)D genes, 496
Gene expression, 384–385
Gene insertion technique, 398
Gene therapy, 404–405
Genes, 290, 301, 361, 372, 373, 377, 380–381, 384–385, 386–388, 400, 402, 403, 496
regulation of, 290, 397–400
Genetic code, 385, 388–389, 389t, 389–390
second, 391
Genetic diseases, 403, 404–405
Genetic engineering, 405–406
Genome, 377
Genomics, 307, 384, 504
Geraniol, 54–55, 56
Geranyl pyrophosphate, 463
Germ cells, immortality of, 374

GH. See Growth hormone (GH)
GILT (gamma-interferon inducible lysosomal thiol) reductase, 493
Glacial acetic acid, 137, 205
β-Globin, biochemical evolution of, 401
Globin chains, in hemoglobin, 307, 309, 401, 401t
Globular proteins, 290, 302–303, 311
Globulins, 511, 511t, 512
Glomeruli of kidneys, 516, 517, 518
Glucagon, 339t, 340, 349, 353, 355
D-Glucitol, 241
Glucoamylase, 249
Glucocerebroside, 271, 272
Glucocorticoid hormones, 280, 503–504
Glucogenic amino acids, 432, 448, 449, 450, 457
Glucokinase, in glycolysis, 433
D-Gluconate anion, 242
Gluconeogenesis, 444, 457, 457
D-Gluconic acid, 243
Glucophage, for diabetes, 354
D-Glucopyranose, 245, 247–248
D-Glucosamine, 234
Glucose, 146, 191, 230, 248t, 430–431, 436–438, 441–443, 449, 455–456, 459, 469, 472, 473, 511, 511t
ATP yield from metabolism of, 437t
diabetes and, 354
in gluconeogenesis, 457
glycolipids and, 265
glycolysis of, 431–435, 433–434, 436–438
oxidation of, 318
from protein catabolism, 444, 448
D-Glucose, 232t, 234, 236, 237–238, 239, 241, 242–243, 244, 248, 249, 250, 473
L-Glucose, 318
Glucose isomerase, 249
Glucose oxidase, 242, 243, 318
Glucose 1-phosphate, 455, 460
Glucose 6-phosphate, 320t, 431, 433–434, 457
D-Glucose 6-phosphate, 244
Glucose transporter molecules (GLUT4), 354
β-Glucosidase, in Gaucher's disease, 272
α-Glucosidases, in glycogen digestion, 250
β-Glucosidases, in cellulose digestion, 250
D-Glucuronic acid, 243–244, 251
Glucuronides, 244
Glues, 107, 108
Glutamate, 320t, 444, 445, 448, 449. See also Glutamic acid, (Glu, E)
Glutamate-oxaloacetate transaminase (GOT), 331
Glutamate-pyruvate transaminase (GPT), 331
Glutamic acid (Glu, E), 291t, 296, 306, 338t, 346, 347, 355, 389t, 400, 419, 449, 464–465
Glutamine (Gln, Q), 291t, 305, 389t, 448, 449, 464
Glutaric acid, 203
Glutathione, 296, 423, 435
Glutathione peroxidase, as cofactor, 475t
Glycation, of proteins, 298
Glyceraldehyde, 231–232, 233, 434
(R)-Glyceraldehyde, 129–130, 232
(S)-Glyceraldehyde, 129–130
Glyceraldehyde dehydrogenase, in glycolysis, 433
Glyceraldehyde 3-phosphate, in glycolysis, 432, 433–434, 437t
D-Glyceraldehyde-3-phosphate dehydrogenase, as cofactor, 475t
Glyceric acid, 205
Glycerin, 94, 95, 108, 263
Glycerol, 95, 108, 259, 261, 263, 264–265, 431, 432, 434, 438, 473. See also Glycerin

Glycerol 1-phosphate, 438. See also Glycerol 3-phosphate
Glycerol 3-phosphate, 461–462. See also Glycerol 1-phosphate
Glycerol 3-phosphate transport, 436–437, 462
Glycerophospholipids, 264–265, 267–269, 271, 462
Glycine, 281, 291t, 292, 294, 346–347, 389t, 447, 449
Glycocholate, 282
Glycogen, 230, 250, 425, 430–431, 434, 438, 455–456, 472, 473
Glycogenesis, 438, 460
Glycogenolysis, 438
Glycogen phosphorylase, signal transduction and, 438
Glycogen synthase, signal transduction and, 438
Glycolic acid, in Lactomer polymerization, 221
Glycolipids, 264–265, 270–273, 462
Glycolysis, 191, 325, 431–438, 432, 433–434, 457, 464
Glycomics, 504
Glycoproteins, 310, 395, 488, 493, 498, 499, 500, 504
Glycosaminoglycans, 251, 310
Glycosides, formation of, 240–241
Glycosidic bonds, 240–241, 245, 247, 248, 249, 250, 310, 362–363
Glycosylase, in base excision repair, 378, 379
Glyset, for diabetes, 354
GMP. See Guanosine 5'-monophosphate (GMP)
Goiter, iodine therapy for, 81
Goldstein, Joseph, 275
Golgi bodies, 411, 492
Gonadotropic hormones, action of, 340
GOT. See Glutamate-oxaloacetate transaminase (GOT)
gp41 glycoprotein, in HIV, 395, 396
gp120 glycoprotein, in HIV, 395
G-protein, 348, 349, 438
G-protein–adenylate cyclase cascade, 349–351, 355
GPT. See Glutamate-pyruvate transaminase (GPT)
Grain alcohol. See Ethanol
Granulocytes, 510
Grapes, 91
Grape sugar, 234
Groups. See Functional groups
Growth hormone (GH), action of, 340
Guanine (G), 361–362, 363, 364, 364t, 365–366, 367–369, 368, 388
Guanosine, 363, 364t
Guanosine diphosphate (GDP), 416, 418
Guanosine 5'-monophosphate (GMP), 364t
Guanosine triphosphate (GTP), 348, 349, 364, 416, 418
Guillain-Barré syndrome, 270
D-Gulose, 232t
Gyrase, during DNA replication, 376

H_1 and H_2 histamine receptors, 352, 509
HAART (highly active antiretroviral treatment) drug cocktail, 395
Haloalkanes, 37–39
Halogenation, 37–38, 81, 105
Halogens, 2
Hard water, soaps and, 264
Haworth, Walter N., 235–236
Haworth projections, 235–240
Hayfever, treatment of, 177
Hb. See Hemoglobin (Hb)

HbF and HbS hemoglobin, in sickle cell anemia, 301
HDL. *See* High-density lipoprotein (HDL)
HDPE. *See* High-density polyethylene (HDPE)
Headaches, nitric oxide and, 352
Health
  basal caloric requirement and, 472
  chemical communication and, 336–337
  immune system and, 486–488
  nutrients and, 469
  vitamins and minerals and, 475–482, 476–479t
Heart, 276, 284, 289, 472, *490*, *514*, 519–520
Heart attack, 331, 423, 435
Heartburn, treating, 145
Heat, 311, 312, 426, 470
Heat-resistant polymerases, in cloning, 381
"Heat shock" protein, *502*
Heavy chains, in immunoglobulins, 493, *494*, *495*, *496*
Heavy metal ions, 312, 423
Heavy metal poisoning, *312*, 326, 327
Helicase, 373, 374t, 376, 385–*386*
Helicobacter pylori, 323, 352
Helium (He), *221*
Helix (helices), *116*. *See also* α-Helix; Double helix; Extended helix; Triple helix
α-Helix
  denaturation of, 311
  of interleukins, 500–501
  in protein secondary structure, 301–304
Helix-turn-helix transcription factor, in gene regulation, 398
Heme, 309, 321, *444*, 450–451, 465, *513*
Hemiacetals, **192**–194, 235–241
Hemlock, poison, 169
Hemodialysis, 515, *516*
Hemoglobin (Hb), 290, 297, 300, 301, *306*, *307*–309, 321, 401, 435, 447, 450–451, 513–514, 515
Hemolytic anemia, 377
Hemophilia, 360, 511t
Henderson-Hasselbalch equation, 158–160
Henle's loop, of kidneys, 516, *517*
Heparin, 251, 512
Hepatitis, 331, 394, 395, 451
Heptane, 21, 22, 35t, 40
Herb tea, for Alzheimer's disease, 345
Herceptin, 497
Hereditary diseases, 361, 377, 449. *See also* Genetic diseases
Heredity, 361, 369
Heroin, 338t
Herpes virus, *394*
Heterocyclic aliphatic amines, **168**
Heterocyclic aromatic amines, **168**, 174, 175
Hexadecanoic acid, 204t
Hexamethylenediamine, 220
Hexanal, 101, 184, 188
Hexane, 20, 21–23, 35, 35t, 36
1,6-Hexanediamine, 170, 220
Hexanediamide, 211–212
Hexanedioic acid, 203, 220
Hexanoic acid, 101, 188, 203, 204t
1-Hexanol, 100–101
2-Hexanol, 100–101
2-Hexanone, 101, 185
3-Hexanone, 185
trans-3-Hexene, 51
Hexoaminadase A, in Tay-Sachs disease, 273
Hexokinase, in glycolysis, *433*
Hexoses, 234, 238–239, 240, 437–438, 457–460, 504
2-Hexyne, 49, 50

HFC-134a, 39
HFC-141b, 39
Hibernation, mitochondria during, 422
High blood pressure, diets for, 469
High-density lipoprotein (HDL), 274, 276–277
High-density polyethylene (HDPE), 66–67, 68t
High-energy phosphate anhydride bond, 413–414
High-fructose corn syrup, 249
Hinckley, Robert, *105*
Hippuric acid, 517
Histamine, 338t, 347, 351–352
Histidine (His, H), 291t, *306*, 338t, 351–352, 389t, 449
Histone acetylase, during DNA replication, 375–376
Histones, 361, 369, *370*, 375–376
HIV. *See* Human immunodeficiency virus (HIV)
HIV-1 protease, inhibition of, 398
HMG CoA. *See* Hydroxymethylglutaryl CoA (HMG CoA)
HMG-CoA reductase, in cholesterol synthesis, 276–277, 463
Hog insulin, primary structure of, 299
Homeostasis, of body fluids, 509
Homosalate, 216
Homozygotes, sickle cell anemia and, 301
Honey, 248, 248t
Hormones, 259, 274, 277, 279–281, 290, 299, 300, **337**, 338–341, 339t, 347–353, 353–355, 356–357, *444*, 511t, 515, 520
Human chromosomes, *360*
Human Genome Project, 377
Human growth hormone, 290
Human immunodeficiency virus (HIV), 395, 499–500
Human insulin, 299–300. *See also* Insulin
Humans, 366t, 373, 374, 375, 401, 401t
Huntington's disease, creatine and, 482
Huperzine A, for Alzheimer's disease, 345
Hyaluronic acid, 251
Hybridization, 380–**381**
Hybridoma cells, *497*
Hydration, 57t, **60**–62, 98–99, 439, *440*
Hydration–dehydration reactions, 98–99
Hydrazinohistidine, 338t
Hydride ion (H⁻), in aldehyde reduction, 190
Hydrobromic acid, 136t
Hydrocarbons, **18**–19, 27, 37, 46–47, 75–87
  aliphatic, **18**
  saturated, **18**
  unsaturated, 18, 46
Hydrochloric acid, 136t, 137
Hydrochlorides, of drugs, 176
Hydrochlorination, of alkenes, 57t, 58, 59
Hydrofluorocarbons (HFCs), 39
Hydrogen (H_2), 2, 7, 139–140. *See also* Hydride ion (H⁻); Hydrogen gas (H_2)
  in autoxidation, 85–86
  in catalyzed reaction with cyclohexanone, 190
  in common catabolic pathway, *410*
  in DNA, *366*
  in Earth's crust, *3*
  in proteins, *302*
Hydrogenation, 57t, 64, 263
Hydrogen bonds, 95–*96*, 103–104, 106, *172*–173, 187, 188, 205, 235, 301–*302*, 304, *305*, 311, *367*–369, 372, 398, 493–495
Hydrogen bromide, 58, 59
Hydrogen chloride, 58, 59, 135, 138, 140, *144*–145, 145–146, 155, *156*, 157, 218, 473
Hydrogen cyanide, 7t
Hydrogen halides, 58–60

Hydrogen iodide, 58, 59, 60
Hydrogen ions (H⁺), 134–135, *410*, 414, 417–419, 420–425, 514, 518. *See also* pH
Hydrogen peroxide, 243, 423, 497
Hydrolases, action of, 319, 320t
Hydrolysis, 214–219, 263, 318, *413*–414, 425–426, 446–447
  of acetyl-CoA, 416, *432*
  of carbohydrates, 231, 249, 472–473
  in fat metabolism, 261, 442, 473
  of glycosides, 240
  in protein catabolism, *444*
Hydronium ion, **134**–135, 136, 138, 140, 143, 145–146, 147–150, 152–153, 154–158
Hydroperoxide group, in autoxidation, 85–87
Hydroperoxy radical, in autoxidation, 85–86
Hydrophilic substances, 264, 265–267, 282, 290–292
Hydrophobic interactions, protein tertiary structure and, 304, *305*
Hydrophobic pocket, in pyruvate kinase, 325
Hydrophobic substances, 259, 264, 265–267, 274, 282, 290–292, 367, *498*–499
Hydroquinone, 84
Hydrothermal vent, *317*, 322
Hydroxide ion, 134, 135–136, 147–148, 151, 152–153, 154–158, 215, 293
Hydroxyacyl CoA, in stearic acid metabolism, 441t
Hydroxyacyldehydrogenase, in fatty acid β-oxidation, *440*
Hydroxyapatite. *See* Calcium hydroxyapatite
4-Hydroxybenzoic acid, 189, 204–205
β-Hydroxybutyrate, 442–443
Hydroxyl group, **9**–11, 9t, 60–62, 186, 235
4-Hydroxy-3-methoxybenzaldhyde, 84
Hydroxymethylglutaryl CoA (HMG CoA), 463
Hydroxyproline, 304
2-Hydroxypropanoic acid, 118, 204. *See also* Lactic acid
(R)-2-Hydroxypropanoic acid, 204–205
Hydroxyurea, 301
Hypercholesterolemia, familial, 276
Hyperglycemia, diabetes and, 354
Hyperphosphorylation, in Alzheimer's disease, 345
Hypertension, 328, 472, 520
Hyperthermophile bacteria, 322, 381
Hypoglycemia, 300, 354, 511t
Hypothalamus, *340*, 355
Hypothyroidism, 81
Hypoxia, 402
Hytrin, 338t, 519

Ibuprofen, 119–120, 122–123, 126, 130, 214, 282, 284
D-Idose, 232t
IgA, IgD, IgE, IgG, and IgM antibodies, 493, 493t, *494*
Ignarro, Louis, 352
Imidazole, molecular structure of, 168
Imines, formation of, 298
Imirapin, 338t
Immune system, 395, 486, *487*, 488, 489–491, 501–504
Immunity, 486–488, 491, 499, 501–502
  innate, **486**–487, 502–503
Immunization, 394, 499
Immunochemistry, 486–504
Immunogens, antigens as, 491
Immunoglobulins, 290, 310, 488, 493–497, 493t, *494*, *496*, 498, 504, 511
Impotence, nitric oxide treatment of, 352
Impurities, in organic compounds, 5–6

Inderal, 338*t*

Indigo, in blue diaper syndrome, 465

Indinavir, in AIDS treatment, 395, *396*, 398

Indomethacin, 282, 284

Induced-fit model, of enzyme action, **323**–326

Infection, 443, 500, 501, 511*t*

Infectious hepatitis, 331, 451

Inflammation, as response to injury, 487

Inflammatory bowel disease, treatment of, 503

Influenza, 394

Information transfer, 361, 369, 373, 384–385

Inhibition, of enzymes, **321**, 324–326, 327

Inhibitors, **321**, 324–326, 352, 398

Inhibitory neurotransmitters, 346–347

Inhibitory receptors, innate immunity and, 503

Initiation, of protein synthesis, 390, 390*t*, 391–392

Initiation codon, 389*t*, 390

Initiation factors, in protein synthesis, 391, 392

Initiation signal, in transcription, 386, 388

Initiator, in transcription, 386–387

Injury, responses to, 500, 501, 511, 512

Innate immunity, **486**–*487*, 502–503

Inorganic chemistry, 2

Inorganic compounds, 3–4, 4*t*, 511*t*

Inositol-1,4-diphosphate (I-1,4-P), 355

Inositol-1-phosphate (I-1-P), 355

Insecticides, pyrethrins as, 210

Insulin, 243, 290, *299*–300, 339*t*, 340, 353, 354, 360, 404, 443

Insulin-dependent diabetes, 354, 503

Integral membrane proteins, 310

Integral proteins, 265, *267*

Integrin, in leukocyte mobilization, 500

Interferon, 395, 490*t*, 491

Interleukins (ILs), 395, 500–501

Intermembrane space, of mitochondria, 412

Internal innate immunity, 486, *487*, 489–491, 490*t*

International Union of Pure and Applied Chemistry (IUPAC)

Interstitial fluid, **508**, 509, 519

Intestine, 431, 451

Intracellular fluid, *489*, 508

Intramolecular hydrogen bonds, in proteins, 301–*302*

Introns, **372**, *386*

Iodide ion, goiter and, 81

Iodine (I₂), 2, 7, 479*t*

Ion channels, 343, 346, 347. *See also* Calcium channels; Sodium ion channels; Transmembrane channels

Ionic compounds, 292–293

Ionizing radiation, 401, 402

β-Ionone, 186

Ion product of water (K_w), 147, 149

Ions, 266, 398, 517

Iproniazide, 338*t*

Iron (Fe), *3*, 290, 451, 475*t*, 479*t*

Iron(II) ion, in heme, 307, 450, 513

Iron–sulfur clusters, 420, 421, 458

Irreversible inhibition, 321, 326, 346

Ischemia reperfusion, 423

Isobutane, 27

Isobutyl alcohol, 92

Isobutylene, 50, 97

Isobutyl group, 25*t*

Isobutyl mercaptan, 105

Isocitrate, 320*t*, *416*, 417

Isocitrate dehydrogenase, 417, 419

Isoelectric point (pI), 291*t*, **293**, 297

Isoenzymes, enzyme regulation via, **330**

Isoflurane, 105

Isoleucine, 291*t*, 389*t*, *449*

Isomerases, action of, 319, 320*t*

Isomers, 21, 23–26, 35, 36, 77–78, *115*

chirality in, 114–130, **117**

cis-trans, **32**–34, **48**, 50–52, 53–55, *115*

constitutional, **21**–23, *115*

Isooctane, in gasoline, 40

Isopentane, 27

Isopentenyl pyrophosphate, 463

Iso prefix, in alkane names, 27

Isoprene, 55–56, 463, 464

Isoprene units, 56

Isopropyl alcohol, 10–11, 92, 108, 215–216

Isopropylamine, 171

Isopropyl group, 25*t*

2-Isopropyl-5-methylcyclohexanol, 100, 127

2-Isopropyl-5-methylcyclohexanone, 100

2-Isopropyl-5-methylphenol, 84

(*R*)-Isoproterenol, 177

Isovaleric acid, 203

Isozymes, enzyme regulation via, **330**

IUPAC. *See* International Union of Pure and Applied Chemistry (IUPAC)

Jackson, Charles, 105

Jaundice, 451

Jenner, Edward, 499

Keflex, 211

Kekulé, Friedrich August, 3, 76

Kekulé structures for benzene, 76–77

Keratin, 289, 290, 303, 311–312

Ketoacidosis, in diabetes, 443

Ketoacids, in amino acid catabolism, 431, 444

Ketoacyl CoA, in stearic acid metabolism, 441*t*

Keto-enol tautomerism, 194–195, 279

Ketogenic amino acids, *432*, 449, 450

α-Ketoglutarate, 320*t*, *416*, 418, 419, 444, 445, 448, *449*, 464

α-Ketoglutarate dehydrogenase, 419

α-Ketoglutaric acid, 431, 444, *449*, 465

D-2-Ketohexoses, 233*t*

Ketone bodies, 441–**443**, *444*, 450, 473, 517

Ketone group, 231

Ketones, 9*t*, **12**–13, 183–195, 298

D-2-Ketopentoses, 233*t*

Ketoprofen, anti-inflammatory drugs and, 282

Ketoses, **231**, 233–234

2-Ketoses, as reducing sugars, 242

Ketosis, 473. *See also* Ketone bodies

D-2-Ketotetroses, 233*t*

β-Ketothiolase, in fatty acid β-oxidation, *440*

Ketotrioses, 231

Kevlar, 220

Kidney failure, 276, 298

Kidneys, 300, 339*t*, 449, 510, 515–*517*, 518, 520

Kidney tubules, *340*, *517*

Killer T cells, in immune system, *487*, 490*t*, 491

Kilocalorie (kcal), 471–472

Kinases, 329, 355

Kinesin, 309

Knocking, in engines, 40

Knoop, Franz, 438–439

Kohler, Georg, 497

Koshland, Daniel, 323

Krabbe's leukodystrophy, 272

Krebs cycle, 413, *416*. *See also* Citric acid cycle

Krebs, Hans, 416, 448

Kwashiorkor, 475

β-Lactam, in penicillin, 211

Lactams **210**, 211, 218

Lactate, 159, 191, *433*–*434*, 435, *457*. *See also* Lactic acid

Lactate dehydrogenase (LDH), 319, 320*t*, 330, 331, 331*t*, *433*

Lactic acid, *118*, 129, 204, 207–208, 221, 486–487. *See also* Lactate

Lactomer, 221

Lactones, **209**, 210

Lactose, 248, 248*t*, 235, 245, 472

Lagging strand, of DNA, *373*, 374

L-Amino acids, 292

Landsteiner, Karl, 246

Lanolin, 262

Lanoxicaps, 338*t*

Laser surgery, protein denaturation and, 312

LASIK (laser in situ keratomileusis), 312

"Laughing gas," 105

Lauric acid, 204*t*, 260*t*, 262*t*, 263

LDL. *See* Low-density lipoprotein (LDL)

LDL-receptor–mediated pathway, 274–275

LDPE. *See* Low-density polyethylene (LDPE)

Le Chatelier's principle, 98–99, 455–456

Leading strand, of DNA, *373*, 374

Lead poisoning, 327

Lecithins, 267–268, *462*

Lemon grass oils, 186

Leptin, 355

Lethal mutations, 400

Leucine, 291*t*, *306*, 389*t*, 397, *449*, 450

Leucine-tRNA synthetase, 395

Leucine zipper, in gene regulation, 398

Leukemia, *486*, 511*t*

Leukocytes, 284, 423, *487*, 500, *510*, **511**

Leukotriene B4, 284

Leukotriene receptors (LTRs), 284

Leukotrienes, 282, 284

Levonogestrel, as contraceptive, 281

Levorotary compounds, **128**

Lewis dot structures, 6, 7–8, 10–11, 76–77, 135

Lexan, synthesis of, 222

Librium, molecular structure of, 172

Ligand-binding G-protein/adenylate cyclase cascade, signal transduction and, 438

Ligand-gated ion channels, 346, 347

Ligands, 340

Ligases, 319, 320*t*, 374*t*, 377, *378*, 379

Ligation step, in DNA replication, 377

Light, 127–129

Light chains, in immunoglobulins, 493, *494*

Light reaction, in photosynthesis, 459

Limonene, 56

Line-angle formulas, **19**–20

Linoleic acid, 260*t*, 262, 262*t*, 263, 460, 473

Linolenic acid, 260*t*, 262, 262*t*, 263, 267, 460, 473

Lipases, 318, 331, 431, *432*, 473

Lipicor, 276

Lipid bilayers, 265–267, *268*, 298, 352, 463

Lipids, 258–284, *265*, *268*, **274**, 410, 417–419, *432*, 443, 461–464, 469, 473, 511. *See also* Fats

Lipid storage diseases, 272–273

Lipomics, 443

Lipopolysaccharides, innate immunity and, 503

Lipoproteins, 86, **274**–277, *275*, 275*t*

5-Lipoxygenase, inhibition of, 284

"Liquid protein" diets, 474

Liquid soaps, 264

Lithium hydroxide, 136*t*

Litmus paper, *137*, 149, 150*t*

Liver, 244, 274–275, 276, 279, 342, *411*, 442–443, 451, 510–511

Lock-and-key model, of enzyme action, **323**

London dispersion forces, 95, 173, 260, 493–495

Loops, DNA molecules arranged in, *370*, 399

Lopressor, 338*t*

Low-density lipoprotein (LDL), 274–275, 276–277, 462
Low-density polyethylene (LDPE), 66, 67, 68t
Lowry, Thomas, 138
Low-sodium diets, 469
Luciferase, in luminescent tobacco plant, 405
Luminescent tobacco plant, 405
Lungs, 509, 513, 514, 515
Luteinizing hormone (LH), 279, 339t, 340, 349, 355
Lyases, action of, 319, 320t
Lye, 137
Lymph, 489, 490, 508, 509
Lymphatic system, 490
Lymphatic vessels, 490, 491
Lymph capillaries, 489, 490
Lymph nodes, in immune system, 489, 490, 491
Lymphocytes, 395, 396, 487, 490, 499–500, 501, 510. See also T cell entries; T lymphocytes
Lymphoid organs, 489, 490
Lymphoma, antisense drugs for, 387
Lysine, 291t, 295, 306, 318, 389t, 449, 474
Lysosomes, 318, 411, 493
Lysozyme, 318, 495
D-Lyxose, 232t

Macrolide drugs, for autoimmune diseases, 504
Macrophages, 487, 488, 489, 490, 490t, 491, 492–493, 500–501, 502–503
Magnesium (Mg), 3, 144, 475t, 479t
Magnesium hydroxide, 137
Magnesium ions, 264, 347, 433, 440, 458, 459, 517
Major groove, in DNA molecule, 367
Major histocompatibility complex (MHC), 487, 488, 491, 492–493, 498, 502–503
Malaria, sickle cell anemia and, 301
Malate, 416, 418–419, 441, 457
Malate dehydrogenase, 419
Malathion, 338t
Malignant tumors, oncogenes and, 402
Malonic acid, 203
Malonyl-ACP, in fatty acid biosynthesis, 461
Maltose, 247–248, 248t, 472, 473
Mammotropin, action of, 339t
Manganese (Mn), 475t, 479t
D-Mannitol, 241–242
D-Mannosamine, 234
D-Mannose, 232t, 246
MAOs. See Monoamine oxidases (MAOs)
Marasmus, 472
Margarine, production of, 263
Markovnikov, Vladimir, 58
Markovnikov's rule, 58, 59, 60, 61, 99
Matrix, of mitochondria, 412
Measles, 394, 487
Mechanical energy, 425–426
Medicine, 94, 114, 119–120, 330–331, 337, 487, 517. See also Disease; Drugs; Therapy
Melting point
    of alkanes, 34–35
    of amino acids, 293
    of fatty acids, 259–260, 260t
    of waxes, 262
Memory cells, 486, 487, 490t, 491, 499
Menstrual cycle, 279, 280
Mental retardation, from PKU, 449
Menthol, 56, 100, 126–127
Menthone, 100
Meperidine, 338t
Mercaptans, 104–106
Mercaptoethylamine, 415, 416
Mercury ion, mercaptans and, 104
Mercury poisoning, 327

Messenger RNA (mRNA), 371, 372, 387, 388, 389, 390t, 391, 392, 393, 397, 401
    in transcription and translation, 385, 386–388
Messengers, drugs affecting, 337, 338t
Metabolic acidosis and alkalosis of blood, 159
Metabolism, 298, 409, 430–451, 432, 433–434, 440, 441t
Metal-binding fingers, in gene regulation, 398, 399
Metal catalysts, for alkene hydrogenation, 64
Metal hydroxides, 135, 136, 136t, 144–145
Metallic ions, as coenzymes, 321
Metal oxides, reactions with acids, 145
Metals, 144
Metaminopeptidase, methionine removal by, 400
Metastasis, 402
Metformin, for diabetes, 354
Methadone hydrochloride, 176
(S)-Methamphetamine, 168
Methanal, 183–184. See also Formaldehyde
Methanamine, 173t. See also Methylamine
Methandienone, 278
Methane, 19, 20t, 35t, 37
Methanethiol, 104, 107t
Methanoic acid, 204t. See also Formic acid
Methanol, 7t, 96t, 107, 107t, 173t
Methedrine, 168
Methenolone, 278
Methicillin, 211
Methionine (Met, M), 291t, 389t, 390, 400, 449, 458, 474
Methionine enkephalin, 353
Methoxycyclohexane, 102
Methoxy group, 102
N-Methylacetamide, 210
Methylacetylene, 50
Methylamine, 7t, 11, 145–146, 167, 173. See also Methanamine
Methylammonium hydroxide, 173
2-Methylaniline, 243
3-Methylaniline, 170
N-Methylaniline, 168, 170
N-Methyl-D-aspartate (NMDA), 338t, 347, 355
Methylation, 386, 388
Methyl benzoate, 213
2-Methyl-1,3-butadiene, 56
Methyl chloride, 37
1-Methylcyclohexanol, 61, 62
trans-2-Methylcyclohexanol, 93
Methylcyclohexane, 31–32
Methyldopa, 338t
Methylene chloride, 37
Methylenecyclohexane, 60, 62
Methylene group, 19–20
Methyleneimine, 7t
Methyl ethyl ketone (MEK), 187
Methyl α-D-glucoside, 240–241
Methyl β-D-glucoside, 240–241
Methyl group, 25t
Methyl GTP, in protein synthesis initiation, 391
1-Methylguanosine, 372
2-Methylhistamine, 338t
Methyl isovalerate, 209
Methyl methacrylate, 65t
Methyl 3-methylbutanoate, 209
Methyl orange, as acid–base indicator, 150
Methylphenidate, 338t, 350
3-Methylphenol, 84
2-Methylpropane, 21, 24
2-Methyl-1-propanethiol, 105
2-Methyl-1-propanol, 92
2-Methyl-2-propanol, 60, 92, 97

2-Methylpropene, 50, 60, 97
Methyl β-D-riboside, 241
Metoclopramide, 338t, 350
Metoprolol, 338t
Metyrosine, 338t
Mevalonate, 463
Mevalonic acid, 205
MHC. See Major histocompatibility complex (MHC)
Micelles, of soap and dirt, 264
Microfilaments, in cytoskeletons, 308
Microsatellites, in DNA, 373
Microtubules, 5, 308–309, 411
Mifepristone, 280, 281
Migitol, for diabetes, 354
Milk of magnesia, 137. See also Magnesium hydroxide
Milk sugar, 248t. See also Lactose
Milstein, Cesar, 497
Mineralocorticoid hormones, 279
Minerals, as nutrients, 469, 475–482, 475t, 476–479t
Minibands, DNA molecules arranged in, 370
Minipress, 338t
Minisatellites, in DNA, 373
Minor groove, in DNA molecule, 367
Mirror-image molecules, 115–119
Mirror images, 117
Mirrors, silvering, 189
Mitchell, Peter, 422
Mitochondria, 318, 336, 389t, 411–412, 421–424, 431, 435, 436, 439, 440, 444–445, 457, 458, 460
Mitochondrial DNA and RNA, 389t
Mitochondrion membrane protein, 310
Mixed anhydrides, 208
Mixtures, racemic, 118
Modulation, enzyme regulation via, 329
Molecular cloning, 380–381
Molecular weight (MW), 173t, 187–188, 187t, 205–206, 206t, 262, 493, 493t
Molecules, 487, 488
    achiral, 118
    chiral, 114–130, 117, 292
    functional groups in, 8–14, 9t, 202–203
    mirror-image, 115–119
    with multiple stereocenters, 123–127
Molina, Mario, 39
Molybdenum (Mo), 475t, 479t
Monoamine messengers, 347, 348
Monoamine oxidases (MAOs), 350, 351, 352–353
Monoamines, oxidation of, 351
Monoclonal antibodies, 497
Monocyte chemotactic proteins (MCPs), 501
Monocytes, 501, 510
Monoglycerides. 261
Monomers, 64, 65, 65t, 458
Monophosphoric esters, of monosaccharides, 244
Monoprotic acids, 139, 140
Monosaccharides, 231–244, 246–247, 361, 362–363, 416, 431–435, 432, 456, 472, 473
    mutarotation of, 240
"Morning after pill," 281
Morphine, 338t, 353–355
Morpholine, 175
Morton, William T. G., 105
Mucins, cystic fibrosis and, 447
Mucus, external innate immunity of, 487
Mullis, Kary B., 380
Multiple sclerosis, 270, 503
Mumps, 394
Murad, Ferid, 352

Muscle, 278, 289, 319, *336–338*, 342, 344, 352, 425, *426*, 437, 438, 503, 513, 514, 519
Muscle relaxants, enzyme specificity and, 319
Muscone, molecular structure of, 186
Muscular dystrophy, creatine and, 482
Mushrooms, thiols in, *105*
Mutagens, 401
Mutarotation, **240**
Mutations, 373, 377–380, 400–403, 464, *496*, 499
Myasthenia gravis, 503
Myelin, composition of, 269
Myelin sheaths, 269–270
Mylar, 78, 221
Myocardial infarction, 276, 331
Myosin, 289, 425, *426*
Myrcene, 56
Myristic acid, 204*t*, 260*t*, 262*t*

NAD⁺. *See* Nicotinamide adenine dinucleotide (NAD⁺)
NADH. *See* Reduced NAD⁺ (NADH)
NADP⁺. *See* Nicotinamide adenine dinucleotide phosphate (NADP⁺)
NADPH. *See* Reduced NADP⁺ (NADPH)
Na⁺, K⁺ ATPase, 266
Naloxone, 338*t*
Nandrolone decanoate, 278
Naphthalene , 80
Naproxen, 126, 130, 214
Narcan, 338*t*
Natural gas, *18*, 39, 91, 107
Natural killer (NK) cells, *486*, 487, 491, 502–503
Near-sightedness, laser surgery to correct, 312
Necrosis, 379
Negative modulation, enzyme regulation via, 329
Nelfinavir, in AIDS treatment, 395
Nembutal, 217
Neonatal herpes, 394
Neo-Synephrine, 338*t*
Neotame, 481
Nephritis, 517
Nephrons, 515, *517*
Nerve axons, 269–270
Nerve gases, 338*t*, 343
Nerves, 319, 344
Nerve transmission. *See* Neurotransmitters
Nervous system, in blood pressure regulation, 520
Neuralgia, capsaicin and, 85
Neurofibrillar tangles, in Alzheimer's disease, 345
Neurons, 309, *336*, 337, 338–339, 342, 343, 345, 350
Neuropeptide Y, 355
Neurotransmitters, 319, **337**, 338–341, 346–355, 356–357, *444*
  adrenergic, 338*t*, 340, 347–353
  cholinergic, 338*t*, 340, 341–346
  peptidergic, 338*t*, 340, 353–355
Neutral salts, **151**
Neutral solutions, 144, 148–150
Neutralization, acid–base, 144
Neutrophils, 501, *510*
Nevirapine, in AIDS treatment, 395
Niacin, 475*t*, 477*t*, 480, 481
Niacinamide, as nutrient, 481
Nickel (Ni), 64, 475*t*
Nicotinamide, 414, *415*
Nicotinamide adenine dinucleotide (NAD⁺), 188, 190–191, *410*, 414, *415*, 416, 417–419, 420, 422, 424, *434*, 438, 439, *440*, 442, 475*t*

Nicotinamide adenine dinucleotide phosphate (NADP⁺), 436, 459, 461
Nicotine, 338*t*, 341, 346
Nicotinic acid, 477*t*
Nicotinic receptor, 341
Niemann-Pick disease, 273
Nifedipine, 338*t*, 520
Nirenberg, Marshall, 389
Nitrate reductase, 475*t*
Nitration, of arenes, 81–82
Nitric acid, 81, 136*t*, 137
Nitric acid test, for proteins, 137
Nitric oxide, 352, 489–491, 490*t*
Nitric oxide synthase, 352, 490*t*, 491
p-Nitroaniline, 170
p-Nitrobenzoic acid, 82
Nitrobenzene, 81
Nitrogen (N₂), 2, 7, 11–12, 167, 169, 209–210, *302*, 431, 443–448, 511*t*
Nitrogenous waste, in urine, 516–517
Nitroglycerin, 82, 94, 108
Nitro group, 82
Nitrous oxide, as anesthetic, 105
N-linked saccharides, 310
Nobel, Alfred, *94*
Nolvadex, treating breast cancer with, 356
Nomenclature
  of alcohols, 92–95
  of aldehydes and ketones, 184–187
  of alditols, 241
  of alkanes, 23–27
  of amides, 209–212
  of amines, 170–172
  of amino acids, 290, 291*t*
  of anhydrides, 208
  of arenes, 77–80
  of carboxylic acids, 203–205
  of carboxylic acid salts, 207–208
  of cycloalkanes, 27–28
  of enzymes, 319, 320*t*
  of esters, 208–209
  of ethers, 102
  of glycosides, 240–241
  of halogenated alkanes, 38
  of monosaccharides, 231, 233–234
  of nucleic acid bases, 361–362
  of nucleotides, 363–364, 364*t*
  of peptides, 294, 296
  of phenols, 83–84
  of polymers, 65, 65*t*
  of proteins, 296, 299
  of thiols, 104–106
Non-insulin-dependent diabetes, 354
Non-nucleoside inhibitor drugs, for AIDS, 395
Nonane, physical properties of, 35*t*
Nonapeptides, 300
Noncompetitive inhibition, **321**, 324–326, 327
Nonessential amino acids, biosynthesis of, 464
Nonheritable errors, in protein transcription, 400
Nonpolar amino acids, 290, 291*t*
Nonsteroidal anti-inflammatory drugs (NSAIDs), 214, 282, 283, 284, 330
Nonsuperposable mirror images, 116
Norepinephrine, 338*t*, 339*t*, 340, 348, *349*
Norethindone, 281
Norethynodrel, 281
Novocain, 176
NSAIDs. *See* Nonsteroidal anti-inflammatory drugs (NSAIDs)
N-terminal amino acid in peptides, 294, 296, 298, 299, *302*
Nuclear membrane, *411*

Nucleic acids, 230, 238, 361–364, **365**, 394, 491. *See also* DNA; RNA
Nucleoside analog drugs, for AIDS, 395
Nucleosides, **362**–363
Nucleosomes, DNA molecules arranged in, 369, *370*
Nucleotide excision repair (NER), 380, 401
Nucleotides, 348, 361, 363–364, 364*t*, 376, 380–381, *444*, 504
Nucleus. *See* Cell nucleus
Nutrasweet, 481
Nutrients, 412, 468–471, 482, *509*, 510
Nutrition, 468–483, 476–479*t*
Nylon-66, 221

Obesity, 354, 469–470, 472, 480
O blood type, 246–247
Octadecanoic acid, 204*t*
Octanal, from oxidation, 100
Octane, 35*t*, 36
Octane antiknock ratings, 40
Octanoic acid, 100, 204*t*
1-Octanol, 100
1-Octene, 49
Octyl *p*-methoxycinnamate, 216
Odors
  of aldehydes and ketones, 188
  of amines, 172
  of carboxylic acids, 206
  of thiols, 104
Oil of almonds, 186
Oil of cinnamon, 186
Oils, **261**–264, 262*t*, *471*, 473. *See also* Essential oils
Okazaki, Reiji, 377
Okazaki, Tuneko, 377
Okazaki fragments, 373, **376**, 377
Oleic acid, 260*t*, 262*t*, 263
Olestra, 481
Oligonucleotide drugs, 387
Oligosaccharides, 231, **245**–248, 273, 457–460, 472–473
O-linked saccharides, 310
Oncogenes, 402, 447, 464
Ondansetrone, 338*t*, 350
Onions, thiols in, *105*
Oocytes, in menstrual cycle, *280*
Opening step, in DNA replication, 375–376
Opiate, 338*t*
Opsin, in retina, 54
Optical activity, of chiral molecules, **127**–129
Oral contraceptives, 281
Orexic agents, 355
Organelles, 265, 309, 318, 336–337, *411*, 412, *431*
Organic chemistry, 2–14
Organic compounds, 3–8, 4*t*
Organic fertilizers, 470
Organisms, chemical communication within, 337
Organophosphorous hydroxylase, nerve gas detoxification via, 343
Origins of replication, 373
Orinase, for diabetes, 354
Ornithine, 445, *446*, 447
L-Ornithine, 318
Osmotic pressure, 512, 519
Osteoarthritis, treatment of, 282
Osteoporosis, aging and, 519
Ovalbumin, 290
Ovarian follicle, *280*
Ovarian phases, in menstrual cycle, *280*
Ovulation, hormones in, 279, *280*
Oxalic acid, *203*, 204

Oxaloacetate, 320*t*, *416*, 417, 419, 442, *449*, 450, *457*
Oxaloacetic acid, 431, *449*
Oxalosuccinate, 417–418
Oxidation, 99–101, 106–107, 176, 188–189, 242–244, 261, 294, 318, 351, 417, 437*t*
β-Oxidation, 431, *432*
  fatty acids and, 438–441, *440*, 460–461
Oxidation–reduction reactions, 458, 459. *See also* Redox reactions
Oxidative damage, 423, 435
Oxidative deamination, 431, *444*–**445**, 448, 464
Oxidative decarboxylation, 435, 437*t*
Oxidative phosphorylation, 436, 437*t*, *410*, 413–414, *420*, 424, *432*. *See also* Electron transport chain
Oxidizing agents, 296
Oxidoreductases, 319, 320*t*, 321
Oxonium ion, **61**, 62
Oxygen, 2, 3, 7, 106–107, 176, 276, *302*, 309, 420, 421, 424–425, 487, 511*t*
  aldehyde reactions with, 188
  autoxidation and, 84–87
  blood and, 243, 509, 513–514, 520
  in detoxification, 423
  lactate accumulation and, 435
  reactions of alkanes with, 37
Oxygen deprivation, 423
Oxygen dissociation curve, 513
Oxytocin, 290, 300, 339*t*, *340*, 353
Ozone layer, CFCs and, 39

p53 protein, as tumor suppressor, 402
PABA. *See p*-Aminobenzoic acid (PABA)
Paclitaxel, 2, 5
Padimate A, 216
PAHs. *See* Polynuclear aromatic hydrocarbons (PAHs)
Pain relievers, 214, 330
Pain transmission, 355
Palladium (Pd), as hydrogenation catalyst, 64
Palmitic acid, 204*t*, 260*t*, 262*t*
Palmitoleic acid, 260*t*
Pancreas, in diabetes, 354
Pancreatic islets, 339*t*, 353, 354
Pantothenic acid, *415*, 416, 475*t*, 478*t*
Paraffin waxes, 34, 262
Parathion, 343
Parathyroid gland, 339*t*
Parathyroid hormone, 339*t*
Parenteral nutrition, 469
Parkinson's disease, 350, 482
Partial pressure, 514, 515
Passive transport, across cell membranes, *266*
Pasteur, Louis, 499
Pathogens, immunity against, 486–487
Pauling, Linus, 76, 301, 367
PCR. *See* Polymerase chain reaction (PCR)
PCR techniques, 375, 380–381
Pellagra, 475
Penicillamine, in treating cystinuria, 449
Penicillin, 126, 210, 211, 326
Penicillin G, 211
Pentaerythritol tetranitrate (PETN), 82
Pentagastin, 338*t*
Pentanal, 190, 206*t*
Pentane, 20, 27, 35*t*, 96*t*, 187*t*
1,5-Pentanediamine, 170–171
Pentanedioic acid, 203
1-Pentanethiol, 105–106
Pentanoic acid, 204*t*, 189, 206
2-Pentanone, 193–194
Pentapeptides, 296
Pentobarbital, 217

Pentose phosphate pathway, **436**
Pentyl mercaptan, 105–106
PEP. *See* Phosphoenol pyruvate (PEP)
Peppers, capsaicin in, 85
Pep pills, 168
Pepsin, 319, 320, 431, 474
Peptide backbone, *296*, 302
Peptide bond, **294**–296, 393
Peptidergic neurotransmitters, 338*t*, 340, 353–355
Peptides, 294–298, 345, 474, *492*
Peptidoglycans, innate immunity and, 503
Peptidyl transferase, *392*, 393
Perchloric acid, 136*t*
Perforins, macrophages and, 491
Peripheral proteins, 265, *267*
Permanent wave, protein denaturation and, *311*–312
Permethrin, 210
Peroxidase, 243, 330
Peroxidase activity, 330
Peroxides, 66, 296
PET. *See* Poly(ethylene terephthalate) (PET)
PET scan, *345*
Petrolatum, 34
Petroleum, *18*, 39–40, 47
Petroleum refinery, *18*, *39*, 40
PGHS. *See* Prostaglandin endoperoxide synthase (PGHS)
pH, 96, 148–152, 150*t*, 154–160, 208, 219, 293, 297, 321, 322, 323, 366, 487, 510, 514, 516, 518. *See also* Hydrogen ions (H⁺)
Phagocytes, 450–451, 501
Phagocytosis, 493, 501, 511
Pharmacogenomics, 377
Phenanthrene, 80, 273–274
Phencyclidine (PCP), 338*t*, 347
Phenergan, 338*t*, 350
Phenobarbital, 217
Phenol, 77, 84, 140
Phenolphthalein, 150*t*, 153
Phenols, **83**–87, 96–97
Phenteramine, 168
N-Phenylacetamide, 218
Phenylalanine, 291*t*, 389, 389*t*, *449*, 450
Phenylalanine hydroxylase, 449
1-Phenylcyclohexene, 77
p-Phenylenediamine, 220
Phenylephrine, 338*t*
Phenyl group, **77**
Phenylketonuria (PKU), 449
(*s*)-1-Phenyl-2-propanamine, 170–171
Phenylpyruvate, hereditary diseases involving, 449
Pheromones, of European corn borer, 52
PHI. *See* Phosphohexose isomerase (PHI)
pH indicators, 149–150
pH meter, 150, *151*, 152
Phosgene, 221–222
Phosphatase, 329, 352–353, 355, 502
Phosphate buffer solution, 155, 156, 158, 159–160
Phosphate esters, in lipid biosynthesis, 462
Phosphate group, 267–268, 325, 329, 363–364, 365–*366*, 413–414
Phosphate ion, in urine, 517
Phosphates, 343, 361, 363–364
Phosphatidates, in lipid biosynthesis, 462
Phosphatidylcholines, 267–268, 462
Phosphatidylethanolamine, 266, 268–269
Phosphatidyl inositol (PI), 355
Phosphatidylinositol-4,5-biphosphates (PIP2), 268
Phosphatidylserine, 266, 268

Phosphatylcholine, 266
Phosphocreatine, 482
Phosphodiesterase, nitric oxide and, 352
Phosphoenol pyruvate (PEP), 325, 432, *433*–434, 442, *449*, 450, *457*
Phosphofructokinase, *433*
2-Phosphoglycerate, *433*–434
3-Phosphoglycerate, 432, *433*–434
Phosphoglycerate kinase, *433*
Phosphoglycerate mutase, *433*
Phosphohexose isomerase (PHI), 320*t*, 331*t*, *433*
Phosphoinositoldiphosphate (PIP2), 348
Phospholipase, 282, 355
Phospholipids, 264–*265*, 274–275
Phosphonates, 343, 346
Phosphoric acid, 97, 137, 219, 244
Phosphoric anhydride bond, 413–414
Phosphoric anhydrides, 219
Phosphoric esters, 219, 244
Phosphorothioate moieties, in antisense drugs, 387
Phosphorus (P), 2, 478*t*
Phosphoryl group, 219
Phosphorylase, 425, 438, 455–456
Phosphorylated ADP, in coenzyme A, *415*
Phosphorylated protein kinase, in gene regulation, 399
Phosphorylation, 244, 329, 348, 349, 354, 355, 387, 388, 397, 399, *410*, 413–414, *420*, 422–424, 431–*432*, *433*–434, 438, 463
Phosphoserine, 329
Phosphotriose isomerase, in glycolysis, *433*
Photofrin, 312
Photorefractive keratectomy (PRK), 312
Photosynthesis, 436, **456**, 458–459
Photosystem I, 458–**459**
Photosystem II, 458–459
Physical properties
  of alcohols, 95–96
  of alkanes, 34–36
  of alkenes and alkynes, 55
  of amines, 172–173
  of amino acids, 293
  of carboxylic acids, 205–206
  of cellulose, 250
  of fatty acids, 259–260, 260*t*
  of lipoproteins, 275*t*
  of monosaccharides, 235
Picric acid, 422
Pigs, hemoglobin of, 401, 401*t*
Pilocarpine, 338*t*
Piperidine, 168, 169
Pituitary gland, 339*t*, *340*
p*K*ₐ, 96, 106, 143–144, 156–157, 158–160, 173–174, 173*t*, 206–207, 208
PKU. *See* Phenylketonuria (PKU)
Plane-polarized light, 127–**128**
Plant proteins, 474–475
Plants, *116*, 210, 405–406, 456, 458–459
Plant waxes, 262
Plaque, cholesterol and, 276
Plasma, 276, 510, 511–512, 511*t*, 515
Plasma cells, 246, *487*, 490*t*, 491, 495, 499
Plasma membrane, 274, *411*
Plasmids, in recombinant DNA techniques, *403*–404
Plastics, recycling of, 68
Platelets, 284, *508*, *510*, **511**, 511*t*, 512
Platinum (Pt), as hydrogenation catalyst, 64
β-Pleated sheet, 301–304, *302*, 311, 345, 501
PM mutant. *See* Poor metabolism (PM) mutant
Pneumocystis carinii, AIDS and, 395
pOH, 148, 149
Poison hemlock, 169

Poison ivy, *84*

Polar amino acids, 290, 291*t*, 292–293

Polar compounds, 103–104, 172–173, 266, 267

Polar covalent bonds, 95

Polarimeter, 128–129, *128*

Polarity
  of aldehydes and ketones, 187
  of amino acids, 290–292
  of carboxylic acids, 205–206
  of cell-membrane lipids, 265–267
  of esters, 209
  of lipids, 259, 261
  of methanol bonds, *95*
  of monosaccharides, 235

Poliomyelitis virus, *394*

Polio vaccines, 499

Pollen, as allergen, 487, 503

Poly A addition, in transcription, *386*

Polyamides, 219, **220**

Polycarbonates, 219, **221**–222

Polycythemia, 511*t*

Polyenes, 53–55

Polyesters, 219, **220**–221

Polyethylene, 64–65, 66–67

Polyethylene films, *66*

Poly(ethylene terephthalate) (PET), 68, 78, 108, 220–221

Poly II. *See* RNA polymerase II

Polymerase chain reaction (PCR), 380–381

Polymerases, *378*, 379, 380–381, 386–388

Polymerization, 47, 64–68

Polymers, **64**, 65, 65*t*. *See also* Polymerization
  DNA and RNA as, 361, 365

Polynuclear aromatic hydrocarbons (PAHs), **80**

Polynucleotide chains, in DNA molecule, *367–369, 370*

Polypeptides, 296, 395, 491, 492

Polypropylene (PP), *66*, 68*t*

Polysaccharides, 230, 231, **245**, 249–251, 430–431, 457–460, 472–473, 491

Polystyrene (PS), *66*, 68*t*

Polysubstituted alkylbenzenes, 78–80

Polyubiquitin, 447

Poly(vinyl chloride) (PVC), 65, 68*t*

Poor metabolism (PM) mutant, of cytochrome P-450 enzyme, 377

Positive modulation, enzyme regulation via, 329

Posterior pituitary gland, hormones secreted by, 339*t*, 340

Postsynaptic membrane, *349*

Postsynaptic nerve endings, 337, 339

Postsynaptic receptor site, acetylcholine at, 343

Post-translational controls, in gene regulation, 400

Potassium (K), *3*, 475*t*, 478*t*

Potassium chloride, 144–145

Potassium dichromate, 101, 188

Potassium dihydrogen phosphate, 145

Potassium hydroxide, 136*t*, 137, 144–145, 207

Potassium ions), 266, 341, 343, 347, 425, 443

Powders, 210

P (peptide) site, *392–393*

P-protein, action of, 349

Pralidoxime chloride, nerve gas detoxification via, 343

Prandin, for diabetes, 354

Prazosin, 338*t*

Precipitates, 297

Pre-initiation complex, in protein synthesis, *391–392*

Premarin, 519

Prenylation, 463, 464

Preservatives, carboxylic acid salts as, *207*

Presynaptic cleft, 342–343

Presynaptic nerve endings, *337*, 339

Preven, as contraceptive, 281

Priestly, Joseph, 105

Primary (1°) alcohols, 10, 93–94, 97, 99–101

Primary (1°) amides, 210

Primary (1°) amines, 11–12, 167, 168, 170

Primary (1°) carbocations, 59

Primary structure
  of DNA, *365–366*
  of nucleic acids, 365–*366*
  of proteins, 297, *298–301, 306*

Primases, *373*, 374*t*, 376

Primatine Mist, 177

Primers, *373*, 376, 380–381

Prion protein, 303

Priorities, in R,S system, 120–123, 121*t*

Procaine hydrochloride, 176

Proenzymes, enzyme regulation via, **327**–328

Progesterone, 277, 279, *280–281*, 339*t*, 356, 357, 519

Prohormones, testosterone and, 278

Prokaryotes, 376, 385, 391, 393, 397

Prolactin, action of, 339*t*, *340*

Proline, 290, 291*t*, 389*t*, *449*

Promethazine, 338*t*, 350

Promoters, 386, 397, 398, *399*

Propanal, 206*t*

2-Propanamine, 170

Propane, 20, 35*t*, 37, 96*t*

Propanedioic acid, 203

1,2-Propanediol, 95

1,2,3-Propanetriol, 94, 95, 263

1,2,3-Propanetriol trinitrate, 94. *See also* Nitroglycerin

Propanoic acid, 187*t*, 188, 204*t*

1-Propanol, 92, 96*t*, 206*t*

2-Propanol, 10–11, 60, 92, 118, 215–216

Propanone, 184. *See also* Acetone

2-Propanone, 184

Propantheline, 338*t*

2-Propenal, 184

Propene, 8, *48*, 50, 58, 60, 61, 108

Propionic acid, 204*t*

Propofol, 244

Propranolol, 338*t*, 352

Propyl alcohol, 92

Propylamine, 171

Propylene, 50, 65*t*

Propylene glycol, 95

Propyl group, 25*t*

Propyne, 49, 50

Prostaglandin A (PGA), 283, 520

Prostaglandin B (PGB), 283

Prostaglandin E (PGE), 283, 520

Prostaglandin endoperoxide synthase (PGHS), 330

Prostaglandin F (PGF), 283

Prostaglandins, 214, 259, 282–283, 330

Prostate cancer, 519

Prostate gland, 283, 519

Prostate-specific antigen (PSA), 519

Prosthetic group, in proteins, 309

Proteases, 328, 400, 447

Proteasomes, 400, 447

Protein coats, of viruses, 394

Protein complementation, 474–475

Protein deficiency diseases, 360, 475

Protein kinase, *349*

Protein kinase II, 342

Protein modification, enzyme regulation via, 329

Protein/peptide-conformation-dependent diseases, 303

Proteins, 137, 265, *266*, 267, **289**–313, *298, 302, 306*, 317, 323, 340, 342, 345, 360–361, 379, 400, 430, 431, *432*, 491–492
  amino acids in, **274**, **290**–292, 291*t*, 292–301, 318, 360–361
  chirality of, 129–130
  in cholesterol transport, 274–275
  in citric acid cycle, *416*
  complete, 474
  degradation of misfolded, 400, 447
  digestion of, 443–*444*, 474
  diseases caused by production of excess, 387
  genes and, 361, 384–385
  genetic code and, 385
  incomplete, 474, 475
  in leukocyte mobilization, 500
  as nutrients, 410, 469, 474–475
  peptide bonds in, **294**–296
  in photosynthesis, 458–459
  in plasma, 511
  as receptors, 337
  in ribosomes, 372
  in serum, 511*t*
  steroid hormones and, 356–357
  in T cell receptor complex, 498–499
  transmembrane, 341
  as transporters, **347**
  tRNA and, 371
  ubiquitin and, 447
  in urine, 517

Protein synthesis, *372*, 388–390, **389**, 389*t*, 390–397, 399–400

Proteoglycans, as glycoproteins, 310

Proteolysis, 447

Proteolytic enzymes, of macrophages, 493

Proteomics, 307, *384*, 504

Prothrombin, in clotting, 512

Proton channel, in electron and hydrogen ion transport, 423–424

Proton donors and acceptors, 138–140, 292–293

Proton gradient, 422–424, 458

Proton transfer, in acids and bases, 135

Proton translocating ATPase, 422–424

Protonophores, 422

Proto-oncogenes, 402

Protozoa, AIDS and, 395

Proventil, *167*, 177

Provera, 519

Proximal tubules, of kidneys, 516, *517*

Prozac, 338*t*

Prusiner, Stanley, 303

D-Psicose, 233*t*

Psoriasis, as autoimmune disease, 503

Puffer fish, poison of, 31

Purine, molecular structure of, 168

Purines, as nucleic acid bases, 361–363, *367–369, 368*, 465

Pyran, 236

Pyranose form, of monosaccharides, 235, **236**–241

Pyrethrin I and II, 210

Pyrethrum powder, 210

Pyridine, 168, 169, 175

Pyridinium acetate, 176

Pyridoxal, *219*, 475*t*, 477*t*

Pyridoxal phosphate (PLP), 219, 475*t*

Pyrimidine, 168

Pyrimidines, as nucleic acid bases, 361–363, *367–369, 368*, 465

Pyrophosphate ion (PP$_i$), 219, 348, *349*, 391, *446*

Pyrophosphoric acid, 219

Pyrrolidine, 168

Pyruvate, 191, 320*t*, 325, *432*, *433–434*, 435, 437*t*, *444*, 448, *449*, 457, 465

Pyruvate dehydrogenase, 435, 475*t*
Pyruvate kinase (PK), 325, 329, 432, *433*, 475*t*
Pyruvic acid, 431, *449*. *See also* Pyruvate

Quarter-stagger arrangement, of collagen
    fibers, 308, *309*
Quaternary molecular structure, of proteins,
    298, 304–313, *306*, 323

Rabies, 394, 499
Racemic mixtures, **118**, 126
Radiation, 401
Radicals, in autoxidation, 85–87
Rancidity, 86, 261
Random coils, in protein secondary structure,
    302–303
Ranitidine, 338*t*, 352
Rapamycin, for autoimmune diseases, 504
Ras protein, prenylation of, 464
Rat liver cell, *411*
RDA. *See* Recommended Daily Allowances
    (RDA)
RDX. *See* Cyclonite (RDX)
Reaction mechanisms, *59–60*
Reaction rates
    of alkene addition reactions, 57–58
    of enzyme activity, 321–323
Receptors, 337, 338*t*, 341, 347, 352, 354, *487*,
    488, 503
Recessive genes, 403
Recombinant DNA, 403–406
Recommended Daily Allowances (RDA), 469,
    476–479*t*, 480, 481–482
Recycling, of plastics, 68
Red blood cells (RBC), 246–247, 266, 274, *301*,
    309, 421, 435, 450–451, *508*, 511, 512. *See
    also* Erythrocytes
Red cabbage juice, as acid–base indicator, *151*
Red litmus paper, *137*, 149
Red No. 40, 83
Redox reactions, 144
Reduced FAD⁺ (FADH₂), *410*, 414, *416*,
    418–419, *420*, 421, 422, 424, 439, *440*,
    441*t*
Reduced NAD⁺ (NADH), *410*, 414, *416*,
    417–419, 420, 422, 424, *434*, 435,
    436–437, 437*t*, 438, 439–441, *440*, 442,
    445
Reduced NADP⁺ (NADPH), 435, 436, 459, 461
Reducing sugars, **242**, 243, 245, 248
5-α-Reductase, aging and, 519
Reduction
    of aldehydes and ketones, 189–191
    of alkenes, 64
    of cystine, 294
    of monosaccharides, 241–242
Refining of petroleum, 39–40
Regioselective reactions, **58**
Reglan, 338*t*, 350
Regulation
    of acetylcholinesterase, 346
    of B cells and T cells, 501–502
    of blood pressure, 519–520
    of enzymes, 327–330
    of genes, **397**–400
    as protein function, 290
Regulators, allosteric enzymes and, 329
Regulatory genes, in transcription, 386
Regulatory sites, allosteric enzymes and, 329
Relaxation step, in DNA replication, 376
Release factors, 393, 399–400
Reminyl, for Alzheimer's disease, 345
Renin, action of, 339*t*
Repaglinide, for diabetes, 354

Repair, of DNA, 377–380
Replication, **373**, *373*–377
Replication fork, in DNA, *373*–374, 376
Replication origins, on DNA molecule, 373
Replisomes, 374
Reserpine, 338*t*
Resins, 107, *108*
Resonance, 76
Resonance hybrid, **76**–77
Resorcinol, 84
Respiration, as redox reaction, 112
Respiratory acidosis and alkalosis, 159
Response elements, in gene regulation, 399
Restriction endonucleases, 403–404
Restriction enzymes, in DNA
    fingerprinting, 375
Retina, 54, 461
Retinitis, cytomegalovirus-induced, 387
Retinol, 54, *55*
Retinopathy, diabetic, 298
Retroviruses, 395, 402
Reuptake, of glutamic acid, 347
Reverse transcriptase, in AIDS, 395, **396**
Reversible denaturation, 313
Reversible inhibition, 321, 346
Rheumatoid arthritis, 282, 501, 503
Rhodopsin, 54, *310*
Ribavirin, in hepatitis treatment, 395
Ribitol, in FAD molecular structure, *415*
Riboflavin, 414, *415*, 475–481, 475*t*, 476*t*
Ribonucleoproteins, telomerase as, 374
Ribonucleoside triphosphates, during DNA
    replication, 376
Ribose, *415*, 436
D-Ribose, 230, 232*t*, 362–363, 364, 413
α-D-Ribose, 238
β-D-Ribose, 362–363
Ribose 5-phosphate, 436
Ribose reductase, as cofactor, 475*t*
Ribosomal complex, in protein synthesis
    initiation, 392
Ribosomal RNA (rRNA), **372**, **385**, 386, 388
Ribosomes, 371, **372**, 388–389, *390–393*
Ribozymes, 317, 372, 392
D-Ribulose, 233*t*
Ribulose 1,5-bisphosphate carboxylase-
    oxygenase
    (RuBisCO), 459
Ribulose 5-phosphate, 436
Rice, incomplete proteins in, 474, 475
Ritalin, 338*t*, 350
Ritonavir, in AIDS treatment, 395
Rituxan, 497
Rivastigmine, for Alzheimer's disease, 345
RNA (ribonucleic acid), 361–364, 364*t*, **385**–388,
    389*t*, 397, 398, 399, 400
    heterocyclic aromatic amines in, 168
    molecular structure of, 365, 370–371,
    371–372
    NADPH and, 436
    in ribozymes, 317
    translation from, **385**, 388–389, *390–393*
    types of, *371*–**372**
    in viruses, 394, 395
RNA polymerase I, 386
RNA polymerase II, 386, 387–388, *399*
RNA polymerase III, 386
RNA polymerases, *386*–388, 397
RNA viruses, 394, 395
Roberts, Richard J., 372
Rod cells, in retina, 54
Rodbell, Martin, 348
"Rotor engine," proton translocating
    ATPase as, 423

Rough endoplasmic reticulum, *411*
Rowland, Sherwood, 39
R (*rectus*) configuration, in R,S system,
    120–123, **121**
R,S system, 119–123, **120**, 232
RU486, 280, 281
Rubbing alcohol, 11. *See also* Isopropyl alcohol
RuBisCO. *See* Ribulose 1,5-bisphosphate
    carboxylase-oxygenase (RuBisCO)
Ruthenium (Ru), as hydrogenation
    catalyst, 64

Sabin polio vaccine, 499
Saccharides, 231, 310
Saccharin, 248*t*, 480
Salicin, 214
Salicylic acid, aspirin and, 214
Saline solution, *234*
Salk polio vaccine, 499
Salmeterol, 178
Salt balance, in blood and kidneys, 518
Salt bridge, in protein tertiary structure,
    304, *305*
Salts, 144–146, 511*t*
    acidic and basic, 150–152, **151**
    amino acids as, 292–293
    of carboxylic acids, 207–208
    neutral, **151**
Sanger, Frederick, 300
Saponification, **215**, 263–264
Saquinavir, in AIDS treatment, 395
Saran Wrap, *66*
Sarcoplasmic reticulum (SR), 266, 348
Sarin, 343
Satellites, in DNA, 372–373
Saturated fatty acids, 259–260, 260*t*, 469
Saturated hydrocarbons, **18**
Schwann cells, 270. *See also* Sheath cells
Scurvy, 469, 475
SDS. *See* Sodium 4-dodecylbenzenesulfonate
    (SDS)
Secondary (2°) alcohols, 10, 11, 93–94, 97,
    99–101
Secondary (2°) amides, 210
Secondary (2°) amines, 11, 167, 168, 170
Secondary (2°) carbocations, 59
Secondary messengers, 337, 340, 342, 348, 349,
    352–353, 355, 489–491
Secondary structure
    of DNA, 367–369, *370*
    of nucleic acids, 365
    of proteins, 297–*298*, 301–304, *306*, 310–313
    of RNA, *371*
Secondary sexual characteristics, 279
Second genetic code, 391
Secretases, in Alzheimer's disease, 345
Sedatives, 172, 217
Selectins, in leukocyte mobilization, 500
Selegiline, 338*t*
Selenium (Se), 475*t*, 479*t*
Self-ionization of water, 147–148
Self recognition, 501–504
Semen, 282–283, 375, 519. *See also* Sperm
Semiconservative replication, 374
Sensor proteins, in gene regulation, 399
Sequences
    of amino acids in proteins, 299–300, 301,
    318, 360–361
    of bases in DNA fingerprinting, 375
    in DNA transcription, 386–387
    genome as, 377
    of nucleotides in nucleic acids, 361–362,
    365–366, 367–369, *372–373*
Serevent, 178

Serine (Ser, S), 268, 291*t*, 295, 305, 310, 389*t*, *449*

Serotonin, 338*t*, 347, 350

Serum, 276, 330–331, 331*t*, 495, 497, 511*t*, **512**, 519

Sex hormones, 277, 279–281, *280*, 519

Sexual reproduction, 361

Sharp, Philip A., 372

Sheath cells, *337. See also* Myelin sheaths; Schwann cells

Sheep clones, *360*

Sheep insulin, primary structure of, 299

Shells, monosaccharides in, 235

Shortenings, production of, 263

Shunt, pentose, 436

Sickle cell anemia, 300, 301, 373, 400

Side chains
  of nucleic acids, *365–366*
  of proteins, 296–298, 304, 305, 307

Sideroblastic anemia, 421

Sieving portions of AARS, in gene regulation, 399

Signal transduction, **348**, 438
  during DNA replication, 375–376
  in immune system, 498–499, 501–502
  in self recognition, 502–503

Silicon (Si), 3

Silk, secondary structure of, 303. *See also* Spider silk

Silver chloride, 3

Silver cyanate, 3

Silvering of mirrors, 189

Silver-mirror test, for aldehydes, *189*

Silver nitrate, in Tollens' reagent, 189

Simple lipids, 259–*265*

Simvastatin, 276

Sinus, of lymph node, *490*

60S ribosome, 388, 390*t*, *392*

Skin, external innate immunity of, 486–487

Skin cancer, xeroderma pigmentosa and, 380

Skunk, thiols in scent of, *104*

Smallpox, 394, 499

Smooth endoplasmic reticulum, *411*

Snake venom, effects on acetylcholinesterase regulation, 346

Soaps, 215, 263, *264*

Sobrero, Ascanio, 94

Sodium (Na), *3*, 221–222, 478*t*

Sodium acetate, 150–151, 154–155, 157, 207, 215–216, 218

Sodium benzoate, *207*, 215–216

Sodium bicarbonate, 84, *140*, 145, 207, 208

Sodium borohydride, 190, 241

Sodium butanoate, 207

Sodium carbonate, 94, 207

Sodium citrate, 512

Sodium D line, in polarimetry, 128–129

Sodium dodecylbenzenesulfonate, 264

Sodium 4-dodecylbenzenesulfonate (SDS), 82–83

Sodium dodecyl sulfate, 264

Sodium ethanethiolate, 106

Sodium ethoxide, 217

Sodium hydroxide (NaOH), 84, 106, 136*t*, 137, 141, 152–154, 155, *156*, 207, 215, 216, 218, 219

Sodium ion channels, 31, 341, 343, 347

Sodium ions, *140*, 341, 343, 425, 443, 517, 518, 520

Sodium lactate, 207–208

Sodium lauryl sulfate, 264

Sodium pentobarbital, 217

Sodium phenoxide, 84, 97

Sodium propanoate, 207

Sodium soaps, 263, 264

Solatol, 338*t*

Solenoid, DNA molecules arranged in, **369**, *370*

Solid shortenings, production of, 263

Solid soaps, 264

Solubility
  of alcohols, 96*t*
  of aldehydes and ketones, 188
  of alkanes, 35–36
  of amine salts, 175
  of carboxylic acids, 205–206
  of ethers, 104
  of lipids, 258, 259
  of monosaccharides, 235
  of proteins, 290, 297
  of soaps, 264

Solutions
  acidic and basic, 148–152, 154–160
  buffering of, 154–158, 158–160
  neutral, 144, 148–150

Solvents, 38–39

Soman, 343

Somatic cells, telomers in, 374

Somatic hypermutation (SHM), of V(J)D genes, 496

D-Sorbitol, formation of, 241

D-Sorbose, 233*t*

Sour taste, of acids, 144, 206

Soybeans, incomplete proteins in, 474, 475

Specificity
  of enzyme action, **318**, 319
  of immune system, 487

Specific rotation, 128–129, 232

Sperm, zygote from, 373

Spermaceti, 262

Sphingolipids, 264–*265*, 269–270, 462

Sphingomyelinase, in Niemann-Pick disease, 273

Sphingomyelins, 266, 269–270, 273, 462

Sphingosine, 264–*265*, 269, 473

Spider silk, *289*, 303

Spiral, left-handed, *114*

Spleen, 451, 489, *490*, 510–511

Splicing, of RNA molecules, 372, *386*

Spongiform encephalitis, 303

Spotted skunk, thiols in scent of, *104*

S (*sinister*) configuration, in R,S system, 120–123, **121**

Starch, 230, 249, 456, 469, 472–473

Starch-like plaques, 303

Start codon, 389*t*, 390

Starvation, 442, 447, 472

Statin drugs, cholesterol and, 276–277, 463

Stearic acid, 204*t*, 260*t*, 262*t*, 263, 267, 269, 439–441, *440*, 441*t*

Stearyl-CoA, in stearic acid metabolism, 441*t*

Stem cells, 374

Step-growth polymerization, 219–222

Stereocenters, **118**, 119–127, 231–235, 292

Stereoisomers, **33**, 114–130
  of amino acids, 292
  with anomeric carbons, 235, **236**, 238, 239, 240, 241
  of monosaccharides, 231–235

Stereospecificity, of enzymes, 318

Steroid hormones, 277–281, 356–357

Steroids, 273–282, 340, 356–357
  anabolic, 278, 482
  as lipids, 259

Stoichiometry, 152, 154

Stomach acid, in protein digestion, 474

Stomach fluid, excess acids in, 145

Stomach inflammation, by aspirin, 214

Stomach ulcers, treatment of, 145

Stop codons, 389*t*, 390, 393, 399–400

Stress, on chemical reactions, 200–206

Stroke, 276, 352, 423

Stroma, *458*

Strong acids, **136**–137, 140–142, 143, 145–146, 155, *156*, 157, 293

Strong bases, 106, **136**–137, 140–142, 155, *156*, 157, 207–208, 293

Structural formulas, 6–8, 47–48
  condensed, 10, 11, 20*t*

Structural genes, in transcription, 386

Structural isomers, 21

Structural proteins, 289, 290

Styrene, 65*t*, 77

Substance(s)
  hydrophilic, 264, 265–267, 282, 290–292
  hydrophobic, 259, 264, 265–267, 282, 290–292

Substance P, 355

Substitutions
  alkyl, 77–80
  aromatic, 81–83

Substrate
  of an enzyme, 321, 323–326
  of pyruvate kinase, 325

Substrate concentration, 321–322

Subunits
  in collagen, 309
  of proteins, 307

Succinate , *416*, 418

Succinate dehydrogenase, 418, 475*t*

Succinic acid, 203, 213

Succinylcholine, 319, 338*t*, 346

Succinyl-CoA, *416*, *449*

Sucralose, 481

Sucrose, 230, 235, 238, 245, 248*t*, 472

Sugar, 300

Sugar beets and sugar cane, sucrose from, 245

"Sugar-free" products, alditols as, *241*, 242

"Sugarless" candy, xylitol in, 242

Sugars, 248, 248*t*, *471*
  glycation of proteins by, 298
  in glycomics, 504
  in nucleic acids, 362–363, 370
  reducing, **242**, 243, 245, 248

Sulfa drugs, 326

Sulfanilamide, 326

Sulfapyridine, 326

Sulfathiazole, 326

Sulfhydryl group, in thiols, 104–107

Sulfonated polysaccharides, heparin as, 251

Sulfonation, of arenes, 82–83

Sulfonyl urea compounds, for diabetes, 354

Sulfur (S), 2, 294, 328, 421, 458

Sulfuric acid, 82–83, 97, 101, 136*t*, 137, 152–154

Sulfuric esters, acidic polysaccharides as, 250

Sun, in photosynthesis, 456, 458–459

Sun protection factor (SPF), 216

Sunscreens, 482

Sunset Yellow (Yellow No. 6), 83

Superimposable molecules. See Superposable molecules

Superoxide, 377, 423

Superoxide dismutase, in detoxification, 423

Superposable molecules, 116

Sutures, biodegradable, 221

Svedberg unit (S), 388

Sweet taste, 235, 240, 248*t*

Sweeteners. *See* Artificial sweeteners

Symmetrical anhydrides, 208

Synapses, *337*, **339**, 342–343

Synergistic effects, of alcohol and barbiturates, 217
Synovial fluid, 508
Synthetases, 319, 320*t*, 391
Synthetic fibers, creation of, 221
Synthetic polypeptides, in AIDS treatment, 395

Table sugar, 230, 245, 248*t*
Tacrine, 338*t*, 345
Tagamet, 338*t*, 352
D-Tagatose, 233*t*
D-Talose, 232*t*
Tamoxifen, breast cancer and, 356
TaqI restriction endonuclease, 404–405
TATA box, 386–387, 397
Tatum, Edward, 361
Tau proteins, in Alzheimer's disease, 345
Taurine, 281, 346–347
Taurocholate, 282
Tautomers, 194–**195**
Taxol, 5
Tay-Sachs disease, 272, 273
T cell receptor complex, formation of, *498–500*
T cell receptors (TcR), *487*, 488, 492, *498–500*, *502–503*, 504
T cells, *487*, 488, 489, 490*t*, 491, 492, *498–500*, *501–502*
Teflon, *66*
Telomerase and telomers, 374
Temperature (*T*), effect on enzyme activity, 321, 322
Terazosin, 338*t*, 519
Terbutaline, 177, 178
Terephthalic acid, 108, 220–221
Termination, of transcription, 390, 390*t*, 393
Termination sequence, in transcription, 387
Termination site, in transcription, *386*, 387
Terpenes, **55–57**
Tertiary (3°) alcohols, 10, 93–94, 97, 100
Tertiary (3°) amides, 210
Tertiary (3°) amines, 11, 167, 168, 170
Tertiary (3°) carbocations, 59
Tertiary structure
    of proteins, *298*, 304–313, *306*, 400
    of RNA, *371*, 372
Testes, 339*t*, *340*, 519
Testosterone, 277, 278, 279, 281, 339*t*, 519
Tetrachloromethane. *See* Carbon tetrachloride
Tetracycline, 323
11-Tetradecenyl acetate, 52
Tetradecanoic acid , 204*t*
Tetrafluoroethylene, polymers of, 65*t*
Tetrahydrofolate, 326
Tetrahydrofuran (THF), 103
Tetraodontidae, 31
Tetrapeptides, 295–296
Tetrasaccharides, blood type and, 246
Tetrodotoxin, 31
Tetroses, 231
Therapy. *See also* Chemotherapy; Drugs; Medicine
    anticancer, 365
    gene, 404–405
    iodine, 81
    laser, 312
    monoclonal antibody, 497
Thermometers, mercury poisoning from, 327
Thiamine, 475*t*, 476*t*
Thiamine pyrophosphate (TPP), as coenzyme, 475*t*
Thiazides, in hypertension treatment, 520
Thioesters, in fatty acid biosynthesis, 460–461
Thioether linkage, in prenylation, 464
Thiokinase, in fatty acid *β*-oxidation, *440*

Thiolase, in fatty acid *β*-oxidation, 439, *440*
Thiolate salts, 106
Thiols, 91, **104**–107
Thiophenol, mercury and, 104
30S ribosome, 391, *393*
3TC. *See* (−)-2-Deoxy-3′-thiacytidine (3TC)
Threonine (Thr, T), 291*t*, 389*t*, *449*, 474
Threose, 123–124
D-Threose, 232*t*, 234
L-Threose, 234
Thrombin, in clotting, 512
Thrombocytes, *510. See also* Platelets
Thromboplastin, in clotting, 512
Thrombosis, 512
Thromboxanes, 259, 282, 284
Thudichum, Johann, 269
Thylakoids, *458*
Thymine, 361–362, 364*t*, 365–*366*, *367–369*, *368*, 386
Thymol, molecular structure of, 84
Thymus, in immune system, 489, *490*
Thyroid gland, 339*t*, *340*
Thyroid hormones, action of, 356
Thyroid-stimulating hormone (TSH), *340*
Thyroxine, 81, 339*t*
Tincture, 92
Titanium (Ti), in Earth's crust, *3*
Titanium dioxide (TiO₂), 216
Titration, **152**–154
T lymphocytes, AIDS and, 395, *396*, 499–500. *See also* T cells
TNT (2,4,5-trinitrotoluene), 82, 422
Tofranil, 338*t*
Tolbutamine, for diabetes, 354
Tollens' reagent, aldehydes and, 188–189
Toluene, 77
Toluidine, 170, 175
Tonsils, 489, *490*
Tooth enamel, 308
Topoisomerases, during DNA replication, 376
Toxicity
    of alkaloids, 169
    of ammonia and ammonium ion, 445, 448
    of benzene, 76
    of botulin, 344
    of ethylene oxide, 103
    of monoclonal antibody therapy, 497
    of nerve gases, 343
    of nitric oxide, 352, 489
Toxins, from *Vibrio cholerae*, 349–351
Trace elements, 475*t*, 478–479*t*
Tranquilizers, amines in, 172
Transaminase, in amino acid catabolism, 444
Transamination, in amino acid catabolism, **444**
Transcription, **385**–388, *386*
Transcriptional level gene regulation, 397–399
Transcription factors, 387, 397–399
Transenoyl CoA, in stearic acid metabolism, 441*t*
Transferases, action of, 319, 320*t*
Transfer RNA (tRNA), 320*t*, **371**–372, *371*, 388–*389*, 390–397, 399, 400
    in transcription and translation, *385*, 388
Transgenic mice, 398
Transition metal catalysts, in aldehyde reduction, 190
Translation, **385**, 388–389, 390–397
Translational level gene regulation, 397, 399–400
Translocation, in protein elongation, 393
Translocator inner membrane (TIM) complexes, 412
Translocator outer membrane (TOM) channels, 422

Transmembrane channel proteins, *267*, 341
Transmembrane channels, 266
Transpeptidase, 326
Transplant rejection, 511*t*
Transport
    of chemicals across cell membranes, 265–267
    via microtubules, 309
    proteins for, 290
Transporters, **347**, 350
Transthyretin, 303
2,4,6-Tribromophenol, 78
Tricarboxylic acid cycle, 413. *See also* Citric acid cycle
Trichloroacetic acid, 206
Trichlorofluoromethane, 38
Trichloromethane, 37
Trienes, 53
Triethylamine, 171
Triethylammonium chloride, 172
Triglycerides, 259–264, 274–275, 473
2,3,4-Trihydroxybutanal, 123–124
Triiodothyronine, 339*t*
Trimethylamine, 11, 167
2,2,4-Trimethylpentane, 36, 40
Trinitroglycerin, 82. *See also* Nitroglycerin
Triols, 95
Triose phosphates, in photosynthesis, 459
Trioses, 231
Tripeptides, 295–296, 299
Triphosphate ion, 219
Triple bond, 6, 7, 18, 19, 46
Triple helix, *304*, 308, 309, 311
Triprotic acids, **139**
Trisaccharides, 245, 246
tRNAᵢ^Met transfer RNA, 391, *392*
Tropocollagen, 309
Trypsin, 318, 319, 320, 327–328, 431, 474
Trypsinogen, 327–328
Tryptophan (Trp, W), 291*t*, 389*t*, *449*, 465, 474
Tubules, of kidneys, 516, *517*
α-Tubulin, in microtubules, 308
β-Tubulin, in microtubules, 308
Tumor cells, 497, 501
Tumor necrosis factor (TNF), 500
Tumors, oncogenes and, 402. *See also* Cancer
Tumor suppressor genes, 402
12:13 dEpoB, 499
Tylenol, 330
Type I and type II diabetes, 354
Typhoid fever, 511*t*
Tyrosine, 291*t*, *306*, 389*t*, *449*
L-Tyrosine, 320*t*
Tyrosine kinase, in B cell and T cell regulation, 502
Tyrosine-tRNA synthetase, 320*t*
L-TyrosyltRNA, 320*t*

Ubiquinone, 420
Ubiquitin, 447
Ubiquitinylation, 447
UDP-glucose, 456, 457–460
Ulcers, 323, 352
Ultra-extensive metabolism (UEM) mutant of cytochrome P-450 enzyme, 377
Ultraviolet (UV) radiation, 39, 216, 261, 377, 401. *See also* UVA and UVB radiation
UMP. *See* Uridine 5′-monophosphate (UMP)
Unsaturated fatty acids, 259–260, 260*t*, 262–263, 265–267, 439, 461
Unsaturated hydrocarbons, 18, 46
Unwinding step, in DNA replication, 376
Uracil, 361–362, 364*t*, 386, 389, 459, 462
Urea, 3, 5, 318, *432*, *444*, 445, *446*, 447–448
    protein denaturation by, 311

in serum, 511*t*
in urine, 516–517
Urea cycle, 352, 431, *432*, *444*, 445–448, *446*
Urease, 323, 475*t*
Uric acid, nitrogen excretion via, 445
Uridine, 364*t*
Uridine diphosphate (UDP), 456, 457–460
Uridine 5′-monophosphate (UMP), 364*t*
Urine, 508, 515, 516–517
Urine tests
for glucose, 243
for anabolic steroids, 278
Urobilin, heme catabolism and, 451
Uronic acids, 243–244
Urushiol, 84
UVA and UVB radiation, 216

Vaccination (vaccines), 394, 395, 499
Valence-shell electron-pair repulsion (VSEPR)
model, 6–8, 28, 47–48
Valeric acid, 204*t*
Valine (Val, V), 291*t*, 389*t*, 400, *449*
Valium, 172
Vanilla beans, *84*, 186
Vanillin, 84, 186
Variola, 499
Variolation, 499
Vaseline, 34
Vasopressin, 300, 339*t*, 340, 349, 353, 355, 518
V domains, in T cell receptor complex, *498*
Vegetable oils, 259, *261*, 262*t*
Vegetarian diets, 470, 474–475
Venom, effects on acetylcholinesterase
regulation, 346
Venous blood, *514*
Very-low-density lipoprotein (VLDL), 274–275
Viagra, nitric oxide and, 352

Vibrio cholerae, G-protein and, 349–351
Vidarabine, 394
Vinegar, acetic acid in, 137
Vinyl chloride, 65, 65*t*, *66*, 401
VIOXX, 282
Vira-A, molecular structure of, 394
Viruses, 394, 402, 405, 447, 486, 487, 488, 489,
490*t*, *492*–493
AIDS, 395–396, 398
Vision, *cis-trans* isomerism in, 54
Visual purple, 54
Vitamin A, 54, 55, 57, 303, 476*t*, 480
Vitamin A aldehyde, 54
Vitamin B$_1$, 475*t*, 476*t*
Vitamin B$_2$, 475–481, 475*t*, 476*t*
Vitamin B$_6$, 219, 475*t*, 477*t*
Vitamin B$_{12}$, 475*t*, 477*t*
Vitamin B complex, 321
Vitamin C, 6, 237, 468, 475, 478*t*, 480
Vitamin D, *475*, 476*t*, 480
Vitamin E, 86–87, 475, 476*t*, 480
Vitamin K, 476*t*, 512
Vitamins, 469, 475–482, 475*t*, 476–479*t*, 511*t*
Vitravene, 387
V(J)D (variable-join-diverse)
recombination, *496*
VLDL. *See* Very-low-density lipoprotein (VLDL)
von Baeyer, Adolph, 217
von Euler, Ulf S., 282
VSEPR model

Water (H$_2$O), 134–136, 143, 150–152, 154–*156*,
173, 230–231, *410*, 421–422, 424, 508. *See
also* Aqueous solution
Water balance, in blood and kidneys, 518
Water-soluble vitamins, 476–478*t*
Watson, James, 367, *368*

Waxes, 34, 259, 262
Weak acids, 84, 96–97, 106–107, **136**–137,
140–142, 143, 143*t*, 154–155, 157,
158–160, 206–207
Weak bases, 106, **136**–137, 140–142, 154–155,
157, 173–175
White blood cells, 284, 423, 488, *508*, 511. *See
also* B cells; Leukocytes; Lymphocytes;
Macrophages; Natural killer (NK) cells;
T cells
White matter, in brain and spinal cord, 270
Whole blood, 510, 511*t*
Wilkins, Maurice, 367
Wöhler, Friedrich, 3, 5
Wood alcohol, 107. *See also* Methanol

Xeroderma pigmentosa (XP), 380
X-ray crystallography, of glyceraldehyde, 233
X-ray diffraction studies, 367, 393
X rays, 401, 402
Xylene, 78
Xylitol, 241–242
D-Xylose, 232*t*

Zafirlukast, 284
Zalcitabone, in AIDS treatment, *396*
Zantac, 338*t*, 352
Ziegler, Karl, 66
Zileuton, 284
Zinc (Zn), 398, *399*, 475*t*, 479*t*
Zinc fingers, in gene regulation, 398, *399*
Zwitterions, 292–293, 297, 444
Zymogens, **327**–328, 379

Standard Atomic Weights of the Elements 1997   Based on Relative Atomic Mass of $^{12}C = 12$

Name	Symbol	Atomic Number	Atomic Weight	Name	Symbol	Atomic Number	Atomic Weight
Actinium*	Ac	89	(227)	Neodymium	Nd	60	144.24
Aluminum	Al	13	26.981538	Neon	Ne	10	20.1797
Americium*	Am	95	(243)	Neptunium*	Np	93	(237)
Antimony	Sb	51	121.760	Nickel	Ni	28	58.6934
Argon	Ar	18	39.948	Niobium	Nb	41	92.90638
Arsenic	As	33	74.92160	Nitrogen	N	7	14.00674
Astatine*	At	85	(210)	Nobelium*	No	102	(259)
Barium	Ba	56	137.327	Osmium	Os	76	190.23
Berkelium*	Bk	97	(247)	Oxygen	O	8	15.9994
Beryllium	Be	4	9.012182	Palladium	Pd	46	106.42
Bismuth	Bi	83	208.98038	Phosphorus	P	15	30.973762
Bohrium*	Bh	107	(264)	Platinum	Pt	78	195.078
Boron	B	5	10.811	Plutonium*	Pu	94	(244)
Bromine	Br	35	79.904	Polonium*	Po	84	(210)
Cadmium	Cd	48	112.411	Potassium	K	19	39.0983
Calcium	Ca	20	40.078	Praseodymium	Pr	59	140.90765
Californium*	Cf	98	(251)	Promethium*	Pm	61	(145)
Carbon	C	6	12.0107	Protactinium*	Pa	91	231.03588
Cerium	Ce	58	140.116	Radium*	Ra	88	(226)
Cesium	Cs	55	132.90545	Radon*	Rn	86	(222)
Chlorine	Cl	17	35.4527	Rhenium	Re	75	186.207
Chromium	Cr	24	51.9961	Rhodium	Rh	45	102.90550
Cobalt	Co	27	58.933200	Rubidium	Rb	37	85.4678
Copper	Cu	29	63.546	Ruthenium	Ru	44	101.07
Curium*	Cm	96	(247)	Rutherfordium*	Rf	104	(261)
Dubnium*	Db	105	(262)	Samarium	Sm	62	150.36
Dysprosium	Dy	66	162.50	Scandium	Sc	21	44.955910
Einsteinium*	Es	99	(252)	Seaborgium*	Sg	106	(266)
Erbium	Er	68	167.26	Selenium	Se	34	78.96
Europium	Eu	63	151.964	Silicon	Si	14	28.0855
Fermium*	Fm	100	(257)	Silver	Ag	47	107.8682
Fluorine	F	9	18.9984032	Sodium	Na	11	22.989770
Francium*	Fr	87	(223)	Strontium	Sr	38	87.62
Gadolinium	Gd	64	157.25	Sulfur	S	16	32.066
Gallium	Ga	31	69.723	Tantalum	Ta	73	180.9479
Germanium	Ge	32	72.61	Technetium*	Tc	43	(98)
Gold	Au	79	196.96655	Tellurium	Te	52	127.60
Hafnium	Hf	72	178.49	Terbium	Tb	65	158.92534
Hassium*	Hs	108	(269)	Thallium	Tl	81	204.3833
Helium	He	2	4.002602	Thorium*	Th	90	232.0381
Holmium	Ho	67	164.93032	Thulium	Tm	69	168.93421
Hydrogen	H	1	1.00794	Tin	Sn	50	118.710
Indium	In	49	114.818	Titanium	Ti	22	47.867
Iodine	I	53	126.90447	Tungsten	W	74	183.84
Iridium	Ir	77	192.217	Uranium	U	92	238.0289
Iron	Fe	26	55.845	Vanadium	V	23	50.9415
Krypton	Kr	36	83.80	Xenon	Xe	54	131.29
Lanthanum	La	57	138.9055	Ytterbium	Yb	70	173.04
Lawrencium*	Lr	103	(262)	Yttrium	Y	39	88.90585
Lead	Pb	82	207.2	Zinc	Zn	30	65.39
Lithium	Li	3	6.941	Zirconium	Zr	40	91.224
Lutetium	Lu	71	174.967	—†		110	(269)
Magnesium	Mg	12	24.3050	—†		111	(272)
Manganese	Mn	25	54.938049	—†		112	(277)
Meitnerium*	Mt	109	(268)	—†		114	[285]
Mendelevium*	Md	101	(258)	—†		116	[289]
Mercury	Hg	80	200.59				
Molybdenum	Mo	42	95.94				

* Elements with no stable nuclide; the value given in parentheses is the atomic mass number of the isotope of longest known half-life. However, three such elements (Th, Pa, and U) have a characteristic terrestrial isotopic composition, and the atomic weight is tabulated for these.

† Not yet named.